HANDBOOK OF RESEARCH
ON SCIENCE TEACHING AND LEARNING

HANDBOOK OF RESEARCH ON SCIENCE TEACHING AND LEARNING

A Project of the National Science Teachers Association

EDITED BY

Dorothy L. Gabel

MACMILLAN LIBRARY REFERENCE USA
SIMON & SCHUSTER MACMILLAN
NEW YORK

SIMON & SCHUSTER AND PRENTICE HALL INTERNATIONAL
LONDON · MEXICO CITY · NEW DELHI · SINGAPORE · SYDNEY · TORONTO

Simon & Schuster Macmillan
Macmillan Library Reference
1633 Broadway, New York, NY 10019-6785

Library of Congress Catalog Card Number: 93-17119

Printed in the United States of America

Printing number
      3   4   5   6   7   8   9   10

Library of Congress Cataloging-in-Publication Data

Handbook of research on science teaching and learning/edited by
   Dorothy L. Gabel.
      p.   cm.
      "A project of the National Science Teachers Association."
      Includes bibliographical references and index.
      ISBN 0-02-897005-5 (alk. paper)
      1. Science—Study and teaching—Research—Handbooks, manuals, etc.
   I. Gabel, Dorothy.   II. National Science Teachers Association.
   Q181.A1H35   1993
   507'.1—dc20   93-17119
                                                 CIP

The paper used in this publication meets the minimum requirements of American
National Standard for Information Sciences—Permanence of Paper for Printed Library
Materials.
ANSI Z39.48-1984.⊗™

CONTENTS

Part I
TEACHING   1

Part II
LEARNING   129

Part
V

CONTEXT 491

PREFACE

Much research on science teaching and learning has occurred during the past 70 years. Science education researchers are fortunate to have reviews of this research available to them, beginning in 1927 with the publication of the digests by Curtis and followed by those of Boenig, Swift, and Lawlor, which summarize research until 1957 and were republished by the Teachers College Press in 1971 in their six-volume set. Research summaries since 1957 have appeared as chapters in each successive edition of the *Handbook of Research on Teaching* and through annual and biannual reports sponsored by the National Association for Research and Science Teaching (NARST). These were begun by the U.S. Office of Education and were later published by the ERIC Science, Mathematics, and Environmental Clearing House at The Ohio State University and by John Wiley & Sons, Inc., as special issues of *Science Education*. The publication of this edition of the *Handbook of Research on Science Teaching and Learning* marks the first time that a prolific number of research studies have been synthesized in a comprehensive manner over an extensive time period. The number and the quality of the entries are testimony and tribute to the productivity of science education researchers in the past decade. By having this large body of research reviewed in one volume and by comparing it with the research summaries of the past, the reader is able to determine the "roots" of science education research, how it has changed in both content and methodology during the past 70 years, and its future direction. Although the primary goals for the *Handbook* are to synthesize and reconceptualize research of the past, the authors were asked to include a chapter that focuses on future research. This volume should therefore serve as a springboard for science education research into the twenty-first century.

Audience

The primary audience of the *Handbook* is science education researchers and others who are doing scholarly work in science education. Included in this group are college and university faculty members; graduate students; members of federal, state, and local agencies who conduct or fund science education research projects; and investigators at science education research and development centers. The *Handbook* will be useful as a resource or textbook for graduate-level seminars in science education and for individual use by graduate students to help them identify and focus on their dissertation topics.

The secondary (but by no means less important) audience for this *Handbook* is curriculum developers, instructional leaders, policymakers, and teachers who desire to base curriculum and teaching decisions on research findings. Although the authors were not asked to make specific recommendations for curriculum developers and teachers, the authors of several chapters have made specific recommendations for practice.

Scope

The *Handbook* is intended to be a comprehensive review of science education research studies. In reality, although the scope of the volume is broad, the authors of some chapters are less comprehensive than others. In addition, some authors give more detailed information about particular studies whereas others summarize across studies.

The same research report may also be referenced in more than one chapter. This overlap in coverage shows the relatedness of the chapters and the research in the field. It also provides the reader with multiple perspectives of a given issue. Although an additional chapter was originally intended to be included in the *Handbook,* its content is maintained in other chapters in a less coherent manner.

Structure

The structure of the *Handbook* is the result of brainstorming by the Editorial Advisory Board. Although it is impossible to separate teaching from learning, separate sections on these topics were proposed. Because of the large number of research studies and emphasis on problem solving in the sciences during the last decade, and the firm belief that research in this area is science-discipline specific, an entire section on problem solving was to be included. In addition, a review of research reports on generic topics that cut across the curriculum and generic issues relating to classroom climate, gender, and cultural diversity were proposed.

As a result, the *Handbook* contains 19 chapters and is divided into 5 major sections. The section entitled "Teaching" contains 3 chapters, including research on teacher education, instructional strategies, and laboratory instruction. The second section, "Learning," examines the epistemological foundations of cognition, alternative conceptions in science, and the affective dimension of learning. The authors in this section were deliberately selected to provide the reader with contrasting points of view. The third section, "Problem Solving," examines problem-solving research in each of the major science disciplines: biology, chemistry, physics, and earth science. Although most of the reviewed discipline-centered research reports focus on mathematical problem solving, the chapters on problem solving at the elementary and middle school levels present a broader view. Section IV, "Curriculum," contains 4 chapters that span all the science disciplines. Included are chapters on the goals of the science curriculum, assessment, history and philosophy, and the uses of technology. Finally, the fifth section, "Context," contains 3 chapters on research studies that are perhaps less specific to science but are important in teaching and learning science. These include research on school climate, gender, and cultural diversity.

Development of the Handbook

The *Handbook* is a project of the National Science Teachers Association (NSTA). Although it had been proposed earlier, Hans Andersen, during his tenure as president of NSTA, asked me if I would be willing to serve as editor. Not realizing how enormous the task would become, I accepted his invitation. After consultation with me, Hans Andersen appointed a Handbook Task Force that later served as the Editorial Advisory Board. The primary task of the board was to assist me in determining the *Handbook* structure, potential authors, and reviewers. At the initial meeting in the spring of 1989, the board began its work by identifying the sections that would be included, the chapters in each section, and a brief outline of the content of each chapter. Board members also made recommendations about possible authors and reviewers for each chapter. Members of the Editorial Advisory Board were: Patricia Blosser, The Ohio State University; David P. Butts, The University of Georgia; Jane Butler Kahle, Miami University; Robert D. Sherwood, Vanderbilt University; and John R. Staver, Kansas State University. All are prominent science educator researchers who are active in both the NSTA as well as in the NARST. Through their continuing efforts, they were most helpful in making this *Handbook* become a reality. In the spring of 1992, when it became apparent that I needed additional support for the final editorial work required on the *Handbook,* I asked Patricia Blosser and David Butts to assist me. They were gracious in their response. In addition to giving me sound advice about several of the manuscripts that had been reviewed and working with particular authors during final revisions, they met with me in Columbus, Ohio, during the summer of 1992 to make the final decisions on the chapter titles and order. At that time, we read all the manuscripts in our possession (17 out of 19) and made final recommendations to the authors.

The authors recommended by the Editorial Advisory Board are outstanding science education researchers. For 16 of the 19 chapters, they were the board's first choice. Only one potential author was unable to complete the manuscript on time to be included in the *Handbook*. Each author contacted was asked to select coauthors if he or she so desired, but not to rely on the work of doctoral students. Each chapter was reviewed by one or two reviewers, the editor, a member of the editorial board, and the editorial staff of the NSTA.

Acknowledgments

The preparation of a manuscript of this magnitude requires much more work than I ever imagined. It is only through the tireless effort of many excellent people that it came to fruition. I am deeply indebted to Hans Andersen for inviting me to become the *Handbook* editor; to the Executive Committee of the NSTA for entrusting me with the editorship; and in particular to Phyllis Marcuccio, Assistant Executive for Publications, for the unending support and encouragement during the past 4 years. In addition, the work of Sheila Marshall and the publication staff at NSTA is greatly appreciated.

The quality of a work of this nature is largely dependent on the chapter authors and the reviewers. I am grateful to each of them and, in particular, to those who were able to send their outlines, manuscripts, and reviews back at the approximate time that they were requested. I also want to especially thank those authors who persevered even though their reviews and revisions were due at an inconvenient time.

I am grateful to the members of the editorial board who were instrumental in providing direction for the *Handbook* and who offered their personal support. In particular, I thank Patricia Blosser and David Butts for their additional work on the *Handbook* and their concern about its progress.

The production of this volume would not have been possible without the continuing effort of my colleagues at Indiana University. I am appreciative of the understanding given to me by our graduate students, whose own work was sometimes delayed because of my busy schedule. I am most indebted to

Vicki Trier, an undergraduate student, who was efficient in keeping track of the manuscripts, reviews, correspondence, and so forth. The enormous task of Vicki and Cheri Trier in checking each reference in every manuscript for accuracy is greatly appreciated, as is the work of Jane Masters in providing direction and support to me and others working in the office at Indiana University.

Finally, I wish to thank Lloyd Chilton, Executive Editor at Macmillan Reference, with whom I began this work, and Philip Friedman, Publisher of Macmillan Reference, with whom I am now working.

References

Gage, N. L. (Ed.). (1963). *Handbook of research on teaching.* Chicago: Rand McNally.

Jacobson, W. J. (Ed.). (1969). *Reviews of research in science education* (6 vols.). New York: Teachers College Press.

Travers, M. W. (Ed.). (1973). *Second handbook on research on teaching.* Chicago: Rand McNally.

Wittrock, M. C. (Ed.). (1986). *Handbook of research on teaching.* New York: Macmillan.

Dorothy L. Gabel

ABOUT THE CONTRIBUTORS

Ronald D. Anderson, Professor of Education at the University of Colorado, is a former president of both the National Association for Research in Science Teaching and the Association for the Education of Teachers in Science. Currently, as director of a study of curriculum reform in science and mathematics education, funded by the U.S. Department of Education, his research interests are primarily in science education policy studies. Recently, he directed an experimental teacher education program at the University of Colorado.

Mary M. Atwater is an associate professor in the Department of Science Education at the University of Georgia. She holds a B.S. in chemistry from Methodist College, an M.A. in chemistry from the University of North Carolina, and a Ph.D. in science education from North Carolina State University. At the University of Georgia, her teaching and research efforts have included 17 federal and state grants. She is a coprincipal investigator of a Howard Hughes grant to encourage science students from underrepresented groups to pursue careers in biomedical research and the project director of a grant to publish a monograph entitled *Multicultural Education: Inclusion of All.* She has published in *School Science and Mathematics, Journal of Research in Science Teaching, Journal of Science Teacher Education, Science Teacher,* and *Science Activities.* She is one of the coauthors of the Macmillan/McGraw-Hill Science program and has been a frequent presenter at national meetings on multicultural science education.

Charles R. Ault, Jr., Associate Professor of Education at Lewis and Clark College, received a Ph.D. in science and environmental education from Cornell University in 1980. Since 1987, he has directed the Science M.A.T. Programs for both preservice and inservice secondary teachers and taught courses in elementary school science and mathematics. Dr. Ault's current research interest builds on the notion of students as "philosophers of their own thinking" and follows several years of interview-based studies of student conceptions of natural phenomena. In addition, he has studied problem solving in the earth science curriculum and the

evaluation of prototype exhibits in museum settings. He has published in *Science Education, The Journal of College Science Teaching, The Journal of Geological Education, Science and Children,* and *Thinking: The Journal of Philosophy for Children.*

Bonnie B. Barr received her Ph.D. in science education from Cornell University. As a professor of science education at the State University of New York at Cortland, she works with K–8 preservice and inservice teachers. Her experience in teaching science includes every grade level from kindergarten through graduate school. Dr. Barr has been the project director for eight federally funded teacher enhancement projects and has served as science consultant for numerous school districts in the United States and Europe. She was an author on the Addison-Wesley Junior High Science series and has written numerous articles and curriculum packages on elementary science. She has served in leadership roles in national organizations including the National Science Teachers Association, the Council for Elementary Science International, and the American Nature Study Society.

Sharolyn J. Belzer is a Ph.D. candidate in biology and education at the University of Michigan. She received her B.S. and M.S. in biology from Washington University and University of Michigan. Her dissertation research focuses on using a hypermedia-based environment, consisting of three components (a database, games, and a communications arena), to facilitate the nonscience undergraduate students' conceptual understanding of evolution and natural selection.

Carl F. Berger is a professor of science education and the director of instructional technology at the University of Michigan. He is a former dean of the School of Education and has been a professor of science education at the university for 20 years. His study of how students use technology dates from the 1960s, when he was a coauthor of the Science Curriculum Improvement Study at the University of California where he received his doctorate. He was a National Science Foundation (NSF) Scholar to the Republic of China in the mid-1970s and president of the National Association for Research in Science Teaching in the 1980s. Dr. Berger has more than 100 publications including 23 books, 12 multimedia packages, and 2 patents.

Diane M. Bunce, an associate professor of chemistry at The Catholic University of America, received a Ph.D. in chemical education from the University of Maryland. Her research interests center on the learning and teaching of chemistry problem solving and chemical concepts. She is a coauthor of both *CHEMCOM* and *Chemistry in Context,* two curriculum projects developed by the American Chemical Society. At present, she is the Director of the Center for Minority High School Chemistry Teachers and Teachers of Minority Students of the Institute for Chemical Education (ICE).

Rodger W. Bybee is the associate director of the Biological Sciences Curriculum Study (BSCS), The Colorado College, Colorado Springs, Colorado. Since 1986, Dr. Bybee has been principal investigator for three new National Science Foundation (NSF) programs. They are: an elementary school program, "Science for Life and Living: Integrating Science, Technology, and Health"; a middle school program, "Middle School Science and Technology"; and a high school program, "Biological Science: A Human Approach." Prior to joining BSCS in August 1985, Dr. Bybee was professor of education at Carleton College in Northfield, Minnesota. He received his Ph.D. in science education and psychology from New York University and his B.A. and M.A. from the University of Northern Colorado. He is coauthor of a leading textbook entitled *Teaching Secondary School Science: Developing Scientific Literacy.* His most recent book is *Reforming Science Education: Social Perspectives and Personal Reflections.*

Frank E. Crawley III is a member of the Science Education Center at The University of Texas at Austin, where he works extensively preparing secondary science teachers to teach majority and minority students in both rural and urban settings. His research has examined the relationship between attitudes and the science-related behaviors of inservice and preservice secondary science teachers, their students, and secondary science students in several foreign countries. Most recent studies have tested strategies designed to encourage more Hispanic students to pursue advanced science study at the precollege level. Dr. Crawley is active in the National Association for Research in Science Teaching, the National Science Teachers Association, the Association for the Education of Teachers in Science, and the American Association of Physics Teachers. He has written more than 60 articles, reports, reviews, and abstracts, as well as one book and three book chapters.

George E. DeBoer is a professor of education at Colgate University in Hamilton, New York, where he has been since 1974. He has a B.A. in biology from Hope College in Holland, Michigan, an M.A.T. in science teaching from the University of Iowa, and a Ph.D. in science education from Northwestern University. His research has focused on gender differences in science course-taking patterns and, most recently, his book, *A History of Ideas in Science Education,* was published by Columbia University Teachers College Press.

Rodney L. Doran, Professor of Science Education at the University of Buffalo (SUNY/Buffalo), has earned a B.S. degree in physics from the University of Minnesota in 1961, an M.S.T. in physics from Cornell University in 1965, and a Ph.D. in science education from the University of Wisconsin in 1969. Dr. Doran's major research interest has been in the assessment of science programs' outcomes, particularly in the cognition, alternative, and laboratory domains. He helped develop and analyze the process skills tests used in the Second International Science Study and was involved with the development, trial testing, and implementation of a manipulative science skills test used statewide in New York at the grade four level. As part of a National Science Foundation (NSF) supported project, Doran helped to develop science performance tests appropriate for high school science courses. Professor Doran has published articles in many journals, several monographs, and the book *Measurement and Evaluation of Science Instruction* (National Science Teachers Association, 1981).

Richard A. Duschl is an associate professor and member of the graduate faculty at the University of Pittsburgh. He holds a primary academic appointment to the Department of Instruction and Learning and affiliated appointments to the Department of History and Philosophy of Science and the Center for Philosophy of Science. Recent publications on incorporating history and philosophy of science in science education include *Restructuring Science Education: The Role of Theories and Their Importance* (Teachers College Press, 1990) and, with Richard Hamilton, an edited volume titled *Philosophy of Science, Cognitive Psychology, and Educational Theory and Practice* (SUNY Press, 1992).

Barry J. Fraser is a professor of education and the director of the National Key Centre for School Science and Mathematics at Curtin University of Technology in Perth, Australia. Currently, he is the executive director of the International Academy of Education, member of the Board of Directors of the National Association of Research in Science Teaching, External Examiner at the University of Brunei Darussalam, and a coeditor of the *International Journal of Educational Research* and the *South Pacific Journal of Teacher Education.* He is author of more than 200 articles and has written 48 book chapters. His best-selling work includes *Classroom Environment, Windows into Science Classrooms* (with K. Tobin), and *Educational Environments: Evaluation, Antecedents and Consequences* (with H. Walberg).

Dorothy L. Gabel, Professor of Science Education at Indiana University, received her M.S. and Ph.D. degree at Purdue University. She was a director of the Teacher Preparation Program at the National Science Foundation in 1987/1988 and is currently involved in a program to improve the science preparation of prospective elementary teachers at Indiana University. Dr. Gabel's major research interest is in chemistry problem solving and concept learning. She is a coauthor of the Prentice Hall high school chemistry text *Chemistry: The Study of Matter,* the editor of the NSTA monograph on *What Research Says to the Chemistry Teacher About Problem Solving,* and director of the *SourceView* videotape project on exemplary chemistry teaching.

Alejandro José Gallard received his Ph.D. from Michigan State University. He is an assistant professor of science education at Florida State University and has taught science to linguistically diverse inner-city students in elementary, middle, and high schools. He conducted research in Latin America on centralized science curricula and secondary science teaching, and collaborated to improve science teacher education programs and classroom-based research on science teaching and learning. Dr. Gallard's research interests focus on classroom and policy issues related to the teaching and learning of science for students with limited English proficiency both in and out of the United States.

Robert Hafner is an assistant professor of science education in the Department of Science Studies at Western Michigan University. He received his Ph.D in curriculum and instruction from the University of Wisconsin–Madison. He teaches biology for prospective teachers, science methods, and graduate science education offerings. His research interests include the role of models in learning science, extending the use of problem solving in biology classrooms, and the application of issues in the history and philosophy of science to science teaching and learning. His publications have appeared in *Science Education, Journal of Research in Science Teaching, Educational Psychologist,* and *American Biology Teacher*.

Stanley L. Helgeson is a professor of science education and the Coordinator of Focus Area 4: New Technologies for the National Center for Science Teaching and Learning at The Ohio State University. Dr. Helgeson was associate director for science education for the ERIC Clearinghouse for Science, Mathematics, and Environmental Education for 22 years. He has published a variety of reports, research summaries, and information analyses, and served as production editor for scores of ERIC publications, including the National Association for Research and Science Teaching (NARST)/ERIC annual review of research in science education, the abstracts of presented papers for the NARST annual meetings, and the Yearbook series for the Association for the Education of Teachers in Science.

Jane Butler Kahle, Condit Professor of Science Education at Miami University, works extensively in the preparation of biology teachers as well as in the assessment of the special needs and issues surrounding women and minorities in science. Professor Kahle has evaluated science education programs and projects in the Philippines, Thailand, Indonesia, Australia, China, New Zealand, Britain, Norway, Sweden, Denmark, Israel, and the United Arab Emirates. She has been chairman of the board of directors of the Gender and Science and Technology Association and has served as president of the National Biology Teachers Association and the National Association for Research in Science Teaching. Dr. Kahle is the author of more than 82 articles, 4 monographs, 13 book chapters, and 6 books. She has served as a reviewer for 6 scholarly journals and has directed more than 20 funded projects.

Thomas R. Koballa, Jr. is an associate professor of science education at the University of Georgia. He holds a B.S. in biology and an M.A. in science education from East Carolina University, and a Ph.D. in curriculum and instruction/science education from The Pennsylvania State University. He is the recipient of research and teacher training grants from the National Science Foundation, the University Research Institute of The University of Texas, and the National Science Teachers Association. He has authored and coauthored more than 40 articles and book reviews in *Journal of Research in Science Teaching, Science Education,* and *School Science and Mathematics,* and has served as editor of the *Attitude Research in Science Education Newsletter.* Dr. Koballa's research interests in persuasive communication design and attitude change are reflected in his contribution to the National Association of Research in Science Teaching (NARST) *Research Matters . . . To the Science Teacher* series entitled "Changing and Measuring Attitudes in the Science Classroom."

Frances Lawrenz is a professor of science education and evaluation at the University of Minnesota. She has published more than 35 articles in national research journals, written chapters for 5 books, and has conducted more than 50 science program evaluations. Dr. Lawrenz has worked as an evaluation specialist for both the National Science Foundation and the U.S. Department of Education as well as serving as a board member for the National Association for Research in Science Teaching (NARST) and School Science and Mathematics Association (SSMA). Presently, she is working on a grant with elementary school districts and teachers to investigate authentic assessment.

Anton E. Lawson is a professor in the Department of Zoology at Arizona State University, where he teaches biology and methods of teaching biology. His research interests center on methods of teaching that promote the development of thinking skills. Professor Lawson received his B.S. degree in biology from the University of Arizona in 1967, his M.S. degree in zoology from the University of Oregon in 1969, and his Ph.D. degree in science education from the University of Oklahoma in 1973. In 1976, 1985, and 1987, Dr. Lawson received the National Association for Research in Science Teaching's (NARST's) JRST Award and, in 1986, its NARST Award for distinguished contributions to science education research.

Reuven Lazarowitz is an associate professor of science education in biology at the Technion—Israel Institute of Technology. He received his M.S. degree in botany, zoology and biochemistry and a Certificate of Teaching from the Hebrew University in Jerusalem. He obtained his Ph.D. in science education in biology at the University of Texas at Austin. Professor Lazarowitz was a member of the Israeli group of scientists and educators who translated the BSCS yellow version and adapted it for Israeli high schools. He was also a member of the Biology Curriculum Committee and the Board of Matriculation Exams in Biology appointed by the Israeli Ministry of Education and Culture. His research interest is in the structure of school biology curriculum; teaching and learning biology concepts and principles in individualized, cooperative small groups; and computer-assisted learning settings.

Casey R. Lu is a Ph.D. candidate in education and biology at the University of Michigan, where she received her B.S. and M.S. degrees in biology. She is a former high school biology teacher. Her research interests include examining the effects of using computers on the thinking and learning processes specific to science.

David P. Maloney is an associate professor of physics at Indiana University–Purdue University, Fort Wayne. After completing an interdisciplinary Ph.D. in physics, geology and education at Ohio University, Dr. Maloney taught at Wesleyan College for 3 years and Creighton University for 10 years. Dr. Maloney's main area of research is students' alternative conceptions and problem-solving difficulties in physics. He has published 18 articles in physics and science education and has presented a number of workshops on physics education research at meetings of the American Association of Physics Teachers (AAPT). Professor Maloney has served two terms on the Research in Physics Education Committee of the AAPT and was chairman of the committee for 2 years.

Judith Meece is an associate professor of education at the University of North Carolina–Chapel Hill. She received her doctoral degree in 1981 from the University of Michigan and has held faculty positions at the University of Pittsburgh and Purdue University. She is a leading researcher in the field of gender differences in motivation and achievement and has published more than 15 articles on that topic. Her 1982 *Psychological Bulletin* article (with Jacquelynne Eccles) presented a new motivational model of mathematics achievement, which served as the basis for her research over the next 10 years. She recently coedited a volume on student motivation and learning. Meece was recently awarded a university fellowship to complete an oral history project focusing on the motivation and development of career teachers.

Joel J. Mintzes, Professor of Biological Sciences at the University of North Carolina at Wilmington, received his B.S. and M.S. in biology at the University of Illinois at Chicago and his Ph.D. in science education at Northwestern University. For the past 20 years, he has focused his research on problems of conceptual development and cognitive processes in the life sciences, and has served as coordinator of introductory biology at several institutions. His publications have appeared in numerous journals, including the *Journal of Research in Science Technology* (JRST), *Science Education, Science and Children,* and *School Science and Mathematics.* He currently serves on the editorial board of the *JRST.*

Carole P. Mitchener is an assistant professor in the School of Education at DePaul University and codirects a partnership between DePaul University and the Institute for Psychoanalysis in Chicago. Her teaching and research combine her interests in teacher development and science education. Her work focuses on understanding how to facilitate preservice teachers' understanding of themselves in teaching-learning situations in ways that promote teacher and student learning. In science education, she studies what types of curriculum and clinical experiences are needed to help preservice teachers make sense of their own science teaching-learning experiences, so they can enhance student learning in science.

Joseph D. Novak did undergraduate and graduate studies at the University of Minnesota, completing his Ph.D. in science education and biology in 1958. He taught biology at Kansas State Teachers College and Purdue University and worked with teacher preparation programs. He moved to Cornell University in 1967 as Professor of Science Education and holds an appointment as Professor of Biological Sciences. His current research interests are in refinement and application of learning theory and theory of knowledge to the improvement of teaching and learning in any educational context. He has authored and coauthored 17 books and more than 100 scholarly papers and book chapters.

J. Steve Oliver is currently a member of the Science Education Department at the University of Georgia. After completing a master's degree from North Carolina State University and a Ph.D. in science education at the University of Georgia, he accepted a faculty position at Kansas State University, where he was science teacher educator and educational researcher until his return to the University of Georgia in 1990. His primary research focus is on the role of attitudes in science learning. He is the author of a variety of articles and has presented papers in this area.

Ronald D. Simpson is director of the Office of Instructional Development at the University of Georgia, where he also is a professor of science education and higher education. Former head of the Department of Mathematics and Science Education at North Carolina State University, he holds degrees from the University of Tennessee and the University of Georgia. The author of more than 100 scholarly publications, Simpson has taught at all levels in higher education. He currently serves as editor of *Innovative Higher Education.*

James Stewart, Professor in the Departments of Curriculum and Instruction (Science Education) and Environmental Studies at the University of Wisconsin–Madison, received his Ph.D. in science education from Cornell University. His teaching includes undergraduate science methods courses for prospective elementary teachers and graduate courses in science education. The focus of Dr. Stewart's research on learning and problem solving in biology, particularly genetics, has been supported by numerous grants over the past 12 years, and has resulted in more than 15 publications in this area. As a member of the University of Wisconsin's Center for Biology Education, Dr. Stewart has been involved in the development of summer inservice offerings that attract approximately 300 K–12 science teachers each year.

Pinchas Tamir received a master's degree in agriculture from Hebrew University of Jerusalem and a Ph.D. in science education, educational psychology and measurement from Cornell University. He taught high school biology and agriculture for 15 years, and was director of the Israel High School biology project. Pinchas Tamir's research interests vary, including various aspects of biological education curriculum development and evaluation, learning in the laboratory, cognitive references, innovative testing, and teacher education. He has published more than 50 books and chapters and nearly 200 articles in professional journals in the United States, Canada, Europe, and Great Britain, and Australia. Pinchas Tamir has also received several awards, among them the prestigious NARST award for life-long contribution to research in science education. In addition, he participates regularly in professional conferences and is frequently involved in keynote lectures. He has served as a visiting professor to 20 universities around the world.

Deborah J. Tippins received a Ph.D. in science education/curriculum and instruction from Texas A&M University in 1989. A veteran elementary and middle school science teacher, she is currently an assistant professor in science education at the University of Georgia. Her research interests center on science teaching and learning in the context of reform, teacher beliefs, and ethical and sociocultural dimensions of science teaching. Dr. Tippins was a recipient of the American Association of Colleges and Schools of Teacher Education Outstanding Scholarship Award in 1991 and the Lilly Teaching/Scholar Fellowship in 1992. She has recently coauthored a book entitled *The Stigma of Genius: Einstein and Beyond Modern Education,* which explores postmodernism as a way of understanding science teaching and learning.

Kenneth Tobin is a professor of science education at Florida State University. For 10 years, he taught physics, chemistry, biology, and general science in Australia and England, and produced curriculum resource materials for teachers of high school science. He has been a science educator in universities in Australia and the United States for 19 years. Tobin is the author of 3 books, more than 180 published journal articles, chapters, and reports, and more than 180 papers presented at international, national, and regional meetings. Dr. Tobin's research interests focus on reform of science curricula, teacher learning and change, and science teacher education.

Burton E. Voss taught junior and senior high school science and then served in the U.S. Army Chemical Corps as a biological sciences research assistant. He completed the Ph.D. in science education at the University of Iowa in 1958. He taught at The Pennsylvania State University from 1958 to 1963 and then joined the faculty at the University of Michigan. His research interests have focused on classroom interaction analysis, inservice and preservice teacher education, and science assessment. He has received numerous grants from the National Science Foundation, the Department of Energy, and the State of Michigan. In addition, he has served in various offices in the National Science Teachers Association, the National Association of Biology Teachers, the School Science and Mathematics Association, and the Michigan Science Teachers Association. He has received distinguished service awards to science education from the National Science Teachers Association and the Michigan Science Teachers Association.

James H. Wandersee is an associate professor of biology education in the Department of Curriculum and Instruction at Louisiana State University. He currently serves as the associate editor of the *Journal of Research in Science Teaching* and as an editorial board member for *The American Biology Teacher*. As a high school biology teacher, college biology professor, and university science educator, he has published more than 70 articles in scientific and science education journals. The focus of his theory-driven research is the graphic representation of scientific knowledge and its effects on science learning.

TEACHING

RESEARCH ON SCIENCE TEACHER EDUCATION

Ronald D. Anderson

UNIVERSITY OF COLORADO

Carole P. Mitchener

DEPAUL UNIVERSITY

In the past decade, when the advocates of educational reform trumpeted their calls for improvement in science education, they usually targeted teacher education as the starting point for correction and change. As a result, the rhetoric about revamping teacher education has grown louder and louder. Not surprisingly, the political maneuvering concerning change in teacher education has matched the rhetoric in intensity and pervasiveness.

What kind of thinking and understanding of science teacher education has produced so much rhetoric and political maneuvering? Unfortunately, much of the talk has been based on simple assumptions and beliefs while little has been heard about the results of disciplined inquiry into science teaching and science teacher education. The more important questions are these: What are the results of serious research into science teacher education? What do these results portend for the future of science teacher education?

The purpose of this chapter is to answer these questions throroughly and in an orderly manner. We do so for the benefit of practitioners who develop and operate science teacher education programs and the researchers who study such enterprises. In this section, we review the results of the most important research on science teacher education. Our purpose is to synthesize these research results within a conceptual framework that reflects differing theoretical perspectives. We offer some insights into the past research and into projected future directions of this research. Having completed this review and reflected on the implications of the research for both the individual teacher and the school as a learning organization, we pose some speculations about the merits of possible alternative paths for future research.

An adequate examination of this research will require investigating science teacher education broadly, including its purposes, alternative theoretical frameworks, instructional content, modes of instruction, institutional arrangements and the people who enter the programs. The purposes of science teacher education, for example, are many and the relative emphasis among these several purposes varies considerably from one program to another. In some programs the primary emphasis is on transmitting to teachers knowledge of learning and instructional practices; in others, the emphasis is on the acquisition of instructional skills and the ability to utilize various instructional strategies. In another case, the program focus may be helping teachers to reconceptualize their role, while in yet another program the focus may be developing a conception of and facility in engaging in productive collegial relationships. Program orientations are many and varied, with many degrees of emphasis among the several purposes found in programs. A review of the research on science teacher education must take account of these varied purposes.

There are obviously many dimensions of science teacher education to consider. It includes the teacher's liberal arts background, instruction in the content field of science, preservice professional education, various means of induction into the profession and inservice education. All must be considered, although the amount of research on the several dimensions varies as does the attention that will be allocated to each in this text.

Appreciation is expressed to Susan Loucks-Horsley (The Network), Linda DeTure (Rollins College), and Nancy Brickhouse (University of Delaware) for their careful reviews and insightful comments.

An adequate understanding of what scholarship says about science teacher education—or any other facet of education—requires that it be considered from a multiplicity of perspectives. Psychological perspectives have been predominant in past research in this arena, but they obviously are not the only perspectives needed for understanding. Sociocultural perspectives (the term is used broadly here for work conducted within several research traditions including sociology and anthropology) also are essential if one is to begin to understand the topic at hand. Philosophical perspectives are essential as well. A synthesis of the results of empirical research from the former two perspectives is strikingly inadequate without philosophical analysis that approaches the matters at hand in an integrative and synoptic manner. Finally, the perspectives offered by the subject matter itself, namely the natural sciences in this case must be considered. The character of science knowledge, the modes of inquiry used in acquiring it, and the applications of science in contemporary society all have implications. Science teacher education has many dimensions and is conducted for many purposes; an adequate understanding of it will require viewing it from a multiplicity of scholarly perspectives.

Other reviews of teacher-education research have appeared in recent years and the reader may wonder how the coverage of this one relates to others. The excellent third edition of the *Handbook of Research on Teaching* (Wittrock, 1986) had a chapter devoted to research on teacher education (Lanier & Little, 1986). In addition to the obvious difference in scope—teacher education in general versus science teacher education—there are differences in areas selected for emphasis. Lanier and Little addressed four main areas, the first of which was teacher educators themselves—a topic not addressed here. Their second topical area, the students of teaching, is one that is addressed here but in considerably less detail. Their third area, the curriculum for teaching, is of direct interest in this review as well, although here the emphasis is on the science-related dimensions. Their final area, the important one of the milieu in which teacher education takes place, is not a direct focus of this review. It is clear that choices must be made, given the size of the review; the review cannot be comprehensive. Lanier and Little's need to make choices was greater and they systematically excluded a number of areas, given the breadth of their topic. Being more limited in scope, the current review can be comprehensive, even though it also does not cover all facets of the topic.

A second publication is the *Handbook of Research on Teacher Education* (Houston, 1990). The fact that its length (925 pages) approaches that of the previously mentioned *Handbook of Research on Teaching* gives some sense of the extensiveness of the literature on teacher education. Within the *Handbook of Research on Teacher Education* is a chapter devoted to science teacher education (Yager & Penick, 1990). Although the length of that chapter is considerably shorter than the current one, its scope is similar and a comparison of their respective coverages is in order. Yager and Penick addressed three major areas: the history of science teacher education, current issues in the field and promising practices, (i.e., current programs that have merit). The current review gives relatively little attention to the first and third of these areas. The focus of the second of their areas—current issues in the field—has a lot in common with the current review, although the current one is considerably more extensive.

Within this comparative context, two characteristics of the review at hand should be noted. In spite of the inadequacies of the space available, an attempt has been made to be fairly comprehensive in coverage of the many facets of science teacher education. Second, an attempt has been made to view the topic at hand at least partially from various theoretical perspectives. Both of these characteristics are visible in the organization of the chapter.

The chapter begins with an exploration of various theoretical perspectives (i.e., alternative conceptual frameworks for describing, conducting, and evaluating science teacher education endeavors). Attention then is turned to the people entering science teacher education, including teacher supply and demand and selection of entrants to science teacher education programs. Standards for programs and their graduates, along with certification processes and teacher assessment systems, also are addressed. Attention then is turned to the curriculum of preservice science teacher education programs, including the dimensions of general education, the science subject matter, and professional education. The means by which this curriculum is delivered also is considered, with attention to various instructional approaches and overall systems. The induction phase of teacher education then is addressed before going to staff development and inservice education. Finally, the chapter offers an overall assessment of what research on science teacher education tells us and what orientations to future research have the most promise.

As is apparent from this brief description of its organization, this chapter has not been organized according to level of education (i.e., elementary, middle/junior high, and senior high school science teacher education). Other ways of categorizing the research seem more fruitful. It is acknowledged, however, that these distinctions are important—particularly that between elementary and the higher levels—because their preparations are quite different. This distinction is discussed where relevant.

What has been the nature of reform efforts? Most educational reform efforts overall can be characterized as "being guided by the rhetoric of a conservation idealogy" (Kyle, 1991, p. 408). The conservative ideology Kyle cites is driven by corporate self-interests and economic growth, and the desire to preserve an ethnocentric Western culture. Schools today, including those of teacher education, continue to reinforce the sociocultural and political contexts in which they function, as opposed to serving as "the sites of counterhegemony" (Kyle, 1991, p. 408). Indeed, the conservative ideology frames educational reform as efforts toward standardization and reproduction, emphasizing mastery and accountability.

Most education reform proposals continue to ignore the clinical nature of teaching (Sarason, Davidson, & Blatt, 1962, 1986). Although phrases such as "teacher as researcher," are becoming more common in the reform literature, true understanding of the complexity of more clinically-oriented practice is absent from most of the debate. For example, Field (1993) argues that the imagined school presented in the "Schools for the 21st Century—A Scenario" section of the Carnegie Report

(1986) is a utopian scenario of schooling, which clearly ignores the reality of urban classrooms. She charges that the Carnegie scenario assumes "that both teacher and student manage to check their emotional baggage at the classroom door" (p. 147). In general, most reform proposals continue to advocate the historical depersonalized, top-down model for change, which reduces the teacher to technician. This view, similar to efforts to make curriculum "teacher-proof," continues to reinforce the assumption that theories for guiding teachers only originate outside of classroom practice. It is worth noting, however, that some efforts exist in developing the clinical nature of teacher and the subjective dimension of teacher education, as noted in "The Reform Agenda" (Stengel, 1992). An example is a program at Brooklyn College under the direction of Dean Madeline Grumet. This program uses the metaphor "Teacher As Artist," emphasizing the moral, caring and spiritual dimensions of the teacher in the construction of its teacher education curriculum.

Educational-reform efforts in teacher education can also be characterized, according to Zeichner and Liston (1990), as lacking a deep understanding of the assumptions and goals of the specific proposals, and the resulting implications. They write:

One of the most notable characteristics of the current reform movement in U.S. teacher education is its lack of historical consciousness.... One consequence of the historical amnesia in the current teacher education reform movement is a lack of clarity about the theoretical and political commitments underlying specific reform proposals. (p. 3)

Since its beginning, science teacher education has relied heavily on the academic tradition—a liberal arts model—for educating teachers. Even with its long history and widespread use, few science teacher educators address the traditional model in their work. Seldom is the model discussed in terms of its theoretical underpinnings and its historical roots. Possibly for many, the academic tradition is only globally understood.

Change in teacher education must be justifiable. Decisions are not justifiable if they are based on inadequate knowledge or processed in a manner that was not deliberative. Teacher educators in general, and science teacher educators in particular, must possess a deep understanding of what it is they are doing and why they are doing it. Without such knowledge, there is little lasting and meaningful reform.

THEORETICAL PERSPECTIVES

When preparing to take his comprehensive examination nearly three decades ago, a Ph.D. candidate in science education asked his major professor if he could expect anything on the exam related to teacher education. The response was, "No, there is nothing but opinion in that area." While there certainly has been extensive research since then that justifies preparing a chapter on research on science teacher education, nonetheless there is a sense that things are not all that different. Granted, there is no definitive picture of science teacher education that scholars in the field can claim is the one dictated by the research literature. There is a variety of approaches to the many aspects of teacher education, and a variety of ways in which

such components are put together to make a "program." But while the more reflective of the people behind these teacher education efforts may well point to a research-based rationale for their efforts, there is considerable diversity among such thoughtful people as to what constitutes the "ideal" program. This diversity of opinion is smaller, of course, than that among the public, politicians and self-appointed experts who are convinced that their personal experience and assumptions provide the needed answers, but nevertheless, the diversity is there.

An indicator of this diversity—and undoubtedly one of the reasons for it—is the lack of a consensus theoretical perspective from which to examine this arena. Nowhere is this diversity of perspectives more apparent than in reviewing the recent *Handbook of Research on Teacher Education* (1990). The chapter on structural and conceptual alternatives (Feiman-Nemser, 1990) describes five conceptual orientations found in teacher education. But the lack of consensus becomes even more apparent when one moves on to other chapters in this large reference book. Chapters prepared by other well-recognized scholars contain yet additional ways of describing alternative conceptualizations of the field. For example, one (Kennedy, 1990) discusses the strategies used by various professions (e.g., medicine, law, architecture, social work) for educating their members. She identifies two well-accepted, but strikingly different, approaches; (1) teaching an extensive body of knowledge and (2) teaching the students to think. The professions of medicine and law provide excellent contrasting examples for her. Other professions, including teacher education, are described as ambivalent. "They are unable to agree on a requisite knowledge base or an appropriate pedagogy for their profession. Educators in these professions lack strong cultural norms regarding what counts in their fields" (Kennedy, 1990, p. 819).

It is within this context of diverse perspectives that science teacher education research is being reviewed. Any thought of a single, obvious conceptual framework for examining this research is quickly dismissed. The beginning point for this review must be an exploration of various theoretical orientations, after which the potential value of such orientations for giving direction to research can be addressed.

ALTERNATIVE PERSPECTIVES

Given the variety of theoretical perspectives—and variety of ways of categorizing and presenting them—it is unrealistic to expect that a discussion of alternative perspectives will be exhaustive. Such a discussion, however, can be illustrative of this variety.

In her review of structural and conceptual alternatives, Feiman-Nemser (1990) surveyed five conceptual orientations for teacher education: (1) academic, (2) practical, (3) technological, (4) personal, and (5) critical/social. She defined orientation as "a set of ideas about the goals of teacher preparation and the means for achieving them." She notes further that such an orientation ideally will include "a view of teaching and learning and a theory about learning to teach." And even though these conceptual orientations are not synonymous with particular forms of a teacher education endeavor, she expects that "such ideas should give direction to the practical activities of teacher preparation such as program planning, course development, instruction, supervision, and evaluation. (p. 220)." A description of these five conceptual orientations is instructive.

The *academic orientation* focuses on transmitting knowledge and developing understanding. This orientation is not uncommon in science education circles because of its traditional association with liberal arts education, subject-matter specialization, and the teacher as intellectual leader. It tends toward emphasizing the subject-matter background of the teacher and favoring images of teaching that include didactic instruction, teaching how to think, inquiry, and the structure of the discipline. Teacher education is seen more in terms of developing a strong subject-matter background than in learning pedagogical skills.

The academic orientation recently has acquired an enhanced formulation from the work of Shulman (1986a; 1986b). His focus on *pedagogical content knowledge* in some ways is a blending of pedagogy and content knowledge. He says that content knowledge, as learned by the prospective teacher in undergraduate science classes, is not adequate for teaching. Prospective teachers must acquire an understanding of alternative ways of conceptualizing science knowledge as well as what makes it easy or difficult for various types of students to learn particular science concepts.

The *practical orientation* focuses on the craft aspect of teaching, (i.e., the techniques, artistry and skills of teaching). Just as the academic orientation tended to emphasize the subject matter preparation of the teacher, this orientation tends to focus on experience in the classroom as the source of learning to be a teacher. Thus, it commonly is associated with various forms of apprenticeship systems of teacher education. The novice learns what works in the "real world" and gains the needed practical skills by working with a master teacher. It is sometimes criticized for its tendency to perpetuate the status quo rather than foster reflection. The risk is that the beginning teacher will simply imitate unimaginative practices in the schools and not reflect on what is experienced there, although the practical orientation need not be this limited. For example, Schon (1983) advocates professional education that includes a reflective encounter with practice.

The primary goal of teacher education with a *technological orientation* is to produce "teachers who can carry out the tasks of teaching with proficiency. Learning to teach involves the acquisition of principles and practices derived from the scientific study of teaching. Competence is defined in terms of performance" (p. 223). This orientation draws heavily on the results of the research on effective teaching that came to the fore in the 1980s.

The manner in which the technological orientation is put into practice varies considerably. Some teacher educators with this orientation, for example, use specific teaching behaviors—identified by the research on effective teaching—both as the content of their program and the performance criteria for assessing success. This orientation obviously includes the competency-based teacher education endeavors, which gained recognition a generation ago and are getting renewed attention in some of the current educational reform efforts.

The *personal orientation* focuses on the teacher-learner as the center of the educational process. Learning to teach is construed as a process of learning to understand, develop, and use oneself effectively. The teacher's own personal development is a central part of teacher preparation . . . . Advocates of the personal orientation favor classrooms in which learning derives from students' interests and takes the form of active, self-directed exploration. They emphasize concepts like readiness and personal meaning and appreciate the interconnections of thinking and feeling. Different versions of the personal orientation draw their rationale and guiding principles from developmental, humanistic, and perceptual psychology (p. 225).

The nature of the personal interactions among the teacher-learners, teacher educators and students is at the core of developing a quality program of this orientation.

The *critical/social orientation* combines a radical theory of teaching and learning with a vision for a new social order. Both educator and political activist, the teacher sought in programs of this orientation is one who works to remove social inequities, promote democratic values in the classroom, and foster group problem solving among students. Practices found within such teacher-education programs are quite varied and not unique to these programs alone. What ties them to this orientation are the purposes for which they are used in these teacher preparation endeavors—preparing teachers to change society.

These five conceptual orientations of Feiman-Nemser (1990) are but one way of describing alternative theoretical perspectives on teacher education. Another example is the dichotomy made by Eisenhart, Behm, and Romagnano (1991) between learning to teach as developing expertise and learning to teach as a rite of passage. Based on their examination of the teacher education literature, they concluded that two kinds of theories were "particularly prevalent." The one, developing expertise, bears considerable resemblance to Feiman-Nemser's technological orientation. The other, rite of passage, somewhat resembles Feiman-Nemser's practical orientation with its emphasis on the craft of teaching, but appears to give more attention to the total culture of the institution into which the novice-teachers are moving.

Another dichotomous way of looking at alternative approaches is the previously mentioned distinction made by Kennedy (1990) between two general strategies of professional education (i.e., strategies of educating individuals in some form of professional practice). One strategy is "to develop, codify, and give to students as much knowledge as possible"; medicine and engineering are the archetypical examples. In the other strategy, the goal is "to prepare students to think on their feet, giving them both reasoning skills and strategies for analyzing and interpreting new situation, until they are sufficiently flexible and adaptable to accommodate the variety of situations they are likely to encounter" (Kennedy, 1990, p. 813) The leading examples of professions that emphasize this goal are law and architecture.

While there is a strong consensus in the fields just mentioned as to the strategy that will be followed in the education of new members of the profession, Kennedy identifies teacher preparation as one of the fields in which there is no consensus. There is ambivalence as to what this education should be. "Educators in these professions lack strong cultural norms regarding what counts in their fields." Examples can be found of programs of teacher education which have remarkably different strategies. Kennedy argues that this ambivalence is not productive and results in endless debates and a tendency toward "overly abstract and idealized characterizations of the work" (p. 821).

A final example of a theoretical conceptualization of teacher education is the set of five models of staff development used by Sparks and Loucks-Horsley (1990) in discussing inservice education and the related educational improvement being sought:

Individually guided staff development is a process through which teachers plan for and pursue activities they believe will promote their own learning. The *observation/assessment* model provides teachers with objective data and feedback regarding their classroom performance. This information may be used to select areas for growth. Involvement in a *development/improvement process* engages teachers in developing curriculum, designing programs, or engaging in a school-improvement process to solve general or particular problems. The *training* model, which could be synonymous with staff development in the minds of many educators, involves teachers in acquiring knowledge or skills through appropriate individual or group instruction. The *inquiry* model requires teachers to identify an area of instructional interest, collect data, and make changes in their instruction on the basis of an interpretation of those data. (p. 235)

Not surprisingly, one can see similarities between some of the categories here and those of other authors mentioned earlier.

In spite of such similarities, however, it is apparent that there is no consensus as to the categories that should be used to describe alternative theoretical orientations for teacher education programs. No discussion of alternative theoretical perspectives can be exhaustive; it can only be illustrative of the ways in which different scholars view the situation. Feiman-Nemser (1990) argues that "a plurality of orientations and approaches exists because people hold different expectations for schools and teachers and because, in any complex human endeavor, there are always more goals to strive for than one can achieve at the same time" (p. 227).

There probably are additional reasons as well. Different scholars have different intellectual perspectives, or ways of "looking at the world." Some have a psychological perspective, (i.e., they see the world through the eyes of a psychologist). Others have a sociocultural perspective; they interpret a given situation from a social and cultural perspective. Others have a philosophical perspective. Still others have a subject-matter perspective, (i.e., they look at the situation from the standpoint of the subject matter itself, in this case, science). While there is a need for scholars to examine a given topic from a multiplicity of intellectual perspectives, the reality is that in this age of specialization, the tendency is for most scholarly analysis to be narrow rather than broad and synoptic.

Whatever the reasons in a given situation, it is clear that there is no consensus, various category systems are overlapping, and many theoretical perspectives on teacher education will demand our attention at times. Having acknowledged this lack of consensus, one still is faced with the question of what set of categories is most useful and just how they are of value to the person studying the field or attempting to develop or modify a functioning program.

From the standpoint of program development or improvement, it may be well to accept any such orientation as viable, assuming the persons responsible for the program have an internally consistent perspective. After all, any program that is truly a program—not, for example, just a collection of courses—has certain goals and means of reaching them. A description of its conceptual orientation is an expression of what these goals and means are. All that is being asked for here is a recognition that a particular orientation exists, a request that it be used for communication purposes and a suggestion that it be used to aid reflection and progress. Given the lack of consensus on theoretical perspectives, no expression of orientation is a definitive expression of what a teacher education program should be. Making the expression explicit, however, should highlight the issues that should be considered and aid analysis and deliberation.

This benefit applies in guiding and interpreting research as well as in developing and improving programs. But once again the variety of conceptual orientations must be taken into account. In the discussion of research that follows, a variety of such perspectives is needed. It will not be possible to adopt one theoretical orientation for interpreting the research on science teacher education.

Given our interest here in a specific academic field, (i.e., the natural sciences), it may seem that we should adopt the academic orientation as described by Feiman-Nemser as our framework for analyzing the research on science teacher education. Such an approach, however, would be exceedingly shortsighted. It offers some important insights, especially from the standpoint of pedagogical content knowledge, as Shulman defines it. But as a total framework for viewing this body of research, it is completely inadequate. In addition to looking at science teacher education from the standpoint of the subject matter, we also must look at it from psychological, sociocultural, and philosophical perspectives, among others. Thus, our look at this body of research probably will be more complex than simple.

THE PARTICIPANTS IN SCIENCE TEACHER EDUCATION

The educational reform endeavors of the last decade focused attention on teachers and the quality of their work. In their attempts to improve education, reformers identify the quality of teachers entering the profession as a major consideration, if not *the* major consideration. Thus, it is not surprising that close attention has been turned to the supply and demand for science teachers, the characteristics of people entering the profession, and the standards they are to meet after completing preparation programs. These interrelated matters shape the discussion here of the participants in programs of science teacher preparation.

Science Teacher Supply and Demand

The supply of and demand for science teachers has been an active area of investigation in recent years. Most of this work has involved use of rather extensive sets of data in established data bases, such as those of the National Center for Education Statistics, and in the records of such agencies as state teacher certification offices. A few case studies of a qualitative nature provide a valuable context for interpreting the more quantitative data.

A large proportion of this work has been done by (1) Richard Murnane and his colleagues, as reported in a number of journal publications and most recently in their book, *Who Will Teach?*; (2) the Panel on Statistics on Supply and Demand for Precollege Science and Mathematics Teachers of the National Research Council (NRC); (3) the National Center for Education Statistics (NCES) with data from some long-standing data bases, some commissioned papers, and, most significantly, the development of a new data base known as the Schools and Staffing Surveys (SASS) initiated in 1988; and (4) Linda Darling-Hammond and others in various studies conducted at the RAND Corporation. The work of these groups tends to be interrelated in that, for example, Murnane is a member of the NRC panel and the panel has focused a considerable amount of attention on the data bases maintained by the NCES.

A reading of the works in this area quickly shows that the data is not definitive; in fact, a conflicting picture emerges. It is apparent that this matter is complex and requires more study and interpretation.

For more information, the following three sources are recommended: (1) *Who Will Teach?* (1991) by Murnane et al., (2) the final monograph of the NRC study panel entitled *Precollege Science and Mathematics Teachers: Monitoring Supply, Demand, and Quality* (1990), and (3) the 1990 *Review of Research in Education* chapter by Darling-Hammond and Hudson with the same title as the NRC monograph. In addition to reporting original data, each provides interpretations of research in the context of a broad body of cited literature. Together, the three give a good picture of this arena.

Understanding Science Teacher Supply and Demand

To determine and understand teacher *supply*, we look at the rate at which teachers leave the profession and their reasons for leaving; the number of people completing teacher-preparation programs; and the number of people reentering the profession after an absence, among other factors. To determine and understand the *demand* for science teachers, we look at high school graduation requirements, the number of students selecting various courses, pupil-teacher ratios, and teacher attrition rates. Darling-Hammond and Hudson (1990) note that different data

sources give differing pictures of supply and demand. For example, college placement officers and state offices have been reporting shortages of science teachers while the National Center for Education Statistics and others have published data indicating negligible shortages. Different definitions of what constitutes a shortage and differing interpretations and assumptions, with respect to projections, account in part for this conflict.

In sorting out this conflicting picture, two important distinctions must be drawn in any consideration of the supply of and demand for science teachers. The first is a distinction between biology teachers on the one hand and physics and chemistry teachers on the other. In the literature, "shortage" of science teachers usually means a shortage of physics and chemistry teachers, not biology teachers. The second needed distinction involves the *quality* of teachers. Clearly, the question is not "can a science teacher be found?" but rather is a *qualified* teacher available? What then are the measures of quality that apply to science teachers? As the NRC report (1990) notes, we do not have a very definitive picture of what constitutes quality with respect to teachers or teaching. The NRC Committee argues that quality would best be defined in terms of what enhances student learning outcomes. However, it notes that assessing quality in these terms is extraordinarily difficult. Meta-analyses of research (Anderson, 1983; Druva & Anderson, 1983) show a positive relationship between student learning and such teacher variables as the amount of training in both science and professional education, but the magnitude of these correlations is very low. Nevertheless, these correlations are higher than those dealing with all other teacher characteristics. The quality measures that generally have been examined in the previously mentioned studies of teacher supply and demand are subject-matter background, certification status, college major, and intellectual ability as measured by test scores.

These measures of quality must be applied to not only people leaving teacher education programs, but also persons in the *reserve pool*. In this group are persons returning to teaching after interrupting their teaching careers or who completed their preparation program at some time in the past but did not enter the profession immediately. Murnane et al. (1991) cite a number of studies indicating that the ratio of teachers hired who come from the reserve pool is in the neighborhood of three out of four (p. 78), a strikingly higher figure than conventional wisdom.

What are the factors determining demand? Although not the only ones, 3 factors stand out: (1) pupil-teacher ratios, (2) student enrollments, and (3) teacher attrition rates. Each of these factors is subject to many influences. In a time of tight budgets, for example, pupil-teacher ratios can be expected to increase somewhat with a resulting reduction in demand for teachers. Projecting, or even tracking, such changes in 15,000 school districts is no small task and it illustrates the difficulty of studying the supply and demand of science teachers. The second factor, student enrollments, appears somewhat easier to project; good demographic data on the size of each school class exists even before these students enter school. But other influences can affect projections, such as the generally increasing high school graduation requirements in science during the 1980s. The third factor, the rate of teacher attrition, is influenced by many variables that are related to each other in a complex manner. Murnane et al. (1991) summarize the predictors of career duration as follows:

- Teachers are most likely to leave the profession during their early years in the classroom, the first year being the most risky. Teachers who survive the early period are likely to continue to teach for many years.
- Once the characteristics of the school districts into which they are hired are taken into account, white teachers are more likely to leave than black teachers.
- On average, secondary school teachers leave earlier than elementary school teachers. Among secondary school disciplines, chemistry and physics teachers are most at risk.
- Teachers with high scores on standardized tests have considerably shorter teaching careers than those with lower scores.
- Teachers who are paid the least leave the most quickly; salary in the first years on the job matters the most." (pp. 59–60)

Murnane et al. note that another important influence on teacher attrition is working conditions, a factor that does not show up in the quantitative analyses on which the above statements are based. They cite Susan Johnson's research and analysis in *Teachers at Work* (1990) as the basis for a convincing case for the importance of working conditions in the retention and recruitment of teachers.

Teacher supply and demand are closely related to the standards to which incoming teachers are held, teacher assessment and certification, and the requirements of teacher-preparation programs. Attention is turned to these matters below.

RESEARCH ON SCIENCE TEACHER SELECTION

Research on selecting science teachers is approached here from three perspectives: (1) the teacher characteristic correlates of successful teaching, (2) characteristics of the current entrants to the profession, and (3) the many facets of the processes and standards by which the flow of people into the profession is controlled.

Research on Science Teacher Characteristics

Research on teacher characteristics has a long history. It began with the optimistic expectation that certain types of people make better teachers than other types. Although the research long ago dashed this notion, there is reason to explore some facets of the topic.

A 1983 meta-analysis of research on science teachers (Druva & Anderson, 1983) included 65 studies and synthesized the research available at the time. Because further research on this topic is exceedingly sparse, that meta-analysis remains the source for an overall picture on science teacher characteristics.

The studies in this meta-analysis address characteristics (e.g., gender, coursework, IQ, and personality variables) as the independent variables and classroom teaching behaviors and student outcomes (e.g., achievement, attitudes) as the dependent variables. The correlations found were low, but those pertaining to student outcomes were even lower than those dealing with teacher behaviors; only six out of 242 cells in the matrix relating teacher characteristics and student outcomes had correlation coefficients exceeding 0.5.

In keeping with the view of the NRC panel on teacher supply and demand referred to earlier—where student outcomes were viewed as the only valuable indicator of teacher quality—the portion of the meta-analysis dealing with student outcomes is attended to in this chapter. The NRC panel found the following, though it is well to note again that the correlation coefficients involved are very low.

1. Student outcomes and teacher age are positively associated.
2. Student outcomes are positively related to the amount of preparation of the teacher, especially their science preparation, but also in professional education.
3. Certain affective student outcomes are associated with certain affective characteristics of teachers (National Research Council, 1990, pp. 477–478).

Based on the research on teacher characteristics, the following assertions are offered:

1. In terms of the *student outcomes* of K–12 students, it is difficult to predict who should be selected for entrance to a science teacher education program on the basis of any teacher characteristics. Although some teacher characteristics correlate to some degree with certain *teacher behaviors*, often thought to be desired outcomes of teacher education programs, these correlations are small and do not provide much of a basis for selection of people for admission.
2. In terms of student outcomes, certain broad characteristics of *programs* are important (e.g., the number of courses in science and education).
3. This research does not indicate much about what the nature of the programs themselves should be.

Other Teacher Factors

These rather uninspiring results, however, are not a basis for saying that certain characteristics might not be related to certain *specific* facets of teaching. Two examples of teacher attributes having such potential are locus of control and cognitive development. In general, in locus of control research related to preservice teachers, internally oriented individuals exhibited an enhanced capacity for applying scientific inquiry skills (Horak & Slobodzian, 1980) and a greater capacity for predicting an understanding of the nature of science (Scharmann, 1988).

In studies on cognitive development, Lawson, Nordland, and DeVito (1975) found that formal reasoning was significantly related to preservice teachers' achievement, aptitude, attitude, and knowledge of the science processes. Sunal and Sunal (1985) found that formal reasoning preservice teachers were significantly higher in performing model classroom teaching behaviors during a field-based science methods course than their transitional or concrete counterparts. As a result of these findings, and in light of the reports that a substantial number of college students do not demonstrate formal reasoning abilities (Lawson et al., 1975; Sunal & Sunal, 1985), some science teacher educators advocate preservice programs which include strategies to facilitate growth in formal reasoning (Lawson, 1982; Sunal & Sunal, 1985; Wright, 1979). So, even though these factors may not provide a basis for teacher selection, research on them may have some promise for understanding and guiding teacher education.

Characteristics of Persons Currently Entering Teaching

Numerous questions in the policy arena have focused attention on certain characteristics of persons entering teaching today. Because of assumptions about the importance of these characteristics, many policymakers are concerned about such characteristics as age, gender, ethnicity, intellectual ability, and subject-matter background. As a result, data has been collected that describes the persons entering the profession today.

Age. Based on data from two states, Murnane et al., (1991) point out that the percentage of older persons entering the teaching profession has increased during the last decade (pp. 24–27). In both states they found an approximate doubling of the percentage of new licensees who were older than 30 years.

Gender. While data indicates that more teachers are women than men, the picture for science teachers is somewhat different. Weiss (1987) reports that 56 percent of science teachers who teach grades 7 through 9 are men and 68 percent of grade 10–12 science teachers are men.

Ethnicity. The percentage of new entrants to the teaching profession who are ethnic minorities has declined during the past two decades. At one time a higher percentage of black college graduates, for example, entered teaching than the percentage of white college graduates; this description no longer fits. Weiss (1987) reports that the percentage of minorities among secondary school science teachers is 8 percent while the percentage of minorities among the student population is approximately 30 percent. The trends just described lead Murnane et al. (1991) to note that "if this trend persists in the 1990s, the nation's teaching force will contain an ever-declining percentage of minority teachers at a time when minority student enrollments are rising dramatically" (p. 33).

Intellectual ability. Historical data indicate that college graduates with higher academic ability have been less likely to enter teaching than their peers of lesser ability. Murnane et al. (1991), based on national data, found that this pattern persisted, and in fact was accentuated, over a 14-year period ending in 1980. They found that by 1980, "a graduate with an IQ score of 100 was more than four times as likely to enter teaching than was a graduate with a score of 130" (pp. 35–36). The pattern of teach-

ers tending to be drawn increasingly from those with lesser ability held across gender and ethnicity lines. This area is one in which the situation may be changing, however; during the 1980s, many states raised grade point average requirements and other criteria for admission to teacher education programs and for certification.

Subject-Matter Background. Indicators of the subject-matter background of science teachers are simply that: indicators, not absolute measures. Among the indicators available are certification status, college major, and number of courses taken in science. One indicator is state certification in the subject taught, a criterion that varies in validity depending on the state, but nevertheless an indicator. Darling-Hammond and Hudson (1990) report that "it appears that significantly more senior high than junior high teachers are certified. In ballpark terms, it appears that roughly 80 to 90% of high school teachers are certified in their subjects, compared to 65 to 75% of junior high teachers." (p. 245)

Possibly the best descriptive information on science background is provided by the national survey done by Weiss (1987). Using the standard of the National Science Teachers Association that a junior high school science teacher should have a minimum of 36 semester hours of college science credit and a senior high school science teacher should have a minimum of 50 hours, Weiss concluded that 43 percent of U.S. junior high science teachers met the standard and 57 percent of senior high science teachers met the standard at that level.

Selection Procedures and Standards

In the political climate of the past decade, much attention has been focused on questions concerning the quality of persons entering the teaching profession. As a result, various agencies have taken specific actions designed to raise this quality. Along with this rush to action, we need to bring to bear the power of systematic inquiry and careful analysis. We need to explore questions about the criteria imposed, how the selection process functions, and who controls it. These questions about science teacher selection must be addressed in the broader context of the general selection of teachers of all fields. Changes in selection criteria and processes generally have been applied across the board to all subject areas and levels.

Selection criteria can be applied either at the beginning of teacher education (entrance requirements), at the end (exit requirements) or at both points. These teacher education exit requirements, of course, also could be thought of as requirements for entrance to the teaching profession.

Many of the teacher characteristics identified earlier (i.e., academic ability, basic skills, subject matter background, and ethnicity) can be addressed as entrance requirements because in most teacher education programs, they are taken as givens, not as something the program is likely to change. The exception is the subject-matter background of the teacher because in many cases, it is a program goal. In other instances, of course, it is expected that this background will be essentially complete at entrance to the program.

Exit criteria generally have more to do with what is learned in the teacher-preparation program. Specifically, what those criteria are and how they are assessed are major issues and lead to consideration of questions of such matters as licensure, certification, and professional assessment approaches. These topics are beyond the scope of this review, but the reader interested in these matters is referred to Roth and Pipho (1990) and Andrews and Barnes (1990).

Who Selects?

Selection criteria are applied by local hiring officials, of course, but much of the selection is exercised by agencies of state government and to some extent by national accrediting groups. Recent changes in practice largely have occurred within state agencies.

Although state control has increased in the last decade, it has a history that reaches back to the beginning of this century. This history, along with detailed attention to changes initiated in the 1980s, is well described in the chapter on "Teacher Education Standards" by Roth and Pipho in the *Handbook of Research on Teacher Education* (1990). State processes evolved into ones in which a teaching certificate generally was awarded on the basis of completing an approved program of teacher education. The educational reform movement of the 1980s, however, brought heightened attention to competency tests, increases in minimum grade point average requirements, and changes in the coursework required in various programs.

These changes have been accompanied by state promotion of alternative certification programs, which generally move some of the control toward the local level. In addition, national accreditation standards established by the National Council for Accreditation of Teacher Education (NCATE), along with cooperating professional associations, have an influence; and the recently established National Board for Professional Teaching Standards has the potential of playing a significant role.

What Are the Selection Criteria?

Possible selection criteria are many, including general intellectual ability, subject-matter background, pedagogical subject-matter competence (Shulman, 1986a), professional knowledge, and professional skills.

Although a relationship between general intellectual ability and teaching proficiency has not been clearly documented (Roth & Pipho, 1990), efforts have been made in the past decade to select teachers with greater intellectual ability. Higher standards have been set including higher minimum college grade point averages and scores on standardized tests. Although these higher standards are often promoted on the basis of getting teachers with greater subject-matter knowledge, the increased attention to the general intellectual ability of teachers (e.g., Murnane et al., 1991, pp. 35–36) would indicate that greater intellectual ability is an important reason for establishing these higher standards.

Greater subject-matter competence was sought in the 1980s through higher minimum grade point averages, new course

requirements for certification, and various testing programs. While some reviewers of the general teacher education research literature assert that there is little evidence to support that knowing one's subject in depth makes one a better teacher, some of the evidence cited earlier indicates that greater knowledge of science makes one a better science teacher (Anderson, 1983; Druva & Anderson, 1983). Although the correlations found in this research were admittedly low, they lend some credence to the attempts to increase subject-matter background of science teachers. On the other hand, limiting selection to such criteria seems short-sighted at best and more sophisticated understandings of subject-matter knowledge, [e.g., pedagogical subject matter (Shulman, 1986a)], are of increasing importance.

Selection on the basis of the amount of professional training has waned in the 1980s with (1) the promotion of alternative teacher certification, (2) restrictions in some states on the number of education courses in teacher preparation programs, and (3) the requirement in some places of a major in a subject field rather than education for elementary teaching. Nonetheless, research does provide some support for professional training for science teachers in the form of positive correlations between the number of hours of professional education and teaching effectiveness (Anderson, 1983; Druva & Anderson, 1983). Teacher education in general has such supporting evidence (e.g., Roth & Pipho, 1990) as well. However the correlations are quite low as in the case of most other correlates of teaching effectiveness.

Teacher Performance Assessment

One approach to selection of teachers is the use of some assessment of actual teaching performance. An example of a performance assessment approach is one used in Connecticut. While the centerpiece of their licensing process is performance assessment, it begins with initial screening that includes an essential skills test and a measure of subject-matter knowledge, namely the appropriate National Teacher Examination (NTE) specialty area test. In the first year of teaching, classroom observations are made six times by professionals from outside the employing district using an instrument encompassing 10 specific teaching competencies. In addition, the teacher undergoes a semistructured interview in an assessment center that probes a number of areas of teaching skills (Murnane et al., 1991, pp. 110–111).

Consideration of assessment systems also focuses attention on the different theoretical perspectives on teacher education discussed earlier. While it is clear that the Connecticut example given above, with its attention to 10 specific teaching competencies, reflects a particular orientation to both teacher education and assessment, a more holistic approach to assessment of performance would also be viable if one had such an orientation to teacher education. The important questions about reliability and validity of the assessments would still be present, even though they would take a different form in the context of a different theoretical perspective on teacher education.

An extensive exploration of the topic of performance assessment of teaching is beyond the scope of this chapter. For a thorough exploration of the research on this topic, the reader is referred to the chapter on "Assessment of Teaching" by Andrews and Barnes in the 1990 *Handbook of Research on Teacher Education*. As they note, there are a number of assessment programs currently in use or development including the work of the National Board for Professional Teaching Standards.

Alternative Teacher Education Programs

A concern about the shortage of qualified teachers—particularly in the areas of science and mathematics—in the educational reform climate of the 1980s, led to the initiation of numerous alternative certification programs. They are addressed here briefly because of their relationship to assessment programs. Now found in approximately one-half of the states (Hawley, 1990; Roth & Pipho, 1990), these programs generally are designed for persons having a bachelor's degree with a major in the field in which they intend to teach. These programs provide a limited amount of professional preparation before the person takes a regular teaching position. These programs supply some training and on-the-job support. They do not yield full certification until after some considerable time teaching (Feiman-Nemser, 1990; Roth & Pipho, 1990). While overall, the number of teachers entering the profession by this route is relatively small, there are substantial numbers in a few urban areas and the numbers are growing (Hawley, 1990).

There is an obvious relationship between such alternative programs and performance assessment, in that the latter potentially is a means of determining the degree of success of such alternative preparation. Some evaluations of the success of these alternative programs of teacher preparation have been completed, and there is some basis for making judgments about their success.

Murnane et al., (1991) cite data indicating that, on the previously mentioned performance assessment in Connecticut—75 percent of teachers in alternative programs passed all 10 dimensions of the Connecticut performance assessment, compared to 88 percent of the graduates of regular teacher education programs. While the difference favors the conventional programs, the difference is not dramatic. This comparison is not one of teacher education versus no teacher education, of course; it is a comparison of conventional teacher education versus an alternative form that is shorter and of a somewhat different character. In their review of research studies that compare teachers having regular teacher preparation with no teacher preparation, Roth and Pipho (1990) claim that the studies "have shown on a variety of measures that those who have teacher education *are* better prepared to teach" (p. 132). They go on to claim that the studies done of alternative certification programs in many states provided a basis for judging that their teachers were "as effective in their initial teaching as were teachers prepared in conventional programs" (p. 143).

Murnane et al. (1991) argue that given the similarity in performance of teachers coming from alternative and conventional programs, judgments about the effectiveness of alternative programs need to be made on additional grounds as well. They believe, for example, that given the somewhat higher quality of people entering alternative programs (on measures of intellectual ability and academic preparation in the data available to them) and the substantially higher percentage of minority teachers entering teaching by this route, that these alternatives serve an important role and should remain as *one* of the routes into the profession.

Other Reform Efforts

The educational reform efforts of the last decade have included considerable attention to teacher education and the selection of people for these programs. The Holmes Group (1986) recommendations appear to be motivated partially by a desire to upgrade the quality of people entering the profession, although the means of doing so largely has been through changing program requirements rather than selection criteria *per se*. The National Board for Professional Teaching Standards (1989), of course, is focused more directly on standards and the means of selecting people for the profession that meet appropriate and high-level standards. In a certain sense program requirements set a standard for entry into the profession and are part of the selection process. In this regard, accrediting groups such as the National Council for the Accreditation of Teacher Education (NCATE) serve a gatekeeper function which limits teacher education graduates to those having certain training in professional education and in their subject field. The specific requirements of subject field groups (e.g., the National Science Teachers Association, 1984) have been incorporated into the NCATE standards and are exerting influence as well.

The selection of people entering the teaching profession is an important consideration, as is the teacher education received by these persons. It is that topic to which attention is now turned.

PRESERVICE SCIENCE TEACHER EDUCATION: THE CURRICULUM

Having addressed a variety of theoretical perspectives and characteristics of people entering science teaching, we turn our attention to the programs which provide the initial preparation of people for the profession. The content and structure of such programs will be addressed first. Attention then is directed to the research on the instruction that takes place in such programs and the resulting student outcomes. In both cases, the focus is on preservice education—whether done as undergraduate, postbaccalaureate nondegree, or graduate work. Inservice education is addressed in a later section entitled "Inservice Education."

Much discussion surrounds defining the word "curriculum" and many definitions exist (see Jackson, 1992; Zumwalt, 1989). In dictionaries, curriculum is simply defined as "a course of study" offered at an institution. In current practice, curriculum commonly refers to a list of courses and topics to be taught, often with instruction recommendations. In this text, however, curriculum, is viewed from a broader perspective. Although program characteristics, such as a teacher education model and a course sequence, are addressed in this discussion, curriculum as it is used here, speaks to more of a "curricular vision" of education and teacher education (Zumwalt, 1989). It is an orientation that derives from an understanding of the complexity of classroom practice (see Schwab, 1983). This understanding views curriculum in a comprehensive manner that acknowledges the interactive, contextual, and evolving nature of teaching and learning, as well as the identities of the individuals involved.

Teacher education is discussed here in the context of the traditional ubiquitous model of an undergraduate curriculum the model which has served as the framework for teacher education in the United States since the turn of the century. This starting point reflects the belief that understanding the broader area of teacher education will enrich one's understanding of the more specific area of science teacher education.

This section has five subsections. The first defines and elaborates on the traditional preservice model used most commonly in preparing science teachers. The second addresses studies of such programs (i.e., evaluations of them), including consideration of alternative theoretical perspectives. The third section critiques the traditional model of science teacher preparation with particular attention to alternative theoretical perspectives. In the fourth section, some examples of traditional programs are provided with consideration of their theoretical orientation. In the last section, some breaks from the traditional models are explored along with consideration of the extent to which they come from different theoretical perspectives.

The Traditional Model

Teacher education in the United States, in general, has a reputation for doing a poor job in preparing teachers. In particular, a fair amount of this criticism has been targeted at the traditional model. This review does not provide evidence to deny this popular assessment. At the same time, it does not support claims that the traditional model itself is the cause for such a condition. In any case, all aspects of the traditional teacher-education model need to be examined further and possibly revised. But it is also the case that this traditional model represents years of both scholarly inquiry and adjustments to the constraints of "real-world" practice. Therefore, although the word "traditional" sometimes connotes a sense of inflexible and outmoded, the word as used in this context could also connote "venerable."

The word "model" as used here refers to the structure and organization of this traditional approach to teacher education, without addressing directly the conceptual or theoretical orientation of such programs. In fact, programs can be operated within this traditional framework while having quite different theoretical orientations. As stated earlier, building an entire "program" on a clearly identifiable theoretical orientation is not commonplace.

Howey (1983) claims that teacher-education programs across the nation are quite similar:

Initial training or teacher preparation programs across the country tend to appear quite similar at least in terms of the number and general types of experiences they afford students and the structure and framework in which these are organized .... The number of hours future teachers spend in professional studies and related foundational work has not significantly increased. Neither has the basic format for study. (p. 11)

We begin our discussion of preservice science teacher education by focusing on this "basic format for study"—the traditional curriculum for the preparation of all teachers nationwide. After providing a brief history, the model is discussed in terms of its three strands: general education, subject-matter preparation, and professional education.

A Brief History. Structured teacher education began in America with the introduction of common schools in the early part of the nineteenth century in New England. Formalized education for students in these common schools created a need for formalized education for the teachers of those schools. This need led to what became known as the normal school where teachers were prepared to teach. The first public normal school was begun in 1893 in Lexington, Massachusetts (Urban, 1990).

Thus, around the turn of the century, normal schools became the primary mode of educating elementary teachers. The curriculum was directed at preparing elementary teachers for classroom work through a 2-year course of study (Feiman-Nemser, 1990). Urban (1990) quotes a reference on the normal school's curriculum for teacher education:

Work in academic subjects at normal schools was largely in subjects that made up the curriculum of the elementary school. Students needed to show a "mastery of reading, writing, spelling, geography, grammar, and arithmetic for admission to the regular professional courses" (Pangburn, 1932, p. 14).

Students who did not demonstrate a mastery would study those subjects at the normal school until they had mastered them. Work in the professional sequence "consisted of thirteen weeks in the History of Education, twenty-seven weeks in the Science of Education, thirty-one weeks in Methods in the Elementary Branches, and twenty weeks in Mental Science, a week being defined as forty-five minutes a day for five days" (Pangburn, 1932, p. 14). Practice teaching in a school, preceded by significant amounts of observation, was also standard in the normal schools. (p. 62)

In general, the normal school curriculum focused on technical teaching skills and mastering content at the elementary school level. While normal schools were educating elementary teachers, universities became involved with preparing secondary school teachers in a manner more in keeping with the liberal arts tradition. Historically, secondary teachers were the products of their university majors.

When did the traditional model originate? Feiman-Nemser (1990) says that according to Cremin (1978), the traditional model for undergraduate teacher education dates back to around the turn of the century and consisted of 4 strands: (1) general culture, (2) special scholarship, (3) professional knowledge, and (4) technical skill. Today, the same general structure has remained, and the essence of those four strands continues, but in a somewhat modified form. In what is now the traditional model, the four strands have become three: (1) general education, (2) subject-matter concentration, and (3) professional education.

General Education. General education, that body of knowledge considered to be at the core of an educated person, is typically the common foundation of study for most college students. Historically known as the seven liberal arts (logic, grammar, rhetoric, arithmetic, geometry, astronomy, and music) and later referred to as general education by the Harvard Committee on General Education (1945), this area of study commonly comprises three general categories: the social sciences, the natural sciences, and the humanities, although various groups and individuals would categorize the area in somewhat different ways.

While the general education component of a student's education is regarded as the guarantee that he or she is becoming a liberally educated person, most professional educators agree that general education is too vaguely defined to live up to such a claim (Conant, 1963). Feiman-Nemser (1990) cites:

The dominance of a general-education sequence reflects widespread agreement about its importance. In practice, however, general education is more like a supermarket where students make independent choices from a wide array of offerings. Rarely does it provide broad cultural knowledge or deep flexible subject-matter understanding. (p. 217)

The general-education portion of the teacher-education program is often criticized. There is little research, however, on how the general-education requirements relate to educating teachers.

Subject Matter Preparation. The importance of teachers knowing their subject is at the center of teacher education. Teachers' knowledge of their subject is critical in shaping their curriculum and pedagogical decisions (Grossman, Wilson, & Shulman, 1989; Hashweh, 1987). A teacher's own knowledge of a subject will enhance or limit the opportunities a student has to learn that subject (McDiarmid, Ball, & Anderson, 1989). Shulman (1986a) lists three categories of content knowledge for teachers: (1) subject matter content knowledge, (2) pedagogical content knowledge, and (3) curricular content knowledge. Most references in the literature to content knowledge address Shulman's first category and it is in that sense that the label is used in most of what follows.

Teachers' subject-matter preparation begins at the precollege level, expands in undergraduate school, and continues during the early teaching years (Feiman-Nemser, 1983). At the undergraduate level, most subject matter preparation occurs outside of the school of education. Beyond general education requirements, students preparing to be secondary teachers spend a large portion of their studies in their particular departments within colleges of liberal arts and sciences. They follow a course of study that, in most cases, is no different from the requirements for non-education students majoring in that discipline.

Although elementary education students generally take more education courses than their secondary counterparts, their situation is similar. Much of their subject matter education also occurs in the academic disciplines in the university, but without the depth provided in the secondary program. Elementary education students' subject-matter preparation usually consists of a composite of academic minors in subjects directly related to the elementary school curriculum. Typically, their studies consist of a collection of introductory subject-area courses designed for field majors or survey courses designed for non-majors.

There are many criticisms of the science course requirements for prospective teachers. At the elementary level, the haphazard collection of science courses is considered a problem. In some colleges and universities, special content courses are developed and taught specifically for elementary majors, but often these courses are said to be watered down and inadequate. At the secondary level, the science courses taken by prospective science teachers are sometimes said to convey an image of scientific inquiry that is not consistent with reality.

Historically, subject-matter knowledge has been equated with the completion of an undergraduate degree that showed evidence of accumulated course credits within an appropriate discipline. Subject matter knowledge has consisted of mastery of fact-dominated information, as evidenced by objective testing. Subject matter preparation of preservice science teachers has been unquestionably the responsibility of the liberal arts and sciences faculty.

New Understandings of Subject Matter. Since the late 1980s, many programs have emphasized broader and deeper understanding of subject matter knowledge and its relationship to teacher education. McDiarmid et al. (1989) explain that the revised goal of educating teachers is for "flexible subject matter understanding":

Flexible understanding of a subject entails the ability to draw relationships within the subject as well as across disciplinary fields and to make connections to the world outside of school . . . . Flexible understanding also involves knowing about the discipline: What experts in the field do, how knowledge evolves, what the standards of evidence are . . . . Finally, flexible understanding means knowing a subject well enough to increase one's understanding of and thereby power within one's environment. (pp. 193, 194)

Grossman et al. (1989) refer to four dimensions of subject matter knowledge that affect teaching and learning: content knowledge, substantive knowledge, syntactic knowledge, and beliefs about the subject matter. The first category is the traditional one commonly thought of when speaking of studying a subject. It includes that knowledge generally specific to a certain field, complete with competing theories and continual change. The substantive and syntactic categories expand upon knowledge about the subject. Grossman et al. (1989) describe substantive knowledge:

The substantive structures of a discipline include the explanatory frameworks or paradigms that are used both to guide inquiry in the field and to make sense of data (Schwab, 1978). In some disciplines, like physics and chemistry, a dominant structure may prevail at any one time, as is reflected in Kuhn's (1970) discussion of normal science and the nature of scientific revolutions. In other disciplines, however, multiple, competing substantive structures may exist at the same time. (p. 29)

In reference to the third category, syntactic knowledge, the same authors write:

Academic disciplines do not simply consist of concepts and organizing frameworks. Disciplinary knowledge includes knowledge of the ways in which new knowledge is brought into the field. The study of physics involves scientific inquiry, the study of literature involves literary analysis, the study of art involves aesthetics and art criticism. Schwab (1978) defines this type of knowledge as knowledge of syntactic structures. In literary criticism, for example, a variety of substantive structures can exist concurrently, such as, for example, the critical traditions of New Criticism and semiotics. (p. 29)

Schwab, himself, made an appeal for syntactic knowledge to the National Association for Research in Science Teaching in his paper presentation, "Decisions and Choice: The Coming Duty of Science Teaching," at the 1974 annual meeting in Chicago. He called for an expansion in the way science educators conceptualized the field of science:

Five terms define a science, that is, bound it and indicate its necessary constituents . . . the set of five terms which define a science: a subject matter of enquiry; the field of potential enquiry from which it comes; the discipline which is brought to bear; the knowledge which results; and the community of enquirers who shape and reshape subject matter, and search and research it. (Schwab, 1974, p. 312)

Efforts to expand ways of defining subject matter continue, as do efforts to expand how these multiple definitions are presented in curriculum (Mitchener, 1991; Pereira, 1990).

The fourth dimension referenced in Grossman et al.'s (1989) model addresses teachers' beliefs about subject matter. Teachers' beliefs strong affect their teaching. In this instance, teachers' beliefs are divided into two categories: (1) beliefs teachers hold about their content and (2) the orientation teachers have toward teaching their subject matter. In the first category, the reference is to how teachers view science and what is important to teach related to it. The second category, however, is a much deeper look at how the teacher approaches science, in terms of how one comes to know it.

CONNECTIONS TO EDUCATIONAL REFORM. These beliefs have a great deal to do with attempts to bring about changes in how science is taught. The many attempts to bring about educational reform in this country require changes on the part of teachers, including changing many of their beliefs about subject matter (Anderson et al., 1992). Belief changes are at the heart of attempts to reform education; teacher education in this context cannot ignore beliefs (Anderson et al., 1992).

Historically, subject-matter preparation was the responsibility of the departments of arts and sciences. That tradition is being challenged. Grossman et al. (1989) write:

Subject matter knowledge has provoked more controversy than study. As teacher educators, we have been satisfied to leave this crucial piece of teachers' knowledge behind its opaque velvet curtain, remanding responsibility for its transmission to departments of arts and sciences . . . . Given the central role that subject matter plays in teaching, we must reexamine our assumption that the subject matter knowledge required for teaching can be acquired solely through courses taken in the appropriate university department. The preliminary results of the growing body of research exploring the relationship between pedagogy and subject matter knowledge indicate that there are several dimensions of subject matter knowledge that are particularly important to the task of teaching. As teacher educators, we must consider how best to introduce this knowledge into programs of teacher education. (p. 24)

Subject-matter preparation, now more than ever, is an area that needs to be addressed in all arenas of teacher preparation. This need has been particularly evident in efforts to broaden how subject-matter is viewed through a variety of interdisciplinary efforts by faculty from both education and the sciences. For examples of such curriculum efforts focusing on the interrelationships between science, technology, and society see Fensham's (1992) "Science and Technology." For an example of the integration of biology, chemistry, physics, and education see a model project for teaching interdisciplinary science to preservice teachers developed at Rollins College (Gregory & DeTure, 1992). The amount of research in this area is growing but still remains sparse. Little is known about what students really learn in their subject area courses, which is particularly interesting in light of the fact that students in teacher education spend the majority of their academic time taking courses in the departments of arts and sciences.

Professional Education. This section deals with the third strand in the traditional teacher-education model: professional education, or pedagogical study—that knowledge needed to practice the specific profession of teaching. It is this area of teacher education that has a history of receiving heaps of criticism from practically every sector (Bestor, 1954; Conant, 1963; Koerner, 1963). For example, in 1963 Conant attacked professional education classes because he considered them a contradiction in terms: they claim to transmit a body of knowledge specific to the teaching profession, yet "professors of education have not yet agreed upon a common body of knowledge that they all feel should be held by school teachers before the student takes his first full-time job" (p. 141). Others add that not only are teachers ill prepared, but the relevance of teachers' professional preparation to the realities and dynamics they face in the classroom is going relatively unstudied (Sarason et al. 1962). The criticisms continue today.

CHANGING NATURE OF PROFESSIONAL KNOWLEDGE. Currently, the professional education strand is changing in terms of how professional knowledge is conceptualized and defined. The education community has taken action on defining the "what" of professional knowledge, addressing Conant's call for "a common body of knowledge." Yet, Tom and Valli (1990) caution against neglecting the more substantial issues related to the "how" to think about professional knowledge:

The term "knowledge base" has become an extremely popular concept in teacher education, even to the point of having a cluster of knowledge-base standards included in the most recent revision of accreditation standards [Gideonse, 1988; National Council for Accreditation of Teacher Education (NCATE), 1987]. The presumption in these accreditation standards is that a knowledge base, actually knowledge bases, exists, and that every faculty ought to attend carefully to the knowledge bases that underlie its teacher education program (NCATE, 1987, p. 37). (p. 389)

The concern is that much of this pedagogical knowledge-base talk is "simplistic" (Tom, 1986a) in that it fails to address the underlying epistemological and theory-practice relationship issues of professional knowledge (see Tom & Valli, 1990). Although there have been moves to identify a professional knowledge base, the real change now occurring is in the deliberation about the nature of teaching and the implications for teacher-education practice. There needs to be attention to the theoretical orientations to teacher education at the heart of these deliberations. Much remains to be done to develop defensible approaches to science teacher education that take into account this emerging knowledge base.

The image of teachers, and science teachers in particular, has been changing as well. This changing image is reflected in some reforms sought in science teacher education (e.g., Roberts & Chastko, 1990; Russell, 1989). Within the last decade, the image of the teacher has shifted from that of technical expert to one of reflective practitioner (Schon, 1983, 1987). With a newer understanding of the limited nature of professional knowledge to solve every problem in real world practice, Schon (1987) cites the need for professional education to develop reflective practitioners. A reflective practitioner is one who develops a capacity to reflect in and on his or her actions in a way that "goes beyond statable rules—not only by devising new methods of reasoning, but also by constructing and testing new categories of understanding, strategies of action, and ways of framing problems" (p. 39). In this way Schon redefines the practitioner as one who constructs his or her practice: "Underlying this view of the practitioner's reflection-in-action is a *constructivist* view of the reality with which the practitioner deals" (p. 36).

Doyle (1990) talks about the reflective practitioner, specifically referring to teachers and teacher education:

Professionally trained teachers, in other words, should first and foremost be able to inquire into teaching and think critically about their work. The knowledge base for the preparation of reflective professionals includes personal knowledge, the craft knowledge of skilled practitioners, and propositional knowledge from classroom research and from the social and behavioral sciences. Within this framework, research and theory do not produce rules or prescriptions for classroom applications but rather knowledge and methods of inquiry useful in deliberating about teaching problems and practices. (p. 6)

Although there is some confusion surrounding the meaning and types of reflection (Zeichner & Liston, 1987), Feiman-Nemser (1990) cites "reflective-teaching as a generic professional disposition" (p. 221) that relates to different theoretical orientations to teacher education. In other words, advocates of many different theoretical orientations have adopted the reflective-teaching label. It appears to be a useful conception for people coming from many different orientations.

This reflective-practitioner image of teaching, however, is not totally new. Around the turn of the century, Dewey (1904) began arguing against teacher education in which "the immediate skill may be got at the cost of the power to go on growing" (p. 8). Dewey spoke of the need to educate prospective teachers in ways of thinking about teaching, as opposed to the technical aspects of teaching. This creates a knowledge derived from their own experiences in classrooms. What is new about Schon's way of thinking about teachers is that it reframes teaching as problematic in nature (Tom, 1985). Cuban (1992) takes this reframing even further by stating that teaching is filled with dilemmas, not merely problems, where the "messier" nature is more characteristic of actual schooling. In both cases, the complexity of teaching is brought into the center of the deliberation. Both cases also define knowledge, in this case knowledge about teaching, as tentative and specific to the context of a situation. And these debates redefine the image of a teacher as one who not only solves problems, but more importantly, as one who manages dilemmas (Lampert, 1984, 1985). This is quite different from the previous image of teachers as decision makers with an emphasis on effective practice.

These changes in professional education, however, are still in the deliberation and early research stages, and have not had a significant impact on the realities of actual practice. The traditional curriculum in the professional education strand remains relatively intact. It consists of university education coursework and education-related fieldwork. In terms of design, there is little variation between teacher-education programs from institution to institution (Clark & Marker, 1975; Feiman-Nemser, 1990; Howey, 1983).

Although some changes may be occurring, recent data indicate that for elementary education students this coursework makes up an average of two-fifths of their college course work, while secondary education students take only one-fifth of their credit hours in education (AACTE, 1988). The professional education curriculum for the secondary school teacher, outside of student teaching, typically includes three or four courses: an introduction to education, a course in educational psychology or psychology of the adolescent, a general methods course, and a special subject-area methods course. The elementary education student takes anywhere from seven to nine courses in the professional educational sequence prior to student teaching. These courses typically include introduction to education, educational psychology or child psychology, and 5 to 7 special subject-area methods courses (e.g., science, math). At either level, the introduction to education course is frequently omitted. And at the elementary level, special methods courses often address more than one subject area in one course (e.g., science and mathematics).

In general, professional education courses fall into three categories: educational foundations, methods courses, and student teaching, a category which includes field and laboratory experiences. Each is addressed below.

EDUCATIONAL FOUNDATIONS. Educational foundations courses serve as the formal interdisciplinary tie between education, the humanities, and the social and behavioral sciences. Historically, this category has included courses in the historical, sociological, philosophical, and psychological aspects of education. More recently, the nature of educational foundations has become more diverse in purpose and content. Educational psychology, for example, is being redefined with a new emphasis on teachers' learning and thinking (Peterson, Clark, & Dickson, 1990). Educational foundations have become more involved with equity issues and issues related to social problems, as evidenced by the growing presence of courses such as multicultural education, now second only to the philosophy of education in its frequency of offerings as an educational foundations course in the undergraduate curriculum (Borman, 1990).

In spite of the broader dimensions of educational foundations, it is clear that a psychological orientation clearly predominates, not just in the foundations courses provided but in the perspectives found in teacher-education courses generally. Psychological perspectives in teacher education programs clearly have more influence than sociocultural or philosophical ones. This psychological orientation would seem to need reconsideration, however, not just because of the many sociocultural aspects of education generally but because of the prominent role of social matters in fully understood constructivist learning and teaching.

Historically, many have viewed the educational foundations courses as the essential core of professional teacher education. Many see these courses as providing teachers with the framework to shape their professional identity. In spite of the changing nature of educational foundations courses and their historical function, they still hold a low prestige position, even in the education community. This has been evident from Conant's (1963) label of educational foundations courses as "pathetic" (p. 127) to Sirotnik's (1990) more recent description of the "serious erosion, decay, decline, or demise of foundations studies in teacher education programs" (p. 715). It is not surprising that educational-foundation courses account for few of the course credits required in teacher-education programs. Indeed, requiring more than two foundations courses in a teacher-education program is rare.

Educational foundations are being reexamined and redefined. The situation remains complex. The complexity lies in understanding how this area relates to classroom knowledge, the foundation for teacher research and practice (Doyle, 1990).

METHODS COURSES. Methods courses, which include both general and subject-specific methods, have been the most criticized component of teacher education. Conant (1963), similar to many other academic critics since, argued that general methods courses lacked intellectual rigor and duplicated material covered in other classes. Although he did see the need for special methods content, he thought this knowledge was best taught in practice teaching. In either case and for different reasons, Conant, like others, was equally negative about both types of methods courses.

Science methods courses act as the bridge between many areas of the teacher education curriculum, as well as between education and studies in the science departments. Methods courses help prospective teachers integrate knowledge and gain experience in applying this integrated learning in actual school settings with real students or in simulated environments with peers. Traditionally, these courses primarily have been the responsibility of schools of education in elementary education. In secondary education, science methods courses often are taught in science departments instead of in education. Where the methods course is taught may well have an impact on the portrayal of the teacher's role and the emphasis on subject matter.

In some cases, methods courses portray the teacher as a content expert and focus on techniques to improve the delivery of that content. On the other hand, some science methods courses emphasize the teacher as a facilitator of learning and focus on making learners active participants in the learning process.

The nature of courses primarily devoted to pedagogy (e.g., methods courses) is currently undergoing considerable rethinking. Shulman's (1986a) research is challenging the appropriateness of the knowledge bases acquired in college science classes and thus assumed for students entering science methods courses. Through the Knowledge Growth in a Professional Project at Stanford University, Shulman's work (1986a) has produced three categories of content knowledge: subject-matter content knowledge, pedagogical content knowledge and curricular content knowledge. Subject matter content knowledge was already addressed under subject-matter preparation.

Shulman's (1986a) two domains of pedagogical content knowledge and curricular knowledge are central to methods course instruction. Pedagogical content knowledge, as he defines it, is "the particular form of content knowledge that embodies the aspects of content most germane to its teachability" (p. 9). He sees pedagogical content knowledge as both an area of specialization specific to teachers and as a bridge between the traditional areas of subject matter and pedagogy. Content knowledge includes identification of the most commonly taught topics and most useful forms of representation, as well as understanding conceptions and preconceptions that students have of particular topics.

Reforms efforts are under way to reframe teacher education programs according to Shulman's research. Stengel (1992) describes Project 30, a Carnegie Foundation initiative to fund 32 institutions' efforts to reform teacher education in ways that engage the faculty of the liberal arts and sciences with that of education. She outlines how three institutions—the University of Northern Colorado, San Diego State University, and Millersville University—have incorporated the concept of pedagogical content knowledge into their teacher-education programs. It is the malleability of the concept that Stengel cites as its greatest value. Her discussion of these three institutions emphasizes that malleability.

It should be noted here that the definition of pedagogical content knowledge has been evolving over time. Some say it is unclear not only to its users but also to its developers, as well (Tom, 1992). Tom claims that Shulman himself has begun to abandon using the term "pedagogical content knowledge" and instead has been referring to the four dimensions of subject-matter knowledge (see Grossman et al., 1989, p. 24) previously outlined in the subject-matter preparation section. Others (e.g., Wilson, 1991), as Tom points out, still use the term "pedagogical content knowledge" and its original definition (Shulman, 1986a).

Another concern that has been expressed over Shulman's work is his distinction between content knowledge and pedagogical content knowledge. McEwan and Bull (1991) believe that the distinction of teachers' knowledge from scholars' knowledge is an untenable dualism and creates complications in that "it is more fruitful to view teachers and scholars as members of a community, bound together by their common intellectual vision and communicative purpose" (p. 332). Yet most agree that this attention to research on teacher knowledge and thinking and its implications for improving teaching has significant potential for teacher education.

Tom (1992) claims that Shulman's pedagogical content knowledge distinction is largely political because such a distinction disarms the old debate on whether teachers should be concerned with knowledge or pedagogy. He asserts that by combining them both into one concept, Shulman has produced a "value-added" idea. Tom (1992) also states that Shulman's concept fuels the recent resurgence of the academic-reform tradition in teacher education by building on the previous structural approach to the study of the subjects associated with the curriculum projects of the late 1950s and 1960s (Bruner, 1960). Advocating this view of teacher-education reform as it is presented, Tom believes, is in opposition to a desperately needed pedagogies "that give primacy to social purpose . . . grounded in interdisciplinary and moral considerations as well as in the disciplines" (p. 28). Stengel (1992) notes that although several schools have begun reforming their teacher-education programs around the concept of pedagogical content knowledge, other schools have been quite successful in focusing on alternative themes as means to restructuring. She particularly cites Brooklyn College's recent restructuring, led by Madeline Grumet, around the metaphor of teacher as artist, with an emphasis on the moral, caring and spiritual dimensions of teaching.

Curricular knowledge, Shulman's third category, educates the prospective teacher in understanding how to work with a particular subject in terms of aims, design, and instructional materials. This type of knowledge, one Shulman cites as being largely neglected in practice, helps teachers understand their subject in arenas larger than day-to-day teaching. It helps prospective teachers understand their subject vertically through the grade levels as well as laterally across disciplines. The work of Project 2061, as reflected in *Science for All Americans* (Rutherford & Ahlgren, 1990), and the work of the National Science Teachers Association's Scope, Sequence and Coordination Project, as reflected in *The Content Core* (1992), are examples of serious attention to this curricular knowledge.

Another way that methods courses—and some programs— are undergoing rethinking has to do with the increasing interest in teaching to promote conceptual understanding. This interest, stemming largely from a growing emphasis on constructivist learning, is evident in many areas in education: learning research, curriculum development, and teacher education. An example of an innovative program developed on the research on conceptual understanding is the Academic Learning program at Michigan State University (Feiman-Nemser,

1990). This program builds on the liberal arts model and operates within an academic theoretical orientation (Feiman-Nemser, 1990). Students graduating from this program have a deep understanding of their subject area, how students learn the subject, and multiple ways of teaching key ideas in that subject.

New understandings of content knowledge, along with new constructivist understandings of teaching and learning, have made the past decade a time of serious rethinking of what constitutes an appropriate science methods course. In fact, most of the research on preservice science teacher education has been done in or relates to science methods courses. Rather than listing all of that research in this section, that research is categorized by topic (e.g., attitude change, instructional strategies, methods course curriculum, etc.) and incorporated in upcoming sections of this review.

FIELD EXPERIENCES AND STUDENT TEACHING. Of all of the areas of the professional education curriculum, this area has received the highest endorsement as the foundation of a strong teacher-education program. Field experiences and student teaching have a long history in teacher education, dating back to 1926 and the first normal school in Lexington, Massachusetts (Cruickshank, 1984). Conant (1963) expressed strong support for this aspect of teacher education:

Let me return now to the term "laboratory experience," which refers to both the observation of children and the practical activity in the classroom carried on in conjunction with professional instruction . . . . I conclude that the effectiveness of education courses is substantially increased when accompanied by "laboratory experiences." I would argue that all education courses for elementary teachers . . . be accompanied by "laboratory experiences" providing for observation and teaching of children. (p. 161)

Lortie (1975), in his landmark sociological study, stressed the importance of preservice field experiences for learning needed practical skills as well as acquiring significant socialization.

Field experience in schools is an opportunity for students to try out who they are as professional educators and what it is they are learning in their classes. In addition, it serves as a general socialization to school contexts. Although Conant referred to field experiences as laboratory experiences, the term generally is used differently. Professionally speaking, laboratory experience (e.g., microteaching, simulations, and protocol materials) is similar to field experience, except it is usually conducted in college settings using preservice peers as students. Laboratory experiences, which were quite popular in teacher education in the 1970s and are still strongly advocated by some (Berliner, 1985), are now largely augmented and/or replaced by a greater emphasis on prospective teachers working in actual schools.

The dilemma science teacher educators face here is whether or not the experiences in the field have the focus they desire. Is it a setting in which prospective teachers can practice the new approaches they are learning in their teacher education program, or is it a setting in which they become socialized to entrenched ineffective practices? Eisenhart et al. (1991) explore this issue in a sensitive manner.

Most students go through a series of school placements in order to give them experiences with a variety of students from a variety of backgrounds in a variety of contexts. Field experiences include many levels of involvement, which range from students simply observing teachers to practice teaching. The trend in recent years has been to increase the number of hours prospective students work in schools with children, prior to actual student teaching. With many programs focusing on an average of 100 clinical hours, some education students—more at the elementary level than secondary—spend as many as 300 contact hours (McIntyre, 1983).

A major emphasis with field experience is on early involvement in schools, sometimes beginning with the prospective teacher's freshman year. Guyton and McIntyre (1990) summarize key research points related to early field experiences:

Applegate (1987) explains that many introductory education courses include early field experiences as a means of developing career commitment. Although it is not clear how early field experiences contribute to self-knowledge, it is clear that many students view knowledge about themselves as a primary outcome of early field experiences. A study by Sunal (1980) indicated that increased involvement in early field experiences improves a preservice teacher's performance of the specific behaviors modeled in the methods course. Denton (1982) found that early field experiences seem to have an effect on subsequent course achievement, rather than on courses of which the field experiences are a part. (p. 516)

On the other end of the continuum that begins with early field experiences is student teaching. Student teaching lasts an average of 10 to 12 weeks, and in most cases, it is the final component of the preservice program. Students begin with observation and eventually take over most, if not all, of the cooperating teacher's classroom responsibilities. Although student teaching is regarded as one of the most valued components of teacher education, Guyton and McIntyre (1990) emphasize that it is still "criticized for lacking a theoretical and conceptual framework, for lacking commonly espoused goals, and for not fulfilling its potential" (p. 515). Cogan (1975) writes:

student teaching generally resembles the training of journeymen preparing for a trade more than it does a supervised internship designed for professionals. (p. 209)

He cites the "lack of intense, systematic and professionally sophisticated supervision" (p. 209) as the major part of this problem. Glickman and Bey (1990) report that research on preservice supervision is substantially less than that on inservice supervision, and the research that is present is sporadic, with many areas still uncovered. They add, however, that the clinical supervision model developed by Cogan (1973) continues to be an ongoing framework for supervision in preservice teacher education.

Research on preservice laboratory and clinical experiences is generally characterized by a "lack of systematic study" (Haberman, 1983, p. 106). There are many substantial areas that require additional research, beyond administrative issues. Two examples of such areas are how to best connect classroom learning with field experiences (Allender, 1991), and how to

best avoid three common "pitfalls of experience" (i.e., the familiarity pitfall, the two-worlds pitfall, and the cross-purposes pitfall) in preservice teacher education (Feiman-Nemser & Buchmann, 1985).

Looking back, this three-pronged traditional model of preservice teacher education has survived relatively intact since its birth in the normal school. There are a lot of reasons why it probably will persist for a long time, not the least of which is that such long-term pervasive practices almost certainly are reinforced by many long-standing socioeconomic forces that resist change. The challenge facing science teacher educators today is this: how will you address in a coherent, comprehensive manner such emerging issues as new views of content knowledge, constructivist approaches to teaching and learning, and a reflective disposition to educating teachers. In addition, thoughtful science teacher educators need to attend to the theoretical orientation of their programs and how important professional issues are addressed within these orientations.

Program-Evaluation Studies

This section focuses on diverse efforts to determine the overall status and effectiveness of preservice science teacher education programs. It primarily reflects Schwab's (1974) first level of research, descriptive studies, thus including such diverse works as scholarly writings, committee reports, and descriptive and survey research.

For the first 60 years of the twentieth century, the information on science teacher education programs and their status was primarily in the form of reports issued by committees of nationally recognized science education scholars. It wasn't until around the mid-1960s that the first survey research was done in preservice science teacher education, a study which examined the status of preservice science teacher education nationally. During the 1970s and even more so in the 1980s, program evaluation studies grew in number and sophistication. This overview looks at three periods of research and highlights major themes that have emerged over the years.

Early History. The first 60 years proved to be very influential in setting standards for preservice science teacher education. Specifically, these standards include (1) a liberal arts education as the basis of science teacher preparation, (2) an overview of the sciences and their interrelationship as the subject-matter preparation, and (3) an emphasis on the laboratory as central to understanding and teaching the nature of science.

Over the years, recommendations were made to educate prospective teachers with greater insights into the nature of science. Use of the laboratory and experimentation in learning and teaching science were viewed as the primary means of offering insight into the nature of science. The laboratory was actually thrust into a central role in the United States when, in the 1880s, Harvard University decided to make laboratory chemistry an admission requirement (see Klainin, 1988). Ever since then documentation has supported and reconfirmed this central role of the laboratory in science teaching (Klainin, 1988; NSSE 1932, 1947, 1960; Shulman & Tamir, 1973; White & Tisher,

1986). Most changes designed to increase the range of science offerings in the preservice curriculum have specifically addressed offering more laboratory courses.

These 60 years, however, were filled more with recommendations than with actual changes in practice; criticism of science teacher education continued (see DeBoer, 1991). Programs to prepare science teachers continued to be cited as being narrow in focus and a haphazard collection of college credits (NSSE, 1947). With the launching of Sputnik in 1957, the criticisms intensified.

The ROSES Report. Prior to the mid-1960s, few statistics for research purposes were collected on the nature of preservice science teacher education in the United States. In 1968, the Research on Science Education Survey (ROSES) became the first major research effort to provide baseline data on the education of science teachers (Bethel, 1984; Yager & Penick, 1990).

In summarizing their report on the status of science teacher education, Newton and Watson (1968) highlighted 3 general trends. First, they conclude that the diversity among the programs was very great, no conceivable pattern was evident. Second, there was a general lack of objective evidence to support program and course decisions. And third, science educators seemed to be isolated from their colleagues at other institutions, and consequently, there was little communication about teacher preparation. Newton and Watson declared that this "chaos in the profession" was due to the inability of science educators to agree upon their goals in preparing science teachers.

A major theme that emerges from the ROSES Report is a perception that education courses generally are irrelevant and a waste of time by preservice teachers, and in many respects by their instructors as well (Bethel, 1984; Craven, 1977; Helgeson, Blosser, & Howe, 1977). Both the preservice teachers and their instructors gave a low priority rating to science methods course topics dealing with professional knowledge, learning theory, and curriculum studies. It is not surprising that the students and their instructors all rated science content and knowledge of that content as the top priority in preservice teacher education.

These perception problems precipitated some changes in practices at universities and colleges. New preservice science education programs (e.g., New Elementary Program, University of Florida; Cooperative Teacher Education Program, University of Illinois) were developed with the aim of making professional study a more meaningful part of teacher preparation (Bethel, 1984; Helgeson et al., 1977). Of the several preservice programs that were funded, Helgeson et al., (1977) note that no significant impact from these programs has been reported. In furthering research, however, the impact of the ROSES Report has been minimal. It is difficult to find literature, beyond brief reviews (Bethel, 1984; Craven, 1977; Helgeson et al., 1977; Yager & Penick, 1990), that addresses this study and its implications in a comprehensive manner. It is also difficult to find subsequent studies that utilize and build on this research or do more than parallel it.

The 1970s and 1980s. Program evaluation investigations in preservice science teacher education during the 1970s were not as high a priority as treatment studies research on inservice education. Although this changed somewhat through the 1980s,

program assessment in general does not have a commanding presence in the literature.

Yager, Lunetta, and Tamir (1979) cite two analyses that summarize science teacher education between 1967 and 1977. The first is Atwood's (1973) paper summarizing a study by the ERIC Information Analysis Center for Science, Mathematics, and Environmental Education (1973). The ERIC study identified promising practices in science teacher education. The second was Tamir's (1976) report on various international patterns in science teacher education.

Similar to the ROSES report, Atwood's (1973) findings reveal a continued preoccupation with the science methods course, science content, and the science teacher. Atwood reported six significant characteristics of science teacher education during the 1970s: (1) a near absence of a theoretical and/or programmatic framework; (2) a noted difference between valued practices and actual practices in teacher-education programs; (3) an inadequate integration of science with education and with other curricula; (4) a lack of congruence between practices advocated in teacher-education institutions and practices occurring in public schools; (5) an attempt to expand field experiences into the total teacher-education program; and (6) an apparent lack of attention to the future development of teacher education.

Atwood's reference to "a near absence of a theoretical and/or programmatic framework" is especially noteworthy. It is yet another indicator of the previously mentioned failure of science teacher educators to attend to the theoretical basis of their programs or to move beyond a conception of a program being a collection of courses to a coherent program with a sound theoretical basis.

The second analysis, reported in the Yager et al., study (1979), is a technical report done by Tamir (1976) on patterns in 1975–1976 teacher education models in the United Kingdom, Australia, Israel, and the United States. Tamir, Lunetta and Yager (1978) developed a self-assessment tool, The Science Teacher Education Inventory, which addresses these trends. Among other trends Tamir noted were more evidence in practice of factors Atwood described earlier as desirable, such as expanded fieldwork and a stronger emphasis on process skills. Tamir's analysis also noted a number of newer curriculum areas that had not been previously addressed: curriculum studies, communication skills, clinical supervision, students as researchers, a science-related social problems view of science, inquiry teaching, and educational technology.

It is important to note the major influence during this period, beginning in the 1960s, as a conceptual orientation to science teaching. Inquiry teaching was not the product of research so much as an intellectual idea promoted by thinkers in the field (e.g., Schwab, 1962). At the beginning of this period, it is only mentioned tangentially in studies of science teacher education (Newton & Watson, 1968; Tamir, 1976). Later science educators began to recognize the importance of inquiry teaching in understanding the nature of science and teaching it, in spite of its limited existence in school practices (Helgeson et al., 1977; Stake & Easley, 1978; Weiss, 1978). Its importance is evidenced by its incorporation into Project Synthesis (Harms & Yager, 1978) as the desired state for teaching (Herron, 1971; Tamir, 1983; Welch, Klopfer, Aikenhead, & Robinson, 1981).

In the 1980s NSTA became active in efforts to make an impact on science teacher preparation because of growing dissatisfaction with science teaching (DeRose, Lockard, & Paldy, 1979; Helgeson et al., 1977; Stake & Easley, 1978; Weiss, 1978). In 1981, an NSTA Committee on Teacher Education was charged with building a data base for the development of standards for science teacher preparation (NSTA, 1984, 1987). Three separate surveys were administered: preservice preparation, by Mechling, certification by Stedman and Dowling; and elementary teachers' perceptions of their preparation for teaching science, by Donnellan (Mechling, Stedman, & Donnellan, 1982). Mechling's study is addressed here.

Of 45 institutions surveyed during the 1979–1980 academic year, Mechling (1982) found that most (98 percent) required prospective elementary teachers to complete science courses and most (93 percent) required 1 or 2 courses in the methods of teaching science. In addition, he found that the typical program was substantially tilted toward process, methods, and techniques, rather than science content. In fact, the survey population saw science content as the greatest need at the middle school and junior high levels. Mechling found that only about 30 percent of the institutions offered special preparations for teaching at either of these levels, with only 5 percent having a specially designed methods course. He reported a difference between middle school and junior high preparation: middle school programs required an average of 30 semester hours of science while the typical junior high program averaged 42 semester hours.

Other research efforts have sought more detailed data on select aspects of science teacher education. Recognizing a need for more specific regional data, Barrow (1987) found that in New England the traditional requirements for preservice elementary education were less than were noted in the Mechling national study. In the Big Eight states, the data closely resembled the national results (Barrow, 1991). Bybee questioned whether these trend studies, conducted primarily in larger institutions, could be applied to smaller four-year colleges, which produce one-third of the science teacher population yearly. Yager and Bybee's (1991) study to address this question was inconclusive.

A focus on practical matters in teacher education is a common desire of teachers. In terms of science methods courses, Stronck's (1985) survey of British Columbia science teachers' perception of their preservice teacher education revealed a desire for the programs to be more pragmatic, and to reflect a greater emphasis on such topics as daily teaching concerns and laboratory safety. Tolman and Campbell's (1989) study of the most preferred instructional topics in an elementary science methods course reflected the same need for pragmatic topics. Inquiry teaching and learning theory, on the other hand, were ranked by educators as having medium educational importance in their course, with women in science and STS (science/technology/society) as low priorities. The use of educational technology in science teacher preparation became an important part of everyday operations during the 1970s and 1980s. This use was prompted largely by the increased use of the microcomputer in schools during this time (see Baird, 1989). Focusing on the use of computers in teacher preparation, Weiss (1987) found that one in four science teachers had no training in how to incorporate computers into their instruction. In addition, two-thirds of the teachers surveyed felt that they were either "totally or somewhat unprepared." Lehman (1985) found that 75 percent of the science teachers who use computers took the equivalent of one college course in microcomputers, usually through inservice education. A year later, Lehman (1986) found that most institutions do indeed provide microcomputer coursework opportunities for preservice teachers, but typically do not require these courses for certification. He found that most of the coursework was centered around content-independent applications. Lehman saw this as a problem in educating science teachers, who need to learn specific techniques for using computers in teaching inquiry science.

Dominant Themes. The last few sections have highlighted descriptive studies of curriculum programs for preservice science teacher education, from the turn of the century to the 1990s. During this time six dominant themes emerge from the research, themes that play a major role in shaping the character of science teacher education. They are as follows:

ESTABLISHED PRESERVICE MODEL. As described earlier, the traditional model of teacher education has been the structure for educating science teachers since early in the twentieth century. Although modified over the years and varied somewhat within different contexts, it still basically consists of the same three strands: general education, subject-matter preparation, and professional education. Most science-education research has been directed at the last strand.

INADEQUATE SUBJECT MATTER PREPARATION. Present-day concerns about subject-matter preparation for elementary and secondary teachers are not new or significantly different from what has been noted over the years (DeBoer, 1991; NSSE, 1932, 1947, 1960). Common criticisms of science teachers, especially at the elementary level, are the following: a lack of science content knowledge, especially in physical and earth sciences (Bethel, 1984; Heikkinen, 1988; Newton & Watson, 1968); an inadequate understanding of the nature of science (NSSE, 1947; Tamir, 1983); and the need for a broader, more interrelated view of science (Atwood, 1973; NSSE, 1947; Tolman & Campbell, 1989). Teachers, especially at the elementary level, also perceive themselves as lacking the necessary knowledge to teach science (DeRose et al., 1979; Donnellan, 1982; Weiss, 1978).

There are several possible explanations that have been offered to address this lack of subject matter expertise: the mix of science-education students with science majors in class coursework (NSTA, 1984), the incongruence between university science coursework and precollege science courses found in schools (Hurd, 1983; NSSE, 1932, 1947), and the need to improve the teaching of college-level science classes (Helgeson et al., 1977; Hurd, 1983; Stake & Easley, 1978; Tamir, 1983). Numerous efforts continue to be made by organizations to address these subject-matter preparation concerns (AAAS, 1989; NSTA, 1984, 1987; Penick, 1987).

HAPHAZARD EDUCATION PREPARATION. The professional education of science teachers has been cited for many years as haphazard (Atwood, 1973; NSSE, 1947; Newton & Watson, 1968). First, the professional education strand in science education has been characterized as irrelevant, lacking a theoretical framework, and lacking attention to planning and development (Atwood, 1973; Newton & Watson, 1968), especially in middle-level preparation (Mechling, 1982). Second, preservice programs have been characterized by a major discrepancy between intended research-based practice and actual teaching practice (Atwood, 1973; Helgeson et al., 1977). And third, preservice science teacher education has been characterized as incongruent with actual school practices (Atwood, 1973; Donnellan, 1982; Hurd, 1983; Stake & Easley, 1978).

This is not to say that there have not been important developments in preservice science teacher education over the years. First, there has been a major increase in the number and types of field experiences incorporated into science teacher education programs (Atwood, 1973; Tamir, 1976), from early involvement (Sunal, 1980) to induction support (Butts, 1989). Second, science teacher education has an increased emphasis on process skills of different kinds (Mechling, 1982; NSSE, 1947; Tamir, 1976). And third, preparing science teachers now includes a heavy emphasis on communication skills and interpersonal relations (Tamir, 1976; Tolman & Campbell, 1989). This last area covers a wide spectrum of topics including learning about cooperative learning (Johnson & Johnson, 1979) and addressing classroom climate (Fraser, 1989).

Although these important changes in practice are related in some ways, they have developed in a more or less isolated fashion. In most cases, changes were specific to small aspects of programs, and usually were of a more structural nature. Too few changes have resulted in furthering deliberations about the knowledge base of science teacher education or in making recommendations for improving the preservice curriculum.

IMPORTANCE OF INQUIRY.   Since the early 1960s, there has been a continued emphasis on inquiry teaching as a means for understanding the nature of science (Schwab, 1962). Science teachers who do not understand the nature of science will have difficulty representing it to their students in a viable manner. Furthermore, studies show that an inquiry curriculum has a significant positive effect on student performance (Shymansky, Kyle, & Alport, 1983), as has the use of instructional strategies associated with teaching by inquiry (Wise & Okey, 1983).

Discrepancies exist, however, between actual practice in inquiry teaching (Stake & Easley, 1978) and that which is desired (Welch et al., 1981). Teachers themselves even report being ill-prepared to guide students in inquiry science lessons, and many attribute this to poor preservice training (DeRose et al., 1979; Donnellan, 1982; Weiss, 1978). Most teachers have never been exposed to science courses that employ inquiry teaching (Helgeson et al., 1977; Stake & Easley, 1978). Even in light of these apparent discrepancies, the importance of inquiry remains strong as a foundation for teaching science, at least in theory, if not in practice.

RELIANCE ON THE LABORATORY.   In 1969, Ramsey and Howe stated that the case for the laboratory in science education was too obvious to argue. Klainin (1988) and White and Tisher (1986) have expressed similar sentiments. Shulman and Tamir (1973) highlight how the role of the laboratory has changed over time in science education:

The laboratory has always been the most distinctive feature of science instruction . . . . With the advent of the new curricula, important changes have taken place in the role assigned to the laboratory . . . . The laboratory acquired a central role not as a means for demonstration, but rather as the core of the science learning process . . . . Ausubel (1968) believes . . . it should typically carry the burden of conveying the methods and spirit of science. (pp. 1118–1119)

How the nature of science is portrayed in the laboratory as well as how and when the laboratory is used have been topics of discussion and concern for years (see DeBoer, 1991). Tamir (1989) discusses the concern in terms of teacher preparation; he writes:

the teacher is the key to effective learning in the laboratory . . . . There is no doubt that improvement in the effectiveness of learning in the laboratory can be achieved only through substantial improvement in teacher preparation. (pp. 59, 60)

He focuses the discussion on methods of educating teachers on how to make laboratory lessons more meaningful and useful.

VALUED EDUCATIONAL TECHNOLOGIES.   Rather recently, the value of educational technologies in preparing science teachers has begun to receive substantial attention. Earlier mention of using educational technology to prepare teachers (Tamir, 1976) has led to attention to areas such as telecommunications and optical storage technologies (see Ellis, 1989). This focus almost certainly will increase with time. To date, the majority of the research on educational technology in preservice programs addresses computer-based learning, microcomputer-based laboratories, computer literacy, and attitudes toward computers (see Ellis, 1989).

Criticism of the Traditional Model

The history of preservice education could be written in terms of promising innovations, but somehow the promises have really not been fulfilled. Progress, yes, but macro-improvements and breakthroughs, no . . . . It is reasonable to propose that the recent history of preservice education has been much more strongly influenced by the rise and fall of educational "styles" and by widely publicized public rancors against teachers than by any tight-knit, closed reasoned rationales of reformers . . . . (Cogan, 1975, p. 209)

The traditional undergraduate model for teacher education is no stranger to criticism. It has been under strong attack almost since the time of its formation, with claims that it lacks substance, depth, coordination, and relevance to actual classroom practice.

Although all strands of the model have received criticism, the majority of complaints, even by those in the field, have been aimed at the professional or pedagogical sequence. Feiman-Nemser (1990) summarizes the criticism of the traditional model of teacher education. She writes:

In the professional sequence, the balance between what Borrowman (1956) calls the liberal and technical seems tilted in the direction of the technical. Methods courses dominate, taught not by master teachers but by university professors. Foundational knowledge comes mostly from educational psychology. The development of technical skills is limited to a brief stint of student teaching. In short, the typical undergraduate program seems more like an organizational compromise, the offspring of an unhappy union between the normal school and the liberal arts tradition. (p. 217)

She explains that the professional sequence overall is a product of not just compromised aims but also of conflict between aims and practice.

Lanier and Little (1986) claim that although teacher educators have been criticized for years for not having agreed on a core body of professional knowledge that all beginning teachers must possess, the research in the other two strands of teacher education—general education and subject-matter preparation—is even more scarce and undefined. Tom (1986a) says that if teacher education is to remain a viable undergraduate program option, a major emphasis needs to be put on rethinking the general education and subject-matter preparation components of teacher education. He says that the problem with general education is its quality and coherence. He says the subject-matter preparation needs to focus "more on core disciplinary ideas and inquiry processes" (p. 31). In addition, there is some argument about the roles of these two strands in the teacher-education program. Historically, a liberal arts education has been advocated and practiced as the foundation for teacher preparation (Scheffler, 1973). More recently, however, arguments have been advanced for emphasizing subject-matter preparation as the curricular priority in educating teachers (Anderson, 1988).

Despite the passing of years and the persistence of debate, the traditional model for teacher education remains remarkably intact, but not without some change. In the 1970s and early 1980s, there were some general pattern shifts, for better or worse, in the teacher-education programs. Lanier and Little (1986) cite the following general trends: (1) field-based experience has increased (Zeichner, 1981); (2) teacher education has increased its reliance on a more technical and vocational-oriented program (Beyer & Zeichner, 1982); and (3) social and philosophical course requirements in educational foundations have been sacrificed to accommodate the changes cited in 1 and 2 (Finkelstein, 1982; Warren, 1982).

More importantly, however, since the mid-1980s there has been significant change in the way teacher education is conceptualized and discussed. For years, an effectiveness movement has dominated research on teaching, a movement which relies on a science-based approach as a means of reforming teacher education. Commonly known as process-product research, it is described as follows by Cochran-Smith and Lytle (1990): For the past 15 years, researchers have been exploring effective teaching by correlating particular processes, or teacher behaviors, with particular products, usually defined as student achievement as measured by standardized tests (see, for example, Brophy & Good, 1985; Denham & Lieberman, 1980; Dunkin & Biddle, 1974). Underlying this research is a view of teaching as a primarily linear activity wherein teacher behaviors are considered "causes" and student learning is regarded as "effects" (p. 2).

As the years progressed, however, many educators grew dissatisfied with this approach, particularly in regards to its portrayal of the teacher as a generic technician and the manner in which it deprofessionalized the teacher (Apple, 1986). Lanier and Little (1986) summarize Lortie's concerns about the impact of such messages on teacher education. They said he feared that these messages "encourage intellectual dependency and discourage professional development and adaptation to change. Lortie recommended, therefore, as many scholars have before him, a strengthening of liberal-professional studies for teachers" (p. 549).

Some programs that have altered the traditional science teacher curriculum model will be addressed later in this section.

Science Teacher Education Programs

Science teacher education programs have been in existence in the United States since around the turn of the century. Over a thousand U.S. higher education institutions operate preservice science teacher education programs. Considering the longevity and volume of such efforts, one would expect a review of preservice science teacher education programs to portray a rich landscape, complete with diverse views, cohesive images, and defined detail. Research on these programs, however, is neither accessible nor diverse.

Indeed, there is a dearth of literature describing preservice science teacher education programs. Although there are many references to these programs, these brief mentions are usually in the context of reporting related research. Actual portrayals of comprehensive programs—including conceptual and structural components—are rare. The limited literature that does exist of this type is primarily from two sources: NSTA publications and National Science Foundation (NSF) reports. This literature, however, primarily addresses structural components; discussions of program ideologies are often absent or vague.

A formal and informal review of programs revealed limited diversity across preservice science teacher education programs, as a whole. Differences that do exist among programs are most often found at the course level. Innovative efforts in reforming science teacher preparation usually are directed at changing one or two isolated components within a program, as opposed to the program as a whole. Such changes, as described earlier, are often seen in creating interdisciplinary subject-matter courses, reworking science-methods courses, and adding field experiences. An extensive study of preservice science teacher education programs is underway at The Network, Inc. In reviewing nearly 100 programs, the results to-date characterize preservice science teacher education programs as "mostly traditional," with innovative components (A. Michaelson personal conversation (1992)). The project is developing a framework for categorizing preservice programs by innovative features. For example, one category might be museum-based science teacher education. Michaelson noted, however, that this review may also be revealing in terms of highlighting key problems in educating science teachers, problems he links to larger reform issues (see Kyle, 1991).

Preservice science teacher education programs across the United States have relied and do rely on the traditional model as a framework for educating teachers. The sections that follow describe selected examples of the traditional model of preservice science teacher education—some with more variations than others—including programs at three levels: elementary, middle/junior high, and senior high.

Elementary Model. The Competency-Based Teacher Education Program (CBTE) at the University of Toledo is representative of an orientation to science teacher preparation which received considerable attention during the 1970s and early 1980s. The program was one of seven exemplary preservice elementary teacher education programs identified by NSTA (see Penick, 1987). As a CBTE program, it has an orientation which Feinman-Nemser (1990) describes as technical, emphasizing the scientific view of teacher.

DeBruin, Gress, and Underfer (1987) describe the CBTE program, which was designed in the late 1960s, as "exacting, flexible, and practical" (p. 9). Each prospective teacher must demonstrate mastery of specific competencies, as stated in behavioral and measurable objectives, within 10 broad educational goals. Instruction is divided into modules around the objectives, with flexible time frames and means for reaching mastery. The program has a major emphasis on clinical experiences beginning as early as the freshman year in the program and totaling 300 hours of field-based experience prior to student teaching.

DeBruin et al. (1987) describe students' coursework in the CBTE program as a combination of general studies, a professional sequence, and an area of specialization. In the general studies component every elementary-education major must also have an area of specialization. If science is that area, the expectation is a minimum of 20 hours of coursework. In the professional sequence, the science-methods course consists of eight modules: (1) unit planning and implementation, (2) teaching science, (3) critiquing and improving faulty test items, (4) classroom management techniques, (5) problem solving, (6) concept lessons, (7) inquiry teaching, and (8) questioning lessons. Pretests and posttests are given to students to measure changes in attitude toward science and science teaching, as well as changes in the mastery of science content.

Middle and Junior High Models. Since the mid-1970s, there have been numerous calls for training middle-level science teachers in separate programs from either elementary or secondary teachers. (Blosser, 1986; NSTA, 1984; Padilla, 1983). Three examples are provided here.

The state of Georgia was one of the first to focus on middle-level education as a separate level of teacher preparation (Riley & Padilla, 1987). In the mid-1970s, Georgia changed to a revised three-tier model of certification: early childhood (grades K–4), middle (grades 4–8), and secondary (grades 7–12). This new middle-level certification is designed to acquaint the prospective teacher with the nature of the middle school grades and students, but it does not certify these teachers in specific subject areas.

In 1986, the NSF funded nine projects to develop models for educating middle school teachers to teach science or mathematics. Doody and Robinson (1988) highlight the key characteristics of two of these NSF middle school models: State University of New York College (SUNY) at Potsdam and Hampton University in Virginia. The key difference between these newly funded models and those developed years earlier is rethinking which has gone into the subject matter preparation for middle-level teachers. Doody and Robinson (1988) write:

Middle school science education spans all of the natural sciences. Traditional college science programs either focus on a series of courses in one domain of science, or provide isolated courses in general science which do not provide for continual increase in rigor and depth through a series of related courses. A model college program to prepare teachers of multiple sciences and interdisciplinary science should include a strong background in interdisciplinary science. (p. 1)

This approach is in contrast to programs developed years ago in which the emphasis was on paralleling a middle school science curriculum with study each year of a different science (i.e. life science, physical science, earth science).

The SUNY Potsdam interdisciplinary science curriculum is divided into two levels. The first level consists of four sequential interdisciplinary science courses, each six semester hours, taken during the freshman and sophomore years. The second level is taken during the junior year and consists of 12 semester hours of science electives in a chosen field. Instruction in these science courses models appropriate science-teaching strategies for students to emulate and experiment with themselves as the semester progresses. Overall, the SUNY program has a heavy emphasis on laboratory activities and the use of microcomputers. The only significant difference between this program and the Hampton University program from earlier models is that it adds a science/technology/society seminar.

Secondary Model. Most of the available information about preservice secondary science teacher education models pertains to the UPSTEP (Undergraduate Preservice Science Teacher Education Programs) programs funded by the National Science Foundation during the 1970s (e.g. Purdue, Iowa, Nebraska).

Krajcik and Penick (1989) summarize the intent of the Iowa-UPSTEP program as follows:

The ultimate goal of Iowa-UPSTEP has been to develop science teachers who have a research-based rationale for teaching science, and the ability to apply that rationale in the science classroom. (p. 796)

Penick and Yager (1988) cite the program's specific goals as highlighting broader perceptions and applications of science, research on effective teaching, and clinical experiences in schools. Included in the course requirements are (1) a two-semester sequence in the history, philosophy, and sociology of science, (2) two applied science courses, (3) three science methods courses, and (4) three field experiences prior to student teaching.

In all, the Iowa UPSTEP students take eight courses in science education over a 2-year period. Students choose two of five applied science courses: earth science, biology, physics, chemistry, and environmental sciences. These science courses focus on science-related social issues specific to particular disciplines. They take three methods courses that cover the span of K–12 schooling, each with a corresponding clinical experience to "attempt to narrow the gap between goals and practices in science teaching and in science teacher education" (p. 63). The first methods course focuses on elementary school science teaching, and students primarily observe other teachers; the second targets working as a tutor with adolescents at the middle school level; and the third methods course gets students involved in planning and teaching secondary school science classes. Krajcik and Penick (1989) found that the graduates of this Iowa UPSTEP program had characteristics similar to a very select national group of outstanding teachers.

Different Models of Preservice Teacher Education

In response to criticism of traditional 4-year teacher education programs, variations have arisen. In the main, these variations do not reflect new theoretical orientations; they are largely structural alternatives. Feiman-Nemser (1990) groups these alternatives in three categories: 5-year or extended program, graduate-level program, and alternative certification program.

The 5-year developmental program is most similar to the traditional preservice model. Students complete 5 years of study at the same institution, typically graduating with a master's degree and certification. Students begin their professional studies as undergraduates, with full-time attention to teacher education in their fifth year. This curriculum, like the traditional model, provides students with a developmental span of time to pursue their professional studies. Possibly because this form of teacher education is only a moderate revision of the traditional form, it has many supporters, including the AACTE. Howey and Zimpher (1986) note that at the time of their study, these 5-year programs were composed largely of structural changes that added more of the same, with little evidence of alternative conceptual frameworks. Many of these programs incorporated the induction year as their fifth year of study.

For example, Austin College has a 5-year developmental liberal arts program (Cobern & White, 1988). In this program, the student's professional education is quite different from the traditional model. Before the fifth year, students take all of their professional preparation within the science coursework, with the exception of a sequence of education "labs." There are four undergraduate education labs that target specific teaching performance skills, such as lesson planning, and include progressively more involved fieldwork in the schools. The graduate study, the fifth year, is made up of a semester of student teaching, two methods and curriculum courses, and/or further academic electives.

The second different form of preservice teacher education is the graduate-level program. This program begins to move toward the inservice style of teacher education. Students enter the program with completed undergraduate degrees and, in most cases, have the needed course requirements to teach in their subject area. Feiman-Nemser (1990) identifies two types of graduate programs that fit this category: (1) an academic model, emphasizing academic knowledge and practical experience; and (2) a professional model, combining professional studies with guided practice.

This first program is best explained as the Master of Arts in Teaching (MAT) degree, a program that emphasizes advanced study in a discipline with complementary professional seminars and an internship. The focus here is largely on the discipline of study, in this case, the sciences. There are a number of MAT programs that prepare science teachers. One example is the Harvard Graduate School MidCareer Math and Science Program, which began in 1983 (Merseth, 1985). Full-time students can complete this program in 9 months: one semester of classes and one semester of student teaching.

The other type of graduate program is described as the professional model. Today, this form of teacher education is best exemplified by the model advocated by the Holmes Group in their initial report, "Tomorrow's Teachers" (1986). In this approach, teacher education becomes a professional-school activity, concentrated in one year—a fifth year—of study and resulting in a graduate degree. This model emphasizes concentrated study in professional education as opposed to the MAT program which devotes more emphasis to the subject area. Typically in the professional model, professional knowledge is defined with a heavy emphasis on scientific research. While this approach has received considerable attention, not all observers are enamored with it. Tom (1986b), for example, characterizes the Holmes recommendation as a "sophisticated analysis followed by simplistic solutions" (p. 44).

The third different model of teacher education, and the farthest from the traditional 4-year program, is alternative certification. This approach is similar to on-the-job training. It relies on students entering the program with a solid general education and subject-matter background. It places high priority on practical, "how-to" types of knowledge. These programs have become popular due to the perceived increasing teacher shortage; approximately half of the states have some form of approved alternative certification (Feistritzer, 1986).

Most alternative certification programs rely on partnerships between universities and school districts. Two such programs operate at DePaul University: one with a suburban school system (Lakebrink, 1991) and the other with the urban city school system (McCormick, 1991). In the suburban program, students receive a teaching certificate, a master's degree, and teaching experience while receiving a salary, all over a 3-year period of time. Students receive all of their training on-site from university faculty and experienced teachers while working in the classroom. The urban program is similar, except the students complete the work in a shorter period of time and attend weekend classes.

These different models cover a wide spectrum. They vary in terms of administrative concerns such as time and expense, and they reflect philosophical differences, such as assumptions about preservice students' knowledge and needs in preparing to teach and the best way to meet those needs. Regardless of the model, it is evident that more in-depth research is needed profiling all types of preservice teacher education (see Howey & Zimpher, 1989).

Concluding Thoughts

Although different theoretical orientations are acknowledged or noted in the literature on science teacher education, it is striking that the program variations described and researched are far more often structural than theoretical. Different theoretical orientations are recognized—such as subject-matter orientations and outcome-based orientations—but they are less prominent in the science teacher education literature than in the general teacher education literature where structural variations still predominate.

There is little theory-based research on innovative science teacher education programs. More is needed. It does not seem particularly venturesome to claim that any other kind probably is not worth the effort at this point in time. New conceptions of subject-matter knowledge create a demand for new orientations to science teacher education and accompanying research.

The growth of understanding of how learners construct meaning and its implications for the classroom create an even more pressing need for disciplined inquiry. As ongoing and future research provides greater understanding of what constitutes constructivist *teaching*, researchers will need to address directly questions about how to educate teachers to successfully teach in this manner. It may well be the most needed—and potentially fruitful—area of research in science teacher education. This generalization is made both with respect to the research on the content of teacher education just discussed and the research on instruction within science teacher education which follows.

PRESERVICE SCIENCE TEACHER EDUCATION: INSTRUCTION

Having considered the content of science teacher education programs and various structural alternatives for organizing it, we turn our attention to the instruction that takes place in such programs. Curriculum and instruction are highly interrelated in this context, as in others, but these categories are a useful basis for organizing the material at hand. This section provides a general overview of research on specific components of science teacher preparation programs—in contrast to research on overall programs as addressed above—and in particular addresses research on the instructional aspects of science teacher education.

Nature of the Research

Even with a long history of educating science teachers, the amount of research on the subject is relatively light. Consequently, there are few reviews of research on the subject. In a discussion of research on teacher education, Yeany (1991) writes: "There is an urgent need to intensify our research efforts in science education to better understand the science teacher" (p. 3).

Research on science teacher education is presented here in two broad categories according to research type, with dominant themes being emphasized within each section. The first category is treatment studies aimed at changing science teachers' knowledge, skills, and attitudes. In the second category, we discuss the more recently emerging research on understanding science teachers, directed at a more meaningful science teacher education.

Experimental Studies of Teacher Change

The curriculum-reform efforts of a generation ago with their emphasis on inquiry teaching spawned a substantial number of studies, mainly between the 1960s and the mid–1980s, of attempts to change science teachers. With reference to the types of change proposed as a result of the 1960s curriculum reform, Sunal (1982) indicated the:

dominant themes were found to include an increased emphasis on a teacher's (1) attitude toward nature and teaching science, (2) ability to use basic skills of scientific investigation, (3) skill in purposefully planning, sequencing and evaluating science instruction, and (4) use of questioning and instructional strategies stressing higher order cognitive interaction. (p. 167)

His listing bears some resemblance to the categories used here to describe the research conducted on such teacher-change endeavors, which are: changing attitudes, changing process skills, and changing teaching behaviors.

Changing Attitudes. A belief that teacher attitudes are important to their effectiveness as teachers led to a surge of studies, especially during the 1970s. These studies were aimed at measuring and altering preservice teachers' attitudes toward science and science teaching through some treatment, generally school-based experiences or specific training strategies. These studies relied on the use of attitude measurement instruments, the Likert scale being the most common form. The instruments included, among others, the Inquiry Science Teaching Strategies Instrument (Lazarowitz & Lee, 1976) and the Revised Science Attitude Scale (Thompson & Shrigley, 1986), specifically for preservice teachers.

These studies show mixed results as is evident, for example, in the research on measuring attitude change resulting from school-based experiences. Some researchers report that field experience and practice teaching positively affect prospective science teachers' attitudes toward science (DeBruin, 1977; Gabel & Rubba, 1979; Piper, 1977; Piper & Hough, 1979; Sparks & McCallon, 1974; Strawitz & Malone, 1986). Others, however, report a lack of change in altering prospective teachers' attitudes toward science during field experiences or practice teaching (Lawrenz & Cohen, 1985; Shrigley, 1974; Sunal, 1982; Weaver, Hounshell, & Coble, 1979). And still others (Shrigley, 1974; Sunal, 1982) note that it is not attitudes toward science itself that field experiences positively impact, but rather it is the attitudes toward the teaching and learning of science that significantly change.

Training strategies have also been used to affect teacher attitudes. Researchers report improvement in attitude toward inquiry teaching as a result of training in inquiry-teaching strategies (Barufaldi, Huntsberger, & Lazarowitz, 1976; Lazarowitz, Barufaldi, & Huntsberger, 1978). Koballa (1992) claims that effective communication training strategies, in particular persuasion messages, can change elementary teachers' attitudes toward teaching science, as long as the teacher remembers the arguments presented in the message. However, "measuring preservice teachers' attitudes toward science cannot adequately predict nor provide a satisfactory explanation of their science teaching behaviors" (Koballa, 1986, p. 501).

The results from attitude studies are not the only point of ambiguity. Understanding the relationships in categories of attitude studies can be equally complicated (White & Tisher, 1986). Shrigley, Koballa, and Simpson (1988) note the evolving definition of attitude within the literature and its confused relationship to belief, value, and opinion.

In the literature, there are numerous claims that significant attitude changes toward science and science teaching took place in prospective teachers. These claims, however, become less clear as one takes a closer look at the literature (Morrisey, 1981; White & Tisher, 1986). At best, the literature says that prospective teachers' attitudes toward science teaching and learning are indeed affected, but the impact upon attitudes toward science itself is unclear. And the relationship is uncertain between these efforts to change teachers' attitudes and actual changes in classroom practice. In general, "no clear principles emerge from this body of research" (White & Tisher, 1986, p. 892). And as Lawrenz and Cohen (1985) suggested, maybe a qualitative research approach is needed to extend our understanding in this arena.

Changing Process Skills. It is generally thought that teachers must have a good knowledge of science process skills, if they are to be effective in promoting the development of these process abilities in their students (Lawrenz, 1975). Therefore, there is a body of research directed at enhancing prospective teachers' knowledge and acquisition of process skills. It has been shown that with specific training in process skills in the science methods course, prospective teachers show an improvement in knowledge of process skills (Berkland, 1974; Campbell, 1974; Campbell & Okey, 1977; Riley, 1979), and in plans to teach process skills (Jaus, 1975; Zeitler, 1981).

There is, however, research that appears to make these findings less clear. Lawrenz and Cohen (1985) found that neither instruction in the methods courses nor the practice-teaching experiences had any significant effect on the elementary and secondary students' knowledge of science processes. Sunal (1980) reported that field experience positively affected prospective teachers' performance of process teaching behaviors, but these improvements were consistent with activity-oriented science teaching and not necessarily inquiry teaching.

Another related claim that accompanied the 1960s curriculum projects was that one must possess a thorough understand-

ing of science before he or she can teach inquiry science to children (Karplus & Thier, 1967; Rowe, 1973). Research exists on the relationship between a preservice teacher's science knowledge and inquiry teaching, however, that does not support this claim. Some researchers argued that the better inquiry elementary teachers were actually those teachers who had minimal science backgrounds (Brehm, 1968; Perkes, 1975). Helgeson et al. (1977) found that a teacher's knowledge of science proved to be less important in developing inquiry teaching than participation in the designing of inquiry lessons.

Overall, the Helgeson et al. (1977) review reveals that not much had come of efforts to improve process-skill development at the preservice level up to the late 1970s. They state that even after the curricular reform of the 1960s, traditional or direct teaching remained the predominant practice in preservice classrooms.

Changing Teaching Behaviors. There is a fair amount of research on the effectiveness of specific science teaching behaviors such as, for example, wait-time (DeTure, 1979; Rowe, 1974a, 1974b; Tobin, 1980) and questioning (Lamb, 1977; Riley, 1979, 1981; Wise & Okey, 1983). There is also research on training teachers to use these identified strategies, thus changing their teaching behavior. The most common type of study in this area compares treatments with respect to their effect on teaching behavior.

Yeany and Padilla (1986) summarized a collection of such research and used meta-analysis to integrate the findings and assess the overall and relative magnitude of the effects. They synthesized two decades of practice and research—188 science teacher preparation studies, related to the use of behavior analysis to train science teachers to exhibit proficiency in designated teaching skill areas. These studies, heavily in the behaviorist tradition, relied on the use of strategy-analysis instruments to improve teaching skills. These instruments ranged from general teaching tools, such as the Flanders System of Interaction Analysis, to instruments developed more specifically for focusing on science teaching skills, such as the Teaching Strategies Observation Differential (Anderson, James, & Struthers, 1974).

Yeany and Padilla (1986) begin with an idealized strategy analysis training model consisting of five training method categories: study of an analysis system, employing models (observing and analyzing models), practice of teaching skills, analysis of lessons (self, peer and supervisor feedback and analysis), and reteaching. They summarized their review of studies according to these common treatment types.

Yeany and Padilla found that all five treatments and the general model were effective methodologies for training teachers in appropriate science teaching strategies, although the impact of the treatments varied. The treatments were ranked in the following order from least to most effective: (1) study of an analysis system and self-analysis, (2) observing models, (3) analyzing models, and (4) practice and analysis with feedback. In a certain sense, each treatment in this progression involves each of the ones that come before it in the list; thus it shows the power of the more sophisticated treatments. Based on an extensive number of research studies, this meta-analysis provides solid evidence of the potential that training has and the form that effective treatments must take.

The Need for Change. That portion of the previous research providing the clearest guidance for science teacher preparation is that which reports on teacher-training models. However, when considering the larger collection of this research, it seems to have fallen short of its goals (Helgeson et al., 1977; Stake & Easley, 1978; Weiss, 1978). Many concerns plagued the usefulness of this research in education (see Darling-Hammond & Snyder, 1992). Darling-Hammond and Snyder, for example, note that in reviewing this type of research over the decades, they found that "eventually the importance of curriculum content and context disappeared from view" (p. 64). They cite Doyle's (1978) expressed concern over the implications that this research even goes so far as to promote behaviors that are "context-proof, teacher-proof and even student-proof" (p. 169). Secondly, in an earlier study, Darling-Hammond, Wise, and Pease (1983) reviewed the effectiveness of research on "direct instruction" versus "indirect instruction," and raised a general concern about this research because in actuality it compares teaching models aimed at entirely different curriculum goals, with resulting misleading comparisons.

It is evident that changing the actions of teachers, especially toward using more inquiry-oriented teaching approaches, is definitely more complex than originally thought. Change does not occur by simply altering such variables as how teachers feel and act. Rather, these desired changes require teachers to learn, rethink, and adopt different knowledge, thoughts, and practices related to teaching. The attempts to achieve changes proved to be more fundamental, in individual ways, to each teacher's personal and professional self, something this type of research could not fully address.

Toward More Meaningful Preservice Science Teacher Education Research

Considering its long history, there is a comparatively small amount of research on preservice science teacher education. In addition, it is rather limited in scope and usefulness. Prior to the mid–1980s almost all of this research was focused on evaluating the results of various programs and techniques with little attention to the dynamics of the learning that occurred or critical examination of the content of the instruction. Research has produced little knowledge about how people actually become teachers, in spite of the fact that the problems of science teacher education have been explored in various ways. The literature identifies problems associated with teacher education, but provides relatively little help in solving them.

Woods (1987) summarizes the problems as follows: (1) much of the knowledge deployed in this training seems remote from the practical concerns of teachers; (2) much of the educational literature appears to be critical of teachers; (3) what we know about teachers is partial and distorted; and (4) the essence of teaching is not amenable to scientific methods. Woods (1987) and Lampert (1985) propose that research in teacher education needs to be directed to studies of self rather than on structure or on volition.

Recently the research on learning to teach, both in general and with respect to science teaching, has begun to go in more promising directions with the use of expanded means of study. These changes are beginning to yield more meaningful information about science teachers: how they see themselves, how they think about teaching and learning science, and how they manage classroom practice. Today's research emphasizes acquiring a deeper understanding of science teachers and their development, including a deeper understanding of self by educators. And these changes are leading to deeper understandings of what it means to be a science teacher, how one becomes a science teacher, and even some glimmers of how these understandings may translate into a science teacher education curriculum.

This section highlights expanding modes of inquiry, some emerging viewpoints that shape the substance of research, examples of the research itself and some speculative reflections on future directions. Many of these research practices and emerging viewpoints are both significant departures from the past and highly interrelated, factors that give promise of notable results in future research.

Expanding Modes of Scholarly Inquiry. The newer modes of inquiry being emphasized today are powerful means of addressing educational problems, especially those more complex problems that Schwab (1974) some time ago encouraged researchers to attack. Within the last couple of decades, educators have come to embrace an interpretive approach (see Erickson, 1986) to research. In the late 1970s, interpretive research was showcased in the U.S. science-education literature with the appearance of the Stake and Easley (1978) case studies. Since the early 1980s, many have advocated this new approach to research in the science education literature (Connelly & Clandinin, 1986; Easley, 1982; Gallagher, 1984; Rist, 1982; Roberts, 1982; Smith, 1982; Welch, 1983). As a result, interest and production in interpretive research in science education has grown rapidly during the late 1980s (see Gallagher, 1991).

An increasing emphasis on cases and case studies (Broudy, 1990; Shulman, 1984) has evolved. Not only does this research produce different results, but it may influence practice in a different manner. Shulman (1986b) writes:

Most individuals find specific cases more powerful on their decisions than impersonally presented empirical findings, even though the latter constitute 'better' evidence. Although principles are powerful, cases are memorable, and lodge in memory as the basis for later judgments. (p. 32)

Cases as exemplars of practice and malpractice constitute one of Shulman's (1986b) 5 categories of knowledge of teaching. Collections of exemplars serve to highlight the complexity and specificity of practices in teaching and teacher education and to act as entries to understanding new problematic teaching situations. These exemplars act as baseline knowledge in understanding new situations—"seeing *this* situation, as *that* one" (Schon, 1983, p. 139). Teachers note the similarities and differences between the two, and this leads to thoughtful professional practice.

From a somewhat different perspective, scholars such as Fenstermacher (1986) make the case that alternative modes of inquiry in education require something more than research alone can offer, interpretive or otherwise. Fenstermacher highlights a needed distinction between the production or generation of knowledge, known as research, and the use or application of knowledge, known as practical reasoning. He argues (1978) that educating a teacher is not merely a matter of inculcating a specific set of teaching skills and competencies, as some research has been advocating. Rather, it is a matter of influencing the premises on which a teacher bases practical reasoning about teaching in specific situations.

the benefit of educational research to educational practice is realized in the improvement of practical arguments, not in programs of performance deduced from the findings of research.... Practical arguments, or some similar way of acknowledging purpose, passionate, intuitive, and moral properties of human action, are the methods for transforming what is empirically known and understood into practice. (Fenstermacher, 1986, pp. 43, 44)

The word "practical" has particular significance here. Schwab (1969) called for understanding teaching and teacher education in light of practical problems of curriculum. Pereira (1992) elaborates on Schwab's use of the term:

For Schwab, the practical, like the theoretic, was a mode of inquiry, a way of dealing with certain kinds of problems, which philosophers have called "uncertain" (Gauthier, 1978). The idea that there should be different methods for different kinds of problems goes back to Aristotle (who distinguished three kinds of knowledge: theoretical, practical, and productive). Practical problems arise in specific times and places out of our concerns (perhaps ill-defined) about particular states of affairs; they compel us to adjudicate between conflicting goals and values; the grounds for choice as well as the outcomes of choice are usually uncertain; and the resolution of the question involves a decision to take action (or not to act at all). (pp. 8, 9)

Research and practical reasoning are similar in their mutual intent to improve practice. But, some see them as unequal in their status in achieving this intent. Fenstermacher (1978) writes:

When it is argued that research has benefit for practice, the criterion of benefit should be the improvement of practical arguments in the minds of teachers and other practitioners.... One of the key reasons I argue for a clear distinction between research and practical reasoning is as a means of preserving the value and force of experience, ethics, passion, and so forth in the work of teachers. The findings of research are one among a number of bases for appraising and changing the practical arguments in the minds of teachers. (p. 45)

Both newer modes of inquiry and new ways of using the results of such inquiry offer promise of influencing the practice of science teaching.

Related Emerging Viewpoints. Emerging viewpoints on science content and human learning are providing new perspec-

tives on science teaching and learning and thus on research in these areas.

BROADENING SCIENCE CONTENT. Since the turn of the century, teacher educators have worked toward developing a broader perspective of science-content knowledge in prospective teachers (NSSE, 1932, 1947; Shulman, 1986a; Tamir, 1976). The established importance of inquiry in understanding the nature of science is an example. Within the last two decades, there has been considerable effort devoted to developing an even broader perspective on science-content knowledge. There are at least three areas in which science-content knowledge is being broadened.

First, there is growing interest in broadening the current understanding of science as a discipline, and in the implications that broader understanding has for teaching science. This work, stemming from Schwab (1978), has been furthered by Shulman (1986a, 1987) in his studies of teaching and teacher knowledge (e.g., substantive, syntactic, pedagogical content knowledge). In science education, studies relating specifically to teachers' understanding of these knowledge bases and beliefs are addressed in a later section.

Secondly, the study of science through the social sciences, an effort to broaden how science is viewed from a humanistic perspective, has been endorsed for years (Duschl, 1990; NSSE, 1947; Tamir et al., 1978). Some discuss it as an outcome of a specific curriculum model, such as STS (Aikenhead, 1986); while others assume that it is a natural outcome of science-related coursework. The prevalent assumption that prospective teachers will develop an understanding of the nature of science in science-methods class (Duschl, 1983; Loving, 1989) or in science courses (Trumbull, 1989) is simplistic and probably incorrect. Many have called attention to the need for responsibly developing this understanding of science in preservice teachers. In practice, however, too few have acted on this need (Hodson, 1988).

A third major effort to broaden science knowledge and understanding has come about as a result of coupling science with the technological nature of society. The STS movement has established a presence at the precollege and college levels, although they are developing somewhat independently (Solomon, 1988). In the curriculum for educating teachers in science, this broadening of science to include technology is advocated in science courses, in the science-methods courses (Aikenhead, 1986) and is reflected in the various recommendations of Project 2061 (Rutherford & Ahlgren, 1990). When discussing actual practices, however, science-methods instructors recently ranked STS as a low priority in their instruction (Tolman & Campbell, 1989).

UNDERSTANDING LEARNING AND THE LEARNER. Learning theory has held an important place in science-education research for a long time. Recently learning and the learner have been portrayed in science-education research in a broader manner than in the past. Within the last several years, for example, ideas and beliefs that children bring with them to the learning process have become an important factor in the search for understanding how to teach and learn science. This newer focus on the learner has become associated with an evolving constructivist view of learning (Confrey, 1990; Driver & Bell, 1986; Osborne & Wittrock, 1985; Piaget, 1929; von Glasersfeld, 1984). In general, a constructivist view of learning states that individuals generate their own understandings. This view acknowledges that learning outcomes depend not only on "known" factors such as the teacher, the curriculum, and the environment, but also on the "unknown" ideas that the learner brings to the task. This new focus on studying the learner has generated numerous research projects on conceptual change that resonant with existing research on metacognitive strategies to help students learn how to learn.

There is, however, some confusion and criticism surrounding this view. For example, Gunstone (1988) claims that although Piaget's work is a major antecedent to constructivism, thus making the two similar, they do have significant differences which many fail to distinguish. He elaborates:

The commonality of the two views of learning is then the common focus on constructivism—each position holds that the individual actively constructs his/her own meaning for experiences. The issue of developmental stages is the substantial difference between the two. (p. 85)

Others think that in spite of its wide acceptance, constructivism is not a promising new perspective. Matthews (1992), for example, argues against the new adoption of constructivism on the grounds that it is merely "a case of old, unpalatable, empiricist wine in a new bottle" (p. 11). He claims that although constructivism is discussed as a new and separate epistemology, in actuality it is fundamentally flawed in that it is wedded to the older Aristotelian-empiricist epistemological paradigm, which "the scientific revolution associated with Galileo effectively destroyed" (p. 8).

Emerging Research. New perspectives and approaches to research are leading to promising investigations regarding science teaching and science teachers' thinking.

REFRAMING SCIENCE TEACHING AND THE SCIENCE TEACHER. Historic images of science teaching are changing as scholars lean in the direction of a reflective epistemology—a reflective image of teaching. An example of naturalistic research that illustrates this emphasis on the reflective teacher is a project on a secondary-science model program at Monash University in Australia. A group of researchers (Baird, J.R., 1989; Baird, Fensham, Gunstone, & White, 1991; Gunstone, Slattery, & Baird, 1989) report on the importance of personal and professional reflection for improving science teaching and learning in a constructivist-based teacher-education program.

Gunstone et al. (1989) report that the most striking finding from their study on preservice science teachers was the personal development experienced by the prospective teachers and teacher educators as a result of involvement in the program. Although each individual differed in his or her development, the importance of reflection for intellectual development and the importance of collaboration for fostering this reflection, were common features applicable to all participants. In

addition, the collaboration served not only as a means of exchanging information and resources, but as a support for participants during the change process. A constructivist perspective in training teachers was cited as important for fostering intellectual development of both a cognitive and affective nature. And combined with personal reflection, this model also helped to improve personal practice, as well.

An increasing amount of science-education research is being conducted on educating prospective teachers to become reflective professionals. In preservice programs, interpretive research methodologies are used in the context of science-methods classes and clinical experiences in schools to study means for developing reflective professionals. Through a science-methods class, Tobin (1989) educates preservice students to think of their role of teacher as that of a researcher; while observing and teaching in schools, preservice teachers collect data (e.g., writing field notes, videotaping, and interviewing). Cronin-Jones (1991) reports a similar use of various interpretive research tools in a teacher-education program, including the Repertory Grid Technique, self-reports, reflective logs, peer reports, field notes, interview data, case studies, and profiles of effective teaching.

Understanding Science Teacher Thinking. For several years there has been an increased focus on understanding teachers' thinking. It is evident in Clark and Peterson's (1986) summary of research on teachers' thought processes in teacher planning, teachers' interactive thoughts and decisions, and teachers' theories and beliefs. More recently, there has been a shift in this literature from the more generic nature of earlier studies (Shulman, 1987) to research with an increased emphasis on teacher development and on subject-matter content.

For several years researchers have been studying science teachers' thinking, although its recognition in the literature has been slow (Roberts & Chastko, 1990). Similar to other areas, this research began with the primary focus on general thought processes of experienced teachers. Using Clark and Peterson's (1986) framework, the majority of this research focused on science teachers' theories and beliefs (Aikenhead, 1984; Gallagher, 1989; Martens, 1990; Mitchener & Anderson, 1989; Munby, 1984). Studies on teacher thoughts in the area of planning (Lederman & Gess-Newsome, 1989) can also be found, but to a more limited degree. And a few studies have reported on teachers' interactive thoughts and decisions (Duschl & Wright, 1989; Pratt, 1985).

More recently, however, this area of research in science teacher education is expanding and deepening. More studies are focusing on understanding preservice science teachers, as well as experienced science teachers. There is an increasing emphasis in the literature on the importance of understanding the personal and professional development of the teacher in order to adequately address his or her thought processes. Similarly, there is an increasing emphasis on the role of science subject matter in understanding teacher thoughts.

In research that addresses individual teachers, Trumbull focuses on personal and professional development in becoming a science teacher in two studies: one on three preservice biology teachers (Trumbull, 1991) and the other on her own development as a biology teacher (Trumbull, 1990). In both studies, Trumbull (1991) highlights the complexity involved in making the transition to becoming a science teacher, a complexity that has at its roots "developing and growing as a person" (p.16). This she claims is essential in programs claiming to develop teachers with a constructivist view, a transition she herself made from a more technocratic one.

In particular, Trumbull writes that people need help in exploring their own relationship to certain types of issues, particularly those of authority, epistemology, and learning. In her autobiography, Trumbull (1990) particularly notes her own transition related to the issues of authority and teaching science:

> My initial notion of teacher control assumed that teachers were lecturers, whether giving formal lectures or explaining material to small groups in laboratory. This image changed as I tried to give increased attention to the ethical issues in biology.... Originally I had felt that the teacher must be the subject matter authority, dispensing the accepted view and identifying and correcting students mistakes. Behavioral objectives reinforced this conception of authority.... [Then,] authority was shared more equally and knowledge was presented as more problematic.... And I was beginning to adjust my teaching practices to facilitate students' own making of meaning. (pp. 173, 176)

Trumbull stresses the need to provide prospective science teachers with a model for constructivist learning situations and help them develop practical knowledge, the type experienced expert teachers possess. Such preservice programs, she continues, lay the seeds that will help prospective teachers in lifelong professional growth as science educators.

Increasing the focus on subject-matter knowledge, a number of studies on science teachers' thinking are being done that address specific content areas for investigation. For example, Duschl (1983) found an inquiry science-methods course unsuccessful in changing prospective elementary teachers' understanding of science as learning content, but successful in creating in them an antagonistic dilemma of science as content versus science as process. Similar research has expanded to connect science teachers' thinking with classroom practice, as evidenced by research relating teachers' understanding of the nature of science and their teaching practices (Brickhouse, 1990) and by research relating teachers' science content knowledge and their teaching practices (Carlsen, 1987; Hashweh, 1987). Lantz and Kass's (1987) functional-paradigm research on chemistry teachers is similar, but it takes this work even further. It addresses the interrelated nature of the multiple factors present in recognizing a teacher's science-content knowledge and how it is portrayed in actual practice.

Roberts and Chastko (1990) further emphasize the role of subject matter in studying teachers' thinking. They present a model framework that acknowledges this emphasis within the complex nature of educating practitioners. The aim of the study is to foster reflective capabilities in preservice science teachers through the use of a systematic analytic framework consisting of four commonplaces, one of which is subject matter. Roberts and Chastko's framework serves multiple purposes: as a way to place subject matter in a more substantial position in studies on teacher thinking, as a way to conceptualize science teaching

events, as a way to educate prospective teachers in a science methods course, and as a way to address collaborative partnerships between schools and universities in educating science teachers.

Looking to the Future. Recent research is encouraging in that it has moved to a level that concerns fundamental changes in how we think about teacher development, the teaching and learning process, and teacher education. We are still a long way, however, from research that addresses such issues in the context of total teacher education programs grounded in a theoretical framework. Furthermore, there is a need to address the question of what theoretical frameworks for a teacher-education program are consistent with what research is telling us about teacher thinking and how teachers come to have an outlook and competencies which are consistent with emerging understandings of student learning.

THE INDUCTION PHASE

The dramatic attrition rate among beginning teachers during their initial years of teaching has led to a sharp increase in attention to the transition period between teachers' university educations and the point where they are accomplished professionals. In other words, the focus is on the realities of their first year or two in a teaching position. Traditionally, teacher education has been discussed in terms of two phases: preservice and inservice teacher education. During the past decade, however, with teacher education being viewed more developmentally, it has become more common to recognize an intermediate stage on the teacher development continuum: the induction phase.

> Despite our growing acknowledgement of the complexity of teaching, there still exists the notion that somehow, if teachers just mastered teaching methods in their college or university preparation programs, many of the barriers to delivering quality instruction would be removed. What is ignored in this view is the power of the first years of actual teaching experience to influence a teaching career, or in some cases, to provide sufficient evidence to new teachers that they should leave teaching and seek fulfillment in other fields. (Griffin, 1989, p. 397)

Schlechty and Vance (1983) estimate that within the first 3 years, 40 percent of beginning teachers leave the profession, compared to the 6 percent total teacher force turnover rate. Schlechty and Vance (1981) report that the higher ability beginning teachers leave teaching in greater proportions than those in the lower ability ranges. For example, in 1978 in North Carolina, 37.5 percent of the top two ranks of white females left teaching, as opposed to 24 percent of the bottom two ranks, as measured by National Teachers Examination scores. These studies reveal very little difference in dropout rate between males and females.

Huling-Austin (1990) offers the following four points as rationale for focusing on formalized induction in teacher education: (1) the isolated nature of teaching, (2) the abrupt nature of the transition into teaching, (3) the documented attrition rate of beginning teachers, and (4) the personal and professional well-being of the beginning teachers. Research on the complexities associated with the first years of teaching is growing, especially studies of individual cases (Bullough, 1989).

The importance of induction programs is being recognized in other countries as well. Building on work done in the United Kingdom, the induction of beginning teachers received a great deal of attention in Australia in the early and mid-1970s, a period when large numbers of teachers were being recruited and authorities were concerned about how they were doing on the job (Tisher, 1990). In some other countries (e.g., Canada), attention to the induction phase has developed more recently (Wideen & Holborn, 1990).

The induction phase of teacher education is a time of transition: between student teaching and teaching, between learning to become a teacher and becoming one, and between preservice and inservice education. It is typically defined as one year, although several professionals refer to induction as the first few years (Griffin, 1989). There have been a variety of induction programs designed to meet the special needs of the inexperienced teacher (Veenman, 1984). The design and components of induction programs vary greatly, but they generally rely on partnerships between universities and school districts. They include various combinations of such elements as university graduate coursework, afterschool seminars, weekend workshops, supervisory cycles, and written materials that include handbooks and newsletters.

Huling-Austin (1990) cites four characteristics that are important to any induction program: (1) flexibility in program goals and design, (2) the presence of a mature and experienced mentor teacher, (3) the careful placement of beginning teachers in situations that promote success, and (4) a process for educating others about the importance of induction. Huling-Austin (1986) warns, however, that there are problems that induction programs cannot solve, such as poorly prepared teachers, personal characteristic problems, major school context problems, misplacements, and long range retention issues. She also warns that the success of the induction program will be diminished if (a) the induction program emphasizes "feeling better" over development and improvement of performance, (b) the program focuses on technical requirements rather than the process of induction, or (c) the program focuses on minimums in terms of course requirements and support.

Teacher induction is commonly viewed as a socialization process, socializing teachers into the education profession. Griffin (1989), however, warns against arguing for induction on the basis of socialization into an occupation because it focuses on induction as a solution to a problem. From his work with the Illinois Initial Years of Teaching Program, he recommends focusing on induction as an initiation process into the profession because of its message to policymakers, one that calls for more deliberate and rational action as opposed to expediency in solving a problem.

Veenman (1984) agrees that induction involves more than socialization, but on different grounds. Based on his research synthesizing the problems of beginning teachers, he presents three strands of teachers' development—and socialization is only one of these strands. The three strands that need to be addressed in planning induction programs are teachers' developmental stages of concern (Fuller, 1969), teachers' cognitive

development over the course of their careers and lives (Sprinthall & Theis-Sprinthall, 1983), and teachers' socialization framework (Zeichner & Tabachnick, 1982).

Induction has been a growing movement in the United States for the last 10 years. Huling-Austin (1990) summarizes the growth of such programs:

In 1980, Florida was the only state that had a mandated induction program. The movement toward induction programs grew dramatically in the 1980s, due to the educational reforms that swept this country in the mid–1980s and in anticipation of future severe teacher shortages. By the late 1980s, at least 31 states plus the District of Columbia had either implemented or were piloting/planning teacher induction programs. (p. 536).

Induction practices are also nationally advocated by the Holmes Group report (1986), and the Carnegie Forum report (1986), two of the reports on teacher-education reform.

In science teacher education, there are references in the literature about the need for an induction phase in science teacher education (Butts, 1989; Loucks-Horsley et al., 1989; Sanford, 1988), but documentation of actual practices is almost nonexistent. Beginning studies on induction practices are presently under way at the University of Georgia. In 1986, the NSF funded the development of a middle-level science teacher education program. Currently, case studies are being conducted on the first-year (Nichols, Monroe, Padilla, & Wiggins, 1992), second-year (Monroe, Padilla, Nichols, & Wiggins, 1992), and third-year (Wiggins, Padilla, Nichols, & Monroe, 1992) graduates of this program as data to develop and implement a support model for these beginning teachers during their initial 3 years of science teaching. Preliminary summaries of the three studies build on the work by Odell et al. (1986) and Henry (1988). Odell et al. identified seven categories where beginning teachers need support: instructional, system, resource, emotional, managerial, parental, and disciplinary. Henry identified the need for support for beginning teachers coming from three main sources: a mentor teacher, university personnel, and school administration. In the case of the first-year teacher (Nichols et al., 1992), she relied most heavily on help from the university person, mainly in the areas of instructional, resource, and emotional support. In the case studies on the second- and third-year teachers, the importance of the on-site teacher was perceived as very valuable to these beginning teachers, especially in the areas of instructional and resource support.

The induction year is seen as the missing link in the continuum of professional teacher education. Howey and Zimpher (1990) see induction as hope, hope for helping more teachers feel the necessary support to stay in the teaching profession. They specifically hope that this added support encourages more teachers to try teaching in some of the more challenging educational situations, in urban and remote rural areas and with at-risk students.

INSERVICE EDUCATION

Inservice education takes place in a context with certain purposes that are related to that context. It takes place in a particular place (often a school-district setting). It is conducted by particular persons for a particular purpose (e.g., professional development of individual teachers or implementation of a new curriculum program). And it is conducted in some degree of isolation from or association with students, teachers and/or other professional personnel from the teachers' school.

Research on inservice science teacher education must take account of the given context. In fact, one could argue that helpful research on many kinds of inservice education simply cannot be done independent of its context. If an inservice education program is conducted for the purpose of implementing a new curriculum program, for example, it is difficult to address its results independent of careful attention to that curriculum and the process by which it is being implemented.

Within a given context, there are a variety of theoretical orientations to the inservice education and research on it. The previously mentioned five orientations to staff development (note that it is a subset of inservice education) are worth reflecting on here and serve as an example of alternative theoretical orientations. Sparks and Loucks-Horsley (1990) identified five such orientations.

Individually guided staff development is a process through which teachers plan for and pursue activities they believe will promote their own learning. The *observation/assessment* model provides teachers with objective data and feedback regarding their classroom performance. This information may be used to select areas for growth. Involvement in a *development/improvement process* engages teachers in developing curriculum, designing programs, or engaging in a school-improvement process to solve general or particular problems. The *training* model, which could be synonymous with staff development in the minds of many educators, involves teachers in acquiring knowledge or skills through appropriate individual or group instruction. The *inquiry* model requires teachers to identify an area of instructional interest, collect data, and make changes in their instruction on the basis of an interpretation of those data. (p. 235)

Without attention to both context and theoretical orientation, any research results are difficult to interpret. In general, we know we can teach most anything we want, within reason, given enough resources and determination. Without attention to context and theoretical orientation, we are left with relatively uninteresting questions, such as what is the most efficient or cheapest way to teach a given specific objective. Of critical importance is the matter of what *should* be taught, something that cannot be addressed adequately independently of context and theoretical orientation.

The general research on inservice education (i.e., not limited specifically to science education) has been reviewed well in several places, most recently and specifically in several chapters in the *Handbook of Research on Teacher Education* (Houston, 1990). There is a striking difference between this general literature and the much more limited collection of research conducted within science. The collection of general research on inservice education is (a) extensive enough and (b) has been conducted within a broad conceptual framework often enough that it is possible to interpret the results in a manner that allows generalizations with respect to some of the basic questions about inservice teacher education.

On the other hand, the science-education literature consists mostly of "one-shot" studies that give an answer to the question of whether or not a particular instructional program (i.e., a workshop or course) had a measurable impact; but these studies do not contribute a lot to broader generalizations. The number of these studies that have been developed within a substantial theoretical framework in the manner of the broader literature really is quite limited.

As a result of the science education literature having this character, there is limited value to summarizing the results of it in comprehensive manner. The resulting catalog of conclusions is not likely to produce a rich collection of generalizations. The broader generalizations about inservice education that come from the reviews of the general literature, however, seem worthy of consideration because they provide a framework for considering the research on science inservice education and offer potential guidance to future research to be conducted specifically in science education.

Therefore, what follows is a brief characterization of generalizations drawn from several extant reviews of the general inservice-education research literature, along with *examples* of specific science-education studies which are not included in these reviews but illustrate how science-education studies fit into this broader conceptual landscape. This research literature on inservice education will be considered in the following three categories:

- individual teacher development
- teacher development in a school context
- teacher development to aid curricular change

with definitions and purposes described for each in the sections that follow.

Individual Teacher Development

A substantial amount of inservice education for teachers is conducted in isolation from the teacher's specific school context or particular curricular changes that are being initiated. It is provided for the potential benefit to that teacher's instruction and its degree of success will be judged by competencies that teacher acquires or by improvement in that teacher's teaching. In the main, it is not assessed in terms of its contribution to some overall curricular or instructional direction established for a school or program.

Several recent reviews include attention to inservice education of this orientation. In the review of research on staff development cited earlier, Sparks and Loucks-Horsley (1990) reviewed research pertaining to various models of staff development that had an individual teacher orientation in contrast to those based on district or state plans of teacher development. The five models of staff development cited earlier provided the categories within which they reviewed the literature on staff development.

Burden (1990) reviewed the research on teachers' personal and professional developmental changes with attention to different domains of developmental changes and various means of promoting developmental growth. In contrast to this focus on individual development, Zeichner and Gore (1990) reviewed the research on teacher socialization. While this label may seem to imply that this literature belongs in the next section to be addressed below—teacher development in a school context—socialization operates for individuals in the broader context and must be addressed here with respect to the individual as well.

Of the five models of teacher development addressed by Sparks and Loucks-Horsley, the one having the most attention in the research literature is training. Most prominent in their review of the research on training is the work of Joyce and Showers which is addressed rather comprehensively in their 1988 book. Training as defined by Joyce and Showers includes not only transmitting knowledge about particular teaching skills (or whatever the object of the training is) but providing opportunities to practice the skill and receive coaching on the job by peers or expert trainers in real classroom settings. These processes can be quite time-consuming but it is clear that teachers can learn to use new strategies that are quite different from their former practice. In their summary of this research Sparks and Loucks-Horsley (1990) indicate that:

It is possible to say with some confidence which training elements are required to promote the attainment of specific outcomes. Likewise, research on coaching has demonstrated the importance of in-classroom assistance to teachers (by an expert or by a peer) for the transfer of training to the classroom. (p. 247)

In his review of the research on teacher development, Burden (1990) addresses a number of different theories of adult development including ones familiar to educators in the context of lower-age schooling such as Piaget's (1963) theory of cognitive development and Kohlberg's (1969) theory of value or moral development. In addition, he examines research on the developmental stages of Loevinger (1966) regarding development of the ego or self, Hunt (1971) regarding conceptual development, Perry (1970) with respect to epistemological or ethical development, and Selman (1980) concerning interpersonal development. He goes on to examine the stages of development proposed by other researchers with respect to both preservice and inservice teacher education. Often built to some extent on the theories of adult development just cited, these scholars have proposed theories on the stages of *professional* development of teachers. Four such theories were considered in the realm of preservice teacher education and seven such descriptions of stages of development in the case of inservice education.

In addition to describing these many developmental theories, Burden goes on to review the research on promoting the developmental growth postulated by these theories and to discuss some of the implications of these findings. Among others, this review attends to induction programs for the beginning teacher, staff development programs for teachers, and supervision of teachers on the job. Although many promising practices are addressed, Burden places strong emphasis on the need to recognize that many of these theories "are not fully articulated concepts" (p. 325). More research is needed in this arena and, in addition, training programs initiated for such purposes need to be evaluated on a continuing basis.

Another approach to teacher development considers it from the standpoint of socialization. Zeichner and Gore (1990) reviewed a substantial body of research on teacher socialization and examined competing explanations of this socialization, along with examining issues pertaining to the relationship of this research to teacher education. Although most of their attention is given to socialization in the context of preservice teacher education, they address socialization in the workplace as well. They do not offer prescriptions as to how inservice education should be conducted, but they provide important insights as to how knowledge in this area should influence practice. In particular, they note the importance of this understanding for educational reform endeavors.

The vast majority of research on inservice teacher development in the area of science has not been conceptualized within the broader theoretical frameworks reflected in the previous reviews. These science education studies largely have been focused on measuring the outcomes of various inservice education endeavors, those outcomes mainly being whatever the developers of the inservice education program had decided they intended to pursue. For example, Kimmel and Tomkins (1985) assessed the impact of some energy workshops for elementary teachers by giving questionnaires to participants at the end. This self-report information indicated that the objectives of the workshop had been met. In a similar context, Lawrenz (1985) evaluated the results of some energy workshops for teachers in a somewhat more substantial manner by giving participating teachers pre- and post-measures of beliefs and attitudes about energy and educational practice. Modest changes were found.

Baird, Ellis, and Kuerbis (1989) researched the impact of a program for training teachers in the use of computers in the classroom. They found that the teachers rated the inservice education as effective in meeting its objectives, but when the researchers subsequently measured the level of use of computers in the classroom on the part of these teachers, the results were disappointing. As the authors themselves note, the results are not completely unexpected. They note that the research literature points to the importance of providing a broader base of support in the school context, a point made earlier in this review and one to which we return in the next section where teacher development is addressed in the broader school context. A final example of studies of this type is one by MacDonald and Rogan (1988) that was based on an extensive review of related research literature and evaluated the impact with rather sophisticated observation instruments in the classrooms of participating teachers. They found that instructional changes had occurred.

While occasional studies of this type are well done, there seems to be little reason to do more of them. First, the evidence is fairly clear that inservice education of this nature — independent of the school context and/or broader support for curricular or instructional changes—is unlikely to have a sizable impact on educational practice and thus probably is a poor investment. Secondly, in view of this probable outcome or lacking some more profound theoretical considerations, research on it seems to be a poor investment as well.

Teacher Development in a School Context

In addition to conducting it within a largely individual context, teacher development may be pursued in the context of a given school or even some grouping of schools such as a school district. In this context, there is the potential for both support for *individual* development from peers in the day-to-day work context, as well as the possibility of a direct contribution to *collective* development goals in the school regarding curriculum and/or instruction. In the reviews on teacher development cited earlier, there is attention to this arena as well, although it is comparatively limited. In this section, attention is turned to research that attends to this broader context. Again, attention will be directed to reviews of research from the general body of research literature, followed by examples of science-education studies of this type.

In their review of research on staff development, Sparks and Loucks-Horsley (1990) examined some studies that addressed staff development in the school context and also identified from the staff development literature some of the factors that increase the probability of success. They note that "research points to a common set of attributes of the organizational context without which staff development can have only limited success" (Loucks-Horsley et al., 1987). They claim staff development is most successful in organizations where the following pertain.

1. Staff members have a common, coherent set of goals and objectives that they helped formulate, reflecting high expectations of themselves and their students.
2. Administrators exercise strong leadership by promoting a norm of collegiality, minimizing status differences between their staff members and themselves, promoting informal communication, and reducing their own need to use formal controls to achieve coordination.
3. Administrators and teachers place high priority on staff development and continuous improvement of personal skills, promoting formal training programs, informal sharing of job knowledge, and a norm of continuous improvement applicable to all.
4. Administrators and teachers make heavy use of a variety of formal and informal processes for monitoring progress toward goals, using them to identify obstacles to such progress and ways of overcoming them, rather than using them to make summary judgments regarding the competence of particular staff members (Conley & Bacharach, 1987).
5. Knowledge, expertise, and resources including time are drawn on appropriately, yet liberally, to initiate and support the pursuit of staff-development goals. (p. 245)

As mentioned in the previous section on teacher development, given the knowledge we have today, there is little reason today to pursue inservice education independent of thorough attention to the broader school context and a systemic approach to making educational changes. Sparks and Loucks-Horsley (1990) conclude from their review that abundant research confirms the need for the following:

(a) schools with norms that support collegiality and experimentation; (b) district and building administrators who work with the staff to clarify goals and expectations and actively commit to and support teachers' efforts to change their practice; (c) efforts that are strongly focused on changes in curricular-, instructional-, and classroom-management practices with improved student learning as the goal; and (d) adequate, appropriate staff-development experiences with follow-up assistance that continues long enough for new behaviors to be incorporated into ongoing practice. (p. 247)

Burden's review (1990) of the literature on stages of teacher development also reflects a view of teacher development that is much broader than inservice education per se. He gives attention to a variety of approaches, including collaborative approaches, supervision, and personalized support systems. A review of the research on teacher induction programs and internships (Huling-Austin, 1990) also is of interest here. Again, the focus is not on inservice education in the manner of traditional classes but on mentoring, flexible induction programs, and collaboration.

Supervision often is an important part of teacher development in the school context. Although it also includes attention to preservice education, a review of the research on supervision (Glickman & Bey, 1990) primarily addresses on-the-job supervision. Based on their review, Glickman and Bay are of the persuasion: "That the supervision of instruction is vital to school success appears nondebatable" (p. 549). They focus on *direct supervision*, that is, practices that "focus narrowly on direct, one-to-one instructional work with teachers." (p. 549). They address the following with respect to direct supervision: philosophy and history, purposes and outcomes, survey of existing supervision practices, orientations to supervision (i.e., behaviors, attitudes, and approaches), variables, nature of feedback, training, particular models, and organizational considerations.

An occasional study of inservice education in science gives attention to the broader contextual matters addressed in the reviews of general research. An example of a study that tangentially attends to this concern is one by Bowyer, Ponzio, and Lundholm (1987) in which a well-developed workshop was made available both through school districts and organizational agencies such as teacher centers or professional associations. The workshops were found to be more successful when offered by the school districts, possibly because of "stronger administrator involvement in customizing workshops for their teachers" (p. 816).

In another instance, Lombard, Konicek, and Schultz (1985) evaluated an inservice education program that extended into the school context. With a sound conceptual framework focused on teacher use of a learning cycle, attention to the several stages of a sound training program cited earlier, and attention to the stages of concern of the participants at various points in the program, they began to see changes in the instructional approaches teachers used.

While science inservice education studies that are firmly grounded in the context of a school are relatively uncommon, a considerable number of studies have been conducted in the context of curricular change endeavors, some of which attend to school context. It is to the context of curricular change endeavors that attention is now turned.

Teacher Development to Aid Curricular Change

Ever since the birth of the science curricular reform movement in the late 1950s, a large portion of science teacher education has been connected in some way to attempts to introduce curricular change. Throughout a substantial portion of the time since, the National Science Foundation has supported inservice education for science teachers—often with the designation of teacher institute—which typically was associated in some way with the implementation of curricular programs which were intended to be different from current practice. Looking at inservice education (i.e., efforts to foster teacher development) in this context requires attention to a fairly broad body of research literature on educational change; it is broader than can be addressed fully here, but some prominent aspects will be highlighted.

The definitive review of the literature on educational change is by Fullan and Steigelbauer (1991). Given that inservice education typically is conducted to produce educational change of some type, this literature would seem to provide an essential foundation for thinking about inservice education. Two aspects of this literature seem particularly pertinent here. First, substantial educational change generally is the result of systemic efforts. That is, it is not the result of any one action for improving education, such as inservice education. It is the result of some combination of endeavors that relate to each other and in combination have a significant impact in the context in which change is being sought. Second, even though inservice education by itself probably will not have a dramatic effect in terms of educational improvement, it probably is an essential ingredient in most systemic efforts at educational improvement.

Although not directly focused on inservice education, meta-analyses of science-education research in the early 1980s did touch on the influence of inservice education in curriculum change (Anderson, 1983; Shymansky et al., 1990; Shymansky, Kyle, & Alport, 1983). The results of these meta-analyses are not in opposition to the generalizations from Fullan cited earlier.

While a focus on the classroom impact of science inservice education or on its role in the curricular change is not typical, occasional attention to this broader context has appeared over a long period of time. For example, in 1972 Anderson measured the impact of a science education improvement project that, though heavily dependent on inservice education, was built on a broader program of support for elementary science-education change. Based on systematic analysis of changes in teaching practice, the program was judged to have a significant impact. As has been reported in many other instances, however, informal observation some years later indicated that the curricular changes had not persisted. In contrast, a recent case study of a long-term comprehensive school district elementary school science-improvement endeavor (Anderson, 1990) showed lasting results. In this case, inservice education had an important role to play, but it was but one part of a much more comprehensive program.

CONCLUSIONS

A lengthy review of research such as this one leads to both conclusions that are stated with considerable confidence and topics that clearly are in need of more study.

Where We Stand

Teacher education is an area of obvious high current concern among the public as well as professionals. It also is a complex matter, an understanding not always present among the public and policymakers. The research on this topic is extensive and varied. From this research, it is clear that if one wishes to teach some specific competency, it generally can be done given moderate resources and a specific focus on the competency. When looking at the totality of a teacher-education endeavor, however, the picture is not so clear; such research generally does not have the definitive answers we seek.

Given its complexity and many facets, it becomes apparent that research must attend to science teacher education in a comprehensive way. It must be examined in its broadest context—including preparation in the subject-matter major—and studied over the long term. It is apparent from examining the extant literature that most science teacher education research is narrow, rather than comprehensive, and generally has a fairly limited theoretical foundation.

The complexity of teacher education (including both preservice and inservice), the lengthy period of time involved, and its many interconnections with the process of change in educational practice in the schools lead to the opinion that too much is expected of preservice teacher education by many of the current reformers. Much of the important teacher education must take place in the school context with teachers who have gotten far enough beyond the trauma of initial job survival that they can reflect on their work in deep ways. In this *classroom* context there is the potential of addressing the essence of both constructivist learning and teaching. Teachers' professional growth also can be addressed in the *school* context and changes in the organization itself can be initiated in a systemic manner. Inservice education has the *potential* of moving much beyond what preservice education can accomplish. Although important and influential, improved preservice teacher education will never be the key impetus to educational reform claimed by some leaders in the public arena.

Future Research

Although in the recent past there has been abundant research on student learning having an orientation often called constructivist, the big advances in understanding about student learning have not been matched by equivalent advances in understanding about teaching. How to teach under real world conditions in such a manner as to foster this kind of learning is not as well understood as learning per se. Given this situation, it is not surprising that even less is known about how to help teachers learn to teach in this manner. But given the direction that research and development are taking in both instruction and curriculum, it seems obvious that research in science teacher education needs to move in this direction as well (i.e., researching how to teach teachers to teach in a constructivist manner).

It also seems important to study science teacher education in the broadest possible context. Teacher education has so many facets, is interactive with so many other aspects of education, and extends over such a long portion of a teacher's career that narrow studies on some limited facet of teacher education, independent of a broader context, seem to have limited potential. In particular, much more attention should be given to researching sophisticated inservice education in the school context where the goal is to effect profound organizational changes as well as individual teacher changes. Some of the research in noneducational settings potentially has important insights to offer. The work of Senge (1990), in the business world, with his prime focus on systems thinking and business organizations as "learning organizations," provides an outstanding example. For a discussion of the potential of this organizational theory for educational situations, see the literature review on curriculum reform by Anderson et al. (1992).

Theoretical perspectives need far more attention than they have been given in most science teacher education. Although it has received limited attention there as well, research in the area of general teacher education has attended to theoretical considerations far more than most of the science-specific research. Finally, teacher education could profit from being studied from *multiple* perspectives. Just as curriculum research would benefit from being examined from multiple perspectives (Anderson, 1992), science teacher education needs to be considered from multiple perspectives. For example, if one were studying competency-based science teacher education—which reflects a particular theoretical orientation—it would be helpful to look at it from psychological, sociocultural, philosophical, and subject-matter perspectives. Different perspectives often reflect different research methodologies and conducting such a study with thorough attention to all of these perspectives may not be feasible. It is possible, however—and not too much to expect—to conduct an analysis that defends the particular perspective(s) taken, explains the limitations of the approach taken, and advocates additional ways other researchers may approach the analysis to provide a more complete picture.

Science teacher education research has a major role to play in improving teaching and learning science in schools. It has the potential of yielding much more than it has in the past. With additional interest and considerably improved efforts, such research can narrow the gap between potential and reality.

References

Aikenhead, G. S. (1984). Teacher decision making: The case of Prairie High. *Journal of Research in Science Teaching, 21,* 167–186.

Aikenhead, G. S. (1986). Preparing undergraduate science teachers in S/T/S: A course in epistemology and sociology of science. In R. K. James (Ed.), *Science, technology, and society: Resources for science education: 1985 AETS yearbook* (pp. 56–64). Columbus, OH: Association for the Education of Teachers in Science. (ERIC Document Reproduction Service No. ED 280 673)

Allender, J. (1991). *Imagery in teaching and learning: An autobiography of research in four world views.* New York: Praeger.

American Association for the Advancement of Science. (1989). *Project 2061: Science for all Americans.* Washington, DC: Author.

American Association of Colleges for Teacher Education. (1988). *Teaching teachers: Facts and figures,* Washington, DC: Author.

Anderson, C. (1988). The role of education in academic disciplines in teacher education. In A. Woofolk (Ed.), *Research perspectives on the graduate preparation teachers* (pp. 88–107). Englewood Cliffs, NJ: Prentice-Hall.

Anderson, R. D. (1983). A consolidation and appraisal of science meta-analysis. *Journal of Research in Science Teaching, 20,* 497–509.

Anderson, R. D. (1990). *District-level implementation of elementary school science: A case study of activity-based learning in science.* Paper presented at the annual convention of the National Association for Research in Science Teaching, Atlanta, GA.

Anderson, R. D. (1992). Perspectives on complexity: An essay on curricular reform. *Journal of Research in Science Teaching, 29,* 861–876.

Anderson, R. D., Anderson, B. L., Varanka-Martin, M. A., Romagnano, L., Bielenberg, J., Mieras, B., & Whitworth, J. (1992). *Review of literature pertaining to curriculum reform in science, mathematics and higher order thinking across the disciplines.* Boulder, CO: University of Colorado.

Anderson, R. D., James, H. H., & Struthers, J. A. (1974). The teaching strategies observation differential. In G. Stanford & A. Roark (Eds.), *Human interaction in education.* Boston, Allyn & Bacon.

Andrews, T. E., & Barnes, S. (1990). Assessment of teaching. In W. R. Houston (Ed.), *Handbook of research on teacher education.* New York: Macmillan.

Apple, M. (1986). *Teachers and texts: A political economy of class and gender relations in education.* New York: Routledge & Kegan Paul.

Applegate, J. (1987). Early field experiences: Three viewpoints. In M. Haberman & J. M. Backus (Eds.), *Advances in teacher education* (Vol. 3, pp. 75–93). Norwood, NJ: Ablex.

Atwood, R. K. (1973). *Elements of science teacher education as abstracted from ERIC - AETS: In search of promising practices in science teacher education.* Paper presented at the annual meeting of the Association for the Education of Teachers in Science.

Ausubel, D. (1968). *Educational psychology.* New York: Holt, Rinehart, & Winston.

Baird, J. R., Fensham, P. J., Gunstone, R. F., & White, R. T. (1991). The importance of reflection in improving science teaching and learning. *Journal of Research in Science Teaching, 28,* 163–182.

Baird, W. E. (1989). Status of use: Microcomputers and science teaching. In J. D. Ellis (Ed.), *The 1988 AETS yearbook: Information technology and science education* (pp. 85–104). Columbus, OH: Association for the Education of Teachers in Science. (ERIC Document Reproduction Service No. ED 307 114)

Baird, W. E., Ellis, J. D., & Kuerbis, P. J. (1989). ENLIST micros: Training science teachers to use microcomputers. *Journal of Research in Science Teaching, 26,* 587–598.

Barrow, L. H. (1987). Status of elementary science teacher education in New England. *Science Education, 71*, 229–237.

Barrow, L. H. (1991). Status of preservice elementary science education. *Journal of Elementary Science Education, 3*, 14–25.

Barufaldi, J., Huntsberger, J., & Lazarowitz, R. (1976). Changes in attitude of preservice elementary education majors toward inquiry teaching strategies. *School Science and Mathematics, 76*, 420–424.

Berkland, T. (1974). An investigation of the understanding of science processes and attitudes toward science of prospective elementary teachers from an unstructured science foundations course and non-science students from a structured earth science course. *Dissertation Abstracts International, 34*, 5741A–5742A. (University Microfilms No. 74–7353)

Berliner, D. (1985). Laboratory settings in the study of teacher education. *Journal of Teacher Education, 36*(6), 2–8.

Bestor, A. (1954). How should America's teachers be educated? *Teachers College Record, 56*, 16–19.

Bethel, L. (1984). Science teacher preparation and professional development. In D. Holdzkom & P. Lutz (Eds.), *Research within reach: Science education*. Washington, DC: National Science Teachers Association.

Beyer, L., & Zeichner, K. (1982). Teacher training and educational foundations: A plea for discontent. *Journal of Teacher Education, 33*(3), 18–23.

Blosser, P. (1986). *The preparation of teachers for middle school science*. Paper presented at the United States–Japan seminar on Science Education, Honolulu, HI. (ERIC Document Reproduction Service No. ED 278 568)

Borman, K. M. (1990). Foundations of education in teacher education. In W. R. Houston (Ed.), *Handbook of Research on Teacher Education* (pp. 393–402). New York: Macmillan.

Borrowman, M. (1956). *The liberal and technical in teacher education: A historical survey of American thought*. New York: Columbia University Teachers College Bureau of Publication.

Bowyer, J., Ponzio, R., & Lundholm, G. (1987). Staff development and science teaching: An investigation of selected delivery variables. *Journal of Research in Science Teaching, 24*, 807–819.

Brehm, S. A. (1968). The role of education methods courses instructors in the preparation of science teachers. *School Science and Mathematics, 68*, 72–77.

Brickhouse, N. (1990). Teachers' beliefs about the nature of science and their relationship to classroom practice. *Journal of Teacher Education, 41*(3), 53–62.

Brophy, J. E., & Good, T. L. (1985). Teacher behaviors and student achievement. In M. C. Wittrock (Ed.), *Handbook of research on teaching* (3rd ed.). New York: Macmillan.

Broudy, H. S. (1990). Case studies—why and how. *Teachers College Record, 91*, 449–459.

Bruner, J. S. (1960). *The process of education*. Cambridge, MA: Harvard University Press.

Bullough, R. V., Jr. (1989). *First-year teacher: A case study*. New York: Teachers College Press.

Burden, P. R. (1990). Teacher development. In W. R. Houston (Ed.), *Handbook of research on teacher education*. pp. 311–328. New York: Macmillan.

Butts, D. (1989). Strengthening the preservice formation of secondary science teachers: Inside-out not Outside-in. In J. Barufaldi (Ed.), *Improving preservice/inservice science teacher education: Future perspectives. 1987 AETS Yearbook* (pp. 123–130). Columbus, OH: Association for the Education of Teachers in Science. (ERIC Reproduction Service No. ED 309 922)

Campbell, R. (1974). The effects of instruction in the basic science process skills on attitudes, knowledge and lesson planning practices of prospective elementary school teachers. *Dissertation Abstracts International, 34*, 4946A–4947A (University Microfilms No. 74–2628)

Campbell, R. L., & Okey, J. R. (1977). Influencing the planning of teachers with instruction in science process skills. *Journal of Research in Science Teaching, 14*, 231–234.

Carlsen, W. S. (1987, April). *Why do you ask? The effects of science teacher subject-matter knowledge on teacher questioning and classroom discourse*. Paper presented at the annual meeting of the American Educational Research Association. (ERIC Document Reproduction Service No. ED 293 181)

Carnegie Task Force on teaching as a profession. (1986). *A nation prepared: Teachers for the 21st century*. Washington, DC: Carnegie Forum on Education and the Economy.

Clark, C., & Peterson, P. (1986). Teachers' thought processes. In M. Wittrock (Ed.), *Handbook of research on teaching* (3rd ed.). New York: Macmillan.

Clark, D. L., & Marker, G. (1975). The institutionalization of teacher education. In K. Ryan (Ed.), *Teacher education (74th yearbook of the National Society for the Study of Education, Part II* (pp. 53–86). Chicago: University of Chicago Press.

Cobern, W., & White, J. (1988, January). *Science teacher training in a five-year developmental liberal arts program*. Paper presented at the annual meeting of the southwest regional division of the Association for the Education of Teachers in Science.

Cochran-Smith, M., & Lytle, S. (1990, March). Research on teaching and teacher research: The issues that divide. *Educational Researcher, 19*, 2–11.

Cogan, M. L. (1973). *Clinical supervision*. Boston: Houghton Mifflin.

Cogan, M. L. (1975). Current issues in the education of teachers. In K. Ryan (Ed.), *Teacher education (74th yearbook of the National Society for the Study of Education, Part II)* (pp. 204–229). Chicago: University of Chicago Press.

Conant, J. B. (1963). *The education of American teachers*. New York: McGraw-Hill.

Confrey, J. (1990). A review of the research on student conceptions in mathematics, science, and programming. *Review of Research in Education, 16*, 3–56.

Conley, S., & Bacharach, S. (1987). The Holmes Group report: Standards, hierarchies, and management. *Teachers College Record, 88*, 340–347.

Connelly, F. M., & Clandinin, D. J. (1986). On narrative method, personal philosophy, and narrative unities in the story of teaching. *Journal of Research in Science Teaching, 23*, 293–310.

Craven, G. (1977). *Science teacher education: Vantage point 1976. 1977 AETS Yearbook*. Columbus OH: ERIC Information Analysis Center for Science, Mathematics, and Environmental Education.

Cremin, L. (1978). *The education of the educating professions (The 19th Charles W. Hunt Lecture)*. Chicago: American Association of Colleges for Teacher Education.

Cronin-Jones, L. (1991). Interpretive research methods as a tool for educating science teachers. In J. Gallagher (Ed.), *Interpretive research in science education* (NARST Monograph No. 4, pp. 217–234). Manhattan, KS: National Association for Research in Science Teaching.

Cruickshank, D. (1984). *Models for the preparation of America's teachers*. Bloomington, IN: The Phi Delta Kappa Educational Foundation.

Cuban, L. (1992). Managing dilemmas while building professional communities. *Educational Researcher, 21*, 4–11.

Darling-Hammond, L., & Hudson, L. (1990). Precollege science and mathematics teachers: Supply, demand and quality. *Review of Research in Education 16*, 223–261.

Darling-Hammond, L., & Snyder, J. (1992). Curriculum studies and the traditions of inquiry: The scientific tradition. In P. Jackson (Ed.), *Handbook of research on curriculum* (pp. 41–78). New York: Macmillan.

Darling-Hammond, L., Wise, A., & Pease, S. (1983). Teacher evaluation in the organizational context: A review of the literature. *Review of Educational Research, 53*, 285–328.

DeBruin, J. (1977). The effect of field-based elementary science teacher education programs on undergraduates. In M. Piper & K. Moore (Eds.), *Attitudes toward science: Investigations*. Columbus, OH: SMEAC Information Reference Center.

DeBruin, J., Gress, J., & Underfer, J. (1987). CBTE: An individualized elementary science teacher education program. In J. Penick (Ed.), *Focus on excellence: Preservice elementary teacher education in science* (Vol. 4, No. 2). Washington, DC: National Science Teachers Association.

DeBoer, G. (1991). *A history of ideas in science education*. New York: Teachers College Press.

Denham, C., & Lieberman, A. (Eds.). (1980). *Time to learn*. Washington, DC: National Institute of Education.

Denton, J. (1982). Early field experience influence on performance in subsequent course work. *Journal of Teacher Education, 33*(2), 19–23.

DeRose, J. V., Lockard, J. D., & Paldy, L. G. (1979). The teacher is the key: A report on three NSF studies. *The Science Teacher, 46*(4), 31–37.

DeTure, L. (1979). Relative effects of modeling on the acquisition of wait-time by preservice elementary teachers and concomitant changes in dialogue patterns. *Journal of Research in Science Teaching, 16*, 553–562.

Dewey, J. (1904). The relation of theory to practice in education. In C. McMurry (Ed.), *The relation of theory to practice in the education of teachers: Third yearbook for the national society for the scientific study of education*: (pp. 9–30) Chicago: University of Chicago Press.

Donnellan, K. (1982). *NSTA elementary teacher survey on preservice preparation of teachers of science at the elementary, middle, and junior high school levels*. Washington, DC: National Science Teachers Association.

Doody, W., & Robinson, D. (1988, April). *National Science Foundation model programs of preparation of middle school science and mathematics teachers*. Paper presented at the annual meeting of the Association for the Education of Teachers of Science, St. Louis, MO.

Doyle, W. (1978). Paradigms for research on teacher effectiveness. In L. Shulman (Ed.), *Review of research in education* (Vol. 5). Itasca, IL: Peacock.

Doyle, W. (1990). Classroom knowledge as a foundation for teaching. *Teachers College Record, 91*, 347–360.

Driver, R., & Bell, B. (1986). Students' thinking and the learning of science: A constructivist view. *School Science Review, 67*, 443–456.

Druva, C. A. & Anderson, R. D. (1983). Science teacher characteristics by teacher behavioral and by student outcome: A meta-analysis. *Journal of Research in Science Teaching, 20*, 467–479.

Dunkin, M., & Biddle, B. (1974). *The study of teaching*. New York: Holt, Rinehart, & Winston.

Duschl, R. A. (1983). The elementary level science methods course: Breeding ground of an apprehension toward science? A case study. *Journal of Research in Science Teaching, 20*, 745–754.

Duschl, R. A. (1990). *Restructuring science education*. New York: Teachers College Press.

Duschl, R. A., & Wright, E. (1989). A case study of high school teachers' decision making models for planning and teaching science. *Journal of Research in Science Teaching, 26*, 467–501.

Easley, J. (1982). Naturalistic case studies exploring social-cognitive mechanisms, and some methodological issues in research on problems of teachers. *Journal of Research in Science Teaching, 19*, 191–203.

Eisenhart, M., Behm, L., & Romagnano, L. (1991). *Learning to teach: Developing expertise or rite of passage? Journal of Education for Teaching, 17*, 51–71.

Ellis, J. D. (Ed.). (1989). *1988 AETS yearbook: Information technology and science education*. Columbus, OH: Association for the Education of Teachers in Science. (ERIC Document Reproduction Service No. ED 307 114)

Erickson, F. (1986). Qualitative methods for research on teaching. In M. Wittrock (Ed.), *Handbook of research on teaching* (3rd ed.). New York: Macmillan.

Feiman-Nemser, S. (1983). Learning to teach. In L. S. Shulman & G. Sykes (Eds.), *Handbook of teaching and policy* (pp. 150–170). New York: Longman.

Feiman-Nemser, S. (1990). Teacher preparation: Structural and conceptual alternatives. In W. R. Houston (Ed.), *Handbook of research on teacher education* (pp. 212–233). New York: Macmillan.

Feiman-Nemser, S., & Buchmann, M. (1985). Pitfalls of experience in teacher preparation. *Teachers College Record, 87*, 53–65.

Feistritzer, E. G. (1986). *Profiles of teachers in the U.S.* Washington, DC: National Center for Education Information.

Fensham, P. (1992). Science and technology. In P. Jackson (Ed.), *Handbook of research on curriculum* (pp. 789–829). New York: Macmillan.

Fenstermacher, G. (1978). A philosophical consideration of recent research on teacher effectiveness. In L. Shulman (Ed.), *Review of research in education* (Vol. 6, pp. 157–185). Itasca, IL: Peacock.

Fenstermacher, G. (1986). Philosophy of research on teaching: Three aspects. In M. Wittrock (Ed.), *Handbook of research on teaching* (3rd ed., pp. 37–49). New York: Macmillan.

Field, K. (1993). Reforming Teacher Education: Two visions. In Field, K., Kaufman, E. and Saltzman, C. (Eds.), *Emotions and Learning Reconsidered: International Perspectives*. New York: Gardner Press, Inc.

Finkelstein, B. (1982). Technicians, mandarins, and witnesses: Searching for professional understanding. *Journal of Teacher Education, 33*(3), 25–27.

Fraser, B. (1989). *Learning environment research in science classrooms: Past progress and future prospects* (NARST Monograph No. 2). Cincinnati, OH: National Association Research in Science Teaching.

Fullan, M. G., & Steigelbauer, S. (1991). *The new meaning of educational change*. New York: Teachers College Press.

Fuller, F. F. (1969). Concerns of teachers: A developmental conceptualization. *American Educational Research Journal, 6*, 207–226.

Gabel, D., & Rubba, P. (1979). Attitude changes of elementary teachers according to the curriculum studied during workshop participation and their role as model science teachers. *Journal of Research in Science Teaching, 16*, 19–24.

Gallagher, J. J. (1984). Qualitative methods for the study of schooling. In B. Fraser & D. Treagust (Eds.), *Looking into classrooms*. Perth: Western Australian Institute of Technology.

Gallagher, J. J. (1989). Research on secondary school science teachers' practices, knowledge and beliefs: A basis for restructuring. In M. Matyas, K. Tobin, & B. Fraser (Eds.), *Looking into windows: Qualitative research in science education* (pp. 43–57). Washington, DC: American Association for the Advancement of Science.

Gallagher, J. J. (Ed.). (1991). *Interpretive research in science education*. National Association for Research in Science Teaching Monograph No. 4.

Gauthier, P. (1978). *Practical reasoning*. Oxford: Oxford University Press.

Gideonse, H. D. (1988). *Relating knowledge to teacher education: Re-

sponding to NCATE's knowledge base and related standards. Washington, DC: American Association of Colleges for Teacher Education.

Glickman, C. D., & Bey, T. M. (1990). Supervision. In W. R. Houston (Ed.), *Handbook of research on teacher education* (pp. 549–566). New York: Macmillan.

Goodland, J. I. (1990, November). Better teachers for our nation's schools. *Phi Delta Kappan, 72.*

Gregory, E., & DeTure, L. (1992, March). *A program to improve elementary teachers' preparation in science, phase II.* Paper presented at the annual meeting of the National Association for Research in Science Teaching, Cambridge, MA.

Griffin, G. A. (1989). A state program for the initial years of teaching. *Elementary School Journal, 89,* 395–403.

Grossman, P. L., Wilson, S. M., & Shulman, L. S. (1989). Teachers of substance: Subject matter knowledge for teaching. In M. Reynolds (Ed.), *Knowledge base for beginning teachers* (pp. 23–36). New York: Pergamon.

Gunstone, R. F. (1988). Learners in science education. In P. Fensham (Ed.), *Development and dilemmas in science education.* New York: Falmer.

Gunstone, R. F., Slattery, M., & Baird, J. R. (1989, March). *Learning about learning to teach: A case study of pre-service teacher education.* Paper presented at the annual meeting of the American Educational Research Association, San Francisco.

Guyton, E., & McIntyre, D. J. (1990). Student teaching and school experiences. In W. R. Houston (Ed.), *Handbook of research on teacher education* (pp. 514–534). New York: Macmillan.

Haberman, M. (1983). Research on preservice laboratory and clinical experiences: Implications for teacher education. In K. Howey & W. Gardner (Eds.), *The education of teachers* (pp. 98–117). New York: Longman.

Harms, N. C., & Yager, R. E. (Eds.). (1978). *What research says to the science teacher* (Vol. 3). Washington, DC: National Science Teachers Association.

Harvard Committee on General Education. (1945). *General education in a free society.* Cambridge, MA: Harvard University Press.

Hashweh, M. (1987). Effects of subject matter knowledge in teaching of biology and physics. *Teaching and Teacher Education, 3,* 109–120.

Hawk, P., & Robards, S. (1987). Statewide teacher induction programs. In D. Brooks (Ed.), *Teacher induction: A new beginning* (pp. 33–44). Reston, VA: Association of Teacher Educators.

Hawley, W. D. (1990). Systematic analysis, public policy-making, and teacher education. In W. R. Houston (Ed.), *Handbook of research on teacher education.* (pp. 136–156). New York: Macmillan.

Heikkinen, M. W. (1988). The academic preparation of Idaho science teachers. *Science Education, 72,* 63–71.

Helgeson, S. L., Blosser, P. E., & Howe, R. W. (1977). *The status of precollege science, mathematics, and social science education. 1955–75. Vol. 1: Science Education.* Columbus, OH: Ohio State University, Center for Science and Mathematics Education.

Henry, M. A. (1988). Multiple support: A successful model for inducting first-year teachers. *Teacher Educator, 24*(2), 7–12.

Herron, M. D. (1971). The nature of scientific enquiry. *School Review, 79*(2), 171–212.

Hodson, D. (1988). Toward a philosophically more valid science curriculum. *Science Education, 72,* 19–40.

Holmes Group. (1986). *Tomorrow's teachers.* East Lansing, MI: Author.

Horak, W., & Slobodzian, K. (1980). Influence of instructional structure and locus of control on achievement of preservice elementary science teachers. *Journal of Research in Science Teaching, 17,* 213–222.

Houston, W. R. (Ed.). (1990). *Handbook of research on teacher education.* New York: Macmillan.

Howey, K. R. (1983). Teacher education: An overview. In K. R. Howey &

W. E. Gardner (Eds.), *The education of teachers* (pp. 6–37). New York: Longman.

Howey, K. R., & Zimpher, N. L. (1986). The current debate on teacher preparation. *Journal of Teacher Education, 37*(5), 41–49.

Howey, K. R., & Zimpher, N. L. (1989). *Profiles of preservice teacher education.* Albany, NY: State University of New York Press.

Howey, K. R., & Zimpher, N. L. (1990). Scholarly inquiry into teacher education in the United States. In R. P. Tisher & M. F. Wideen (Eds.), *Research in teacher education: International perspectives* (pp. 163–190). New York: Falmer.

Huling-Austin, L. (1986). What can and cannot reasonably be expected from teacher induction programs. *Journal of Teacher Education, 37,* 2–5.

Huling-Austin, L. (1990). Teacher induction programs and internships. In W. R. Houston (Ed.), *Handbook of research on teacher education* (pp. 535–548). New York: Macmillan.

Hunt, D. E. (1971). *Matching models in education.* Toronto: Ontario Institute for Studies in Education.

Hurd, P. (1983). State of precollege education in mathematics and sciences. *Science Education, 67,* 57–67.

Jackson, P. (1992). Conceptions of curriculum and curriculum specialists. In P. Jackson (Ed.), *Handbook of research on curriculum* (pp. 3–40). New York: Macmillan.

Jaus, H. (1975). The effects of integrated science process skill instruction on changing teacher achievement and planning practices. *Journal of Research in Science Teaching, 12,* 439–447.

Johnson, D. W., & Johnson, R. T. (1979). Conflict in the classroom: Controversy and learning. *Review of Educational Research, 49,* 51–69.

Johnson, S. M. (1990). *Teachers at work: Achieving excellence in our schools.* New York: Basic Books.

Karplus, R., & Thier, H. (1967). *A new look at elementary school science.* Chicago: Rand-McNally.

Kennedy, M. M. (1990). Choosing a goal for professional education. In W. R. Houston (Ed.), *Handbook of research on teacher education.* (pp. 813–825). New York: Macmillan.

Kimmel, H., & Tomkins, R. P. T. (1985). Energy education workshops for elementary school teachers. *School Science and Mathematics, 85,* 595–602.

Klainin, S. (1988). Practical work and science education I. In P. Fensham (Ed.), *Development and dilemmas in science education* (pp. 169–188). New York: Falmer.

Koballa, T. R. (1986). Teaching hands-on science activities: Variables that moderate attitude-behavior consistency. *Journal of Research in Science Teaching, 23,* 493–502.

Koballa, T. (1992). Persuasion and attitude change in science education. *Journal of Research in Science Teaching, 29,* 63–80.

Koerner, J. (1963). *The miseducation of American teachers.* Boston: Houghton Mifflin.

Kohlberg, L. (1969). Stage and sequence: The cognitive development approach to socialization. In D. Goslin (Ed.), *Handbook of socialization theory and research* (pp. 347–380). Chicago: Rand-McNally.

Krajcik, J., & Penick, J. (1989). Evaluation of a model science teacher education program. *Journal of Research in Science Teaching, 26,* 795–810.

Kuhn, T. (1970). *The structure of scientific revolutions.* Chicago: University of Chicago Press. (Original work published 1962)

Kyle, W. C., Jr. (1991). The reform agenda and science education: Hegemonic control vs. counterhegemony, *Science Education, 75,* 403–411.

Lakebrink, J. (1991). *Teacher preparation in professional development schools.* Paper presented at the annual meeting of the American Association of Colleges of Teacher Education, Atlanta, GA.

Lamb, W. G. (1977). Evaluation of a self-instruction module for training science teachers to ask a wide cognitive variety of questions. *Science Teacher, 61*, 29–39.

Lampert, M. (1984). Teaching about thinking and thinking about teaching. *Journal of Curriculum Studies, 16*, 1–18.

Lampert, M. (1985). How do teachers manage to teach? Perspectives on problems in practice. *Harvard Educational Review, 55*, 178–194.

Lanier, J. E., & Little, J. W. (1986). Research on teacher education. In M. C. Wittrock (Ed.), *Handbook of research on teaching* (3rd ed.). (pp. 527–569). New York: MacMillan.

Lantz, O., & Kass, H. (1987). Chemistry teachers' functional paradigms. *Science Education, 71*, 117–134.

Lawrenz, F. (1975). The relationship between science teacher characteristics and student achievement and attitude. *Journal of Research in Science Teaching, 12*, 433–437.

Lawrenz, F. (1985). Impact on a five week energy education program on teacher beliefs and attitudes. *School Science and Mathematics, 85*, 27–36.

Lawrenz, F., & Cohen, H. (1985). The effect of methods classes and practice teaching on student attitudes toward science and knowledge of science processes. *Science Education, 69*, 105–113.

Lawson, A. (1982). The reality of general cognitive operations. *Science Education, 66*, 229–241.

Lawson, A. E., Nordland, F. H., & DeVito, A. (1975). Relationship of formal reasoning to achievement, aptitudes, and attitudes in preservice teachers. *Journal of Research in Science Teaching, 12*, 423–431.

Lazarowitz, R., & Lee, E. (1976). Measuring inquiry attitudes of secondary science teachers. *Journal of Research in Science Teaching, 13*, 455–460.

Lazarowitz, R., Barufaldi, P. J., & Huntsberger, P. J. (1978). Student teacher's characteristics and favorable attitudes toward inquiry. *Journal of Research in Science Teaching, 15*, 559–566.

Lederman, N. G., & Gess-Newsome, J. (1989, March). *A qualitative analysis of the effects of a microteaching course on preservice science teachers' instructional decisions and beliefs about teaching.* Paper presented at the annual meeting of the National Association for Research in Science Teaching, San Francisco.

Lehman, J. (1985). Survey of microcomputer use in the science classroom. *School Science and Mathematics, 85*, 578–583.

Lehman, J. R. (1986). Microcomputer offerings in science teacher training. *School Science and Mathematics, 86*, 119–125.

Loevinger, J. (1966). The meaning and measurement of ego development. *American Psychologist, 21*, 195–206.

Lombard, A. S., Konicek, R. D., & Schultz, K. (1985). Description and evaluation of an inservice model for implementation of a learning cycle approach in the secondary science classroom. *Science Education, 69*, 491–500.

Lortie, D. (1975). *Schoolteacher: A sociological study.* Chicago: University of Chicago Press.

Loucks-Horsley, S., Harding, C., Arbuckle, M., Murray, L., Dubea, C., & Williams, M. (1987). *Continuing to learn: A guidebook for teacher development.* Andover, MA: Regional Laboratory for Educational Improvement of the Northeast and Islands/National Staff Development Council.

Loucks-Horsley, S., Carlson, M. O., Brink, L. H., Horwitz, P., Marsh, D. D., Pratt, H., Roy, K. R., & Worth, K. (1989). *Developing and supporting teachers for elementary school science education.* Andover, MA: The NETWORK. (ERIC Document Reproduction Service No. ED 314 235).

Loving, C. (1989, March). *Current models in philosophy of science: Their place in science teacher education.* Paper presented at the annual meeting of National Association for Research in Science Teaching, San Francisco.

MacDonald, M. A., & Rogan, J. M. (1988). Innovation in South African science education (part I): Science teaching observed. *Science Education, 72*, 225–236.

Martens, M. L. (1990). *Either/or = neither/nor: Case study of a teacher in change.* Paper presented at the annual meeting of the National Association for Research in Science Teaching, Atlanta, GA.

Matthews, M. R. (1992, March). *Old wine in new bottles: A problem with constructivist epistemology.* Paper presented at the annual meeting of the National Association for Research in Science Teaching, Boston.

McCormick, J. (1991, December 23). A class act for the ghetto. *Newsweek*, pp. 62, 63.

McDiarmid, G. W., Ball, D. L., & Anderson, C. W. (1989). Why staying one chapter ahead doesn't really work: Subject-specific pedagogy. In M. Reynolds (Ed.), *Knowledge base for beginning teachers* (pp. 193–205). New York: Pergamon.

McEwan, H., & Bull, B. (1991). The pedagogic nature of subject matter knowledge. *American Educational Research Journal, 28*, 316–334.

McIntyre, D. J. (1983). *Field experiences in teacher education: From student to teacher.* Washington, DC: Foundation for Excellence in Teacher Education and the ERIC Clearinghouse on Teacher Education.

Mechling, K. (1982). *Survey results: Preservice programs of teachers of science at the elementary, middle, and junior high levels.* Washington, DC: National Science Teachers Association.

Mechling, K., Stedman, C., & Donnellan, K. (1982). Preparing and certifying science teachers. *Science and Children, 20*(2), 9–14.

Merseth, K. (1985). Harvard graduate school midcareer math and science programs. *Journal of College Science Teaching, 14*, 239–241.

Mitchener, C. P. (1991, April). *Understanding and acting on the complexity of subject matter: Schwab and STS education.* Paper presentation at the annual meeting of the National Association of Research for Science Teaching, Lake Geneva, WI.

Mitchener, C. P., & Anderson, R. D. (1989). Teachers' perspective: Developing and implementing an STS curriculum. *Journal of Research in Science Teaching, 26*, 351–369.

Monroe, L. M., Padilla, M. J., Nichols, B. K., & Wiggins, J. R. (1992, March). *Induction: A case study of a second year teacher—Susan.* Paper presented at the annual meeting of the National Association for Research in Science Teaching, Cambridge, MA.

Morrisey, J. (1981). An analysis of studies on changing the attitude of elementary student teachers toward science and science teaching. *Science Education, 65*, 157–177.

Munby, H. (1984). A qualitative study approach to the study of a teacher's beliefs. *Journal of Research in Science Teaching, 21*, 27–38.

Murnane, R. J., Singer, J. D., Willett, J. B., Kemple, J. J., & Olsen, R. J. (1991). *Who will teach?: Policies that matter.* Cambridge, MA: Harvard University Press.

National Board for Professional Teaching Standards. (1989) *Toward high and rigorous standards for the teaching profession.* Washington, DC: Author.

National Council for Accreditation of Teacher Education. (1987). *NCATE standards, procedures, and policies for the accreditation of professional education units: The accreditation of professional education units for the preparation of professional school personnel at basic and advanced levels.* Washington, DC: Author.

National Research Council. (1990) *Precollege science and mathematics teachers: Monitoring supply, demand, and quality.* Washington, DC: National Academy Press.

National Science Teachers Association. (1984). *Standards for the preparation and certification of teachers of science, K–12.* Washington, DC: Author.

National Science Teachers Association. (1987). *Standards for the preparation and certification of teachers of science, K–12.* Washington, DC: Author.

National Science Teachers Association. (1992) *The content core.* Washington, DC: Author.

National Society for the Study of Education. (1932). *A program for teaching science: Thirty-first yearbook of the NSSE.* Chicago: University of Chicago Press.

National Society for the Study of Education. (1947). *Science education in American schools: Forty-sixth yearbook of the NSSE.* Chicago: University of Chicago Press.

National Society for the Study of Education. (1960). *Rethinking science education: Fifty-ninth yearbook of the NSSE.* Chicago: University of Chicago Press.

Newton, D. E., & Watson, F. G. (1968). *Research on science education survey: The status of teacher education programs in the sciences, 1965–67.* Cambridge, MA: Authors.

Nichols, B. K., Monroe, L. M., Padilla, M. J., & Wiggins, J. R. (1992, March). *Induction: A case study of a first year teacher—Melinda.* Paper presented at the annual meeting of the National Association for Research in Science Teaching, Cambridge, MA.

Odell, S., Loughlin, C., & Ferraro, D. (1986–87). Functional approach to identification of new teacher needs in an induction program. *Action in Teacher Education, 8*(4), 51–57.

Osborne, R., & Freyberg, P. (1985). *Learning in science: The implications of children's science.* Portsmouth, NH: Heinemann.

Osborne, R. J., & Wittrock, M. C. (1983). Learning science: A generative process. *Science Education, 67,* 489–508.

Padilla, M. (1983). *Science and the early adolescent.* Washington, DC: National Science Teachers Association.

Pangburn, J. M. (1932). *The evolution of the American teachers college.* New York: Columbia University, Teachers College, Bureau of Publications.

Penick, J. E. (Ed.). (1987). Preservice elementary teacher education in science. *Focus on excellence.* Washington DC: National Science Teachers Association (ERIC Document Reproduction Service No. ED 281 724)

Penick, J. E., & Yager, R. E. (1988). Science teacher education: A program with a theoretical and pragmatic rationale. *Journal of Teacher Education, 39*(6), 59–64.

Pereira, P. (1990, October). *Facets of subject matter.* Paper presented at the Institute for Educational Research at the University of Oslo, Oslo, Norway.

Pereira, P. (1992). *An introduction to curriculum deliberation.* Manuscript submitted for publication.

Perkes, V. (1975). Relationship between a teacher's background and sensed adequacy to teach elementary science. *Journal of Research in Science Teaching, 12,* 85–88.

Perry, W. (1970). *Forms of intellectual and ethical development during the college years.* New York: Holt, Rinehart & Winston.

Peterson, P., Clark, C. M., & Dickerson, W. P. (1990). Educational psychology as a foundation in teacher education: Reforming an old nation. *Teachers College Record, 91,* 322–346.

Piaget, J. (1929). *The child' conception of the world.* London: Routledge & Kegan Paul.

Piaget, J. (1963). *Psychology of intelligence.* Totowa, NJ: Littlefield, Adams.

Piper, M. (1977). Investigation of attitude changes of elementary preservice teachers in a competency-based, field-oriented, science methods course. In M. Piper & K. Moore (Eds.), *Attitudes toward science: Investigations.* Columbus: SMEAC Information Reference Center.

Piper, M., & Hough, L. (1979). Attitudes and open mindedness of undergraduate students enrolled in a science methods course and a freshman physics course. *Journal of Research in Science Teaching, 16,* 193–197.

Pratt, H. (1985). *Factors that affect teachers' decisions on content and process in science courses.* Unpublished manuscript.

Ramsey, G., & Howe, R. (1969). An analysis of research on instructional procedures in secondary school science: Part II. *The Science Teacher, 36*(4), 72–81.

Riley, J. P. (1979). Influence of hands on science process training on preservice teachers' acquisition of process skills and attitude toward science and science teaching. *Journal of Research in Science Teaching, 16,* 373–384.

Riley, J. P. (1980). *The influences of systematic analysis of self and peer taught videotaped lessons on teacher instructional behavior.* Paper presented at the annual meeting of the Association for the Education of Teachers in Science, Anaheim, CA.

Riley, J. P. (1981). The effects of preservice teacher's cognitive questioning level and redirecting on student science achievement. *Journal of Research in Science Teaching, 18,* 303–309.

Riley, J. P., & Padilla, M. (1987). Early childhood and middle grades science programs. In J. Penick (Ed.), *Focus on excellence: Preservice elementary teacher education in science* (Vol. 4, No. 2, pp. 21–26). Washington, DC: National Science Teachers Association.

Rist, R. (1982). On the application of ethnographic inquiry to education: Procedures and possibilities. *Journal of Research in Science Teaching, 19,* 439–450.

Roberts, D. A. (1982). The place of qualitative research in science education. *Journal of Research in Science Teaching, 19,* 277–292.

Roberts, D. A., & Chastko, A. M. (1990). Absorption, refraction, reflection: An exploration of beginning science teacher thinking. *Science Education, 74,* 197–224.

Roth, R. A., & Pipho, C. (1990). Teacher education standards. In W. R. Houston (Ed.), *Handbook of research on teacher education.* (pp. 119–135). New York: Macmillan.

Rowe, M. B. (1973). *Teaching science as continuous inquiry* (2nd ed.). New York: McGraw-Hill.

Rowe, M. B. (1974a). Wait-time and rewards as instructional variables, their influence on language, logic and fate control; part one—wait time. *Journal of Research in Science Teaching, 11,* 81–94.

Rowe, M. B. (1974b). Relation of wait-time and rewards to the development of language, logic, and fate control: Part 2. Rewards. *Journal of Research in Science Teaching, 11,* 291–308.

Russell, T. (1989). Defective, effective, reflective: Can we improve science teacher education programs by attending to our images of teachers at work? In J. Barufaldi (Ed.), *Improving preservice/inservice science teacher education: Future perspectives* (pp. 161–168). Columbus, OH: AETS and ERIC. (ERIC Document Reproduction Service No. ED 309 922)

Rutherford, F. J., & Ahlgren, A. (1990). *Science for all Americans.* Oxford: Oxford University Press.

Sanford, J. P. (1988). Learning on the job: Conditions for professional development of beginning science teachers. *Science Education, 72,* 615–624.

Sarason, S., Davidson, K., & Blatt, B. (1962). *The preparation of teachers: An unstudied problem in education.* New York: Wiley.

Sarason, S. B., Davidson, K. S., & Blatt, B. (1986). *The preparation of teachers: An unstudied problem in education* (rev. ed.). Cambridge, MA: Brookline Books.

Scharmann, L. (1988). Locus of control: A discriminator of the ability to foster an understanding of the nature of science among preservice elementary teachers. *Science Education, 72,* 453–465.

Scheffler, I. (1973). *Reason and teaching.* New York: Bobbs-Merrill.

Schlechty, P. C., & Vance, C. (1981). Do academically able teachers leave education? The North Carolina case. *Phi Delta Kappan, 63,* 106–112.

Schlechty, P. C., & Vance, C. (1983). Recruitment, selection and retention: The shape of the teaching force. *Elementary School Journal, 83,* 469–487.

Schon, D. A. (1983). *The reflective practitioner.* New York: Basic Books.

Schon, D. A. (1987). *Educating the reflective practitioner.* San Francisco: Jossey-Bass.

Schwab, J. J. (1962). The teaching of science as enquiry. In J. Schwab & P. Brandwein (Eds.), *The teaching of science* (pp. 1–103). Cambridge, MA: Harvard University Press.

Schwab, J. J. (1969). The practical: A language for curriculum. *School Review, 78,* 1–23.

Schwab, J. J. (1974). Decisions and choice: The coming duty of science teaching. *Journal of Research in Science Teaching, 11,* 309–317. (Originally presented in plenary session at the 1974 annual meeting of the National Association for Research in Science Teaching in Chicago)

Schwab, J. J. (1978). Education and the structure of the disciplines. In I. W. Westbury & N. J. Wilkof (Eds.), *Science, curriculum, and liberal education* (pp. 229–272). Chicago: University of Chicago Press. (Original work published 1961)

Schwab, J. J., (1983). The practical 4: Something for curriculum professors to do. *Curriculum Inquiry, 13,* 239–265.

Selman, R. L. (1980). *The growth of interpersonal understanding.* New York: Academic Press.

Senge, P. (1990). *The fifth discipline: The art and practice of the learning organization.* New York: Doubleday.

Shrigley, R. L. (1974). The attitude of pre-service elementary teachers toward science. *School Science and Mathematics, 74,* 243–250.

Shrigley, R. L., Koballa, T. R., & Simpson, R. D. (1988). Defining attitudes for science educators. *Journal of Research in Science Teaching, 25,* 659–678.

Shulman, L. S. (1984). The practical and the eclectic: A deliberation on teaching and educational research. *Curriculum Inquiry, 14,* 183–200.

Shulman, L. S. (1986a). Those who understand: Knowledge growth in teaching. *Educational Researcher, 15*(2), 4–14.

Shulman, L. S. (1986b). Paradigms and research programs in the study of teaching: A contemporary perspective. In M. C. Wittrock (Ed.), *Handbook of research on teaching* (3rd ed., pp. 3–36). New York: Macmillan.

Shulman, L. S. (1987). Knowledge and teaching: Foundations of new reform. *Harvard Educational Review, 57,* 1–22.

Shulman, L. S., & Tamir, P. (1973). Research on teaching in the natural sciences. In R. Travers, (Ed.), *Second handbook of research on teaching.* (pp. 1098–1148) Chicago, IL: Rand McNally.

Shymansky, J. A., Hedges, L. V., & Woodworth, G. (1990). A reassessment of the effects of inquiry-based science curricula of the 60's on student performance. *Journal of Research in Science Teaching, 27,* 127–144.

Shymansky, J. A., Kyle, W. C., & Alport, J. M. (1983). The effects of new science curricula on student performance. *Journal of Research in Teaching, 20,* 387–404.

Sirotnik, K. A. (1990). On the eroding foundations of teacher education. *Phi Delta Kappan, 71,* 710–716.

Smith, M. L. (1982). Benefits of naturalistic methods in research in science education. *Journal of Research in Science Teaching, 19,* 627–638.

Solomon, J. (1988). The dilemma of science, technology, and society education. In P. Fensham (Ed.), *Development and dilemmas in science education.* New York: Falmer.

Sparks, D., & Loucks-Horsley, S. (1990). Models of staff development. In W. R. Houston (Ed.), *Handbook of research on teacher education.* (pp. 234–250). New York: Macmillan.

Sparks, R., & McCallon, E. (1974). Microteaching: Its effect on student attitudes in an elementary science methods course. *Science Education, 58,* 483–487.

Sprinthal, N. A., & Theis-Sprinthall, L. (1983). The teacher as an adult learner: A cognitive developmental view. In G. A. Griffin (Ed.), *Staff development* (Eighty-second Yearbook of the National Society for the Study of Education). Chicago: University of Chicago Press.

Stake, R. E., & Easley, J. A. (1978). *Case studies in science education.* Washington, DC: U.S. Government Printing Office.

Stengel, B. S. (1992, April). *The reform agenda.* Paper presented at the annual American Educational Research Association meeting, San Francisco.

Strawitz, B., & Malone, M. (1986). The influence of field experiences on stages of concern and attitudes of preservice teachers toward science and science teaching. *Journal of Research in Science Teaching, 23,* 311–320.

Stronck, D. R. (1985). Recommendations of British Columbia science teachers for revising teacher education programs. *Science Education, 69,* 247–257.

Sunal, D. W. (1980). Effect of field experiences during elementary methods courses on preservice teacher behavior. *Journal of Research in Science Teaching, 17,* 17–23.

Sunal, D. W. (1982). Affective predictors of preservice science teaching behavior. *Journal of Research in Science Teaching, 19,* 167–175.

Sunal, D. W., & Sunal, C. (1985). Teacher cognitive functioning as a factor in observed variety and level of classroom teaching behavior. *Journal of Research in Science Teaching, 22,* 631–648.

Tamir, P. (1976). The Iowa-UPSTEP program in international perspective. *Technical report series* (No. 9). Iowa City: University of Iowa, Science Education Center.

Tamir, P. (1983). Inquiry and the science teacher. *Science Education, 67,* 657–672.

Tamir, P. (1989). Training teachers to teach effectively in the laboratory. *Science Teacher Education, 73,* 59–69.

Tamir, P., Lunetta, V., & Yager, R. (1978). Science teacher education: An assessment inventory. *Science Education, 62,* 85–94.

Thompson, C. L., & Shrigley, R. L. (1986). What research says: Revising the science attitude scale. *School Science and Mathematics, 86,* 331–343.

Tisher, R. P. (1990). One and a half decades of research on teacher education in Australia. In R. P. Tisher & M. F. Wideen (Eds.), *Research in teacher education: International perspectives* (pp. 67–88). New York: Falmer.

Tobin, K. G. (1980). The effect of an extended teacher wait-time on science achievement. *Journal of Research in Science Teaching, 17,* 469–475.

Tobin, K. G. (1989). Teachers as researchers: Expanding the knowledge base of teaching and learning. In M. Matyas, K. Tobin, & B. Fraser (Eds.), *Looking into windows: Qualitative research in science education* (pp. 1–12). Washington, DC: American Association for the Advancement of Science.

Tolman, M. N., & Campbell, M. K. (1989). What are we teaching the teachers of tomorrow? *Science and Children, 27*(3), 56–59.

Tom, A. R. (1985). Inquiring into inquiry-oriented teacher education. *Journal of Teacher Education, 36*(5), 35–44.

Tom, A. R. (1986a). *The case for maintaining teacher education at the undergraduate level.* St. Louis: Washington University, Coalition of Teacher Education Programs. (ERIC Document Reproduction Service No. ED 267 067)

Tom, A. R. (1986b). The Holmes Report: Sophisticated analysis, simplistic solutions. *Journal of Teacher Education, 37*(4), 44–46.

Tom, A. R. (1992, April). *Pedagogical content knowledge: Why is it popular? Whose interests does it serve, and What does it conceal?* Paper presented at the annual American Educational Research Association meeting, San Francisco.

Tom, A. R., & Valli, R. (1990). Professional knowledge for teachers. In W. R. Houston (Ed.), *Handbook of research on teacher education* (pp. 373–392). New York: Macmillan.

Trumball, D. (1989, March). *University researchers' inchoate critiques of science teaching: A case for the study of education.* Paper presented at the annual meeting of the National Association for Research in Science Teaching, San Francisco.

Trumball, D. (1990). Evolving conceptions of teaching: Reflections of one teacher. *Curriculum Inquiry, 20*(2), 161–182.

Trumball, D. (1991, April). *What's a teacher to know: What's a teacher to be.* Paper presented at the annual meeting of the National Association for Research in Science Teaching, Lake Geneva, WI.

Urban, W. J. (1990). Historical studies of teacher education. In W. R. Houston (Ed.), *Handbook of research on teacher education* (pp. 59–71). New York: Macmillan.

Veenman, S. (1984). Perceived problems of beginning teachers. *Review of Educational Research, 54,* 143–178.

von Glasersfeld, E. (1984). An introduction to radical constructivism. In P. Watzlawick (Ed.), *The invented reality* (pp. 17–41). New York: W.W. Norton.

Warren, D. (1982). What went wrong with the foundations and other off-center questions. *Journal of Teacher Education, 33*(3), 28–30.

Weaver, H., Hounshell, P., & Coble, C. (1979). Effects of science methods courses with and without field experiences on attitudes of preservice elementary teachers. *Science Education, 63,* 655–664.

Weiss, I. R. (1978). *Report of the 1977 national survey of science, mathematics, and social studies education.* Washington, DC: U.S. Government Printing Office.

Weiss, I. R. (1987). *Report of the 1985–1986 national survey of science and mathematics education.* Washington, DC: National Science Foundation.

Welch, W. W. (1983). Experimental inquiry and naturalistic inquiry. *Journal of Research in Science Teaching, 20,* 95–103.

Welch, W. W., Klopfer, L. E., Aikenhead, G. S., & Robinson, J. T. (1981). The role of inquiry in science education: Analysis and recommendations. *Science Education, 65,* 33–50.

White, R. T., & Tisher, R. P. (1986). Research on natural sciences. In M. Wittrock (Ed.), *Handbook of research on teaching,* (3rd ed., pp. 874–905). New York: Macmillan.

Wideen, M. F., & Holborn, P. (1990). Teacher education in Canada: A research review. In R. P. Tisher & M. F. Wideen (Eds.), *Research in teacher education: International perspectives* (pp. 11–32). New York: Falmer.

Wiggins, J. R., Padilla, M. J., Nichols, B. K., & Monroe, L. M. (1992, March). *Induction: A case study of a third year middle school science teacher.* Paper presented at the annual meeting of the National Association for Research in Science Teaching, Cambridge, MA.

Wilson, S. M. (1991). Parades of facts, stories of the past: What do novice history teachers need to know. In M. M. Kennedy (Ed.), *Teaching academic subjects to diverse learners* (pp. 99–116). New York: Teachers College Press.

Wise, K. C., & Okey, J. R. (1983). A meta-analysis of the effects of various science teaching strategies on achievement. *Journal of Research in Science Teaching, 20,* 419–436.

Wittrock, M. C. (Ed.). (1986). *Handbook of research on teaching* (3rd ed.). New York: Macmillan.

Woods, P. (1987). Life histories and teacher knowledge. In J. Smyth (Ed.), *Educating teachers: Changing the nature of pedagogical knowledge* (pp. 121–136). New York: Falmer.

Wright, E. (1979). Effect of intensive instruction in cue attendance on solving formal operational tasks. *Science Education, 63,* 381–393.

Yager, R. E., & Bybee, R. W. (1991, Winter). Science teacher education in four-year colleges: 1960–1985. *Journal of Science Teacher Education, 2*(1), 9–15.

Yager, R., & Penick, J. (1990). Science teacher education. In R. Houston (Ed.), *Handbook of research on teacher education* (pp. 657–673). Reston, VA: Association of Teacher Educators and MacMillan Publishing Company.

Yager, R., Lunetta, V., & Tamir, R. (1979). Trends in science teacher education. *School Science and Mathematics, 79,* 308–312.

Yeany, R. H. (1991). Research and teacher education. *NARST NEWS, 33*(3), 1–3.

Yeany, R. H., & Padilla, M. J. (1986). Training science teachers to utilize better teaching strategies: A research synthesis. *Journal of Research in Science Teaching, 23,* 85–95.

Zeichner, K. M. (1981). Reflective teaching and field-based experience in teacher education. *Interchange, 12*(4), 1–22.

Zeichner, K. M., & Gore, J. M. (1990). Teacher socialization. In W. R. Houston (Ed.), *Handbook of research on teacher education.* (pp. 329–348). New York: Macmillan.

Zeichner, K. M., & Liston, D. P. (1987). Teaching student teachers to reflect. *Harvard Educational Review, 57,* 23–48.

Zeichner, K. M., & Liston, D. P. (1990). Theme: Restructuring teacher education. *Journal of Teacher Education, 41,* 3–20.

Zeichner, K. M., & Tabachnick, B. R. (1982). The belief systems of university supervisors in the elementary student teaching programs. *Journal of Education of Teaching, 81,* 34–54.

Zeitler, W. (1981). The influence of the type of practice in acquiring process skills. *Journal of Research in Science Teaching, 18,* 189–197.

Zumwalt, K. (1989). Beginning professional teachers: The need for a curricular vision of teaching. In M. Reynolds (Ed.), *Knowledge base for beginning teachers* (pp. 173–184). Washington, DC: American Association of Colleges for Teacher Education.

· 2 ·

RESEARCH ON INSTRUCTIONAL STRATEGIES FOR TEACHING SCIENCE

Kenneth Tobin
FLORIDA STATE UNIVERSITY

Deborah J. Tippins
UNIVERSITY OF GEORGIA

Alejandro José Gallard
FLORIDA STATE UNIVERSITY

The critical question is not "What does the teacher do?" Basically we're talking about teaching as an act of leadership. Teaching is essentially helping people get excited in a subject area, which leads them to engage in big ideas, the cultural ideas. (Holmes Group, 1990, p. 10)

The preceding excerpt from a recent report of 30 leading educators in the United States begins with a statement that the most important questions related to teaching should not focus on the role of the teacher. On the contrary, the role of the teacher is to be facilitative, to help learners get excited about learning. The excerpt frames the ideas on which to build a vision of what a teacher might do in tomorrow's schools. Although questions about the teacher's role might not be considered as important as questions about the roles of learners by the Holmes Group, our view is that teaching and learning occur in a culture in which the actions of teachers and students are inextricably linked. Thus, it does not seem prudent to question learners and learner roles without, at the same time, questioning teachers and teaching roles. If tomorrow's classrooms are to be transformed to be compatible with the Holmes Group's vision, we have no choice but to begin with what is happening now. This requires the introduction of reforms intended to (1) create schools that have mediation of learning as their primary goal

and (2) produce teachers who, as professionals, can lead, guide, and mediate the learning process so that (3) learning becomes a way of life and schools become communities of learning. The reform we envision does not lend itself to a recipelike list of procedures. Rather, creativity and sophisticated thinking will play critical roles in the educational change.

Calls for reform of science education are frequently heard. Some investigators, such as Marsh and Odden (1991), have used a wave metaphor to describe the efforts at reform. The first wave of educational reform, which is anchored in *A Nation at Risk* (National Commission on Excellence in Education, 1983), involved increased graduation requirements in terms of traditional academic courses, such as requiring more mathematics and science in order to graduate from high school. The second wave of reform involved the preparation of better courses, new model curriculum standards, improved textbooks, advocacy of improved teaching, and the development of indicators of program quality. The third wave focused on radical curriculum reform, integration of the curriculum across content areas, advocacy of higher-order thinking skills, and applications of technology to improve learning. The fourth wave shifted its focus to professionalism, teacher decision making, career levels, and restructuring of schools.

The authors thank reviewer James J. Gallagher (Michigan State University).

A close look at suggested reforms in science education suggests a high level of agreement on what ought to be done. However, despite repeated calls for reform, there is little evidence of significant change. Essentially, the science education community would like to see students learn science as a process and learn about science in a way that is meaningful and enriching, enabling citizens to interpret their world in terms of science. The National Science Teachers Association (NSTA), through its report *Essential Changes in Secondary Science: Scope, Sequence and Coordination* (1989), advocated a curriculum that was integrated in nature, beginning in a formal sense at grade 7 and becoming increasingly abstract in the later high school years. The emphasis was on understanding science. Traditional practices stand in stark contrast to the recommendations of this and other reports. Research suggests that the tasks in which students engage in science classes have low cognitive demand and emphasize the learning of facts and memorization of algorithms to solve problems without necessarily understanding why the algorithms work (Doyle, 1983; Stake & Easley, 1978; Tobin & Gallagher, 1987a). How will traditional science classrooms such as those described in the cited studies be transformed to fit the vision of NSTA's *Scope, Sequence and Coordination*? We believe that change is a systemic process requiring the exposure of tacit infrastructures of culture that often hinder social and educational reform. Because of their traditionally central role in the curriculum process, teachers are viewed as key components in the current endeavors to reform science curricula. First, it is essential to begin the process with an understanding of what is happening now in science classrooms and an understanding of the teachers' rationales for maintaining traditional practices. Concomitantly, opportunities to construct new visions of science education should be provided for teachers and other participants in science curriculum. Finally, in an ever-tortuous and sometimes frustratingly slow manner, change can proceed from what is happening in classrooms toward new visions encapsulated in documents such as NSTA's *Scope, Sequence and Coordination*. A key ingredient of the success of reform efforts is to be found not in teacher training, no matter how well planned and implemented such programs might be, but in soundly conceived and implemented educational programs that reach out to all who have roles in science education and take for granted the professionalism of teachers.

OVERVIEW OF THE CHAPTER

Thirteen sections follow this overview. The first provides a theoretical rationale that is a significant part of the chapter because it provides a context for understanding how we view research syntheses. The second section presents an overview of constructivism, which is an important component of the interpretive framework that was applied to the review of articles and to the synthesis. Subsequent sections address views of traditional classrooms, exemplary teaching practices, teacher beliefs, pedagogical content knowledge, use of textbooks, mediating learning through verbal interaction, the learning cycle, cooperative learning, laboratory activities, and assessment practices. A conclusion reexamines the role of teachers and relates teaching to the present calls for reform.

Our intention in writing this review is to go beyond a summary of all the research on science teaching. Instead, we set out to relate what we learned about science teaching from a review of research. Many more articles could have been included in the review; however, selection was based on the following procedures. The authors met as a group to ascertain which research areas were most salient to the teaching of science. Approximately 16 sections emerged as a consensus. A reduced number was included in the chapter because some, such as conceptual change and learning cycle, overlapped considerably in terms of the implications for teaching. Furthermore, space limitations precluded the inclusion of all areas. Papers within sections were identified through the use of data bases, examination of journals and programs of annual meetings of professional associations, and papers that were readily accessible because of our extensive involvement in research on teaching.

RATIONALE

Although constructivism is not a new way of thinking, it has not been accepted as the prevalent way to conceptualize knowledge by the community of educators. Modern science and education have rested on the separation of the knower and the known, a central tenet in the Cartesian-Newtonian way of organizing the world. Over the past 20 years, however, a few persons, such as Ernst von Glasersfeld and Les Steffe, have endeavored to make sense of educational problems from a constructivist perspective. Despite being in a minority, they have constructed the groundwork for applying constructivist ideas to education and have endeavored to educate colleagues to understand education in terms of constructivism (e.g., Steffe, von Glasersfeld, Richards, & Cobb, 1983). Whether it was due to the efforts of such people in whole or in part or to the need for new ways to examine the persistent problems in science and mathematics education, the educational community now seems to be involved in a process in many ways similar to what Kuhn called a paradigm shift (Kuhn, 1962). Whereas constructivism was not widely accepted just a decade ago, there are now efforts to understand constructivism in all areas of education. This cognitive revolution is challenging and exciting, as well as confusing and frightening, as new meanings for educational inquiry emerge. The essential elements of constructivism as they apply to science teaching and learning are described in the next section.

Research on science teaching published in the past several years shows a marked shift in assumptions from those embodying process-product models for research, reflecting the dominant Cartesian-Newtonian paradigm of the time. A Cartesian-Newtonian way of seeing the world is one in which the uncertainties of science teaching are replaced by verified empirical knowledge about the act of teaching. During the previous 20 years, research on science teaching was shaped by the fundamental assumption that "for research to be valid and publishable, ... it had somehow to mirror 'scientific' paradigms" (Russell & Munby, 1992). This assumption has strongly influenced research on science teaching and the struggle to conceptualize science teacher education. In this chapter, we analyze research as it relates to science teaching from the perspective of broad changes that have paralleled a shift in assumptions about the nature of research.

CONSTRUCTIVISM

Constructivism: A Way of Knowing

Constructivism is a set of beliefs about knowledge that begin with the assumption that a reality exists but cannot be known as a set of truths because of the fallibility of human experience (von Glasersfeld, 1989). From a constructivist perspective, learning is a social process of making sense of experience in terms of extant knowledge. Persons interact with objects and events through their senses, which are inextricably associated with extant knowledge that includes beliefs and images. Knowledge exists only in the minds of cognizing beings. Objectivity, as it has been envisioned traditionally, is not possible. The categories of the mind, which are shaped by human experience and mediated by a sociocultural milieu, make no claim to represent the universe in a real way. By extension, teaching, as framed within a constructivist rationale, is characterized by an almost total absence of truths or unimpeachably "correct" answers to the most important issues (Kagan, 1992).

All knowledge is constructed. Accordingly, social and cultural phenomena are also personal constructs. However, an individual is born into a social and cultural environment in which all of the objects and events that are encountered have particular meanings that were also constructed. Through interactions with the living and nonliving components of the milieu in which an individual is raised, all learning is socially mediated. Accordingly, knowledge has a social component and cannot be seen as constructed by an individual acting alone. For example, the use of language or mathematics in thought is social because both language and mathematics are social constructions. Because it is not possible for an individual to separate individual thought from the sociocultural components of what is known, constructivism can be thought of as a set of beliefs that incorporate the social and the cultural among the most important postulates. Any study of teaching and learning should focus, therefore, not only on the manner in which an individual attempts to make sense of phenomena but also on the role of *the social* in the mediation of meaning.

Constructivism: A Method or a Referent?

Some authors (e.g., Fosnot, 1988) use constructivism to represent a method of teaching whereby the teacher bases what happens on beliefs that are consistent with constructivism. By this they mean that constructivism has been used as a referent to build a classroom that maximizes student learning. For example, the teacher would typically take account of what students know, maximize social interactions between learners so that they can negotiate meaning and provide a variety of sensory experiences from which learning is built. Another example of this is Russell's (1992) description of teaching practices, such as lecturing. Russell regards lecturing as having little value compared with alternatives such as small-group learning or interactive discussions that are constructivist in nature. Such positions, although understandable, reduce constructivism to a set of methods and diminish its power as a set of intellectual referents for making decisions in relation to actions. Just as constructivism can be used to explain how students make sense of experience in interactive discussions or in small-group problem-solving activities, so too it can be used to explain why learning occurs in lectures and how lectures can be adapted to improve the quality of learning.

Wheatley's (1991) approach to curriculum has been carefully built with constructivism as a referent. In this approach, known as problem-centered learning, students work together in small groups making meaning of tasks and setting out to solve problems that are perplexing. The teacher in such classes has an important mediating role, ascertaining what students know and structuring tasks so that they can build knowledge structures that are commensurate with knowledge of the discipline. According to Wheatley, students negotiate meaning in small-group situations and then negotiate consensus in whole-class settings. In the elementary classes, we have observed that a period of 45 minutes of small-group problem solving is followed by 45 minutes of whole-class discussion, usually led by students. The teacher's role is to monitor student understandings and guide discussions so that all students have opportunities to express their understandings in language and engage in activities such as clarifying, elaborating, justifying, and evaluating alternative points of view. Such visions of classroom learning environments are appealing alternatives to those so often reported as traditional in studies of learning in classrooms (e.g., Tobin & Gallagher, 1987a). However, there are many acceptable ways to implement a problem-centered approach to learning. The best management system for organizing activities and bringing students and tasks together will depend on the cultural milieu in which the curriculum is embedded. Our vision of problem-centered learning begins with students who are motivated to learn and teachers who see their roles in terms of facilitating learning. It begins with teachers understanding the rationale for implementing the curriculum as they do and reflecting as they implement the curriculum. Teachers negotiate with students, asking questions to elicit thinking about the viability of knowledge representations, arranging students together so that they can argue toward consensus, and pointing the way to additional learning resources. To the students, the teacher is a mediator, guide, provocateur, friend, and colearner. Constructivism is a referent, a set of beliefs about knowledge, a set of reflective tools that enable the science teacher to consider the optimal ways to mediate the process of learning science.

To be a viable theory of knowing, constructivism must have explanatory power in all situations in which knowledge is constructed or cognizing beings are deemed to know. Thus, constructivism, as a set of beliefs about knowing and knowledge, can be used as a referent to analyze the learning potential of any situation. For example, constructivism can be used to explain why certain students have been successful in learning science in contexts in which teachers lecture day after day and students listen and copy notes until their hands ache. Similarly, constructivism can be useful in explaining how any given set of circumstances might be changed to improve the learning opportunities of individuals. For example, if a biology department has as policy that all classes will contain at least 200 students, it is probably not feasible to think in terms of Wheatley's problem-centered learning. However, from a constructivist perspective, learning can be thought of as a social process of making sense of experience in terms of what is known. To improve learning, therefore, a teacher might consider how to improve the quality of experiences given the constraint of 200 or more learners. Similarly, in countries, such as Taiwan, where class sizes of more than 40 or even 50 students are common, teachers can think in terms of improving the quality of social interactions, providing a range of meaningful experiences to each learner, and making it possible for each student to become aware of his or her relevant prior knowledge and apply it to the process of learning. Constructivism, when used as a reflective tool, empowers teachers and enables them to fashion learning activities to the circumstances in which they find themselves. Thus, teachers can focus planning and implementation strategies on the needs of learners as they understand them from a constructivist perspective.

In contrast to the use of constructivism as a method is the notion of constructivism as a tool for critical reflection. As such, constructivism can act as a referent for deciding which teacher and learner roles are likely to be more productive in given circumstances. When used in this way, it is possible to plan and implement activities that are believed to facilitate learning. Constructivism tends to sophisticate our thinking about educational problems, leading to unique questions that open the door to new recognitions. For example, in teacher education a question that needs to be answered is: How can prospective and practicing teachers learn to teach science? Research findings based on the gathering of more and more data have been regarded as contributors to this body of knowledge, yet they have not increased our insight into how best to prepare prospective and practicing teachers. Accordingly, solutions to the problem are framed within a set of assumptions that a body of knowledge exists "out there" and learners (in this case practicing and prospective teachers) can come to know this body of knowledge by interacting with it. The focus is on the interaction between disciplinary knowledge and the learner. Such thinking stands in marked contrast to two questions that are of significance from a constructivist point of view: What does the learner already know about teaching and learning and how can this knowledge be represented? An associated question is related to the process of making sense of experience: What experiences should teachers be provided to enable them to build an understanding of teaching and learning? Finally, teachers need time to make sense of their experiences. That is, they need time to think things through and then time for such cognitive activities as to clarify, elaborate, justify, and consider the merits of alternatives. From a constructivist point of view, the emphasis is on the teacher as a learner, a person who will experience teaching and learning situations and give personal meaning to those experiences through reflection, at which time extant knowledge is connected to new understandings built from experience and social interaction with peers and teacher educators.

Constructivism and the Curriculum

Many science educators and teachers have an image or model of the curriculum as a set of printed materials that can be transported from place to place. In this sense, curriculum is perceived as something that exists apart from teachers, students, and other aspects of the learning environment. By contrast, curriculum theorists such as Grundy (1987) have a conceptual view of curriculum as the set of all learning experiences. Although broad, this definition of curriculum is consistent with constructivism because it does not regard knowledge as separate from the knower and the culture in which learning is to occur. In the past, objectivist ways of thinking about a body of knowledge existing out there to be learned resulted in curricula that were thought to be transferrable to sites in greatly differing contexts. When teachers and learners adapted the resources to fit local needs, the designers of the materials viewed this as implementation infidelity. From a constructivist way of thinking, a curriculum is embedded in culture and cannot be separated from culture that includes other learners and the shared taken-for-granted knowledge of the culture, myths, customs, taboos, and history. Curriculum takes into account the possibility of multiple frames of reference and different ways of viewing the world. From a constructivist perspective, curriculum includes the larger historical, social-political-economic context, and the influence of others such as parents, administrators, and teachers.

Grundy's research provides a theoretical framework for curriculum that encompasses three knowledge-constitutive interests: technical, practical, and emancipatory interests. When technical interests are foremost, classrooms are oriented toward controlling and managing the learning environment. Control is manifest through rule-following actions often associated with state or local laws, curriculum guides, and standardized tests. Where practical interests are evident, the emphasis in classrooms is on the development of understanding and interpretation through processes of consensus building and negotiation. Emancipatory interests are visible in classrooms where both teachers and students are empowered to the extent that they are able to engage in autonomous actions that arise from reflection.

A learner has to make sense of science through an existing conceptual structure. The resulting knowledge of science is an interpretation of experience in terms of extant knowledge. Accordingly, two questions become fundamental to considerations of a science curriculum: What experiences should be provided to the learner in order to facilitate learning, and how can the learner represent what is already known to give meaning to these experiences?

Accordingly, by thinking of science teaching in a constructivist terms, science educators shift perspectives and are able to reconceptualize the nature of the curriculum. Because all knowledge must be individually constructed, it makes no sense to begin by thinking solely about the disciplines of science in the absence of learners or in isolation from each other. For example, the familiar debate in science education over whether to emphasize concepts or processes has little meaning from a constructivist point of view. Making sense of science is a dialectical process involving both content and process. The two can never be meaningfully separated. The process skills can be thought of as thinking processes such as using the senses to experience; representing knowledge through language, diagrams, mathematics, and other symbolic modes; clarification; elaboration; comparison; justification; generation of alternatives; and selection of viable solutions to problems.

Teacher as Mediator

The teacher's role is to mediate the learning of students. This assertion should have an important influence on the way teachers think about teaching. To begin with, the focus ought to be on learners rather than the discipline. This is in contrast to Tobin and Gallagher's (1987a) findings in a case study of a group of Australian teachers to determine what happened when they implemented the science curriculum. Content coverage was one of the teachers' highest priorities. They planned with this in mind and stuck to the plan to the best of their abilities. They changed topics at the scheduled time regardless of the extent to which students had learned what had been covered. If teachers see themselves as mediating students' learning, two critical components of their role are to monitor learning and concentrate on providing constraints so that student thinking is channeled in productive directions. To undertake such a role, it will be important for teachers to interact with students to a greater extent, to ascertain what they know and what they are thinking. The interactions can begin with the goals of the curriculum, which should take into account the goals of the students. Too often there is a wide discrepancy between the teacher's goals and those of the students. Study after study has revealed a less than optimal level of student engagement in science classes. In very few classrooms have the majority of students been motivated to learn science for the majority of the time. With few exceptions, the students do not appear to be in class to learn science. Perhaps this problem can be overcome, to a degree at least, if teachers begin to negotiate the goals of the curriculum with students.

As a mediator, the teacher must ensure that students are given opportunities for quality learning experiences to provide a solid base for learning with understanding. This requires that teachers constantly learn about which learning experiences would be the most appropriate for their students. A metaphor we would use to describe this role of the teacher is one of student/learner. From a constructivist perspective, this suggests that a teacher can facilitate learning in the ways already mentioned and by constraining experiences so that students have opportunities to build on extant knowledge.

Other metaphors, such as the teacher as provocateur, can bring into focus the variety of productive roles available to teachers. The provocateur is seen as a master fencer engaged in a struggle with students, thrusting at times, defending at other times, but always making visible the skills of a master fencer. Thrust, defend, defend, thrust; the master and student engage in a struggle intent on attaining the goal of becoming a more expert fencer. In the process, the teacher assumes a questioning, challenging role and creates situations in which students can challenge the official "facts."

The teacher contributes a great deal and is exhausted by the effort. Similarly, the student is stretched to the limit in a manner in which no one is hurt and mutual satisfaction is attained. The gap between the current skill level and what is needed to be successful is always managed in such a way that it is elusively beyond the student. In the process, learners create perturbations as they attempt to give meaning to particular experiences through the imaginative use of existing knowledge. The resolution of these perturbations leads to an equilibrium state in which new knowledge has been constructed to cohere with a particular experience and prior knowledge. Because of the importance of students testing the viability of knowledge claims, it is necessary for teachers to consider how to provide opportunities for testing the viability of what is learned through negotiations with students and by providing opportunities for problem solving.

Discussion

In our experience, it is unusual to find teachers who require students to generate questions and seek answers to them. Far too often, questions are regarded as a threat to authority. Constructing questions might be one way for students to build conceptual conflict, and seeking answers to them might begin the process of resolving the conflict. Group discussions can play a significant role in students' learning by providing time for interaction with peers to answer student-generated questions, clarify understandings of specific science content, identify and resolve differences in understanding, raise new questions, design investigations, and solve problems. Group interactions also provide a milieu in which students can negotiate differences of opinion and seek consensus. Perhaps even more important, opportunities for students to generate questions and interact with one another develop their ability to speak out, unafraid, in order to take a moral stand. Our awareness of the social context of science teaching and learning can provide the impetus for a vision of science teaching and learning that is grounded in the search for meaningful experiences.

LIFE IN SCIENCE CLASSROOMS: TRADITIONAL VIEWS

This introductory section, which portrays life in traditional science classrooms, relies mainly on a review of three studies by one of the authors of this chapter (Tobin & Espinet, 1989; Tobin, Espinet, Byrd, & Adams, 1988; Tobin & Gallagher, 1987a, 1987b). Some of the more important insights gained from these studies are described.

Reducing Higher Cognitive Demands

On occasions, when teachers emphasized high-level cognitive outcomes, teachers and students interacted to reduce the cognitive demands of academic tasks. A frequent example of this occurred when teachers asked a question at a relatively high cognitive level and then rephrased it before students could answer. The question to which students finally responded often required a dichotomous selection such as yes or no. Students attempted to renegotiate assigned tasks, students in small groups copied from one another, and, in open-book tests, students could copy information that had previously been copied from another source (e.g., the chalkboard or another student's writing). Other examples of "safety nets" (strategies used to reduce the risk of students feeling embarrassed as a result of engaging in tasks) included group assignments; peer assistance; using easy or familiar content on tests; assigning more credit for memory or procedural components of a task than for high-level cognitive components; allowing students to revise products after they had submitted them; peer review of products before submission; teacher assistance, prompts, and cues; extra-credit assignments; less exacting grading for low-achieving students; grading on completion rather than accuracy; and last-minute review of key content before a test.

The emphasis on quantitative aspects of science in all grade levels was on learning how to use procedures and computational efficiency. Teachers provided students with type examples, algorithms, and rules of thumb to reduce the cognitive demands of the work and enable students to cope with it in a procedural manner. For example, students were given up to 10 examples of the same type of problem to enable them to recognize the type if it occurred in an examination. The general approach to solving this type of problem was routinized as a result of working a relatively large number of similar problems. As a consequence, students could obtain the correct answer without necessarily understanding the science or mathematics involved. Student interviews indicated that many students were able to solve complex problems by applying a formula that had been learned by rote. Discussion of the problems indicated that students had scant knowledge of scientific or mathematical principles but had learned approaches to obtaining correct answers.

Target students, defined as those who dominated interactions with the teacher, were evident in most science classes in the studies discussed in this section. In nearly all science classes a small group of target students dominated class interactions by answering most of the questions, asking the teacher most questions, and assisting in teacher-directed activities in numerous ways, such as working problems on the chalkboard. These actions of target students reduced the cognitive demand of the work for their classmates. For example, in one grade 11 physics lesson the teacher worked a difficult problem on the chalkboard and called for students to contribute to the solution. The work at a higher cognitive level was carried out by the teacher as he worked the physics problem step by step on the chalkboard and by the more able students as they responded to the teacher's questions. Most students watched and listened to the teacher and target students respond to questions. The higher-cognitive-level work was done for them; their role was to memorize correct answers and procedures to follow when solving similar problems.

Classroom Management

The activity setting used by teachers has considerable bearing on the way students engage. Two major patterns of activity organization were evident. Most teachers taught using predominantly whole-class activities. Teachers used lectures to introduce information, whole-class interactive activities to involve students, and around-the-class reading to cover material in the textbook. Students were also involved in individualized activities and to a lesser extent in small-group activities. Similar activities were used in heterogeneous and tracked classes. In most instances, teachers had difficulty in maintaining discipline. The main problems stemmed from low teacher expectations for student performance and failure to apply rules for standards of behavior and performance in a consistent manner.

In all classes taught by these teachers, a relatively small group of students distracted others in the class. Interviews with a number of disruptive students indicated that some lacked motivation mainly because of lack of success and others were successful but did not find the content stimulating. Disruptive behavior of a socially unacceptable type (e.g., physically interacting with others) was a problem in many classes, and disruptive behavior of a type that is socially acceptable (e.g., quietly talking with a peer) was evident in most classes. Dealing with management problems robbed teachers of time in which to teach for understanding. As a consequence, teachers with management difficulties also had problems with instruction.

Throughout the studies three types of monitoring were evident. When management was a problem, the teacher monitored student misbehavior and circulated in the room in an endeavor to maintain order. The second type of monitoring occurred when the teacher observed whether or not students were engaged in a task. The third type of monitoring occurred when teachers were able to probe the understanding of individuals. In classes with few management problems there was a greater proportion of the last two types of monitoring. When management was a problem, teachers were unable to monitor for understanding to an appreciable extent.

The results of these studies support the view that the reward structure in classes was an important force that needed attention before the effects of other instructional variables could be fully understood. Teachers constantly motivated students by referring to tests and examinations. This teaching style focused student attention on the content to be tested. If the teacher emphasized a specific type of problem as being important for the test, students tended to concentrate more than when the teacher gave no such emphasis. Two trends were evident. Work identified as likely to be tested had higher status than work not so identified, and activities assumed importance only if they were associated with learning for an assessment or were directly assessed. Consequently, learning for the test, copying notes from the chalkboard or from the textbook, completing laboratory reports, and completing homework had high status in most classrooms.

Textbooks, Laboratory Activities, and Assessment

Teachers in all of the studies used textbooks as a source of student activities. In some instances, the activities emphasized higher-cognitive-level learning, and in other cases they emphasized learning facts and algorithms. Mainly because of the emphasis given to discussion of questions from the textbook, the majority of class time was devoted to discussion of questions students had attempted at home. Consequently, most new content was introduced for the first time in the textbook. Teachers used class time to elaborate on student responses and to extend and clarify concepts from the textbook.

In contrast, many textbooks focused on presenting science content, and the questions at the end of each chapter required students to locate specific science content in the chapter. Few questions necessitated higher-cognitive-level processes. Teachers often asked students to make summaries from the textbook and to answer end-of-chapter questions. Such activities made low cognitive demands, usually involving a brief search through the text for relevant information and transcription of the information into notebooks.

When planning science programs, all teachers partitioned the year's work into topics to be covered each term and subsequently planned the content to be covered each week and in each lesson. Each chapter from the textbook was scheduled to be completed in a specified time. In most instances the work was covered in the scheduled time and students completed a test before commencing a new topic or chapter. On occasions, when additional time was necessary, teachers allocated less time to the next topic or chapter.

The focus of the implemented curriculum tended to be on covering planned content rather than ensuring that students developed depth of understanding. All teachers stated that it was important to complete the topics and chapters in the scheduled time so that students could take tests and examinations. A consequence of this concern with covering content was that teacher explanations dealt with surface features of the content and frequently did not probe to the depth necessary for student understanding. In many cases, complex content and algorithms were presented for students to learn by rote, and there was little probing of the underlying logic.

Laboratory activities tended to be of a "cookbook" type, with strong emphasis on following procedures in order to collect data. There was little emphasis on planning an investigation or on interpreting results. Teachers provided the procedures to be followed and a table in which to record data. Recipes for most experiments were in the textbook, in a manual, or on the chalkboard. The typical pattern was for data-collecting activities to continue until approximately 5 minutes of the lesson remained. At this time the teacher issued instructions to stop work, restore the equipment, and write the conclusions. Throughout these studies, students were continually motivated by assessment in the form of homework, tests, and examinations. Rarely was the importance of topics related to the world outside the classroom. The focus was on completing the work in a manner that allowed students to succeed on a test, examination paper, or assessment activity. Rewards were pervasive and constituted the most common form of extrinsic motivation in all classes. Although many of the teachers in this study recognized deleterious effects of the assessment system, they did not feel they had the power to change the form or frequency of assessments. Teachers stated that responsibility for changing the curriculum and the assessment system was vested in the head of department, school administrators, and state education authorities. Heads of department involved in the studies also accepted little responsibility for changing the type of assessment.

Interviews with teachers revealed that most accepted responsibility for student learning. When questioned about the most effective learning activities, teachers usually cited clear and logical explanations of content. They appeared to believe that learning could be enhanced to the greatest extent by presenting the material in a more appropriate manner. Seldom was the form of student engagement mentioned as an important variable. Furthermore, teachers viewed the curriculum in terms of content to be covered, and student learning was demonstrated by success on achievement tests. It is easy to see why teachers broke down the content so that students could observe and learn correct answers. It is also easy to understand why teachers helped students to succeed in conceptually difficult subjects like chemistry and physics by teaching them procedures for arriving at correct answers to problems.

The teachers' perceptions that external factors had a strong influence on the curriculum were supported by our observations. For example, the state department of education provided guidelines on what should be taught and teaching strategies to be used. In addition, all teachers were heavily reliant on textbook materials and little adaptation occurred. When deviations from the textbook were observed, they usually involved use of information or materials from another textbook. Consequently, adopting appropriate textbooks is a means of improving the quality of student engagement. This was observed in grade 11 and 12 biology classes in which the end-of-chapter questions did not lead to search-and-find activities, but required students to understand the science from the chapter and to construct high-level cognitive responses.

Discussion

The results of the studies reviewed in this section suggest that teachers' knowledge and beliefs are potent forces associated with the implemented curriculum. Beliefs about how students learn and what they ought to learn appear to have the greatest impact. The results are consistent with the interpretation that teachers' knowledge of student learning is based mainly on the teachers' own style of learning. The intuitive knowledge that drives teachers' behavior seems to be related to the hours of time they spent in classrooms during their own elementary and secondary education. We observed numerous examples that accord with the maxim that teachers teach the way they learned. In most cases, one or more rote learning procedures were preferred and the content that was valued consisted of facts and procedures that could be learned in this manner. As a consequence, teachers felt comfortable with what was happening in their classes. Students were learning what the teacher intended, using methods the teacher had used to learn similar content. The tendency in the past has been to look at traditional practices in an oversimplified way. It is tempting to suggest that this "cyclic" state of affairs could be remedied by placing the teacher in an environment in which new content could be learned in a manner facilitating understanding. The implication is that teachers' beliefs are causing the curriculum to be the way it is. In our view, which is supported by research reviewed later in this chapter, the curriculum is a product of many dialectical relationships between sets of interrelated variables. Whereas teachers' beliefs influence the manner in which the curriculum is implemented, so too does the curriculum influence teachers' beliefs. If changes are to occur in the curriculum, it is unlikely that changing one variable alone will be sufficient to overcome the buffering effect of the multitude of dialectical relationships that tend to restore the traditional equilibria.

LIFE IN SCIENCE CLASSROOMS: EXEMPLARY PRACTICES

What is it about educational research and educational policy reports that inevitably leads to conclusions that the quality of education generally, and science education in particular, needs to improve substantially? Is it the nature of educational research to highlight aspects of teaching and learning that must be improved? Or is it that researchers and report writers typically focus on what needs to be improved rather than what is already being accomplished? Possibly a great deal could be learned by focusing on the good things that teachers do and in particular by investigating exemplary teachers and identifying what makes them so good. This approach has been successful in the past. More than 20 years ago Mary Budd Rowe (1969) identified the wait-time variable by examining several anomalous audiotapes of science lessons in which students were demonstrating high levels of inquiry. The value of using an average wait time of 3 to 5 seconds to promote student inquiry was revealed because Rowe investigated the few classes in which student inquiry was high rather than several hundred in which it was low. That is not to deny that a great deal could have been learned by intensive investigations of the low-inquiry classes but to emphasize that Rowe's seminal work on wait time was a result of changing focus from the majority of teachers to a few successful teachers. The change in focus was fruitful, and her research findings captured the imagination of educators throughout the world. To many educators, the appeal of Rowe's work on wait time was that there were no surprises. The results made sense, and with a few moments of thought almost any educator could understand how the use of an extended wait time could lead to

improvements. How many other variables like wait time are there? What more can we learn from investigating the practices of exemplary teachers? The negative tenor of so many educational research and policy reports suggests that it is time to refocus on exemplary teaching practices and see what can be learned. Of course, there is no guarantee that such research will provide grounds for greater optimism. Exemplary teachers simply provide a new data base from which implications for the practice of science education can be derived.

Some optimistic research endeavors have highlighted educational accomplishments and paved the way for improvements in schooling. The effective-schools movement (e.g., Bickel, 1983; Cohen, 1982) is based on the assumption that successful schools do exist and that other schools could be improved by adopting some of the practices found in them. Penick (1983) advocated studies with a focus on successful science education as holding hope for improving practice. These ideas were incorporated in a project known as the Search for Excellence, sponsored by the NSTA, the Council of State Supervisors, the National Science Supervisors Association, and the National Science Board (Bonnstetter, Penick, & Yager, 1983; Yager, 1984). Because the focus of the Search for Excellence was on programs, the initial output included case studies of over 50 excellent science programs published as several volumes by the National Science Teachers Association (e.g., Penick, 1983, 1988; Penick & Bonnstetter, 1983; Penick & Lunetta, 1984; Penick & Minehard-Pellens, 1984; Penick & Yager, 1986a). Six programs were investigated closely to identify "the critical ingredients of the program, the critical factors in the emergence of the program, and the critical forces required for its maintenance" (Yager, 1984, p. 5).

Just as investigations of exemplary programs provided directions for school districts and schools to organize and implement science programs, investigations of exemplary teachers might provide models for other teachers to emulate in order to enhance science learning of students. This philosophy was adopted by a team of Australian researchers who conducted a number of case studies of exemplary science and mathematics teachers (Tobin & Fraser, 1987). Many of the studies selected for review in this section concern Australian science teachers identified as exemplary. The studies are reported in a book (Tobin & Fraser, 1987) and in numerous journal articles referenced in this chapter.

The findings of the review are presented in terms of (1) classroom management, (2) teaching for understanding, and (3) maintaining a learning environment that is conducive to learning.

Classroom Management

Tobin, Capie, and Bettencourt (1988) reviewed the research on classroom management and highlighted implications for science teaching. Teachers were urged to anticipate decisions about who does what, where, when, and for how long. An important planning role was to provide work for students who completed activities such as laboratories or worksheets before others. Communicating expectations to students was important for a variety of factors such as hand raising, calling out, moving about the room, and seating plans. Monitoring student engagement was crucial so that teachers could ensure that their expectations were translated into appropriate classroom behavior and engagement. Specifically, teachers should be aware that off-task behavior or passive engagement is a precursor to more severe disruptions so that minor infractions can be addressed in a relatively unobtrusive manner. The importance of these elements was supported in a study in which differences in classroom management accounted for more than one-third of between-class differences in science process-skill achievement (Padilla, Capie, & Cronin, 1986). Implicit in the Tobin, Capie, and Bettencourt review are beliefs that the teacher should be the controller of students with a view to maximizing learning in what might be described as technical circumstances. Why should a teacher intervene if a student is passively engaged? Presumably, the student is contemplating, making connections, deciding what to do next, and reflecting in practice. If a teacher constructs learners as being autonomous, it makes sense to think of them taking time out to reflect on what has been happening and what needs to happen next. Passive engagement is a positive sign when emancipatory interests are emphasized and a negative sign when technical interests based on teacher control predominate.

Gallagher and Tobin (1987) described an interaction between classroom management and the cognitive demands of the tasks in high school science activities. When the cognitive demands of the tasks were high, students demonstrated task-avoidance behaviors that taxed the teacher's managerial effectiveness. In the case of a low-ability class of students, the teacher was able to retain control of the class by changing the activity to one in which students could engage without risk of failure. In a higher-ability class, students returned to the task after a period of restlessness in which most of the class was off task. These examples suggest that students may resist teacher attempts to have them engage for sustained periods when cognitive demands are high.

An implication of these findings is that teachers should plan for cognitive demand and for management. Because students become restless after a period of high cognitive demand, plans should provide for high-cognitive-level tasks to be interspersed with low-cognitive-demand tasks. In addition, teachers might try to reduce the risks associated with engagement in higher-cognitive-level tasks. The lower-ability class in the Gallagher and Tobin study may have remained on task if the activity had been less public and students could have engaged without the fear of failing before their peers. Small-group and individualized modes of engagement may be of more use for higher-cognitive-level tasks because engagement of individuals is relatively more private.

A distinctive feature of the classes of exemplary science teachers was the high level of managerial efficiency. The teachers actively monitored student behavior by moving around the room and speaking with individuals from time to time and also maintained control at a distance over the entire class. In order to monitor understanding successfully, it was necessary for students to be well behaved and cooperative. Little evidence of student misbehavior was noted. In most classes taught by exemplary science teachers, students demonstrated a capacity to work together if problems arose, to seek help from a peer, or to wait for the teacher to provide assistance. Students were able to work independently and cooperatively in groups. Consequently, teachers were not under pressure to maintain order, nor were they rushing from one student to another at the behest of students experiencing difficulties. Rather, the teachers had time to consider what to do next and to reflect on the lesson as it progressed. Interestingly, many of the teachers listed the development of autonomy and independence among the goals of their curriculum. Although a rule structure was firmly in place in these classes and students worked within the rules, there was no need for constant enforcement of the rules. Students knew what to do and appeared to enjoy working in the classroom. The evidence suggests that most teachers monitored student engagement and understanding in a thoughtful, systematic, and routine manner.

A variety of styles was used by different teachers in establishing and maintaining an environment conducive to learning. For example, the two chemistry teachers in Tobin and Garnett's study (1988) had different styles of teaching. The first exemplary teacher emphasized small-group work, whereas the second used mainly whole-class interactive activities. As with the exemplary teachers in other case studies, both of these teachers managed their classes effectively and were able to monitor for understanding and consistent student engagement rather than for misbehavior.

Enhancing Student Understanding of Science

A finding that applied to most teachers in the exemplary practice studies was that they had a concern for assisting students to learn with understanding. All of the exemplary teachers used activities in which students could have overt involvement in the academic tasks. In elementary grades, the activities were based on the use of materials to solve problems, and in high school grades teachers often used concrete exemplars for abstract concepts. The key to teaching with understanding was verbal interaction that enabled teachers to monitor students' understanding of science concepts.

The exemplary teachers were effective in a range of verbal strategies, including asking questions to stimulate thinking, probing student responses for clarification and elaboration, and offering explanations to provide additional information. For example, in Goodrum's (1987) study, three exemplary elementary science teachers taught with a materials-centered emphasis that encouraged formulation and testing of predictions. The teachers were concerned with students understanding the methods of science, developing scientific concepts that could be used to interpret the environment, and acquiring an attitude of scientific inquiry. Students were allowed to work together with materials and discuss their findings with peers and the teacher. Each teacher adopted a problem-solving orientation in which a problem was outlined, data were collected, and solutions were discussed. The key to students developing understanding about science was the quality of the teachers' questions in small-group and whole-class activities. The teachers knew what questions to ask to facilitate important understandings about science.

The exemplary high school teachers used laboratory activities in a manner that facilitated learning of science. For example, Tobin, Treagust, and Fraser (1988) reported that an exemplary biology teacher emphasized inquiry rather than verification of facts and principles and that the teacher was a model inquiry teacher, not only in the way that he asked questions but also in the way he fostered student independence and curiosity. At the beginning of each week, the teacher ensured that students knew what was to be accomplished during the week and part of the following week. The students knew when laboratory activities were scheduled and were expected to prepare for them before the class. In most instances, laboratory activities commenced without a prelaboratory discussion. Thus, students were involved in the laboratory activity at the outset of each lesson. They knew what to do as a result of their preparation at home or work in previous lessons. Laboratory activities were completed with a whole-class discussion of results and conclusions in which the findings of the laboratory activity were related to content in the textbook and to prior learning.

Maintaining a Favorable Classroom Learning Environment

A synthesis of the exemplary-practice studies (Fraser & Tobin, 1989) indicated that students of exemplary teachers perceived the environment as conducive to learning. Fraser and Tobin reported that students in both classes of an exemplary biology teacher perceived their actual classroom climate more favorably than students taught by nonexemplary teachers viewed science. The biggest differences for both classes were related to involvement, teacher support, order, and organization. Also, in classes taught by exemplary teachers there was a surprising congruence between the psychosocial environment preferred by students and the classroom environment they perceived to be provided by their teachers. These findings are in accord with earlier research which suggested that achievement is higher when the preferred and perceived learning environments match (e.g., Fraser, 1986).

Discussion

What we learned from the synthesis of nine case studies of exemplary practice was not surprising, nor did it provide grounds for total optimism. The exemplary teachers managed their classes well, taught with student understanding as a focus, maintained a learning environment that was conducive to learning, and had a sound grasp of the science content students were to learn. None of the factors alone was sufficient for effective teaching. All of the factors applied to each of the teachers who matched up to the researchers' constructivist view of exemplary practice. The findings emphasize the complex nature of teaching and learning in classrooms. To be an effective teacher requires much more than presenting content from a textbook and effectively managing a class of learners. The synthesis of exemplary-practice studies showed that effective practice derives from beliefs about teaching and learning and a reservoir of discipline-specific pedagogical knowledge (i.e., the knowledge needed to teach specific science content) (Shulman, 1986).

It was apparent in the synthesis of exemplary-teacher case studies that teachers' questions were used skillfully to focus student engagement and to probe for misunderstandings. When explanations were given, they were clear and appropriate. Concrete examples were often used to illustrate abstract concepts, and analogies and examples from outside the classroom were used frequently to facilitate understanding. In addition, teachers appeared to anticipate areas of content likely to give students problems. At the conclusion of a lesson, the main points were highlighted and reinforced. Clearly, exemplary teachers had extensive knowledge of how students learned as well as what to teach and how best to teach it. The findings are a salient reminder that teaching is a demanding profession. Without the necessary content and pedagogical knowledge, teachers can expect to flounder. Those who are experiencing difficulties can anticipate continuing problems unless they attain mastery over what they are teaching and how to teach it.

The important role of content knowledge was demonstrated when teachers taught outside their field of expertise in general science classes in the secondary school and in some elementary science classes. Because teachers did not have the content knowledge, errors of fact were made and opportunities to elaborate on student understandings and to diagnose misunderstandings were missed. In some instances, flaws were evident in attempts to explain concepts with which students were having difficulty. In other cases, analogies were selected that compounded students' problems in understanding the concepts. The result of teachers' lack of content knowledge in high school classes was an emphasis on learning facts and a sowing of seeds for the development or reinforcement of misconceptions. These instances of teachers having less than optimal backgrounds in the content to be taught occurred in classes of teachers who had been nominated as exemplary. Such problems are likely to be of greater significance in classes of nonexemplary science teachers. There is considerable need for greater research into the extent to which science teachers have the knowledge needed to teach science effectively. Recent studies of the relationships between pedagogical content knowledge, instruction, and student knowledge are described later in this chapter.

Administrators should be loath to schedule teachers for out-of-field teaching assignments. Willing and dedicated teachers can expect to encounter problems if they are required to teach in areas in which they do not have the discipline-specific pedagogical knowledge necessary to sustain an appropriate learning environment. Policymakers often feel they have few alternatives because suitably qualified science teachers are in short supply. As a consequence, teacher educators should reexamine courses and ensure that professional science teacher education programs provide the knowledge needed to teach general science as well as areas of specialization such as biology, physics, and chemistry.

This review of exemplary practice provides models of teaching that all science teachers can seek to emulate. In this sense, the review provides grounds for optimism about the future of science education. In addition, the analysis of exemplary teaching reveals an Achilles' heel that needs the close attention of all science educators. Other resources are probably needed to help teachers obtain the knowledge needed to teach specific science content. Providing these resources to all science teachers and convincing them that knowledge limitations might be inhibiting *their* science teaching effectiveness are substantial problems for all science teacher educators.

TEACHER BELIEFS

The Concept of Belief

Research on teacher beliefs has been fundamental to the teacher-education rhetoric of recent years (Calderhead, 1988; Crow, 1987; Eisenhart, Shrum, Harding, & Cuthbert, 1988; Feiman-Nemser & Buchmann, 1987; Floden, 1985; Koehler, 1988; Taylor, 1990; von Glasersfeld, 1989; Weinstein, 1989). Science educators, as well, have begun to recognize the importance of examining teachers' pedagogical beliefs about science teaching and learning. Various studies have examined the beliefs held by prospective and practicing science teachers about students and classrooms (Abell & Smith, 1992; Haggerty, 1992; Hewson & Hewson, 1989; Jasalavich, 1992; Lorsbach, 1992). Other studies (Ball & McDiarmid, 1987; Brousseau & Freeman, 1988) have explored the role of teacher beliefs in teacher-education programs. Brousseau and Freeman, for example, have probed the way in which teacher-education faculty members define desirable teacher beliefs. This review, however, focuses more exclusively on studies that emphasize inquiry into the beliefs of practicing teachers, many of whom are secondary science teachers.

The central importance of the concept of belief to many science-education researchers is apparent in a recent study of science educators' definitions of belief and the sources of their definitions (Oliver & Koballa, 1992). Eight categories of definitions of belief emerged from data obtained through questionnaires and telephone interviews of selected science educators. These categories included responses that: (1) equate belief with knowledge; (2) consider belief as a functional representation of knowledge; (3) view belief as antecedents of attitude, motivation, and behavior; (4) describe belief as "a relationship between object and attribute that a person holds true"; (5) define belief as "personal convictions that may or may not be based on observation or logical reasoning"; (6) reflect a definition of "belief as a vision oriented by an epistemic system"; (7) explain belief as "the acceptance or rejection of a proposition"; and (8) specify a dictionary definition of belief.

The respondents' multiple definitions of belief suggest that science educators have different conceptions about belief. Oliver and Koballa submit that, nevertheless, the following elements are common to the varying definitions of belief: a relationship between belief and knowledge, the idea that beliefs are acquired through communication, the concept that beliefs prompt action, and a continuum that reflects a range of beliefs from factual to evaluative.

There appears to be no common use of the concept of belief in teacher-education research. However, the importance of new efforts to engage in such conversations that confront the meaning of *belief, voice, negotiation*, and other words used frequently in research on science teaching cannot be overlooked. For example, Nespor (1987), in his study of the beliefs of eight middle school teachers, makes a distinction between beliefs and knowledge. He describes six structural features of beliefs that distinguish them from other forms of knowledge. An example is the heavier reliance of belief systems than of knowledge systems on affective and evaluative components. In developing a model for understanding teacher beliefs, Nespor's fundamental assumptions about beliefs stand in marked contrast to those guiding our own research. Nespor argues that because beliefs have structural features they cannot be shown to be unreasonable or propositional in nature.

In our current research, the concept of belief has shifted in meaning as we have come to understand the nature of referents as a form of belief. The viability of an individual's scientific knowledge became more important as we began to think about beliefs as referents as opposed to a form of professional knowledge to which language had been assigned. Referents (i.e., constructivism, objectivism, emancipatory interests, control, time, social expectations) act as organizers of teacher knowledge in the form of beliefs and images. They are useful in determining whether particular actions are legitimate in the culture of the science classrooms in which they operate. The viability of knowledge became a more important gauge than whether knowledge was propositional or whether it was represented symbolically as language. The issue of viability was useful in that it enabled us, as individuals, to make qualitative sense of experience. If an image was used as a referent for a particular action, it seemed that the referent was regarded by the actor, in the particular context of action, as having viability. Accordingly, we redefined belief as a form of knowledge that is personally viable in the sense that it enables a person to meet his or her goals. Viability is a significant part of the definition because it implies a test of the belief against experience. This can be undertaken only in a social milieu. From our perspective, all beliefs have a social component even though they are personal constructs.

Beliefs associated with world-view theory have become an area of interest for science-education researchers in recent years (Cobern, 1991; Proper, Wideen, & Ivany, 1988). Cobern makes a distinction between world-view presupposition as belief and what he refers to as "ordinary belief." This distinction is rooted in the idea that the fundamental difference between world view beliefs and ordinary beliefs is really a matter of degree. Proper et al. (1988) focused on science teachers' classroom discourse in terms of identifying expressed world-view beliefs . Data sources for this study were 54 segments of teacher talk taken from 42 lessons taught by secondary science teachers. The transcripts were examined using Kilbourn's (1980–81) analytical scheme, which describes six different sets of world-view beliefs. Teachers in this study projected a limited set of world-view beliefs, which were categorized as formism, mechanism, contextualism, and organicism. As science educators and researchers struggle with understanding the complexities of teacher belief, it is useful to recollect Cobern's words: "the exact meaning of belief is determined by the context of usage."

Metaphor and Teacher Belief

One promising line of research on teacher beliefs recognizes that metaphors can be useful in our attempt to understand science teaching and learning, teachers' roles, and the complex relationships between belief and action. The relationships between metaphor and understanding have been explored in numerous contexts (e.g., Howard, 1989; Jarolimek, 1991; Muscari, 1988). Howard, for example, graphically illustrated how a change in metaphor can lead to enhanced understandings of phenomena. At one time, he indicated, the heart was conceptualized as a furnace that heats the blood. As studies of the structure of the heart matured, however, the furnance metaphor was preempted by more salient metaphors such as the heart as a pump. Howard cautioned that not all aspects of a concept can be transferred to another, and it is often forgotten that a metaphor is being used to make sense of a phenomenon. When teachers use metaphor to facilitate understanding of science concepts, it is important for students to learn the ways in which the metaphor does and does not apply. It is implied that students should be reflective about their understandings of phenomena and aware of the limitations of their knowledge. For example, Roth (1990, 1991), who is a teacher researcher, describes teaching and learning roles in terms of cognitive apprentice, scaffolding, and coach. He does not suggest that every aspect of each of the three metaphors can be used to form an understanding of teaching and learning. Rather, meaning is associated with dynamic images such as a young learner apprentice with a wise master teacher, a sturdy scaffold reaching into new territory but firmly attached to extant knowledge, and a thoughtful diagnostician contemplating how to get improved performance from a hard-working athlete.

Similarly, Grant (1991) described learning in terms of a metaphor of climbing Mount Everest. The metaphor was used to assist students in making sense of their roles in the literature classroom and to better understand assessment as rewarding production rather than reproduction of knowledge. However, not all aspects of climbing the mountain can be considered productively in relation to learning. Hard work, effort, strength, perseverance, courage, teamwork, and accomplishing goals in stages are images of a successful assault on the summit of Everest that seem relevant and appropriate to enhancing learning. In contrast, hunger, poor health, fatigue, and safety hazards are possible consequences of climbing Everest that ought not be considered inevitable, even in a metaphorical sense, in the process of learning.

Metaphors can constrain the manner in which events, problems, and solutions are framed (Drake, 1991; Gozzi, 1991; Grandin, 1989). Drake used the metaphor of a journey to describe learning as a lifelong process containing significant events such as the call to adventure, the decision to act, disequilibrium, anxiety, and cognitive dissonance. The final stage of the journey, the actualization of human potential, involves sharing the products of learning with others to benefit the community

as a whole. Gozzi addressed the problems of metaphors constraining thinking. Using the example of mind as computer and anthropomorphic metaphors such as intelligence, memory, and user friendly, Gozzi noted that use of the mind-as-computer metaphor has led to exaggerated claims for the potential of computers. Whereas this is seen as a problem in relation to the way we think about the value of computers, the use of a computer as a metaphor for the brain (e.g., Leong, 1988) may have even greater consequences for the way we think of learning of science and the education of science teachers. In such cases, the computer, which we do understand, is used to explain the functioning of the brain and the learning process. Accordingly, the widespread availability of personal computers has made it possible to explain such things associated with learning as cognitive overload, information processing, and short- and long-term memory. Although in some contexts the brain-as-computer metaphor might be useful, in others it might restrict the range of possible approaches to learning in ways that are deleterious to the process of learning. Consequently, one goal for reflection is to identify metaphors and ascertain how their use might constrain actions in given contexts.

Grandin (1989) advocated the use of metaphors to empower students. In her discussion of the efficacy of three metaphors—conflict, midwife, and web—Grandin implies that metaphors have universal applicability. She indicated that despite the positive benefits of struggle, conflict is seen as self-defeating in many situations. Although the teacher as midwife is seen as leading to a connected view of teaching and learning, Grandin suggested that males might not readily identify with the midwife metaphor and the teaching role implied by the metaphor is relatively passive. Grandin advised that teachers must be active to assist students in grappling with positive aspects of science learning. According to Grandin, the web metaphor views writing as a process and suggests a final product that is potentially empowering for students, and, compared with the midwife metaphor, it is more likely to appeal to male teachers. Grandin seems to overlook the personal aspect of making sense of concepts with metaphors. What works in the sense-making process for one person might not work for another.

Gilbert (1989) also does not appear to appreciate that it is for the learner to construct and make sense of metaphorical relationships. In his study of 201 ninth- and tenth-grade general-biology students, analogies and metaphors were inserted into text to be used by a treatment group while a control group used "literal" text. The study indicated that analogies and metaphors inserted in text do not promote achievement or improve attitudes. However, Gilbert appears to overlook the significance of the analogies and metaphors to learners. Indeed other studies suggest that a metaphor that becomes a heuristic for one learner might be of little value to another (e.g., Tobin & Ulerick, 1989).

Numerous authors have conceptualized teacher and learner roles in terms of metaphors (e.g., Gurney, 1990; Philion, 1990; Power, 1990; Provenzo, McCloskey, Kottkamp, & Cohn, 1989; Tobin, 1991). Philion (1990), for example, described the metaphors used by three student teachers in English classrooms. The metaphors were teacher as classroom manager, referee, canvasser, and the nurturer who was able to talk with students rather than to them and who remarked that fertile soil is necessary for successful student-teacher relationships. Power (1990) used metaphors to argue for a blurring of the roles of the teachers, students, and researchers. She examined the metaphors and models associated with instruction and showed how new blurred roles for teachers and acceptance of a range of methods can be applied in working with prospective teachers.

When teachers act in the classroom, they do what makes sense to them in the circumstances. Criteria that guide a teacher's selection of appropriate practices include what has worked in the past and seems likely to be effective in a given context. However, decision making is usually not conscious and is not focused on each of the myriad behaviors associated with the various roles to be adopted in any lesson. Research (e.g., Tobin & Ulerick, 1989) suggests that metaphors used to conceptualize particular teaching roles guide many of the practices adopted by teachers. The manner in which metaphors are related to teaching practices is exemplified in the following examples of science teaching.

Gary did not have many problems with discipline even though misbehavior of students was a widespread problem in the school in which he taught. Gary, a martial-arts instructor and holder of a black belt in karate, adopted a metaphor of *teacher as intimidator*. The images he projected in class were frequently carried over from his martial-arts training and knowledge. His posture, movement around the class, and intent stare at potential troublemakers could easily have belonged in a karate contest. These images, together with proximity to students who were off task, were a deterrent to most students in the class who might have contemplated misbehavior. Gary was not physically intimidated by students in his class and was not prepared to accept unruly behavior from them. Gary was an authority figure who demanded the respect of students because of his managerial style (Tobin & Gallagher, 1987b).

Jonathon used a metaphor of *teacher as preacher* to make sense of teaching (Tobin & Espinet, 1989). In his life outside the classroom Jonathon was a preacher. As a teacher, he lectured from the front of the class and set seatwork tasks from the textbook. His lectures had many of the characteristics of a sermon, the textbook was his bible, and his role in the classroom was consistent with the roles he fulfilled as a preacher.

Sandra, a high school science teacher, was a *resource* to her students. She allowed them to learn together in groups or to complete tasks independently. The resource metaphor appeared to define Sandra's role and constrain her from behaving in certain ways. For example, few whole-class activities were conducted in 10 weeks of teaching and, when they did occur, they were of short duration and intended to clarify the schedule or provide details related to the administration of the program. Sandra was untiring in her efforts to share the teacher resource among the student consumers. To the extent that she was free to do so, Sandra responded to student needs by answering questions, providing explanations, and generally assisting students to remain cognitively active. Even when Sandra visited a group she usually interacted with two or three of the students at the table on an individual basis. Few visits to groups exceeded 30 seconds.

Diana, an elementary teacher, used three metaphors to describe her teaching role in different contexts (Tobin, 1991). She usually managed her class as a *policeman*, but in some circumstances she was a *mother hen* and on other occasions she was an *entertainer*. Her mode of behavior (i.e., the metaphor she used to drive behavior) depended on the context in which learning was to occur. Each conceptualization of her role as manager was associated with a discrete set of beliefs.

Provenzo et al. (1989) demonstrated that teachers used a variety of metaphors to describe their roles. They concluded that a systematic investigation of teacher metaphors associated with the critical events of teaching can provide valued insights into the practical wisdom of teachers and lead to knowledge that can provide a foundation for the education of prospective and practicing teachers. Provenzo et al. suggested that

exploration of teacher metaphors on a larger scale can lead to the identification of critical events in teaching and how the teacher experiences them. Such events and experiences, if introduced into teacher education programs, can provide a dialectic for current programs in order to expand the repertoire of the teacher-to-be and illuminate his or her entry into the reality of the profession, as it is experienced. (p. 571)

Munby (1990) assumes that the metaphors teachers use when they talk about their work represent their professional realities. For example, many of the traditional practices reported in science classrooms can be explained in terms of metaphors that are commensurable with the conduit metaphor. Munby (1986) explained how a teacher used a metaphor associated with movement to conceptualize the curriculum. The metaphor was apparent throughout the teacher's oral accounts of teaching and learning. Other metaphors identified in the study were knowledge as an entity to be picked up and the student's mind as a container. Conceptualizations such as these pervaded the language and presumably the actions of teachers in classrooms. Mosenthal (1988) advocated a change from the conduit metaphor to one that takes account of the user's rather than an authority's point of view. This call for a change of metaphor acknowledges the association of knowledge, power, and authority. If teachers are to alter the balances of power in classrooms, changes may be needed in the metaphors used to conceptualize teaching and learning roles.

In the remainder of this section, explanations are provided for the use of metaphors to make sense of teachers' roles, how belief sets are associated with specific roles and metaphors, and how new metaphors can be constructed to help teachers reconceptualize teaching roles and change instructional practices.

Metaphors and Teacher Change

Bullough (1991) explored the use of metaphors by three prospective teachers as they moved through a teacher-education program. He concluded that metaphors were not always useful as a focus for reflection because in some instances the construction of new metaphors did not represent a thoughtful reexamination of teaching and learning. The metaphors discussed range from teacher as husband, teacher as butterfly, and teacher as policewoman to complex metaphors such as teacher as chameleon ("always changing and trying to find a spot where I am most comfortable," p. 49). The label (husband, chameleon) was not the characteristic that made the metaphor simple or complex but the explanation of how teaching was related to being a husband or a chameleon, whichever the case might be. Munby and Russell (1989) suggest that it is productive for all teachers to become students of their own metaphors. Careful attention to how one describes the world gives insights into how one constructs it. Accordingly, the language of teachers provides an image of their professional knowledge and the ways in which that knowledge might be represented and constrained. Bullough's study highlights the important point that metaphor is just one tool for thinking about teaching. Use of metaphors does not guarantee that the quality of reflection will increase; that depends on the person constructing the metaphor as a heuristic for knowledge of teaching. In a study of middle school science teaching, learning, and curriculum reform, Tobin, Tippins, and Hook (1992) reported that initially the teacher in their study did not use metaphors to represent his knowledge of teaching. However, once he knew about metaphor as a way to think about teaching, he began to describe his roles in terms such as prison warden and air-traffic controller. He then used a metaphor to reconceptualize his teaching roles in terms of being a nurturing gardener. In this case his thinking was thoughtful and elaborate. Indeed, the gardener metaphor became a heuristic for teacher and student roles and sets of beliefs about teaching and learning. In this instance, the teacher learned to use metaphor to build new knowledge of teaching, to guide implementation of the curriculum, and to reflect *on* and *in* practice. In this sense, metaphor was a powerful tool for thinking about teaching and learning. As the research team identified metaphors and specific actions consistent with these metaphors, they were able to deconstruct classroom routines, exposing many of the tacit beliefs underlying given actions.

Elias (1990) also explored a growth and cultivation metaphor, noting that although plants have the power to grow, they grow best under the proper conditions, to which a gardener can attend. He also explored metaphors of the teacher as doctor, the person in charge with the medical cure, in relation to a metaphor that combines the roles of doctor and nurse (i.e., cure and care). This metaphor introduces the notions of expertise, power, and care.

Not all doctors would fit Elias's image of cure prescribers, nor would all nurses be as caring. For him, these characteristics had greatest salience when he related the two sets of roles to teaching. His suggestions are a reminder that metaphors are not universally appropriate or inappropriate. On the contrary, meaning depends on the manner in which an individual constructs the metaphor and relates it to educational situations. Elias had undergone heart surgery and had encountered doctors and nurses of particular types and, as a consequence, had reflected on his own roles as a teacher. For him it made sense to alter his teaching roles in terms of images of cure and care that had personal meaning to him. Each of the metaphors advocated by Elias incorporated care giving to the patient, the plant, or the student. The concern for the care of the recipient is consistent with a constructivist perspective of the learning of students.

Fawcett (1992) showed that changes in teaching are possible if roles are reconceptualized in terms of new metaphors. Fawcett, a teacher researcher, described her experience in becoming a colearner in terms of moving from the sanctity of the big teacher's desk to the small desks of the students. She described her change process as slow and painful. Instrumental in her change was a shift in metaphor from commander of the big desk to teacher as lifelong learner. Underlying the change in metaphor is a change in the teacher's beliefs about control of students. Behind the desk, the teacher maintains distance from learners, the big desk signifying control and power of the teacher; in the colearner situation, control and power are more equitably distributed between teacher and students as the teacher interacts at eye level with students in their small desks.

Knowles (1990) described his change from a "really good teacher" who was a transmitter of content and a controller of students to one who was a facilitator of learning. The stimulus for his change was a course taught by a professor who perceived himself as a facilitator. The experience as a learner in the class enabled the author to reflect on his own teaching practices and construct a commitment to change and images on which to build changes in his own classroom practices. Tobin and Jakubowski (1991) also reinforced the importance to curricular change of teachers' commitment to personal change and images of a reformed curriculum.

These studies and others illustrate how metaphors that reveal beliefs and orientations about teaching are useful in conceptualizing experience and may provide an alternative means of evaluating teacher cognition. Briscoe (1991) examined the way in which a secondary science teacher, Brad, made sense of his roles in terms of metaphors that served as organizers of important belief systems. While Brad expressed a commitment to change, certain beliefs about social expectations and classroom management constrained his actions. Briscoe underscores the importance of opportunities for reflection as teachers struggle to "construct new knowledge which is consistent with the role metaphors they use to make sense of changes in their practice."

The initial insights into the potential power of metaphors as a "master switch" to changing belief sets came in a study conducted in Australia (Tobin, Kahle, & Fraser, 1990). One teacher in this study, Peter, conceptualized his management role in terms of being captain of the ship. When the context was right, Peter became the captain of the ship and his students were regarded as the crew. How students and the teacher were expected to behave in activities was defined in terms of the metaphor he used to understand management. The metaphor became a filter for formulating beliefs associated with management in the contexts in which it was deemed relevant to managing the class in this way. Peter did not believe it was always appropriate to be captain of the ship. Some contexts required different management styles. In such contexts Peter believed he should be an entertainer.

Observations of teaching and interviews suggest that the captain of the ship and entertainer metaphors influenced the way Peter conceptualized his roles and the way he taught. In a particular activity, Peter used the metaphor that made most sense to him in the circumstances. When he was entertaining the class he was humorous, interactive, and amenable to student noise and risqué behavior. As captain of the ship, Peter was assertive and businesslike. He was in charge of the class and emphasized whole-class activities in order to maintain control of a teacher-centered and -paced learning environment. He was particularly severe on students who stepped out of line and often scolded them in a strong voice. In this mode he called on non volunteers and ensured that all students listened and participated in an appropriate manner. Whole-class activities were appropriate for both metaphors. The captain of the ship gave orders and explanations to the entire crew, and the entertainer performed to the whole audience. In both contexts the teacher was in charge (i.e., controlled the students), just as the captain manages the ship and the entertainer manages the audience.

Peter's beliefs about student learning all involved the teacher in some way. Peter liked to interact with students in small groups or on a one-to-one basis. Consequently, the particular management style he employed became relevant as he interacted with students as the captain of the ship or the entertainer. Although Peter's beliefs about his management style constrained the way he facilitated learning, he did not appear to reflect *on* or *in* action to any great extent. There was no evidence that Peter critically reviewed his beliefs about management or his management metaphors in relation to his beliefs about facilitating learning.

What was so interesting in Peter's teaching was the distinct teaching style associated with each metaphor for managing student behavior. As Peter switched metaphors, a great many variables changed as well. This finding suggests that helping teachers to acquire new metaphors for specific teaching roles might help them improve their classroom learning environment. Peter's ability to manage the class in distinctly different ways suggests that he might be able to improve his teaching by using different metaphors to conceptualize his teaching roles. For example, if Peter understood teaching in terms of a gardener nurturing new seedlings, would he tend individually to the learning needs of his students? Would students then perceive the class to be more personalized? These questions have interesting implications for teacher education. If teachers conceptualize their teaching roles in terms of metaphors, it is possible that the process of teacher change can be initiated by introducing a variety of metaphors and reflecting on their viability in the classroom.

The findings reviewed to this point suggest that significant changes in classroom practice are possible if teachers are assisted to understand their teaching roles in terms of new metaphors. Further insights into the importance of metaphors in conceptualizing teaching roles were obtained in a study of another teacher, Marsha (Tobin & Ulerick, 1989). When the study commenced, Marsha had been teaching two classes for one semester and patterns of behavior and interaction were well established.

Marsha had major problems with classroom management. Because of the importance of beliefs about teaching and learning in earlier research, Marsha's beliefs were examined in relation to each of three salient roles she identified (i.e., facilitator, manager, assessor). It was conjectured that Marsha's beliefs were contributing to the problems she was experiencing. Interviews with Marsha soon revealed a puzzle. Although she used a number of metaphors in understanding her roles as a teacher, she described each of her teaching roles in terms of a distance metaphor that implied contradictory actions. Marsha believed that as a facilitator of learning, she should be close to students; to be an effective manager, she should be distant from them; and as an assessor, she should not get too close to them. Until Marsha began to reflect on her practices in terms of beliefs and metaphors, these contradictions were not apparent to her. Once she identified them, these contradictions became a perturbation that Marsha wanted to resolve, in a theoretical sense and in terms of her classroom practices.

The context of learning and teaching is a mediator for beliefs (i.e., beliefs are always anchored to context). In a specific context one set of beliefs might apply to a given role, but in another a different belief set might "drive" behavior. Thus, a teacher (e.g., Marsha) might hold a particular belief and not allow it to influence her actions because contextual factors made it seem inappropriate. What emerged as being more important was the manner in which Marsha conceptualized her roles. In the context of the classes she was teaching, the metaphors used to make sense of the three roles seemed incompatible. For example, the principal metaphors associated with Marsha's understanding of her roles as manager and assessor seemed inconsistent with how she would facilitate learning in her ideal class.

Although Marsha believed her role as facilitator should have highest priority, her inability to manage her classes effectively required that greater priority be given to her management role. The main metaphor Marsha used to conceptualize management was *teacher as comedian*. Teaching behaviors associated with the metaphor seemed to elicit aggressive student behavior; students took advantage of Marsha, did not cooperate with her, and the environments in her classes were not conducive to learning.

Marsha received assistance from a research team to learn and change. Discussions with team members focused on constructivism (von Glasersfeld, 1987), and Marsha's reflections on teaching focused on students having opportunities to learn. Eventually, Marsha decided to reconceptualize her role as manager in terms of being a *social director*. She understood what this meant because she had been a social director on many occasions in the past. Her application of this role to teaching was metaphorical and resulted in rejection of many of her previous beliefs about managing a class. Her social-director metaphor was associated with beliefs that were compatible with constructivism. The social director managed the class in a manner that supported learning. According to the metaphor, the teacher invites students to the party of learning. Students decide whether or not to come, and the teacher's role is to create opportunities for them to learn. If students decide not to accept the invitation, the teacher has the responsibility to make the invitation more attractive. Only two rules applied: guests (i.e., students) should be courteous to their host (i.e., the teacher) and to one another, and guests should not disrupt the fun (i.e., learning) of others. Guests who violated these rules would be invited to leave the party.

When Marsha perceived the conditions to be right, she implemented her new metaphor and the associated teaching role. Student misbehavior, which was previously widespread, almost disappeared, and management became less of an issue. Although disruptive behavior diminished considerably, many students exhibited latent hostility. An additional change that produced almost immediate results grew from the suggestion that Marsha view her role as an assessor in terms of the metaphor of "*a window into the student's mind.*" An assessment would allow the teacher to see what a student knew or permit a student to show what he or she had learned. Marsha realized that so many students need not fail science, changed her procedure for assigning grades, communicated the new system to her classes, and endeavored to create an expectation of success in science. Students who had regarded her assessment procedures as unreasonable responded with enthusiasm to the new approach to assessment.

This research suggests that a teacher's personal epistemology acts as a constraint to the way he or she thinks in relation to teaching. Marsha used constructivism as a referent for thinking about science teaching and developed new metaphors for conceptualizing the roles she considered to have greatest salience for teaching. She reflected on possible ways to conceptualize roles and used constructivism as a cognitive filter to decide whether there was a fit between what she believed about learning and teacher and student roles that were conceptualized within the boundaries of a metaphor. Only when the fit was determined to be good did the teacher accept the metaphor as viable. For Marsha, constructivism became the main cognitive referent for acting in relation to teaching science. As new metaphors were implemented, Marsha reflected on her actions and the practices of students in terms of constructivism, and anything that did not fit became a focus for change. For Marsha, the test of viability of the actions embedded within the curriculum became constructivism. In contrast, when Marsha began her teaching career, her principal referent for action was control. Her beliefs about science and the teaching and learning of science were grounded in the belief that the teacher should maintain control of students.

The metaphors used to make sense of roles and the belief sets associated with particular actions might be productive foci for reflection. This is consistent with research by Kottkamp (1990), who also explored the significance of metaphor in the process of reflection. Teachers can identify the salient metaphors for specific teaching roles and consider whether alternatives would lead to improvements in the classroom. If teachers decide to alter the metaphors they use to understand particular roles, beliefs previously associated with the roles might be perceived to be no longer relevant. Beliefs consistent with the new metaphor can then be deemed relevant and influence what teachers do as they plan and implement the curriculum.

The dialectical nature of the relationships between ways in which teacher knowledge can be represented and can thereby influence the implemented curriculum was demonstrated by Tobin et al. (1992). The authors concluded that teaching occurs in the context of a complex multitask environment in which an effective teacher acts in accordance with multiple referents that take the form of images, metaphors, verbal propositions, and overt actions. The referents associated with particular actions are context specific. Tobin et al. reported that the teacher in their study was a professional who acted in accordance with extensive knowledge of science teaching. At times he was deliberative, using beliefs for which he had language as referents in deciding what to do; at the same time, he handled routines by acting in accordance with knowledge that took the form of images. Sometimes there was a high level of compatibility between referents used in a deliberative way, and on other occasions there were conflicts that led to dilemmas to be resolved during a lesson. A more common situation, involving competition between alternative referents, occurred when the teacher discussed his teaching practices and attempted to justify what was done. During discussions, conflicts became apparent between referents the teacher would like to use and the referents he appeared to use while teaching. Situations such as these provided a context for changes to occur in specific actions associated with teaching and espoused beliefs that did not seem to affect the way he taught.

Over an extended period, the teacher constructed a language which was linked to his practices as a teacher. Gradually the language became a tool for describing experience, for framing descriptions of what happened in his classroom, for raising questions, and ultimately for providing solutions and plans for future teaching. Notable examples were the teacher's use of constructivism, emancipatory interests, and metaphors as referents for actions. Initially the teacher was resistant to the idea of constructivism and was not convinced that it had anything to do with how he should teach. In essence, constructivism challenged his conventional thinking about the curriculum, the roles of teachers and students, assessment of students, and his beliefs about the nature of science, learning, and knowledge. Acceptance of constructivism necessitated too much change, and the need for change was not apparent to him. However, constructivism was being discussed among his colleagues in the school and university educators who visited the school. Accordingly, he learned about constructivism by participating in discussions and debates. Initially at least, constructivism was not a referent for changing other knowledge forms. However, the teacher was committed to changing his role in relation to controlling students.

From the time he was a student in school, the teacher in the Tobin et al. study believed that teachers should exercise less control over students and allow them to make decisions in relation to their own learning. Although his initial years of teaching were influenced by the traditional culture of schools, he was not satisfied with the way he was teaching and was motivated to change. He lacked a vision of how to give students more autonomy in the classroom but had a commitment to personal change. His opportunity to change arose through sharing a classroom with colleagues and participating in a study with prospective teachers in which he was required to answer questions about his classroom practices. The teacher began to think about his teaching in relation to his beliefs about how he should teach, and his reading of Grundy's (1987) typology of cognitive interests provided a theoretical framework to support change. The teacher implemented change because he had a commitment to personal change, a vision of what the curriculum could be like in his classroom, and the cognitive tools to reflect on his practices. In addition, prospective teachers were undertaking collaborative research in his classroom, providing a context in which he could discuss his teaching practices and reasons for teaching as he did. Many of his colleagues in the school were endeavoring to change their curricula, as well. Conversations centered on topics such as concept maps, cooperative learning, alternative assessment, constructivism, and Grundy's emancipatory interests. For almost any position that was stated there were advocates and dissenters. Accordingly, there was a lively and ongoing debate about teaching and the science curriculum.

The study of Tobin et al. suggests that the teacher began to see his emphasis on providing students with emancipatory interests as a way of teaching that was consistent with constructivism. That is, constructivism and the closely associated emancipatory interests became referents for action. Over time, these two ideas became differentiated as the teacher reflected on his actions in terms of both referents.

Initially, the teacher did not use metaphors to a notable extent in describing his teaching or the roles of learners. However, the research team was interested in metaphors and the teacher became aware of them as a tool for conceptualizing teaching and learning. Gradually, the teacher began to describe teaching and learning in terms of metaphor. This represented a more holistic way of describing teaching. Rather than focusing on the myriad behaviors that occur in a classroom, the teacher was able to capture the essence of his roles in terms of metaphors such as *prison warden, air-traffic controller, symphony conductor*, and *gardener*. These metaphors were holistic descriptions of teaching and learning to which language could be assigned if necessary. Metaphors seemed to provide a convenient way to plan teacher and learner roles and to reflect on action after teaching. Thus, in an out-of-classroom context the teacher could think about teaching and learning in metaphorical terms, relating metaphors to valued referents such as constructivism, control, and constraints and to activities such as assessment, cooperative learning, laboratories, and projects. The study suggests that metaphor is a powerful cognitive tool for categorizing experience and organizing actions.

Appreciating how teachers can construct and reconstruct knowledge suggests different ways to structure learning environments for practicing and prospective teachers. For example, teachers can be given opportunities to experience teaching and learning and to give meaning to those experiences in their own ways. However, their own ways of experiencing depend on their beliefs and the values they ascribe to those beliefs. In addition, teachers need opportunities to test their descriptions with peers. Through such discussions they can defend their ways of knowing and resolve perturbations that arise from considering the differences between the perspectives of those in their peer group. Books, articles, and the persuasions of teachers also have a significant role to play in the debate that results in learning. Different theoretical models for describing experience can lead to additional learning as teachers make sense of what is being advocated and apply it to their own experiences.

The teacher, who in this case is the learner, is at the center of the learning process but learns in a sociocultural setting where the beliefs of others mediate the process of experiencing and learning.

The Relationship Between Belief and Practice

Research on teacher thinking has led teacher educators (Clandinin & Connelly, 1986; Elbaz, 1983) and, more recently, science educators, to ask questions about the relationship between beliefs and action. Studies have emphasized the ways in which teachers' beliefs about the teaching-learning process play a meaningful part in science classrooms. Emerging research explores how teachers' beliefs are pervasive in the classroom and influence the nature of teacher roles, planning and decision-making processes, and ultimately the curriculum. O'Loughlin (1990) cautions researchers that "teacher beliefs should not be conceptualized as simple cohesive systems which have a direct causal bearing on teachers' actions."

Using a naturalistic case-study approach, Cornett, Yeotis, and Terwillinger (1990) investigated the relationship between teacher beliefs and action in the science classroom. In this study of a first-year middle school science teacher, the researchers inferred seven personal theories of the teacher that appeared to guide her practice: (1) visual learning, (2) talking in kids' terms, (3) science learning as fun, (4) higher-level learning, (5) very disciplined class, (6) reinforced concepts, and (7) help students save face. The interaction of these theories became an important focus of the study, as it became apparent to the researchers that certain theories were more dominant or less dominant at certain times and in specific situations. Similarly, in our own attempts to understand the way in which a middle school science teacher made sense of his thoughts and actions as related to assessment or safety, we observed the relation of the actions to possible referents in the form of beliefs, images, and metaphors. Not all referents were of equal value or appropriateness. Several different, sometimes competing referents that were context specific influenced the teacher's actions (Tippins et al., 1992).

Cornett, Yeotis, and Terwillinger recommend additional cases that begin to focus on the evolution of teacher beliefs or personal theories over time. Kagan and Tippins (1993) analyzed case studies written by prospective teachers at the beginning, middle, and end of their field experiences for signs that teachers' beliefs about pupils and classrooms had changed over the course of the student-teaching experience. Cases written by secondary teachers reflected little change over time, as they continued to focus on beliefs associated with academic achievement and the potential for disruptive behavior. In contrast, cases written by elementary teachers reflected beliefs about teaching that involved coming to understand complex interactions between family and school life, the multifaceted nature of children, and the thick personal histories of children.

Cronin-Jones and Shaw (1992) also explored the relationship between belief and practice in the context of elementary and secondary science education methods courses. Their study was designed to analyze and compare the belief structures of prospective elementary and secondary teachers before and after participation in methods courses. A repertory grid technique was used to ascertain student responses in four major categories: student concerns, task concerns, broad concerns, and changes in concerns. The beliefs of elementary and secondary teachers did not appear to change much as a result of course participation. However, elementary teachers appeared to develop more narrow foci of beliefs and secondary teachers demonstrated broader and more diverse foci of beliefs, in contrast to the findings of Kagan and Tippins. The repertory grid technique used in this study was a fruitful research tool. However, because many of teachers' beliefs are implicit and tacitly held, they cannot always be ascertained by asking the teachers about their beliefs in relation to given situations. Narratives, case studies, metaphors, and imagery hold great potential for further insight into the belief structures of teachers.

Teacher Beliefs about the Nature of Science

Science teachers also have beliefs that influence their personal thinking about the nature and philosophy of science and the construction of a personal image of science. Several studies have provided insight into the process by which teachers formulate beliefs about the nature of science.

Gallagher (1991) inquired into the role of secondary school science teaching in framing teachers' beliefs and understandings of the nature of science and scientific knowledge. Using science textbooks and studies of classroom practice, he examined teachers' perceptions of science. He concluded that many of the secondary textbooks gave only perfunctory attention to the nature of science, usually in the first two chapters of the book. Science and scientific knowledge were usually conveyed as truth. During instruction, more attention was given to concepts and principles of science than to the process by which scientific knowledge is formulated. Gallagher found that secondary science teachers responded to textbooks in a way that contributed to a transmission model of science teaching. That is, all 25 practicing science teachers in this study portrayed science as an objective body of knowledge. Even two teachers who had a strong background in the history and philosophy of science and understood its importance in teaching held deeply entrenched beliefs about the objectivity of scientific knowledge. That science teachers have preexisting beliefs about the objective nature of science is not surprising. Their beliefs have been shaped by thousands of hours spent in college classrooms internalizing objective models of science. Many science teacher education programs are now using self-reflection to help both practicing and prospective teachers expose and confront their already well-developed beliefs about the nature of science.

In a study of four teachers participating in the BioTAP component of the Teacher Assessment Project, Collins (1989) used portfolios and simulation exercises to explore teacher beliefs about the nature of science. Like Gallagher, Collins reasoned that teachers do not understand the nature of science. She emphasized the difficulties in making inferences about the tacit knowledge of teachers from the responses obtained with the alternative forms of assessment.

As part of an interpretive study of 10 secondary science teachers participating in an authentic research experience in collaboration with university research laboratories, Arora and Kean (1992) looked closely at participants' beliefs about the nature of science. Using journals, responses to journals, transcriptions of group meetings and seminars, formal evaluations, and informal communications as data sources, the authors concluded that the teachers' beliefs about the nature of science and scientific work that were "consistent with some current philosophical views of the nature of science but inconsistent with how science is generally taught in schools."

Benson (1989) was also interested in the relationship between the epistemological beliefs of science teachers and their approach to teaching science in the classroom. A case-study approach was used to investigate the beliefs of secondary biology teachers. The gap between what teachers say they believe about the nature of science and what they do in practice was again apparent. Although broad biological concepts were mentioned to students as being important, all three teachers stressed the learning of detailed information about specific organisms consistent with an objectivist view of science. When teachers were questioned about the apparent contradictions between their beliefs and practice, they cited external constraints as a major factor influencing their practice.

Constraints and Beliefs

Why do teachers teach as they do? Like Benson, we have found that constraints can be obstacles to change in teachers' practices. When constraints act as myths for a culture (i.e., time, scarce resources, control, social expectations), they may suppress any changes considered, even when teachers are strongly committed to personal change (Tobin, 1991; Tobin et al., 1992).

Science teachers often move through 15 to 20 years of schooling without ever being induced to think about their own beliefs about the nature of science and scientific knowledge and that which has shaped them (Gallagher, 1991). Often they are not educated to think in terms of exposing the tacit assumptions embedded in teaching practices and conventions. Research on teacher beliefs about the nature of science has provided the groundwork for viewing beliefs and actions in a larger historical and philosophical context. Understanding teacher beliefs can lead to redefinitions of our images of science and the way it is taught.

Teacher Beliefs About Science, Technology, and Society (STS)

STS has become one of the central organizers for science teaching during the past decade, representing an appropriate learning context for all students. Studies have shown that learning science in an STS context enhances creativity, improves attitudes, increases academic achievement, and expands the use of science in daily life (Bybee, 1987; Bybee & Mau, 1986; Penick & Yager, 1986a; Yager, 1988). Science-education researchers have also focused on understanding STS implementation

(Aikenhead, 1992; Ellis, Bybee, Giese, & Parisi, 1992), but how this might be accomplished remains unclear. Studies have examined the way in which teachers' beliefs influence the STS implementation process.

Teacher's beliefs about STS-oriented issues and their personal STS literacy have been of foremost concern in recent studies. Zoller, Dunn, Wild, and Beckett (1991) examined the beliefs of 49 STS and 134 non-STS teachers and compared them with the beliefs of 302 grade 11 students enrolled in an STS course and 205 non-STS students. A questionnaire comprising six representative statements selected from the VOSTS inventory form (Aikenhead, 1987) was used to elicit teacher and student beliefs. The aim was to establish baseline data on the teachers' and students' STS beliefs, determine the STS literacy of teachers and students, and consider the ethical issues associated with STS education, or what the authors refer to as "indoctrination." The authors, emphasizing the need for research on teacher and student STS beliefs, state that "assessment and monitoring of the STS beliefs and positions on STS-related topics and issues of both teachers and students who may or may not have been exposed to an STS course, is essential if a valid research-based response to the education versus 'indoctrination' issue within the teaching reality of STS courses is sought." They found that the desired STS literacy among both teachers and students has yet to be realized. They described significant differences between the STS beliefs of teachers and students, even when students have been exposed to an extensive STS course with a heavy emphasis on values and beliefs.

Investigating the way in which teachers' beliefs influenced the acceptance, rejection, or alteration of the STS TOPICS curriculum, Mitchener and Anderson (1987) conducted a study of 14 teachers and 2000 students. Data collected through observations, interviews, and document analysis were used to develop a profile of teachers to provide insight into the acceptance, rejection, or alteration of STS curriculum. The three types of profiles generated indicated that teachers who felt the STS curriculum captured students' interest in real-life situations and who valued decision-making and group skills were more accepting of the curriculum; teachers who disliked the inclusion of social-studies content and the lack of real science topics and did not value activity, decision-making, or group skills rejected the STS curriculum; teachers who were concerned about the time and energy involved in implementing the TOPICS curriculum altered it in various ways. It is clear from these studies that teachers' beliefs do influence the way the curriculum is interpreted and implemented.

Self-efficacy Beliefs

Self-efficacy theory, in its original form, was most frequently used to explain coping behaviors in situations containing elements of fear. It has since been redefined, employed, and studied extensively in other contexts, including science teacher education. Self-efficacy theory originates from Bandura's (1986) social cognitive theory, which includes psychological phenomena as important aspects of learning theory. The original theory postulated by Bandura (1977) suggested that a person's behav-

ior and behavior change are mediated by the person's belief about his or her ability to perform certain tasks or behaviors. These beliefs are referred to as self-efficacy beliefs, expectations, or judgments. Self-efficacy beliefs were thought to be useful in examining whether behavior would be initiated, the effort expended on the behavior, and the duration of the effort (Lent & Hackett, 1987) and thus were considered mediating variables in relation to a person's behavioral intentions.

Two basic components of self-efficacy theory are the self-efficacy beliefs referred to as expectations or judgments and outcome expectations. Lent and Hackett (1987) described self-efficacy beliefs as concern about the ability to perform a given task and outcome expectations as the expected consequences of performing the tasks. Bandura (1986) argued that self-efficacy expectations associated with the perception of performance capabilities play a more important role in determining behavior than outcome expectations.

A factor important to the conceptualization of self-efficacy theory, with particular implications for science teacher education, is the hypothesized domain-specific nature of self-efficacy beliefs. Self-efficacy is believed to correspond to specified domains, necessitating the construction of task-specific measures in various domains. Self-efficacy theory has been useful for investigations of behavior and behavior change in teacher education (Guskey, 1988; Woolfolk, Rosoff, & Hoy, 1990), mathematics and computer domains (Miura, 1986; Schunk, 1987), and, more recently, science education (Czerniak & Waldon, 1991; Enochs & Riggs, 1990; Lent, Brown, & Larkin, 1987; Post-Kammer & Smith, 1986; Riggs, 1991). In science education, self-efficacy theory was initially useful in examining career-relevant behaviors associated with occupational preferences for scientific and technological careers. For example, Lent et al. (1987) developed the Self-Efficacy for Technical/Scientific Fields Educational Requirements and Self-Efficacy for Academic Milestones instrument for the purpose of predicting grades and persistence in scientific and technologic fields. More recently, science educators have focused their research efforts on efficacy of teachers, particularly elementary science teachers. Enochs and Riggs (1990) concentrated on the development of an elementary science teaching-efficacy-belief instrument. Using this instrument with both practicing and prospective teachers, they found significant gender differences in self-efficacy for science teaching.

Self-efficacy theory has been positively correlated with internal locus of control (Anderson & Schneier, 1978; Gist, 1987). Fate control is closely associated with self-efficacy theory and is directly related to the locus-of-control variable in social learning theory. The notion of fate control was originally employed by Rowe (1974a), who used the analogy of bowlers and crapshooters to emphasize the importance of chance or skill for influencing destiny. Haury (1988, 1989) extended the study of locus of control in relation to science teaching. In particular, he examined the relationship of teaching strategies to locus of control and factors that might predict science locus of control.

Qualitative methods of inquiry have provided new insight into self-efficacy beliefs and science teaching. Czerniak and Waldon (1991) compared quantitative science teaching efficacy measures with qualitative measures of teaching success. They found no relationship between science efficacy scores and teachers' rating of the success of actual science lessons.

The self-efficacy model is used to explain causal links between self-efficacy and science-related behaviors. Thus, it represents research shaped by a Newtonian perspective of a world that is never changing and "out there." Science education researchers operating within the Cartesian-Newtonian boundaries have examined only small pieces of the world of science teaching and learning and associated teacher beliefs.

With the emergence of ethnographic and qualitative approaches to research and the engagement of teachers in action research, different epistemological perspectives implicit in research on science teaching become visible. New and different assumptions guiding research on teacher beliefs become central: assumptions about what yields valid research information, the importance of context in making sense of science teaching and learning, and assumptions about the complex relationships between power and knowledge, the learner, the teacher, and other aspects of the learning environment. In this spirit, Connelly and Clandinin (1988) pioneered the use of teachers' narratives as primary sources of teacher beliefs. Over the past few years, researchers have developed a number of qualitative methods for inferring teacher beliefs and used many of them in making sense of beliefs associated with science teaching.

Qualitative Methods of Inferring Teacher Belief

Teachers' narratives can serve as sources of valuable information about their beliefs. Kagan and Tippins (1993) have described a number of approaches used to analyze narratives to infer underlying belief: (1) categorizing and computing the incidence of different kinds of perceptions (Zeichner, Tabachnick, & Densmore, 1987) or judgments (Thomas, 1990) about students and classrooms, (2) comparing a teacher's problem-solving strategy with some ideal model (Gliessman, Grillo, & Archer, 1989; Manning & Payne, 1989; Pugach & Johnson, 1989); (3) classifying statements in terms of levels of reflective thought (Simmons et al., 1989); (4) categorizing a case as descriptive, problem setting, or problem solving (Laboskey & Wilson, 1987); and (5) examining the metaphors used in a narrative (Munby, 1987; Russell, 1987; Tobin, 1990a).

In the search for a system of analysis to evaluate teachers' narratives, Kagan and Tippins (1991) examined narratives (classroom cases) written by science teachers in terms of structural and content features. A system that separated content and structural features of narratives did not reveal underlying belief. However, three themes that emerged from the study have implications for researchers interested in using classroom cases to infer underlying beliefs to assess professional growth: internal conflicts that provoke problems in a teacher, the long-term evolutionary nature of problems, and ethical overtones.

Teacher Beliefs and Educational Reform

Within the broader education community, a number of studies have recognized the importance of teacher beliefs in the process of school reform. Brousseau, Book, and Byers (1988) studied the impact of teacher beliefs on the process and culture of teaching. They attempted to define components of a "teaching culture" and concluded that the experience of working in such a culture has a significant impact on shaping beliefs that cross the boundaries of school setting and gender. Beliefs and orientations of 382 experienced classroom teachers and 332 prospective teachers were compared using data from a survey that included an Educational Beliefs Inventory. Experienced teachers expressed beliefs noticeably different from those of inexperienced and prospective teachers. The authors stress the need for studies that go beyond identifying teacher beliefs by considering how beliefs change and the factors that influence change.

Eisenhart, Shrum, Harding, and Cuthbert (1988) looked at the way in which teacher beliefs influence the implementation of educational policy by examining naturalistic studies of teacher belief published during the past 20 years. They began their study with an assumption that educational reform policies should be designed to fit with teacher beliefs. They found that "teachers appear to become less positive about activities as particular conditions associated with the activities change," notably with regard to activities outside the classroom. Consequently, although any attempt at reform must seriously consider the beliefs of teachers, if policies are designed to fit with these beliefs, teachers may stay uninvolved in reform taking place outside the school at policy levels. They iterate the need for teachers to be engaged in the development and monitoring of standards for the profession to guarantee that implementation of policy is compatible with teacher beliefs.

In the past decade dramatic changes in the practice of science education have occurred, reflecting broad educational reform and the use of constructivism as a basis for thinking about research and practice. *Project 2061: Science for All Americans* (American Association for the Advancement of Science, 1989) and *The Content Core: A Guide for Curriculum Designers* (NSTA, 1992) are among the many national reports setting new directions for science education. It is important for researchers, teacher educators, and policymakers to consider the nature of local reform activities in response to these national reforms.

Discussion

Future research should seek to enhance our understanding of the relationships between teacher beliefs and science education reform. Many of the reform attempts of the past have ignored the role of teacher beliefs in sustaining the status quo. The studies reviewed in this section suggest that teacher beliefs are a critical ingredient in the factors that determine what happens in classrooms. However, taking account of teacher beliefs is a necessary but insufficient condition for sustaining changes in science classrooms. Student beliefs are important too. Teachers should not expect to make significant changes in the classroom without the cooperation of students. Similarly, the beliefs of others in the educational community (e.g., parents, policymakers) might have to be considered if the quality of science learning is to be improved.

PEDAGOGICAL CONTENT KNOWLEDGE

Researchers have described an intrinsic connection between content knowledge and the "methods" and "strategies" of teaching. This type of teacher knowledge, an integration of subject matter with pedagogy, is referred to as pedagogical content knowledge in the literature. According to Hashweh (1985), teacher subject-matter pedagogical knowledge is more than a knowledge of the concepts, principles, and topics in a discipline—it includes knowledge of how to teach particular topics. Shulman's (1986) work in this area is familiar. He describes three categories of content knowledge—subject-matter content knowledge, curricular content knowledge, and pedagogical content knowledge. Pedagogical content knowledge differs from subject-matter knowledge in having the added dimension of subject-matter knowledge for teaching.

The Dilemmas of Content and Pedagogy

Shulman's differentiation of content from pedagogical content knowledge has been challenged on epistemological grounds. McEwan and Bull (1991) contend that "all subject matter is pedagogical." Accordingly, viewing pedagogical content knowledge as a category of teacher expertise separate from content knowledge supports an objectivist view of the world. By contrast, Shulman (1986) distinguishes between pedagogical content knowledge and content knowledge on scholarly grounds by describing "150 ways of pedagogical knowing" that differ from the ways in which scholars represent subject matter. Like Shulman, Cochran (1992) distinguishes between the ways a teacher and a scientist organize knowledge. She states that "an experienced teacher's knowledge of science is organized from a teaching perspective and is used as a basis for helping students to understand specific concepts. A scientist's knowledge, on the other hand, is organized from a research perspective and is used as a basis for developing new knowledge in the field" (p. 4). McEwan and Bull discuss the implications of Shulman's view of knowledge, which they say suggests that "because scholars' representations are an accurate reflection of the world as it really is, they are objectively true."

The knower cannot be separated from the known when constructivism is used as a referent to think about teaching and learning. Like McEwan and Bull, we would assert that content, pedagogy, and pedagogical content knowledge are inextricably linked. Another question growing out of this dilemma, with even broader implications for science teaching, concerns whether an epistemological distinction can be made between teaching and research. How we view the notion of methods is a reflection of this debate. Too often, instructional efforts have concentrated on showing practicing and prospective teachers new methods or tricks that presuppose one body of knowledge and classrooms that exist in a politically and socially neutral world. This is of particular concern given the wide diversity of learners in classrooms, especially those who have limited English proficiency (Carroll & Gallard, 1993; Gallard, 1992; Gallard & Tippins, 1992). However, science is not a body of knowledge that is politically and socially neutral and ahistorical; science classrooms are not simply places where neutral instruction takes place. Rather, they are "contested cultural sites" (Giroux, 1989) where certain types of speaking and particular forms of knowledge are legitimate and others are not. A rethinking of the notion of methods requires us to examine the way in which we view content, pedagogy, and pedagogical content as forms of knowledge. It compels us to consider the relationship between scholarship and teaching. The process of engaging teachers as researchers collaborating in inquiry has enabled science educators to begin to confront these complex issues.

Different aspects of science teacher knowledge are currently receiving attention from science education researchers. Our examination of these studies revealed a clear distinction between research pedagogical content knowledge and research on subject-matter knowledge. This dichotomy is highlighted in the NETWORK (1989, p. 17) document on *Developing and Supporting Teachers for Elementary School Science Education*, which states that: "teachers need to know both science content and pedagogy to teach science well. As Shulman (1986) argues, it is not enough to have good generic teaching skills; rather, each discipline requires its own teaching strategies. Teachers' content knowledge as well as their 'pedagogical content knowledge' are both of concern." For organizational purposes in this section, we will distinguish between these two categories, mirrored in much of the research on science teaching. Although this chapter does not focus on teacher preparation, several studies have investigated pedagogical knowledge in this context (Marks, 1991; Powell, 1991–1992; Rovegno, 1992; Tamir, 1988). For example, Tamir (1988) has outlined a framework of teacher knowledge consisting of six major categories: general liberal education, personal performance, subject matter, general pedagogical, subject matter–specific pedagogical, and foundations of the teaching profession. He has provided suggestions for addressing subject-matter knowledge, general pedagogical knowledge, and subject matter–specific pedagogical knowledge within science teacher education programs, particularly those for prospective teachers. Marks (1991) raises questions about the role of pedagogical content knowledge in teacher-education programs. He questions the extent to which pedagogical content knowledge should or should not be taught, the way in which prospective teachers derive pedagogical content knowledge, and how and where it should be emphasized in teacher-education programs.

The Development of Pedagogical Content Knowledge in Science Teachers

Several studies have focused on how pedagogical content knowledge is *developed* in teachers. Tobin and Garnett (1988) compared the science knowledge of two high school chemistry teachers and two primary teachers from the exemplary-practice studies. The two high school chemistry teachers had degrees in science and were able to use their knowledge to stimulate overt learner engagement on high-level cognitive objectives. Students in their classes were involved in thinking about science content in an overt way. This was not the case to the same extent in either of the primary teachers' classes. Of course, some of the differences could be explained in terms of the relative ages of the students. Perhaps older students could be expected to engage in high-level cognitive tasks more often and for longer periods than younger students. Although this might be the case, Tobin and Garnett postulated that the two high school teachers facilitated overt student engagement through the strategies they used. For example, one teacher used questions directed to individuals to promote thinking about content in materials-centered environments. This was possible because he had a firm grasp of the content he wanted to teach and he knew how to sustain student engagement. His monitoring was effective because he was in the right place at the right time and had the content knowledge to maintain the momentum of the lesson. He could ask students what they were thinking and could provide feedback on the substance of their thoughts.

In contrast, the two primary teachers were certified to teach primary grades and were not specialist science teachers. Their academic preparation was based on professional courses in a teacher-education degree and graduate work in science education. The focus of all their science-related courses was on teaching science. The emphasis was on curriculum resources and types of activities appropriate for teaching and learning science. Although both teachers were convinced of the value of primary students learning science in a "hands-on" manner, neither possessed sufficient knowledge of science to implement materials-centered activities successfully. Both teachers knew how to arrange students in groups and how to get the activities started, but they were unable to help the students develop science content from the activities. For example, one teacher attempted to focus student thinking on concepts associated with the properties of water. He knew how to monitor student engagement and how to probe student thinking, but he could not ask the crucial questions to focus student thinking on what was to be learned. In fact, there was no evidence that the teacher knew what students were supposed to learn from the activity.

Having sufficient content knowledge to teach science in an adequate manner is not just a problem for primary science teachers. Happs (1987) reported that an exemplary high school teacher made many errors while teaching a general science topic that was out of his field. The knowledge limitations were evident in the observed lessons and most likely resulted in students developing or reinforcing misconceptions. Tobin (1987) also reported that an experienced and "well qualified" high school teacher made numerous errors while teaching grade 10 general science. Although the teacher had an undergraduate major in anatomy and a graduate qualification in science, his knowledge of nuclear energy was extremely limited and his teaching reflected this. These findings highlight the importance of content knowledge and its impact on science teaching at all educational levels.

The pedagogical content knowledge of a high school physics teacher was the focus of an ethnographic study by Treagust, Wilkinson, Leggett, and Glasson (1988). This 6-month study of a teacher's pedagogical content knowledge used the concept of teacher as researcher to identify the process in which content knowledge in the curriculum is transformed by means of the teacher's content-specific pedagogy into a form of knowledge understood by students. The classroom teacher was a member of the research team, and this approach allowed him to reflect on his practices and implement changes as the research progressed.

Espinet (1988) examined the development of pedagogical content knowledge by two high school chemistry teachers in Spain. She was interested in determining the manner in which the teachers would develop pedagogical understandings to correspond to their content knowledge. Both teachers demonstrated a high level of content knowledge but had not experienced ideas of pedagogical knowledge as part of their education. Her findings led Espinet to conclude that pedagogical content knowledge merges with chemistry content knowledge when the teaching environment is structured and student centered, providing an opportunity for the teacher to learn about students' learning and thinking processes.

Pedagogical Content Knowledge and Instruction

Research in teacher education has shown that teachers' pedagogical content knowledge has an influence on the content and process of instruction (Grossman, 1988). Studies of science teachers have also provided insight into the ways in which pedagogical content knowledge influences instruction. Yager, Hidayat, and Penick (1988) contend that a strong content knowledge of science has little relationship to a person's understanding of science and his or her ability to communicate this understanding. They cite the classic example of university professors strongly versed in science content, yet widely reputed to be poor teachers. They suggest that the traditional notion of content should be expanded to include a pedagogy based on "historical and philosophical application of science."

Mason's (1988) research has revealed that beginning and prospective teachers considered to have a good knowledge of their content area are unable to apply this content information to teaching methodologies. They had difficulty in conceptually organizing their knowledge of science and the relationships of major concepts and consequently experienced difficulties in implementing their knowledge together with effective teaching strategies. Mason concludes that there is a need to show teachers how to apply information covered in educational courses to their content area, or to merge content and pedagogy.

An examination of the content knowledge of six experienced secondary school physics and biology teachers revealed a relationship between the way they modified the subject-matter content of the textbook and their subject matter knowledge. The interaction between subject-matter knowledge and pedagogy was also evident in the way teachers used explanatory representations (Hashweh, 1987).

Knowledge of subject matter influences not only the way teachers modify the textbook and utilize explanations but also the way teachers sequence content and interact with students (Gallagher & Tobin, 1985). Results of a study of 15 high school science teachers indicated that teachers behave differently when they are teaching outside their content expertise. The manner in which teachers sequenced content and the quality of the interactions with students both appeared to be influenced by the teachers' relative lack of knowledge when teaching outside their content expertise.

In a study of teacher decision making, Duschl and Wright (1989) used an ethnographic approach to examine the relationship between decision making and teachers' knowledge of the nature and structure of science and content knowledge. The 13 high school science teachers participating in the study made decisions that were prompted by lack of knowledge about the structure of science and content knowledge. Student developmental needs, curriculum objectives, and pressures of accountability were more likely to have a bearing on teacher decision making.

Pedagogical Content Knowledge and Learning

Although most of the studies we reviewed were concerned with the relationship between pedagogical content knowledge and instruction, researchers have more recently inquired into the relationship between pedagogical knowledge and student knowledge. Magnusson, Borko, Krajcik, and Layman (1992) were interested in the relationship between teacher content and pedagogical knowledge and student knowledge in the context of a microcomputer-based laboratory experience designed to investigate heat energy and temperature. The six experienced eighth-grade teachers and 22 students participating in this study were interviewed using semi-structured protocols. Transcripts were analyzed in terms of (1) content knowledge of both teachers and students (i.e., statements concerning temperature, heat energy, or their relationship), and (2) pedagogical content knowledge (as represented by teachers' statements concerning what students were likely to know, what they might find difficult, what was important for them to know, etc.). Interviews were used to derive information related to alternative frameworks, student misconceptions or understandings, and pedagogical strategies. The authors concluded that there was evidence of a relationship between teacher and student knowledge, and they summarized the patterns that emerged in the following way: "If teacher knowledge was strong but the opportunities to exhibit it under certain conditions were few, student knowledge did not change. If teacher knowledge was strong and opportunities to exhibit it were many, student knowledge did improve. If teacher knowledge was not strong and opportunities were many, student knowledge did not improve substantially, and in some cases became less correct." A similar study examined the genetics, pedagogical knowledge, and pedagogical content knowledge of four high school biology teachers in relation to student knowledge (Bellamy, Borko, & Lockard, 1992). Teachers' content knowledge of genetics was found to be inadequate in facilitating student understanding of genetics concepts. The authors found that mental processes involving pedagogical knowledge and pedagogical content knowledge are necessary for advancing student understanding of science concepts.

Subject Matter Knowledge

Much of the research on teacher thinking has focused on describing process rather than content. The interest in content that is currently being expressed seems to reflect the notion of the "missing paradigm problem," or what is viewed as an absence of focus on subject matter in teaching (Shulman, 1986). A listing of essential knowledge for beginning teachers developed for a meeting of the American Association of Colleges for Teacher Education (AACTE) included such topics as planning, management, evaluation of teacher learning, and influence of context. Subject-matter knowledge was not included in this list of essential knowledge (Buchmann, 1983). Research on science teaching during the past several years reflects a growing concern about the specific subject-matter knowledge of teachers and its effect on problem solving (Hashweh, 1987; Tamir, 1991).

Subject-matter knowledge has been widely recognized as an important aspect of science teaching. It is often defined as the teacher's personal knowledge of the content, such as the declarative, procedural, and conceptual understanding of physics (Leinhardt & Smith, 1985). Only recently, however, have science education researchers attended more fully to the nature of subject-matter knowledge and the way in which it develops in teachers.

Subject Matter Knowledge and Conceptual Change

Research has focused primarily on the subject-matter learning by secondary science teachers. Smith and Neale (1989), however, have studied the development of subject-matter knowledge among 10 elementary teachers enrolled in a summer program that concentrated on conceptual-change teaching in science. The purpose of their study was to document conceptual change in the teachers' content knowledge. Data consisted of videotapes of science lessons taught by participating teachers before and after the summer program, interviews focused on teachers' knowledge of the physics of light and shadows and of their students' understandings in that area, and demographic questionnaires. Before the summer program, teachers appeared to have limited and fragmented content knowledge, with evidence of "conceptual flaws." Their knowledge of the content changed dramatically as a result of participating in the program, which emphasized making sense of specific science content and developing pedagogical strategies for translating the content in teaching. In discussing implications of this research, the authors suggest the need for inservice teacher-education models that provide opportunities for teachers to construct case knowledge of specific science content.

Extending this line of research, Smith (1991) presented a model of teaching intended to address a problem that has attracted considerable attention in science education. His work is cited here in lieu of a comprehensive review of the literature on conceptual change, which would command more space than is available in this chapter. Smith makes it clear that what he means by conceptual change is much more than telling a student he or she is wrong and informing him or her of the correct answer. The student must be convinced of the viability of one answer compared to another. Smith discusses the conditions associated with conceptual change in terms of four criteria used by Posner, Strike, Hewson, and Gertzog (1982) and Hewson and Hewson (1984). These conditions, none of which are sufficient and all of which are necessary for conceptual change, are discussed briefly here.

Smith's criteria for conceptual change are consistent with the idea that knowledge constructions are perceived as viable as long as they enable a person to meet his or her personal goals. Accordingly, the challenge for a teacher wishing to mediate the learning process is to establish and maintain contexts in which given understandings are seen by students to be inadequate to meet particular goals. Conceptual change can occur when a learner is dissatisfied with present understandings and has identified intelligible alternatives that are coherent with other understandings and enable him or her to meet personal goals. Smith adopts the metaphor of learners as cognitive apprentices (Collins, Brown, & Newman, 1989) to describe how teachers can facilitate conceptual change in science. Essential ingredients in a model of teaching are modeling, coaching, and scaffolding. Initially, a teacher takes an active role in showing students how to act in certain situations and assists them to overcome limitations. Over time, the teacher's role in the learning process becomes less prominent and greater autonomy for learning and use of knowledge is assumed by the learner. The teaching role that Smith describes is reminiscent of that of a provocateur who requires students, through questions and other verbal challenges, to examine the viability of their ideas.

Subject-Matter Knowledge and the New Curriculum

Research suggests that teachers' subject-matter knowledge may also influence curricular processes. In a year-long study involving four beginning biology teachers, Carlsen (1991) examined the relationship between the teachers' subject-matter knowledge and the discourse strategies that were used. Data consisted of 12 to 16 biology lessons (per teacher) that were audiotaped and transcribed, interviews with teachers' sources of knowledge, and information obtained through a "card-sort task of 15 biological topics by self-reported subject matter knowledge." Discourse in the biology lessons was analyzed by constructing coding scheme based on activity types (ActTypes). An ActType was a class of activities such as a laboratory exercise, announcements simulation, discussion, recitation, student seatwork, movie, or quiz. These activity codings were returned to teachers and modified in response to their comments. Carlsen (1991) found that knowledge-related differences influenced teachers' use of instructional strategies and ultimately the nature of the curriculum. For example, teachers who had a strong grasp of their subject matter tended to employ whole-class instructional strategies to introduce new material or review previous material. In contrast, student-centered instructional formats were favored by teachers who did not understand their subject well. Accordingly, teachers' decisions about instructional formats had a bearing on the opportunities students had to communicate in their science classrooms; student-centered instructional formats provided greater opportunities for students to interact with teachers and peers. Because this study examined the subject matter knowledge and discourse strategies of experienced teachers, Carlsen advises that the findings cannot necessarily be used to characterize the subject-matter knowledge of novice teachers.

Tomanek (1992) used an action research approach to look at the role of teachers' subject-matter knowledge in the curricular process. As a teacher researcher, she was interested in examining the interaction between teacher knowledge, content and its associated meanings, and the experienced curriculum in a secondary environmental-science course. After taping and transcribing 22 lessons and examining lesson plans and notes, Tomanek identified five topics that appeared to be significant content components of the experienced curriculum. She then wrote "case histories" for three of the topics (niche, entropy, and population change) and analyzed them for patterns within and across the cases. One group of patterns with particular relevance to this review of subject-matter knowledge was associated with teacher subject-matter knowledge. Tomanek found that the meanings associated with the three curricular topics evolved over time and were closely associated with her subject-matter knowledge and her knowledge of the students. For example, her restricted knowledge about a particular topic acted as a curriculum "halt" in terms of facilitating multiple meanings of the topic. In addition, her level of comfort with a particular topic was reflected in her approach to the curriculum process. When she felt more comfortable, she was more likely to expose students to the content in diverse ways. Tomaneck's study reminds us of the complexities and uncertainties of science teaching and leads us to think about the way we define a curriculum, as well as curricular pedagogical interests. How we think about teachers' subject-matter knowledge in relation to the curriculum is important because it frames assumptions made about knowledge, teaching, learning, students, and teachers.

Gess-Newsome (1992) suggests that many of the studies that provide the basis for our current understanding of subject-matter knowledge have been limited to laboratory simulations of classroom situations, have been restricted in scope by overreliance on the use of concept or cognitive maps, or fail to consider that the subject-matter structure (or conceptual schema for a teacher's content) may vary with level of expertise. She examined the nature and sources of teacher's subject-matter structures and the relationship of their subject-matter structures to classroom practice. Data collected from five male, secondary biology teachers consisted of pre- and postobservation interviews, questionnaires, classroom observations, documents, and anecdotal information. Gess-Newsome found six variables that most strongly influence the translation of subject-matter structures into practice: teacher intentions, content knowledge, pedagogical knowledge, students, teacher autonomy, and time. However, her findings challenge the results of many earlier studies (which assumed that subject-matter structures could be directly translated into practice) by elucidating some of the ways that teachers' subject-matter structures are mediated by context-specific factors in particular classrooms. A second finding in this study concerns the sources of teachers' subject-matter structures. Teachers reported that their college-level content courses and actual teaching experience were primary influences on subject-matter structures. Gess-Newsome suggests that the findings of this study and others should be considered in current efforts to reform science education, particularly in relation to the nature and amount of science content that should be included in science teacher education programs.

Discussion

Efforts to analyze and understand teachers' pedagogical knowledge and subject-matter knowledge have important implications for both the way we teach and the types of research questions we pursue. Different orientations that are apparent in contemporary research lead to different conclusions about the way in which teachers' pedagogical knowledge and subject knowledge influence the nature of the curriculum and the quality of instruction. Future research in this area should consider the implications of translating theory into practice. In examining how teacher knowledge and beliefs influence student views, researchers must consider the teaching context, the curriculum, and the complex relationships between the teacher, learner, and other aspects of the learning environment. Shulman (1990) has stated that "how you learn a subject in college affects how you teach it." This statement is indicative of the need for research on the role of teacher knowledge in our university science-education programs.

TEXTBOOKS: RESEARCH ON SCIENCE TEACHING

Textbooks have long been considered a major factor in shaping science-education instructional programs. Efforts to improve the quality of textbooks and reform the curriculum stem, in part, from the widespread belief that students have difficulty in learning from science textbooks. Students encounter obstacles to learning from textbooks because of fundamental contradictions "between the nature of science textbooks and the educational goals that texts are meant to serve" (Finley, 1991). In many ways, the design and use of science textbooks mirror the current structure and function of schools. As Alvermann and Hinchman (1991) stated, "teachers' use of textbooks and other printed materials seems tied to their position at either end of the process-product continuum."

Efforts to reform science education at all levels must embody new images of the science textbook. The changing role of the textbook in the context of an increasingly technological society is an important gap in the textbook literature. Many of the studies we reviewed have taken a first step in examining alternative ways of thinking about the role of textbooks in today's science classrooms. Pervasive themes in these studies include the nature of science as portrayed in textbooks, the inclusion of STS issues in science textbooks, student comprehension of written text, the characteristics of textbooks, metacognitive strategies in textbook comprehension, textbook questioning strategies, the role of the textbook in the curriculum, textbook

comprehension monitoring strategies and student attitudes to the textbook (Tippins & Tobin, 1992). In spite of the wealth of studies emphasizing the role of the textbook in science classrooms, much room remains for research on how science textbooks can become facilitators of thinking for both students and teachers, rather than passive warehouses of accumulated knowledge. Just as knowledge cannot be transferred from the teacher's head to the empty vessel of the student's mind, knowledge cannot be siphoned from a textbook. In addition, we were struck by the paucity of research on the mediational role of the teacher in relation to the use of textbooks.

Kyle and Gottfried's (1992) ethnographic study, attempting to fill this gap, focused on the use of textbooks in six high school biology teachers' classrooms. To understand how teachers' use of textbooks influenced the Project Synthesis "desired" and "actual" state criteria for biology education the researchers collected data during classroom observations from which they constructed profiles of three textbook-centered (TC) teachers and three multiple-reference (MR) teachers. Project Synthesis descriptors equate actual state classrooms with traditional textbook-centered learning environments. At the same time, the role of the textbook in desired state learning environments remains unclear. The researchers sought to examine relationships that might reveal the role of textbooks in such classrooms. The six biology teachers who participated in the study were initially identified through their response to a questionnaire, as part of a purposeful sampling technique. Teacher interviews and classroom observations of one general biology class taught by each of the six teachers took place over an 8-week period. The resulting data were coded according to textbook use, desired and actual program characteristics, and teacher characteristics and analyzed using a data base that collated information as representative of teacher-centered or multiple-reference classrooms. The composite profiles of the TC and MR classrooms revealed that textbook-centered biology classrooms were aligned with actual state criteria in over 95% of the data entries, whereas multiple-resource classrooms were about equally aligned with actual state and desired state criteria. Some additional factors that appeared to be related to textbook-use alignment with the desired state of biology education included curricular goals expressed by individual teachers, the instructional strategies selected in implementing the curriculum, and the individual commitment of teachers to curriculum and professional development activities. For example, the findings suggest that use of a broad repertoire of classroom activities and professional involvement in curriculum development and science projects influence the extent to which biology teachers' classrooms exemplify desired state classrooms.

Considering the importance of textbooks in science classrooms, it is surprising that much of the research on science text properties has been limited to isolated studies of short duration focusing on specific properties such as readability, textbook helps, sexism, vocabulary, and concepts. Meyer, Crummey, and Greer's (1988) study focused on an in-depth analysis of three elementary science programs and the implications for teachers using these textbook programs: Merrill and Silver-Burdett (grades 1–5) and one level of the Holt science program and the McGraw-Hill science program. The study took place in the context of a larger longitudinal study of science learning in three school districts that were using the selected textbooks. The researchers determined the content and general properties of the text materials and compared selected segments of text representing several content domains across textbooks. They found striking differences between the programs in terms of content and pedagogy. Programs characterized by the greatest amount of text also had more hands-on and teacher-directed activities. The authors suggest that the emphasis on teacher-directed activities and the associated reduction of activities using student materials could be interpreted as a positive finding, because teacher-directed activities using readily available materials may be more likely to be done than activities found only in student materials. The authors raise a number of issues that should be considered as new textbooks are selected and introduced in the classrooms—the teachers' probable use of textbooks, the long-term effects of students' exposure to various textbooks, and the importance of detailed analyses of the quality of science textbooks.

Discussion

Textbooks remain an integral part of many science classrooms and in many cases define the curriculum. Thus, when textbooks place a major emphasis on isolated bits of knowledge and technical vocabulary, science learning is reduced to obsession with details and the ability to memorize and regurgitate vocabulary words and pieces of information. We can anticipate that the textbook industry, like other industrial counterparts, will continue to be driven by the free-enterprise market—thus, changes will occur slowly. Accordingly, the role of the textbook in relation to both teacher and student learning is inextricably linked to policy issues that have important implications for how science and reading should be taught.

Understanding the politics of textbook publishing, adoption, and use is critical to any reform effort focusing on new images of the science textbook for science classrooms. At national, state, and local levels, textbooks have historically been a complex key political issue and continue to be today. The National Society for the Study of Education (1990), in dedicating its eighty-ninth yearbook to the examination of *Textbooks and Schooling in the United States*, raised concern about the political issues associated with the quality of textbooks in the context of the current curriculum reform movement. (Elliot & Woodward, 1990). Whereas, federal funds have supported research leading to reports emphasizing the issues associated with textbook quality, the responsibility for improving textbooks remains at the state and local levels. As Cody (1990) suggested, "the national debate, acknowledging state responsibility for education, challenged the states to address the problems coming up in the research about textbooks."

The extent to which states have undertaken direct political action with regard to textbook reform varies. Generally, northeastern and central states have assumed a more indirect role in the development of policies and procedures for textbook issues. By contrast, southern and western states have become known as "adoption states" because tax revenues are used to provide free textbooks that are selected from a state list of approved textbooks. In keeping with the tradition of local control of schools and what children will learn, textbook adoption states have provided mechanisms for community involvement in the selection process (Cody, 1990). Yet controversy continues to surround the political aspects of selection, reflecting the diverse interests of publishers, policymakers, single-issue pressure groups, and classroom teachers. With regard to science textbooks, market pressures are aimed at producing textbooks that cover the contents of all possible curricula, while special-interest groups argue for the exclusion of sensitive topics. Woodward and Elliot (1990b) suggest that "there need not be a disequilibrium between scholarship, community needs, and the instructional materials market place." Clearly, in science education, in which the quality of textbooks and their role in the classroom have been persistent concerns, reform efforts should strive for such a balance.

MEDIATING LEARNING THROUGH VERBAL INTERACTION

Traditionally, the teacher's role in classrooms has been examined in terms of a knowledge-transmission metaphor, focusing on variables such as classroom management (e.g., Sanford, 1984), clarity (e.g., Land, 1981), and questioning (e.g., Winne, 1979) . From a constructivist perspective, studies in such areas can still be regarded as significant. However, from the perspective of the teacher mediating students' learning, a promising area of research involves the study of the use of silence in verbal interaction. Wait time, or pausing phenomena, has the potential to provide opportunity for reflection in action and to transfer control from one speaker to another. The overwhelming majority of studies of wait time suggested changes that were positive. This prompted a leading educator[1] to remark that wait time to education was like penicillin to medicine, widespread cures and no harmful side effects.

Winne and Marx (1983) stated that classroom researchers should consider internal processes of students and focus on the student as an information-processing learner. They stated that for learning to occur the student must perceive the instructional stimuli, note their occurrence, understand the required cognitive processes, use the processes to create or manipulate information to be stored as learned material, and encode the information for later retrieval. If these criteria are accepted, teaching roles can be defined in terms of maintaining appropriate student task involvement and utilizing cues to stimulate the cognitive processes deemed necessary for learning. The criteria support a mediational role for the teacher. Central to the learning process is the use of relevant extant knowledge, conscious effort to reflect and make sense of experience, and time dedicated to the attainment of specific goals.

If a constructivist perspective on learning is adopted, the importance of providing sufficient time for teacher and student thinking is apparent. If teacher discourse is to influence student learning, each spoken word must be heard and assigned meaning in the context that applies at the time of speaking and hearing. As a consequence, the rate at which words are spoken should fit each hearer's capacity to assign meaning to the spoken text. This involves much more than giving meaning to each individual word, because collections of words and other signs in the culture contribute to the meanings of discourse. If the hearer constructs herself as an active sense maker, she will assign meaning to the context, interpret each word and collection of words, and relate the text of what she is learning to what she already knows. If adequate time is to be provided, teachers should consciously manage the duration of pauses *after solicitations* and provide regular intervals of silence during explanations. The rate of speaking should fit the capacity of learners to make sense, and this can be monitored by students and by teachers by interpreting the verbal and nonverbal signs provided by learners.

The pauses *following student discourse* are also of potential importance. As Rowe (1974a) noted, speech is interspersed with pauses that range from quite short time intervals separating individual words to much longer intervals as a speaker completes a segment of speech and considers what to say next. These time intervals often exceed 3 to 5 seconds. Siegman and Pope (1965) reported that the length of pauses in discourse increased in proportion to the difficulty of the task, Rochester (1973) stated that pauses in speech were related to cognitive processing. Consequently, as a student attempts a complex explanation, greater cognitive activity is called for and longer pauses separate bursts of speech. Longer pauses provide ample opportunity for a teacher to interrupt speech and associated cognitive activity by completing the answer for a student or by asking another question. The student is therefore prevented from developing a complete answer to a question or correcting errors that may have been made. Interruptions of this type disrupt thinking and can inhibit learning. If teachers can refrain from speaking until 3 to 5 seconds have elapsed, a student may continue speaking or another student might initiate discourse.

Rowe (1969) defined two types of wait time: wait time I as the duration of the pause after a teacher utterance and wait time II as the duration of the pause after a student utterance. An extended wait time I or II was defined as an average of 3 to 5 seconds. In most instances, wait time I is related to the pause after a teacher question and wait time II is the pause after a student response to a question. In an endeavor to overcome difficulties encountered in implementing extended wait time I and II, Lake (1973) suggested that wait time should be redefined in terms of the period of silence that precedes teacher talk. Lake defined two types of wait time according to which speaker has primary control over the length of the pause. Teacher wait time was defined by Lake, and supported empirically in a study by Fowler (1975), as the length of the pause preceding teacher discourse. Student wait time was similarly defined as the length of the pause preceding student discourse.

Rowe (1969) reported that average wait time I and II values in science classes throughout the United States were less than 3 seconds and usually less than 1 second. These findings were replicated in the United States (e.g., Swift & Gooding, 1983) and in other parts of the world (e.g., Tobin, 1986b). After studying audiotapes of science lessons in which students displayed high levels of inquiry, Rowe (1974a) noted that features of classroom discourse were related to wait time I and II. Rowe's studies were based on results from elementary and high school classes as well as from teachers working with small groups of students. Rowe conducted studies of teachers using wait time with their regular classes and teachers teaching microgroups. During most of the studies wait time was manipulated to ascertain the effects of using an average wait time between 3 and 5 seconds. The results were a synthesis of studies conducted over a 7-year period.

Use of an extended wait time changed the quality of teacher and student discourse in elementary science classes. Rowe reported that teachers demonstrated greater response flexibility, asked fewer but more appropriate questions, and developed higher expectations for students previously rated as slow learners. These changes in teacher discourse together with more time for thinking may have contributed to the changes observed in student participation.

When the average length of wait time I and II was greater than about 3 seconds, Rowe reported an increase in the length of student responses; an increase in the number of unsolicited, but appropriate, student responses; an increase in the number of responses rated as speculative; a decrease in the number of students failing to respond; an increase in student-to-student comparisons of data; an increase in student inferences supported by evidence; an increase in the number of responses from students rated by the teacher as relatively slow learners; and an increase in the variety of verbal behaviors exhibited by students.

Although these changes occurred in an environment in which teachers endeavored to increase both types of wait time beyond 3 seconds, Rowe reported that wait time II had the greatest effect on the length of student responses, the number of unsolicited but appropriate student responses, and the use of evidence before or after inference statements. When wait time I was extended there was a lower incidence of student failures to respond to teacher solicitations.

Arnold, Atwood, and Rogers (1974) utilized 11 teachers of students in grades 1 to 5 to investigate relationships between wait time I and the cognitive level of questioning. They reported that longer pauses followed analysis questions (4.6 seconds) than questions at other levels of Bloom's taxonomy. Jones (1980) hypothesized that the time required to respond to a question was related to the type and complexity of the question. The type of question was classified as convergent or divergent, and complexity was classified in terms of Piagetian level (concrete or formal). The investigation involved 32 eighth-grade students who were studying a unit on projection of shadows. The students were instructed as a group and questions were posed during interviews with individuals. Student response wait time was used as the unit for analysis. The average time following convergent questions was 2.8 seconds, whereas the average time for divergent questions was 6.9 seconds. The difference between student response wait times after questions classified as concrete and formal was not statistically significant. However, Jones indicated that students were not consistently responding to formal questions in a formal manner. Jones concluded that students being questioned would be better served if the individual asking the question provided time for students to think and then answer.

When teacher wait time was increased to average between 3 and 5 seconds, Tobin (1986b) reported that the number of utterances per unit time was reduced in classes with extended wait time. This trend was balanced by an increase in the average length of utterances. One possibility is that teachers and students used the increased time between utterances to consider what they were going to say and that sufficient time was provided for utterances to be completed. Tobin obtained some support for this assertion with the finding that the number of times teachers interrupted student discourse was reduced in classes with an extended wait time.

One change that has been consistently reported for extended wait-time conditions is a decrease in the number of teacher questions (e.g., DeTure & Miller, 1984; Honea, 1982). Tobin (1986) noted that the proportion of solicitations (the number of solicitations compared to the total number of verbal moves) was greater in extended wait-time classes than in regular wait-time classes. At first sight this appears to be contrary to the results of other wait-time research. However, it should be noted that in each study the actual number of solicitations per unit time decreased in extended wait-time classes. Thus, the consistent pattern observed in extended wait-time classes was that the actual number of questions asked per unit time decreased, but the proportion of solicitations increased. The result is consistent with the interpretation that teachers in extended wait-time classes focused to a greater extent on formulating questions designed to make students think. This assertion is supported by other studies.

Swift and Gooding reported a higher occurrence of evaluative questions and less frequent use of chain questions. In a follow-up study, Gooding, Swift, and Swift (1983) reported that the use of an extended wait time was associated with the use of fewer memory-level questions, fewer rhetorical questions, fewer management questions, and fewer leading questions. Similarly, DeTure and Miller (1984) examined the types of questions and found a reduction in the number of cognitive memory questions and an increase in the numbers of questions classified as requiring divergent thinking in the extended wait-time group.

Differences in teacher discourse after a student utterance show that the teacher does use additional wait time to think about subsequent discourse. DeTure and Miller found a decrease in the number of repeated verbal patterns (e.g., any phrase repeated by the teacher more than five times during an interaction sequence) and in the amount of mimicry in extended wait-time classes. Tobin (1986b) reported that the extended wait-time classes had a lower proportion of teacher mimicry and a lower proportion of low-level teacher reactions than in control-group classes. This trend was associated with a tendency for teachers to probe for additional information or input from students in extended wait-time classes. Anshutz (1975) also reported that teachers in an extended wait-time group asked more probing questions than teachers in a control group.

Wait Time and Student Achievement

Research on wait time indicates that student learning increases when teachers utilize an extended wait time in their lessons. For example, Samiroden (1983) investigated the relationship between higher cognitive-level questions, wait time, and student achievement. Prospective teachers in two experimental groups were trained to use wait times of 1 to 4 seconds or 4 to 7 seconds, respectively. Seventeen prospective teachers each taught a 60-minute lesson to two 11th-grade biology classes. Only eight prospective teachers achieved the desired wait-time lengths. The results indicated that classes receiving the extended wait-time treatment achieved at a significantly higher level than those receiving the short wait-time treatment.

In an experimental study involving 23 teachers of students in grades 5, 6, and 7 (Tobin, 1980), all teachers taught an introductory science topic to allow their wait time to be calculated and prior measures of student achievement to be obtained. In the experimental phase of approximately 10 weeks, a random sample of teachers endeavored to increase wait time to an average of more than 3 seconds. Posttest measures were administered at the end of each of the two instructional units taught during the experimental phase. The outcome measures in the study consisted of items measuring higher cognitive-level science achievement. Class mean achievement was the unit for analysis. When prior differences in achievement were considered, a significant relationship between teacher wait time and science achievement was obtained for the experimental phase.

Anderson (1978) reported an increased apathy toward physics for students in the increased wait-time classes. Females in short wait-time classes found the class more formal but also more satisfying than females in long wait-time classes. Anderson cautioned that the use of long wait time could lead to decreased satisfaction and increased apathy for pupils conditioned to more rapid question-answer interactions. This is the only wait-time study included in the review that used student perceptions of the psychosocial learning environment as a dependent variable. The result highlights the significance of considering students as a significant part of the curriculum. Changes cannot be made in teacher actions without involving students in a fundamental way. If the changes in the learning environment do not fit well with student's beliefs about their roles as learners, there could be a problem, a symptom of which might have been increased apathy of the students toward learning physics. Too many of the studies reviewed in this section considered wait time as an independent variable. From a sociocultural perspective it is evident that such an assumption about classrooms cannot be meaningful. The actions of all participants in a classroom are inextricably linked and confounded.

Tobin and Capie (1982) investigated the effects of increased teacher wait time on student engagement rates and integrated process-skill achievement for students in grades 6, 7, and 8 in an experimental study involving 13 teachers. The design of the study allowed measures of student formal reasoning ability, summative achievement, and retention. The process-skill outcomes were all at a higher cognitive level. When student differences in formal reasoning ability were considered, use of a teacher wait time between 3 and 5 seconds was associated with higher student achievement and retention. Similar results were obtained when the individual student score and the mean class score were used as units of analysis.

In an experimental study involving 20 teachers of students in grades 6 and 7 (Tobin, 1986b), the student outcomes were higher cognitive-level concepts associated with probabilistic reasoning. The design enabled formal-reasoning ability to be measured for all students participating in the study. Class mean achievement scores were used as the units for analysis. When variation in formal-reasoning ability was considered, classes receiving an extended wait time achieved at a higher level than classes receiving a short teacher wait time.

The study also linked student discourse variables to achievement. When between-class variation in formal-reasoning ability was statistically removed, a multiple regression analysis indicated that the average length of student discourse and the proportion of student reacting utterances were each related positively to summative achievement. These results suggest that teachers might concentrate on the average length of student responses and the proportion of student reactions as a possible means of increasing achievement. Use of a longer wait time appears to be one way to induce such changes in pupil discourse.

Riley (1986) reported an interaction between wait time I and cognitive level of questioning on achievement for students in grades 1 to 5. Achievement decreased when wait time was extended from medium to long for low-level questions. In contrast, achievement increased when an average wait time of 3 seconds was used in conjunction with high- and mixed-cognitive-level questioning. Riley suggested that the optimal wait time may depend on the cognitive level of questioning and the cognitive level of the outcomes to be achieved. The results suggest that short wait times may be most appropriate for learning at a low cognitive level. This finding accords with the commonsense notion of many teachers we have spoken to. Indeed, this is precisely what happens in many classes. Teachers use a short wait time because they expect learning, in many cases, to be little more than learning by rote what is said or written.

Another type of wait time also benefits student cognitive processing. Rowe (1983) reported that science achievement increased when instructors provided 2 minutes for thinking after every 8 minutes of instruction. The pause enabled students to read over their notes and clarify concepts with peers as well as to think about what had been taught in the preceding 8 minutes of instruction.

Training Studies

A number of studies suggest that teachers have difficulty utilizing a wait time of 3 seconds or more. The initial approach to this problem was to develop training programs directed at extending wait time. Chewprecha, Gardner, and Sapianchai (1980) compared three training methods for modifying wait time I. The study utilized 77 high school chemistry teachers from Bangkok, Thailand. Four treatment groups were formed. Three groups participated in a 2-hour orientation on the importance of questioning and wait time, and one group was a control. After the orientation, group I teachers studied three different pamphlets each month for a semester; group II teachers listened to three different audio models each month and provided written comments on them; and group III teachers attended a workshop on questioning and undertook a quantitative analysis of an audiotape each month. Wait time I was measured from an audiotape of a chemistry lesson taught at the beginning of semester 2.

The use of instructional pamphlets was found to be more effective than qualitative or quantitative analyses of audiotapes. However, neither of the analyses involved self-analysis. The Chewprecha et al. study is significant for several reasons. In the first place, the lessons were conducted in the Thai language and was the first wait time study conducted in a language other than English; second, each treatment group attained an average wait time close to the criterion; and third, the results are probably generalizable only to teacher education in developing countries. The latter point is well illustrated by the authors, who noted that the success of the pamphlets was probably due to limited access to books and articles on education in Thailand. The pamphlets were read with considerable enthusiasm and the ideas were eagerly implemented.

Rice (1977) randomly selected 10 undergraduate elementary education majors enrolled in a science methods course and randomly assigned them to two groups. All 10 teachers prepared six science lessons to be presented to a class of elementary students. Wait time, number of questions, and cognitive level of questioning were determined from an audiotape of the first lesson. After that, one group of teachers participated in an instructional treatment that consisted of viewing films and reading articles about aspects of questioning. The wait-time component of the instructional treatment consisted of reading an article written by Rowe (1969). The other group of teachers discussed aspects of the presented lesson.

After the instructional treatment, the group of five teachers increased their average wait time from 1.3 seconds to 2.1 seconds. However, only two teachers attained the threshold of approximately 3 seconds. Although one teacher in the instructional treatment group actually decreased wait time from 1.5 seconds to 1.0 seconds, no effort was made to ascertain his or her reasons for so doing. The average wait time for the control group was 1.3 seconds on both occasions.

These results contrast with those reported by Swift and Gooding (1983) for a sample of American teachers. Swift and Gooding reported that questioning training based on the use of pamphlets only marginally increased wait time I and II. Teachers using training pamphlets maintained a mean wait time I of 1.4 seconds and a mean wait time II of 0.7 seconds during a 15-week study. These values were only slightly greater than the average of 1.2 seconds and 0.6 seconds maintained by a control group in the study.

Esquivel, Lashier, and Smith (1978) investigated wait-time extension with a group of 92 prospective elementary teachers.

Each was assigned to teach science to 6 to 10 elementary students in grades 3 through 5. Teachers were assigned to their group in pairs so that they could be assisted to maintain an extended wait time. Esquivel et al. reported that feedback on wait time II did not enable teachers to maintain a 3 second average over a sequence of three science lessons. In this study the average wait time II was 1.2 seconds, much below the criterion of 3 seconds but greater than the 0.5 to 0.9 seconds that is typically reported for wait time II.

Other studies have shown regular feedback to be beneficial. Tobin (1980, 1986b) and Tobin and Capie (1982) reported substantial gains in teacher wait time when performance feedback was regularly provided. In these studies teachers receiving feedback maintained an average teacher wait time above a 3-second criterion.

DeTure (1979) used a factorial design to investigate the effects of feedback on the ability of prospective teachers to implement an extended wait time. The subjects were 52 prospective teachers. Each was randomly assigned to one of four treatment groups to microteach a group of four elementary students from grade 4 or 5. The four treatments were audio model with no feedback, audio model with feedback, video model with no feedback, and video model with feedback. The model consisted of a master teacher and four grade 5 students discussing a discrepant event while using an average wait time I and II above 3-seconds. The use of a videotape followed by feedback from an advisor enabled teachers to attain a mean wait time II of 3.7 seconds, significantly higher than the mean wait time attained by groups using other training techniques. However, no teachers in the study were able to attain an average wait time I above 1.8 seconds.

Swift and Gooding (1983) used a wait timer to signal a 3-second pause to teachers and students. A voice-activated relay system operated a red light that signaled when an appropriate period of silence had elapsed. This system provided teachers and students with an indication of how long they needed to wait. With the aid of this device, teachers and students were able to maintain an average wait time I of 2.6 seconds and an average wait time II of 1.4 seconds. These results indicate that the wait timer was more effective in controlling student discourse than teacher discourse. This may have been attributable to the placement of the device in the classroom. It is possible that students had a good view of the light, whereas teachers may have had their backs to the wait timer or may have been too preoccupied with teaching to concentrate on the light. An alternative explanation is that the teachers did not believe in the appropriateness of a long wait time in the context of the implemented curriculum.

DeTure and Miller (1984) used a written-protocol model to change wait time. A written model was read and participants in a training program were required to transcribe a tape of their teaching and to calculate their wait-time averages for two or three lessons. The written-protocol model required a teaching cycle to be repeated until an average wait time of more than 3 seconds was reached. Feedback was also incorporated in the treatment. Seventy percent of the participants reached a criterion average for wait time I and wait time II after two lessons. The remaining teachers attained criterion after three lessons.

However, the teachers in this study were not required to utilize an extended wait time in a sustained manner with a regular class.

In a follow-up study involving 10 teachers from the Swift and Gooding (1983) study, Swift, Swift, and Gooding (1984) used a supportive intervention technique that was successful in assisting teachers to maintain an extended wait time during their regular teaching assignments. The supportive intervention procedure had many elements in common with coaching (Joyce & Showers, 1983), which has been successful in facilitating sustained teacher change involving the use of other teaching strategies.

The training studies raise questions about the probable effectiveness of teacher-education courses in which teachers are asked or urged to try a particular strategy or are simply provided with a handout. The chance of substantially improving the quality of classroom discourse through such methods appears to be remote. The crucial question to be addressed in training studies is how to sustain an extended wait time. In natural settings, most teachers maintain an average wait time between 0.2 and 0.9 seconds. The main problem appears to be that the magnitude of required change is of the order of 600%. Such a change represents a major departure from normal teaching style, and concomitant changes in teacher and student behavior necessitate new approaches to management of classroom interactions. Although Swift et al. (1984) used positive feedback only, most studies have used a form of feedback that involved discussion of positive and negative features of classroom discourse. It is possible that different types of teachers respond to different forms of feedback.

A salient finding in the wait-time studies related an attempt to increase wait time with anxiety. Honea (1982) reported an increase in teacher anxiety when moving from short wait time to long wait time. This finding has important implications for teachers endeavoring to change the way they teach. If teachers are to sustain an extended wait time in their classes, or for that matter attempt to change any teaching routine, they may need external support to do so. Increased anxiety might be explained in terms of the significant changes in interaction patterns that occur quickly when wait time is extended. If teachers become anxious and discouraged, they could reject the use of an extended wait time and return to their customary pattern of classroom interactions.

Perhaps the biggest omission in the wait-time studies reviewed in this chapter is a failure to find out why teachers used a short wait time when they taught. In other studies reviewed in this chapter, it is clear that traditional approaches to the science curriculum give greatest value to rote learning of facts and algorithms. If, as Riley suggests, the optimal wait time for such curricular goals is short, it is no surprise that teachers would maintain a short wait time. In addition, teachers might not give highest priority to facilitating learning. A teacher who believes that learning cannot occur unless students are well controlled might also believe that use of a long wait time will lead to transfer of control from the teacher to students. In such circumstances, the teacher might believe that learning would be inhibited in long-wait-time classes because of loss of teacher control. Other teachers have emphasized content coverage because of a strong perceived connection between student learning and the amount of content covered. Once again, it might be construed that the use of a long wait time would lead to less content coverage and hence less opportunity to learn. Finally, teachers have argued that it is important to maintain momentum in a lesson. Use of a long wait time can be seen as decreasing momentum, resulting in loss of student attention and thereby lower achievement. Each of these arguments is consistent with an epistemology of objectivism. If teachers believe that knowledge is out there to be learned as truths, it is consistent to have beliefs such as those just mentioned, to support the use of a short wait time. Research on wait time would be improved if researchers took into account why teachers used a short wait time and then provided educational experiences demonstrating that the use of a short wait time in most circumstances makes no sense. The training methods employed in most of the cited studies probably have not been successful because, under the circumstances, it would make little sense to persist in the use of an extended wait time.

Thinking about Thinking: The Importance of Wait Time

The research that has been reviewed indicates that wait time is an important instructional variable when higher-cognitive-level learning is pursued. The studies have been conducted across grades kindergarten through 12. When teachers maintain an average wait time between 3 and 5 seconds achievement is enhanced; however, the increased achievement is probably a result of changes in teacher discourse and subsequent student discourse and thinking. The silence provides teachers with time to think and to formulate and use higher-quality discourse, which then influences the thinking and responding of students.

The findings reported by Tobin (1986b) suggest that similar outcomes can be expected whether an extended wait time is used with whole-class groups or small groups of three to five students. However, Tobin and Gallagher (1987b) reported that a small number of target students monopolized classroom interactions in whole-class settings. In that study teachers utilized a short wait time. Under extended wait time conditions, a greater number of students might be involved in interactions with the teacher. It is also possible that increased thinking time would be of most benefit to the three to five target students in each class.

Although research has indicated that the quality of teacher and student discourse can be improved when an extended wait time is used in whole-class and small-group settings, none of the reviewed studies investigated wait time in dyadic interactions. The optimal wait time for individualized settings is probably dependent on attributes of students and teachers as well as the intended purpose of the interactions. Sufficient time should be provided for student and teacher discourse to be completed and for appropriate cognitive processing to occur. However, for certain types of teachers, students, and intended outcomes, the optimal time may vary from the 3 to 5 seconds normally advocated for whole-class and small-group settings.

An assumption in the use of an extended wait time is that additional time can be used by teachers and students for cognitive processing. However, more time may not always be beneficial. In an experimental setting, lesson plans are often provided in order to control for unwanted variance in the implemented curricula in different classes. In this way, the activities in which students are to engage are specified. As a consequence, an implemented curriculum usually provides opportunities for students to engage in higher-cognitive-level tasks. In a nonexperimental situation, engagement patterns are often quite different. The evaluation system tends to shape the implemented curricula in most classrooms (Doyle, 1983; Tobin, 1985). Teachers plan and implement activities to enable students to pass examinations and tests or to attain a high course grade. As a consequence, many of the interactions that occur are pitched at a low cognitive level because many tests are also at a low cognitive level. Algorithms are introduced and learned in order to allow students to get the right answers.

The use of algorithms tends to reduce the cognitive demands of the work. As a consequence, when teachers implement the regular curriculum, the cognitive level of whole-class interactions is often low. In most lessons, there is often not much to be gained by pausing for 3 to 5 seconds. Unless the implemented curriculum requires students to think, there is little value in providing additional time to think. For example, if the questions contained in topic tests and examinations are at a lower cognitive level, it is likely that the most important classroom interactions will also be at a lower cognitive level. In such circumstances it is difficult to see how a long wait time can facilitate achievement. Consequently, introducing an extended wait time alone may not produce the improvements that research on wait time suggests. Unless the implemented curriculum provides students with opportunities to develop higher-cognitive-level outcomes, there is little point in using an extended wait time.

Wait time probably affects higher-cognitive-level achievement directly, by providing additional time for student cognitive processing, and indirectly, by influencing the quality of teacher and student discourse. This review has indicated that the use of an average wait time between 3 and 5 seconds provides an environment in which substantial changes in teacher and student behavior can occur. Some of these changes are positively associated with higher achievement. Further research is needed to ascertain the relative importance of the direct and indirect effects of wait time on higher-cognitive-level achievement.

Discussion

One final note related to educating teachers to use an extended wait time is offered. We feel that Rowe was extremely creative to identify from her 300 tapes the 3 that were anomalous and from those tapes to select wait time as the most salient variable. The studies that followed Rowe's 7-year study were reductionist in the sense that many more variables were identified as related to wait time and having salience in the classroom. From Rowe's powerful insights and relatively simple conceptual framework emerged a complex interactive maze that was clearly context dependent. It is dubious that years of further research in this line would be fruitful. The training programs needed for teachers to attain and maintain optimal levels on a critical number of these variables would have to be vastly superior to those reviewed in this chapter. Perhaps a more holistic approach offers greater potential for success. For example, the use of a metaphor such as teacher as provocateur or teacher as gardener might result in the use of optimal levels of a great many variables including wait time. In any event, from the research we have reviewed, it is apparent to us that the reductionist approach so clearly typified by research on wait time leads to a research-practice discontinuity not apparent in the research focused on metaphors and beliefs. At the heart of the difference is the use of training and control rather than education and empowerment.

THE LEARNING CYCLE

In the past few years, researchers and teacher educators have become increasingly interested in the learning cycle as a model of instruction and procedure for curriculum development. The learning cycle is not a new concept in science education, its origins stemming from (1) the development of the Science Curriculum Improvement Study (SCIS) program in the 1960s, (2) the work of Atkin and Karplus (1962) related to "guided discovery" teaching, and (3) the research of Lawson, Abraham, and Reuner (1989) leading to the development of a theory of human learning compatible with the basic elements of the learning cycle. It has traditionally been described as a "method" or "theory" of instruction that centers on investigative activity preceding any formal introduction to scientific concepts.

Since its introduction in the 1960s, the learning cycle has been referred to alternatively as exploration-invention-discovery (Karplus & Thier, 1967), exploration-concept introduction-concept application (Karplus et al., 1967), exploration-conceptual invention-expansion of idea (Renner, Abraham, & Birnie, 1985); d) exploration-conceptual invention-conceptual expansion (Abraham & Renner, 1986), and exploration-term introduction-concept application (Lawson, 1988).

Contemporary writers describe the phases of the learning cycle essential to the development of concepts in terms of Piagetian theory: the exploration phase parallels the ideas of assimilation and disequilibrium, the conceptual invention stage is analogous to the principle of accommodation, and the conceptual expansion phase facilitates organization (Renner & Marek, 1988).

Earlier research related to the learning cycle appears under different names and is often couched in behavioristic language, whereas contemporary studies are embedded in cognitive aspects of teaching and allude to the learning cycle as a "constructive process" that facilitates conceptual change. Lawson, Abraham, and Renner (1989), in their monograph entitled *A Theory of Instruction: Using the Learning Cycle to Teach Science Concepts and Thinking Skills*, trace the evolution of research related to the learning cycle, beginning with studies that were

related to the overall effectiveness of the SCIS program (which used the learning cycle as a primary instructional strategy). These studies examined variables associated with student achievement and process learning, the effect of the learning cycle on student attitudes such as self-concept, research centered on teacher variables related to the use of the learning cycle, and studies of the long-term outcomes of using the SCIS model.

Over a period of years, as the learning cycle became incorporated in a variety of science curricula and programs, the focus of research shifted to investigations of attitudes, content achievement, and thinking skills within the context of individual curricula (e.g., Campbell, 1977; Carlson, 1975; Davis, 1977; Lawson & Wollman, 1976; McKinnon & Renner, 1971; Purser & Renner, 1983; Saunders & Shepardson, 1987; Schneider & Renner, 1980; Ward & Herron, 1980).

Essentially, the inherent value of the learning cycle continues to be reflected in studies that regard it as primarily a model for organizing instruction, rather than a tool for reflecting on the what and why of teachers' actions in their science classes. What can we learn, then, from the most recent studies of the learning cycle? The purpose of this section is to describe current research on the learning cycle as it relates to science teaching, discuss the utility of this research in answering fundamental questions about science teaching and learning, and suggest implications for the teaching profession. In the current science-education literature, the learning cycle is most often discussed as an instructional strategy; it is frequently considered in terms of integrating contemporary research involving alternative conceptions and novice-expert problem-solving models.

Scharmann's (1992) descriptive study provides a rationale for use of the learning cycle as an alternative instructional strategy for assessing misconceptions and promoting conceptual change. He examined the utility of the learning cycle in assessing misconceptions associated with a unit on angiosperm reproduction and helping students overcome misconceptions related to the classification of fruits and vegetables. Scharmann stresses a need to think about the exploration phase of the learning cycle as going beyond "hands-on" manipulation of materials to include "minds-on" experiences in such forms as analogies, opinion statements, independent decision making, and analysis of situational contexts. This is an important idea, as far too often good science teaching is equated with "hands-on" experiences with little consideration for the cognitive engagement of learners.

Marek and Methven (1991) examined the relationship of teachers' attitudes and the implementation of workshop-developed learning cycles in their science classes to elementary students' conservation reasoning and the language used to describe properties of objects. The experimental group in this study were 16 teachers who had participated in a summer workshop on the use of the learning cycle in teaching and curriculum development, while a comparison group of teachers taught using traditional expository methods. There were two types of comparison teacher groups. One group provided students with materials to do experiments but the teacher "told" students the concept before or immediately after the experiment; the experiment was for the purpose of verification, and student data and peer discussions were not used to invent the concept. The second group used textbook-driven instruction, with no student experiments and hence no data collected that could be used to invent concepts. A questionnaire was used to assess changes in teacher attitudes about the learning cycle and the degree to which it was implemented in their classes. Teachers were also observed to examine the teaching procedures that were being used and the teacher and student roles in activities. Piagetian conservation tasks and individual student interviews focused on the language students used to describe objects, and these were analyzed both quantitatively and qualitatively. In terms of evaluating the outcomes of the inservice workshop, the authors found that participating teachers valued the learning cycle as an instructional tool and implemented it in their science instruction. Their students demonstrated a much larger percentage of conservation reasoning abilities than students in any comparison-group classrooms.

An assumption underlying the learning-cycle model is that three distinct phases with a definite sequence and structure are necessary for the development of conceptual understanding. Renner et al. (1985) planned a study for the purpose of "testing" the "necessity variable" with students in three secondary school physics classes. Students in one section of the physics course were exposed to instruction in which all three phases of the learning cycle were used in sequence, while students in the other sections had instruction that omitted one or two phases. However, the researchers decided to give all students experience with the missing phases after the necessary research data had been collected. Instruction centered on two experiments designed to facilitate concept development of Ohm's law and the ideas of weight and mass. Student achievement was evaluated with Concept Achievement Tests (CATs) before and after each phase of the learning cycle. Other data consisted of measures of student IQ, intellectual development level, and individual student interviews. In general, the authors conclude that (1) "free exploration" of science concepts is not an efficient way to learn the content and structure of science; (2) although all the phases of the learning cycle are necessary, in-depth laboratory experiences can be substituted for some of the phases; (3) the sequence of phases is important (i.e., conceptual invention needs to follow exploration); and (4) students believe in the necessity of all phases of the learning cycle.

Abraham (1989) reached similar conclusions concerning the necessity for all three phases of the learning cycle. In an investigation of six chemistry classes he compared instruction based on the learning cycle with instruction based on traditional reading-lecture formats. Sequencing variations of the learning cycle were studied, and the author concluded that optimal learning occurs when the normal gathering data, invention, expansion (GIE) sequence of the learning cycle is employed. The research focus on the necessity of each phase of the learning cycle seems to represent what Grundy (1987, p. 12) would describe as a technical orientation to research, curriculum, and the way in which knowledge is generated. That is, the technical interest is manifest in the rigid interpretation of the learning-cycle model

through an emphasis on "controlling pupil learning so that, at the end of the teaching process, the product will conform to the eidos (that is, the intentions or ideas) expressed in the original objectives."

The effectiveness of the learning cycle as an instructional strategy compared with a systematic-modeling approach to instruction was the focus of a study by Rubin and Norman (1989). The 13 teachers participating in this study received "training" in the use of both the learning-cycle and systematic-modeling approaches. As its name suggests, the modeling approach involves introducing students initially to concepts and process skills and then modeling the skills by demonstration and metacognition (e.g., teachers talk aloud about their thoughts as they model specific process skills). The process is repeated, replacing the teacher metacognition step with questions designed to elicit responses from students, followed by opportunities for students to have independent practice. Over a period of 3 months, 15 lessons were taught to 327 sixth- through ninth-grade students using both the learning-cycle and systematic-modeling approaches. Students in the control group were not exposed to either approach. The Group Assessment of Logical Thinking and the Middle Grade Integrated Process Skills Test were used to assess integrated process skills of students. The authors concluded that the modeling approach was more effective than the learning-cycle approach in leading to improvements in achievement and formal reasoning. Parallels can be drawn between this study and previous studies of the effectiveness of different models of concept development that offered little insight into the teacher's role in this process.

Lavoie (1992) pursued a similar track in comparing the effects of a traditional learning-cycle model (LC) with a prediction/discussion-based learning cycle (PDLC). Extending previous research that introduced the importance of prediction and discussion in the development of conceptual understanding (Good, 1989; Lavoie, 1989), Lavoie suggests the need to including a prediction-discussion phase in the learning cycle before the exploration phase. To support this proposition, he conducted a study in which he sought to compare "teachers' impressions and tendencies toward learning-cycle instruction, their students, and themselves; students' attitudes and motivations toward science, learning-cycle instruction, their teacher and peers; and students' development of process skills, logical thinking abilities, and conceptual understanding." As yet, there have been no attempts to make sense of the learning cycle using a predominantly qualitative approach in which teachers are engaged as action researchers—this study begins to do so. Quantitative techniques, however, are used to assess attitudes and differences between groups in students' logical-thinking abilities, science process skill abilities, and conceptual understanding. In this regard, a concern for standardization and generalizability permeates the study, in which five biology teachers, who were already familiar with the learning-cycle, each taught ten learning cycle lessons within one PDLC class and one LC class for 3 months. The teachers were introduced to interpretive research techniques before the study and met each week during the study to observe, discuss, and interpret videotapes of learners. In analyzing the qualitative and quantitative data in this study, Lavoie discovered that when the PDLC version of the learning cycle was used students achieved greater gain scores for science-process skills, logical thinking, and conceptual understanding. Instruction using the PDLC version of the learning cycle also facilitated the organization, augmentation, and structuring of procedural and declarative knowledge. Lavoie suggests that including prediction and discussion in the learning cycle enables students to engage in hypotheticopredictive reasoning that facilitates the way in which they interpret and make sense of phenomena, make explicit their own beliefs, and deal with newly encountered phenomena and cognitive conflict. This study highlights the importance of the social aspects of knowledge construction; further research concerning the role of discussion in negotiation and consensus making is needed. What remains unclear is the teacher researcher's voice in the final interpretation and discussion of the results. As the teacher researcher mindset appears more predominantly in research, it becomes increasingly important for researchers to clarify the nature of the teacher's role and voice in the research.

Science education research is conducted in different contexts and often reflects a tension between what is normative and what is unique to a teacher's practice. In Glasson and Lalik's (1990) study of the learning cycle from a language-learning perspective the focus was on the "unique." Three experienced teachers (elementary, middle, and high school) were interviewed about their personal epistemologies, participated in reflective sessions, taught science lessons using the language-oriented learning-cycle model, and interpreted videotapes of their teaching at the beginning and end of the school year. Qualitative analysis techniques, including discrepant-case analysis and categorical analysis, were used to interpret the way in which the teachers implemented a language-oriented adaptation of the learning-cycle model. The data analyses were used to develop assertions that were eventually organized into case studies of the respective teachers. The authors found that when teachers framed their instruction with the learning cycle, they focused on aspects that were consistent with (and not conflicting with) their preexisting beliefs. However, the emphasis on reflection in relation to the use of the learning-cycle model gave teachers opportunities to reframe problems and make changes in their practice. This study illustrates how a curriculum (which may involve models such as the learning cycle) cannot be meaningfully separated from the teacher, the learner, and other factors associated with implementation. Teachers need opportunities for reflection in order to construct new ways of making sense of experience.

Renner et al. (1985), incorporating the learning-cycle model, compared student achievement and attitudes toward physics in an investigation in which some students collected their own data and others were given secondhand data from which to learn. Students who experienced data-collecting activities liked physics better and achieved at a higher level than students who worked with secondhand data. However, the authors cautioned:

The active experimentation experienced in the learning cycle is *not* the verification laboratory so often encountered. The findings of this research just referred to, therefore must not be interpreted to mean that laboratory experience in general will necessarily increase the permanency of learning. (pp. 322–323)

As well as providing support for use of the learning cycle, the findings of Renner and his colleagues raise numerous questions. Should teachers utilize the learning cycle as a basis for implementing all science lessons? Is the learning cycle the best way to facilitate learning with understanding? What should teachers and students do in activities based on the learning cycle? Can the learning cycle be implemented inappropriately? What teacher and student roles are most salient in laboratory activities incorporating the learning cycle? Why did students fail to learn with understanding in activities involving the use of secondhand data?

Discussion

The learning-cycle model grew from the Piagetian school in the United States. In essence, it is an approach to teaching and learning that ensures that students engage in the kinds of thinking that constructivists would argue is necessary to promote learning. The studies reviewed in this section provide evidence to support the use of the learning cycle in one form or another. What is not studied, however, is important. How do teachers actually implement the learning cycle, and how do their beliefs, roles, and metaphors change with respect to teaching and learning science?

These questions seem critical to us if we are to learn something about teaching and learning from the research on learning cycles. In addition, we know little about the actual engagement of learners and the manner in which they change their beliefs in relation to their roles as learners, the nature of learning and science, and content-specific concepts. Instead, the authors tend to speak of the learning cycle as if it is implemented in a uniform manner from class to class, irrespective of the cultural mores that we know make a significant difference to the curriculum. If the learning cycle is to be a useful heuristic for guiding science teachers in planning and implementing science curricula, there is much more to be done with respect to its use by teacher researchers who can tell us the conditions under which it makes sense to adopt and adapt the learning cycle in their science classrooms. How is the learning cycle represented in science classrooms where teachers, using referents such as constructivism and emancipatory interests for students, implement science curricula? This is a question to be answered by advocates of the learning cycle.

An alternative learning-cycle model that is worth consideration is Gallagher's (1992) model developed to enable teachers to make changes from teaching practices consistent with objectivism to ones more closely associated with constructivism. Gallagher's model contains three elements that center on student learning: acquisition of scientific ideas, integration, and application. The acquisition component of the model assists teachers in understanding the needs of students in terms of conceptual understanding and relevance of science in their daily lives. The integration component emphasizes the sense-making process and the development of connections essential in the understanding of subject matter. Application involves an awareness of how the connectedness of science facilitates a better understanding of additional scientific knowledge and of how scientific knowledge can be used to make sense of experience and solve problems.

Another outgrowth of the learning cycle was developed by Bybee and his colleagues in the Biological Sciences Curriculum Study (Bybee et al., 1989). The BSCS model has five phases that have been incorporated into curriculum resources developed by the group. The phases are engagement, exploration, explanation, elaboration, and evaluation. A feature of each of these phases is the focus on student actions. Bybee et al. define each phase in terms of what students must do in a science activity. For example, the evaluation phase is described in terms of students assessing their knowledge, skills, and abilities. Because of this focus the five-phase model proposed by Bybee et al. has the potential to help teachers plan and implement activities with student learning as a critical focus.

COOPERATIVE LEARNING

Cooperative learning has become an important strand of research on science teaching during the past decade and continues to be of interest to science educators. Against this backdrop of renewed interest in cooperative learning, we found remarkably little empirical research on the teacher's mediational role in cooperative-learning environments, the role of negotiation and consensus building in the collaborative process, or how collaborative learning actually develops. Tobin (1990b) points out that "although studies of cooperative learning in the context of science education abound, . . . the focus of these studies has not been so specifically on the learning process." The majority of the articles we identified fit into two categories: (1) comparative studies of student achievement, attitudes, and associated "variables" across diverse instructional formats, and (2) articles on the application of cooperative-learning research findings in classroom and laboratory settings.

Two reviews provide evidence that cooperative learning has been introduced, to good effect, in numerous studies over many years. In a review of 46 studies of learning in groups, Slavin (1984) reported that 29 showed cooperative learning to have significantly positive effects on student achievement. Fifteen studies showed no difference, and two studies showed significantly higher achievement for a control group than for a cooperative group. More detailed analyses revealed interesting trends in the results. Of 27 studies that used group study *and* group rewards for individual learning, 24 found positive effects and 3 found no differences. In contrast, of nine studies of group methods that did not use group rewards, none found positive effects on achievement. Slavin's analyses indicated that group rewards for individual performance were critical to the effectiveness of cooperative-learning methods. Slavin (1984) stated that attempts to create more effective instruction by improving the quality of instruction, time on task, and so on may be ineffective without considering incentives directed at increasing the students' motivation to learn.

In a review of over a thousand studies dating back to the late 1800s, Johnson and Johnson (1985a, p. 23) reported that cooperative-learning experiences tend to promote more learning than competitive or individualistic learning experiences and are linked with higher self-esteem. Compared with competitive and individualistic activities, cooperative-learning experiences tend to promote higher motivation to learn, produce more positive attitudes toward learning experiences and teachers, and result in stronger perceptions that students care about learning and assisting one another. Johnson and Johnson (1985b) also noted the value of structuring learning activities so that students can resolve controversies during their interactions. The resolution of controversies resulted in students sharing ideas with one another and working together to obtain a final solution to a problem. The authors noted that resolution of controversies was preferred to engaging in debate, and both strategies were superior to working alone.

Of central importance to the discussion of cooperative learning as a teaching-learning strategy in the research articles we reviewed were some of the following themes: comparisons of achievement and/or attitudes across individualistic, competitive, and cooperative-learning environments and diverse settings (Jones, 1989; Okebukola, 1986a, 1986b; Sherman, 1989; Tingle, 1989); the relationship of students' preferred learning styles to achievement in cooperative or competitive groups (Okebukola,1986a, 1986b); the nature of discourse when students are engaged in problem solving in cooperative learning groups (Kempa & Ayob, 1991; Pugh & Lock, 1989; Rogg & Kahle, 1992; Tippins & Kagan, 1992); the relationship of cooperative learning to students' on-task behavior and academic achievement (Lazarowitz, Hertz, Baird, & Bowlden, 1988); the effect of cooperative incentives and heterogeneous grouping in cooperative-learning settings (Watson, 1992); and gender issues associated with physics problem solving in cooperative-learning groups (Heller, 1992). A second category of articles centered on the application of theory to practice through these themes: cooperative-learning strategies (Jones & Steinbrink, 1991; Scott & Heller, 1991; Sharan & Sharan, 1989); cooperative problem solving (Lapp, Flood, & Thrope, 1989); and observation of cooperative-learning classrooms (Furtwengler, 1992).

Few studies in science have investigated the collaborative processes within groups and examined the negotiation of meaning that occurs. Wheatley, Cobb, and their colleagues have undertaken studies of the social construction of knowledge in mathematics classes (e.g., Cobb & Wheatley, 1988; Cobb, Yackel, Wood, Wheatley, & Merkel, 1988) and advocate cooperative learning in groups together with whole-class discussions in which students share what they have learned in small groups. The emphasis in the activities is on negotiating meaning and arriving at consensus. Student's making sense of what they are learning is given highest priority, and rote learning of procedures and facts to obtain correct solutions is not encouraged. In regard to cooperative learning, there are 3 studies in science education (Brody, 1991; Jones & Steinbrink, 1989; Scharmann, 1992), which are reviewed next.

Brody used constructivism as an interpretive framework in a study of 15 experienced elementary, middle, and secondary school teachers who participated in a 9 month cooperative-learning program that focused on reflection. The program utilized journals and an interactive seminar environment to promote reflection with regard to cooperative learning. In this study, Brody (1991, p. 3) interviewed participants to make sense of the way in which teachers modify cooperative learning to fit their existing beliefs about pedagogy; how teachers interact with this innovation to reconstruct their assumptions about teaching and learning; and how these reconstructions affect certain pedagogical themes that emerged from them: conceptions of their role, their sense of authority, and locus of control; their notions of the nature of knowing and knowledge; their conceptions of their decision making; their understanding of cooperative learning; and their resolution of the dilemmas of practice that emerge in the process.

Brody used a model developed by Miller and Seller (1985) to analyze teachers' epistemological orientations along a continuum ranging from "transmission" to "transaction" to "transformation." When this model is used, new and different views of cooperative learning emerge. A transmission view promotes technical interests (Grundy, 1987) whereby cooperative learning is viewed as a highly structured activity in which the teacher's role is that of disseminator of knowledge. A transactional approach to teachers' epistemological orientations emphasizes the nature of the dialogue between teachers and students, students and other students, and learning as a process of negotiation. The teacher's role is to facilitate learning. A transformational approach to cooperative learning, which considers ethical implications of teaching in a framework of broad educational goals for social change and personal actualization, emphasizes emancipatory interests, autonomy of learners, and engagement with peers to negotiate consensus and personal meaning.

Brody found that the majority of the teachers who participated in the study held a similar range of beliefs with regard to the pedagogical themes. Most of these teachers appeared to have transitional orientations. Although they still held some beliefs about knowledge, control, and cooperative learning that were consistent with traditional ways of thinking, they also demonstrated beliefs consistent with transactional value orientations. Brody suggests that teachers need time for reflection and sustained administrative support, that implicit values associated with cooperative learning should be made explicit early in the cooperative-learning training program, and that participation in cooperative-learning programs should be voluntary because of the risks of being unsuccessful. The manner in which teachers implement cooperative learning depend on their beliefs about teaching and learning. To many teachers, cooperative learning is vastly different from their beliefs and thus involves an element of risk. Accordingly, attempts to implement cooperative learning can provide opportunities to describe what is happening in the classroom and ascertain why it is happening. What is learned can provide a rationale for an educational program to consider whether cooperative learning can be used to enhance learning in the contexts in which curricula are to be implemented. The educational program will need to extend beyond teachers to include students, colleagues, administrators, and others, such as parents, who have an interest in science education.

In another study, Jones and Steinbrink (1989) were interested in identifying learning outcomes associated with cooperative learning and examining how teachers felt about the use of cooperative-learning strategies. Fifty teachers participated in a cooperative-learning seminar and then used cooperative-learning strategies with their students throughout the semester. At a second seminar at the end of the semester, teachers analyzed test-score data, wrote reports, and discussed their experiences. Although most teachers reflected positive attitudes about cooperative learning, their concerns were focused on individual achievement (as measured by test scores) when cooperative-learning strategies prevailed. Teachers expressed difficulties associated with integrating cooperative learning "team scores" and "gain scores" into traditional evaluation schemes. The second year of the study became an attempt to refine specific cooperative-learning strategies so that there would be a better "fit" with test scores and evaluation schemes. Using Brody's analysis scheme, teachers in this study appeared to be using a transmission approach to cooperative learning. That is, the teachers had a strong directive role in initiating structured cooperative-learning strategies.

Secondary biology and earth-science teachers' understanding of and ability to apply the nature of science in instruction on evolutionary topics was the subject of a study by Scharmann (1992) centered on teachers participating in the second year of a 2-week NSF-sponsored institute. One aspect of Scharmann's study relevant to cooperative learning was a qualitative assessment of the degree to which mentor and participant teachers would consider the adoption of more student-centered approaches to the teaching of evolution and its applied principles. According to Scharmann, such approaches may promote a learning environment that would be appropriate for dealing with evolution theory and an instructional format that would allow students to discuss the uncertainties of scientific knowledge. The *Stages of Concern* questionnaire (Hall, George, & Rutherford, 1979) was used to examine teachers' views of the use of a peer-discussion instructional format. During the second year of the project, mentor teachers introduced participating teachers to peer-discussion strategies. Scharmann found a significant difference between the concerns expressed by teachers participating in the second year of the project and the concerns of first-year participants. Profiles of these teachers' stages of concerns showed a much higher score with regard to refocusing efforts to implement peer-discussion strategies. The author inferred that the interaction with mentor teachers provided the support and modeling for teachers to become comfortable with and excited about this innovation and ultimately enhanced the potential for adoption of student-centered instruction. The peer-discussion format in this study appeared to reflect ideas consistent with what Brody (1991) described as the transactional approach. Social skills and the belief that knowledge is not fixed or immutable are characteristics of this cooperative-learning approach. The work of Brody (1991) and Grundy (1987) provides a useful framework for thinking about research on cooperative learning in future studies.

Discussion

Although we do not view cooperative learning as a panacea, we consider it as a potentially valuable activity type because of the potential for students clarifying, defending, elaborating, evaluating, and arguing with one another. The negotiation and consensus building that are possible in cooperative groups suggest that teachers should give serious consideration to employing cooperative-learning strategies when they consider it appropriate. The decision must rest with teachers and students because of the interactive manner of classroom processes. It is not possible, a priori, to predict whether cooperative learning will enhance learning. Among the factors that determine whether cooperative learning is likely to be successful are the goals of students, the extent to which students are motivated to learn and cooperate, and the degree to which the teacher believes cooperative learning to be a viable activity for his or her class. Thus, some tasks might lend themselves to cooperative learning while others do not, and it is surely the case that cooperative learning is feasible in some classes and not in others.

THE SCIENCE LABORATORY

To many educators the laboratory is the essence of science, a metonymy (or central defining attribute of a concept) that can be used to make sense of experience in the context of science teaching. From a constructivist perspective, laboratory activities can be seen as a means of allowing learners to pursue learning autonomously, having varied multisensory experiences. However, from the studies reviewed earlier in this chapter, it is clear that in many science curricula laboratory activities are not viewed as "mainstream." Unfortunately, when they are a part of the curriculum they often take the form of cookbook recipes (verification labs) in which students follow prescriptions to obtain predetermined outcomes. Technical interests appear to be the characteristic of laboratory activities when they occur. Furthermore, the evidence suggests that laboratory activities fall short of achieving the potential to enhance student learning with understanding (Hofstein & Lunetta, 1982; Stake & Easley, 1978; Tobin & Gallagher, 1987a).

Laboratory activities promise so much in terms of students being able to solve problems and construct relevant science knowledge. Tamir and Lunetta (1978) indicated that the main purpose of the laboratory in the science curricula of the 1960s was to promote student inquiry and allow students to undertake investigations. This emphasis was in marked contrast to using the laboratory primarily as a place to illustrate, demonstrate, and verify known concepts and laws. According to Hofstein and Lunetta (1982), laboratory activities can be effective in promoting intellectual development, inquiry, and problem-solving skills. Furthermore, laboratory activities have the potential to assist in the development of observational and manipulative skills and in understanding science concepts.

Despite widespread acceptance of the importance of laboratory activities in science curricula, research on teaching and learning in laboratory activities is not substantial. Novak (1988) described the problems graphically:

The science laboratory has always been regarded as the place where students should learn the process of doing science. But summaries of research on the value of laboratory for learning science did not favor laboratory over lecture-demonstration . . . and more recent studies also show an appalling lack of effectiveness of laboratory instruction. . . . Our studies showed that most students in laboratories gained little insight either regarding the key science concepts involved or toward the process of knowledge construction.

In the past 20 years numerous reviews have been equivocal about the purpose and effectiveness of laboratory activities in science. Hofstein and Lunetta (1982) concluded their review with the suggestion that the right questions have probably not been asked to focus research on laboratory activities. That assertion is supported by the following review.

Since the curriculum innovations of the 1960s, the emphasis in laboratory activities has been on providing students with hands-on experiences. Less attention has been directed to the negotiation of meaning. The understandings that emerge from laboratory activities depend on direct and vicarious experiences and negotiations of meaning as learners engage to solve problems and make sense of what they do. Experiences with phenomena and events provide opportunities for students to construct images to which language can be assigned at a later time. Social collaboration enables understandings to be clarified, elaborated, justified, and evaluated. Incorrect understandings can be recognized as such. Time for reflective thinking is crucial, even when psychomotor skills are the main goals of an activity. This assertion was clearly illustrated by Beasley (1979), who found that mental rehearsal of the steps in a procedure enhanced the learning of technical skills required in volumetric analysis. An interesting implication of Beasley's study was the comparable effectiveness of mental and physical practice.

Studies of interactions between teachers and students in laboratory activities can focus on processes such as the assignment of language to concepts, negotiation of meaning, and arriving at consensuses. However, earlier studies of laboratory activities have not been based on the development of meaning central to a radical constructivist epistemology.

Several studies compared student achievement in activities that incorporated pre- and postlaboratory discussions with achievement in activities lacking such discussions. These studies gave little emphasis to the purposes of laboratory activities, the content being developed, and the complex nature of interactions between teachers and students. Studies by Isom and Rowsey (1986), Raghubir (1979), and Nelson and Abraham (1976) provide support for discussion in conjunction with pre- and postlaboratory activities but do not provide insights into the interactive and cooperative strategies that must emerge if discussions are to be effective in facilitating learning. How do students use knowledge from a laboratory investigation to negotiate meaning in a cooperative-learning group? The studies that are most needed involve close examination of the negotiation process in classes in which students know what they are expected to do and are motivated to learn. How students construct and reconstruct ideas, test them with their peers, and transform them as a result of negotiation is not well understood in the context of science education. In addition, it is not only of interest to know that postlaboratory activities are beneficial, we need to know what form such discussions should take if students are to arrive at a negotiated consensus at the whole-class level.

How students engage in laboratory activities also influences how and what they learn. Unfortunately, most studies of classrooms have shown that students do not have many opportunities for direct experience with phenomena. Several studies suggested a pattern of teacher-structured activities with students watching and listening for most of the lesson (Newton & Capie, 1981; Tobin, 1986a; Tobin & Capie, 1982). For example, Tobin and Capie (1982) reported that middle school students were covertly engaged for more than 50% of the allocated time; engaged in overt planning and data-processing tasks for only 2% and 5% of the time, respectively; and off task for approximately 30% of the allocated time. The findings indicate that summative achievement and retention were each related to the proportion of time engaged in planning and data-processing tasks. Whereas students undoubtedly can learn by watching and listening, the results suggest that higher achievement is associated with getting involved in a more overt manner.

Although teachers can use these findings as an argument for increasing the amount of overt engagement for students in laboratory activities, causal arguments are not actually warranted. The results are consistent with the hypothesis that overt engagement in learning tasks enhances learning. However, the results are also consistent with other hypotheses. It is likely that high-achieving students (target students) engage to a greater extent than low-achieving students in laboratory activities (Tobin & Gallagher, 1987b). Detailed investigations of learning in classrooms are necessary to provide insights into the types of laboratory activities that enhance learning with understanding. In conducting such studies, we should expect to find that no one approach will be ideal for all learners.

Many studies have shown that male students become more involved in science laboratories than female students (e.g., Kelly, 1987; Tobin & Garnett, 1988). Kahle (1990) described a relatively small group of male students dominating interactions with equipment in laboratory activities. Not only did males monopolize the use of equipment and opportunities to learn, they also sometimes contaminated reagents and otherwise interfered with equipment and materials, thereby minimizing opportunities for others to learn in the intended manner. Undoubtedly, males and females engage in different ways in laboratory activities because of cultural factors that extend beyond learning in science classrooms. However, such factors cannot be ignored. Science teachers can take active measures to reverse trends associated with differential involvement of males and females in science activities and adopt proactive measures to increase engagement of female students in laboratory activities and facilitate learning.

Studies of learning technical skills also raise more questions than are answered (e.g., Vann, May, & Shugars, 1981; Yager, Engen, & Snider, 1969). Many of these studies compared one approach with another, disregarding many of the contextual factors that make a difference in the teaching and learning of science. Can it be assumed that all laboratory activities are equally conducive to learning psychomotor skills? Is a highly structured environment always better than a nonstructured environment? The studies reviewed did not probe learning environments in laboratory activities, show what worked and what did not, or seriously consider variations needed for learning different skills.

The manner in which students are grouped directly affects their opportunities to learn. Several studies have highlighted the problems teachers face in managing students in laboratory activities (e.g., Gallagher & Tobin, 1987). Teachers have difficulty coping with management of disruptive behavior and the learning needs of approximately 30 students in a laboratory setting. The extent to which students are cooperative and motivated to learn obviously makes a difference to the efficacy of laboratory activities. Do students understand their roles in laboratory activities? Can students communicate with peers to identify a problem, reach agreement on the nature of the problem, design strategies for solving the problem, agree on those strategies, and implement a plan to collect and analyze data? These communication skills, which are craved by business and industry leaders around the United States (e.g., Florida Chamber of Commerce, 1989), are somewhat elusive in most classes. Students cannot be placed in groups with an expectation that productive outcomes will occur. Collaboration is a role most students must learn. Studies of laboratory learning should not assume that students organized in groups are learning cooperatively. In a study of group problem solving, Tippins, Kagan, and Jackson (1992) found that prospective teachers did, indeed, equate the assignment of students to small groups with cooperative learning.

Is it possible for students to work independently of the teacher in laboratory activities? Leonard, Cavana, and Lowery (1981) investigated the importance of providing students with discretion in learning in high school laboratory activities. Students were able to select procedural options commensurate with personal interests, goals, and their own discretionary abilities. Leonard et al. noted:

It appears that tenth-grade biology students . . . are capable of using discretion to a greater extent and with greater rewards than is presently being allowed. Students in this study were found to be able to learn on their own discretion for periods of 10–15 minutes at the beginning of the school year and for at least three class periods later in the school year. . . . When teacher expectations were increased, students adjusted to these expectations. Students also adjusted successfully to seeking teacher help less frequently and only at specific times during laboratory investigations. (p. 503)

The study by Leonard et al. suggests that teachers might be able to ease the management problems and enhance science learning by giving students opportunities to learn independently. If students do pursue solutions to problems independently of the teacher, the teacher is free to become a facilitator of learning and focus on students who are in need of assistance.

There is considerable evidence that assessment can focus learning activities in laboratory contexts (e.g., Stake & Easley,

1978; Tobin & Gallagher, 1987a). How teachers assess students and what they assess have a major impact on implemented curricula. In most instances test-driven systems result in classroom activities that emphasize rote learning of science facts and rote learning of algorithms to solve exercises similar to those included on the test. Alberts, van Beuzekom, and de Roo (1986) also recognized the powerful driving force of examinations when they advocated the separation of the summative assessment of practical achievement from formative assessment. They argued that adherence to such a policy might help reduce the extent to which assessment focuses the curriculum and thereby determines what is valued by students and subsequently learned. Close attention needs to be given to other assessment issues as well. What do students learn from laboratory activities? How can students represent their knowledge? How should knowledge constructed in laboratory activities be assessed?

In ideal circumstances, an assessment scheme should provide students with opportunities to represent what they know about identified aspects of science. At the classroom level teachers are encouraged to use a variety of methods to assess student knowledge acquisition. These methods include traditional pencil-and-paper methods, personal oral interviews, and performance tests. The desirability of using a range of techniques is based on an assumption that much of the knowledge acquired in a hands-on and minds-on science program is tacit and has not been verbalized. Accordingly, although students can apply certain knowledge when they *do science*, they cannot necessarily reproduce that knowledge in verbal form on a pencil-and-paper test or in a discussion with the teacher. A paradox associated with this point was emphasized in a review of practical assessment in science:

The paradox is obvious. Many teachers place great value in science as a practical subject. The *practical* nature of the subject is commonly regarded as an important source of pupil motivation. Science is taught in laboratories and teachers spend a considerable amount of time in supervising practical work. *Yet the bulk of science assessment is traditionally non-practical.* (Bryce & Robertson, 1985, p. 1)

Olsen (1973) highlighted the difficulty of assessing the knowledge gained from a laboratory because of the tacit nature of the skills acquired. Abouseif and Lee (1965) concluded that practical tests revealed abilities different from those required for written tests but that different kinds of practical tests assessed different abilities. Kelly and Lister (1969) reinforced this point with the observation that practical work involves both manual and intellectual abilities that are in some measure distinct from those used in nonpractical work. Kelly and Lister noted: "we have evidence that not only are the abilities measured by written papers different, but that different practical tests produce different measurements." Comber and Keeves (1973) reported that practical tasks measured quite different abilities from those assessed by the more traditional tests, even those designed to assess practical skills without using actual apparatus. Grobman (1970) considered the use of written tests to be highly inappropriate in the assessment of either complex laboratory experiments or basic laboratory tasks. Finally, Tamir (1972) stated that practical testing situations are mandatory for

the practical mode. Tamir noted: "Since practical skills are seen as being valid only in the context of investigation, practical tests should involve real investigations" (p. 181).

A priority for research is the use of alternative assessment tasks (i.e., not multiple-choice, pencil-and-paper items) to determine what students have learned from laboratory activities. What have students learned and how can they apply their knowledge to solve problems? To what extent is knowledge learned in laboratory activities interconnected with prior knowledge? How can practical assessment tasks be incorporated into district- and state-level assessments of science achievement? These questions are just a sample of those that teachers and researchers need to investigate as many states (e.g., Florida, Michigan) endeavor to implement strategies to reform science education. Because of the demonstrated driving force of assessment on the curriculum, it is imperative that methods of assessment be reformed at the same time that approaches to teaching and learning science are being changed.

Discussion

Substantive changes in laboratory learning are unlikely to occur unless teachers and students change the manner in which they conceptualize their roles in science classes. At present, most teachers do not understand the role of laboratory activities as a means of allowing students to solve problems and thereby construct knowledge of science. Verification or technical-skills laboratories are common and usually take the form of cookbook activities. Changing the type of activity encountered in the laboratory will necessitate a change in the culture of the science classroom. Students will have to adopt new roles and make sense of them before they can be expected to be productive learners in problem-solving activities. How can students work together to learn from one another? What is involved in sharing? Negotiating? Forming consensuses? Evaluating assertions from peers? Similarly, teachers will need to identify the salient roles to be undertaken in problem-solving laboratories and formulate a facilitation role. For many teachers, the role will be new. In the past it has been assumed that teachers and students will implement reforms in the manner intended by curriculum designers; however, history has shown that this rarely is the case. More often than not, teachers adapt the curriculum to fit their own beliefs about what ought to be done in science classes. Teachers do what makes sense to them in the given circumstances. If changes are to be sustained, teachers will need assistance in changing their beliefs about what ought to be done in science laboratory activities.

ASSESSMENT PRACTICES

A most significant role of the teacher, from a constructivist perspective, is to assess student learning. In a study of exemplary teachers, Tobin and Fraser found that teachers routinely monitored students in three distinctive ways: they scanned the class for signs of imminent off-task behavior, closely examined the nature of the engagement of students, and investigated the extent to which students understood what they were learning (Tobin & Fraser, 1987). If teachers are to mediate the learning process, it is imperative that they develop ways to assess what students know and how they can represent what they know.

Traditional assessment practices seem to be associated with teaching roles akin to judging and rewarding. The interpretive study of science classrooms undertaken by Tobin and Gallagher (1987a) revealed that teachers used assessment as a motivator of students. Unless an activity was assessed, it was difficult to obtain the active participation and cooperation of students. The effect of this trend was an emphasis on completing products for grades. There seemed to be an implicit bargain in the culture that students would work for grades in much the same way that employees work for pay. Students focused on completing tasks and getting the grade, and learning became a by-product of the main activity in the culture. An inadvertent outcome of this process might have been that negotiation between teacher and student led to overspecification of the tasks, making them clear to students, reducing risk, and making it intellectually safe to do science. From a constructivist point of view, such an environment is not conducive to learning science with understanding. To achieve such a goal students must be perplexed at times and struggle to resolve perturbations that are created in the act of experiencing and trying to figure out puzzles associated with engaging in an intellectually rigorous manner. Numerous studies indicated that the cognitive level of academic tasks was low and the cognitive level of the pencil and paper tests used to assess learning was correspondingly low. The process of education, like the currency of many countries, appeared to have been inflated.

In a study of a beginning middle or high school science teacher, Tobin and Ulerick graphically described the problems associated with assessment practices that are built on a metaphor for assessment of a fair judge (Tobin & Ulerick, 1989). Even though Marsha, the teacher in the study, had gained control of her difficult-to-handle students, they had built negative attitudes in response to the teacher because she was holding the line on what she considered to be legitimate standards of learning, standards built from a consideration of the discipline rather than a consideration of what students knew. The negative attitude toward the teacher was widespread in the class and manifested itself in latent hostility toward the teacher and the subject of science. In terms of the metaphor of bargaining work in exchange for grades, it seemed as if Marsha was trying to negotiate too hard a deal and the workers, the students, were not buying into it. Marsha, like so many teachers we have worked with, had no alternative ways to think about or practice assessment. She had only one conceptualization of assessment, and when it was not working she had no options to fall back on.

Marsha was involved in research on her teaching and had solved some of her management problems by reconceptualizing her role in relation to management by constructing a new metaphor that was consistent with constructivism. The new metaphor resulted in a greater focus on the needs of learners and a system of rules that was simple and emphasized student responsibility. Once the new strategies were implemented, a striking transformation of her classroom occurred. Accordingly, Marsha was open to the idea that further improvements would be possible if she could reconstruct her understandings of as-

sessment. She also had strong beliefs about constructivism as a way of thinking about knowing that opened the possibility of trying different approaches to assessment. Initially, Marsha was inclined to try different approaches advocated by those whom she respected in science education. She tried concept mapping, oral interviews, and problem-centered learning. None seemed to be successful. The essence of the problem seemed to be how she viewed the nature of science and how she thought about assessment. Her beliefs about the nature of science were evolving in relation to her beliefs about constructivism. Because she could not construct a metaphor for assessment, a member of the research team suggested to her that it was like a mirror. Although the metaphor appealed to the researcher (a student looking into a mirror and seeing her knowledge displayed in her mind), it made no sense to Marsha. Subsequently, the researcher suggested a window into the students' minds, an opportunity for them to show the teacher what they knew. Marsha could see how this might work.

Using the metaphor of a window into the students' minds, Marsha rethought her assessment policies and practices. She implemented a new approach in her class and matters began to improve. The curious aspect of the metaphor she built was that it transferred power from the teacher to the students. The students now had the responsibility to make decisions about what they knew, how to represent what they knew, and when to schedule time with the teacher to show what they had learned. To Marsha and the students this seemed like a reasonable approach to assessment. An approach to assessment that emphasized the autonomy of students and learning with understanding evolved over a period of time.

Considerable attention has been given to the idea of creating a portfolio culture as a means of assessing students in alternative ways. For example, Duschl and Gitomer developed an approach to assessment based on students displaying evidence of what they have learned in a portfolio (Duschl & Gitomer, 1991; Duschl, Smith, Kesidou, Gitomer, & Schauble, 1992). Essentially, a portfolio is a container in which artifacts are placed, over a period of time, to represent what learners know. What goes in the container and what artifacts represent tend to vary from study to study. At one level students can have autonomy with respect to what is placed in the portfolio. If this is the case, it is important to provide each artifact with a label, describe what it represents and why it was included, and from time to time to review what is in the portfolio, replacing artifacts with better evidence of learning. In this way the portfolio is dynamic and is updated continuously to best represent what is known in a particular topic area. In such cases the teacher becomes a learner in the classroom, using portfolios to construct a profile of what each student does and does not know and the circumstances under which specific knowledge can be represented. At the other end of a spectrum, learners have less control over what is included in a portfolio; the teacher decides what entries, or artifacts, are included. In both cases the artifacts provide a basis for learners to reflect on their own learning and interact with peers and the teacher. Portfolio assessment has the potential to change the roles of teachers in significant ways. Freundlich, Gitomer, Duschl, and Faux (1992) described the potential of portfolios to change teacher and learner roles:

The portfolios envisioned are part of a new classroom culture where the philosophical and psychological underpinnings of the curricula are based on the field of the history and philosophy of science and the research of cognitive scientists. In the portfolio culture the assessment of student work is not terminal. Rather the purpose is to provide the teacher and the student with information that each can use to facilitate the process of meaningfully learning concepts and to develop reasoning. The classroom culture encourages students to be active rather than passive learners and teachers to be facilitators and role models rather than information givers ... A portfolio culture is based on a growth model of student development rather than a deficit model. This means that the student is viewed in terms of what they possess and how they grow and not on what they lack. We expect that portfolio instruction will become tailored to the individual needs of students. (p. 3)

Duschl and Gitomer (1991) examined the manner in which portfolio entries provide a basis for curriculum decisions. The entries provide teachers with a window into the minds of students, and, in a way that is not often possible, teachers have access to what students know about particular topics and how they can represent what they know. Accordingly, the portfolio entries make it possible for teachers to build learning activities around the students' prior knowledge.

We believe that the use of portfolios can be enhanced by thinking about the process from a constructivist perspective. First, it makes sense to think of a portfolio as a means of enabling students to show what they know. In a sense it is a showcase that provides an interface between the displayer and those who want to assess learning. Second, the student can be given autonomy in deciding what goes into the portfolio and what these artifacts represent. This description and justification process could take place in writing, orally, or a combination of both. Third, because of the difficulty in understanding what an artifact represents, the portfolio would provide a context for discussion in which a teacher and a student could negotiate what the artifacts represent and what progress they had made toward meeting the goals of the course. Interactions of this type would undoubtedly require considerable time on the part of the teacher, but it is also the case presently that assessment often consumes a disproportionate amount of a teacher's time. As Duschl and Gitomer imply with the term *portfolio culture*, the use of portfolios might lead to radically different ways to implement a curriculum and significantly different ways for teachers and students to interact in classrooms.

Discussion

The availability of portfolios on an ongoing basis opens the opportunity for teachers to get in touch with what students know on an ongoing basis. Examination of students' portfolios can provide a basis for a learner-focused dialogue between the students and teacher. Also, portfolios provide opportunities for students to interact with one another about what they know. We foresee a situation in which group members spend some time each week examining the portfolios of peers. Some students could assume a tutoring role toward those in their group who do not appear to have grasped a particular concept, and all group members can provide formative evaluation of the representativeness of artifacts and the extent to which they accurately represent what has been learned. Much of the learning could be centered on portfolios. The use of portfolios is a tangible way to bring teaching, learning, and assessment closer together, and from a constructivist perspective it is easy to see how these interactive processes could result not only in better assessment but also in better learning.

Portfolios are but one example of what has come to be known as alternative assessments, often described as "authentic", "performance-based," or "outcome-based" assessment. Historically, the goal of most assessment practices was equity, but the approach was reductionist, so that assessment may have skewed the curriculum in the direction of easily measurable basic skills and isolated facts. The more recent forms of alernative assessments have been attempts to capture equity in the context of the discipline and the context of the school and learner.

Constructivism provides a different conceptual apparatus from which to view assessment. When alternative assessment systems are rooted epistemologically in social constructivist theory, assessment will involve the representation of knowledge and the social viability of knowledge. From a constructivist perspective, the knower personally participates in all acts of understanding. The emphasis is on determining what the student knows; thus, the point of reference for assessment becomes what the student understands. New questions in relation to assessment assume significance from a constructivist perspective. For example, how can specific knowledge be represented? To what extent should students be given control over the what, how, and when of assessment? Ultimately, as teachers construct new knowledge and attempt to change their practices in relation to assessment, they negotiate with each other. In this way they test new knowledge and meaning about their changed practices in relation to assessment.

CONCLUSION

David Berliner (1986) once used the metaphor of a half-filled glass to describe the state of what we know about teaching from research on teaching. Is the glass half full? Or is it half empty? Our conclusion might well take up where Berliner left off. What we have found in reviewing the research on science teaching is that there is no shortage of studies. Indeed, we could have written a chapter several times as long had we undertaken to review all of the studies that have been done. We noted, however, that many of the studies did not provide results directly related to teaching and the mediational role of teachers. We had difficulty fitting many of the studies into the constructivist framework we were using to make sense of teaching and learning science. From our perspective, the paradigm change that has occurred in educational research is associated with profound differences in the way researchers view teachers, learners, researchers, and research on teaching and learning. Accordingly, problems are conceptualized differently, the questions that arise to be studied require new methodologies, and the purposes of research are distinctive. Rather than undertaking research designed to obtain objectified and generalizable knowledge, many researchers are focusing on learning from research and reporting to others what has been learned and the circumstances under which learning occurred. These studies map closely onto classrooms, teachers, and students. Furthermore, the discontinuity that is so often observed between research and practice is no longer apparent.

Despite the somewhat pessimistic opening to this conclusion, we were able to learn much from these studies, but even these lessons are but simple beginnings. For each study we asked the question: What can we learn about the mediational role of teachers? The result was an overwhelming sense that teachers can make a difference in classrooms. The roles of teachers begin with setting goals for the curriculum. If teachers view learning in terms of understanding, which is associated with connectedness and usefulness in social contexts, a curriculum that begins with student learning rather than content coverage is likely to emerge. This is no simple task. How do we learn to shift perspectives in a way that moves us to understand the complex web of dialectical interactions that constitute a science curriculum?

As the curriculum is implemented, the roles undertaken by the teacher seem to make a difference. How teachers and students view their roles in relation to interacting with one another and what has to be learned is a critical component of the environments in which learning occurs. The roles undertaken by teachers depend on the circumstances in which teachers and learners find themselves. Accordingly, each teacher must be responsive to cues that arise as students undertake learning tasks in a complex social milieu. It is not the case that any teacher variable, like wait time, can be manipulated above a threshold and create conditions to enhance learning by as much as a standard deviation. The optimal wait time depends on the circumstances that prevail in the learning environments. However, the wait-time research is convincing in its overwhelming support for the provision of silence in classrooms. Providing time for teachers and students to think seems much more important than mechanistically increasing the duration of selected types of pauses separating speakers during verbal interaction.

What can we learn from research on teaching? This question has been asked many times in the past decade. The key, we believe, is that knowledge is not objective or generalizable across situations, nor is research an educational quest for certainty. We read each study within our framework for teaching and learning and made sense of it in a context of what we knew already and how the study contributed to a growing picture of teachers mediating the science learning of students. In this sense, our contentions defy traditional notions of the teacher's role by recognizing the complexities and uncertainties inherent in the mediation of science learning. What can teacher educators do to facilitate the professional development of prospective and practicing science teachers? It is clear from the research that a plethora of variables are related in some way to learning and engagement of students. It is also clear that focusing on these in a singular manner can be counterproductive. In contrast, holistic ways of learning about teaching and reflecting on teaching, such as through the use of metaphors and images, appear promising. Can teachers be educated to reflect on practice in terms of metaphors, images, and powerful referents such as beliefs about control of students and beliefs about learning? Can teachers view themselves as learners in their own classrooms? We see the potential of teachers as educated professionals who are able to look into their own classrooms, identify problems, and on reflection construct a variety of alternatives to overcome the problems. Of course there also are times when professionals and learners seek assistance from others. Thus, the sources of professional knowledge, growth, and change are learning from experience in classrooms and, as necessary, reaching out to colleagues and other sources for assistance.

We began this chapter with thoughts on the reform of science curricula. Tried and tested approaches have not been successful, in part, we believe, because of the reformers' beliefs about knowledge, learning, and teaching. Top-down approaches to reform are in many ways the antithesis of empowerment, yet we have endeavored to show that teachers need the autonomy to structure the learning environments of their students as circumstances dictate. As we pause to consider the purpose and direction of research in science teaching, we see reflective and responsive teachers who are able to learn from experience as the missing link in attempts at reform. We see teachers as learners in their own classrooms, continuing in their attempt to identify "what works." As new and different ideas of what it means to be a teacher emerge, we can begin to think of teachers as researchers and reformers actively engaged in the construction of knowledge to change classroom practices and inextricably linked to the efforts of systemic reform. However, as we face the challenge of reconstructing K–12 science and teacher-education curricula, we should not forget that teaching is much more than being a learner. Provocateur, coach, role model, guide, social director . . . are these parts of an essential set of metaphors to be learned by all teachers? We do not believe so. It is for individual teachers to build their own understanding of the possible links between metaphors, teaching, and learning. We do not believe in universals, but by understanding how others have found such metaphors useful in planning and implementing curricula, it is possible that teachers will build their own metaphors, images, and belief sets in a process of reconceptualizing the curriculum and reflecting on classroom practices as it is implemented.

References

Abell, S., & Smith, D. (1992). What is science? Preservice elementary teachers conceptions of the nature of science. In S. Hills (Ed.), *Proceedings of the Secondary International History and Philosophy of Science & Science Teaching Conference* (pp. 11–22). Kingston, Ontario, Canada: Queen's University.

Abouseif, A. A., & Lee, D. M. (1965). The evaluation of certain science practical tests at secondary school level. *British Journal of Educational Psychology, 35,* 41–49.

Abraham, M. R. (1989). Research and teaching: Research on instructional strategies. *Journal of College Science Teaching, 18,* 185–187.

Abraham, M. R., & Renner, J. W. (1986). The sequence of learning cycle activities in high school chemistry. *Journal of Research in Science Teaching, 23*, 121–143.

Aikenhead, G. (1987). *Views on science-technology-society* (Form CDN.MC.4). Saskatchewan: Department of Curriculum Studies, University of Saskatchewan.

Aikenhead, G. (1992). How to teach the epistemology and sociology of science in a historical context. In S. Hills (Ed.), *Proceedings of the Second International Conference on the History and Philosophy of Science and Science Teaching* (pp. 23–34). Kingston, Ontario, Canada: Queen's University.

Alberts, R. V. J., van Beuzekom, P. J., & de Roo, I. (1986). The assessment of practical work: A choice of options. *European Journal of Science Education, 8*, 361–369.

Alvermann, D. E., & Hinchman, K. A. (1991). Science teachers' use of texts: Three case studies. In C. Santa & D. Alvermann (Eds.), *Science learning: Processes and applications.* Newark, DE: International Reading Association.

American Association for the Advancement of Science. (1989). *Project 2061: Science for all Americans.* Washington, DC: Author.

Anderson, B. O. (1978). The effects of long wait-time on high school physics pupils' response length, classroom attitudes and achievement (Doctoral dissertation, University of Minnesota). *Dissertation Abstracts International, 39*, 3493A. (University Microfilms No. 78-23, 871)

Anderson, C. R., & Schneier, C. E. (1978). Locus of control, leader behavior and leader performance among management students. *Academy of Management Journal, 21*, 690–698.

Anshutz, R. J. (1975). An investigation of wait time and questioning techniques as an instructional variable for science methods students microteaching elementary school children (Doctoral dissertation, University of Kansas). *Dissertation Abstracts International, 35*, 5978A. (University Microfilms No. 75-06, 131)

Arnold, D. S., Atwood, R. K., & Rogers, V. M. (1974). Question and response levels and lapse time intervals. *Journal of Experimental Education, 43*, 11–15.

Arora, A., & Kean, E. (1992). Perceptions of doing science: Science teachers' reflections. In S. Hills (Ed.), *Proceedings of the Second International Conference on the History and Philosophy of Science and Science Teaching Conference* (pp. 54–68). Kingston, Ontario: Canada: Queen's University.

Atkin, J. M., & Karplus, R. (1962). Discovery or invention? *Science Teacher, 29*, 45.

Ball, D., & McDiarmid, W. (1987). Understanding how teacher knowledge changes. *Colloquy, 1*, 9–23. East Lansing: Michigan State University, National Center for Research on Teacher Education.

Bandura, A. (1977). Self-efficacy: Toward a unifying theory of behavioral change. *Psychological Review, 84*, 191–215.

Bandura, A. (1986). *Social foundations of thought and action: A social cognitive theory.* Englewood Cliffs, NJ: Prentice Hall.

Beasley, W. F. (1979). The effects of physical and mental practice of psychomotor skills on chemistry student laboratory performance. *Journal of Research in Science Teaching, 16*, 473–479.

Bellamy, M., Borko, H., & Lockard, D. (1992, March). *The roles of three types of teacher knowledge—content knowledge, pedagogical knowledge, and pedagogical content knowledge—in the teaching of high school mendelian genetics.* Paper presented at the annual meeting of the National Association for Research in Science Teaching, Boston.

Benson, G. (1989). Epistemology and science curriculum. *Journal of Curriculum Studies, 21*, 329–344.

Berliner, D. C. (1986). In pursuit of the expert pedagogue. *Educational Researcher, 15*(7), 5–13.

Bickel, W. E. (1983). Effective schools: Knowledge, dissemination, inquiry. *Educational Researcher, 12*(4), 3–5.

Bonnstetter, R. J., Penick, J. E., & Yager, R. E. (1983). *Teachers in exemplary programs: How do they compare?* Washington, DC: National Science Teachers Association.

Briscoe, C. (1991). The dynamic interactions among beliefs, role metaphors, and teaching practices: A case study of teacher change. *Science Education, 75*, 186–199.

Brody, C. (1991, April). *Cooperative learning and teacher beliefs about pedagogy.* Paper presented at the annual meeting of the American Educational Research Association, Chicago, IL.

Brousseau, B. A., & Freeman, D. J. (1988). How do teacher education faculty members define desirable teacher beliefs? *Teaching and Teacher Education, 4*, 267–273.

Brousseau, V., Book, C., & Byers, J. (1988). Teacher beliefs and the cultures of teaching. *Journal of Teacher Education, 39*, 33–39.

Bryce, T. G.K., & Robertson, I. J. (1985). What can they do? A review of practical assessment in science. *Studies in Science Education, 12*, 1–24.

Buchmann, M. (1983). *The priority of knowledge and understanding in teaching* (Occasional Paper No. 61). East Lansing, MI: Institute for Research on Teaching. (ERIC Document Reproduction Service No. ED 237 503)

Bullough, R. V., Jr. (1991). Exploring personal teaching metaphors in preservice teacher education. *Journal of Teacher Education, 42*, 43–51.

Bybee, R. W. (1987). Science education and the science-technology-society (S-T-S) theme. *Science Education, 71*, 667–683.

Bybee, R. W., & Mau, T. (1986). Science and technology related global problems: An international survey of science educators. *Journal of Research in Science Teaching, 23*, 599–618.

Bybee, R. W., Buchwald, C. E., Crissman, S., Heil, D. R., Kuerbis, P. J., Matsumoto, C., & McInerney, J. D. (1989). *Science and technology education for the elementary years: Frameworks for curriculum and instruction.* Washington, DC: National Center for Improving Science Education.

Calderhead, J. (1988). Learning from introductory school experience. *Journal of Education for Teaching, 14*, 75–83.

Campbell, T. C. (1977). An evaluation of a learning cycle intervention strategy for enhancing the use of formal operational thought by beginning college physics students. *Dissertation Abstracts International, 38*(7), 3903A.

Carlsen, W. (1991). Effects of new biology teachers' subject-matter knowledge on curricular planning. *Science Education, 75*, 631–647.

Carlson, D. A. (1975). Training in formal reasoning abilities provided by the inquiry role approach and achievement on the Piagetian formal operational level. *Dissertation Abstracts International, 36*(11), 7368A.

Carroll, P. S., & Gallard, A. J. (1993). Students and teachers talking in the middle school science classroom. What does their discourse mean? In S. Tchudi (Ed.), *Language in science and humanities* (pp. 76–84). Urbana, IL: National Council of Teachers of English.

Chewprecha, T., Gardner, M., & Sapianchai, N. (1980). Comparison of training methods in modifying questioning and wait-time behaviors of Thai high school chemistry teachers. *Journal of Research in Science Teaching, 17*, 191–200.

Clandinin, J., & Connelly, M. (1986). On narrative method, personal philosophy, and narrative unities in the story of teaching. *Journal of Research in Science Teaching, 23*, 293–310.

Cobb, P., & Wheatley, G. (1988). Children's initial understandings of ten. *Focus on Learning Problems in Mathematics, 10*, 1–28.

Cobb, P., Yackel, E., Wood, T., Wheatley, G., & Merkel, G. (1988). Research into practice: Creating a problem solving atmosphere. *Arithmetic Teacher, 36*, 46–47.

Cobern, W. (1991). *World view theory and science education research* (Monograph of the National Association for Research in Science Teaching, No. 3). Manhattan: Kansas State University.

Cochran, K. (1992). *Pedagogical content knowledge: Teachers' transformations of subject matter. Research matters . . . to the science teacher* (Monograph of the National Association for Research in Science Teaching, No. 5). Manhattan: Kansas State University.

Cody, C. (1990). The politics of textbook publishing, adoption and use. In D. L. Elliot & A. Woodward (Eds.), *Textbooks and schooling in the United States.* Chicago: University of Chicago Press.

Cohen, M. (1982). Effective schools, accumulating research findings. *American Education, 18,* 13–16.

Collins, A. (1989). Assessing biology teachers: Understanding the nature of science and its influence on the practice of teaching. In D. Herget (Ed.), *Proceedings of the First International Conference on the History and Philosophy of Science in Science Teaching* (pp. 61–70). Tallahassee: Florida State University.

Collins, A., Brown, J. S., & Newman, S. E. (1989). Cognitive apprenticeship: Teaching the craft of reading, writing, and mathematics. In L. B. Resnick (Ed.), *Knowing, learning, and instruction: Essays in honor of Robert Glaser* (pp. 453–494). Hillsdale, NJ: Lawrence Erlbaum.

Comber, L. C., & Keeves, J. P. (1973). *Science education in 19 countries.* New York: Wiley.

Connelly, F. M., & Clandinin, D. J. (1988). *Teachers as curriculum planners: Narratives of experience.* New York: Teachers College Press.

Cornet, J. W., Yeotis, C., & Terwilliger, L. (1990). Teacher personal practice theories and their influences upon teacher curricular and instructional actions: A case study of a secondary science teacher. *Science Education, 74*(5), 517–529.

Cronin-Jones, L., & Shaw, S. (1992). The influence on the beliefs of preservice elementary and secondary teachers. Preliminary comparative analyses. *School Science and Mathematics, 92,* 14–22.

Crow, N. A. (1987, April). *Preservice teachers' biography: A case study.* Paper presented at the annual meeting of the American Educational Research Association, Washington, DC.

Czerniak, C., & Waldon, M. (1991, April). *A study of science teaching efficacy using qualitative and quantitative research methods.* Paper presented at the annual meeting of the National Association for Research in Science Teaching, Lake Geneva, WI.

Davis, J. O. (1977). The effects of three approaches to science instruction on the science achievement, understanding, and attitudes of selected fifth and sixth grade students. *Dissertation Abstracts International, 39,* 211A.

DeTure, L. R. (1979). Relative effects of modeling on the acquisition of wait-time by preservice elementary teachers and concomitant changes in dialogue patterns. *Journal of Research in Science Teaching, 16,* 553–562.

DeTure, L. R., & Miller, A. P. (1984, April). *The effects of a written protocol model on teacher acquisition of extended wait-time.* Paper presented at the annual meeting of the National Association for Research in Science Teaching, New Orleans.

Doyle, W. (1983). Academic work. *Review of Educational Research, 53,* 159–199.

Drake, S. M. (1991). The journey of the learner: Personal and universal story. *The Educational Forum, 56,* 47–59.

Duschl, R. A., & Gitomer, D. H. (1991). Epistemological perspectives on conceptual change: Implications for educational practice. *Journal of Research in Science Teaching, 28,* 839–858.

Duschl, R. A., & Wright, E. (1989). A case study of high school teachers' decision making models for planning and teaching science. *Journal of Research in Science Teaching, 26,* 467–502.

Duschl, R., Smith, M., Kesidou, S., Gitomer, D., & Schauble, L. (1992, April). *Assessing student explanations for criteria to format conceptual change learning environments.* Paper presented at the annual meeting of the American Educational Research Association, San Francisco.

Eisenhart, M. A., Shrum, J. L., Harding, J. R., & Cuthbert, A. M. (1988). Teacher beliefs: Definitions, findings and directions. *Educational Policy, 2,* 51–70.

Elbaz, F. (1983). *Teacher thinking: A study of practical knowledge.* New York: Nichols.

Elias, J. (1990). Teaching as cure and care: A therapeutic metaphor. *Religious Education, 85,* 436–444.

Elliot, D. L., & Woodward, A. (Eds.). (1990). *Textbooks and Schooling in the United States: National Society for the Study of Education 89th Yearbook.* Chicago: NSSE, Chicago University Press.

Ellis, J., Bybee, R., Giese, J., & Parisi, L. (1992). Teaching about the history and nature of science and technology: Issues in teacher development. In S. Hills (Ed.), *Proceedings of the Second International Conference on the History and Philosophy of Science and Science Teaching.* Kingston, Ontario, Canada: Queen's University.

Enochs, L., & Riggs, I. (1990). Further development of an elementary science teaching efficacy belief instrument: A preservice elementary scale. *School Science and Mathematics, 90,* 694–706.

Espinet, M. (1988, April). *The development of pedagogical content knowledge of beginning high school chemistry teachers with a strong chemistry background but no formal training in education.* Paper presented at the annual meeting of the National Association for Research in Science Teaching, Lake of the Ozarks, MO.

Esquivel, J. M., Lashier, W. S., & Smith, W. S. (1978). Effect of feedback on questioning of preservice teachers in SCIS microteaching. *Science Education, 62,* 209–214.

Fawcett, G. (1992). Moving the big desk. *Language Arts, 69,* 183–185.

Feiman-Nemser, S., & Buchmann, M. (1987). When is student teaching teacher education? *Teaching and Teacher Education, 3,* 255–273.

Finley, F. N. (1991). Why students have trouble learning from science texts. In C. Santa & D. Alvermann (Eds.), *Science learning: Processes and applications.* Newark, DE: International Reading Association.

Floden, R. E. (1985). The role rhetoric in changing teachers' beliefs. *Teaching and Teacher Education, 1,* 19–32.

Florida Chamber of Commerce. (1989). *Rise to the challenge. Improving math and science: Business agenda.* Tallahassee: Florida Chamber of Commerce.

Fosnot, C. T. (1988, January). *The dance of education.* Paper presented at the annual meeting of Association for Educational Communications and Technology, New Orleans.

Fowler, T. W. (1975, March). *An investigation of the teacher behavior of wait-time during an inquiry science lesson.* Paper presented at annual meeting of National Association for Research in Science Teaching, Los Angeles. (ERIC Document Reproduction Service No. ED 108 872)

Fraser, B. J. (1986). *Classroom environment.* London: Croom Helm.

Fraser, B. J., & Tobin, K. (1989). Student perceptions of psychosocial environment in classrooms of exemplary science teachers. *International Journal of Science Education, 11,* 19–34.

Freundlich, J., Gitomer, D., Duschl, R., & Faux, R. (1992, March). *Constructing portfolio assessment in science classrooms.* Paper presented at the annual meeting of the Association of the Education of Teachers of Science, Boston.

Furtwengler, C. (1992). How to observe cooperative learning classrooms. *Educational Leadership, 49*(7), 59–62.

Gallagher, J. (1991). Prospective and practicing secondary school science teachers' knowledge and beliefs about the philosophy of science. *Science Education, 75,* 121–133.

Gallagher, J. J. (1992). Secondary science teachers and costructivist practice. In K. Tobin (Ed.), *The practice of constructivism in science education.* Washington, DC: American Association for the Advancement of Science Press.

Gallagher, J., & Tobin, K. (1985). *Teacher management and student engagement in high school science.* Paper presented at the annual meeting of the National Association for Research in Science Teaching, French Lick Springs, IN.

Gallagher, J., & Tobin, K. (1987). Teacher management and student engagement in high school science. *Science Education, 71,* 535–555.

Gallard, A. J., (1992). Learning science in multicultural classroom. In K. Tobin (Ed.), *Constructivism and the teaching and learning of science and mathematics.* Washington DC: American Association for the Advancement of Science.

Gallard, A. J., & Tippins, D. J. (1992). *Language diversity and science learning: The need for a critical system of meaning.* Unpublished manuscript, University of California at Santa Cruz.

Gess-Newsome, J. (1992, March). *Biology teachers' perceptions of subject matter structure and its relationship to classroom practice.* Paper presented at the annual meeting of the National Association for Research in Science Teaching, Boston.

Gilbert, S. W. (1989). An evaluation of the use of analogy, simile, metaphor in science texts. *Journal of Research in Science Teaching, 26,* 3315–3327.

Giroux, H. A. (1992). *Border crossings: Cultural workers and the politics of education.* New York: Chapman and Hall.

Gist, M. E. (1987). Self-efficacy: Implications for organizational behavior and human resource management. *Academy of Management and Review, 12,* 472–483.

Glasson, G., & Lalik, R. (1990). *Interpreting the learning cycle from a language learning perspective.* Paper presented at the annual meeting of the National Association for Research in Science Teaching, Atlanta.

Gliessman, D. H., Grillo, D. M., & Archer, A. C. (1989, March). *Changes in teacher problem solving: Two studies.* Paper presented at the annual meeting of the American Educational Research Association, San Francisco.

Good, R. (1989, March). *Toward a unified conception of thinking: Prediction with a cognitive science perspective.* Paper presented at the annual meeting of the National Association for Research in Science Teaching, San Francisco.

Gooding, C. T., Swift, P. R., & Swift, J. N. (1983, April). *An analysis of classroom discussion based on teacher success in observing wait time.* Paper presented at annual conference of New England Educational Research Organization, Rockport, ME.

Goodrum, D. (1987). Exemplary teaching in upper primary science classes. In K. Tobin & B. J. Fraser (Eds.), *Exemplary practice in science and mathematics teaching.* Perth: Curtin University of Technology.

Gozzi, R., Jr. (1991, March). *The metaphor of the mind-as computer: Some considerations for teachers.* Paper presented at the 42nd annual meeting of the Conference on College Composition and Communication, Boston.

Grandin, S. L. (1989, March). *English studies and the metaphors we live by.* Paper presented at the 40th annual meeting of the Conference on College Composition and Communication, Seattle.

Grant, G. E. (1991). Ways of constructing classroom meaning: Two stories about knowing and seeing. *Journal of Curriculum Studies, 23,* 397–408.

Grobman, H. (1970). *Developmental curriculum projects: Decision points and processes.* Itasca: F. E. Peacock.

Grossman, P. L. (1988). *A study in contrast: Sources of pedagogical content knowledge for secondary English.* Unpublished doctoral dissertation, Stanford University, Stanford.

Grundy, S. (1987). *Curriculum: Product or praxis?* London: Falmer.

Gurney, B. F. (1990, April). *Tugboats and tennis games: Preconceptions of teaching and learning through metaphor.* Paper presented at the annual meeting of the National Association for Research in Science Teaching, Atlanta.

Guskey, T. R. (1988). Teacher efficacy, self-concept, and attitudes toward the implementation of instructional innovation. *Teaching and Teacher Education, 4,* 63–69.

Haggerty, S. (1992). Student teachers' perceptions of science and science teaching. In S. Hills (Ed.), *Proceedings of the Second International Conference on the History and Philosophy of Science and Science Teaching* (pp. 483–494). Kingston, Ontario, Canada: Queen's University.

Hall, G. E., George, A. A., & Rutherford, W. L. (1979). *Measuring stages of concern about the innovation: A manual for the use of the SoC questionnaire* (2nd ed.). Austin: University of Texas.

Happs, J. (1987). Good teaching of invalid information: Exemplary junior secondary science teachers outside their field of expertise. In K. Tobin & B. J. Fraser (Eds.), *Exemplary practice in science and mathematics teaching.* Perth: Curtin University of Technology.

Hashweh, M. (1985). An exploratory study of teacher knowledge and teaching: The effects of science teachers' knowledge of subject-matter and their conceptions of learning on their teaching. *Dissertation Abstracts International* (AAC 8602482), *46*(12), 3672A.

Hashweh, M. (1987). Effects of subject-matter knowledge in the teaching of biology and physics. *Teaching and Teacher Education, 3*(2), 109–120.

Haury, D. (1988). Evidence that science locus of control orientation can be modified through instruction. *Journal of Research in Science Teaching, 25,* 233–246.

Haury, D. (1989). The contribution of science locus of control orientation to expressions of attitude toward science teaching. *Journal of Research in Science Teaching, 26,* 503–517.

Heller, P. (1992, March). *Teaching physics problem solving through cooperative grouping: Do men perform better than women?* Paper presented at the annual meeting of the National Association for Research in Science Teaching, Boston.

Hewson, P. W., & Hewson, M. G. A. (1984). The role of conceptual conflict in conceptual change and the design of science instruction. *Instructional Science, 13,* 1–13.

Hewson, P. W., & Hewson, M. G. (1989). Analyses and use of a task for identifying conceptions of teaching science. *Journal of Education for Teaching, 15,* 191–209.

Hofstein, A., & Lunetta, V. N. (1982). The role of the laboratory in science teaching: Neglected aspects of research. *Review of Educational Research, 52*(2), 201–217.

Holmes Group. (1990). *Tomorrow's teachers: A report of the Holmes Group.* Lansing, MI: Author.

Honea, J. M. (1982). Wait time as an instructional variable: An influence on teacher and student. *Clearinghouse, 56*(4), 167–170.

Howard, R. (1989). Teaching science with metaphors. *School Science Review, 70*(252), 100–103.

Isom, F. S., & Rowsey, R. E. (1986). The effect of a new prelaboratory procedure on students' achievement in chemistry. *Journal of Research in Science Teaching, 23,* 231–235.

Jarolimek, J. (1991). Focusing on concepts: Teaching for meaningful learning. *Social Studies and the Young Learner, 3*(3), 3–5.

Jasalavich, S. (1992, March). *Preservice elementary teachers' beliefs about science teaching and learning and perceived sources of their beliefs prior to their first formal science teaching experience.* Paper presented at the annual meeting of the National Association for Research in Science Teaching, Boston.

Johnson, D. W., & Johnson, R. T. (1985a). Classroom conflict: Controversy versus debate in learning groups. *American Educational Research Journal, 22,* 237–256.

Johnson, R. T., & Johnson, D. W. (1985b). Student-student interaction: Ignored but powerful. *Journal of Teacher Education, 36*(4), 22–26.

Jones, R. (1989). Cooperative learning in the elementary science methods courses. *Journal of Science Teacher Education, 1,* 1–3.

Jones, R., & Steinbrink, J. (1989). Using cooperative groups in science teaching. *School Science and Mathematics, 89,* 541–551.

Jones, R., & Steinbrink, J. (1991). Home teams: Cooperative learning in elementary science. *School Science and Mathematics, 91*(4), 139–143.

Joyce, B. R., & Showers, B. (1983). *Power in staff development through research on training.* Alexandria, VA: Association for Supervision and Curriculum Development.

Kagan, D. (1992). Implication of research on teacher belief. *Educational Psychologist, 27*(1), 65–90.

Kagan, D. M., & Tippins, D. J. (1991). How teachers' classroom cases express their pedagogical beliefs. *Journal of Teacher Education, 42*(4), 281–291.

Kagan, D., & Tippins, D. (1993). Classroom cases as gauges of professional growth. In *Teacher education yearbook I: Diversity and teaching.* (pp. 98–110. New York: Harcourt, Brace, Jovanovich).

Kahle, J. B. (1990). Real students take chemistry and physics. In K. Tobin, J. B. Kahle, & B. J. Fraser (Eds.), *Windows into science classrooms: Problems associated with high level cognitive learning in science.* London: Falmer.

Karplus, R., & Thier, H. D. (1967). *A new look at elementary school science. Science curriculum improvement study.* Chicago: Rand McNally.

Kelly, A. (Ed.) (1987). *Science for girls?* Bristol, PA: Open University Press.

Kelly, P. J., & Lister, R. E. (1969). Assessing practical ability in Nuffield A-level biology. In J. F. Eggleston & J. F. Kerr, (Eds.) *Studies in assessment.* London: English University Press.

Kempa, R., & Ayob, A. (1991). Learning interactions in group work in science. *International Journal of Science Education, 13,* 341–354.

Kilbourn, B. (1980–81). World views and curriculum. *Interchange, 11*(2), 1–10.

Knowles, M. S. (1990). Appendix E. From teacher to facilitator. In M. S. Knowles (Ed.), *Adult learner: A neglected species* (3rd ed.). Houston: Gulf Publishing.

Koehler, V. (1988). Barriers to the effective supervision of student teaching: A field study. *Journal of Teacher Education, 39*(2), 28–35.

Kottkamp, R. B. (1990). Means for facilitating reflection. *Education and Urban Society, 22*(2), 182–203.

Kuhn, T. S. (1962). *The structure of scientific revolutions.* Chicago: University of Chicago Press.

Kyle, W. C., Jr., & Gottfried, S. (1992). Textbook use and the biology education desired state. *Journal of Research in Science Teaching, 29,* 35–49.

Laboskey, V. K., & Wilson, S. N. (1987, April). *Case writing as a method in preservice teacher education.* Paper presented at the annual meeting of the American Educational Research Association, Washington, DC.

Lake, J. H. (1973). The influence of wait-time on the verbal dimensions of student inquiry behavior (Doctoral dissertation, Rutgers University). *Dissertation Abstracts International, 34*(10), 6476A. (University Microfilm No. 74-08866)

Land, M. L. (1981). Actual and perceived teacher clarity—relations to student achievement in science. *Journal of Research in Science Teaching, 18,* 139–143.

Lapp, D., Flood, J., & Thrope, L. (1989). How to do it: Cooperative problem solving: Enhancing learning in the secondary science classroom. *American Biology Teacher, 51,* 139–143.

Lavoie, D. (1989, April). *Enhancing the learning cycle with prediction and level three interactive videodisc lessons in science.* Paper presented at the annual meeting of the National Association for Research in Science Teaching, San Francisco.

Lavoie, D. (1992, March). *The effects of adding a prediction/discussion phase to a science learning cycle.* Paper presented at the annual meeting of the National Association for Research in Science Teaching, Boston.

Lawson, A. E. (1988). A better way to teach biology. *American Biology Teacher, 50,* 266–278.

Lawson, A. E., & Wollman, W. T. (1976). Encouraging the transition from concrete to formal cognitive functioning—an experiment. *Journal of Research in Science Teaching, 13,* 413–430.

Lawson, A. E., Abraham, M. R., & Renner, J. W. (1989). *A theory of instruction: Using the learning cycle to teaching science concepts and thinking skills* (Monograph of the National Association for Research in Science Teaching, No. 1). Cincinatti, OH: NARST.

Lawson, C. A. (1967). *Brain mechanisms and human learning.* Boston: Houghton Mifflin.

Lazarowitz, R., Hertz, R., Baird, J., & Bowlden, V. (1988). Academic achievement and on-task behavior of high school biology students instructed in a cooperative small investigative group. *Science Education, 72,* 475–487.

Leinhardt, G., & Smith, D. A. (1985). Expertise in mathematics instruction: Subject matter knowledge. *Journal of Educational Psychology, 77,* 241–271.

Lent, R. W., Brown, S. D., & Larkin, K. C. (1987). Comparison of three theoretically diverse variables in predicting career and academic behavior: Self-efficacy, interest congruence and congruence thinking. *Journal of Counseling Psychology, 34*(3), 293–298.

Lent, R. W., & Hackett, G. (1987). Career self-efficacy: Empirical status and future directions. *Journal of Vocational Behavior, 30,* 347–382.

Leonard, W. H., Cavana, G. R., & Lowery, L. F. (1981). An experimental test of an extended discretion approach for high school laboratory investigations. *Journal of Research in Science Teaching, 18,* 497–504.

Leong C. K. (1988). On the computer metaphor for learning. *Canadian Journal of Special Education, 4,* 3–7.

Lorsbach, A. (1992, March). *An interpretive study of prospective teachers' beliefs about the nature of science.* Paper presented at the annual meeting of the National Association for Research in Science Teaching, Boston.

Magnusson, S., Borko, H., Krajcik, J., & Layman, J. (1992, March). *The relationship between teacher content and pedagogical content knowledge and student content knowledge of heat energy and temperature.* Paper presented at the annual meeting of the National Association for Research in Science Teaching, Boston.

Manning, B. H., & Payne, B. D. (1989). A cognitive self-direction model for teacher education. *Teacher Education Quarterly, 18,* 49–54.

Marek, E., & Methven, S. (1991). Effects of the learning cycle upon student and classroom teacher performances. *Journal of Research in Science Teaching, 28,* 41–53.

Marks, R. (1991, April). *When should teachers learn pedagogical content knowledge?* Paper presented at the annual meeting of the American Educational Research Association, Chicago.

Marsh, D., & Odden, A. (1991). Implementation of the California mathematics and science curriculum frameworks. In A. Odden (Ed.), *Educational policy implementation.* Albany: State University of New York Press.

Mason, C. (1988, April). *A collaborative effort to effectively evolve pedagogical content knowledge in preservice teachers.* Paper presented at the annual meeting of the National Association for Research in Science Teaching, Lake of the Ozarks, MO.

McEwan, H., & Bull, B. (1991). The pedagogic nature of subject matter knowledge. *American Educational Research Journal, 28,* 316–334.

McKinnon, J. W., & Renner, J. W. (1971). Are colleges concerned with intellectual development? *American Journal of Physics, 39,* 1047–1052.

Meyer, L. A., Crummey, L., & Greer, E. A. (1988). Elementary science textbooks: Their contents, text characteristics and comprehensibility. *Journal of Research in Science Teaching, 25,* 435–463.

Miller, J. P., & Seller, W. (1985). *Curriculum perspectives and practice*. New York: Longman.

Mitchener, C. P., & Anderson, R. D. (1987). Teachers' perspective: Developing and implementing an STS curriculum. *Journal of Research in Science Teaching, 26*, 351–369.

Miura, I. T. (1986, April). *Computer self-efficacy: A factor in understanding gender differences in computer course enrollment*. Paper presented at the annual meeting of the American Educational Research Association, San Francisco.

Mosenthal, P. B. (1988). The conduit metaphor and the academic goal of reading. *The Reading Teacher, 41*, 448–450.

Munby, H. (1986). Metaphor in the thinking of teachers: An exploratory study. *Journal of Curriculum Studies, 18*, 197–209.

Munby, H. (1987). Metaphor and teachers' knowledge. *Research in the Teaching of English, 21*, 377–397.

Munby, H. (1990). Metaphorical expressions of teachers' practical curriculum knowledge. *Journal of Curriculum and Supervision, 6*(1), 18–30.

Munby, H., & Russell, T. (1989, March). *Metaphor in the study of teachers' professional knowledge*. Paper presented at the annual meeting of the American Educational Research Association, San Francisco.

Muscari, P. G. (1988). The metaphor in science and in science classroom. *Science Education, 72*, 423–431.

National Center for Improving Science Education. (1989). *Developing and supporting teachers for elementary school science education*. Colorado Springs, CO: NETWORK, Inc.

National Commission on Excellence in Education. (1983). *A nation at risk: The imperative for educational reform*. Washington, DC: U.S. Government Printing Office.

National Science Teachers Association (NSTA). (1989). *Essential changes in secondary science scope: Scope, sequence and coordination*. Washington, DC: Author.

National Science Teachers Association (NSTA). (1992). *Scope, sequence, and coordination of secondary school science: Vol. 1. The content core: A guide for curriculum designers*. Washington, DC: Author.

Nelson, M. A., & Abraham, E. C. (1976). Discussion strategies and student cognitive skills. *Science Education, 60*, 13–27.

Nespor, J. (1987). The role of beliefs in the practice of teaching. *Journal of Curriculum Studies, 19*, 317–328.

Newton, R. E., & Capie, W. (1981, April). *Effects of engagement quality on integrated process skill achievement in grade 7 and 8 science students of varying ability levels*. Paper presented at the annual meeting of the National Association for Research in Science Teaching, Grossinger, NY.

Novak, J. D. (1988). Learning science and the science of learning. *Studies in Science Education, 15*, 77–101.

Okebukola, P. (1986a). Cooperative learning and students' attitudes to laboratory work. *School Science and Mathematics, 86*, 582–590.

Okebukola, P. (1986b). The influence of preferred learning styles on cooperative learning in science. *Science Education, 70*, 509–517.

Oliver, S., & Koballa, T. (1992, March). *Science educators use of the concept of belief*. Paper presented at the 65th annual meeting of the National Association for Research in Science Teaching, Boston.

O'Loughlin, M. (1990, April). *Evolving beliefs about teaching and learning: The view from Hofstra University: A perspective on teachers' beliefs and their effects*. Paper presented at the annual meeting of the American Educational Research Association, Boston.

Olsen, D. R. (1973). What is worth knowing and what can be taught? *School Review, 82*, 27–43.

Padilla, M., Capie, W., & Cronin, L. (1986, March). *The relationship of teacher performance to science process skill achievement of students*. Paper presented at the annual meeting of the National Association for Research in Science Teaching, San Francisco.

Penick, J. E. (Ed.). (1983). *Focus on excellence: Science as inquiry*. Washington, DC: National Science Teachers Association.

Penick, J. E. (Ed.). (1988). *Focus on excellence: Elementary science*. Washington, DC: National Science Teachers Association.

Penick, J. E., & Bonnstetter, R. J. (Eds.). (1983). *Focus on excellence: Biology*. Washington, DC: National Science Teachers Association.

Penick, J. E., & Lunetta, V. N. (Eds.). (1984). *Focus on excellence: Physical science*. Washington, DC: National Science Teachers Association.

Penick, J. E., & Meinhard-Pellens, R. (Eds.). (1984). *Focus on excellence: Science/technology/society*. Washington, DC: National Science Teachers Association.

Penick, J. E., & Yager, R. E. (1986a). Science education: New concerns and issues. *Science Education, 70*, 427–431.

Philion, T. (1990). Metaphors from student teaching: Shaping our classroom conversations. *English Journal, 79*(7), 88–89.

Posner, J., Strike, K., Hewson, P., & Gertzog, W. (1982). Accommodation of a science conception: Toward a theory of conceptual change. *Science Education, 66*, 211–227.

Post-Kammer, P., & Smith, P. L. (1986). Sex differences in math and science career self-efficacy among disadvantaged students. *Journal of Vocational Behavior, 29*, 89–101.

Powell, R. R. (1991–1992). Acquisition and use of pedagogical knowledge among career-change preservice teachers. *Action in Teacher Education, 13*(4), 17–23.

Power, B. M. (1990). Research, teaching and all that jazz: New metaphors and models for working with teachers. *English Educator, 22*(3), 179–191.

Proper, H., Wideen, M., & Ivany, G. (1988). World view projected by science teachers: A study of classroom dialogue. *Science Education, 72*, 547–560.

Provenzo, E. F., McCloskey, G. N., Kottkamp, R. B., & Cohn, M. M. (1989). Metaphor and meaning in the language of teachers. *Teachers College Record, 90*, 551–573.

Pugach, M. C., & Johnson, L. J. (1989, March). *Developing reflective practice through structured dialogue*. Paper presented at the meeting of the American Educational Research Association, San Francisco.

Pugh, M., & Lock, R. (1989). Pupil talk in biology practical work—a preliminary study. *Research in Science and Technological Education, 7*, 15–26.

Purser, R. K., & Renner, J. W. (1983). Results of two tenth-grade biology teaching procedures. *Science Education, 67*, 85–98.

Raghubir, K. P. (1979). The laboratory investigative approach to science instruction. *Journal of Research in Science Teaching, 16*, 13–17.

Renner, J., & Marek, E. (1988). *The learning cycle and elementary school science teaching*. Portsmouth, NH: Heinemann.

Renner, J., Abraham, M., & Birnie, H. H. (1985). The importance of the form of student acquisition of data in physics learning cycles. *Journal of Research in Science Teaching, 22*, 303–325.

Rice, D. R. (1977). The effect of question-asking instruction on preservice elementary science teachers. *Journal of Research in Science Teaching, 14*, 353–359.

Riggs, I. (1991, March). *Gender differences in elementary science teacher self-efficacy*. Paper presented at the annual meeting of the American Educational Research Association, Chicago.

Riley, J. P., II. (1986). The effects of teachers' wait-time and knowledge comprehension questioning on pupil science achievement. *Journal of Research in Science Teaching, 23*, 335–342.

Rochester, S. R. (1973). The significance of pauses in spontaneous speech. *Journal of Psycholinguistic Research, 2*, 51–81.

Rogg, S., & Kahle, J. (1992, March). *The characterization of small instructional workgroups in 9th grade biology*. Paper presented at the annual convention of the National Association for Research in Science Teaching, Boston.

Roth, W. M. (1990, April). *Collaboration and constructivism in the science classroom*. Paper presented at the annual meeting of the American Educational Research Association, Boston.

Roth, W. M. (1991, April). *Aspects of cognitive apprenticeship in science teaching*. Paper presented at the annual meeting of the National Association for Research in Science Teaching, Lake Geneva, WI.

Rovegno, I. (1992). Learning to teach in a field-based methods course: The development of pedagogical content knowledge. *Teaching and Teacher Education, 8*(1), 69–82.

Rowe, M. B. (1969). Science, silence and sanctions. *Science and Children, 6*(6), 11–13.

Rowe, M. (1974a). Relation of wait-time and rewards to the development of language, logic, and fate control: Part II—Rewards. *Journal of Research in Science Teaching, 11*, 291–308.

Rowe, M. B. (1974b). Wait time and rewards as instructional variables, their influence on language, logic, and fate control: Part I—Wait time. *Journal of Research in Science Teaching, 11*, 81–94.

Rowe, M. B. (1983). Getting chemistry off the killer course list. *Journal of Chemical Education, 60*, 954–956.

Rubin, R. L., & Norman, J. J. (1989, March/April). *A comparison of the effect of a systematic modeling approach and the learning cycle approach on the achievement of integrated science process skills of urban middle school students*. Paper presented at the annual meeting of the National Association for Research in Science Teaching, San Francisco. (ERIC Document Reproduction Service No. ED 308 838).

Russell, T. (1987, July). *From pre-service teacher education to first year of teaching: A study of theory and practice*. Paper presented at the British Educational Research Association Conference on Teachers' Professional Learning, University of Lancaster, Lancaster, England.

Russell, T. (1992). Learning to teach science: Constructivism, reflection, and learning from experience. In K. Tobin (Ed.), *The practice of constructivism in science education*. Washington, DC: AAAS Press.

Russell, T., & Munby, H. (1992). *Teachers and teaching: From classroom to reflection*. London: Falmer.

Samiroden, W. D. (1983). The effects of higher cognitive level questions wait time ranges by biology student teachers on student achievement and perception of teacher effectiveness (Doctoral Dissertation, Oregon State University). *Dissertation Abstracts International, 43*(10), 3208A.

Sanford, J. P. (1984). Management and organization in science classrooms. *Journal of Research in Science Teaching, 21*, 575–587.

Saunders, W. L., & Shepardson, D. (1987). A comparison of concrete and formal science instruction upon science achievement and reasoning ability of sixth-grade students. *Journal of Research in Science Teaching, 24*, 39–51.

Scharmann, L. (1992, March). *Teaching evolution: The influence of peer instructional modeling*. Paper presented at the annual meeting of the National Association for Research in Science Teaching, Boston.

Schneider, L. S., & Renner, J. W. (1980). Concrete and formal teaching. *Journal of Research in Science Teaching, 17*, 503–517.

Schunk, D. H. (1987, April). *Domain-specific measurement of students' self-regulated learning processes*. Paper presented at the annual meeting of the American Educational Research Association, Washington, DC.

Scott, L., & Heller, P. (1991). Team work: Strategies for integrating women and minorities into the physical sciences. *The Science Teacher, 58*, 24–28.

Sharan, Y., & Sharan, S. (1989). Group investigation expands cooperative learning. *Educational Leadership, 47*(4), 17–21.

Sherman, L. W. (1989). A comparative study of cooperative and competitive achievement in two secondary biology classrooms: The group investigation model versus an individually competitive goal structure. *Journal of Research in Science Teaching, 26*, 55–64.

Shulman, L. (1986). Paradigms and research programs in the study of teaching: A contemporary perspective. In M. C. Wittrock (Ed.), *Third handbook of research on teaching*. New York: Macmillan.

Shulman, L. S. (1986). Those who understand: Knowledge growth in teaching. *Educational Researcher, 15*(2), 4–14.

Shulman, L. (1990). *Aristotle had it right: On knowledge and pedagogy*. (Occasional paper no. 4). East Lansing, MI: The Holmes Group.

Siegman, A. W., & Pope, B. (1965). Effects of question specificity and anxiety producing messages on verbal fluency in the initial interview. *Journal of Personality and Social Psychology, 2*, 522–530.

Simmons, J. M., Sparks, G. M., Starko, A., Pasch, M., Colon, A., & Grinberg, J. (1989, March). *Exploring the structure of reflective pedagogical thinking in novice and expert teachers: The birth of a developmental taxonomy*. Paper presented at the annual meeting of the American Educational Research Association, San Francisco.

Slavin, R. E. (1984). Students motivating students to excel: Cooperative incentives, cooperative tasks, and student achievement. *The Elementary School Journal, 85*, 53–63.

Smith, D., & Neale, D. (1989). The construction of subject matter knowledge in primary science teaching. *Teaching and Teacher Education, 5*, 1–19.

Smith, E. L. (1991). A conceptual change model of learning science. In S. Glynn, R. Yeany, & B. Britton (Eds.), *Psychology of learning science* (pp. 43–63). Hillsdale, NJ: Lawrence Erlbaum.

Stake, R. E., & Easley, J. A. (1978). *Case studies in science education* (Vols. 1 & 2). Urbana: Center for Instructional Research and Curriculum Evaluation and Committee on Culture and Cognition, University of Illinois at Urbana–Champagne.

Steffe, L. P., von Glasersfeld, E., Richards, J., & Cobb, P. (1983). *Children's counting types: Philosophy, theory, and application*. New York: Praeger Scientific.

Swift, J. N., & Gooding, C. T. (1983). Interaction of wait time feedback and questioning instruction on middle school science teaching. *Journal of Research in Science Teaching, 20*, 721–730.

Swift, J. N., Swift, P. R., & Gooding, C. T. (1984, April). *Observed changes in classroom behavior utilizing workshop, wait time feedback and immediate supportive intervention*. Paper presented at the annual conference of the New England Educational Research Organization, Rockport, ME.

Tamir, P. (1972). The practical mode: A distinct mode of performance. *Journal of Biological Education, 6*, 175–182.

Tamir, P. (1988). Subject matter and related pedagogical knowledge in teacher education. *Teaching and Teacher Education, 4*, 99–110.

Tamir, P. (1991). Professional and personal knowledge of teachers and teacher educators. *Teaching and Teacher Education, 7*(3), 263–268.

Tamir, P., & Lunetta, V. N. (1978). An analysis of laboratory inquiries in the BSCS yellow-version. *American Biology Teacher, 40*, 353–357.

Taylor, P. (1990, March). *The influence of teacher beliefs on constructivist teaching practices*. Paper presented at the annual meeting of the American Educational Research Association, Boston.

Thomas, M. (1990, August). *Reflective or reactive pupils and teachers? Special educator or special trainer? Role dilemmas for teachers and education tutors*. Paper presented at the International Special Education Congress, Cardiff, Wales.

Tingle, J. B. (1989). Effect of cooperative grouping on stoichiometric problem solving in high school chemistry (Doctoral dissertation, Louisiana State University Agricultural and Mechanical College, 1988). *Dissertation Abstracts International, 49*, 2097A.

Tippins, D., & Kagan, D. (1992, April). *How preservice teachers translate learning theory into instruction: A study of group problem solving*. Paper presented at the annual meeting of the American Educational Research Association, San Francisco.

Tippins, D., & Tobin, K. (1992). *Using textbooks to enhance science learning* (Technical Report 92-1). Tallahassee: Florida State University.

Tippins, D., Kagan, D., & Jackson, D. (1992). *Translating theory into practice: A study of group problem solving among prospective secondary science teachers*. Manuscript submitted for publication.

Tobin, K. G. (1980). The effect of an extended teacher wait time on science achievement. *Journal of Research in Science Teaching, 17,* 469–475.

Tobin, K. G. (1985). *Academic work in science classes*. Paper presented at the annual meeting of the American Educational Research Association, Chicago.

Tobin, K. (1986a). Student task involvement and achievement in process-oriented science activities. *Science Education, 70,* 61–72.

Tobin, K. G. (1986b). Effects of teacher wait time on discourse characteristics in mathematics and language arts classes. *American Educational Research Journal, 23,* 191–200.

Tobin, K. (1987, April). *Teaching for higher cognitive level learning in science*. Paper presented at the annual meeting of the National Association for Research in Science Teaching, Washington, DC.

Tobin, K. (1990a). Changing metaphors and beliefs: A master switch for teaching? *Theory into Practice, 29*(2), 122–127.

Tobin, K. (1990b). Research on science laboratory activities: In pursuit of better questions and answer to improve learning. *School Science and Mathematics, 90*(5), 403–418.

Tobin, K. (1991, April). *Making sense of science teaching*. Paper presented at the annual meeting of the National Association for Research in Science Teaching, Lake Geneva, WI.

Tobin, K. G., & Capie, W. (1982). Relationships between classroom process variables and middle school science achievement. *Journal of Educational Psychology, 14,* 441–454.

Tobin, K., Capie, W., & Bettencourt, A. (1988). Active teaching for higher cognitive learning in science. *International Journal of Science Education, 10,* 17–27.

Tobin, K., & Espinet, M. (1989). Impediments to change: Application of peer coaching in high school science. *Journal of Research in Science Teaching, 26,* 105–120.

Tobin, K., Espinet, M., Byrd, S. E. & Adams, D. (1988). Alternative perspectives of effective science teaching. *Science Education, 72,* 433–451.

Tobin, K., & Fraser, B. J. (Eds.). (1987). *Exemplary practice in science and mathematics education*. Perth: Curtin University of Technology.

Tobin, K., & Gallagher, J. (1987a). What happens in high school science classrooms? *Journal of Curriculum Studies, 19,* 549–560.

Tobin, K., & Gallagher, J. J. (1987b). The role of target students in the science classroom. *Journal of Research in Science Teaching, 24,* 61–75.

Tobin, K., & Garnett, P. (1988). Exemplary practice in science classrooms. *Science Education, 72,* 197–208.

Tobin, K., & Jakubowski, E. (1991). The cognitive requisites for improving the performance of elementary science and mathematics teaching. In E. W. Ross, J. W. Cornett, & G. McCutcheon (Eds.), *Teacher personalizing, theorizing: Issues, problems and implications* (pp. 161–178). Columbia University Press.

Tobin, K., Kahle, J. B., & Fraser, B. J. (1990). *Windows into science classrooms: Problems associated with high level cognitive learning in science*. London: Falmer.

Tobin, K., Tippins, D., & Hook, K. (1992, April). *The construction and reconstruction of teacher knowledge*. Paper presented at the annual meeting of the American Educational Research Association, San Francisco.

Tobin, K., Treagust, D. F., & Fraser, B. J. (1988). An investigation of exemplary biology teaching. *American Biology Teacher, 50,* 142–147.

Tobin, K., & Ulerick, S. (1989, March). *An interpretation of high school science teaching based on metaphors and beliefs for specific roles*.

Paper presented at the annual meeting of the American Educational Research Association, San Francisco.

Tomanek, D. (1992, April). *Studying content as a part of a curriculum process*. Paper presented at the annual meeting of the American Educational Research Association, San Francisco.

Treagust, D., Wilkinson, W., Leggett, M., & Glasson, P. (1988, April). *Enhancement of physics learning: Teacher knowledge of content and content-specific pedagogy*. Paper presented at the annual meeting of the National Association for Research in Science Teaching, Lake of the Ozarks, MO.

Vann, W., May, K. N., & Shugars, D. A. (1981). Acquisition of psychomotor skills in dentistry: An experimental teaching method. *Journal of Dental Education, 45,* 567–575.

von Glasersfeld, E. (1987). *The construction of knowledge: Contributions to conceptual semantics*. Seaside, CA: Systems Inquiry Series, Intersystems Publication.

von Glasersfeld, E. (1989). Cognition, construction of knowledge, and teaching. *Synthese, 80,* 121–140.

Ward, C. R., & Herron, J. D. (1980). Helping students understand formal chemical concepts. *Journal of Research in Science Teaching, 17,* 387–400.

Watson, S. (1992, March). *Cooperative incentives and heterogeneous arrangement of cooperative learning groups: Effects on achievement of elementary education majors in an introductory science course*. Paper presented at the annual meeting of the National Association for Research in Science Teaching, Boston.

Weinstein, C. S. (1989). Teacher education students' preconceptions of teaching. *Journal of Teacher Education, 40*(2), 53–60.

Wheatley, G. H. (1991). Constructivist perspectives on science and mathematics learning. *Science Education, 75,* 9–21.

Winne, P. H. (1979). Experiments relating teachers' use of higher cognitive questions to student achievement. *Review of Educational Research, 49,* 13–49.

Winne, P. H., & Marx, R. W. (1983). *Students cognitive processes while learning from teaching: Summary of findings* (Occasional paper). Instructional Psychology Research Group, Simon Fraser University, Burnaby, BC, Canada.

Woodward, A., & Elliot, D. L. (1990). Textbooks: Consensus and controversy. In D. L. Elliot & A. Woodward (Eds.), *Textbooks and schooling in the United States*. Chicago: University of Chicago Press.

Woolfolk, A., Rosoff, B., & Hoy, W. (1990). Teachers' sense of efficacy and their beliefs about managing students. *Teaching and Teacher Education, 6*(2), 137–148.

Yager, R. E. (1984, April). *Searching for excellence*. Paper presented at annual meeting of the American Educational Research Association, New Orleans.

Yager, R E. (1988). A new focus for school science: S/T/S. *School Science and Mathematics, 88,* 181–189.

Yager, R. E., Engen, H. B., & Snider, B. C. F. (1969). Effects of the laboratory and demonstration methods upon the outcomes of instruction in secondary biology. *Journal of Research in Science Teaching, 6,* 76–86.

Yager, R., Hidayat, E., & Penick, J. (1988). Features which separate least effective from most effective science teachers. *Journal of Research in Science Teaching, 25,* 165–177.

Zeichner, K. M., Tabachnick, B. R., & Densmore, K. (1987). Individual, institutional, and cultural influences on the development of teachers' craft knowledge. In J. Calderhead (Ed.), *Exploring teachers' thinking* (pp. 21–59). London: Cassell.

Zoller, U., Dunn, S., Wild, R., & Beckett, P. (1991). Students' versus their teachers' beliefs and positions on science/technology/society oriented issues. *International Journal of Science Education, 13,* 25–36.

RESEARCH ON USING LABORATORY INSTRUCTION IN SCIENCE

Reuven Lazarowitz

IIT TECHNION

Pinchas Tamir

HEBREW UNIVERSITY

Science teaching laboratories are places for learning. Almost 200 hundred years ago, Edgeworth and Edgeworth (1811) wrote:

The great difficulty which has been found in attempts to instruct children in science has, we apprehend, arisen from the theoretic manner in which preceptors have proceeded. The knowledge that cannot be immediately applied is quickly forgotten and nothing but disgust connected with useless labor remains in the pupil's mind.... (Pupils) senses should be exercised in experiments, and these experiments should be simple, distinct and applicable to some object in which the pupils are immediately interested. We are not solicitous about the quantity of knowledge that is obtained at any given age, but we are extremely anxious that the desire to learn should continuously increase.... Until children have acquired some knowledge of effects, they cannot inquire into causes. Observation must precede reasoning; and as judgement is nothing more than a perception of the results of comparison, we should never urge our pupils to judge until they have acquired some portion of experience. (pp. 226, 329, 424)

It is readily apparent that the claims made by the Edgeworths so many years ago are still valid. The school laboratory is the setting established by the educational institutions to meet the needs and goals identified in their statement. The term *laboratory work* as used in U.S. literature corresponds to the term *practical work* for the United Kingdom and literature of its previously affiliated countries. For the purpose of this chapter, we follow Hegarty-Hazel's (1990b) definition of laboratory work:

Student laboratory work (is) a form of practical work taking place in a purposely assigned environment where students engage in planned learning experiences ... interact with materials to observe and understand phenomena. (Some forms of practical work such as field trips are thus excluded.) (p. 4)

Perhaps no other area in science education has attracted so many research reviews as learning, teaching, and assessment in the laboratory. Several reasons may account for this unusual interest in conducting and publishing such reviews. An important reason is the distinctive role and unique potential of the laboratory and the difficulty in obtaining convincing data on the effectiveness of learning in the laboratory . The latter is due to the complexity of factors related to the practical work per se and to the use of assessment procedures that have often been inadequate. Hence, as observed by Blosser (1983), educators are "still faced with the problem of defending laboratory activities as an essential component of the science curriculum" (p. 168). They continuously seek for valid research evidence to support their belief in the significant and unique contribution of students' work in the laboratory.

We have located 37 reviews between 1954 and 1990. The distribution of authors of these reviews is interesting; 13 reviews were authored by Americans, 7 each by British and Australians, 5 by Israelis, 2 in collaboration between and American and an Israeli, 2 in collaboration between an Israeli and an Australian, and 1 in collaboration between an Australian and an

The authors wish to thank David P. Butts (University of Georgia) for his help in reviewing earlier drafts of the chapter and for preparing it for publication.

American. Tables 3.1, 3.2, and 3.3 present lists of reviews by their authors, nationalities, and major focus.

The tables divide the reviews into three categories reflecting authorship: (1) the United States and collaborators, (2) the United Kingdom and (3) Israel and Australia. The rationale for this division is that the U.S. authors cite predominantly research published in the United States. English authors have tended to cite mainly British publications, whereas Australian and Israeli

TABLE 3.1 A List of Research Reviews on the Student Laboratory: the United States and Collaborators

Authors	Year and Type of Publication[a]	Focus(es)/Conclusions
Rosen	1954, J	History of the physics laboratory in U.S. public schools.
Pella	1961, J	Goals and degrees of freedom available to the teacher using the laboratory.
Belanger	1969, J	Limitations of studies comparing method X versus method Y. There is no convincing evidence for the superiority of discovery learning.
Ramsey & Howe	1969, J	Multireference laboratory-centered approach produces better understanding of the scientific enterprise and more positive attitudes. How best to integrate the laboratory with more conventional classwork remains an open question.
Novak	1972, B	How to arrange laboratory facilities for effective learning.
Shulman & Tamir[b]	1973, C	Philosophical and psychological background of learning in the laboratory with special reference to inquiry. A critical review of an exemplary study. Concept learning, critical thinking, manual and investigative skills, attitudes—all in relation to the attributes and dynamics of the laboratory.
Bates	1978, C	Laboratory experiences are superior for teaching manipulative skills. Some kinds of inquiry-oriented laboratory activities are better than labs for teaching the processes of inquiry, provided that skilled teachers, time, and guidance are available. Laboratories have a potential for nurturing positive student attitudes and providing a wide variety of students with opportunities to be successful in science.
Hickman	1978, C	Hands-on experience is integral to learning in the BSCS programs. Case studies of "troubleshooting" are presented.
Blosser	1980, T	Laboratory work is used less frequently than science educators would desire. On the average, the level of achievement in science is higher among students who made observations and did experiments.
Hofstein[b] & Lunetta	1982, J	The case for laboratory teaching is not as obvious as it may seem. Improvement in research is needed in terms of using better research designs and more sensitive instruments, attending to characteristics of student samples and the nature of laboratory activities before and during the study, and including more relevant dependent and independent variables.
Igelsrud & Leonard	1988, J	Practical work with concrete materials is effective in promoting the development of formal concepts. The preferred approach is guided inquiry. Microcomputer-based laboratories are helpful in developing concepts and skills.
Lawson, Abraham, & Renner	1989, B	A theory of instruction based on the learning cycle is described, supportive research is presented, and recommendations are offered. All three phases of the learning cycle involve students in observing and exploring phenomena and in testing possible explanations by experimentation most frequently conducted in the laboratory. Further applications often involve laboratory work as well.
Novak	1990, C	The interplay of theory and methodology in laboratory work can be enhanced through Vee diagrams and concept maps.
Klopfer	1990, C	To help students obtain scientific inquiry outcomes, investigative laboratory activities accompanied by reflection are recommended, along with use of special instructional technique-enhancing science process skills. There is a large scope for legitimate variations in desirable inquiry-related outcomes for different students.
Tobin	1990, J	Meaningful learning in the laboratory is possible. It is crucial that students reflect on their findings and consult a range of resources including other students, the teacher, books, and materials. Research is needed on how students engage, construct understandings, and negotiate meaning in cooperative groups and on how to guide teachers in establishing and maintaining environments conductive to learning. Teacher researchers are the logical inquirers in such studies. Collaboration among teachers and researchers is essential.
Gunstone[c] & Champagne	1990, C	Learning is an act of construction when pupils generate new meanings from incoming information by linking it with prior knowledge. This implies less time spent in interacting with apparatus, worksheets, and instructions and more time devoted to reflection and discussion. When this is done, the laboratory can be an effective means or creating conceptual change.

[a]B = book; C = chapter in book; J = journal; T = technical report.

[b]Israeli.

[c]Australian.

TABLE 3.2   A List of Research Reviews on the Student Laboratory: the United Kingdom

Authors	Year and Type of Publication[a]	Focus(es)/Conclusions
Kerr	1964, B	Insufficient laboratory facilities, overuse of laboratories, shortage of qualified teachers, lack of laboratory assistants, and large classes impair the effectiveness of laboratory work. Cognitive and psychomotor objectives are viewed by teachers as more important than affective objectives.
Connel	1971, J	A historical review of opinions that indicates lack of consensus and the existence of conflicting evidence regarding the relative effectiveness of lecture-demonstration compared with student practical work.
Ogborn	1978, C	The review deals with two main questions about student work in the laboratory: what it should be and how its aims can be brought to reality.
Toothacker	1983, J	University students in introductory courses do not benefit from the laboratory work. There are equally effective but less costly and time-consuming ways of teaching the empirical roots of science and promoting positive attitudes. Even the aim of becoming familiar with experimental techniques and equipment is not adequately attained. Because most beginning students come to the laboratory ill-prepared for undirected research, first- and second-year physics laboratory classes should be eliminated.
Woolnough & Allsop	1985, B	Practical work is essential for teaching the processes of science. It is possible to acquire knowledge (albeit shallow and limited) of much scientific information without doing any practical work. Early scientific education should be concerned not with theoretical content, but with developing the processes of science. Later, practical work should not be used as a subservient strategy for teaching scientific concepts and knowledge; there are self-sufficient reasons for doing practical work.
Bryce & Robertson	1985, J	A review of practical-work assessment in science criticizing the present situation, in which the bulk of assessment is nonpractical, and describes ways and procedures for assessing practical work by practical tasks.
Layton	1990, C	A historical philosophical account of the relationship between the processes in which students engage in the laboratory and our understanding of the nature of science. At present, students' laboratory practices have often been reflective of dubious or discarded philosophies of science. There is a need for critical appraisal of the burden of the student laboratory in conveying an authentic image of science.

[a]B = book; C = chapter in book; J = journal.

authors present a more varied list of references. Each list is in chronological order.

The examination of the topics selected by or assigned to particular authors illustrates an interesting pattern. For example, in three of four reviews focusing on assessment (Giddings & Hofstein, 1980; Tamir, 1985; Friedler & Tamir, 1990), the authors are Israeli, and in the remaining (Bryce & Robertson, 1985) they are British. This reflects the importance and prevalence of practical assessment in Israel and the United Kingdom as contrasted with its scarcity in the US.

For each review in Tables 3.1, 3.2, and 3.3, the venue of its publication is indicated. Books, journal articles, and technical reports are usually written at the initiative of the authors, whereas chapters are usually invited. Reviews written at the authors' initiative often attempt to provide empirical support for their previously established opinion. For example, in Woolnough and Allsop (1985), the authors attempted to provide evidence in support of separating content from process and concentrating practical work on the development of process skills. In Lawson, Abraham, and Renner (1989), there is an attempt to support integration of processes and content into one learning cycle. Similarly, Igelsrud and Leonard (1988) and Tamir (1976b) emphasize the distinctive and indispensable role of the laboratory, whereas Toothacker (1983) recommends limiting practical work in introductory college courses. An invitation to write a chapter in a book may be considered an expression of recognition of the author's expertise in a particular field, such as the history of science or assessment.

Reflecting on this history of research related to use of the laboratory in science teaching, we note four points as follows:

- There is little reference in the reviews to science in the elementary school, even though hands-on experience is essential for meaningful learning at this level.
- Certain issues related to the student laboratory, such as separation or integration of content and process, have remained controversial in spite of the research conducted.
- Improving research by formulating more "telling" questions and using better research design has remained a major challenge even in the 1990s.
- It is still necessary to provide sound empirical support for the role of the laboratory and the steps that are needed to ensure that the potential of the laboratory will be realized

In this chapter, we attempt to address these issues while emphasizing the four goals for teaching science through laboratory work that emerge from the review of the literature. These goals are as follows:

- Science laboratories should provide concrete experiences and ways to help students confront their misconceptions.

TABLE 3.3   A List of Research Reviews on the Student Laboratory: Australia and Israel

Authors	Year and Type of Publication[a]	Focus(es)/Conclusions
Tamir[b]	1976, b T	A comprehensive review of the literature on the role of the laboratory in science teaching, including a historical account of opinions and practices, discussion of goals, types of laboratories in terms of guidance and levels of inquiry, individualized and small-group work, variation among laboratories in different science disciplines and possible alternatives to practical work.
Giddings[c] & Hofstein[b]	1980, J	Trends in assessment of laboratory performance in high school science are presented and discussed in terms of what can be evaluated and how it can be done. Written evidence is contrasted with practical examinations and continuous assessment.
Tamir[b]	1985, C	A rationale for using practical examinations is provided, based on the strong impact that assessment has had on learning. Various kinds of practical tests are described and analyzed in terms of their validity and feasibility for student assessment by the teacher and for external examinations.
Boud,[c] Dunn,[c] & Hegarty-Hazel[c]	1986, B	A book aimed at teaching in the laboratory at the tertiary level. Various aspects of the use of the laboratory, its role, its integration within the curriculum, and its effectiveness are discussed.
Hofstein[b]	1988, C	This chapter reviews and redefines goals and teaching practices used in the laboratory. It concludes with two assertions. It is unreasonable to expect the laboratory to teach all goals effectively. However, the laboratory can play an important role in achieving some goals such as logical reasoning, inquiry, and problem-solving skills, and the development of positive attitudes.
Ainley[c]	1990, C	The physical facility is one component that influences the teaching and learning of science and colors the views that students hold of their science education in school.
Atkinson[c]	1990, C	Special instructional strategies are needed in order to enhance knowledge acquisition from laboratory activities. Examples of such strategies are the audiotutorial approach, the Personified System Instruction, and the creation of conceptual conflict, which can be done effectively in the laboratory.
Edwards[c] & Power[c]	1990, C	The role of the laboratory in an individualized inquiry-oriented science curriculum for junior high students, the Australian Science Education Project (ASEP) is described. The laboratory activities frequently failed to attain their goals because of inexperienced teachers and lack of pupils' understanding of the purposes of their activities.
Gardner[c] & Gauld[c]	1990, C	A distinction was made between attitude toward science and scientific attitude. It was concluded that laboratory experiences can improve both. Factors that have an effect on attitudes are laboratory facilities, instructional conditions, time allowed, variety, cognitive challenge, integration of theory with lab work, social interaction, and well-prepared teachers.
Hegarty-Hazel[c]	1990, C	This is the only review on learning technical skills available in the literature, a symptom of the observation that technical skills have been undervalued. It was concluded that student laboratories, especially at the tertiary level, should be restricted to skills that are necessary for scientific inquiries or future learning or vocational use. Assessment of the technical skills attained should be criterion referenced based on direct observations.
Hegarty-Hazel[c]	1990, C	Life in science laboratory classes at the tertiary level has often been described by students as busy, hard work, tedious, demanding, and tiring. However, some laboratory experiences have been regarded very highly. Factors such as classroom transactions, management, the learning environment, and curriculum are discussed. Recommendations regarding desirable patterns of instruction are presented.
Prosser[c] & Tamir[b]	1990, C	Students can learn about the use of the computer in the laboratory as a research tool, as an aid to learning, and as a management tool. In general, learning with the aid of microcomputers does not influence the level of achievement but does save instructional time. The most useful application of computers is simulation. Gradual introduction of computers with careful monitoring is recommended.
Friedler[b] & Tamir[b]	1990, C	Life in science laboratory classrooms at the secondary level is described in terms of intended, perceived, and implemented transactions as well as the subject matter, the teachers, the students, and the curriculum. Suggestions for improvement are offered and the initial results of their application are reported.
Tamir[b]	1990b, C	Various approaches to the evaluation of student learning outcomes in the laboratory are critically reviewed. The impact of including practical examinations in the biology matriculation examination in Israel is described and its implications for policy making are discussed.

[a] T = technical report; J = journal; C = chapter in book; B = book.

[b] Israeli.

[c] Australian.

- Science laboratories should provide opportunities for data manipulation through the use of microcomputers.
- Science laboratories should provide opportunities for developing skills in logical thinking and organization, especially with respect to science, technology, and society (STS) issues.
- Science laboratories should provide opportunities for building values, especially as they relate to the nature of science.

After reviewing the literature in light of these four goals for teaching science through laboratory work, we review factors, such as curricula, resources, teachers, and assessment strategies, that help students accomplish these goals in the laboratory.

PART I: GOALS FOR TEACHING SCIENCE THROUGH THE SCIENCE LABORATORY

According to White (1988),

The laboratory sets science apart from most school subjects. It gives science teaching a special character, providing many teachers and their students liveliness and fun that are hard to obtain in other ways. That character is almost sufficient alone to justify the high capital and recurrent cost of laboratories. (p. 186)

The student science laboratory is indeed unique. A reflection on what are perceived as its goals and objectives is useful. Bryce and Robertson (1985), following Doran (1978), noted that "practical science" is ill-defined because of "lack of an overall conceptual framework" (p. 14). According to Shulman and Tamir (1973) and Friedler and Tamir (1990), a rationale for student laboratory work includes these features:

1. Science involves highly complex and abstract subject matter. Even high school students may fail to grasp science concepts without the concrete props and opportunities for manipulation afforded in the laboratory (e.g., Lawson, 1975; Lawson & Renner, 1975; McNally, 1974). The laboratory also offers unique opportunities to identify student misconceptions (Driver & Bell, 1986; Friedler, 1984).
2. Students' participation in actual investigations that develop their inquiry and intellectual skills is an essential component of the inquiry curriculum (Schwab, 1962). It gives students an opportunity to appreciate the spirit of science (Ausubel, 1968) and promotes understanding the nature of science—for instance, the scientific enterprise, the way in which scientists work, the multiplicity of scientific methods, and the interrelationship between science, technology, and society (Shulman & Tamir, 1973).
3. Laboratory work promotes the development of cognitive abilities such as problem solving, analysis, generalizing (Ausubel, 1968), critical thinking, applying, synthesizing, evaluating, decision making, and creativity (Shulman & Tamir, 1973).
4. Laboratory work is essential for developing skills of various kinds: manipulative, inquiry, investigative, organizational, and communicative (Olson, 1973; Shulman & Tamir, 1973; Tamir 1975). Many authors have offered detailed taxonomies for the various skills—for example: manipulative skills (Bryce & Robertson, 1985; Eglen & Kempa, 1974; Hegarty-Hazel, 1990c; Klopfer, 1990) and inquiry skills (Klopfer, 1990; Tamir, Nussinovitz, & Friedler, 1982).
5. A goal rarely discussed in the literature is understanding the concepts that underlie scientific research, such as the definition of a scientific problem, hypothesis, assumption, prediction, conclusion, and models (e.g., Friedler & Tamir, 1986).
6. An important goal that captures both the cognitive and effective domains is the development of scientific attitudes, such as honesty, readiness to admit failure, critical assessment of the results and their limitations, curiosity, risk taking, objectivity, precision, confidence, perseverance, responsibility, collaboration, and readiness to reach consensus (e.g., Henry, 1975; Shulman & Tamir, 1973).
7. Students usually enjoy practical work in the laboratory and, when offered a chance to experience meaningful, nontrivial but not too difficult experiences, they become motivated and interested not only in their laboratory assignment but also in studying science (Ben-Zvi, Hofstein, Samuel, & Kempa, 1977; Henry, 1975; Selmes, Ashton, Meredith, & Newal, 1969).

Although these seven dimensions of a rationale for student practical work are useful, it is evident that some are related exclusively to laboratory work. However, most of them are relevant to nonlaboratory learning as well. It should be noted that the contribution of practical work exceeds by far that of other instructional settings. For example, Ausubel (1968) maintains that

In dividing the labor of scientific instruction, the laboratory typically carries the burden of conveying the method and the spirit of science whereas the textbook and teachers assume the burden of transmitting subject matter content. (p. 346)

Olson (1973) argues that

While schools aspire to facilitate the acquisition of both knowledge and skills, they are reasonably successful in serving the goals pertaining to the acquisition of knowledge, but they serve poorly the educational goals pertaining to the development of skills. (p. 41)

He maintains that this is so because learning skills require "direct experience," whereas the dominant mode of instruction is through "symbolically coded" transmission, which is not suitable for the development of skills. Tamir (1990b) builds on Olson's observation and emphasizes that

The laboratory is certainly expected to provide for the development of motor and intellectual skills as well as problem solving abilities and affective outcomes (since) the major learning mode is direct experience. (p. 244)

Woolnough and Allsop (1985) argue:

We need to deliberately and consciously separate practical work from the constraint of teaching scientific theory. We must stop using practical work as a subservient strategy for teaching scientific concepts and knowledge. There are self sufficient reasons for doing practical work in science and neither these, nor the aims concerning the teaching and understanding of scientific knowledge are served by continual linking of practical work to the content syllabus of science. (p. 39)

In contrast, Atkinson (1990) maintains that "a purpose of many science laboratory exercises is to promote the learning of particular symbols or facts and concepts" (p. 121). Kreitler and Kreitler (1974) hold an even more extreme position. For them, the laboratory provides direct experiences with concepts that lead to episodes that give the concepts meaning. The laboratory experiences also help students establish the accuracy of their beliefs. However, these authors do not believe that problem solving and arousing interest and curiosity are realistic roles of the laboratory.

Just as researchers do not agree on the relative importance of the laboratory, teachers and scientists also have different perceptions. Surveys conducted by Kerr (1964) among teachers in Britain and by Boud, Dunn, Kennedy, and Thorly (1980) among Australian scientists indicated that in both groups cognitive and psychomotor objectives were regarded as more important than affective objectives. Gauld (1978) found that over a 20-year period, teachers had shifted away from the view of laboratory work as an aid to understanding and toward the view of practical work as a facilitator of interest in the ability to solve problems. Gardner and Gauld (1990) suggest that "another way of considering the question of importance is to identify those objectives regarded as best achieved through lab work" (p. 135). They report the results of Henry (1975), who found that laboratory lessons were most clearly favored for objectives related to psychomotor and cognitive aspects of experimental procedures and social interactions in the class.

Another important consideration in assessing the relative importance of particular objectives is the preference of students. Keightley and Best (1975), who compared the preferences of 2616 Australian grade 11 biology students with the views of their 82 teachers regarding the relative importance of 12 educational objectives, reached the following conclusion:

The discrepancy between what teachers hope to teach and what students hope to learn suggests that students are more instrumental in their approach to biological knowledge than their teachers. The teachers on the other hand are more conceptual in their approach. (p. 65)

Indeed, the students' most preferred objective was "practical skills" (e.g., to use a microscope or to dissect an animal), whereas the teachers' most valued objective was understanding of the biologists' view of the world. In a similar vein, in surveys reported by Osborne (1976) and Boud et al. (1980), students perceived that they learn very little about problem solving in the laboratory.

Based on this review regarding the intended use of laboratory instruments, we conclude that the instructional potential of the laboratory is enormous. For some of the intended outcomes, such as the development of certain skills, the laboratory is the only available setting in school. For the remaining objectives, the decision about whether and to what extent they will

be pursued in the laboratory depends on the particular conditions and context as well as on the values of teachers and the preferences of students. Some of the unique characteristics of practical work discussed in the following may help in making reasonable decisions about when and how to use the student laboratory in science teaching.

First Goal: Science Laboratories Provide Concrete Experiences and Opportunities to Confront Student Misconceptions.

Woolnough and Allsop (1985) devoted their book to arguing for separation between concept learning, which should be the function of nonlaboratory instruction, and process learning, problem solving, getting a "feel" for science, and developing scientific attitudes, which should be done in the laboratory without the constraints of particular contents.

It is our premise that concept learning may be significantly enhanced by laboratory experiences, and for some students (e.g., elementary school students who are at the concrete operational stage) the laboratory is absolutely necessary for meaningful learning of concepts as well as for the development of reasoning skills.

The laboratory is a place in which science students have experiences that interact with their existing conceptions and at the same time develop new concepts. Hence, the laboratory can be used as a means of identifying students' preconceptions, as well as a vehicle for extending or modifying such conceptions. The effectiveness of the laboratory in fulfilling both tasks—diagnosing and affecting conceptual change—depends on how the laboratory is used. By this, we mean the nature of the exercises and investigations, the way students interact with the teacher and with each other, and the role played by pre- and postlaboratory discussion (e.g., Friedler & Tamir, 1990; Hegarty-Hazel, 1990a; Tamir, 1977).

The laboratory may help in identifying student beliefs, alternative frameworks, and misconceptions in two ways: (1) observing, listening, and questioning students during their work reveals their thinking, and (2) conducting structured interviews based on their observations and manipulation of objects and apparatus provides a second source.

Friedler and Tamir (1984) reported results of a combination of structured and unstructured observations of students' work in the laboratory:

The following difficulties appeared quite frequently: a distorted understanding of fundamental concepts (such as: cell, enzyme); inability to link theoretical knowledge to observed phenomena; inability to distinguish the relevant from the irrelevant in the experiment; misleading associations and deficiencies in knowledge, especially of chemistry related to biological processes. (p. 92)

In extensive observations of practical work of 11- to 14-year-old students, Tasker (1981) found several misconceptions related to the processes of scientific investigation. For example, pupils often fail to understand the relationship between the purpose of investigation and the design of the experiment they carry out. Frequently, the students' view of the purpose of the task they perform is different from that of their teachers. Lawson, Abraham, and Renner (1989) found that "the learning cycle provides opportunity for students to reveal alternative beliefs" (p. 89). This learning takes place mainly at the exploration phase, which paves the way for the development of more adequate conceptions.

The potential of laboratory tasks for revealing children's reasoning patterns and for identifying their conceptions led Piaget to construct his clinical interviews around concrete objects, using tasks that served as the basis for the conversation and allowed the researcher to probe deeply into the thinking processes of the child. Researchers studying misconceptions have been quick to recognize that the clinical interview is a major research tool. Indeed, many student misconceptions have been identified through clinical interviews involving manipulation of concrete objects.

Nussbaum and Novak (1976) studied childrens' concepts of the earth by using a globe and a figure of a girl holding a rock. The figure was placed at different places on the globe, and for each position the students were asked to draw on a corresponding diagram the direction in which the rock would fall. Erickson (1979) studied 12-year-old children's conceptions of heat and temperature. In one task, the children had to place sugar, butter, metal, and naphthalene on a hot plate and observe what happened. Tibergheim and Delacote (1979) provided a dry cell, a bulb, and a wire and asked children to arrange them so that the bulb lit up. Many students 7 and 12 years old were unable to complete the task. Rather than their spoken comments, their observed attempts were the main source of information about their concept of electrical current. Another example is the use of an egg in salty water to identify students' conceptions of osmosis and diffusion (Tamir, Amir, & Friedler, 1988). Thus, a major task of science teachers is to find ways to help motivate students to change their conceptions and instructional strategies to effect the desired conceptual change.

Another important implication of the research on student concepts concerns a crucial component of laboratory work—namely, observation. Observation is not the objective, bias-free process assumed by inductivists and still conceived as such by naive people. It has been repeatedly demonstrated that observations are affected by the beliefs of the observer (e.g., Champagne, Klopfer, & Anderson, 1980; Gunstone & White, 1981). Sometimes students hold beliefs so strongly that they adjust the observation to fit them, rather than change their views on the basis of observations. Another important factor is the ability to distinguish between observation and inference. "This is a commonly difficult task but a crucial one if laboratories are to be validly used for learning from observation" (Gunstone & Champagne, 1990, p. 175).

Finally, several researchers have designed special laboratory exercises for the purpose of identifying students' ideas about particular concepts. Clement (1982) suggests that for diagnostic purposes qualitative rather than quantitative experiments should be used, because the qualitative investigation avoids the "noise" associated with formula and "number stuffing" and is

"an effective way of getting students to think about their own preconceptions" (p. 70).

The laboratory may offer the conditions and means for creating conceptual conflict that would pave the way for inducing the desired conceptual change. Nussbaum and Novick (1982) describe a case study that exemplifies the role the laboratory can play in achieving this important goal.

The learning sequence begins with the presentation of a flask containing air and an evacuating hand pump, whose operation is demonstrated. Students actually manipulate the pump and feel the sucking effect. They are told: Let us pretend that each of us can look at the air in the flask through "magic magnifying spectacles." What would you see? Here is a sheet with two drawings of the same flask. Please make a drawing of the air in the flask before operating the pump and another of the air remaining after we push the pump and remove half of it.

The teacher, moving about, selects different types of drawings, which are then reproduced on the chalkboard. Students explain their reasons for the way they have made the drawings. Then a class debate takes place, followed by a class poll about each drawing. The lesson ends by the teacher demonstrating how air can be compressed in a syringe. The question is posed: How can the air in the cylinder be compressed into half of its original space? In the next lesson, students work in small groups trying to reach consensus. Finally, with the appropriate teacher guidance, the class is ready for the teacher's statement that the air is composed of particles and empty spaces around them. The students understand that when the air is compressed the particles get closer to each other (Nussbaum & Novick, 1982, pp. 190–197). In the trials reported by the authors, a high level of understanding has been attained. The crucial role of the laboratory-based "exposing event" is strongly acknowledged.

A kind of misconception that is peculiar to biology is related to what might be regarded as a satisfactory causal explanation of events, phenomena, and processes pertaining to living organisms. A large proportion of high school students have been repeatedly found to confuse causal and teleological explanations (e.g., Bartov, 1978; Jungwirth, 1975; Tamir, 1985; Zohar & Tamir, 1990). Teleological explanations imply means-end relationships such that presumably the purpose is identified as the cause. These often involve anthropomorphism i.e., assuming the possession of conscious, intelligent aiming at desired outcomes), which, according to the accepted scientific view, cannot be attributed to nonhumans. Regardless of whether one supports the use of teleological formulations (see Zohar & Tamir, 1991, for possible justifications for such support) or rejects such formulations altogether, there is a consensus about the importance of being able to distinguish causal from teleological explanations.

Bartov (1978) developed a sequence of five lessons that were found to be highly effective in teaching 600 grade 10 students how to distinguish between causal and teleological explanations. The instructional strategy was to create conceptual conflict by demonstrating responses of organisms that are useless or even harmful. This refutes the hypothesis that certain behaviors are caused by their usefulness, while supporting the alternative hypothesis that these behaviors are brought about by some causal mechanisms. Two experiments were conducted. In the first, electrical, mechanical, or chemical simulation applied to a pithed frog and to isolated frog legs resulted in reflex withdrawal of the legs, even though this was of no use to the dead frog. The second experiment, was performed with twigs of pea seedlings, causing them to bend away from the window or in a direction opposite to the source of light. This deprives the seedling of light, which is essential for photosynthesis. These two experiments, in which students could observe behaviors that work against desirable outcomes, were found much more powerful than any verbal persuasion in helping students understand the difference between causal and teleogical explanations in a variety of test situations.

Lawson et al. (1989) argue that the use of the learning cycle not only provides opportunities for students to reveal their beliefs and conceptions but also, allows "students to test their beliefs thus developing more adequate conceptions and thinking patterns" (p. 89). Moreover, the learning cycle, with its emphasis on laboratory work, develops reasoning and process skills, as well as the concepts underlying scientific research, such as controlled experiment (p. 93), inference (p. 97), hypothesis (p. 99), and prediction (p. 101).

We have already mentioned the importance of explicit teaching of concepts underlying scientific research, such as distinguishing between observation and inference (Gunstone & Champagne, 1990) or between hypothesis and assumption (Friedler & Tamir, 1986). The qualitative experiments suggested by Clement (1982) can be used not only for diagnostic purposes but also to induce conceptual change. Gunstone and Champagne (1990) note:

What is being argued here is that there are times when laboratory focus ... the active processing and interpreting of observations can be more readily fostered by the teacher. This fostering depends on considerably more time than is usually being given to intellectual interaction with observations and their consequences. (p. 178)

To argue for more time in the laboratory to be spent on interaction and reflection is to argue for the use of discussion. This is a teaching strategy which has been widely under used in laboratories (p. 179)

Finally, the conceptions of students, if properly utilized, may enhance positive attitudes toward laboratory work. Frequently, students who are asked to test hypotheses prescribed by the teacher or the text see no reason to test these hypotheses other than to satisfy the teacher. However, if different students in a class put forward alternative explanations (i.e., hypotheses) for a certain phenomenon, the experiments they design to test these hypotheses have a sense of reality because they may indeed determine whose hypothesis is supported and whose has been refuted (Driver, 1983). According to Gunstone and Champagne (1990), it might be expected under these circumstances.

That observations may be uniform through a group and that the interpretations of observations will be even more varied. Active reconstruction of personal views is likely to be required by the students.... Consequently moving from the data to an accepted conclusion is likely to be different, time consuming and to demand considerable effort on the part of the teacher. (p. 1721)

Second Goal: Science Laboratories Provide Opportunities for Data Manipulation Through the Use of Microcomputers

Microcomputers are gradually finding their way into education in general and into science education in particular. The attributes of the computer are its ability to store information, to process the information rapidly according to the user's requirements, to display the information in a variety of representations, and to carry on an interactive dialogue with the user (Walker, 1983). Thus, learning with computers in laboratories is active and interactive.

At present, the research evidence on the effectiveness of using microcomputers in schools is unclear (Becker, 1987). The available research findings indicate that in general there is little difference in achievement between users and nonusers, but the former need less time to reach the same level of performance. So far the positive impact of computers was found to be greater in elementary schools than in secondary schools. Finally, integration of computers with other instructional approaches was more effective than using microcomputers as the sole means of instruction (Roblyer, 1985).

We believe that the microcomputer can make an important contribution to the learning traditionally associated with the student laboratory. Based on our review of the literature, we describe a number of ways in which computers have been used in the context of student learning in the laboratory. Even though research is limited, the examples we cite indicate that the potential is enormous.

In research laboratories, computers are increasingly being interfaced with laboratory equipment to control experiments, collect experimental data, and construct graphical or statistical analyses of data obtained in observations and experiments. Integrating the computer with laboratory equipment has received substantial attention in physics (Millar & Underwood, 1984), biology (Adams & Baker, 1986; Watson, 1984), and chemistry (Warren, 1984). The collection of data is carried through sensors or probe to the computer by an interface that transforms analog data into digital signs. This process enables the researcher to see the measurements taken as the process under investigation is evolving. Moreover, the computer can show the measurements in tables and graphs and perform calculations and statistical analysis promptly. Thus, a microcomputer-based laboratory makes the following contributions to learning (Nachmias, 1989):

1. It provides opportunities for viewing a complete process rather than discrete phases of the process, whereas in an ordinary laboratory the whole picture is drawn through synthesis of separate observations.
2. The phenomenon under investigation is represented by a graph that appears on the screen. This enables the student to construct a bridge between the phenomenon and its formal representation. This association is important because research has shown that delayed data analysis, which is common in an ordinary laboratory, influences adversely the quality of conclusions (Driver, 1982).

3. The computer enables the student to observe the same phenomenon in multiple representations. For example, in an experiment with a moving car, the screen shows the car as it is moving, a graph showing velocity as a function of time, and another graph showing acceleration. By viewing the three representations simultaneously, the student acquires better insight into the phenomenon as well as a better grasp of concepts such as constant velocity and constant acceleration.

4. Because several sensors may be connected simultaneously to the microcomputer, it is possible to take several measurements at the same time. For example, the following measurements were taken in an aquarium during a period of 24 hours: oxygen concentration, pH, water temperature, light intensity, and rate of fish movement.

5. The role played by students in the school laboratory is often similar to that of a researcher. Because most of the technical work in a microcomputer-based laboratory (MBL) is done by the computer, the student is free to think and to solve problems without being overloaded with technicalities.

6. The continuous interaction between the student and the computer during the computer-based investigation may be used as a means for identifying misconceptions and sometimes as an aid in bringing about conceptual change.

7. The very departure from ordinary laboratory work may have a motivating effect on the students and improve their attitudes toward science.

Nachmias (1989) describes a series of six studies carried out over a period of 2 years to study the development of students' graphing skills, content achievement, development of scientific reasoning, evaluation of information presented by the computer, and problem solving in MBLs and ordinary laboratories. Because of its educational significance and exemplary design, we describe in some detail the study dealing with the development of scientific reasoning (Friedler, Nachmias, & Linn, 1990). The study was designed to test the effect of computer-based laboratory activities specially designed to foster the development of observation *or* prediction skills on the performance of eighth-grade students in four classrooms. All the students studied a semester-long program about heat and temperature. However, two classes engaged in activities designed to foster observation skills, while the two other classes engaged in activities designed to develop prediction skills. All classes were taught by the same teacher. The effects of the treatments were measured using the Discover Final Test, based on the Discover game, which measures observation and prediction skills in a subject-matter-free or domain-general context. The Scientific Reasoning Test confronted the students with a novel problem—the variables that affect the cooling rate of a swimming pool. The students were required to predict, justify the predictions, design an experiment, carry out the experiment, make observations, draw conclusions, and apply the findings in the form of practical recommendations for the best solution to the stated problem. This test was considered to be domain specific because it was administered at the end of a semester-long study of temperature and energy. The Subject Matter Test designed by Songer and Linn (1986) assessed students' understanding of the influence of different variables on the heating and cooling processes. In addition, classroom observations and interviews with individual students regarding the instructional program were conducted. The program was successful in the sense that each treatment group performed significantly better than its counterpart on the skills (observation vs. prediction) emphasized by the treatment in both domain-general and domain-specific contexts. At the same time, the two treatment groups did not differ from each other in mastery of subject-matter knowledge and other scientific reasoning skills. The findings demonstrated differences in the use of reasoning skills in general compared with specific contexts and showed that the possession of certain reasoning skills does not guarantee that these skills are applied in problem solving unless specific instructional measures are taken to foster such application.

New, improved courseware is expected to foster this trend of incorporating technological innovations in the laboratory learning environment as catalysts for better, more effective instructional methods. Conducting graphical and statistical analysis of data obtained by observations and experiments and stored in the computer for later analysis is another use of computers in the laboratory.

An example is the analysis of data stored in a spreadsheet to see whether rats fed on different diets gained weight differently (Cooler, 1985). Other examples are described by Arons (1984) with regard to physics and Crovello (1982) with regard to measurements from field studies and data collected in the laboratory about population dynamics. Data bases may be used in various science-related areas. For example, the GEOCOLOGY Project at Oak Ridge National Laboratory makes available information on environmental, climatic, and societal characteristics that can easily be retrieved and utilized by students (Olson, Emerson, & Nungessner, 1980).

An impressive example is a study unit developed in Australia—BIRDS OF ANTARCTICA (Anderson, 1984). In 1982, one task of the Australian National Antarctic Research Expedition was to record all sightings of bird life. The data base comprises all the observations of seabirds made by scientists during the voyage, together with meteorological information, time, date, and ship's position and activity. Interrogation of the data base allows the user to answer questions such as, In what latitude ranges, or in what conditions of density, are cape petrels observed? The student can work as a scientist, building up a picture of the habitat and behavior of seabirds. The information can be displayed in histograms, bar charts, or scattergrams, with the option of several descriptive statistics. Using this computerized research data base with eleventh- and twelfth-grade students, Maor (1990) identified two significant groups of inquiry skills. One is related to manipulating the data base and retrieving information; the other is related specifically to the formulation of questions or hypotheses that are considered as higher-level skills. Her results show that students developed inquiry skills only after they acquired practical skills. The students who mastered some of the practical skills and easily manipulated the data base used this source of information in a more effective way to develop inquiry skills and were able to apply them in other related areas inferred from the data base.

Various instructional strategies involving computers may be

used in the laboratory. Drill and practice is undoubtedly the most common use of microcomputers in school. As observed by Arons (1984),

Efficient and well planned drill, presented on an individual basis with immediate feedback reinforcing correct responses and correcting mistakes is a powerful instructional device. It is important in helping the student build bases of vocabulary and factual knowledge that underlie subsequent thinking, reasoning, studying, and problem solving. (p. 1051)

In the laboratory, drill and practice is useful in developing an overview of technical skill routines as well as skills in processing experimental data (Prosser & Tamir, 1990).

Instructional dialogue, or "Socratic dialogue" is a program that uses "a series of questions to lead a student through lines of reasoning to insights and conclusions" (Arons, 1984, p. 1052). The dialogues are conducted through a sequence of multiple-choice questions that have "yes" or "no" answers. Most questions require response in words, symbols, or numbers chosen by the student. Graphic displays are used throughout, and the student uses a built-in pointer to indicate places in diagrams in answer to some questions or to indicate how to construct appropriate graphs. A correct answer allows the student to continue in the main sequence. An incorrect answer is dealt with in a remedial sequence. From the user's point of view, these dialogues are similar to expert systems that can lead a medical student through symptoms and test results to diagnosis of an illness. The following three examples are described by Arons (1984): HEAT, which teaches the difference between heat and temperature and leads students to an operational definition of transfer of heat; BATTERIES AND BULBS, which involves the student in a series of simulated experiments integrated with a dialogue leading to understanding of current, circuit, resistance, and other related concepts; and OBSERVATION AND INFERENCE, which uses interesting tasks to teach students the difference between observation and inference. Arons asserts that instructional dialogues have a great potential for helping students develop their reasoning skills.

Marin, Friedler, and Tamir (1990) developed an inquiry-oriented test involving factors that affect enzymatic reactions, which resembles the inquiry-oriented laboratory tests used in the matriculation examinations in Israel (Tamir, 1974). The authors report their intention to design an integrated laboratory-microcomputer test, in which certain steps will be performed with the computer and others require hands-on activities.

Simulation, or the use of microcomputers to simulate an operation system, constitutes one of the most powerful potential functions of the computer in science teaching. In addition to their potential for bringing about meaningful learning of concepts, simulations offer unique opportunities for developing and practicing problem-solving and decision-making skills in a variety of contexts. As Anderson (1984) noted, simulations

usually permit users to examine some aspects of the real world under controlled conditions, they enable the study of variables which might otherwise be inaccessible, and they promote a range of educational goals. (p. 62)

Becase of limitations of time and experimental skill, students are rarely engaged in the whole series of repetitive experiments that are conducted by a research scientist. With the computer program controlling the complexity of the system, students can undertake such "real" investigations and learn to appreciate the significance of the various stages of scientific research independent of the complex and often confusing aspects of field or laboratory work. Simulations are particularly useful for laboratory exercises that require sophisticated and expensive equipment or long periods of time.

Whereas many simulations have been designed to replace ordinary laboratories, others are intended to supplement routine laboratory work (Wood, 1984). Butler and Griffin (1979) describe a laboratory course in chemistry in which they integrated a number of simulations. Hegarty-Hazel (1990c) argues that reading and mental rehearsal before physically attempting technical skills, such as pipetting, enhance the acquisition of the skills. This is an example of the sorts of laboratory tasks to which appropriate simulations can make a substantial contribution. Computer simulations can also contribute to the development of observation and inquiry skills, thereby enhancing student investigative and problem-solving competencies in the real laboratory.

CATLAB (Kinnear, 1982a) is an example of a simulated laboratory which offers learning opportunities in that students have an access in an ordinary laboratory setting. CATLAB

simulates the inheritance of coat color and coat pattern in the domestic cat. The program is open-ended and provides genetically valid outcomes from user-selected matings. The problem to be investigated, the starting point, the sequence of the investigation and the finishing point can be defined and controlled by the students. Manipulative errors that sometimes swamp and confound laboratory investigations are not encountered and the time constraints of genetics experiments that preclude repeat matings are absent. (p. 90)

In this program and similar ones, such as BIRDBREED (Kinnear, 1982b), the students' task is qualitatively different from the usual textbook-style problem. Textbook-style problems typically contain a clear statement of the initial state and of the terminal goals. All data required are given as part of the problems. Thus, the students' task is often reduced to deciding which rule should be applied. In contrast, in the microcomputer simulation students are given only a statement of the terminal state. Students need to decide on the required data and define a strategy to generate these data from the microcomputer data base, to judge their adequacy and significance, and, if necessary, to modify their strategy or generate additional data in light of the results obtained. These genetic simulations provide students with experiences that they typically do not have in their regular courses (Barnato & Barrett, 1981). Hounshell and Hill (1989) suggested that simulations allow analysis of genetic characteristics of many generations during a single laboratory period. The simulations allow students to be free from "time-consuming procedures." This time can be used for learning other important concepts in genetics.

Computers have been used for simulation of processes such as diffusion, osmosis, and mitotic division (Hounshell & Hill,

1989), as well as for investigating population-growth curves (Lazarowitz & Huppert, 1988). In their study, Lazarowitz and Huppert (1988) show that while doing computer simulations, students make decisions about what data to enter, what dilutions should be made, and what errors have been made in their simulated experiments. The students, who worked either in groups of two or three or individually, were able to progress at their own pace, an important aspect for the gifted learners as well as for the slow ones.

The role of a simulation is not to replace a laboratory experiment but to give students an opportunity to gain more experience with the manipulation of variables than is possible in real experiments. Students should be active during the simulation by writing down their hypotheses, collecting data through the computer, and presenting the results in graphs or tables (Dennis, 1979; Lazarowitz & Huppert, 1988; Lunneta & Hofstein, 1981; Marks, 1982; Okey, Shaw, & Waugh, 1983; Roberts, Anderson, Dean, Garet, & Shaffer, 1983).

Unfortunately, although the microcomputer may be used to improve problem-solving and logical thinking skills, it is often used for mastery of facts by drill and practice (Graef, 1984). A software program, the "Growth Curve of Microorganisms," developed by Huppert and Lazarowitz (1985) and based on the recommendations of Graef (1984) and the approach of Daley and Hiller (1981), was integrated into the teaching of microorganisms in an existing tenth-grade biology curriculum. This software included simulated experiments in which the impact of several factors on population growth was investigated. Three microcomputers were available in the laboratory, and the students worked individually or in small groups on the simulation program. In the classroom, students learned the characteristics of the microorganisms, such as their structure, the life processes, and other related facts. The students became acquainted with the definition of population, generation time, lag phase, and exponential phase of growth of microorganisms. In the laboratory, students examined yeast cells under the light microscope, observing their reproduction and the budding process, and learning how to count cells on a hemocytometer, how to dilute a yeast cell culture, and how to calculate the number of cells in the sample. Performing single experiments, students were able to investigate separately the impact on the microorganism population growth curve of each independent variable such as temperature, nutrient concentration, and the initial number of cells in the culture. The simulation computer program enabled the students to perform many "experiments" in a short time and to investigate simultaneously the interaction of the factors on the growth curve. The microcomputer simulation was integrated into the sequence of the learning activities in the classroom and the laboratory and was performed at the student's own pace. This software was used in four studies in which the experimental group used the combination of Computer Assisted Lab™ (CAL) with classroom laboratory instruction and the control groups followed the same learning material without the CAL component. The main result was that kibbutz students who were concrete or transitional achieved significantly better in the experimental group. In a second study, the experimental group in a city school achieved signifi-

cantly higher mean scores than the control group. No evidence of differences between scores of boys and girls was found within each group.

In a third study, Lazarowitz and Yaakoby (1989a, 1989b) used the "Growth Curve of Microorganisms" software to investigate the impact of the classroom-laboratory work combined with CAL on students' inquiry skills. They assessed the inquiry skills with the Biology Test of Science Processes, developed by Royce (1979), which measures nine science-inquiry skills. The results indicated that the experimental group performed significantly better on inquiry skills, graph communication, interpreting data, and controlling variables. No evidence of differences was found between the performance scores of the boys and girls in the experimental group, but the mean scores of the girls in the experimental group were significantly higher than those of the girls in the control group on two subscales: interpreting data and controlling variables. The control group performed better on only one subscale, prediction.

The "Growth Curve of Microorganisms" was used by Lazarowitz and Shemesh (1989) in a study of the relationship between students' cognitive operational stage and their academic achievement while learning in a CAL combined with classroom and laboratory work. The results showed that within each group, the higher the students' cognitive operational stage, the higher their academic achievement. But when comparing the mean scores between the groups, the concrete and transitional operational students in the experimental group achieved significantly better than their counterparts in the control group. No evidence for differences in the mean scores of formal operational students in both groups were found. Thus, the evidence indicates that the CAL mode of instruction combined with laboratory work enables students at the concrete and transitional operational stages to achieve significantly better than their counterparts in the control group.

Hounshell and Hill (1989) studied the achievement of students who were exposed to a computer-enhanced biology course. In this course, microcomputers were used to expand, enrich, and supplement the laboratory and the lectures of an introductory course in biology (from cells through microbiology, genetics, plants, nutrition and energy, zoology, human anatomy and physiology, to ecology). The experimental classes used the computers about 70 to 80 percent of the laboratory time for experiments on the topics. The students were assessed with a comprehensive test of basic skills that included the following objectives: recognition, classification, quantification, interpretation of data, prediction of data, hypothesis evaluation, and design analysis. The content areas were botany, zoology, and ecology. Their results showed that the experimental group achieved significantly higher than students in the control group.

Wainwright (1989) reported no evidence that the use of microcomputers contributed to more effective learning of chemistry concepts for non–formal-level students. Formal-level students in both groups also did not differ in their achievement. Wainwright (1989) explains that the inability to learn the concepts effectively through the computer may be due to the "excessive information-load demand" in working memory. Another explanation given is that both groups were assessed with a written achievement test, which was much more similar to worksheets used with the control group than to the computer activities.

In a meta-analysis study, Wise and Okey (1983) found an effect size of +82 on student performance when microcomputers were used in instruction. The use of computer-assisted instruction has been found to improve the academic performance of below-average and average students in middle schools (Becker, 1987). Improvement in achievement associated with the introduction of CALs has been reported by Hallworth and Brehner (1980), Burns and Bozeman (1980), and Lazarowitz and Huppert (1988). Higher achievement by students taught with microcomputer simulations, laboratory activities, and a combination of these two strategies versus conventional classroom instruction was described in a study by Shaw and Okey (1985), who found differences in achievement among students with different levels of logical reasoning ability.

Other studies have shown that computer-simulated experiments were as effective as "hands-on" experiments in terms of performance on achievement tests for both boys and girls and that retention was better in boys than in girls (Soon Choi & Gennaro, 1987). Computer simulations enhanced active involvement in the learning process and helped students meet the learning unit objectives (Rivers & Vockell, 1987). Anderson (1987) reported that girls did better than boys in programming areas in which problems were presented verbally rather than mathematically.

In summary, the CAL method of instruction was shown to be an effective method for providing students with the opportunity to manipulate simultaneously three independent variables in simulated experiments (Lazarowitz & Huppert, 1988; Lazarowitz, Yaakoby, & Huppert, 1990).

Microcomputers can be used as tutoring devices. In this mode, users teach the computer rather than being tutored. Here the user must communicate with the computer to program it in a language it understands. As programmer, the student assumes responsibility for learning, which, according to Papert (1980), makes learning qualitatively different. The best example is the LOGO project developed at the Massachusetts Institute of Technology. De Sessa (1982) describes how LOGO is used in learning certain concepts of elementary physics. In this program students control the motion of the "turtle" by "pushing" it with forces of specified direction and magnitude. The turtle then moves on the screen according to the laws of Newtonian physics as if it were on a frictionless surface. According to De Sessa, students who undergo such experiences develop an intuitive understanding of elementary mechanics that is difficult to obtain in traditional learning environments. Similarly Clement (1985) found that students who had experienced LOGO had a significant gain in reflective thinking and creativity.

Although learning LOGO certainly offers unique experiences, Hartley and Lowell (1977) noted that

it should be noted that the dialog is largely one sided with the computer responding to the instructions of the student and providing feedback on syntax errors, or displaying the output of a satisfactory program. The student is strictly limited to the command and syntax structure of the language and any instruction on formulating the problem or on methods of generalizing or correcting programs has to be supplied by the teacher or by the student himself. (p. 42)

Much more work on development and research is needed in using this approach and finding out about its effects and potential. Some of this has been done with college students. The use of microcomputers in college science laboratory courses, according to Leonard (1990), is in two major areas: (1) direct instruction of laboratory concepts by simulation using traditional computer-assisted instruction or advanced versions of computer-assisted instruction using an interactive video disk system and (2) use of the microcomputer for analysis of data input with laboratory instrumentation interfaced with the microcomputer.

Leonard (1990) reported that students who used computer-simulated experiments in laboratory investigations of kinetics, absorbance spectroscopy, emission spectroscopy, and equilibrium achieved as well as or better than students who performed the laboratory investigation without Computer-assisted instruction on the same topic. The group who used the computers spent less time learning the material. In another study, Curtis (1986) used a software system to teach students how to fit simple response functions to experimental data. Curtis found that students with low and low-average mathematical skills benefited from it more than students with higher skills.

Leonard (1990) found that microcomputers interfaced with laser video disk players permit a high level of activity between the student and the computer and provide high-resolution, lifelike video images of natural phenomena. However, Stevens' (1985) findings indicated no difference in gains between a group who used interactive video disk and a group who had conventional laboratories and teaching of physical principles of standing waves and strings. When students who studied equilibrium in chemistry using simulation with an interactive video disk system were compared with students using conventional laboratory activies, the former group achieved significantly higher on laboratory reports and tests (Waugh, 1987). According to Leonard (1987, 1988), students felt that videodisk instruction gave them more experimental and procedural options and they used the instructional time more efficiently. The interactive videodisk increased their interest, helped them to understand basic principles, improved their achievement, and modified their attitudes toward science.

Interfacing the physiological instruments with the microcomputer, Nicklin (1985) found that many physiological experiments were improved and made more accurate. When the microcomputer-based workstations were interfaced with old kynograph transducers in physiology laboratories, Rhodes (1986) found that they were more functional and successful. Morgan, Markell, and Feller (1987) developed a complete description of interfacing physiology measuring devices with a microcomputer. An guide to constructing interfaces for laboratory instruments such as thermistors, motion timer, pH meters and humidity meters has been described by Vernier (1987).

The advantages of using instruments interfaced with a microcomputer in the laboratory were summarized by Leonard

(1990) as reducing cost, improving effectiveness, saving time, and providing an opportunity to learn to use state-of-the-art scientific instrumentation. Using this approach to laboratory work, data analysis becomes simpler, experimental results are more meaningful and students can perceive more easily the relationship between dependent and independent variables in a study (Leonard, 1988).

In summary, in this section we have described the potential of microcomputers to help the educational enterprise in attaining goals associated with the student laboratory. It is important, however, to realize that microcomputers are no substitute for actual laboratory work. Students need the opportunity to engage directly with materials, organisms, and natural phenomena. An attempt to use field experiments has indicated that, regardless of cognitive achievement, most students would strongly resist replacement of laboratory work by vicarious experiences (Ben Zvi et al., 1977). Stein, Nachmias, and Friedler (1990) compared the learning process and the achievement of students working in ordinary laboratories with those of students who had performed an identical investigation in a microcomputer-based laboratory. They found no evidence of differences in the outcomes, quality of graphs, understanding of concepts, and conclusions drawn. However, students in the ordinary laboratory took twice as much time as those in the microcomputer based lab. Unfortunately, the time saved was not used for reflection and substantive discussion but was wasted on procedures and techniques. Moreover, there were some indications that students who had made their own graphs attended to the various features of the graph more carefully than their counterparts who had used graphs made by the computer.

An interesting and potentially sobering finding is that many students tend to view any information presented by the computer as accurate and consequently fail to recognize flaws in the data (Nachmias, 1989).

While simulation programs can provide illustrations which it is difficult to give by any other means, and although teaching dialogs can allow more extending types of exercises, the student cannot in any real sense place his own construction on the problem and set out his own methods of solution. Practically the program languages which are used do not permit this (Hartley & Lowell, 1977, cited in Walker & Hess, 1984, p. 42).

In order to overcome this limitation, Huppert and Lazarowitz (1985) designed software that enables students to decide by themselves the initial number of cells to start population growth, what nutrient concentrations to work with, and the appropriate range of temperatures to use in the simulated experiments. Unrealistic decisions concerning these variables immediately provide results that do not allow the students to go further; for example, the simulated hemocytometer can contain cells that are so crowded that nobody can count them.

With regard to laboratory applications, Arons (1984) cautioned that if the computer short-circuits insight or if it simply makes available end results for analysis or "confirmation," it is educationally sterile or even deleterious, particularly in introductory courses. Although the computer can relieve students from tedious numerical calculations and its graphic display offers considerable advantages over programmable hand calculators, "where extensive numerical computation is done without direction, students' attention to analysis and interpretation of methods and results, the effort is largely wasted" (Arons, 1984, p. 1051) and may promote rote learning. With regard to simulations, Arons (1984) cautions that:

Some of the effort being devoted to simulations is producing undesirable materials. As a rule, if the phenomena are readily accessible, then it is better to expose the beginning students to the actual phenomena than to simulation thereof. (p. 1052)

This implies that students benefit from computer simulations after they have had some experience with the real phenomena. Continuous learning with computers might cultivate dependence on the reinforcement provided by the computer so that students would not be able to engage in genuine independent learning, ask their own questions, or check their own reasoning for internal consistency.

Further studies of this displacement of laboratories by computers are needed. Does classroom instruction provide contact with distinguished intellects; motivate through human relations; encourage conversation, argument, and discussion; and create a unique social system that can be neither substituted nor simulated by computers?

Third Goal: Science Laboratories Provide Opportunities for Developing Skills in Logical Thinking and Organization, for Example, Through Science-Technology-Society Issues

The science, technology, and society (STS) movement, which began in the 1970s, gained considerable momentum in the 1980s and seems to continue its expansion in the 1990s. It developed because of disappointment in the weak impact of science programs of the 1960s (e.g., Stake & Easley, 1978; Welch, Klopfer, Aikenhead, & Robinson, 1981) and their failure to meet the needs of many students, especially the lower half in terms of intellectual abilities and socioeconomic status. A second reason is recognition of the importance of scientific literacy for living in our technological society.

STS is included in the school curriculum as a complete course extending over a whole semester or even over an entire year—for example, the BSCS (1991) elementary science course "Science for Life and Living." More frequently, however, STS instruction has been confined to specific topic modules that can be incorporated in the regular science program. For example, the Science and Technology in Society curriculum developed under the auspices of the British Association for Science Education includes a large number of resource units designed to be quick and easy to use, occupying about 75 minutes of classroom time. The units are designed to actively involve students by demanding more than just a passive role of reader or listener. Thus, discussion questions, simulations, role-playing, data analysis, and decision-making exercises are involved (Holman, 1986, p. 2).

Note that the laboratory has not been mentioned in this connection. Indeed, Woolnough and Allsop (1985) stated that issues related to science and society would be trivialized if they were tied to the practical work that students were able to do. "An introduction to the way that decisions are made in our democratic society needs to be tackled more directly—through discussion, simulation, case studies and games" (p. 78). Examination of topics included in STS programs suggests that for most of them the claim is valid. For example, the topics include recycling aluminum, test-tube babies, the heart pacemaker, energy from biomass, robots at work, fluoridation of the water supply, noise risks, and nuclear power.

However, STS instruction can also be used to enrich and make laboratory work more relevant and more attractive to the student. Fensham (1990) suggests that science teachers choose real-world problems that may be solved by a combination of science process skills and scientific and technological knowledge. For example "school science could include extensive investigations into the claims and relative merits of products that are relevant to its various age levels" (p. 306). Fensham points out that cooperative activities in the laboratory that have obviously human-related outcomes are being suggested as appropriate types of practical work in secondary schools if girls are to be more attracted to science. Fensham concludes his advocacy for science for all by stating that "practical work is likely to be quite critical if this movement is to have any success" (p. 309).

A number of other science educators advocate that laboratories use common phenomena and materials familiar in everyday life. For example, Gunstone and Champagne (1990) note that one of the most common everyday experiences involving chemical change is cooking. Yet these chemical changes rarely find a place in laboratory investigations. Furthermore, it is uncommon for students even to regard cooking as an example of chemical change (p. 169). White (1988) argues that "if substantial proportion of laboratory investigations used common materials instead of things never encountered elsewhere, it will be possible to meet the aim for science to illuminate people's lives" (p. 190).

Rubin and Tamir (1988) demonstrated that the study of everyday phenomena in the laboratory could serve as an advance organizer for developing junior, middle, and high school students' understanding and ability to apply cognitive strategies and inquiry skills. They found that students who had received the advance-organizer treatment outscored students who had followed the regular laboratory investigations on various achievement and attitude measures. Moreover, the teachers reported that learning sequences with everyday phenomena were most helpful.

Research studies that focus directly on the role and characteristics of practical work in the laboratory in STS-oriented courses are not widely reported. Some indirect evidence comes from the National Science Teachers Association (NSTA) study of exemplary programs, in which "we find students studying issues—often related to energy, population, natural resources, the environment or sociology—collecting data, writing reports and taking action" (Penick, 1986, p. 1).

Research is needed on the relative effectiveness of STS-oriented laboratories and the effectiveness of STS-oriented laboratories with other STS activities. Finally, research is needed to determine the optimal integration of STS-oriented laboratories with other STS instructional strategies.

Fourth Goal: Science Laboratories Provide Opportunities for Building and Communicating Values Concerning the Nature of Science

In a comprehensive review, Layton (1990) stated:

The general thesis is that the philosophy of science has rarely been used in a systematic and deliberate manner as a prime source of objectives for student laboratory work. Instead, it has been resource drawn upon selectivity—raided even—to underwrite purposes and practices which have their origins in considerations remote for philosophy. As a result, student laboratory practice has been reflective of dubious or discarded philosophies of science. (p. 37)

As early as 1973, Shulman and Tamir noted that

... works of general philosophers and historians of science such as Kuhn (1970) are frequently cited ... to clarify the domain that students are to understand.... Philosophical studies of this sort are used to justify a particular choice of educational objective. (p. 1105)

Using content analysis, Jacoby and Spargo (1989) reviewed 16 physics texts and found that "with the exception of one textbook, namely, the Project Physics course, all the texts examined revealed a predominance of an inductivist-empiricist approach" (p. 45).

During the 1960s and 1970s many evaluation studies of the "new" curricula administered tests and inventories designed to assess students' understanding of the nature of science—example, Test on Understanding Science (TOUS), (Cooley & Klopfer, 1961) and the *Science Process* Inventory (SPI) (Welch & Pella, 1967–1968)—and scientific attitudes such as curiosity (e.g., Campbell, 1972). We could locate only one study that attempted to determine the impact of the laboratory on understanding the nature of science, but no evidence of such an impact was reported (Yager, Englen, & Snider, 1969). Many studies compared the effects of different curricula. For example, Ramsey and Howe (1969) concluded in their review that, based on TOUS, a multireference laboratory-centered approach to biology will produce greater student growth in understanding the scientific enterprise. Some other studies (e.g., Jungwirth, 1970; Trent, 1965) found no evidence of differences in performance in TOUS between students in "new" and "conventional" programs. The Science Process Inventory (SPI) was designed with 135 agree or disagree statements that describe assumptions, activities, products, and ethics of science. Using SPI with high school students in Israel, some indirect evidence regarding the positive impact of the laboratory on understanding the nature of science processes was found. The new biology curriculum, which relied heavily on students' work in the laboratory, resulted in higher SPI scores than the less laboratory-oriented courses in chemistry and physics (Tamir, 1972). In a study of more than 3000 students in grades 9 to 12, major misconceptions were found regarding interrelationships among assumptions, hypotheses, and theories; the role of scientists in creating models and classification schemes; the plurality of scientific methods and approaches; the nature of experiment and control; and the relationship between scientific research and practical application.

Although there are differences in opinion about aspects of the nature of science (Hodson, 1988b; Layton, 1990), it appears that, at the high school level, agreement would far exceed the disagreements. Hence, the poor performance on particular items of TOUS and SPI should be taken seriously. Perhaps measures such as concept maps, Vee diagrams (Novak, 1990), and more direct treatment of basic concepts of scientific research (Friedler & Tamir, 1986) would improve the present deficiencies. New emphases such as STS may require some revision and supplementation of instruments such as SPI, but understanding the nature of science continues to be a major goal of science education. This assertion applies to the importance of scientific attitudes, as well. Instruments such as those developed by Moore and Sutman (1970) and Vitrogan (1969) should be reconsidered for both student and curriculum evaluation with special reference to the effect of different kinds of laboratory experiences.

As part of their laboratory experiences, students are often required to write laboratory reports following a formal structure commonly used in textbook and scholarly journals. Medawar (1963) wrote a provocative article entitled "Is the scientific paper a fraud?" explaining why the scientific paper gives a distorted account of the nature of science. A similar view is expressed by Hodson (1988a):

The actual chronology of experiment and theory is rewritten in textbooks. This helps to sustain the myth that the path of science is certain and assigns a simple and clear cut role to experiment thereby assisting the perpetuation of further myths concerning experiments. . . . The *post hoc* description and justification of actions that interpret the experimenter's motives and idealize the decision making events in terms of currently held theories. (p. 57)

It has been suggested that indoctrinating young children in writing such formal reports is even more damaging than having them read such textbooks. According to the London Association for the Teaching of English (1969):

The written language used by the children must be their own expression of observations, ideas, conclusions. Teachers should allow a variety of responses: factual and imaginative; personal and impersonal. As children's thought develops they would be able to *choose* the kind of writing appropriate to the set task. (p. 128)

Providing the opportunity to become familiar with phenomena and the freedom to write laboratory reports in the style preferred by the student would make significant contributions to the development of personal knowledge and the enhancement of positive scientific attitudes, but many believe the best way to achieve these goals is to involve students in individual projects. Individual projects may be used at any level from lower elementary grades to graduate courses at the university and inservice education courses. However, the most widely used are projects at the senior high school level. In some courses in the United Kingdom, such as the Nuffield A-level biology, chemistry, and physics programs, projects are required of all students and special time is allocated for this purpose in the school timetable. In Israel, high school students specializing in biology in grades 11 and 12 are required to do an individual project in outdoor ecology. Although such projects may produce severe logistic problems and place a heavy burden on teachers, they are frequently considered by students as the peak learning experience of the entire school program (Tamir, 1989b). Project work is especially suitable for gifted students, students who have a special interest in science, and potential scientists.

The laboratory blocks of the Biological Sciences Curriculum Study (BSCS) in which the whole class is engaged in a research project illustrate the potential of project work. Metzner (1962) wrote:

It may be that for the gifted student the laboratory block will provide the motivation to engage in a related research problem. An objective of the block is to give students an appreciation of the need for careful quantitative observations, for application of inductive and deductive reasoning, and for experimental controls. They learn that conclusions in science are tentative, and they learn the need for suspended judgement. (p. 35)

Brandwein (1962) observed some 1000 teachers in biology over a period of 10 years. He found that "the science work in those schools that were successful in stimulating individual investigation seemed to have related patterns" (p. 46). Among these patterns, he mentions the following:

1. A teacher who has the energy and desire to work with individual students.
2. A clear emphasis on intellectual attainment in school.
3. Leaning toward the individual rather than group work in the laboratory. Demonstrations do not rob students of the right to discover.
4. An administration that is sympathetic to science and to individual investigations, shows active interest, and takes an active part in planning.

Based on his rich experience, Brandwein provided useful suggestions for how to guide students in their individual investigations from the difficult task of selecting a problem for investigation through routines in management, monitoring the progress of the investigation, and reporting the findings. He presented a concrete example of how students can use *Biological Investigations for High School Students*, a BSCS publication that includes 150 topics suitable for individual projects. Each topic is presented in the form of a prospectus giving the necessary background, suggested questions for investigation, suggested approaches, expected pitfalls, and suggested readings.

In Israel, the Ministry of Education encourages students of good standing in their school studies to replace one matriculation examination with an individual project on a topic of their choice. Students who choose to do such a project in science have the opportunity to benefit from all the aspects just mentioned. The majority of these students would specialize in the university in science or science-related fields, such as medicine or agriculture.

In an interesting study, Bliss (1990) compared students' reactions to undergraduate physics laboratory and project work. Rather than using an agree-disagree type questionnaire, students were interviewed and asked to tell about some time when learning felt particularly good or particularly bad. Of 271 accounts, 47% were about lectures, 16% about laboratories, 15% about individual work, 9.5% about tutorials, 8.5% about project work, and 4% about other areas such as examinations. When the reactions to the different experiences were compared, the projects came out very positive.

Common to both "good" projects and "good" laboratory stories are feelings of independence, freedom and responsibility. [They differ in] the intensity of the satisfactions produced by projects, by contrast with the more muted pleasure and enjoyment associated with "good" experiments. Projects are one of the rare learning situations where students do feel totally involved. . . . Project work presented students with a real challenge and often they experienced a tremendous sense of achievement. (p. 393)

It may be concluded that the goal of enhancing understanding of the nature of science, especially how scientists really work, and the associated goals of developing scientific attitudes and interest in science can be best achieved by project work, provided this work is carefully planned and adequately supervised. The more laboratory activities resemble project work, the higher the probability of achieving these goals.

In a conference on the History and Philosophy of Science in Science Teaching held in Tallahassee, Florida in 1989, the role of the history of science in helping students develop a realistic image of the scientific enterprise was strongly advocated (e.g., Matthews & Winchester, 1989). But how can the history of science be related to the student laboratory? Matthews presents the following example:

A sample lesson on the pendulum can be enhanced if historical matters about Galileo and his dispute with his Aristotelian opponents are introduced. . . . Galileo said of heavy and light pendula (iron and cork) that when set swinging they will remain in synchrony for a thousand oscillations. When students do the experiment, the pendula are out of synchrony within two dozen swings, a matter confirmed in replications of Galileo experiment. What does that tell us about Galileo, about the relationship of evidence to theory in science, about the role of rhetoric in science? . . . Students can only gain by being sensitive to these issues, and by being encouraged to think them through. (p. 11)

This is an example of performing classical experiments and comparing our present interpretation with those offered by scientists in the past. In addition to its contribution to the understanding of natural phenomena, the classical experiment is highly motivating because of the opportunity it offers to "stand on the shoulders of giants" and realize that many of their experiments can be replicated by the students. Thus the distorted image many students have of scientists (unusual persons wearing white gowns, working in isolation, and exhibiting extraordinary behavior) may be discarded, and students may realize that

scientists are ordinary persons (Kaufman, 1989) and that anyone who is creative or interested and ready to undertake the necessary studies may become a scientist.

Because students frequently hold intuitive conceptions that were held by scientists hundreds of years ago (e.g., Driver & Bell, 1986), it is reasonable that the same experiments that helped scientists to change their conceptions in the past, might be useful in effecting conceptual change in present-day students. A good example is that of van Helmont carrying out his famous experiment with a willow tree and concluding on the basis of his results after maintaining the tree for 5 years with just water that "Therefore 164 pounds of wood, bark and roots were formed out of water only" (BSCS, 1973, p. 76). Thus, Matthews and Winchester (1989) stated:

No one is suggesting an identity of individual and historical learning. [However,] understanding the obstacles to development in the history of science can in some measure throw light on problems in individual learning. (p. 9)

Through specific and fairly detailed examples, Arons (1988) illustrates how historical and philosophical elements can be infused into introductory physics courses. Some of his examples bear directly on the laboratory. For instance, in the classic experiment associated with André Ampère, students observe that two parallel wires attract each other when carrying electric current in the same direction. Most teachers are quick to assert that the interaction between wires is electromagnetic without reference to any other possibility. Arons suggests that "this quick and casual assertion destroys a significant learning experience for the student and marks the loss of a valuable pedagogical opportunity" (p. 17). He goes on to show that an alternative explanation could have been offered and how Ampère, who had to consider the two alternatives, arrived at the presently accepted explanation.

It may be concluded that occasional reference to the history of science in relation to laboratory work may enrich the laboratory experiences, add to the integration of theory and practice, and enhance understanding of the nature of science.

PART II: FACTORS THAT FACILITATE SUCCESS IN THE SCIENCE LABORATORY

Curricula, resources, learning environments, teaching effectiveness, and assessment strategies are factors shown by research studies to facilitate successful science laboratory instruction.

A. Curricula

People have different conceptions of what is meant by curriculum. One way to deal with the ambiguity is to identify distinct phases of curriculum, such as the intended, perceived, and implemented. The goals and objectives represent the *intended*. Teachers' and students' views reflect the *perceived* curriculum. Teaching, learning, and the learning environment describe the *implemented* curriculum. In this section we deal with the implemented curriculum, namely instructional materials, selected activities, and integration of the laboratory with other learning experiences.

The nature of classroom transactions is strongly dependent on the curriculum materials, in (1) a laboratory manual consisting of a series of exercises or investigations that may or may not be integrated with nonlaboratory experiences or (2) worksheets or (3) a textbook that includes laboratory exercises. Worksheets are usually prepared by the teacher and consequently permit flexibility and matching with students' capabilities. However, teachers often do not have the time and resources available to authors of textbooks and laboratory manuals, and as a result the quality of teacher-designed exercises leaves much to be desired. It is not easy to design high-quality curriculum materials. It is especially demanding with regard to inquiry-oriented laboratory investigations because of the need to try the experiments to make sure they "work," as well as the importance of having "balanced" exercises in terms of their "cognitive challenge" (Gardner & Gauld, 1990, p. 141). Hazardous activities must be avoided, as well as exercises that may have negative affective connotations, such as dissections and inconsiderate experiments with live animals (Silberstein & Tamir, 1981). Because of these difficulties, teachers rarely invent laboratory exercises; instead, they adopt them from various sources. Because a major characteristic of the "new" curricula of the 1960s was extensive use of investigative laboratories, each of the projects, such as the BSCS, PSSC, and CHEM Study in the United States and the Nuffield projects in the United Kingdom, engaged special groups of scientists and teachers to develop and test laboratory exercises. In addition to the laboratory exercises that accompany the textbooks, the BSCS developed a series of "laboratory blocks" so that interested teachers could set aside 4 to 6 weeks for intensive laboratory-based work on a selected topic. The story of the development of these blocks illustrates the complexity of the task. According to Grobman (1969), the following procedure was used:

The prospective author prepared a rough draft of the Block. This . . . was received by members of Dr. Lee's staff in the Austin laboratory and they made estimate of the practicality of the laboratory work in terms of typical high school laboratory facilities. The manuscript was then criticized by members of the Laboratory Block Committee. . . . After the suggestions of the project associates had been incorporated, the Laboratory Block was tested in a number of schools and feedback information forwarded to the author. The Block was then rewritten and subjected to a second series of reviews and subsequent trials in additional schools. Based upon that experience a manuscript was finally prepared for commercial publication. Each Block was accompanied by a teacher's supplement. (p. 248)

In spite of the heavy investment in preparation of laboratory exercises, content analysis carried out by Herron (1971), Tamir and Lunetta (1978), and Lunetta and Tamir (1981) revealed serious deficiencies in these exercises in terms of opportunities to practice and develop major inquiry skills, such as defining a research problem, formulating hypotheses, planning experiments, and identifying limitations of an experiment. When the BSCS Yellow version was adapted for use in Israel, many of the original exercises were gradually deleted and the new investigations developed for the practical matriculation examinations were substituted (Tamir, 1990b).

In planning a course or a unit, special considerations relate to the laboratory. Two important considerations are time allocation and sequencing. The work in the laboratory is by nature time-consuming. In addition, the higher the level of inquiry or the lower the level of guidance provided for open-ended problem-solving investigations (see Shulman & Tamir, 1973, pp. 1112, 1114), the more time is needed to make the work meaningful. In the inquiry-oriented curricula of the 1960s it was often assumed that at least half of the class time should be spent on activities and laboratory exercises (e.g., Romey, 1969). Recently, the state of Texas legislated that at least 40 percent of class time in science be devoted to laboratory work (James, 1987). Interestingly, observations of laboratory classes indicated that at both the high school level (Friedler & Tamir, 1984) and the university level (Hegarty-Hazel, 1990) the highest inquiry level was associated with more time spent on discussion about management and on arrangements, preparation, and other organizational activities at the expense of discussion of processes and concepts. Hegarty-Hazel (1990a) suggests that "this could perhaps be due to (incorrectly) equating scientific inquiry with the now rather discredited notion of unguided student discovery" (p. 376). A similar variation was reported by Kyle, Penick, and Shymansky (1979). They found, for example, that the time spent on actual laboratory activities was 12 percent of the total time of the laboratory lessons in introductory geology, 36 percent in introductory botany, 34 percent in advanced botany, and 55 percent in advanced chemistry. Blosser (1988) observed that

The teacher has to put in time setting up the laboratory and maintaining it, assisting students in their investigations [and] reading and reviewing laboratory reports. . . . Ideally science teachers would like double class periods for laboratory activities. Such a time frame presents scheduling problems for administrations. (p. 57)

It may be concluded that time allocation and its effective use are major factors to consider in planning curricula for practical work in the laboratory.

There are problems of sequencing of activities within a particular laboratory period and integrating the laboratory with other instructional strategies. Blosser (1988) reminds us that according to Bruner, teachers should organize curriculum and instruction so that students first experience the material (enactive learning), then react to a concrete presentation of it (iconic learning) and finally symbolize it (symbolic learning). This kind of instructional organization fits nicely with the classroom protocol of introducing students to science concepts by having them conduct laboratory activities, discuss their laboratory work, and finally read background material so they can extend and solidify their learning. Such protocols also parallel the learning-cycle approach to science instruction. A laboratory lesson following the recommended protocol is likely to represent a guided-inquiry approach (Igelsrud & Leonard, 1988) rather than the verification approach characterizing many conventional curricula. Equally important is integration of the laboratory with the remaining components of the science course. For example, Igelsrud and Leonard (1988) noted:

> The 1987 edition of BSCS Green Version contains more than 40 laboratory investigations most of which employ the [guided inquiry] approach. Importantly, the investigations in this approach are placed in appropriate sections within each chapter of the textbook thus providing a systematic basis for the development of problem solving and reasoning skills. (p. 306)

Some empirical evidence for the benefits of appropriate integration of laboratory and test comes from the work of Boud et al. (1980), Quinlan (1981), and Zieleniecova (1984).

Learning in the laboratory and achieving its goals may be enhanced and supported by other strategies, such as vicarious experiences through media such as films, slides, and television; dry thought experiments (Reif & St. John, 1979); invitations to inquiry (Schwab, 1963); microcomputers (Fraser, Giddings, Griffith, Hofstein, & Herkle, 1990; Lunetta, 1974); individualized curricula such as the Australian Science Education Project (Edwards & Power, 1990) and personalized system of instruction (Dunn, 1986); and audiotutorial methods (Postlethwaite, Novak, & Murray, 1964).

Thus, the interrelationships between laboratory work and curriculum are clear. To a considerable extent, the laboratory determines the nature of the curriculum and the curriculum has a strong influence on the effectiveness of the laboratory.

B. Resources

Most science teaching from middle schools through higher levels is conducted within a context of special facilities, according to Archenhold, Jenkins, and Wood-Robinson (1978):

> The work that can be done in a laboratory is heavily dependent upon the use that may be made of the accommodation available.... Many of the items of equipment developed for school science teaching are expensive.... Some require regular and skilled maintenance.... The variety of chemical solutions and living organisms employed in school science teaching is much greater than a decade or so ago. (p. ii)

How important are the facilities? According to Showalter (1984), "Research can show that without adequate laboratory facilities and materials, most students cannot learn biology in any meaningful way" (p. 1). Englehard (1968) investigated the relationship between characteristics of the laboratory room and classroom transactions. He found that when all classes took place in the laboratory room, there was greater use of inquiry methods than when laboratory work was carried out in a separate place. Davis (1972) found that provision of equipment and materials had improved patterns of teaching science. In this study, science teachers reported that adequate supply of materials made teaching more convenient and more effective, increased the amount of students' experimental work, and enabled teachers to broaden the science curriculum. Ainley (1978) found that better facilities were associated with what students perceive as an enriched learning environment, namely greater involvement in purposeful activities and more stimulation to study. In contrast, the use of textbooks was independent of the quality of facilities. Similarly, comments by teachers suggested that adequate facilities led to greater use of laboratory work and exploration as a vehicle for learning. Several studies (e.g., Beisenherz & Olstad, 1980) found that lack of material, equipment, and facilities was a major impediment to laboratory activities.

The new curricula of the 1960s emphasized students' investigations in the laboratory. Where new curricula were introduced, there had been concurrent attempts to improve laboratory facilities. In Israel in the mid-1960s, for example, only schools that had adequately equipped laboratories were considered as candidates for adopting the Israel Adaptation of the BSCS program (Tamir, 1976a). Using the BSCS Biology Laboratory Facilities Checklist (Abraham, Novak, & Schaefer, 1965), a very substantial improvement in laboratory facilities was made in Israeli schools between 1966 and 1973 (Tamir, 1978). Similar findings were reported independently by Dreyfus (1979). Grobman (1963) found that in the United States adequacy of laboratory facilities contributed significantly to student achievement. Ainley (1990) concludes his review of laboratory facilities by stating that "it does appear that lack of resources can limit the way in which curricula based on extensive laboratory work are implemented in practice" (p. 237). Conversely, there is some evidence that removal of impediments can promote laboratory work. For example, the provision of easy access to laboratory materials, living organisms, and chemicals through special supply centers has been influential in promoting laboratory work in high school biology in Israel (Tamir, 1976b)

A comprehensive study of science facilities was carried out under the auspices of the NSTA in 140 selected schools throughout the United States (Novak, 1972). Conclusions and recommendations reflect trends toward flexibility, movable furniture, arrangements for individualized work, a large preparation-storage-distribution area, and carrels designed so that they can be used by two students.

An almost totally neglected area in research is the role of laboratory technicians or assistants and their impact on management and learning in the school laboratory. However, In the Second International Science Study (SISS), positive correlations were found in several countries between the extent of available assistance and the frequency of laboratory experiences, the level of inquiry in the laboratory, and students' achievements (Tamir & Doran, 1992).

Other resources for the science laboratory are the instructional materials teachers have. The planned format and procedures of the science laboratory must accommodate the activity and cognitive stage of the student, as well as provide the teachers with a laboratory framework that enables them to be available to individuals and small groups, while still being in control of the entire class. This means that the content and directions must be of the utmost clarity but must be structured so that they are essentially independent of the teacher's directions for most of the laboratory session. Furthermore, in proceeding from simple to complex, from the unknown to the observed, students in the concrete and transitional stages should be able to draw conclusions regarding their laboratory activity within a small group, and in interaction with their peers (Lazarowitz, Baird, Hertz-Lazarowitz, & Jenkins, 1985).

One might hypothesize that the science teacher and the developer of science curricula could facilitate achievement and interest of the students, particularly the concrete and transitional ones, by adapting the learning materials to the cognitive stage and style of the learner. Emphasis should be placed on laboratory work that allows the individual student to work in a cooperative atmosphere and yet affirms the need for individual advancement in understanding science and the use of process skills. This may provide stimuli for learners to develop positive attitudes toward the study of science and technology in high school.

Noelting (1980) noted that both learning and development must be considered when judging the importance of course experience in fostering a particular principle such as first-order proportionality. Various investigators, including Fennema and Sherman (1977, 1979) and Farrell and Farmer (1985), have found a positive relationship between course experience and first-order direct proportional reasoning. One may assume that the same is true for biology topics, like the pH scale and osmosis. First-order proportional reasoning might be considered necessary for a simple understanding of the pH scale, which relates a scale number to the concentration of the H^+ ion, and so would be within the range of understanding of most students. The presentation of the formula $pH = \log_{10}(H^+)$ would not be grasped because the student has not studied logarithms and has not yet incorporated first-order *inverse* proportions in his or her problem-solving strategies. Farell and Farmer (1985) found that even many college-bound adolescents were not able to use inverse proportions. In high school chemistry, students often do not deal with the pH scale and the mole concept until the eleventh grade. Yet the biology student continues to deal with cellular and physiological processes during the tenth grade, requiring at least an elementary understanding of the pH scale. Lazarowitz and Witenoff (1990) drew attention to the fact that certainly for upper-stream students, and for many lower-stream ones, inclusion of the pH scale in the ninth grade curriculum is possible. Lazarowitz and Witenoff (1990) formed that the girls in their sample were working against what might be considered "intellectual handicaps." The basic reasons for their general lower assessment are most complex. In relation to proportional reasoning, Karplus (1977), Farmer, Farrell, Clark, and McDonald (1982), and Farrel and Farmer (1985) found significant gender differences in favor of males in first-order direct proportional reasoning.

Harty and Beall (1984), who studied self-concepts and gender roles in relation to achievement and attitudes in science, found that girls thought of science as incongruent with their gender-role conception and were concerned with their ability to measure up to what is "proper" for a man or a woman. Handley and Morse (1984) found that roots for these attitudes can already be found in the fifth grade, where girls felt that science was useful but boys exhibited more positive attitudes toward science and were more interested in doing scientific experiments. Lazarowitz and Huppert (1981) found that a noncompetitive atmosphere in which girls could advance at their own pace promoted higher achievement.

It is possible that task-directed and more structured materials appeal more to girls and may be related to their cognitive style (Charlton, 1980). There are cognitive styles in which order, objective and clear plans, and working within groups are preferred. These may not be sufficient for real intellectual advancement, but if accompanied by other cognitive styles such as intuitive thinking and being concerned with concepts and values, they could further academic achievement. As Charlton (1980) suggests, every student is a different composite of cognitive styles, and earlier concrete and late concrete operational reasoners have cognitive styles that require hands-on involvement and use of the senses. These students usually have to be coaxed into working with order, being accurate in their observation, and learning to draw conclusions. If they worked with more structured materials during their junior high school science studies, these experiences might have some effect on their cognitive style and serve as a stimulus for their cognitive development.

There is an ongoing discussion in educational circles about whether streaming aids or detracts from the achievement of the student. Lazarowitz and Witenoff (1990) found that the lower-stream classes were far more heterogeneous than the upper-stream classes. This is one reason why teachers find it difficult to deal with lower-stream classes. This difficulty is magnified by the lack of appropriately written texts and laboratory exercises. One should add because of the present emphasis on integration in the school systems, there is a trend toward accepting students from the lower stream into the comprehensive high schools, where on the whole they remain streamed into nonacademic majors. Examination of the present laboratory texts and exercises in science curricula indicates that their major objective is presenting information and technical directions. Not enough thought has been given to structuring the laboratory in terms of the reasoning level and language ability of students. The use of a structured or programmed "discovery" laboratory worksheet integrates and involves the student in many of the reasoning and practical skills that are the aims of the inquiry method.

Students on the level of concrete reasoning often pose a contradiction to Piaget's definition of the vertebrate behavior as being "strongly curious" (Piaget, 1978). Students in streamed classrooms typically lack maturity, interest, and curiosity and show real difficulty in assimilating the material taught in the science class. If one accepts Piaget's view that accommodation is not possible without assimilation, what possibility is there for change and maturation unless this takes place? Piaget emphasized that disequilibrium or conflict is necessary for new accommodation, but that the previous operational level in any subject determines the assimilation of the new experience (Inhelder & Piaget, 1958).

Piaget discriminated between the effect of the environment on the organism and the effect of the organism on the environment (Piaget, 1978). When children feel they have mastered an understanding of some process or principle, they develop a greater feeling of control over their environment, and in a sense this mastery enables them to use the information for understanding that specific environment. This can be considered an emphatic change in behavior. On a higher level of mastery, this leads eventually to what Piaget considered the ultimate possibility of change of the ecological environment by the change in the behavior of the individual organism. The science educator can contribute to this mastery—if given the correct tools.

C. Learning Environment

Learning success is linked to the environment in which learning occurs. The laboratory lesson is often much less formal than nonlaboratory lessons. Students are allowed to talk and move about the room. During long periods they are free to do what they see fit, and they often have opportunities to interact individually or in a small group with the teacher and their peers.

The characteristics of learning environments have been studied with instruments through which students or teachers indicate their level of agreement with statements about what happens in class—for example, "students in the class find the work hard to do" (Anderson & Walberg, 1974) or "my teacher usually tells us step by step what we are to do in the laboratory" (Barnes, 1967). The Learning Environment Inventory designed by Fraser (1989) is widely used. White (1988) stated that practical work usually takes place in a distinct environment, namely a different room with "color, mystery and address of materials and equipment" (p. 186). Tamir added that in the laboratory, students usually cooperate and are expected to help each other.

Studies reporting results based on the Learning Environment Inventory pertaining to the laboratory are rare. In one study, Hofstein, Gluzman, Ben-Zvi, and Samuel (1980) found that students in academic schools perceived their laboratory classes as more satisfying, better organized, and more competitive than students in vocational schools. In both types of school, laboratory work compared with regular lessons was perceived as requiring more effort, involving more active learning, and facilitating sensitivity to others and the ability to work in a group. Among about 50 studies using the classroom environment as an independent variable and about 30 studies using it as the dependent variable, which represent 30 years of research in this area (Fraser, 1989), only one study (Lawrenz & Munch, 1984) focused on the laboratory.

A Science Laboratory Environment Inventory (LEI) designed specifically for studying laboratory classes has been developed (Fraser et al., 1990). This instrument, following revisions recommended by Tamir (1991), may enhance the research that is needed in order to capture the unique features of the learning environment of the laboratory.

Instruments are very useful, but important information may be obtained through structured and unstructured observations as well. For example, Hertz-Lazarowitz, Baird, Webb, and Lazarowitz (1984) found that cooperative and helping behaviors occurred considerably more frequently in laboratory than in recitation lessons. Using unstructured observations, it was found that students in inquiry-oriented laboratories are more active and initiate more ideas than in conventional laboratories. Teachers are less direct; processes of science receive more emphasis; there is more postlaboratory discussion; and teachers give less instruction in front of the class and move around more, checking, probing, and supporting (Eggleston, 1983;

Friedler, 1984). These results are similar to those obtained by Hofstein et al. (1980) using LEI. As more studies of the unique learning environment in the student laboratory are carried out, it may be possible to obtain convergence not only between LEI and observations but also with teachers' self-reports (Tamir, 1983a).

High school classrooms are highly heterogeneous. This is expressed in students' different ethnic origins, gender, levels of academic abilities, cognitive operational stages, different needs, and vast interests. In such a classroom context, teachers face the challenge of finding ways to adapt learning materials to their students and selecting methods of instruction that maximize students' academic success. The traditional frontal teaching cannot reach a heterogeneous population of students. An alternative that received prominence in recent years is cooperative learning.

Cooperative learning has its roots in the social psychological theory concerned with problems of cooperation and competition (Deutsch, 1949). Cooperative learning methods were developed as an antithesis to the competitive individualistic learning structures. The significance of cooperative learning and its advantages and disadvantages in classrooms with a heterogeneous student population were described by Sharan (1980), Slavin (1980), Johnson, Johnson, Maruyama, Nelson, and Skon (1981), and Webb (1982). These investigators reported that the cooperative learning methods enhanced students' academic achievement and self-esteem and improved their classroom learning environment. Studies performed in science classrooms at the junior and senior high school levels showed that when the cooperative learning approach was implemented, students' academic achievement, inquiry skills, and self-esteem increased; their "on task" behavior was higher; and the classroom learning environment became more positive (Johnson, 1976; Humphreys, Johnson, & Johnson, 1982; Lazarowitz, 1991; Lazarowitz et al., 1985; Lazarowitz & Karsenty, 1990). With cooperative learning the boundaries between classroom and laboratory work fade. Learning activities in classroom and laboratory are performed in a sequence of readings, writings, laboratory work, discussions, presentations, and tests.

Cooperative learning in laboratory work is illustrated by peer tutoring in small investigative groups. Okebukola and Ogunniyi (1984) studied the effects of different laboratory modes of instruction and learners' abilities on students' achievement and the acquisition of practical skills. Comparing the effect of cooperative groups, competitive groups, and individualized work, they found that cooperatively based groups did better in cognitive achievement, competitive groups did better in practical and process skills, and both were equal in cognitive achievement with the students who learned in an individualized mode. The cooperative group showed slightly lower achievement in process skills. The mixed-ability cooperative groups did better in cognitive achievement than the mixed-ability competitive groups. The same was true for the high-ability cooperative groups. Their conclusion was that in order to promote laboratory skills, some competitive and individual modes are necessary in addition to the cooperative mode. Peterson and Janicki (1979) also found that high-ability students did better in cooperative learning situations. Skon, Johnson, and Johnson (1981) found no evidence of effect when students were cooperative or competing with peers of equal, lower, or higher ability. Michaels (1977) concluded that competition between individuals is more effective. Miller and Hamblin (1963) found that cooperation is superior when there are cooperative tasks to be carried out but that competitive groups are superior for individual tasks. Sharan (1980), Howe and Durr (1982), and Sharan and Hertz-Lazarowitz (1980) concluded in favor of cooperative learning.

The laboratory experience in most secondary schools is that of small groups of three or four students who share the same equipment and cooperate in performing experiments. Hertz-Lazarowitz et al. (1984) reported that laboratory settings produce the highest frequency of natural interaction and cooperation on task compared with other modes of instruction in science. These results were found in noncooperative classrooms. The sensitive teacher can emphasize the cooperative nature of the learning task and encourage student-student interactions that can help facilitate learning through peer information sharing and peer discussion. This goal can be accomplished when curricular material is planned to encourage such cooperative interaction.

The studies cited earlier concerned different cooperative methods of instruction developed mostly for elementary schools. A method that was developed for high schools—peer tutoring in small investigative groups—was used in several studies in biology by Lazarowitz et al. (1985), Lazarowitz and Karsenty (1990), and Lazarowitz and Galon (1990). This method is a combination of the jigsaw (Aronson, Stephan, Sikes, Blaney, & Snapp, 1978) and group investigation (Sharan & Hertz-Lazarowitz, 1980). Instead of being passive listeners to teachers, students are free to be learners, that is, to study, ask, and seek knowledge and to interact with other students and with teachers. They are responsible for their learning.

In the two studies reported by Lazarowitz and Galon (1990) and Lazarowitz and Karsenty (1990), the cooperative learning was carried out in a sequence of activities in classrooms and laboratories. Students scored significantly higher in their academic achievement, mastered the concepts they had to learn, and mastered several inquiry skills. In the affective domain, students increased their helping behaviors and assumed responsibility for their learning. Students who were in the concrete and transitional cognitive operational stages coped successfully with topics such as mitosis and meiosis (cell division). Those topics are difficult because they require delicate psychomotor skills to use the microscope for higher magnification and involve abstract concepts in cell biology. Girls achieved higher scores in their studies, indicating that the cooperation method may help them overcome social and learning obstacles and achieve according to their full capacity. It appears that cooperative learning enhanced students' learning by facilitating mutual help and cooperation among group members. These characteristics were well supported by the learning-environment findings, which showed that in the experimental classes there was a decrease in the competitive atmosphere during the learning process and an increase in cooperation among students. The learning environment in the cooperative classroom was more flexible than in the frontal classroom. The approach facilitated more interaction among students and teachers and thereby created a positive learning climate. The findings on the learning environment support this interpretation, because students perceived their classroom climate as positive. They were more motivated to participate in reaching group goals because of positive attitudes toward the learning process. The findings that cooperative learning promotes girls' achievement and induces a positive atmosphere in the learning process support Fenshman's (1990) idea that laboratory work performed in a cooperative mode may have human-related outcomes and may be a more appropriate type of practical work in high schools in order that girls be more attracted toward science. The fact that no differences were found between girls and boys in the experimental group is encouraging. Cooperative learning seems to promote equal achievement in science for boys and girls. In a learning environment in which girls do not have to compete or behave according to teachers' and students' social expectations, they are able to fulfill their academic potential. Thus, group instruction can be an avenue by which girls reach higher levels of achievement in science (Haukoos & Penick, 1983; Lazarowitz et al., 1985).

Learning in a cooperative method required students to make measurements, communicate and read about graphs, interpret data, and design experiments. These skills were enhanced by exchange of ideas and discussion in the cooperative groups. Students who studied in chemistry laboratory work groups were also found to achieve higher scores on some inquiry skills than students in the control group (Cohen, 1987; Ztadock, 1983).

In cooperative learning, students examine different sources of information and read and prepare learning material and laboratory work so they are able to teach it to their peers. They check with each other to confirm their knowledge and readiness to teach and consult with each other about the best ways of teaching. In these ways, students practice teaching the learning material. These rehearsals probably helped students to master the learning material more effectively. Similar findings were reported by DeVries and Slavin (1978), Johnson, Johnson, and Scott (1978), Sharan, Kussell, Hertz-Lazarowitz, Bejarano, Raviv, and Sharan (1984), and Lazarowitz et al. (1985).

In his 1990 survey, Tobin reported that studies of cooperative learning show a strong relation between students' higher self-esteem and academic achievement. Cooperative learning promotes higher motivation to learn and "on task behavior." Students' positive attitudes toward the subject matter, learning experiments, and teachers increased, and as a result they cared more about learning and helping one another. Tobin suggested that there should be more studies of cooperative learning in laboratory work because the opportunity to work in groups may illuminate how students collaborate and assist one another. These social learning activities are vital for mastering scientific concepts and inquiry skills, since the group work imitates the teams of scientists who work in research. There is no doubt that if one wants science to be for "all" (Fenshman, 1990), cooperative learning in laboratory work can be a vehicle for obtaining this goal.

D. Teaching Effectiveness

Teachers' attitudes, knowledge, skills, and behaviors can also affect whether the learning in the student laboratory attains its objectives. Teaching in the laboratory requires a high level of skill proficiency and subject-matter knowledge, a special kind of subject matter–specific pedagogical knowledge, certain specific attitudes, and a readiness for risk taking.

Research indicates that many teachers are ill-prepared. They perceive inquiry as too difficult and time-consuming and as including goals that do not match their personal philosophy and are hard to assess (e.g., Welch et al., 1981). Much can be done to upgrade the teaching in the laboratory by improving undergraduate science studies (Tamir, 1989a).

Research and evaluation studies indicate various weaknesses and inadequacies of the laboratory work as currently practiced in schools (Friedler & Tamir, 1986; Hofstein & Lunetta, 1982; Novak & Gowin, 1984; Stake & Easley, 1978). Tobin's (1987) description summarizes the general findings:

Although teachers appeared to value laboratory activities they did not implement it in the manner that facilitated the type of learning that was planned.... For a variety of reasons most teachers appear to avoid laboratory investigations, particularly in classes containing low ability students. When laboratory investigations are implemented they rarely comprise an integral part of the science program. In most cases the laboratory investigation is intended to confirm something that has already been dealt with in an expository type lesson. Students are usually required to follow a recipe in order to arrive at a predetermined conclusion. As a consequence the cognitive demand of the laboratory tends to be low. (p. 210)

Assuming that Tobin's description is accurate, even in the United Kingdom, where practical work has traditionally been given great emphasis, many secondary school students have failed to develop basic practical skills such as observing, estimating quantities, designing experiments, and making inferences (Assessment of Performance Unit, 1984).

Although many factors influence the nature of learning in the student laboratory, the single factor that makes the greatest impact is the teacher. Surprisingly, none of the reviews cited earlier dealt with the development or education of teachers to teach in the laboratory. Because teaching in the laboratory is in many ways different from teaching in an ordinary classroom, one would expect various approaches to teaching in the laboratory to receive much more emphasis in preservice and inservice education than is currently devoted to them in most places. In fact, few teacher education programs offer direct systematic instruction on how to teach in the laboratory. It is often assumed that students' participation in science laboratories in college gives them the knowledge and skills to teach in a school laboratory. A quick look at how laboratories are used in most college courses (e.g., Kyle et al., 1979; Tamir, 1977) shows that this assumption is wrong. There is no doubt that improved effectiveness of learning in the laboratory can be achieved only through substantial improvement in teacher preparation. The developers of the curricula in the 1960s recognized the need for inservice courses to prepare teachers to teach by the approach they advocated. In these courses special attention was given to the laboratory. For example, the BSCS published a special guide entitled *BSCS materials for preparation of inservice teachers in biology* (Andrews, 1964). In the inservice courses introducing teachers to the BSCS, about half of the time was allocated to the necessary procedures, to teaching inquiry-oriented laboratories, and to effective management in the laboratory. Specific laboratory investigations were selected for inclusion in teacher-education courses.

Long before Johnstone and Wham (1962) drew our attention to the cognitive overload in investigative laboratories, Schwab (1963) had developed Invitations to Enquiry. They were designed to engage students in scientific inquiry through analysis and discussion without the burden of manipulating apparatus, materials, and organisms. Even though Invitations to Enquiry was not meant to replace the laboratory, a BSCS teacher-education program would be considered inadequate if it did not give participants opportunities to experience learning and teaching some invitations to inquiry. Indeed, a list of selected invitations to inquiry for teacher-education courses was included in Andrews (1964).

The Biology teachers handbook (Klinckman, 1970) provides ample information and useful examples related to teaching in the laboratory. In the United States, teacher-education materials in the 1960s and 1970s were developed within the framework of particular curricula such as the BSCS. In the United Kingdom, science-education tutors cooperated in devising a variety of student activities for preservice and inservice courses without reference to any particular curriculum. The project, known as the Science Teacher Education Project (STEP), arranged the various activities under 13 topics, about half of which concerned teaching and managing in the laboratory. For example, Methods and Techniques offers activities on planning and designing class experiments. Safety engages student teachers in spotting hazards in the laboratory and trying some experiments that need rehearsal. Laboratory Design and Management provides experience in ordering, storing, and distributing materials; in looking after small animals; and in designing simple apparatus (Hayson & Sutton, 1974). The STEP materials were tried and evaluated in the United Kingdom (e.g., Holford, 1974) and were adapted in several countries such as Israel and Australia. In addition to the collection of Activities and Experiences, which served as a bank of ideas for professional courses, STEP published a number of resource books including readings, case studies of problems and events, and an annotated collection of pupils' writings in science.

This package of materials was intended to help teacher educators relate theory to practice. Even though no information is available on their current use, there is no doubt that STEP has had a strong impact on science teacher educators and on the kind of activities and experiences offered to student teachers.

A unit on planning and development of learning material focusing on the laboratory was designed by Rasmussen and reported by Lee (1969). The unit begins with a comparative analysis of a laboratory exercise and a laboratory investigation and ends with students each choosing a topic with which they feel comfortable and designing their own laboratory exercise and an investigation on this topic. One of the innovations introduced by the BSCS was the Laboratory Block, designed to engage a biology class for 6 weeks in an in-depth study of a particular topic. Lee, Lehman, and Peterson (1967) present detailed suggestions on how the Laboratory Blocks can be used as a basis for teacher-education courses lasting 1 to 8 weeks. The only evidence is the testimony of the BSCS director that the "Laboratory Block Program has been successful beyond our original expectations" and that this program can "serve as a significant aid to the planning of flexible programs emphasizing laboratory work" (Mayer, 1967, p. iv).

A special module entitled "Basic concepts of scientific research" (Friedler & Tamir, 1986) was developed to teach high school students about the nature and development of scientific knowledge. Basic concepts such as research problem, hypothesis, assumption, control, and experiment are constructed by the students. Various activities guide the students toward understanding the processes and concepts well enough to make their laboratory investigations meaningful. Teachers who used the module (which is accompanied by a teacher guide) reported that it helped them not only to teach their students but also to better understand the nature of scientific research themselves.

Shadmi (1983) describes a physical science course for preservice elementary teachers in which

the backbone of the whole learning process is the individual activity of the student. All happens in the physics laboratory.... The students work on various assignments planning and carrying out experiments, drawing conclusions and solving problems.... During the activities individual teacher/student interaction constantly takes place.... Homework complements class activities, class discussions clarify certain points.... Some problems are left open [and] serve as a starting point for the following assignment.... One cannot over-emphasize the importance of the student's individual activities and his interaction with the teacher. This is the only guarantee that individual weaknesses will be detected. (p. 121)

The program was accepted enthusiastically, as evidenced by informal conversations and responses to questionnaires. "The objectives of creating positive motivation and improving the self confidence of primary school teachers have been achieved" (p. 122). The course requires 90 hours, which correspond to half of the time usually allocated to the study of physical science in teacher colleges. It has been taught successfully by instructors whose specialized fields are physics, chemistry, biology and geography, provided they themselves had worked through the course as students. Most of the participants who scored poorly on the pretest achieved mastery of the major concepts and skills that are the objectives of the program.

Dreyfus (1983) describes a preservice laboratory course for biology and agriculture secondary school teachers. Every 2 weeks a pair of student teachers teach, in the presence of their peers, a laboratory lesson to high school pupils who are brought to the university campus. The student teachers' peers, who are unaware of the planning of the lesson, have to observe the class, identify the objectives and main decisions made by the teacher, evaluate the decisions, and propose alternatives.

At the following meeting, school pupils not present, the declaration of intentions of the teacher was confronted with the perceptions of observers.... The discussions were expected to yield two main outcomes. Firstly, they served to sum up and clarify the potential of laboratory lessons.... Secondly they demonstrated ... civilized discussion between members of a team of teachers. (Dreyfus, 1983, p. 213)

A fifth example is the teacher-preparation course developed by Life and Agricultural Science (LAS) in Israel. Feinstein and Blum (1983) described its thrust as follows:

Instead of letting teachers do all experiments systematically, stations were set up where the teachers could do those activities for which they thought they needed practice.... There were 2 to 3 more stations than teacher groups so that all teachers could work without having to wait.... Experienced teachers served 3–4 stations each ready to answer questions and discuss technical or didactic problems. In order to encourage teachers to engage in lengthy experiments (which are typical to agricultural research) some of the experiments were presented, broken down into parts.... The station system proved to have some considerable advantages: Teachers did only those experiments for which they needed training.... More experienced teachers finished more quickly than others. Thus 50% of time could be saved.... The different stations were used simultaneously and therefore fewer units of apparatus were needed.... Teachers realized that different groups can work simultaneously on different experiments.... Teachers had to choose among alternatives.... Teachers could solve many problems by themselves. (pp. 380–381)

Considering the need to prepare prospective teachers to teach in the laboratory and help inservice teachers realize the great potential of the student laboratory, we have described examples of structured units and courses that have been used in different teacher-education programs. Yet, teacher educators use different approaches to guide their student teachers and acquaint them with strategies that might improve meaningful learning in the laboratory.

One approach is content analysis of laboratory manuals. Teachers may offer their students laboratory exercises taken from commercial textbooks without really considering the nature of these exercises and what the students are actually learning from them. Yet a simple, inexpensive, and effective way is readily available to help teachers choose laboratory experiences that meet their objectives and modify them to meet the needs of their students. Two content-analysis schemes are useful for this purpose. The first was suggested by Schwab (1962) and elaborated by Herron (1971) regarding the level of guidance or "openness and permissiveness" in Schwab's (1962) terms:

The manual can pose problems and describe ways and means by which the student can discover relations he does not already know from his books. At a second level, problems are posed by the manual but methods as well as answers are left open. At a third level, problem as well as answer and method are open: the student is confronted by the raw phenomenon. (p. 55)

Herron added the most highly guided dimension where all three levels are given. The usefulness of this simple level analysis is easily seen when one compares the results of this analysis with lofty claims of curriculum designers. For example, Herron's analysis indicated that about three-quarters of the laboratory exercises in the BSCS and PSSC manuals were at level zero, 20 percent at level 1, 5 percent at level 2, and none at level 3. Tamir and Lunetta (1978) designed the Laboratory Analysis Inventory (LAI), which, when used in content analysis of laboratory manuals, yielded much more detailed information.

Another skill for the teacher is efficient management in the laboratory. According to Gallagher and Tobin (1987),

The teacher's most important role is to facilitate learning by maintaining an environment in which students . . . receive challenges and assistance as required . . . most teachers seem to be preoccupied with management in laboratory activities. (p. 535)

Capie and Tobin (1982) reported that middle school students were covertly engaged 50 percent of the allocated time; engaged in overt planning and data-processing tasks for only 2 percent and 5 percent of the time, respectively; and were off task for approximately 30 percent of the time. Because achievement and retention were found to be related to the proportion of time engaged in planning and data processing, it follows that improvement in management to decrease off-task time and increase covert engagement is highly needed.

Berliner (1983) presented a review of the research that is pertinent to training teachers for their executive functions. Eight of the nine executive functions he identified are especially relevant to the management of learning in the laboratory. Berliner (1983) argued that

Mastery of the requisite subject matter areas, together with the managerial skill to meet the demands of complex and dynamic classroom environments may constitute both the necessary and sufficient conditions for effective teaching to take place. (p. 559)

The executive functions are particularly applicable to the student laboratory, and teacher educators need to attend to them in their program.

A third skill is the use of concept mapping as a tool. Careful planning of laboratory lessons is essential if their learning potential is to be realized. Novak and Gowin (1984) observed that

Often students enter into a laboratory or field setting wondering what they are supposed to do or see; and their confusion is so great that they may not get as far as asking what regularities in events or objects they are to observe, or what relationships between concepts are significant. As a result, they proceed blindly to make records or manipulate apparatus with little purpose and little consequent enrichment of their understanding of the relationships they are observing and manipulating. Concept maps can be used to help students identify key concepts and relationships, which in turn will help them to interpret the events and objects they are observing. (pp. 47–48)

The Vee heuristic, developed by Novak and Gowin (1984), is a device designed to help students understand the structure and process of knowledge construction. It is especially useful for laboratory instruction. First, when used in an analysis of laboratory manuals, it reveals conceptual gaps such as key concepts that are missing or failure to fit observations and experiments into an appropriate and theoretical background. Second, it helps students recognize what objects or events they are observing, what concepts they already know that relate to the events or objects, and what records are worth making. Finally, it helps students make sense of their data and test their conclusions against those that relevant concepts or principles suggest.

But do these skills in desired teacher behavior help in the laboratory? In order to assess teacher-education activities related to teaching in the laboratory, two questions are asked. First, do different components of a teacher-education program affect teachers' performance in the laboratory? Second, do laboratory-based experiences in the teacher-education program affect the performance of teachers in general and their performance while teaching in the laboratory in particular?

The desired outcome of teacher education, namely the teachers' performance, is mediated by acquisition of knowledge, skills, and positive attitudes. Because the assessment of teacher performance is complex and long term, most studies of the impact of teacher-education courses report changes in these mediating variables. Often, the level of satisfaction of student teachers with the course is reported as well.

Piper and Moore (1977) collected 12 studies focusing on factors that affect the development of positive attitudes as a result of preservice science-education courses for elementary teachers. Their analysis led to the following conclusions:

1. Special courses for elementary teachers designed and taught through collaboration between science departments and science educators would improve attitudes toward science.
2. Teachers, inservice and preservice, continue to ask *how* to teach—not *what* to teach.
3. A structured training program provides foundational structure that builds confidence and positive attitudes without stifling creativity and originality.
4. Instructors should model the teaching strategies that a teacher can use in the classroom.
5. One of the primary needs of teachers teaching hands on activities is classroom management.
6. Telling teachers about strategies is not enough. Teachers need to see and be part of these strategies so that they in turn can put them in practice with children.
7. Peer teaching and tutoring small groups of children do not give teachers realistic practice in classroom management techniques, such as grouping, passing around materials, etc. Therefore teachers need to practice with children in classroom situations.
8. Teachers indicate they need practical suggestions as to: a) how to begin an activity; b) how to ask open-ended questions; c) how to group children; d) how to disperse materials; e) how to lead children in synthesizing the data from the activity; and g) how to evaluate a lesson.

Welch (1983) reported results of a portion of a need-assessment study of research in science education that dealt with topics relevant to teacher education. According to a meta-analysis of 153 studies located and analyzed by Sweitzer (1982), a grand mean effect size for all treatments was 0.77 standard deviation, which indicates a substantial effect of teacher-education courses and activities. (According to Cohen, 1969, a value of 0.4 to 0.6 indicates a moderate effect, 0.6 to 0.8 a large effect, and greater than 0.8 a very large effect). Common experiences and the effect size of their impact are as follows: questioning analysis (1.38); teachers' science courses (0.97); student self-directed study (0.81); methods course (0.79); use of laboratory (0.75); workshop (0.73); inquiry instruction (0.63); and field-based program (0.35). Preservice courses had a somewhat higher effect size than inservice courses (0.78 and 0.72, respectively). Finally, different outcomes had different effect sizes as follows (in descending order): science process (1.08); indirect verbal behavior (0.72); questioning level (0.70); science knowledge (0.52); and attitude toward science (0.39).

It may be concluded that "teachers who received the various training treatments tended to outperform the comparison groups on measures of science knowledge, process, attitude and desired teaching behaviors" (Welch, 1983, p. 460). Compared with science-curriculum materials, the mean effect size for teacher-education efforts as reported by Welch (1983) was twice as large: 0.77 and 0.35, respectively. On the basis of the reported data, it was concluded that

Teacher education efforts are quite effective, at least in the short term, for influencing teaching performance.... Whether this change is permanent and in turn affects student performance is unknown at present. Little research was found on this topic. (pp. 460–461)

Shymansky, Hedges, and Woodworth (1990) reported on a meta-analysis of the effects of inquiry-based science curricula. The composite effect size based on 320 studies was 0.26. They analyzed the effect size by various categories, one of which is related to the impact of the teacher inservice training:

Students in the new programs taught by teachers having received special inservice training on the use of the materials significantly outperformed students in traditional programs. Where teachers received no inservice training the results, though positive, are not statistically significant. (p. 141)

The effect sizes were 0.27 and 0.23, respectively. The authors do not discuss the small difference between the two effect sizes. They stress that their data "indicate that student performance is significantly enhanced when special teacher inservice accompanies implementations" (p. 142). They observed that only 33 percent of the studies analyzed reported whether the teachers had participated in special preparation, even though this could have influenced the size of the effect of any curriculum on student learning.

It is commonly believed that teacher characteristics such as gender, age, training in science, experience, personality, and attitudes have significant effects on student learning. Druva (1982) found very low correlations between these characteristics and student learning and attitudes. His meta-analysis involved 300 studies and the mean correlations were 0.05 and 0.04 for cognitive and affective achievement, respectively. One

wonders whether similar results will be found in relation to teaching in the laboratory. Wise and Okey (1983) reported the results of meta-analyses of 411 cases comparing the impact of various teaching techniques on student learning. The total effect size for 12 techniques was 0.34. The two techniques related to the laboratory were "manipulative learning" and "inquiry-discovery," with effect sizes of 0.57 and 0.32, respectively.

Of 13 instructional systems reviewed by Willett, Yamashita, and Anderson (1983) 2 involved student-guided individual study with a substantial laboratory component. Welch (1983) concluded, on the basis of Druva and Wise and Okey's studies, that "teacher characteristics and behaviors have only slight influence on student learning" (p. 464). On the other hand, students' learning behavior, classroom environments, and exposure to instructional resources all had effect sizes of 0.50 or greater. Hence "teachers must learn techniques to stimulate student responsible behaviors in science learning.... [These] include ... cooperative learning, learning contracts, personalized systems of instruction and mastery learning" (p. 464). Because student behaviors have always been the basis of learning in the laboratory, Welch's recommendation to change the direction of teacher education toward student-responsible behaviors should be especially apt for training teachers to teach effectively in the laboratory.

E. Student-Assessment Strategies

Learning in the laboratory is indeed a unique experience and "practical work involves abilities both manual and intellectual which are in some measures distinct from those used in nonpractical work" (Kelly & Lister, 1969). It has been shown repeatedly that the practical mode of performance (this term was first used by Tamir, 1972) is only weakly correlated with performance on paper-and-pencil tests (Ben-Zvi et al., 1977; Comber & Keeves, 1973; Robinson, 1969; Tamir, 1972, 1975). Even when seemingly same skills are assessed (e.g., planning experiments), achievement on a laboratory examination differs from achievement on paper-and-pencil tests (Assessment of Performance Unit, 1984; Tamir, 1990b). Yeany, Larussa, and Hale (1989) concluded that when physical objects are present, tests seem to provide a more valid measure of both logical reasoning and process skills than corresponding paper-and-pencil tests.

Recently 6 of the 24 countries that participated in the SISS administered a paper-and-pencil, multiple-choice test as well as a practical laboratory test to large samples of 10- and 14-year-old students. The paper-and-pencil test included both "content" and "process" items. Three process skills were identified in the practical test: performing (observing, measuring), investigating (hypothesizing, designing experiments), and reasoning (drawing conclusions, explaining). The items of the process paper-and-pencil subtest were classified into the same three skills. A comparison of the results obtained on the two tests and the various subtests indicated that across all six countries and two age levels, the "process" and "content" paper-and-pencil subtest results (both mean scores and correlations) were much more similar to each other than to the practical test results. In all six countries, 10-year-olds performed significantly better in the practical test, as might be expected from children in a concrete stage of development. No such difference was found in most countries among 14-year-olds. A most interesting finding was the position of 14-year-old Japanese students, who achieved the highest mean score on the paper-and-pencil test and the lowest in the practical test. Finally, whereas Singapore males exhibited higher achievement than females on both tests, in the remaining countries females had lower achievement on the paper-and-pencil test, but on the practical test they achieved as well as or better than the males (Tamir & Doran, 1992).

The implications of these findings are clear. Paper-and-pencil tests are inadequate for assessing student performance in the laboratory. "Yet the bulk of science assessment is traditionally non-practical" (Bryce & Robertson, 1985, p. 1). This conclusion seems remarkable because excellent means of practical assessment are available (see reviews by Hofstein, 1988; Tamir, 1990b).

Because the kinds of tests used in class have a strong influence on what and how students learn and where they invest their efforts and paper-and-pencil tests can never do justice to the performance in laboratory investigations, teachers who are interested in providing learning in the laboratory should become familiar with innovative practical laboratory tests. Inquiry-oriented laboratory tests have been used successfully in high school biology classes in Israel for more than 15 years. Details of their construction and use may be found in Tamir (1974).

In order to standardize the assessment of the students' written responses, a Practical Tests Assessment Inventory was designed. This has been routinely employed in assessing the laboratory tests that constitute one component of the biology matriculation examination in Israel (Tamir, Nussinovitz, & Friedler, 1982). It contains 21 categories, beginning with problem formulation and concluding with application of knowledge discovered in the investigation.

In Israel, where practical laboratory tests are an important component of the matriculation examination in biology, novel problems are designed every year. After their use in the matriculation examination, these tests are published and become part of the curriculum, often replacing less inquiry-oriented exercises. Also, preservice teachers perform selected inquiry-oriented investigations and analyze them as a regular activity in their methods course. Inservice meetings are held every year for analysis and discussion of the results of the matriculation examination.

Many teachers (about 120 every year) serve as evaluators of students' performance on the practical laboratory tests—a golden opportunity to become familiar with learning deficiencies and misconceptions. Indeed, with this rich potential for learning about student reasoning patterns and difficulties, special diagnostic exercises have been developed for particular questions in the laboratory tests. Each exercise focuses on one inquiry skill, such as problem identification and formulation, hypothesis formulation, identifying dependent and independent variables, or designing experiments. These published exercises (Tamir & Nussinovitz, 1979) proved to be very useful in "helping prospective teachers identify the deeper attributes of students' learning, understanding, conceptions and achievements" (Tamir, 1983b, p. 618).

Especially in the United Kingdom, there has been a tendency to replace external practical tests with continuous internal assessment by the teachers. Continuous assessment takes place on special occasions during the year. Normally, only a few students are observed in each lesson. Assessment follows detailed instructions along predetermined categories, all provided by the examination board. Ganiel and Hofstein (1982) carried out a study in which they trained 25 high school physics teachers in Israel to evaluate their students' performance in the laboratory. They prepared videos in each of which a student was shown performing a physics experiment in a school laboratory. Three familiar experiments were selected from the current curriculum because they contained many elements that could be assessed. They found that as a result of their training, "the assessment of laboratory performance had been greatly improved both in precision and in objectivity" (p. 598).

Remarks

The major characteristics of learning in the laboratory are that it is based on concrete experiences, involves hands-on activities, and makes use of the senses. This kind of learning may be designated as "experiential learning," which is consistent with the constructivist views of the learning process. We have chosen to review two popular learning models to illustrate how the potentially unique contribution of practical work may operate within their frameworks: the generative-learning model and learning cycles.

According to Osborne and Wittrock (1985),

The generative learning model is centrally concerned with the influence of existing ideas on what sensory input is selected and given attention, the links that are generated between stimuli and aspects of memory store, the construction of meaning from sensory input and information retrieved from long-term memory, and finally the evaluation and possible subsumption of constructed meaning. (p. 64)

The underlined words represent the learning activities. For each activity we suggest examples of the potential contribution of the laboratory. The laboratory is the major source of sensory input. Both selection and attention are regularly practiced in practical work. Links between stored information and the sensory input have a high probability of being recorded in episodic rather than semantic memory (Tulving, 1972). Thus, the laboratory can be source of episodes and images that give meaning to propositions the students have already acquired or will acquire (White, 1988). These episodes and images are actively constructed by the learner. Finally, the laboratory offers ample opportunities for the learner to test the constructed meaning against meanings constructed from other sensory input. Because students work individually or in small groups, they usually accept major responsibility for their own learning as advocated by the generative-learning mode. As Osborne and

Wittrock (1985) observed, quite often students fail to generate the links expected by the teacher but instead form links between the sensory input and existing ideas that would be considered irrelevant by the teacher. Hence, the laboratory is an effective setting for identifying student misconceptions.

As defined by Lawson et al. (1989), the learning cycle consists of three phases: exploration, term introduction, and concept application. During exploration, students explore new materials and ideas with minimal guidance. The new experiences should raise questions or complexities. The second phase begins when the teacher introduces a new term or terms that refer to the patterns (concepts) discovered during the exploration. In the last phase the students apply the new terms to additional examples in order to extend and generalize the concept's meaning. Quite frequently, phases 1 and 3 involve some laboratory work. This is the case in all the examples given by Lawson et al. The learning cycle exemplifies one of the principles of experiential learning: that to facilitate meaningful learning, concrete experiences should precede the presentation of concept or theory. Lawson et al. provide ample evidence for the effectiveness of the learning-cycle model in which practical work plays a major role.

The actual learning in the laboratory depends on the nature of the tasks assigned as well as the opportunities offered to students to learn. The nature of the assigned tasks is often determined by textbooks. Following Herron (1971), who analyzed the level of inquiry in laboratory manuals, Tamir and Lunetta (1978) designed the LAI to identify the opportunities to learn offered by a particular laboratory exercise. Using the LAI it was found, for example, that neither the PSSC nor the BSCS editions of the early 1970s provide opportunities to formulate a research problem, to design and carry out an experimental procedure, to discuss assumptions and limitations underlying experiments, or to share their efforts with others (Tamir & Lunetta, 1978, 1981). This kind of content analysis explains why frequently little inquiry is found in the science classroom (Welch et al., 1981) and why the enormous potential of practical work is actualized to only a limited extent.

Another indicator of the nature of learning that takes place in the laboratory is the extent to which prelaboratory and postlaboratory discussions take place. Using an observation instrument designed by Smith (1971), Tamir (1977) found a high positive correlation between inquiry orientation and postlaboratory discussion in university laboratories. Summarizing the findings of several studies, Friedler and Tamir (1984) concluded that in laboratory lessons teachers talk much less than in recitation lessons (40 percent as compared with 75 percent of the time), students receive more individual assistance, and students talk on the average 13 percent of the time and are off task only 6 percent of the time. They also found that in inquiry experiments, compared with verifying exercises, planning and other processes of science receive more emphasis, there is more postlaboratory discussion, students are more active, and they initiate ideas more frequently.

At the university level, Reif and St. John (1979) found that only 25 percent of the students in traditional physics laboratories were able to account coherently for what they had been doing, compared with 80 percent in laboratories specially designed to enhance "thinking like a physicist." It may be concluded that the student laboratory can offer unique learning experiences. Whether and to what extent students actually engage in the learning activities depend on the teacher and on local conditions.

CONCLUSIONS AND RECOMMENDATIONS

Pickering (1980) questioned whether laboratory courses were a waste of time. In their review, Hofstein and Lunetta (1982) concluded that laboratory activities can promote inquiry, intellectual development, mastery of problem-solving skills, and manipulative skills that can lead to formation of concepts in science. Yet their findings indicate that laboratory activities, as they are currently implemented, do not enhance students' learning or understanding of science. These findings are supported by Stake and Easley (1978) and Tobin and Gallagher (1987).

What are the reasons for this unsatisfactory situation? A short historical review of the practical work in schools by Lock (1988) gave some reasons. Lock mentioned that to the end of the nineteenth century, the role of the practical work in laboratories was to confirm the theory that had already been taught. It did not include any genuine experiments, and the procedures and expected results were printed in the textbooks. By the end of the nineteenth century, practical work was separated from teachers' demonstrations. Also toward the end of the nineteenth century, according to Jenkins (1979), the role of investigation became central in practical work and the inquiry procedures were expected to lead to understanding of the theory. Practical work declined in both the United States and the United Kingdom at the beginning of the twentieth century, and doubts were expressed about its role in teaching science to students with different abilities. In addition, practical work had the role of confirming what was taught and learned in the classrooms.

Only after the 1950s and the new science curricula of the 1960s were new courses developed reemphasizing the role of discovery learning through practical work. These new courses embodied the idea that practical work would lead to the formation and learning of concepts and principles, while leaving teachers to provide a learning environment in which they guide students who are seeking knowledge. Although the goals of the new science curricula were well stated, the textbooks and laboratory manuals did not always reflect these goals. In the United States there was still a separation between the textbook material to be learned in the classroom and the manual containing the experiments to be performed in the laboratories.

In the United Kingdom, the Nuffield Project in Biology adapted a format in which the textbook included the topics in a continuous sequence of readings and experiments to be performed in such a way that none of them could be ignored. Only the results of a performed experiment enabled the students to continue with a new topic to be read. This format made certain that experiments were performed, results were discussed, and conclusions were drawn. This kind of procedure provided experiments that resulted in meaningful learning. Only with discussions, interpretations, inferences, and conclusions do experiments play a meaningful role in learning (Millar 1987; Pickering 1987; Tobin, 1990).

In the Nuffield Project, textbooks include supplementary experiments and the teacher decides whether they can be performed. This textbook format has enhanced the spirit of inquiry. Practical work followed by questions and discussion led to the conceptualization process and mastery of concepts and principles. Thus one of the conditions for practical work to be performed in a spirit of inquiry appears to be the format in which the learning material and the experiments are presented to the teachers and students. Separating the reading material and the experiments into a textbook and lab book may contribute to lack of inquiry while performing the practical work or permit the laboratories to be ignored altogether.

If it is a curriculum goal that practical work will lead to concept formation and formulations of theories, then each experiment should start from an existing theory and knowledge base and identify a problem that will lead to hypothesizing, planning, and executing an experiment. The results of the experiment should be discussed and interpreted. The conclusions obtained should lead to a new problem.

Two more components of the application of science curricula and use of practical work in schools have been addressed: teachers and students. Teachers are expected to teach in an inquiry approach and to use a scientific method of investigation while their students are performing experiments. In order to create an investigative laboratory environment, teachers must exhibit behaviors that can be acquired only through past experience in doing some kind of research themselves. It is presumptuous to assume that by using learning materials of an inquiry nature, teachers will be able to exhibit and use investigative behaviors. Studies have shown that science teachers who have college research experience have more positive attitudes toward teaching science through an inquiry approach than other teachers (Lazarowitz & Lee, 1976). It is well known that positive attitudes toward a phenomenon, object, or events are prior conditions to behavioral changes. Therefore, serious consideration must be given to providing science teachers with an investigative experience so that they will have positive attitudes toward the inquiry approach. Only then may one expect teachers to create an investigative.environment while students are performing experiments. Tobin (1990) noted the advantage of having teachers who can conduct research in their own classrooms in collaboration with university researchers:

By conducting research in classrooms, it is hypothesized that teachers produce knowledge about learning in a manner which facilitates utilization of the findings of the research conducted on-site and similar research published in the literature. (p. 413)

One of the most common findings in the research studies is that science teachers do not use any practical work in their teaching or use it as a way to confirm what was already taught. This situation can be changed in two ways. One way is to encourage teachers to use internal assessment of practical work in their laboratories as a means of enhancing the performance of

experiments. Second, because external assessment of practical work has been found to be an important factor in emphasizing the role of experiments in learning science and teachers' behaviors (Lock, 1988; Tamir, 1983a), external practical assessment may be used to influence teachers' instructional behaviors in a manner expected by the curriculum goals.

Yet another element is important in improving laboratory instruction. Science teachers must have a continuous supportive system of inservice training courses. In the Israeli school system, a weekly inservice day for teachers is offered on the same day in different locations by the Israeli Center for Science Teaching, which includes Science Education Centers from different institutions in collaboration with the Ministry of Education and Culture. The course includes (1) lectures by professors aimed at updating teachers' knowledge in subjects that are on the cutting edge of the biology research, (2) pedagogical and didactical issues raised by teachers and discussed in collaboration with science educators, (3) and new experiments performed by teachers and discussion of logistic and methodology issues. We strongly believe that the combination of these three elements, textbooks written from a specific pedagogical approach, external practical assessment, and the development of a continuous supportive system for teachers will enhance the proper use of practical work in school laboratories and facilitate the expected academic achievements (Tamir, 1975, 1976a).

Students' learning styles, cognitive preferences, cognitive stages, and interests, needs, and attitudes have been shown to be important in laboratory learning. Teachers need to develop the ability to adapt suitable methods of teaching and learning materials to meet a student's needs. Research in science education has already provided the knowledge needed to move in this direction. This includes information about the effectiveness of individualized and tutorial approaches, and the integration of computer-assisted learning, microcomputer-based laboratories, cooperative learning, assessment, and evaluation methods.

Pickering (1987), who stated that learning and student enthusiasm can be fostered by laboratory activities that are geared to students' needs and give students the satisfaction of finding out that they can overcome challenges, suggests that experiments in laboratories should be written according to several criteria. First, an experiment should be structured so that it presents students with what he calls a "puzzle" and not with an illustration of what they already know. Second, experiments should be written with "carefully defined procedures" so that each student in the class can carry them out. Third, experiments in science should also include topics for which current knowledge is incomplete or not understood even by scientists. Fourth, students should be required to prepare in advance, in their notebooks, a plan for how to proceed with the experiments, rather than using manuals and written instructions in laboratory work. Fifth, students should be required to write reports using a very flexible format.

That learning material designed to enhance the mastery of inquiry skills and other cognitive and affective outcomes can result in reaching the goals is well supported by research. For example, when BSCS programs were used, science process skills were mastered, more positive attitudes toward science were developed, students understood biological concepts better, and scientific thinking skills improved (Igelsrud & Leonard, 1988; Shymansky, 1984; Tamir & Jungwirth, 1975). Research findings also confirm that correlations exist between students' cognitive stages, learning strategies, laboratory experiments performed, instructional worksheet format, and achievement (Lawson & Blake, 1976; Lawson & Renner, 1975; Lawson & Wollman, 1976; Lazarowitz & Witenoff, 1990).

Pickering (1987) asks whether the role of laboratory work is to train future scientists or to crystallize what students were already taught and know. Is the role of the laboratory to teach science concepts and principles and to master science process skills? Can we assume that by practicing science process skills, students will develop scientific skills? Can we hypothesize that this kind of thinking can help our students use these skills in approaching the problems of daily life? Can we expect that by collecting data and information and organizing their thoughts in objective and logical ways, students will sub-stitute these skills for more prejudiced and subjective approaches?

To answer these questions, we need more research. But this research is complex and should be conducted in a more structured manner in which science-education research teams focus on one particular issue from different points of view. Studies should be repeated in order to increase reliability. If researchers use similar approaches, such as Tamir and Amir's (1987) model of factors of process skills for planning laboratory practical work, Igelsrud and Leonard's (1988) model for guided inquiry, or Lazarowitz and Witenoff's (1990) format for laboratory instruction, results can be obtained from different sources. The mix of studies performed at the present time will not lead us into the future. Well-structured studies by science-education research teams studying specific topics and using common methods and procedures will provide the needed answers about the role of laboratory work in science instruction.

References

Abraham, N., Novak, A., & Schaefer, J. R. (1965). 1965 revised BSCS biology laboratory facilities checklist. *BSCS Newsletter, 24,* 3–8.

Adams, D., & Baker, R. (1986). Science technology and human values: An interdisciplinary approach to science education. *Journal of College Science Teaching, 15,* 254–258.

Ainley, J. (1978). *An evaluation of the Australian Science Facilities Program and its effect on science education in Australian schools.* Unpublished Ph. D. thesis, University of Melbourne.

Ainley, J. (1990). School laboratory work: Some issues of policy. In E. Hegarty-Hazel (Ed.), *The student laboratory and the science curriculum* (pp. 223–241). London: Routledge.

Anderson, G. J., & Walberg, H. J. (1974). Learning environments. In H. J. Walberg (Ed.), *Evaluating educational performance.* Berkeley, CA: McCutchan.

Anderson, J. (1984). Computing in schools. *Australian Educational Review, 21.* Hawthorn: Australian Council of Educational Research.

Anderson, R. E. (1987). Females surpass males in computer problem solving: Findings from the Minnesota Computer Literacy Assessment. *Journal of Education Computing Research, 3,* 39–51.

Andrews, T. F. (1964). *BSCS materials for preparation of inservice teachers of biology, special publication* (No. 3). Boulder, CO: Biological Sciences Curriculum Study.

Archenhold, W. F. A., Jenkins, E. W., & Wood-Robinson, C. W. (1978). *School science laboratories: A handbook of design and management.* London: John Murray.

Arons, A. B. (1984). Computer based instructional dialogs in science courses. *Science, 224,* 1051–1056.

Arons, A. B. (1988). Historical and philosophical perspectives attainable in introductory physics courses. *Educational Philosophy and Theory, 20*(2), 13–23.

Aronson, E., Stephan, C., Sikes, J., Blaney, N., & Snapp, M. (1978). *The jigsaw classroom.* Beverly Hills: Sage.

Assessment of Performance Unit. (1984). *The assessment framework of science at age 13 and 15 APU Science Report for Teachers, 2.* London: Department of Education and Science.

Atkinson, E. T. (1990). Learning scientific knowledge in the student laboratory. In E. Hegarty-Hazel (Ed.), *The student laboratory and the science curriculum* (pp. 119–131). London: Routledge.

Ausubel, D. P. (1968). *Educational psychology: A cognitive view.* New York: Holt, Reinhart & Winston.

Barnato, C., & Barrett, K. (1981). Microcomputing in biology inquiry. *American Biology Teacher, 43,* 372–378.

Barnes, L. W. (1967). The development of student checklist to determine laboratory practices in high school biology. In A. E. Lee (Ed.), *Research and curriculum development in science education* (pp. 90–96). Austin: University of Texas Publication.

Bartov, H. (1978). Can students be taught to distinguish between teleological and causal explanations? *Journal of Research in Science Teaching, 15,* 567–572.

Bates, G. R. (1978). The role of laboratory in secondary school science programs. In M. B. Rowe (Ed.), *What research says to the science teacher.* Washington, DC: National Science Teachers' Association.

Becker, H. J. (1987). *The impact of computer use in children's learning. What research has shown and what it has not* (Report No. 18). Baltimore: Johns Hopkins University Center for Research in Elementary and Middle Schools.

Beisenherz, P. C., & Olstad, R. G. (1980). The use of laboratory instruction in high school biology. *American Biology Teacher, 42,* 166–168.

Belanger, M. (1969). Learning studies in science education. *Review of Educational Research, 39,* 377–395.

Ben-Zvi, R., Hofstein, A., Samuel, D., & Kempa, R. F. (1977). Modes of instruction in high school chemistry. *Journal of Research in Science Teaching, 14,* 433–439.

Berliner, D. (1983). Training teachers for their executive functions. In P. Tamir, A. Hofstein, & M. Ben Peretz (Eds.), *Preservice and inservice education of science teachers* (pp. 545–564). Rehovot: Balaban Interscience.

Biological Sciences Curriculum Study (BSCS). (1973). *Biological sciences: An inquiry into life* (3rd ed.). New York: Harcourt, Brace, Jovanovitz.

Biological Sciences Curriculum Studies (BSCS). (1975). *An inquiry into life* (Yellow Version; Unity) Israeli Adaptation. Jerusalem: Israeli Science Teaching Center, Hebrew University.

Biological Sciences Curriculum Study (BSCS). (1991). *Basic genetics: A human approach* (2nd ed.). Dubuque: Kendall/Hunt.

Bliss, J. (1990). Students' reactions to undergraduate science laboratory and project work. In E. Hegarty-Hazel (Ed.), *The student laboratory and the science curriculum* (pp. 383–396). London: Rutledge.

Blosser, P. (1980). *A critical review of the role of the laboratory in science teaching.* Columbus, OH: Center for Science and Mathematics Education.

Blosser, P. (1983, May). The role of the laboratory in science teaching. *School Science and Mathematics, 83,* 165–169.

Blosser, P. (1988). Labs—are they really as valuable as teachers think they are? *Science Teacher, 55,* 57–59.

Boud, D., Dunn, J., & Hegarty-Hazel, E. (1986). *Teaching in laboratories.* Guildford: Society for Research in Higher Education, Nelson. National Foundation for Educational Research (*NFER*) in England.

Boud, D. J., Dunn, J., Kennedy, T., & Thorly, R. (1980). The aims of science laboratory courses: A survey of students, graduates and practicing scientists. *European Journal of Science Education, 2,* 415–428.

Brandwein, P. (1962). Beginnings in developing an art of investigation. In P. Brandwein, J. Metzner, E. Morholt, A. Roe, & W. Rosen (Eds.), *Teaching high school biology: A guide to working with potential biologists,* BSCS Bulletin No. 20, pp. 43–60. Washington, DC: American Institute of Biological Sciences.

Bryce, T. G. K., & Robertson, I. J. (1985). What can they do? A review of practical assessment in science. *Studies in Science Education, 12,* 1–24.

Burns, P. K., & Bozeman, W. C. (1980). Computer-assisted instruction and mathematics achievement: Is there a relationship? *Educational Technology, 20,* 32–39.

Butler, W., & Griffin, H. (1979). Simulations in general chemistry laboratory with microcomputers. *Journal of Chemical Education, 56,* 543–545.

Campbell, J. R. (1972). Is scientific curiosity a viable outcome in today's secondary school science programs? *School Science & Mathematics, 52,* 139–147.

Capie, W., & Tobin, K. G. (1982). Pupils engagement in learning tasks. *Journal of Research in Science Teaching, 18,* 409–418.

Champagne, A. B., Klopfer, L. E., & Anderson, J. H. (1980). Factors of influencing the learning of classical mechanics. *American Journal of Physics, 48,* 1074–1079.

Charlton, R. E. (1980). Teacher-to-teacher cognitive style considerations for the improvement of biology education. *American Biology Teacher, 42,* 244–247.

Clement, D. H. (1985). Logo programing: Can it change how children think? *Electronic Learning, 4*(4), 28.

Clement, J. (1982). Students' preconceptions in introductory mechanics. *American Journal of Physics, 50,* 66–71.

Cohen, I. (1987). *Various strategies for the teaching of the topics of chemical energy and chemical equilibrium: Development, implementation and evaluation.* Thesis for doctor of philosophy degree, Weizman Institute of Science, Rehovot, Israel.

Cohen, J. (1969). *Statistical power and analysis for the behavioral sciences.* New York: Academic Press.

Connel, L. (1971). Demonstration and individual practical work in science teaching: A review of opinions. *School Science Review, 52,* 692–702.

Cooler, D. (1985). The biologist toolbox: Anova with Super Cale. *Bioscience, 35,* 306–307.

Cooley, W. W., & Klopfer, L. E. (1961). Test on understanding science (Form W). Princeton, NJ: Educational Testing Service.

Comber, L. C., & Keeves, J. (1973). *Science education in nineteen countries.* New York: Wiley.

Crovello, T. J. (1982). Computers in biological education. *American Biology Teacher, 44,* 476–483.

Curtis, J. B. (1986). Teaching college biology students the simple linear regression model using an interactive microcomputer graphics software package. *Dissertation Abstracts International, 46,* 7a.

Daley, M., & Hiller, D. (1981). Computer simulation of the population growth (*Schizosaccharomyces pombe*) experiment. *Journal of Biology Education, 15,* 266–268.

Davis J. (1972). *An assessment of change matriculation and science facilities initiated by NDEA Title III used for high school science in Tennessee between 1965–1970.* Unpublished Ed. D. thesis, University of Tennessee.

Dennis, J. (1979). *The Illinois series on educational applications of computers.* Urbana, IL: College of Education.

De Sessa, A. (1982). Unlearning Aristotelian physics: A study of knowledge based learning. *Cognitive Science, 6,* 37–75.

Deutsch, M. A. (1949). A theory of cooperation and competition. *Human Relations, 2,* 129–151.

De Vries, D. L., & Slavin, R. E. (1978). Teams-Games-Tournament (TGT): Review of ten classroom experiments. *Journal of Research and Development in Education, 12,* 28–38.

Doran, R. L. (1978). Assessing the outcomes of science laboratory activities. *Science Education, 62,* 401–409.

Dreyfus, A. (1979). *Evaluation of the impact of the BSCS on laboratory facilities in schools.* Jerusalem: Israel Science Teaching Center, Hebrew University (in Hebrew).

Dreyfus, A. (1983). The devil's advocate as a teachers trainer. In P. Tamir, A. Hofstein, & M. Ben Peretz (Eds.), *Preservice and inservice education of science teachers* (pp. 211–214). Rehovot: Balaban Interscience.

Dreyfus, A. (1986). Manipulating and diversifying the levels of the difficulty and task sophistication of one and the same laboratory exercise. *European Journal of Science Education, 8,* 17–25.

Driver, R. (1982). *The pupil as a scientist?* Milton Keynes: Open University Press.

Driver, R., & Bell, B. (1986). Students' thinking and the learning of science: A constructivist view. *School Sciences Review, 67,* 443–456.

Druva, C. A. (1982). *Science meta-analysis: Teacher characteristics by teacher behavior and student outcomes.* Paper delivered at the 1992 annual meeting of the National Association for Research in Science Teaching, April 7, Lake Geneva, WI.

Dunn, J. (1986). Teaching strategies. In D. Boud, J. Dunn, & E. Hegarty-Hazel (Eds.), *Teaching in laboratories* (pp. 35–36). Guildford: Society of Higher Education, NFER.

Edgeworth, R. L., & Edgeworth, M. (1811). *Essays on practical education* (3rd ed.). London: Johnson.

Edwards, J., & Power, C. (1990). Role of laboratory in a national junior secondary science project: Australian Science Education Project (ASEP). In E. Hegarty-Hazel (Ed.), *The student laboratory and the curriculum* (pp. 315–336). London: Rutledge.

Eggleston, J. (1983). Teacher pupil interactions in science lessons: Explorations and theory. In P. Tamir, A. Hofstein, & M. Ben Peretz (Eds.), *Preservice and inservice education of science teachers* (pp. 519–536). Rehovot: Balaban International Science Services.

Eglen, J. R., & Kempa, R. F. (1974). Assessing manipulative skills in practical chemistry. *School Science Review, 56,* 261–273.

Engelhard, D. F. (1968). *Aspects of spatial influence on science teaching methods.* Unpublished Ed. D. thesis, Harvard University, School of Education.

Erickson, G. L. (1979). Children's conceptions of heat and temperature. *Science Education, 63,* 221–230.

Farmer, W. A., Farrell, M. A., Clark, R. M., & McDonald, J. (1982). A validity study of two paper-pencil tests of concrete and formal operations. *Journal of Research in Science Teaching, 19,* 475–483.

Farrell, M. A., & Farmer, W. A. (1985). Adolescents performance on a sequence of proportional reasoning tasks. *Journal of Research in Science Teaching, 22,* 503–518.

Feinstein, B., & Blum, A. (1983). Adaptation of teacher inservice training to a change in curriculum style and teachers' experience. In P.

Tamir, A. Hofstein, & M. Ben-Peretz (Eds.), *Preservice and inservice education of science teachers* (pp. 379–381). Rehovot: Balaban Interscience.

Fennema, E., & Sherman, J. (1977). Sex-related differences in mathematics achievement, spatial visualization and affective factors. *American Educational Research Journal, 14*, 51–71.

Fennema, E., & Sherman, J. (1979). Sex-related differences in mathematics achievement and related factors: A further study. *Journal for Research in Mathematics Education, 10*, 189–201.

Fensham, P. J. (1990). Practical work and the laboratory in science for all. In E. Hegarty-Hazel (Ed.), *The student laboratory and the science curriculum* (pp. 291–311). London and New York: Routledge.

Fraser, B. (1989). *Learning environment research in science classrooms: Past progress and future prospects* (NARST Monograph No. 2). Washington, DC: National Association of Research in Science Teaching.

Fraser, B. J., Giddings, G. J., Giffith, A. K., Hofstein, A., & Herkle, D. G. (1990, April). *A cross-national study of science laboratory classroom environments*. Symposium presented at the annual meeting of National Association of Research in Science Teaching, Atlanta, GA.

Friedler, Y. (1984). *Problems and processes in learning and teaching in the biology laboratory in Israeli high schools*. Unpublished Ph. D. thesis, Hebrew University, Jerusalem (in Hebrew with English summary).

Friedler, Y., Nachmias, R., & Linn, M. C. (1990). Learning scientific reasoning skills in micro-computer-based laboratories. *Journal of Research in Science Teaching, 27*, 173–192.

Friedler, Y., & Tamir, P. (1984). Teaching and learning in high school laboratory in Israel. *Research in Science Education, 15*, 89–96.

Friedler, Y., & Tamir, P. (1986). Teaching basic concepts of scientific research to high school students. *Journal of Biological Education, 24*, 263–269.

Friedler, Y., & Tamir, P. (1990). Life in science laboratory classroom at secondary level. In E. Hegarty-Hazel (Ed.), *The student laboratory and the curriculum* (pp. 337–354). London: Rutledge.

Gallagher, J. J., & Tobin, K. (1987). Teacher management and student engagement in high school science. *Science Education, 71*, 535–555.

Ganiel, U., & Hofstein, A. (1982). Objective and continuous assessment of student performance in the physics laboratory. *Science Education, 66*, 581–591.

Gardner, P., & Gauld, C. (1990). Labwork and students' attitudes. In E. Hegarty-Hazel (Ed.), *The student laboratory and the curriculum* (pp. 132–158). London: Rutledge.

Gauld, C. D. (1978). Practical work in sixth-form biology. *Journal of Biological Education, 12*, 33–38.

Giddings, G., & Hofstein, A. (1980). Trends in the assessment of laboratory performance in high school science instruction. *Australian Science Teacher, 26*(3), 57–69.

Graef, J. (1984). Teaching science with computers. *Physics Teacher, 22*, 430–436.

Grobman, A. (1969). *The changing classroom*. New York: Doubleday.

Grobman, H. (1963). Some comments on the evaluation program findings and their implications. *BSCS Newsletter, 19*, 25–29.

Gunstone, R. F., & Champagne, A. B. (1990). Promoting conceptual change in the laboratory. In E. Hegarty-Hazel (Ed.), *The student and the curriculum* (pp. 159–182). London: Rutledge.

Gunstone, R. F., & White, R. T. (1981). Understanding of gravity. *Science Education, 65*, 291–299.

Hallworth, H. J., & Brehner, A. (1980). *Computer assisted instruction in schools: Achievement, present development and projections for the future*. Calgary, Canada: ERIC Document Reproduction Service No. ED 200 18.

Handley, H. M., & Morse, L. W. (1984). Two-year study rating adolescents' self-concept and gender role perceptions to achievement towards science. *Journal of Research in Science Teaching, 21*, 599–607.

Hartley, J., & Lowell, K. (1977). The psychological principle underlying the design of computer-based instructional systems. In D. Walker & R. Hess (Eds.), *Instructional Software*. Belmont, CA: Wadsworth.

Harty, H., & Beall, D. (1984). Attitudes towards science of gifted and non-gifted 5th-graders. *Journal of Research in Science Teaching, 21*, 438–488.

Haukoos, G. A., & Penick, J. E. (1983). The influence of classroom climate on science process and content achievement of community college students. *Journal of Research in Science Teaching, 20*, 629–637.

Hayson, J. T., & Sutton, C. R. (1974). *Science teachers education project: Activities and experiences*. London: McGraw-Hill.

Hegarty-Hazel, E. (Ed.). (1990a). *The student laboratory and the science curriculum*. London: Rutledge.

Hegarty-Hazel, E. (1990b). Overview. In E. Hegarty-Hazel (Ed.), *The student laboratory and the science curriculum* (pp. 3–26). London: Rutledge.

Hegarty-Hazel, E. (1990c). Learning technical skills in the student laboratory. In E. Hegarty-Hazel (Ed.), *The student laboratory and the curriculum* (pp. 75–94). London: Rutledge.

Hegarty-Hazel, E. (1990d). Life in science laboratory classrooms at tertiary level. In E. Hegarty-Hazel (Ed.), *The student laboratory and the curriculum* (pp. 357–382). London: Rutledge.

Henry, N. W. (1975). Objectives for laboratory work. In P. L. Gardner (Ed.), *The structure of science education* (pp. 61–75). Hawthorn, Victoria: Longman.

Herron, M. D. (1971). The nature of scientific inquiry. *School Review, 79*, 171–212.

Hertz-Lazarowitz, R., Baird, H. J., Webb, C. D., & Lazarowitz, R. (1984). Student-student interaction in science classrooms: A naturalistic study. *Science Education, 68*, 603–619.

Hickman, M. F. (1978). Teaching strategies and styles. In W. V. Mayer (Ed.), *Biology Teachers' Handbook* (3rd ed.) (pp. 101–298). New York: Wiley & Sons.

Hodson, D. (1988a). Experiments in science and science teaching. *Educational Philosophy and Theory, 20*(2), 53–66.

Hodson, D. (1988b). Philosophy of science, science and science education. *Studies in Science Education, 12*, 25–57.

Hofstein, A. (1988). Practical work and science education II. In P. Fensham (Ed.), *Development and dilemmas in science education* (pp. 189–216). London: Falmer.

Hofstein, A., Gluzman, R., Ben-Zvi, R., & Samuel, D. (1980). A comparative study of chemistry students' perception of the learning environment in high schools and vocational schools. *Journal of Research in Science Teaching, 17*, 547–552.

Hofstein, A., & Lunetta, V. N. (1982). The role of the laboratory in science teaching: Neglected aspects of research. *Review of Educational Research, 52*, 201–217.

Holford, D. G. (1974). Case studies in the evaluation of particular curriculum units. In J. T. Haysom & C. R. Sutton (Eds.), *Innovation in teacher education* (pp. 94–109). London: McGraw-Hill.

Holman, J. (1986). SATIS—Another STS initiative from Britain Reporter. *Science Through Science Technology and Society, 2*(3), 1–3.

Hounshell, P. B., & Hill, S. R., Jr. (1989). The microcomputer and achievement and attitudes in high school biology. *Journal of Research in Science Teaching, 26*, 543–549.

Howe, A. C., & Durr, B. P. (1982). Analysis of an instructional unit for level of cognitive demand. *Journal of Research in Science Teaching, 19*, 217–224.

Humphreys, B., Johnson, R. T., & Johnson, D. W. (1982). Effects of cooperative, competitive and individualistic learning on students' achievement in science class. *Journal of Research in Science Teaching, 19,* 351–356.

Huppert, J., & Lazarowitz, R. (1985). *The growth curve of the microorganisms: Computer assisted learning units.* Israel: Biology, TALGAR, Granot Educational Computer System. (Software and Learning Unit in English).

Igelsrud, D., & Leonard, W. H. (1988). Labs: What research says about biology laboratory instruction. *American Biology Teacher, 50,* 303–306.

Inhelder, R., & Piaget, J. (1958). *The growth of logical thinking from childhood to adolescence.* New York: Basic Books.

Jacoby, B. A., & Spargor, P. E. (1989). Ptolemy revived? Instrumentalism in selected British, American and South African high school physical science textbooks. *Interchange, 20,* 33–80.

James, R. K. (1987, April). *Follow up study of the concerns of Texas science teachers about the forty percent lab time role.* Paper presented at the annual meeting of the National Association for Research in Science Teaching, Washington, DC.

Jenkins, E. W. (1979). *From Armstrong to Nuffield.* London: John Murray.

Johnson, D. W., Johnson, R. T., Maruyama, G., Nelson, D., & Skon, L. (1981). Effects of cooperative competitive and individualistic goal structures on achievement: A metanalysis. *Psychological Bulletin, 89,* 47–62.

Johnson, D. W., Johnson, R. T., & Scott, L. (1978). The effects of cooperative and individualized instruction on student attitudes and achievement. *Journal of Social Psychology, 104,* 207–216.

Johnson, R. T. (1976). The relationship between cooperation and inquiry in science classrooms. *Journal of Research in Science Teaching, 13,* 55–63.

Johnstone, A. H., & Wham, A. J. B. (1962). The demands of practical work. *Education in Chemistry, 19,* 71–73.

Jungwirth, E. (1970). An evaluation of the attained development of the intellectual skills needed to understanding the nature of scientific inquiry by BSCS pupils in Israel. *Journal of Research in Science Teaching, 7,* 141–151.

Jungwirth, E. (1975). The problem of teleology in biology as a problem of biology teacher education. *Journal of Biological Education, 9,* 243–246.

Karplus, R. (1977). Science teaching and the development of reasoning. *Journal of Research in Science Teaching, 14,* 169–175.

Kaufman, G. B. (1989). History in the chemistry curriculum. *Interchange, 20,* 81–98.

Keightley, J. V., & Best, E. D. (1975). Student preferences for year 11 biology classes in some south Australian schools. *Research in Science Education, 5,* 57–67.

Kelly, P. J., & Lister, R. (1969). Assessing practical ability in Nuffield A level biology. In J. F. Eggleston & J. F. Kerr (Eds.), *Studies of assessment* (pp. 129–142). London: English Universities Press.

Kerr, J. F. (1964). *Practical work in school science.* Leicester: Leicester University Press.

Kinnear, J. F. (1982a). Computer simulation and concept development in students of genetics. *Research in Science Education, 12,* 89–96.

Kinnear, J. F. (1982b). *Birdbreed, program and demonstration.* Melbourne, Australia: Massa Educational Technology.

Klinckman, E., ed. (1970). *Biology teacher handbook* (2nd ed.). New York: Wiley.

Klopfer, L. E. (1990). Learning scientific inquiry in the school laboratory. In E. Hegarty-Hazel (Ed.), *The student laboratory and the science curriculum* (pp. 95–118). London: Rutledge.

Kreitler, H., & Kreitler, S. (1974). The role of the experiment in science education. *Instructional Science, 3,* 75–88.

Kuhn, T. S. (1970). *The structure of scientific revolutions* (2nd ed.). Chicago: University of Chicago Press.

Kyle, W. C., Penick, J. E., & Shymansky, J. A. (1979). Assessing and analyzing the performance of students in college science laboratories. *Journal of Research in Science Teaching, 16,* 545–551.

Layton, D. (1990). Student laboratory practice and the history and philosophy of science. In E. Hegarty-Hazel (Ed.), *The student laboratory and the science curriculum* (pp. 37–59). London: Rutledge.

Lawson, A. E. (1975). Developing formal thought through biology teaching. *American Biology Teacher, 37,* 411–419, 429.

Lawson, A. E., Abraham, M. R., & Renner, J. W. (1989). *A theory of instruction* (NARST Monograph No. 1). National Association of Research in Science Teaching.

Lawson, A. E., & Blake, A. J. D. (1976). Concrete and formal thinking abilities in high school biology students as measured by three separate instruments. *Journal of Research in Science Teaching, 13,* 227–235.

Lawson, A. E., & Renner, J. W. (1975). Piagetian theory and biology teaching. *American Biology Teacher, 37,* 336–343.

Lawson, A. E., & Wollman, W. T. (1976). Encouraging transition from concrete to formal cognitive functioning an experiment. *Journal of Research in Science Teaching, 13,* 413–430.

Lawrenz, F., & Munch, T. (1984). The effect of grouping of laboratory students on selected educational outcomes. *Journal of Research in Science Teaching, 21,* 699–708.

Lazarowitz, R. (1991). Learning biology cooperatively: An Israeli junior high school study. *Cooperative Learning, 11,* 18–20.

Lazarowitz, R., Baird, J. H., Hertz-Lazarowitz, R., & Jenkins, J. (1985). The effects of modified jigsaw on achievement classroom social climate, and self-esteem in high school science classes. In R. Slavin, S. Sharan, S. Kagan, R. Hertz-Lazarowitz, C. Webb, & R. Schmuck (Eds.), *Learning to cooperate, cooperating to learn* (pp. 231–253). New York: Plenum.

Lazarowitz, R., & Galon, M. (1990, July). *Learning biology in a cooperative setting: Ninth grade students' achievement and cognitive reasoning stages.* International Convention on Cooperative Learning, Baltimore Convention Center, Baltimore.

Lazarowitz, R., & Huppert, Y. (1981). The development of individualized audio-visual learning units in junior high school biology and students' achievements. *European Journal of Science Education, 3,* 195–204.

Lazarowitz, R., & Huppert, J. (1988). Computer assisted learning in biology; student achievement by gender and cognitive levels. In F. Lovis & E. D. Tagg (Eds.), *Computer in education* (pp. 589–597). Amsterdam: Elsevier Science Publications, North Holland.

Lazarowitz, R., & Karsenty, G. (1990). Cooperative learning and student academic achievement, process skills, learning environment and self-esteem in tenth grade biology classrooms. In S. Sharan (Ed.), *Cooperative learning, theory and research* (pp. 123–149). New York: Praeger.

Lazarowitz, R., & Lee, A. E. (1976). Measuring inquiry attitudes of secondary science teachers. *Journal of Research in Science Teaching, 13,* 455–460.

Lazarowitz, R., & Shemesh, M. (1989, July). *Students, achievement and cognitive operational levels in a computer assisted learning in microbiology* (pp. 9–13). Paper presented at the tenth biennial meetings, International Society for the Study of Behavioral Development (ISSBD), Jyvaskyla, Finland.

Lazarowitz, R., & Witenoff, S. (1990). Teaching the pH concept to nonformal operational students in the ninth grade biology laboratory. *Research in Education, 44,* 21–38.

Lazarowitz, R., & Yaakoby, J. (1989a, March). *Inquiry skills of tenth grade biology students in a computer assisted learning setting.* Paper presented at the annual meeting of the National Association of Research in Science Teaching (NARST), San Francisco.

Lazarowitz, R., & Yaakoby, J. (1989b, September). *The integration of computer assisted learning in existing curriculum in 10th grade biology: Students, achievements by method and gender.* Paper presented at the Third European Conference for Research on Learning and Instruction, Madrid.

Lazarowitz, R. Yaakoby, J., & Huppert J. (1990, April). *The integration of computer assisted learning in existing curriculum in 10th grade biology: Students, academic achievement, inquiry skills and cognitive operational stages.* Paper presented at the seventh annual meeting of the Israeli Association for Computers in Education, Kfar Hamacabia, Israel.

Lee, A. E., ed. (1969). *The preservice preparation of secondary school biology teachers* (Publication No. 25). Washington, DC: U. S. Commission on Undergraduate Education in the Biological Sciences.

Lee, A. E., Lehman, D. L., & Peterson, G. E. (1967). *Laboratory blocks in teaching biology* (Special Publication No. 5). Boulder, CO: Biological Science Curriculum Study.

Leonard, W. H. (1987). Interactive videodisc: Computer instruction of the future? *Collegiate Microcomputer, 5,* 197–201.

Leonard, W. H. (1988). Interfacing in the biology laboratory: State of art. *American Biology Teacher, 50*(8), 523–526.

Leonard, W. H. (1990). Computer-based technology in college science laboratory courses. *Journal of College Science Teaching, 19,* 210–211.

Lock, R. (1988). A history of practical work in school science and its assessment, 1860–1986. *School Science Review, 70,* 115–119.

London Association for the Teaching of English. (1969). *Language, the learner and the school* (pp. 125–128). London: Penguin.

Lunetta, V. N. (1974). Computer based dialogs: A supplement to the physics curriculum. *Physics Teacher, 12,* 355–356.

Lunetta, V. N., & Tamir, P. (1981). An analysis of laboratory activities: Project Physics and PSSC. *School Science & Mathematics, 81,* 230–236.

Lunneta, V., & Hofstein, A. (1981). Simulation in science education. *Science Education, 65,* 243–252.

Marin, O., Friedler, Y., & Tamir, P. (1990). *The design and validation of an inquiry microcomputer based test.* Paper presented in the annual meeting of the Association for Teacher Education in Europe, Limerick, Ireland.

Marks, G. H. (1982). Computer simulations in science teaching: An introduction. *Journal of Computers in Mathematics and Science Teaching, 1,* 18–20.

Maor, D. (1990, July). *Development of student inquiry skills in a computerized classroom environment.* Paper presented at the 21st annual conference of Australian Science Education Research Association (ASERA), Curtin University, Perth, Western Australia.

Matthews, M. R., & Winchester, I. (1989). History, science and science teaching. *Interchange, 20,* 1–15.

May, P. M., Murray, K., & Williams, D. (1985). CAPE: A computer program to assist with practical assessment. *Journal of Chemical Education, 62,* 310–312.

Mayer, W. V. (1967). Foreword. In A. E. Lee, D. L. Lehman, & G. E. Peterson (Eds.), *Laboratory in teaching biology* (Special Publication No. 25, p. iv). Boulder CO: Biological Science Curriculum Study.

McNally, D. W. (1974). *Piaget education and teaching.* Sydney: Argus & Robertson.

Medawa, P. B. (1963). Is the scientific paper a fraud? *The Listener* (BPC publication), Sept. 12, 377–378.

Metzner, J. (1962). Promising practices with capable students. In P. Brandwein, J. Metzner, E. Morholt, A. Roe, & W. Rosen (Eds.), *Teaching high school biology: A guide to working with potential biologists* (Bulletin No. 2, pp. 13–42). Washington, DC: American Institute of Biological Sciences.

Michaels, J. W. (1977). Classroom reward structures and academic performance. *Review of Educational Research, 47,* 87–98.

Millar, R. (1987). Towards a role for experiment in the science teaching laboratory. *Studies in Science Education, 14,* 109–118.

Millar, R. H., & Underwood, C. I. (1984). Using the analogue input port of the BBC microcomputer: Some general principles and a specific example. *School Science Review, 66,* 270–279.

Miller, L. K., & Hamblin, R. L. (1963). Interdependence, differential rewarding and production. *American Sociological Review, 28,* 768–778.

Moore, R. W., & Sutman, F. X. (1970). The development field test and validation of an inventory of scientific attitudes. *Journal of Research in Science Teaching, 7,* 85–94.

Morgan, R. M., Markell, C. S., & Feller, R. F. (1987). A microcomputer exercise on muscle physiology. *Journal of College Science Teaching, 17,* 23–27.

Nachmias, R. (1989). The microcomputer based laboratory: Theory and practice. *Megamot Behavioral Science Quarterly, 32,* 245–261 (in Hebrew).

Nicklin, R. C. (1985). The computer as a lab partner. *Journal of College Science Teaching, 15,* 31–35.

Noelting, G. (1980). The development of proportional reasoning and the ratio concept: Part II. Problem structural at successive stages: Problem-solving strategies and the mechanism of adaptive restructuring. *Educational Studies in Mathematics, 11,* 331–363.

Novak, J. D. (1972). *Facilities for secondary school science teaching: Evolving patterns in facilities and programs.* Washington, DC: National Science Teachers Association.

Novak, J. D. (1990). The interplay of theory and methodology. In E. Hegarty-Hazel (Ed.), *The student laboratory and science learning* (pp. 60–70). London: Rutledge.

Novak, J. D., & Gowin, B. (1984). *Learning how to learn.* Cambridge: Cambridge University Press.

Nussbaum, J., & Novak, J. O. (1976). An assessment of children's concepts of the earth utilizing structured interviews. *Science Education, 60,* 535–550.

Nussbaum, J., & Novick, S. (1982). Alternative frameworks conceptual conflict and accommodation: Toward a principled teaching strategy. *Instructional Science, 11,* 183–200.

Ogborn, J. (1978). Aims and organization of laboratory work. In J. G. Jones & J. L. Lewis (Eds.), *The role of the laboratory in physics education* (pp. 1–64). Birmingham: Goodman.

Okebukola, P. A. O., & Ogunniyi, M. B. (1984). Cooperative, competitive and individualistic laboratory interaction patterns: Effects on students' performance and acquisition of practical skills. *Journal of Research in Science Teaching, 21,* 875–884.

Okey, J. R., Shaw, E. L., & Waugh, M. L. (1983, April). *Development and evaluation of lesson plans for computer simulations.* Paper presented at the annual meeting at the National Association for Research in Science Teaching, Dallas.

Olson, D. R. (1973). What is worth knowing and what can be taught. *School Review, 82,* 27–43.

Olson, J. R., Emerson, C. J., & Nungessner, M. K. (1980). *Geoecology: A country level environmental data base for the continental United States.* Oak Ridge, TN: Oak Ridge National Laboratory.

Osborne, R. J. (1976). Using student attitudes to modify instruction in physics. *Journal of Research in Science Teaching, 13,* 525–531.

Osborne, R., & Wittrock, M. (1985). The generative learning model and its implications for science education. *Studies in Science Education, 12,* 59–87.

Pappert, S. (1980). *Midstorms.* New York: Basic Books.

Pella, M. O. (1961). The laboratory and science teaching. *Science Teacher, 28,* 20–31.

Penick, J. (1986). The STS curriculum: Notes on exemplary programs. *Reporter: Science Through Science Technology and Society, 2,* 1–5.

Peterson, P. L., & Janicki, T. C. (1979). Individual characteristics and children's learning in large-group and small group approaches. *Journal of Educational Psychology, 71*, 677–687.

Piaget, J. (1978). *Behavior and evolution*. New York: Pantheon.

Pickering, M. (1980). Are lab courses a waste of time? *Chronicle of Higher Education, 19*, 44–50.

Pickering, M. (1987). Laboratory education as a problem in organization. *Journal of College Science Teaching, 16*, 187–189.

Piper, M. K., & Moore, K. D. (1977). *Attitudes toward science: Investigations*. Columbus, OH: SMEAC Information Reference Center.

Polyani, M. (1969). *Knowing and being known*. London: Rutledge & Kegan Paul.

Postlethwaite, S. N., Novak, J., & Murray, H. (1964). *An integrated experience approach to learning*. Minneapolis: Burgess.

Prosser, M. T., & Tamir, P. (1990). Developing and improving the role of computers in students laboratories. In E. Hegarty-Hazel (Ed.), *The student laboratory and the science curriculum* (pp. 267–290). London: Rutledge.

Quinlan, C. (1981). Project Physics in N. S.W. *Research in Science Education, 11*, 86–87.

Ramsey, G. A., & Howe, R. W. (1969). An analysis of research in instructional procedures in secondary school science, part II. *Science Teacher, 36*, 72–81.

Reif, F., & St. John, M. (1979). Teaching physicists thinking skills in the laboratory. *American Journal of Physics, 47*, 750–757.

Rhodes, S. B. (1986). A microcomputer kymograph. *Journal of College Science Teaching, 15*, 523–527.

Rivers, R. H., & Vockell, E. (1987). Computer simulations to stimulate scientific problem solving. *Journal of Research in Science Teaching, 24*, 403–415.

Roberts, N., Anderson, D., Dean, R., Garet, M., & Shaffer, W. (1983). *Introduction to computer simulations*. Reading, MA: Addison-Wesley.

Robinson, J. (1969). Evaluating laboratory work in high school biology. *American Biology Teacher, 31*, 236–240.

Roblyer, M. D. (1985). *Measuring the impact of computers in construction: A non-technical review of research for educators*. Washington, DC: AEDS.

Romey, W. O. (1969). *Inquiry techniques for teaching science*. Englewood Cliffs, NJ: Prentice-Hall.

Rosen, S. A. (1954). History of the physics laboratory in the American public schools (to 1910). *American Journal of Physics, 22*, 194–204.

Royce, G. K. (1979). *The development and validation of a diagnostic criterion-referenced test of science processes*. Ph. D. dissertation, University of Nebraska.

Rubin, A., & Tamir, P. (1988). Meaningful learning in the school laboratory. *American Biology Teacher, 50*, 477–482.

Schwab, J. J. (1962). The teaching of science inquiry. In J. J. Schwab & P. F. Brandwein (Eds.), *The teaching of science*. Cambridge MA: Harvard University Press.

Schwab, J. J. (1963). *Biology teachers handbook*. New York: Wiley.

Selmes, C., Ashton, B. G., Meredith, H. M., & Newal, A. (1969). Attitudes to science and scientists. *School Science Review, 51*, 7–22.

Shadmi, Y. (1983). Training primary school teachers. In P. Tamir, A. Hofstein, & M. Ben Peretz (Eds.), *Preservice and inservice education of science teachers* (pp. 117–124). Philadelphia: Rehovot Balaban International Science Services.

Sharan, S. (1980). Cooperative learning in small groups: Recent methods and effects on achievements attitudes and ethnic relations. *Review of Educational Research, 50*, 241–271.

Sharan, S., & Hertz-Lazarowitz, R. (1980). A group investigation method of cooperative learning in the classroom. In S. Sharan, P. Hare, D. D. Webb, & R. Hertz-Lazarowitz (Eds.), *Cooperation in education* (pp. 14–46). Provo, UT: Brigham Young University Press.

Sharan, S., Kussell, P., Hertz-Lazarowitz, R., Bejarano, Y., Raviv, S., &

Sharan, Y. (1984). *Cooperative learning in the classroom*. Hillsdale, NJ: Lawrence Erlbaum.

Shaw, E. L., & Okey, J. R. (1985, April). *Effects of microcomputer simulations on achievement and attitudes of middle school students*. Paper presented at the annual meeting of the National Association for Research in Science Teaching, French Lick Springs, IN.

Showalter, B. (1984). What is unified science education? *Prism, 11*, 1(2). Columbus: Center for Unified Science Education, Ohio State University.

Shulman, L. S., & Tamir, P. (1973). Research on teaching in the natural sciences. In R. M. W. Travers (Ed.), *Second handbook of research on teaching* (pp. 1098–1140). Chicago: Rand McNally.

Shymansky, J. A., Hedges, L. V., & Woodworth, G. (1990). A reassessment of the effects of inquiry-based science curricula of the 60's on student performance. *Journal of Research in Science Teaching, 27*, 127–144.

Silberstein, M., & Tamir, P. (1981). Factors which affect students' attitudes toward the use of living animals in learning biology science. *Science Education, 65*, 119–130.

Skon, L., Johnson, D. W., & Johnson, R. T. (1981). Cooperative peer interaction versus individual competition and individualistic efforts: Effects on the acquisition of cognitive reasoning strategies. *Journal of Educational Psychology, 73*, 83–92.

Slavin, R. E. (1980). Cooperative learning. *Review of Educational Research, 50*, 315–342.

Smith, J. P. (1971). The development of a classroom observation inventory relevant to the Earth Science Curriculum Project. *Journal of Research in Science Teaching, 8*, 231–235.

Songer, N., & Linn, M. C. (1986). *Linking laboratory science and practical science*. Unpublished manuscript. Berkeley: University of California Education in Mathematics Science and Technology.

Soon Choi, B., & Gennaro, E. (1987). The effectiveness of using computer simulated experiments on junior high students, understanding of the volume displacement concept. *Journal of Research in Science Teaching, 24*, 539–552.

Sparkes, R. A. (1982). Microcomputers in science teaching. *School Science Review, 63*, 442–452.

Stake, R. E., & Easley, J. (1978). *Case studies in science education*. Urbana-Champaign. University of Illinois, Center for instructional and Curriculum Evaluation.

Stein, J. S., Nachmias, R., & Friedler, Y. (1990). An experimental comparison of two science laboratory environments: Traditional and microcomputer-based. *Journal of Education Computing Research, 6*(2) 183–202.

Stevens, S. M. (1985, April). *Interactive computer/videodisc lessons and their effect on students', understanding of science*. Paper presented to the annual meeting of the National Association for Research in Science Teaching, French Lick Springs, IN.

Tamir, P. (1972). The practical mode of performance in biology: A distinct mode. *Journal of Biological Education, 6*, 175–182.

Tamir, P. (1974). An inquiry oriented laboratory examination. *Journal of Educational Measurement, 11*, 25–33.

Tamir, P. (1975). Nurturing the practical mode in schools. *The School Review, 83*, 499–506.

Tamir, P. (1976a). The Israeli High School biology project: A case of adaptation. *Curriculum Theory Network, 5*, 305–315.

Tamir, P. (1976b). *The role of the laboratory in science teaching* (Technical report no. 10). Iowa City: University of Iowa Science Education Center.

Tamir, P. (1977). How are the laboratories used? *Journal of Research in Science Teaching, 14*, 311–316.

Tamir, P. (1978). The impact of BSCS on laboratory facilities for biology students in Israeli high schools. *International Evaluation Journal*. Princeton, NJ: Educational Testing Service.

Tamir, P. (1983a). Teachers' self reports: As an alternative approach for the study of classroom transactions. *Journal of Research in Science Teaching, 20,* 815–823.

Tamir, P. (1983b). External examinations as a means for teacher education. In P. Tamir, A. Hofstein, & M. Ben Peretz (Eds.), *Preservice and inservice of science teachers* (pp. 615–621). Rehovot: Balaban Interscience Services.

Tamir, P. (1985). Practical examinations. In T. Husen & N. Postlethwite (Eds.), *International encyclopedia of education* (Vol. 7, pp. 4011–4017). Oxford: Pergamon.

Tamir, P. (1989a). Training teachers to teach effectively in the laboratory. *Science Education, 73,* 56–69.

Tamir, P. (1989b). The ecological project as an example of bridging between central and school based curriculum development. In M. Silberstein & A. Lewy (Eds.), Theory and practice in curriculum development. *Hallacha Lemaseeh, 5,* 25–38 (in Hebrew).

Tamir, P. (1990a). *Teaching for meaningful active inquiry.* Unpublished paper, Israel Science Teaching Center, Hebrew University, Jerusalem.

Tamir, P. (1990b). Evaluation of student work and its role in developing policy. In E. Hegarty-Hazel (Ed.), *The student laboratory and the science curriculum* (pp. 242–266). London: Rutledge.

Tamir, P. (1991). Practical work in school science: An analysis of current practice. In B. Woolnough (Ed.), *Practical science* (pp. 13–20). Celtic Court: Open University Press.

Tamir, P., & Amir, R. (1987). Interrelationships among laboratories process skills in biology. *Journal of Research in Science Teaching, 24,* 137–144.

Tamir, P., Amir, R., & Friedler, Y. (1988). High school students' difficulties in understanding osmosis. *International Journal of Science Education, 9,* 541–551.

Tamir, P., & Doran, R. (1992). *Conclusions and disscusion of findings related to practical skills testing in science. Studies in Educational Evaluation, 18,* 393–408.

Tamir, P., & Jungwirth, E. (1975). Students, growth as result of studying BSCS biology for several years. *Journal of Research in Science Teaching, 12,* 263–279.

Tamir, P., & Lunetta, V. N. (1978). An analysis of laboratory activities in the BSCS Yellow Version. *American Biology Teacher, 40,* 353–357.

Tamir, P., & Lunetta, V. N. (1981). Inquiry related tasks in high school science laboratory handbooks. *Science Education, 65,* 477–484.

Tamir. P., & Nussinovitz, R. (1979). *Analysis of students' answers to questions in the biology practical laboratory examinations.* Jerusalem, Israel Science Teaching Center, Hebrew University Jerusalem (in Hebrew).

Tamir, P., Nussinovitz, R., & Friedler, Y. (1982). The design and use of practical tests assessment inventory. *Journal of Biological Education, 16,* 42–50.

Tasker, R. (1981). Children's views and classroom experiences. *Australian Science Teachers Journal, 27,* 33–37.

Tiberghein, A., & Delacote, G. (1979). Manipulation et représentations de circuits électriques simples par des enfants de 7 a 12 ans. *Revue Francaise de Pédagogie, 34,* 32–47.

Tobin, K. (1987). Secondary science laboratory activities. *European Journal of Science Education, 8,* 199–211.

Tobin, K. (1990). Research on laboratory activities: In pursuit of better questions and answers to improve learning. *School Science and Mathematics, 90,* 403–418.

Tobin, K., & Gallagher, J. J. (1987). What happens in high school science classrooms? *Journal of Curriculum Studies, 19,* 549–560.

Toothacker, W. S. (1983). A critical look at introductory laboratory instruction. *American Journal of Physics, 51,* 516–520.

Trent, J. H. (1965). The attainment of the concept "understanding science" using contrasting physics courses. Unpublished Ph.D. diss., Stanford University, Stanford, CA.

Tulving, E. (1972). Episodic and semantic memory. In E. Tulving & W. Donaldson (Eds.), *Organization of memory.* New York: Academic Press.

Vernier, D. L. (1987). *How to build better mousetrap.* Portland, OR: Vernier Software.

Vitrogan, D. (1969). Characteristics of a generalized attitude toward science. *School Science and Mathematics, 69,* 150–158.

Wainwright, C. L. (1989). The effectiveness of computer assisted instruction package in high school chemistry. *Journal of Research in Science Teaching, 26,* 275–290.

Walker, D. F. (1983). Reflections on the educational potential and limitations of microcomputers. *Phi Delta Kappan, 65,* 103–107.

Walker, D. F., & Hess, R. D. (1984). *Instructional software.* Belmont, CA: Wadsworth.

Warren, J. (1984). Calorimetric analysis, curve fitting and the BBC microcomputer. *School Science Review, 66,* 257–269.

Watson. J. (1984). Using a microcomputer to monitor temperature and light. *Journal of Biological Education, 18,* 57–64.

Waugh, M. (1987, April). *The influence of interactive videodisc simulations on student achievement in an introductory college chemistry course.* Paper presented to the annual meeting of the National Association for Research in Science Teaching, Washington, DC.

Webb, N. W. (1982). Student interaction and learning in small groups. *Review of Educational Research, 53,* 421–445.

Welch, W. W. (1983). Research on science teacher training: Summary and implications. In P. Tamir, A. Hofstein, & M. Ben Peretz (Eds.), *Preservice and inservice in education for science teachers* (pp. 459–466). Rehovot: Balaban Interscience Service.

Welch, W. W., Klopfer, L. E., Aikenhead, G. S., & Robinson, J. T. (1981). The role of inquiry in science education: Analysis and recommendations. *Science Education, 65,* 33–50.

Welch, W. W., & Pella, M. D. (1967–1968). The development of an instrument for inventory knowledge of the processes of science. *Journal of Research in Science Teaching, 5,* 64–68.

White, R. T. (1988). *Learning science.* Oxford: Basil Blackwell.

Willett, J., Yamashita, J., & Anderson, R. (1983). A meta-analysis of instructional systems applied in science teaching. *Journal of Research in Science Teaching, 20,* 405–417.

Wise, K. C., & Okey, J. R. (1983, April). *The impact of microcomputer based instruction on student achievement.* Paper presented at the annual meeting of the National Association for Research in Science Teaching, Dallas.

Wood, P. (1984). Use of computer simulations in microbial and molecular genetics. *Journal of Biological Education, 18,* 309–312.

Woolnough, B., & Allsop, T. (1985). *Practical work in science.* London: Cambridge University Press.

Yager, B. E., Englen, H. B., & Snider, B. C. (1969). Effects of the laboratory and demonstration methods upon the outcomes of instruction in secondary biology. *Journal of Research in Science Teaching, 6,* 76–86.

Yeany, K. H., Larussa, A. A., & Hale, M. L. (1989, March). *A comparison of performance based versus paper and pencil measures of science processes and reasoning skills as influenced by gender and reading ability.* Paper presented at the annual meeting of the National Association of Research in Science Teaching, San Francisco.

Ztadock, N. (1983). *The learning of the concept "chemical energy" in inquiry groups: Evaluation of effectiveness* (in Hebrew). Thesis for the degree of Master in Science, Weizman Institute of Science, Rehovot, Israel.

Zieleniecova, F. (1984). *Interests in science and technology in Czechoslovakia.* National Report to the 12th IPN Symposium, University of Kiel.

Zohar, A., & Tamir, P. (1991). Anthropomorphism and teleology in reasoning about biological phenomena. *Science Education, 75*(1), 57–67.

LEARNING

· 4 ·

RESEARCH ON THE ACQUISITION OF SCIENCE KNOWLEDGE: EPISTEMOLOGICAL FOUNDATIONS OF COGNITION

Anton E. Lawson

ARIZONA STATE UNIVERSITY

THE ISSUE

How do people acquire knowledge? Although this question has occupied some of the world's best minds for more than 2000 years, it is fair to say that anything resembling a "scientific" answer has emerged only during the past few decades. The likely reason for this is that the question is difficult in the sense that alternative theories of knowledge acquisition have not lent themselves to scientific test. Also, it has been philosophers who have traditionally been interested in this question and philosophers, by definition, are persons who have a consistent interest in armchair speculation and an equally consistent disinterest in putting their ideas to scientific test. Indeed, if they did put their ideas to the test they would no longer be philosophers. They would be scientists. Very few philosophers in the past have been interested in or able to make this obviously necessary (to a scientist) transformation. Likewise, few scientists before the twentieth century have been interested in asking epistemological questions.

One of the first scientists to ask epistemological questions was the Swiss psychologist Jean Piaget. As a precocious teenager growing up in Switzerland in the early twentieth century, Piaget became deeply interested in the question of how people acquire knowledge. Piaget was clearly influenced by the German philosopher Immanuel Kant. Kant viewed knowledge acquisition as a complex interaction between our sensory impressions of things and a priori, or rational, judgments that do not depend on experience. Although Piaget was taken by the epistemological question and was sympathetic to many of Kant's views, he quickly became frustrated (to the point of a nervous

breakdown) over philosophy's general disdain for empirical test. To remedy this obvious (to him) shortcoming of philosophy, he embarked on a scientific research program to discover how humans acquire knowledge by studying children as they learn about their world. Hence, the field of genetic (developmental) epistemology was born.

The primary purpose of this chapter is to follow in Piaget's footsteps in an attempt to provide a scientific answer to the question, "How is knowledge acquired?" In this chapter, we take a close look at Piaget's work and his process of knowledge acquisition called equilibration, but we do not stop at that. Rather we consider alternatives to Piaget's equilibration theory and go beyond his work to consider possible neurological mechanisms involved in learning and knowing. Modern work in neural physiology and the neural modeling principles of mathematician Steven Grossberg serve as the major theoretical perspective for this. We also consider the role that logic plays (or does not play) in reasoning and learning and then take a look at the nature of concepts, conceptual systems, and conceptual change in both scientists and science students. We then consider the very important role that analogy plays in scientific insight, learning, memory, and cognition and propose a new stage theory of development. Along the way, we explore the implications of our findings for teaching in a variety of areas of science and mathematics.

Empiricism and Nativism

Before launching into a lengthy and complex discussion of how knowledge is acquired, it is helpful to discuss briefly two early answers to the question, namely empiricism and nativism. Empiricism is the doctrine that all knowledge is derived from sensory experience. Although alternative forms of empiricism have been espoused by philosophers such as the Greek Aristotle; Berkeley, Hume, and Locke of Great Britain; and Ernst Mach and the logical positivists of Austria, the critical point of the empiricist doctrine appears to be that the ultimate source of knowledge is the external world. To the empiricist, the essence of knowledge acquisition is the incorporation of representations of the external world gained primarily through keen observation of that world.

Nativism in its various forms stands in stark opposition to empiricism. Its basic claim is that the source of knowledge is from within. Plato, for example, argued for the existence of innate ideas that unfold during the intellectual development of the child. Chomsky and Foder (in Piattelli-Palmerini, 1980) are modern proponents of a nativist-inspired view. What are we to make of these opposite and extreme positions? Following Piaget's advice to resist mere speculation, we should start by investigating organisms as they are engaged in knowledge acquisition. Consequently, let us consider the following examples.

Van Senden (in Hebb, 1949) reported work with congenitally blind persons who had gained sight after surgery. Initially these persons could not distinguish a key from a book when both lay on a table in front of them. Also, they were unable to report seeing any difference between a square and a circle. Only after considerable experience with the objects, including touching and holding them, were they able to "see" the obvious (to others) differences. Observations such as these strongly suggest that visual experience alone is insufficient for knowledge acquisition, at least in this type of situation.

In an experiment described by Von Foerster (1984), microelectrodes were inserted into a cat's brain. The cat was then placed in a cage with a lever that would dispense food when pressed, but only when a tone of 1000 Hz was produced repeatedly. In other words, to obtain food the cat had to learn to press the lever while the tone was sounding. Initially the electrodes indicated no neural activity associated with the tone. Eventually, however, the cat learned to press the lever at the correct time, and from that point on the microelectrodes showed significant neural activity when the tone was produced. Apparently the tone was not even perceived until it was of some consequence to the cat. This result is reminiscent of Piaget's contention that a stimulus cannot be a stimulus until some "mental structure" has been acquired that allows its assimilation.

Another experiment showed that cats reared in normal environments have cells in the visual center that become electrically active when the cats are shown objects with vertical lines. However, the corresponding cells of cats reared to the same age in a modified environment with no vertical lines show no response to the identical objects. Thus, contrary to the nativist's position, it seems clear that mere passage of time is not sufficient for these brain cells to develop fully.

Finally, consider a problem faced by my son when he was a 14-month-old child playing with the Shape-Sorter shown in Figure 4.1. Typically the child would pick up the cylinder sitting at the left on top of the Shape-Sorter. He would then hunt for a

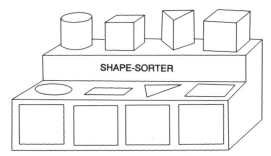

FIGURE 4.1. The Fisher-Price Shape-Sorter.

hole into which he could drop the cylinder. Initially he was unable to locate the correct hole even though it was directly below the spot where he had picked up the cylinder. Even, if by chance, he happened to pick the correct hole, he was unable to orient the cylinder correctly to make it fit. Nevertheless, with my physical assistance he was eventually able to succeed. When he placed the cylinder above the correct hole, I would gently push the object to orient it correctly so that it fit. Then, when he let go of the cylinder, it would drop out of sight. He was delighted—success! But when he picked up the rectangular solid next to the spot where the cylinder had been, which hole do you think he tried to put it in? The one below the rectangular solid? Wrong. He did not even consider that hole, even though it is clearly (to us) the correct choice. Instead, he tried repeatedly to put it in the round hole. Presumably this is because it was that behavior, the act of placing an object above the round hole and letting it go, that previously led to success. In other words, he would attempt to assimilate the new situation to the previously successful behavior. In general, successful behavior will be repeated until contradicted. Of course, when the rectangular object was placed over the round hole it would not fit, so the behavior was contradicted. But only after numerous contradictions was the child willing to try another hole. I tried showing him which holes the various objects would actually go into, but this was to no avail. He had to behave himself. In other words, the child learned from his behaviors. Only after repeated behaviors did he find the correct holes. It was the coordination of behaviors that ultimately led to consistent success.

These examples make it clear that knowledge acquisition is not merely a matter of direct recording of sensory impressions, nor is the mere passage of time sufficient for innate processes to unfold. Rather, knowledge acquisition, much as Kant suggested, appears to involve a complex interaction among sensory impressions, properties of the organism's developing brain, and the organism's behavior in a dynamic and changing environment.

ALTERNATIVE THEORIES OF KNOWLEDGE ACQUISITION

Because Piaget was one of the first and foremost investigators to attempt to answer epistemological questions by scientific means, his theory of knowledge acquisition and its relationships to various alternative theories of knowledge acquisition deserve special consideration. Piaget began his professional studies as a biologist. His psychological theory of knowledge acquisition was inspired by biological theory, particularly theories of embryology, development, and evolution. In fact, Piaget's thinking was firmly grounded in the assumption that intelligence is itself a biological adaptation. Therefore, the same developmental principles should apply to both processes of knowledge acquisition and biological evolution. As Piaget put it, "Intelligence is an adaptation to the external environment just like every other biological adaptation" (Bringuier 1980, p. 114). In other words, Piaget's basic assumption is that the acquisition of knowledge during childhood and adolescence can be understood in the same, or analogous, terms as the evolutionary acquisition of a hard protective shell, strong leg muscles, or keen vision.

Biological and Psychological Theories

In Piaget's view, at least five biological theories are available to explain the evolutionary development of species; hence, by analogy, there are five psychological theories to explain knowledge acquisition in individuals. The biological theories can be termed (1) vitalism, (2) preformism, (3) Lamarckianism, (4) neo-Darwinism, and (5) genetic assimilationism. Piaget referred to the analogous psychological theories as (1) intellectualism, (2) apriorism, (3) associationism, (4) pragmatism, and (5) equilibrationism, respectively. The following discussion of each pair draws heavily on Piaget's analysis in the introduction of *The Origins of Intelligence in Children* (Piaget, 1952) and Lawson (1982).

Vitalism/Intellectualism. Vitalism holds that living organisms can arise spontaneously from nonliving substances. Likewise, new evolutionary structures can arise spontaneously as a result of changes in the environment. By analogy, intellectualism explains intellectual development by endowing the child with an innate ability to construct knowledge by mere consideration of his or her own activity. In other words, knowledge is derived from activity and reason is the final arbitrator of what is reality. Descartes, in his famous dictum of the 1600s "I think, therefore, I am," thus adopted the principle that anything that people can clearly and distinctly think about must be true.

In spite of the seemingly spontaneous appearance of maggots in rotting flesh and microorganisms in stagnant broth, vitalism was defeated long ago by the classical experiments of Redi, Spallanzani, and Pasteur. These experiments showed convincingly that life arises only from prior life. Thus, following Piaget's basic assumption, it follows that vitalism is not a viable theory on which to build sound epistemological theory.

Preformism/Apriorism. The theory of preformism, which gained popularity among seventeenth- and eighteenth-century biologists, asserts that if a structure seems to appear during embryological development, it must have been there all along but in an invisible or microscopic form. Biologists know that leaves and, in some cases, parts of flowers can be seen folded up inside buds long before they start growing and spreading.

Furthermore, in the pupal stage all the parts of the butterfly can be discovered curled up inside. The preformists believed that something like this existed curled up in the egg (or in the sperm) and that embryological development was merely the unfolding and growing of tiny preformed parts. Preformism, when applied to evolutionary development, holds that new structures have a purely internal origin and become observable simply by coming in contact with new environments.

Apriorism (or nativism), the psychological analogue of preformism, considers new ideas to be innate, thus prior to experience. Experience gives the ideas the opportunity to manifest themselves but in no way contributes to their form. Although apriorism has had illustrious proponents, no less than Plato himself, its time has come and for the most part gone in psychology, as has the theory of preformism in embryology. More recent microscopic views of egg and sperm cells fail to reveal tiny preformed people curled up inside.

Lamarckianism/Associationism. The third evolutionary theory, Lamarckianism, holds that organisms acquire new characteristics in response to environmental pressures and these new characteristics become hereditarily fixed, thus passed from parent to offspring. The process has been termed the "inheritance of acquired characteristics" (Figure 4.2).

The analogous psychological theory, referred to as associationism, claims that intellectual development results from the direct acquisition of habits without mediation by internal mental structures. In other words, new mental structures result directly from the incorporation of habits acquired through experience. Thus, mental structures have a purely external origin. Of course, Lamarck's theory (at least as generally understood) has fallen by the wayside, as have vitalism and preformism. This

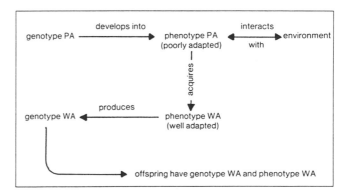

FIGURE 4.2. Lamarck's theory of the inheritance of acquired characteristics. An organism with a specified genotype (PA) develops into an organism with a corresponding phenotype PA. The organism then interacts with the environment, which reveals the initial phenotype to be poorly adaptive; thus the organism acquires the well-adapted phenotype WA (e.g., strong muscles). In some unspecified way the new well-adapted phenotype then acts to modify the initial genotype PA to produce a new genotype WA, which is transmitted to the next-generation offspring and produces the WA or well-adapted phenotype.

fact alone may not cause one to reject associationism as a viable theory of intellectual development, yet if we hold, as Piaget did, to the basic assumption that knowledge acquisition should be explained in analogous terms to evolutionary development, then it must be rejected.

Neo-Darwinism/Pragmatism. The fourth theory, known as neo-Darwinism (neo- because Darwin knew nothing of the mechanics of genetics or mutations at the time he wrote *Origin of Species*), is held by biologists who believe that evolution occurs through a natural selection of already-existing genetic variations initially produced by spontaneous mutation. Mutations in the genome cause changes in observable characteristics that are then selectively evaluated by the environment (Figure 4.3).

The psychological analogue of neo-Darwinism is pragmatism, in which new mental structures arise by random, nondirectional changes in previous structures that are then tested through behavior. Behavior is either successful and retained or unsuccessful and relinquished. Thus, mental structures are a consequence of trial and error and selection after the fact. New mental structures are internal in origin, but the environment plays an active role by selecting only the appropriate structures for retention.

The validity of neo-Darwinian theory is undisputed among modern biologists, yet many acknowledge that after-the-fact natural selection is by no means the final word. A number of instances of biological adaptation cannot be explained solely in

FIGURE 4.4. Snails of the genus *Limnaea* studied by Piaget. (A) The elongated form, L. *stagnalis*, found in calm water; (B) the contracted form, L. *bodamica*, found near wave-battered shorelines. (From Waddington, 1975.)

terms of neo-Darwinian theory. Piaget himself investigated the adaptation of a variety of aquatic snails to wave-pounded and calm environments in which changes in shell shape cannot be explained solely by after-the-fact natural selection (Piaget, 1929a, 1929b). Let us consider these data in some detail, for they clearly show a limitation of neo-Darwinian theory and the need to consider the fifth biological theory, genetic assimilationism. Only then was Piaget in a position to advance a psychological theory to account for knowledge acquisition.

Assimilationism/Equilibrationism. Snails of the genus *Limnaea* are found in almost all European lakes, including those in Switzerland, where Piaget made his initial observations. These snails are famous for their variability in shell shape. Those that live in calm waters have an elongated shape, whereas those that live on wave-battered shorelines develop a contracted, more globular shape (Figure 4.4).

Piaget was able to demonstrate that offspring of the elongated forms, when reared in laboratory conditions simulating the wave-battered shoreline, develop into the contracted form. The contracted form is due to a contraction of the columellar muscle to hold the snails more firmly to the bottom whenever a wave threatens to dislodge them. As a consequence of muscle contraction, the shell develops a contracted form as it grows. Thus, the contracted form is a phenotypic adaptation without genotypic change. Interestingly, the phenotypic adaptation has become genetically fixed in naturally occurring populations. This was discovered by taking contracted forms into the laboratory and rearing their eggs in calm conditions. The offspring retained contracted phenotypes through many generations. This is an excellent example of a characteristic acquired in the course of a lifetime that has become genetically fixed. It clearly appears to be support for Lamarckianism.

Can neo-Darwinian theory adequately explain the phenomenon? I think not. The key argument (Piaget, 1952, 1975, 1978) rests on the fact that the elongated snails, when placed in a wave-battered environment, are able to acquire the contracted form. Thus, the contracted form is an acquired (nonhereditary) adaptation. In the past, when the elongated forms moved into wave-battered environments, there would be no need for natu-

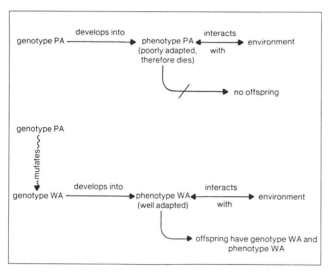

FIGURE 4.3. Neo-Darwinian theory of evolution through mutation and natural selection. Within a population of organisms, genotypes vary by genetic recombination of initially spontaneous and random nondirectional mutations. Consider genotype PA. It develops into phenotype PA, which interacts with its environment. Because phenotype PA is poorly adapted, it dies and fails to leave offspring. However, genotype PA may mutate to produce genotype WA, which develops into phenotype WA. Phenotype WA interacts successfully with the environment and leaves offspring, also with the well-adapted genotype WA.

ral selection for the contracted form to make it a hereditary genotypic trait. In fact, natural selection for snails having the contracted genotype would seem to be impossible because there would have been nothing to select. All of the snails with either genotype would be contracted! How, then, did the contracted phenotype become incorporated into the genome?

Genetic Assimilation.   The generally accepted answer to this question among evolutionary biologists draws heavily on the work of Waddington and his theory of genetic assimilation. It should be pointed out that, although Waddington's theory allows for the assimilation of genes ensuring the inheritance of initially acquired characteristics, it does so through natural selection, but not of the relatively simple sort envisioned by Darwin. In reality, the theory of genetic assimilation represents a further differentiation of neo-Darwinism rather than a contradiction of it.

Genetic assimilation involves natural selection of individuals with a tendency to develop certain beneficial characteristics. It is a widely accepted and valid model of gene modification that appears in modern textbooks of evolutionary biology (e.g., Ehrlich, Holm, & Parnell, 1974; Futuyma, 1979). In order to understand Waddington's theory of genetic assimilation, it is necessary to consider embryological development and Waddington's concept of canalization.

Canalization.   The fertilized egg is but a single cell. As it divides, the resulting cells differentiate into the myriad of cell types such as skin, brain, and muscle cells that constitute the newborn. The developing embryo has a remarkable ability to buffer itself against environmental disturbances to ensure that "correct" types of cells are produced. This is evidenced even before the first cell divides. For example, many eggs contain particular types of cytoplasm arranged in definite places. When such an egg is centrifuged, the types of cytoplasm are displaced. But if the egg is then left alone, the types of cytoplasm gradually move back to their original places. This self-righting tendency is also found in eggs that are cut in half. Identical human twins are produced by one egg that divides in an abnormal way so that each twin arises from what one would expect to produce only half of a normal individual.

The term Waddington gave to the developing organism's ability to withstand perturbations to the normal course of development was "canalization." As Waddington (1966) described it,

The region of an early egg that develops into a brain or a limb or any other organ follows some particular pathway of change. What we have found now is that these pathways are 'canalized', in the sense that the developing system has an inbuilt tendency to stick to the path, and is quite difficult to divert from it by any influence, whether an external one like an abnormal temperature or an internal one like the presence of a few abnormal genes. Even if the developing system is forcibly made abnormal—for instance, by cutting part of it away—it still tends to get itself back onto the canalized pathway and finish up as a normal adult. (p. 48)

Waddington went on to point out that canalization is not complete. The developing system does not always end up as a properly formed adult. The important point is that it has the tendency toward self-regulation, toward a final end product, even in the face of considerable variance in the paths taken.

Waddington likened canalization to a ball rolling downhill with several radiating canals opening up in front of it (Figure 4.5). As the ball rolls downhill, certain internal (genetic) or external (environmental) factors can deflect it into one or another canal with the ball ending up at the bottom at one of a finite number of predetermined points. The roll does not terminate halfway up the side of one canal. Waddington called the system of radiating canals the "epigenetic" landscape. To describe the development of an entire organism, a number of epigenetic landscapes would be required—one for each characteristic.

Suppose, for example, an epigenetic landscape were constructed to represent the development of the sex of an individual. The landscape would contain two canals, thus would dictate one of two final end points—male or female. Genetic factors would then operate to deflect the ball into one of the two canals and the normal adult would end up male or female (but not somewhere in between) despite intrusions which, at intermediate points, might cause the ball to roll part way up the side of one of the canals. The environment might also cause the ball to be deflected into one or the other canal.

An excellent example of this occurs in the marine worm *Bonellia*. The environment determines the individual's sex, but canalization usually ensures a male or female—not an intersex. Figure 4.6 shows the female and male *Bonellia* worms. The larvae are free-swimming. If the larva settles down alone, it develops into a female. If, however, it lands on the proboscis of a female, it develops into a dwarf male.

According to Waddington, organisms vary in their ability to respond to environmental pressures because of differences in their epigenetic landscapes, that is, their degree of canalization, the heights of thresholds, and number of alternative canals. Some individuals have well-canalized landscapes with few alternatives, hence are relatively unresponsive to environmental pressures. Compare the two epigenetic landscapes shown for the two first-generation individuals in Figure 4.7a and b. Both

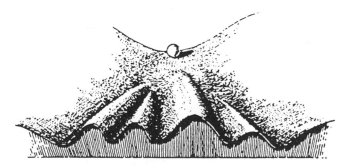

FIGURE 4.5. The epigenetic landscape of the developing individual. The various parts of a developing embryo have a number of possible canals of development open to them. Genetic or environmental pressures may switch the course of development from one canal to the next, resulting in different observable characteristics. (From Waddington, 1966.)

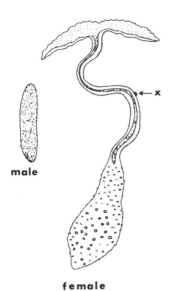

male

female

FIGURE 4.6. The marine worm *Bonellia* showing the dwarf male and female forms. The arrow marked X points to the dwarf male (actual size) on the female proboscis. (From MacGinitie & MacGinitie, 1968.)

individuals have well-canalized landscapes and have two alternatives, yet the threshold in early development of landscape H is higher than that in landscape L. Hence, an environmental pressure, depicted by the white arrow, will most likely fail to force the ball across the high threshold in H to produce the developmental modification (WA). On the other hand, in landscape L with its lower threshold, the same environmental pressure is more likely to push the ball over the threshold into another canal and produce the developmental modification.

Because of such differences, individuals vary in their ability to respond to environmental pressures. Thus, some may acquire beneficial modifications, while others may acquire non-beneficial modifications, and some may not be easily changed. Of course, those that acquire the beneficial modifications have a better chance for survival and leave more offspring. The poor responders die out. Hence landscape L and the characteristic of being able to respond in a certain beneficial way are selected.

As shown in the figure, the population becomes one in which all members have landscape L. At this point, only the slightest genetic mutation (black arrow) can now push the ball over the threshold into the new canal. Once this happens, the organism develops the well-adapted phenotype WA with or without the environmental pressure. In a sense, the selection for landscape L has put the developmental machine on hair trigger. Any number of gene mutations, which can be considered random at the level of nucleic acid structure, are likely to produce the present well-adapted phenotype. Therefore, the mutations are not random in their adaptive effect but may produce modifications in the genome that are positive in their direction. The end result is that beneficial characteristics initially acquired in response to specific environmental pressures are assimilated into the genome. This result appears neo-Lamarckian, yet it involves mutation and natural selection for individuals able to acquire the desired characteristics.

Although Waddington (1975) stated that Piaget's studies of *Limnaea* represent one of the most thorough and interesting examples of genetic assimilation in naturally occurring populations, the biological literature is replete with additional natural and experimental demonstrations of its validity. The now classical experiments of Clausen, Keck, and Hiesey (1948) suggest genetic assimilation at work in the development of the plant *Achillea lanulosa* at various altitudes in the Sierra Nevada of California. Low-altitude plants are taller than those found at higher altitudes, but when the seeds of low-altitude plants are planted at high altitudes, the plants become dwarfed just like the ones naturally occurring in the high altitudes. Thus, dwarfing is initially an environmentally acquired characteristic. When they planted the seeds of naturally occurring high-altitude dwarfs in the gardens at low altitudes, Clausen et al. found that the seeds developed into dwarfed plants even though the high-altitude environmental pressure had been removed; the dwarfed form had been genetically assimilated.

Waddington (1959) studied genetic assimilation in the laboratory through a series of natural selections of fruit flies raised on mediums with high concentrations of salt. He was able to produce a population of fruit flies in which genes for large anal papillae (known to assist the regulation of osmotic pressure of body fluids) were assimilated. The development of large anal papillae, like the contracted shell shape and dwarfed plant height in the previous examples, was shown to be a phenotypic adaptation in response to direct environmental pressure; with selection for genes ensuring this phenotypic adaptation, it became genetically assimilated. Additional examples can be found in Futuyma (1979) and Rendel (1967).

Psychological Equilibrationism

[The following discussion of psychological equilibration differs in subtle ways from Piaget's conception of the process, which is based on Piaget's concept of biological phenocopy (see Bringuier, 1980, p. 113; Piaget, 1975, pp. 216–217; Piaget, 1978, pp. 78–83). As far as I am aware, Piaget's concept of phenocopy has failed to receive favor among modern biologists; therefore, the present discussion of psychological equilibration is confined to its relationship to the generally accepted theory of genetic assimilation.]

Piaget referred to the psychological counterpart of genetic assimilation as psychological equilibration. Figure 4.8 explicates the equilibration process as analogous to Waddington's theory of genetic assimilation. Again, the psychological analogue of the changing genotype during evolution is the developing mental structure during one's lifetime. The epigenetic landscape (itself shaped by the genes) corresponds to one's predisposition to acquire new behaviors determined by what Piaget (1971b) called "assimilation schemata" (p. 22). The phenotype corresponds to overt behaviors. Figure 4.8a corresponds to a situation in which the individual with assimilation schemata H simply does not respond to pressures imposed by experience and does not develop a new mental structure (WA). Interaction with the environment does not produce "disequilibrium" or subsequent mental accommodation. The individual is not "developmentally ready" because the assimilation sche-

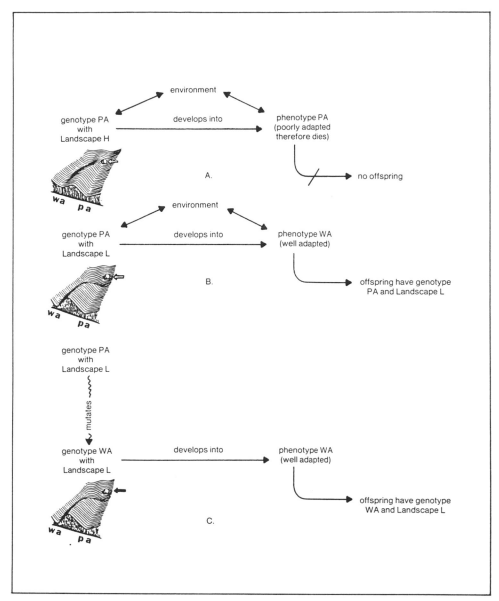

FIGURE 4.7. Waddington's theory of genetic assimilation. (A) An organism with a poorly adapted genotype in an epigenetic landscape nonresponsive to environmental pressure (white arrow) will develop into a poorly adapted phenotype that is selected against and eliminated. (B) An organism with genotype PA in a responsive landscape L acquires a well-adapted phenotype as the environmental pressure (white arrow) is able to push the course of development across the threshold into the WA canal. However, the well-adapted phenotype has yet to be assimilated into the genome, and the offspring will exhibit the new phenotype only in the presence of continued environmental pressure. (C) Genotype PA with landscape L mutates spontaneously (black arrow) to produce new genotype WA, which produces the phenotype WA even without continued environmental pressure (i. e., the initially acquired characteristic has been genetically assimilated).

mata available to the individual are not adequate to assimilate the new experience. The available assimilation schemata are built up by the interplay between the individual's powers of coordination and the data of experience. Note that such an explanation dictates incremental and sequential development.

Figure 4.8b on the other hand, indicates that the individual with assimilation schemata L is able to respond to environmental pressures and acquire a new behavior. But the newly acquired behavior has not yet been assimilated into his or her mental structure (i.e., the mental structure remains PA). The new behavior and the person's previous ways of thinking have not been integrated. The result is mental disequilibrium. With removal of environmental pressure, the individual is apt to revert to previous inappropriate behaviors just as the offspring of genetically elongated but phenotypically contracted snails

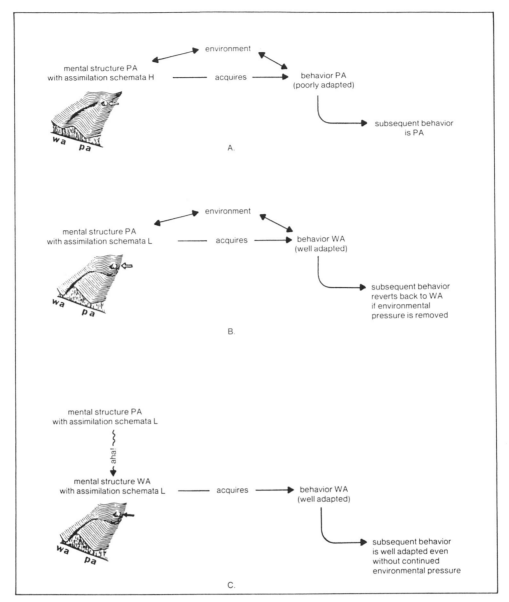

FIGURE 4.8. Psychological equilibration seen as analogous to Waddington's theory of genetic assimilation. (A) An individual with mental structure PA and inadequate assimilation schemata H interacts with the environment and continues to exhibit poorly adapted behavior PA. (B) An individual with mental structure PA and adequate assimilation schemata L interacts with the environment and acquires a new well-adapted behavior WA. However, the new behavior has not been assimilated into the present mental structure and will be relinquished without continued environmental pressure. Disequilibrium exists. (C) The individual with mental structure PA and schemata L undergoes an "aha" experience (the new environmental input finally makes sense to the individual in light of what he or she already knows). The mental structure is accommodated to allow complete assimilation of the input. Subsequent behavior is well adapted even without continued environmental pressure.

develop into the elongated form if reared in a calm environment.

In the classroom, students may be able to solve a proportions problem correctly if the teacher suggests the procedure or if the problem is similar enough to ones previously solved. But if left to their own devices, use of the proportions strategy may never occur to the students because they have failed to comprehend why it was successful in the first place (i.e., it has never been integrated with previous thinking). Thus, Figure 4.8b can be said to represent a state of disequilibrium because a mismatch exists between the poorly adapted mental structure and the sometimes successful behavior.

Finally, Figure 4.8c represents the restoration of equilibrium through a spontaneous, internal, yet directional reorganization of mental structure that allows complete assimilation of the new behavior pattern into the accommodated mental structure of the individual. Thus, psychological assimilation corresponds to the entire process of the incorporation of new well-adapted behavior patterns (phenotypes) into one's mental structure (the genome) by way of a spontaneous accommodation of mental structure (i.e., the mutation). Hence, one does not have assimilation without accommodation. Piaget was fond of quoting the child who, when asked about the number of checkers in two rows of unequal length, responded correctly and reported that "Once you know, you know forever." Here is a child with an accommodated mental structure who had completely assimilated the notion of conservation of number.

Equilibration Theory and Instruction

The importance for instruction of the psychological theory of equilibration can be simply stated. If one adopts the vitalism/intellectualism or preformism/apriorism positions, then the teacher is superfluous to the student's intellectual development. The teacher can do nothing but wait for the proper mental structures to develop. On the other hand, if one adopts the Lamarckian/associationism position, all one would need to do is force students to work by rote problems that required the appropriate behaviors for solution. The mere habit of correctly working these problems would ensure the development of the corresponding appropriate mental structures. Of course, teachers know that such a procedure simply does not work. As Ausubel (1979) correctly pointed out, faculty psychology or the "formal discipline" approach to education has long been discredited.

If one adopts the neo-Darwinian/pragmatism approach to education, then one is forced to wait until spontaneous and nondirectional reorganizations of mental structures occur before intellectual development takes place. Again, the entire process is internal and not amenable to environmental-instructional shaping. The teacher is relegated to the relatively unimportant position of telling a student when ideas are right or wrong and is in no way able to shape the direction of the student's thinking.

But if one adopts the psychological analogue of genetic assimilation theory, the teacher is not placed in a position of sitting idly by, waiting for intellectual development to occur. Rather, the teacher knowledgeable of developmental pathways can produce the environmental pressures that enable students to reorganize their thinking spontaneously along the path toward more complex and better-adapted thought processes. The teacher can be an instigator of disequilibrium and can provide pieces of the intellectual puzzle for the students to put together. Of course, the ultimate mental reorganization must be accomplished by the students, but the teacher is far from passive. He or she can set the process on hair trigger just as the directional natural selection of Waddington sets the genome on hair trigger to await the slightest mutation.

The key point is that external knowledge (that presented by the teacher) can become internalized if the teacher accepts the notion that the equilibration process is the route to that internalization. This means that students (1) must be prompted to engage their previous ways of thinking about the situation to discover how they are inadequate to assimilate the new situation and (2) must then be given ample opportunity to think through the situation to allow the appropriate mental reorganization (accommodation), which in turn allows successful assimilation of the new situation (see Karplus, 1979; Lawson, Abraham, & Renner, 1989).

Consider the well-known case of the high school and college students who employ an additive strategy to solve a proportionality problem. Given two plastic cylinders equal in height but unequal in diameter, the students note that water from the wide cylinder at the fourth mark rises to the sixth mark when poured into the narrow cylinder. When asked to predict how high water at the sixth mark in the wide cylinder will rise when poured into the narrow cylinder, the students respond by predicting mark 8, "because it rose 2 marks last time so it will rise 2 marks again."

How can these students be taught to use a proportions strategy? According to equilibration theory, they must first discover the error of their previous thinking. In this case, this is quite simple to arrange. Simply pour the water into the narrow cylinder and let the students note the rise to mark 9. Even without pouring, the error can be discovered by considering the process of pouring from the narrow cylinder to the wide one. The students predict that water at 6 in the narrow will rise to 4 when poured into the wide. They predict that water at 4 in the narrow will rise to 2 in the wide. They predict that water at 2 in the narrow will rise to 0 in the wide. Aha, the water disappears! Of course, the students see the absurdity of the situation and are forced into mental disequilibrium.

At this point, the students are prepared for step 2, the introduction of the "correct" way to think through the problem. Keep in mind, however, that according to the analogy, it is the students themselves who must undergo mental reorganization to appreciate your suggestions and assimilate the new strategy. This will *not* happen immediately. Our experience suggests that it requires considerable time and repeated experience with the same strategy in a number of novel contexts (see Lawson & Lawson, 1979; Wollman & Lawson, 1978). The fact that the use of a variety of novel contexts is helpful (perhaps even necessary) is an argument in favor of breaking down some of the traditional subject-matter distinctions. For example, in a biology course one should not hesitate to present problems that involve proportions in comparing prices at the supermarket, altering recipes in cooking, comparing the rotations of coupled gears, balancing weights on a balance beam, estimating the frog population size in a pond, comparing the relative rates of diffusion of chemicals, and estimating gas mileage. If the range of problem types was confined to traditional biology subject matter, many students would fail to undergo the necessary mental reorganization to internalize the proportions strategy.

Although the previous example dealt with learning to use a proportions strategy (an aspect of logicomathematical knowledge), equilibration theory does not deal only with the acquisition of logicomathematical knowledge. As Piaget (1975) pointed out, "Now it is essential to note that this tendency to replace exogenous knowledge by endogenous reconstructions is not confined to the logico-mathematical realm but is found throughout the development of physical causality" (p. 212).

A lovely classroom example of using equilibration theory to help students develop physical understanding was reported by Minstrell (1980). He was interested in teaching his high school physics students about the forces that keep a book "at rest" on a table. Before simply telling the students that the book remains at rest because of the equal and opposite forces of gravity (downward) and the table (upward), Minstrell asked his students what forces they thought were acting on the book. Many of the students believed that air pressing in from all sides kept the book from moving. Others imagined a combination of gravity and air pressure pushing downward. A few students also thought that wind or wind currents "probably from the side" could affect the book. The most significant omission seemed to be the students' failure to anticipate the table's upward force. Although some students did anticipate both downward and upward forces, most believed that the downward force must total more than the upward force "or the object would float away."

After the crucial first step of identifying the students' initial misconceptions, Minstrell took the class through a carefully planned sequence of demonstrations and discussions designed to provoke disequilibrium and initial mental reorganization, stopping along the way to poll the students for their current views. The key demonstrations included piling one book after another on a student's outstretched arm and hanging a book from a spring. The student's obvious expenditure of energy to keep the books up led some students to admit the upward force. When students lifted the book already supported by the spring, the initial response was surprise at the ease at which it could be raised. "Oh my gosh!" "There is definitely a force by the spring." Although Minstrell admits that the series of demonstrations was not convincing to all, in the end about 90 percent of his students voiced the belief that there must be an upward force to keep the book at rest. Of course, instruction did not stop here. Nevertheless, the majority of Minstrell's students were well on the way to the appropriate mental reorganization.

In short, the teacher who takes equilibration theory to heart becomes an asker of questions, a provider of materials, a laboratory participant, a class chairperson and secretary. He or she gathers the class together and solicits the data they have gathered and their meaning. Most important, this teacher is not a teller. He or she is a facilitator and director of learning. If materials are well chosen, questions are posed, and students are prompted to think through data and problems, much can be done to encourage the acquisition of more adaptive mental structures.

A NEUROLOGICAL ACCOUNT OF EQUILIBRATION

Although Piaget's theory of equilibration has won numerous advocates and to most seems intuitively reasonable, it has a serious theoretical weakness. The psychological theory, although based on a valid biological analogue (namely genetic assimilation), remains essentially that, an analogy. Obviously, to

move our understanding beyond analogical arguments of this sort, we must actually understand how the brain functions, because ultimately it is the brain that records and stores our knowledge. Considerable progress has been made during the past 20 years or so in the related fields of neural physiology and neural modeling. We now turn our attention to that work with a brief introduction to general brain anatomy and neurological signaling. This will be followed by a neurological explanation of the psychological process of equilibration.

Basic Neurological Principles

General Brain Anatomy. Figure 4.9 is a side view of the human brain showing the spinal cord, brain stem, and cerebral cortex. In general, the cortex is divided into a frontal portion containing neurons that control motor output and a rear portion that receives sensory input.

The thalamus is a center at the top of the brain stem that relays signals from the sense receptors to the sensory cortex. All sensory input, with the exception of smell, passes through one of 29 thalamic regions on its way to the cortex. One of the most important nuclei is the lateral geniculate nucleus, the relay station of the optic tract from the retina to the visual cortex (see Figure 4.10).

At the center of the brain stem from just below the thalamus down to the medulla (lowest segment of the brain stem) is the reticular formation. The reticular formation is believed to play a key role in nerve networks by serving as a source of nonspecific arousal. Located in the inner surface of the deep cleft between the two brain hemispheres lies the hypothalamus. It appears to be the source of specific drive dipoles such as fear-relief and hunger-satisfaction, which also play a key role in the neural models developed.

Neuronal Signals. The basic unit of the functioning nervous system is the nerve cell or neuron. There are many types of

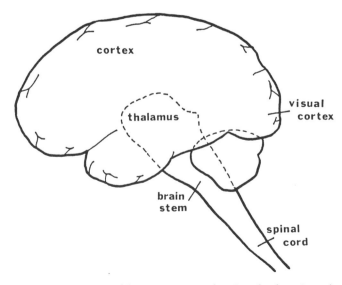

FIGURE 4.9. General brain anatomy showing the location of key structures involved in neural modeling.

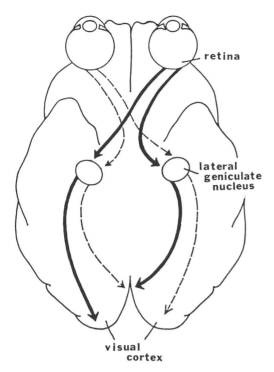

FIGURE 4.10. The optic tract showing the path of sensory input from both retinas to the lateral geniculate nuclei near the center of the brain and to the visual cortex at the rear of the brain.

Excitation of retinal cells fires signals along the optic nerve to a layer of neurons located in the lateral geniculate nucleus (LGN). Cells of the LGN process the signals and in turn relay signals to a third layer of cells in the visual cortex at the back of the brain (see Figure 4.10). From the visual cortex, signals are transmitted back to the LGN and to additional layers of neurons for further processing. The signals that are sent back to the LGN play a significant role by allowing the system to compare incoming signals with prior expectations acquired from previous learning. More will be said about this later (also see Grossberg, 1982, especially pp. 8–15). An excellent discussion of the neural anatomy relevant to learning and memory can also be found in Mishkin and Appenzeller (1987).

General Principles of Network Modeling

Table 4.1 (after Hestenes, 1983) lists crucial components and variables of neurons and layers of neurons as well as their physiological and psychological interpretation in Grossberg's theory. Consider the ith neuron in a collection of interacting neurons, but they all share characteristics exemplified by pyramid cells in the cerebral cortex (shown in Figure 4.11).

Pyramid cells consist of four basic parts: a cell body, a set of dendrites, an axon or axons, and a set of terminal knobs. The dendrites and the cell body are capable of receiving electrical signals from axons of other neurons. The axons (which may branch) conduct signals away from the cell body. When stimulated by incoming signals, the terminal knobs open small packets of chemical transmitter, which, if released in sufficient quantity, cause the signal to pass across the gap or synapse to the next neuron.

A nonfiring neuron has a slightly negative potential across the cell membrane (approximately -70 mV), which is termed its resting potential. Incoming signals (which can be either excitatory or inhibitory) modify the resting potential in an additive fashion and induce what is referred to as the cell's generating potential. When the generating potential exceeds a certain threshold, a "spike" or action potential is generated in the cell body and travels down the axons.

The action potential travels at a constant velocity with an amplitude of up to about 50 mV. Signals are emitted in bursts of varying frequency, depending on the amount of depolarization of the neuron. It is believed that all of the information in the signal is dependent on the pulse frequency of the bursts.

Neurons appear to be arranged in distinct layers or slabs within various substructures of the brain. Consider, for example, the arrangement of neurons in the visual system. Light is received by an initial layer of photoreceptors in the retina.

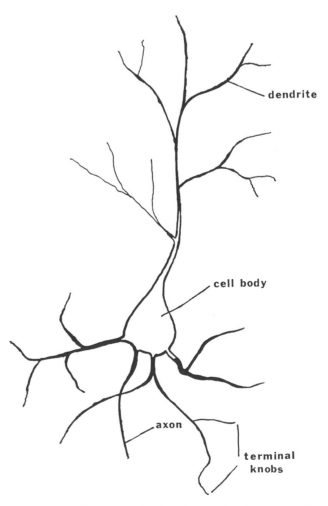

FIGURE 4.11. The pyramid cell of the cerebral cortex and its major parts.

TABLE 4.1. Neuron Components and Variables

Components	Node v_i	Directed axon pathway e_{ij}	Synaptic knob $N_{ij} v_i$
Variables	X_i	S_{ij}	$Z_{ij} X_i$

Name/Variable	Physiological Interpretation	Psychological Interpretation
Activity X_i	Average generating potential	Stimulus trace or STM trace
Signal S_{ij}	Average firing frequency	Sampling or performance signal
Synpatic strength Z_{ij}	Transmitter release rate	LTM trace

Source: Hestenes, 1983.

neurons. The average generating potential of the *i*th neuron at node V_i is denoted by X_i, the stimulus or short-term memory (STM) trace. This activity can be sustained by a feedback loop; thus Grossberg makes STM the property of any neuron in which activity is sustained for a specific period of time. STM is not a single undetermined location in the brain into which a limited amount of information can be input and temporarily stored, as is the common interpretation in cognitive psychology.

The signal that propagates along the axon or axons, e_{ij}, from node V_i to synapse knob N_{ij}, is signified by S_{ij}. The signal is, of course, a function of the activity X_i at node V_i. The final extremely important neural variable is the synaptic strength, Z_{ij}, of knob N_{ij}. The tentative physiological interpretation of Z_{ij}, is the average rate of transmitter release at the synapse. Z_{ij}, in effect, represents the ease with which signals down e_{ij} can cause V_j to fire. If signals can get the knob to release a great deal of transmitter (a large value for Z_{ij}), they are sent across the synapse and the next cell in line (V_j) fires. If signals are not able to get the knob to release much transmitter (a small value for Z_{ij}), they do not cause V_j to fire. Increases in Z_{ij} represent modification of knobs to allow transmission of signals among neurons. Thus Z_{ij} becomes the location of long-term changes in systems of neurons, or the long-term memory (LTM) of the system. Learning is thus considered to be a biochemical modification of synaptic strengths. As was the case for STM, LTM is a property of neuron connections rather than a specific single location in the brain.

Equations of Variable Interactions. Grossberg has proposed equations describing the basic interaction of the variables just mentioned. Of particular significance are equations describing changes in X_i and Z_{ij} (i.e., changes in short-term memory and long-term memory). In general, these equations, for a network with *n* nodes, are of the form

$$\dot{X}_i = -A_i X_i + \sum_{k=1}^{n} S_{ki} Z_{ki} - \sum_{k=1}^{n} C_{ki} + I_i(t) \qquad (1)$$

$$\dot{Z}_{ij} = -B_{ij} Z_{ij} + S'_{ij} [X_j]^+ \qquad (2)$$

Where the overdot represents a time derivative and $i, j = 1, 2, \ldots, n$.

The equations describe what is necessary to drive a change in activity of V_i (i.e., X_i) and a change in rate of transmitter release at knob N_{ij} (i.e., Z_{ij}). Equation (1) is referred to as the activity equation of node V_i because it describes factors that influence changes in STM; equation (2) is referred to as the learning equation because it describes elements necessary to cause changes in LTM.

Consider first the terms in equation (1). As mentioned, X_i represents the initial level of activity of nodes V_i. The term A_i represents a passive decay constant inherent in any dissipative system. The sign is negative, indicating a drop in activity of V_i across time due to the product of A_i and X_i. In other words, if V_i receives no additional input or feedback from itself, the activity stops. The term $S_{ki} Z_{ki}$ represents inputs (S_{ki}) to the nodes V_i from prior cells in the system mediated by their respective synaptic strengths (Z_{ki}). The positive sign indicates that these signals stimulate or increase the activity of cells V_i. These inputs are additive; hence their summation is called for. The C_{ki} term represents inhibitory node-node interactions of the network, hence the negative sign. Recall that inputs to neurons can be excitatory or inhibitory. The final term, I_i, represents inputs to V_i, from sources outside the network (i.e., neurons other than those in slab V_k).

Equation (2), the learning equation, describes conditions necessary to modify the synaptic strength of knobs N_{ij}. The term Z_{ij} represents initial synaptic strength. B_{ij} is a decay constant; thus the term $B_{ij} Z_{ij}$ is a forgetting or decay term. $S'_{ij}[X_j]^+$ can be considered the learning term, as it drives increases in Z_{ij}. S'_{ij} is the signal that has passed from node V_i to knob N_{ij}. The prime reflects the fact that the initial signal, S_{ij}, may be slightly altered as it passes down e_{ij}. The term $[X_j]^+$ represents the level of activity at postsynaptic nodes V_j that exceed firing threshold. Only activity above threshold can cause changes in Z_{ij}. In short, the learning term indicates that for information to be stored in LTM, two events must occur simultaneously: signals must be received at N_{ij}, and nodes V_j must receive inputs from other sources that cause them to fire. When these two events drive activity at N_{ij} above a specified constant of decay, the Z_{ij}s increase; that is, the network learns.

Learning in a Simple Neural Circuit: Classical Conditioning

Learning in Grossberg's theory occurs when synaptic strengths (Z_{ij}s) increase, that is, when the transmitter release rate increases and makes transmission of signals from one neuron to the next easier. Hence learning is, in effect, an increase in the number of connections among neurons "operative" until the transmitter release rates at certain terminal knobs increase above a specified threshold. Thus, in order to have a "mental structure" become more complex, transmitter release rates must increase at a number of knobs so that the signals can easily be transmitted across synapses that were previously there but inoperative. This view reveals the sense in which the nativists are "correct." If one equates mental structures with already present, but inoperative, synapses, then the mental structures are present before any specific experience. The view also reveals how the empiricists are "correct." For the synapses to become operative, experience is necessary to strengthen some of the connections. Let us see how experience can do this.

Consider Pavlov's classical conditioning experiment in which a dog is stimulated to salivate by the sound of a bell. As you may recall, when Pavlov first rang a bell the dog, as expected, did not salivate. However, after repeated simultaneous presentation of food (which did initially cause salivation) and bell ringing, the ringing alone was eventually able to cause the dog to salivate.

In Pavlovian terms, the food is the unconditioned stimulus (US). The salivation on presentation of the food is the unconditioned response (UCR), and the bell is the conditioned stimulus (CS). In general terms, Pavlov's experiment showed that when a conditioned stimulus (e.g., a bell) is repeatedly presented with an unconditioned stimulus (e.g., food), it will eventually evoke the unconditioned response (e.g., salivation) by itself.

How can the unconditioned stimulus do this? Figure 4.12 shows how Grossberg uses his principles to construct a simple neural network capable of explaining classical conditioning. The network can be depicted as just three cells, A, B, and C, each of which actually represents many neurons of type A, B, and C.

Initial presentation of food causes cell C to fire. This creates a signal down its axon that, because of prior learning (i.e., a relatively large Z_{cb}), causes the signal to be transmitted to cell B. Cell B fires and the dog salivates. At the outset, ringing the bell causes cell A to fire and send signals toward cell B; however, when the signal reaches knob N_{AB}, its synaptic strength Z_{AB} is not initially large enough to cause B to fire. The dog does not salivate. However, when the bell and the food are paired, cell A can learn to fire cell B according to equation (2). Firing cell A results in a large S'_{AB}, and the appearance of food results in a large $\Sigma[X_B]^+$. Thus the product $S'_{AB}[X_B]^+$ is sufficiently large to drive an increase in Z_{AB} to the point at which it alone can cause

node V_B to fire and evoke salivation. Food is no longer needed. The dog has learned to salivate at the ringing of a bell. The key point in terms of neural modeling theory is that learning is driven by simultaneous activity of pre- and postsynaptic neurons, in this case activity of cells A and B.

Learning in Humans: A More Complex Network

The Basic Pattern of Knowledge Acquisition. How might Grossberg's principles be applied to account for learning in humans? Consider the simple behavior of a human infant learning to orient a bottle to suck milk. Again, Grossberg's principles will be used to deal with the increase in complexity. Such a step will provide us with a model of simple learning through which we can begin to visualize neural events that may be involved in equilibration and in higher-order cognition.

The behavior of interest was observed by Jean Piaget in his son Laurent (from 7 to 9 months of age). Piaget (1954) reports as follows:

From 0:7 (0) until 0:9 (4) Laurent is subjected to a series of tests, either before the meal or at any other time, to see if he can turn the bottle over and find the nipple when he does not see it. The experiment yields absolutely constant results; if Laurent sees the nipple he brings it to his mouth, but if he does not see it he makes no attempt to turn the bottle over. The object, therefore, has no reverse side or, to put it differently, it is not three dimensional. Nevertheless Laurent expects to see the nipple appear and evidently in this hope he assiduously sucks the wrong end of the bottle. (p. 31)

Laurent's initial behavior consisted of lifting and sucking whether the nipple was oriented properly or not. Apparently Laurent did not notice the differences between the bottom of the bottle and the top and/or he did not know how to modify his behavior to account for the presentation of the bottom. Thanks to his father, Laurent had a problem.

In order to construct a neural model of Laurent's behavior, we need to be clear about just what new behavior (knowledge) Laurent must acquire. At the outset he knew how to flip the bottle to orient it properly for sucking provided the nipple was visible. He also knew how to bring the bottle to his mouth and suck. What he lacked was the ability to flip the bottle *before* lifting and sucking when only the bottom was visible. This was the behavior that needed to be acquired. How was it acquired?

Again, let's return to Piaget's experiment. On the sixth day of the experiment, when the bottom of the bottle was given to Laurent, "he looks at it, sucks it (hence tries to suck glass!), rejects it, examines it again, sucks it again, etc., four or five times in succession" (Piaget, 1954, p. 127). Piaget then held the bottle out in front of Laurent to allow him to look at both ends simultaneously. Although his glare oscillated between the bottle's top and bottom, when the bottom was again presented to Laurent he still tried to suck the wrong end.

The bottom of the bottle was given to Laurent on the eleventh, seventeenth, and twenty-first days of the experiment. Each time Laurent simply lifted and began sucking the wrong end. But by the thirtieth day he "no longer tries to suck the glass as before, but pushes the bottle away, crying" (Piaget, 1954, p.

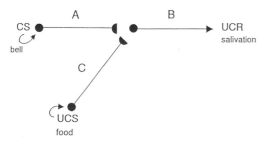

FIGURE 4.12. Classical conditioning in a simple neural network. Cells A, B, and C represent layers of neurons.

128). Interestingly, when the bottle was moved a little farther away, "he looks at both ends very attentively and stops crying" (p. 128).

Finally, 2 months and 10 days after the start of the experiment, when the bottom of the bottle was given to Laurent he was successful in first flipping it over: "he immediately displaces the wrong end with a quick stroke of the hand, while *looking beforehand* in the direction of the nipple. He therefore obviously knows that the extremity he seeks is at the reverse end of the object" (Piaget, 1954, pp. 163–164).

Laurent's learning behavior, although relatively simple, follows a pattern. That pattern consists of an initially successful behavior driven in part by a response to an external stimulus and in part by an internal drive (in this case hunger). That initially successful behavior is contradicted when it is misapplied beyond the situation in which it was acquired. This leads to frustration (reminiscent of Piaget's concept of disequilibrium) and, in neural modeling terms, an eventual shutting down of the internal drive coupled with a nonspecific arousal that causes the child to stop the incorrect behavior and attend more closely to the external stimulus that initially provoked the behavior. Attention, once aroused, allows the child to notice previously ignored cues and/or relationships among the cues, which in turn allows him to couple those cues with modified behavior to deal successfully with the new situation. Hence, a new procedure has been actively acquired. How does this happen at the level of neurons?

The Neural Network. Figure 4.13 depicts a hypothetical neural network (after Grossberg, 1982, chapter 6) that might drive this learning. In general, CS_i represents the ith conditioned stimulus among possible stimuli that excites cell population U_{i1} in the sensory cortex. Input to U_{i1} has already passed through prior slabs of neurons, specifically the retina and the lateral geniculate nucleus, as in this specific case CS_1 represents the

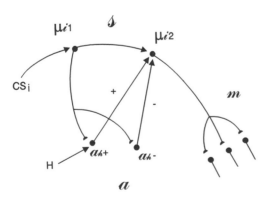

FIGURE 4.13. A minimal network capable of modeling the initiation of the lifting and sucking response when cued by the appearance of the bottle (CS). Cell populations in the sensory cortex (μ^+) send excitatory input to excitatory arousal cells (Ah^+) and inhibitory arousal cells (Ah^-). The hunger drive (H) coupled with excitation from CS causes μ_{i2} to fire and initiate the motor response. S represents the sensory system, A the arousal system, and M the motor control system.

undifferentiated pattern of visual inputs from Laurent's bottle (i.e., either the top or the bottom). In response to CS_i, U_{i1} sends signals to another slab of neurons in the motor cortex, U_{i2} (Brodmann area 4; Albus, 1981, pp. 89–90), as well as to all populations of arousal cells for specific drives (probably located in the hypothalamus, Grossman, 1967). Because in this case hunger is the drive of interest, CS_i will generally be limited to arousal of the cell populations that increase the hunger drive, Ab^+, and those that decrease the hunger drive, Ab^- (see Grossberg, 1982, pp. 259–262, for a discussion of the pairing of cue with appropriate drives). Populations Ab^+ and Ab^- then send signals to U_{i2}. Finally, and only when excited by signals from U_{i1} and by excitatory signals from Ab^+, U_{i2} will fire and send signals to M (the motor cells controlling the behavioral response) that release the conditioned response—the lifting and sucking of the bottle.

Notice that this sequence of events causes the synaptic weights at the Ab^+ layer and at the U_{i2} layer to increase because pre- and postsynaptic activity occurs at both layers, thus conditioning the behavior of lifting and sucking to the appearance of the bottle when the child is hungry. Therefore, this minimal network can explain the initiation of Laurent's behavior. How might it explain the behavior's termination upon satisfaction of the hunger drive?

The Rebound from Hunger to Satisfaction. Intake of food gradually reduces the activity of Ab^+ cells, which in turn causes a "rebound" or activation of Ab^- cells, which in turn inhibits activity at U_{i2}, thereby stopping the motor response. Precisely how does the satisfaction of hunger at Ab^+ generate a rebound of activity at Ab^-? The simplest version of the neural rebound mechanism, referred to as a dipole, is shown in Figure 4.14.

An internally generated and persistent input, I, stimulates both the Ab^+ and the Ab^- channels. This input will drive the rebound at Ab^- when the hunger-derived input H is shut off. When Laurent is hungry, inputs I and H sum and create a signal along e_{13} (i.e., from V_1 to V_3). A smaller signal is also set along e_{24} by I alone. At synaptic knobs N_{13} (the knob connecting V_1 to V_3) and N_{24} (the knob connecting V_2 to V_4), transmitter is produced at a fixed rate but is, of course, used more rapidly at N_{13} than at N_{24}. Signals emitted by V_3 exceed those emitted by V_4. Because these signals compete subtractively at V_5 and V_6, only the output from V_5 is positive, hence produces a positive incentive motivation to drive feeding behavior.

When hunger is reduced and the hunger drive stops, the network exhibits a rebound due to the relative depletion of transmitter at N_{13}. This occurs because input I to both V_1 and V_3 is the same but signals leaving N_{24} are now stronger than those leaving N_{13} (due to varying levels of transmitter). Thus, the subtractive effect causes firing of V_6, which, through its inhibitory effect on U_{i2}, causes feeding behavior to stop.

Stopping Feeding Behavior Due to Frustration. A mechanism to stop feeding while the hunger drive persists is also needed. Again, the dipole is involved. A nonspecific orienting arousal source (OA) is also required (see Figure 4.14). In general, unexpected feedback to the sense receptors causes a decrease in input to U_{i1}, to Ab^+, and to Ab^- to the point at which

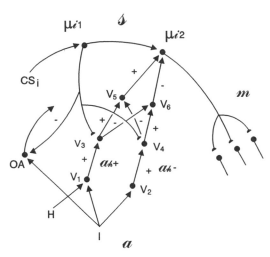

FIGURE 4.14. A hypothesized network capable of explaining the stopping of feeding behavior when the hunger drive, H, is reduced. I represents a tonic, persistent input from the arousal system. Cell populations V_1 through V_6 represent the dipole system responsible for the rebound from hunger to satisfaction when H is reduced. OA is a source of nonspecific orienting arousal.

activity at U_{i2} falls below threshold and the motor behavior is stopped. A decrease in activity at U_{i1} will also cause a decrease in inhibitory output to the nonspecific orienting arousal cells, OA (probably located in the reticular formation). With inhibition shut down, the orienting arousal cells fire and provoke a motor response of cue search. Simply put, unexpected events are arousing. When the maladaptive behavior is extinguished, attention can be focused on the situation and the problem solver, Laurent in this case, is free to attend to previously ignored cues. Recall Laurent's behavior on the thirtieth day of Piaget's experiment. Laurent "no longer tries to suck the glass as before, but pushes the bottle away, crying" (Piaget, 1954, p. 128). Furthermore, when the bottle was moved a little farther away, "he looks at both ends *very attentively* and stops crying" (Piaget, 1954, p. 128). A key question then is, how do unexpected events cause a decrease in input to U_{i1}?

Match and Mismatch of Input with Expectations: Adaptive Resonance. A detailed answer to the previous question lies beyond the scope of the present chapter. The reader is referred to Grossberg (1982, pp. 8–14 and 262–265) for details. In general, however, it can be shown that the suppression of specific input and the activation of nonspecific arousal depend on the layer-like configuration of neurons and feedback expectancies. Consider a pattern of sensory representations to the visual system (i.e., the retina). The retina consists of a layer of retinal cells, $V_1, V_2, \ldots, V_n$, each of which has an activity, $X_1(t), X_2(t), \ldots, X_n(t)$, at every time t due to inputs $I_i(t)$ from an external source. At every time t, the input drives a pattern of activity $X(t) = X_1(t), X_2(t), \ldots, X_n(t)$ across the layer. From the retina the pattern of activity is sent to the LGN, where it excites another layer of cells $V_1, V_2, \ldots, V_n$ and also sends inhibitory

signals to the nonspecific arousal source (see Figure 4.15). Thus, nonspecific arousal is initially turned off by the input.

Following Grossberg, the field of excitation in the LGN is referred to as $F^{(1)}$. Now suppose that, because of prior experience, the pattern of activity, X_1, at $F^{(1)}$ causes firing of another pattern X_2 at $F^{(2)}$ where X_2 may be the next pattern to follow X_1 in a sequence of events previously recorded and $F^{(2)}$ is another layer of cells, which, in this case, would be in the visual cortex. X_2 then constitutes an expectation of what will occur when X_1 excites cells in the LGN. Suppose further that pattern X_2 at $F^{(2)}$ is now fed back to the LGN to be compared with the retinal input following X_1. This would allow the two patterns to be compared. The present is, in effect, compared with an expectation. If the two patterns match, you see what you expect to see. This allows uninterrupted processing of input and continued quenching of nonspecific arousal. Grossberg refers to the match of input with expectations as an *adaptive resonance*.

But suppose the new input to $F^{(1)}$ does not match the expected pattern X_2 from $F^{(2)}$. Mismatch occurs, and this causes activity at $F^{(1)}$ to be turned off, which in turn shuts off the inhibitory output to the nonspecific arousal source. This turns on nonspecific arousal and initiates an internal search for a new pattern at $F^{(2)}$ that will match X_1. If no match can be found, new cells are used to record the new neural sensory input to which the subject is now attending.

The previous discussion can explain the shutting down of maladaptive behavior even in the presence of a continued hunger drive. But why did it take so many contradictions to extinguish that behavior? The answer lies partially in the fact that, on

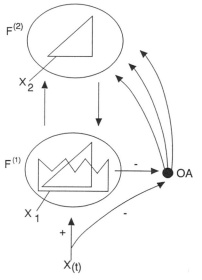

FIGURE 4.15. Match and mismatch of patterns of activity on successive layers of neurons in the brain (after Grossberg, 1982). Input X(t) excites a pattern of activity at $F^{(1)}$ and inhibits nonspecific orienting arousal (OA). Pattern X_1 at $F^{(1)}$ excites pattern X_2 at $F^{(2)}$, which feeds back to $F^{(1)}$. A mismatch causes quenching of activity at $F^{(1)}$ and shuts off inhibition of OA. OA is then free to excite $F^{(2)}$ to search for another pattern to match input.

many trials, Laurent's behavior was not maladaptive. It was successful before the start of Piaget's experiment. Furthermore, it was successful on the many trials during the experiment when Laurent was allowed to feed in his normal way. On these trials the synaptic strengths of Ab^+ and U_{i2} continued to be increased while fired by C_{i1} (the top of the bottle). On the other hand, when his behavior was frustrated, the sensory representation of the bottom of the bottle (C_{i2}) active at U_{i1} during the Ab^- rebound is being conditioned to Ab^-. The *net* feedback from Ab^+ to U_{i1} (either directly to U_{i2} or via U_{i2}) is smaller than when behavior is always successful. As Piaget's experiment continued, the C_{i2} projections to Ab^- became stronger until they finally, on the thirtieth day, dominated the C_{i1}-Ab^+ projections and Laurent stopped attempting to lift and suck the bottle when the bottom was presented. At last he is free to build new connections. His incorrect behavior has been extinguished, and nonspecific arousal cells are firing to allow him to search for important sensory cues previously ignored.

What Laurent must learn is to flip the bottle when its bottom is visible. The bottom is the important cue to be linked with flipping. According to the theory, to provoke this learning the neural activity in the cells responsible for recognizing the bottom of the bottle must be sustained in STM while the motor act of flipping occurs. Nonspecific arousal serves as the source drive to provoke a variety of behaviors (e.g., turning and flipping the bottle); thus when Laurent hits on the act of flipping while he is paying attention to the relevant visual cues or they are stored in STM, the situation exists to allow the required learning to take place. Note the similarity of this neurological account to the biological process of genetic assimilation.

Again consider Figure 4.14. In this case let U_{i1} represent the pattern of excitation in the sensory cortex provoked by sight of the bottom of the bottle. If this pattern remains active in STM during the act of flipping (see Grossberg, 1982, pp. 247–250, for mechanisms), the synaptic strengths of the sensory pattern playing at the nonspecific orienting arousal center (firing due to nonspecific arousal) are increased. The sensory pattern from U_{i1} plus the nonspecific arousal provides pre- and postsynaptic activity that drives increases in the Z_{ij}s. This in turn fires the O_A-U_{i2} pathway. Thus U_{i2}, the cells responsible for the flipping of the bottle, receive inputs from U_{i1} (the bottom of the bottle) and from the orienting arousal source, both of which cause increases in synaptic strengths. In other words, the network allows the child to link, or condition, the sight of the bottom of the bottle with the behavioral response of flipping. Flipping the bottle when the bottom is seen was the behavior that had to be acquired. When the behavior was performed, it resulted in the sight of the nipple, which had previously been conditioned to the act of lifting and sucking. Thus, flipping became linked to lifting and sucking. With each repetition of the sequence, the appropriate synaptic strengths increase until the act takes place with considerable ease.

Thus, Laurent has actively solved his problem and the network has become more complex by the strengthening of synaptic connections. As with Pavlov's dog, an increase in complexity of the neural networks (mental structures) has resulted. This increase is not due to the unfolding of some innate ideas, unless one considers the presence of prewired synaptic connections as innate ideas.

Extension of Network Characteristics to Higher Levels of Knowledge Acquisition

Data on the deployment of advanced reasoning indicate that the percentage of students who successfully use advanced strategies increases gradually with age (see Lawson, Adi, & Karplus, 1979). These advances cannot be attributed to direct teaching because either the increases often come much later than the direct teaching that could have lead to success (e.g., proportional reasoning) or they come without the benefit of any direct teaching (e.g., correlational reasoning) (Lawson & Bealer, 1984). For simplicity, let us restrict our discussion to problems of proportional reasoning because they seem to evoke the most consistent and smallest class of student responses. Let us further restrict the discussion to just one problem of proportional reasoning, the Suarez and Rhonheimer (1974) Pouring Water Task.

The Pouring Water Task, as adapted by Lawson et al. (1979), requires the student to predict how high water will rise when poured from one cylinder into another. Students are shown that water at mark 4 in a wide cylinder rises to mark 6 when poured into a narrow cylinder. They must predict how high water at mark 6 in the wide cylinder will rise when poured into the narrow cylinder. Student responses vary but typically fall into one of four categories:

Additive strategy, e.g., water rose from 4 to 6 (4 + 2 = 6); therefore, it will rise from 6 to 8 (6 + 2 = 8).
Qualitative guess, e.g., the water will rise to about 10.
Additive proportions strategy, e.g., the water will rise to 9 because the ratio is 2 to 3 and 2 + 2 + 2 = 6 in the narrow cylinder, while 3 + 3 + 3 = 9 in the wide one.
Proportions strategy, e.g., 2/3 = 4/6 = 6/x, x = 9.

Again, for simplicity the discussion will be limited to consideration of just two of these strategies, the additive strategy and the proportions strategy, as they appear to reflect the most naive and sophisticated strategies, respectively.

Given the typical naive response of the child to the task is 8, by use of the additive strategy, and the typical sophisticated response of the adolescent is 9, by use of the proportions strategy, the central question becomes: How does the shift from use of the additive strategy to use of the proportions strategy during adolescence come about? The present hypothesis is that it comes about in basically the same manner as Laurent learned to flip his bottle. The child who responds to the Pouring Water Task with the additive strategy is like Laurent when he responds to the bottom of the bottle by lifting and sucking. The neural modeling problem is one of modeling the shift from an additive to a proportions strategy, just as the shift from immediate lifting and sucking to a modified response of flipping and only then lifting and sucking was modeled. To the naive problem solver, the Pouring Water Task presents problem cues, just as the bottle presented cues to Laurent. The difficulty is that these cues set off the wrong response—the additive response. In other words, by analogy with the neural network involved in Laurent's shift in behavior, we assume that the problem cues (CS_1), plus some internal drive to solve the task, combine to evoke the UCR, the additive response controlled by some U_{i2} in the motor cortex.

For children, say below age 10 to 11 years, responding to quantitative problems by use of addition and/or subtraction is indeed a common strategy and, in many instances, leads to success. They do, of course, know how to multiply and divide and many can solve textbook proportions problems, yet use of the proportions strategy (which involves multiplication and/or division) seldom occurs to them, just as it did not occur to Laurent to flip the bottle before lifting and sucking (at least during the first 39 days of Piaget's experiment). Laurent's motor behavior had been successful in the past, and he had no reason to believe it would not continue to be successful. Indeed, many children who use the additive strategy are certain that they have solved the problem correctly.

How, then, do additive reasoners come to recognize the limitations of their thinking? And once they do, how do they learn to deploy the correct proportions strategy? The steps in that process are as follows: (1) indiscriminate use of the additive strategy to solve additive *and* multiplicative problems; (2) contradictory and unexpected (i.e., negative) feedback after use of the additive strategy to solve multiplicative problems, which eventually leads to termination of its use in a knee-jerk fashion; (3) initiation of nonspecific orienting arousal that provokes an external search for problem cues and an internal search through memory for successful strategies that can be linked to those cues; (4) selection of cues and the discovery of a new strategy that appears successful; (5) repeated positive feedback from the successful use of the discovered strategy; and (6) acquisition of an internal system of strategy monitoring to allow a check for consistency or contradiction, which allows the matching of problem cues with appropriate strategies in future situations. Let us consider each step in turn.

Initiation and Termination of Problem-Solving Behavior. A hypothesized minimal neural network (analogous to that previously derived to account for Laurent's behavior) that may account for some of the characteristics of the problem-solving behavior in question is shown in Figure 4.16. The figure is based on the assumption that some problem-solving drive (P) exists and functions to stimulate arousal cells Ap^+. Although the physical basis of specific drives such as hunger and fear are well known, the very existence of a "problem-solving drive" is speculative. Let CS_i represent problem cues from the Pouring Water Task that initially evoke use of the additive strategy. Specific problem cues of the task fire U_{i1} cells in the sensory cortex, which in turn send that pattern of activity to arousal cells, which are also being stimulated by the hypothesized problem-solving drive P to arousal cells Ap^+. Because of prior conditioning, this activity feeds to U_{i2} (also activated by U_{i1}), which in this case represents neural activity to initiate the motor response of addition.

Thus, problem cues (CS_i) from the Pouring Water Task are initially conditioned to additive behavior. The key cue is, perhaps, that the water previously rose "2 more" marks when poured into the narrow cylinder (an absolute difference). Other cues, such as the relative difference of the water levels (narrow cylinder = 1 ½ × the height of the wide cylinder) are ignored. Just as Laurent's feeding behavior was terminated by

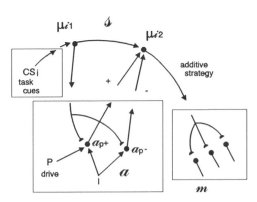

FIGURE 4.16. Components of a hypothesized neural network combining sensory cues from the Pouring Water Task (CS) with an internal problem-solving drive (P) from the arousal system A to fire activity at μ_{i2}, which initiates use of the additive strategy in the motor control system M.

satisfaction of the hunger drive, the student's problem-solving behavior is terminated by reduction of the problem-solving drive P when a solution has been generated. When input from P stops, the tonic input I to both Ap^+ and Ap^- causes a rebound at Ap^- that quickly inhibits U_{i2} activity to stop problem-solving behavior.

Stopping Use of the Additive Strategy Because of Contradiction. A student using the additive strategy to generate a response of 8 to the Pouring Water Task fully expects that the answer is correct, just as Laurent expected to get milk when he sucked the bottom of the bottle. As in the case of Laurent, the unexpected feedback of an incorrect answer will eventually result in stopping the conditioned motor response in similar situations and turning on nonspecific arousal, which will lead to closer inspection of the problem cues and a search for a more effective strategy.

What form can this contradiction take? Certainly, missing problems on a math test could be one form. Yet that feedback normally occurs well after the act of problem-solving has been terminated and would most likely lose some effectiveness in focusing attention on the problem. A seemingly more effective source of feedback, at least on the Pouring Water Task, would be to have the student actually pour the water after predicting a value of 8. The rise of water to mark 9 presents immediate contradictory feedback that could produce the desired effect. However, single contradiction, no matter what its source, is probably not sufficient to shut down the additive strategy. Recall that it took Laurent many trials before he stopped bringing the bottom of the bottle to his mouth. This is probably because the student's use of the additive strategy does not always lead to contradiction. In many problem situations, addition and subtraction are the correct operations. Even if they are not correct, the student may not discover that they are wrong for many days, if ever.

Thus, sensory task cues from additive problems (C_{i1}) to the Ap^+ channel (via U_{i1}) linked to the additive strategy (U_{i2}) would continue to be strengthened in some situations. In situations in which use of the additive strategy leads to contradictions, the sensory task cues from proportions problems (C_{i2}) lead to Ap^- (via U_{i1}); thus the $C_{i2} \rightarrow Ap^-$ channel would be conditioned. As students who use the additive strategy meet continued contradictions, the C_{i2} projections to Ap^- would become stronger than the C_{i1} projections to Ap^+, until they eventually dominate and the student no longer responds unthinkingly with an additive strategy to quantitative problems of this sort.

Orienting Arousal and the Search for a New Strategy.   Only when the unthinking use of the additive strategy has been extinguished and nonspecific arousal is sufficient can the sort of problem inspection and strategy search occur that will lead to successful conditioning of the C_{i2} input to the proportions strategy. How might this occur?

Again consider the example of Laurent learning to flip his bottle before lifting and sucking as shown in Figure 4.14. What seems to be required in the case of proportional reasoning is to link C_{i2} input (multiplicative cues from proportions problems) to the U_{i2} motor response of the proportion strategy. In other words, input C_{i2} at $F^{(1)}$ must match $F^{(2)}$ feedback. This does not appear difficult; it seems to require simply that it occur to the student to use multiplication and division instead of addition and subtraction in the presence of C_{i2} input and nonspecific arousal (see Figure 4.14). But this is not the entire story. A student so conditioned may respond to additive problems with a proportions strategy if the student is not sufficiently aware of the problem cues that suggest which strategy to use!

FEEDBACK AND INTERNAL MONITORING OF PROBLEM SOLVING. How, then, do we resolve the problem of reliably matching cues with reasoning strategies? This is a central question. The proposed answer is as follows. When confronted with a quantitative problem, certain key words or concrete referents are conditioned to strategies (as stated earlier). For example, the word "twice" suggests a multiplicative relationship. The word "more" suggests an additive relationship. However, because these cues are not always reliable, the problem solver must initiate the use of a strategy to determine its consequences, or probable consequences if carried out completely, and then compare those consequences with other known information about the problem situation. If this leads to an internal contradiction, that strategy must be incorrect and another strategy must be tried. Internal contradiction in this case would mean that an adaptive resonance is not found between input and expectations. This would lead to immediate termination of the $S \rightarrow Ap^+$ input, which drives a rebound at Ap^- to terminate that strategy and provoke excitation of nonspecific arousal and a search through LTM for another strategy.

For example, in the Pouring Water Task, use of the additive strategy would lead one to predict that water at mark 2 in the narrow cylinder would rise to mark 0 when poured into the wide cylinder. The water disappears! This is, of course, impossible; therefore, the additive strategy must be wrong as it leads to contradictory feedback. Or consider the following problem: John is 6 years old and his sister Linda is 8 years old. When John is twice as old as he is now, how old will Linda be? The word twice may suggest that you multiply $6 \times 2 = 12$ to get John's age, so you should multiply $8 \times 2 = 16$ to get Linda's age (many

students do this after a lesson on proportions). But this is wrong, as we all know that Linda cannot age faster than John! Therefore, if one internally monitors his or her tentative solution, an internal contradiction results that shuts down S input to U_{i1}, Ap^+, and U_{i2} supporting the proportions strategy. This strategy is given up because it fails to obtain an adaptive resonance, and a new one is searched for in LTM, is found, and is tried until solution expectancies no longer generate internal contradictions. Thus, internal reasoning and internal contradiction are used to match problem cues to problem strategies. This qualitative novel thinking process presumably takes place when students have learned a variety of problem cues and a variety of solution strategies and are left on their own to match cues with strategies.

To summarize, advanced problem solvers appear to have at their disposal the memory record of a variety of problem cues, a variety of problem strategies, and a general mode of operation that says "try the various strategies available until you get one that does not produce internal contradictions." Consequently, the key difference between the additive-reasoning child and the proportional-reasoning adolescent is not the strategies they possess. Both are capable of using both types of strategies. Rather, the key difference appears to be that the child unthinkingly initiates a strategy and then fails to check its consequences internally for consistency with other known data (e.g., water does disappear when poured from one container to another), whereas the adolescent unthinkingly initiates a strategy and (this is new) checks its results for possible contradictions. If contradictions are found (if an adaptive resonance is not found), a new strategy is tried until no contradictions are discovered. This key difference may arise because adolescents have gradually become aware of the fallacy of automatically "jumping to conclusions," while their younger counterparts have not. Thus, a novel behavior has emerged not by direct assimilation of environmental input, nor has it emerged by the maturation of innate structures. Rather, it has emerged by the novel combination of already present, but previously unlinked, problem-solving behaviors and problem cues.

Relation of the Neural Theory to Piagetian Theory and Educational Practice

The proposed theory of neural processing and equilibration shows why the normal curriculum is insufficient to provoke many students to acquire the skills needed to deal successfully with problems of the Pouring Water type. Students learn algorithmic strategies, but they are seldom confronted with the diversity of problems needed to provoke the close inspection of problem cues necessary to link cues with strategies and tentative results with implied consequences. Of course, students must be taught to add, subtract, multiply, and divide. There is little reason to suppose, however, that they need to know, for example, that "the product of the means equals the product of the extremes."

In short, what is acquired in school lessons is insufficient. This statement is reminiscent of Piaget's position regarding the role of teaching in intellectual development. Piaget long insisted that normal teaching practices are insufficient because they seldom, if ever, afford the possibility for equilibration (Piaget, 1964). Unfortunately, as previously mentioned, Piaget's model of psychological equilibration is based on evolutionary rather than neurological models. The present theory, although certainly too simplistic to account for the details of advanced problem solving, nevertheless suggests neurological mechanisms that may be involved in some aspects of the equilibration and knowledge-acquisition process.

Consider the child's initial use of the additive strategy in a knee-jerk fashion as an instance of the immediate assimilation and processing of input by previously acquired mental structures (strategies). This is the Piagetian state of equilibrium. The individual is satisfied by his or her response and not intellectually aroused. But suppose repeated attempts at using that strategy lead to contradiction. At the neurological level, this could be interpreted speculatively as weakening of the $S \rightarrow Ap^+$ channel and strengthening of the $S \rightarrow Ap^-$ channel until it dominates, and nonspecific orienting arousal is turned on and searching behavior initiated to acquire an appropriate response to solve the problem. This is the state of disequilibrium. Finally, through the internal trial-and-error search behavior (see Grossberg, 1982, pp. 14–15) and/or closer inspection of the phenomena, a successful behavior pattern is acquired; that is, new neural connections are formed by increases in the synaptic strengths of the pathways from the input stimulus to the output response. This constitutes an accommodation of mental structures, the acquisition of more complex behavior, and resolution of the problem. In Piagetian terms, it restores equilibrium but at a more sophisticated, emergent level.

Having suggested a sequence of events involved in the successful emergence of proportional reasoning, it is possible to identify why some students never acquire that ability. First, if prerequisite strategies and knowledge are not in place, they cannot be utilized. By analogy, Laurent already knew how to flip his bottle. That was not the problem. Rather, the problem was to connect the flipping with the appearance of the bottom of the bottle. Likewise, the problem for most adolescents is not that they do not know how to multiply and divide or have not "learned" that the "product of the means equals the product of the extremes." Rather, the problem is that they have failed to link the appropriate operations with the appropriate problem cues. Second, the student must be confronted with many diverse problem-solving opportunities that provide the necessary contradictions to his or her use of the additive strategy. Without feedback and contradiction, the necessary arousal will not occur. Therefore, even if students are told to use "proportions" to solve the problems, they are likely to fail to do so in transfer situations because use of the old incorrect strategy has not been extinguished.

The previous discussion, although related most directly to the gradual acquisition of proportional reasoning, does not necessarily preclude its direct teaching. For example, with respect to the Pouring Water Task, the additive strategy can be contradicted directly simply by pouring the water from the wide to the narrow cylinder and noting the rise to mark 9 instead of mark 8. Other problems with similar contradictory feedback can be utilized. One would expect this type of instruction to be very effective, yet teachers and curriculum developers must continue to remind themselves of the remaining limiting factor—namely the student. No matter how potentially interesting the material may seem to the teacher, it is the student who must be aroused by the contradictory feedback to relinquish an incorrect strategy and begin the search for a new one! Sufficient arousal may be difficult to achieve in the impersonal classroom setting, particularly if the problems bear little resemblance to problems of personal importance. Furthermore, short-term direct teaching is probably insufficient to promote the development of the internal monitoring system needed to match problem cues to solution strategies in novel situations. Long-term efforts are needed to accomplish this.

EQUILIBRATION, CONSTRUCTIVISM, AND THE LEARNING PARADOX

The process of equilibration as described by Piaget and as previously detailed at the level of neural models clearly implies that knowledge acquisition is a process in which the learner actively constructs his or her knowledge. Such a view has become known as *constructivism* (Piaget, 1977a, 1977b; Resnick, 1980; Wittrock, 1974). Constructivism, as discussed, stands in marked opposition to the empiricist doctrine that complex knowledge is assimilated directly from the environment. It also stands in opposition to the nativist doctrine, which assumes the existence of innate structures that unfold spontaneously with maturation (Chomsky & Fodor in Piattelli-Palmerini, 1980). According to Bereiter (1985), however, a fundamental theoretical problem exists with constructivism. As Bereiter sees it, constructivist theory offers no satisfactory account of how the learner, if guided only by simple cognitive structures, can generate more complex structures, which allow new behaviors with distinctive properties. Bereiter asks, "How can a structure generate another structure more complex than itself?" (p. 204). In his view, it would not seem possible for a simple self-generating system to become more complex without some external ladder or rope to climb on, which presumably does not exist. Hence, we are left with what has been called the *learning paradox* (Pascual-Leone, 1976, 1980).

How can the learning paradox be resolved? Without reverting to empiricism or nativism, how can simple structures guided by simple cognitive structures generate more complex structures with novel properties not present in the initial structures? Recently, Lawson and Staver (1989) proposed a solution to the learning paradox based in part on the concept of emergent properties and in part on the neural modeling principles just discussed. What follows is recapitulation of the Lawson and Staver proposal. Although their proposal is neural in nature, it may be helpful to precede the discussion with a nonneural analogy to provide a sense of how the constructivist perspective differs from its alternatives.

Suppose you happen upon a snow-covered lawn with a rectangular sidewalk and two piles of snow heaped up either end of the rectangle on the inside of the sidewalk. What could have caused the snow to be piled up like this? Perhaps two tiny clouds passed over the lawn and deposited the snow at either end. Perhaps some children had been playing in the area, and they piled up the snow. Neither of these two explanations is constructivist in nature, because in both cases the snow "structure" was present in the environment (either in the clouds or in the children's minds) before its emergence in the lawn. The constructivist explanation for the piles is that the snow was shoveled off the sidewalk by someone moving counterclockwise and always tossing the snow to the left. Consequently, the amount of snow on the inside of either end of the rectangle would be greater than in the middle and the snow would form two piles. This explanation is contructivist because the resulting structure was nowhere before the behavior. Instead, the behavior coupled with the geometry of the landscape created the new structure. In analogous fashion, the following argument suggests that it is the geometry of neural connections and the behavior of the individual that create new structures that were not previously present in the organism or in the environment. We begin with a brief discussion of emergent properties.

Emergent Properties in the Natural Sciences

Emergent properties are defined as qualitatively unique properties of an object or a system of interacting objects that are derived from a unique combination or configuration of the system's component parts and are not easily predictable from knowledge of the parts. Consider, for example, graphite and diamond, two substances composed only of carbon atoms. In graphite, the carbon atoms are arranged in layers that slide past one another easily, giving graphite a soft, greasy feel and an opaque, black appearance. Diamond can be produced from graphite under conditions of extremely high temperature and pressure, which alter the arrangement of the carbon atoms. In diamond, each carbon atom is bonded to four others in a three-dimensional structure. This three-dimensional array contains no layers and makes diamond a very hard, brittle, crystalline material. Thus the properties of diamond are emergent from those of graphite. The key point as far as resolution of the learning paradox is concerned is that those emergent properties do *not* arise from novel parts. Rather, they arise from a novel arrangement of the *same* parts.

Several examples of emergent properties exist in the biological sciences. Consider evolution. The properties of all living things, from the simplest, single-celled bacterium to the most complex multicellular organism, depend on deoxyribonucleic acid (DNA) molecules. Bacterial DNA and human DNA differ not in kind but only in the amount and arrangement of their constituent parts (nucleotides). DNA provides the blueprints for the construction of all living things. Evolution has progressed from the single-celled bacteria-like organisms of the primordial soup to today's complex multicellular organisms. Those organisms have qualitatively unique, emergent properties (e.g., bilateral symmetry, limbs, brains, leaves) and behavior (e.g., conducting photosynthesis, capturing prey, speaking, reasoning) that have developed from the unique combination of the component parts of DNA molecules, *not* from novel components!

Developmental biology offers numerous examples of emergent properties. Although preformism (the idea that if we see something develop from the egg, it must have been there all the time but in an invisible form) was a popular theory in the seventeenth and eighteenth centuries, it has long been abandoned. Initially, the human infant is but a single fertilized egg cell. During embryological development, the zygote divides to form billions of cells in novel combinations with novel shapes, sizes, and functions that together bestow on the fetus new structures such as a heart, lungs, a brain, eyes, and legs. These structures allow the organism to carry out novel behaviors, not the least of which is to begin to sustain life independent of its mother.

Emergent Properties in Cognition

Suppose that the development of more complex behaviors with novel properties takes place in a way analogous to these physical and biological examples of emergent properties. Perhaps the learner initially possesses simple innate structures (Chomsky & Fodor in Piatelli-Palmerini, 1980), or perhaps the learner possesses innate functions (Piaget in Piatelli-Palmerini, 1980). Selection of either position leads to the same end, as innate functions must be guided by innate structures. The essential task of the learner, then, is to actively construct qualitatively different emergent behaviors out of unique combinations of those initially given parts. Viewed in this way, qualitatively novel "developments" arise from combinations of prior simple structures.

Thus, in the Lawson and Staver view, the learner has innate capabilities that allow initial learning to take place but nativism alone does not result in distinctive behaviors. Rather, the learner actively constructs novel combinations, thereby giving rise to new levels of behavior. The key idea is that properties of thought emerge that are qualitatively different from those that came before. How might this occur on a neural level? To answer this question, one must consider the individual parts and how they combine. In the case of cognition, the individual parts are neurons. To resolve the learning paradox, two conditions must be met. First, it must be shown that the individual neurons can and do organize themselves into more complex combinations given only normal environmental stimulation and their inherent functional and structural properties. Second, it must be shown that cognitive development consists of the acquisition of new behaviors with properties not previously exhibited, that is, with emergent properties. Having shown these two conditions to hold, the learning paradox is, in theory at least, resolved at a general level. For a more convincing argument, however, it must also be shown how specific novel combinations of neurons combine to produce specific behaviors with specific emergent properties. Of course, our present knowledge of neural anatomy and cognition does not allow such specificity. At present, we shall be satisfied if the first and second points can be explicated, leaving the specific details for future research.

In fact, the solution of the learning paradox rests primarily on the first factor, as no one would seriously argue that cognitive development does not involve the acquisition of behaviors with emergent properties. For example, the newborn is unable to move about in his or her environment and eat or drink without considerable parental assistance. Yet during the first 1 to 2 years of life, effective sensory motor behaviors to do so are acquired. Next, the ability to use sounds to describe and to communicate with others about what is seen, heard, smelled, and felt is acquired. Still later, the ability to use language to invent and test ideas about unseen causes of perceived events is acquired. All of these behaviors were not present in prior stages of development; thus, they constitute behaviors with emergent properties. In essence, the Lawson and Staver argument is that all of these behaviors depend on neural activity and all new neural activity involves not new sorts of neurons, but novel combinations of the same sorts of neurons—just as changing graphite to diamond involves transforming one arrangement of carbon atoms into another.

Given normal environmental stimulation, do individual neurons, like individual atoms of carbon or individual cells of an embryo, join spontaneously with one another to form more complex organizations that allow the organism to perform behaviors with emergent properties? Lawson and Staver believe the answer to this question is yes. To explain how they arrived at this answer, we must return to the basic principles of neural modeling theory.

A Return to Classical Conditioning

Recall that learning in Grossberg's theory occurs when synaptic strengths (Z_{ij}s) increase, that is, when the transmitter release rate increases and makes transmission of signals from one neuron to the next easier. Hence learning is, in effect, an increase in the number of connections among neurons, or an increase in complexity. Presumably the structural connections already exist, but they are not functional until the transmitter release rates at certain terminal knobs increase above a specified threshold. Thus, in order to have a "mental structure" become more complex, transmitter release rates must increase at a number of knobs so that signals can easily be transmitted across synapses that were previously there but inoperative. This view of the situation reveals the sense in which the nativists are correct. If one equates mental structures with already present, but inoperative, synapses, then the mental structures are present before any specific experience. This view also reveals how the constructivists are correct. For the synapses to become functional, experience is necessary to strengthen some of the connections. Let us return to Pavlov's classical conditioning experiment to see just how experience can do this.

When Pavlov first rang a bell, the dog, as expected, did not salivate. However, after repeated simultaneous presentation of food and bell ringing, the ringing alone was able to cause the dog to salivate. Pavlov's experiment showed that when a conditioned stimulus (e.g., a bell) is repeatedly presented with an unconditioned stimulus (e.g., food), it will eventually evoke the unconditioned response (e.g., salivation) by itself.

Figure 4.12 showed a simple neural network capable of explaining classical conditioning based on Grossberg's princi-

ples. Initial presentation of food caused cell C to fire, which created a signal down its axon that, because of prior learning, caused the signal to be transmitted to cell B. Cell B fired and the dog salivated. At the outset, ringing of the bell caused cell A to fire and send signals toward cell B. However, when the signal reached knob N_{AB}, its synaptic strength Z_{AB} was not initially large enough to cause B to fire. The dog did not salivate. However, when the bell and the food were paired, cell A learned to fire cell B according to Grossberg's learning equation. The firing of cell A resulted in a large S'_{AB}, and the appearance of food resulted in a large $[X_B]^+$. Thus the product $S'_{AB}[X_B]^+$ was sufficiently large to drive an increase in Z_{AB} to the point at which it alone caused node N_B to fire and evoke salivation. Food was no longer needed. The dog had learned to salivate at the ringing of a bell. The key point in terms of neural modeling theory is that learning is driven by simultaneous activity of pre- and postsynaptic neurons, in this case activity of cells A and B. In terms of constructivism, the key point is that learning results from novel combinations of the basic elements of the brain, not from the production of new elements.

There is no homunculus that controls this increase in complexity (i.e., new connections between cell A and cell B); rather, the increase results from the simultaneous ringing of the bell and appearance of the food. The bell evokes presynaptic activity at cell A and the food evokes postsynaptic activity at cell B, so transmitter release rate at synapse A-B increases. A new functional connection has been made. The system has become functionally more complex because of its own "nature" *and* because of experience.

Lawson and Staver argue that they have shown in principle how neural networks spontaneously become functionally more complex. Resolution of the learning paradox requires only that these spontaneous increases in neural network complexity allow for novel behaviors with emergent properties. It remains for future research to determine the specific details of which specific new functional connections lead to which specific behaviors with which specific emergent properties.

THE ROLE OF LOGIC IN KNOWLEDGE CONSTRUCTION

Clearly, a faithful recording of external events, even if possible, is not sufficient to account for all of human cognition. If one notes, for example, that John is older than Mike and Mike is older than Beth, then it seems "logical" that John *must be* older than both. We can quickly draw this obvious conclusion in spite of the fact that the relative ages of John and Beth are not among the information given. Rather, we were able to deduce the information using our powers of reason. Without those powers of reason—that is, our inferential ability—knowledge acquisition would be severely restricted. For many people, including Piaget, logic and reasoning amount to one and the same thing. Indeed, early logicians studied logic because it was supposed to be the discipline involved in discovering effective patterns of reasoning. Piaget made one of the strongest statements about the role of propositional logic in advanced reasoning: "Reasoning is nothing more than the propositional calculus itself" (Inhelder & Piaget, 1958, p. 305). Other statements by Piaget reflect his belief that logical operations, once acquired, allow one to reason with the logical form of propositions regardless of presumably irrelevant contexts; for example, "concrete operations fail to constitute a formal logic; they are incompletely formalized since form has not *yet* been divorced from subject matter" (Piaget, 1957, p. 17).

Similarly, Inhelder and Piaget (1958) argued that the formal stage is reached when coordination is achieved "between a set of operations of diverse kinds and that the form be liberated from particular contents" (pp. 331–332). However, a long line of research beginning with the work of Wason (1966) and others (Byrne, 1989; Cheng & Holyoak, 1985; Evans, 1982; Griggs & Cox, 1982; Johnson-Laird, 1983; Markovits, 1984, 1985; Overton, Ward, Black, Noveck, & O'Brien, 1987; Wason & Johnson-Laird, 1972) makes it clear that form is not completely liberated from content in "logical" reasoning. Even Piaget admitted that content might play a significant role in formal reasoning (Piaget, 1972). However, he did not suggest how his theory would have to be modified to account for such a result. More recently, Cheng and Holyoak (1985) argued that people possess general schemes of permission, causation, obligation, and so forth and that these relate to particular classes of events from experience. Johnson-Laird (1983) saw reasoning as even more tied to specific events and argued that reasoning involves constructing representations of particular events described by sets of specific premises. He used the phrase "mental models" to refer to these representations.

In basic agreement with more recent authors, I would argue that rather than logic, the essential feature of advanced reasoning is the tendency of the reasoner to initiate reasoning with more than one specific antecedent condition (Markovits, 1984, 1985). If the reasoner is unable to imagine more than one antecedent condition, at least he or she is aware that more than one is possible, and conclusions that are drawn are tempered by this possibility. Presumably, knowledge of specific antecedent conditions (multiple hypotheses) arises from previously constructed mental models of specific situations, as proposed by Johnson-Laird (1983), and the general tendency to search for multiple hypotheses arises by generalization from a variety of particular situations in which this has turned out to be the case. This proposed view of advanced reasoning will be referred to as the multiple-hypothesis theory.

Two Common Forms of Logic

Although the multiple-hypothesis theory of advanced reasoning is not based on the use of abstract forms of logic, it is necessary to begin with a discussion of two basic logical forms to more fully explicate what the multiple-hypothesis theory is, how it differs from its alternatives, and how it has been tested. The two abstract forms of logic are those of basic conditional (if . . . then) and biconditional (if and only if) logic. Specific forms of conditional logic are as follows:

1. $p \supset q, p \therefore q$ (this form of logic is referred to as *modus ponens*);
2. $p \supset q, \bar{q} \therefore \bar{p}$ (this form of logic is referred to as *modus tollens*).

Two presumed misapplications of conditional logic are

3. $p \supset q, \bar{p} \therefore \bar{q}$ (the conclusion of $\bar{q}$ is referred to as the fallacy of denying the antecedent).
4. $p \supset q, q \therefore p$ (the conclusion of p is referred to as the logical fallacy of affirming the consequent).

The corresponding forms of biconditional logic are as follows: (1) $p \equiv q, p \therefore q$; (2) $p \equiv q, \bar{q} \therefore \bar{p}$; (3) $p \equiv q, q \therefore p$; and (4) $p \equiv q, \bar{p} \therefore \bar{q}$. Let us consider different types of situations in which these forms might be employed. In his logic textbook, Copi (1972) identified six types of conditional statements, each of which he believes corresponds to a different sense of if . . . then; thus each asserts a different type of implication. The following statements represent the six types and all follow precisely the same "if . . . then" form.

1. *If* a drawing is a square, *then* it has four sides.
2. *If* a person has successfully completed high school degree requirements, *then* he or she is a high school graduate.
3. *If* the car is running, *then* there is gas in the tank.
4. *If* a person has a valid driver's license, *then* he or she may legally drive a car.
5. *If* the ASU football team plays in the Rose Bowl game, *then* John will be in Pasadena on New Year's Day.
6. *If* Hitler was a great man, *then* I am a monkey's uncle.

Clearly, the six conditional statements, are of the same logical form, yet they vary in context. In statement 1, the consequent follows logically from the antecedent by virtue of the fact that squares are a subclass of a class of drawings that have four sides and must have properties of that class. Another subclass called rectangles also has four sides but they are unequal. In statement 2, on the other hand, the consequent follows from its antecedent by the definition of the term graduate, which means one who has completed the degree requirements. The consequent of 3 does not follow from its antecedent either by logic alone or by definition. Here the connection is an empirical one. The connection is one of a necessary but not sufficient cause. The car cannot run without gas in the tank, but gas alone is not sufficient to get the car running. Statement 4 also involves necessary conditions; however, here the condition of having a valid driver's license is not only necessary but also sufficient to *allow* one to drive a car (assuming the person is not drunk or otherwise impaired). Statement 5 does not involve causal relationships of the type implied in 3 or 4 because it simply reports a decision on the part of John to behave a certain way under certain circumstances. Finally, statement 6 differs from the previous five statements in that no "real" connection between the antecedent and consequent is suggested. Rather, this sort of conditional is often used as an emphatic or humorous way to deny the antecedent via *modus tollens*. Clearly, the speaker is not a monkey's uncle; therefore Hitler must not have been a great man. These six conditional statements will be referred to respectively as (1) class membership, (2) definitional, (3) necessary but not sufficient cause, (4) necessary and sufficient cause, (5) decisional, and (6) material. Although these conditional statements appear similar to Cheng and Holyoak's (1985) pragmatic reasoning schemes, we make no claim that they actually constitute six different reasoning schemes.

A seventh type of if . . . then statement has appeared in the psychological literature that does not fit any of these categories. This type was first investigated by Wason (1966). Wason's if . . . then statement, which is part of his Four Card Task, reads as follows: "*If* a card has a vowel on one side, *then* it has an even number on the other side." Because there is no nonarbitrary reason for this connection between vowels and even numbers to be asserted, this seventh form will be referred to as arbitrary. This is an important point and could well explain why the Four Card Task elicits such a wide range of responses from adults. In other words, because the task does not contain any nonarbitrary information about the nature of the connection between the antecedent and the consequent, the subject may be unable to decide what sort of connection (e.g., causal, decisional, definitional) it is supposed to hold.

Seven "logic" tasks were designed that involved these seven types of conditional statements (Lawson, 1992). In each task, the subject (S) was given four sentences and asked to draw an inference and then write that inference in spaces provided. The four sentences corresponded to the four possible conditions (i.e., p, $\bar{p}$, q, $\bar{q}$). For example, in the task designed to elicit inferences from the conditional statement "If a drawing is a square, then it has four sides," the following four questions were asked:

1. If a drawing is a square, then it has four sides. This drawing is a square, therefore. . .
2. If a drawing is a square, then it has four sides. This drawing is not a square, therefore. . .
3. If a drawing is a square, then it has four sides. This drawing has four sides, therefore. . .
4. If a drawing is a square, then it has four sides. This drawing does not have four sides, therefore. . .

If one uses the standard rules of conditional logic (*modus ponens, modus tollens*) to guide reasoning on each of the seven tasks, the conclusions reached in all tasks are identical. For question 1 of the Squares Task, for example, one would reason with the form $p \supset q, p \therefore q$ and conclude that "the drawing has four sides." For question 2, the reasoning would follow the form $p \supset q, \bar{p} \therefore q$ or $\bar{q}$, and the subject would conclude that "no conclusion can be reached." Question 3 should lead to the same conclusion as question 2 when one follows reasoning of the form $p \supset q, q \therefore p$ or $\bar{p}$, and the subject would conclude that "no conclusion can be reached." Finally, question 4 reasoning would be of the form $p \supset q, \bar{q} \therefore \bar{p}$, and the subject would conclude that "the drawing is not a square."

Therefore, Piaget's position that adult reasoning follows logical form regardless of context would lead to the prediction that

persons identified as formal operational by performance on Piagetian tests of formal reasoning would draw the same inferences on all seven tasks because the form of each task is identical. However, previous research and further analysis of the seven tasks allow one to anticipate different results of such a test. Consider, for instance, application of the conditional rules of inference to the definitional High School Graduate Task and the necessary and sufficient Driver's License Task. Although these two tasks follow the same logical form as the others, it seems that the context dictates that the reader draw a different set of conclusions. Use of *modus ponens* in question 1 on the High School Graduate Task, for example, would lead appropriately to the conclusion that "John is a high school graduate." Also, use of *modus tollens* in question 4 would lead appropriately to the conclusion that "Joe has not successfully completed high school degree requirements." These conclusions are of the same form as in the other tasks. But questions 2 and 3 could, I think, also be answered "correctly" with a biconditional interpretation of the relationships between p and q (i.e., $p \equiv q$) because the relationship is a definitional one: one who has successfully completed high school degree requirements *is* a high school graduate and vice versa. Therefore question 2 states that Linda has not successfully completed high school degree requirements, so "she is not a high school graduate." Is this conclusion unreasonable? I think not. Of course, it is possible to think of situations that could render this conclusion false (e.g., Linda did not complete the requirements; her parents paid the principal so the principal awarded her the degree), but these situations are beyond the meaning normally associated with the phrase "high school graduate." In other words, the correct conclusions in the definitional situation and in the necessary and sufficient conditions of the Driver's License Task could well be of the biconditional form. Thus, context clearly appears to affect reasoning and the view that a formal-operational person reasons with the form of arguments regardless of their context appears wrong, at least when we are dealing with these types of implications.

What is it about the context that cause one's inferences to shift from a conditional to a biconditional form? As mentioned previously, the multiple-hypothesis theory would argue that the answer lies in one's ability to conceive of alternative antecedent conditions (or alternative hypotheses). Given $p \supset q$, can the thinker imagine other antecedent conditions that could reasonably lead to q? If the thinker can, his or her inferences appear like those of someone using conditional logic. If not, the inferences appear like those of someone using biconditional logic. Again consider the High School Graduate Task. Can you think of antecedent conditions other than "completing high school requirements" that can lead to becoming a high school graduate? Possibly, as just discussed, but you must make a considerable effort to do so. Therefore, for most people p implies q and q implies p. Contrast this to the causal Gas Tank Task. Given $p \supset q$ (if the car is running, there is gas in the tank), can you think of other occasions (conditions) in which there may be gas in the tank? Suppose, for example, there is gas in the tank, but the ignition has not been turned on. This would prevent the car from running. Thus, when confronted with the incomplete argument, "If the car is running, then there is gas in the tank. The car is not running, therefore . . . ," one can easily avoid the incorrect conclusion that there is no gas in the tank by simply imagining that the ignition has not been turned on.

More specifically, the context of the task should affect the pattern of inferences one draws (i.e., conditional or biconditional) to answer a question, because some contexts make it easier (or more difficult) for the thinker to imagine alternative antecedents. When alternative antecedents are easy to imagine, one uses conditional reasoning. When antecedents are difficult to imagine, biconditional reasoning will more likely be used.

More specifically for the seven tasks, the following questions regarding alternative antecedent conditions could be asked (note that the questions are listed, by and large, in order of ease of generation of alternatives):

1. Can I imagine a reason, other than no gas in the tank, that could keep the car from running? Answer: Yes. The ignition is not turned on.
2. Can I imagine drawings, other than squares, that have four sides? Answer: Yes. Rectangles are not squares, but they have four sides.
3. Can I imagine a reason, other than ASU playing in the Rose Bowl game, that could get John to Pasadena on New Year's Day? Answer: Yes. He may be there to visit some friends.
4. Can I imagine letters, other than Es, that could have 4s on their back side? Answer: Yes. There may be an A on the front side.
5. Can I imagine a situation, other than Hitler being a great man, that could cause me to be a monkey's uncle? Answer: This question is difficult to answer because it is not really a "serious" question. Hitler being a great man could not really cause someone to be a monkey's uncle in the first place. But if you are willing to accept this, you may be willing to accept alternatives as well.
6. Can I imagine conditions, other than having a valid driver's license, that would enable me to drive a car legally? Answer: Possibly there are some, but I cannot think of any; therefore, I will answer no.
7. Are there conditions, other than completing high school degree requirements, that would enable me to become a high school graduate? Answer: Possible there are some, but I cannot think of any; therefore, I will answer no.

Presumably, as one progresses from question 1 to question 7, it becomes more and more difficult to imagine alternative antecedent conditions. Alternatives seem to come easily to mind in the gas tank context, with considerable difficulty in the monkey's uncle context, and perhaps not at all in the driver's license and high school graduation contexts. Therefore, subjects should respond to the earlier tasks with an inference pattern like that of a person using conditional logic, but should respond to the later tasks with an inference pattern like that of a person using biconditional logic. This result, therefore, supports the view that *the context of the task influences the inference patterns one employs in its solution, and the context effect is due to the extent to which the subject is able to use his or her knowledge of the context to generate alternative hypotheses or*

antecedent conditions that could lead to the imagined consequent. Note that this position differs from the popular familiarity hypothesis, which suggests that the subject's familiarity with the context is crucial in determining his or her reasoning patterns (e.g., Griggs & Cox, 1982). The assumption is made that all normal adults are familiar with the contexts used in all seven of the tasks but will nevertheless draw different inferences, depending on whether the context facilitates or restricts alternative-hypothesis generation. Also, the multiple-hypothesis theory goes beyond the Johnson-Laird mental-model hypothesis in that, although it argues that reasoning is based on the use of specific mental representations (not the use of abstract forms of logic), it also argues that a key factor that determines the ultimate form of the inferences (i.e., conditional or biconditional) is a person's knowledge of alternative antecedent conditions. The multiple-hypothesis theory goes beyond the Cheng and Holyoak (1985) pragmatic-reasoning scheme hypothesis for the same reason.

Key Elements of the Multiple-Hypothesis Theory of Advanced Reasoning

Table 4.2 and Figure 4.17 summarize key elements of the multiple-hypothesis theory. In Table 4.2, two types of mental representations are depicted. In the first type, only one antecedent condition p (depicted by the solid circle) is linked through past experience (depicted by the solid line) with one consequent q (depicted by another solid circle). The inferences that result from use of such a mental representation follow the form of the biconditional. Note that the argument is not that the reasoner is using biconditional reasoning. In agreement with Johnson-Laird (1983), the argument is that the reasoner is reasoning about the specific mental representation. The form of the reasoning is that of the biconditional *because* of the nature of the representation. For example, if p represents the antecedent condition of "completing high school degree requirements," q represents the status of being "a high school graduate," and no other antecedent conditions can be imagined, it follows that if one completes high school degree requirements then one is a high school graduate ($p \therefore q$) and if one is a high school graduate then one must have completed the degree requirements ($q \therefore p$). Likewise, if one has not completed the requirements, one is not a graduate ($\bar{p} \therefore \bar{q}$) and vice versa ($\bar{q} \therefore \bar{p}$).

In the second type of representation, two antecedent conditions (o and p) are linked to one consequent q. Because of the nature of this more complex representation, the pattern of inferences drawn is that of the conditional. For example, suppose o represents the alternative antecedent condition that "one has not completed high school degree requirements, but instead has done special favors for the school principal." Suppose that p represents the condition of "completing high school degree requirements." In this case, both conditions are linked to the consequent q of "being a high school graduate." Now there are two ways to become a high school graduate; therefore, it follows that if $o \therefore q$ or if $p \therefore q$. However, if one is told that Sue is a high school graduate (q), it is not clear how she became one;

TABLE 4.2. Mental Representations of Specific Situations with One and Two Antecedent Conditions and the Inferences that Result from Each Type

Mental Representations	Resulting Inferences	
One antecedent condition		
$p \qquad q$	$p \therefore q$ $q \therefore p$	$\bar{p} \therefore \bar{q}$ $\bar{q} \therefore \bar{p}$
Two antecedent conditions		
o q p	$o \therefore q$ $p \therefore q$ $q \therefore$ no conclusion	$\bar{o} \therefore$ no conclusion $\bar{p} \therefore$ no conclusion $\bar{q} \therefore \bar{p} \cdot \bar{o}$

Source: Based in part on Lawson (1992).

therefore, given q, no conclusion regarding o or p can be drawn. Likewise, given $\bar{o}$, q, or $\bar{q}$ may follow because we are not informed about p (i.e., $\bar{o} \therefore$ no conclusion). Finally, if one is told that Sue is not a graduate ($\bar{q}$), one may conclude that she has not completed the requirements, nor has she done special favors for the principal (i.e., $\bar{q} \therefore \bar{p} \cdot \bar{o}$).

As another example, consider the conditional statement that "if the car is running, then there is gas in the tank." Given the statement that "the car is not running" ($\bar{p}$), what conclusions should be drawn? If one's mental representation merely associates gas ($\bar{q}$) with running (p) and no gas ($\bar{q}$) with not running ($\bar{p}$), one would incorrectly conclude that there is no gas in the

tank ($\bar{p} \therefore \bar{q}$). On the other hand, if one can imagine at least two reasons why the car may not be running, such as no gas ($\bar{q}$) and ignition not turned on ($\bar{o}$), then reasoning takes the form of the conditional and the conclusion is that no conclusion follows.

Figure 4.17 shows that three levels of intellectual development can arise as a consequence of the nature of one's mental representations and from a generalization from the more complex type. At level 1 only mental representations with one antecedent condition exist; therefore all inferences take the form of the biconditional. At level 2 the child's increased experiences have produced many types of mental representations that vary from situation to situation; therefore the resulting form of inferences (biconditional or conditional) varies from situation to situation depending on whether or not alternative antecedent conditions are imagined. Finally, at level 3, the thinker has acquired a more general representation that can be applied to all situations, which allows him or her to argue that "even though I may be able to think of only one antecedent condition, I know from past experience that in most situations other possibilities exist, so I had better look for some before I jump to a hasty conclusion." Therefore, at level 3, most reasoning takes the form of the conditional.

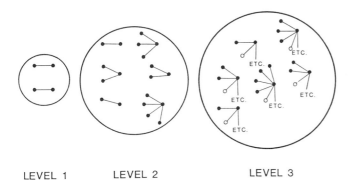

LEVEL 1   LEVEL 2   LEVEL 3

FIGURE 4.17. Levels of intellectual development in terms of types of mental representations. At Level 1 only mental representations of specific situations with one antecedent condition exist; therefore, inferences take the form of the biconditional. At level 2 mental representations exist for specific situations with different numbers of antecedents; therefore, the pattern of inferences one draws (biconditional or conditional) depends on the specific situation. At level 3 the reasoner has acquired the knowledge that alternatives probably exist, even when not easily imagined (designated by the open circles); therefore reasoning becomes more reflective as the person seeks alternatives and, if they are found, the inferences drawn take the form of the conditional.

Testing the Alternatives

As a test of the multiple-hypothesis theory of advanced reasoning contrasted with Piaget's theory that adult formal-operational persons reason with the form of conditional statements regardless of their context, 922 college students were administered a test that allowed them to be classified as "concrete," "transitional," and "formal" thinkers as defined by Piaget's theory. The seven "logic" tasks were then randomly distributed to the students so that each student received just one task (Lawson, 1992). Under these conditions, Piagetian theory led to the prediction that formal-operational subjects would respond similarly using conditional logic to each task because of their similar logical form. The multiple-hypothesis theory leads to the prediction that responses should vary from task to task depending on the extent to which the context suggests alternative antecedent conditions. More specifically the multiple-hypothesis theory predicted that

- The percentage of conditional responses should significantly decrease from task 1 to task 7. (See the prior seven questions for the predicted task order.)
- The percentage of biconditional responses should significantly increase from task 1 to task 7.
- Because the multiple-hypothesis theory argues that Piagetian tests of formal-operational reasoning actually measure whether a person has acquired a general approach to reasoning that tells him or her to seek alternative antecedent conditions even when none may be obvious, subjects who have been classified as formal operational should respond with a greater proportion of inferences of the conditional form than subjects classified as transitional or concrete operational on each of the logic tasks.

Figure 4.18 shows the percent of Ss responding to the seven tasks with the conditional form for all questions (i.e., question 1 $\therefore q$; question 2 no conclusion; question 3 no conclusion; question 4 $\therefore \bar{p}$). The percentages ranged from 56.7% on the Gas Tank Task to 1.9% on the Driver's License Task.

Figure 4.19 shows the percent of Ss who responded to the seven logic tasks with the biconditional form for all questions (i.e., question 1 $\therefore q$; question 2 $\therefore \bar{q}$; question 3 $\therefore p$; question 4 $\therefore \bar{p}$). Again, the percentages varied widely from task to task. Only 17.1% of the Ss' responses followed the biconditional pattern on the Gas Tank Task, but 76.4% of responses did so on the Driver's License Task. Results fit the pattern predicted by the multiple-hypothesis theory, that task context influences the thinking used because of the context's ability to evoke alternative antecedents. The task for which alternatives were assumed to be the easiest to generate (the Gas Tank Task) elicited the greatest percentage of inferences of the conditional form (and the

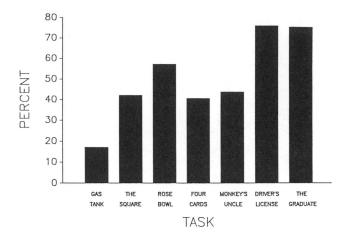

FIGURE 4.19. Percentage of subjects drawing inferences with the biconditional logic pattern on each task.

least percentage of the biconditional form), and the tasks that were assumed to be the most difficult to imagine alternatives (the Driver's License and the Graduate tasks) elicited the least percentage of inferences of the conditional form (and the greatest use of the biconditional form).

Figure 4.20 shows the percentage of Ss at each Piagetian developmental level (concrete, transitional, formal) who drew inferences with the conditional pattern on each task. Two trends should be noted. First, as predicted by the multiple-hypothesis theory, the formal Ss tended to respond more frequently with the conditional pattern than the other two groups. Group differences reached significance ($p < .05$) on two of the tasks (The Square, $X^2 = 10.08, p = .0065$; Monkey's Uncle, $X^2 = 6.6, p = .035$). For the Gas Tank, Rose Bowl, Four Cards, Driv-

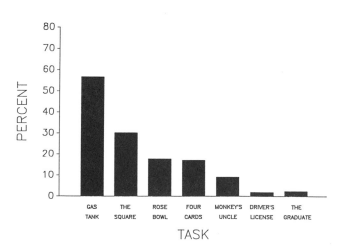

FIGURE 4.18. Percentage of subjects drawing inferences with the conditional logic pattern on each task.

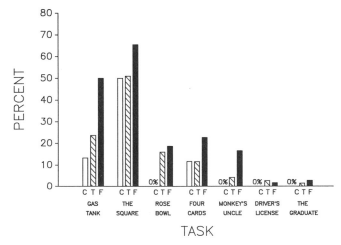

FIGURE 4.20. Percentage of subjects at each Piagetian reasoning level (C = concrete, T = transitional, F = formal) drawing inferences with the conditional logic pattern on each task.

er's License, and Graduate tasks, the respective X^2 values and probability levels were $X^2 = 4.49, p = .11; X^2 = 3.06, p = .22; X^2 = 3.5, p = .17; X^2 = 3.39, p = .18; X^2 = .41, p = .81$. Second, the formal operational Ss did not consistently draw inferences that followed the conditional pattern from task to task (the percentage varied from 65.5% on the Gas Tank Task to 1.7% on the Driver's License Task). This indicates a very strong context effect. Thus, students identified as formal operational do *not* appear to be reasoning with context-free rules of conditional logic.

Conclusions and Recommendations

The results lead to two major conclusions. First, specific contexts influence the pattern of inferences one draws (conditional or biconditional), presumably because they facilitate or restrict one's ability to imagine alternative antecedent conditions. When one is able to imagine reasonable alternatives, the pattern of inferences drawn is similar in form to that of conditional logic. When one is unable to imagine reasonable alternatives, the pattern of inferences is similar in form to that of biconditional logic (Markovits, 1984, 1985). Second, the more skilled reasoners (as identified by performance on the Piagetian reasoning items) responded more often with the conditional response pattern than their less skilled peers. Hence, the study's general conclusion is that task context influences the pattern of inferences drawn (conditional or biconditional) but that a general (context-free) reasoning ability exists that consists primarily of a propensity to consider alternatives. That is, a general rule to guide reasoning appears to exist in some persons, but instead of being a logical one, it is of the sort that asks, "Before I jump to a hasty conclusion (i.e., $p \therefore q, \bar{p} \therefore \bar{q}, q \therefore p, \bar{q} \therefore \bar{p}$), are there alternative antecedent conditions that would render a hasty conclusion false? If so, do not draw the conclusion." In other words, the context influences one's conclusions based on one's ability to imagine alternatives within specific contexts, given the general propensity to look for them.

Rather than the acquisition of a new set of logical operations, adolescent intellectual development can be viewed primarily as the acquisition of a general disposition to consider alternative possibilities and the acquisition of accompanying hypothesis-testing schemes that allow one to process evidence and to choose among the alternatives (e.g., control of variables, correlational reasoning, probabilistic reasoning). In causal contexts, Lawson and Hegebush (1985), for instance, found that young children showed little interest in or awareness of cause-effect relationships. Older children were aware of causal relationships but tended to restrict their thinking to one cause for one effect (hence, a frequent drawing of inferences of the biconditional form), and adolescents were more likely to generate multiple possible causes. It appears then, that some adolescents and adults move beyond this restricted view and become aware of multiple causes for specific effects in specific contexts and, indeed, acquire a general disposition (a habit of mind) that leads them to probe deeply for alternatives even when the alternatives are not thrust on them by the circumstances (hence

a frequent avoidance of drawing inferences of the biconditional form).

THE ROLE OF REASONING IN THE CONSTRUCTION OF CONCEPTS

Knowledge acquisition and reasoning, according to the view presented here, are not primarily a matter of applying logical rules or merely abstracting external features. Rather, knowledge acquisition is seen as a constructive process that involves actively generating and testing alternative possibilities. To be more specific, knowledge acquisition is analogous to the working of the scientist as he or she generates and tests alternative hypotheses. Let us consider how the process is involved in the construction of descriptive concepts and theoretical concepts.

How Are Descriptive Concepts Constructed?

To acquire a sense of how the construction of descriptive concepts takes place, consider the drawings in Figure 4.21. The first row of Figure 4.21 contains five "creatures" called Mellinarks (Elementary Science Study, 1974). None of the creatures in the second row is a Mellinark. From this information, try to decide which of the creatures in the third row are Mellinarks.

If you correctly identified the first, second, and sixth figures as Mellinarks, you have acquired a concept of the term Mellinarks. Note that the term "concept" is being defined as some pattern of regularity identified by the learner plus a term (in this case Mellinark) that is used to refer to that pattern. Assuming that you correctly discovered the pattern referred to by the term Mellinark, how did you do it? Outdated theories of abstraction (Hume, 1739; Locke, 1690) would claim that you "induced" a set of specific characteristics and generalized it to other instances. The multiple-hypothesis theory, on the other

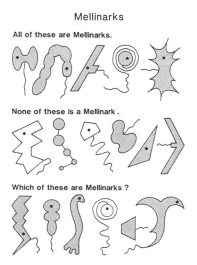

FIGURE 4.21. Imaginary creatures called Mellinarks. (From Elementary Science Study, 1974.)

hand, emphasizes the importance of procedural knowledge and generating and testing alternatives (Bolton, 1977; Holland, Holyoak, Nisbett, & Thagard, 1986; Mayer, 1983).

Let us consider a solution employing this generative process of concept construction. A glance at row one reveals several features of the Mellinarks. They have tails. They contain one large dot and several smaller dots. They have an enclosed cell-like membrane that may have curved or straight sides. If we assume that features such as these are crucial, then which ones? The nature of the membrane (curved or straight) can be eliminated immediately because both membrane types exist in row one. The importance of the other three features can be tested easily starting with some alternatives as follows. Mellinarks consist of creatures with

1. one large dot only
2. several small dots only
3. one tail only
4. one large dot and several small dots
5. one large dot and one tail
6. several small dots and one tail
7. one large dot and several small dots and one tail

According to alternative 1, all the creatures of row one and none of the creatures in row two should contain one large dot. Because this is not the case, the prediction is disconfirmed and the proposition that Mellinarks are distinguished solely by the presence of one large dot is also disconfirmed. The same pattern of deductive reasoning leads one to disconfirm alternatives 2 through 6, leaving proposition 7, that Mellinarks are defined by the presence of all three features, as correct. Thus, only the first, second, and sixth creatures in row three are Mellinarks.

Concept construction, seen in this light, is not a purely abstractive process but depends on the ability to generate and test alternative propositions. In this sense, one's conceptual knowledge (an aspect of declarative knowledge) depends on one's procedural knowledge. As one gains skill in using these deductive procedures, concept formation becomes easier. In the case of the Mellinarks, the concept acquired is a descriptive one, as its defining attributes are directly perceptible. An interesting question that was not addressed is where the alternative propositions came from in the first place. The answer appears to be that they came directly from an inspection of the creatures themselves. This process is perhaps best labeled by the term *induction*, as the learner notes *specific* features on specific creatures and induces, that is, generates the *general* propositions that those features may be critical to *all* Mellinarks. In this sense the "alternatives" are really inductively generated general propositions. We can, therefore, refer to this generative process of concept construction as *inductive-deductive* in nature.

The Role of Chunking in Higher-Order Concept Construction

At any one moment, the human mind can mentally integrate or process only a limited amount of information. Miller (1956) introduced the term "chunks" to refer to the discrete units of information that could be consciously held in working memory and transformed or integrated. He cited considerable evidence that the maximum number of these discrete chunks was approximately seven.

Clearly, however, we all acquire concepts that contain far more information than seven units. The term ecosystem subsumes a far greater number of discrete units or chunks than seven. Furthermore, the term ecosystem itself is a concept; thus, it probably occupies but one chunk in conscious memory. This implies that a mental process must occur in which previously unrelated parts—that is, chunks of information (a maximum of about seven chunks)—are assembled by the mind into one higher-order chunk or unit of thought. This implied process is known as "chunking" (Simon, 1974).

The result of higher-order concept acquisition (chunking) is extremely important. It reduces the load on mental capacity and simultaneously opens up additional mental capacity that can be occupied by additional concepts. This in turn allows one to acquire still more complex and inclusive concepts (i.e., concepts that subsume greater numbers of subordinate concepts). To turn back to our initial example, once we all know what a Mellinark is, we no longer have to refer to them as "creatures within an enclosed membrane that may be curved or straight, which have one large dot and several smaller dots inside and one tail." Use of the term Mellinark to subsume all of this information greatly facilitates thinking *and* communication when both parties have acquired the concept. See Ausubel (1963) and Ausubel, Novak, and Hanesian (1968) for details of the assumption process.

How Are Theoretical Concepts Constructed?

The preceding discussion of descriptive concept construction leaves two important issues unresolved. How does concept acquisition take place when the defining attributes are not directly perceptible, that is, when the concept in question is a theoretical one? What takes place when the theoretical concept to be acquired contradicts a previously construction concept?

Again, let us consider these issues through the use of an example. The example is that of Charles Darwin as he changed his view from that of a creationist to that of an evolutionist. Furthermore, he invented a satisfactory theory of evolution through a process referred to as natural selection. Note that the concepts of creationism, evolution, and natural selection are all theoretical, in the sense that their defining attributes are not directly perceptive.

First, let us consider the process of conceptual change. How are inappropriate theoretical concepts modified or discarded in favor of more appropriate theoretical concepts? This is a difficult question to answer, primarily because the process takes place inside people's heads away from the observer and often at a subconscious level. Thus it is not only hidden from the researcher but is often hidden from the subject as well (Finley, 1986).

Conceptual Change. Gruber and Barrett (1974) analyzed Darwin's thinking during the period 1831 to 1838, when he underwent a conceptual change from a creationist theory of the world (a misconception in today's scientific thinking) to that of an evolutionist (a currently valid scientific conception). Fortunately for Gruber and Barrett and for us, Darwin left a record of much of his thinking during this period in copious diaries. Figure 4.22 highlights the major changes in his theoretical conceptual system during this time.

Darwin's theory in 1831 has been described by Gruber and Barrett (1974) as one in which the creator made an organic world (O) and a physical world (P) and the organic world was perfectly adapted to the physical world (see A of Figure 4.22). This view of the world served Darwin well, and his thoughts and behavior were consistent with it.

Although Darwin was certainly a creationist in 1831, he was aware of evolutionary views. In fact, his grandfather, Erasmus Darwin, published a work entitled *Zoonomia: or the laws of organic life* that contained speculative ideas about evolution and its possible mechanism. Nevertheless, on the day in 1831 when Darwin boarded the H.M.S. *Beagle* as the ship's naturalist, he was seeking an adventure—not a theory of evolution.

During the first 2 years of the voyage on the *Beagle*, Darwin read some persuasive ideas about the modification of the physical environment through time in Charles Lyell's two-volume work *Principles of geology*. At each new place Darwin visited, he found examples and important extensions of Lyell's ideas. Darwin was becoming increasingly convinced that the physical world was not static—it changed through time. This new concept stood in opposition to his earlier beliefs, and it created a contradiction. If the organic world and the physical world are perfectly adapted and the physical world changes, then the

organic world must also change. This, of course, is the logical extension of the argument. Its conclusion, however, was the opposite of Darwin's original theory that organisms did not evolve.

This contradiction put Darwin into a state of mental disequilibrium, because he did not immediately accept its implication and conclude that organisms must also change. It was not until 1837, after Darwin returned to England, that he was converted to the idea of evolution of species (Greene, 1959). In hindsight, it seems unlikely that it would require this amount of time for Darwin to assimilate the implication of the situation, but the fact is that in the 2000 pages of geological and biological notes made during the voyage, there is little discussion of the evolution of organisms. What little there is opposes the idea.

Precisely how and why Darwin changed his view are not known. Figure 4.22, however, appears to be a fairly accurate summary of his changing world view. Smith and Millman (1987) have also carefully examined Darwin's notebook (particularly the B notebook) and characterized Darwin's mind as in a state of "exploratory thinking," meaning that, rather than accepting any particular theory, Darwin was considering various views (alternative hypotheses) to explain the situation as he saw it. If we assume that the weight of accumulating evidence forced a rejection of special creation (e.g., physical change, intermediate "forms" of organisms, untold diversity of species and more than could reasonably be held on Noah's ark), then this exploratory thinking was aimed primarily at explaining evolution. Figure 4.22e thus represents the partial restoration of mental equilibrium as it eliminates the contradiction implied in Figure 4.22b.

Recall that Piaget refers to the process of moving from a mental state of equilibrium to disequilibrium and back to equilibrium as equilibration. Therefore, an initial answer to the question of how conceptual change occurs is through the process of equilibration. The conditions necessary for conceptual equilibrium to take place appear to be (1) data that are inconsistent with prior ways of thinking; (2) the presence of alternative conceptions or hypotheses (the hypothesis of evolution); and (3) sufficient time, motivation, and thinking skills to compare the alternative hypotheses and their predicted consequences with the evidence (Anderson & Smith, 1986; Hewson & Hewson, 1984; Lawson & Thompson, 1988; Posner, Strike, Hewson & Gertzog, 1982).

The Use of Analogy. Once Darwin accepted the alternative hypothesis that organisms evolve, the question of "how" immediately arose. Darwin's answer was, through a process called natural selection. Thus, natural selection was a theoretical concept employed by Darwin. Unlike the descriptive concept of Mellinarks, the defining attributes of the concept of natural selection are not visible. By what intellectual process did Darwin arrive at the concept of natural selection? How, in general, are theoretical concepts acquired?

According to the record (e.g., Greene, 1959; Gruber & Barrett, 1974; Smith & Millman, 1987), Darwin's search for a theory to explain the evolution of organisms involved a number of initially unsuccessful trials and a good deal of groping until September 1838, when a key event occurred. Darwin read

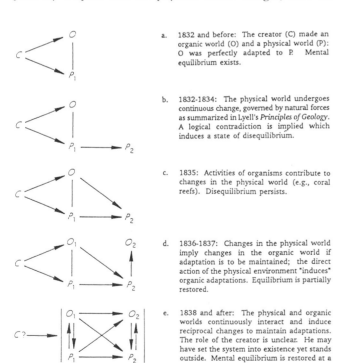

a. 1832 and before: The creator (C) made an organic world (O) and a physical world (P): O was perfectly adapted to P. Mental equilibrium exists.

b. 1832-1834: The physical world undergoes continuous change, governed by natural forces as summarized in Lyell's *Principles of Geology*. A logical contradiction is implied which induces a state of disequilibrium.

c. 1835: Activities of organisms contribute to changes in the physical world (e.g., coral reefs). Disequilibrium persists.

d. 1836-1837: Changes in the physical world imply changes in the organic world if adaptation is to be maintained; the direct action of the physical environment "induces" organic adaptations. Equilibrium is partially restored.

e. 1838 and after: The physical and organic worlds continuously interact and induce reciprocal changes to maintain adaptations. The role of the creator is unclear. He may have set the system into existence yet stands outside. Mental equilibrium is restored at a higher more complex plane.

FIGURE 4.22. Charles Darwin's changing world view from 1832 to 1838 as an example of mental equilibration. (After Gruber & Barrett, 1974.)

Thomas Malthus' *Essay on population*. Darwin wrote, "I came to the conclusion that selection was the principle of change from the study of domesticated productions; and then reading Malthus, I saw at once how to apply this principle" (Greene, 1959, pp. 257–258). Darwin saw in Malthus' writing a key idea that he could *borrow* and use to explain evolution. That key idea was that *artificial* selection of domesticated plants and animals was *analogous* to what presumably occurs in nature and could account for a change or evolution of species. As Gruber (in Gruber & Barrett, 1974, pp. 118–119) pointed out, Darwin had read Malthus before, but it was not until this reading that he became conscious of the import of artificial selection. Once it had been assimilated, Darwin turned to the task of marshaling the evidence for his theory of descent with modification. He turned to the facts known about plant and animal breeding, to the evidence that had first led him to doubt the fixity of species, namely the facts concerning the geographic distribution of organic forms, and to the creatures of the Galápagos Islands. He discovered support for his ideas in the geological, anatomical, ecological, and embryological records of the time, and by the year 1842 he was ready to commit a rough draft of his entire theory to paper (Greene, 1959).

The example of Darwin's use of the analogous process of artificial selection suggests that analogy plays a central role in theoretical concept acquisition. The "idea" or pattern that allowed Darwin to make sense of his data was analogous to the pattern in the process of artificial selection. Hanson (1947) referred to this process of borrowing old ideas and applying them in new situation as "abduction." Others have referred to the process as analogical reasoning (Karplus, 1979; Lawson & Lawson, 1979) or analogical transfer (Holland et al., 1986).

Examples of abduction are numerous in history of science. Kepler borrowed the idea of the ellipse from Apollonius to describe planetary orbits. Mendel borrowed patterns of algebra to explain heredity. Kekulé borrowed the idea of snakes eating their tails (in a dream!) to determine the molecular structure of benzene, and Coulomb borrowed Newton's ideas of gravitational attraction to describe the electrical forces that exist at the level of atomic particles.

Abduction, the use of analogy to apply old ideas in new situations to invent new concepts and new explanations, is all-pervasive. According to Pierce (quoted in Hanson, 1947),

All the ideas of science come to it by way of Abduction. Abduction consists in studying the facts and devising a theory to explain them. Its only justification is that if we are ever to understand things at all, it must be that way. Abductive and inductive reasoning are utterly irreducible, either to the other or to Deduction, or Deduction to either of them. (p. 85)

Therefore, theoretical concepts are constructed by applying a previously acquired pattern from the world of observable objects and events to explain unobservable events. The scientist must discover the analogy, but the student in the classroom can be assisted by having the teacher point out the relevant analogy.

The General Pattern of Concept Construction and Conceptual Change

Upon reflection, we can identify a general pattern that exists in both processes of concept formation (whether one is construction of descriptive concepts via induction or theoretical concepts via abduction) and conceptual change. The pattern exists in both, because the processes of concept construction and conceptual change are not really different but are two ends

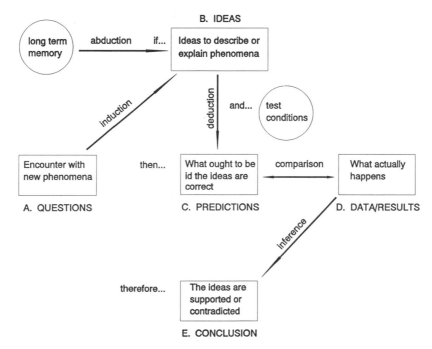

FIGURE 4.23. The basic pattern of deductive thinking.

of the same continuum. As Piaget reminds us, every act of assimilation to a cognitive structure is accompanied by some accommodation of that structure. No two experiences are ever identical; therefore, pure assimilation is not possible. Likewise, pure accommodation presumably does not take place because it would imply a cognitive reorganization without any input from the environment. Thus, at the concept-construction end of the continuum, we have the dominance of assimilation over accommodation, and at the conceptual-change end we have dominance of accommodation over assimilation.

The general pattern is shown in Figure 4.23. Box A represents a question that has been raised because of some novel experience that cannot be adequately assimilated to prior structures (e.g., what is a Mellinark? How did the diversity of species arise?). Box B represents alternative propositions that have arisen by either selection of perceptible features of the problem situations (induction) or analogical reasoning (abduction) from one's own memory or that of others (e.g., in books). The use of analogical reasoning is an important component of what is often referred to as creative thinking. The subconscious mind plays an important role in the generation of alternative explanations.

To test the alternatives, some experimental or correlational situation must be imagined that allows deduction of the propositions' consequences (box C). The implied consequences (predictions) are then compared with the results of the test, which are represented by box D. If the predicted results and the actual results are essentially the same, support for the proposition has been obtained. If not, the proposition has been weakened and others should be generated and tested until reasonable agreement is obtained. Note how the words *if, and, then,* and *therefore* tie the elements of the process together into a reasonable argument for or against any particular proposition or set of alternatives.

The acquisition of declarative knowledge is very much a constructive process that makes implicit or explicit use of procedural knowledge. Of course, students can memorize aspects of declarative knowledge, but such learning by rote will not assist in improving the procedural knowledge. The pedagogical task is to teach in such a way that students participate in the constructive process, because doing so improves meaningfulness and retention of the declarative knowledge and increases consciousness and generalizability of the procedural knowledge.

THE NATURE OF PROCEDURAL KNOWLEDGE

Figure 4.23 depicted the way in which concept acquisition occurs, that is, the way people learn about their world. The result of this learning process is conceptual or declarative knowledge. The procedures one uses to generate that declarative knowledge are collectively known as procedural knowledge. Thus, the boxes of the figure represent various aspects of declarative knowledge (e.g., questions, hypotheses, predictions, results, and conclusions) and the arrows (from box to box) represent various procedures (abduction, induction, deduction, and inference). Various reasoning patterns (cognitive

strategies), such as combinatorial reasoning (generation of combinations of alternative propositions), control of variables (experimenting with only one independent variable), and correlational reasoning (comparing ratios of confirming to disconfirming events), are embedded in the process.

Because of the central importance of procedural knowledge in science and in creative and critical thinking in general, psychologists and educators alike have attempted to identify its components with as much precision as possible. One of the early attempts to do so concerned eight central skills and several subskills (Burmester, 1952). A modified list of those skills grouped in seven categories relatable to the general pattern of thinking depicted in Figure 4.23 follows:

1. Skill in accurately describing nature
2. Skill in sensing and stating causal questions about nature
3. Skill in recognizing, generating, and stating alternative hypotheses and theories
4. Skill in generating logical predictions
5. Skill in planning and conducting controlled experiments to test hypotheses
6. Skill in collecting, organizing, and analyzing relevant experimental and correlational data
7. Skill in drawing and applying reasonable conclusions

Some of these skills are creative, while others are critical. Still others involve both creative and critical aspects of scientific thinking. A skill is defined as the ability to do something well. Skilled performance includes knowing what to do, when to do it, and how to do it. In other words, being skilled at something involves knowing a set of procedures, knowing when to apply those procedures, and being proficient at executing the procedures. The seven general skills can be expanded into the following subskills:

1.00 Skill in accurately describing nature
    1.10 Skill in describing objects in terms of observable characteristics
1.20 Skill in seriating objects in terms of observable characteristics
1.30 Skill in classifying objects in terms of observable characteristics
1.40 Skill in describing, seriating, classifying, and measuring objects in terms of variables such as amount, length, area, weight, volume, and density
1.50 Skill in identifying variable and constant characteristics of groups of objects
    1.51 Skill in identifying continuous and discontinuous variable characteristics and naming specific values of those characteristics
    1.52 Skill in measuring, recording, and graphing the frequency of occurrence of certain values of characteristics in a sample of objects
    1.53 Skill in determining the average, median, and modal values of the frequency distribution in 1.52
1.60 Skill in recognizing the difference between a sample and a population and identifying ways to obtain a random (unbiased) sample

1.61 Skill in making predictions concerning the probability of occurrence of specific population characteristics based on the frequency of occurrence of those characteristics in a random sample

2.00 Skill in sensing and stating causal questions about nature

2.10 Skill in recognizing a causal question from observation of nature or in the context of a paragraph or article

2.20 Skill in distinguishing between an observation and a question

2.30 Skill in recognizing a question even when it is stated in expository form rather than interrogatory form

2.40 Skill in distinguishing a question from a possible answer to a causal question (hypothesis) even when the hypothesis is presented in interrogatory form

2.50 Skill in distinguishing between descriptive and causal questions

3.00 Skill in recognizing, generating, and stating alternative hypotheses (causal explanations) and theories

3.10 Skill in distinguishing a hypothesis from a question

3.20 Skill in differentiating between a statement that describes an observation or generalizes from the observation and a statement that is a hypothesis (causal explanation) for the observation

3.30 Skill in recognizing the tentativeness of a hypothesis or theory

3.40 Skill in distinguishing between a tentative explanation for a phenomenon (hypothesis) and a term used merely to label the phenomenon

3.50 Skill in systematically generating all possible combinations of generated hypotheses

4.00 Skill in generating and stating logical predictions based on the assumed truth of hypotheses and imagined experimental or correlational conditions

4.10 Skill in differentiating between hypotheses and predictions

5.00 Skill in planning and conducting controlled experiments to test alternative hypotheses

5.10 Skill in selecting reasonable alternative hypotheses to test

5.20 Skill in differentiating between an uncontrolled observation and an experiment involving controls

5.30 Skill in recognizing that only one independent factor in an experiment should be variable

5.31 Skill in recognizing the independent variable factor and the dependent variable factor(s)

5.32 Skill in recognizing the factors being held constant in the partial controls

5.40 Skill in recognizing experimental and technical problems inherent in experimental designs

5.50 Skill in criticizing faulty experiments when

5.51 The experimental design was such that it could not yield an answer to the question

5.52 The experiment was not designed to test the specific hypotheses stated

5.53 The method of collecting the data was unreliable

5.54 The data were not accurate

5.55 The data were insufficient in number

5.56 Proper controls were not included

6.00 Skill in collecting, organizing and analyzing relevant experimental and correlational data

6.10 Skill in recognizing existence of errors in measurement

6.20 Skill in recognizing when the precision of measurement given is warranted by the nature of the question

6.30 Skill in organizing and analyzing data

6.31 Skill in constructing tables and frequency graphs

6.32 Skill in measuring, recording, and graphing the values of two variables on a single graph

6.33 Skill in constructing a contingency table of discontinuous variables

6.40 Skill in seeing elements in common to several items of data

6.50 Skill in recognizing prevailing tendencies and trends in data and extrapolating and interpolating

6.60 Skill in applying quantitative notions of probability, proportion, percent, and correlation to natural phenomena; recognizing when variables are related additively or multiplicatively; and setting up simple quantitative equations describing these relationships

6.61 Skill in recognizing direct, inverse, or no relationship between variables

6.62 Skill in recognizing that when two things vary together, the relationship may be coincidental, not causal

6.63 Skill in recognizing additional evidence needed to establish cause and effect (see 6.62 above)

7.00 Skill in drawing and applying reasonable conclusions

7.10 Skill in evaluating the relevance of data and drawing conclusions through a comparison of actual results with predicted results

7.11 Skill in differentiating between direct and indirect evidence

7.12 Skill in recognizing data that are unrelated to the hypotheses

7.13 Skill in recognizing data that support a hypothesis

7.14 Skill in recognizing data that do not support a hypothesis

7.15 Skill in combining supportive and contradicting evidence from a variety of sources to weigh the likely truth or falsity of hypotheses

7.16 Skill in postponing judgment if no evidence or insufficient evidence exists

7.17 Skill in recognizing the tentativeness inherent in all scientific conclusions

7.20 Skill in applying conclusions to new situations

7.21 Skill in refraining from applying conclusions to new situations that are not closely analogous to the experimental situation

7.22 Skill in being aware of the tentativeness of conclusions about new situations even when there is a close parallel between the two situations

7.23 Skill in recognizing the assumptions that must be made in applying a conclusion to a new situation

These skills function in concert in the mind of the creative and critical thinker as he or she learns about the world. They include key steps and the key words *if, and, then,* and *therefore* as depicted in Figure 4.23. The skills are, in essence, learning tools essential for success and even for survival. Hence, if teachers help students improve their use of these creative and critical thinking skills, they have helped them become more intelligent and helped them "learn how to learn."

Stages in the Development of Procedural Knowledge

A great deal has been written about the development of procedural or operative knowledge within the Piagetian tradition (e.g., Collea et al., 1975; Collette & Chiappetta, 1986; Inhelder & Piaget, 1958; Karplus et al., 1977). Piaget's sensory-motor, preoperational, concrete operational, and formal operational stages are well known. There is little argument about the validity of the notion of levels or phases in the development of procedural knowledge, but considerable controversy exists regarding the details.

In Piaget's theory, the child at birth is in the *sensory-motor* stage. During this stage, which lasts for about 18 months, the child acquires such practical knowledge as the fact that objects continue to exist even when they are out of view (object permanence). The name of the second stage describes the characteristics of the child: *preoperational*—the stage of intellectual development before mental operations appear. In this stage, which persists until around 7 years of age, the child is extremely egocentric; centers attention only on particular aspects of given objects, events, or situations; and does not demonstrate conservation reasoning. In other words, the child's thinking is very rigid. The major achievement during this stage is the acquisition of language.

At about 7 years of age, the thinking processes of children begin to "thaw out"; they show less rigidity. This stage, called *concrete operational*, is marked by the development of operations. Concrete operations are defined by Piaget as mentally internalized and reversible systems of thought based on manipulations of classes, relations, and quantities of objects. Children can now perform what Piaget calls mental experiments; they can assimilate data from a concrete experience and arrange and rearrange them in their heads. Concrete operational children have much greater mobility of thought than when they were younger.

The name of this stage of development is representative of the type of thinking of this type of learner. As Piaget explains this stage, "The operations involved . . . are called 'concrete' because they relate directly to objects and not yet to verbally stated hypotheses" (Piaget & Inhelder 1969, p. 100). In other words, the mental operations performed at this stage are "object bound"—tied to objects.

The potential for the development of what Piaget calls *formal operational* thought develops between 11 and 15 years of age. For Piaget, the stage of formal operations constitutes the highest level in the development of mental structures. A person who has entered that stage of formal thought "is an individual who thinks beyond the present and forms theories about everything, delighting especially in considerations of that which is not" (Piaget, 1966, p. 148).

Presumably, there is nothing genetically predetermined in this sequence of development of mental structures. Rather, as Inhelder and Piaget (1958) state, "maturation of the nervous system can do no more than determine the totality of possibilities and impossibilities at a given stage. A particular social environment remains indispensable for the realization of these possibilities" (p. 337). Piaget chose the name formal operational for the highest stage of thought because he believed that thinking patterns are isomorphic with rules of formal propositional logic (Piaget, 1957). This is perhaps the most problematic position in Piaget's theory. As previously discussed, a long line of research indicates that, although advances in reasoning performance do occur during adolescence, no one, even professional logicians, reasons with logical rules divorced from the subject matter. For this reason it may be preferable to abandon Piaget's terminology and adopt new terminology for these differences in childlike and adultlike thought. I have suggested use of the terms intuitive and reflective respectively (Lawson et al., 1989) as they seem to capture better the key differences. That is, before adolescence, children may raise causal questions and generate answers, yet they have no systematic means of asking themselves whether those answers are correct. They must rely on others for this, so when left on their own they simply generate ideas and for the most part use them for better or worse. Without such a reflective ability, children confronted with complex tasks simply choose the most obvious solution that pops into their heads and conclude that it is correct without considering alternatives and arguments in their favor or disfavor.

Kuhn, Amsel, and O'Loughlin (1988) reached a similar conclusion about the differences between childlike and adultlike thinking. They identified three key abilities acquired by some adults. First is the ability to think *about* a theory rather than thinking only with a theory. In other words, the reflective adult is able to consider alternative theories and ask which is the most acceptable. On the other hand, the intuitive thinker does not consider the relative merits and demerits of alternative theories (hypotheses); he or she merely has a "theory" and behaves as if it were true. Chamberlain (1897) referred to these as ruling theories.

Second is the ability to consider the evidence to be evaluated as distinct from the theories. For the child, evidence and theory are indistinguishable. In our experience, perhaps the most difficult distinction to be made in the classroom is that between the words *hypothesis, prediction,* and *evidence* (Lawson, Lawson, & Lawson, 1984). The words are essentially meaningless if one has never tried to decide between two or more alternative explanations, thus has never before considered the role played by predictions and evidence. Third is the ability to set aside one's own acceptance (or rejection) of a theory and evaluate it objectively in light of its predictions and the evidence.

A New View of Stage Theory. Although Piaget's characterization of the thinking processes that dominate his stages has been found wanting, a strong case could be made that the stages are real and in need of reinterpretation. I would like to suggest a reinterpretation, based on a look at how patterns of deductive reasoning are utilized across age. Its basic premise is that deductive reasoning (at least in some form) is present virtually at birth; thus, development does not amount to changes in this reasoning pattern with age but instead involves changes in what the reasoning pattern can be applied to. In this sense, my view of development is opposite that of Piaget's, in that I see horizontal décalage as the rule rather than the exception. Let's see how this might work with respect to deductive reasoning.

Stage 1 (Birth to Age 18 Months).   Of course children during the first 18 months of life are not able to generate verbal arguments of the *if-and-then* form. Nevertheless, experimental evidence suggests that their behavior is consistent with the hypothesis that their preverbal behavior follows this pattern. Consider, for example, Piaget's famous object permanence task in which an experimenter, in full view of the infant, hides a ball under one of two covers. Diamond (1990) has shown that infants as young as 5 months of age will reach under the cover for the hidden ball, indicating that they have retained a mental representation of the ball even though it is out of sight. Furthermore, I argue that such behavior indicates that the infant is reasoning in the deductive form as follows:

If . . . the ball is still where he or she put it (even though I can no longer see it)
and . . I reach under the cover where it was hidden
then . . I will find the ball.

In agreement with Meltzoff (1990), I term the infant's representation of the ball an *empirical representation* because it is a representation of an event that the infant has experienced empirically. That is, the infant actually saw the ball hidden under the cover.

Stage 2 (18 Months to 7 Years).   Although infants younger than 18 months are able to solve the simple object-permanence task just described, it is not until 18 months that they can solve one in which they must represent what they have not experienced. Piaget (1954) invented a hiding task, which he called "serial invisible displacement," to tap this higher-order skill. In this task the adult hides a ball in his or her hand in view of the infant and then moves the hand under a series of three occluders, dropping the ball under one of them. The infant is not given any indication when the ball was dropped. Instead, he or she only sees the empty hand emerge at the other side of the occluders. Consequently, the infant will look in the last place he or she saw the ball (the hand) and then must deduce that because the ball is not there, but must exist somewhere, it must

be in one of the places the hand traveled to after the ball was hidden.

In this experiment, children younger than 18 months look in the hand and look no further. They are stumped. However, children older than 18 months are able to make the correct deduction and find the ball. Again, I argue that the thinking that leads to successfully locating the ball takes on the *if-and-then* deductive pattern:

If . . . the ball is hidden behind one of the occluders
and . . . I lift each in turn
then . . . I will eventually find the ball.

Consequently, what separates the first and second stages of development is not this *if-and-then* pattern of thinking. Rather, it is the context in which the pattern can be applied. Notice that in this second, more difficult task the child must create and initiate reasoning with a *hypothetical* as opposed to an *empirical* mental representation. In other words, in this case the representation is of something that the infant has *not* experienced (the ball being dropped off behind one of the occluders). The infant must generate this hypothetical representation from the task conditions and then use it as a starting point for thinking (Meltzoff, 1990). In Meltzoff's words,

By 18 months of age there has been the growth of a kind of second-order representational system and a capacity for hypothetical representations. This enables the child to wonder "what if," to contemplate "as if," and to deduce "what must have been" in advance of, and often without, the perceptual evidence. (1990, p.22)

Notice that language is not necessary in arguments of either type (i.e., those of stage 1 that involve empirical representations and those of stage 2 that involve hypothetical representations).

Stage 3 (7 Years to Early Adolescence).   During much of stage 2 the child becomes proficient at using language to name and describe the objects, events, and situations in his or her environment. Acquisition of language allows the deductive pattern to be applied at a new level—the level of naming, describing, and classifying.

In a recent study by Lawson (1993), the Mellinark task and similar tasks were administered to children from 6 to 8 years of age. Carefully sequenced instruction was designed to teach the children how to use the correct pattern of reasoning to discover the relevant attributes of the creatures. As discussed previously, the pattern of reasoning was inductive-deductive in nature:

If . . . tiny spots are the key features that make a creature a Mellinark
and . . . all of these creatures in row 1 are Mellinarks
then . . . they all should have tiny spots.

Interestingly, none of the 30 six-year-olds were able to generate and/or comprehend this pattern of reasoning, whereas 15 of the 30 seven-year-olds and virtually all of the eight-year-olds (29 of 30) were. Work reviewed by Dempster (1992) and by

Levine and Prueitt (1989) indicates that the younger childrens' failure is most likely related to relatively late maturation of the prefrontal cortex. Levine and Prueitt even present a Grossberg-inspired neural model of the shift.

My position is that stage 3, which beings at age 7, involves use of the inductive-deductive pattern to name, describe, and classify the objects, events, and situations in the child's environment, all mediated by language. However, this type of reasoning is limited. Notice that reasoning (like that in stage 1) is initiated with what the child directly perceives in his or her environment; for example, the child is able to see the tiny spots on the creatures in the Mellinark task. In this sense, the representations the child uses to initiate reasoning are empirical in origin and have been derived via *induction*, that is, generalization of specific observations to general classes of objects, events, or situations.

STAGE 4 (EARLY ADOLESCENCE AND OLDER). At roughly age 11 to 12 years, some children become increasingly able to use language to apply the deductive pattern of reasoning to "hypothetical" rather than empirical representations. I use the term hypothetical in much the same way as when it was applied to stage 2 reasoning. The key difference between stage 4 and stage 2 is that whereas stage 4 reasoning is language mediated, stage 2 reasoning is not.

Consider, for example, the question; How are salmon able to locate the stream of their birth before spawning? Because the salmon cannot tell you how they do it, nor can you find out by watching the salmon as they head upstream, answering a question of this sort requires generating and testing alternative hypotheses about how the salmon might accomplish the task. Such a hypothesis generation process requires the use of abduction, not induction. The relevant abductive-deductive argument might go as follows:

If . . . salmon navigate by using their eyes
and . . . I blindfold some of the returning salmon
then . . . they should not be as successful at finding their home stream as those that are not blindfolded

Again, the deductive pattern of the argument is the same as in the prior three stages. The key difference between this (stage 4) reasoning and the prior (stage 3) reasoning is not the pattern but what initiates the pattern. Stage 3 reasoning was initiated by empirical representation, by induction (the direct perception of environmental stimuli). Stage 4 reasoning, on the other hand, is initialized by hypothetical representation, by abduction (the imagined process of the salmon looking at remembered landmarks much as you would when hunting for a house not visited for years).

The importance of such a shift in thinking cannot be overemphasized. Whereas thinking at stage 3 is primarily a response to environmental encounters (i.e., reactive), reasoning at stage 4 is reflective, self-contained, and proactive. In this sense stage 4 is the only stage that is truly scientific. The epistemology of the stage 3 reasoner is one of observation. What causes events? To find the answer, you observe the events. The epistemology of the stage 4 reasoner is vastly different. What causes events?

To find the answer, you must generate several possible causes, deduce their implied consequences, and then observe the results of experimental manipulations to support or reject the generated possibilities. In this sense, Piaget was right in his claim that childrens' reasoning begins with the *real* whereas adult reasoning begins with the *hypothetical*.

How Procedural Knowledge Develops

Lawson et al. (1984) hypothesized that the transition to stage 4 thinking (reflecting on the correctness of one's abductively generated hypotheses) is a consequence of the internalization of patterns of external argumentation. Presumably this internalization occurs when alternative hypotheses are proposed and their relative merits and demerits are debated. This hypothesis appears to be in essential agreement with Piaget's earlier thinking. Piaget (1928) proposed that the development of advanced reasoning was a consequence of "the shock of our thoughts coming into contact with others, which produces doubt and the desire to prove" (p. 204). Piaget went on to state:

The social need to share the thought of others and to communicate our own with success is at the root of our need for verification. Proof is the outcome of argument. . . . Argument is therefore, the backbone of verification. Logical reasoning is an argument which we have with ourselves, and which produces internally the features of a real argument. (p. 204)

In other words, the growing awareness of and ability to use the pattern of hypotheticodeductive thought during adolescence (defined as the ability to ask questions of oneself, generate alternative answers, deduce predictions based on the assumed truth of those alternative answers, and then sort through the available evidence to verify or reject those alternatives, all inside one's own head) is a consequence of attempting to engage in arguments of the same sort with other persons and listening to arguments of others in which alternative hypotheses are put forward and accepted or rejected as the basis of evidence and reason as opposed to authority or emotion.

This position also seems consistent with that of Vygotsky (1962), who viewed speech as social in origin and only with time coming to have self-directive properties that eventually result in internalized, verbalized thought. The position is also similar to that of Luria. According to Luria (1961), the progressive differentiation of language to regulate behavior occurs in four steps: (1) the child learns the meaning of words, (2) language can serve to activate behavior but not limit it, (3) language can control behavior through activation or inhibition via communication from an external source, and (4) the internalization of language can serve a self-regulating function through instructions to oneself.

Piaget (1976) proposed a similar three-level theory of procedural knowledge development. The first level (sensory-motor) is one in which language plays little or no role, as it has yet to be acquired. The child learns primarily through sensory-motor activity and knowledge is that of action. The second level is characterized by the acquisition of language. The child is able to respond to spoken language and acquire knowledge transmitted from adults who speak the same language. To learn, the child is able to raise questions and have adults respond verbally to those questions. Of course, not all adult responses are understood; nonetheless, a new and powerful mode of learning is available to the child. The essential limitation of this level is that the use of language as a tool for reflection and as an internal guide to behavior is poorly developed and limited to use in the perceptible world. Thus, reasoning at this level is essentially intuitive. The final level begins at the moment at which the individual begins to ask causal questions, not of others but of him-, or herself, and through gradual "internalization" of elements of the language of argumentation acquires the ability to "talk to himself," which constitutes the essence of reflective thought and makes it possible to internally test alternative hypotheses and arrive at internally reasoned decisions to answer those questions.

Voss, Greene, Post, and Penner (1983) have characterized advanced thinking in the social sciences as largely a matter of constructing proposals for action that conform to many of the classical principles of rhetorical argumentation. Likewise, Lawson and Kral (1985) view the process of literary criticism as mainly one using classical forms of argumentation.

As mentioned, this view of the development of procedural knowledge suggests that the terms *intuitive* and *reflective thought* are more descriptive of the intellectual changes that take place during adolescence than Piaget's terms *concrete* and *formal thought*. The childlike thinker is not conscious of the hypotheticodeductive nature of his or her thought processes; therefore thinking is dominated by context-dependent cues and intuitions. The adultlike thinker, on the other hand, has become conscious of his or her thought patterns and has internalized a powerful pattern of argumentation that allows conscious reflection on the adequacy or inadequacy of alternative hypotheses before action. Reflective thinking is probably not based on formal logic as Piaget claimed, but on alternative hypotheses, predictions, evidence, and arguments, all mediated by language.

Essential Elements of Instruction

The argument has been made that the constructive process results in acquisition and/or change of declarative knowledge, which resides in conceptual systems of various degrees of complexity and abstractness. Furthermore, conscious awareness of the procedures involved in constructing such knowledge "develops" when arguments with others force one to reflect on the adequacy or inadequacy of one's procedures. Conscious verbal rules to guide behavior develop from such encounters, which serve as "anticipatory schemes" to guide behavior in new situations. Thus, development extends the range of effective performance from familiar to novel situations. We now come to the final issue of this chapter. How can instruction be designed and carried out to help students construct and retain useful declarative knowledge and develop a conscious awareness of effective procedural rules with general applicability?

Our previous discussion suggests that the following elements must be included in lessons designed to improve both declarative and procedural knowledge:

1. Questions should be raised or problems should be posed that require students to act on the basis of prior beliefs (concepts and conceptual systems) or prior procedures.
2. Those actions should lead to results that are ambiguous or can be challenged or contradicted. This forces students to reflect back on the prior beliefs or procedures used to generate the results.
3. Alternative beliefs or more effective procedures should be proposed by students and the teacher.
4. Alternative beliefs or the more effective procedures should now be utilized to generate new predictions or new data to allow either the change of old beliefs or the acquisition of a new belief (concept).

Suppose, for example, that students in a biology class are asked to use their prior declarative knowledge (beliefs) to predict the salinity in which brine shrimp eggs will hatch best and to design and conduct an experiment to test their prediction. If students work in teams of two or three, about 10 to 15 sets of data will be generated. These data can be displayed on the board. Because no specific procedures were given to the groups, the results will vary considerably. This variation in results allows students to question one another about the procedures used to generate the results. It also provokes in some students the cognitive state of disequilibrium, as their results contradict their expectations. A list of differences in procedures can then be generated. For example,

- *The hatching vials contained different amounts of water.*
- *Some vials were capped, others not capped. The amounts of eggs varied from vial to vial and group to group.*
- *Some eggs were stirred, others not stirred.*
- *Some groups used distilled water, others tap water, and so on.*

Once this list is generated, it becomes clear to the students that these factors should not vary. Thus, a better procedure can be suggested. All the groups can then follow the same procedure (that is, variables will be controlled). When this is done, the "real" effect of various concentrations of salt can be separated from the spurious effects of the other variables. Finally, once the new data are obtained, the results are clear, allowing students to see whose predictions were correct and whose were not. Moreover, the results permit the teacher to introduce the term *optimum range* for the pattern of hatching that was discovered. For some students this will help restore equilibrium; for other students additional activities may be necessary.

The main thesis is that situations that allow students to examine the adequacy of prior beliefs (conceptions) force them to argue about and test those beliefs. This in turn can provoke disequilibrium when those beliefs are contradicted and can provide the opportunity to acquire more appropriate concepts and become increasingly skilled in the procedures used in concept acquisition (i.e., reasoning patterns or forms of argumentation). The current educational literature contains a number of examples of how instruction can be designed to be consistent with this constructivist approach to knowledge acquisition.

A NEUROLOGICAL EXPLANATION OF ANALOGICAL REASONING

As we have seen, the process of abduction or analogical reasoning is essential for scientific insight and for theoretical concept acquisition. Consequently, the current interest in the psychology of analogical transfer—the use of analogies to solve novel problems and to form generalized rules or concepts (e.g., Anderson & Thompson 1989; Carbonell, 1986; Catrambone & Holyoak, 1985; Gentner, 1987, 1989; Gick & Holyoak, 1983; Holland et al., 1986; Holyoak, 1985; Holyoak, Junn, & Billman, 1984; Holyoak & Thagard, 1989)—appears well justified. In this section we present a neurological-level theory of memory and analogical transfer, based in part on Grossberg's neural modeling principles presented earlier. Although Grossberg has not directly confronted the issue of analogical transfer, his principles of neural networks and theory of human memory can provide a framework for understanding how analogical transfer occurs and why it plays a central role in human learning and memory.

The central question is this: What causes some experiences to find their way into long-term memory while other experiences do not? Grossberg's answer to this question is contained in a general way in his learning equation, which was discussed earlier in the context of the process of equilibration. We will return to the learning equation again, as it provides a key to understanding memory in general and, more specifically, the role of analogy in facilitating information transfer from short-term to long-term memory. I begin the discussion by describing an experience of mine while visiting a Japanese science classroom. That experience will more sharply focus the issue.

During my visit to a Japanese elementary school, a teacher and his students were discussing the results of an experiment on the role of a variety of factors on the growth of seeds. The teacher organized the students' comments in words and diagrams on the chalkboard. The students were very enthusiastic and the teacher wrote very clearly on the board. The experiment was familiar, but my inability to understand spoken or written Japanese made it difficult to understand much of the discussion.

At the conclusion of the lesson, our group of visitors adjourned to the school principal's office for a traditional cup of tea. At that time it occurred to me that I had observed a very good lesson and should attempt to make a few notes, including a record of what the teacher had written on the board. I was able to reconstruct some, but not all, of what had been written. Interestingly, recalling the relative position of the major items on the board was easy. The diagram of the seeds and their container, the numbers 1 and 2, and letters A and B, and the question mark were all easily recalled. But recalling the shapes of the Japanese or Chinese symbols and words was impossible. To be more specific, a symbol shaped like this "?" was recalled, but a symbol shaped like this "才" was not. Why?

Readers may be saying to themselves that this is not the least bit surprising. Familiar English language elements were recalled, while unfamiliar Japanese language elements were forgotten. Because the observer does not speak or write in Japa-

nese, this is predictable. Agreed! This can clearly be predicted from our past experiences in trying to remember familiar and unfamiliar items. But how can it be explained at the neurological level? After all, all of the stimuli on the board were clear and all could have been copied at the time. Why does one remember items that are familiar and forget items that are not? Precisely what does "familiar" mean in neurological terms? How does familiarity facilitate transfer into long-term memory?

As you will recall, Grossberg's learning equation, $Z_{ij} = -B_{ij}Z_{ij} + S'_{ij}[X_j]^+$, describes conditions necessary to modify the synaptic strength of knobs N_{ij}. The term Z_{ij} represents initial synaptic strength. B_{ij} is a decay constant; thus the term $B_{ij}Z_{ij}$ is the forgetting term. $S'_{ij}[X_j]^+$ is the learning term as it drives increases in Z_{ij}. S'_{ij} is the strength of the signal that has passed from node V_i to knob N_{ij}. The term $[X_j]^+$ represents the level of activity at postsynaptic nodes V_j that exceeds the firing threshold. The learning term indicates that for information to be stored in LTM, two events must occur simultaneously. First, signals must be received at N_{ij}. Second, nodes V_j must receive inputs that cause them to fire. When these two events drive activity at N_{ij} above a specified constant of decay, the Z_{ij}'s increase.

Grossberg proposed a process called adaption resonance, which is a crucial component of transferring information to LTM. The following is a brief review of that process. As we know, the brain is able to process a continuous stream of changing stimuli and constantly modify behavior accordingly. This implies that a mechanism exists for matching input with expectations from prior experience and selecting alternative expectations when a mismatch occurs. Grossberg's model for such a mechanism was shown in Figure 4.15.

Following Grossberg, the field of excitation on a particular slab of neurons was referred to as $F^{(1)}$ and the field at the next slab as $F^{(2)}$. Suppose that, because of prior experience, a pattern of activity, P_1, at $F^{(1)}$ causes firing of a pattern P_2 at $F^{(2)}$, where P_2 could be a single neuron. P_2 then excites a pattern P on $F^{(1)}$. The pattern P is compared with the input following P_1. P is the expectation. P will be P_1 in a static scene and the pattern to follow P_1 in a temporal sequence. If the two patterns match, you see what you expect to see. This allows uninterrupted processing of input and continued quenching of nonspecific arousal. Grossberg refers to this match of input with expectations as an *adaptive resonance*. One is aware only of patterns that enter the resonant state. Unless resonance occurs, no coding in LTM takes place, because only in the resonant state is there both pre- and postsynaptic excitation of the cells at $F^{(1)}$.

Now suppose the new input to $F^{(1)}$ does not match the expected pattern P from $F^{(2)}$. Mismatch occurs, and this causes activity at $F^{(1)}$ to be turned off by lateral inhibition, which in turn shuts off the inhibitory output to the nonspecific arousal source. This turns on nonspecific arousal and initiates an internal search for a new pattern at $F^{(2)}$ that will match P_1.

Such a series of events explains how information is processed across time. The important point is that stimuli are considered familiar if a memory record of them exists at $F^{(2)}$ so that the pattern of excitation sent back to $F^{(1)}$ matches the incoming pattern. If they do not match, the incoming stimuli are unfamiliar and orienting arousal is turned on to allow an unconscious search for another pattern to attempt the match. If no such match is obtained (as in the case of looking at an unfamiliar Japanese symbol), no coding in LTM will take place unless attention is directed more closely at the object in question. Directing one's attention at the unfamiliar object will boost presynaptic activity high enough so that the sensory input is recorded in a set of previously uncommitted cells. [For details see Grossberg (1982), especially pages 29–31, and Lawson (1986).]

Grossberg's theory of adaptive resonance is a general explanation of how input patterns find their way into LTM. That theory must be extended at this point. In general, the theory of analogical transfer that we propose describes specific neural processes that greatly facilitate coding of new experiences in LTM. However, before discussing the role of analogy, we need to take a closer look at the way neurons function to recall and reproduce patterns.

Outstars and Instars: The Fundamental Units.

Grossberg identified the outstar as the underlying neural mechanism for reproduction and recall of patterns. An outstar is a neuron whose cell body lies in one slab of interconnected neurons with a set of synaptic knobs that connect it to a set of cell bodies embedded in a lower slab of neurons (Figure 4.24). The outstar is a fundamental functional unit able to learn and reproduce a pattern. Understanding how outstars accomplish this is central to understanding how appropriate analogies can greatly enhance learning.

The outstar shown in Figure 4.24 is actively firing impulses down its axon to a lower slab of neurons that is simultaneously receiving a pattern of input from a still lower slab of neurons, or perhaps from the environment (e.g., a pattern of visual input on the retina). In the figure, the darkened neurons represent active neurons and the more the cell body is darkened, the more active it is (i.e., the more input it is receiving, hence the more frequently it is firing). When outstar N_0 is firing and signals S_{0j} are reaching the slab as the pattern is, the synaptic strengths Z_{0j} will grow. An important consequence of this

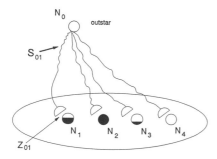

FIGURE 4.24. The darkened neurons represent active neurons; the more the cell is darkened, the more active it is. If the outstar is sampling the slab during this activity, the synaptic strengths of each synaptic knob will tend to mirror the activity of the cell (as shown in Figure 4.25). The more active the cell, the more the synaptic strength will grow.

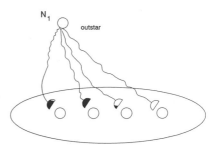

FIGURE 4.25. The darkened synaptic knobs represent the synaptic strength after the outstar has sampled the slab long enough to learn the pattern. The synapses shown have learned the pattern they have sampled in Figure 4.24.

change in synaptic strengths is that when the pattern of activity on the slab is gone, the outstar is able to *reproduce* the pattern whenever it fires again. When the outstar fires repeatedly, synapses with high synaptic strength cause their associated cells to become very active, and cells with low synaptic strength are less active, just as they were when the slab was being sampled. In this manner the pattern will reappear (Figure 4.25).

Not only are slabs of neurons connected via axons from higher slabs, as depicted in Figures 4.24 and 4.25, but also the neuron cell bodies on the input slab have axons that connect them to the cell bodies of higher slabs. As depicted in Figure 4.26, a pattern of activity on a lower slab is mirrored by the rate of transmitter release in the synapses leading to the cell bodies on the higher slab. Thus, when the pattern is active on the lower slab and the cell body on the higher slab (the outstar) is active, these synaptic strengths increase in a fashion that mirrors that pattern of activity. Consequently, if the pattern appears again, the outstar will fire. In this sense, the outstar has remembered the pattern. Note that a "sufficient" period of time is needed for the outstar (the neuron on the higher slab) to learn the pattern. We shall see later that analogy can play a key role by reducing this period of time, making learning possible when it would otherwise not occur.

The activity depicted in Figure 4.26 is of such functional importance that Grossberg has given it a name, the instar. The instar is actually the set of synaptic weights associated with the synaptic knobs connected to a neuron. If a pattern fires a neuron repeatedly, that pattern will reappear as part of the instar of that neuron.

To summarize, the synaptic strengths of an outstar align themselves to an input. The outstar is then able to reproduce the input. If an outstar or a collection of outstars are not aligned to an input, that input cannot be reproduced unless it is presented again. Thus it will not be remembered. The important point that if an outstar is not present, a pattern cannot be reproduced and thus is not remembered. In the initial example of the Japanese classroom, the Japanese symbols could not be reproduced because the outstar (or outstars) necessary to reproduce them did not exist. We conclude, therefore, that outstars must be present for recall to occur.

Having said this, one must keep in mind that actual input patterns, such as those on the retina, are never exactly the same.

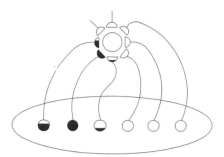

FIGURE 4.26. An instar that has learned the pattern that fired it. The darkened cells are active cells, the more darkened, the more active. If the cell above is fired, then according to Grossberg's learning equation, the synaptic strengths will grow. The synaptic strengths of other cells decay. In this fashion the synapses (the instar) learn the pattern that has fired the cell they impinge on. If this pattern appears again as part of an input, the synaptic strengths of the instar will cause the cell to fire.

We do not look the same when we awake as we remember looking before going to bed, but we recognize ourselves nevertheless. Grossberg's theory of adaptive resonance shows how this can be accomplished. Adaptive resonance is a method by which slabs of neurons interact with each other to find a best fit. Suppose that slab 1 is presented with an image. That image has several features; it consists of several patterns. Each pattern (or feature) that has been learned triggers an outstar in slab 2. The outstars triggered in slab 2 in turn fire on slab 1. Patterns that have been correctly triggered reproduce the pattern that triggered them. If all is correct, or "close," a best fit has been found. If a pattern does not match, it causes a search for a different outstar associated with a different pattern. Grossberg's theory of adaptive resonance includes a detailed description of this search for a "best fit."

THE OUTSTAR AS A MECHANISM FOR CHUNKING. Before discussing the role of analogy in facilitating learning, one more neurological mechanism needs to be in place—that for the well-known psychological phenomenon of chunking. Chunking, which was referred to earlier in the context of concept formation, has an interesting place in the literature of psychology. Miller's magic number 7, plus or minus 2, refers to the fact that, almost universally, people can remember only seven unrelated units of data (if they do not resort to memory tricks or aids) (Miller, 1956). It is sometimes said that this is why telephone numbers are seven digits long. Clearly, however, we all form concepts (patterns) that contain far more information than seven "units." Thus, a mental process must occur in which previously unrelated pieces of input are grouped or "chunked" together to produce higher-order chunks (units of thought). As previously mentioned, this implied process is known as chunking (Simon, 1974).

Grossberg hypothesized that the outstar is the anatomical or functional unit that makes chunking possible. An outstar sampling a slab can group a set of neurons that are firing at the same time. To do this, it merely fires at a high enough rate to allow its synaptic strengths to mirror the activity of the neurons being grouped (or chunked). For adaptive resonance to occur, the neurons being chunked must fire the outstar. In this sense, the purpose of an outstar is to form a chunk and later to identify the chunk when it has been formed.

An architecture called on-center, off-surround (OCOS) plays an important part in chunk formation. OCOS is sometimes referred to as winner-take-all architecture. In an OCOS architecture, active cells excite nearby cells and inhibit those that are farther away. Because OCOS cells excite nearby cells, cells close together will excite each other. "Hot spots" of active cells close together result, which inhibit cells farther away. The cells in the hot spots sample the slab and become the outstars that learn the pattern. Thus, the cells in a pattern in slab 1 become those that will excite a hot spot, and the cells in the hot spot become the outstars that learn the pattern.

Chunking can be either temporal or spatial. For example, a spoken word is the sequential chunk of neural activity required to produce the word and, a heard word is the sequential pattern of sounds that have been identified as that word. The OCOS architecture can force a winner (a hot spot) on the sampling slab and thus force chunking to occur in either the spatial or the temporal case.

If the outstar is indeed the biological mechanism for chunking, then Miller's magic number 7 must have some physical relationship to the outstar architecture. What might this be? There are probably physical limits associated with the activity of cells, their rates of decay, and the spread of axonal trees. This is purely speculation, but only so many hot spots can exist on a slab, so some limit must exist, and an excited neuron can continue to fire for only a certain length of time. Constraints such as these should force a physical limit on the size of a chunk.

The Neural Basis For Analogy

An analogy consists of objects, events, or situations that have features (or patterns) in common; that is, they are similar in one or more ways. We shall show that shared features have a significant neural impact. We implicitly assume that *similar* features will have the same sort of impact. The heart of the argument will be the discovery that chunks with similar *or* shared features reinforce each other significantly by forming feedback loops. These feedback loops cause the activity of the sampling outstars (i.e., the codes that are sampling the new, to-be-learned patterns) to grow exponentially as the feedback loop forms. Such an exponential increase in cell activity is significant for two reasons. First, it causes rapid sampling, and rapid sampling means fast learning (or learning period, as slow learning and no learning are often synonymous). Second, cells on an OCOS slab compete with each other and those that become active first are able to quench less active cells. A specific example of analogical transfer will be used to explicate these points.

Analogy Facilitates Learning: An Example. When I was in seventh grade, my mathematics teacher introduced the word *perpendicular* and the symbol ⊥ to refer to two lines that intersect at a 90-degree angle. The teacher wanted us to re-member the word and the symbol (and of course their mean-ing), so when he introduced the word and the symbol, he also introduced the words "pup-in-da-cooler." The teacher believed intuitively that introducing these similar-sounding words and the images they would evoke in our minds would aid in recall. The words not only brought out a few laughs from the class but also worked extremely well. To this day, I cannot think of the word perpendicular or the symbol ⊥ without "pup-in-da-cooler" following close behind. The words perpendicular and pup-in-da-cooler are very similar. The letters are similar, and of course, and so are the chunks. We shall focus on each of these facts as, they make the example analogous to analogies that share similarities at different levels.

Assume that the word "pup" is an already acquired chunk in LTM and the word "perp" represents a chunk to be learned. Having pup as an active chunk in LTM will speed the learning of perp. Similarly, hearing the words pup-in-da-cooler will speed the learning of perp-en-dic-ular because they share features in common.

Basic auditory features (or patterns) are called phonemes. However, it will simplify the discussion to assume that letters are the basic auditory patterns, the phenomes. The reader must merely replace "letter" with "phoneme" to provide a more technically correct version of the following discussion. In brief, when "perpendicular" is spoken, much of the neural activity in the STM related to the words pup-in-da-cooler remains because the two words sound much the same. The shared features re-main active and cause chunking to occur, which makes it possi-ble to learn quickly the word perpendicular.

To explain how shared features cause chunking of new in-put to occur, consider the neural model depicted in Figure 4.27. The word "pup" is heard. The sound presents input to slab 1 as shown and in turn activates the chunk for "pup" already stored in LTM on slab 2. In other words, the sound "perp" creates neural activity on slab 1 and a hot spot of neural activity on slab 2. We thus have the outstar representing the chunk "pup" (on slab 2) feeding the beginning and ending letters "p," which remain active on slab 1. These letters begin to form a portion of an instar connected to the hot spot on slab 2 that will

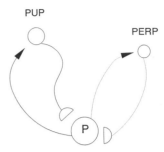

FIGURE 4.28. A feedback loop forming. The arrows indicate portions of instars. A signal travels from pup to p (as part of an outstar), then along the dotted arrow from p to perp, back to p along the dotted line, and from there along the arrow back to pup. The dotted lines indicate an instar (from p to perp) and an outstar (from perp to p) during formation.

chunk "perp." This neural activity will create a feedback loop from "pup" on slab 2 to "p" on slab 1, then to "perp" on slab 2 and back again (see Figure 4.28). This feedback loop will greatly increase the activity of the slab 2 neuron "perp," and this increased activity will make it much easier for the chunk (the outstar) "perp" to form.

On slab 2 there are other neurons that chunk each of the syllables pup, in, da, coo, and ler. There is also a slab 3, on which there is a single neuron that will chunk the five syllables pup-in-da-coo-ler. Again, feedback loops are involved. As shown in Figure 4.29 multiple feedback loops connect the neu-rons on slab 3 chunking the words "pupindacooler" and "per-pendicular" with each other via neurons on lower slabs (e.g., "pupindacooler" on slab 3 to "in" on slab 2 to "perpendicular" on slab 3, then back to "in" on slab 2 and from there back to "pupindacooler" on slab 3). Another feedback loop forms from

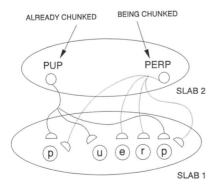

FIGURE 4.27. Chunks sharing features. Pup chunks the fea-tures (the phenomes) beginning p sound, u sound, and end-ing p sound. Perp chunks the same beginning and ending p sounds.

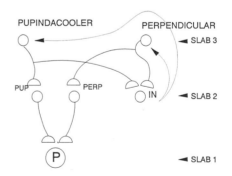

FIGURE 4.29. Multiple feedback loops connect the neurons on slab 3 chunking the words "pupindacooler" and "perpen-dicular." One feedback loop begins at "pupindacooler," then goes to "in" to "perpendicular," back to "in" and from there back to "pupindacooler." Another feedback loop goes from "pupindacooler" to "pup" to "p," then to "perp" and from there to "perpendicular," back to "perp," back to "p," to "pup" and finally to "pupindacooler." Others exist as well. The instars from "p" to "perp," from "perp" to "perpendicu-lar," and from "p" to "pup" and "pup" to "pupindacooler" are not shown.

"pupindacooler" on slab 3, to "pup" on slab 2, to "p" on slab 1, then to "perp" on slab 2, from there to "perpendicular" on slab 3, back to "perp" on slab 2, back to "p" on slab 1, to "pup" on slab 2, and finally to "pupindacooler" on slab 3. Other similar feedback loops are also formed.

An Emergent, Self-Organizing Neural Control System

How the symbol $\perp$ is learned has not yet been explained. Consequently, this section shows that the shared features that exist between pup-in-da-cooler and perpendicular are responsible for the creation of an emergent neural control system that can greatly increase the speed at which the symbol $\perp$ is learned.

The section begins with two simple alternative configurations to the emergent neural control system and then introduces the control system itself. Why the neural control system is such an improvement over the alternatives and why it can greatly increase the rate of learning are explained.

Figure 4.30 depicts the first simple alternative, in which perp represents the word perpendicular that is to be associated with recall of the symbol $\perp$. A and C each represent a neuron or, in the OCOS architecture, a small group of neurons that mutually excite each other. A is the neuron or group of neurons that are active in the auditory neural subsystem when the word perpendicular is spoken. This group of neurons either chunks or will chunk this word. C is a neuron that will sample the area of the visual system that contains the symbol $\perp$ and will chunk this symbol if the learning is successful.

Excitation of neurons A and C will result in association of the word perpendicular with the symbol $\perp$. If the association is successful, the word "perpendicular" will cause recall of the symbol. The activity of A could be considered chunking enhancement. This is because C is a sampling cell, and its activity will result in the formation of a chunk, a set of features that will be grouped together. Activity of cell A will help to increase the sampling rate of C and thus the ability of C to chunk (or form concepts). The problem with the preceding configuration is that the activity of cell C depends solely on the activity of A. Thus, unless A is extremely active or repeated many times, the learning that C is attempting will not take place.

Figure 4.31 shows an extension of Figure 4.30. In this configuration the two words pup and perp activate neuron C. Thus, C will be excited twice as much as in the previous case (provided $X_A = X_B$ and $z_{AC} = z_{BC}$). This configuration is an improvement over that shown in Figure 4.30, but it still does not allow for large-scale boosts in neural activity.

Figure 4.32 shows an emergent self-organizing neural control system that can cause large-scale boosts in activity. In fact, it

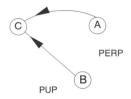

FIGURE 4.31. An extension: two words, PUP and PERP, activate neuron C. C will be excited twice as much as in Figure 4.29 provided $X_A = S_B$ and $z_{AC} = z_{BC}$.

can cause the sampling rate of C to increase exponentially. As shown, neurons A and B form a feedback loop. Presumably this is the neural mechanism that an analogy produces. Because neurons A and B of the control structure form a feedback loop, as A and B fire each will increase the rate at which the other fires. A increases the firing rate S_{BA} of B, and B increases the firing rate S_{AB} of A and the sampling rate of C will initially grow exponentially.

In the architecture presented in Figure 4.32, signal S_{AB}, which travels down the axon leaving A, travels to both B and C. Thus, $S_{AB} = S_{AC}$. In the same fashion $S_{AB} = S_{BC}$. Thus, an increase in S_{AB} and S_{BA} will result in an increase in S_{AC} and S_{BC} as well. The signals to C from A and B are by-products of this architecture. The feedback loop from A to B and back again to A (see Figure 4.33) emerges when the organism is presented with data that cause A and B to fire at the same time. As a by-product, other regions are also flooded with neural excitation. The neurons chunking "pup" and "perp" are such an A and B. If the region they flood is a winner-take-all region, a C will emerge, become strongly excited, and be able to learn the symbol $\perp$. "Perp" will be associated with $\perp$, and if C's axonal tree also reaches the auditory cortex, then $\perp$ will be associated with perp.

THE CONTROL SYSTEM DRIVES THE LEARNING OF THE SYMBOL $\perp$. To say that the control system drives the learning of the symbol $\perp$ means that the control system determines the rate at which the learning will occur. To understand why the analogy controls the rate of learning, notice in the sampling rate equation that even if z_{AC} is small, if $S_{BA} \cdot z_{BA}$ is large then $d/dt(X_C)$, the sampling rate of C, will be large. This is interesting because $S_{BA} \cdot z_{BA}$ will be large if the association between A and B is strong. Thus, the analogy, the signals between A and B, drives the learning of the symbol $\perp$.

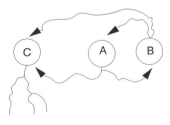

FIGURE 4.32. A self-organizing neural control system that can cause the sampling rate of C to increase exponentially. C's activity increases exponentially because neurons A and B form a lateral feedback loop.

FIGURE 4.30. Associating the word "perp" with the symbol $\perp$.

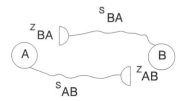

FIGURE 4.33. The components of the feedback loop.

The shared features within the input data "pup-in-da-cooler" and "perpendicular" caused the control system (neurons A and B and the feedback loop they form) to arise. A represents the neuron chunking the word "pup-in-da-cooler" and B represents the neuron chunking "perpendicular." The system arose because the shared features caused the chunking to occur. As mentioned before, a major reason chunking occurred was that the shared features caused an exponential growth in neural activity. This rapid rise in activity allowed the chunks to form. Thus the input data, the neural ability to chunk, and the exponential growth associated with feedback all cause the control system to emerge. [For mathematical demonstrations of these points, see Lawson and Lawson (in press).]

A Return to the Japanese Classroom

In summary, the example of learning the word perpendicular and its mathematical symbol has a natural explanation within hierarchical neural networks. A network to explain why the Japanese symbol could not easily be recalled would also consist of a hierarchy of slabs. The first slab would be a slab of neurons activated by line segments tilted (or oriented) in a specific direction. The second slab would consist of cells (outstars) that chunk neurons in the first slab. If there is a neuron on the second slab that has chunked the neurons on the first that are activated by the Japanese symbol, then the symbol can be recalled by activating that neuron. If no neuron has previously chunked the Japanese symbol, the symbol cannot be recalled unless presynaptic activity associated with receives a considerable boost.

The neural models that have been presented suggest that analogies can greatly facilitate learning and memory. How could an analogy be used to help recall a symbol such as the Japanese symbol 木? According to the theory, the correct approach would be to try to imagine something "like" the symbol. For example, the symbol might remind you of the three crosses at Christ's crucifixion, with the middle cross taller than the other two. As demonstrated, activation of these similar images already stored in LTM greatly increases postsynaptic neural activity; this, according to Grossberg's learning equation, allows storage of the new input in LTM. Of course, one may not be able to generate a satisfactory analogy, image, or set of images, in which case one would have to resort to the more tedious task of describing the symbol: it has three vertical lines attached to a horizontal line, and so forth. Provided patterns for these terms exist in memory at $F^{(2)}$, this procedure will work, but it requires considerable effort to describe all the relevant variables. This

effort is, in fact, a method for maintaining relevant portions of the image in STM so that chunking can occur. Thus, we have a neurological account of why "a picture is worth a thousand words." The images are far superior because the relevant features are already linked in the analogous image. Of course, if irrelevant features are linked, the analogy will be misleading. The pup-in-da-cooler, perpendicular example was meant to demonstrate the manner in which analogy can speed learning by incorporating relevant features shared by new ideas and those already known.

CONCLUDING REMARKS: INTEGRATING PHILOSOPHY, NEURAL MODELING, SCIENTIFIC INSIGHT, AND INSTRUCTION

This chapter concludes with an attempt to integrate specific aspects of philosophy, neural modeling, scientific insight, and instruction. The chapter began with a discussion of two opposing philosophical views of knowledge acquisition; namely, empiricism and nativism. At this point, review of those two philosophical views should prove instructive. Recall that Grossberg's learning equation was stated as follows:

$$\dot{Z}_{ij} = -B_{ij}Z_{ij} + S_{ij}[X_j]^+$$

The equation claims that novel experiences will make their way into LTM when the product of neural activity due to incoming stimuli (i.e., the presynaptic activity S_{ij}) times the neural activity due to already established connections in the brain (i.e., the postsynaptic activity $[X_j]^+$) exceeds present synaptic strength (Z_{ij}) times a decay constant ($-B_{ij}$). This implies that there are at least two different ways to get novel information into LTM. The first way is to boost the incoming signal (the S_{ij}) so that even very little postsynaptic activity ($[X_j]^+$) will drive the product of S_{ij} and $[X_j]^+$ high enough to increase the Z_{ij}s significantly. We have all had the experience of trying to memorize information, such as the names of the states and their capitals, by simply repeating the words over and over again. This repetition presumably boosts the presynaptic incoming signals high enough for the connections to be eventually made (i.e., for the Z_{ij}s to increase sufficiently). Thus some experiences *can* make their way *directly* into LTM. Highly emotionally charged situations also boost presynaptic activity, so that "one-shot" learning is possible in some circumstances. In this sense, the empiricist doctrine is correct.

But recall our discussion of trying to remember the question mark and the Japanese symbol after the visit to the Japanese science classroom. The question mark was easily recalled, while the Japanese symbol was just as easily forgotten. Presumably this was because the incoming signals from the question mark (the presynaptic signals, S_{ij}) were greeted in the brain by postsynaptic activity ($[X_j]^+$) due to a memory record of question marks seen in the past; thus, the product of S_{ij} and $[X_j]^+$ was great enough to increase the transmitter release rates of the relevant neurons to the extent that a new set of functional connections was formed. The incoming signals from the Japanese symbol were not greeted by comparable postsynaptic activity because no prior memory record of the Japanese symbol existed. Therefore, no change in transmitter release rate of those neurons occurred. Hence, when attention was directed away from the board and the incoming signals from the Japanese symbol decayed, the symbol was forgotten.

In this sense, Piaget's statement that one must have a prior "mental structure" in order to "assimilate" experience is correct. The nativist position is also partially correct in the sense that prior "ideas" must be present for the experiences to make their way into the mind. Of course, the nativists incorrectly conclude that these prior ideas must be innate because they could not come from experience. But even though the empiricists and the nativists can claim to be partially correct, neither has recognized both key terms of Grossberg's learning equation, namely the S_{ij} and the $[X_{ij}]^+$, and neither has properly included the role of behavior and the process of adaptive resonance.

Of course, to get new experiences into LTM, we need not always rely on boosting presynaptic activity. Instead, we can use clever mnemonic techniques that boost postsynaptic activity. Suppose, for example, you have just met someone for the first time, and he tells you his name is Bill. One way to recall his name would be to repeat the name Bill over and over again, hoping to boost the S_{ij}s to the extent that a permanent record is stored in LTM. Although this might work, a much easier way is to ask yourself who else you know with the name Bill. Most likely it will take only a moment's reflection to recall another Bill, at which point a host of memories linked to this other Bill will be triggered. You recall his last name, his face, his wife, and a variety of experiences you have shared with this other Bill. The net effect of this reminiscence is that the postsynaptic activity associated with the neuron or neurons associated with the name Bill is given a huge boost of activity (i.e., the $[X_{ij}]^+$s associated with "Bill" increase). Thus, when you are looking at the new Bill, the S_{ij}s that are active in STM due to this experience are greeted by the large $[X_{ij}]^+$s due to the postsynaptic activity associated with the name Bill and all of those prior experiences from LTM. Consequently, the transmitter release rates (Z_{ij}s) increase and a new functional connection is made so that the next time you see the new Bill, the name Bill will be recalled from LTM.

The use of analogy to boost postsynaptic activity amounts to much the same thing. Instead of using the same name and associated memories from LTM, an analogous word or experience can be used to boost the postsynaptic activity. As we have seen, this allows feedback loops to form that lead to an exponential increase in postsynaptic activity. This idea gives us a neural-level explanation for scientific insight, and it also suggests an extremely valuable teaching device.

Recall that the term abduction was introduced in our discussion of Charles Darwin and his use of analogy to invent the process of natural selection. Abduction was contrasted with induction and deduction. Induction moves from particular to general rules, whereas deduction proceeds from general rules to their logically implied specific consequences. Abduction, on the other hand, goes neither from specific to general nor from general to specific but proceeds from one context to another by way of analogy. Abduction amounts to borrowing an idea from one area and applying it in a new area because the two areas are seen to be alike or analogous. One could use the phrase analogical reasoning in place of abduction. As we saw, Darwin used analogical reasoning when he explained the evolution of species as a process *like* the artificial selection of domestic plants and animals. He used the term natural selection to label this hypothetical analogous process in nature. The scientific inventions of others such as Kepler, Mendel, Coulomb, and Kekulé can also be viewed as the products of analogical reasoning or abduction. The list of analogies in science is long, for as Pierce stated, perhaps not overzealously, "all the ideas of science some to it by way of Abduction" (Hanson, 1947, p. 85).

Our discussion of neural models reveals why this is so. Consider the words of Charles Darwin from his autobiography, as quoted in the editor's preface of Darwin's *Voyage of the Beagle* (1962):

Innumerable well observed facts were stored in the minds of naturalists ready to take their proper places as soon as any theory which would receive them was sufficiently explained. (p. xxii)

Two points need to be made. First, the key event for Darwin in his invention of natural selection as the mechanism for evolution occurred when he "saw" that the evolutionary process in nature was analogous to the process of artificial selection in domestic breeding. In neural terms, for Darwin to invent natural selection, he needed the postsynaptic neural activity of his experiences with artificial selection boosted at the same time that he was contemplating these "innumerable observed facts" that were also stored in his memory. When these two sets of memory records were boosted simultaneously—that is, when both presynaptic activity, the S_{ij}s, and postsynaptic activity, the $[X_{ij}]^+$s were boosted—a new set of connections was formed. When Darwin became conscious of these new connections he felt the moment of insight, the Eureka experience. Second, as reflected in Darwin's words, other informed naturalists will also easily put innumerable facts in their proper places once the relevant pattern is presented to them.

Of course, Darwin was interested in teaching other naturalists about the process of natural selection. But suppose we wish to teach students about the great theoretical ideas of science, such as natural selection. How can students come to understand these ideas? The answer, predictably, is to use the same process that led to their invention in the first place, that is, analogy. In this case the appropriate analogy to use is that of artificial selection. But there is a problem. Unfortunately, most students have not had personal experiences with artificial selection, so a verbal presentation of examples of artificial selection may be ineffective. Instead of relying on verbal examples, a simulation activity can be conducted in which the students take an active part in the selection process (e.g., Stebbins & Allen, 1975). In the simulation the students play the role of predatory birds capturing and eating mice (paper disks of various colors). Over three generations of this selection process, the color of the mouse population changes to "fit" the environment in which they live (a patterned piece of fabric). So the students actively simulate the process of selection in the classroom and establish an LTM record of the process. Consequently, when the teacher introduces the terms natural selection to refer to this process in nature, the students will have the experiences necessary to allow the appropriate assimilation. In neurological terms, the simulation activity provides connections to LTM and sets off in the students' brains the postsynaptic pattern of activity, the boosted $[X_{ij}]^+$s, that can be matched with presynaptic S_{ij}s activated when the teacher introduces the words natural selection. This allows new connections to be made. Without such a match, no assimilation will occur; transfer to LTM will not take place. Here assimilation is defined neurologically as the resonant state that occurs when the pattern driven from LTM matches that derived from the environment.

One final point should be made. The preceding discussion has omitted one possibly crucial element. Recall that the young Charles Darwin began his explorations as a believer in special creation. Before he could even begin to wonder about the causes of evolution, he had to become convinced that his initial belief in special creation was wrong and that evolution, in fact, does occur. When old ideas must be changed before new ones can be assimilated, the more complex and time-consuming process of equilibration must occur. This implies that the truly effective teacher must not only know the subject matter well enough to be able to present all of the puzzle pieces and the patterns that will enable the students to put those pieces together, but must also know the contradictory patterns that the students may bring to class and how to open their minds so that they become willing and able to wrestle successfully with the alternatives and the evidence. Only then will students gain scientifically appropriate conceptual knowledge and the procedural skills to evaluate old and perhaps inappropriate ideas and allow the process of equilibration to proceed. In other words, the primary goal of instruction is not merely to teach concepts and conceptual systems but to teach students how to change old concepts and conceptual systems and invent new ones when it is appropriate to do so.

References

Albus, J. S. (1981). *Brains, behavior, and robotics*. Peterborough, NH: BYTE Books.

Anderson, C. W., & Smith, E. R. (1986). Teaching science. In V. Koehler (Ed.), *The educator's handbook: A research perspective*. New York: Longman.

Anderson, J. R., & Thompson, R. (1989). Use of analogy in a production system architecture. In S. Vosniadou & A. Ortony (Eds.), *Similarity and analogical reasoning*. London: Cambridge University Press.

Ausubel, D. P. (1963). *The psychology of meaningful verbal learning*. New York: Grune & Stratton.

Ausubel, D. P. (1979). Education for rational thinking: A critique. In A. E. Lawson (Ed.), *The psychology of teaching for thinking and creativity*. AETS 1980 Yearbook. Columbus, OH: ERIC/SMEAC.

Ausubel, D. P., Novak, J. D., & Hanesian, H. (1968). *Educational psychology: A cognitive view* (2nd ed.). New York: Holt, Rinehart & Winston.

Bereiter, C. (1985). Toward a solution of the learning paradox. *Review of Educational Research, 55*(2), 201–226.

Bolton, N. (1977). *Concept formation*. Oxford: Pergamon.

Bringuier, J. (1980). *Conversations with Jean Piaget*. Chicago: University of Chicago Press.

Burmester, M. A. (1952). Behavior involved in critical aspects of scientific thinking. *Science Education, 36*(5), 259–263.

Byrne, R. M. J. (1989). Suppressing valid inferences with conditionals. *Cognition, 31*, 61–83.

Carbonell, J. G. (1986). Learning by analogy: Formulating and generalizing plans from past experience. In R. Michalski, J. G. Carbonell, & T. M. Mitchell (Eds.), *Machine learning: An artificial intelligence approach*. Palo Alto, CA: Tioga Press.

Catrambone, R., & Holyoak, K. J. (1985). *The function of schemas in analogical problem solving*. Poster presented at the meeting of the American Psychological Association, Los Angeles, CA.

Chamberlain, T. C. (1897). The method of multiple working hypotheses. *Science, 148*, 754–759, 1965.

Cheng, P. W., & Holyoak, K. J. (1985). Pragmatic reasoning schemas. *Cognitive Psychology, 17*, 391–416.

Clausen, J., Keck, D., & Hiesey, W. (1948). *Carnegie Institution of Washington Publication, 581*.

Collea, F. P., Fuller, R. G., Karplus, R., Paldy, L. G., & Renner, J. W. (1975). *Physics teaching and the development of reasoning*. Stony Brook, NY: American Association of Physics Teachers.

Collette, A. T., & Chiappetta, E. L. (1986). *Science instruction in the middle and secondary schools*. Columbus, OH: Merrill.

Copi, I. M. (1972). *Introduction to logic* (4th ed.). New York: Macmillan.

Darwin, C. (1962). *The voyage of the beagle*. (L. Engel, Ed.). Garden City, NY: American Museum of Natural History and Doubleday.

Dempster, F. N. (1992). Resistance to interference: Developmental changes in a basic processing mechanism. *Developmental Review, 12*, 45–57.

Diamond, A. (1990). The development and neural bases of inhibitory control in reaching in human infants and infant monkeys. In A. Diamond (Ed.), *The development and neural basis of higher cognitive functions*. New York: New York Academy of Sciences.

Ehrlich, P. R., Holm, R. W., & Parnell, D. R. (1974). *The process of evolution* (2nd ed.). New York: McGraw-Hill.

Elementary Science Study. (1974). *Attribute games and problems*. New York: McGraw-Hill.

Evans, J. St. B. T. (1982). *The psychology of deductive reasoning*. London: Routledge and Kegan Paul.

Finley, F. N. (1986). Evaluating instruction: The complementary use of clinical interviews. *Journal of Research in Science Teaching, 23*(7), 635–650.

Futuyma, D. J. 1979. *Evolutionary biology*. Sunderland, MA.: Sinauer.

Gentner, D. (1987). Structure-mapping: A theoretical framework for analogy. *Cognitive Science, 7*, 155–170.

Gentner, D. (1989). The mechanisms of analogical reasoning. In S. Vosniadou & A. Ortony (Eds.), *Similarity and analogical reasoning*. London: Cambridge University Press.

Gick, J. L., & Holyoak, R. J. (1983). Schema induction and analogical transfer. *Cognitive Psychology, 15*, 1–38.

Green, J. C. (1959). *The death of Adam*. Ames: Iowa State University Press.

Griggs, R. A., & Cox, J. R. (1982). The elusive thematic materials effect in Wason's selection task. *British Journal of Psychology, 73*, 407–420.

Grossberg, S. (1982). *Studies of mind and brain*. Dordrecht, Holland: D. Reidel.

Grossman, S. P. (1967). *A textbook of physiological psychology*. New York: Wiley.

Gruber, H. E., & Barrett, P. H. (1974). *Darwin on man.* New York: Dutton.

Hanson, N. R. (1947). *Patterns of discovery.* London: Cambridge University Press.

Hebb, D. O. (1949). *The organization of behavior.* New York: Wiley.

Hestenes, D. (1983). *How the brain works: The next great scientific revolution.* Paper presented at the third workshop on maximum entropy and Bayesian methods in applied statistics, University of Wyoming.

Hewson, P. W., & Hewson, M. G. A. (1984). The role of conceptual conflict in conceptual change and the design of science instruction. *Instructional Science, 13,* 1–13.

Holland, J., Holyoak, K., Nisbett, R., & Thagard, P. (1986). *Induction: Processes of inference, learning, and discovery.* Cambridge, MA: MIT Press.

Holyoak, K. J. (1985). The pragmatics of analogical transfer. *Psychology of Learning and Motivation, 19,* 59–87.

Holyoak, K. J., Junn, E. N., & Billman, D. O. (1984). Development of analogical problem-solving skill. *Development Psychology, 55,* 2024–2055.

Holyoak, K. J., & Thagard, P. R. (1989). Analogical mapping by constraint satisfaction. *Cognitive Science, 13,* 295–355.

Hume, P. (1739). *A treatise of human nature.* Oxford: Clarendon Press, 1896.

Inhelder, B., & Piaget, J. (1958). *The growth of logical thinking from childhood to adolescence.* New York: Basic Books.

Johnson-Laird, P.N. (1983). *Mental models.* Cambridge, MA: Harvard University Press.

Karplus, R. (1979). Teaching for the development of reasoning. In A. E. Lawson (Ed.), *1980 AETS yearbook: The psychology of teaching for thinking and creativity.* Columbus, OH: ERIC/SMEAC.

Karplus, R., Lawson, A. E., Wollman, W., Appel, M., Bernoff, R., Howe, A., Rusch, J. J., & Sullivan, F. (1977). *Science teaching and the development of reasoning: A workshop.* Berkeley: Regents of the University of California.

Kuhn, D., Amsel, E., & O'Loughlin, M. (1988). *The development of scientific thinking skills.* San Diego: Academic Press.

Lawson, A. E. (1982). Evolution, equilibration, and instruction. *The American Biology Teacher, 44*(7), 394–405.

Lawson, A. E. (1986). A neurological model of sensory-motor problem solving with possible implications of higher-order cognition and instruction. *Journal of Research in Science Teaching, 23*(16), 503–522.

Lawson, A. E. (1992). What do tests of "formal" reasoning actually measure? *Journal of Research in Science Teaching, 29*(9), 965–983.

Lawson, A. E. (1993). Deductive reasoning, brain maturation, and science concept acquisition. *Journal of Research in Science Teaching, 30*(9), 1029–1051.

Lawson, A. E., Abraham, M. R., & Renner, J. W. (1989). *A theory of instruction: Using the learning cycle to teach science concepts and thinking skills.* Cincinnati, OH: National Association for Research in Science Teaching.

Lawson, A. E., Adi, H., & Karplus, R. (1979). The acquisition of propositional logic and formal operational schemata during the secondary school years. *Journal of Research in Science Teaching, 15*(6), 465–478.

Lawson, A. E., & Bealer, J. M. (1984). The acquisition of basic quantitative reasoning skills during adolescence: Learning or development? *Journal of Research in Science Teaching, 21*(4), 417–423.

Lawson, A. E., & Hegebush, W. (1985). A survey of hypothesis testing strategies: Kindergarten through grade twelve. *The American Biology Teacher, 47*(6), 348–355.

Lawson, A. E., & Kral, E. A. (1985). Developing formal reasoning through the study of English. *The Educational Forum, 49*(2), 211–226.

Lawson, A. E., & Lawson, C. A. (1979). A theory of teaching for conceptual understanding, rational thought and creativity. In A. E. Lawson (Ed.), *The psychology of teaching for thinking and creativity.* AETS Yearbook. Columbus, OH: ERIC/SMEAC.

Lawson, A. E., Lawson, D. I., & Lawson, C. A. (1984). Proportional reasoning and the linguistic abilities required for hypothetico-deductive reasoning. *Journal of Research in Science Teaching, 21*(2), 119–131.

Lawson, A. E., & Staver, J. R. (1989). Toward a solution of the learning paradox: Emergent properties and neurological principles of constructivism. *Instructional Science, 180,* 169–177.

Lawson, A. E., & Thompson, L. D. (1988). Formal reasoning ability and misconceptions concerning genetics and natural selection. *Journal of Research in Science Teaching, 25*(9), 733–746.

Lawson, D. I., & Lawson, A. E. (1993). Neural principles of memory and a neural theory of analogical transfer. *Journal of Research in Science Teaching,* in press.

Levine, D. S., & Prueitt, P. S. (1989). Modeling some effects of frontal lobe damage: Novelty and perseveration. *Neural Networks, 2,* 103–116.

Locke, J. (1690). *Essay on human understanding.* Oxford: Clarendon Press, 1924.

Luria, A. R. (1961). *The role of speech in the regulation of normal and abnormal behavior.* (J. Tizard, Ed.). Oxford: Pergamon.

MacGinitie, G. E., & MacGinitie, N. (1968). *Natural history of marine animals.* (2nd ed.). New York: McGraw-Hill.

Markovits, H. (1984). Awareness of the "possible" as a mediator of formal thinking in conditional reasoning problems. *British Journal of Psychology, 75,* 367–376.

Markovits, H. (1985). Incorrect conditional reasoning among adults: Competence or performance? *British Journal of Psychology, 76,* 241–247.

Mayer, R. E. (1983). *Thinking, problem solving, cognition.* New York: W. H. Freeman.

Meltzoff, A. N. (1990). Towards a developmental cognitive science. In A. Diamond (Ed.), *The Development and Neural Basis of Higher Cognitive Functions.* New York: New York Academy of Sciences.

Miller, G. A. (1956). The magical number seven, plus or minus two: Some limits on our capacity for processing information. *The Psychological Review, 63*(2), 81–97.

Minstrell, J. (1980). *Conceptual development of physics students and identification of influencing factors.* Unpublished research report, Mercer Island School District, Washington.

Mishkin, M., & Appenzeller, T. (1987). The anatomy of memory. *Scientific American, 256*(6), 80–89.

Overton, W. F., Ward, S. L., Black, J., Noveck, I. A., & O'Brien, D. P. (1987). Form and content in the development of deductive reasoning. *Developmental Psychology, 23*(1), 22–30.

Pascual-Leone, J. (1976). A view of cognition from a formalist's perspective. In K. F. Riegel & J. A. Meacham (Eds.), *The developing individual in a changing world: Vol 1. Historical and cultural issues, 89–110.* The Hague, The Netherlands: Mouton.

Pascual-Leone, J. (1980). Constructive problems for constructive theories: The current relevance of Piaget's work and critique of information processing simulation psychology. In R. H. Kluwe & H. Spada (Eds.), *Developmental models of thinking.* New York: Academic Press.

Piaget, J. (1928). *Judgment and reasoning in the child.* New York: Harcourt-Brace.

Piaget, J. (1929a). Les races lacustres de la *Limnaea stagnalis* and recherches sur la rapports de l'adaptation héréditaire avec la milieu. *Bulletin biologique de la France et de la Belgique, 62,* 424.

Piaget, J. (1929b). Adaptation de la *Limnaea stagnalis* aux milieux lacustres de la Suisse romande. *Revue Suisse de Zoologie, 36,* 263.

Piaget, J. (1952). *The origins of intelligence in children*. New York: International Universities Press.

Piaget, J. (1954). *The construction of reality in the child*. New York: Basic Books.

Piaget, J. (1957). *Logic and psychology*. New York: Basic Books.

Piaget, J. (1964). Cognitive development in children: Development and learning. *Journal of Research in Science Teaching, 2*(2), 176–186.

Piaget, J. (1966). *Psychology of intelligence*. Totowa, NJ: Littlefield Adams.

Piaget, J. (1971a). *Biology and knowledge*. Chicago: University of Chicago Press.

Piaget, J. (1971b, May 26). *Problems of equilibration*. In C. F. Nodine, J. M. Gallagher, & R. H. Humphreys (Eds.), *Piaget and Inhelder: On equilibration*. Proceedings of the First Annual Symposium of the Jean Piaget Society, Philadelphia: The Jean Piaget Society.

Piaget, J. (1972). Intellectual evolution from adolescence to adulthood. *Human Development, 15*, 1–12.

Piaget, J. (1975). From noise to order: The psychological development of knowledge and phenocopy in biology. *The Urban Review, 8*(3), 209.

Piaget, J. (1976). *Piaget's theory*. In B. Inhelder & H. H. Chipman (Eds.), *Piaget and his school*. New York: Springer-Verlag.

Piaget, J. (1977a). *The development of thought: Equilibrium of cognitive structures*. New York: Viking.

Piaget, J. (1977b). *The grasp of consciousness*. London: Routledge and Kegan Paul.

Piaget, J. (1978). *Behavior and evolution*. New York: Random House.

Piaget, J., & Inhelder, B. (1969). *The psychology of the child*. New York: Basic Books.

Piattelli-Palmerini, M. (Ed.). (1980). *Language and learning: The debate between Jean Piaget and Noam Chomsky*. Cambridge, MA: Harvard University Press.

Posner, G. J., Strike, K. A., Hewson, P. W., & Gertzog, W. A. (1982). Accommodation of a scientific conception: Toward a theory of conceptual change. *Science Education, 66*(2), 211–227.

Rendel, J. M. (1967). *Canalization and gene control*. London: Logos.

Resnick, L. B. (1980). The role of invention in the development of mathematical competence. In R. H. Kluwe & H. Spada (Eds.), *Developmental models of thinking*. New York: Academic Press.

Simon, H. A. (1974). How big is a chunk? *Science, 183*, 482–488.

Smith, C. L., & Millman, A. B. (1987). Understanding conceptual structures: A case study of Darwin's early thinking. In D. N. Perkins, J. Lochhead, & J. C. Bishop (Eds.), *Thinking: The second international conference*. Hillsdale, NJ: Lawrence Erlbaum.

Stebbins, R. C., & Allen, B. (1975). Simulating evolution. *The American Biology Teacher, 37*(4), 206.

Suarez, A., & Rhonheimer, M. (1974). *Lineare function*. Zurich: Limmat Stiftung.

Von Foerster, H. (1984). On constructing a reality. In P. Watzlawick (Ed.), *The invented reality: How do we know what we believe we know?* New York: Norton.

Voss, J. F., Greene, T. R., Post, T. A., & Penner, D. C. (1983). Problem-solving in the social sciences. In G. H. Bowen (Ed.), *The psychology of learning and motivation: Advances in research theory*. New York: Academic Press.

Vygotsky, L. S. (1962). *Thought and language*. Cambridge, MA: MIT Press.

Waddington, C. H. (1959). Canalization of development and genetic assimilation of acquired characters. *Nature, 183*(4676), 1654.

Waddington, C. H. (1966). *Principles of development and differentiation*. New York: Macmillan.

Waddington, C. H. (1975). *The evolution of an evolutionist*. Ithaca, NY: Cornell University Press.

Wason, P. C. (1966). Reasoning. In B. M. Foss (Ed.), *New horizons in psychology*. Harmondsworth: Penguin.

Wason, P. C., & Johnson-Laird, P. N. (1972). *Psychology of reasoning: Structure and content*. Cambridge, MA: Harvard University Press.

Wittrock, M. C. (1974). Learning as a generative process. *Educational Psychologist, 11*, 87–95.

Wollman, W. T., & Lawson, A. E. (1978). The influence of instruction on proportional reasoning in seventh graders. *Journal of Research in Science Teaching, 15*(3), 227.

RESEARCH ON ALTERNATIVE
CONCEPTIONS IN SCIENCE

James H. Wandersee
LOUISIANA STATE UNIVERSITY

Joel J. Mintzes
HARVARD UNIVERSITY

Joseph D. Novak
CORNELL UNIVERSITY

The purpose of this chapter is to introduce the reader to the large body of literature on alternative conceptions in science education and to offer a selective overview of that literature. We see our primary audience as science education researchers whose research expertise lies outside this domain and our secondary audience as practicing science teachers. It is our intention to serve both groups. Thus, in preparing this chapter, we chose not to write a conventional scholarly review of the literature but to focus on the research issues we consider most salient for the person who is just beginning to explore this knowledge base. We make no claims of total objectivity, although we have endeavored to provide an original, balanced, and useful (but not encyclopedic) view.

We are experienced science teachers, veterans of the high school and college science classroom, who have become science educators. Our research in this field was based on the meaningful learning theory of David Ausubel. Although our thinking has moved beyond Ausubel's original work, it is still our foundation and, therefore, our bias. Thus, we do not claim to have approached this topic in a totally detached and clinical fashion or to pretend we believe the literature speaks for itself. We also offer our own interpretations and opinions about the issues involved, signaling the reader when we do so.

Later in this chapter we explain our method of selecting the studies we discuss. We think that the studies we present here are representative of the research literature but acknowledge that another set of authors might have selected a somewhat different sample. To ensure intellectual honesty and offer divergent perspectives, we cite other helpful overviews of the literature and assume the reader will want to compare what we say with what they say.

To make this chapter as reader friendly as possible, we have constructed a concept map (see Figure 5.1) that illustrates its basic organization. As you examine the map, note that we see the complex language that researchers use (i.e., specialized terms) as the novice's initial impediment to understanding the various research orientations of the alternative conceptions literature. After gaining such background knowledge, we think the novice is in a better position to understand how its research claims and its researchers' views interact to affect the field's research agenda.

Is the alternative conceptions literature actually more complex than this? Most certainly. However, past experience has shown that our students are less likely to "get lost" when they use a concept map on their intellectual journeys and that concept maps which attempt to include too much detail can hinder

The authors thank reviewers Audrey B. Champagne (State University of New York at Albany) and Patricia E. Blosser (The Ohio State University).

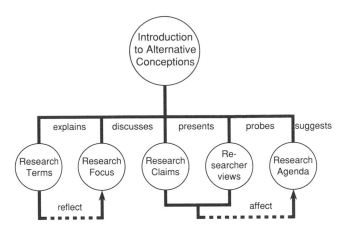

FIGURE 5.1. A concept map of this review: Chapter 5, basic organization.

rather than promote a learner's progress (see Novak & Wandersee, 1990).

MEANING, THOUGHT, AND LANGUAGE

A Plethora of Terms

The words researchers use may block the novice's path to understanding the alternative conceptions literature. We see our first task as helping the novice blaze a trail through the thicket of terminology. Francis Bacon, whom some have called the "father of experimental science" (Wandersee, 1987b), perhaps said it best in his *Novum Organum* of 1620, wherein he wrote: "the ill and unfit choice of words wonderfully obstructs the understanding" (Mackay, 1977, p. 13). Few who know the alternative conceptions literature well would disagree! Because it is a relatively new and rapidly evolving research area in science education, many new terms have been proposed by researchers (including this chapter's authors).

As a result, try to read some of the research and you may soon be mired in a morass of terms capable of dissuading all but the most persistent reader from pressing forward. In this swamp you will encounter a host of perplexing terms, such as naive beliefs (Caramazza, McCloskey, & Green, 1981); erroneous ideas (Fisher, 1983); preconceptions (Hashweh, 1988); multiple private versions of science (McClelland, 1984); underlying sources of error (Fisher & Lipson, 1986); personal models of reality (Champagne, Gunstone, & Klopfer, 1985); spontaneous reasoning (Viennot, 1979); and persistent pitfalls (Meyer, 1987)—to name just a few. This is not to imply that these terms are without merit, but only to demonstrate that the literature is replete with terms that have slightly (or, at times, greatly) different meanings. Many of the nuances in meaning can easily elude the novice researcher.

As Carmichael et al. (1990) aptly observe, "One difference between an expert and a novice lies in the coherence and structure of their understandings.... Novices' attempts to understand the natural world may be both insightful and shallow,

too particular and too general, holistic and disjointed, coherent and confused, logical and illogical, speculative and experience-based" (p. 1). Although this intellectual ungainliness may be normal in learning science, perhaps there are ways of teaching science that can help novices anticipate or avoid some of the conceptual difficulties others have previously encountered. Science teachers may not intend to produce students who are content experts, but they do want their students to know the current scientific explanations for natural phenomena and how scientists have arrived at these explanations. They want their students to see the power of scientific theories—not only in science class, but in everyday life. In order to teach science in an effective manner, teachers must understand not only what their students currently know about the topics they wish to teach but also how students learn science and which teaching strategies are most appropriate for their conceptual development. Perhaps by studying these young scholastic wayfarers at different points along the paths they choose to follow toward scientific understanding, one can understand how some are able to reach their destination without delay, while others get lost along the way. This is, in essence, the implicit assumption of what Millar (1989) has called the "Alternative Conceptions Movement (or ACM for short)."

What is an alternative conception, and why is that term now preferred by many researchers to the originally dominant term misconception? Abimbola (1988), Gilbert and Swift (1985), and other researchers have built a strong case for using the term alternative conception. Not only does it refer to experience-based explanations constructed by a learner to make a range of natural phenomena and objects intelligible, but also it confers intellectual respect on the learner who holds those ideas—because it implies that alternative conceptions are contextually valid and rational and can lead to even more fruitful conceptions (e.g., scientific conceptions). Because a conceptual framework or knowledge structure can often be inferred (validly or invalidly) after one knows the "self-reported" alternative conceptions a student holds about a particular science topic, the meaning of the term alternative conception is more specific than that of a related term (alternative framework) and lies one transformation level closer to the actual cognitive state of the learner than the latter term does. As Novak and Gowin (1984) point out, with each transformation of initially valid raw data comes the possiblity that the original "truth" may be lost in the process.

Here is how one science educator views the self-report data gathered by many of the researchers in the ACM: "The major strength I see is that we are giving students an opportunity to tell us what they think, rather than impose on them our view of the world" (D. P. Butts, personal communication, August 5, 1991). Thus, "what they say they think" may indicate their alternative conceptions; "what we think they think," based on their set of alternative conceptions, may yield a knowledge structure called an alternative framework. (The framework is thus an additional transformation of the data, based on self-report data generated by students from which researchers first infer alternative conceptions and then alternative frameworks. Such reductionist practices may be helpful, but they introduce the possibility that the resultant frameworks may distort the learners' views.) The idea of a framework is also suggestive of a relatively stable pattern of knowledge (compare "framework" to a carpenter's scaffold), one that a larger group of learners share. Not all researchers would accept such assertions of stability.

Because the initially preferred and closely related term, misconception, means a vague, imperfect, or mistaken understanding of something—one that is commonly held by learners, is difficult to "teach away," and is at variance with current scientific knowledge—an increasing majority of researchers contend that its use not only contradicts constructivist views of knowledge but also erroneously implies that such ideas have a negative value, serve no cognitive purpose for the learner, and should be quickly eradicated.

Not all researchers, however, agree that the term misconception should be discarded. The term already has meaning to the layperson and, in a science education context, readily conveys the concept of an idea at variance with current scientific thought. Gowin (1983) has proposed simplifying terms even more by changing misconception to mistake. Mistake, however, has an even stronger and more specific negative connotation than misconception, for it often implies bad judgment or disregard of a key rule or principle. For that reason, misconception seems to be more appropriate than mistake (or error).

Most science teachers are, to some extent, "laypersons" who find the science education literature to be relatively arcane, and the fact that the term misconception is easily understood by both practicing teachers and the general public is no small advantage for a discipline that is already known for its jargon. Nevertheless, as Good (1991) points out, precision does have its virtues; he maintains that "it is important to select words carefully in our efforts to convey our intended meanings" (p. 387).

One of this chapter's authors, learning theorist Joseph Novak, contends that a technical term which he introduced at the first International Seminar on Misconceptions in Science and Mathematics at Cornell University in 1983 actually has more explanatory power than the term misconception. Novak prefers to speak of LIPHS—an acronym for limited or inappropriate propositional hierarchies—because that term is based on (Ausubelian) learning theory and points to the underlying structural and functional weaknesses in the learner's cognitive structure that are relevant to a given science topic. It also underscores the conceptual reorganization required of the learner in order to form valid conceptions and is thus compatible with constructivism.

Offering another perspective, Good (1991) prefers the term prescientific conception because it is less negative than misconception, is specific to science, and emphasizes that the learner's ideas may eventually lead him or her to the current scientific conception about a topic. Good recognizes that "words tend to oversimplify by setting up dichotomies where none exist" (p. 387) and that "levels of science understanding fall along a continuum" (p. 387), but he thinks that the term prescientific conception best conveys science education researchers' intended meanings.

As researchers learn more about alternative conceptions, the dominant terms used in the literature may indeed change, but we see an *e pluribus unum* trend indicating convergence (as the field matures) to fewer terms—terms that not only communicate the key insights embedded in the larger population of extant terms but also convey meaning to science teachers, whom most researchers ultimately hope to influence.

The latter issue is no small concern. The German alternative conceptions researcher Reinders Duit has written: "So far, the amount of research data is becoming bigger and bigger, and the gulf between teachers and researchers increases accordingly. Maybe they do not hear us at all anymore, shouting out the wonderful new findings we have" (R. Duit, personal communication, July 2, 1991). We might add, "Maybe they are trying to understand, but our 'shouting' is unintelligible to them because of the terms we use."

It may seem ironic for us to suggest that a single term can be applied to a research program that argues for tolerance of existing conceptions (cf. Gunstone, 1989), but until a new consensus emerges, alternative conception appears to be the "term of choice" for the majority of researchers in this field and, fortunately, it does not seem too difficult for diverse audiences to understand. Remember, however, that no vote was actually taken and that this field is a dynamic one.

A Term-Categorization Scheme Based on Meaning

In his book *Explication*, Chaffee (1991) recommends two approaches to the process of analyzing the meaning of a concept—distillation and listing. He notes, "In either case, it is helpful to organize our thinking into lower-order and higher-order concepts" (p. 25). He describes how preparation of a listing hierarchy can lead to the production of a "definition tree." We have built upon both his suggestions and those of Gauld (1987) in preparing this overview. The differences in the terms used by members of the ACM are not insignificant; they often reflect distinctly different epistemological and theoretical commitments (details of which lie beyond the purview of this chapter).

Nomothetic and Idiographic Terms. Driver and Easley (1978) pointed out that alternative conception studies can be divided into two groups. In one group, the nomothetic studies, knowledge is evaluated by its conformity to (or deviation from) a standard knowledge base (e.g., accepted scientific knowledge). The word nomothetic literally means "founded upon or derived from custom or law." The pupil's understanding is to be assessed according to the congruence between the pupil's responses and current scientific thought. Terms such as errors, naive conceptions, erroneous ideas, misunderstandings, preinstructional ideas, persistent pitfalls, classroom mismatches, conceptual difficulties, children's learning problems, preconceptions, limited or inappropriate propositional hierarchies, superstitious beliefs, student difficulties, prescientific conceptions, naive theories, incorrect generalizations, conceptual disorders, differential uptake of science, conflicting schemas, unfounded beliefs, mistakes, underlying sources of error, and misconceptions can be placed in this category. Such studies are also more likely to be experimental—often using pencil-and-paper tests, quantification, and inferential statistics.

In the other group, the idiographic studies, the student's understanding of natural objects and events is probed, studied, and analyzed on its own terms, much as an anthropologist might study another culture. The word idiographic literally means "self-written," and such studies involve the explication of individual cases to (possibly) uncover common features and at least "raise awareness of possible perspectives pupils may bring and difficulties they may have, and hence enable more effective communication to take place" (Driver & Easley, 1978, pp. 79–80). Terms such as personal models of reality, pupil's ideas, alternative conceptions, spontaneous ways of reasoning, alternative frameworks, multiple private versions of science, developing conceptions, children's science, commonsense theories, schoolchildren's criteria, children's views, personal constructs, children's understanding, children's knowledge, intuitive beliefs, and everyday physical and chemical conceptions can be placed in this category. Such investigations are more likely to be naturalistic, study fewer students in greater depth, use student's self-report data, and employ thick description.

However, there are exceptions to the preceding generalizations. Some studies use both qualitative and quantitative methods. In addition, Gunstone (1989) warns that the temptation to place a particular researcher into either the nomothetic or the idiographic camp on the basis of terms used by the researcher in a particular publication should be resisted because "the epistemological issues underlying a researcher's choice of terminology are more complex" (p. 643). Often, researchers' viewpoints evolve as the field does.

Some terms may emphasize the developmental (e.g., qualifiers such as naive, pre-, developing), the chronological (e.g., terms such as first notions, intermediate understandings, young children's conceptions), or the contextual (e.g., qualifiers such as everyday, classroom, folk, school, world) aspects of students' understanding. However, Millar (1989) contends that "the history of the field is, by and large, of a movement from the 'misconceptions' language towards the 'alternative conceptions' and 'children's science' end of the continuum" (p. 590). Thus, we think that the nomothetic-idiographic distinction may be the most useful one for understanding the ACM.

We would be remiss if we did not point out that the historical "movement of the field," as identified by Millar (1989), is not entirely positive. Those of us in science education who work with practicing scientists on a daily basis realize that they are deeply concerned about any shift in instructional focus from "authentic, basic, current science" to what they perceive (albeit erroneously) to be the "whatever-works-for-you" or "whatever-you're-interested-in" school science of the current constructivism and science-technology-society (STS) movements. In philosophical terms, most of the scientists we work with are realists, if not positivists; they become relativists only during times of controversy in their discipline, and once consensus is reached, they revert to being realists.

We believe that any effort to reform science education must have the intellectual support of scientists in order to succeed. Therefore, we caution against wholesale abandonment of the ACM's nomothetic (science-centered) research dimension, even though we are sympathetic to the reasons behind the shift toward the idiographic (personal knowledge–centered) re-

search dimension. Thus, we see a place for both kinds of research and we support an emerging synergy.

We are therefore encouraged by statements that appear in such school science reform documents as the American Association for the Advancement of Science (AAAS) *Science for All Americans* (1989). For example, "A fundamental premise of Project 2061 is that the schools do not need to be asked to teach more and more content, but rather to focus on what is essential to scientific literacy and to teach it more effectively" (AAAS, 1989, p. 4). Such statements indicate that scientists' views on science education are not only beginning to converge with those of science educators, they are actually coalescing to support the shared goal of fostering scientific literacy. It seems reasonable to assume that alternative conceptions research findings will be an invaluable resource in the pursuit of scientific literacy for all students.

EMERGENT KNOWLEDGE CLAIMS

Beginning with the work of Piaget in the 1920s and continuing to the present, an enormous effort has been concentrated on understanding the ways in which learners view the natural world and, consequently, what teachers need to do to encourage conceptual change in the physical and life sciences. In recent years, the volume of this research has grown rapidly. The field's leading comprehensive bibliography (Pfundt & Duit, 1985, 1988, 1991) listed 700 references in its first edition, 1400 in its second, and approximately 2000 in its latest version.

Out of this alternative conceptions research has emerged a wealth of potentially valuable knowledge—knowledge which, if modified for use in the science classroom, could have a profound impact on the quality of learning achieved there. Unfortunately, the sheer volume of this work, the format in which it has been reported, and its virtual inaccessibility have made its products largely unavailable to the classroom teacher. Even those active or experienced in research might be intimidated by the large quantity of published studies to date, in view of the effort required to wade through and make sense of them.

In this section we provide an overview of several of the most significant "knowledge claims" (Novak & Gowin, 1984) emerging from the research on students' alternative conceptions, placing emphasis on some seminal studies that have been published during the past 20 years in the science education literature. Our purpose is to provide an introduction to the field for both the seasoned science education researcher whose research expertise lies outside this domain and the novice researcher who wishes to contribute to its growing body of knowledge. We make no claim of comprehensiveness, for to do so would require considerably more research time and publication space than has been allotted to us. Instead, we have attempted to identify and cite the studies that, in our judgment, have had the greatest impact on the direction taken by subsequent workers in the field.

As a point of departure, we initially chose to focus on efforts by the "most published" scholars in the field and then to expand our citations by reviewing papers cited by those authors as central to their own thinking. Using the latest edition of the

Pfundt and Duit (1991) Bibliography, as well as the ERIC data base, we identified approximately 400 papers in this manner. This represents about 20 percent of the entire literature corpus and about 40 percent of the published work in the English language. The Pareto principle, which holds that the significant items in a group normally constitute about 20 percent of the total items in the group, also supports such a selection strategy.

Perhaps more than any other recent research endeavor in science education, the alternative conceptions program has benefited by substantial international cooperation and communication, with important contributions by researchers from nearly every continent and from a host of countries. As the products of this effort accumulate, a pattern of repeated knowledge claims has begun to emerge in the literature. In this section we attempt to winnow these claims and, ultimately, we identify eight of the most frequently repeated assertions which appear to signal potentially important changes in how science educators view teaching and learning in the natural sciences.

In attempting to evaluate these assertions, we have focused on providing a balance of some of the best available evidence in our field—the products of some of the most able minds in science education. However, in such a brief chapter we are able to describe only a representative sample of the alternative conceptions research literature; this means some key studies and important researchers may not be included.

Claim 1: LEARNERS COME TO FORMAL SCIENCE INSTRUCTION WITH A DIVERSE SET OF ALTERNATIVE CONCEPTIONS CONCERNING NATURAL OBJECTS AND EVENTS.

This assertion is clearly the cornerstone of the research program and continues to be the focus of the most intense activity in the field. It is important to note that the word "diverse" does not mean "infinite"; most researchers believe that the set of common alternative conceptions for a given science topic is relatively small. The latest version of the Pfundt and Duit (1991) *Bibliography* lists over 1000 studies that specifically address the issue of students' understandings of scientific concepts. Of this number, about two-thirds fall within the domain of physics (including earth and space sciences); about 200 studies have examined issues in biology, and about 125 focused on chemistry.

Representative Findings—Physics

Of some 700 studies within the school subject of physics, about 300 have been devoted to concepts in mechanics (including force and motion, gravity, velocity, and acceleration), about 159 to electricity, and about 70 each to concepts of heat, optics, the particulate nature of matter, and energy. The earth and space sciences have sparked some 35 studies and "modern physics" (physics based on relativity and quantum theory) about 10.

Possibly the most influential studies to date, if citation frequency is a good indicator, have been those devoted to the exploration of intuitive ideas held by learners in the topic area

of simple mechanics (Bliss, Ogborn, & Whitlock, 1989; Caramazza, McCloskey, & Green 1980; Champagne, Klopfer, & Anderson, 1980; Clement, 1982; diSessa, 1982; Gunstone & White, 1981; McCloskey, 1983; McDermott, 1984; Minstrell, 1982; Viennot, 1979; Watts, 1983). When asked to describe the forces acting on a moving body, the overwhelming majority of students seem to hold to a commonsense, everyday notion that has elements (in varying proportions) of both an Aristotelian and an impetus theory of motion. Among the propositions that students frequently embrace are the following: (1) when a force is applied to an object, it produces motion in the direction of the force; (2) under the influence of a constant force, objects move with constant velocity; (3) the velocity of an object is proportional to the magnitude of the applied force; and (4) in the absence of a force, objects are either at rest or, if moving, are slowing down (Hashweh, 1988; Lythcott, 1985).

Clement (1983) proposes that the basic model most students use is quite simple: motion implies a force. A number of studies have revealed that learners harbor an array of conceptual problems concerning the potential relationships of force, velocity, and acceleration. In one set of preinstructional interviews involving inservice science teachers and college physics students, fewer than 25 percent of the subjects demonstrated a sufficient qualitative understanding of acceleration to be able to apply the concept in a real-world situation (Trowbridge & McDermott, 1981).

Several studies have asked students to predict the course of a moving body by applying the principles of basic Newtonian dynamics. Working with a small group of sixth-grade children in suburban Boston, diSessa (1982) studied the strategies children used to hit a target with a moving "dynaturtle"—a computer graphics entity controlled at a keyboard. The children's initial efforts were described by diSessa as the collision of fundamentally Aristotelian physics with a Newtonian reality. Similar work with both high school and college students revealed that about 60 percent were unable to describe (accurately) the trajectory of a projectile dropped from an airplane in flight (Fishbein, Stavy, & Ma-Naim, 1989; McCloskey, 1983), and a similar proportion were unsuccessful in predicting the path of a moving rocket after a midcourse correction (Clement, 1982). In a related problem of vector kinematics, Aguirre (1988) found that a large number of tenth-grade prephysics students had difficulty in predicting the path taken by a powerboat while traversing a flowing river.

The problem of curvilinear motion has also received substantial attention. When 50 undergraduate students at Johns Hopkins University were asked to sketch the trajectory of a moving ball emerging from a spiral tube, fully 36 percent of the drawings were of a curvilinear path (Caramazza et al., 1980). A similar set of predictions emerged when students were asked to describe the path of a pendulum bob cut from its string at the bottom of its arc (Caramazza et al., 1981).

The velocity of falling bodies under the influence of gravity and the relationship of position to relative weight of objects pose additional difficulties for many students. Gunstone and White (1981) confronted a group of first-year physics students at Monash University in Australia with a problem involving metal and plastic spheres of comparable size and asked them to predict the time each would take to fall to the floor when dropped from a given height. Nearly 25 percent of them believed the metal ball would fall faster. Similar results were found in a study by Watts (1982). In another task, a bucket and a weight—tied at opposite ends of a rope—were slung over a pulley and equilibrated at the same height above the floor. The experimenter then pulled the bucket down and asked students to predict what would happen when the bucket was released. Over 30 percent of the subjects thought the system would return to its initial state.

Results such as these seem to support the conclusion that "students do not merely lack knowledge (of mechanisms); they espouse 'laws of motion' that are at variance with formal physical laws" (Caramazza et al., 1980). Or, as expressed elsewhere, students have a "rich accumulation of interrelated ideas that constitute a personal system of commonsense beliefs about motion" (Champagne et al., 1980).

A large number of studies have examined students' understanding of electricity, especially related to the workings of simple circuits involving a battery, bulbs, and a set of wires (Bauman & Adams, 1990; Black & Solomon, 1987; Cohen, Eylon, & Ganiel, 1983; Fredette & Clement, 1981; Fredette & Lochhead, 1980; Heller, 1987; Johsua, 1984; Osborne, 1981, 1983; Peters, 1982; Shipstone, 1984). Taken together, the work of Osborne (1983) and Shipstone (1984) is especially enlightening, because the subjects of these studies span an age range from 9 to 18 years.

Five distinct models of a simple circuit were employed by these students. The "single-wire" notion suggests that current leaves the battery and travels through one wire to a bulb, which serves as a kind of electricity "sink." In the "clashing-currents" model, electricity leaves the battery from both terminals and travels toward the bulb, where it is "used up." In addition to these ideas, three kinds of "unidirectional models" were identified: (1) "unidirectional without conservation," in which current flows through a circuit, becoming gradually weaker as it encounters successive components (such as bulbs); (2) "unidirectional with sharing," wherein the current is distributed to and consumed equally by all components of the circuit, with all bulbs achieving the same brightness; and (3) "unidirectional with conservation," which is the scientifically acceptable view.

The single-wire model apparently predominates in younger children, more than 50 percent of whom set up such a "circuit" when asked to "get the bulb to glow." The clashing-current model seems to appeal to about 35 to 40 percent of the children in their middle years, and it is gradually replaced by "unidirectional" models—with the notion of conservation slowly becoming evident in about 10 percent of the subjects at about age 12 and rising to about 60 percent acceptance by age 18.

Students' ideas of heat and temperature have occupied the attention of a number of researchers (Albert, 1978; Andersson, 1986; Erickson, 1979, 1980; Erickson & Tiberghien, 1985; Hewson & Hamlyn, 1984; Johnstone, McDonald, & Webb, 1977; Nachmias, Stavy, & Avrams, 1990; Shayer & Wylam, 1981; Triplett, 1973). Whereas Shayer and Wylam (1981) attempted to map the development of heat and temperature concepts onto the classical Piagetian stages, other researchers sought to explore children's ideas on their own terms, without comparison

with any fixed external standard. We think Erickson's work (1979, 1980) is especially notable and insightful in this regard.

In a pair of studies conducted in British Columbia, Erickson interviewed 10 children (ages 10 to 12) and discovered a vast store of ideas about thermal phenomena. To illustrate: (1) "heat makes things rise," (2) "heat and cold are material substances that can be transferred from one thing to another," and (3) "heat accumulates in some areas and flows to other areas." Some of these ideas seemed to be unique constructions (i.e., the "children's viewpoint"); others were reminiscent of the old "caloric theory" of heat held by some eighteenth-century scientists. Still others represented reasonable approximations of the currently accepted "kinetic-molecular" explanation. In a follow-up investigation, Erickson administered a forced-choice instrument to more than 250 students in grades 5, 7, and 9. He found, through principal component factor analysis, that students responded consistently to the three viewpoints represented. Common views among the children were that "large ice cubes take longer to melt than small ones because they are colder" and that "wax melts when heated because it is soft."

Students' notions of light, vision, and optical phenomena have also received considerable attention (Andersson & Karqvist, 1983; Eaton, Anderson, & Smith, 1983, 1984; Feher, 1990; Feher & Rice, 1987, 1988; Goldberg & McDermott, 1986; Guesne, 1985; Jung, 1987; Mohapatra, 1988; Rice & Feher, 1987; Shapiro, 1989; Watts, 1985). Whereas early studies (Stead & Osborne, 1980) focused on issues such as relative distance traveled by light from sources of different sizes and at different times of day, subsequent work has addressed more fundamental problems: (1) How do people see? (2) What is a shadow? (3) What is a mirror image? and (4) Why is light necessary for life?

Perhaps the most persistent problem documented in this work is the relationship of light to vision and the significance of reflection, refraction, and absorption. It appears that the overwhelming majority of learners insist that light "brightens or illuminates" objects, which enables them to be seen, or that the eyes play an active role in "reaching out and grabbing" images (Eaton et al., 1983). The absence of a conceptual model of "light traveling in all directions from a source" apparently leads to faulty predictions about shadow and image formation and also results in scientifically unacceptable explanations of such phenomena as rainbows and the workings of optical instruments (Smith, 1987).

How learners view the nature and states of matter (especially their understanding of the particulate theory of matter) has held a central place in the alternative conceptions research tradition from its inception (Andersson, 1990; Ault, Novak, & Gowin, 1984; Gabel, Samuel, & Hunn, 1987; Hibbard & Novak, 1975; Novick & Nussbaum, 1978, 1981; Nussbaum, 1985; Scott, 1987; Stavy, 1988, 1989; Wandersee, 1983c). Two studies by Novick and Nussbaum (1978, 1981) are typical of the work in this area. Eighth-grade Israeli students were interviewed about the particulate nature of matter after a year of instruction in physical science. Responses to the interviews formed the basis of a paper-and-pencil instrument that was administered to more than 500 pupils in elementary, junior high, senior high, and college classrooms. Three aspects of the theory posed special problems for students: (1) that gas molecules are in constant motion (only about 20 percent of elementary and 50 percent of college students agreed), (2) that heating and cooling cause a change in particle motion (10 to 40 percent agreed), and (3) that empty space is found between particles (10 to 40 percent agreed).

The last proposition is apparently a universal stumbling block and a recurring problem revealed in studies of this issue. It seems that students "abhor nothingness" and are willing to fill it with the almost anything, including "pure air, other gases, and muck" (Scott, 1987). The notion that matter is continuous, rather than particulate, demonstrates a kind of "raisin cake" model of matter commonly held by students of virtually any age.

When young children are asked to describe their understanding of the concept of energy, the most frequent association they make is with living things, especially humans. Typically, these youngsters mention growth, physical fitness, exercise, and food. Energy seems to be an anthropomorphic notion that is heavily laden with images of movement or activity. By the early teenage years, these ideas seem to expand to include such nonliving entities as power stations, the sun, and fire (Bliss & Ogborn, 1985; Brook, 1987; Carr, Kirkwood, Newman, & Birdwhistel, 1987). Watts (1983) identified a number of alternative conceptions of energy used by secondary school pupils in addition to the human-centered and activity-based models. These conceptions consider energy as a "source of force" that is expendable or rechargeable, as an ingredient that is dormant and must be triggered, and as a fuel associated with processes that make life comfortable.

Solomon (1982, 1983, 1984) has suggested that learners develop two separate domains or knowledge systems for dealing with energy phenomena: a "physics" system, which is used in school to solve problems posed by teachers and textbooks, and a "life-world" system, to which even successful students retreat when confronted by novel or difficult situations. Apparently many students are aware of the differences between systems, and the ablest ones demonstrate greater facility at switching back and forth at appropriate times. Duit and Haussler (1992) have reviewed the research on learning and teaching energy and argue that exergy (the amount of useful energy in a system) is a more precise term to use in many real-world teaching situations than energy. As they observe, "this point of view has important consequences for drawing conclusions on how to introduce the energy concept in school science" (p. 10).

Studies of students' understandings of earth and space concepts have generated a wealth of information, especially on the earth as a cosmic body, the nature of gravitational attraction, and the relationship between astronomical objects or events and common everyday occurrences (Baxter, 1989; Finegold & Pundak, 1990; Jones, Lynch, & Reesink, 1987; Klein, 1982; Mali & Howe, 1979; Nussbaum, 1979; Nussbaum & Novak, 1976; Nussbaum & Sharodini-Dagan, 1983; Sneider & Pulos, 1983; Vosniadou, 1990). The seminal work in this area was done by Nussbaum and Novak (1976) at Cornell University. This research study was subsequently expanded by Nussbaum and others to include several cross-age and cross-cultural investigations involving populations in the United States, Israel, Nepal, and Nigeria.

The original study (Nussbaum & Novak, 1976) involved a set of interviews with 26 second-grade children who had received six audio-tutorial lessons on the earth as a cosmic body. Five "notions" of the earth emerged from these interviews: (1) the earth is flat; (2) the earth is round but limited in space by a "ceiling" or "floor," and "things" fall "down" relative to an outside observer; (3) the earth is round, bounded by limitless space, and on it things fall "down"; (4) the earth is round, bounded by limitless space and things fall toward the earth but not toward its center; and (5) the earth is round, bounded by limitless space, and things fall toward the earth's center.

In a follow-up study in Israel, Nussbaum (1979) interviewed students in grades 4 to 8 and discovered a number of interesting developmental trends. For example, notion 1 is most prevalent at grade 4 (some 25 percent of Israeli children subscribe to it) and it drops to under 5 percent by the time children reach grade 8. At the same grade levels, the frequency of notion 5 rises from under 5 percent to about 30 percent. Another study, done in San Francisco (Sneider & Pulos, 1983), obtained comparable results—with most children through age 10 holding notions 1, 2, and 3 and most older children (ages 13 and 14) preferring notion 4 or 5.

At the Harvard-Smithsonian Center for Astrophysics, Project STAR (Science Teaching through its Astronomical Roots) is worthy of note. This curriculum project, funded with seed money from the National Science Foundation, seeks not only to disabuse high school students of their astronomical misconceptions but also to use astronomy as a vehicle to teach basic scientific principles (Allis, 1989). To accomplish these goals, Philip M. Sadler and his colleagues have conducted extensive research on students' cognitive difficulties in understanding basic astronomical concepts and facts. Although the project is not yet complete, innovative curriculum modules and materials aimed at providing experiences that overcome misconceptions have already been produced.

Of special note is the award-winning 18-minute videotape created by Project STAR called "A Private Universe." This program reveals the persistent, fundamental misunderstandings that even Harvard graduates (still in their robes) harbor about the solar system in which they live. Nobel laureate (physics, 1988) Leon Lederman saw the video and commented that this is "a most dramatic example of the failure of even our most prestigious colleges to remove the misconceptions developed in childhood" (Pyramid Film & Video, 1988, p. 1).

Representative Findings—Biology

Research on students' biological concepts has expanded significantly (Mintzes, Trowbridge, Arnaudin, & Wandersee, 1991). A content analysis through mid-1986 listed more than 100 studies (Wandersee, Mintzes, & Arnaudin, 1989), and a recent count suggests that the number has doubled since that time. Studies in biology have been categorized into five topical areas: students' concepts of life, animals and plants, the human body, continuity (including reproduction, genetics, and evolution), and other biological phenomena (ranging from cells to food webs).

Work on children's understanding of the life concept spans some 60 years, from Piaget's (1929) formative efforts to more recent studies growing out of conceptual change theory (Carey, 1985) and the alternative conceptions research program (Brumby, 1982; Stepans, 1985; Tamir, Gal-Choppin, & Nussinovitz, 1981).

Piaget proposed a four-stage theory that chronicles the emergence of the life concept. Accordingly, children ascribe life to (1) objects exhibiting activity or usefulness (ages 3 to 7), (2) movement of any kind (ages 7 and 8), (3) spontaneous movement (ages 9 to 11), and finally (4) (at ages 11 and 12) to plants and animals. Subsequent studies produced a set of anomalous results that cast doubt on the stage theory, and efforts by Carey (1987) suggest that the acquisition of the life concept may involve a fundamental reorganization of relevant knowledge structures between the ages of 7 to 10—replacing naive psychological notions with an "intuitive biology."

The meanings that learners assign to the concepts of animal and plant have been reasonably well documented, with the former receiving considerably more attention than the latter (Bell, 1981a, 1981b; Bell & Barker, 1982; Jungwirth, 1971; Lazarowitz, 1981; Ryman, 1974; Stavy & Wax, 1989; Tema, 1989; Trowbridge & Mintzes, 1985, 1988; Wandersee, 1986b). It appears that learners of all ages subscribe to a highly restricted understanding of animals—applying the label almost exclusively to vertebrates, especially to the land mammals commonly found in the home, farm, or zoo. When asked, "What is an animal?", learners' typical responses are that animals "are alive," "have legs," "move," "have hair or fur," and "live outside or in the woods." Characteristics of plants are even more poorly understood by science students. Stavy and Wax (1989) found that only 57 percent of the 11- to 12-year old Israeli children in their study considered plants to be alive, only 66 percent thought that plants reproduce, and 88 percent thought that plants must eat.

Research on students' understanding of anatomical and physiological concepts reveals that even young children have well-developed ideas about the human body (Arnaudin & Mintzes, 1985, 1986; Gallert, 1962; Mintzes, 1984; Porter, 1974). Many children (as young as 10 years) depict a vast array of internal structures when asked to draw the inside of their body. Common structures depicted by children include the stomach, heart, brain, muscles, bones, lungs, kidneys, and veins. The location, size, and shape of the organs that children depict and the functions they assign to them are fairly consistent across studies. For example, the heart is often drawn in the shape of a "valentine," and many children assert that it cleans, filters, or manufactures the blood. A large proportion of students at all levels seem to believe that a separate system of tubes carries air to the heart and other body structures.

In addition to their notions about the human body, many learners have a store of prescientific knowledge about reproduction (Kreitler & Kreitler, 1966; Moore & Kendall, 1971), heredity (Engel-Clough & Wood-Robinson, 1985; Deadman & Kelly, 1978; Kargbo, Hobbs, & Erickson, 1980), and evolutionary change (Bishop & Anderson, 1990; Brumby, 1979, 1984; Hallden, 1988; Jungwirth, 1975). Young children seem to subscribe to a set of sequential ideas about origins. First they deny birth; then they acknowledge it but deny the parental role; subsequently, they accept the role of the mother and, much later, that of the father. When asked about the transmission of hereditary traits, children often ascribe some traits (e.g., eye and hair color) to the mother and others (e.g., height and weight) to the father. Many individuals of all ages subscribe to a type of pre-Mendelian blending of parental traits in offspring. The notion that hereditary traits change over time is often acknowledged, but the change is sometimes seen as a response to environmental variation—an idea that Jungwirth (1975) has called "inverted evolution" and "preconceived adaptation."

The structure of cells and cellular phenomena, including cell division, protein synthesis, respiration, and photosynthesis, have received growing attention (Anderson, Sheldon, & Dubay, 1990; Barker & Carr, 1989a, 1989b; Bell, 1985; Dreyfus & Jungwirth, 1988, 1989; Eisen & Stavy, 1988, 1989; Fisher, 1985; Roth, Smith, & Anderson, 1983; Stavy, Eisen, & Yaakobi, 1987; Stewart, Hafner, & Dale, 1990; Wandersee, 1983a, 1984, 1986a). The notion that green plants synthesize their own food intracellularly seems to pose an almost insurmountable problem for many students. When asked about plant nutrition, a large proportion of individuals, including those who have taken several previous biology courses, insist that plants obtain food from the soil.

Representative Findings—Chemistry

In addition to studies of students' conceptions of heat, temperature, and the particulate nature of matter, work in chemistry has addressed notions of covalent bonding, electrochemistry, transformation of matter, chemical reactions, chemical equilibrium, stoichiometry, and molar measurement (Andersson, 1986, 1990; Cachapuz & Martins, 1987; Cros, Chastrette, & Fayol, 1988; Garnett, Garnett, & Treagust, 1990; Gorodetsky & Gussarsky, 1986; Gussarsky & Gorodetsky, 1990; Hackling & Garnett, 1985; Novick & Menis, 1976; Peterson & Treagust, 1989; Schmidt, 1987).

Using a "two-tier" multiple-choice instrument in which students were asked about the nature of covalent bonds and required to justify their responses, Peterson and Treagust (1989) identified a number of potentially important conceptual problems among grade 12 chemistry students in Australia. About one-fourth of the students subscribed to the notions that equal pairing of electrons occurs in all covalent bonds and that bond polarity determines the shape of a molecule. More than one-third were confused about the relationship between relative electronegativity of atoms and the tendency to form polar molecules.

But many conceptual problems concerning atoms and molecules are considerably more basic than these. In a study of 12- to 16-year-olds in Sweden, Andersson (1990) found that many pupils thought that atoms vary in shape (e.g., square, rectangular). He also discovered that learners often transfer macroproperties of elements to the micro world—suggesting, for example, that phosphorus atoms are yellow, copper atoms are maleable, naphthalene molecules smell, and water molecules are made of small drops.

Andersson (1990) found, too, that students invent chemical systems to avoid explaining phenomena they don't understand. For example, particles are often visualized as tightly packed units with no space between them, in order to avoid the difficult issue of "nothingness."

Even when concepts of atoms and molecules are seemingly well understood, the nature of the interactions between them poses further problems. Hackling and Garnett (1985) found a substantial number of high school chemistry students who had fundamental misunderstandings about the nature of chemical equilibrium. Many seemed to think equilibrium is reached when the concentrations of reactants and products are equal. Often students do not recognize the dynamic nature of systems at equilibrium, sometimes failing to conceive of a mixture as a single entity and attempting to "balance" the right and left sides of an equation independently (Gussarsky & Gorodetsky, 1990).

Australian chemistry educator Peter Fensham (1992) reviewed chemistry teaching responses to the alternative conceptions literature related to chemical substances, structure, and reactions. He found evidence to support Millar's (1990) contention that kinetic-molecular theory is so difficult to teach at the beginning of an introductory chemistry course because it is of virtually no use to the learners then. Fensham also questions the conventional wisdom that the states of matter ought to be taught early in a chemistry course to set the stage for kinetic-molecular theory, noting that many alternative conceptions research studies show learners have great difficulty understanding the particulate nature of matter—especially with respect to gases. He advises that when students have many more experiences of chemical phenomena and are familiar with "chemical descriptions on both the continuous and discontinuous levels, they may be in a much better position to construct into their own thinking the powerful explanatory concepts that atomic structure and kinetic-molecular theory provide" (Fensham, 1992, p. 24).

Generalizations Across Disciplines

Claim 2: THE ALTERNATIVE CONCEPTIONS THAT LEARNERS BRING TO FORMAL SCIENCE INSTRUCTION CUT ACROSS AGE, ABILITY, GENDER, AND CULTURAL BOUNDARIES.

A claim frequently encountered in the alternative conceptions literature is that students' notions of natural phenomena are robust with respect to such factors as age, ability, gender, and culture. For example, Champagne, Gunstone, and Klopfer (1983) claim that "the naive descriptive and explanatory systems show remarkable consistency across diverse populations, irrespective of age, ability, or nationality" (p. 173). Ogunniyi (1987) reports a variety of traditional and scientific explanations of cosmological phenomena among literate and nonliterate Nigerians "regardless of level of education, sex, age, religion, locality or tribe" (p. 113). Peters (1982) suggests that "even honors students have conceptual difficulties with physics" (p. 501).

What evidence do we have to support these claims? Of the four factors mentioned, the effect of age has received the most

attention by researchers, although it has usually been secondary to the issue of identifying and describing students' notions. The age factor itself is problematic, however, and is often confounded with other variables, notably the effects of varying levels and quality of instruction. Nonetheless, an examination of studies that include subjects across a broad range of age groups suggests, not surprisingly, that alternative conceptions are characteristic of students at all age levels. However, the frequencies of these ideas vary considerably by domain of knowledge and by level and quality of prior instruction (Arnaudin & Mintzes, 1985; Novick & Nussbaum, 1978, 1981; Trowbridge & Mintzes, 1988; Wandersee, 1983a).

In Wandersee's (1983a) cross-age study of photosynthesis involving some 1400 subjects throughout the United States in grades 5, 8, 11, and 14, he asked, "Where does most of the food of plants come from?" More than 60 percent of the elementary and secondary school students and about half of the college students said "the soil." Arnaudin and Mintzes (1985) found more than 66 percent of fifth-graders and about 50 percent of college students subscribe to an "open" cardiovascular system with blood leaving the vessels and circulating into or around the cells. Apparently such a conception is even common in first-year medical students (Patel, Kaufman, & Magder, 1991). Novick and Nussbaum (1981) reported the persistence of the "continuous" model of matter, which characterized some 90 percent of elementary school children and about 60 percent of university students.

It is clear that the frequencies of some alternative conceptions change little over time, while others change dramatically. It also seems that some domains harbor more intransigent alternative conceptions than others (Engel-Clough & Driver, 1986; Engel-Clough, Driver, & Wood-Robinson, 1987). In seeking explanations for these phenomena, we must continually remind ourselves that the explanatory tools of meaningful learning theory, conceptual-change theory, and alternative-conceptions research are relatively new (Carey, 1986; Novak, 1987; Posner, Strike, Hewson, & Gertzog, 1982).

Issues of ability and gender have received surprisingly little attention in well-controlled studies. Peters (1982) reported the frequencies of errors made by students enrolled in an honors-level introductory physics course at the University of Washington. Many of these students (who ranked in the top 3 percent of their classes) displayed a wide array of conceptual difficulties in working problems in kinematics, dynamics, electricity, and magnetism. In a study comparing "gifted" (mean IQ = 146) and "non-gifted" (mean IQ = 116) high school physics students, Placek (1987) found, surprisingly, no significant differences between those groups in the ability to solve problems involving the motions of ordinary objects encountered in everyday life. Indeed, both groups had common misconceptions about moving objects.

In an era in which gender issues are becoming increasingly important, few well-designed studies have examined the incidence of alternative conceptions among males and females. Where differences have been reported, our impression is that they generally suggest that males have fewer alternative conceptions than females, but we realize that this may be a subtle result of study focus and design. Examples that seem to support the generalization include studies on projectile motion (Maloney, 1988), astronomy (Jones et al., 1987; Lightman, Miller, & Leadbeater, 1987), biological classification (Lazarowitz, 1981; Ryman, 1977), natural selection (Jimenez Aleixandre & Fernandez Perez, 1987), geometrical optics (Bouwens, 1987), and pressure, weight, and gravity (Mayer, 1987).

Perhaps because of the international nature of the alternative-conceptions research effort, cross-cultural studies are not uncommon. Studies that compare populations across advanced Western countries seem to find few differences of statistical significance. An example is found in the work of Viennot (1979), who compared French, British, and Belgian university students on intuitive notions in basic dynamics. In contrast, studies that search for differences in substantially divergent cultures often find an "overlay" of traditional views that are quite distinct from explanations offered by contemporary science. Examples include Nepali notions of the earth and gravity (Mali & Howe, 1979), Nigerian ideas about cosmology (Ogunniyi, 1987) and growth (Ross & Sutton, 1982), views about air and air pressure held by primary school teachers in Swaziland (Rollnick & Rutherford, 1990), conceptions of force and movement among secondary and university students in Zimbabwe (Thijs, 1987), and understandings of velocity in Japanese and Thai children (Mori, Kojima, & Tadang, 1976). We think it is important to note that these "traditional" ideas often constitute a second layer of explanations that may contribute to a mosaic of alternative conceptions that includes so-called intuitive conceptions found in Western countries. This population mosaic, in turn, may reflect the individual conceptual-propositional mosaics.

Claim 3: ALTERNATIVE CONCEPTIONS ARE TENACIOUS AND RESISTANT TO EXTINCTION BY CONVENTIONAL TEACHING STRATEGIES.

Perhaps the most frequently quoted aphorism in the alternative conceptions literature is Ausubel, Novak, and Hanesian's (1978) warning, "These preconceptions are amazingly tenacious and resistant to extinction" (pp. 372–373). Although resistance to change has been noted in a variety of knowledge domains, it is interesting to note, and of potential significance, that the phenomenon has been reported more often in the physical than in the life sciences and within the physical sciences is most often identified with counterintuitive phenomena such as those involving moving bodies. Much has been made of the "resistance" issue, but we think great care must be taken to differentiate between resistance in the face of sustained, high-quality, conventional instruction involving able, highly motivated, well-prepared students and simple failure to learn over a short period of time under substandard conditions. We believe this distinction has profound implications for those committed to the development of intervention strategies.

A number of studies satisfy the criteria of high-quality instruction and well-prepared students. DiSessa (1982) reported a case study of a freshman student at Massachusetts Institute of Technology (MIT) who "after a year of high school physics and essentially all the Newtonian mechanics in the freshman curriculum" still encounters many of the learning [problems] seen in sixth-grade children attempting the "dynaturtle" task mentioned earlier in this chapter.

After a physics course for preengineering students at the University of Massachusetts, a full 75 percent of those tested were unable to predict the trajectory of a rocket after a midcourse engine burn, compared with 88 percent who failed this question at the beginning of the course (Clement, 1982). Somewhat comparable results were reported by Gunstone (1987), who surveyed more than 5000 high school physics students on an end-of-course examination in Australia.

Trowbridge and McDermott (1981) found that approximately 66 percent of the students enrolled in a calculus-based physics course at the University of Washington were unable, on a postcourse assessment, to apply adequately the concept of acceleration. Similarly, you may recall that Caramazza et al., (1980) reported that more than a third of the freshmen admitted to Johns Hopkins University who had taken a year of high school physics could not accurately predict the path of a ball emerging from a spiral tube.

In contrast, Bell and Barker (1982) reported a successful attempt to modify students' ideas about the concept of animal by using small-group methods and multiple examples and nonexamples, an approach that might arguably be called "conventional," at least by comparison with recently proposed strategies. One group of 14-year-olds was reluctant, on a pretest, to apply the label animal to such organisms as a worm, a moth, and a spider (only about 25 percent were successful). After instruction, virtually all of the subjects included such invertebrates within the animal domain.

We think the message to teachers and would-be instructional developers seems clear: Not all alternative conceptions are tenacious. It is important to differentiate between the concepts that might require high-powered conceptual change strategies and those that are equally likely to yield to well-planned, conventional methods. Accordingly, we think a working knowledge of discipline-specific alternative conceptions research findings might well be considered basic to the professional preparation of master science teachers.

Claim 4: ALTERNATIVE CONCEPTIONS OFTEN PARALLEL EXPLANATIONS OF NATURAL PHENOMENA OFFERED BY PREVIOUS GENERATIONS OF SCIENTISTS AND PHILOSOPHERS.

This theme appears repeatedly in the alternative conceptions literature, spanning a continuum of notions from those asserting that the conceptual development of children recapitulates significant ideas in the history of science to those that merely point to similarities in these domains. Piaget (1970) based his work squarely on this idea: "The fundamental hypothesis of genetic epistemology is that there is a parallelism between progress made in the logical and rational organization of knowledge and the corresponding psychological processes." In their study of students' understanding of classical mechanics, Champagne et al. (1980) concluded that "the rules of the belief system [employed by college physics students] parallel the descriptive aspects of Aristotelian physics."

In contrast, there have been others who deny such a proposition (McClelland, 1984): It is tempting to draw parallels between descriptions of phenomena in so-called primitive societies and in ancient Greek schools with those given by young children, and to suppose that the historical development of ideas is reproduced by the individual. I believe this to be both a snare and a delusion ... it may encourage attempts to teach ideas in an historical sequence, when the evidence upon which present day ideas are elaborated is freely available. (p. 3)

Nevertheless, substantial effort has gone into an analysis of potential parallels (Clement, 1983; Fishbein et al., 1989; Furio Mas, Hernandez-Perez, & Harris, 1987; Lythcott, 1985; Matthews, 1987; McCloskey, 1983; Nussbaum, 1983; Preece, 1984; Shanon, 1976; Wandersee, 1986a; Wiser & Carey, 1983). While several studies have examined parallel historical concepts in biology and chemistry, the overwhelming majority have been within the domain of physics, especially the area of kinematics (Hashweh, 1988).

Speculation about the forces acting on moving bodies predates the ancient Greeks, but the Aristotelian (circa 350 B.C.) and medieval impetus (circa A.D. 1320) theories held sway over much Western thought until the time of Newton. Aristotle held that objects move only when they are under the influence of a force and then only while in direct contact with it. To account for the movement of projectiles that are not in direct contact with any observable "mover," Aristotle suggested that the medium itself (air or a water current moving rapidly backward) pushes the objects forward. One exception to this explanation is the movement inherent in falling bodies, which occurs in the absence of any force and is said to be the product of an object seeking its "natural place." Objects are considered to be at rest when they reach their natural place, and the weight of an object is related to the rapidity with which it reaches its natural place. Heavy things fall faster than light ones, and gases rise because they lack mass.

Finding the Aristotelian explanation of movement unsatisfactory for projectiles and falling bodies, several medieval thinkers, notably Franciscus de Marchia, Jean Buridan, and Nicole Oresme, proposed the concept of "impetus." Impetus was conceived as an entity impressed on a moving body by a "mover" that causes the body to continue its motion in the same direction until the impetus dissipates (according to Marchia) or the object is acted on by another force (according to Buridan and Oresme).

In contrast to these notions, Isaac Newton demonstrated that objects at rest remain at rest and those in motion remain in motion at constant velocity unless acted on by an external force. When acted on by such a force, the object accelerates or decelerates

To what extent are these notions reflected in the ideas learners bring to the classroom? It is clear from what has been said previously (see the exposition of Claim 1) that students subscribe to a variety of notions about moving bodies. First, many believe that when a force is applied to an object, it causes motion. This is consistent with both Aristotelian and impetus theories. Second, under the influence of constant force, objects move with constant velocity, and the magnitude of the velocity is proportional to the magnitude of the force. Here, historian

A. R. Hall (1954) comments that "though Aristotle never explicitly formulates the proposition that the application of a constant force gives a body a constant velocity, it is implied in the whole of pre-Galilean mechanics." Third, the notion that heavy things fall faster than light ones is certainly characteristic of many students' alternative conceptions and is a prominent feature in Aristotle's "sublunary" realm. Finally, students typically believe that, in the absence of a force, objects are either at rest or slowing down. This is clearly inconsistent with the Aristotelian explanation of falling bodies but is encompassed by impetus theory.

It has also been suggested that biology students' views about organic change in living organisms imply a kind of Lamarckian view of evolution. Jean Baptiste Pierre Antoine de Lamarck (1774–1829) provided a comprehensive explanation of evolution in his *Philosophie Zoologique* (1809), which rested largely on three basic principles: those of "needs," "use and disuse," and "inheritance of acquired characteristics." How similar are these principles to those held by students?

The first of these (Lamarck, 1809) suggests that "great alterations in the environment of animals lead to great alterations in their needs" and "if the new needs become permanent, the animals then adopt new habits which last as long as the needs that evoked them." Bishop and Anderson (1990) and Brumby (1984) have cited several examples of this kind of thinking in college biology students. In one problem, students asserted that "cheetahs needed to run fast for food, so nature allowed them to develop faster running skills." In another, it was suggested that dark skin develops in Africans because they need to be protected from intense solar radiation.

The second of Lamarck's (1809) principles holds, that "a more frequent and sustained use of an organ, gradually strengthens that organ, develops it, enlarges it ... while the constant lack of use of such an organ, insensibly weakens it, worsens it ... and ends by causing it to disappear." Again, Bishop and Anderson (1990) provide an example of this in college students who claim that blind cave bats develop small nonfunctional eyes through nonuse of these organs. Many students confuse hypertrophy of muscle groups in athletes as an instance of this principle.

The third principle (Lamarck, 1809) asserts, "All that nature has caused individuals to acquire or lose by the influence of the environment ... is preserved by reproduction for the new individuals which arise therefrom, provided that the modifications acquired are common to both sexes." The notion of inheritance of acquired characteristics seems to have a great deal of intuitive appeal, especially in the context of a problem that extends over a lengthy period of time (Engel-Clough & Wood-Robinson, 1985). In one problem, 12- to 16-year-olds were presented with a situation involving a married couple "who trained hard to become good athletes." The students were asked, "If the children of this family practiced hard over several generations— would you automatically get fast runners in about 200 years?" Forty-eight percent agreed. Some 44 percent also thought that baby mice whose ancestors had their tails removed for several generations would be born with shorter tails.

Parallels between the historical development of chemical concepts and students' alternative conceptions have not been studied as extensively as parallels in physics and biology (Furio Mas, Hernandez-Perez, & Harris, 1987; Gil-Perez & Carrascosa, 1987). Nonetheless, several studies have reported ideas about the material nature of gases that appear consistent with pre–seventeenth-century thinking, especially that of Aristotle (Stavy, 1988). Aristotle recognized just four elements (air, fire, earth, and water) and held that only the latter two were material, in the sense that they possessed mass and occupied space. For Aristotle and subsequent natural philosophers, there was no clear distinction between "air" (gases) and spirit, mist, vapor, or breath. All these entities were thought to be condensation products of a substance that was alternately labeled "pneuma" or "spiritus" (Taylor, 1963).

Studies in Spain and Israel with students ranging in age from 9 to 18 suggest that many pupils believe gases lack both substance and mass and have a natural tendency to rise. In one set of questions, Spanish pupils were presented with an experiment involving the total vaporization of a liquid in a hermetically sealed container and were asked about conservation of substance, mass, weight, and the tendency of the products to rise (Furio Mas et al., 1987). More than 80 percent of the 12- to 13-year-olds and about 50 percent of the 17- to 18-year-olds were found to be nonconservers. Similarly, nearly 90 percent of the younger students and 60 percent of the older ones reported that gas is "something that rises."

Israeli students were shown a glass of soda water and asked to predict the relative weight of the glass before and after the escape of carbon dioxide (Stavy, 1988). More than 50 percent of the younger children (grades 4–6) claimed that the glass would maintain the same weight ("air has no weight"). This response dropped to about 20 percent in the older students (grades 8 and 9). Interestingly, a substantial number of younger children (over 20 percent in grade 5) suggested that the glass would actually gain weight as the gas escaped ("air adds lightness to water"). Clearly, these ideas bear some resemblance to Aristotle's understanding of the nonmaterial nature of gases.

At one of the plenary sessions of the first International Seminar on Misconceptions in Science and Mathematics held at Cornell University in 1983, J. Nussbaum and J. Wandersee (see Helm & Novak, 1983, p. 3) requested that probing the uses of the history of science for understanding students' misconceptions be placed on the alternative conceptions community's research agenda. In his historically based alternative conception study of photosynthesis, Wandersee (1986a) concluded that the history of science can serve as a valuable heuristic device for science teaching by suggesting (to both students and teachers) misconceptions about a particular science topic that may still be present today (albeit in modified form). Although he found no substantial evidence that students recapitulate the history of science in learning the photosynthesis concept, Wandersee proposed that the history of science could be used to encourage and assist the science student in discovering his or her own conceptual weaknesses.

Claim 5: ALTERNATIVE CONCEPTIONS HAVE THEIR ORIGINS IN A DIVERSE SET OF PERSONAL EXPERIENCES INCLUDING DIRECT OBSERVATION AND PERCEPTION, PEER CULTURE, AND LANGUAGE, AS WELL AS IN TEACHERS' EXPLANATIONS AND INSTRUCTIONAL MATERIALS.

An overview of the origins of alternative conceptions must, by its nature, remain speculative. The evidence for origins is often inferential at best, and certainly such origins are difficult to document. This is especially true of the alternative conceptions derived from direct observation and perception, where the primary data are often statements of self-report by the subjects involved. Nonetheless, a number of studies have yielded useful insights.

Several workers have suggested that many intuitive notions in simple mechanics are products of repeated interactions between children and physical objects or events in their immediate environment. "The students' macroschema for motion derives from years of experience with moving objects and serves the students satisfactorily in describing the world" (Champagne et al., 1983, p. 173). On the other hand, McClelland (1984) says that "for a young child, interactions with a mother, parents, siblings, other adults and peers will be the most salient events and it is in these areas that we might expect the earliest and most elaborated theories . . . . Physical phenomena may be expected to be much less salient (p. 3).

These two assertions are not mutually exclusive, but they focus on a somewhat controversial issue: the origins of children's alternative conceptions in the physical sciences as compared with the biological sciences (Bloom & Borstad, 1990; Lawson, 1988, 1991a, 1991b; Lythcott & Duschl, 1990; Mintzes, 1989). If Champagne et al. (1983) are right, one would expect to find young children arriving at school with established notions about moving bodies. McClelland's position, in contrast, would seem better in predicting the child's early speculation on human behavior (Carey, 1987) and possibly the nature of such basic biological functions as ingestion, digestion, respiration, and excretion. Perhaps it is the relative proportion of influence (direct, physical experience versus socialization) that is the key to understanding such findings.

It appears that alternative conceptions about moving bodies are among the earliest and most persistent of children's science problems (Clement, 1983; diSessa, 1982) and that many children bring to their first school experiences well-entrenched ideas about gravity and velocity that develop in a real world in which factors such as friction and air resistance are important. On the other hand, there is good evidence that young children also have firmly established ideas about bodily functions (Mintzes, 1984).

Perhaps these findings are not surprising when one considers the constructivist assertion that learners build meanings around the objects and events with which they come into immediate and direct contact (Driver & Bell, 1986; Osborne & Wittrock, 1985; Pope & Gilbert, 1983). In this light, one would predict early ideas about simple phenomena in both the physical and biological domains, but would no more expect early theorizing about cellular respiration or mitotic cell division than about quantum mechanics or relativity. Accordingly, at present the assertion that alternative conceptions in the physical and biological sciences derive from different sources (Lawson, 1988) does not appear to be well founded and needs to be well documented if it is to stand.

The role of peer culture and language in the development of scientific understandings has received considerable attention (Adeniyi, 1985; Bell & Freyberg, 1985; Duit, 1981, 1985;

Hawkins & Pea, 1987; Hewson, 1985; Hewson & Hamlyn, 1984; Kenealy, 1987; Lin, 1983; Mayer, 1987; Solomon, 1983, 1984, 1987; Solomon, Black, Oldham, & Stuart, 1985; Stavy & Wax, 1989; Stenhouse, 1986; Watts & Gilbert, 1983). The evidence strongly suggests that the meanings learners construct in their everyday usage of "technical" terms have a significant impact on their understanding of those terms in a scientific context.

Solomon's work demonstrates that learners often utilize two separate knowledge systems that share common elements of terminology. In order to function efficiently in the world of science and in the everyday "life world," students learn prompts and cues that help them discriminate between the two systems. It appears that better students are more aware of these two domains and are more adept in switching at appropriate times. In one study of 14- to 15-year-olds (Solomon, 1984), it was shown that students who are aware of the multiple meanings of the concept energy are better able to provide an acceptable definition of it on a written examination.

Commenting on students' multiple uses of the word "energy," Duit (1981) suggested that the principle of conservation was not a part of its meaning in common usage. "Quite the reverse; 'energy' in everyday language is a quantity which can be produced and consumed, but not conserved." To demonstrate his point, he presented the results of a conservation-of-energy task in which students were asked to predict the height reached by a ball on its return trip along a parabolic surface. More than half of the students were unsuccessful in this basic exercise.

In a revealing study of the role of peer culture and language in conceptual development in science, Hewson and Hamlyn (1984) interviewed both children (mean age = 15) and adults of the Sotho tribe inhabiting the interior plateau of southern Africa on their understanding of heat and temperature phenomena. The investigators discovered that the culture embraces a "powerful metaphor concerning heat [which] pervades many aspects of life [such that] heat is bad and coolness is good." As a result, the designation "hot" is given to individuals who are exhausted, impatient, angry, sick or bereaved and to women who are pregnant or menstruating. Furthermore, physical explanations of heat generally do not include the "caloric" notion that is often a part of the explanatory framework used in Western cultures.

There is now substantial evidence that textbooks and teachers inadvertently provide another avenue for the introduction of alternative conceptions (Andersson, 1990; Barras, 1984; Bauman & Adams, 1990; Cho, Kahle, & Nordland, 1985; Garnett et al., 1990; Heller, 1987; Veiga, Costa Pereira, & Maskill, 1989; Wandersee, 1984).

Cho et al. (1985) examined the three most commonly used American high school biology textbooks as potential sources of misconceptions in genetics. Among the problems they identified were those related to content sequencing, establishment of conceptual relationships, use of terminology, and introduction of mathematical elements. All three books presented the topics of meiotic cell division and genetics in separate chapters. None of the books clearly related the central concepts of chromosomal separation and DNA replication. The terms "gene" and "allele" were used interchangeably, and the Punnett square was treated as a graphic algorithm for solving genetics problems

without relating it to segregation and independent assortment of chromosomes.

Barras (1984) identified 15 misconceptions in the written work of students that he claims are frequently perpetuated by teachers and textbooks of biology. Among these errors are the esoteric (neurons of *Hydra* are located in the mesoglea) and the seemingly trivial (the parts of a tapeworm are the head and the tape), as well as some that appear to be serious conceptual problems (green plants photosynthesize during the day and respire at night; plant cells have a cell wall but no cell membrane).

Andersson (1990) has pointed out that diagrams and pictures encountered in textbooks are another potential source of conceptual difficulty. For example, a Swedish chemistry book depicts (at the molecular levels) a flask containing potassium and chlorine gas which suggests that the distance between particles in a solid and in a gas is the same. Another picture shows a crystal of salt with five to six atomic diameters between particles, suggesting that "a salt is like a gas with molecules in perfect order."

Teachers (unfortunately) serve as another major source of alternative conceptions. In some cases, the problem apparently resides in conceptual errors or misinformation held by the teachers themselves (Claim 6); in others, the language that teachers employ in the classroom results in a set of unintended learning outcomes (Claim 7).

Claim 6: TEACHERS OFTEN SUBSCRIBE TO THE SAME ALTERNATIVE CONCEPTIONS AS THEIR STUDENTS.

The assertion that teachers hold a substantial array of alternative conceptions in the domain of natural science should not be particularly surprising, especially to teachers themselves, and is surely not news to those engaged in teacher-education programs. In the defense of teachers, the persistence of their alternative conceptions may be an effect of poor college science textbook writing or poorly taught college science courses. Nonetheless, this finding may be profoundly disturbing to some and has significant implications for selecting, educating, and employing the next generation of science teachers.

The number of studies focusing on teachers' understandings of scientific concepts has grown significantly during the past decade (Ameh, 1987; Ameh & Gunstone, 1985; Bloom, 1989; Enochs & Gabel, 1984; Feher & Rice, 1987; Heller, 1987; Kruger, 1990; Lawrenz, 1986; Mohapatra & Bhattacharyya, 1989; Ogunniyi & Pella, 1980; Rollnick & Rutherford, 1990; Smith, 1987; Veiga et al., 1989). Of the studies conducted to date, a large proportion have examined the views of elementary school teachers and those enrolled in elementary teacher education programs. This focus probably reflects a concern among science educators about the adequacy of the preparation these teachers receive and their consequent abilities to assist young children in understanding complicated scientific concepts.

At an inservice workshop in physical science, Lawrenz (1986) administered 31 multiple-choice items taken from the National Assessment for Educational Progress (NAEP) physical science test to more than 300 elementary school teachers in Arizona. The mean score, 19 out of 31, was comparable to the 48th percentile for all 17-year-olds taking the examination. One question on conservation of mass posed the following problem: "Iron combines with oxygen to form rust. One should therefore find that rust weighs (the same, more, less) than the iron it came from." Only 36 percent of teachers (about 3 percent more than we would expect due to chance) selected the correct answer (weighs more). Forty-six percent thought the iron would lose weight.

Reminiscent of earlier studies on students' understanding of basic Newtonian mechanics, Mohapatra and Bhattacharyya (1989) asked elementary and secondary school teachers in India about the forces acting on two moving bodies, A and B. Bodies A and B have equal mass, and A is moving on a frictionless surface with a uniform velocity of $2V$, while B is moving with uniform velocity V. The questions were: (1) Is there any force acting on A and B in the horizontal direction? and (2) if yes, is the horizontal force acting on A greater than that acting on B? Over 75 percent of the teachers believed that a force is acting in the direction of movement, and more than half thought the magnitude of the force is proportional to the velocity of the moving object.

Several studies of teachers' understandings of light and optical phenomena reveal the essential similarity of their alternative conceptions and those of their students (Feher & Rice, 1987). Smith (1987) interviewed 10 elementary school teachers before and after a 4-week summer program on light, shadows, and other phenomena. The teachers had an opportunity to read about children's alternative conceptions, to explore their own ideas on the topics, to perform a variety of relevant experiments, and to interview children. At the beginning of the institute, the teachers "provided evidence of conceptual flaws similar to children's misconceptions," holding a model of light as something that "illuminates" objects and providing examples (i.e., sun, bulb) much as children do. These teachers lacked the general conceptual model of light traveling in all directions from a source and, as a result, were unable to provide satisfactory explanations of shadows, colors, and refraction. By the end of the program, most of the teachers had begun to replace their alternative conceptions, but they "had a difficult time giving them up, and used their memories of activities and discussions to support their *prior* [emphasis added] conceptions."

Heller (1987) reported similar conclusions in a study of elementary and middle school teachers' conceptions of electric circuits. In a written test, the teachers were asked to predict what would happen if a change were made in the circuit and to compare the brightness of two bulbs in the same and different circuits. The majority (10 out of 14) subscribed to a "sequential model" of electric circuits: a fixed current flows out of a battery until it reaches a bulb; bulbs use up current; the brightness of a bulb depends on the amount of current reaching it; when a circuit has more than one bulb, each bulb reduces the available current so that all bulbs receive less. Heller suggests that this model is virtually identical to one used by many children and, furthermore, that the misconception "energy flows from the battery through empty conductors" is reinforced in most elementary and middle school science books, the principal source of basic science information for many teachers.

In a departure from these studies, Linder and Erickson (1989) studied the understanding of sound and auditory phe-

nomena by 10 postgraduate secondary teacher-education candidates who held bachelor degrees in physics from a Canadian university. The students were interviewed about their interpretations of sounds emanating from a bursting balloon, a tuning fork, and a hollow glass tube. They were also asked to explain sound using schematic, graphical representations and to describe factors affecting the speed of sound in air. These would-be secondary school physics teachers suggested that sound is an entity that is "carried or transferred by individual molecules through a medium" and that sound is a "bounded substance with impetus in the form of flowing air."

Claim 7: LEARNERS' PRIOR KNOWLEDGE INTERACTS WITH KNOWLEDGE PRESENTED IN FORMAL INSTRUCTION, RESULTING IN A DIVERSE SET OF UNINTENDED LEARNING OUTCOMES.

As if the problem of teachers' alternative conceptions weren't enough, the difficulty is compounded when the teacher's knowledge interacts with the prescientific knowledge students bring to the classroom. Gilbert, Osborne, and Fensham (1982) and Osborne, Bell, and Gilbert (1983) have presented convincing evidence for a variety of unintended instructional outcomes that they call the "consequences of children's science." These consequences are products of an interaction between two epistemologically and conceptually divergent "world views"—those of the teacher and those of the student. In reference to these two studies, transcripts from dozens of interviews with children (ages 10–17) in New Zealand suggest five broad categories of unanticipated instructional outcomes that occur frequently but often go unrecognized by teachers.

The "Undisturbed Children's Science Outcome" is apparently typical of much instruction in basic Newtonian mechanics in which the learner's views remain substantially unchanged by the interaction. Evidence for this "tenacity" phenomenon is overwhelming, possibly the most frequently reported finding in the field (Claim 3).

In the "Two Perspectives Outcome" the learner acquires one conceptual framework that is readily applied to school science but retains another set of explanations for real-world phenomena. The cognitive outcome is characterized by lack of integration of concepts and is typical of much rote mode learning in school science (Ausubel et al., 1978). Examples include: $F = MA$ (for use in school) versus "motion implies force" (for use in the real world) (Clement, 1983); natural selection versus creationism (Bloom, 1989); and the kinetic-molecular theory versus the caloric theory of heat (Erickson, 1979, 1980).

Learners often misinterpret knowledge presented by their science teachers and use it instead to support their own preconceptions—a result referred to as the "Reinforced Outcome." Osborne (1983) reported an attempt to teach electric circuits to elementary school children using an analogy of the human circulatory system; the wires were said to be analogous to blood vessels, the electric current to blood flow, the battery to the heart and central body, the glowing bulb to a finger gaining heat, and battery depletion to aging of the heart. Apparently, for one male student, the teacher's analogy reinforced his preconception that positive current flows to a bulb, where it is changed to a negative current. The child explained, "I think it [the analogy] is right because blood changes also ... when it gets oxygen it turns red ... so it is different kinds ... different currents."

The "Mixed Outcome" is a hybrid product that results from an attempt to reconcile two seemingly incompatible views. Biology students learning that blood contains cells must reconcile this school-based knowledge with their perceptual knowledge that blood is a red liquid. As a result, they construct a variety of idiosyncratic hybrids in which blood cells float in a red liquid. When asked to identify the liquid, it is frequently labeled "blood" (Arnaudin & Mintzes, 1985, 1986). This is similar to the problem of learners who "fill" the space between gas particles with air (Novick & Nussbaum, 1978, 1981).

In a hypothetical "Unified Scientific Outcome," both the teacher's understanding and that of the learner are modified and extended, producing a "coherent scientific perspective." Although it is epistemologically enticing, to our knowledge evidence for this type of outcome is meager. However, with efforts under way throughout the world to develop constructivist science teaching strategies, we anticipate advances in science education here.

In addition to the work of the New Zealand group, several other researchers have attempted to document the unanticipated outcomes resulting from classroom interactions of teachers and students (Happs, 1985; Eaton et al., 1983, 1984; Smith & Anderson, 1984). The Michigan State group (Eaton et al., 1984) investigated children's interpretations of light after conventional instruction on the topic.

In one study (Eaton et al., 1984), two veteran teachers were observed as they taught a 13-lesson unit on light to several classes of fifth-grade children. One lesson focused on the speed of light and assumed the children understood that we see objects that reflect light to our eyes. The intended outcome of the lesson was learning that "light brightens things so we can see them." Postinstruction interviews revealed that the children's interpretation ("light travels quickly to the things it shines on") was very different from that anticipated by these experienced teachers.

In a companion study (Smith & Anderson, 1984), Ms. Howe, another experienced fifth-grade teacher, engaged her students in growing grass under dark and under illuminated conditions. The plants growing in the dark were yellow and spindly; those grown in light were green and tall. The children concluded that "plants grown in the dark are not healthy"—a reasonable conclusion if one begins with the assumption that plants get their food from the soil and the light in order to stay healthy. Ms. Howe, of course, began with the assumption that plants need light to make food and those grown in the dark are starving to death. For Ms. Howe, this lesson was the culmination of a 5-week unit on plant nutrition and the students' unintended conclusions constituted a "surprising, if not shocking" outcome.

In another study of unanticipated learning outcomes, Happs (1985) interviewed 14-year-old students after they had 3 hours of instruction about rocks and minerals. Based on the respected "generative learning model" (Osborne & Wittrock, 1983, 1985), the researchers assumed that long-term learning depends on

the meanings that learners bring with them into the classroom. Accordingly, students were given several samples including one of marble and were asked to classify them as "rock" ("a solid, natural material found on or under the surface of the earth") or "not rock." The intention was that students would observe marble to be a solid, natural material and conclude that it is a rock. Instead, the students focused on the polished surface of the marble slab and, believing that natural means "untouched by humans," they concluded that marble is not natural and is therefore not a rock. Thus, the problem did not lie in in the generative learning model but in its application by the researchers.

Claim 8: INSTRUCTIONAL APPROACHES THAT FACILITATE CONCEPTUAL CHANGE CAN BE EFFECTIVE CLASSROOM TOOLS.

During the past 15 years, a number of novel intervention strategies have been developed with the aim of assisting students in the transition toward scientifically acceptable understandings of natural phenomena. These so-called conceptual-change techniques include a diverse and somewhat eclectic set of approaches that can only loosely be referred to as a method. They share a common epistemological assumption—students come to new learning situations with a fund of prior knowledge and the teacher's primary role is that of meaning negotiator or change agent. Several of the techniques grow out of a theory of conceptual change (Posner et al., 1982) that is grounded in contemporary philosophy of science (Kuhn, 1970; Lakatos, 1970; Toulmin, 1972) and a general constructivist view of learning (Osborne & Wittrock, 1983, 1985); others emerge from an Ausubelian (or neo-Ausubelian) assimilation theory of learning (Ausubel et al., 1978; Novak, 1977; Novak & Gowin, 1984).

Among the conceptual change techniques are those that focus on the functions of externalizing and modifying the learner's knowledge structure and others that address the need for self-monitoring and controlling the events of learning. Although these techniques are in an embryonic stage of development and implementation, a number of studies have attempted to assess their effectiveness in classroom situations. We think it is likely that the most successful conceptual change strategies will ultimately rely on a wide range of theory-driven, basic techniques used in various combinations as the needs of the learner require.

Externalizing and Modifying Cognitive Structure

It has been suggested that conceptual change strategies are founded, in part, on the premise that science and science learning are best promoted in the light of public scrutiny amidst the clash of opposing viewpoints (Gowin, 1981; Mintzes et al., 1991). The notion is that conceptual change typically involves hard intellectual work (often fraught with emotion) that necessitates restructuring of relationships among existing concepts and often requires the acquisition of entirely new concepts. Accomplishing these goals may be facilitated by a combination of individual, small-group, and whole-class activities in which alternative explanations and descriptions of scientific phenomena are verbalized, justified, debated, tested, and applied to new situations.

Among the most frequently used techniques for externalizing existing ideas are clinical interviews (Osborne & Gilbert, 1980; Pines, Novak, Posner, & Van Kirk, 1978), concept maps (Novak & Gowin, 1984; Novak & Wandersee, 1990, Trowbridge & Wandersee, 1992) and related diagrammatics (Gunstone & White, 1986; Stewart, 1980; Wandersee, 1987a), open-ended and multiple-choice response items (Tamir, 1989; Treagust, 1988), problem sets (Browning & Lehman, 1988), computer simulations (Nachmias et al., 1990; Zietsman & Hewson, 1986), sorting and word-association tasks (Cachapuz & Maskill, 1987; Champagne et al., 1985; Jungwirth, 1988), classroom discussions (Nussbaum & Novick, 1982), and journal writing (Kuhn & Aquirre, 1987). The "workhorse" in this area is the clinical interview (Wandersee et al., 1989).

Once students' ideas have been identified, the task of modifying those ideas begins. To this end, several research groups have experimented with a wide range of modification techniques with varying levels of success. Research efforts such as those by Anderson and Smith (1983), Champagne et al., (1985), Clement (1987), Driver (1987), Driver, Guesne, and Tiberghien (1985), Hewson and Hewson (1983), Novak and Gowin (1984), and Nussbaum and Novick (1982), are especially noteworthy. Conceptual change strategies and cooperative learning approaches have been combined with some success (Basili & Sanford, 1991).

A brief word of caution about the status of research on conceptual change seems in order. Much of this work is relatively recent in origin and, although promising, is probably best described as exploratory in nature. Many of the studies have relied on small sample sizes, untested methods, anecdotal records, and relatively nonrigorous research designs lacking control-group comparisons. Virtually none of the studies has been replicated. However, purely qualitative research continues to improve as research design keeps pace with advances in methods. So, even with the aforementioned caveats, we remain impressed by the relative success some researchers have achieved to date.

Research by the Children's Learning in Science (CLIS) project at the University of Leeds, United Kingdom, has been among the most far-reaching efforts (Brook, 1987; Scott, 1987). The project has been guided by a general constructivist learning model (Driver, 1981, 1989; Driver & Oldham, 1986) and has involved the significant collaboration of university personnel and public school teachers. Working in three groups of about 10 individuals each, the initial effort concentrated on developing instructional materials in the school science topic areas of energy, structure of matter, and plant nutrition. Units developed by working groups consist of 6 to 12 lessons and are organized around a general scheme with four phases: an orientation activity, elicitation of prior knowledge, a restructuring component, and a reflective or metacognitive exercise. The restructuring component consists of several "teaching maneuvers" that focus on broadening and differentiating existing concepts, building experimental bridges, unpacking conceptual problems, importing new models or analogies, and progressively shaping students' ideas (Driver, 1987).

Anyone who plans to do research on alternative conceptions should become familiar with the work of Rosalind Driver and her colleagues at CLIS. A good place to start might be the CLIS publication *Research on Students' Conceptions in Science: A Bibliography* (Carmichael et al., 1990).

In one study using the CLIS materials, 12- to 14-year-old pupils were engaged in a series of 10 lessons aimed at broadening their understanding of energy (Brook, 1987). Pretests revealed that most of the pupils subscribed to an anthropomorphic notion of energy and associated it strongly with movement, growth, fitness, exercise, and food. As a result of instruction, substantial change was detected in the extent of pupil agreement with several propositions about energy. For example, the majority of pupils agreed with the suggestions that machines use up energy and that cars use up energy as they consume gasoline. Unfortunately, several other alternative conceptions were apparently unaffected by instruction. Most of the pupils continued to believe that "plants get their food energy from the soil through their roots" and most continued to disbelieve that "we need energy all the time even when we are sleeping."

In a companion study (Scott, 1987), high school students were given 13 lessons on the particulate nature of matter with the intent of encouraging change from a continuous model to a discrete model. During a period of 2½ months, one 14-year-old girl, Sharron, apparently experienced significant movement toward the scientifically accepted view. However, even at the end of that time, Sharron was not entirely committed to the notion of an intermolecular vacuum. These findings are consistent with those of other researchers, including Nussbaum (1985).

A substantial number of studies have grown out of a conceptual-change model that proposes the use of conflict or confrontation strategies in the accommodation of novel scientific concepts (Posner et al., 1982). The model focuses on, among other things, the conditions thought necessary for the replacement of one central concept with another. Primary among those conditions is that learners become dissatisfied with existing conceptions—finding them insufficiently powerful to explain central problems in the domain and resulting in a set of accumulated puzzles or anomalies. Also, the new concept to be learned must be seen as both intelligible and plausible, as well as have the potential to explain a broad range of new phenomena (i.e., it must be fruitful).

Among the studies that explore conflict strategies as a means of effecting conceptual change are several that involve heat and temperature (Erickson & Tiberghien, 1985; Rogan, 1988; Stavy & Berkovitz, 1980), electricity (Arnold & Millar, 1987; Gauld, 1988; Licht, 1987; Osborne, 1983; Thorley & Treagust, 1987), mass and volume (Hewson & Hewson, 1983; Rowell & Dawson, 1981), velocity (Zeitsman & Hewson, 1986), acids and bases (Hand & Treagust, 1988), and the particulate nature of matter (Nussbaum & Novick, 1982). Studies involving the nature of electric circuits are especially enlightening because the range of alternative conceptions has been well documented and the relative simplicity of the problem has enabled researchers to devise unambiguous conflict situations.

In one study, a sequence of six 70-minute lessons was presented to 17 pupils (ages 11 and 12) at an inner-city middle school in England (Arnold & Millar, 1987). The lessons encouraged students to formulate theories about electrical circuits and then, through the introduction of conceptual conflict, to reconceptualize. Preinterviews revealed that 15 of the students had no understanding of the bipolar nature of a circuit and employed monopolar solutions in attempting to light a bulb. After subsequently lighting the bulb, the majority of pupils (13) employed a "clashing currents" model to explain the phenomenon. Following instruction, 14 pupils demonstrated an understanding of bipolarity and the rule of circuit closure and 12 used a "circulation" model of current flow. However, it appeared that a major stumbling block was the tenacity with which pupils held to the belief that current is consumed by electrical components in a circuit.

In a similarly well-designed study, Licht (1987) employed confrontation techniques to teach concepts of electric circuits to secondary school students (mean age 14.6 years) in Holland. Subjects enrolled in the experimental group ($n = 70$) were given 15 50-minute lessons that introduced conflicts with their initial conceptions that current is consumed by elements in an electric circuit and that batteries produce a constant current. Several control groups ($n = 74$) that received "traditional electricity education" provided a source of comparison. Pretests, immediate posttests, and delayed (by 5 months) posttests were administered. Before instruction, about 50 percent of the students subscribed to a current-consumption theory, a frequency that dropped in the experimental group to 11 percent at posttest and remained at 14 percent in the delayed posttest. The comparable frequencies for the control treatment were 37 percent at posttest and 43 percent after 5 months. The concept that batteries supply a constant current was held by a third of the experimental students before instruction and dropped at posttest to 11 percent (immediate) and 14 percent (delayed). Comparable data for the control groups showed a drop from 40 to about 30 percent at posttest times.

In addition to confrontation strategies, several researchers have reported varying levels of success with the use of analogies (Clement, 1987; Clement, Brown, & Zietsman, 1989; Glynn, Yeany, & Britton, 1991; Dupin & Johsua, 1989; Johsua & Dupin, 1987). Clement especially has made a persuasive case in his pursuit of "anchoring conceptions," which he defines as "intuitive knowledge structures that are in rough agreement with physical theory." Lessons he has developed focus on convincing students that the target problem is analogous to some commonly understood physical phenomenon. To do this, he sometimes employs an intermediary "bridging analogy."

For example, to convince students that a table exerts an upward force on a book lying on it, he uses the analogy of the force exerted by a spring on a hand that compresses it. The bridging analogy is the force exerted by a book resting on a flexible table.

Similar analogies, in combination with conflict strategies, have been used by Johsua and Dupin (1987) and Dupin and Johsua (1989) to overcome the persistent misconception that separate currents from the positive and negative poles of a battery clash inside a bulb, causing it to light up. In one study (Johsua & Dupin, 1987) more than 75 percent of French sixth- and eighth-graders were found to adhere to the clashing-currents model. The classroom procedure (consisting of five lessons) elicited student viewpoints, presented a set of conflict

situations, encouraged extensive discussion, and introduced a "train analogy." The two experiments performed by the students were intended to eliminate both the clashing currents and current-consumption models. Students observed that directional compasses placed on wires leading to and from a bulb deflected in opposite directions as the current flowed, demonstrating that the currents travel in different directions, and an ammeter placed on both wires demonstrated that current is not used up by the bulb.

Despite these efforts, virtually none of the students embraced the scientifically accepted view of current flow. The compass experiment caused many students to jettison the clashing-current model in favor of a "circulatory with consumption" model. However, the ammeter experiment posed a seemingly insoluble dilemma: "current flows in a circle and the battery wears down, but current is not consumed by the bulb." The result was an epistemological obstacle. The students rejected the ammeter results—in fact, "rejection was total, harsh, and definitive." Alas, in the end, most students subscribed to a simple circulatory-with-consumption viewpoint.

Although most studies investigating conceptual-change techniques have focused on concepts in the physical sciences, several have addressed issues in biology (Barker & Carr, 1989a, 1989b; Bell & Barker, 1982; Eisen & Stavy, 1989; Happs & Scherpenzeel, 1987).

The work of Barker and Carr (1989a) addresses a persistent and well-documented misconception—that plants obtain their food from the soil. In so doing, it departs from previous approaches and instead focuses on photosynthesis as the production of familiar, energy-rich organic materials (wood), using the students' prior knowledge of wood as a point of departure. The instructional package is composed of several practical investigations that focus students' attention on their own prior knowledge and help them formulate theories about photosynthesis. A brief, self-teaching booklet introduces new concepts, and a set of application exercises relate already familiar situations to the new concepts.

Twenty-eight fourth-formers (14-year-olds) in New Zealand who had had no previous instruction in photosynthesis received 12 1-hour lessons using the Barker and Carr (1989a) materials. Prior to instruction only two students held ideas that were consistent with the scientifically accepted view. This number rose to 14 immediately after instruction, and a delayed-retention test revealed that 11 students retained these views 12 weeks later.

Monitoring and Controlling Learning

In addition to the wide range of domain-specific approaches currently used to effect conceptual change in the natural sciences, several generic strategies have been developed to help students "learn how to learn" (Novak & Gowin, 1984). The rationale behind these so-called metacognitive strategies is the assumption that students who know how to monitor and control their learning are empowered to engage in more purposeful "meaning making" (Novak, 1987) and, as a result, can be expected to recognize and attempt to correct inconsistencies in their own thinking.

Among the strongest lines of evidence supporting the use of metacognitive strategies is that which addresses differences in the structure and use of knowledge by experts and novices (Chi, Feltovich, & Glaser, 1981). Dozens of studies that compare individuals with varying degrees of expertise in such disparate areas as chess playing, physics, genetics problem solving, medical diagnosis, and computer programming reveal that successful people have learned how to integrate their knowledge by reflecting on its meaning and identifying and resolving contradictions in its structure. It has been suggested that: "If metalearning can be taught, then the problem of how to bring about conceptual change may be solved" (White & Gunstone, 1989).

This may be a somewhat optimistic projection, and a good amount of empirical work needs to be done to substantiate it. Nonetheless, studies consistently reveal that poor students have difficulty in (1) judging when they understand something, (2) determining the amount of time and effort required to reach their goals, and (3) deciding on appropriate learning strategies. In contrast, able students (1) repeatedly use self-questioning strategies to probe their conceptions, (2) often employ reviewing techniques, (3) typically reflect on their performance, and (4) consider how their knowledge may be of use in future situations (Alexander & Judy, 1988; Garner & Alexander, 1989, Wandersee, 1988). Baird (1986) has gone so far as to suggest that: "Only minor improvements in learning will come about through a search for new styles of teaching . . . substantial improvements depend on a fundamental shift from teacher to student in responsibility for, and control of learning."

Several important questions have been posed by those who advocate metacognitive approaches. Can students learn to use these strategies? If so, will students use them? Finally, what kinds of conceptual change can be expected if the answers to the first two questions are in the affirmative? No definitive answers to these questions are yet available—in part because of the complexity of the issues involved, but also because the research program itself is still relatively new. Although studies on metacognition have occupied researchers in the fields of reading (Garner, 1987; Paris & Oka, 1986), special education (Slife, Weiss, & Bell, 1985), and mathematics education (Schoenfeld, 1987) for some time, science educators are relatively underrepresented in the area. The most well-established research centers on metacognition in science are those at Cornell University (Novak & Gowin, 1984); the University of California, Los Angeles (Wittrock, 1992); and Monash University in Melbourne, Australia (White & Gunstone, 1989).

In Melbourne, the Project to Enhance Effective Learning (PEEL) has brought together teachers and university personnel in an attempt to implement metalearning strategies in a variety of science-related contexts. In one 6-month study, the effects of metalearning materials on ninth-grade science and eleventh-grade biology students were investigated (Baird, 1986). The students completed checklists, evaluation notebooks, and workbooks that required them to reflect on their own learning. The materials focused student attention on a variety of important questions: What do I know about this topic? How do I feel about it? What are the most important points? What is required to complete this exercise? How does this new knowledge compare with what I used to think? How am I doing? Have I com-

pleted the task? Do I fully understand this? How can I make sure I remember this? What use will I make of it? By the end of the 6-month period, 60 percent of the ninth-graders and 75 percent of the eleventh-graders were regularly asking themselves such questions.

White and Gunstone (1989) summarized several insights about implementing metalearning techniques based on their experience with the PEEL project. Several of these suggestions are in accord with those offered by Garner and Alexander (1989). Among them are the following:

1. Attention must be paid to the match between the student's short-term goals of performing well on classroom assessments and the teacher's long-term objective of facilitating meaningful learning; understanding must be rewarded.
2. Context is also important; the success of metalearning strategies in science depends on their implementation within and across disciplines during an extended period of time.
3. Metalearning and conceptual change require effort; they are endothermal processes or nonspontaneous events that must be fueled; variation in strategies may provide the necessary fuel.
4. Sustained commitment to metalearning strategies depends on continued support of colleagues, administrators, parents, and peers, and perhaps the most important factor is the pupil's self-discipline.

Much of what we have learned about effective teaching and learning does not yield quicker results with less effort by the parties involved. Quite the contrary, there is no "royal road" to knowledge. If we have learned anything, it is that more effective instruction takes more time to implement. In the science classroom, instructional emphasis on speed learning or on "coverage" (a curriculum concept that is the bane of meaningful learning) is misplaced.

In Ithaca, New York, at Cornell University, Novak and his colleagues have developed, implemented, and tested several metalearning strategies that grew out of the Ausubelian learning tradition, including concept mapping, Vee diagramming, and concept-circle diagramming (Heinze-Fry & Novak, 1990; Novak & Gowin, 1984; Novak & Wandersee, 1990).

The guiding principle has always been Ausubel's prior-knowledge dictum—that the most important factor influencing learning is the student's own prior knowledge (Ausubel, 1963, 1968). Wandersee (1986a) has put it this way, "The most important things students bring to their science classes are their concepts (p. 581)."

Novak and his colleagues developed concept mapping and concept-circle diagramming to represent students' knowledge structures in graphic form. Novak's colleague D. B. Gowin developed Vee diagramming as a graphic way to represent what students understand about the nature of scientific knowledge and knowledge construction. Taken together, these tools (and clinical interviewing) have allowed Novak's research group to make progress in theory building via their alternative conceptions program.

The tools have also given classroom science teachers ways of representing science content and scientific thinking in a graphic form accessible to both students and teachers. The Cornell research group has been fine-tuning these tools for more than a decade (with modifications based on emerging relevant theoretical principles and research findings). The value of the tools resides chiefly in their active, rather than passive, usage by the learner.

Closing Comments on Knowledge Claims

We have attempted to cite representative research from across the school science subjects and from various perspectives, without regard for epistemological or theoretical stance. However, because of our selection based on frequency of citation, some perspectives (e.g., radical constructivist and neo-Piagetian views) may be underrepresented in the resulting analysis. If that is the case, we hope the other ACM review sources that we have cited will allow the reader to fill such lacunae. We have employed one of many possible selection strategies and have interpreted only the literature that our search flagged.

Summary of Information Presented Thus Far

Figure 5.2 depicts the portion of the review that has now been completed. It consists of a portion of the original chapter concept map that was presented in Figure 5.1. This abbreviated concept map emphasizes how the information we have just considered is interrelated. We think it will be useful to retrace visually the linkages developed in the foregoing chapter sections, especially those between the purpose and design of this review that were set forth in the introduction, the terminology barriers the novice will encounter in the ACM literature, the two basic types of research foci reflected in the terms used by ACM researchers, and the fundamental research claims emerging from both types of ACM studies, spanning the spectrum of school science subjects.

To assist the reader in considering the ACM knowledge claims as a coherent set, the following list shows them, not as isolated assertions, but as an integrated whole—aiding the process of summarization:

Claim 1: LEARNERS COME TO FORMAL SCIENCE INSTRUCTION WITH A DIVERSE SET OF ALTERNATIVE CONCEPTIONS CONCERNING NATURAL OBJECTS AND EVENTS.

Claim 2: THE ALTERNATIVE CONCEPTIONS THAT LEARNERS BRING TO FORMAL SCIENCE INSTRUCTION CUT ACROSS AGE, ABILITY, GENDER, AND CULTURAL BOUNDARIES.

Claim 3: ALTERNATIVE CONCEPTIONS ARE TENACIOUS AND RESISTANT TO EXTINCTION BY CONVENTIONAL TEACHING STRATEGIES.

Claim 4: ALTERNATIVE CONCEPTIONS OFTEN PARALLEL EXPLANATIONS OF NATURAL PHENOMENA OFFERED BY PREVIOUS GENERATIONS OF SCIENTISTS AND PHILOSOPHERS.

Claim 5: ALTERNATIVE CONCEPTIONS HAVE THEIR ORIGINS IN A DIVERSE SET OF PERSONAL EXPERIENCES INCLUDING DIRECT OBSERVATION AND PERCEPTION, PEER CULTURE AND LANGUAGE, AS WELL AS IN TEACHERS' EXPLANATIONS AND INSTRUCTIONAL MATERIALS.

Claim 6: TEACHERS OFTEN SUBSCRIBE TO THE SAME ALTERNATIVE CONCEPTIONS AS THEIR STUDENTS.

Claim 7: LEARNERS' PRIOR KNOWLEDGE INTERACTS WITH KNOWLEDGE PRESENTED IN FORMAL INSTRUCTION, RESULTING IN A DIVERSE SET OF UNINTENDED LEARNING OUTCOMES.

Claim 8: INSTRUCTIONAL APPROACHES THAT FACILITATE CONCEPTUAL CHANGE CAN BE EFFECTIVE CLASSROOM TOOLS.

Taken together, these eight knowledge claims represent a large number of research studies worldwide and reflect what we think is an emerging consensus of the ACM.

Figure 5.3 presents a flowchart to help the reader trace the main path the review has taken up to this juncture. In order to be both concise and sequential, we have depicted the principal flow of ideas—highlighting the points we consider most important. We invite the reader to take a visual excursion through the chart and to recall the story line. Of necessity, much of the chapter text we wish to summarize consists of salient examples of important research designed to communicate the nature and dimensions of typical alternative conceptions studies. At this point, we think that the reader now has a basic understanding of (1) the evolution of the alternative conceptions movement (drawing on research studies published primarily from 1970 to mid-1991), (2) some current knowledge claims and hypotheses of the field, and (3) how ACM research relates to conceptual change research. Until now, we have focused on the past and

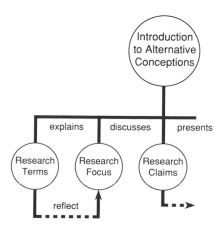

FIGURE 5.2. A concept map of the portion of the review that has now been completed.

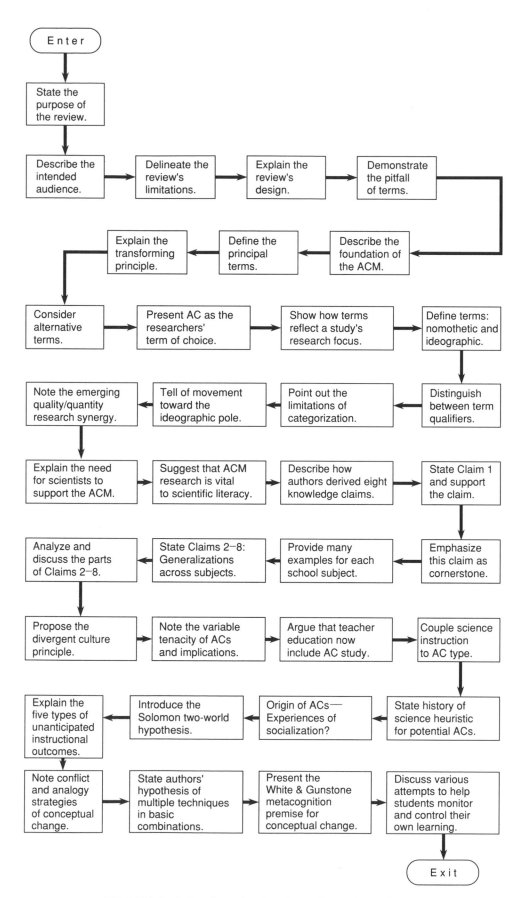

FIGURE 5.3. A flowchart showing the major pathway this review has taken up to this point.

the present. The next two sections lead from the present to the future, exploring the current views of key ACM researchers and discussing some issues that affect tomorrow's ACM research agenda.

THE VIEWS OF LEADING RESEARCHERS

As part of our background research in writing this chapter, we conducted an international survey of influential and/or promising science-education researchers who have done work related to alternative conceptions in science. One of this chapter's authors (J.D.N.) was not involved in designing or conducting the survey, because he was one of the experts our (J.H.W., J.J.M.) science education colleagues had suggested we query. We asked each expert to answer three questions:

1. Which alternative conceptions study do you consider most influential?
2. What strengths and weaknesses do you see in the published alternative conceptions research?
3. Where should we go from here?

The intent of the question series was to identify the bases of research decision making and assess the progress of the ACM. First we used citation frequency and peer recommendation selection criteria to compile a list of 100 names of influential or promising science education researchers; then we designed a purposive, stratified sample that we drew from this list. Stratification dichotomies included (a) male and female ACM researchers, (b) established and promising ACM researchers, (c) United States-based and international ACM researchers, (d) physical science and life science ACM researchers, and (e) ACM leader-experts and science education leader-experts in other specialty areas. (Leader-experts were defined as published researchers who have also served in key organizational leadership positions.)

We mailed 30 questionnaires to the subjects composing our sample in the spring (Northern Hemisphere) of 1991. We included a small number of promising but as yet relatively unpublished researchers ($n = 10$) on our master list in order to include new Ph.D.s who had just been immersed in the literature while completing their own alternative conceptions dissertations in science education.

A cover letter explained the primary purpose of our questionnaire in preparing this NSTA *Handbook of Research*. We asked the subjects whom we queried to answer the aforementioned questions (in free-response form) and advised them explicitly that their responses might well be quoted in the *Handbook*. We explained that, if included, their responses would be cited as "personal communications," following the practices recommended in the third edition of the *Publication Manual of the American Psychological Association* (1983).

The final response rate for the questionnaire was 70 percent. A few experts also sent books and papers, not only to corroborate their answers but also to help ensure that we would be able to represent the state of the field without distortion. As a result, we feel confident that the conclusions we have drawn in this chapter have external and ecological validity.

Responses to Question 1 Concerning Influential Studies. In the first question, we asked the researchers we queried to choose the single alternative-conceptions study or publication that had had the greatest influence on their own research and then asked them to explain their answer. Most researchers offered specific reasons for their final selections—chosen from the more than 2400 published studies.

Upon careful analysis of all the responses we received, we constructed the following generalizations: (1) there is a small group of leading researchers whose multiple references (even among the experts' responses) testify to their authority (e.g., Driver, Erickson, Novak, Nussbaum, Viennot); (2) beyond the work of the handful of acknowledged leaders in the field, the experts seemed to be influenced most by alternative conceptions publications that were either related to their own science content area or helped them gain a panoramic overview of or new perspective on the field; and (3) the analytical comments by the experts reflected their ability to read widely, evaluate critically, and synthesize research results with mental agility. Interestingly, sometimes a study was cited by an expert as having influence on him or her because it evoked a negative response and thus served as a stimulus to marshall evidence against a position taken in the article. Many studies were described as a springboard to the expert's research area or approach—showing that exemplary studies can have far-reaching intellectual effects.

Responses to Question 2 Concerning Strengths and Weaknesses. In the second question, we asked the experts to specify some important strengths and weaknesses they saw in the published ACM research. To provide a context for their comments, they were asked (hypothetically) what advice they would give one of their own graduate students who was contemplating a dissertation project involving research on alternative conceptions. What should the doctoral student be sure to do and what should he or she avoid doing? The responses we received were varied and insightful. Analysis of all the responses resulted in the following generalizations: (1) it is difficult, even for experts, to characterize a corpus of 2400+ studies, because even the experts have read only a subset of them; (2) there is a need for more replication studies and for meta-analytic studies of the research to date; (3) the field needs more and better studies of teachers' alternative conceptions and of the roles they should play in the conceptual change process; (4) qualitative or synergistic approaches to research should be employed more often than in the past; (5) researchers should investigate the philosophical and historical background behind the science concepts they intend to study; (6) all studies should be based on pilot studies; (7) ACM terms should be defined very carefully; (8) alternative conceptions may be considered as natural intermediates of the learning process and the students who hold them should be treated with intellectual respect; (9) the alternative conceptions that are found during a study should be analyzed as to their sources, their prevalence, and the degree to which they impede further learning; (10) researchers should use research tools that provide maximum latitude of expression for their subjects; (11) researchers should recognize that for a student to change a scientific conception requires a major conceptual reorganization; and (12) clinical interviews and concept maps appear to be the current research tools of choice.

Responses to Question 3 Concerning Future Research. In the third question, we asked the experts to assess the current status of science education's knowledge about alternative conceptions and to comment on the future they foresaw for such a research program (i.e., the ACM). To provide a context for their remarks, they were asked whether they thought there is still much to be done or whether we have reached a point of diminishing returns.

Upon analysis of all the responses offered, we constructed the following generalizations: (1) the research is moving from description of alternative conceptions to understanding the process of conceptual change; (2) there is a need for a theoretical foundation that can describe, predict, and explain alternative conceptions, but the field has not reached a consensus on which theory to use; (3) some researchers think that looking for the cause of alternative conceptions is more promising than looking for their "cure"; (4) the questions of how many more and what kinds of studies are needed remain open; (5) the cultural dimensions of the alternative conceptions phenomenon appear ready for productive exploration; (6) several experts saw the need for integration of research findings about alternative conceptions in science with those of related domains (e.g., problem-solving research, mathematics education research, and research in the cognitive sciences); (7) some experts would like to establish a moratorium on alternative conceptions studies (especially those of the identification and cataloguing genre) so we can analyze, integrate, and assimilate the ACM findings we now have; (8) a number of researchers were skeptical about conceptual-change theory but saw it as a good starting point; (9) some experts pointed out that alternative conceptions were "not all of the same cloth" and that a new categorization system was needed; (10) it was often suggested that alternative conceptions be viewed as a natural concomitant of learning and be considered less harmful to the student than present ACM studies portray; (11) several experts pointed out a need to translate existing alternative conceptions research findings into practical lessons and activities that teachers can use in their classrooms; (12) a research sequence was proposed by one expert to ensure that alternative conceptions findings ultimately affect classroom practice; and (13) the majority of experts saw the alternative conceptions research field as having the potential to yield many more important findings—if future studies are carefully designed to be more analytically sophisticated than past studies have been.

EVALUATION OF THE RESEARCH AGENDA

The ACM and Changes in Science Education

At this time, it appears that even the most committed supporters of the so-called stage theory of cognitive development see evidence that many of its central claims are no longer tenable (Lawson, 1991a, 1991b). In its place has emerged a con-

structivist synthesis that draws largely on the work of Ausubel (1963), Kelly (1955), Vygotsky (1962), Novak & Gowin (1984), Posner et al. (1982), von Glaserfield (1989), and others. The new synthesis has emerged rapidly—during the last 5 or 6 years—and may provide a strong theoretical framework for further research in the alternative conceptions tradition. Based on recent discussions at the annual meetings of the National Association for Research in Science Teaching, the emerging constructivist synthesis appears to be gaining wide support and may become the first broad-based research perspective in the field. It has already begun to be "translated" for science teachers and is now affecting classroom practice (see Yager, 1991).

Along with progress in theory construction (which has been influenced by theoreticians such as Piaget, Ausubel, Kelly, Novak, and Vygotsky and by philosophers such as Gowin, Hanson, Kuhn, Feyerabend, Lakatos, and Toulmin), equally impressive empirical and methodological progress has occurred. Accordingly, we agree with Confrey (1990), who, in her compelling cross-disciplinary review, characterized the alternative conceptions research program as "progressive" in the Lakatosian sense. Compare the following descriptions of the alternative conceptions research program, which are separated by a time span of just 6 years:

> The present situation in which workers exploring this approach find themselves may be compared analogously to Kuhn's "preparadigmatic" phase in science, viz. so far, the efforts of individuals have tended to be uncoordinated: terminology has not been agreed upon, a common methodology not shared, ultimate aims not stated, and classroom implications not explored. (Gilbert & Swift, 1985, p. 683)

After 6 years of accumulated research and a coalescence of viewpoints from theoretical liquidity, Duit (1991) tells us:

> The trends are, in my view, really exciting. Research started from a very important, but nevertheless limited, emphasis on investigations of students' conceptual frameworks. The scope is now much broader. The constructivist view of learning has proven to be a powerful and valuable driving force of research. (p. 83)

In many ways, progress seen in theory construction has been matched by advances in methodology. It now appears that science education is beginning to metamorphose to a discipline with a distinctive perspective on conceptual change—one that is uniquely grounded in history, sociology, philosophy of science, and cognitive psychology and offers a unique contribution to understanding science learning. At the center of these changes is the ACM.

Gone, for the most part, are the rigid experimental designs borrowed from the agricultural sciences and the heavy reliance on psychometric techniques and multivariate analytic approaches of a decade ago. Reflecting a newfound willingness to experiment with research methods (and even to develop new ones), various discipline-based techniques have been employed by science education researchers (e.g., clinical interviewing, science problem-solving tasks with think-aloud protocols, concept mapping, Vee diagramming). Such techniques have enabled researchers to ask and answer questions heretofore unamenable to investigation. The study of familiar classroom events and the use of relatively direct data analyses (with few data transformations) have generated "believable" results that many teachers find both plausible and useful in their work.

There has been another key advance. We have long understood that science teachers' subject matter knowledge and pedagogical knowledge were both important. We now know that a third area is also vital to teaching success. Pedagogical content knowledge (PCK)—knowing how best to teach particular science concepts and principles to a given group of students—distinguishes the science educator and the science teacher from the scientist. It draws on and integrates science content knowledge, science education research, science teaching experience, and principles of pedagogy. The alternative conceptions research findings dovetail with Shulman's (1986) efforts to illuminate what has been called the "missing paradigm." Surely, teachers must understand their students' prevalent alternative conceptions about a particular science topic if they are to tailor their lessons to fit the students they serve. Cochran (1991) has prepared a concise explanation of PCK and its transformative characteristics, which we recommend.

Central to PCK is knowing how to teach the subject-specific and perennially difficult science concepts that are part of every science curriculum and to teach them well. Here is where alternative conceptions research may be valuable beyond estimation (see Ganiel & Idar, 1985; Viennot, 1985).

Equally encouraging is the extent to which our discipline has begun to contribute to, rather than merely borrow from, closely related fields. Many researchers in allied areas are reading our journals, trying our techniques, and applying our findings. Collaboration between science educators and specialists in the natural sciences, cognitive sciences, applied sciences, and the history, sociology, and philosophy of science, mathematics, and technology has resulted in the establishment of interdisciplinary research centers—from Boston to San Diego to Melbourne to Utrecht, to name just a few. Combined with the globalization of science education research, this has resulted in a diversification of the field, shifting much activity away from traditional schools of education and strengthening the other schools of education that are now adapting, transforming, and collaborating.

It seems to us that the relatively young research movement (ACM) that began by probing students' "wrong ideas" has opened a vast frontier of immense promise and, in so doing, has forever changed the landscape of science education. Before considering where this program might lead, we will summarize briefly what we perceive to be the state of present knowledge and the tools that help construct it.

Knowledge Claims

Based on our terminological analysis of the ACM, our survey of ACM experts, and our citation-based thematic analysis of some of the seminal studies on alternative conceptions that have appeared in the science education research literature over the past 20 years, we believe there is now sufficient evidence to support the following research claims:

1. Learners bring a diverse array of ideas about natural objects and events to their science classes, and these ideas are often at variance with the scientifically accepted views in physics, biology, and chemistry. Without knowledge of these student ideas, the science teacher is at a great disadvantage.

2. Learners of both genders and of virtually all ages, ability levels, and cultural backgrounds subscribe to alternative conceptions. However, the frequency of occurrence for a given alternative conception may vary considerably—based on such factors as the level and quality of prior instruction—along with personal and social experiences beyond the classroom walls.

3. The tenacity with which alternative conceptions are held is not a myth, but it does vary substantially. The extent of possible conceptual change is probably affected by an imposing grouping of epistemological factors (including degree of current satisfaction, intelligibility, plausibility, fruitfulness) that may reflect underlying psychological issues (e.g., the degree of novelty or counterintuitiveness embodied by the new conception, the opportunities to observe regularities in natural objects or events that are central to understanding a science topic, and the degree to which science instruction challenges a given pupil's existing alternative conceptions).

4. There are obvious similarities between the explanations of natural phenomena suggested by previous generations of scientists and the explanations associated with today's students' alternative conceptions. However, the proposition that learners recapitulate the conceptual development pattern chronicled by the history of science in arriving at currently accepted scientific conceptions has found little empirical support. Research suggests that the use of the history of science as a heuristic device for science teaching and learning shows much greater promise.

5. Tracing the origins of alternative conceptions is, at present, a speculative intellectual enterprise. The roots of such conceptions are often hidden and therefore difficult to study using empirical methods. The conceptual history of the individual learner is idiosyncratic and difficult to trace (Wandersee [1992] has called this time-linked, experience-based phenomenon "the historicality of cognition"). However, the widespread occurrence of many alternative conceptions across diverse populations and cultures suggests that they reflect common cultural experiences involving direct observation of nature, the use of everyday language, the influence of mass media, and experiences due to instructional practices.

6. Teachers, as well as students, may (and often do) hold alternative conceptions and harbor misunderstandings about current science. Because, at least in the past, elementary teachers took fewer and less challenging science courses than secondary teachers, the alternative conceptions that elementary teachers subscribe to are not substantially different from those of the population at large—including their students.

7. Learning frequently takes an unforeseen path, and the outcomes of any planned intervention are often inconsistent with the science teacher's original instructional intentions. It appears possible to anticipate and circumvent some, but not

all, of this deviation from the intended learning outcomes. Some factors (e.g., social and economic) appear to lie beyond the science teacher's sphere of intervention.

8. Several of the instructional approaches designed to foster conceptual change appear promising, including those that rely on use of conceptual conflict, analogies, and metacognitive strategies. However, more research is needed before such approaches can be considered for widespread adoption. In addition, science educators have yet to learn how to convince a majority of science students to modify their science concepts or to take advantage of the metacognitive strategies that are available. We and a sizable number of other researchers see an emphasis on promoting meaningful learning (in marked contrast to rote learning) as crucial to progress in this area. Others, for example, prefer to emphasize the mastery of critical thinking skills.

The Research Tools

If a recent content analysis we performed within the biological alternative conceptions literature is a good indicator (Wandersee et al., 1989), interviewing is the most common research tool in alternative conception studies (see Pines, Novak, Posner, & Van Kirk [1978] for details on clinical interviewing). In the sample of 103 ACM studies that we examined, 46 percent used clinical interviews, 20 percent used multiple-choice tests, 19 percent used sorting tasks, 9 percent used student drawings, 8 percent used questionnaires, 7 percent used open-ended tests, 6 percent used classification tasks, 6 percent used association tasks, 6 percent used essay writing, 5 percent used identification tasks, 4 percent used concept mapping, 4 percent used problem-solving tests, 4 percent used Piagetian tasks, and 19 percent used unique approaches. (Note: the percentages do not sum to 100 percent because some studies used several methods.) Various researchers have tried other innovative approaches. Gilbert, Watts, and Osborne (1985) have used an interview-about-instances technique; Champagne et al. (1985) have used a DOE (demonstrate, observe, explain task) interview; and Novak and Gowin (1984) recommend a concept map–based clinical interviewing technique.

Jean Piaget, of course, perfected the interview as a way of probing children's cognitive capabilities, and science educators owe him a debt of gratitude for all the interviewing strategies his work spawned. Now, however, most (but not all) ACM researchers focus on changes in personal knowledge structures, rather than on general reasoning ability, and on the social and experiential bases of cognition rather than on the genetic and developmental ones. Considering the subject matter content knowledge involved in cognition was relatively unimportant to Piaget's research; it is very important to many ACM researchers.

Whereas initially, paper-and-pencil tests using multiple-choice items were commonly used by ACM researchers, concept maps, POE (prediction, observation, and explanation) probes, concept circle diagrams, question production, Vee diagrams, and other new knowledge probes are becoming more and more common. Novak and Gowin (1984) see concept maps constructed from taped interviews as an excellent way to multi-

ply the power of two tools (interviews and concept maps), facilitating both identification of the learner's alternative conceptions and subsequent graphic representation of the learner's relevant cognitive structures. White and Gunstone (1992) have published a book entitled *Probing Understanding*, which we recommend to both researchers and teachers seeking "a range of practical methods of probing their students' understanding" (p. vii).

We think the work of the late John E. Feldsine, Jr. (1987) is particularly noteworthy with respect to probing science student's understanding. Feldsine was a chemistry educator who used student-constructed concept maps as the heart of his general chemistry course for more than 6 years. His doctoral dissertation and related ACM seminar papers demonstrate how much can be accomplished with concept mapping—especially as a tool for revealing a greater portion of student's knowledge of a content domain than conventional course examinations can. Like Feldsine, we contend that concept maps are powerful evaluation tools that have yet to be fully exploited by the ACM (Wandersee, 1990). Not all researchers, however, share this view, and the search for other new research tools continues.

A NEW RESEARCH AGENDA: FROM ALTERNATIVE CONCEPTIONS TO CONCEPTUAL CHANGE

In this section, we offer a proposal for the future focus of science education, showing how alternative conceptions research can play a vital role in such a reconceptualization. We also give specific examples of how tomorrow's researchers might be prepared and what research might be like if such a proposal is adopted. The section's sole purpose is to stimulate thought and explore possibilities.

There has been some debate about what, if anything, constitutes the unique domain of professional activity to which science educators can rightfully lay claim (Good, Herron, Lawson, & Renner, 1985; Wandersee, 1983b; Yager, 1984). Some have suggested a broadly expanded role for science educators at the interface of science, technology, and society (STS). Others see many practical problems with the implementation of such a proposal and recommend that science educators concentrate solely on the teaching and learning of science. The result has been a partial polarization of the field—with those who have not yet taken a position vacillating in the middle. The issue turns on the classic curriculum question, which Novak (1977, p. 129) rephrased as, "what knowledge is of most worth for most students?"

We think there is value in reconceptualizing our professional domain in a way that eliminates the aforementioned impasse and is consonant with the research findings of the ACM. We suggest that science educators concern themselves more broadly with the goal of understanding and promoting conceptual change in the teaching and learning of the natural sciences. This is not to imply that the science learner's concepts are always totally changed—they may be deleted, replaced, augmented, paralleled, exchanged, or refined—but that conceptual restructuring (or more simply put, change) lies at the heart of science teaching and learning, whether the curriculum intends to produce scientifically literate citizens, prize-winning physicists, or qualified physicians.

Such a focus would encourage research collaboration with scientists in the natural sciences, especially biology, chemistry, and physics (the primary source disciplines for the school subjects); with practicing science teachers; and with others whose work can help us understand cognitive restructuring at various levels and in similar situations (e.g., historians, sociologists, and philosophers of science; cognitive scientists; ethnographers; and anthropologists).

Some two decades of research on alternative conceptions have yielded a wealth of knowledge and the recognition that ACM researchers' interests and those of the aforementioned specialists are converging. We suggest a more panoramic view of our professional domain and recommend close collaboration with science faculty members in the schools and with relevant academic disciplines on issues of conceptual change in teaching and learning the natural sciences. The following are representative of the problems such a research program might engage.

Critical Junctures in Learning

Studies on the understanding of critical concepts in the context of existing course structures could be enlightening. For example, such studies might have researchers going "head first" into classrooms, literally "following students around" in the school laboratory and attending lectures and discussions with them. The purpose would be to observe directly the kinds of conceptual problems that arise during the course of instruction. Rather than merely isolating potentially troublesome concepts, such an ethnographic approach might enable science education researchers to identify the propositional relationships between key clusters of concepts as they occur in vivo. Perhaps then we could begin to pinpoint how conceptual difficulties in one domain affect learning in another.

The design of such a study might require regular videotaping of classroom events and multiple assays of student understanding before and after the lessons taught during the term. Ultimately, a log of potentially difficult "critical junctures in learning" (a term proposed by ACM researcher J. J. Mintzes) might be compiled—each juncture representing a challenging cognitive event in the course of the term (see Trowbridge & Wandersee, 1992). This kind of longitudinal look at the conceptual difficulties arising within existing course structures might be particularly useful to teachers in reflecting on and rethinking teaching approaches or course scope and sequence.

Comparative Knowledge Structures

Studies of the ways in which knowledge is structured and applied by individuals teaching and learning the natural sciences could prove especially helpful to science teachers, science teacher-educators, and curriculum developers. We see concept mapping, as well as the SemNet™ semantic networking software developed by Kathleen Fisher and her associates at San Diego State University, as possible ways of doing this.

A number of studies have explored differences in the structure and use of knowledge by experts and novices in various of science-related domains (see Smith, 1991). We believe this work will continue to become more sophisticated and expansive, as it also becomes more relevant to understanding conceptual change. It would also be useful to explore variations in knowledge structures within disciplines (e.g., within biology: how do molecular geneticists and evolutionary biologists differ in their understanding and application of the gene concept?).

A related area that has received some research attention and should be elaborated is the restructuring of domain-specific and pedagogical knowledge by preservice, novice, and experienced teachers. Ultimately, knowledge of topic-specific PCK for a given school subject would be useful to those engaged in innovative teacher education programs such as the Holmes Plan.

Cognitive-Affective Relationships

Many research programs in science education address either the cognitive or the affective domain, but few have developed a sustained interest in the relationship between the two. We now have good evidence that individuals who successfully construct complex and elaborate conceptual structures in a given domain of knowledge also tend to form positive attitudes about that domain and, in tandem, raise their own self-esteem. It appears that conceptual change is intimately linked to affective change, and we think that is an area worthy of concentrated investigation. Such new knowledge might help us attract larger numbers of individuals, especially women and minorities, to the study of the natural sciences. The "fluent integration of thinking, feeling, and acting" is a Gowinian goal that has yet to be realized in science education.

Intervention

The number of research efforts that focus on intervention strategies has been growing rapidly. However, the majority have been primarily exploratory. As proposals for conceptual-change interventions increase in number, science educators must examine the effectiveness of these widely promoted strategies in rigorous, well-controlled studies. Attempts to assess the quality of these strategies must avoid the pitfalls that confronted earlier researchers—including uncritical use of existing standardized tests, even though irrelevant to the instructional goals of the intervention.

It may be that no single concept has had as pernicious an effect on educational improvement as the one called achievement. This was typically defined (operationally) as the difference between pretest and posttest scores, often after a planned intervention of relatively short duration. Thus, a great deal of meaning has been tied to a single score and to an established (but conceptually inadequate) technology (we might say "tyranny") of testing.

Studies designed to measure the conceptual change resulting from an intervention need to employ assessment techniques sensitive to subtle changes in student's understanding. Fortunately, several such techniques have been developed, including interviews about instances and events, Vee diagramming, computer-based diagnostic systems, concept circle diagramming, constructing computerized semantic networks (e.g., SemNet™), and concept mapping. New ones are sure to follow.

Metacognition

Research on the use of metacognitive strategies (strategies that help students reflect on "what and how they know" in order to monitor their own learning) to enhance conceptual change has been a source of optimism within the ACM. However, it appears that these techniques cannot provide a "quick fix" or make learning easy. Rather, we are finding out that quality learning takes much longer to occur than was previously thought. Ideally, a metacognitive strategy would be studied longitudinally to assay its effects on learning over time. This kind of research program requires extensive collaboration between universities, school personnel, students, and parents. To ensure a valid trial, the benefits of cooperation must be manifest to all. Such studies might begin in the early elementary grades with simple metacognitive strategies and gradually implement increasingly sophisticated strategies as students move through the school system. For example, Wandersee (1987a), building on the work of Novak and Gowin (1984), proposed a sequence of graphic metacognitive strategies from constructing simple concept-circle diagrams to concept maps to Gowin's Vee diagrams through grades 1 to 12. Other researchers might use a different sequence of age-appropriate metacognitive strategies, such as models of comprehension in science (Wittrock, 1992).

EPILOGUE: ON EDUCATING FUTURE RESEARCHERS

As investigators reflect on the extensive research base on alternative conceptions in science, they may note that the accumulated efforts of the past 20 years offer some important lessons on the education of future researchers in science education. The most significant may be that the next generation of researchers must be more broadly educated than in the past, especially in relevant areas outside science education proper.

Because we proposed to focus on the integrated study of conceptual change and cross-disciplinary integration requires the student to have at least one strong knowledge base from which to make connections (to the new ones being learned), we shall assume an already strong science content background (depth) upon entry to the graduate program in science education—before the student attempts the integration with relevant disciplines (breadth).

Graduate education at the doctoral level (in science education) has typically involved graduate-level natural science courses, offerings in research design and statistics, science education content and methods courses, courses in educational psychology, and possibly some work in educational foundations (e.g., educational history, philosophy, and sociology). Although such preparation may be appropriate for those seeking applied research positions in public schools and related agencies, we think it is generally inadequate for college- and university-based researchers because it does not provide a strong theory base or the conceptual framework needed to formulate worthwhile research questions.

We suggest that those who seek the doctorate in science education and who wish to contribute to the ACM or to science education's understanding of conceptual change should have (1) at least 5 years of experience teaching science (this is the minimum time required for the prospective science educator to achieve science teaching proficiency, according to Berliner's [1989] model, which focuses on the cognition that underlies teacher behaviors); (2) both a solid content foundation and some research experience in at least one physical or life science (preferably at the master's level or beyond); and (3) coursework in a science education sequence involving the research literature, curricula, instruction, and current topics; the history, sociology, and philosophy of science; selected cognitive sciences; both qualitative and quantitative research methods; and writing for publication. This coursework should be complemented by relevant professional experiences, from collaboration with practicing researchers to scholarly paper presentations to grant writing to course design.

We see the doctoral student and his or her adviser developing an individualized sequence of coursework and professional experiences using the preceding criteria as benchmarks. We endorse early (but thoughtful) identification of a research problem by the student so that most of his or her interdisciplinary coursework and reading can be done with that problem in mind.

To understand how people think in a scientific domain and what constitutes scientific literacy requires in-depth knowledge of that domain, broad appreciation of the ways in which its central concepts evolved, and awareness of both the habits of mind and the epistemological commitments of experts in that discipline. This kind of knowledge is not made available to those who opt to pursue more parochial training, rather than interdisciplinary education at the graduate level. We must move beyond *training* science educators. We need persons who can ask good new questions, not parrot pat answers to old questions.

In addition, learning to do quality research does not occur in a vacuum, nor is it born of coursework alone. It is nurtured by intensive, regular, and intellectually stimulating interaction between the novice, his or her peers, and at least one experienced researcher who maintains a high level of direct involvement in a coordinated, goal-directed, theory-driven research program.

We see the graduate student proposing an original research problem that falls somewhere within the research program of the doctoral adviser. The benefits to both student and adviser far outweigh the disadvantages. Such a nurturing environment is possible, not only at a nucleus of large research centers but also at smaller institutions that can offer the resources and the established researchers necessary for guidance and success.

An Ending and a Beginning

In conclusion, we have provided an overview of the ACM, its major research findings, and its experts' thoughts on progress in the field. We have concluded with our own examination of the research agenda, and we have proposed new directions for both the education of researchers and the research they might do. In an attempt to make a unique contribution to the literature, we did not want to duplicate the important contributions made by others.

For example, the bibliographies by Pfundt and Duit (1985, 1988, 1991) and Carmichael et al. (1990) are unparalleled. The latter also concisely addresses some epistemological, psychological, and methodological issues within ACM that we have not raised.

The book edited by Glynn et al. (1991), *The Psychology of Learning Science,* bridges the gap between state-of-the-art ACM research and classroom teaching practice. For example, in the chapter on "Children's Biology" (Mintzes et al., 1991), we review some of our own research findings and suggest ways to use ACM research findings in teaching the life sciences. Another chapter, by Duit (1991), examines the consequences of students' conceptions for learning science and should be of interest to most readers.

A feature article in the *Journal of NIH Research* by Hooper (1990) may also be of interest in view of its unique slant—it was written especially for practicing research scientists in order to explain "what science is learning about learning science," and it discusses the ACM in the context of other advances in science education. Driver's (1989) article on students' conceptions and science learning is also noteworthy and leads off an issue of the *International Journal of Science Education* devoted to research on students' conceptions. West and Pines (1985) have edited a fine volume on ways to represent cognitive structure and effect conceptual change. There is certainly no shortage of literature for those who want to learn more about the alternative conceptions movement.

References

Abimbola, I. O. (1988). The problem of terminology in the study of student conceptions in science. *Science Education, 72,* 175–184.

Adeniyi, E. O. (1985). Misconceptions of selected ecological concepts held by Nigerian students. *Journal of Biological Education, 19,* 311–316.

Aguirre, J. M. (1988). Student preconceptions about vector kinematics. *Physics Teacher, 26,* 212–216.

Albert, E. (1978). Development of the concept of heat in children. *Science Education, 63,* 389–399.

Alexander, P. A., & Judy, J. (1988). The interaction of domain-specific and strategic knowledge in academic performance. *Review of Educational Research, 58,* 375–404.

Allis, S. (1989, January 9). Lessons from on high. *Time,* 65–66.

Ameh, C. (1987). An analysis of teachers' and their students' views of the concept "gravity." *Research in Science Education, 17,* 212–219.

Ameh, C., & Gunstone, R. (1985). Teachers' concepts in science. *Research in Science Education, 15,* 151–157.

American Association for the Advancement of Science. (1989). *Science for all Americans.* Washington, DC: Author.

American Psychological Association. (1983). *Publication manual of the American Psychological Association* (3rd ed.). Washington, DC: Author.

Anderson, C., Sheldon, T., & Dubay, J. (1990). The effects of instruction on college nonmajors' conceptions of respiration and photosynthesis. *Journal of Research in Science Teaching, 27,* 761–776.

Anderson, C., & Smith, E. (1983). *Children's conceptions of light and color: Understanding the concept of unseen rays.* East Lansing: Michigan State University.

Andersson, B. (1986). The experiential gestalt of causation: A common core to pupils' preconceptions in science. *European Journal of Science Education, 2,* 155–171.

Andersson, B. (1990). Pupils' conceptions of matter and its transformations. *Studies in Science Education, 18,* 53–85.

Andersson, B., & Karqvist, C. (1983). How Swedish pupils, aged 12–15 years, understand light and its properties. *European Journal of Science Education, 5,* 387–402.

Arnaudin, M. W., & Mintzes, J. J. (1985). Students' alternative conceptions of the circulatory system: A cross-age study. *Science Education, 69,* 721–733.

Arnaudin, M., & Mintzes, J. J. (1986). The cardiovascular system: Children's conceptions and misconceptions. *Science and Children, 23,* 48–51.

Arnold, M., & Millar, R. (1987). Being constructive: An alternative approach to the teaching of introductory ideas in electricity. *International Journal of Science Education, 9,* 553–563.

Ault, C. R., Novak, J. D., & Gowin, D. B. (1984). Constructing Vee maps for clinical interviews on molecule concepts. *Science Education, 68,* 441–462.

Ausubel, D. P. (1963). *The psychology of meaningful verbal learning.* New York: Grune & Stratton.

Ausubel, D. P. (1968). *Educational psychology: A cognitive view.* New York: Holt, Rinehart, & Winston.

Ausubel, D. P., Novak, J. D., & Hanesian, H. (1978). *Educational psychology: A cognitive view.* New York: Holt, Rinehart, & Winston.

Baird, J. R. (1986). Improving learning through enhanced metacognition: A classroom study. *European Journal of Science Education, 8,* 263–282.

Barker, M., & Carr, M. (1989a). Teaching and learning about photosynthesis. Part I: An assessment in terms of students' prior knowledge. *International Journal of Science Education, 11,* 49–56.

Barker, M., & Carr, M. (1989b). Teaching and learning about photosynthesis. Part II: A generative learning strategy. *International Journal of Science Education, 11,* 141–152.

Barras, R. (1984). Some misconceptions and misunderstandings perpetuated by teachers and textbooks of biology. *Journal of Biological Education, 18,* 201–206.

Basili, P., & Sanford, J. (1991). Conceptual change strategies and cooperative group work in chemistry. *Journal of Research in Science Teaching, 28,* 293–304.

Bauman, R., & Adams, S. (1990). Misunderstandings of electric current. *Physics Teacher, 28,* 334.

Baxter, J. (1989). Children's understanding of familiar astronomical events. *International Journal of Science Education, 11,* 502–513.

Bell, B. (1981a). What is a plant?: Some children's ideas. *New Zealand Science Teacher, 31,* 10–14.

Bell, B. (1981b). When is an animal not an animal? *Journal of Biological Education, 15,* 213–218.

Bell, B. (1985). Students' ideas about plant nutrition: What are they? *Journal of Biological Education, 19*, 213–218.

Bell, B., & Barker, M. (1982). Towards a scientific concept of "animal." *Journal of Biological Education, 16*(3), 197–200.

Bell, B., & Freyberg, P. (1985). Language in the science classroom. In R. Osborne & P. Freyberg (Eds.), *Learning in science. The implications of children's science* (pp. 9–40). Auckland, NZ: Heinemann.

Berliner, D. C. (1989). Implications of studies on expertise in pedagogy for teacher education and evaluation. In *New directions for teacher assessment* (pp. 39–68). Proceedings of the 1988 ETS Invitational Conference. Princeton, NJ: Educational Testing Service.

Bishop, B., & Anderson, C. (1990). Student conceptions of natural selection and its role in evolution. *Journal of Research in Science Teaching, 27*, 415–428.

Black, D., & Solomon, J. (1987). Can pupils use taught analogies for electric current? *School Science Review, 68*, 249–254.

Bliss, J., & Ogborn, J. (1985). Children's choices of uses of energy. *European Journal of Science Education, 7*, 195–203.

Bliss, J., Ogborn, J., & Whitlock, D. (1989). Secondary school pupils commonsense theories of motion. *International Journal of Science Education, 11*, 261–272.

Bloom, J. (1989). Preservice elementary teachers' conceptions of science, theories and evolution. *International Journal of Science Education, 11*, 401–415.

Bloom, J., & Borstad, J. (1990). Comments on "The acquisition of biological knowledge during childhood: Cognitive conflict or tabula rasa?" *Journal of Research in Science Teaching, 27*, 399–403.

Bouwens, R. (1987). Misconceptions among pupils regarding geometrical optics. In J. D. Novak (Ed.), *Proceedings of the Second International Seminar on Misconceptions and Educational Strategies in Science and Mathematics* (Vol. III, pp. 97–107). Ithaca, NY: Department of Education, Cornell University.

Brook, A. (1987). Designing experiences to take account of the development of children's ideas: An example from the teaching and learning of energy. In J. D. Novak (Ed.), *Proceedings of the Second International Seminar on Misconceptions and Educational Strategies in Science and Mathematics* (Vol. II, pp.49–64). Ithaca, NY: Department of Education, Cornell University.

Browning, M., & Lehman, J. (1988). Identification of student misconceptions in genetics problem solving via computer program. *Journal of Research in Science Teaching, 25*, 747- 761.

Brumby, M. (1979). Problems in learning the concept of natural selection. *Journal of Biological Education, 13*, 119–122.

Brumby, M. (1982). Students' conceptions of the life concept. *Science Education, 66*, 613–622.

Brumby, M. (1984). Misconceptions about the concept of natural selection by medical biology students. *Science Education, 68*, 493–503.

Cachapuz, A., & Martins, I. (1987). High school students' ideas about energy of chemical reactions. In J. D. Novak (Ed.), *Proceedings of the Second International Seminar on Misconceptions and Educational Strategies in Science and Mathematics* (Vol. III, pp. 60–68). Ithaca, NY: Department of Education, Cornell University.

Cachapuz, A., & Maskill, R. (1987). Detecting changes with the learning in the organization of knowledge: Use of word association tests to follow the learning of collision theory. *International Journal of Science Education, 9*, 491–504.

Caramazza, A., McCloskey, M., & Green, B. (1980). Curvilinear motion in the absence of external forces: Naive beliefs about the motion of objects. *Science, 210*, 1139–1141.

Caramazza, A., McCloskey, M., & Green, B. (1981). Naive beliefs in 'sophisticated' subjects: Misconceptions about the trajectories of objects. *Cognition, 9*, 117–123.

Carey, S. (1987). *Conceptual change in childhood*. Cambridge, MA: MIT Press.

Carey, S. (1986). Cognitive science and science education. *American Psychologist, 41*, 1123–1130.

Carmichael, P., Watts, M., Driver, R., Holding, B., Phillips, I., & Twigger, D. (1990). *Research on students' conceptions in science: A bibliography*. Leeds, Eng: Children's Learning in Science Research Group, University of Leeds.

Carr, M., Kirkwood, V., Newman, B., & Birdwhistel, R. (1987). Energy in three New Zealand secondary school junior science classrooms. *Research in Science Education, 17*, 117–128.

Chaffee, S. H. (1991). *Explication*. Newbury Park, CA: Sage.

Champagne, A., Gunstone, R., & Klopfer, L. (1983). Naive knowledge and science learning. *Research in Science and Technological Education, 1*, 173–183.

Champagne, A., Gunstone, R., & Klopfer, L. (1985). Effecting changes in cognitive structures among physics students. In L. H. T. West & A. L. Pines (Eds.), *Cognitive structure and conceptual change* (pp. 163–187). New York: Academic Press.

Champagne, A., Klopfer, L., & Anderson, J. (1980). Factors influencing the learning of classical mechanics. *American Journal of Physics, 48*, 1074–1079.

Chi, M., Feltovich, P., & Glaser, R. (1981). Categorization and representation of physics problems by experts and novices. *Cognitive Science, 5*, 121–152.

Cho, H., Kahle, J., & Nordland, F. (1985). An investigation of high school biology textbooks as sources of misconceptions and difficulties in genetics and some suggestions for teaching genetics. *Science Education, 69*, 707–719.

Clement, J. (1982). Students' preconceptions in introductory mechanics. *American Journal of Physics, 50*, 66–71.

Clement, J. (1983). A conceptual model discussed by Galileo and used intuitively by physics students. In D. Gentner & A. Stevens (Eds.), *Mental models* (pp. 325–339). Hillsdale, NJ: Lawrence Erlbaum.

Clement, J. (1987). Overcoming students' misconceptions in physics: The role of anchoring intuitions and analogical validity. In J. D. Novak (Ed.), *Proceedings of the Second International Seminar on Misconceptions and Educational Strategies in Science and Mathematics* (Vol. III, pp. 84–97). Ithaca, NY: Department of Education, Cornell University.

Clement, J., Brown, D., & Zeitsman, A. (1989). Not all preconceptions are misconceptions: Finding "anchoring" conceptions for grounding instruction on students' intuitions. *International Journal of Science Education, 11*, 554–565.

Cochran, K. F. (1991). Pedagogical content knowledge: Teachers' transformations of subject matter. *NARST News, 33*, 7–9.

Cohen, R., Eylon, B., & Ganiel, M. (1983). Potential difference and current in simple electric circuits: A study of students' concepts. *American Journal of Physics, 51*, 407–412.

Confrey, J. (1990). A review of the research on student conceptions in mathematics, science, and programming. In C. Cazden (Ed.), *Review of research in education* (pp. 3–56). Washington, DC: American Educational Research Association.

Cros, D., Chastrette, M., & Fayol, M. (1988). Conceptions of second year university students of some fundamental notions in chemistry. *International Journal of Science Education, 10*, 331–336.

Deadman, J., & Kelly, P. (1978). What do secondary school boys understand about evolution and heredity before they are taught the topics? *Journal of Biological Education, 12*, 7–15.

diSessa, A. (1982). Unlearning Aristotelian physics: A study of knowledge based learning. *Cognitive Science, 6*, 37–75.

Dreyfus, A., & Jungwirth, E. (1988). The cell concept of 10th graders: Curricular expectations and reality. *International Journal of Science Education, 10*, 221–229.

Dreyfus, A., & Jungwirth, E. (1989). The pupil and the living cell: A taxonomy of dysfunctional ideas about an abstract idea. *Journal of Biological Education, 23*, 49–55.

Driver, R. (1981). Pupil's alternative frameworks in science. *European Journal of Science Education, 3*, 93–101.

Driver, R. (1983). *The pupil as scientist?* London: Milton Keynes.

Driver, R. (1987). Promoting conceptual change in classroom settings: The experience of the Children's Learning in Science Project. In J. D. Novak (Ed.), *Proceedings of the Second International Seminar on Misconceptions and Educational Strategies in Science and Mathematics* (Vol. II, pp. 97–107). Ithaca, NY: Department of Education, Cornell University.

Driver, R. (1989). Students' conceptions and the learning of science. *International Journal of Science Education, 11*, 481–490.

Driver, R., & Bell, B. (1986). Students' thinking and the learning of science: A constructivist view. *School Science Review, 67*, 443–456.

Driver, R., & Easley, J. (1978). Pupils and paradigms: A review of the literature related to concept development in adolescent science students. *Studies in Science Education, 5*, 61–84.

Driver, R., Guesne, E., & Tiberghien, A. (Eds.). (1985). *Children's ideas in science*. London: Milton Keynes.

Driver, R., & Oldham, V. (1986). A constructivist approach to curriculum development in science. *Studies in Science Education, 13*, 105–122.

Duit, R. (1981). Understanding energy as a conserved quantity—remarks on the article by R. U. Sexl. *European Journal of Science Education, 3*, 291–301.

Duit, R. (1985). The meaning of current and voltage in everyday language and its consequences for understanding the physical concepts of the electric circuit. In R. Duit, W. Jung, & C. von Rhoneck (Eds.), *Aspects of understanding electricity* (pp. 205–214). Kiel, Ger: Schmidt & Klaunig.

Duit, R. (1991). Students' conceptual frameworks: Consequences for learning science. In S. Glynn, R. Yeany, & B. Britton (Eds.), *The psychology of learning science* (pp. 65–85). Hillsdale, NJ: Lawrence Erlbaum.

Duit, R., & Haussler, P. (1992, July). *Learning and teaching energy: Combining aspects of constructivist, girl suited and STS approaches.* Paper presented at the Content Pedagogy Writing Sorkshop, Centre for Science, Mathematics and Technology, Monash University, Melbourne, Australia.

Dupin, J., & Johsua, S. (1989). Analogies and "modeling analogies" in teaching: Some examples in basic electricity. *Science Education, 73*, 207–224.

Eaton, A., Anderson, C. , & Smith, E. (1983). When students don't know they don't know. *Science and Children, 20*, 7–9.

Eaton, A., Anderson, C. , & Smith, E. (1984). Students' misconceptions interfere with science learning: Case studies of fifth grade students. *Elementary School Journal, 84*, 365–379.

Eisen, Y., & Stavy, R. (1988). Students' understanding of photosynthesis. *American Biology Teacher, 50*, 208- 212.

Eisen, Y., & Stavy, R. (1989). Development of a new science study unit following research on students' ideas about photosynthesis: A case study. In P. Adey (Ed.), *Adolescent development and school science* (pp. 295–302). London: Falmer.

Engel-Clough, E., & Driver, R. (1986). A study of consistency in the use of students' conceptual frameworks across different task contexts. *Science Education, 70*, 473–496.

Engel-Clough, E., Driver, R., & Wood-Robinson, C. (1987). How do children's scientific ideas change over time? *School Science Review, 69*, 255–267.

Engel-Clough, E., & Wood-Robinson, C. (1985). Children's understanding of inheritance. *Journal of Biological Education, 19*, 304–310.

Enochs, L., & Gabel, D. (1984). Preservice elementary teachers' conceptions of volume. *School Science and Mathematics, 84*, 670–680.

Erickson, G. (1979). Children's conceptions of heat and temperature. *Science Education, 63*, 221–230.

Erickson, G. (1980). Children's viewpoints of heat: A second look. *Science Education, 64*, 323–336.

Erickson, G., & Tiberghien, A. (1985). Heat and temperature. In R. Driver, E. Guesne, & A. Tiberghien (Eds.), *Children's ideas in science* (pp. 52–84). London: Milton Keynes.

Erlwanger, S. H. (1973). Benny's conceptions of rules and answers in IPI mathematics. *Journal of Children's Mathematical Behavior, 1*, 7–26.

Feher, E. (1990). Interactive museum exhibits as tools for learning: Explorations with light. *International Journal of Science Education, 12*, 35–49.

Feher, E., & Rice, K. (1987). A comparison of teacher-student conceptions in optics. In J. D. Novak (Ed.), *Proceedings of the Second International Seminar on Misconceptions and Educational Strategies in Science and Mathematics* (Vol. II, pp. 108–117). Ithaca, NY: Department of Education, Cornell University.

Feher, E., & Rice, K. (1988). Shadows and anti-images: Children's conceptions of light and vision II. *Science Education, 72*(5), 637–649.

Feldsine, J. E. (1987). Distinguishing student misconceptions from alternative conceptual frameworks through construction of concept maps. In J. D. Novak (Ed.), *Proceedings of the Second International Seminar on Misconceptions and Educational Strategies in Science and Mathematics* (Vol. I, pp. 177–181). Ithaca, NY: Department of Education, Cornell University.

Fensham, P. (1992, July). *Beginning to teach chemistry*. Paper presented at the Content Pedagogy Writing Workshop, Monash University Centre for Science, Mathematics, and Technology, Melbourne, Australia.

Finegold, M., & Pundak, D. (1990). Students' conceptual frameworks in astronomy. *Australian Science Teachers Journal, 36*, 76–83.

Fishbein, E., Stavy, R., & Ma-Naim, H. (1989). The psychological structure of naive impetus conceptions. *International Journal of Science Education, 11*, 71–81.

Fisher, K. (1983). Amino acids and translation: A misconception in biology. In H. Helm & J. Novak (Eds.), *Proceedings of the International Seminar on Misconceptions in Science and Mathematics* (pp. 407–419). Ithaca, NY: Department of Education, Cornell University.

Fisher, K. (1985). A misconception in biology: Amino acids and translation. *Journal of Research in Science Teaching, 21*, 53–62.

Fisher, K., & Lipson, J. (1986). Twenty questions about student errors. *Journal of Research in Science Teaching, 23*, 783–803.

Fredette, N., & Clement, J. (1981). Student misconceptions of an electric current: What do they mean? *Journal of College Science Teaching, 10*, 280–285.

Fredette, N., & Lochhead, J. (1980). Student conceptions of simple circuits. *Physics Teacher, 18*, 194–198.

Furio Mas, C., Hernandez-Perez, J., & Harris, H. (1987). Parallels between adolescents' conception of gases and the history of science. *Journal of Chemical Education, 64*, 616–618.

Gabel, D., Samuel, K., & Hunn, D. (1987). Understanding the particulate nature of matter. *Journal of Chemical Education, 64*, 695–697.

Gallert, E. (1962). Children's conceptions of the content and structure of the human body. *Genetic Psychology Monographs, 65*, 293–405.

Ganiel, U., & Idar, J. (1985). Student misconceptions in science—how can computers help? *Journal of Computer, Mathematics and Science Teaching, 4*, 14–19.

Garner, R. (1987). *Metacognition and reading comprehension*. Norwood, NJ: Ablex.

Garner, R., & Alexander, P. (1989). Metacognition: Answered and unanswered questions. *Educational Psychologist, 24*, 143–158.

Garnett, D., Garnett, P., & Treagust, D. (1990). Implications of research on students' understanding of electrochemistry for improving science curricula and classroom practice. *International Journal of Science Education, 12*, 147–156.

Gauld, C. (1987). Student beliefs and cognitive structure. *Research in Science Education, 17*, 87–93.

Gauld, C. (1988). The cognitive context of pupils' alternative frameworks. *International Journal of Science Education, 10*, 267–274.

Gilbert, J., Osborne, R., & Fensham, P. (1982). Children's science and its consequences for teaching. *Science Education, 66*, 623–633.

Gilbert, J., & Swift, D. (1985). Towards a Lakatosian analysis of the Piagetian and alternative conceptions research programs. *Science Education, 69*, 681–696.

Gilbert, J., Watts, D., & Osborne, R. (1985). Eliciting student views using an interview-about-instances technique. In L. H. T. West & A. L. Pines (Eds.), *Cognitive structure and conceptual change* (pp. 11–27). Orlando, FL: Academic Press.

Gil-Perez, D., & Carrascosa, A. (1987). What to do for science misconceptions. In J. D. Novak (Ed.), *Proceedings of the Second International Seminar on Misconceptions and Educational Strategies in Science and Mathematics* (Vol. II, pp. 149–157). Ithaca, NY: Department of Education, Cornell University.

Glaserfeld, E. von (1989). Cognition, construction of knowledge, and teaching. *Synthese, 80*, 121–140.

Glynn, S., Yeany, R., & Britton, B. (Eds.). (1991). *The psychology of learning science.* Hillsdale, NJ: Lawrence Earlbaum.

Goldberg, F., & McDermott, L. (1986). Student difficulties in understanding image formation by a plane mirror. *Physics Teacher, 24*, 472–480.

Good, R. (1991). Editorial. *Journal of Research in Science Teaching, 28*(5), 387.

Good, R., Herron, J. D., Lawson, A., & Renner, J. (1985). The domain of science education. *Science Education, 69*, 139–141.

Gorodetsky, M., & Gussarsky, E. (1986). Misconceptualization of the chemical equilibrium concept as revealed by different evaluation methods. *European Journal of Science Education, 8*, 427–441.

Gowin, D. B. (1981). *Educating.* Ithaca, NY: Cornell University Press.

Gowin, D. B. (1983). Misconceptions, metaphors, and conceptual change: Once more with feeling. In H. Helm & J. D. Novak (Eds.), *Proceedings of the International Seminar on Misconceptions in Science and Mathematics* (pp. 39–41). Ithaca, NY: Department of Education, Cornell University.

Guesne, E. (1985) Light. In R. Driver, E. Guesne, & A. Tiberghien (Eds.), *Children's ideas in science* (pp. 10–32). London: Milton Keynes.

Gunstone, R. (1987). Student understanding in mechanics: A large population survey. *American Journal of Physics, 55*, 691–696.

Gunstone, R. (1989). A comment on 'the problem of terminology in the study of student conceptions in science.' *Science Education, 73*, 643–646.

Gunstone, R., & White, R. (1981). Understanding of gravity. *Science Education, 65*, 291–299.

Gunstone, R., & White, R. (1986). Assessing understanding by means of Venn diagrams. *Science Education, 70*, 151–158.

Gussarsky, E., & Gorodetsky, M. (1990). On the concept "chemical equilibrium": The associative framework. *Journal of Research in Science Teaching, 27*, 197–204.

Hackling, M., & Garnett, D. (1985). Misconceptions of chemical equilibrium. *European Journal of Science Education, 7*, 205–214.

Hall, A. R. (1954). *The scientific revolution, 1500–1800: The formation of the modern scientific attitude.* London: Longmans, Green.

Hallden, O. (1988). The evolution of the species: Pupil perspectives and school perspectives. *International Journal of Science Education, 10*, 541–552.

Hand, B., & Treagust, D. (1988). Application of a conceptual conflict teaching strategy to enhance student learning of acids and bases. *Research in Science Education, 18*, 53–63.

Happs, J. (1985). Cognitive learning theory and classroom complexity. *Research in Science and Technological Education, 3*, 157–174.

Happs, J., & Scherpenzeel, L. (1987). Achieving long term conceptual change using the learner's prior knowledge and a novel teaching setting. In J. D. Novak (Ed.), *Proceedings of the Second International Seminar on Misconceptions and Educational Strategies in Science and Mathematics* (Vol. II, pp. 172–181). Ithaca, NY: Department of Education, Cornell University.

Hashweh, M. (1988). Descriptive studies of students' conceptions in science. *Journal of Research in Science Teaching, 25*, 121–134.

Hawkins, J., & Pea, R. (1987). Tools for bridging the cultures of everyday and scientific thinking. *Journal of Research in Science Teaching, 24*, 291–307.

Heinze-Fry, J., & Novak, J. D. (1990). Concept mapping brings long-term movement toward meaningful learning. *Science Education, 74*, 461–472.

Heller, P. (1987). Use of core propositions in solving current electricity problems. In J. D. Novak (Ed.), *Proceedings of the Second International Seminar on Misconceptions and Educational Strategies in Science and Mathematics* (Vol. III, pp. 225–235). Ithaca, NY: Department of Education, Cornell University.

Helm, H., & Novak, J. D. (Eds.). (1983). *Proceedings of the International Seminar on Misconceptions in Science and Mathematics.* Ithaca, NY: Department of Education, Cornell University.

Hewson, M. (1985). The role of intellectual environment in the origin of conceptions: An exploratory study. In L. H. T. West & A. L. Pines (Eds.), *Cognitive structure and conceptual change* (pp. 152–161). Orlando, FL: Academic Press.

Hewson, M., & Hamlyn, D. (1984). The influence of intellectual environment on conceptions of heat. *European Journal of Science Education, 6*, 245–262.

Hewson, M., & Hewson, P. (1983). Effect of instruction using students' prior knowledge and conceptual change strategies in science learning. *Journal of Research in Science Teaching, 20*, 731–743.

Hibbard, K. M., & Novak, J. D. (1975). Audio-tutorial elementary school science instruction as a method for study of children's concept learning: Particulate nature of matter. *Science Education, 59*, 559–570.

Hooper, C. (1990). What science is learning about learning science. *Journal of NIH Research, 2*, 75–81.

Jimenez, Aleixandre, M., & Fernandez Perez, J. (1987). Selection or adjustment? Explanations of university biology students for natural selections problems. In J. D. Novak (Ed.), *Proceedings of the Second International Seminar on Misconceptions and Educational Strategies in Science and Mathematics* (Vol. II, pp. 224–238). Ithaca, NY: Department of Education, Cornell University.

Johnstone, A., McDonald, J., & Webb, G. (1977). Misconceptions in school thermodynamics. *Physics Education, 12*, 248–251.

Johsua, S. (1984). Students' interpretation of simple electrical diagrams. *European Journal of Science Education, 6*(3), 271- 275.

Johsua, S., & Dupin, J. (1987). Taking into account student conceptions in instructional strategy: An example in physics. *Cognition and Instruction, 4*, 117–135.

Jones, B., Lynch, P., & Reesink, C. (1987). Children's conceptions of the earth, sun and moon. *International Journal of Science Education, 9*, 43–53.

Jung, W. (1987). Understanding students' understanding: The case of elementary optics. In J. D. Novak (Ed.), *Proceedings of the Second International Seminar on Misconceptions and Educational Strategies in Science and Mathematics* (Vol. III, pp. 268–277). Ithaca, NY: Department of Education, Cornell University.

Jungwirth, E. (1971). A comparison of the acquisition of taxonomic concepts by BSCS and non-BSCS pupils. *Australian Science Teachers Journal, 17*(4), 80–82.

Jungwirth, E. (1975). Preconceived adaptation and inverted evolution (a case study of distorted concept formation in high school biology). *Australian Science Teachers Journal, 21*, 95–100.

Jungwirth, E. (1988). The associative field as a diagnostic instrument in assessing the breadth of multicontextual concepts: The concept "development." *International Journal of Science Education, 10,* 571–579.

Kargbo, D., Hobbs, E., & Erickson, G. (1980). Children's beliefs about inherited characteristics. *Journal of Biological Education, 14,* 137–146.

Karmiloff-Smith, A., & Inhelder, B. (1975). If you want to get ahead, get a theory. *Cognition, 3,* 195–212.

Kelly, G. (1955). *The psychology of personal constructs.* New York: Norton.

Kenealy, P. (1987). A syntactic source of a common "misconception" about acceleration. In J. D. Novak (Ed.), *Proceedings of the Second International Seminar on Misconceptions and Educational Strategies in Science and Mathematics* (Vol. III, pp. 278–292). Ithaca, NY: Department of Education, Cornell University.

Klein, C. (1982). Children's concepts of the earth and the sun: A cross-cultural study. *Science Education, 65,* 95–107.

Kreitler, H., & Kreitler, S. (1966). Children's concepts of sexuality and birth. *Child Development, 37,* 363–378.

Kruger, C. (1990). Some primary teachers' ideas about energy. *Physics Education, 25,* 86–91.

Kuhn, C., & Aguirre, J. (1987). A case study of the "journal method," a method designed to enable the implementation of constructivist teaching in the classroom. In J. D. Novak (Ed.), *Proceedings of the Second International Seminar on Misconceptions and Educational Strategies in Science and Mathematics* (Vol. II, pp. 262–274). Ithaca, NY: Department of Education, Cornell University.

Kuhn, T. (1970). *The structure of scientific revolutions.* Chicago: University of Chicago Press.

Lakatos, I. (1970). Falsification and the methodology of scientific research programmes. In I. Lakatos & A. Musgrave (Eds.), *Criticism and the growth of knowledge.* New York: Cambridge University Press.

Lamarck, J. B. (1809). *Philosphie zoologique, ou exposition des considerations relatives a l'histoire naturelle des animaux.* Paris. [Reprinted in English by University of Chicago Press, 1984.]

Lawrenz, F. (1986). Misconceptions of physical science concepts among elementary school teachers. *School Science and Mathematics, 86,* 654–660.

Lawson, A. (1988). The acquisition of biological knowledge during childhood: Cognitive conflict or tabula rasa? *Journal of Research in Science Teaching, 25,* 185–199.

Lawson, A. (1991a). Constructivism and domains of scientific knowledge: A reply to Lythcott and Duschl. *Science Education, 75,* 481–488.

Lawson, A. (1991b). Is Piaget's epistemic subject dead? *Journal of Research in Science Teaching, 28,* 581–592.

Lazarowitz, R. (1981). Correlations of junior high school students' age, gender and intelligence with ability to construct classification in biology. *Journal of Research in Science Teaching, 18,* 15–22.

Licht, P. (1987). A strategy to deal with conceptual and reasoning problems in introductory electricity education. In J. D. Novak (Ed.), *Proceedings of the Second International Seminar on Misconceptions and Educational Strategies in Science and Mathematics* (Vol. II, pp. 275–284). Ithaca, NY: Department of Education, Cornell University.

Lightman, A., Miller, J., & Leadbeater, B. (1987). Contemporary cosmological beliefs. In J. D. Novak (Ed.), *Proceedings of the Second International Seminar on Misconceptions and Educational Strategies in Science and Mathematics* (Vol. III, pp. 309–321). Ithaca, NY: Department of Education, Cornell University.

Lin, H. (1983). A "cultural" look at physics students and physics classrooms—an example of anthropological work in science education. In H. Helm & J. D. Novak (Eds.), *Proceedings of the International Seminar on Misconceptions in Science and Mathematics* (pp. 194–213). Ithaca, NY: Department of Education, Cornell University.

Linder, C., & Erickson, G. (1989). A study of tertiary physics students' conceptualizations of sound. *International Journal of Science Education, 11,* 491–501.

Lythcott, J. (1985). "Aristotelian" was given as the answer, but what was the question? *American Journal of Physics, 53,* 428–432.

Lythcott, J., & Duschl, R. (1990). Qualitative research: From methods to conclusions. *Science Education, 74,* 445–460.

Mackay, A. (Ed.). (1977). *The harvest of a quiet eye: A selection of scientific quotations.* New York: Crane, Russak.

Mali, G., & Howe, A. (1979). Development of earth and gravity concepts among Nepali children. *Science Education, 63,* 685–691.

Maloney, D. (1988). Novice rules for projective motion. *Science Education, 72,* 501–513.

Matthews, M. (1987). Experiment as the objectification of theory: Galileo's revolution. In J. D. Novak (Ed.), *Proceedings of the Second International Seminar on Misconceptions and Educational Strategies in Science and Mathematics* (Vol. I, pp. 289–298). Ithaca, NY: Department of Education, Cornell University.

Mayer, M. (1987). Common sense knowledge vs. scientific knowledge: The case of pressure, weight and gravity. In J. D. Novak (Ed.), *Proceedings of the Second International Seminar on Misconceptions and Educational Strategies in Science and Mathematics* (Vol. I, pp. 299–310). Ithaca, NY: Department of Education, Cornell University.

McClelland, J. (1984). Alternative frameworks: Interpretation of evidence. *European Journal of Science Education, 6,* 1–6.

McCloskey, M. (1983). Naive theories of motion. In D. Gentner & A. Stevens (Eds.), *Mental models* (pp. 299–324). Hillsdale, NJ: Lawrence Erlbaum.

McDermott, L. (1984). Research on conceptual understanding in mechanics. *Physics Today, 37,* 24–32.

Meyer, E. (1987). Thermodynamics of mixing ideal gases: A persistent pitfall. *Journal of Chemical Education, 64,* 676.

Millar, R. (1989). Constructive criticisms. *International Journal of Science Education, 1,* 587–596.

Millar, R. (1990). Making sense: What use are particle ideas to children? In P. L. Lijnse, P. Licht, W. de Vos, & A. J. Waarlo (Eds.), *Relating macroscopic phenomena to microscopic particles* (Proceedings of a seminar). Utrecht, Netherlands: Centre for Science and Mathematics Education, University of Utrecht.

Minstrell, J. (1982). Explaining the "at rest" condition of an object. *Physics Teacher, 20,* 10–14.

Mintzes, J. J. (1984). Naive theories in biology: Children's concepts of the human body. *School Science and Mathematics, 84,* 548–555.

Mintzes, J. J. (1989). The acquisition of biological knowledge during childhood: An alternative conception. *Journal of Research in Science Teaching, 26,* 823–824.

Mintzes, J. J., Trowbridge, J. E., Arnaudin, M., & Wandersee, J. H. (1991). Children's biology: Studies on conceptual development in the life sciences. In S. Glynn, R. Yeany, & B. Britton (Eds.), *The psychology of learning science* (pp. 179–202). Hillsdale, NJ: Lawrence Erlbaum.

Mohapatra, J. (1988). Induced incorrect generalizations leading to misconceptions—an exploratory investigation about the laws of reflection of light. *Journal of Research in Science Teaching, 25,* 777–784.

Mohapatra, J., & Bhattacharyya, S. (1989). Pupils [sic] teachers [sic], induced incorrect generalization and the concept of "force." *International Journal of Science Education, 11,* 429–436.

Moore, J., & Kendall, D. (1971). Children's concepts of reproduction. *Journal of Sex Research, 7,* 42–61.

Mori, I., Kojima, M., & Tadang, N. (1976). The effect of language on a child's conception of speed: A comparative study on Japanese and Thai children. *Science Education, 60,* 531–534.

Nachmias, R., Stavy, R., & Avrams, R. (1990). A microcomputer-based diagnostic system for identifying students' conceptions of heat and temperature. *International Journal of Science Education, 12,* 123–132.

Novak, J. D. (1977). *A theory of education*. Ithaca, NY: Cornell University Press.

Novak, J. D. (1987). Human constructivism: Toward a unity of psychological and epistemological meaning making. In J. D. Novak (Ed.), *Proceedings of the Second International Seminar on Misconceptions and Educational Strategies in Science and Mathematics* (Vol. I, pp. 349–360). Ithaca, NY: Department of Education, Cornell University.

Novak, J. D., & Gowin, D. B. (1984). *Learning how to learn*. New York: Cambridge University Press.

Novak, J. D., & Wandersee, J. H. (Eds.). (1990). Perspectives on concept mapping [Special issue]. *Journal of Research in Science Teaching, 27*, 922–1079.

Novick, S., & Menis, J. (1976). A study of student perceptions of the mole concept. *Journal of Chemical Education, 53*, 720–722.

Novick, S., & Nussbaum, J. (1978). Junior high school pupil's understanding of the particulate nature of matter: An interview study. *Science Education, 62*, 273–281.

Novick, S., & Nussbaum, J. (1981). Pupils' understanding of the particulate nature of matter: A cross-age study. *Science Education, 65*, 187–196.

Nussbaum, J. (1979). Children's conception of the earth as a cosmic body: A cross-age study. *Science Education, 63*, 83–93.

Nussbaum, J. (1983). Classroom conceptual change: The lesson to be learned from the history of science. In H. Helm & J. D. Novak (Eds.), *Proceedings of the International Seminar on Misconceptions in Science and Mathematics* (pp. 272–281). Ithaca, NY: Department of Education, Cornell University.

Nussbaum, J. (1985). The particulate nature of matter in the gaseous phase. In R. Driver, E. Guesne, & A. Tiberghien (Eds.), *Children's ideas in science* (pp. 124–144). London: Milton Keynes.

Nussbaum, J., & Novak, J. D. (1976). An assessment of children's concepts of the earth utilizing structured interviews. *Science Education, 60*, 535–550.

Nussbaum, J., & Novick, S. (1982). Alternative frameworks, conceptual conflict and accommodation: Toward a principled teaching strategy. *Instructional Science, 11*, 183–200.

Nussbaum, J., & Sharodini-Dagan, N. (1983). Changes in second grade children's preconceptions about the earth as a cosmic body resulting from a short series of audio-tutorial lessons. *Science Education, 67*, 99–114.

Ogunniyi, M. (1987). Conceptions of traditional cosmological ideas among literate and nonliterate Nigerians. *Journal of Research in Science Teaching, 24*, 107–117.

Ogunniyi, M. B., & Pella, M. O. (1980). Conceptualizations of scientific concepts, laws and theories held by Kwara State, Nigeria secondary school science teachers. *Science Education, 64*, 591–599.

Osborne, R. (1981). Children's ideas about electric current. *New Zealand Science Teacher, 29*, 12–19.

Osborne, R. (1983). Towards modifying children's ideas about electric current. *Research in Teaching and Technological Education, 1*, 73–82.

Osborne, R. J., Bell, B., & Gilbert, J. K. (1983). Science teaching and children's views of the world. *European Journal of Science Education, 5*, 1–14.

Osborne, R. J., & Gilbert, J. K. (1980). A method for investigating concept understanding in science. *European Journal of Science Education, 2*, 311–321.

Osborne, R., & Wittrock, M. (1983). Learning science: A generative process. *Science Education, 67*, 489–508.

Osborne, R., & Wittrock, M. (1985). The generative learning model and its implications for science education. *Studies in Science Education, 12*, 59–87.

Paris, S., & Oka, E. (1986). Children's reading strategies, metacognition, and motivation. *Developmental Review, 6*, 25–56.

Patel, V., Kaufman, D., & Magder, S. (1991). Causal explanation of complex physiological concepts by medical students. *International Journal of Science Education, 13*, 171–185.

Peters, P. (1982). Even honor students have conceptual difficulties with physics. *American Journal of Physics, 50*, 501–508.

Peterson, R., & Treagust, D. (1989). Grade 12 students' misconceptions of covalent bonding and structure. *Journal of Chemical Education, 66*, 459–460.

Pfundt, H., & Duit, R. (1985). *Bibliography: Students' alternative frameworks and science education*. Kiel, Ger: University of Kiel Institute for Science Education (Institut für die Padagogik der Naturwissenschaften).

Pfundt, H., & Duit, R. (1988). *Bibliography: Students' alternative frameworks and science education*. Kiel, Ger: University of Kiel Institute for Science Education (Institut für die Padagogik der Naturwissenschaften).

Pfundt, H., & Duit, R. (1991). *Bibliography: Students' alternative frameworks and science education*. Kiel, Ger: University of Kiel Institute for Science Education (Institut für die Padagogik der Naturwissenschaften).

Piaget, J. (1929). *The child's conception of the world*. New York: Harcourt Brace.

Piaget, J. (1970). *Genetic epistemology*. New York: Columbia University Press.

Pines, A., Novak, J. D., Posner, G., & Van Kirk, J. (1978). *The clinical interview: A method for evaluating cognitive structure* (Research Report No. 6). Ithaca, NY: Department of Education, Cornell University.

Placek, W. (1987). Preconceived knowledge of certain Newtonian concepts among gifted and non-gifted eleventh grade physics students. In J. D. Novak (Ed.), *Proceedings of the Second International Seminar on Misconceptions and Educational Strategies in Science and Mathematics* (Vol. III, pp. 386–391). Ithaca, NY: Department of Education, Cornell University.

Pope, M., & Gilbert, J. (1983). Personal experience and the construction of knowledge in science. *Science Education, 67*, 193–203.

Porter, C. (1974). Grade school children's perceptions of their internal body parts. *Nursing Research, 23*, 384–391.

Posner, G., Strike, K., Hewson, P., & Gertzog, W. (1982). Accommodation of a scientific conception: Toward a theory of conceptual change. *Science Education, 66*, 211–227.

Preece, P. (1984). Intuitive science: Learned or triggered? *European Journal of Science Education, 6*, 7–10.

Pyramid Film & Video. (1988). *A private universe. An insightful lesson on how we learn* [Flier]. Santa Monica, CA: Author.

Rice, K., & Feher, E. (1987). Pinholes and images: Children's conceptions of light and vision I. *Science Education, 71*, 629–639.

Rogan, J. (1988). Development of a conceptual framework of heat. *Science Education, 72*, 103–113.

Rollnick, M., & Rutherford, M. (1990). African primary school teachers—what ideas do they hold on air and air pressure? *International Journal of Science Education, 12*, 101–113.

Ross, K., & Sutton, C. (1982). Concept profiles and the cultural context. *European Journal of Science Education, 4*, 311–323.

Roth, K., Smith, E., & Anderson, C. (1983). *Students' conceptions of photosynthesis and food for plants*. East Lansing: Institute for Research on Teaching, Michigan State University.

Rowell, J., & Dawson, C. (1981). Volume, conservation and instruction: A classroom based Solomon four group study of conflict. *Journal of Research in Science Teaching, 18*, 533–546.

Ryman, D. (1974). Children's understanding of classification of living organisms. *Journal of Biological Education, 8*, 140-144.

Ryman, D. (1977). Teaching method, intelligence and gender factors in pupil achievement on a classification task. *Journal of Research in Science Teaching, 14,* 401–409.

Schmidt, H. (1987). Secondary school students' learning difficulties in stoichiometry. In J. D. Novak (Ed.), *Proceedings of the Second International Seminar on Misconceptions and Educational Strategies in Science and Mathematics* (Vol. I, pp. 396–404). Ithaca, NY: Department of Education, Cornell University.

Schoenfeld, A. (1987). What's all the fuss about metacognition? In A. Schoenfeld (Ed.), *Cognitive science and mathematics education* (pp. 189–215). Hillsdale, NJ: Lawrence Erlbaum.

Scott, P. (1987). The process of conceptual change in science: A case study of the development of a secondary pupil's ideas relating to matter. In J. D. Novak (Ed.), *Proceedings of the Second International Seminar on Misconceptions and Educational Strategies in Science and Mathematics* (Vol. II, pp. 404–419). Ithaca, NY: Department of Education, Cornell University.

Shanon, B. (1976). Aristotelianism, Newtonianism, and the physics of the layman. *Perception, 5,* 241–243.

Shapiro, B. (1989). What children bring to light: Giving high status to learner's views and actions in science. *Science Education, 73,* 711–733.

Shayer, M., & Wylam, H. (1981). The development of the concepts of heat and temperature in 10–13-year-olds. *Journal of Research in Science Teaching, 18,* 419–434.

Shipstone, D. (1984). A study of children's understanding of electricity in simple DC circuits. *European Journal of Science Education, 6,* 185–198.

Shulman, L. S. (1986). Those who understand: Knowledge growth in teaching. *Educational Researcher, 15,* 4–14.

Slife, B., Weiss, J., & Bell, T. (1985). Separability of metacognition and cognition: Problem solving in learning disabled and regular students. *Journal of Educational Psychology, 77,* 437–445.

Smith, D. C. (1987). Primary teachers' misconceptions about light and shadows. In J. D. Novak (Ed.), *Proceedings of the Second International Seminar on Misconceptions and Educational Strategies in Science and Mathematics* (Vol. II, pp. 461–476). Ithaca, NY: Department of Education, Cornell University.

Smith, E., & Anderson C. (1984). Plants as producers: A case study of elementary science teaching. *Journal of Research in Science Teaching, 21,* 685–698.

Smith, M. U. (1991). *Toward a unified theory of problem solving: Views from the content domains.* Hillsdale, NJ: Lawrence Erlbaum.

Sneider, C., & Pulos, S. (1983). Children's cosmographies: Understanding the earth's shape and gravity. *Science Education, 67,* 205–221.

Solomon, J. (1982). How children learn about energy or does the first law come first? *School Science Review, 63,* 415–422.

Solomon, J. (1983). Learning about energy: How pupils think in two domains. *European Journal of Science Education, 5,* 49–59.

Solomon, J. (1984). Prompts, cues and discrimination: The utilization of two separate knowledge systems. *European Journal of Science Education, 6,* 63–82.

Solomon, J. (1987). Social influences on the construction of pupils' understanding of science. *Studies in Science Education, 14,* 63–82.

Solomon, J., Black, P., Oldham, V., & Stuart, H. (1985). The pupils' view of electricity. *European Journal of Science Education, 7,* 281–294.

Stavy, R. (1988). Children's conception of gas. *International Journal of Science Education, 10,* 553–560.

Stavy, R. (1989). Students' conceptions of matter. In P. Adey (Ed.), *Adolescent development and school science* (pp. 273–282). London: Falmer.

Stavy, R., & Berkovitz, B. (1990). Cognitive conflict as a basis for teaching quantitative aspects of the concept of temperature. *Science Education, 64,* 679–692.

Stavy, R., Eisen, Y., & Yaakobi, D. (1987). How students aged 13–15 understand photosynthesis. *International Journal of Science Education, 9,* 105–115.

Stavy, R., & Wax, N. (1989). Children's conceptions of plants as living things. *Human Development, 32,* 635–647.

Stead, B., & Osborne, R. (1980). Exploring students' concepts of light. *Australian Science Teachers Journal, 26,* 84–90.

Stenhouse, D. (1986). Conceptual change in science education: Paradigms and language-games. *Science Education, 70,* 413–425.

Stepans, J. (1985). Biology in elementary schools: Children's conceptions of life. *American Biology Teacher, 47,* 222–225.

Stewart, J. T. (1980). Techniques for assessing and representing information in cognitive structure. *Science Education, 64,* 223–235.

Stewart, J., Hafner, B., & Dale, M. (1990). Students' alternative views of meiosis. *American Biology Teacher, 52,* 228–232.

Tamir, P. (1989). Some issues related to the use of justifications to multiple-choice answers. *Journal of Biological Education, 23,* 285–292.

Tamir, P., Gal-Choppin, R., & Nussinovitz, R. (1981). How do intermediate and junior high school students conceptualize living and nonliving? *Journal of Research in Science Teaching, 18,* 241–248.

Taylor, R. (1963). *A short history of science and scientific thought.* New York: Norton.

Tema, B. (1989). Rural and urban African pupils' alternative conceptions of "animal." *Journal of Biological Education, 23,* 199–207.

Thijs, G. (1987). Conceptions of force and movement. Intuitive ideas of pupils in Zimbabwe in comparison with findings from other countries. In J. D. Novak (Ed.), *Proceedings of the Second International Seminar on Misconceptions and Educational Strategies in Science and Mathematics* (Vol. III, pp. 501–513). Ithaca, NY: Department of Education, Cornell University.

Thorley, N., & Treagust, D. (1987). Conflict within dyadic interactions as a stimulant for conceptual change in physics. *International Journal of Science Education, 9,* 203–216.

Toulmin, S. (1972). *Human understanding: The collective use and evolution of concepts.* Princeton, NJ: Princeton University Press.

Treagust, D. (1988). Development and use of diagnostic tests to evaluate students' misconceptions in science. *International Journal of Science Education, 10,* 159–169.

Triplett, G. (1973). Research on heat and temperature in cognitive development. *Journal of Children's Mathematical Behavior, 1,* 27–43.

Trowbridge, D. E., & McDermott, L. (1981). Investigation of student understanding of the concept of acceleration in one dimension. *American Journal of Physics, 49,* 242–253.

Trowbridge, J. E., & Mintzes, J. J. (1985). Students' alternative conceptions of animals and animal classification. *School Science and Mathematics, 85,* 304–316.

Trowbridge, J. E., & Mintzes, J. J. (1988). Alternative conceptions in animal classification: A cross-age study. *Journal of Research in Science Teaching, 25,* 547–571.

Trowbridge, J. E., & Wandersee, J. H. (in press). *Concept mapping in a college course on evolution: Identifying critical junctures in learning. Journal of Research in Science Teaching.*

Veiga, M., Costa Pereira, D., & Maskill, R. (1989). Teachers' language and pupils' ideas in science lessons: Can teachers avoid reinforcing wrong ideas? *International Journal of Science Education, 11,* 465–479.

Viennot, L. (1979). Spontaneous reasoning in elementary dynamics. *European Journal of Science Education, 1,* 205–221.

Viennot, L. (1985). Analyzing students' reasoning in science: A pragmatic view of theoretical problems. *European Journal of Science Education, 7,* 151–162.

Vosniadou, S. (1990). Conceptual development in astronomy. In

S. Glynn, R. Yeany, & B. Britton (Eds.), *The psychology of learning science* (pp. 149–177). Hillsdale, NJ: Lawrence Erlbaum.

Vygotsky, L. (1962). *Thought and language*. Cambridge, MA: MIT Press.

Wandersee, J. H. (1983a). Students' misconceptions about photosynthesis: A cross-age study. In H. Helm & J. D. Novak (Eds.), *Proceedings of the International Seminar on Misconceptions in Science and Mathematics* (pp. 441–466). Ithaca, NY: Department of Education, Cornell University.

Wandersee, J. H. (1983b). Suppose a world without science educators. *Journal of Research in Science Teaching, 20*, 711–712.

Wandersee, J. H. (1983c). What research says: The concept of "away." *Science and Children, 21*(2), 47–49.

Wandersee, J. H. (1984). Why can't they understand how plants make food?: Students' misconceptions about photosynthesis. *Adaptation, 6*(3), 13, 17.

Wandersee, J. H. (1986a). Can the history of science help science educators anticipate students' misconceptions? *Journal of Research in Science Teaching, 23*, 581–597.

Wandersee, J. H. (1986b). Plants or animals: Which do junior high school students prefer to study? *Journal of Research in Science Teaching, 23*, 415–426.

Wandersee, J. H. (1987a). Drawing concept circles: A new way to teach and test your students. *Science Activities, 24*(4), 9–20.

Wandersee, J. H. (1987b). Francis Bacon: Mastermind of experimental science. *Journal of College Science Teaching, 17*(2), 120–123.

Wandersee, J. H. (1988). Ways students read texts. *Journal of Research in Science Teaching, 25*, 69–84.

Wandersee, J. H. (1990). Concept mapping and the cartography of cognition. *Journal of Research in Science Teaching, 27*, 923–936.

Wandersee, J. H. (1992). The historicality of cognition: Implications for science education research. *Journal of Research in Science Teaching, 29*, 423–434.

Wandersee, J. H., Mintzes, J. J., & Arnaudin, M. (1989). Biology from the learners' viewpoint: A content analysis of the research literature. *School Science and Mathematics, 89*, 654–668.

Watts, D. M. (1982). Gravity—don't take it for granted! *Physics Education, 17*, 116–121.

Watts, D. M. (1983). A study of school children's alternative frameworks of the concept of force. *European Journal of Science Education, 5*, 217–230.

Watts, D. M. (1985). Students' conceptions of light—a case study. *Physics Education, 20*, 183–187.

Watts, D. M., & Gilbert, J. (1983). Enigmas in school science: Students' conceptions for scientifically associated words. *Research in Science and Technological Education, 1*, 161–171.

West, L. H. T., & Pines, L. (Eds.). (1985). *Cognitive structure and conceptual change*. Orlando, FL: Academic Press.

White, R., & Gunstone, R. (1989). Metalearning and conceptual change. *International Journal of Science Education, 11*, 577–586.

White, R., & Gunstone, R. (1992). *Probing understanding*. New York: Falmer.

Wiser, M., & Carey, S. (1983). When heat and temperature were one. In D. Gentner & A. L. Stevens (Eds.), *Mental models* (pp. 267–297). Hillsdale, NJ: Lawrence Erlbaum.

Wittrock, M. (1992, July). *Generative science teaching*. Paper presented at the Content Pedagogy Writing Workshop, Centre for Science, Mathematics and Technology, Monash University, Melbourne, Australia.

Yager, R. (1984). Defining the discipline of science education. *Science Education, 68*, 35–37.

Yager, R. (1991). The constructivist learning model: Towards real reform in science education. *Science Teacher, 58*, 52–57.

Zietsman, A., & Hewson, P. (1986). Effect of instruction using microcomputer simulations and conceptual change strategies on science learning. *Journal of Research in Science Teaching, 23*, 27–39.

· 6 ·

RESEARCH ON THE AFFECTIVE DIMENSION
OF SCIENCE LEARNING

Ronald D. Simpson, Thomas R. Koballa, Jr., and J. Steve Oliver
THE UNIVERSITY OF GEORGIA

Frank E. Crawley III
THE UNIVERSITY OF TEXAS AT AUSTIN

Science is the human attempt to command more of nature's hidden potential. Focusing on the outcomes of a few key discoveries over the past century reveals the meaning of this concept. Life has been enhanced by the products of science and technology, and the United States is an example of what can be accomplished by humankind through the scientific enterprise.

It is possible to think about the hidden potential of science teaching in much the same way. That is, what is it about our work as teachers that holds perhaps the greatest hope or potential for the future? What is it about an educational experience that significantly changes a young person? Why do some graduates become intelligent consumers of science and some do not? Why do some people immerse themselves in science while others shun the scientific enterprise? Why do some students seem to "click" into place while others wander outside the mainstream of science? Answers to these questions are important to science educators. Also, these questions have become increasingly important to researchers in psychology, sociology, and education. Furthermore, these are questions whose solutions may help us solve some of the tough problems the United States now faces within the education enterprise.

Those who teach science have come to a universal observation: Students' behavior is influenced by the values they hold, the motivation they possess, the beliefs they bring from home to the classroom, and the myriad attitudes they have formulated about school, science, and life in general. The key to success in education often depends on how a student feels toward home, self, and school.

A broad overview of how the affective domain relates to science education is developed in this chapter. We attempt to deal with ideas that are relevant and useful to practicing teachers. The landscape of this somewhat new area of science education is sketched. Attitudes, values, motivation, and other important concepts are identified and defined. Research on this topic is summarized and discussed. The results of social-psychological research in the context of science education are addressed. Finally, the future implications of affective research are discussed. This chapter should provide science educators with important ideas, possible strategies, and suggested plans of action associated with the hidden potential of the affective domain.

ESSENTIAL INGREDIENTS OF THE AFFECTIVE DOMAIN

The field of science education brings learners, science, and society together so that growth occurs in all three educational domains: cognitive, psychomotor, and affective. The cognitive domain involves the acquisition of facts and concepts, along with the development of problem-solving and reasoning skills. The psychomotor domain involves the development of physical and dexterity skills. The affective domain (from the Latin *affectus*, meaning "feelings") includes a host of constructs, such as attitudes, values, beliefs, opinions, interests, and motivation.

The authors thank reviewer Robert L. Shrigley (Pennsylvania State University).

The concept of attitude, for example, can be subdivided into three dimensions: feeling, cognition, and behavior. Attitude is commonly defined as a predisposition to respond positively or negatively to things, people, places, events, or ideas. A related concept, value, normally represents more abstract objects. For example, we have attitudes toward such objects as "teacher," "mathematics," or "school," whereas values tend to focus on abstract ideas such as "democracy," "love," or "freedom." As the affective domain becomes more attractive for study and discourse, clearer definitions of important terms continue to emerge. The following is a simple comparison of four terms and how they differ with respect to psychological objects and the components of affect, cognition, and behavior:

Term	Typical Object	Major Components
Attitudes	Things, people, places	Contains affect, cognition, and behavior
Values	Abstract ideas such as love, democracy, freedom	More emphasis on affect and cognition and less on immediate behavior
Beliefs	The general acceptance or rejection of basic ideas	More emphasis on cognitive acceptance or rejection
Motivation	Focused more on the desire to act or not act	More emphasis on the behavior component

SCIENTIFIC ATTITUDES AND VALUES

Because attitude research in science education includes the student-science-society triangle, it is only natural that some confusion would emerge between the terms "scientific attitudes" and "attitudes toward science." Scientific attitudes have come to be known as ways in which scientists believe in and conduct their work. For example, a classic report written in 1966 called *Education and the Spirit of Science* was published by the Educational Policies Commission 1966 and contained values that were deemed important by scientists and educators as "modes of thought" important to humans and their pursuit of science and technology. Some of the values outlined in that publication are still relevant today:

- *Longing to know and understand.* A conviction that knowledge is desirable and that inquiry directed toward its generation is a worthy investment of time and other resources.

- *Questioning of all things.* A belief that all things, including authoritarian statements and "self-evident" truths, are open to question. All questions are prized, although some are of greater value than others because they lead to further understanding through scientific inquiry.

- *Search for data and their meaning.* Prizing the acquisition and ordering of data because they are the basis for theories, which, in turn, are worthwhile because they can be used to explain many things and events. In some cases these data have immediate practical applications of value to humans, as when data enable one to assess accurately the severity of a problem in society or the effects of policies directed to improving such situations.

- *Demand for verification.* A high regard for requests that supporting data be made public and that new empirical tests be invented or conducted to assess the validity or accuracy of a finding or assertion.

- *Respect for logic.* Esteem for chains of inference that lead from raw data to conclusion according to some logical scheme and insistence that conclusions or actions not based on such chains be subject to doubt.

- *Consideration of premises.* Prizing frequent review of the basic external and internal assumptions from which a line of inquiry has arisen, especially when they are used as a basis for determining further action.

- *Consideration of consequence.* A belief that frequent and thoughtful review of the direct and indirect effects of pursuing a given line of inquiry or action is worthwhile and that a decision to continue or abort the inquiry or action will be made in terms of the consequences.

More recently, the National Academy of Sciences (1989) has published a monograph entitled *On Being a Scientist.* Written by a committee of prestigious scientists and scholars, the publication addresses professional as well as personal issues that are at the core of scientific values and ethics. The following excerpt is an example of how scientists view their work in terms of human values and attitudes:

When methods are defined as all of the techniques and principles that scientists apply in their work, it is easier to see how they can be influenced by human values. As with hypotheses, human values cannot be eliminated from science, and they can subtly influence scientific investigations.

The influence of values is especially apparent during the formulation or judgment of hypotheses. At any given time, several competing hypotheses may explain the available facts equally well, and each may suggest an alternate route for further research. How should one select among them?

Scientists and philosophers have proposed several criteria by which promising scientific hypotheses can be distinguished from less fruitful ones. Hypotheses should be internally consistent, so that they do not generate contradictory conclusions. Their ability to provide accurate predictions, sometimes in areas far removed from the original domain of the hypothesis, is viewed with great favor. With disciplines in which prediction is less straightforward, such as geology or astronomy, good hypotheses should be able to unify disparate observations. Also highly prized are simplicity and its more refined cousin, elegance.

The above values relate to the epistemological, or knowledge-based, criteria applied to hypotheses. But values of a different kind can also come into play in science. Historians, sociologists, and other students of science have shown that social and personal values unrelated to epistemological criteria—including philosophical, religious, cultural, political, and economic values—can shape scientific judgment in fundamental ways. For instance, in the nineteenth century the geologist Charles Lyell championed the concept of uniformitarianism in geology, arguing that incremental changes operating over long periods of time have produced the Earth's geological features, not large-scale catastrophes. However, Lyell's preference for this still important idea may have depended as much on his religious convictions as on his geological observations. He favored the notion of a God who is an unmoved mover and does not intervene in His creation. Such a God, thought Lyell, would produce a world where the same causes and effects keep cycling eternally, producing a uniform geological history.

The obvious question is whether holding such values can harm a person's science. In many cases the answer has to be yes. The history of science offers many episodes in which social or personal values led to the promulgation of wrong-headed ideas. For instance, past investigators produced "scientific" evidence for overtly racist views, evidence that we now know to be wholly erroneous. Yet at the time the evidence was widely accepted and contributed to repressive social policies.

Attitudes regarding the sexes also can lead to flaws in scientific judgments. For instance, some investigators who have sought to document the existence or absence of a relationship between gender and scientific abilities have allowed personal biases to distort the design of their studies or the interpretation of their findings. Such biases can contribute to institutional policies that have caused females and minorities to be underrepresented in science, with a consequent loss of scientific talent and diversity.

Conflicts of interest caused by financial considerations are yet another source of values that can harm science. With the rapid decrease in time between fundamental discovery and commercial application, private industry is subsidizing a considerable amount of cutting-edge research. This commercial involvement may bring researchers into conflict with industrial managers—for instance, over the publication of discoveries—or it may bias investigations in the direction of personal gain.

The above examples are valuable reminders of the danger of letting values intrude into research. But it does not follow that social and personal values necessarily harm science. The desire to do accurate work is a social value. So is the belief that knowledge will ultimately benefit rather than harm humankind. One simply must acknowledge that values do contribute to the motivations and conceptual outlook of scientists. The danger comes when scientists allow values to introduce biases into their work that distort the results of scientific investigations.

The social mechanisms of science discussed later act to minimize the distorting influences of social and personal values. But individual scientists can avoid pitfalls by trying to identify their own values and the effects those values have on their science. One of the best ways to do this is by studying the history, philosophy, and sociology of science. Human values change very slowly, and the lessons of the past remain of great relevance today. (pp. 6–8)

It is clear from this excerpt that scientists are becoming increasingly concerned about social and ethical issues. Relationships between science and society and how they are interwoven with the values and beliefs of people will no doubt become a more important part of the science curriculum.

ATTITUDES TOWARD SCIENCE AND RELATED OBJECTS

"Scientific attitudes" should not be confused with "attitudes toward science." Attitudes or feelings toward science refer to a person's positive or negative response to the enterprise of science. Put another way, they refer specifically to whether a person likes or dislikes science. A seven-item subscale designed by Simpson and Troost (1982) and Simpson and Oliver (1990) measures *student attitude* (specific feelings) *toward science*:

- Science is fun.
- I have good feelings toward science.
- I enjoy science courses.
- I really like science.

- I would enjoy being a scientist.
- I think scientists are neat people.
- Everyone should learn about science.

Included in the scale are other dimensions related to attitude toward science. One example is *motivation to achieve in science*. Items from this subscale include:

- I always try hard, no matter how difficult the work.
- When I fail that makes me try that much harder.
- I always try to do my best in school.
- I try hard to do well in science.

Another dimension related to attitude toward science is *science anxiety*, the negative pole of the attitude concept. Items in this subscale include:

- Science makes me feel as though I'm lost in a jungle of numbers and words.
- My mind goes blank when I am doing science.
- Science tests make me nervous.
- I would probably not do well in science if I took it in college.

The science attitude of students can include *attitude toward science teacher*. Examples of items representing this dimension include:

- My science teacher encourages me to learn more science.
- I enjoy talking to my science teacher after class.
- My science teacher makes good plans for us.
- Sometimes my science teacher makes me feel dumb.
- My science teacher expects me to make good grades.

Attitude toward science curriculum is another important affective variable. Examples of statements representing feelings are:

- We do a lot of fun activities in science class.
- We learn about important things in science class.
- We cover interesting topics in science class.
- I like our science textbook.

AFFECT AS A CHARACTERISTIC OF PEOPLE

In addition to the aforementioned dimensions of attitude, there are other important variables related to science education and the affective domain. These are to be discussed in this section.

Self-Concept

Considerable research in social psychology and education focuses on self-concept and how it is related to school learning. Many earlier researchers, including Bloom (1976) and

Brookover, Thomas, and Paterson (1964), posited that the way students feel about themselves may be the most important variable in the process of education. Bloom even reported that our schools may be turning out increasingly larger numbers of students who view themselves as "failures" or "rejects" in society. He believes that 12 or 13 years of negative feedback to some individuals may lead to serious mental health problems later in life.

Fate Control

Several years ago, Rowe (1974a) called attention to a phenomenon known as "fate control." Fate control is defined as the belief that events that occur to us are under our own control. People who rate high in fate control are those who believe that skill and hard work are the major reasons for success or failure. Those low in fate control believe that their future lies in the hands of other powerful people and that they have little influence over their plight.

Locus of Control

Fate control is closely related to another variable known as "locus of control." This term describes the degree to which other individuals believe that reinforcement is contingent on their own behavior. In other words, some people believe their deeds influence what eventually happens to them, whereas others think "whim" or "luck" is largely behind the reward or punishment they receive.

Cultural Background

In addition to a plethora of self-related variables associated with the affective domain, a host of cultural values and social belief systems influence strongly how people (and students) behave. A classic study by Spilka (1970) in a school of 753 Sioux Indians and 455 white secondary school pupils showed how a school program may conflict with the cultural heritage of some who attend and thus alienate certain students from both their school peers and family. For instance, we have learned from observing science teaching in other countries patterned after American curriculum models that if certain values embodied in science instruction conflict with cultural values held by students or their parents, the educational goals of the program are not adequately realized.

Kluckhohn (1956), in early discussions of value orientations, suggested that human beings relate to nature in three basic ways: "Man Subjugated to Nature," "Man in Nature," and "Man over Nature." (These were the terms she used at that time.) According to Kluckhohn, the philosophy of "Man Subjugated to Nature" implies a fatalistic attitude and is held by many Spanish-Americans. "If it is the Lord's will that I die, then I shall die" represents this position. The "Man in Nature" philosophy regards all natural elements, including humans, as parts of one harmonious whole. This attitude has been dominant in China in past centuries. "Man over Nature" is characteristic of many Americans. In this view, natural forces are to be overcome and used for human purposes. In many Western civilizations, sci-

ence and technology have been the primary tools for enacting this philosophy. It is obvious that science instruction for students from different cultural backgrounds needs to be planned with these different perspectives in mind.

Belief Systems

Harvey, Prather, White, and Hoffmeister (1968) and Harvey (1970) found that beliefs of most adults fall into one of four systems. These belief systems influence the way people learn, develop new skills, cope with stress, and relate to others. *System 1* people are characterized by a strong belief in supernaturalism, positive attitudes toward tradition and authority, and absolute adherence to rules and roles. They tend to think in concrete terms and to view things as black or white, subject to little change. Members of this group are generally dogmatic and hold fairly rigid beliefs about the world around them.

System 2 people are only slightly less dogmatic and inflexible than those of System 1. However, they tend to have strong negative attitudes toward tradition and authority. System 2 people have the lowest self-esteem and the greatest cynicism. Paradoxically, they want and need to rely on other people, but they fear loss of personal control and power. When members of this belief system lack power, they denounce authority; but when they are in possession of power, they frequently abuse it.

System 3 reflects a strong emphasis on friendship, interpersonal harmony, and dependency relationships. Members of this system exhibit strong needs to help others, sometimes controlling others by establishing strong dependency bonds.

System 4 members are the most abstract and openminded—they tend to be creative, flexible, pragmatic, and utilitarian in their problem-solving styles. They respond with moderation to rules and regulations, not seeming to need much structure or dependency for themselves but recognizing that they are frequently necessary for others.

In studying these and other belief systems of students and educators, Harvey (1970) found that 75 percent of the elementary school teachers tested and 90 percent of school superintendents studied belong to System 1. Only about 7 percent of all people tested by Harvey belonged to System 4. Although Harvey's research did not include children, he suggested that science teachers should help System 1 children by initially providing needed structure, then gradually encouraging students to shift to a more independent style of observing, thinking, and problem solving. System 2 students need structure, coupled with warmth and fairness. Teachers should be sure that the rules and regulations of the classroom or school are explained reasonably and logically. System 3 students need external reinforcement (to meet their need for dependency), but should be encouraged to become more independent. Students who are members of System 4 have the greatest need for academic freedom and flexibility. Too many rules and regulations stifle the creative individuals in this group. Harvey et al. (1968) theorize that a child's ultimate belief system may be determined to a large extent by the freedom he or she has to explore values and to evolve and internalize rules on the basis of pragmatic outcomes. Thus, teaching science as open-ended inquiry would appear to encourage more flexible behavior in children, but some students may need to be phased gradually into less structured teaching formats.

Social Values

For decades, the famous researcher Milton Rokeach (1973) studied the basic values of American and other societies. He concluded that there are 18 important terminal values that we share and 18 significant instrumental values. The *terminal values*, the cherished outcomes that we uphold and strive for as a society, are:

- A comfortable life (a prosperous life)
- An exciting life (a stimulating, active life)
- A sense of accomplishment (lasting contribution)
- A world at peace (free of war and conflict)
- A world of beauty (beauty of nature and the arts)
- Equality (brotherhood, equal opportunity for all)
- Family security (independence, free choice)
- Happiness (contentedness)
- Inner harmony (freedom from inner conflict)
- Mature love (sexual and spiritual intimacy)
- National security (protection from attack)
- Pleasure (an enjoyable, leisurely life)
- Salvation (saved, eternal life)
- Self-respect (self-esteem)
- Social recognition (respect, admiration)
- True friendship (close companionship)
- Wisdom (a mature understanding of life)

Instrumental values are the enablers that help us accomplish the terminal values. These values, as established by Rokeach, are:

- Ambitious (hard-working, aspiring)
- Broadminded (open-minded)
- Capable (competent, effective)
- Cheerful (lighthearted, joyful)
- Clean (neat, tidy)
- Courageous (standing up for your beliefs)
- Forgiving (willing to pardon others)
- Helpful (working for the welfare of others)
- Honest (sincere, truthful)
- Imaginative (daring, creative)
- Independent (self-reliant, self-sufficient)
- Intellectual (intelligent, reflective)
- Logical (consistent, rational)
- Loving (affectionate, tender)
- Obedient (dutiful, respectful)
- Polite (courteous, well-mannered)
- Responsible (dependable, reliable)
- Self-controlled (restrained, self-disciplined)

It is, therefore, important for science educators to understand that science is not value free, nor is it taught to individuals lacking attitudes, beliefs, and values. Often, schools and teachers represent the values of the American middle class, while large numbers of students receiving instruction adhere to different value systems. The society and culture within which students, science, and teachers are placed influence dramatically the goals, strategies, and outcomes of science education.

FOCUSING ATTITUDE RESEARCH IN SCIENCE EDUCATION

In attempting to show the directions and usefulness of attitude research in today's schools and with today's students, it is necessary to reexamine the means by which this research has traditionally been conducted. Gardner (1975) summed up the state of attitude research for the late 1960s and early 1970s when he concluded: "What is needed is the professional expertise and will to move the results of research into our schools. The ultimate goal of all these endeavors is simply stated: to stimulate joy, wonder, satisfaction and delight in children as a result of their encounters with science" (p. 33). To this end, he noted, as have others, teachers of science place importance on the study of attitudes as important characteristics of students.

Simpson and Troost (1982) supported the need for the study of attitudes by saying, "We use the term 'commitment to science' to include the interests, attitudes, values and other affective behaviors of students being addressed in this study. Commitment to science is not merely defined in terms of student desire to major in science, but rather in terms of the desire to take more science courses, to continue reading about science, to explore new scientific topics, and to be involved in science-related social issues. These and other indicators suggest lifelong learning patterns associated with science in the broadest sense" (p. 765). These and other authors believed that the study of attitudes as characteristic of students is important, but also they believed that enhancement of attitudes is related to a long-term commitment to science.

Early Research in Attitudes

As with the enterprise of science itself, teachers of science, administrators, and parents can learn more about effective science instruction through research. Attitude research began in 1928, when Thurstone published an article entitled "Attitudes Can Be Measured." Of course, the concept of attitude is much older than that (see Shrigley, Koballa, & Simpson, 1988), but Thurstone opened the door for research on attitudes, and since that time several techniques have emerged for measuring the construct of attitude.

The field of science education began as a "specialty area" only in the late 1950s and early 1960s, and research in the affective domain in science education is relatively new. During the 1960s and 1970s researchers in this area were interested in more descriptive and fundamental areas of student attitudes toward science and how these and other feelings related to the learning of science. As this became a serious topic for research, theoretical underpinnings from social psychologists and other behavioral and social scientists began to form a basis for new directions of research concerning more specific components of attitude development and communication theory.

Research in attitudes during the 1960s, 1970s, and 1980s tended to focus on three issues, which may be summed up by three questions: (1) What is the magnitude of science-related affect in school students? (2) What change in attitudes can be detected in response to research treatments or changes in school science practices? (3) What relationships can be supported between attitude and school science-related behaviors?

It is not trivial to ask what aspect of the study of attitudes teachers find important. Explication of the relationship between attitudes and achievement might be high on the list. However, the teacher may also see a need for positive student attitudes whether or not they are related to achievment. Positive attitude is thus a variable related to classroom environment or student satisfaction. One author wondered how much success that one can expect to have in stimulating positive attitudes toward science (Schibeci, 1984); "How many students can we reasonably expect to feel the sense of wonder and excitement which is felt by professional scientists?" (p. 47).

What follows is a brief summary of what the research community has said about attitude in relation to science education in the 1960s, 1970s, and 1980s. The primary sources of information are reviews of the literature on attitude and science education. These should allow more efficient follow-up for the interested reader. Two particularly comprehensive studies, those of Gardner (1975) and Schibeci (1984), help provide focus. By examining the results of these parallel reviews and supplementing the information with individual studies, it is hoped that a clearer picture of the results and needs of attitude research may evolve.

Types of Attitudes

Gardner (1975) began the review of the literature of attitudes in science education by contrasting the "two broad categories" of attitude toward science and scientific attitudes. Attitude toward science encompasses interest in science, attitudes about scientists, and attitudes toward the use of science, while scientific attitudes include the characteristics of scientists that are believed to be desirable in students of science (such as open-mindedness and objectiveness). This is still one of the major distinctions made in attitude research. Koballa and Crawley (1985) summed up the difference: "the term attitude toward science should be used to refer to a general and enduring positive or negative feeling about science. It should not be confused with scientific attitude, which may be aptly labeled scientific attributes" (p. 223).

Other researchers have identified attitudes in terms of more specific subgroups of these broad categories. Simpson and Troost (1982) identified 15 of these categories, including science affect, science self-concept, general self-esteem, locus of control, achievement motivation, science anxiety, emotional climate of the science class, physical environment of the science class, other students in the science class, science teacher, science curriculum, family—general, family—science, friends and science, and school. These categories are typical of much of the research in attitudes.

Traditional Measurement of Attitudes

In the 1970s attitudes were measured by use of a variety of techniques, but two received the most notice in the 1975 review by Gardner. The first is the Likert scale method. In a typical Likert scale instrument, respondents read each item and then respond along a range of categories from strongly agree to strongly disagree. For instance, an item might read, "I like science." The respondent would indicate the degree to which he or she agrees or disagrees with this statement.

The second method is known as interest inventories. Gardner (1975) described this measurement technique as follows: "Interest inventories typically contain a list of careers, topics, or activities; the respondent indicates those in which he is interested" (p. 7). In some cases these interest inventories allowed the respondent to indicate a degree of interest and thus were scored in a manner similar to the Likert scale.

Two other methods for assessing attitudes that were mentioned by Gardner have come to the fore in recent years as strong contenders for the most popular methods. In the quantitative realm, the semantic differential has been used by a significant number of researchers. The instrument consists of statements that are responded to through pairs of adjectives that form scale end points. The scale often consists of five or seven points between the adjective pairs. An advantage of this system of measurement is the association of multiple pairs of adjectives with a given attitude object. For instance, if "my science teacher" is the attitude object, then adjective pairs such as "understanding - uncaring", "organized - disorganized", and "interesting - boring" might be used.

Emerging qualitative research methodologies were referred to by Gardner as "clinical" or "anthropological". In practice, these methods involved interviews with students, nonparticipating observation of class sessions, and participating observation of classes. Today, qualitative methodologies in educational research are rivaling quantitative methods as the most popular approach for collecting data and evaluating school settings.

Attitude, Ability, and Behavior

One of the questions that is most on the minds of either teachers or attitude researchers is this: If teachers could instill in students a sense of wonder and excitement (to use Gardner's terminology) about the study of science, would it result in greater achievement or participation by those students? Or perhaps more appropriately, do students who demonstrate more positive attitudes achieve at higher levels? Many teachers have an intuitive belief in this connection, but can it be demonstrated?

Gardner's review of this literature in 1975 offered little evidence of strong attitude-achievement or attitude-ability relationships. Six of 10 studies with specific goals of examining these relationships found either no evidence of a relationship or a negative relationship. All four of the remaining studies reported positive relationships, but Gardner questioned the validity of the instrument used to measure the attitudes of students in two of these. In the last two, positive correlations between attitude and achievement were reported in the range of 0.20 to 0.35, with higher correlations being reported for older students. The picture of the relationship between attitude and achievement was not a particularly inviting one in 1975.

The picture of this relationship as painted by Schibeci in 1984 is at least somewhat more promising. He cited a variety of studies that reported a positive link between attitude and achievement. Many of the correlations were in the range of 0.3 to 0.5, with one study (Simpson & Wasik, 1978) reporting a correlation between an affective behavior checklist and achievement in biology of 0.84. On the other hand, he also reported studies in which there was no link at all. In this vein, Schibeci describes Fraser's view: "Many studies of the attitude-achievement association are motivated by the assumption that attitudes and achievement are strongly related, if only on the grounds that 'the best milk comes from contented cows' as Fraser (1982) has stated! Fraser, in fact, argued that there was no empirical support for this seemingly commonsense assumption.... Fraser concluded ... that if teachers want to improve achievement, they would be well advised to concentrate on achievement *per se* instead of trying to improve achievement scores by improving attitudes" (p. 30). Schibeci and many other attitude researchers have not accepted this position.

Shrigley (1990), in a synthesis entitled "Attitude and Behavior are Correlates," returned to this point. He asked, "Can science attitude scores be expected to reflect science behaviors in the classroom?" (p. 97). In examining this issue and coming to the conclusion that attitudes can be predictors of behavior in the classroom, Shrigley examined five perspectives of the issue: (1) attitude precedes behavior, (2) attitude is behavior, (3) attitude is not directly related to behavior, (4) attitude follows behavior, and (5) attitude and behavior are reciprocal. His conclusions are twofold: "The premise that attitudes precede behavior dominates current attitude research" (p. 107) and "The research generated since 1976 suggests that science attitude scores can be expected to correlate moderately, at least, with the behavior of teachers and students in the science classroom" (pp. 108–109). There is some disagreement with Shrigley's first statement regarding the direction of the relationship between attitudes and behavior. For example, Punch and Rennie (1989) concluded that affect is more strongly related to previous than to subsequent achievement.

Relationships with Other Variables

Gender. Gardner (1975) expressed the general intuitive response of many educators regarding the relationship of gender and attitudes in science: "Sex is probably the single most important variable related to pupils' attitude to science" (p. 22). Gardner went on to say that sex differences in regard to attitude arise early in life, but he identified some evidence that these

differences diminish as children get older. Schibeci (1984) concurs with Gardner's assertion concerning the importance of gender differences but reports examples from studies that support virtually any conclusion, from a much greater positive attitude by boys to no difference at all. Haladyna and Shaughnessy (1982), in reporting a large meta-analytic study, concluded that "while boys were found to like science more than girls like science, this difference is consistently small and variable from study to study and from grade to grade" (p. 555). Ormerod and Duckworth (1975) point to an attitude gender difference related to the content of science courses. Females may exhibit a more positive attitude toward the biological sciences than the physical sciences, thus confounding a measure of general attitude toward science. In contrast, Simpson and Oliver (1985) found a clear and consistent distinction between males' and females' attitudes toward science across grades 6 through 10. Males had significantly more positive attitudes than females at each level except grade 9. Participants in this study took classes in biological science in the seventh and tenth grades and received physical science instruction in grades 8 and 9.

Structural Variables. Other variables that may be related to attitude toward science are structural variables. They are defined as "demographic variables such as geographic location and socio-economic status, and home variables of various kinds" (Gardner, 1975, p. 24). Additional variables in this list might include grade level, religious background, and cultural background.

Differences in attitude related to geographic location are found in the research literature, but not too commonly. Gardner (1975) reported differences in scientific interest in both studies that examined this issue during his 1975 review. Gender, however, was a mediating variable of these differences. Lin and Crawley (1987) found that in Taiwan, "metropolitan students did report higher attitudes toward science, but they also reported higher levels of speed, friction, favoritism, difficulty, cliqueness, and competitiveness in classes, indicating differences in classroom climates" (cited in Staver et al., 1989, p. 256).

Considerably more studies have examined the relationship between socioeconomic status and attitude toward science. The research literature seems to indicate consistently that there is no relationship between these variables.

Grade level is another structural variable that is commonly discussed in relation to interest in science. In most cases, it is believed that attitude toward science declines across grade levels beginning in middle school. Schibeci (1984) sums up these beliefs by simply saying, "Many studies have reported a decline in attitudes with increasing grade level" (p. 35) and citing five examples. Simpson and Oliver (1985) found this pattern of decline across the grades 6 through 10 in a study of 4000 students in North Carolina during a single school year. In addition, attitude reports in the beginning of the school year follow a consistent pattern of decline within the school year. In the Simpson and Oliver study, for example, attitude toward science within each grade except ninth became markedly less positive by midyear.

School Variables. The most consistent finding reported in the traditional research literature of the 1970s and 1980s is a lack of studies dealing with school variables and affect related to science. This was reported by Gardner (1975), Simpson and Troost (1982), Schibeci (1984), and others.

According to these authors, many of the school variables studied involve the relationship of what is commonly called "classroom climate" and attitude toward science. As might be expected, this relationship tends to be positively correlated, but Gardner (1975) cited a study by Walberg (1969) which indicated that "the classroom characteristics associated with higher achievement were different from those associated with better attitudes" (p. 25).

Curriculum and Instructional Variables. Perhaps the largest number of attitude research studies conducted in science classrooms have focused on the use of new and innovative science curriculum materials and subsequent changes in attitude toward science. Many new science curricula were introduced in the 1960s and early 1970s, creating many opportunities for study. Unfortunately, the findings of this research as a whole indicate that the effect of a new curriculum on the attitude of students was relatively small. As Gardner (1975) stated, "In most cases, a particular treatment group's mean attitudes have been shifted a fraction of a standard deviation along the attitude continuum" (p. 29). In other words, there was then little reason to believe that curriculum materials have a positive effect on student attitudes. At that time, "second-generation" curriculum projects were coming out and there was hope that these projects would lead to greater attitude gains than were evident in the earlier works.

Nearly 10 years later, the relationship between instructional strategies and the attitudes of school students was no clearer than before. In Schibeci's literature of 1984, nearly equal numbers of innovations are reported to increase student attitudes (field-based science, use of literature, "optional retesting contingencies," structured films, student-managed diagnosis, and unstructured museum tours) and to effect no change (laboratory work, self-pacing, modified mastery learning strategy, values clarification, and others). No pattern emerges to distinguish innovations that do and do not positively affect attitude. Perhaps the issue lies not with the curriculum or the teachers, but in the definitions and theoretical underpinning of attitude that have been chosen by researchers.

PRELUDE TO A NEW PARADIGM

Clear and distinct definition of terms has been recognized as a problem by several authors evaluating the attitude research. Klopfer (1976) stated, "Although the affective domain taxonomy has been available to science teachers and researchers for quite some time, it has not had much impact on either practice or research in science education. A primary reason for this neglect is the lack of clarity and prevalence of confusion regarding the phenomena of concern in science education that the student might have feelings or attitudes about" (p. 301). Krynowsky (1988) summed up the situation in the following way:

The lack of conceptual clarity in attitudinal assessment in science education is associated to the broader problem of being able to explain what an attitude is and how it can be defined, measured, and related to behavior. Other disciplines, especially social-psychology, have been grappling with this problem for at least the last century.

What science education assessments have lacked, however, and what Chapter 6 intends to demonstrate, is that the provision of some theoretical foundation can assist in the development of instruments which could provide more readily interpretable and comparable results. (p. 577)

It is to this issue that we now turn.

SOCIAL PSYCHOLOGY–BASED RESEARCH AND SCIENCE EDUCATION

Decision making is of considerable interest to science teachers at all educational levels, particularly as it pertains to the choices students make concerning the study of science in high school and college and the pursuit of science-related careers. From their experience in the classroom, science teachers realize that some adolescents willingly tackle the decisions they face, whereas others are reluctant to take charge. Moreover, some choices that young people make at times appear to be both logical and in their best interest, whereas at other times their choices seem irrational, even self-defeating. The long-term effectiveness of science instruction could be greatly improved if we knew more about the motivational bases for behavior and how we might interest more students in tackling the important decisions in their lives with forethought. It is important for students to choose courses of action that expand rather than restrict the many personal, social, educational, and career opportunities available to them in science.

Track, Course, and Career Choices

The U.S. Congress, Office of Technology Assessment (1988) uses a pipeline model to visualize the path students follow and critical junctures that exist along the way from high school into college then on to careers in one of the many science and engineering professions. The route from high school to one of these professions is anything but direct for many students, particularly females and non-Asian minorities. There is considerable evidence that the flow is disrupted at several key points along the way.

Track choice offers the first opportunity for the pipeline to spring a leak. The talent flow at this critical point diverges sharply, as 39 percent of high school students follow an academic track, 37 percent a general track, and 24 percent a vocational track, with about equal numbers of males and females pursuing an academic track but slightly more females than males entering a vocational track. Greater percentages of black and Hispanic students enter the general track (35 percent and 42 percent, respectively), followed by vocational (33 percent and 31 percent, respectively) and academic (32 percent and 27 percent, respectively) tracks. Students enrolled in nonacademic tracks complete fewer and less challenging science and mathematics courses than academic-track students and are less prepared to enter science and engineering programs offered in most four-year colleges. Data further indicate that by the senior year academic-track students planning to major in either science or engineering outnumber students of similar interest in nonacademic tracks by nearly 2 to 1, a ratio that increases to 3 to 1 by the college sophomore year.

Further leaks occur in the science and engineering pipeline as high school students face key decisions about course choice. Students who complete algebra II, calculus, biology, chemistry, and physics courses are more likely to remain in the science and engineering pipeline than students who do not. Completion of courses beyond those required for graduation provides students with more opportunities to major in science or engineering fields in college. College students also find it easier to switch to science or engineering if they have completed one or more elective science and mathematics courses. Nonetheless, compared with their science and engineering counterparts, transfer students are more likely to have completed fewer science and mathematics courses and are more likely to be black, Hispanic, or female. All told, 85 percent of those enrolled in science and engineering in college in 1988 were white, compared with 10 percent black and 5 percent Hispanic (National Science Board, 1989).

At the end of the pipeline, white males dominate the science and engineering professions. In 1986, 87 percent of the 3.9 million scientists and engineers in the work force were males; 91 percent were white, 2 percent black, 5 percent Asian, and 2 percent "other"—a rather precipitous drop in black participation in these professions in light of their 32 percent representation in the academic track in high school.

Attitude appears to be a crucial factor in charting a course through the academic pipeline into one of the many science and engineering professions. Having an interest in science-related hobbies during childhood is strongly related to interest in high school science courses, as are holding early aspirations of being a scientist and receiving encouragement to major in science from parents, teachers, and peers (Thomas, 1986). Interest in science and science careers among ninth-graders translates into greater interest in the study of subsequent high school science courses (Stallings & Robertson, 1979).

Volitional Behavior and Situational Constraints

Students face continuously numerous school-related decisions about tasks to complete. Teachers make frequent in-class and out-of-class assignments, ranging from daily readings and quizzes to extended laboratory reports and current-events projects. Some students take the initiative and tackle an assigned task right away; some take no action. At first glance, personal motivation alone appears to separate students who begin planning the current-events project, for example, from others who either work on an assignment for another class, begin talking, or do nothing. Both groups of students may be genuinely interested in completing the assigned task, but the latter group may believe that task completion depends primarily on the availability of needed resources (reference materials), opportunities (time), other people (a ride to the library), and personal knowledge (how to write reports). Completion of the current-events paper by the assigned date, therefore, seems impossible for students who believe that factors beyond their control regulate their behavior. In the absence of controls, real or imagined, behavior is self-regulatory or volitional, and personal intention alone directs personal action.

If availability of resources, provision of opportunities, and possession of the requisite skill affect performance of a relatively simple, discrete behavior such as completion of a current-events project, it is easy to see how these factors might influence high school students' track, course, or career choices. Research has shown that numerous barriers impede female and minority students' access to careers in science and engineering. Female students, for example, believe that science is for males only and is an inappropriate and unfeminine activity, beliefs that may be unconsciously reinforced by teachers, peers, and parents (Hill, Pettus, & Hedin, 1990). Moreover, girls and minorities lack many of the math- and science-related experiences in and out of school that are provided to boys (Thomas, 1986).

External controls can facilitate as well as inhibit behavior. Intervention programs designed to increase the participation of women and minorities in science and engineering have utilized facilitative factors and met with considerable success (Clewell, 1987). Scientists who were female, minorities, or both have visited schools and served as mentors to students in an effort to refute the belief that science is the exclusive domain of the white male. Successful intervention programs for blacks and Hispanics have included career counseling for high school students to call attention to the many opportunities available to persons who study science or engineering (Mulkey & Ellis, 1990). Removing barriers to access and providing opportunities to participate in science and engineering programs ease the grip that external controls exert on students' interest in the study of science and mathematics.

Students gain a greater sense of control over their lives when they carefully consider all information related to educational and career decisions. Only then are they likely to be willing to take risks and pursue educational programs that lead to careers in science or engineering.

Forced Compliance, Mindlessness, and Reason

To what extent are students rational actors? Do they carefully frame, examine, and weigh each alternative before making a decision about what courses to take or what careers to pursue? Perhaps the nation's need for science and engineering talent would be better served if high school students were required to complete 4 years of science and mathematics courses. This would better prepare all students for careers in science and engineering, but it would be difficult to justify requiring 4 years of science and mathematics and not history, foreign language, and art. What is more, the requirement may not lead to increased numbers of scientists and engineers as intended. Instead, departure from the science and engineering pipeline might only be delayed until completion of the senior year, when graduates unhappy with science and engineering would opt for nonscience jobs or postsecondary education in nonscience fields.

Forced-Compliance. Social psychologists have used three prevailing mechanisms to explain the effects of forced or induced compliance: dissonance, self-perception, and impression management (Tesser & Shaffer, 1990). The theory of cognitive dissonance was developed by Festinger in 1957. Dissonance is an uncomfortable psychological (and physiological) state that arises when a person holds two inconsistent cognitions; cognitions take the form of thoughts, attitudes, beliefs, or behaviors. The presence of dissonance gives rise to pressures to return to a pleasant state, consonance, which can be achieved through several means. One approach to dissonance reduction is to decrease the number and importance of the dissonant elements, increase the number and importance of consonant elements, or both. Another is to change one of the dissonant elements. To see dissonance and its reduction in action, imagine what takes place in the mind of a high school student who, on the first day of the junior year, finds that he is enrolled in chemistry against his will. Dissonance is aroused because one cognition, disliking chemistry, is inconsistent with a second cognition, attending chemistry class. Dissonance can be reduced by downplaying the importance of the unfavorable beliefs associated with chemistry (e.g., "There really can't be but so much homework!") or enhancing the favorable ones (e.g., "Some of the people in the class are quite nice."). Two other routes to dissonance reduction are also possible. The student may be able to drop chemistry and add physics, which perhaps is taught by one of the student's favorite teachers. Or the student may add the cognition that chemistry is mandatory, which is consonant with the cognition of attending class, and suppress a negative attitude for the time being, only to exit the science and engineering pipeline upon graduating from high school. This latter option casts doubt on the efficacy of mandating 4 years of high school science and mathematics.

Self-perception accounts for processes that take place when initial opinions are weak or behaviors lie within what social psychologists call one's latitude of acceptance. Behaviors that lie outside this psychological realm and are undergirded by well-formed opinions are explained by social psychologists using one of the routes to dissonance reduction. Self-perception theory was proposed by Bem in 1972. Attitudes are not reflections of what humans are predisposed to do, according to the theory, but are inferred from behaviors. The phrase "Try it, you'll like it" best summarizes the behavior-attitude relationship according to self-perception theory. Let us return to the example of mandated chemistry enrollment to see the theory in action. The student who has a neutral attitude toward chemistry and, much to her surprise, finds that she is enrolled in the course on the first day of school forms an attitude toward chemistry that is dependent on the favorableness of her initial experiences.

The third model offered to explain forced-compliance phenomena is impression management (Tedeschi, 1981). According to the theory, people are concerned about the impressions that others have of them. Therefore, situations in which individuals find themselves accountable to others for engaging in counterattitudinal behaviors generate great need to present an image that is consistent, rational, and predictable. Considerable thought is devoted to the counterattitudinal behavior, and a publicly defensible position is formed for any discrepant behavior. By contrast, people held accountable only to themselves quickly change their attitudes to be consistent with their counterattitudinal behavior and thereby reaffirm a consistent self-image. The motivation for attitude change in impression management is external, namely the desire to appear consistent to other people. In the example of mandated chemistry enrollment, students who are concerned primarily with projecting a consistent self-image will openly defend enrollment in chemistry and adopt attitudes consistent with their public stance.

Mindlessness and Mindfulness. Social and cognitive psychologists today realize that humans perform many seemingly complex tasks (and simple ones) with minimal information processing, or in a state that has been referred to as mindlessness (Langer, 1989). In contrast to a state of reduced attention during mindlessness, mindfulness is a condition of alertness and lively awareness in which information is actively processed and categories and distinctions are created. Behavior during a state of mindlessness is rigid and rule governed, whereas mindful behavior is rule guided. The rigidity associated with mindlessness occurs on both the cognitive and emotional levels, but by default rather than by design. Research has shown that mindfulness is not more effortful than mindlessness, may even enhance one's health, and is associated with increased longevity. More important, mindfulness can be taught, according to Langer. Someone instructed to view situations as conditional and events as having multiple causes learns to be mindful. When mindful, a person is in a position to notice more in the environment, to see opportunities for control that the mindless person would overlook, and to engage in greater risk-taking behavior.

The goal of education is to teach mindfulness. Students who learn to engage in mindful behavior are more likely to see the many opportunities available to them in the science and engineering professions, gain a greater sense of control over their educational and career plans, and be willing to take the risks necessary to remain in the talent pool. By contrast, forced compliance undermines personal decision making and mindfulness. Mandated enrollment in the traditionally elective science courses such as chemistry and physics appears to be gaining support nationwide, but the nature of these courses is changing. Science courses currently being proposed and developed will integrate the science fields rather than teach them as separate disciplines. It remains to be seen whether new curricula will bring about the intended result of improving scientific literacy and increasing the number of traditionally underrepresented black and Hispanic students who are prepared to pursue postsecondary science education and possible careers in science and engineering. In view of the results of research on the mindlessness-mindfulness paradigm, mandated science enrollment throughout high school may serve only to delay departure from the science and engineering pipeline until high school graduation.

Reason and Choice Framing*.* Numerous decisions are made each day without advance knowledge of their consequences. Because the consequences of each decision depend on uncertain events, the choice of one option over another may be construed as the acceptance of a gamble that can yield various outcomes with different probabilities. The way in which choices are framed has rather dramatic effects on the decisions that people make. According to prospect theory, people avoid risk when considering possible gains and seek risk when considering possible losses, even when the consequences of both decisions are mathematically identical (Kahneman & Tversky, 1984).

Kahneman and Tversky assert that an analysis of rational choice incorporates two principles: invariance and dominance. The principle of invariance stipulates that the preference order between prospects should not depend on the manner in which they are described. The dominance principle requires that if prospect A is at least as good as prospect B in every respect and better than B in at least one respect, then A should be preferred to B. Problems presented to people reveal that they make choices that clearly violate these two principles of rational choice and illustrate the extent to which choice framing determines which options ordinary people select. Two representative problems help to illustrate the extent to which the invariance principle is violated. (The total number of respondents in each problem is denoted by *N*, and the percentage who chose each option is indicated in parentheses.)

Problem 1 (N = 152): Imagine that the United States is preparing for the outbreak of an unusual Asian disease, which is expected to kill 600 people. Two alternative programs to combat the disease have been proposed. Assume that the exact scientific estimates of the consequences of the programs are as follows:

- *If Program A is adopted, 200 people will be saved. (72%)*

- *If Program B is adopted, there is a one-third probability that 600 people will be saved and a two-thirds probability that no people will be saved. (28%)*

Which of the two programs would you favor (Kahneman & Tversky, p. 343)?

Both outcomes have the same mathematical expectation, which is determined by multiplying each outcome by its probability of occurrence and summing over all possible outcomes. Under Program A 200 people are certain to be saved (200 × 1 = 200). Under Program B the expectation is computed as 600 × 1/3 + 0 × 2/3, which gives the identical expectation of 200 lives saved. The reference point adopted for the problem is one in which the disease is allowed to take its toll of 600 lives. The outcomes include that reference point and two possible gains. The fact that 72 percent of the people selected Program A clearly indicates a risk-aversive preference in the face of gain; a majority of respondents prefer saving 200 lives for sure to a gamble that offers a one-third chance of saving 600 lives.

Consider a second version of the same problem in which the options are framed differently.

Problem 2 (N = 155):

- *If Program C is adopted, 400 people will die. (22%)*

- *If Program D is adopted, there is a one-third probability that nobody will die and a two-thirds probability that 600 people will die. (78%)*

Once again, both options have the same mathematical expectation, namely that 400 people die. Furthermore, the outcomes of Programs A and C are identical, as are those of Programs B and D, but the preferred option in Problem 1 turns out to be less preferred in Problem 2. The second problem assumes a reference state in which no one dies of the disease and the choices are framed in terms of certain or likely deaths. In the Problem 2 version of choice framing, people clearly prefer to seek risk to avoid loss rather than choose an option with a certain loss of 400 lives.

Violations of the dominance principle are also widespread. People choose one option over its alternative in a simple decision but choose the alternative when presented with the same problem cast in the form of a set of concurrent decisions. Even decisions made with a great deal of confidence have been reversed when the same problem is framed differently.

Rules of rational decision making appear to be psychologically unfeasible for most people to handle in the course of the daily choices that they make. It is unlikely, as Kahneman and Tversky show, that the average person can engage in unbiased reasoning without devoting considerable time and effort to recasting decisions into equivalent representations and computing the mathematical expectation of each outcome rather than relying on each outcome's subjective value.

Application of prospect theory in science education may help researchers and teachers gain a better understanding of students' course-choice decisions. It is clear from the work of Kahneman and Tversky that the majority of students apply subjective criteria rather than logical analysis when faced with risky decisions such as enrolling in elective science or mathematics courses. Unknowingly, teachers and counselors may dissuade students from enrolling in elective science courses by framing course-choice decisions in terms of increased educational and career opportunities available to persons who successfully complete these courses. Faced with possible gain, students may actively seek to avoid what they see to be the many risks associated with enrolling in chemistry or physics. On the other hand, framing course-choice decisions in terms of lost educational and career opportunities for students who avoid chemistry or physics may achieve the desired outcome of increased enrollment. Confronted with lost opportunities, students may become risk seekers and willingly enroll in one or both of these key elective science courses.

Reasoned Action and Planned Behavior

Attitude has been the subject of investigation by social psychologists for decades. One reason for this interest is the belief that people make evaluative judgments about a wide variety of targets and rely on these judgments, or attitudes, in deciding among several possible courses of action. The attitude-behavior link is so intuitively appealing that it should need no further elaboration. After all, are not people who like foreign cars, for example, more likely to buy one?

What intuition suggests, however, research has found difficult to support. Why has attitude-behavior consistency been so difficult to demonstrate? One possible explanation lies in the complexity of the definition of attitude that has dominated the social psychology literature. Since antiquity, attitude has been viewed as a latent variable that manifests itself in the form of one or more of three types of responses: cognitive, affective, and conative. Cognitive responses reflect perceptions of and information about the object of an attitude, called attitude object, which can be a person, an event, or a real object. Affective responses include the feelings one holds toward an attitude object. Behavioral inclinations and actions with respect to an attitude object constitute conative responses. Tests of the attitude-behavior relationship have typically involved examination of what people say versus what they do, of their verbal and nonverbal responses of either a cognitive, affective, or conative nature.

Fishbein and Ajzen proposed a theoretical framework that structured attitude in a different way (1975; Ajzen, 1989; Ajzen & Fishbein, 1980). According to this model, the cognitive, affective, and conative aspects of attitude are not merely related to one another, they interact in a causal manner. Attitude is viewed as the personal evaluation of some highly specific behavior, which in turn is defined in terms of action, target, context, and time components. This personal evaluation of a specific behavior (the *affective* component), called attitude toward the behavior (AB), gives rise to a commitment or an intention to engage in the behavior, called behavioral intention (BI, the *conative* component). Personal beliefs (the *cognitive* component) give rise to attitudes. Each belief about the behavior links the behavior with a specific attribute (a characteristic, outcome, or event). The strength of the link between an attribute and the object (called behavior belief, b) is weighted by the attribute's subjective utility (called outcome evaluation, e) through the expectancy-value theorem. A belief-based estimate of attitude toward the behavior (AB) is obtained by summing the product of each behavioral belief by its outcome evaluation.

A normative factor also contributes to the formation of behavioral intention. Called subjective norm (SN), this component represents the extent to which the individual perceives that there is support from important referents to engage in the behavior. Beliefs also form the subjective norm. In this case, the extent to which each referent supports enactment of the behavior (called normative belief, nb) is weighted by motivation to comply with the referent (mc). A belief-based or expectancy-value estimate of subjective norm is obtained by summing the product of each normative belief by its motivation to comply. Behavior (B), behavioral intention (BI), attitude toward the behavior (AB), subjective norm (SN), and beliefs are related in a causal chain, known as the theory of reasoned action, as follows:

$$B \sim BI = w_1 AB + w_2 SN,$$

where w_1 and w_2 represent the relative contributions of AB and SN to the determination of BI, respectively, and AB and SN are formed by their belief-based estimates, $\Sigma_i \, b_i e_i$ for i personal beliefs and $\Sigma_k \, (nb)_k (mc)_k$ for k normative beliefs. (For more information on this equation, see Ajzen and Fishbein, 1980.) The utility of the theory of reasoned action rests on the assumption that the behavior of interest is volitional, free of any external constraints.

The theory of reasoned action has been used successfully to explain numerous science-related behaviors. Koballa (1988) examined the determinants of female junior high school students' intentions to enroll in at least one elective physical science course in high school and found that external variables traditionally associated with course enrollment interest (ability group, science grades, and attitude toward science were unrelated to female students' intentions. Attitude toward and social support for course enrollment (AB and SN, respectively, in the preceding equation) were found to determine female students' intentions (BI). Female students' personal attitude, more than social support, gave rise to greater commitment to enroll in an elective physical science course.

Crawley and Coe (1990) investigated middle school students' intentions to enroll in a high school science course the next year, assuming enrollment was optional. The external variables sex, ethnicity, general ability, and science ability were found to be unrelated to course enrollment intentions. Attitude toward enrolling in a high school science course and social support for doing so were predictive of course enrollment intentions. The relative contributions of attitude and subjective norm were found to vary depending on students' sex, ethnicity, general ability, and science ability, in rather unexpected ways. Attitude was more important for male students, but subjective norm (social support) was more important for female students. Attitude alone was predictive of minority students' course enrollment intentions. Among students with high general and science ability, course enrollment intentions were a product of social support rather than personal attitude. This influence shifted progressively toward the attitude component as the general and science abilities of student groups shifted from high to medium to low.

The explanatory value of the theory of reasoned action rests on the assumption that the individual can decide at will whether to perform the behavior of interest. This assumption of volition greatly restricts the range of applicability of the personal decision-making model. Individuals faced with behavioral decisions encounter barriers, personal limitations, and lack of resources—real or imagined—at one time or another that control the extent to which they feel free to act.

In response to this limitation, Ajzen (1985) proposed and validated a more inclusive form of the theory of reasoned action, called the theory of planned behavior. Added to the original model was a third component, perceived behavioral control (PBC), which was hypothesized to have a direct impact on the formation of behavioral intention, independent of the contributions of attitude and subjective norm. Perceived behavioral control represents the extent to which the individual believes that behavioral performance is complicated by internal factors (inadequate information, skill, or ability) and external factors (lack of resources, opportunity, or the cooperation of others).

Like attitude and subjective norm, perceived behavioral control has belief-based antecedents, namely control beliefs (*cb*) and evaluation of controls (*ec*). The causal relation among the variables in the theory of planned behavior is represented in the following equation:

$$B \sim BI = w_1AB + w_2SN + w_3PBC,$$

where w_1, w_2, and w_3 represent the relative contributions of *AB*, *SN*, and *PBC* to the determination of *BI*, respectively, and *AB*, *SN*, and *PBC* are formed by their belief-based estimates, $\Sigma_i b_i e_i$ for *i* personal beliefs, $\Sigma_k (nb)_k (mc)_k$ for *k* normative beliefs, and $\Sigma_r (cb)_r (ec)_r$ for *r* control beliefs. The theory of planned behavior reduces to the theory of reasoned action when the behavior of interest is volitional. (For more information, see Ajzen and Madden, 1986, and Crawley and Koballa, 1992.)

Crawley (1990) used the theory of planned behavior to examine the determinants of science teachers' intentions to use investigative teaching methods. Teachers identified availability of materials, facilities, and equipment along with time for preparation, class size, student ability and interest, and course objectives as controls that could influence use of investigative teaching methods with the students in their classes. Perceived behavioral control and subjective norm, however, were found to be unrelated to teachers' intentions to use investigative teaching methods. Teachers' attitude toward use of the methods proved to be the sole factor contributing to the formation of their intentions.

Use of the theory of planned behavior in science education permits teachers and researchers to identify the key beliefs underlying personal action and demonstrates the pivotal role that these beliefs play for the individual in constructing a model of reality. People consult their specific model of reality when faced with the need to make personal decisions. Once determined, salient beliefs can be reinforced, downplayed, and supplemented with additional information relevant to specific behavioral decisions, whether the decision is to select an elective science course or to plan for a career in one of the many science or engineering professions.

PERSUASIVE COMMUNICATION

On the last Friday in May, a swimsuit-clad lifeguard and a well-dressed dermatologist visited neighboring Florida high schools. It was the first time either person had addressed students in a professional capacity. The lifeguard's message, presented to the entire student body near the end of a Spring awards ceremony, concerned the harmful effects of overexposure to the sun and called for all students to limit their time outdoors between noon and 3 P.M. during the summer months. The dermatologist's message, which was broadcast by closed-circuit television to all classrooms during the first 10 minutes of the day, also concerned the harmful effects of overexposure to the sun but encouraged the copious use of sunblock and sunscreen when outdoors. How effective were these presentations? What factors would most contribute to their effectiveness? Would it be the arguments put forth in the message or the attractiveness and credibility of the speaker? If the messages produced attitude changes, would these changes be long lasting? Would the changes in attitude lead to behavioral changes?

Such questions as these have been the concern of social psychologists since the turn of the century. The study of persuasion and attitude change has also been of interest to science educators, but not for quite so long. Investigating the effect of interventions on science-related attitudes and behaviors began in earnest in the 1950s, coinciding with the emergence of science education as a discipline and the identification of the affective domain as one of three domains of human learning by Benjamin Bloom and his colleagues (Shrigley, 1990). Today most science teachers recognize the need to persuade students to accept the truth of certain ideas, such as the importance of wearing safety goggles in chemistry lab. Science teachers are also concerned about the persuasion techniques used by other people, because they believe a scientifically literate citizen should be able to distinguish fact from opinion with respect to issues of science and technology.

The science-education literature contains hundreds if not thousands of reports of interventions designed to change attitudes. Development of programs to influence the likelihood of certain science-related attitudes is important because it is assumed that changes in attitude will result in changes in behavior. Unfortunately, few simple and straightforward generalizations can be made about how and why science-related attitudes change. Favorable attitude changes are reported for semester-long science courses, teacher-training institutes, computer-based instruction, museum tours, and many other treatments. Moreover, a number of researchers have found negative shifts in attitudes or no changes at all (e.g., Lucas & Dooley, 1982; Vockell & Fitzgerald, 1982). Science-education researchers overall seem not to be bothered by the inconsistent findings, explaining away the results that do not show favorable shifts in attitude by inadequacies in measurement or treatments of too short duration.

Undergirding the development of programs to change science-related attitudes is the assumption that, when provided with the appropriate information, subjects will be induced to develop socially desirable attitudes (Shrigley, 1983). However, little attention has been given to identifying the instructional elements that constitute the turnkey variables in the many intervention programs. More often than not, the interventions are broad, convenient variables that in many ways resemble treatments abandoned by social psychologists researching attitude change in the 1930s (Shrigley, 1983). As Koballa and Shrigley (1983) point out, because the programs designed to change science-related attitudes and those to which they were compared are often more alike than they are different, nonsignificant differences between treatments should not be a surprise but expected. Moreover, because of their broad nature it was impossible to determine what aspects of an intervention were responsible for the attitude change when significant differences were found (Shrigley, 1983). These problems make transfer of research findings to the classroom virtually impossible and serve only to frustrate teachers seeking useful suggestions about how to improve students' science-related attitudes.

At the heart of these programs designed to change science-related attitudes are one or more persuasive messages (Koballa, 1992). Social scientists have been about the business of constructing theoretic models of persuasion, designing persuasive messages, and testing their effectiveness for nearly 50 years, and their efforts are well chronicled. The quantity and quality of the literature amassed by social psychologists prompted Shrigley (1983) to suggest that science educators look to the social psychology literature for guidance in developing interventions to change attitudes.

Persuasion: A Mode of Social Influence

Persuasion is one mode of social influence that has been conceptualized in a number of ways. It has evolved from a means of winning in the Greek courts of law to a concept affected by our mass-media society (Larson, 1986). In the minds of some educators, persuasion connotes unethical processes. Keeping teachers from consciously using persuasion may be the tendency to associate such unsavory forms of social influence as indoctrination and brainwashing with teaching to achieve affective outcomes (Bloom, Hastings & Madaus, 1971). However, Petty and Cacioppo (1981) view persuasion simply as an "instance in which an active attempt is made to change a person's mind" (p. 4). Compared with other forms of social influence, persuasion appears to be "neither inherently good nor bad" (Myers, 1983, p. 260).

A situation in which the source consciously intends to influence and both the source and receiver function as active agents in the persuasion process better reflects today's conception of persuasion. Koballa (1992) defines persuasion as "a conscious attempt to bring about a jointly developed product common to both source and receiver through the use of symbolic cues." Reflecting current social-psychological thought, this definition stresses the need for cooperation between source and receiver in the act of persuasion and portrays the essence of the process by which paradigmatic shifts occur in the scientific community (Miller, 1987). Central to the persuasion process is the formation or modification of beliefs that are supported by evidence and good reason.

Other forms of social influence are often confused with persuasion, including propaganda, coercion, indoctrination, brainwashing, and instruction. Science educators and teachers have good reason to confuse persuasion with these other forms of social influence. Little has been written about the distinguishing characteristics of persuasion in the education literature, and, as is true for persuasion, the conceptualizations of forms of social influence have changed over time. Revealing how persuasion is related to these other forms of social influence should assist science educators and teachers in deciding on its appropriate uses.

Propaganda, according to Kecskemeti (1973), is a type of persuasion directed toward a mass audience. Central here are the audience and means of communication. Another conception presents propaganda in an unfavorable light, as an intentional process to disrupt rational decision making through the use of half-truths and emotional appeals (Chase, 1956). Propaganda in this more popular view is inherently unethical.

Coercion relies on reinforcement control, whereas persuasion is prompted by information (Smith, 1982). Recipients of a persuasive message are able to accept or reject the appeal; the same cannot be said for coercion. French and Raven (1960) concede that mild coercion is an integral and often necessary aspect of education. Actions taken by teachers and principals to govern student misbehavior are often coercive in nature. Further distinguishing coercion from persuasion is that, when functioning under duress, people will say or do things in public that they do not believe or privately endorse to avoid punishment or obtain reward (Kelman, 1958). Statements made by American and European hostages held by Iraqi troops in Kuwait that were later renounced upon their release illustrate this aspect of coercion.

Both indoctrination and persuasion are concerned with establishing and changing certain beliefs, but their measures of success are very different. The emphasis placed on the reasons for holding certain beliefs as opposed to the content of beliefs is the basic difference between these two forms of social influence. According to Green (1971), the purpose of indoctrination is to "inculcate the right answer, but not for the right reasons or even for good reason" (p. 31). In contrast, inculcating certain beliefs so that they are based in argument and evidence is the aim of persuasion. From this perspective, the measure of successful persuasion is changes in belief based on convincing argument and evidence.

Journalist Edward Hunter (1951) invented the term brainwashing shortly after the Korean War to describe the techniques used by the North Korean military to obtain the cooperation and compliance of captured Allied soldiers. In sharp contrast to persuasion, physical brutality, intensive interrogation, and psychological pressure are the tools of brainwashing. Even when it is used with prisoners of war, the testimonies of soldiers captured during both the Korean and Vietnam wars reveal that brainwashing is not completely successful (Striker, 1984).

The activities of persuasion resemble classroom instruction in several ways. Communications that include argument and evidence are the vehicles of both persuasion and instruction. Like instruction, persuasion is intended "to shape someone's belief or behavior by helping him to see that the belief is reasonable and the behavior is justified" (Green, 1971, p. 29). For either persuasion or instruction to be successful, a message must be attended to and comprehended by the person to whom it is directed. Both instruction and persuasion have been influenced by the theoretic viewpoint of constructivism. Accordingly, recipients of both engage in conscious cognitive activity and respond in terms of their prior knowledge and beliefs. Nevertheless, people who are unfamiliar with the similarities between these two forms of social influence often label influence attempts they disapprove of as persuasion.

Inducing Persuasion

Numerous models of persuasion have been developed by social psychologists over the last 40 years. From these many models, four general theoretical paradigms have been identified: learning, consistency, perceptual, and functional (McGuire, 1976). Although each paradigm specializes in explaining certain outcomes, Petty and Cacioppo (1981) explain that the research methodologies they employ and the themes they address often overlap. Of the four paradigms, the learning-theory paradigm of persuasion is the most thoroughly researched and often facilitates research based on the other theoretical paradigms. Educators should take comfort in knowing that the most prominent persuasion paradigm is based squarely on many of the theories of learning that dictate instructional strategy in the cognitive domain.

Information Processing. Information processing directed much of the persuasion research conducted during the 1950s under the auspices of the Yale Communication and Attitude Change Program. A major force in shaping contemporary persuasion research, the Yale Program did not present a systematic theory of persuasive communication. As documented in Hovland, Janis, and Kelley's book *Communication and Persuasion* (1953), the program's focus was empirical research organized around the question, "Who says what to whom with what effect?" (Smith, Lasswell, & Casey, 1946). Specifically, Hovland and his colleagues and students studied how attitude is affected by the message (what), its source (who), and the audience of the message (whom). In addition, the channel used to deliver the message and the persistence of attitude change were investigated. The program operated under the basic assumption that learning new information in a persuasive message will change beliefs, the knowledge base of attitude, and that attitude change is maintained only when the information presented in a persuasive message is remembered (Hovland et al., 1953). Persuasion research in science education can be traced directly to the approach used by the Yale researchers (Koballa, 1992).

The persuasive messages developed by science educators during the 1970s and 1980s were based on the Yale group's model of persuasion. Factors related to the message's content, its organization, and the channel through which the message is transmitted were manipulated to test their effects on attitude change. Messages were developed by selecting arguments based on the investigators' knowledge of the audience, and then the arguments were combined with supportive evidence. Arguments selected were directed at both positive and negative consequences of the desired science-related attitude or behavior. These consequences were often couched in terms of their agreement with societal and cultural values. Although efforts to change science-related attitudes using messages built in this way have achieved moderate success, it has been argued that the Yale group's model does not deal with the content of persuasive messages (Ajzen & Fishbein, 1980), the thoughts people generate when listening to messages, and how those thoughts affect their attitudes (Petty & Cacioppo, 1986).

Based on principles of the theory of reasoned action, Fisbein and Ajzen (1981) argue that "only when the message brings about a shift in the summed products across a total set of underlying beliefs can it be expected to influence attitudes or subjective norms [or perceived controls] and, by extension, intentions and behavior" (p. 344). From this perspective, a persuasive message consists of a set of arguments and supporting evidence that addresses behavioral, normative, and control beliefs that are salient for the target audience regarding the behavior of interest. For a persuasive message to be effective, arguments and supporting evidence must attack beliefs salient for the audience or the evaluation of those beliefs which underlie the behavior chosen for modification. Salient beliefs, you will recall, may be identified by questioning a sample of the target audience about reasons for engaging or not engaging in the behavior, referents who would or would not support their engaging in the behavior, and factors that would facilitate or inhibit their doing so.

In line with the theory of reasoned action, three conditions must be met for changes in beliefs salient for an audience to result in behavioral change. First, only when a sufficient number of the behavioral, normative, and control beliefs are changed will attitude, subjective norm, and perceived behavioral control be changed. Second, changes in the beliefs that are the antecedents of attitude, subjective norm, and perceived behavioral control will affect intention only to the extent that they have significant weight in the prediction of intention. Third, the strength of the attitude-behavior relation will determine the degree of correspondence between change in intention and change in behavior.

Working from Fishbein and Ajzen's model, Stutman and Newell (1984) propose a systematic procedure for constructing persuasive messages. Central to their procedure is the notion that specific beliefs rather than societal and cultural values are the critical elements of persuasive appeals. Stutman and Newell (1984) contend that values are generalized beliefs that function most admirably in persuasive appeals directed at general attitudes (e.g., attitude toward science) or general behaviors (e.g., teaching science or conserving energy). Specific beliefs, they continue, tend to express how a person actually sees a value manifested in a specific context and should be used when appeals are made to specific behaviors. An example may clarify the usefulness of specific beliefs in successful persuasion. All teachers in a single school may value discipline, yet argue over policies regarding punishment for student misbehavior. In this situation, differences in belief are likely for the following positions: (1) students should be allowed to select their punishment from among options, (2) paddling is an acceptable form of punishment, and (3) psychological or physical harm is a likely consequence of corporal punishment. In this example, an appeal to values alone seems inappropriate if the formulation of a school discipline policy is the desired outcome. As one can see, even among a seemingly homogeneous population— teachers at a single school— beliefs about how a value will manifest itself in a concrete situation may vary considerably. Thus, appeals to specific beliefs, rather than values, seem to be appropriate when attempting to persuade persons to engage in specific science-related behaviors (Stutman & Newell, 1984).

The message-construction approach based on the theory of reasoned action and Stutman and Newell's (1984) suggestions has been used to develop appeals to persuade adolescents and adults to engage in science-related behaviors. Behaviors that are the focus of messages constructed using this approach include enrolling in an elective science course (Black, 1990; Crawley & Koballa, 1990), becoming the laboratory partner of a classmate who has AIDS (Warden, 1991), and teaching children about the environment (Chen, 1988). The steps followed in the construction of these messages mirror the message construction sequence recommended by Ajzen and Fishbein (1980) and Stutman and Newell (1984):

1. The target audience is defined as precisely as possible and the thesis of the message is identified in terms of action, target, context, and time. In Black's (1990) study, for example, the target audience was Hispanic students enrolled in three San Antonio, Texas high schools. The focus of her appeal was students enrolling in (action) chemistry (target) in high school (context) in September 1990 (time).

2. From a sample representative of the target audience, beliefs are obtained about one's personal interest in engaging in the behavior, referents who could influence one's decision to engage in the behavior, and factors likely to facilitate or inhibit one's doing so. Open-ended questionnaires modeled after ones shown by Ajzen and Fishbein (1980) were used by all investigators to collect information specific to the three areas of belief. Questions found on the open-ended instruments used by Crawley and Koballa (1990) include. "What do you see to be the advantages, for you, of signing-up to take chemistry in September, 1990?" "Are there any people who would disapprove of your signing-up to take chemistry in September, 1990?" "What things outside of your control could happen to make it difficult for you to sign-up to take chemistry in September, 1990?"

3. Behavioral, normative, and control beliefs considered salient for the target audience are generated using information obtained from the sample on the open-ended questionnaire. Among the beliefs considered salient for Hispanic students in the Rio Grande Valley of Texas when asked about enrolling in high school chemistry are: "help them reach their educational goal" (behavioral/advantage), "doing more work at home and at school" (behavioral/disadvantage), "parents and/or guardians" (referent/approve or disapprove), "working in an after-school job that would interfere with study time" (control/inhibiting), and "if the chemistry course is offered at 7:00 a.m." (control/facilitating) (Crawley & Koballa, 1990).

4. Salient beliefs considered to be based on inaccurate information, easily argued against, or likely to reinforce the desired behavior are selected for inclusion in the message. Then arguments and evidence are chosen to support favorable salient beliefs or discredit unfavorable ones to produce a message to encourage engagement in the desired behavior. In constructing her message, Black (1990) chose to support favorable beliefs by using concrete examples and to discredit unfavorable beliefs by either giving contradictory information or downplaying the belief's importance. For the

unfavorable salient belief "enrolling in chemistry will hurt my grade point average," Black (1990) first identified the belief in the message and then downplayed its importance and counterargued: "Grades? Don't worry about grades! No one wants to get bad grades. Usually this only happens if you treat chemistry like a no effort course instead of a challenging experience. If you use good study skills and plan your time you can work it out and make good grades. The benefits are well worth the extra time and effort" (pp. 156–157).

Messages constructed by science educators using this process are typically no more than five pages long and can be read or presented via videotape or audiotape in less than 30 minutes. The study found that there were significant differences in behavioral intention, subjective norm, and perceived behavioral control when both students and parents received the belief-based message, but not when only the students received the message.

Cognitive Processing. Paralleling advances in persuasive-message design were advances in the way persuasion is conceptualized. Greenwald's (1968) cognitive-response model was seen by science educators as an improvement over the Yale group's model of information processing. Central to the model is the notion that people relate their existing knowledge to the information presented in a persuasive message. Unlike the Yale group's model, which assumes a stringent correspondence between remembering specific arguments presented in a persuasive message and attitude change, the cognitive-response model postulates that attitude change and its maintenance are a result of people remembering the thoughts they generate in response to a persuasive message (Petty, Ostrom, & Brock, 1981). Thus, attitude change occurs only to the extent that thoughts in agreement with the message are generated, and changed attitude persists as long as the favorable thoughts are remembered. Of obvious importance to the cognitive response model is the notion that self-generated arguments are more salient than those generated by someone else (Petty et al., 1981). That persuasion is due to the cognitive responses generated as a result of listening to a science-related persuasive message is supported by a study by Koballa (1985). He found that preservice elementary teachers' attitudes toward teaching about energy conservation correlated significantly with cognitive responses elicited immediately after the persuasive communication and cognitive responses recalled 3 weeks later, but not with the recall of arguments presented in the message.

Elaboration Likelihood. A refinement of the cognitive-response approach to persuasion is Petty and Cacioppo's (1981, 1986) Elaboration Likelihood Model (ELM). The ELM proposes a continuum of persuasion, anchored at one end by a central route and at the other end by a peripheral route to message processing. The central route corresponds to Greenwald's (1968) conception of persuasion and emphasizes "thoughtful consideration" of arguments presented in a persuasive message. The peripheral route accounts for situations in which people may be persuaded without giving much thought to the arguments of the message. The importance of the ELM to science-education researchers and teachers rests on the insights it can provide about how persuasive strategies influence the science-related decisions that people make.

Important to the ELM are the ideas that (1) people are motivated to hold correct attitudes and (2) considerable variability exists regarding the kind and amount of issue-relevant elaboration people are willing to engage in to evaluate a persuasive message. Attitudes are judged as correct to the degree that they are considered beneficial, with context determining the standards of correctness. Elaboration refers to the "extent to which a person scrutinizes the issue-relevant arguments contained in a persuasive message" (Greenwald, 1968, p. 7).

The ELM purports that a message recipient can objectively or in a biased manner process the arguments presented in a persuasive message or rely on peripheral cues in the persuasive context. The degree to which people who are presented with a persuasive message scrutinize issue-relevant arguments and thereby follow the central route to persuasion is determined by several motivational and ability variables (Petty & Cacioppo, 1986). We now turn to how one's motivation and ability to think about the arguments proffered in a persuasive message determine the impact of message, source, and recipient factors on attitude change.

Message Factors. Several message variables influence one's motivation and ability to engage in issue-relevant thinking and thereby follow the central route to processing. Messages that rely heavily on numbers and statistical data, Petty and Cacioppo (1986) contend, reduce a person's motivation to process a message. Messages that present vague information and use emotionally charged language and those that are not self-paced (e.g., those presented via radio or television) are also likely to interfere with the ability to process issue-relevant arguments (Petty & Cacioppo, 1986). Other message variables noted by McGuire (1976) that are likely to affect message processing include the use of humor, positive versus negative appeals, whether conclusions are drawn or left to the receiver to draw, and whether potent arguments are presented early or late in the message.

Experimental manipulations to encourage issue-relevant processing have been used in one science-education study. Warden (1991) made a message about becoming a laboratory partner of a classmate who has AIDS more appealing to seventh-grade students by encouraging them to think about a hypothetical personal encounter with the disease before listening to the message. Her results support the use of an orally presented focusing task to enhance student motivation and thereby increase issue-relevant processing for students with little knowledge about AIDS.

Source Factors. Tests of the ELM indicate that quality arguments alone are not sufficient to induce persuasion in all circumstances. Argument quality is the most important persuader for messages considered highly relevant, but for messages viewed as irrelevant or moderate in relevance, source characteristics may affect their persuasiveness (Petty & Cacioppo,

1986). Concerning science-related issues, it is most likely that personal motivation to attend to issues such as suntanning, dissecting animals, and teaching the theory of evolution in science classes will differ from one person to the next. For recipients not highly motivated to engage in issue-relevant thinking, what is believed about the source seems to have a decided effect on the persuasiveness of the appeal (Petty & Cacioppo, 1986).

Credibility, attractiveness, and power are the qualities of the source described by Kelman (1961) that function in the persuasion process. Credibility stems from a source's perceived expertise and trustworthiness. A credible source is one who is viewed as knowledgeable about the topic and seems to have the best interest of the recipient at heart. Internalization is the mechanism by which highly credible sources are thought to change attitudes (Kelman, 1961). Internalization occurs when a new attitude is integrated into one's belief and value system, according to Hass (1981). Changes in attitude through internalization are likely to be long lasting.

Attractive sources change attitudes through identification. Identification is the message recipient's attempt to establish a psychological bond with the message source (Hass, 1981). Various factors seem to account for why a source is considered attractive and therefore more persuasive, including physical appearance (Chaiken, 1979), repeated exposure (Zajonc, 1968), and similarity in appearance or attitude (Byrne, 1971).

Powerful sources alter attitude by means of compliance (Hass, 1981). Power may be derived from the ability to dispense rewards or punishment. Power also comes from legitimate authority, often extended by social institutions such as school, church, and family. Changes in attitude by compliance are often short lived because they tend to represent public adoption of the source's position rather than private acceptance (Hass, 1981).

A number of persuasive skills can be used by science teachers to enhance their credibility, attractiveness, or power. Describing one's past science accomplishments, raising anticipated objections when lecturing on a controversial topic, and providing specific evidence to support theory-based claims are just a few ways to establish credibility in science class. Although changing one's physical appearance is often difficult, attractiveness may be enhanced by communicating liking for the audience through verbal and nonverbal cues such as attentive posture, head nodding, voice tone, and using jargon (Barak, Patkin, & Dell, 1982; Claiborn, 1979). Sharing information about one's background and attitudes may be used to establish teacher-student similarity and thereby boost attractiveness and persuasion (Hass, 1981). It is usually advantageous for a source to emphasize power. Legitimate power can be conveyed by using such words as "should," "ought," and "oblige" when making requests and phrasing requests to emphasize reward over coercion (Raven & Rubin, 1976).

The source is not the only factor that affects the persuasiveness of an appeal, but knowledge of the source is important to an understanding of persuasion. Persuasion skills that enhance credibility, attractiveness, and power are suspected to be of particular value when recipients are not highly motivated to scrutinize issue-relevant arguments.

Audience Factors. Several motivation and ability factors specific to the message recipient are suspected to affect message processing. Personal relevance is one factor known to affect motivation to attend to the arguments addressed in a persuasive message (Petty & Cacioppo, 1986). Personal relevance is likely to be heightened when people sense that the issue addressed in the message will significantly affect them in the not too distant future. Numerous studies support the notion that motivation to scrutinize the issue-relevant arguments is increased when message relevance is high (e.g., Chaiken, 1980; Petty & Cacioppo, 1979). Motivation to process the arguments presented in persuasive messages is also enhanced when arguments are presented by multiple sources (Harkins & Petty, 1981). Festinger's (1957) dissonance effect—that people desire to hold the majority view because it is considered correct—is one explanation for this phenomenon. Other studies indicate that increasing the number of persons responsible for message evaluation reduces one's motivation to process the arguments in a persuasive message (e.g., Brickner, Harkins, & Ostrom, 1986; Weldon & Gargano, 1985). Petty and Cacioppo (1986) use the term "social loafing" to describe the reduced productivity observed in groups compared with predicted individual performance.

Another important audience factor that affects motivation to attend to message arguments is the construct of need for cognition. Need for cognition is the desire a person has to "experience an integrated and meaningful world" (Cohen, Stotland, & Wolf, 1955) and embraces behaviors that, according to Petty and Cacioppo (1986), cannot be explained as drives or instincts. The construct differs from locus of control, self-esteem, cognitive style, self-efficacy, and other motivational variables often used to detect individual differences in persuadability in that it focuses exclusively on people's motivation to engage in cognitive rather than physical endeavors. Using versions of Petty and Cacioppo's Need for Cognition Scale, researchers found that persons high who have a high need for cognition are more likely to base their attitudes on thoughtful analysis of relevant information, whereas persons who have a low need for cognition are not (e.g., Petty & Cacioppo, 1984).

If a person is highly motivated to scrutinize a message thoughtfully but lacks the ability to do so, little issue-relevant thinking is likely to occur (Petty & Cacioppo, 1986). Indeed, the ability to evaluate carefully the arguments in a persuasive message is needed for the central route to persuasion to be followed. An important ability variable that affects message processing is prior knowledge (Petty & Cacioppo, 1986). In general, the results of several studies reported by Petty and Cacioppo (1986) suggest that when prior knowledge is low, simple cues in the persuasive context, such as the length of the message or the credibility of the source, influence message persuasiveness. When prior knowledge is high, recipients are likely to engage in issue-relevant processing.

In a study designed to compare the effects of issue-relevant process versus peripheral cues, Cacioppo & Petty (1979) tested the hypothesis that males and females should be equally able to evaluate factual and verifiable comments about fashions and football, domains for which knowledge differences were found between men and women. The results revealed that the subjects' prior knowledge affected issue-relevant thinking and acceptance of the messages presented, with men and women being less accepting of inaccurate characterizations of fashions and football tackles, respectively. These findings seem particularly relevant to science education because science is often viewed as a masculine enterprise.

Prior knowledge about the topic of a message may lead to objective processing and strengthening of the proffered attitude position or result in biased evaluation of the persuasive message (Fisk & Taylor, 1984). The tendency toward biased processing reflects the fact that "stored knowledge tends to be biased in favor of an initial opinion" (Petty & Cacioppo, 1986, p. 111). Variables found to increase motivation to engage in biased processing include forewarning of message content or of persuasive intent and excessive message repetition (Petty & Cacioppo, 1979).

Also suspected of affecting the ability to process a persuasive message is education. Over the years it has been reported that people with little education are more easily persuaded than well-educated people (see review by Bettinghaus, 1968). An explanation for this finding consistent with the ELM is that those well educated with regard to the issue are better able to process the arguments in the message. However, issue-relevant processing may occur in one of two ways. Argument strength or weakness may be assessed on logic alone, suggesting objective processing, or biased by prior knowledge. Well-educated people, according to Petty and Cacioppo (1986), may also focus on peripheral cues not attended to by less well-educated people. Presentation of both sides of an issue in a persuasive message is paramount among the cues suspected of capturing the attention of well-educated people.

In summary, research based on the ELM suggests that the message recipient is the key to persuasion. The recipient can objectively or in a biased manner process the arguments presented in a persuasive message or rely on peripheral cues in the persuasive context. Any variable that tends to reduce one's motivation or ability to process issue-relevant arguments tends to increase the importance of peripheral cues associated with the message, source, recipient, or persuasive context. Conversely, when one's motivation or ability to think about issue-relevant arguments is high, peripheral cues become less important determinants of persuasion. The benefits of persuasive situations that foster issue-relevant thinking are greater persistence of attitudes and resistance to counterarguments and improved relations between attitude and behavior (Petty & Cacioppo, 1986).

Assessing the Effects of Persuasive Interventions

Inadequacies in the design of attitude scales are often blamed for the lack of consistent research findings regarding science-related attitudes (Munby, 1983). Because attitude is a construct that must be measured indirectly, usually through self-report, it is imperative that instruments used to assess attitudes be both reliable (i.e., produce consistent results) and valid (i.e., measure what you want to measure). Statistical tests to determine instrument reliability are much more prevalent in the development of attitude scales than are plans for establishing instrument validity. Establishing validity is a process that involves human judgment in combination with statistical procedures, according to Abdel-Gaid, Trueblood, and Shrigley (1986). With attention to issues of reliability and validity, a number of Likert and semantic differential scales were developed by science educators.

The development of semantic differential scales by science educators stems from the use of Fishbein and Ajzen's (1975) theory of reasoned action to investigate science-related attitudes and decision making. As described in some detail earlier, the theory of reasoned action intimates that attitude measures should focus on a person's attitude toward a behavior. The role of specificity in the model is operationalized by the deliberate inclusion of the elements action, target, context, and time. Fishbein and Ajzen (1975) argued that the correlation between attitude and behavior is determined in part by the degree of correspondence between the elements composing the attitudinal and behavioral variables.

Fishbein and Ajzen (1975) and Ajzen (1985) also identify other variables, operationalized in a similar manner, that should be measured along with attitude toward the behavior to facilitate behavioral prediction. The variables are subjective norm and perceived behavioral control. You will remember from the preceding discussion that subjective norm refers to a person's beliefs about whether or not significant others think he or she should engage in the behavior and perceived behavioral control specifies factors that could facilitate or inhibit engagement in the behavior. Derived from a combination of the attitude, subjective norm, and perceived behavioral control scores is a behavioral intention score, considered the best predictor of actual behavior. Studies conducted within the framework of this theory have had a substantial impact on the field of attitude research in science education since the mid–1980s.

Measuring intention, attitude, subjective norm, and perceived behavioral control with respect to the behavior requires the use of semantic differential items similar to the one modeled below:

I intend to read my biology textbook during study hall for 15 minutes every day throughout the school year.

likely ____ : ____ : ____ : ____ : ____ : ____ : ____ unlikely
extremely quite slightly neither slightly quite extremely
3   2   1   0   −1   −2   −3

Semantic differential scales have been developed that measure the antecedents of several science-related behaviors. To date, scales have been constructed that measure adolescents' intentions to enroll in an elective chemistry course (Black, 1990; Crawley & Koballa, 1990) and to become the laboratory partner of a classmate who has AIDS (Warden, 1991). Instruments that measure preservice teachers' intentions to teach about the environment (Chen, 1988) and teachers' intentions to use activities and investigations completed during a summer science improvement program with their students (Crawley, 1990) are also available.

Instruments developed to measure variables specified in the theory of reasoned action offer several advantages over traditional measures of attitude. Their development and use is based on a systematic theory of human behavior, the goal of which is to predict and understand behavior. Clear distinctions are made between belief, attitude, intention, and behavior. Attitude is assumed to be a function of all salient beliefs and the evaluation of the beliefs about the attitude object. As a result, refinement of measures by means of item analysis and assurance of scale unidimensionality by means of confirmatory factor analysis are not required (Ajzen & Fishbein, 1980). Furthermore, scales based on the theory of reasoned action are closely linked to the development of belief-based, persuasive messages.

ATTITUDE: AN ESSENTIAL INDICATOR OF THE QUALITY OF SCIENCE EDUCATION

According to Shrigley and Koballa (1992), attitude research in science education is undergoing a revolution. Evidence of this change is described in this chapter. In early sections, research that mirrors Kuhn's (1962) definition of "normal science" is chronicled. The numerous efforts to develop measurement tools, to study the effects of interventions on attitude change, and to investigate the associations between attitude and a host of school, home, and societal variables provided few useful conclusions. In later sections, the transformation in thinking that depicts the onset of a revolution is unveiled. The implications of this overthrow of traditional lines of research are many. One implication is illustrated in the description of the indicators of the quality of science education.

The National Research Council established the Committee on Indicators of Precollege Science and Mathematics Education in 1983 to develop a means for monitoring the success of the reform efforts under way to improve the teaching and learning of precollege mathematics and science (Raizen & Jones, 1985). The educational system was modeled in terms of inputs, processes, and outputs. The Committee's first task was to specify the most important outcomes desired of mathematics and science instruction, then select the schooling variables (inputs and processes) that are related to these outcomes, and finally identify appropriate measures for the specified variables in order to assess current conditions and monitor changes.

Several outcomes were considered as possible indicators of the condition of precollege science and mathematics education. Variables that were examined included achievement, attitude, college attendance, choice of college majors, choice of careers, later career paths, life income, and job satisfaction (Raizen & Jones, 1985, pp. 29–31). College attendance, choice of majors, choice of careers, life income, and job satisfaction were rejected as feasible outcomes because of their dependence on economic conditions and perceptions of future employability. Attitude was also excluded, over the objections of one Committee member (Raizen & Jones, 1985). As justification for its decision, the Committee cited uncertainties about the significance of favorable attitudes toward a particular field of study, about some of the measures used, and about the relationship of attitude to student achievement or to later life outcomes. It recommended reconsideration of the use of student attitudes toward science and mathematics in any further work on indicators. Student achievement was the sole variable to survive the Committee's scrutiny. The acquisition of knowledge was singled out as the main reason for the existence of formal education, and achievement was deemed to be the primary indicator of the condition of precollege science and mathematics education.

The authors of this chapter take exception to this omission. We believe the data on track, course, and career choices presented at the beginning of this chapter show that the dramatic underrepresentation of women and non-Asian minorities in the science and engineering pipeline is attributable as much to a lack of interest in pursuing either of these academic routes as it is to inferior science and mathematics achievement.

Attitude has been a key outcome examined in major national and international assessments of the condition of science and mathematics education. A sample of 9-, 13-, and 17-year-old students nationwide were assessed in 1977, 1981, and 1986 as part of the National Assessment of Educational Progress (NAEP) (Hueftle, Rakow, & Welch, 1983; Mullis & Jenkins, 1988). One or more of the three national assessments and the first International Assessment of Educational Progress (IAEP) (Lapointe, Mead, & Phillips, 1989) have included a substantial number of items that probed student attitudes toward science classes, science teachers, science careers, applications of science, professional ethics in science, and the value of science. An examina-

tion of the content of a representative sample of these items from several of the attitude-related categories reveals the nature of the nation's expectations for students regarding affective outcomes of precollege science (and mathematics) education.

The content of items related to the categories "attitudes toward science classes" and "attitudes toward science careers" refers to students' perceptions of the present or future consequences of studying science. For example, several items ask students whether their current science class is boring, fun, interesting, easy, stimulating, or makes them feel stupid, curious, successful, or confident. Items such as these that target the impact of science learning reflect students' beliefs about current science enrollment (behavioral beliefs), but these beliefs may not be salient to enrolling in science courses. These beliefs offer starting points for developing attitudinal antecedents but are not acceptable indicators of the condition of science education.

Attitude-related items that probe future consequences of science enrollment are more easily revised to represent target behaviors and thus acceptable indicators of the condition of science education. Before attitudinal antecedents can be developed, the target behaviors to which these antecedents allude must first be specified in terms of four components: action, target, context, and time (Ajzen & Fishbein, 1980). Attitude items probing student interest in science-related career preparation serve as better indicators of the condition of science education in high school courses, perhaps, than in the elementary or middle school grades.

Science Classes. Sample items taken from prior NAEP administrations regarding the usefulness of science classes illustrate how beliefs about the consequences of studying science may be transformed into target behaviors. Once these target behaviors are specified, their attitudinal antecedents can be developed. Sample NAEP Items—Much of what you learn in science classes:

is useful in everyday life.
will be useful in the future.
is related to the real world.

Target Behaviors

1. I am learning science that I can use in everyday life.
2. I am learning science that I can use in the future.
3. I am learning science that is related to the real world.

Attitudinal Antecedent

Learning science that I can use in everyday life is:

good	extremely	quite	slightly	neither	slightly	quite extremely	bad
valuable	extremely	quite	slightly	neither	slightly	quite extremely	worthless
harmful	extremely	quite	slightly	neither	slightly	quite extremely	beneficial
pleasant	extremely	quite	slightly	neither	slightly	quite extremely	unpleasant

In a similar manner, students' attitudes toward the target behaviors "I am learning science that I can use in the future" and "I am learning science that is related to the real world" can be assessed. The task for science teachers, educators, and policymakers is to decide the extent to which these three target behaviors related to science classes, for example, represent important outcomes for instruction.

Science Careers. The 1981 NAEP questionnaire included several items that addressed student interest in pursuing additional education in science and possibly a science-related career. Most of the items required that students think about long-term aspirations, but a few of them asked students about things they could do now in preparation for the future. Six items questioned students about their attitudes toward a science-related career or career preparation. The original items taken from the 1981 assessment were posed as questions to which students responded using a five-point scale. For purposes of illustration, each item has been revised to reflect a behavioral outcome, instead of posing a question. Once transformed in this way, each behavior represents an appropriate indicator of the quality of science education in the domain of science career and career preparation.

Target Behaviors

1. I plan to stop taking science courses after I have completed as many as my school requires for graduation.
2. I would like to work at some job that lets me use what I know about science.
3. I want to work with scientists in an effort to solve problems.
4. I would like to visit a scientist at work.
5. I want to learn about science-related jobs that I can do.
6. I would like to learn more about jobs in a science or engineering field.

To assess students' attitude toward taking more science courses, for example (target behavior 1), each statement of the target behavior would be transformed into a verb phrase and followed by the four bipolar, adjectival scales used in assessing students' attitudes, namely good-bad, valuable-worthless, harmful-beneficial, and pleasant-unpleasant. Use of these behaviors as outcomes of science instruction, it seems to us, offers science teachers and researchers an opportunity to monitor students' attitudes within each grade and as they progress through the upper elementary and secondary school grades. The ultimate task of selecting appropriate target behaviors to be used as indicators of the condition of science education, we realize, requires the collaboration of science teachers, educators, parents, and policymakers.

Science Literacy. Science, technology, and their applications are a ubiquitous part of the world marketplace. In response to the United States' need to remain competitive in a global economy, schools are called on to provide programs in science that engender scientific literacy in all students. The traditional meaning of scientific literacy—an understanding of the norms of science and knowledge of major scientific constructs—is no longer adequate. Students today should be aware of the impact of science and technology on society and the policy choices that must inevitably emerge; that is, they must acquire favorable attitudes toward organized science knowledge. According to the best available evidence, only 7 percent of adults meet these standards (Miller, 1983).

Knowledge acquisition reduces illiteracy, but efforts to reduce illiteracy may fall short of their mark. Becoming literate does not guarantee widespread participation in the scientific enterprise. Programmatic efforts to reduce illiteracy through knowledge acquisition alone may do little more than produce persons who are scientifically aliterate; that is, they possess knowledge but choose not to act on their knowledge. For the general public to be attentive but indecisive is ominous for a nation likely to become increasingly dependent on public support to sustain an economy committed to science and technology.

To produce scientifically literate graduates—persons who understand the scientific approach, basic scientific constructs, and science policy issues and exercise their civic responsibility—science teachers, educators, and policymakers must insist on the development of favorable attitudes toward the use of science alongwith the acquisition of knowledge as benchmarks against which the condition of science education is assessed.

Implications of Attitude Research for Science Education

Policymakers. Which science-related attitudes merit consideration as indicators of the condition of science education? The answer to this question lies in the specification of desirable target behaviors toward which favorable attitudes are to be developed. Only then is the causal link between attitude and behavior enhanced. Acceptable science-related behavioral goals represent outcomes that maximize the likelihood that all students will be prepared to become active participants in the professional, service, and civic aspects of the scientific enterprise. Representative behaviors include:

1. Learn science that is useful, now and in the future.
2. Use the methods and tools of science to address out-of-school problems.
3. Participate in community conservation or environmental efforts.
4. Support the application of science to solve global problems.
5. Promote the ethical use of scientific research.
6. Study precollege science beyond the courses that are required for graduation.
7. Work in a science- or engineering-related service or professional field.

These behaviors represent outcomes about which all students studying science should develop favorable attitudes. Behaviors 6 and 7, however, reflect outcomes more appropriate for middle and secondary school students than for students in the elementary grades. It seems reasonable that all students in the secondary grades should acquire favorable attitudes toward the study of elective science courses (behavior 6), regardless of their future career or educational aspirations. Moreover, it seems equitable to expect that as a result of their science study all secondary school students should acquire favorable attitudes toward entering a science or engineering profession (behavior 7), particularly women and non-Asian minorities, who are underrepresented in these professions, even though the social support, opportunity, and ability may be lacking. Unless policymakers are willing to set standards and monitor performance on attitudinal as well as achievement outcomes, students are unlikely to attain the level of scientific literacy necessary to become a functional member of society.

Science Teachers.   The theories of reasoned action and planned behavior detail the formative role that information plays in ultimately directing personal action. Information that is salient to the performance of a specific behavior, it was shown in the theory of planned behavior, shapes perceptions of three types:

1. Personal consequences—the advantages and disadvantages associated with personal action
2. Social support—encouragement and discouragement from others regarding personal action
3. Self-efficacy—perceptions of one's ability, the barriers, resources, and opportunities related to personal action

Schools, in the view of the authors, are best qualified to provide students with salient information related to the consequences of personal, science-related action (or inaction), as well as to assist them with the development of social support systems and the acquisition of a sense of self-control regarding personal action.

References

Abdel-Gaid, S., Trueblood, C. S., & Shrigley, R. L. (1986). A systematic procedure for constructing a valid microcomputer attitude scale. *Journal of Research in Science Teaching, 23*(9), 823–839.

Ajzen, I. (1985). From intentions to actions: A theory of planned behavior. In J. Kuhl & J. Beckmann (Eds.), *Action control: From cognition to behavior.* New York: Springer-Verlag.

Ajzen, I. (1989). Attitude structure and behavior. In A. R. Pratkanis, S. J. Breckler, & A. G. Greenwald (Eds.), *Attitude structure and function.* Hillsdale, NJ: Lawrence Erlbaum.

Ajzen, I., & Fishbein, M. (1980). *Understanding attitudes and predicting social behavior.* Englewood Cliffs, NJ: Prentice Hall.

Ajzen, I., & Madden, T. J. (1986). Prediction of goal directed behavior: Attitude, intentions, and perceived behavioral control. *Journal of Experimental Social Psychology, 22,* 453–474.

Barak, A., Patkin, J., & Dell, D. M. (1982). Effects of certain counselor behaviors on perceived expertness and attractiveness. *Journal of Counseling Psychology, 29*(3), 261–267.

Bem, D. J. (1972). Self-perception theory. *Advances in Experimental Social Psychology, 6,* 1–62.

Bettinghaus, E. P. (1968). *Persuasive communication.* New York: Holt, Rinehart & Winston.

Black, C. B. (1990). Effects of student and parental messages on Hispanic students' intention to enroll in high school chemistry: An application of the theory of planned behavior and the elaboration likelihood model of persuasion. *Dissertation Abstracts International, 52,* 124A.

Bloom, B. S. (1976). *Human characteristics and school learning.* New York: McGraw-Hill.

Bloom, B. S., Hastings, J. T., & Madaus, G. F. (Eds.). (1971). *Handbook of formative and summative evaluation of student learning.* New York: McGraw-Hill.

Brickner, M. A., Harkins, S. G., & Ostrom, T. M. (1986). The effects of personal involvement: Thought provoking implications for social loafing. *Journal of Personality and Social Psychology, 51,* 763–769.

Brookover, W. B., Thomas, S., & Paterson, A. (1964). Self-concept of ability and school achievement. *Sociology of Education, 37*(3), 271–278.

Byrne, D. E. (1971). *The attraction paradigm.* New York: Academic Press.

Cacioppo, J. T., & Petty, R. E. (1979). Sex differences in influenceability: Toward specifying the underlying process. *Personality and Social Psychology Bulletin, 6,* 651–656.

Chaiken, S. (1979). Communicator physical attractiveness and persuasion. *Journal of Personality and Social Psychology, 37,* 1387–1397.

Chaiken, S. (1980). Heuristic versus systematic information processing and the use of source versus message cues in persuasion. *Journal of Personality and Social Psychology, 39*(5), 752–766.

Chase, S. (1956). *Guides to straight thinking, with 13 common fallacies.* New York: Harper & Row.

Chen, C. S. (1988). Cognitive style and intention change: The effects of anecdotal and data-summary persuasive strategies on teachers' intentions to teach about the environment. *Dissertation Abstracts International, 49,* 1311A.

Claiborn, C. D. (1979). Counselor verbal intervention, nonverbal behavior, and social power. *Journal of Counseling Psychology, 26*(5), 378–383.

Clewell, B. C. (1987). What works and why: Research and theoretical bases of intervention programs in math and science for minority and female middle school students. In A. B. Champagne & L. E. Hornig (Eds.), *Papers from the 1987 National Forum for School Science: Students and Science Learning.* Washington, DC: American Association for the Advancement of Science.

Cohen, A. R., Stotland, E., & Wolfe, D. M. (1955). An experimental investigation of need for cognition. *Journal of Abnormal and Social Psychology, 51,* 291–294.

Crawley, F. E. (1990). Intentions of science teachers to use investigative teaching methods: A test of the theory of planned behavior. *Journal of Research in Science Teaching, 27*(7), 685–697.

Crawley, F. E., & Coe, A. E. (1990). Determinants of middle school students' intention to enroll in a high school science course: An application of the theory of reasoned action. *Journal of Research in Science Teaching, 27*(5), 461–476.

Crawley, F. E., & Koballa, T. R. (1990). *Hispanic students' interest in chemistry project: Final report for the Pharr-SanJuan-Alamo Independent School District.* Unpublished report, Science Education Center, University of Texas, Austin.

Crawley, F. E., & Koballa, T. R. (1992). Hispanic-American students' attitudes toward enrolling in high school chemistry: A study of planned behavior and belief-based change. *Hispanic Journal of Behavioral Sciences, 14*(4), 469–486.

Educational Policies Commission. (1962). *Education and the spirit of science.* Washington, DC: Author.

Festinger, L. A. (1957). *A theory of cognitive dissonance.* Stanford, CA: Stanford University Press.

Fishbein, M., & Ajzen, I. (1975). *Belief, attitude, intention and behavior. An introduction to theory and research.* Reading, MA: Addison-Wesley.

Fishbein, M., & Ajzen, I. (1981). Acceptance, yielding and impact: Cognitive processes in persuasion. In R. Petty, T. Ostrom, & T. Brock (Eds.), *Cognitive responses in persuasion* (pp. 339–359). Hillsdale, NJ: Lawrence Erlbaum.

Fisk, S. T., & Taylor, S. E. (1984). *Social cognition.* Reading MA: Addison-Wesley.

Fraser, B. J. (1982). How strongly are attitude and achievement related? *School Science Review, 63*(224), 557–559.

Gardner, P. L. (1975). Attitudes to science: A review. *Studies in Science Education, 2,* 1–41.

Green, T. F. (1971). *The activities of teaching.* New York: McGraw-Hill.

Greenwald, A. G. (1968). Cognitive learning, cognitive response to persuasion and attitude change. In A. G. Greenwald, T. C. Brock, & T. M. Ostrom (Eds.), *Psychological foundations of attitudes.* New York: Academic Press.

Haladyna, T., & Shaughnessy, J. (1982). Attitudes toward science: A quantitative synthesis. *Science Education, 66*(4), 547–563.

Harkins, S. G., & Petty, R. E. (1981). Effects of source magnification of cognitive effort on attitudes: An information processing view. *Journal of Personality and Social Psychology, 40*(3), 401–413.

Harvey, O. J. (1970). Beliefs and behavior: Some implications for education. *Science Teacher, 37*(9), 10–14.

Harvey, O. J., Prather, M., White, B. J., & Hoffmeister, J. K. (1968). Teachers' beliefs, classroom atmosphere and student behavior. *American Educational Research Journal, 5*(2), 151–166.

Hass, R. G. (1981). Effects of source characteristics on cognitive response and persuasion. In R. E. Petty, T. M. Ostrom, & T. C. Brock (Eds.), *Cognitive responses in persuasion.* Hillsdale, NJ: Lawrence Erlbaum.

Hill, O. W., Pettus, W. C., & Hedin, B. A. (1990). Three studies of factors affecting the attitudes of blacks and females toward the pursuit of science and science-related careers. *Journal of Research in Science Teaching, 27*(4), 289–314.

Hovland, C. I., Janis, I. L., & Kelley, H. H. (1953). *Communication and persuasion: Psychological studies of opinion change.* New Haven, CT: Yale University Press.

Hueftle, S. J., Rakow, S. J., & Welch, W. W. (1983). *Images of science: A summary of results from the 1981–82 national assessment in science.* Minneapolis: Minnesota Research and Evaluation Center.

Hunter, E. (1951). *Brain-washing in Red China.* New York: Vanguard.

Kahneman, D., & Tversky, A. (1984). Choices, values, and frames. *American Psychologist, 39*(4), 341–350.

Kecskemeti, P. (1973). Propaganda. In I. Pool, W. Schramm, F. W. Frey, N. Maccoby, & E. B. Parker (Eds.), *Handbook of communication* (pp. 844–870). Chicago: Rand-McNally.

Kelman, H. C. (1958). Compliance, identification, and internalization: Three processes of attitude change. *Journal of Conflict Resolution, 2,* 51.

Kelman, H. C. (1961). Process of opinion change. *Public Opinion Quarterly, 25,* 57–78.

Klopfer, L. E. (1976). A structure for the affective domain in relation to science education. *Science Education, 60*(3), 299–312.

Kluckhohn, F. R. (1956). Dominant and variant value orientations. In F. R. Kluckhohn & H. A. Murray (Eds.), *Nature, Society, and Culture.* New York: Alfred A. Knopf.

Koballa, T. R. (1985). The effect of cognitive response in attitudes of preservice elementary teachers toward energy conservation. *Journal of Research in Science Teaching, 22*(6), 555–564.

Koballa, T. R. (1988). The determinants of female junior high school students' intentions to enroll in elective physical science courses in high school: Testing the applicability of the theory of reasoned action. *Journal of Research in Science Teaching, 25*(6), 479–492.

Koballa, T. R. (1992). Persuasion and attitude change in science education. *Journal of Research in Science Teaching, 29,* 63–80.

Koballa, T. R., & Crawley, F. E. (1985). The influence of attitude on science teaching and learning. *School Science and Mathematics, 85*(3), 222–232.

Koballa, T. R., & Shrigley, R. L. (1983). Credibility and persuasion: A sociopsychological approach to changing that attitudes toward energy conservation of preservice elementary teachers. *Journal of Research in Science Teaching, 20*(7), 683–696.

Krynowsky, B. A. (1988). Science education assessment instruments: Problems in assessing student attitude in science education: A partial solution. *Science Education, 72*(5), 575–584.

Kuhn, T. S. (1962). *The structure of scientific revolutions.* Chicago: University of Chicago Press.

Langer, E. J. (1989). Minding matters: The consequences of mindlessness-mindfulness. *Advances in Experimental Social Psychology, 22,* 137–173.

Lapointe, A. E., Mead, N. A., & Phillips, G. W. (1989). *A world of differences: An international assessment of mathematics and science.* Princeton, NJ: Educational Testing Service.

Larson, C. U. (1986). *Persuasion: Reception and responsibility.* Belmont, CA: Wadsworth.

Lin, B., & Crawley, F. E. (1987). Classroom climate and science related attitudes of junior high school students in Taiwan. *Journal of Research in Science Teaching, 24*(6), 579–591.

Lucas, K. B., & Dooley, J. H. (1982). Student teachers' attitudes toward science and science teaching. *Journal of Research in Science Teaching, 19*(9), 805–809.

McGuire, W. J. (1976). Attitude change and the information-processing paradigm. In E. P. Hollander & R. G. Hunt (Eds.), *Current perspectives in social psychology.* New York: University Press.

Miller, C. M. (1987, May). *Paradigmatic and presumptive shifts: Thomas Kuhn and Richard Whately in tandem.* Paper presented at the Annual Meeting of the Eastern Communication Association, Syracuse, NY. (ERIC Document Reproduction Service No. ED 286 223)

Miller, J. D. (1983). Scientific literacy: A conceptual and empirical review. *Daedalus, 112*(2), 29–48.

Mulkey, L. M., & Ellis, R. S. (1990). Social stratification and science education: A longitudinal analysis, 1981–1986, of minorities' integration into the scientific talent pool. *Journal of Research in Science Teaching, 27*(3), 205–217.

Mullis, I., & Jenkins, L. B. (1988). *The science report card: Elements of risk and recovery.* Princeton, NJ: Educational Testing Service.

Munby, H. (1983). Thirty studies involving the 'Scientific Attitude Inventory': What confidence can we have in this instrument? *Journal of Research in Science Teaching, 20*(2), 141–162.

Myers, D. G. (1983/1987). *Social psychology.* New York: McGraw-Hill.

National Academy of Sciences. (1989). *On being a scientist.* Washington, DC: National Academy Press.

National Science Board. (1989). *Science and engineering indicators—1989 (NSB 89–1).* Washington, DC: U. S. Government Printing Office.

Ormerod, M. B., & Duckworth, D. (1975). *Pupils' attitudes to science: A review of research.* Berkshire: NFER.

Petty, R. E., & Cacioppo, J. T. (1979). Issue involvement can increase or decrease persuasion by enhancing message-relevant cognitive responses. *Journal of Personality and Social Psychology, 37*(10), 1915–1926.

Petty, R. E., & Cacioppo, J. T. (1981). *Attitudes and persuasion: Classic and contemporary approaches.* Dubuque, IA: Brown.

Petty, R. E., & Cacioppo, J. T. (1984). The effects of involvement on responses to argument quantity and quality: Central and peripheral routes to persuasion. *Journal of Personality and Social Psychology, 46,* 69–81.

Petty, R. E., & Cacioppo, J. T. (1986). *Communication and persuasion: Central and peripheral routes to attitude change.* New York: Springer-Verlag.

Petty, R. E., Ostrom T. M., & Brock, T. C. (1981). Historical foundations of the cognitive response approach to attitudes and persuasion. In R. E. Petty, T. M. Ostrom, & T. C. Brock (Eds.), *Cognitive responses in persuasion.* Hillsdale, NJ: Lawrence Erlbaum.

Punch, K. F., & Rennie, L. J. (1989, March). *Clarification of the direction of the affect-achievement relationship in science.* Paper presented at the annual meeting of the National Association for Research in Science Teaching meeting, San Francisco. (ERIC Document Reproduction Service No. ED 306 119)

Raizen, S. A., & Jones, L. V. (Eds.). (1985). *Indicators of precollege education in science and mathematics.* Committee on Indicators of Precollege Science and Mathematics Education, National Research Council. Washington, DC: National Academy Press.

Raven, B. H., & Rubin, J. Z. (1976). *Social psychology: People in groups.* New York: Wiley.

Rokeach, M. (1973). *The nature of human values.* New York: Free Press.

Rowe, M. B. (1974a). Relation of wait-time and rewards to the development of language, logic, and fate control: Part II–Rewards. *Journal of Research in Science Teaching, 11*(4), 291–308.

Rowe, M. B. (1974b). Wait-time and rewards as instructional variables, their influence on language, logic, and fate control: Part I–Wait-time. *Journal of Research in Science Teaching, 11*(2), 81–94.

Schibeci, R. A. (1984). Attitudes to science: An update. *Studies in Science Education, 11,* 26–59.

Shrigley, R. L. (1983). The attitude concept and science teaching. *Science Education, 67*(4), 425–442.

Shrigley, R. L. (1990). Attitude and behavior are correlates. *Journal of Research in Science Teaching, 27*(2), 97–113.

Shrigley, R. L., & Koballa, T. R. (1992). A decade of attitude research based on Hovland's learning theory model. *Science Education, 76,* 17–42.

Shrigley, R. L., Koballa, T. R., & Simpson, R. D. (1988). Defining attitude for science educators. *Journal of Research in Science Teaching, 25*(8), 659–678.

Simpson, R. D., & Oliver, J. S. (1985). Attitude toward science and achievement motivation profiles of male and female science students in grades six through ten. *Science Education, 69*(4), 511–526.

Simpson, R. D., & Oliver, J. S. (1990). A summary of major influences on attitude toward and achievement in science among adolescent students. *Science Education, 74,* 1–18.

Simpson, R. D., & Troost, K. M. (1982). Influences on commitment to learning of science among adolescent students. *Science Education, 66*(5), 763–781.

Simpson, R. D., & Wasik, J. L. (1978). Correlation of selected affective behaviors with cognitive performance in a biology course for elementary teachers. *Journal of Research in Science Teaching, 15,* 65–71.

Smith, B. L., Lasswell, H. D., & Casey, R. D. (1946). *Propaganda, communication, and public opinion.* Princeton, NJ: Princeton University Press.

Smith, M. J. (1982). *Persuasion and human action: A review and critique of social influence theories.* Belmont, CA: Wadsworth.

Spilka, B. (1970, August). Alienation and achievement among Oglala Sioux secondary school students. *Resources in Education* (ERIC Report No. ED 045225). A final report to the National Institute of Mental Health, Bethesda, MD.

Stallings, J., & Robertson, A. (1979). *Factors affecting women's decisions to enroll in advanced mathematics courses.* Menlo Park, CA: SRI International.

Staver, J. R., Enochs, L. G., Koeppe, O. J., McGrath, D., McLellan, H., Oliver, J. S., Scharmann, L. C., & Wright, E. L. (1989). A summary of research in science education—1987. *Science Education, 73*(3), 243–389.

Striker, L. D. (1984). *Mind-bending.* Garden City, NY: Doubleday.

Stutman, R. K., & Newell, S. E. (1984). Beliefs versus values: Salient beliefs in designing a persuasive message. *Western Journal of Speech Communication, 48*(4), 362–372.

Tedeschi, J. T. (1981). *Impression management theory and social psychological research.* New York: Academic Press.

Tesser, A., & Shaffer, D. R. (1990). Attitudes and attitude change. In M. R. Rosenzweig & L. W. Porter (Eds.), *Annual Review of Psychology, 41,* 479–523.

Thomas, G. E. (1986). Cultivating the interest of women and minorities in high school mathematics and science. *Science Education, 70,* 31–43.

Thurstone, L. L. (1928). Attitudes can be measured. *American Journal of Sociology, 33*(4), 529–554.

U. S. Congress, Office of Technology Assessment. (1988, December). *Elementary and secondary education for science and engineering–A technical memorandum* (OTA-TM-SET–41). Washington, DC: U. S. Government Printing Office.

Vockell, E. L., & Fitzgerald, T. (1982). Developing favorable attitudes toward animal life among elementary school pupils. *Hoosier Science Teacher, 8*(2), 53–59.

Walberg, H. J. (1969). Social environment as a mediator of classroom learning. *Journal of Educational Psychology, 60,* 443–448.

Warden, M. A. (1991). The effect of two videotaped persuasive messages on 7th grade science students' intentions to perform an AIDS-related laboratory behavior. *Dissertation Abstracts International, 52,* 126A.

Weldon, E., & Gargano, G. M. (1985). Cognitive effort in additive task groups: The effects of shared responsibility on the quality of multiattribute judgment. *Organizational Behavior and Human Decision Processes, 36,* 348–361.

Zajone, R. B. (1968). Attitudinal effects of mere exposure. *Journal of Personality and Social Psychology Monograph Supplement, 9*(2), 1–27.

Part

◆ III ◆

PROBLEM SOLVING

·7·

RESEARCH ON PROBLEM SOLVING: ELEMENTARY SCHOOL

Bonnie B. Barr

SUNY CORTLAND

"Problem solving through reflective thinking should be both the method and valued outcome of science instruction in America's schools."
—John Dewey, 1916, Method in Science Teaching, *General Science Quarterly*.

Researchers of science teaching and learning today continue to regard problem solving as a valued outcome of science education. Numerous reports over the past decade have called for reform in education, particularly curriculum and instruction in mathematics and science. In a paper entitled "Science Education and the Nation's Economy," Hurd (1989) summarized the findings of 26 national reports. Eighteen of the reports specifically describe problem solving in science as an educational objective. *Science for All Americans* (Rutherford & Ahlgren, 1990) states that "knowledge should be understood in ways that will enable it to be used in solving problems" (p. 175) and that "in the science classroom wondering should be as highly valued as knowing" (p. 191). Champagne and Klopfer (1977) wrote, "On one aspect of science teaching, there is a remarkable degree of agreement in the professed beliefs of most science educators—the belief that problem solving and reflective thinking play an important role in children's learning of science in school" (p. 437).

Belief in problem solving as a goal of elementary science instruction is pervasive. However, there is a disparity of some magnitude between belief and practice. Despite the calls for restructuring and rethinking of the mission of science education, elementary science instruction in many classrooms has remained unchanged since the 1950s. This review addresses the research that most directly provides insights into what educators can do to change this scenario. The research reviewed deals with student learning, curriculum development, and classroom instructional strategies as they relate to problem solving.

Because only moderate progress has been made in implementing John Dewey's problem-solving philosophy in the instructional strategies of today's classrooms (Champagne & Klopfer, 1977), a historical approach to the research is in order. The historical approach reveals that ideas on problem-solving education that are considered novel today were contemporary in the 1950s. An outcome of the historical review may be the notion that to produce a population literate in the skills of science will require much broader organizational restructuring in our educational system than has occurred in the past.

ATTITUDE OR PROCESS: DEFINING THE ATTRIBUTES OF PROBLEM SOLVING

Dewey (1913) used terms such as problem solving, reflective thinking, inquiry, and scientific attitude interchangeably. In the hope of reducing ambiguity in terminology, research from the 1930s through 1950 focused on defining the attributes of

The author thanks reviewers Doris A. Trojacak (University of Missouri) and David P. Butts (University of Georgia).

the problem-solving process. It soon became apparent that the problem-solving process had many dimensions. Some researchers concentrated on the nature of the scientific enterprise and the attitudes and behaviors required of successful practitioners. Lists of behaviors consistent with a scientific attitude were developed. Behaviors included such things as a desire to try things out experimentally, willingness to change one's opinion or conclusion when confronted with new evidence, determination to be objective in judgment, and unwillingness to base a conclusion on one or a few observations. Many tests were developed to measure one or more elements of the scientific attitude (Davis, 1935; Edwards & Robertson, 1939; Keeslar, 1956; Baumel & Berger, 1965; Kozlow & Nay, 1976). The complexities of these scientific attitudes are so great, however, that encouraging them has proved to be an enormous undertaking. Other than stimulants to intellectual discourse, the "lists" have had only minimal impact on the instructional strategies employed in contemporary classrooms. This does not mean that scientific attitudes are not essential to problem solving. It implies that future research must address the barriers preventing the inclusion of scientific attitudes as valued educational objectives.

Research in the late 1960s and early 1970s illustrates the trend toward applying psychological theories of development and learning to problem solving. This resulted in a better definition of the component skills or behaviors of inquiry. Shulman and Keislar (1966) classified the component skills into four categories:

1. Problem sensing. A discrepant event or an apparent incongruity stimulates the awareness of a problem.
2. Problem formulating. An attempt to define or clarify the problem is made. Solutions to the problem are anticipated.
3. Searching. Questions about the problem are raised. Information is gathered. Hypothesizes are formulated and alternative solutions are explored.
4. Problem resolving. The incongruity or disequilibrium is removed and the problem is resolved to the satisfaction of the learner.

Instructional psychologists call these behaviors the *process skills*. Looking at problem solving as a patterned thought process or a series of specific behavioral steps has been supported by numerous researchers who have articulated similar steps, but with varying emphasis (Obourn, 1956; Gagne, 1970). The delineation of specific process skills has served as a springboard for research on problem solving. Much of the resulting research on process skills is prescriptive in its attempt to validate specific instructional strategies and curricular approaches that enhance the problem-solving abilities of the elementary student.

THE ESSENTIAL ELEMENTS: CONTENT, ATTITUDES, AND THE PROCESS SKILLS

Suchman, as early as 1960, expounded on the futility of teaching process skills outside a content context. The whole nature of problem solving is content embedded. Scientific problems are formulated as the result of a cognitive dissonance (Festinger, 1962). Clarification, problem analysis, and problem resolution depend on the relevant concepts stored in memory (Greeno, 1978; Novak, 1977; Norton & Butts, 1973). Strategies to promote meaningful conceptual learning constitute another body of research and are not dealt with in this review. However, bear in mind that content knowledge that can be retrieved and applied to a problem is a critical factor in any successful problem-solving strategy (Finley, 1983; Pizzini, Shepardson, & Abell, 1989).

As described earlier, attitudes that are consistent with the nature of science are necessary for successful problem resolution. Scientific attitudes are nurtured when the problem solver experiences ownership of the problem to be solved. The problem must be derived from the students' own conceptual framework. The student needs to see the problem as meaningful and relevant. The problem must be generated by the student and not imposed by the teacher. Pizzini et al. (1989) found that "student ownership of the problem is one of the most essential variables resulting in successful problem solving" (p. 527).

Once applicable concepts are stored in memory and ownership of the problem has been achieved, problem-solving abilities can be enhanced through instruction on logical reasoning strategies or process skills. One of the first projects to study the effects of inquiry training on the problem-solving skills of elementary science students was that done by Suchman in 1960. Suchman developed the Inquiry Training curriculum, which was used in a 3-year study with fifth-grade students. The curriculum utilized inquiry strategies to motivate learning of science concepts, while, at the same time, stimulating the growth of scientific problem-solving skills. The Inquiry Training curriculum consisted of a series of single-problem film loops, a handbook of teacher demonstrations of discrepant events, and a student handbook with simple illustrations of science discrepancies.

The curriculum is implemented by exposing the children to a problem via either a film loop or a teacher demonstration. The students are directed to ask questions about the problem that elicit a "yes" or "no" response from the teacher. After about 30 minutes of information gathering, the children identify variables and formulate hypotheses of cause-and-effect relationships. The purpose of this protocol is to teach children to probe in an objective, systematic manner and to reason productively on their own to discover meaningful patterns. The protocol ends when causal relationships are discovered. As new problems are introduced, the children receive guidance in how to approach them. They are introduced to a three-stage plan. In the first stage, the children identify the objects and systems embedded in the problem, conditions that exist, and changes that occur during the episode. In the second stage, the children focus on which conditions are really necessary to produce the event. In the third stage, children formulate imaginary tests to verify perceived relationships. After each episode, questions are critiqued, the stage to which each question belongs is identified, and the type of information that was gained from each question is clarified. Suchman stated that the questions must be answerable by yes or no to discourage children from transferring control of the thinking process to the teacher. Suchman found that the inquiry skills of the fifth-grade students improved when they were exposed to the Inquiry Training curriculum. The children became more productive in their design and use of verification in experimentation and were fairly consistent in their ability to transfer the inquiry thinking pattern to new problems. The importance to the problem-solving process of student-generated questions has been confirmed by Winne and Marx (1977), Anderson and Smith (1981), and Zoller (1987).

Neal (1961) provided instruction in the following process skills to students in grades 1 through 6 in a laboratory school setting: identifying and stating problems, selecting pertinent and adequate data, formulating and evaluating a hypothesis, generalizing and forming a conclusion, and applying concept or seeing relationships. The ability to recognize and state problems was developed through techniques including the use of discrepant observations obtained from current events, field trips, demonstrations, and questions posed by the instructor. The ability to select pertinent and adequate data was developed by use of question captions on illustrations and demonstration materials, cause-effect experimentation, and hypothetical problem episodes posed by the teacher. Techniques used to develop the ability to formulate and evaluate hypotheses included the use of discrepant observations made during field trips on the school site, tasks that helped students differentiate between facts and assumptions, and probing questions that asked students to expand their explanations of causal variables in an experiment. Acquisition of the process skills was assessed through (1) analysis of student record books, that is, voluntary statements of problems, plans for solving problems, analysis of data, generalization and applications of learning; (2) records of the child's creative work: demonstrations, exhibits, visual materials, and student-constructed evaluation forms used to document achievement; (3) anecdotal records; and (4) objective tests. Neal found that students in all age groups developed the ability to use methods of scientific inquiry. The greatest growth in all five process skills occurred between grades 4 and 6.

The hypothesis that instruction on process skills can enhance students' problem-solving abilities was further verified by research conducted by Butts and Jones (1966). The study involved 109 sixth-grade students. Thirty- to sixty-minute training sessions were held 5 days a week for 3 weeks. The training sessions followed the Suchman Inquiry Training protocol. Each problem-solving episode was followed by a critique session. The Tab Inventory of Science Processes (TISP) was used to assess problem-solving skills. Students involved in the training demonstrated significant growth in behavior patterns indicative of effective problem solvers.

Problem-solving models rank hypothesis formation high among the skills necessary for problem solving. Some researchers believe that skill in hypothesis formation is intuitive and cannot be taught (Ramsey & Howe, 1969). However, this view is in direct conflict with earlier research by Atkin (1958), who found that in classrooms in which science discrepancies were presented and inquiry was encouraged, children offered more hypotheses as solutions to science problems. This view was confirmed in the research of Quinn and George (1975) on

whether children's hypotheses could be improved. A hypothesis was defined as a "testable explanation of an empirical relationship among variables in a given problem situation" (p. 289). Students in four sixth-grade classes participated in the study. The students were shown film clips of a single physical science problem. After a discussion in which the problem was verified, the students were encouraged to gather information about the problem by asking the teacher questions that could be answered with a yes or no. The film was reshown. Any additional questions were answered before the students were asked to write as many hypotheses about the question as they could. Upon completion of this task, the teacher shared with the students the Hypothesis Quality Scale, which they then used to judge their own hypotheses and to improve on them. After every two films (total of 12) there was a discussion session focused on formulating good hypotheses. Students discriminated between empirical statements and nonempirical statements, such as "Increased air pressure caused the bottle to move ten units" and "Magic did it." Students listed observations from the film in sequential order, made inferences about their observations, and proposed tests for their hypotheses. Conclusions drawn from the study indicate that hypothesis formation can be taught. Furthermore, the gain in skill in forming hypotheses is unrelated to socioeconomic level or demographics. The key ingredient in teaching children to formulate hypotheses is a classroom environment that encourages reflection on cause-effect relationships to problems.

Ability to control variables in a science experiment is a process skill necessary for successful problem solving. Wollman and Chen (1982) found that social-interaction episodes before hands-on problem solving situations significantly enhanced the students' ability to control variables. Two classrooms of fifth-grade students were involved in the study. Each classroom received a different instructional treatment. The Social Interaction/Hands-on Treatment consisted of two parts. At the beginning of the lesson the teacher demonstrated a problem event. The students were asked to respond to the following questions: What caused the event? What evidence do you have for your response? What alternative explanations might account for the event? What explanation best describes the event? The discussion lasted about 20 minutes in each of eight 45-minute sessions. The remainder of each session was devoted to small-group work with kits of materials that posed problems requiring control of variables. The Hands-on Treatments was identical to the first approach except that the initial social interaction was omitted. The Hands-on Treatment group had more time to work with the problems in the kit. In both an immediate and delayed posttest, the students in the Social Interaction/Hands-on Treatment group scored significantly higher on ability to control variables than students in the Hands-on treatment group. The researchers concluded that this type of regular social interaction associated with hands-on science problems should substantially raise the level of inquiry for all elementary students. It appears that it is not enough to "do"—reflection on the process is important to the acquisition of process skills.

As mentioned, control of variables is a sophisticated process skill essential to problem solving. Elementary students frequently find the steps in experimental design confusing and approach experimentation in a trial-and-error manner without regard for causal relationships. Numerous research studies have demonstrated that mnemonics can enhance recall of detailed information. Ross (1990) conducted a study to see if mnemonics could be used to help students attend to specific steps when designing an experiment. A peg-word mnemonic was used in the study. The first letter in each of the words of the mnemonic was designed to trigger recognition of a specific science process involved in experimentation. The mnemonic used in the study is as follows:

COWS	MOO	SOFTLY
Change something.	Measure something.	Something stays the same.

Two treatment groups comprising of fifth-grade students were involved in the 4-week study. In the first treatment group the teacher demonstrated a design for a real-life experiment and then applied the mnemonic to the design. The students were then given seven paper-pencil tasks in which they were asked to design, using the mnemonic, experiments to test real-life problems such as "What dog food does your dog like best?" In the second treatment group, the teacher demonstrated the real-life experiment and then applied the mnemonic and an organized set of rules for designing an experiment to the problem. The rules were as follows:

1. Decide what will be changed.
2. Pick different amounts of the change to test.
3. Decide on the main result of the change.
4. Find a way to measure it.
5. Think of things that might make a difference to the result.
6. Keep things the same during the test.

The students in the second treatment group were then given the seven problems and asked to use the organized rules to design an experiment to test each problem. Both treatment groups were substantially more likely to control variables than the equivalent control group. The treatment group receiving the organized rules scored somewhat better than those receiving only the mnemonic.

Application of the Research to the Classroom

1. Problem solving is supported by discrepant events accompanied by student-generated questions.
2. Problem solving is supported by providing students with guidance on how to ask productive questions.
3. Problem solving is supported by activities and experiences that cause children to reflect on the strategies used to resolve a problem.
4. Problem solving is supported by providing students with standards that help them evaluate their own hypotheses.
5. Problem solving is supported by social interaction about a problem before experimentation.

6. Problem solving is supported by mnemonics that help students memorize the steps in a controlled experiment.

PROCESS-ORIENTED CURRICULA: DO THEY ENHANCE PROBLEM SOLVING?

Based on the substantial body of research affirming that instruction supports the development of process skills, three major federally funded K–6 elementary science curricula, SAPA, SCIS, and ESS, were developed in the 1960s. The SAPA (Science, A Process Approach) curriculum (Gagne, 1963, 1965) is developed around an elaborate hierarchy of process skills (Gagne, 1963). Acquisition of these skills is viewed as a linear relationship in which basic skills (observation, inference, classification, predicting, collecting and recording data, and measurement) are learned before the integrated skills (controlling variables, interpreting data, defining operationally, formulating hypotheses, and experimentation). Instructional emphasis in the primary grades is, therefore, focused on the development of the basic skills. Each lesson is presented in a detailed instructional module that articulates both the prerequisite and subsuming process skills for the lesson.

The SCIS (Science Curriculum Improvement Study) curriculum integrates process skills into a conceptual approach. In SCIS, Piaget's research on cognitive development is translated into instructional strategies via the learning-cycle model. In the SCIS Learning Cycle, students are involved in an exploration phase, followed by a concept-introduction and a concept-application phase (SCIS, 1968; Karplus, 1964). Wholistic conceptual science schemes are developed in a K–6 physical and life science strand. The Elementary Science Study (ESS) curriculum is designed to help students construct their own investigations on a variety of real-world science experiences. Students are encouraged to explore novel relationships among variables, formulate hypotheses, and conduct investigations. Emphasis of the hands-on program is on knowledge acquisition through student discovery of causal relationships among variables (Elementary Science Study, 1971).

Extensive research on the effect of these curricula on the development of student attitudes, process skills, and content knowledge has been conducted. Renner, Strafford, Coffia, Kellogg, and Weber (1973) measured the development of science process skills of students in SCIS programs and those in textbook-oriented classrooms. The experimental group was composed of 46 children who had 5 years of SCIS experience. In every identifiable respect, the inquiry science program was the only difference between the experimental and the control group. The process skills evaluated were observation, classification, measurement, experimentation, interpretation, and prediction. A 17-task assessment was constructed to measure the growth in process skills. Each task focused on a performance problem that required the use of a specific process skill. Following is an example of one of the process tasks (Renner et al., 1973, p. 298):

Materials: A rubber band, a small piece of stiff wire, a support stand, a ruler, graph paper, string, and four hardware washers.

Administrative Procedure: Give the materials to the students. *Instructions*: You have four hardware washers here. How far would eight of these washers stretch this rubber band?

Scoring: 1. Give one point if the child determines how far the four washers will stretch the rubber band. 2. Give one point if, based on his/her measured data, the child gives an answer for the stretch that would be caused by eight washers.

The instrument was individually administered to each of the 60 students (45 experimental, 15 control) in the study. Results of the study indicate that the SCIS program made a significant contribution to the development of process skills. The inquiry-based manipulative program proved superior to a textbook program in aiding the development of all process skills tested.

Davis (1979) reported on a districtwide effort to measure the achievement and creativity of students (grades 1–6) enrolled in the process-based SAPA curriculum and those in a traditional textbook program. No significant difference was found in the achievement scores of the two groups. The SAPA students, however, scored higher than students in control groups on a subtest of the Torrance Test of Creativity. SAPA students produced more and a greater variety of ideas or questions than traditional students. From this research, it appears that a process-oriented curriculum fosters skills necessary for divergent thinking and problem solving.

One of the first problems encountered in assessing problem-solving outcomes of the activity-based, process-oriented curricula was lack of available instruments to measure process and problem-solving skill. This deficit resulted in an avalanche of test development. Most tests were designed to validate specific curricula. For example, Walbeeser and Carter (1970), McLeod, Berkheimer, Fyffe, and Robinson (1975), and Ludemann (1974) all developed tests to measure the integrated process skills defined by the SAPA curriculum. Shaw (1983), in an attempt to validate the effect on problem solving of process-oriented curricula in general, developed a content-free test to measure the integrated process skills of interpreting data, controlling variables, defining operationally, and formulating hypotheses (Objective Referenced Evaluation; Shaw, 1983). The test was used to determine differences in problem-solving ability between students in two classrooms receiving SAPA II instruction and students in two classrooms, similar in all other measures except that they received traditional instruction (textbook), on the same content. Test results at the end of 24 weeks of instruction indicated that the SAPA II students are better able to apply problem-solving strategies to content outside the curriculum than students in the control group.

After a decade of use, the activity-based, process-oriented curricula were scrutinized from a national perspective. Many small, local studies were synthesized into more global reports. Linn and Thier (1975) conducted a nationwide survey of fifth-grade students in seven states living in both rural and urban settings to determine whether SCIS instructional materials affect reasoning involved in controlling variables. The experimental group received instruction through the SCIS module Energy Sources. The students in the control group received instruction on the same concepts but through traditional textbook-lecture methods. Testing procedures consisted of showing both groups of students a film of a cart rolling down an inclined plane. In subsequent interviews students were asked to describe experiments that might be conducted to determine the causal variables involved in the cart's rate of movement. Scores of SCIS students in both rural and nonrural settings were substantially superior to those of students who received traditional instruction. These results suggest that instructional materials used in diverse settings by regular classroom teachers can significantly assist in enabling children to acquire important reasoning patterns.

All three activity-based, process-oriented curricula (ESS, SAPA, and SCIS) were included in a national synthesis of research reports by Bredderman (1982). Bredderman reviewed more than 60 studies reported over a 15-year period encompassing the late 1960s and 1970s. All studies involved classrooms using one of the activity-based elementary science curricula described earlier. Each study reviewed included a control group that received comparable content instruction from a textbook. Over 1000 classrooms and 13,000 students were involved in the study. The pattern of results of these studies shows that children in activity-based classrooms outperformed in all categories, including process skills, creativity, attitudes, logic, and science content, those in textbook-oriented classrooms. This was true regardless of sample size, same or different teachers, the length of time the program was used, or the grade level at which the assessment was done.

Students who were academically or economically disadvantaged made the greatest gains in both content and process from the activity-based programs. SAPA students tended to make the greatest gains in process skills, whereas SCIS students gained most in content and logical reasoning. Too few studies involved ESS students to warrant their inclusion in these comparisons.

A quantitative synthesis of over 25 years of primary research on the effect on student performance of activity-based, process-oriented elementary science programs was conducted by Shymansky, Kyle, and Alport (1983). The meta-analysis procedures developed by Glass, McGraw, and Smith (1981) were used in the interpretation of data. Each criterion was analyzed by specific curriculum, grade level, gender, socioeconomic status, and length and validity of study and method of testing. Results of the synthesis indicate that the average student in the process-based curricula exceeded the performance of 63 percent of the students in traditional classrooms. Students exposed to the process-based programs showed the greatest gains in process skill development, attitude to science, and achievement. Using more sophisticated statistical methods for meta-analysis, Shymansky, Hedges, and Woodworth (1990) refined their synthesis. In the new review, the overall effectiveness of the activity-based, process-oriented elementary science programs is confirmed. The process-based curricula had a positive impact on student performance across all measures including academic achievement. This is an important finding, in that some teachers and parents felt that an inquiry-based curriculum would compromise content attainment for gain in process skills. This study alleviates those fears. In the primary grades (K–3) the greatest attainment was made in achievement and process-skill development, whereas in the intermediate grades (4–6) students' perceptions of science were also significantly affected by the programs.

Application of the Research to the Classroom

1. Problem solving is supported through the acquisition of process skills by activity-based, process-oriented elementary science curricula.
2. Confronted with the back-to-basics movement, some educators feared that process-based curricula sacrificed content for the sake of inquiry. Research clearly indicates that this is not the case. Elementary students enrolled in process-oriented classrooms had higher academic achievement scores than students in textbook-oriented classrooms.
3. Regardless of local or teacher experience, students in activity-based elementary science classrooms outperform students in traditionally taught classrooms.
4. Academically or economically disadvantaged students appear to gain the most from activity-based elementary science programs.
5. In many elementary science classrooms across the country, the textbook is still the predominant instructional vehicle and process is deemphasized. It seems apparent that better communication of the research is warranted in order to reverse this trend.

THE CLASSROOM ATMOSPHERE: WHAT CAN BE DONE TO MAKE IT CONDUCIVE TO PROBLEM SOLVING?

The research reviewed to this point shows that instruction can indeed improve the problem-solving skills of students. Process-based curricula have been developed that significantly enhance the acquisition of not only process skills but also content. Research also indicates that the instructional environment, starting with the expectations of the teacher, can enhance problem solving.

Teachers' expectations of students and of themselves have been shown to play a vital role in creating a classroom atmosphere in which students become productive problem solvers. Cobb, Yackel, Wood, Wheatley, and Merket (1988) describe a research study involving 23 second-grade teachers who were involved in teaching mathematics through small-group problem solving and whole-class discussion. Teacher expectations that positively supported a problem-solving atmosphere are listed below. In whole-class discussions the child will be able to (Cobb et al., 1988, p. 47):

- Explain how he or she understood and attempted to solve a problem that the group has completed.

- Listen to, and try to make sense of, explanations given by other children.
- Indicate his or her agreement or disagreement with solutions given by other children.
- In the event of conflicting solutions, attempt to justify a solution and question alternatives and thus work toward the achievement of a consensus.

The researchers found that teachers holding these expectations for children had to accept certain expectations for themselves.

- They had to accept children's explanations in a nonevaluative way.
- They must communicate with the children that they are genuinely interested in their thinking and that they can learn from errors.
- They must view the children's solution attempts as their thinking which should be treated with respect rather than as an example of faulty thinking that needs to be corrected.
- They must allow children to work at their own rates and demonstrate that they expect the children to persist and figure problems out for themselves.
- They must be practical reasoners using their experiences of interacting with children to make wise decisions that foster student confidence and competence at problem solving.

Although the research described was done in the context of mathematics, the importance of teacher expectations in creating a problem-solving atmosphere has been substantiated in elementary science by research done by Martens (1988) and Tobin, Espinet, Byrd, and Adams (1988).

Children's expectations are also important. If children perceive a problem to be too difficult, they tend not to search for solutions. Success nurtures risk taking, and with the mastery of each new problem the student grows in confidence in his or her problem-solving ability. The expectation of success stimulates the search to resolve the problem. Tyler (1958) found that expectancy of successful completion of a problem had an effect on the individual's performance. If students viewed a problem as "doable," they were more likely to solve the problem.

External elements and processes within the instructional strategies of the classroom appear to effect students' acquisition of problem-solving skills. According to Wilson and Koran (1976), "The mental processing performed by the learner during inquiry is sensitive to external elements within the instructional situation" (p. 486). Elements found to promote appropriate mental processing are the selection of equipment provided to the students for exploration and the use of science experiences that create conceptual or cognitive conflict. Cognitive conflict was found by Palmer (1965) to accelerate concept acquisition by stimulating the inquiry processes. Thompson (1989) found that the use in elementary science classrooms of discrepant events, materials, or experiences that produce cognitive conflict not only motivates children to learn basic science principles but also improves students' questioning skills, which are basic to problem solving.

Teacher demeanor, questioning protocol, and reward system are additional environmental elements that contribute to students' successful use of process skills and problem-solving behaviors. Rowe (1973) analyzed more than 800 tape recordings of science lessons taught by teachers in a variety of settings. In an analysis of the teachers' questioning protocols, Rowe found that the average teacher asks questions at a rate of two or three per minute, with as much as 20 percent of the teacher's talk during a lesson consisting of evaluative responses, such as "Good," "Fine,", "OK," "Very nice," "All right," and "You know better." Because inquiry requires a significant amount of cognitive processing and is driven intrinsically by conceptual conflict rather than by extrinsic rewards, Rowe predicted that a higher order of questioning, more wait time between questions, and a reduced external reward system should improve student inquiry behaviors. Fifty elementary teachers received training on question-asking strategies that utilized increased wait time between questions. The teachers also learned techniques to facilitate discussion during question sessions so that they were regarded by students not as mediators but as partners in the process of inquiry. The teachers were taught to refrain from verbal rewards during the inquiry sessions.

After the training institute, the teachers returned to their classrooms to implement the inquiry strategies with their students. First, they introduced their students to a meaningful inquiry experience. Next, they asked a greater variety of questions. They increased their wait time between questions. They practiced a demeanor that allowed them to become facilitators of group discussion, and finally they minimized the verbal rewards they used during the inquiry session. Using this protocol, they found that (Rowe, 1973, pp. 258–260):

- The length of students' responses increased.
- The number of unsolicited but appropriate responses by students increased.
- Failures to respond to questions decreased.
- Confidence, as reflected in fewer inflected responses (Is that what you want?), increased.
- The incidence of speculative thinking increased.
- Teacher-centered show-and-tell decreased and child-child comparing increased.
- The number of questions asked by children increased, and the number of experiments they proposed increased.
- Contributions by "slow" students increased.
- Disciplinary moves decreased.

The "simple" protocol seemed to empower students and make them less dependent on the teacher and more willing to assume initiative for resolving the problems on their own.

Journaling, another external element that seems to promote mental processing, was studied in 1976 by Wilson and Koran. Wilson and Koran hypothesized that "hunch generation," describing a possible cause for an observed situation, would enable elementary students to be better processors of information. Forty-five children between the ages of 9 and 11 were involved in this study. The students were first shown a discrepant event in which four wooden blocks, each painted a different color with an equal-length string and weight attached, were placed at the starting gate of a slide and then released. In every race, the same block always won and the same blocks always lost. To force the discrepancy, Wilson and Koran required the students to note that although the blocks looked the same except for their color, they slid at different rates and for different distances. After viewing the discrepant event, each student received a copy of one of three forms. Form one asked students to write some hunches that might explain the discrepancy. Form two asked students to read a list of possible hunches. The direction on form three was to proceed to the next page for further instructions. All students then performed 15 experiments, each of which investigated a discrete variable related to the discrepant event. The posttest consisted of a drawing of an inclined plane and a block with a string and weight attached. Children were instructed to make any necessary change in order to make the block the "winner of the sliding block race." Students were also given a list of brief statements of procedures and asked to identify procedures as "useful" or "not useful" in finding a solution to the situation. Children who were asked to write out hunches received higher scores on recognition of causal variables on the posttest. The interesting point of this research is that differences appear to exist as a result of the mental processing that occurred when students were asked to write down the hunches. This is consistent with research by Novak, Ring, and Tamir (1971) on advance organizers. The research validates the use in elementary classrooms of journals in which students record observation, inferences, hypotheses (hunches), and interpretations of data to support the development of problem solving skills.

Classroom environments that provide opportunities for small groups of children to work together to solve problems tend to foster the development of problem solving. Socialization is seen by Piaget (1959) as significant in the development of reasoning in the concrete-operational child. Piaget found that children become much less egocentric when they discuss things with one another. From such discussions, they become aware of the different views of a problem held by their peers. This confrontation of mental theories requires the student to seek new information and formulate new hypotheses. Participation in small-group problem solving tends to stimulate cognitive disequilibrium in such a way that over time new structures of thinking arise (Haste, 1987). Noddings and Shore (1984) attempted to explain this phenomenon by suggesting that through class discussions students are forced to construct explanations of their reasoning. In the telling, the students are forced to clarify their thinking, which leads to a deeper understanding of the concept. Kamii and DeVries (1978) found that to stimulate growth in problem-solving skills, the cooperative groups must work on tasks that they are able to resolve with their own actions. The result of the child's action on the task should be immediate and clearly visible. The task must be of intrinsic interest to each child in the group if dialogue is to be sustained and action pursued. Research indicates that teachers play a significant role in the success of cooperative group problem solving by selecting challenging materials and structuring group composition (Johnson & Johnson, 1987).

Application of the Research to the Classroom

1. Problem solving is supported when teachers expect students to be able to describe how they solved a problem, critique problem solutions of other students, and cooperate with others to seek alternative solutions.
2. Problem solving is supported when a nonevaluative, flexible, nurturing classroom environment is created.
3. Problem solving is supported when teachers ask a variety of different types of questions (i.e., recall, higher order, probing, divergent), wait longer for student responses between questions, encourage class dialogue rather than teacher-mediated discussion, and withhold extrinsic rewards during problem-solving episodes.
4. Problem solving is supported when developmentally appropriate and intrinsically rewarding discrepant events are presented to students.
5. Problem solving is supported when appropriately structured small groups work together on the resolution of a problem.

MODES OF EXPERIMENTS: HOW DO SCIENTIFIC AND ENGINEERING EXPERIMENTS DIFFER?

In recent years there has been a significant amount of research related to the science-technology-society type of problem (Yager & Hofstein, 1986; Hungerford, Volk, & Ramsey, 1990). The new vintage (since 1985) of federally funded elementary science curricula reflects this direction. CHEM (Chemicals, Health, Environment and Me, 1991) focuses on environmental, real-world problems related to chemistry. GEMS: Great Explorations in Math and Science (1991a, b) addresses environmental issues such as acid rain and global warming but also has a technology strand reflected in modules such as Soap Bubbles and OOblick. FOSS (Full Option Science System, 1992) combines a discovery, technology approach with conceptual understandings. Some contemporary elementary science programs have incorporated a design technology strand in which students build equipment to resolve problems such as "Build a car that turns as it rolls down an incline plane" and "Build a wind-up machine that makes a noise every two minutes" (Johnsey, 1986, 1991; Dunn & Larson, 1990).

The new curricula are all activity based and process oriented. In some cases, however, they reflect a different type of experimental mode than the elementary science activity-based curricula of the 1960s. In the new programs more engineering experiments are used to engage the interest of the students, whereas in the 1960s curricula science experiments received greater emphasis. Science experiments are defined as those in which the primary purpose is to identify cause-effect relationships. Engineering problems, on the other hand, have as the primary purpose a product or a desired outcome. Recent research indicates that initial emphasis on engineering experiments with transfer to science experiments may be an appropriate developmental direction for contemporary elementary science curricula.

When confronted with a science problem, researchers noted, children often bypass the science experiment mode and move directly into an engineering mode. For example, when asked to explore which of several chemicals was responsible for turning a mixture pink, the children moved directly to trying to produce the pink color (Kuhn & Phelps, 1982). When asked to determine which design features of a car affect its speed, the children moved directly to producing the fastest car (Schauble, 1990). Dewey (1913) suggested that interest in science experiments is likely to begin on the practical side, but he predicted that given enough opportunities to explore science problems, children will move on their own from a goal (engineering mode) to a search for causal relationships (science mode) of experimentation. Schauble, Klopfer, and Raghavan (1991) conducted research to determine the validity of Dewey's prediction and determine if and when children adopt the engineering or science mode in their approach to experiments. Students were randomly assigned to two treatment groups. One group started with a problem that they would likely perceive as an engineering problem ("How should water canals be designed to optimize boat speed?") and then proceeded to a science experiment (Lower an object suspended from a spring into a fluid, to observe the effects of buoyant force on objects of different mass and volume). The second group started with the science experiment first and then proceeded to the engineering problem. Approximately the same number of investigations were conducted by both groups. The students who started with the canal problem (engineering mode) tended to perform fewer different investigations but ran some tests over and over. Students starting with the buoyancy task performed a greater variety of experiments. The greatest improvement in strategic performance was achieved by the students who started with the canal task. This result seems consistent with Dewey's prediction. It appears that students move to new and specialized forms of thinking by building on more familiar, everyday ways of thinking. The results of this investigation support the emphasis placed on engineering problems in contemporary elementary science curricula.

BARRIERS TO THE IMPLEMENTATION OF PROBLEM SOLVING IN ELEMENTARY SCIENCE CLASSROOMS

Even though much is known about what strategies nurture productive problem-solving behavior, there is evidence that many of the strategies are not being implemented into the majority of elementary classrooms. In many elementary classrooms, there still seems to be a great emphasis on coverage of factual material. Newmann (1988) stated that "the addiction to coverage fosters the delusion that human beings are able to master everything that is worth knowing" (p. 346). Ninety-five percent of science teachers tend to use a textbook 90 percent of the time (Stake & Easley, 1978). Brandwein (1981) found that most science students do not conduct one experiment in which the solution is unknown throughout the academic year. In the remaining paragraphs of this review several of the barriers preventing a greater emphasis on problem solving in elementary schools are discussed.

Barrier One: Assessment. There is little reward for changing teaching strategies and classroom organization to emphasize high-level cognitive learning and problem solving if systemwide, state, and national assessments continue to evaluate the recall of facts. If assessment directly parallels academic work, there is little encouragement for students to take risks and think divergently. New and better avenues of communication between all stakeholders of the educational endeavor (students, parents, community leaders, businesses, schools) need to be established. Alternative approaches to assessment need to be developed and researched. An atmosphere conducive to problem solving will prevail in our classrooms when evaluative instruments value engagement of the mind rather than completion of tasks.

Barrier Two: Teacher Preparation Programs. The philosophical nature of science is often overlooked in preservice courses for elementary teachers. Too often, science is portrayed at the college level as a body of knowledge. Preteachers need to be exposed to science experiences that nurture their curiosity, develop their process skills, and help them expand their conceptual schemes. Preservice experiences in science should link with pedagogy in order to enable teachers to become flexible in their thinking, receptive to change and innovation, questioning in their outlook, aware of their own perceptions and assumptions, open to a wide range of alternatives, tolerant of ambiguity, and reflective in their thinking. Education programs should help preteachers view problems as a challenge and believe in their own ability to solve problems.

Barrier Three: Differing Perspectives on the Purpose of Education. "Actions speak louder than words" is a truism that describes to some extent the dichotomy between what schools write in mission statements and what occurs in the school community. To nurture the acquisition of problem-solving skills and science attitudes, a paradigm shift must occur from perceiving the teacher as the giver or imparter of knowledge to perceiving the teacher as the mentor or facilitator of learning. Perspectives of effective teaching often vary (Tobin et al., 1988). More philosophical congruence on the value of problem solving is needed before atmospheres conducive to cognitive engagement of students are the norm in contemporary classrooms.

Barrier Four: Societal Values. Reforms called for in national assessments of education do not necessarily reflect the values of all the stakeholders of the educational process. Perhaps site-based management, community-education partnerships, novel student-teacher incentives, and greater national commitment will bring about system change that values problem solving and promotes congruence between theory, research, and practice.

References

Anderson, C. , & Smith, E. (1981). Patterns in the use of elementary school science program materials: An observational study. *National Association for Research in Science Teaching Abstracts*. Columbus, OH: ERIC.

Atkin, M. J. (1958). A study of formulating and suggesting tests for hypotheses in elementary school science learning experiences. *Science Education, 42*(5), 414–422.

Baumel, H. B., & Berger, J. J. (1965). An attempt to measure scientific attitudes. *Science Education, 49(3)*, 267–269.

Brandwein, P. F. (1981). *Memorandum; On renewing schooling and education*. New York: Harcourt Brace Jovanovich.

Bredderman, T. (1982) What research says: Activity science—the evidence shows it matters. *Science and Children, 20*, 39–41.

Butts, D., & Jones, H. (1966). Inquiry training and problem solving in elementary school children. *Journal of Research in Science Teaching, 4*, 21–27.

Champagne, A. B., & Klopfer, L. E. (1977). A sixty year perspective on three issues in science education: I. Whose ideas are dominant? II. Representation of women. III. Reflective thinking and problem solving. *Science Education, 61*(4), 431–52.

CHEM. (1991). *Chemicals, health, environment and me*. Chemical Education for Public Understanding Program. Berkeley: Lawrence Hall of Science, University of California.

Cobb, P., Yackel, E., Wood, T., Wheatley, G., & Merket, G. (1988). Creating a problem-solving atmosphere. *Arithmetic Teacher, 36*, 46–47.

Davis, I. C. (1935). Measurement of scientific attitude. *Science Education, 19*(3), 117–122.

Davis, M. (1979). The effectiveness of guided-inquiry discovery approach in an elementary school science curriculum. (Doctoral dissertation, University of Southern California, 1978). *Dissertation Abstracts International, 39*, 4164A.

Dewey, J. (1913). *Interest and effort in education*. Boston: Houghton Mifflin.

Dewey, J. (1916). Method in science teaching. *General Science Quarterly, 1*, 3–9.

Dunn, S., & Larson, R. (1990). *Design technology: Children's engineering*. New York: Falmer.

Edwards, L. E., & Robertson, M. (1939). The construction of a scale for the determination of the scientific attitude sensitive curiosity. *Science Education, 23*(4), 198–206.

Elementary Science Study: ESS. (1971). *Batteries and bulbs. II: An electrical gadget suggestion book*. New York: Webster–McGraw Hill.

Festinger, L. (1962). Cognitive dissonance. *Scientific American, 207*, 93–106.

Finley, F. N. (1983). Science processes. *Journal of Research in Science Teaching, 20*, 47–54.

FOSS. (1992). *Full option science system*. Berkeley, CA: Lawrence Hall of Science.

Gagne, R. M. (1963). The learning requirements for enquiry. *Journal of Research in Science Teaching, 1*, 144–153.

Gagne, R. M. (1965). *The psychological basis of science—a process approach* (AAAS Miscellaneous Publication, 65–68). Washington, DC: American Association for the Advancement of Science.

Gagne, R. M. (1970). *The conditions of learning*. New York: Holt Rinehart & Winston.

GEMS. (1991a). *Great explorations in math and science*. Berkeley: Lawrence Hall of Science, University of California.

GEMS. (1991b). *To build a house*. Berkeley: Lawrence Hall of Science, University of California.

Glass, G. V., McGraw, B., & Smith, M. L. (1981). *Meta-analysis in social research*. Beverly Hills, CA: Sage.

Greeno, J. G. (1978). Natures of problem-solving abilities. In W. K. Estes (Ed.), *Handbook of learning and cognitive process* (Vol. 1). Hillsdale, NJ: Lawrence Erlbaum.

Haste, H. (1987). Growing into rules. In J. Bruner & H. Haste (Eds.), *Making sense: The child's construction of the world*. New York: Methuen.

Hungerford, H. R., Volk, T. L., & Ramsey, J. M. (1990). *Science-technology-society: Investigating and evaluating STS issues and solutions*. Champaign, IL: Stipes.

Hurd, P. D. (1989). *Science education and the nation's economy*. Paper presented at the AAAS Symposium on Science Literacy.

Johnsey, R. (1986). *Problem solving in school science*. London, UK: MacDonald Educational. (Available in the U. S. from Teachers Laboratory, Brattleboro, VT.)

Johnsey, R. (1991). *Design and technology through problem solving*. New York: Simon & Schuster.

Johnson, D. W., & Johnson, R. T. (1987). *Learning together and alone: Cooperative, competitive, and individualistic learning*. Englewood Cliffs, NJ: Prentice-Hall.

Kamii, C. (1985). *Young children reinvent arithmetic: Implications of Piaget's theory*. New York: Teachers College Press.

Kamii, C., & DeVries, R. (1978). *Physical knowledge in preschool education: Implications of Piaget's theories*. Englewood Cliffs, NJ: Prentice-Hall.

Karplus, R. (1964). *Theoretical background of the science curriculum improvement study*. Berkeley: Science Curriculum Improvement Study, University of California.

Keeslar, O. (1956). The science teacher and problem solving. *Science Education, 23*, 13–14.

Kozlow, M. J., & Nay, M. A. (1976). An approach to measuring scientific attitudes. *Science Education, 60*(12), 147–172.

Kuhn, D., & Phelps, E. (1982). The development of problem-solving strategies. In H. Reese (Ed.), *Advances in Child Development and Behavior, 17*, 1–44.

Linn, M. C., & Thier, H. D. (1975). The effect of experiential science on development of logical thinking in children. *Journal of Research in Science Teaching, 12*, 49–62.

Ludemann, R. R. (1974). Development of the science processes test (TSPT) (Doctoral dissertation, Michigan State University). *Dissertation Abstracts International, 36*, 203A.

Martens, M. L. (1988). Implementation of a problem solving curriculum for elementary science: Case studies of teachers in change. *Dissertation Abstracts International, 49*, 716A.

McLeod, R. J., Berkheimer, G. D., Fyffe, D. W., & Robison, R. W. (1975). The development of criterion-validated test items for four integrated science processes. *Journal of Research in Science Teaching, 12*(4), 415–421.

Neal, L. (1961). Techniques for developing methods of scientific inquiry in children in grades one through six. *Science Education, 45*(4), 313–320.

Newmann, F. M. (1988). Can depth replace coverage in the high school curriculum? *Phi Delta Kappan, 69*(5), 345–348.

Noddings, N., & Shore, P. J. (1984). *Awakening the inner eye: Intuition in education*. New York: Teachers College Press.

Norton, R., & Butts, D. (1973). A developmental study in assessing children's ability to solve problems in science. *National Association for Research in Science Teaching Abstracts*. Columbus, OH: ERIC.

Novak, J. D. (1977). *A theory of education*. Ithaca, NY: Cornell University Press.

Novak, J. D., Ring, D. G., & Tamir, P. (1971). Interpretation of research findings in terms of Ausubel's theory and implications for science education. *Science Education, 55*(4), 483–526.

Obourn, E. S. (1956). Analysis and check list on the problem-solving objective. *Science Education, 40*, 338–392.

Palmer, E. L. (1965). Accelerating the child's cognitive attainments through inducement of cognitive conflict: An interpretation of the Piagetian position. *Journal of Research in Science Teaching, 3*(4), 318–325.

Piaget, J. (1959). *The language and thought of the child* (3rd. ed.). London: Routledge & Kegan Paul. (Original work published 1923)

Pizzini, E. L., Shepardson, D. P., & Abell, S. K. (1989). A rationale for and the development of a problem solving model of instruction in science education. *Science Education, 73*(5), 523–534.

Quinn, M. E., & Geroge, K. D. (1975). Teaching hypothesis formation. *Science Education, 59*(3), 289–296.

Ramsey, G. A., & Howe, R. W. (1969). An analysis of research related to instructional procedures in elementary school science. *Science and Children, 6*(7), 25–26.

Renner, J. W., Strafford, D. C., Coffia, W. J., Kellogg, D. H., & Weber, M. C. (1973). An evaluation of science curriculum improvement study. *School Science and Mathematics, 73*(4), 291–318.

Ross, J. A. (1990). Learning to control variables: Main effect and aptitude treatment interactions of two rule-governed approaches to instruction. *Journal of Research in Science Teaching, 27*(6), 523–539.

Rowe, M. B. (1973). *Teaching science as continuous inquiry*. New York: McGraw-Hill.

Rutherford, F. J., & Ahlgren, A. (1990). *Science for all Americans*. New York: Oxford University Press.

Schauble, L. (1990). Belief revision in children: The role of prior knowledge and strategies for generating evidence. *Journal of Experimental Child Psychology, 49*, 31–57.

Schauble, L., Klopfer, L. E., & Raghavan, K. (1991). Students' transition from an engineering model to a science model of experimentation. *Journal of Research in Science Teaching, 28*(9), 859–882.

SCIS. (1968). *SCIS Elementary science sourcebook*. Science Curriculum Improvement Study. Berkeley: Lawrence Hall of Science, University of California.

Shaw, T. J. (1983). The effect of a process-oriented science curriculum upon problem-solving ability. *Science Education, 67*(5), 615–623.

Shulman, L. S., & Keislar, E. R. (Eds.). (1966). *Learning by discovery: A critical appraisal*. Chicago: Rand-McNally.

Shymansky, J. A., Hedges, L. V., & Woodworth, G. (1990). A reassessment of the effects of inquiry-based science curricula of the 60's on student performance. *Journal of Research in Science Teaching, 27*(2), 127–144.

Shymansky, J. A., Kyle, W. C., & Alport, J. M. (1983). The effects of new science curricula on student performance. *Journal of Research in Science Teaching, 20*(5), 387–404.

Stake, R. E., & Easley, J. A. (1978). *Case studies in science education*. Urbana, IL: Center for Instructional Research and Curriculum Evaluation, University of Illinois.

Suchman, J. R. (1960). Inquiry training in the elementary school. *Science Teacher, 27*, 42–47.

Thompson, E. L. (1989). Discrepant events: What happens to those who watch? *School Science and Mathematics, 89*, 26–29.

Tobin, K., Espinet, M., Byrd, S. E., & Adams, D. (1988). Alternative perspectives of effective science teaching. *Science Education, 72*(4), 433–451.

Tyler, B. B. (1958). Expectancy for eventual success as a factor in problem solving behavior. *Journal of Educational Psychology, 49*(3), 166–172.

Walbeeser, H. H., & Carter, H. L. (1970). The effects of test results of change in task and response format required by altering test administration from an individual to a group form. *Jounal of Research in Science Teaching, 7*, 1–8.

Wilson, J. T., & Koran, J. J. (1976). Effect of generating hunches on subsequent search activity when learning by inquiry. *Journal of Research in Science Teaching, 13*(6), 479–488.

Winne, P. H., & Marx, R. W. (1977). Reconceptualizing research on teaching. *Journal of Educational Psychology, 69*(6), 668–678.

Wollman, W. T., & Chen, B. (1982). Effects of structured social interaction on learning to control variables: A classroom training study. *Science Education, 66*(5), 717–730.

Yager, R. E., & Hofstein, A. (1986). Features of a quality curriculum for school science. *Journal of Curriculum Studies, 18*(2), 133–146.

Zoller, U. (1987). The fostering of question-asking capability: A meaningful aspect of problem-solving in chemistry. *Journal of Chemical Education, 64*(6), 510–512.

· 8 ·

RESEARCH ON PROBLEM SOLVING:
MIDDLE SCHOOL

Stanley L. Helgeson
THE OHIO STATE UNIVERSITY

Problem solving has long been a concern in science education. In examining 60 years of literature as represented by articles published in *Science Education*, Champagne and Klopfer (1977) noted that the first article in the first volume of the journal (then named *General Science Quarterly*) was written by John Dewey, whose position was that "the method of science—problem solving through reflective thinking—should be both the method and valued outcome of science instruction in America's schools" (p. 438). A clear link between science and problem solving is noted by Simon (1981): "scientific discovery is a form of problem solving, and ... the processes whereby science is carried on can be explained in terms that have been used to explain the processes of problem solving" (p. 48).

Approximately 50 percent of the students in the United States take no science beyond grade 10 (Helgeson, Blosser, & Howe, 1977). For these students, science in middle or junior high school may constitute the major portion of science they will encounter and the last in the physical sciences. Problem solving thus becomes a crucial issue in the middle grades. The importance of problem solving in middle or junior high school science is reflected in the results of a study reported by Rakow (1985) in which a group of middle or junior high school science teachers reviewed reports on the status of science education to determine criteria for excellence at that level. They concluded that an exemplary middle or junior high school science program is one that, among other things, develops students' problem-solving skills. Rakow notes that the objectives identified in this study were substantially the same as those identified by the National Science Teachers Association (NSTA) Task Force on Excellence in Middle/Junior High School Programs (Reynolds, Pitotti, Rakow, Thompson, & Wohl, 1984).

Because many of the studies of problem solving in science extend beyond the middle or junior high school level, some of the research reported here includes subjects from both the upper elementary and senior high school levels. In view of necessary space limitations, the studies are greatly abbreviated in this report; more extensive summaries are reported elsewhere (Helgeson, 1992).

DEFINITIONS OF PROBLEM SOLVING

Although science educators agree substantially on the importance of the role of problem solving in school science, there is much confusion and inconsistency in the use of terminology related to problem solving. Instead of defining problem solving, science educators have often tried to categorize and describe the process, using such terms as scientific method, scientific thinking, critical thinking, inquiry skills, and science processes (Champagne & Klopfer, 1981a).

The relationship between problem solving and science processes can be seen in the development of Science—A Process Approach (SAPA), which represented "an attempt to establish the specific competencies in students which will make it possible for them to solve problems, to make discoveries, and more generally think critically about science" (Gagné, 1965, p. 7). Gagné (1970) noted that the processes underlying SAPA are equivalent to these intellectual skills and can be categorized under the general names of observing, classifying, measuring, using space-time relations, using numbers, communicating, and inferring. Integrated processes include formulating hypotheses, defining operationally, manipulating variables, inter-

The author thanks reviewer Glenn Markle (University of Cinncinnati).

preting data, drawing conclusions, and, the most complex activity of all, experimenting (Gagné, 1970, pp. 260–266). Gagné (1977) later says:

Problem solving may be viewed as a process by which the learner discovers a combination of previously learned rules which can be applied to achieve a solution for a novel problem situation.
Problem solving is not simply a matter of application of previously learned rules, however. It is also a process that yields new learning. (p. 155)

More recently, Shaw (1983) defined problem solving skills to include the four integrated processes of interpreting data, controlling variables, defining operationally, and formulating hypotheses.

ASSESSMENT OF PROBLEM SOLVING

Assessment of problem solving reflects some of the ambiguity noted in the various definitions of problem solving. However, the close relationship between problem solving and science process skills is apparent even in the most cursory examination of the science-education literature. In most cases problem solving is, in effect, defined by what is measured or assessed, and most often this includes at least some aspect of the processes of science. In an overview of measurement instruments in science, Mayer and Richmond (1982) note the increased attention paid to science process skills and problem-solving behavior in the curriculum development activities of the 1960s. Much of the effort in assessing problem solving appears to derive from these curriculum-development activities. The assessment instruments reviewed here fall generally into four categories: (1) those including defined behaviors or strategies of problem solving, (2) those dealing with science process skills in general, (3) those restricted to the integrated science process skills, and (4) assessment of logical thinking.

Problem Solving Behaviors

Four instruments were identified that were designed to evaluate problem solving. In each of these cases, problem solving was defined by specified strategies or behaviors. One of the earliest of these instruments was reported by Butts (1964), Jones (1966), and Butts and Jones (1966). Originally called the X–35 Test of Problem Solving, the name evolved to the Tab Inventory of Science Processes (TISP) and later became better known as the Tab Science Test. In this instrument, the child is presented with a discrepant event via a film loop and asked to select an explanation from a given list. The student is then provided data by clue questions whose answers are covered by tabs. The student removes the tab from each desired question and places it on an answer sheet. Under each tab is a "yes" or "no" indicating whether the explanation is correct. The student continues processing data until a correct explanation is found.

The TAB Science Puzzler, derived from the Tab Science Test, was designed by Norton (1971) as a measure of problem-solv-

ing performance. The five subtasks of the test coincide with steps of a problem-solving model: (1) problem orientation, (2) problem identification, (3) problem solution, (4) data analysis, and (5) problem verification.

As a part of the evaluation of the Unified Science and Mathematics for Elementary School (USMES) program, Shann (1976) developed the Test of Problem Solving Skills (TOPSS), a group-administered, paper-and-pencil, multiple-choice instrument. Criteria for test items were that they (1) measured understanding of the processes of science, (2) drew from real-life experience, and (3) were written for elementary school children.

Ross and Maynes (1983a) developed a multiple-choice test for seven problem-solving skills in science using learning hierarchies based on expert-novice differences. The seven skills were (1) developing a focus (hypothesis formulation), (2) developing a framework (designing an experiment), (3) judging the adequacy of data collected, (4) recording information, (5) observing relationships in data, (6) drawing conclusions, and (7) generalizing. Learning hierarchies were constructed for each skill by contrasting the cognitive behavior of novices with the cognitive behavior of sophisticated problem solvers.

Science Processes

Another group of four instruments assessed students' science process skills, including the basic skills, and a fifth instrument incorporated enquiry skills. The Test of Science Processes (TOSP), was constructed by Tannenbaum (1969) to assess the ability of junior high school students to use the processes of observing, comparing, classifying, quantifying, measuring, inferring, experimenting, and predicting.

The Science Process Skills Test (SPST), developed by Moliter and George (1976), was designed to evaluate the performance of children in grades 4, 5, and 6 on the inquiry skills of inference and verification. The items were designed to be as content free as possible, and items and distractors were presented as illustrations to avoid reading problems. The final form of the instrument consisted of nine inference and nine verification items.

The development of a science-processes test using external criterion-referenced validation and an objective method of item selection was reported by Ludeman (1975). A pool of items intended to assess students' ability to use the processes of science was generated and a subset of the Individual Competency Measures from the SAPA program was used as the external criterion. The author concluded that The Science Processes Test (TSPT) was useful for assessing students' ability to use the processes of science.

A multiple-choice test of Basic Process Skills in Science (BAPS) appropriate for students in grades 3 through 8 was developed by Cronin, Twiest and Padilla (1985) to assess six basic science-process skills: observation, inference, prediction, measurement, communication, and classification. Criteria for development included (1) an emphasis on the six most widely used basic science-process skills; (2) a multiple-choice, four-option format; (3) an emphasis on pictures and drawings to clarify and enhance items; (4) average test readability below the fourth-grade level; (5) test length permitting completion within one class period (45 minutes or less); (6) wide range of difficulty of items addressing each process skill; and (7) content-free test items. Twiest and Twiest (1989) modified the BAPS to include new graphics, corrected some items, and validated the instrument with parallel station tests using actual demonstrations (BAPSST) and interviews (BAPSIT). All three tests appear to be valid and reliable methods for measuring basic science-process skills.

Integrated Science Processes

The integrated science process skills served as the focus of seven instrument-development efforts. Fyffe (1972) and Robison (1974) collaborated on developing instruments to measure process-skill development for four integrated process skills: controlling variables, interpreting data, formulating hypotheses, and defining operationally.

As part of a study of the effect of a process-oriented curriculum (SAPA II) on problem-solving skills, Shaw (1982) developed a process-skills test, using a paper-and-pencil, multiple-choice, group testing format. Problem-solving skills were defined as including the integrated process skills of interpreting data, controlling variables, defining operationally, and formulating hypotheses. Because the purpose of the test was to evaluate the transfer of problem-solving skills to new content, the items were designed not to include science content covered during the treatment. The 60-item Objective Referenced Evaluation in Science (ORES) was developed for sixth-grade students, but the author indicates that it would also be appropriate for seventh- and eighth-graders.

The Test of Enquiry Skills (TOES) developed by Fraser (1979) measures nine separate enquiry skills that fall in three major groups. The first group of scales measures skills related to using reference materials; the second measures skills related to reading and processing information; the third measures three critical thinking in science skills. A pool of items was developed for enquiry skills considered important on the basis of a search of the literature and was field tested and revised. A final version was administered to students in year 7 to year 10. The resulting scores showed that mean performance at a given grade level varied markedly from skill to skill; however, the average performance on all TOES scales increased with grade level. Standard deviation tended to decrease as year (grade) level increased. Mean reliability (KR–20) values ranged from 0.72 for year 7 scores to a low of 0.59 for year 10 scores. Test-retest reliabilities for samples of year 7 students ranged from 0.65 to 0.82, with a mean of 0.73 for the nine scales. Thus, the reliability of the instrument was deemed satisfactory.

The next four instruments in this group effectively constitute a series with the first test as the basis from which the others derive. The Test of Integrated Process Skills (TIPS), designed by Dillashaw and Okey (1980), is a 36-item multiple-choice test for middle and secondary school students covering content from all science areas. The items have four possible responses and are related to five integrated process skills: hypothesizing, identifying variables, operationally defining, designing investigations, and graphing and interpreting data.

Because of the continued need for and relative scarcity of process-skills tests for middle and high school grades, Burns, Okey, and Wise (1985) developed another set of test items aimed at the same set of science-process skills assessed by the TIPS instrument. The new instrument, TIPS II, is also composed of 36 multiple-choice items and can be administered in a normal class period. The authors determined that the TIPS II is a reliable instrument for measuring science process skill achievement and that it increases the available item pool for measuring these skills.

A project to develop a middle-grades integrated process skills test (MIPT) was reported by Cronin and Padilla (1986). The criteria for the test were (1) an emphasis on the skills associated with experimenting; (2) multiple-choice, four-option format; (3) average test readability below the seventh-grade level; (4) test length to permit completion within one class period (45 minutes); (5) a wide range of difficulty of items addressing each identified skill; and (6) content-free test items. Based on the results of field testing, the investigators concluded that most items were functioning well, making the MIPT useful for evaluating the integrated process-skill performance of middle school students.

The Performance of Process Skills (POPS) test, based on refined and modified items of the MIPT, was developed by Mattheis and Nakayama (1988) in an attempt to construct a valid and reliable non-curriculum-specific measure of integrated science-process skills intended for use with middle school students. Six process-skill objectives were identified to form a basis for the POPS test items: identifying experimental questions, identifying variables, formulating hypotheses, designing investigations, graphing data, and interpreting data. From the pool of 40 MIPT items, the 21 items judged to be the best measures of the identified process skills were selected and modified for inclusion in the POPS test. The authors consider the POPS test to be a reliable and valid instrument for diagnostic or summative assessment in science classes or research studies, useful in evaluation of instruction and learning, curriculum validation, and assessing process-skills competence of middle school students.

Although not part of the same line of development as the preceding four instruments, the last test in this group is closely related in orientation. Based on the belief that students should be involved in predicting relationships between variables and attempting to quantify these relationships, McKenzie and Padilla (1986) developed the Test of Graphing Skills (TOGS). Nine objectives encompassing the skills associated with constructing and interpreting line graphs were written and refined. The content of the objectives included selecting appropriate axes, locating points on a graph, drawing lines of best fit, interpolating, extrapolating, describing relationships between variables, and interrelating data displayed on two graphs. The authors note that with the increased emphasis on multiple sources of data for decision making and the availability of large amounts of data, ways of condensing and interpreting data are becoming more important. Because experimenting and other data-collecting activities provide an opportunity to practice such skills in the science classroom, it seems natural that teaching line-graph construction and interpretation skills should occur within the study of science.

Logical Thinking

The final group of instruments to be reviewed reflects the relationship of logical thinking, science-process skills, and problem solving. The need to adapt curricula to the developmental levels of learners motivated a search by Lawson (1978) for a reliable and valid classroom test of formal reasoning. The items selected for inclusion in the test required isolation and control of variables, combinatorial reasoning, probabilistic reasoning, and proportional reasoning. No items directly measuring correlational reasoning were readily available at the time the test was constructed. In addition, one item involving conservation of weight and one involving displaced volume were included. The 15 items each involved a demonstration using some physical materials or apparatus. The test was administered to 513 students in eighth, ninth, and tenth grades. Questions requiring a prediction were posed, followed by possible answers. Students selected the best answer and wrote explanations for their choices. Principal-components analysis indicated that three factors were present: formal reasoning, concrete reasoning, and "early formal" reasoning. Lawson concluded that the same psychological parameters measured by classical Piagetian interviews were measured by the classroom test with a fairly high degree of reliability.

Tobin and Capie (1980, 1981) selected items from the test developed by Lawson to construct the Test of Logical Thinking (TOLT), a 10-item test for students of middle school age and older. The instrument contains two items related to each of five modes of logical thought: identifying and controlling variables and proportional, correlational, probabilistic, and combinatorial reasoning. The test uses a double multiple-choice format in which the student is presented with a problem and asked to select a correct answer from among five responses and then to select a reason for the answer from among five choices. To be correct, the student must choose both the right answer and the correct reason. This minimizes the effect of guessing and results in high test reliability with relatively few test items. The TOLT was validated with both college and secondary school students by correlating its items with performance on tasks presented using traditional Piagetian interviews. The evidence supports the TOLT as a valid, reliable measure of formal reasoning ability.

Because the existing tests of cognitive development measured no more than five modes of reasoning and most were influenced by students' reading and writing ability, Roadrangka, Yeany, and Padilla (1983) developed the Group Test of Logical Thinking (GALT) to measure six logical operations: conservation, proportional reasoning, controlling variables, probabilistic reasoning, correlational reasoning, and combinatorial logic. Items were constructed that presented a problem (with pictorial representations of real objects) to the student, who then selected the best answer and a justification for that answer from multiple choices. The FOG Index was used to adjust sentences to produce a written test at or above the sixth-grade level. In final form, the 21-item GALT was administered to 450 subjects in grades 6 through college. The results indicate that the GALT is a reliable instrument that adequately measures the six logical operations.

Summary

Assessment of problem solving at the middle or junior high school level has evolved from evaluation of defined problem-solving behaviors to measurement of science-process skills to evaluation of the integrated science-process skills. All of the instruments are paper-and-pencil, multiple-option tests and were either developed in conjunction with evaluation of a specific curriculum or clearly derived from curricular emphases. Most recent efforts (TIPS, TIPS II, MIPT, POPS) have been concerned with the integrated science-process skills identified in SAPA. The Classroom Test of Formal Reasoning (Lawson), TOLT, and GALT represent an approach based on our increased knowledge of the learning process and the relationships among logical thinking, science-process skills and problem solving. None of the instruments can be considered the complete solution to the difficulty of assessing problem solving. All, however, have some claim to legitimacy in measuring some aspects of problem solving. The most recent instruments, in particular, are reliable and report reasonable validity claims. In brief, instruments capable of assessing elements of problem solving are available, justifying further attempts to measure and to understand the process of problem solving. It should be noted that paper-and-pencil instruments are not the only method of assessment. A protocol based on differences in the ways experts and novices organize knowledge (see Pirkle & Pallrand, 1988) and various think-aloud techniques also provide means for examining problem solving. However, such studies reported in the middle or junior high school science literature tend to involve special cases in which a procedure is designed for a particular study and not readily generalizable. Another promising approach involves performance assessment in which students are evaluated while doing a science task. Another section of this book deals with this topic.

PROBLEM-SOLVING STRATEGIES AND BEHAVIORS

Several studies dealt with the kinds of strategies or behaviors students exhibited in problem solving. The first six studies reviewed deal with such strategies in relatively structured or guided situations. The final study in this section, however, examines a general problem-solving strategy.

Structured Strategies

Schmiess (1971) attempted to determine whether sixth-grade students ($n = 34$) could successfully engage in scientific investigation. The sixth-graders' ability to solve problems was measured by comparing the students' investigative data with homologous professional scientific data. Schmiess found that (1) the question of whether sixth-graders had the ability for scientific investigation could not be answered statistically; however, student data indicated that 50 percent of the students were accurate on 78 percent of their investigations; (2) the posttest scores on science interest and on solving new problems were significantly higher than pretest scores; (3) high and average achievers were significantly higher than were the low achievers on ability to solve new problems; and (4) no significant differences were obtained among high, average, and low achievers on interest, nor were there significant differences between boys and girls on solving selected science problems, interest in science, or ability to solve new problems.

Mandell (1980) conducted a study to identify common problem-solving behaviors and strategies used by sixth-graders ($n = 25$) who had been classified by their teachers as superior problem solvers. A secondary purpose was to determine whether the students so classified were, in fact, superior problem solvers. A series of six problems were presented in the same order to each student in individual, audiotaped interview sessions and the students were divided into successful and nonsuccessful problem solvers for each problem. Forty-eight percent of the sample (labeled the successful problem-solver group) solved one-half or more of the six problems; the other subjects constituted the nonsuccessful problem solvers group. The successful problem-solving group means on IQ, SRA Math, and SRA Science subtests were significantly higher than the nonsuccessful problem-solving group means on these same measures. All 10 of the students multiply nominated (selected by two or three teachers) as superior problem solvers were in the successful problem-solving group; none of the multiply nominated students were in the nonsuccessful problem-solving group. Mandell found that the students in the successful problem-solving group had certain characteristics in common: (1) they were quick to identify the nature of the problem, (2) they were able to use all four abilities in Piaget's INRC group (identify, negation, reciprocity, and correlativity), (3) they were not dependent on physical manipulations or calculations in solving most problems, (4) they used rough tables or matrices if calculations were needed, and (5) they expressed their reasoning and procedures with ease.

Wilson (1973) investigated the effect of generating hunches on subsequent search activities in problem-solving situations by 45 students, 9–11 years of age. The students, divided into three groups, were assigned to observe a contradictory stimulus. The first group was asked to write hunches, the second group was allowed to read hunches, and the third group had no hunch activities. Wilson found evidence of direct influences of generating hunches on search behaviors and quality of solutions. The findings suggest that structuring stimulus events and arranging learning conditions so that hunches are generated can influence search activities and the quality of the solutions generated. In effect, the students become more efficient as problem solvers.

Berger (1982) and Berger and Pintrich (1986) reported studies to examine students' attainment of skill in using science processes. A computer simulation presented a vertical line or "wall" on the right side of the screen with a circle representing a balloon touching the wall at a predetermined vertical distance. Students were to estimate the height of the balloon. As they entered their estimates into the computer, an arrow appeared on the left side of the screen, traveled to the right, and stuck in the wall at the height of the estimate. If the estimate was correct, the balloon "popped." Three basic strategies were identified: random, ladder (moving up or down systematically), and bracket. The researchers found, among other things, that (1) the microcomputer provided a useful and powerful tool to gather strategy data, (2) students used effective strategies and improved their estimation skill quickly, and (3) age and amount of information presented in the task affected performance. Younger students did not perform as well as older students, a finding consistent with other studies of developmental differences in learning and memory. As the amount of information available increased, students showed a decrease in reaction time, indicating that there was less demand on short-term memory. With less demand on short-term memory, performance improved.

Rudnitsky and Hunt (1986) conducted a study to examine and describe approaches and strategies used by children to solve a complex problem involving or discovering a set of cause-effect relationships. The students, from grades 4 through 6, were told that a dot of light on the computer represented a vehicle that could be moved by colored keys on the keyboard. The problem was to determine what effect each colored key had on the vehicle. The students had to make two or three keystrokes at a time to cause the vehicle to move; the program would not accept fewer than two or more than three keystrokes. The researchers identified four different types of move sequences: (1) Exploration segments occurred when the subjects were trying combinations of moves freely, often forgetting or ignoring the built-in restrictions on number of keystrokes. (2) Patterns involved the repetition of a particular two- or three-keystroke sequence. (3) Focusing appeared to represent a transition and seemed to involve the application of a strategy, such as putting together various combinations of keystrokes or pursuing a hunch about the effects of a particular key. Focusing was systematic and often generated information that could be transformed into a theory or testable hypothesis. (4) Hypothesis testing involved making and testing a prediction. Not all hypotheses were perfectly correct or testable and not all hypothesis testing led to a correct, or any, answer. Typically, however, one or two predictions that were borne out constituted sufficient evidence for students to draw conclusions.

The results of this study support the position that theories do not exist in nature, waiting to "be gotten." Rather, theories must be constructed. Thus, if we are to aid students in developing problem-solving skills, we must provide for the kind of theory-generating activity implied by these results.

Champagne and Klopfer (1981b) investigated the interaction between semantic knowledge and process skills in eighth-grade students' ($n = 27$) performance in solving two types of problems (analogies and set membership) of inducing structure from physical geology concepts. It was hypothesized that, with performance adjusted for differences in IQ and science knowledge, (1) students with high ratings in structuring process skills and students with low ratings would perform equally well in solving analogies problems involving the same science concepts, and (2) students with high ratings in structuring process skills would perform better in solving set-membership problems involving the same science concepts than would students with low ratings. Champagne and Klopfer concluded that, as they had hypothesized, successful performance on analogies problems cannot be attributed to differences in the students' structuring process skills. For set-membership problems, however, students rated high in processing skills performed better in problem solving both before and after instruction, although not in a comparison 1 year-later.

General Problem Solving Strategy

Finally, Ronning and McCurdy (1982) reported on (1) a study of problem-solving processes of 150 junior high school students and (2) the results of a training program in problem-solving processes for junior high school students. Each student was given a set of six problems selected to resemble problems in junior high school science textbooks. Before attempting the problem set, junior high school students in the treatment group were exposed to a 4-hour general problem-solving program that attempted to develop skills in defining, attacking, and solving science problems. The control group, composed of students from the same junior high school, attempted the problems without the training program. The "junior high school students evinced not even rudimentary general problem attack skills" (p. 31). The evidence suggested that junior high school students are developmentally unable to profit from a general problem-solving strategy. It was further suggested that a hands-on approach to teaching science using tasks to pique the curiosity may help students approach problems more skillfully and solve them more successfully.

Summary

The evidence suggests that teachers can correctly distinguish between students with superior problem-solving skills and those without. Within the relatively structured settings of the studies reviewed, successful problem solvers were able to identify the nature of the problem, employ at least some appropriate logical thinking skills, and express their reasoning procedures. Using computer-simulated problems, students quickly developed effective strategies to improve estimation and hypothesis-formulation skills. Similarly, generating hunches in written format seemed to improve student's problem-solving effectiveness. Performance generally improved with age, which is probably a matter of increased maturation. Performance also improved with increased availability of information because the demand on short-term memory was decreased. However, although performance improved, students did not necessarily generate correct or even testable hypotheses; hypotheses were often based on insufficient evidence. The case for teaching general problem-solving skill did not fare well. The evidence suggests that junior high students do not have even rudimentary

general problem attack skills, furthermore, it appears that they are not developmentally able to profit from a general problem-solving strategy.

COGNITIVE STYLE AND PROBLEM SOLVING

Five investigations focused on students' field dependence or field independence in relationship to problem solving. Three other studies considered other student characteristics.

Field Dependence/Field Independence

Ronning, McCurdy, and Ballinger (1984) assert that a viable theory of problem solving must consider at least three dimensions: domain knowledge, problem-solving methods, and characteristics of problem solvers. Therefore, they selected a single characteristic and examined the relationship of problem-solving success and field independence or field dependence of junior high school students ($n = 150$). Five females and five males were randomly selected from each of three grades of five randomly selected junior high schools. The Group Embedded Figures Test (GEFT) was used as the measure of field independence, and problem-solving protocols were analyzed for six science problems presented to the students. It was found that (1) junior high school students have difficulty with problem solving, especially with problems involving proportional reasoning or separation and control of variables, (2) field-independent students significantly outperformed field-dependent students on the problems, and (3) there was no sex difference for problem solving.

Based on research evidence that novice and expert problem solvers represent and organize knowledge differently, Pirkle and Pallrand (1988) examined the way in which field dependence or field independence affected information perception and processing of junior high school students ($n = 39$). The study protocol elicited from the subjects (1) their intuitive mental models of horizontal, vertical, and projectile motion; (2) their intermediate mental models of projectile motion; and (3) their postexperimental models of projectile motion. The students, identified as either field dependent or field independent by means of the GEFT, were individually questioned about their understanding of the effect of gravity on vertical, horizontal, and projectile motion. They were then given the opportunity to compare or verify their responses with information presented graphically on a computer monitor. It was found that erroneous intuitive knowledge representation tended to persist for some subjects after information to the contrary was presented graphically on the computer monitor. The results indicated that the field-independent subjects experienced greater success in solving the problems presented than did the relatively field-dependent adolescents.

Lawson and Wollman (1977) involved 54 sixth-graders (28 males, 26 females) in a study to answer three questions: (1) What is the relationship between subject performance on Inhelder and Piaget's bending rods and balancing beam tasks? (2) What is the relationship between performance on these tasks and ability to make critical value judgments in social contexts? (3) What is the relationship between these abilities and degree of field dependence or field independence (measured by the GEFT)? Lawson and Wollman found that (1) the bending rods and balance beam tasks were significantly correlated, (2) the tasks were significantly correlated with the value questions, and (3) high correlations were found between the GEFT and the bending rods and balance beam tasks. The authors suggest that the high correlation between scores on the bending rods and balance beam tasks supports the Piagetian position that these tasks measure the same psychological parameters and that the ability to abstract formal reasoning patterns from their concrete content seems to be restricted by the factors responsible for a high degree of field dependence. This agrees with other studies in which degree of field independence correlated highly with success on certain types of problems. The authors indicate that these findings imply that if we wish to enhance success in problem solving, science instruction should attempt to foster autonomy by allowing students to investigate phenomena freely. They further suggest that inquiry-discovery methods that allow students autonomy and initiative may foster field independence and, therefore, cognitive development. Science classrooms should thus provide a variety of increasingly complex and repeated experiences. It appears likely that such experiences will occur most readily when students investigate real science phenomena with direct, hands-on activities.

Two other studies included field dependence and independence as factors. Squires (1977) analyzed sex differences and cognitive styles (field dependent/independent) of 13-year-olds in science problem-solving situations and found that field-independent students scored higher on problem solving than field-dependent students. Stuessy (1988), in constructing a model for the development of reasoning ability in adolescents, found a significant direct effect of field dependence or independence and an indirect effect of locus of control through the field dependence-independence variable on scientific reasoning.

Other Student Characteristics

The influence of the reflective-impulsive dimension on problem-solving skills was investigated by Jacknicke and Pearson (1979). Of the 184 sixth-graders involved, 68 were classified as reflective in disposition, 70 as impulsive, and the remainder as neither reflective nor impulsive. The problem solving tasks required either a guided-discovery approach or an open-ended approach for completion. Analysis of the data showed that (1) reflectives and impulsives performed about equally well in selecting, generating, and evaluating observations and hypotheses. (2) Reflectives and impulsives asked similar types and quantities of questions during the problem solving process. (3) Reflectives' performance was superior to that of the impulsives in selecting and generating hypotheses with content and modes of presentation different from those in the guided-discovery and open-ended tasks. (4) Gagné's (1965) claim that observing and hypothesizing abilities exhibit high intertask generality was only partially supported by this study (that is, the abilities appear to be more task specific than Gagné suggests).

Dunlop and Fazio (1975) studied abstract preferences in problem-solving tasks and their relationship to abstract ability and formal thought. An assumption underlying this study was that the level of abstract reasoning used by a student when solving problems is often substantially below the student's capacity. Randomly selected students ($n = 329$) from grades 8, 9, 12, 13, and 16 were given the Shipley Test of Abstract Reasoning. All students were presented with 18 written problem-solving tasks and asked to state their preferences concerning methods for arriving at a solution for the tasks. As expected, older groups demonstrated greater abstract reasoning ability as well as a greater percentage of students in the formal-operational stage of development; however, no significant differences were found between grade levels with respect to abstract preference scores. This preference was independent of abstraction ability and the development of formal operational thought. It was concluded that an individual's preference for a concrete algorithm to a problem-solving situation was not dependent on his or her abstraction ability. No significant correlations were found between abstract reasoning ability and abstract preference, supporting the independence of these two variables. The assumption that a student's level of reasoning is often below his or her capacity is supported by the results of this study; that is, preference for a specific solution may be partly responsible for a student's below-capacity functioning.

In a 5-year study, Scott (1973) examined the longitudinal effects of the inquiry strategy method on students' styles of categorization. Ninety-two students were included in the study; 42 (experimentals) received 2 to 3 years of inquiry strategy exposure in their science classes during their later elementary or junior high school years; the remaining 50 (comparisons) received conventional science teaching during their elementary, junior, and senior high school years. The longitudinal groups (16 experimentals and 16 comparisons) were tested twice, once in 1966, when the students were ending seventh grade, and again in 1971, before graduation from high school. The results of the longitudinal testing indicate that the inquiry process had a persistent enough effect on students' analytical behavior that they maintained a significant advantage over the comparison students for 6 years.

Summary

It is apparent from the evidence that middle and junior high school students find problem solving difficult. This is particularly true when the problems involve separation and control of variables. The research on cognitive style also clearly indicates that middle and junior high school students who are field independent enjoy a significant advantage over field-dependent students in solving science problems. A comparison of students characterized as reflective or impulsive indicates no significant difference in their ability to solve problems; however, the reflective students appeared better able to transfer their skills as long as the new problems were not too different from the familiar ones. Although students' abstract reasoning ability appears to increase, within limits, with age, a student's preference for a concrete approach to problem solving may result in below-capacity reasoning. Finally, inquiry strategy seems to have long-term effects on student analytical behavior, with benefits persisting at least 6 years.

REASONING ABILITY AND COGNITIVE DEVELOPMENT

Several studies examined various aspects of reasoning ability and its relationship to cognitive development. Reasoning is usually considered to include such skills as identifying and controlling variables, proportional reasoning, probabilistic reasoning, and correlational reasoning—skills that are also related to problem solving.

Reasoning Ability

Linn and Levine (1976) investigated the development of the ability to control variables by students ($n = 120$) in a sample composed of 40 students, half females and half males, from each of three age groups: 12, 14, and 16 years-old. Physics problems involving either familiar or unfamiliar variables were presented to the students in three different informational formats. It was found that (1) there were no consistent sex differences across problems for any question, (2) both familiarity with the variable and format of the question influenced success, and (3) there was a qualitative change in ability to control variables between ages 12 and 16. The investigators concluded that there was some evidence to support the hypothesis that subjects try to solve new problems by drawing on apparently relevant past experience.

The effects of problem format and number of independent variables in the problem on the responses of eighth-grade students ($n = 548$) to a control-of-variables reasoning task (the bending rods task) was investigated by Staver (1986). The number of independent variables refers to the number of ways in which the rods differed (thickness, materials, weight, length). The task was presented in two separate formats: (1) completion answer followed by essay justification and (2) completion answer followed by multiple-choice justification. The results showed that (1) task format had no effect on subjects' scores, (2) the differences between subjects' mean scores on the 2- to 3-variable essay versions and the 4- to 5-variable essay versions were significantly greater than the mean scores on the corresponding multiple-choice versions of the task, which were rather uniform, and (3) the 2- to 3-variable forms together were significantly less difficult than the 4- to 5-variable forms together. The results indicate that adding independent variables to a control-of-variables reasoning problem leads to an overload of working memory, which affects performance. The results also suggest that the effect of format on reasoning assessment is connected to the degree of working-memory overload that occurs during such evaluation. Science teachers must pay close attention to the demands placed on working memory during both instruction and evaluation.

The purposes of a study by Saunders and Jesunathadas (1988) were to examine the effect of familiar and unfamiliar task content on the proportional-reasoning abilities of ninth-grade students ($n = 96$) and to determine whether students' problem-solving abilities would generalize across specific subject-matter domains. The researchers found that student performance was higher on proportional-reasoning problems involving familiar content than those with unfamiliar content but that content familiarity interacted with difficulty. When the proportional-reasoning problems involved simple ratios, the mean score with familiar content was greater than with unfamiliar content. But when the problems involved difficult ratios, familiarity with content did not have a significant effect on the students' problem solving. When level of difficulty was considered separately, the students' mean score for problems with simple ratios was significantly higher than their scores for problems with difficult ratios. Independent of content familiarity and level of difficulty, the mean score of male students was significantly higher than that of female students.

Recognizing that proportionality is one of the most ancient and fundamental connections between mathematics and science, Heller, Ahlgren, Post, Behr, and Lesh (1989) investigated the effects of two context variables, ratio type and problem setting, on the performance of seventh-grade students ($n = 254$) on a qualitative and numerical proportional-reasoning task. Analysis of the data indicated that qualitative reasoning is a necessary but not sufficient condition for success on proportional-reasoning problems. Some student success with proportional reasoning could be attributed to memorization of procedures and skill with rational numbers even in the absence of quantitative reasoning skills. Some students were also successful on proportional reasoning even though they lacked skill in rational numbers, but skill with rational numbers ensured success in proportional reasoning. A hierarchy of difficulty for problem settings was found, with the least familiar being the most difficult and the most familiar the least difficult.

Bady (1977) studied the logical-reasoning ability of 55 male students, of whom 20 were ninth-graders, 20 eleventh-graders, and 15 college freshmen. Three tasks were presented in clinical interviews to investigate adolescents' ability to see correlations in data and to test hypotheses. One task was to determine whether the student tested a hypothesis by finding confirming instances or, logically, by finding the disconfirming instance. Two tasks asked the student to find a relationship between two variables. Significant differences in scores were found for each task across age; however, the scores were relatively low in general. The results indicated that logical reasoning abilities develop with age, but that the students frequently lacked the ability to deal with correlated data in problems.

A series of four related investigations into patterns of reasoning were reported by Capie, Newton, and Tobin (1981), McKenzie and Padilla (1981), Newton, Capie, and Tobin (1981), and Tobin, Capie, and Newton (1981). The study attempted to determine if developmental patterns in the ability to control variables, in correlational reasoning, in probabilistic reasoning, and in proportional reasoning (1) could be identified in a large, diverse sample; (2) were similar regardless of problem context; and (3) were similar for subjects of similar developmental levels regardless of educational level. Data for 2282 subjects in grades 6–13 were collected using two forms of the Test of Logical Thinking (TOLT). In general, the response patterns were similar regardless of educational level or the context of the problem. That is, concrete students tended to respond in the same ways regardless of their educational level.

The relationship between integrated process skill and formal thinking abilities of middle and high school students ($n = 492$) was examined by Padilla, Okey, and Dillashaw (1983). Approximately 80 students from each of the seventh through twelfth grades were selected to provide a full range of ability levels. The TIPS was used to measure the five integrated skills of hypothesizing, identifying variables, operationally defining, designing an investigation, and graphing and interpreting data. Formal thinking ability was measured by the TOLT. Correlational analysis showed a strong relationship ($r = 0.73$) between achievement on the two measures and all subtests of the measures; factor analysis data corroborated the correlational evidence. The evidence thus shows that science process skill ability is strongly associated with logical thinking. Further research is needed to determine whether process skill ability influences logical thinking ability, or whether the converse is the case, and to determine whether teaching for one kind of ability will have an effect on the other.

A series of studies of logical-reasoning abilities of students in grades 6 through 12 conducted by Bitner (1987, 1988, 1989, 1990) involved a total of 592 students in several school districts. The students' logical thinking abilities were assessed using the Group Assessment of Logical Thinking (GALT). Among other things, she found that (1) the majority of students were not functioning at the formal operational level; (2) there was a significant movement from concrete to transitional operational reasoning at the end of grade 7; (3) a plateau effect occurred between grades 8 and 9; (4) there were no significant gender differences for total GALT score, although some differences were found for individual items; (5) the five formal operational modes on the GALT were predictors of critical thinking as measured by the Watson-Glaser Critical Thinking Appraisal and the Ross Test of Higher Cognitive Abilities and were also predictors of academic success; and (6) incorporation of logical, critical, and creative thinking skills into instruction resulted in significantly higher student performances in controlling variables, correlational reasoning, and total GALT scores.

A comparison study involving middle grade students in North Carolina and Japan was reported by Coble (1986), Spooner (1986), and Mattheis, Coble, and Spooner (1986). The primary purpose was to measure the reasoning skills and integrated process skills, and their relationship, of students in grades 7, 8, and 9 in the two countries. A total of 3291 students in North Carolina and 4397 Japanese students participated in the study. Instruments used were the Attitude Toward Science Scale (ATSS), the GALT, and the TIPS II. North Carolina students had more positive attitudes than their Japanese counterparts at all grade levels, but there was a decrease in scores from the seventh to the ninth grade. For the North Carolina students, no

significant differences in TIPS II mean scores were found between male and female students. There was a progression in scores from the seventh to the ninth grade. The most difficult area for the North Carolina students was identifying variables. In comparison, Japanese students scored highest on this subscale. A significant relationship was found between the integrated process skills scores on the TIPS II and logical thinking scores as measured by the GALT ($r = .64$; $p < .0001$). This relationship is consistent with findings in other studies using these tests. The GALT results indicated that only 10 percent of the North Carolina students were functioning at the formal operational stage. In comparison, 32 percent of the Japanese students were functioning in the formal mode.

The Burney Logical Reasoning Test and the Science subtest of the Stanford Achievement Test were used by Dozier (1986) to investigate the relationships between objective measures of logical reasoning abilities and science achievement. All students in seventh-grade life science ($n = 39$), eighth-grade general science ($n = 35$), and ninth-grade physical science ($n = 33$) classes were tested with both instruments. Significant correlations were found between the sets of raw scores for the seventh grade ($r = .541$), eighth grade ($r = .702$), and ninth grade ($r = .386$); for total population raw scores, $r = .553$. The author concluded that a significant relationship existed between the objective measures of logical reasoning abilities and science achievement.

Development Patterns/Models

Stuessy (1988) developed and tested a model for the development of reasoning abilities in adolescents. A battery of assessments for locus of control, field dependence-independence, IQ, rigidity-flexibility, and reasoning was given to middle school ($n = 101$) and high school ($n = 89$) students. Student characteristics of age, gender, and experience with science-related activities were also included. The model was tested and revised using path analysis. Significant path coefficients were found between scientific reasoning abilities for age (.54), experience (.11), IQ (.49), and field dependence-independence (.15). A path from locus of control (.29) through field dependence-independence to reasoning was also significant. These five variables accounted for 61 percent of the variance in scientific reasoning abilities. Support for an indirect effect of locus of control on scientific reasoning through the field dependence-independence variable suggests that experiences which encourage a shift from external toward internal locus of control may ultimately affect the development of scientific reasoning abilities. Support for strong direct effects of IQ and field dependence-independence on scientific reasoning suggests individualizing instruction as a way to meet the individual needs of children exhibiting differences in aptitudes and abilities. Finally, this study supported the increase of scientific reasoning abilities with age. Age, in fact, was the strongest predictor of scientific reasoning in the model, with IQ being the other strong predictor. This suggests that acquisition of scientific reasoning abilities in adolescents is a developmentally complex process.

A search for a learning hierarchy among skills comprising formal operations and the integrated science process skills was reported by Yeany, Yap, and Padilla (1986). Ordering theoretic and probabilistic latent structure analysis methods were used to analyze data on five process skills and six logical thinking skills. Intact classes including 741 science students in grades 7 to 12 from three schools participated in the study. The GALT was used to measure the performance of students on the six Piagetian cognitive modes: conservation reasoning, proportional reasoning, controlling variables, probabilistic reasoning, correlational reasoning, and combinatorial reasoning. The TIPS II was used to measure student performance on the five integrated process skills: identifying variables, identifying hypotheses, operational defining, designing experiments, and graphing and interpreting data. The data analysis indicated that the skills of conservation reasoning, combinatorial reasoning, and designing experiments formed the base of the hierarchy. That is, students who had mastered these skills were more likely to have the skills at the next higher level. At the next level were the skills of graphing and interpreting data, operationally defining, controlling variables, and propositional reasoning. These skills appeared to be superordinate to the three skills at the base and subordinate to the remaining four skills. After the second level, the hierarchical relationships became more complex. Proportional reasoning was prerequisite to probabilistic reasoning which in turn was prerequisite to both correlational reasoning and identifying variables. Identifying variables had the most complex set of underlying prerequisite skills. In addition to having probabilistic reasoning, most students who had skill in identifying variables had also acquired the skills of hypothesizing, graphing, and interpreting data as well as operationally defining. Thus, the integrated process skills appear inextricably entangled with the Piagetian modes of reasoning. Students may not be able to acquire certain scientific process abilities until prerequisite cognitive skills are developed. For example, according to the results of this study, the probability of a student being able to formulate hypotheses before the abilities to conserve, use proportional reasoning, control variables, and use combinatorial logic develop is quite low. This implies that process skill–based curriculum activities need to be developed and presented with a structure that reflects the super- and subordinate relationships of the skills.

Reviews of Research Related to Reasoning Ability

Two reports summarized research related to reasoning ability, particularly that implied by formal thought, which is the ability to apply principles of problem solving to any problem. Linn and Levine (1977) reviewed the research on scientific reasoning and found that changes did take place during adolescence but that they were not as complete as suggested by earlier descriptions of formal thought. They found that performance related to scientific reasoning appeared to be influenced by (1) the number of variables to be considered, (2) familiarity of the variables to the student, (3) previous experience with variables, (4) method of presenting information about the task, (5) procedure for interacting with the apparatus (e.g., free or constrained), and (6) subject matter of the problem (e.g., physics, biology). Of importance for instruction were the findings that (1) only a small number of adolescents could effectively control variables in familiar situations; (2) relatively few adolescents reach the level of formal operations, hence concrete experience is a valuable aid to learning at all stages of adolescence; (3) programs that offer a choice of mode of learning or provide several approaches for teaching a principle are probably more useful than programs that adhere to a particular theory; (4) programs that encourage learners to manipulate materials have great potential benefit; and (5) the method that learners use to organize information is important; emphasis should be placed on solution strategies that optimize the organization of the problem.

Staver (1984) examined research on formal reasoning patterns in science education and identified a number of implications for science teachers: (1) Adolescents and young adults frequently do not use formal reasoning patterns when such thinking is needed to comprehend fully a segment of learning. (2) An adolescent may utilize control-of-variables reasoning on one task before another; a second student may reverse the order of tasks. (3) Achievement in science includes both conceptual knowledge and reasoning skills. (4) The conclusion that children are unable to comprehend certain concepts until they reach a certain stage in their reasoning development overlooks the idea (called constructivism) that knowledge may be acquired by formation within the brain through interactions with the environment as opposed to being internalized from the outside environment (an empiricist model). (5) Science teachers should design and carry out instruction that does not overtax working memory. These situations should involve cognitive conflicts in which difficulty is controlled and success is virtually assured at frequent points. This can be accomplished by minimizing the number of items of information, maximizing familiarity of information, and highlighting cues.

Summary

The evidence strongly associates science process skills with logical thinking. It also suggests that both cognitive development and problem-solving skills are related to maturation. Ability to control variables tends to increase (within limits) with age; however, control of variables may vary from one student to another and from one task to another by the same student. Most adolescents do not operate at the formal-operations level. To maximize opportunities for both cognitive development and problem solving, we need to limit the number of variables involved in problems, to draw on familiar examples when possible, and to allow freedom of interaction with equipment and materials in direct, concrete, hands-on experience (Lawson & Wollman, 1977).

PROBLEM SOLVING AND INSTRUCTION

The largest segment of the research dealing with problem solving in middle school science is related to the effects of various aspects of instruction on students' problem-solving ability.

Modes of Instruction

Bowyer, Chen, and Thier (1976) studied the effects of a free-choice environment on sixth graders' ($n = 90$) ability to control variables. Half of the students were randomly assigned to a treatment group that had access to a Science Enrichment Center in which they could select a science activity of their choice along with their choice of materials. It was found that (1) more than 60 percent of the students within all subgroups did not control variables, less than 32 percent controlled variables inconsistently, and an insignificant number controlled consistently, and (2) the experimental group differed significantly from the control group. Approximately one-third of the experimental population moved from a state in which they were unable to recognize the need for controlling variables to one in which they recognized the necessity to control experiments. Seventy-eight percent of the experimental group understood the necessity for controlling variables, particularly when the variables were familiar ones such as weight and length.

McKee (1978) investigated the effects of two contrasting teacher behavioral patterns on science achievement, problem-solving ability, confidence, and classroom behavior for students ($n = 100$) in sixth-grade science. The teaching patterns were student-structured learning in science, which minimized restrictions on students, and teacher-structured learning in science, which was moderately restrictive. The data analysis indicated, among other things, that (1) students in the student-structured learning group performed significantly better on problem-solving tasks; (2) students in the student-structured group had significantly higher confidence levels; and (3) there were significant correlations for problem-solving ability and confidence level, for problem-solving ability and IQ, and for confidence level and IQ. McKee concluded that student-structured learning strategy functioned better to improve problem-solving skills of black students while working just as well for white students and was the obvious choice for teachers interested in promoting a student's ability to solve problems while improving self-confidence.

Creativity and the group problem-solving process was studied by Foster (1982) to determine whether cooperative small groups would stimulate the creativity of fifth- and sixth-grade students ($n = 111$) more than an individualized learning environment. Creativity was defined as becoming sensitive to problems, identifying the difficulty, searching for solutions, testing and retesting hypotheses, and communicating the results. Half of the population (control) worked by themselves, while the other half (experimental) worked together in groups of four or five. Each half worked in a student-structured environment on the same science activity, which involved creating as many different types of electrical circuits as possible from a given set of batteries and bulbs. The same trained teacher guided students in both the individual and small-group settings. Among other things, it was found that both fifth- and sixth-graders in small groups did significantly better on creating electrical circuits

than individuals working alone. Understanding of electrical circuits was not differentially affected by the treatment conditions, although the scores indicated that both groups of students had a better understanding of electrical circuits from being in the study than from previous experience. Analysis by gender indicated that boys did better in the experimental group situation and girls did better in the control group situation. Foster concluded that, for fifth- and sixth-grade students, working in cooperative small groups was more effective than working alone.

The effectiveness of mastery instruction was compared with that of an equivalent nonmastery mode of instruction for improving students' learning and retention of selected science process skills by Brooks (1982). Ninety average and above-average junior high school students were involved in the study. No significant differences were found in levels of achievement between mastery- and non-mastery-instructed students. Above-average mastery-instructed students scored significantly higher than their non-mastery counterparts in retention of low-level process skills, but there was no significant difference in retention for the average students. Both average and above-average mastery-instructed students scored higher on retention of the higher process skills than did the non–mastery-instructed students. Although mastery instruction did not result in greater achievement gains, the evidence suggests that it may result in a more permanent mastery of the higher process skills than an equivalent time with a non-mastery instructional strategy.

Cox (1980) studied early adolescent use of selected problem-solving skills using microcomputers. The subjects included 66 seventh- and eighth-grade students in four urban junior high schools; 48 were males, 18 were females. The students were randomly assigned to work alone or in groups of two, three, or five for three 50-minute sessions. Cox concluded, among other things, that (1) students can improve in problem-solving skills in a short time using a microcomputer, (2) the training program on organizing data in a matrix was successful, (3) individuals worked better in teams than alone, (4) the influence of group interaction enabled subjects of all abilities to participate successfully and solve problems, (5) all subjects adapted quickly and easily to use of the microcomputers, and (6) microcomputers can be considered a viable motivating aid for the development of some problem-solving skills of early adolescents.

Friedler, Nachmias, and Songer (1989) reported on a study that was part of an investigation of the educational potential of Microcomputer-Based Laboratories (MBL). This study focused on the development and implementation of a scientific reasoning Skill Development Module, which was a series of computerized and non-computerized activities geared toward the instruction of general scientific skills such as experimental planning, prediction formation and revision, control of variables, and formation of discriminating observations. The module was implemented over a period of 16 weeks in eight eighth-grade classes ($n = 250$). Evaluation of the modules showed an improvement in student abilities to plan, to control variables while designing an experiment, to make careful observations, to distinguish between observations and inferences, to make detailed predictions, and to justify them on the basis of their previous experiments. Efforts to teach explicitly strategies for general problem-solving skills "were largely successful." However, the data suggested that students had difficulty transferring general problem-solving skills from one context to another. These results support the contention that basic scientific reasoning skills should be taught in a variety of contexts.

Pirkle and Correll (1990) were interested in the performance of early adolescent students on the "Flight Protocol," with particular attention to interpretation of the graphs of paths of projectiles, as well as the comparative effects of the teachers' use of verbal intervention strategies. Eighth-grade students ($n = 30$) volunteered to participate and were assigned to one of three groups (half were boys and half were girls in each group), each of which received a different intervention. There was no significant difference in performance on the Flight Protocol by the three groups of subjects. Neither cognitive modeling nor probing questions appeared to enhance the subjects' performance significantly. The subjects performed poorly on this activity regardless of the verbal strategies employed by the interviewer. Inspection of the students' drawings revealed that their ability to restructure and transform the information graphically presented, a skill associated with inherent perceptual style, appeared to have a more consistent impact on success in this task than the use of verbal intervention strategy.

Novak, Gowin, and Johansen (1983) explored the use of concept mapping and knowledge Vee mapping with seventh- and eighth-grade science students. In general, students of any ability could be successful in concept mapping and other factors (e.g., motivation) were more important. Both seventh- and eighth-graders could acquire an understanding of the Vee heuristic and apply the learning tool in regular junior high school science. An important finding was that the experimental classes demonstrated superiority in problem-solving performance on novel problems after less than 6 months of instruction with these strategies. The data suggested that concept mapping and Vee mapping were helpful with respect to changes in student knowledge about science and problem-solving skills and performance on novel problem-solving tests. The data also indicated that effective use of the Vee heuristic takes time for students to acquire and that 2 or more years may be required for students to achieve high competence.

The relative effects of hands-on and teacher demonstration laboratory procedures on ninth graders' declarative knowledge (factual and conceptual) and procedural knowledge (problem-solving) achievement were studied by Glasson (1987). The students ($n = 54$) were randomly assigned to classes taught by either a hands-on or a teacher demonstration laboratory method. He found that (1) the two instructional methods resulted in equal declarative knowledge (factual and conceptual) achievement; (2) all students in the hands-on laboratory class performed significantly better, regardless of reasoning ability, on the procedural knowledge (problem-solving) test than did students in the teacher demonstration class; (3) prior knowledge significantly predicted performance on the declarative knowledge test; and (4) reasoning ability, prior knowledge, and teaching method (in that order) had significant effects for procedural knowledge. These three variables accounted for 53 percent of the variance in problem solving. Glasson concluded that the ability to solve problems is enhanced if students have ap-

plied reasoning strategies in the process of actively performing an experiment.

Problem-Solving Instruction

Based on the premise that hypothesis formation is an important step in problem solving, Quinn and George (1975) and Quinn and Kessler (1980) evaluated a method for teaching hypothesis formation to sixth-grade children. It was found, among other things, that (1) the treatment groups did significantly better than the control groups in hypothesis formation; (2) there was no significant difference in the performance of children in upper or lower socioeconomic groups; (3) bilingual children performed better than monolingual children; and (4) the ability to hypothesize was correlated with intelligence, overall grade point average, reading ability, and sex (with boys doing better). It was suggested that students educated in more than one language are better problem solvers than their monolingual peers, which would imply the need for closer working relationships between science and language educators.

Egolf (1979) examined the effects of two modes of instruction on seventh- and eighth-grade students' abilities to solve quantitative word problems in science. The treatment group included 420 students, the control group 105 students. Researcher-developed booklets were used to teach the students how to solve a specific type (density) of word problem and to teach a general method for solving word problems. Among the conclusions drawn were: (1) the booklets could be effectively used to teach students how to solve density problems, (2) no booklet successfully taught the students a general problem-solving process, (3) significant grade-level effects existed regarding word problem-solving ability, and (4) student gender was not related to word problem-solving ability.

An instructional program based on expert-novice differences in experimental problem-solving performance was taught to sixth-grade students ($n = 265$) by Ross and Maynes (1983b). Based on a chronological account of what successful scientists do when designing an experiment, the domain of problem solving was broken into a set of seven skills. The first two skills—(1) developing a focus for the investigation (hypothesis formulation) and (2) establishing a framework for the investigation (including control of variables)—were included in the instructional design. Performance was measured by tests of specific transfer; that is, the same skills were to be performed in a different experimental setting using different apparatus. It was found that treatment-condition students consistently outperformed control group students. The investigators concluded that the instructional program had a beneficial effect.

Lawsiripaiboon (1983) examined the effects of a problem-solving strategy on ninth-grade students' ($n = 423$) ability to apply and analyze physical science subject matter. The problem-solving group received problem-solving laboratory activities and classroom discussions emphasizing the application and analysis levels. The conventional group received laboratory activities and discussions focusing on the knowledge and comprehension levels. The problem-solving group significantly outperformed the conventional group, and teachers in the problem-solving group asked significantly more high-level questions. The investigator concluded that the problem-solving strategy used was an effective means for improving overall achievement, particularly achievement at the application and analysis levels. Problem-solving laboratory activities and teacher-initiated questions at the application and analysis levels thus appeared to be practical strategies.

A similar study was reported by Russell (1979), Russell and Chiappetta (1980, 1981), and Chiappetta and Russell (1982). The purpose was to improve eighth-graders' ability to apply and analyze earth science subject matter. Fourteen sections of students ($n = 287$) were randomly assigned to seven experimental and seven control groups taught by four teachers. The experimental groups received a problem-solving form of instruction that included reading, problem-solving tasks, and discussion and laboratory exercises emphasizing application and analysis levels. The control groups received a traditional textbook-oriented form of instruction that included reading, discussion, and laboratory activities. The experimental group significantly outperformed the control group. A comparison of the median questioning level of the teachers showed that the teachers in the experimental group employed a significantly higher level of questions. The investigators concluded that an instructional program using a problem-solving approach will significantly increase overall achievement, particularly at the application and knowledge levels, and that such an approach should include written problem-solving activities and teacher-directed questions that emphasize application of knowledge.

An investigation of the effects of intensive instruction on the ability of ninth-grade students ($n = 205$) to generate written hypotheses and ask questions about variables pertaining to discrepant scientific events was reported by Pouler (1976) and Pouler and Wright (1977). The students were randomly assigned to a control and four experimental groups of 41 students each. The experimental instruction involved watching a discrepant event (a Suchman film loop) until six acceptable hypotheses were written. It was found that reinforcement was essential for producing a greater quantity of written hypotheses. For higher-quality hypotheses, reinforcement plus knowledge of the criteria was superior to no instruction.

In a related study, Wright (1978) examined the feasibility of intensive instruction in either observation of details or hypothesis generation, using a discrepant-event film loop, as a model for improving the open exploration skills of ninth-graders ($n = 120$). In general, the treatment groups were superior to the control group in number of details reported, number and quality of hypotheses generated, and in number and diversity of questions asked. No significant differences were found between treatment groups in the number and quality of hypotheses or the number and diversity of questions generated. Wright concluded that both treatments were equally effective in improving the open exploration skills of the students, with the exception of the number of observed details, for which the subjects who had been instructed in observing details exhibited superior performance.

McKenzie and Padilla (1984) investigated the effects of three instructional strategies and student entry characteristics on student engagement and acquisition of skills necessary for the construction and interpretation of line graphs. The strategies examined were an activity-based approach, a written simulation-based approach, and a combination of activity- and written-simulation instruction. Students' ($n = 101$) entry characteristics examined included level of cognitive development and spatial ability. The results indicated that (1) no single instructional strategy of those examined appeared to be superior to the others in regard to level of graphing achievement attained by students; (2) instructional strategies consisting of activities and written simulations resulted in higher levels of engagement than those consisting only of written simulations; (3) students classified as transitional or formal tended to score higher than concrete-operational students on the graphing-achievement measure; (4) spatial scanning ability was minimally related to graphing achievement; and (5) although differences in engagement across the three treatment groups accounted for a portion of the variance in graphing achievement, the treatments themselves accounted for a unique portion of the variance; this was also true across levels of development.

Inquiry Instruction

Butts and Jones (1966) studied the effectiveness of inquiry training in improving problem-solving behavior in sixth-grade students ($n = 109$). Approximately half of the students were involved in a planned guidance program based on that devised by Suchman as inquiry training. The remaining students served as the controls and did not receive the inquiry training. There was a significant relationship between inquiry training and changes in problem-solving behaviors as measured by the TAB Inventory of Science Processes (TISP). No relationship was found between inquiry training and changes in concept transfer as measured by TISP, or between inquiry training and recall of factual science knowledge as measured by STEP and the Elementary Science Test. No relationship was found between changes in problem-solving behaviors that occurred in conjunction with inquiry training and IQ, chronological age, science factual knowledge, or sex. The investigators concluded that students could benefit from directed instruction in problem-solving behaviors and that age, IQ, sex, and science factual knowledge were not significant factors in students' benefitting from inquiry training. The assertion that meaningful concept development results from inquiry training could not be supported; children who were successful problem solvers on TISP could not apply the concept to a different situation.

Jones (1973) investigated what effect acknowledging successful autonomous discovery had on seventh-grade students' ($n = 49$) problem-solving abilities, concept development, science achievement, and self-concept as learners when they were exposed to the Inquiry Development Program for a semester. Data analyses revealed that (1) there was no significant difference in changes in self-concept, problem-solving behavior, or science concept development between the experimental and control groups; (2) there was no significant change in self-concept or science concept development for either the experimental or control group; (3) the control group showed a significant gain in science achievement but the experimental group did not; and (4) there was a significant positive change in problem-solving behavior for both the experimental and control groups, suggesting that the act of discovery was more important than whether the discovery was acknowledged by the teacher.

Davis (1979) examined the effects on achievement of using two approaches to science instruction: (1) an expository-text approach and (2) a guided inquiry-discovery approach. In the expository-text approach, the students ($n = 52$) received direct presentation of information and concepts from the text and teachers. In the guided inquiry-discovery approach, the students ($n = 51$), guided by the materials and teachers, engaged in investigations involving inquiry processes structured to develop information and concepts. It was found that (1) the guided inquiry-discovery approach was significantly more effective than the expository-text approach in achievement of knowledge and information of content contained in the science units; (2) achievement in understanding of science inquiry and processes was slightly, but not significantly, higher for the guided inquiry-discovery group, suggesting that the two methods were equally effective for these students; and (3) students receiving the guided inquiry-discovery instruction expressed significantly more positive attitudes than the expository-text group. In conclusion, it appeared that the guided inquiry-discovery method of instruction provided a means of combining the products and processes of science while enhancing positive attitudes.

Thomas (1968) analyzed the effects of two instructional methods, didactic and guided discovery, on eighth-grader's ($n = 143$) understanding of the scientific enterprise, understanding of scientists, understanding of the methods and aims of science, achievement of factual-conceptual understanding, use of critical thinking skills, and use of problem-solving skills. Possible interactions of methods with the learner variables of sex, intelligence, creativity, interest in science, general scholastic achievement, and achievement in science were also included. For the three areas of understanding science, the only difference that could be attributed to method was for understanding of the scientific enterprise, in which the high scholastic achievement group was superior when instructed by guided discovery. For factual-conceptual achievement of content, the didactic method was superior for all pupil variables. For the acquisition of inquiry skills, critical thinking skills, and problem solving, interactions of method with levels of ability showed guided discovery to be better for the high groups, didactic better for the low groups, and neither superior for the middle groups.

Using different patterns and amounts of instruction on planning experiments with sixth- and eighth-grade students, Padilla, Okey, and Garrard (1984) examined the effects of instruction on integrated science process skill achievement. Treatment One ($n = 168$) involved a 2-week introductory unit emphasizing designing and carrying out experiments; subsequent content units had approximately one period-long process skill activity per week integrated into the regular curriculum. Treatment Two ($n = 85$) involved only the same 2-week introductory unit emphasizing experiments; subsequent instruction was primarily content oriented with little process emphasis. Treatment Three ($n = 76$) was a control treatment with the same content-oriented instruction as Treatment Two but no direct process skill experience. All groups showed increases in process skills achievement and logical thinking over the course of the treatment. No significant process skill (TIPS) differences were found among the three treatments for the sixth-graders. Among eighth-graders, however, the extended process skill group (Treatment One) scores were significantly higher than scores for either the 2-week process skill group (Treatment Two) or the control group (Treatment Three). On the logical thinking (TOLT) variable, no significant differences were found with either grade level; scores for all groups in both grades increased from pre- to posttest, but no differences due to treatment were found. To more closely examine specific skills and the degree to which instruction influenced them, the process skills test was divided into three subtests, the logical thinking test was divided into five subtests, and the data were analyzed for these scales. A significant difference due to treatment was found for both sixth- and eighth-graders on the hypothesizing and identifying variables subtest. Treatment One was higher than the control among sixth-graders and both Treatments One and Two were higher than the control for eighth-graders. No differences for either grade were found for the other two process skills subtests or for any of the logical thinking subtests. Although the results were statistically significant, relatively little of the sixth- and eighth-graders' achievement on hypothesizing and identifying variables was accounted for by the treatment. Logical thinking skills were also not much affected in the time span of the treatment. It appeared that affecting one ability had little influence on the other over the 14-week period. However, greater benefit to students seemed to result from integrating science content and process instruction over a longer period of time.

Jordan (1987), using information processing as a theoretical base, sought to determine whether specific cognitive strategies (memory skills) could be taught and learned. Four tasks were designed as follows: (1) summarizing a reading, (2) self-testing vocabulary, (3) writing questions, and (4) writing answers. Ninth-grade science students ($n = 311$) were assigned to an experimental group that was assigned the tasks described and a comparison group that was assigned predetermined tasks from the text. The results showed statistically significant differences in favor of the comparison group on recall and the experimental group on problem solving. Further analysis indicated that although the results were statistically significant, mean score differences were very small, suggesting that they might also be educationally questionable.

Linn, Clement, Pulos, and Sullivan (1989) assessed the role of science topic instruction combined with logical reasoning (control of variables) instruction in teaching adolescents about blood-pressure problems. Four major findings emerged from this study: (1) all participants acquired science topic knowledge about blood pressure, (2) those receiving strategy instruction acquired knowledge of the controlling-variables strategy, (3) acquisition of science topic knowledge influenced reasoning performance, and (4) the combination of science topic knowledge and strategy instruction produced more generalized reasoning performance than did science topic knowledge alone. Based on these results, both science topic knowledge and strategy instruction increase the likelihood of successful performance by helping students bring relevant information to bear on problem solving.

Farrell (1988) sought to determine whether skills in proportional reasoning taught to eighth-graders ($n = 115$) before they received instruction in three physical science problems would promote transfer of learning among the problems. The treatment materials were self-instructional packets that included explanations, diagrams, drawings, and problems on the balance beam, the inclined plane, and the hydraulic lift. The subjects were pre- and posttested, with a random half of each treatment group receiving additional instruction in fractional proportions. Students who received instruction in proportionality exhibited greater learning than uninstructed subjects. In each of the tasks, there was greater transfer of learning to new problems for the students receiving prior instruction in proportionality. It did not appear that transfer was task specific in the context of this study.

Finally, Curbelo (1984) conducted a meta-analysis of 68 experimental studies, producing a pooled sample size of 10,629 students, to determine the effects of problem-solving instruction on science and mathematics student achievement. The author concluded that (1) when groups of students were given instruction in problem solving, their achievement exceeded that of students not provided with instruction in problem solving by an average of 0.54 standard deviation; (2) the duration of instruction in problem solving is positively correlated with performance on problem-solving measures; (3) the most effective duration for instruction in problem solving appears to be 5 to 10 weeks; (4) problem solving can be taught effectively in any topical area in science and mathematics; and (5) the inquiry method seems to be one of the most effective strategies for teaching problem solving (p. 78).

Summary

Most middle and junior high school students are probably developmentally unable to benefit from instruction in a general problem-solving approach. However, the research shows that students do benefit from instruction in problem solving; they can and do learn to use integrated science process skills. They are also able to transfer these skills to new problems if the new problems are not too dissimilar from those with which they have had experience. The most effective approach to teaching science appears to integrate science process skills and science content over several weeks, using hands-on, inquiry activities concentrating on specific problem-solving skills. Moreover, students who receive such instruction tend to learn more science and to develop more positive attitudes toward science and more self-confidence in their own abilities.

SCIENCE CURRICULA AND PROBLEM SOLVING

In tracing the historical development of problem solving as a curriculum goal, White (1978) noted that it received attention as a major curriculum strand from the Dewey School of 1896 to the 1970s. As the current effort indicates, science curriculum developers' concern with problem solving has persisted into the 1990s.

Junior high school science curriculum developments in the 1960s departed from traditional methods as well as traditional objectives. The new objectives were directed toward development of the ability to solve problems in a logical manner. The relationship between inquiry teaching, as represented by the then newly developed science curricula, and intellectual development was studied by Friot (1971). Tasks described by Piaget and Inhelder were used to evaluate the development of interpropositional logic (i.e., formal operations). The study showed that logical thinking processes could be evaluated using the Piagetian tasks and that some curricula were effective at some grade levels but not others. The Time, Space, and Matter (TSM) curriculum was significantly more effective in enhancing the development of formal operations than either the Introductory Physical Science (IPS) or the Earth Science Curriculum Project (ESCP) materials. The IPS curriculum was significantly better than the ESCP curriculum. At the ninth-grade level, both the ESCP and IPS curricula were significantly better than the control and the IPS curriculum was significantly better than the ESCP curriculum. It was also found that sex and IQ were not significantly related to gains in logical thinking ability.

Schlenker (1971) wished to determine whether a physical science inquiry development program, based on materials developed at the University of Illinois, achieved results significantly different from those of a traditionally taught program with similar science content when taught to pupils ($n = 582$) in grades 5 through 8. Schlenker concluded that children who studied under the inquiry development program developed significantly greater understanding of science and scientists and fluency and productivity in using the skills of inquiry or critical thinking than did children who studied under the traditional program. In addition, there was no consistently significant difference in the mastery or retention of the usual content of elementary school science between the two groups of students.

Butzow and Sewell (1971) administered the Test of Science Processes as a pre- and posttest to 92 eighth-graders studying the first five chapters of the IPS materials to determine whether process learning was significantly changed. The students were assigned to four homogeneously grouped classes with placement based on IQ, teacher recommendation, interviews, and achievement records in English and mathematics. There were significant improvements in the students' ability to observe, compare, classify, quantify, measure, experiment, and infer during the course, but not in the ability to predict. It appeared that there was a relationship between intelligence and process skills, with the more intelligent groups at the eighth-grade level farther along a continuum of difficulty from observing to experimenting at the onset of instruction. Consequently, classes with students having lower ability levels showed the greatest change in process learning during IPS instruction.

Selected aspects of two seventh-grade science programs—Interaction of Man and the Biosphere (IMB), an experimental inquiry program, and Science Is Explaining, the control program—were compared by Gudaitis (1971). The experimental group and the control group each consisted of 48 boys and 48 girls. From pretest to posttest on attitude toward science, students in the experimental program showed no significant attitude change and students in the control group showed a significant decrease in attitude; there was no significant difference in mean gain scores between the two groups. For science process skills, the mean gain scores for students in both programs increased significantly but there was no significant difference in mean gain scores between the groups. For critical thinking skills, the mean gain scores for students in both programs increased significantly and students in the experimental group made significantly greater gains than students in the control group.

Hill (1982) compared the effect of a Human Science Program module and traditional science classes on pupils' logical thinking skills and attitudes toward their science course. No significant difference was found between treatment groups in pupils' logical thinking scores; however, a significant teacher-by-treatment interaction was found. No difference was found between treatment groups on the total attitude instrument; however, a significant difference in favor of Human Sciences was found on the activity subscale. A significant sex-by-treatment interaction was also found. It appeared that the Human Sciences Program, when compared to traditional science classes, was partially successful in promoting more positive attitudes; however, it did no more to promote logical thinking skills than the traditional science classes.

Stallings (1973) and Stallings and Snyder (1977) compared the inquiry behavior of ISCS and non-ISCS students as measured by the Tab Science Test, which was administered to 178 seventh-grade, 164 eighth-grade, and 104 ninth-grade ISCS students and to 165 seventh-grade, 113 eighth-grade, and 68 ninth-grade non-ISCS students. No significant differences were found between ISCS and non-ISCS groups in the seventh or eighth grades, but a difference in favor of the non-ISCS group was found in the ninth grade. No differences in patterns of clue questions selected were found between the ISCS and non-ISCS groups in seventh or eighth grade. Differences in clue question selection were found in the ninth grade, with the ISCS students exhibiting fewer inefficient patterns and more efficient patterns of inquiry than the non-ISCS group. The ISCS program as used by teachers involved in this study did not result in clear gains in inquiry skills; however, the reliability of the Tab Test was found to be low (0.556 for ninth grade, 0.455 in eighth grade, and 0.434 in the seventh grade), suggesting caution in interpreting the results.

Evaluation of the Unified Science and Mathematics for Elementary Schools (USMES) program was the basis for studies reported by Shapiro (1973), Shann (1975), and Shann, Reali, Bender, Aiello, and Hench (1975). A primary objective of the USMES program was "the enhancement of elementary school student's abilities in real, complex problem solving" (Shann et al., 1975, p. 127). Shapiro (1973) reported that "the USMES experience had, irrespective of units and teachers involved, a marked and positive effect on the students' problem solving behavior" (p. 13). However, the other studies reported mixed

results, with few significant differences favoring the USMES program, although there was some evidence that USMES enhanced basic skill development in mathematics, science, and social science as measured by SAT tests (Shann et al., 1975). Despite these findings, interviews disclosed that teachers perceived children in the USMES program to be more responsible for their own learning and to show growth in data collection abilities, graphing skills, hypothesis testing, and communication with their peers. Although not supported by data, these perceptions were consistent and persistent among teachers from all geographic and demographic areas involved in the program (Shann, 1975). Concern with the validity and reliability of the problem-solving instruments led to the development of a separate instrument described in the section on assessment of problem solving (Shann, 1976).

Bullock (1973) reported a study to determine the relative effectiveness of three different types of elementary school science curricula in the development of selected problem-solving skills of sixth-grade students. The three curricula were Science—A Process Approach (SAPA), the Laidlaw textbook series, and the Environmental Studies (ES) project materials. A total of 27 teachers and 512 students participated in the study. The TAB Science Puzzler Form B was administered as a pretest and the TAB Science Puzzler Form C as a posttest 6 months later. It was found that (1) there was no significant difference in the improvement of the problem-solving skills of students using the Laidlaw textbook series and those using the SAPA curriculum materials; (2) there was a significant difference in the improvement of problem-solving skills of students using the Laidlaw textbook series compared with students using the ES materials; (3) there was a significant improvement in the problem-solving skills of the students exposed to the SAPA materials compared with those exposed to the ES materials; and (4) significant improvement in problem-solving skills was attained with both the SAPA program and the textbook series.

In a related study, the long-term effectiveness of SAPA in the development of problem-solving skills in fifth- and sixth-graders was assessed by Breit and Bullock (1974). This was a comparison of certain problem-solving skills of children who had been in classrooms using the SAPA program for at least 4 years with the performance of children who had been in classrooms not using SAPA. A random sample of 100 students from each group was selected for the study. The TAB Science Puzzler, Form B, was administered to both groups. A significant difference in problem-solving skills was found in favor of the children who had been using the SAPA materials.

Shaw (1978, 1983) was concerned with the effect of the process-oriented science curriculum, SAPA II, on the ability of sixth-graders to utilize problem-solving skills, which were defined as the integrated process skills of controlling variables, forming hypotheses, interpreting data, and defining operationally. Other areas investigated included (1) determining whether problem solving learned in science would transfer to social studies, (2) testing models concerning problem-solving skills to determine whether there was evidence for a hierarchy of problem-solving skills, and (3) determining if training in problem-solving skills would increase students' proficiency in basic skills such as observing, inferring, and predicting. The treatment group scored higher than the control group on the problem-solving skills portions of both the science and social studies instruments, indicating that problem-solving skills can be taught by the process-oriented science curriculum and will transfer to a social studies content. No significant differences between the groups were found on either instrument for basic process skills, but the treatment group scored higher than the control group for the process of classifying on the social studies instrument but not on the science instrument. Evidence supported the hierarchy model of process skills, suggesting that mastery of the basic skills is a prerequisite to proficiency in the problem-solving skills.

An analysis of inquiry level in junior high textbooks, supplemental activity guides, and discipline was undertaken by Pizzini, Shepardson, and Abell (1991) using a scheme that classifies the inquiry level of activities as confirmation, structured-inquiry, guided-inquiry, and open-inquiry. The Merrill Focus series and Principle series and the Scott, Foresman series were selected for analysis because they accounted for 44 percent of the science textbooks used at the junior high level. The findings of this study indicate that activities in both textbooks and supplemental activity guides and across all disciplines are restricted to the confirmation and structured-inquiry levels, with confirmation level activities dominating. No guided-inquiry or open-inquiry activities were found in any of the textbooks or supplemental activity guides. The authors concluded that commercially published activities are unlikely to contribute to open-inquiry science instruction because of their emphasis on confirmation and structured-level inquiry. Thus, the use of commercially published materials will do little or nothing to contribute to the development of problem-solving skills.

The following three reports are all based on meta-analyses of other studies and include findings related to other topics as well as problem solving. Studies of the effects of "new" (inquiry-oriented) science curricula on student performance were synthesized by Shymansky, Hedges, and Woodworth (1990) in a reanalysis of an earlier meta-analysis (Shymansky, Kyle, & Alport, 1983). Six criterion clusters were used as dependent variables in this meta-analysis: science achievement, student perceptions, process skills, problem solving, related skills (reading, math computation, writing), and other performance areas (involving mostly studies of Piagetian task performance). "The impact of the new curriculum in terms of an effect-size estimate is the difference between the mean criterion scores for the new and traditional curricula expressed in standard deviation units" (p. 132). For example, an effect size of 1.0 indicates that a criterion score that would be at the 50th percentile under the new curriculum would have been at the 84th percentile under the traditional curriculum; in effect, the new curriculum raises "C" pupils almost to "A" status (p. 133). The analysis showed that student performance across all curricula at the junior high school level was significantly affected ($p < .05$) for composite (Mean ES = .33), science achievement (Mean ES = .39), process skills (Mean ES = .39), and problem solving (Mean ES = .23). The results of this reanalysis generally support the conclusions of the earlier study:

the new science curricula of the 1960s and 1970s were more effective than the traditional textbook programs of the time.

Science curriculum effects in the secondary school were examined by Weinstein, Boulanger, and Walberg (1980) in a meta-analysis of 33 studies involving more than 19,000 students in the United States, Great Britain, and Israel. The studies involved 13 different curricula, 8 at the senior high school level and 5 at the junior high school level. Outcomes considered were (1) conceptual learning, (2) inquiry skills, (3) attitudinal development, (4) laboratory performance, and (5) concrete skills. Weinstein et al. (1980) reported a ratio of approximately 4:1 in favor of outcomes related to the use of innovative curricula and concluded that:

The post-Sputnik (1958) curricula produced beneficial effects in science learning that extended across science subjects in secondary schools, types of students, various types of cognitive and affective outcomes, and experimental rigor of the research. . . . The present analysis shows a moderate 12 point percentile advantage (effect size = 0.31) on all learning measures of average student performance in the innovative courses. (pp. 518–519)

Bredderman (1983) synthesized the research on the effectiveness of three activity-based elementary school science curricula (ESS, SAPA, and SCIS). Outcomes were measured in 57 studies, including more than 900 classrooms. "The mean effect size for all outcome areas was 0.35. This indicates about a 14 percentile improvement for the average student as a result of being in the activity-based program" (p. 504). Gains for the activity-based curricula were found for science content, science processes tests, and affective outcomes. Gains were also reported, on the average, in creativity, intelligence, language, and mathematics. Disadvantaged students derived greater benefits than other students. The effects of the particular programs reflected their relative curricular emphases (p. 499). Bredderman concluded that:

The accumulating evidence on the science curriculum reform efforts of the past two or three decades consistently suggests that the more activity-process-based approaches to teaching science result in gains over traditional methods in a wide range of student outcomes at all grade levels. (p. 513)

Summary

In most, but not all, cases, using inquiry-oriented curricula resulted in significant gains in problem-solving skills, as well as gains in achievement and in attitudes toward science. These gains vary, however, from one curriculum to another and from one grade level to another. Yet, when we consider the total picture, we find convincing evidence that the curriculum does make a difference. In short, using a curriculum designed to promote an inquiry approach will result in enhanced problem-solving abilities, as well as gains in other outcomes.

REFERENCES

Bady, R. (1977, March). *Logical reasoning abilities in male high school science students.* Paper presented at the 50th annual meeting of the National Association for Research in Science Teaching, Cincinnati. (ERIC Document Reproduction Service No. ED 138 444)

Berger, C. F. (1982). Attainment of skill using science processes: I. Instrumentation, metholodology and analysis. *Journal of Research in Science Teaching, 19*(3), 249–260.

Berger, C. F., & Pintrich, P. R. (1986). Attainment of skill in using science processes. II: Grade and task effects. *Journal of Research in Science Teaching, 23*(8), 739–748.

Bitner, B. L. (1990, April). *Thinking processes model: Effect on longitudinal reasoning abilities of students in grade six through twelve.* Paper presented at the 63rd annual meeting of the National Association for Research in Science Teaching, Atlanta. (ERIC Document Reproduction Service No. ED 322 009)

Bitner-Corvin, B. L. (1987, April). *The GALT: A measure of logical thinking ability of 7th through 12th grade students.* Paper presented at the 60th annual meeting of the National Association for Research in Science Teaching, Washington, DC. (ERIC Document Reproduction Service No. ED 293 717)

Bitner-Corvin, B. L. (1988, April). *Logical and critical thinking abilities of sixth through twelfth grade students and formal reasoning modes as predictors of critical thinking abilities and academic achievement.* Paper presented at the 61st annual meeting of the National Association for Research in Science Teaching, Lake of the Ozarks, MO. (ERIC Document Reproduction Service No. ED 293 715)

Bitner-Corvin, B. L. (1989, March). *Developmental patterns in logical reasoning of students in grades six through ten: Increments and plateaus.* Paper presented at the 62nd annual meeting of the National Association for Research in Science Teaching, San Francisco. (ERIC Document Reproduction Service No. ED 306 092)

Bowyer, J., Chen, B., & Thier, H. D. (1976, September). *A free-choice environment: learning without instruction. Advancing education through science-oriented programs.* Berkeley: Lawrence Hall of Science, University of California, (ERIC Document Reproduction Service No. ED 182 166)

Bredderman, T. (1983). Effects of activity-based elementary science on student outcomes: A quantitative synthesis. *Review of Educational Research, 53*(4), 499–518.

Breit, F. D., & Bullock, J. T. (1974, April). *The effectiveness of science—A process approach in the development of problem-solving skills in fifth and sixth grade students.* Paper presented at the 47th annual meeting of the National Association for Research in Science Teaching, Chicago.

Brooks, E. T. (1982). The effects of mastery instruction on the learning and retention of science process skills (Indiana University, 1982). *Dissertation Abstracts International, 43*(4), 1103A.

Bullock, J. T. (1973). A comparison of the relative effectiveness of three types of elementary school science curricula in the development of problem-solving skills (University of Florida, 1972). *Dissertation Abstracts International, 34*, 185A.

Burns, J. C., Okey, J. R., & Wise, K. C. (1985). Development of an integrated process skill test: TIPS II. *Journal of Research in Science Teaching, 22*(2), 169–177.

Butts, D. P. (1964). The evaluation of problem solving in science. *Journal of Research in Science Teaching, 2*(2), 116–122.

Butts, D. P., & Jones, H. L. (1966). Inquiry training and problem solving in elementary school children. *Journal of Research in Science Teaching, 4*, 21–27.

Butzow, J. W., & Sewell, L. E. (1971, March). *The process learning components of introductory physical science: A pilot study.* Paper presented at the 44th annual meeting of the National Association for Research in Science Teaching, Silver Spring, MD.

Capie, W., Newton, R., & Tobin, K. G. (1981, April). *Patterns of reasoning: controlling variables.* Paper presented at the 54th annual meeting of the National Association for Research in Science Teaching, Grossinger's in the Catskills, NY.

Champagne, A. B., & Klopfer, L. E. (1977). A sixty year perspective on three issues in science education: I. Whose ideas are dominant? II. Representation of women. III. Reflective thinking and problem solving. *Science Education, 61*(4), 431–452.

Champagne, A. B., & Klopfer, L. E. (1981a). Problem solving as outcome and method in science teaching: Insights from 60 years of experience. *School Science and Mathematics, 81*, 3–8.

Champagne, A. B., & Klopfer, L. E. (1981b). Structuring process skills and the solution of verbal problems involving science concepts. *Science Education, 65*(5), 493–511.

Chiappetta, E. L., & Russell, J. M. (1982). The relationship among logical thinking, problem solving instruction, and knowledge and application of earth science subject matter. *Science Education, 66*, 85–93.

Coble, C. R. (1986, March). *A cooperative study of science attitudes and involvement in science activities of U.S. and Japanese middle grades students.* Paper presented at the 59th annual meeting of the National Association for Research in Science Teaching, San Francisco.

Cox, D. A. H. (1980). Early adolescent use of selected problem-solving skills using microcomputers (University of Michigan, 1980). *Dissertation Abstracts International, 41*(9), 3855A.

Cronin, L. L., & Padilla, M. J. (1986, March). *The development of a middle grades integrated science process skills test.* Paper presented at the 59th annual meeting of the National Association for Research in Science Teaching, San Francisco.

Cronin, L. L., Twiest, M. M., & Padilla, M. J. (1985, April). *The development of a test of basic process skills in science.* Paper presented at the 58th annual meeting of the National Association for Research in Science Teaching, French Lick Springs, IN.

Curbelo, J. (1984). The effects of teaching problem-solving on students' achievement in science and mathematics. In J. J. Gallagher & George Dawson (Eds.), *Science education and cultural environments in the Americas.* (Report of the Inter-American Seminar on Science Education. Panama City, Panama.) Washington, DC: National Science Teachers Association.

Davis, M. (1979). The effectiveness of a guided-inquiry discovery approach in an elementary school science curriculum (University of Southern California, 1978). *Dissertation Abstracts International, 39*(7), 4164A.

Dillashaw, F. G., & Okey, J. R. (1980). Test of the integrated science process skills for secondary science students. *Science Education, 64*(5), 601–608.

Dozier, J. L., Jr. (1986). Relationships between objective measures of logical reasoning abilities and science achievement of students in a nonpublic junior high school in South Carolina (University of South Carolina, 1985). *Dissertation Abstracts International, 46*(10), 2986A.

Dunlop, D. L., & Fazio, F. (1975, March). *A study of abstract preferences in problem solving tasks and their relationship to abstract ability and formal thought.* Paper presented at the 48th annual meeting of the National Association for Research in Science Teaching, Los Angeles.

Egolf, K. L. (1979). The effects of two modes of instruction on students' abilities to solve quantitative word problems in science (University of Maryland, 1978). *Dissertation Abstracts International, 40*(2), 778A.

Farrell, E. (1988). How teaching proportionality affects transfer of learning: Science and math teachers need each other. *School Science and Mathematics*, *88*(8), 688–695.

Foster, G. W. (1982, April). *Creativity and the group problem solving process*. Paper presented at the 55th annual meeting of the National Association for Research in Science Teaching, Lake Geneva, WI.

Fraser, B. J. (1979). *Test of enquiry skills handbook*. Hawthorne, Victoria: Australian Council for Educational Research.

Friedler, Y., Nachmias, R., & Songer, N. B. (1989). Teaching scientific reasoning skills: A case study of a microcomputer-based curriculum. *School Science and Mathematics*, *89*, 58–67.

Friot, F. E. (1971). The relationship between an inquiry teaching approach and intellectual development (University of Oklahoma, 1970). *Dissertation Abstracts International*, *31*(11), 5872A.

Fyffe, D. W. (1972). The development of test items for the integrated science processes: Formulating hypotheses and defining operationally (Michigan State University, 1971). *Dissertation Abstracts International*, *32*(12), 6823A.

Gagné, R. M. (1965). Psychological issues in *Science—a process approach*. In *The psychological basis of science—a process approach*. Washington, DC: Commission on Science Education, American Association for the Advancement of Science.

Gagné, R. M. (1970). *The conditions of learning* (2nd ed.). New York: Holt, Rinehart & Winston.

Gagné, R. M. (1977). *The conditions of learning* (3rd ed.). New York: Holt, Rinehart & Winston.

Glasson, G. E. (1989). The effects of hands-on and teacher demonstration laboratory methods on science achievement in relation to reasoning ability and prior knowledge. *Journal of Research in Science Teaching*, *26*(2), 121–131.

Gudaitis, D. J. (1971). The effects of two seventh grade science programs, *Interaction of man and the biosphere* and *Science is explaining* on student attitude, science processes, and critical thinking (New York University, 1971). *Dissertation Abstracts International*, *32*(3), 1259A.

Helgeson, S. L. (1992). *Problem solving research in middle/junior high school science education*. The Ohio State University, Columbus: ERIC Clearinghouse for Science, Mathematics, and Environmental Education.

Helgeson, S. L., Blosser, P. E., & Howe, R. W. (1977). *The status of precollege science, mathematics, and social science education: 1955–1975* (Vol. 1, Science Education). The Ohio State University, Columbus: Center for Science and Mathematics Education. (ERIC Document Reproduction Service No. ED 153 876)

Heller, P. M., Ahlgren, A., Post, T., Behr, M., & Lesh, R. (1989). Proportional reasoning: The effect of two context variables, rate type, and problem setting. *Journal of Research in Science Teaching*, *26*(3), 205–220.

Hill, A. E. (1982). A comparative study of the effects of a human sciences program module and traditional science classes on pupils' logical thinking skills and attitudes toward their science course (University of Colorado at Boulder, 1981). *Dissertation Abstracts International*, *42*(8), 3533A.

Jacknicke, K. G., & Pearson, D. A. (1979, March). *The influence of the reflective/impulsive dimension on problem-solving skills in elementary school science*. Paper presented at the 52nd annual meeting of the National Association for Research in Science Teaching, Atlanta.

Jones, H. L. (1966). The development of a test of scientific inquiry, using the tab format, and an analysis of its relationship to selected student behaviors and abilities (University of Texas, 1966). *Dissertation Abstracts International*, *27*(2), 415A.

Jones, W. W. (1973). An investigation of the effect of acknowledging successful autonomous discovery by seventh grade students exposed to the inquiry development program (University of Northern Colorado, 1972). *Dissertation Abstracts International*, *33*(7), 3425A.

Jordan, E. A. (1987). The effect of teaching learning-strategies on achievement in ninth grade (University of British Columbia, 1986). *Dissertation Abstracts International*, *48*, 96A.

Lawsiripaiboon, P. (1983). The effects of a problem solving strategy on ninth grade students' ability to apply and analyze physical science subject matter (University of Houston, 1983). *Dissertation Abstracts International*, *44*(5), 1409A.

Lawson, A. E. (1978). The development and validation of a classroom test of formal reasoning. *Journal of Research in Science Teaching*, *15*, 11–24.

Lawson, A. E., & Wollman, W. T. (1977). Cognitive level, cognitive style, and value judgment. *Science Education*, *61*(3), 397–407.

Linn, M. C., Clement, C., Pulos, S., & Sullivan, P. (1989). Scientific reasoning during adolescence: The influence of instruction in science knowledge and reasoning strategies. *Journal of Research in Science Teaching*, *26*(2), 171–187.

Linn, M. C., & Levine, D. I. (1976). *Adolescent reasoning: The development of ability to control variables*. Berkeley: Lawrence Hall of Science, University of California. (ERIC Document Reproduction Service No. ED 182 164)

Linn, M. C., & Levine, D. I. (1977). Scientific reasoning ability in adolescence: Theoretical viewpoints and educational implications. *Journal of Research in Science Teaching*, *14*(4), 371–384.

Ludeman, R. R. (1975, March). *Development of the science processes test (TSPT)*. Paper presented at the 48th annual meeting of the National Association for Research in Science Teaching, Los Angeles. (ERIC Document Reproduction Service No. ED 108 898)

Mandell, A. (1980). Problem-solving strategies of sixth-grade students who are superior problem solvers. *Science Education*, *64*(2), 203–211.

Mattheis, F. E., Coble, C. R., & Spooner, W. E. (1986, March). *A study of the logical thinking skills of junior high school students in North Carolina and Japan*. Paper presented at the 59th annual meeting of the National Association for Research in Science Teaching, San Francisco.

Mattheis, F. E., & Nakayama, G. (1988, September). *Development of the performance of process skills (POPS) test for middle grade students*. Greenville, NC: East Carolina University. (ERIC Document Reproduction Service No. ED 305 252)

Mayer, V. J., & Richmond, J. M. (1982). An overview of assessment instruments in science. *Science Education*, *66*, 49–66.

McKee, D. J. (1978, March). *A comparative study of problem-solving ability and confidence for sixth grade science students exposed to two contrasting strategies*. Paper presented at the 51st annual meeting of the National Association for Research in Science Teaching, Ontario.

McKenzie, D. L., & Padilla, M. (1981, April). *Patterns of reasoning: Correlational thinking*. Paper presented at the 54th annual meeting of the National Association for Research in Science Teaching, Grossinger's in the Catskills, NY. (ERIC Document Reproduction Service No. ED 201 487)

McKenzie, D. L., & Padilla, M. J. (1984, April). *Effects of laboratory activities and written simulations on the acquisition of graphing skills by eighth grade students*. Paper presented at the 57th annual meeting of the National Association for Research in Science Teaching, New Orleans. (ERIC Document Reproduction Service No. ED 244 780)

McKenzie, D. L., & Padilla, M. J. (1986). The construction and validation of the test of graphing in science (TOGS). *Journal of Research in Science Teaching*, *23*(7), 571–579.

Moliter, L. L., & George, K. D. (1976). Development of a test of science processes skill. *Journal of Research in Science Teaching*, *13*(5), 405–412.

Newton, R., Capie, W., & Tobin, K. G. (1981, April). *Patterns of reason-

ing: Proportional reasoning. Paper presented at the 54th annual meeting of the National Association for Research in Science Teaching, Grossinger's in the Catskills, NY. (ERIC Document Reproduction Service No. ED 202 733)

Norton, R. E. (1971). A developmental study in assessing children's ability to solve problems in science (University of Texas at Austin, 1971). *Dissertation Abstracts International, 33,* 204A.

Novak, J. D., Gowin, B., & Johansen, G. T. (1983). The use of concept mapping and knowledge Vee mapping with junior high school students. *Science Education, 67*(5), 625–645.

Padilla, M. J., Okey, J. R., & Dillashaw, F. G. (1983). The relationship between science process skill and formal thinking abilities. *Journal of Research in Science Teaching, 20*(3), 239–246.

Padilla, M. J., Okey, J. R., & Garrard, K. (1984). The effects of instruction on integrated science process skill achievement. *Journal of Research in Science Teaching, 21*(3), 277–287.

Pirkle, S. F., & Correll, J. (1990, April). *Computer technology: The dynamic problem presentation interfaced with three instructional modes*. Paper presented at the 63rd annual meeting of the National Association for Research in Science Teaching, Atlanta.

Pirkle, S. F., & Pallrand, G. J. (1988, April). *Knowledge representation about projectile motion in junior high school students*. Paper presented at the 61st annual meeting of the National Association for Research in Science Teaching, Lake of the Ozarks, MO. (ERIC Document Reproduction Service ED 294 739)

Pizzini, E. L., Shepardson, D. P., & Abell, S. K. (1991). The inquiry level of junior high activities: Implications to science teaching. *Journal of Research in Science Teaching, 28*(2), 111–121.

Pouler, C. A. (1976). *The effect of intensive instruction in hypothesis generation upon the quantity and quality of hypotheses and the quantity and diversity of information search questions contributed by ninth grade students* (Ph. D. dissertation, University of Maryland). (ERIC Document Reproduction Service No. ED 128 225)

Pouler, C. A., & Wright, E. (1977, March). *The effect of intensive instruction in hypothesis generation upon hypothesis forming and questioning behaviors of ninth grade students*. Paper presented at the 50th annual meeting of the National Association for Research in Science Teaching, Cincinnati. (ERIC Document Reproduction Service No. ED 135 661)

Quinn, M. E., & George, K. D. (1975). Teaching hypothesis formation. *Science Education, 59*(3) 289–296.

Quinn, M. E., & Kessler, C. (1980, April). *Science education and bilingualism*. Paper presented at the 53rd annual meeting of the National Association for Research in Science Teaching, Boston. (ERIC Document Reproduction Service No. ED 193 024)

Rakow, S. J. (1985). Excellence in school science. *School Science and Mathematics, 85*(8), 631–634.

Reynolds, K. E., Pitoti, N. W., Rakow, S. J., Thompson, T., & Wohl, S. (1984). *Excellence in middle/junior high school science programs*. Washington, DC: National Science Teachers Association.

Roadrangka, V., Yeany, R. H., & Padilla, M. J. (1983, April). *The construction of a group assessment of logical thinking (GALT)*. Paper presented at the 56th annual meeting of the National Association for Research in Science Teaching, Dallas.

Robison, R. W. (1974). The development of items which assess the processes of controlling variables and interpreting data (Michigan State University). *Dissertation Abstracts International, 35*(3), 1522A.

Ronning, R., & McCurdy, D. W. (1982). The role of instruction in the development of problem solving skills in science. In R. E. Yager (Ed.), *What research says to the science teacher* (Vol. 4, pp. 31–41). Washington, DC: National Science Teachers Association.

Ronning, R. R., McCurdy, D., & Ballinger, R. (1984). Individual differences: A third component in problem-solving instruction. *Journal of Research in Science Teaching, 21,* 71–82.

Ross, J. A., & Maynes, F. J. (1983a). Development of a test of experimental problem-solving skills. *Journal of Research in Science Teaching, 20,* 63–75.

Ross, J. A., & Maynes, F. J. (1983b). Experimental problem solving: An instructional improvement field experiment. *Journal of Research in Science Teaching, 20*(6), 543–556.

Rudnitsky, A. N., & Hunt, C. R. (1986). Children's strategies for discovering cause-effect relationships. *Journal of Research in Science Teaching, 23*(5), 451–464.

Russell, J. M. (1979). The effects of problem solving on junior high school students' ability to apply and analyze earth science subject matter (University of Houston). *Dissertation Abstracts International, 40*(3), 1386A.

Russell, J. M., & Chiappetta, E. L. (1980, April). *The relationship between level of cognitive development and achievement at four levels of Bloom's Taxonomy for junior high school students*. Paper presented at the 53rd annual meeting of the National Association for Research in Science Teaching, Boston.

Russell, J. M., & Chiappetta, E. L. (1981). The effects of a problem solving strategy on the achievement of earth science students. *Journal of Research in Science Teaching, 18*(4), 295–301.

Saunders, W. L., & Jesunathadas, J. (1988). The effect of task content upon proportional reasoning. *Journal of Research in Science Teaching, 25,* 59–67.

Scott, N. C., Jr. (1973). Cognitive style and inquiry strategy: A five-year study. *Journal of Research in Science Teaching, 10*(4), 323–330.

Schlenker, G. C. (1971). The effects of an inquiry development program on elementary school children's science learnings (New York University, 1970). *Dissertation Abstracts International, 32,* 104A.

Schmiess, E. G. (1971). An investigative approach to elementary school science teaching (University of North Dakota, 1970). *Dissertation Abstracts International, 31*(9), 4391A.

Shann, M. H. (1975). *An evaluation of unified science and mathematics for elementary schools (USMES) during the 1973–74 school year*. Boston: Boston University. (ERIC Document Reproduction Service No. ED 135 861)

Shann, M. H. (1976). *Measuring problem solving skills and processes in elementary school children*. Boston: Boston University, School of Education. (ERIC Document Reproduction Service No. ED 135 807)

Shann, M. H., Reali, N. C., Bender, H., Aiello, T., & Hench, L. (1975). *Student effects of an interdisciplinary curriculum for real problem solving: The 1974–75 USMES evaluation. Final report*. Boston: Boston University. (ERIC Document Reproduction Service No. ED 135 864)

Shapiro, B. (1973). *The notebook problem. Report on observations of problem solving activity in USMES and control classrooms*. Newton, MA: Education Development Center. (ERIC Document Reproduction Service No. ED 116 917)

Shaw, T. J. (1978). The effects of problem solving training in science upon utilization of problem solving skills in science and social studies (Oklahoma State University, 1977). *Dissertation Abstracts International, 38*(9), 5227A.

Shaw, T. J. (1982). *ORES—Objective Referenced Evaluation in Science*. Manhattan: Kansas State University. (ERIC Document Reproduction Service No. ED 216 876)

Shaw T. J. (1983). The effect of a process-oriented science curriculum upon problem-solving ability. *Science Education, 67*(5), 615–623.

Shymansky, J. A., Hedges, L. V., & Woodworth, G. (1990). A reassessment of the effects of inquiry-based science curricula of the 60's on student performance. *Journal of Research in Science Teaching, 27*(2), 127–144.

Shymansky, J. A., Kyle, W. C., & Alport, J. M. (1983). The effects of new science curricula on student performance. *Journal of Research in Science Teaching, 20*(5), 387–404.

Simon, H. S. (1981). The psychology of scientific problem solving. In R. D. Tweney, M. E. Doherty, & C. R. Mynalt (Eds.), *On scientific thinking* (pp. 41–54). New York: Columbia University Press.

Spooner, W. E. (1986, March). *An assessment of integrated process skills of junior high school students*. Paper presented at the 59th annual meeting of the National Association for Research in Science Teaching, San Francisco.

Squires, F. H. (1977). An analysis of sex differences and cognitive styles on science problem solving situations (The Ohio State University). *Dissertation Abstracts International, 38*(5), 2688A.

Stallings, E. S. (1973). A comparison of the inquiry behavior of ISCS and non-ISCS science students as measured by the TAB science test (Florida State University). *Dissertation Abstracts International, 34*(3), 1149A.

Stallings, E. S., & Snyder, W. R. (1977). The comparison of the inquiry behavior of ISCS and non-ISCS science students as measured by the TAB science test. *Journal of Research in Science Teaching, 14*, 39–44.

Staver, J. R. (1984). Research on formal reasoning patterns in science education: Some messages for science teachers. *School Science and Mathematics, 84*(7), 573–589.

Staver, J. R. (1986). The effects of problem format, number of independent variables, and their interaction on student performance on a control of variables reasoning problem. *Journal of Research in Science Teaching, 23*(6), 533–542.

Stuessy, C. L. (1988). Path analysis: A model for the development of scientific reasoning abilities in adolescents. *Journal of Research in Science Teaching, 26*, 41–53.

Tannenbaum, R. S. (1969). The development of *Test of science processes* (Columbia University, 1968). *Dissertation Abstracts International, 29*(7), 2159A.

Thomas, B. S. (1968). An analysis of the effects of instructional methods upon selected outcomes of instruction in an interdisciplinary science unit (University of Iowa). *Dissertation Abstracts International, 29*(6), 1830A.

Tobin, K. G., & Capie, W. (1980, April). *The test of logical thinking: Development and applications*. Paper presented at the 53rd annual meeting of the National Association for Research in Science Teaching, Boston. (ERIC Document Reproduction Service No. ED 188 891)

Tobin, K. G., & Capie, W. (1981). The development and validation of a group test of logical thinking. *Educational and Psychological Measurement, 41*(2), 413–423.

Tobin, K. G., Capie, W., & Newton, R. (1981, April). *Patterns of formal reasoning: Probabilistic reasoning*. Paper presented at the 54th annual meeting of the National Association for Research in Science Teaching, Grossinger's in the Catskills, NY.

Twiest, M. M., & Twiest, M. G. (1989, March). *Modification and validation of a test of basic process skills using different methods of test administration*. Paper presented at the 62nd annual meeting of the National Association for Research in Science Teaching, San Francisco.

Weinstein, T., Boulanger, F. D., & Walberg, H. J. (1980). Science curriculum effects in high school: A quantitative synthesis. *Journal of Research in Science Teaching, 19*(6), 511–522.

White, E. P. (1978). Problem solving: Its history as a focus in curriculum development. *School Science and Mathematics, 78*(3), 183–188.

Wilson, J. T. (1973, March). *An investigation into the effects of generating hunches upon subsequent search activities in problem solving situations*. Paper presented at the 46th annual meeting of the National Association for Research in Science Teaching, Detroit. (ERIC Document Reproduction Service No. ED 079 064)

Wright, E. (1978). The influence of intensive instruction upon the open exploration behavior of ninth grade students. *Journal of Research in Science Teaching, 15*(6), 535–541.

Yeany, R. H., Yap, K., & Padilla, M. J. (1986). Analyzing hierarchical relationships among modes of cognitive reasoning and integrated science process skills. *Journal of Research in Science Teaching, 23*(4), 277–291.

·9·

RESEARCH ON PROBLEM SOLVING:
EARTH SCIENCE

Charles R. Ault, Jr.
LEWIS & CLARK COLLEGE

THE IMPORTANCE OF EDUCATING PROBLEM SOLVERS IN EARTH SCIENCE

Our children must understand ideas from earth science and the ideas earth science shares with other scientific disciplines. Without that understanding, we cannot hope to increase scientific literacy or to safeguard the Earth. Nothing less than the survival of our children and our planet is at stake today. (National Center for Earth Science Education [NCESE] and American Geological Institute [AGI], 1991a, p. 1)

Despite the evident hyperbole, this quotation expresses two rationales for the importance of the earth science curriculum: the interdisciplinary nature of scientific literacy and the essential role of this understanding in promoting the habitability of Earth. Cursory reflection suggests that earth science is simply a domain in which problem solvers apply knowledge from chemistry, physics, and biology as well as procedures from mathematics. The issue is not so simple, for Earth is perhaps best characterized as a complex system in its own right, an object of study within an emerging discipline of earth system science.

Scientific research continues to yield fundamental new knowledge about Earth. Studies of the continents, oceans, atmosphere, biosphere, and ice cover over the past 30 years have revealed that these are components of a far more dynamic and complex world than could have been imagined a few generations ago. These investigations have also delineated with increasing clarity the complex interactions among Earth's components and the profound effects of these interactions on earth history and evolution. We can now, for example, incorporate the global effects of atmospheric wind stress into models of oceanic circulation, study volcanic activity as a link between convection in Earth's mantle and worldwide atmospheric properties, and trace the global carbon cycle through the many transformations of this vital element by terrestrial and ocean biota, atmospheric chemistry, and the weathering of Earth's solid surface and soils.

Our new knowledge is providing us with deeper insight into the Earth as a system. This insight has set the stage for a more complete and unified approach to its study, Earth System Science. (Earth System Sciences Committee and the National Aeronautics and Space Administration [NASA], 1986, p. 4)

THE NATURE OF EARTH SCIENCE PROBLEMS IN SCHOOL CURRICULA

Evaluation Studies of Project Curricula

The *Activity Sourcebook for Earth Science* (edited by Mayer, Champlin, Christman, & Krockover, 1980; produced by the Educational Resources Information Center's [ERIC] Science Mathematics and Environmental Education Clearinghouse [SMEAC] with funding from the National Institute of Education [NIE] in cooperation with the Office of Educational Research and Improvement [OERI], U.S. Department of Education) provides an excellent guide to what science educators consider appropriate problems for school earth science (see the review by Mayer, 1982). This volume contains selections from several curriculum projects created by national organizations in the 1960s and 1970s. Prominent among these are the Earth Science Curriculum Project (ESCP; supported by the National Science Founda-

The author thanks reviewers Victor J. Mayer (The Ohio State University) and Andy I. Verdon, Jr. (American Geological Institute).

tion [NSF] and sponsored by the American Geological Institute [AGI]) and the Crustal Evolution Education Project (CEEP; supported by NSF and sponsored by the National Association of Geology Teachers [NAGT]).

The most recent collection of such problems or exercises may be found in the AGI's *Earth Science Investigations* (Oosterman & Schmidt, 1990). The problems in this volume are in an important sense more "conservative" than those in the earlier ERIC publication. The AGI collection attempts to supplement the norms of instruction that are based on adoptions of commercial programs; the NAGT selections are very close, if not identical, to the problems framed in keeping with an attempt to reform traditional science teaching from an inquiry approach. In either case, these collections are the most widely accepted examples of what scientists and science teachers through their professional organizations have felt important to use in teaching earth science problem solving.

Five years after ESCP instructional materials entered secondary classrooms, Champlin (1970) reviewed the research evaluating the program. These studies included analyses of student knowledge in astronomy as well as geology. Champlin concluded that ESCP had made a "substantial contribution to secondary science education" and that its successful implementation depended on the "combination of the right curriculum with the appropriate teacher" (p. 31). Most of the studies reviewed pertained to student achievement. Several of these used the ESCP "comprehensive final" as a measure of achievement. Generally, Champlin reported studies to show comparable or favorable achievement in ESCP classes when matched with traditional ones. One study found that ESCP students were better prepared than others for introductory-level college geology (Paull, Larson, & VandenAvond, 1969) as measured by a test of factual knowledge, application of principles, and drawing inferences from data tables. None of the studies reviewed by Champlin examined student cognition or problem solving, although ESCP placed a strong emphasis on fieldwork. Instead, research at that time dealt primarily with instructional variables (authoritarian versus inquiry approach, for example) and teacher preparation. Project developers had assumed that emphasis on inquiry processes in the curriculum would promote problem solving.

More recently, in a similar format using achievement as the dependent variable (in terms of knowledge, application, explanation, and experimental techniques), Tunley (1986) found that high school geology compares favorably with other secondary science subjects in promoting student knowledge of basic science. He tested 42 geology students the year before and then immediately after completion of an elective geology course. A *t*-test of the mean pre and post scores on the Houghton Mifflin Test of Achievement and Proficiency in Science (TAPS) was significant at the .05 level (alpha). His paper argued that geology could fulfill "unified" science requirements because the gain in science knowledge on the TAPS was consistent with expectations for achievement in school science. Again, problem solving was not addressed.

The ESCP staff developed its project materials before plate-tectonic theory had radically transformed the field of geology. Six CEEP centers completed development of this project by the end of the 1970s to ensure that earth science instructors had a valid representation of the change in geological thinking that evolved so rapidly in less than 20 years. This project prompted another pulse of evaluative research (Mayer, 1979, 1984; Mayer & Stoever, 1978, 1979; Rojas & Mayer, 1979).

CEEP modules (64 in all) guide students through simulations of problems whose solutions proved critical to the construction and acceptance of plate-tectonic theory. Research scientists aided in the design of these modules. The best and most comprehensive summary of the evaluation is that of Mayer (1984). One objective of evaluation was to ascertain whether the modules were "as interesting" as "normal" classroom procedures. The other major objective was to ascertain whether "each module successfully taught the concepts that were outlined in its objectives" (p. 9). More than 20 teachers and 1000 students tested each module.

Demographic and IQ data on students in grades 8 through 11 were collected. An 11-item, multiple-choice Cognitive Level Test (adapted from the Australian Science Education Project by Tisher & Dale, cited in Mayer, 1984) was administered at the end of each CEEP unit. General achievement was assessed with a 40-item Concept Process Test (Disinger & Mayer, 1974).

The evaluation found little difference in attitudes toward regular class and crustal-evolution modules and documented achievement gains for all modules. The modules "successfully taught concepts related to plate tectonics in a way that was as interesting to students as the regularly used materials" (Mayer, 1984, p. 25). This summative evaluation had as primary aims to describe teachers and students and the achievement of cognitive objectives. It achieved these aims but did not contribute significantly to a theoretical understanding of student problem solving in the context of crustal-evolution problems.

Evaluators did use an activity checklist to monitor instructional practices and obtained background information on teachers. None of these procedures conformed to protocols for studying problem solving or cognition (i.e., think aloud, task-centered interviews, etc.) that have emerged in the past decade or so, partially in response to useful distinctions among procedural knowledge (knowing how to do X), declarative knowledge (knowing that X is the case or relation of importance), and structural knowledge, the kind of knowledge needed to coordinate the first two (Barba & Rubba, 1992b). Nor did evaluation of CEEP consider issues surrounding the relation between skills for problem solving in well-defined contexts and the acquisition of more general problem-solving heuristics. Summative evaluation of CEEP did at least consider achievement in problem solving from the perspective of applying science process skills as measured by the Concept Process Test.

Comment on Trends in the Assessment of Earth Science Problem Solving

Assessment of science learning in schools rarely emphasizes problem-solving performance in the context of highly integrated or interdisciplinary science teaching. Evaluation studies, perhaps the broadest attempts to examine student thinking in earth science, cannot be expected to illuminate problem-solving outcomes unless school evaluation of students places high value on problem-solving performance itself, not to mention problem solving in interdisciplinary contexts presupposed by the definition of an "earth" science. Revisions of curriculum under way at present may encourage change in assessment goals and procedures. Still, little is known about the nature of student problem solving in earth science, not to mention appropriate instruction (and curriculum packages) for promoting such proficiency. New tests are available or forthcoming: (1) the New York State Regents syllabus for earth science is being revised and will be accompanied by a new test rewritten each year (the test historically used five hands-on ESCP tasks now being incorporated into statewide performance assessment in science at the fourth-grade level) Agrusso, personal communication, July 20, 1991) and (2) the AGI/NSTA Earth Science Examination (Callister & Mayer, 1988). The task force that authored the AGI/NSTA test claimed that "The test was specifically designed to include questions that stress more than just facts and concepts. Questions assess students' problem-solving abilities, understanding of relationship between science and technology, and understanding of the nature of science. More than half of the questions require an understanding of an ability to apply concepts, many by asking for interpretations of maps, diagrams, and tables" (Callister & Mayer, 1988, p. 34).

The AGI/NSTA Test

The questions on the AGI/NSTA test are multiple choice in format. The choices are fixed and one is presumably correct, a structure fundamentally incongruous with problem solving in complex domains. The task force of educators and scientists who authored AGI's planning guide for earth science education has called for something more ambitious in the way of assessment:

Today, even when we claim we want students to learn to describe, explain, and predict, they are all too often asked on tests to fill in a blank in a sentence, match a word with a definition, label diagrams, or choose a correct answer from one of four or five alternatives. These types of tests usually can be mastered by students who have learned by rote unconnected facts, definitions, and principles. These types of tests provide little insight regarding how well integrated and truly functional a student's knowledge might be.

New methods of assessment that require students to present descriptions, explanations, predictions, decisions, problems solutions, and their reasons or justifications for their work must be developed. Essay writing, solving simulated, and, whenever possible, real scientific problems, solving laboratory-based problems, interpreting real data, developing predictions based on patterns of change, and writing descriptions and explanations should be included in all new test instruments. (NCESE & AGI, 1991b, p. 19)

The NCESE and AGI now distribute two publications intended to assist in improving precollege earth science education: *Earth Science Content Guidelines* (NCESE & AGI, 1991b) and *Earth Science Education for the 21st Century: A Planning Guide* (NCESE & AGI, 1991a). Educators and scientists prepared these guides through a series of conferences and with NSF funding (Mayer & Armstrong, 1990, reported on the conclusions from the first of these conferences). The *Content Guidelines* are organized according to "essential questions," an approach conducive to thinking of problem-solving activities in the curriculum.

Earth Systems Education Program

Somewhat in competition with, but ultimately complementing, the AGI publications is the Earth Systems Education program being developed at The Ohio State University. This program distributes its own newsletter titled *PLESE*, an acronym for the Program for Leadership in Earth Systems Education, and conducts staff development.

The point of departure for *PLESE* is the importance of the systems concept for modern problem solving about Earth. In *PLESE*, seven understandings are used to build this systems perspective: Earth is rare, beautiful, and unique; collectively and individually, human activities are having a serious impact on Earth; technology and science help us to utilize Earth and space; interacting subsystems of water, land, ice, air, and life comprise the earth system; systems have evolved over 4 billion years; Earth is a subsystem in a vast and ancient universe; and many careers are devoted to the study of Earth's origin, processes, and evolution. In summary, the time has never been better to undertake studies of problem solving in earth science education, for the curriculum has matured to incorporate research perspectives, environmental concerns, and systems thinking. "The time appears to be ripe for the first total restructuring of the science curriculum since the Committee of Ten established the current high-school sequence in the late 1800s" (Mayer et al., 1992, p. 72).

Regrettably, there is insufficient research into the nature of learning and problem solving in this seemingly intractable context. There are some notably interesting studies, however, on how students reason about Earth as a body in space (e.g., Nussbaum, 1979; Sneider & Pulos, 1983), infer geological relationships (e.g., Ault, 1982), and apply the hydrologic cycle (e.g., Stevens & Collins, 1980). In addition, educational theorists have examined earth science learning simply because it is a subject commonly taught to adolescent students. Their interest has been in instructional variables (e.g., Russell & Chiapetta, 1981), application of philosophy of science to curriculum structure (e.g., Finley, 1981, 1982), the disparity in meaning between children's earth science notions and scientific concepts (e.g., Happs, 1984), and the nature of cognitive processes (e.g., Champagne & Klopfer, 1981; Novak, Gowin, & Johansen, 1983). On an optimistic note, work on earth science problem solving within clearly established paradigms of problem solving in science education has commenced (Barba, 1990; Niedelman, 1990), with due consideration to questions about the role of domain-specific knowledge and the nature of transfer.

VISUAL REASONING

Significance of Visual Aptitude and Geometric Representation

Instructors often claim that students must develop visualization skill in order to solve earth science problems—using, for example, geometric representations of rock strata (cross-sectional or block diagrams) and contour maps to depict everything from topography to air masses. What makes visualization difficult is the need to scale a set of incomplete observations into a workable diagram where problem-solving conventions apply—for example, from weather station data into a regional weather map composed of isobars. As diverse as the earth sciences are, their problem domains are characterized by the concepts of scale, incomplete (or indirect) observation of complex systems, and conventions for visually representing the problem situation.

The attempt to connect processes that change the features of the earth on various time scales with patterns in those features found in the present is the fundamental characteristic of geological reasoning. Such reasoning encompasses a vast array of events, yet very often presupposes skill in visualizing "structures" or geometric patterns in the rock record of geologic processes. Thus, the study of structural geology is an essential component in the preparation of a geologist. Typically, a course in structural geology includes many problem-solving exercises requiring visualization skill.

In a preliminary effort to characterize how students of structural geology reason, Ault (1986b) focused on "thickness problems." (Thickness problems present the student with the task of inferring the true thickness of a rock unit from measurement of its expression at the surface of the earth.) A case-study approach led to think-aloud, problem-solving sessions with four students. Each student attempted to work through a self-generated hierarchy of thickness problems. Increase in the difficulty of thickness problems conformed to the degree and number of departures from perpendicular intersections among three planes and one line of interest: the bed surface, the slope surface, an imaginary horizontal surface (map plane), and the traverse line across the bed exposure. Poor grasp of the utility of problem-solving conventions (e.g., stereonet plotting), perhaps restricted by the implicit associations of curving contour lines with hills on topographic maps, and ambiguous comprehension of apparent versus true dips (meaning angles from the horizontal to a bearing on a surface) constrained structural geology students' performance on these problems, as suggested by a preliminary, qualitative analysis of the audio-recorded transcripts. Students might do better if they described a goal state diagram and labeled the initial problem drawing and each subproblem phase with as complete a set of verbal phrases as possible (e.g., "This line is the bearing of the line of intersection of the bed surface with the slope surface in the map view or horizontal plane").

Griffin (1983) posed a series of map-interpretation tasks to adult geography students, asking them (in a manner similar to the thickness problem study) to make a written report of their methods. Griffin's objective was to learn how users solved visual problems involving cartographic material. In this task, a "cartogram" radically altered space (voting districts) while keeping contiguous relationships among districts accurate. The cartogram showed each voting district greatly distorted, its area proportional to population. The task challenged subjects to identify corresponding units on the map and the cartogram. Success rates approached 80 percent (112 subjects; 1344

trial tasks) on some problems. Griffin conceptualized subject thinking as "procedures" for solving subproblems and "strategies" combining these procedures into solution paths. A visual reference system was needed to succeed with the task. Subjects had to invent some way to identify reference points on both displays, a problem-solving step not unlike those taken by geology students moving from one planar surface to another when projecting lines in thickness problems.

Components of the ESCP attempted to capture the sense of inquiry experienced when gathering data in the field for the purpose of constructing geological maps. Making or interpreting geological maps is fundamental to much of geological problem solving, and instructional materials emphasizing problem solving with maps are common. The *Journal of Geological Education* frequently publishes papers on instructional aspects of helping students construct and interpret geological maps as well as related exercises: thickness problems, "three-point" problems (i.e., determining the orientation of a plane in space), projection of surfaces from block diagrams, and topographic map construction (e.g., Bart, 1991). These instructor-authored "how-to" articles fill the void created by an absence of a research tradition explicating the relationship between geological reasoning and visuospatial aptitude.

Visual Frame of Reference: Reasoning About The Earth in Space

Perhaps no problem area better illustrates the demands placed on visualization skill than transforming how we directly perceive the cosmos—a big, rotating bowl above a horizontal and circular horizon—into an orbital model of motion accounting for the seasons, solar and lunar eclipses, phases of the moon or Venus, retrograde motion of the planets, stellar parallax, and so forth.

Most students believe that Earth orbits the sun. They seldom recognize patterns in the change of position of celestial objects on an hourly, daily, weekly, monthly, or annual basis. Even less often do they grasp how to interpret these changes in terms of an orbital model—one involving a movable observer on a spinning sphere (Earth) that circles another (the sun). Simultaneously, other spheres (the planets) orbit the central one, at different rates proportional to their distance from the center. Other objects (the stars) are so distant that their position appears unchanged on the scale of human lifetimes and the resolution of everyday observation. The belief in an orbital model and facility with language about it often belie the haunting presence of an intuitive understanding of a flat earth, even for adolescent students (Nussbaum, 1979).

Research on understanding both spherical-earth and gravity concepts began with Nussbaum and Novak's (1976) interview study of learning by second-graders. Interviewers probed children's notions of the path a falling object might take when dropped from different locations on the earth. Props included picture cards of the earth seen from space and globes. Even children who viewed the earth as spherical with people everywhere clung to a notion of "absolute down." According to this

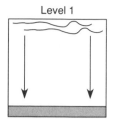 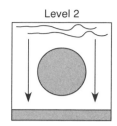 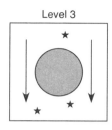 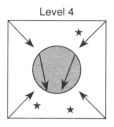 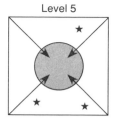

FIGURE 9.1.

Source: Figure redrawn from Sneider and Pulos' diagram (1983), which was published in *Science Education*, 67(2), p. 206; that figure was in turn based on one by J. Nussbaum printed in the same journal, Vol. 63(1), p. 83. Redrawn figure copyright © 1979, 1983, appears by permission of John Wiley & Sons.

notion, if objects could keep falling through the earth they would end up at the south pole, not the center of the earth.

Nussbaum (1979) reported that the intuitive notions of absolute down held by Israeli students persisted through grade 8. Klein (1982) found common levels of naive thinking about the earth in space across Mexican-American and Anglo-American cultural groups. Mali and Howe (1979) concluded from interviews with Nepalese children that cultural differences strongly influence performance on tasks designed to test understanding of Earth in space. However, in a comprehensive summary and rigorous replication of this research tradition, Sneider and Pulos (1983) reexamined the cross-cultural data, finding little support for the hypothesis that cultural differences affect earth concept development—differences within age groups were more substantial than those between cultural groups. The Sneider and Pulos work validated the basic Nussbaum scheme of levels for the earth notion. However, they distinguished more carefully between notions of Earth's shape and understanding the direction of gravitational acceleration (see Figure 9.1.).

Ault (1986a) critiqued the Sneider and Pulos paper and pointed out additional unanswered questions raised by this line of inquiry. For example, do individuals hold incompatible views from two levels simultaneously? In what way is gravitational up and down in particular related to comprehending changing frames of reference in general?

With a good mental model of orbital motion—an "earth concept" at level five in the Nussbaum scheme—students can attack questions such as: "Will a Brazilian see the same moon phase as a Canadian tonight? When does the sun shine on the south wall of a home in Sydney, Australia? How can shadows be used to prove which direction is north for someone in Ohio?" However, even on entering college, they are probably not ready to do so. Kelsey (1980) determined that a majority of students in a general, introductory astronomy course could not reason properly about projective space. Kelsey presented six problems to the students: (1) coordinating perspectives of mountains, (2) coordinating perspectives of stars in the sky, (3) recognizing distortions due to planar rotations and projection, (4) interpreting lunar phases from a person-centered frame of reference, (5) interpreting lunar phases from an Earth-centered model, and (6) interpreting lunar phases from a two-dimensional

Earth, moon, sun diagram. Too few students passed the sixth problem to include it as a category separate from number five.

Kelsey's work and the Nussbaum tradition suggest how tenaciously students may hold to intuitive notions gleaned from everyday perspectives to the detriment of reasoning about Earth in space. The questions generated by this research provided the impetus for Project STAR (Science Teaching through its Astronomical Roots) at the Harvard-Smithsonian Center for Astrophysics. Two notable productions from Project STAR are the provocative videotape, "A Private Universe," which features interviews about seasons and lunar phases with students at a Harvard graduation and local middle school and high school students, and *Where We Are in Space and Time*, a compendium of 21, secondary-level activities. Interviews in the film conform to a think-aloud problem-solving protocol; the exercises in the activity book call for imaginative construction of devices for indirect measurement.

TRADITIONS OF RESEARCH ABOUT STUDENT THINKING IN EARTH SCIENCE

Contrasting Student and Scientist Views of Descriptive Categories in Earth Science

Using data from the Learning in Science Project (Osborne & Freyberg, 1985), Happs (1982a, 1982b, 1982c, 1982d, 1984, 1985; Happs & Stead, 1989) reported exhaustively on student's views of earth science phenomena. He drew implications for teaching about rocks and minerals, glaciers, soils, and mountains from middle through secondary school levels.

From interviews and sorting tasks, Happs (1982a) concluded that children's approach to constructing categories for rock and mineral samples contrasted sharply with the approach likely for earth scientists. For example, students of ages 11 through 17 were asked to sort a collection of rocks and minerals according to common properties. In discussing their decisions, they rarely used the term "mineral," preferring to call all objects "rocks." "Stones" referred to small pieces of rock, a major subcategory. They used "crystal" and "pebble" frequently but in nonscientific ways. Weight and shape commonly emerged as critical attributes for identification. The students made little connection between minerals and the fabric of rocks. Older students jointly used everyday and scientific schemes for classifying specimens: "sedimentary, ordinary, exotic," for example.

Children in the Learning in Science Project described what they noticed about pictures of glaciated landforms (Happs, 1982b). Interviews probed children's ideas concerning the processes forming these landforms. They were unaware that glaciers tended to move, were bodies of ice, and produced major erosional features. Instead, they imagined glaciers as "always there" and made no association between fluctuating climate and glacial movement, if they had an awareness of movement at all. They made no link between U-shaped valleys, fjords, moraines, and the erosional work of glaciers. Some thought the valleys resulted from old volcanoes.

In reference to students' grasp of mountain-building processes, Happs (1982c) found that two-thirds of his sample of 37 had no awareness that a dormant volcano might erupt again. Eighty percent could not relate mountain building in New Zealand to plate tectonics. Mountain building—as well as glaciers—is especially important to New Zealanders, who live in an active, plate margin zone. Happs judged the absence of any appreciation of tectonic concepts among a majority of adolescents a disturbing situation.

With regard to soils, Happs (1982d, 1984) concluded that students had little sense of soil-formation processes or dynamic, environmental factors related to soil change with time. According to Happs, teachers should emphasize time-dependent properties of soils. Many adolescents in the Happs study held the view of simple cycling between soil and clay: soil changes to clay, clay to rock, rock to soil.

Common to all of the Happs examples is the basically good thinking of children according to the classification schemes for earth science phenomena that suit their purposes. Difficulties arise when the meanings their teachers ascribe to geological terms differ from those held by children—and the children have yet to recognize the alternative purposes served by the scientific meaning. An example is given by Happs (1985). A teacher may expect a child to infer that marble is a rock from the observation that it is a solid, natural material. The child, however, observes that the marble sample is highly polished—a change induced by humans. Since the marble is not an example of "things untouched by people"—the everyday sense of "natural" to Happs' subjects—it is not natural and cannot be a rock. The child's logic is impeccable, although the conclusion is at odds with the teacher's intentions. The child and the teacher are simply not "talking science" (Lemke, 1990) according to the same pattern of meaning. Researchers label this mismatch a "misconception" (cf. Philips, 1991).

Cohen (1968) used a "mudpile mountain" to probe upper elementary grade children's grasp of fluvial processes in a microgeology context. Water running down the dirt pile created gullies. When blocked, small lakes formed. Cohen asked children to predict the future of the lake and justify their predictions in terms of processes making the lake larger or smaller, deeper or shallower. In addition, Cohen asked how runoff water might sort dirt at the bottom of the gully. Before instruction, predictions varied widely. Afterwards, nearly all grasped a simple principle of transport: water carries smaller particles farther from the slope than larger ones.

Cohen's work, which antedated the Happs study by more than a decade, underscores the value of listening to how children explain in their own terms events in their everyday environment. Happs has reminded us to presume very little about preinstructional, commonly shared earth science knowledge grasped by adolescent students. Research with young children suggests that logical reasoning improves when the question makes intentional sense to the child (Donaldson, 1982). That is, the reason for asking the question and the question itself fit purposes that have meaning to the child. There is no reason to discount the importance of either the Happs view or the Donaldson conclusion when teaching students of any age. Without shared purpose (or without cognizance of another's purpose), shared meaning of concepts is not likely.

Curriculum materials should illustrate abstract ideas with everyday objects. When reasoning about temporal order and duration in terms of everyday experience, for instance, children's logic parallels that of geologists attempting to solve problems about the relative ages of rocks. Children's concepts about time present no barrier to understanding the geologic past; however, their lack of familiarity with and knowledge about geologic objects and events leads to confused, contradictory thinking (Ault, 1982). When provided with clear, plastic tubes containing layers of trash, sampled from different positions within an imaginary garbage pile, children from grades 4 and 6 had no difficulty inferring an overall sequence of layers from top to bottom. They simply matched or correlated portions of one tube with another, allowing for possible thinning or thickening of a layer. Some even surmised that grass layers repeated on a weekly basis, reflecting weekend yardwork. In this context, children easily isolated age cues: position indicated how long ago, relatively speaking, a type of garbage was disposed of. Perhaps, as some suggested, the stuff was "already old"; torn pieces of last month's newspaper, for example. When presented with diagrams of rock columns from local roadsides and park gorges, the children were unable to reason consistently about relative ages of rocks. They used cues such as crumbliness, color, compaction, and hardness to evaluate the possible age of a rock independent of its position in the column. Consider the analogous confusion for the geologist: very ancient igneous rock redeposited as a conglomerate above younger (older?) rock. Students who solve problems involving reasoning about relative age must attend carefully to precise meanings of "age." They must sort the relative time age cues from aging process cues (Ault, 1982).

Understanding Models of Earth Science Phenomena

Mayer (1989) has summarized research on the efficacy for understanding in science of pictorial or diagrammatic models. He found that models "improve recall of conceptual information, decrease verbatim retention, and increase creative solutions on transfer problems. . . . Models can help lower aptitude learners to think systematically about the scientific material they study" (p. 43). For example, students who studied a conceptual model version of the nitrogen cycle solved 42 percent more transfer problems than students who read a 670-word passage on the same topic (Mayer, Dyck, & Cook, cited in Mayer, 1989).

Perhaps the most interesting and everyday problem for many earth science students is weather prediction. They can match their skills against the evening television weather forecaster. Weather prediction involves data gathering and interpretation on several different scales. First, there is local observation of temperature, humidity, air pressure, cloud cover (and type), visibility, wind speed and direction (at the surface and above), dew point temperature, and thermal inversions. Next, there is the need to collect similar information from many other localities. Although keen observation of cloud type, wind direction, humidity, and short-term temperature change might eventually enable a student to make reliable forecasts, local observation cannot explain weather change.

Gathering data on a large scale permits mapping of pressure and temperature contours, wind circulation, and air mass distribution. Cloud cover, storms, wind direction change, temperature, and humidity patterns can then be seen as the effects of interacting masses of air with characteristics based on where they originally came into relative equilibrium with the earth's surface (polar land, tropical sea, etc.). In these problems the data are straightforward and simple but the system being represented is complex beyond complete prediction—a situation typifying problems in earth science.

Comprehending complex systems requires elaborate, even multiple, conceptual models (Stevens & Collins, 1980). Teaching aimed at bringing about such understanding involves probing student "surface errors" for possible deeper misconceptions, then guiding instruction according to "correction strategies." According to Stevens and Collins, such diagnosis and correction require:

1. Knowledge of common student errors and the relationship of such errors to misconceptions.
2. Understanding of the types of "real-world, experiential knowledge" students use to comprehend novel problems.
3. Understanding of various ways to apply such real-world knowledge.

In their study of student understanding of rainfall, Stevens and Collins (1980) pursued these issues. They characterized learning as a process of refining mental models into better correspondence with the world. For rainfall, they identified four models used by people or explained in textbooks: a "simulation" model of evaporation, a "functional" model of evaporation, a water-cycle model, and a climate model.

The simulation model is based on the imagery of water as discrete, moving, colliding, and attracting-repelling particles. This model accounts for evaporation in terms of particle escape from the liquid phase.

In a functional model, evaporation is associated with the amount of heat. Escape rate is the focus of attention and a function of water temperature in sophisticated versions of this model.

The water-cycle model connects evaporation and precipitation. In its sophisticated version, air masses play critical roles; in its rudimentary form, clouds.

Paralleling the water-cycle model is the climate model, similar in most respects except that it represents a large-scale, geographic perspective. This last model incorporates water and air currents that interact with land masses around the world.

Stevens and Collins (1980) found the educational implications of their analyses "profound":

This view suggests that multiple models should be taught explicitly as alternative points of view about a topic. The emphasis should be on the kinds of situations and problems for which each model is applicable, and on how to apply them to solve different types of novel problems. At the same time, students should learn the limitations of each model and how to test out a solution derived from one model against another. Students might also be taught how various distortions of a model lead to different misconceptions, and how any model can be systematically refined to increase its predictive accuracy. (p. 196)

Stevens and Collins' (1980) work on conceptual models grew out of their attempts to build computer tutors according to expert-systems programming conventions (e.g., numerous if-then rules as subroutines). At a less formal level of description, Stepans and Kuehn (1985) conducted interviews with children in grades 2 and 5 on such weather phenomena as wind, clouds, thunder, lightning, rain, snow, and rainbows. Children in both grades, whatever their level of knowledge, failed to use true causal explanations to account for these phenomena. In general, children who experienced hands-on science instruction gave true causal answers more frequently than those who learned school science from textbooks only.

Sometimes textbooks are misleading—as in the case of cross-sectional diagrams depicting underground water as a stripe of pure blue. Such an illustration reinforces naive conceptions of underground water. Children tend to have images of subterranean lakes and caverns filled with clear water. This may be true in regions with carbonate bedrock but is highly misleading for most aquifers, saturated ground, buried gravel beds, and permeable bedrock. In addition, I have inferred through informal conversations that students rarely consider the relationship among water table, ground water discharge, and stream or lake surface levels. They are likely to imagine water-filled ditches and basins as perched far above the level of the water table. Notions of continuity between surface runoff and ground water movement do not arise spontaneously, but problems involving aquifer recharge, well drawdown, and stream discharge when surface runoff stops make no sense to someone reasoning from a naive conception of underground water.

Direct observation of the phenomena of interest is seldom possible for underground water. Small-scale experiments may reveal what happens on true scales, but only by analogy. Analogies that resolve dilemmas may prove reliable. Adding water to buckets of gravel to measure "empty spaces" may help to teach the concept of pore space and resolve the dilemma, "What holds the rock above the water table up? If the water as an underground lake supports the rock above, then why doesn't the pressure of the rock above squirt the water back up through the cracks it came down by? Perhaps the grains of the rock support some of the load." The idea of the water being under pressure is a good one and can account for the flow of some springs. Solutions to complex problems involving underground water must integrate the concept of hydrostatic pressure with rock permeability and porosity. Images of underground lakes are a poor place to begin.

Meteorology, oceanography, and hydrology are excellent earth science subjects for model development based on fundamental physical science principles extrapolated to complex, large-scale systems. Density currents, fluid behaviors, and percolation phenomena all occur on small scales that can be studied as analogues of larger ones. Thus, the goal of "integration across scales" applies in these contexts as well as in geological ones.

In an analysis of how to integrate mathematical formulas, model building, and social concern, Calkins (1989) explained how science attempts to make sense of ocean motion. In his approach, students are taught the value of models according to the pragmatic philosophy of William James.

The essential principle of James' philosophy of pragmatism is that the test of truth in a proposition is not in a proposition's source, but in how the proposition works. Students of science should understand that a model predicting an outcome is only 'true' in the particular instance when the model works and that scientific knowledge is not based on absolute "truth." (p. 30)

Recall from the earlier discussion of Toulmin's philosophy the importance accorded pragmatic methods responsive to the particular problem of interest. Calkins has shown how to cast earth science problems as ones with "right" solutions, "right" being circumscribed by the pragmatic value of the model used to frame the problem.

Weather problems are also amenable to mathematical treatment. Gibb (1984a, 1984b) considers the possibilities so abundant that he recommends that mathematics instructors use meteorology as a rich source of applications problems. These problems might illustrate applications of laws (energy radiation rate as a function of the fourth power of temperature), development of mathematical models (to determine the difference in receipt of solar energy per square meter of surface on the solstice for a given location), or scaling and estimation (e.g., finding that the daily transport of heat across the 40th parallel is nearly equal to burning 1.4×10^{10} metric tons of coal).

Earth science textbooks often introduce students to the idea that winter comes because solar radiation is dispersed over Earth's surface due to the inclination of its rotational axis with respect to the plane of its orbit. They also recognize that the sun shines for fewer hours during winter months. Which factor, for San Francisco, has the greater effect on seasonal cooling? This is a remarkably difficult mathematical problem, but one that secondary students may work to model in a computing environment. Simplifying assumptions are needed, even for the basic step of solving for the energy received per square meter at noon on the solstices (ignore absorption and reflection). Pro-

jecting a square perpendicular to the sun's rays onto the plane of Earth is a necessary first step (and a challenging application of geometry or trigonometry in a purposeful context). The solution, according to Gibbs, yields a factor of 2. Twice the energy (in joules per square meter per second) reaches Earth's surface on June 21 as on December 21 at this latitude at noon. However, the angle of the sun with respect to the horizontal surface at a given latitude changes continuously; how does one sum up an infinite number of "noonlike" solutions to get the total energy received? Such problems have the potential to establish the need for a mathematical procedure before learning it—to place purpose (and intelligibility) ahead of technique.

Again, earth science contexts for doing so are legion. However, the Gibbs example reduces problem-solving behavior in earth science to that found in physics texts. As interesting and valuable as problem-solving skill in this context might be, it remains distinct from the study of problem solving from a holistic or systems perspective.

Instructional Variables and the Promotion of Problem Solving in Earth Science Teaching

Research about early adolescent learning frequently addresses earth science because it is a common eighth-grade subject. The most significant study of student earth science learning at this level from a problem-solving perspective was done by Chiapetta and Russell (1982). They modeled problem solving in three parts: (1) problem presentation, (2) collection of relevant information, and (3) analysis of information and production of a solution. They assigned 287 students randomly to 14 sections of earth science. Seven sections received conventional instruction and seven experienced modifications emphasizing problem solving. The investigators randomly chose 70 students from each group for study over a 6-week period.

Teachers of the conventional sections asked questions almost entirely at the knowledge and comprehension levels of Bloom's *Taxonomy* (Bloom, 1956) and led laboratory exercises intended to confirm classroom concepts. Teachers of the problem-solving sections asked questions at the application and analysis levels of the taxonomy. They conducted laboratory exercises using the researchers' three-part model. First, they presented a problem, such as a situation in which a farmer was puzzled that apparently similar fields differed in irrigation needs and flooding tendencies. Students next performed soil tests for porosity and permeability. In step 3 they assumed the roles of agricultural consultants and analyzed the data in order to solve the flooding and irrigating puzzle.

Russell and Chiapetta (1981) reached two important conclusions. First, logical thinking (or "intellectual competence" as measured by a 10-item Piagetian test) had a greater influence on achievement than did instructional style. However, after accounting for intellectual competence among all students, the problem-solving treatment did lead to improved achievement, especially at the higher levels of the Bloom taxonomy (as measured by a 20-item test).

Even studies that attempt no instructional intervention can inform teaching practice. Solarte (1984) attempted to model cognitive complexity in order to generate a guide for matching subject matter to cognitive ability. His study involved 433 Regents earth science students in western New York. He concluded that such a scheme is feasible and, when used, would encourage teachers to adjust instruction.

Wiley (1984) contrasted abstract geologic conceptual development among ninth-grade students according to field-trip activities. Students who participated in a process-oriented field trip were given a guidebook that posed questions requiring inquiry in order to make a response. Other students who participated in a content-oriented field trip received a guidebook containing direct information. Results showed that the process treatment developed long-lasting concrete concepts. Also, the "order of concept revelation" greatly influenced the development of lasting abstract concepts.

Wiley's results echoed the findings of Thomas (1968). From research with 143 eighth-grade earth science students, Thomas found a "guided discovery teaching method" superior to "didactic techniques" for helping high- and low-ability students acquire inquiry, critical-thinking, and problem-solving skills. However, didactic teaching exceeded guided discovery in helping students achieve factual and conceptual knowledge.

Monk and Stallings (1975), in contrast to Thomas, failed to find any difference in student course scores in response to receiving factually oriented quizzes versus quizzes that stressed application and analysis. Here the subjects were 200 students in an undergraduate physical geography course. Because these subjects were so radically different from the earth science eighth-graders, no conclusions by comparison should be drawn. Moreover, finding differences in achievement based on instructional treatment may become progressively more difficult with older, more capable learners, such as college students.

Monk (1983) attempted to distinguish the "daily achievement patterns" between students grouped according to formal versus concrete cognitive tendency while learning the abstract concept of plate tectonics. He found that formals outperformed concretes as expected. The plan to examine daily achievement patterns mirrored the time-series design of research in meteorology and oceanography. This paradigm has been ignored for the most part in studying learning, a complex system of interacting variables with changes developing (perhaps) through time in analogy with pattern development in atmospheric and oceanic systems. At least intensive sampling of changes in time makes sense in both contexts. However, the use, in effect, of daily testing raises the problem of whether such frequent evaluation of learning might invalidate claims about instructional effectiveness.

In response to this concern, Mayer and Rojas (1982) determined that the frequency of data collection had no measurable effect on classroom achievement or student attitudes toward learning science. Subsequently, Farnsworth and Mayer (1984) concluded that intensive time-series designs for classroom investigations could help researchers distinguish differences in learning between formal and concrete students.

Mayer and Kozlow (1980) conducted a feasibility study of the time-series design. They determined that daily measures of student learning of the crustal-evolution concept by eighth-grade students did not interfere with established classroom routines. Kwon and Mayer (1985), who assumed the validity of the intensive time-series design, reported a postintervention increase in achievement. Instead of a decline in understanding after instruction on the abstract concept of plate tectonics was over, they found significant increase in understanding (or at least maintenance of understanding) for a period of time. They termed this finding a "momentum effect" and found its strength to be dependent on the level of student understanding tested and the cognitive level of the learner.

Ulerick (1982) emphasized the dependence of introductory geology students on their prior knowledge as a means of integrating information from lectures, readings, and laboratories. Ulerick tested student performance at three levels: literal knowledge, integration of knowledge, and transfer of knowledge. Students performed well only on literal knowledge. Students with relevant prior knowledge specific to higher-level questions did the best on integration and transfer. Teacher directions for student study before a test emphasizing synthesis had no effect.

Even when an innovative instructional approach intended to engage students in inquiry or problem-solving processes has no greater effects on concept learning than does conventional instruction, student attitudes toward the subject may improve (Kern & Carpenter, 1984). Field-oriented, on-site problems are natural to the earth science curriculum, but their use demands extra teaching effort, according to Kern and Carpenter. Not surprisingly, students who elect to take additional earth science courses beyond a required introductory offering tend to have low anxiety about the subject (Westerback, Gonzalez, & Primavera, 1984).

Philosophy of Science, Learning Theory, and Cognitive Process Perspectives on Earth Science Problem Solving

Finley (1981, 1982) has studied student mastery of geologic classification from the perspective of philosophy of science and an interest in geology subject matter in its own right. He found that classification schemes are context bound to content. A classification scheme has superordinate and subordinate classes and criteria for defining membership—whether the elements are objects, events, or processes. Characteristics for defining membership may be classificatory (characteristic present or not), comparative (ordinal judgments of more or less of an attribute), quantitative (real number value assigned to each element), or some combination of these three. In some respects, Finley's work resembles that of Stevens and Collins (1980) on multiple models of the water cycle, for he has found psychological profit in conceptual analysis of content.

Geologic classification uses mineral composition as a primary characteristic of igneous rocks. Finley contrasted a simple scheme with a more sophisticated one, finding that the former used comparative characteristics and the latter, quantitative ones. In addition, the simpler scheme lacked defining values, which prevented completion of rules for class membership. Finley hypothesized that the ambiguity in the supposedly simpler scheme could create problems for students using it. To remedy such problems, an instructor would have to teach students how to conduct the measurement procedures needed to assign membership to a category. Simpler systems are not so simple if the operational definition of a category involves potentially ambiguous criteria for making comparative judgments. However, ambiguity itself may play an essential role in constructing classification systems for reasoning about similarities and differences.

Finley and Smith (1980a) have argued for guidance from the teacher in the initial solution of problems, even when the task is one of using a single characteristic classification scheme. They argue that "task-specific strategies" are a type of knowledge. The task specific strategies they analyzed in depth pertain to rock sorting according to relative grain size or amount of light-colored grains and placing rock samples in proper matrix cells according to column only or row and column descriptors. This work treats concepts learned (e.g., rock properties) as information needed for decision making. Finally, Finley and Smith (1980b) stress that learning "conceptually related tasks" (i.e., all pertaining to igneous rock classification in particular, not classification in general) contributes to achieving success on a complex, related task.

In keeping with an explanatory ideal similar to that pursued by Finley and Smith, Champagne and Klopfer (1981) attempted to explain the distinctive contributions of semantic knowledge (what concepts mean) and processing skills (strategies or procedures to follow) to the successful solution of problems. Champagne and Klopfer examined student performance on two types of verbal problems: analogies and set membership. The problems shared subject matter but differed in processing skills needed for solution. In the analogies problem, students picked a word from a short list (e.g., metamorphic, marble, sandstone, sedimentary) in order to complete an analogy (e.g., shale is to slate as limestone is to _____). In the set-membership problem students crossed out a word they believed did not belong with the other three (e.g., "volcano" in the list "volcano, igneous, metamorphic, sedimentary"). These problems had in common the need for students to induce a structure relating the parts.

Champagne and Klopfer (1981) hypothesized that students with greater processing skills on structuring tasks would do better on the set-membership problems than those with lesser structuring skills. They assessed structuring skills with the ConSAT (Concept Structuring Analysis) technique (Champagne, Klopfer, Squires, & Desena, 1981). Students observed by means of ConSAT guidelines arranged a small set of concept cards according to how they thought about the words. They explained their arrangements to an interviewer. In this instance the two tasks were ROCK and MINERAL. Each consisted of a few more than a dozen cards labeled with a single word such as slate, magma, graphite, and calcite. Each set included general terms such as rock, mineral, inorganic solid substances, and sedimentary. Indeed, students rated high in structuring process skill—those able to order semantic knowledge—scored high on set-membership problems after instruction in rock classification (this correlation could not be found a year later). Processing skill did not contribute significantly to analogy problem solving. Champagne and Klopfer (1981) concluded:

The cognitive analysis of the problems led us to hypothesize some explicit differences in the process skills necessary for the solution of each problem type. This suggests that instruction to produce successful solution of geology analogies and geology set-membership problems will differ along the dimension of process instruction. It also suggests that the processes should be explicitly incorporated into the statements of instructional objectives. (pp. 508–509)

Records such as those obtained through ConSAT expose preinstructional knowledge as well as changes in conceptual understanding after instruction. Researchers at Cornell have invented similar procedures for apprehending and invoking change in student cognitive structure (Novak & Gowin, 1984).

Novak, Gowin, and Johansen (1983) claimed that "concept mapping" and "Vee mapping" were helpful strategies for improving student knowledge achievement and problem-solving performance. Their work capped a decade-long effort at understanding variables within cognitive structure according to Ausubelian theory. The point of mapping exercises is to guide students in making their own knowledge explicit. Mapping practice helps students organize what they know for efficient application, find gaps in their understanding, and raise questions about how to improve their understanding (Ault, 1985). The Cornell tradition is predicated on the value of having students take their own knowledge as an object of study in its own right and is therefore closely linked to traditions examining naive conceptions, alternative frameworks, and preinstructional knowledge.

Like Champagne and Klopfer, the Cornell researchers have been concerned with how learning processes for structuring knowledge may transfer to problem-solving performance. Novak (1977) believed that effective use of knowledge presupposes its organization hierarchically as displayed on a concept map. Whether the topic is oceans, moon phases, or continents, maps have a focus or superordinate concept, propositional or prepositional relationships to other concepts, and concrete examples of key concepts. Maps depict layers of abstractions.

The desire to compare students' transfer of learning for problem solving led Novak, Gowin, and Johansen (1983) to devise a conceptual problem in which measures other than concept maps could be used to evaluate performance. They reported a "Winebottle Test." Four classes of eighth-grade students received instruction in solids, liquids, gases, and gas laws. Each student attempted to answer in essay form why a cork popped out of an empty wine bottle that was placed in sunlight on a windowsill after a night in a cold refrigerator. Students in two control classes answered only in essay form. Students in two experimental classes constructed maps of the concepts from their original paragraphs and then rewrote the concept map in paragraph form. Reviewers scored answers according to the number of valid relationships (e.g., "air expands when warmed"). Students in the experimental group composed more than twice as many valid relationships as control students.

Another team (Rollins, Denton, & Jake, 1983), working with many of the same assumptions about learning held by the Cornell group, returned to the Janke and Pella (1972) earth science concept list for grades K–12. From the list they distilled a nucleus of essential earth science concepts to use in assessing the attainment of environmental, earth, and astronomical concepts by Texas high school seniors. They investigated how well students could select an example of a given concept and whether students could identify appropriate supraordinate and subordinate concepts for day-night, changing earth, seasons, solar energy, and environment. Seasons had a supraordinate category of orbital motion and a subordinate one of earth-axis angle of inclination, for example. Rollins et al. (1983) did not investigate transfer of learning or problem solving directly.

Expert-Novice and Transfer-of-Learning Paradigms

Barba (1990, 1992a) investigated expert and novice problem solving in earth science by comparing the abilities of 30 "accomplished" teachers (30 graduate hours of study, at least 3 years of experience, and not less than 12 course hours in geological sciences) with those of 30 "neophytes" (preservice, undergraduate, student teachers with less than 7 course hours in geological sciences). All participants volunteered. Her study assumed subjects were aware of, had access to, and could verbalize their mental processes. She looked for activation of knowledge structures and interactions among "declarative, procedural, and structural knowledge." She recorded for each subject measures of general mental ability (Otis Lennon test), declarative knowledge in earth science (AGI/NSTA test), "Gagnéan levels" of declarative knowledge (fact, rule, concept, and problem-solving levels of the AGI/NSTA test), and use of procedural knowledge while engaged in solving tasks typical of earth and space science school curricula (five ESCP tasks). Audiotapes of subjects engaged in problem-solving activities were utilized in a "protocol analysis."

Barba anchored her study in the conception of an expert as someone who adopts one of five strategies when confronted with an unfamiliar problem (Clement, cited in Barba, 1990): (1) construct a simpler problem of the same type, (2) investigate extreme cases (push problem parameters to infinity or zero), (3) appeal to mental models based on visual intuition, (4) seek analogy with systems already understood, or (5) search for potential misanalogies in the one chosen. However, the problems in her study could not be considered particularly unfamiliar to the expert cohort.

In summarizing the findings about research on problem solving in biology, physics, and chemistry, Barba concludes that, in contrast to experts, novices struggle to categorize problems as a type amenable to a solution procedure and do not engage in the flexible reformulation of the problem typified by Clement's five strategies. Experts use more factual knowledge and more knowledge at higher Gagnéan levels than novices, break problems into subproblems, recognize relevant principles governing a solution, find multiple solution paths, test more hypotheses, and reflect on the reasonableness of their solutions—all in contrast to how novices function.

Despite a professed intention to "better understand the cognitive processes inherent" in the discipline of earth science through the expert-novice paradigm, Barba did not proceed from an analysis of features of reasoning in geology or astronomy. Instead, she accepted the sample materials of school science as valid representations of earth science problems. Each problem-solving task—mineral hardness, vertical profile from a topographic map, radioactive decay calculation, wave speed, and "capture/recapture" (crabs captured, marked, released; second capture characterized as proportion of marked to unmarked and total population calculated by proportional reasoning)—had an unambiguous solution but allowed for multiple solution pathways.

All null hypotheses were rejected in this study, including, surprisingly, that of no difference in general mental ability (experts' higher scores were viewed as a reflection of maturation, in contrast to the hypothesis of static IQ). There were significant differences in declarative knowledge, Gagnéan levels of declarative knowledge, and procedural knowledge. Barba concluded that experts brought more knowledge to the problem-solving situations and used this knowledge far more than the novices did. Prior knowledge helped experts to make judgments about solutions.

As summarized by Barba and Rubba (1992), experts solved problems in fewer steps than novices, made fewer nonproductive steps, and appeared to look ahead by joining pieces of procedural knowledge and recognizing nonproductive solution paths. Experts used subroutines; novices tended not to. Knowledge of principles by experts aided in these choices.

Lacking prior experience with the task, novices were "forced to formulate schema while solving the tasks (Barba & Rubba, 1992). They did not move efficiently between declarative and procedural knowledge. They lacked the structural knowledge needed to organize information for completing the tasks. Alarmingly, the tasks they could not perform were typical of ESCP problems posed for middle and secondary school students. To repeat, these were not open-ended, inquiry-oriented tasks.

Apart from the minimal earth science training prospective teachers in this study received, the challenge for teacher preparation appears to be that of posing the kind of problems for novices that best promote the initial development of schemata. Barba admits that much more study is needed in earth science in order to plan training designed to help novices become experts.

Niedelman (1990) approached the study of problem solving in earth science from the perspective of testing the hypothesis of "low road" transfer of a high-level thinking skill. By "low road" Niedelman meant spontaneous activation in a new context of a behavior practiced until nearly automatic in the initial one. By "higher-level thinking" he meant, in this study, the skill of "seeking out the relevant data" (sorting relevant from irrelevant information and deciding whether sufficient information existed to solve the problem and, if not, what additional information would be necessary).

Niedelman chose earth science problems based primarily on applications of physical science concepts as the training program and used a video disk program for instruction. His design used "underspecified problems" in which students had to decide (1) that there was insufficient data to solve a problem and (2) what data were missing. He looked for transfer in the context of solving algebra problems. On all measures of student decision making about data, results were promising.

Forty-seven eighth-grade students in two separate classes participated in the experimental study; 122 students studying from a traditional earth science text served as the control group. The experimental sections were taught by an acknowledged expert teacher. Niedelman convincingly presented a case for gains in earth science knowledge and problem solving; the video disk format and training using distributed practice worked well together in teaching students to evaluate information.

The claim of transfer to mathematics is less compelling, because it is possible to attribute the gains in math problem solving to the fact that all students took eighth-grade math concurrently with earth science. Defining earth science problems as situations calling for applications of physical science principles, such as density-driven currents, Niedelman's dissertation represents an exemplary integration of instructional technology with training in earth science problem solving at both domain-specific and higher-order levels of thinking. His conceptualization of transfer deserves incorporation into future studies as an alternative to the expert-novice paradigm.

Transfer of problem-solving skill is elusive and the interaction of domain-specific knowledge with strategic knowledge in problem solving is complex, differing, perhaps, from discipline to discipline in substantial ways. In a synthesis of the literature on this topic, Alexander and Judy (1988) conclude that domain knowledge is needed to develop strategic knowledge and that strategic knowledge helps not only to utilize but also to acquire domain-specific knowledge. Their relationship is reflexive. Moreover, increasing domain knowledge alters strategic processing, and strategy strength varies with its appropriateness to the task. Alexander and Judy feel there is a need for research "explicitly designed to address the question of the interaction of domain specific knowledge and strategic knowledge on academic performance" (p. 397). They also call for increased understanding of how competence develops. A problem-solving context for learning is critical: "All programs [of instructional research] advocate learning in the context of working on specific problems.... Useful knowledge is not acquired as a set of general propositions, but by active application, during problem solving in the context of specific goals" (Glaser, 1990, p. 37).

What is problem solving and how do these two dissertation studies define it? Barba views problem solving as a continuum from simple computational tasks to questions involving analysis, synthesis, and evaluation as defined on Bloom's taxonomy of educational objectives (1956). (Ironically, mathematics educators hesitate to label "computation" as "simple" in the context of learning.) Barba expresses interest in how experts differ from novices in access to information and generation of routines for solving problems. She characterizes knowledge (after Gagné, Briggs, & Wager, 1988) as being of three distinct types: declarative, procedural, and structural. Structural knowledge means knowledge of how to coordinate the first two—the key to accessing information and generating solution strategies, habits of thought synonymous with expertise.

Niedelman (1990), working from a mathematics tradition, has defined problem solving as two levels of thinking skills: domain-specific knowledge and generic strategies for making decisions about problem-solving procedures. Niedelman values teaching for the transfer of heuristic reasoning from a familiar situation to a novel one, expediting the learning of problem-solving strategies in the new domain. Recognizing correspondences between the familiar and novel contexts automatically and without deliberate reflection is the major step in low-road transfer. Niedelman concludes that explicit heuristics for solving problems are best acquired in the context of "distributed practice" across many examples and levels of problem difficulty. In their briefest sense, heuristics are what problem solvers choose to do when they are uncertain about what to do. Niedelman defined one of these heuristics as the evaluation of given data: Is it sufficient to solve the problem?

For both Barba and Niedelman, earth science problems provided convenient grist for their research mills. Barba looked for a fit between the findings of the expert-novice paradigm as followed in research about problem solving in physics, chemistry, and biology. She defined earth problem-solving expertise operationally in terms of what kinds of problem-solving behaviors were carried out by the experts in her study. The experts were experienced teachers, not research scientists.

Niedelman happened to choose earth science as a training context for the transfer of a data-evaluation heuristic to algebra. Both defined earth science problems as applications of physical science principles to situations in which these principles could yield deductively unambiguous solutions. Despite the merits of their inquiries, the products do not contribute to the task of defining the nature of earth science problems or to the goal of specifying a model of rationality for reasoning in earth science.

CONCLUSION

From almost any research perspective, educational theorists argue that learning happens best in the context of solving problems—best if real, all right if simulated. However, earth science teachers who challenge students to solve problems have received little guidance from educational researchers. In the broadest sense, the study of problem solving in earth science education has the potential to:

1. Expand our perspective of the nature of reasoning in science.
2. Examine the cognition governing problem solving in complex contexts.
3. Test hypotheses of transfer and the interaction of domain specific learning with strategic knowledge.

The problem-solving studies reviewed in this chapter remain essentially silent on the issue of reasoning about complex systems with contingent histories and the consequent difficulties of extrapolation—the essence of earth system science in the introductory quotation. Studies of earth science education have not treated science principles in context—either a broad earth system context or an even broader sociopolitical one. What we seem to know least about is the kind of understanding of science most important to people in general.

This scenario is hardly new to earth science education. Hagner and Henderson (1959) alluded, in effect, to the centrality of complex systems in claiming that geology "requires the ability to think in terms of several variables at once, to recognize problems and ways of dealing with them" (p. 36). Hagner and Henderson (1959) also observed that geology was absent "in the great majority of high school curricula" (p. 39).

Also writing in the middle of the century, Holmes (1956) argued for the importance of the study of Earth in general education. He felt that more valuable than the inclusion of principles from other sciences in geology was the "essence of geology itself: history of the Earth and evolution of its inhabitants" (p. 37). For Holmes, "Of all the physical sciences, geology touches the most broadly and vitally upon the life and destiny of Man" (p. 37). These are the same themes as found in the opening quotation of this chapter.

Hill's (1891) provocative essay of a century ago still rings true as well. He called geology—the precursor of earth science—"vaguely understood" and "indefinitely represented." Geology, according to Hill, "is a grand study of the present structure of the earth"—the earth system in today's parlance. In 1891, Hill acknowledged the connection between the study of Earth and the kind of scientific knowledge people are most likely to encounter. He claimed quite bluntly, "The cultural aspects of civilization are due to geologic structure" (Hill, 1981, p. 42).

A century after Hill, earth science remains neglected in the school curriculum. Research about problem solving in earth science education cannot proceed unless we "get our epistemology straight." Geology and the other earth sciences are not immature, merely descriptive sciences. They are fields of study with distinctive methods created in response to the characteristics of compelling problems, united to other fields of science by acknowledging the need for and difficulty of understanding the operation of complex systems on grand scales.

Earth (and space) sciences have distinctive characteristics—the aspect of scale, for example. As enormous as a thundercloud may seem, the frontal system of which it is a part can overwhelm the common scale of observation. Whether crustal history, stellar evolution, or heat transfer through ocean currents, the scale of change for earth science phenomena presents an obstacle to meaningful student solutions to problems.

Scale interacts with a second theme, the need to visualize. Incomplete observation complicates the task of visualization. Making matters worse is the fact that phenomena of interest are complex systems. Several conventions are applicable to a wide range of problems from different domains of earth science. Among these are contour mapping and geometric models—inferences of large-scale objects' patterns and simplified three-dimensional structures, respectively. With maps and models, whether of bedrock and landscapes, air masses, ocean circulation, or celestial motion, earth scientists predict and explain events on the grandest scales of space, time, and energy. Some of these events or the earth science perspectives of them have significance for public, environmental policy.

Two factors mitigate the seemingly insurmountable difficulty of teaching earth science to naive learners. First, nearly all earth science instruction can tap a wealth of everyday experience and imagery. Second, earth science knowledge can help students make unsuspected connections among their everyday experiences, often with relevance for understanding policy issues debated publicly (karst terrain, underground water, and hazardous waste disposal, for example).

Science educators have paid relatively little attention to students' difficulties in understanding the earth sciences. We know too little about why some students succeed with problems requiring visualization skill while others fail. Nor do we understand how people come to grasp the scale of earth science phenomena or how failure to appreciate scale inhibits meaningful learning and problem solving. Lastly, we have yet to begin in earnest studies of how students solve complex problems embedded in value contexts

Ultimately, research will provide no one best way to teach or measure student learning. Teachers ought to synchronize their teaching to the kind of knowledge they expect from their students. Rote learning may overwhelm efforts to teach problem solving if students become addicted to "right answer" science. Problem solvers seem to tolerate ambiguity or at least multiple pathways of reasoning. They need to know what they know, how to access information, and when to apply which parts. Often the role of the teacher is to make students uncomfortable with their present state of understanding.

The general wisdom is to learn to apply knowledge in many contexts and to organize meaning with precision and hierarchy. Quality research literature on the earth science curricula is sparse, but the findings from other domains probably transfer in cases in which earth science involves chemistry, physics, biology, and mathematics.

The complexity of teaching and learning earth science vastly exceeds the ability of research to offer prescriptive advice. Earth science fascinates us because if we solve its problems, it takes us well beyond our own experiences. At the same time, earth and space science is the science of things right beneath our feet and above our heads—everywhere, always. In tandem with ecology, it teaches where we belong, how we fit in, and what we can do—and must do—to try to maintain the habitability of our globe.

References

Alexander, P. A., & Judy, J. E. (1988). The interaction of domain-specific and strategic knowledge in academic performance. *Review of Educational Research, 58*(4), 375–404.

Ault, C. R., Jr. (1982). Time in geological explanations as perceived by elementary school students. *Journal of Geological Education, 30*(5), 304–309.

Ault, C. R., Jr. (1985). Concept mapping as a study strategy in earth science. *Journal of College Science Teaching, 15*, 38–44.

Ault, C. R., Jr. (1986a). Expanded abstract of "Children's cosmographies: Understanding the earth's shape and gravity" by C. Sneider & S. Pulos. *Science Education, 67*(2), 205–221, 1983. Prepared for *Investigations in Science Education, 12*, 72–80.

Ault, C. R., Jr. (1986b, October). *Spatial vs. conceptual reasoning in the "first" structural geology lab*. Paper presented at the Association for Educators of Teachers in Science (AETS) regional meeting, Indianapolis, IN.

Barba, R. H. (1990). A comparison of expert and novice earth and space science teachers' problem solving abilities. *Dissertation Abstracts International, 52*(12), 4078A. (University Microfilms No. AAC91–04849)

Barba, R. H., & Rubba, P. A. (1992a). A comparison of preservice and inservice earth and space science teachers' general mental abilities, content knowledge, and problem-solving skills. *Journal of Research in Science Teaching, 29*(8), 520–535.

Barba, R. H., & Rubba, P. A. (1992b). Procedural task analysis: A tool for science education problem-solving research. *School Science and Mathematics, 92*(4), 188–192.

Barba, R. H., & Rubba, P. A. (1993). Expert and novice earth and space science teachers' declarative, procedural, and structural knowledge. *International Journal of Science Education, 15*(3), 273–282.

Bart, H. A. (1991). A hands-on approach to understanding topographic maps and their construction. *Journal of Geological Education, 39*(4), 303–305.

Bloom, B. S. (Ed.). (1956). *Taxonomy of educational objectives. Handbook I: Cognitive domain*. New York: McKay.

Calkins, J. R. (1989). Modeling the ocean in motion. *Science Teacher, 56*, 28–30.

Callister, J. C., & Mayer, V. J. (1988). NSTA's new earth science test. *Science Teacher, 55*(4), 32, 34.

Champagne, A. B., & Klopfer, L. E. (1981). Structuring process skills and the solution of verbal problems involving science concepts. *Science Education, 65*(5), 493–511.

Champagne, A. B., Klopfer, L. E., Squires, D. A., & Desena, A. T. (1981). Structural representations of student's knowledge before and after science instruction. *Journal of Research in Science Teaching, 18*(2), 97–111.

Champlin, R. F. (1970). A review of research related to ESCP. *Journal of Geological Education, 18*, 31–39.

Chiapetta, E. L., & Russell, J. M. (1982). The relationship among logical thinking, problem solving instruction and knowledge and application of earth science subject matter. *Science Education, 66*, 85–93.

Cohen, M. R. (1968). *The effect of small scale geologic features on concepts of fluvial geology among fifth and sixth grade children*. Unpublished doctoral dissertation, Cornell University, Ithaca, NY.

Disinger, J. F., & Mayer, V. J. (1974). Student development in junior high school science. *Journal of Research in Science Teaching, 11*(2), 149–155.

Donaldson, M. (1982). Conservation: What is the question? In P. Bryant (Ed.), *Piaget: Issues and experiments*. Cambridge, Eng: University Press.

Earth System Sciences Committee & the National Aeronautics and Space Administration [NASA]. (1986). *Earth system science overview: A program for global change*. Washington, DC: NASA.

Farnsworth, C. H., & Mayer, V. J. (1984). An assessment of the validity and discrimination of the intensive time-series design by monitoring learning differences between students with different cognitive tendencies. *Journal of Research in Science Teaching, 21*(4), 345–355.

Finley, F. N. (1981). A philosophical approach to describing science content: An example from geologic classification. *Science Education, 65*(5), 513–519.

Finley, F. N. (1982). An empirical determination of concepts contributing to successful performance of a science process: A study in mineral classification. *Journal of Research in Science Teaching, 19*(8), 689–696.

Finley, F. N., & Smith, E. L. (1980a). Effects of strategy instruction on the learning, use, and vertical transfer of strategies. *Science Education, 64*(3), 367–375.

Finley, F. N., & Smith, E. L. (1980b). Student performance resulting from strategy-based instruction in a sequence of conceptually related tasks. *Journal of Research in Science Teaching, 17*(6), 583–593.

Gagne, R. M., Briggs, L. J., & Wager, W. W. (1988). *Principles of instructional design* (3rd ed.). New York: Holt, Rinehart, & Winston.

Gibb, A. A. (1984a). Talking about the weather, part I. *Mathematics and Computer Education, 18*, 13–20.

Gibb, A. A. (1984b). Talking about the weather, part II. *Mathematics and Computer Education, 18*(2), 100–106.

Glaser, R. (1990). The reemergence of learning theory within instructional research. *American Psychologist, 45*, 29–39.

Griffin, T. L. C. (1983). Problem solving on maps—the importance of user strategies. *Cartographic Journal, 20*(2), 101–109.

Hagner, A. F., & Henderson, D. M. (1959). Problems in geologic education: The elementary course. *Journal of Geological Education, 7*, 36–39.

Happs, J. C. (1982a). *Rocks and minerals* (Working Paper No. 204). Hamilton, NZ: University of Waikato, Science Education Research Unit. (ERIC Document Reproduction Service No. ED 236 034)

Happs, J. C. (1982b). *Glaciers* (Working Paper No. 203). Hamilton, NZ: University of Waikato, Science Education Research Unit. (ERIC Document Reproduction Service No. ED 236 033)

Happs, J. C. (1982c). *Mountains* (Working Paper No. 202). Hamilton, NZ: University of Waikato, Science Education Research Unit. (ERIC Document Reproduction Service No. ED 236 032)

Happs, J. C. (1982d). *Soils* (Working Paper No. 201). Hamilton, NZ: University of Waikato, Science Education Research Unit. (ERIC Document Reproduction Service No. ED 236 031)

Happs, J. C. (1984). Soil genesis and development: Views held by New Zealand students. *Journal of Geography, 83*(4), 177–180.

Happs, J. C. (1985). Cognitive learning theory and classroom complexity. *Research in Science & Technological Education, 3*(2), 159–174.

Happs, J. C., & Stead, K. (1989). Using the repertory grid as a complementary probe in eliciting student understanding and attitudes toward science. *Research in Science & Technological Education, 7*(2), 207–220.

Hill, R. T. (1891). Do we teach geology? *Popular Science Monthly, 40*(4), 41–43.

Holmes, C. D. (1956). Key to a more efficient and effective elementary geology course. *Journal of Geological Education, 4*(2), 37–42.

Janke, D. L., & Pella, M. O. (1972). Earth science concept list for grades K–12 curriculum construction and evaluation. *Journal of Research in Science Teaching, 9*(3), 223–230.

Kelsey, L. J. G. (1980). The performance of college astronomy students on two of Piaget's projective infralogical grouping tasks and their relationship to problems dealing with phases of the moon. *Dissertation Abstracts International, 41*(6), 2539A.

Kern, E. L., & Carpenter, J. R. (1984). Enhancement of student values, interests and attitudes in earth science through a field-oriented approach. *Journal of Geological Education, 32*(5), 299–305.

Klein, C. A. (1982). Children's concepts of the earth and the sun: A cross cultural study. *Science Education, 65*, 95–107.

Kwon, J. S., & Mayer, V. J. (1985). Identification and descriptions of the momentum effect in studies of learning: An abstract science concept. *Journal of Research in Science Teaching, 22*(3), 253–259.

Lemke, J. L. (1990). *Talking science: Language, learning, and values*. Norwood, NJ: Ablex.

Mali, G. B., & Howe, A. (1979). Development of earth and gravity concepts among Nepali children. *Science Education, 63*(5), 685–691.

Mayer, R. E. (1989). Models for understanding. *Review of Educational Research, 59*, 43–64.

Mayer, V. J. (1979). The role of scope and evaluation in the development of CEEP modules. *Journal of Geological Education, 27*(4), 157–159.

Mayer, V. J. (1982). The NAGT earth sciences activities book. *The Journal of Geological Education, 30*(5), 312–314.

Mayer, V. J. (1984). Crustal evolution education project materials: National evaluation. *School Science and Mathematics, 84,* 7–26.

Mayer, V. J., & Armstrong, R. E. (1990). What every 17-year-old should know about planet Earth: The report of a conference of educators and geoscientists. *Science Education, 74*(2), 155–165.

Mayer, V. J., Armstrong, R. E., Barrow, L. H., Brown, S. M., Crowder, J. N., Fortner, R. W., Graham, M., Hoyt, W. H., Humphris, S. E., Jax, D. W., Shay, E. L., & Shropshire, K. L. (1992). The role of planet earth in the new science curriculum. *Journal of Geological Education, 40,* 66–73.

Mayer, V. J., Champlin, R. A., Christman, R. A., & Krockover, G. H. (1980). *Activity sourcebook for earth science.* Columbus, OH: ERIC/SMEAC.

Mayer, V. J., & Kozlow, M. J. (1980). An evaluation of a time series single-subject design used in an intensive study of concept understanding. *Journal of Research in Science Teaching, 17*(5), 445–461.

Mayer, V. J., & Rojas, C. A. (1982). The effect of frequency of testing upon the measurement of achievement in an intensive time-series design. *Journal of Research in Science Teaching, 19*(7), 543–551.

Mayer, V. J., & Stoever, E. C., Jr. (1978). NAGT crustal evolution education project: A unique model for science curriculum materials development and evaluation. *Science Education, 62*(2), 173–179.

Mayer, V. J., & Stoever, E. C., Jr. (1979). The role and scope of evaluation in the development of CEEP modules. *Journal of Geological Education, 27*(4), 157–159.

Monk, J. S. (1983). Using an intensive time series design to examine daily achievement and attitude of eighth and ninth grade earth science students grouped by cognitive tendency, sex, and IQ. *Dissertation Abstracts International, 44*(9), 2727A.

Monk, J. J., & Stallings, W. M. (1975). Classroom tests and achievement in problem solving in physical geography. *Journal of Research in Science Teaching, 12*(2), 133–138.

National Center for Earth Science Education [NCESE] & the American Geological Institute [AGI]. (1991a). *Earth science education for the 21st century: A planning guide.* Alexandria, VA: Author.

National Center for Earth Science Education [NCESE] & the American Geological Institute [AGI]. (1991b). *Earth science content guidelines grades K–12.* Alexandria, VA: Author.

Niedelman, M. S. (1990). An investigation of transfer to mathematics of a problem-solving strategy learned in earth science. *Dissertation Abstracts International, 51*(11), 3622A. (University Microfilms No. AAC91-11130)

Novak, J. D. (1977). *A theory of education.* Ithaca, NY: Cornell University Press.

Novak, J. D., & Gowin, D. B. (1984). *Learning how to learn.* Cambridge, MA: Cambridge University Press.

Novak, J. D., Gowin, D. B., & Johansen, G. T. (1983). The use of concept mapping and knowledge Vee mapping with junior high school science students. *Science Education, 67*(5), 625–645.

Nussbaum, J. (1979). Children's conceptions of Earth as a cosmic body: A cross-age study. *Science Education, 63,* 83–93.

Nussbaum, J., & Novak, J. D. (1976). An assessment of children's concepts of the Earth utilizing structured interviews. *Science Education, 60*(4), 535–550.

Oosterman, M. A., & Schmidt, M. T. (Eds.). (1990). *Earth science investigations.* Alexandria, VA: American Geological Institute.

Osborne, R., & Freyberg, P. (1985). *Learning in science: The implications of children's science.* Portsmouth, NH: Heineman.

Paull, R. A., Larson, A. C., & VandenAvond, R. (1969). Predicting the effect of ESCP on introductory college geology courses. *Journal of Geological Education, 17*(2), 47–53.

Philips, W. C. (1991). Earth science misconceptions. *Science Teacher, 58*(2), 21–23.

Rojas, C. A., & Mayer, V. J. (1979). Characteristics of teachers and students in the second stage of CEEP evaluation. *Journal of Geological Education, 27*(3), 107–110.

Rollins, M. M., Denton, J. J., &. Jake, D. L. (1983). Attainment of selected earth science concepts by Texas high school seniors. *Journal of Educational Research, 77*(2), 81–88.

Russell, J. M., & Chiapetta, E. L. (1981). The effects of a problem solving strategy on the achievement of earth science students. *Journal of Research in Science Teaching, 18*(4), 295–301.

Sneider, C., & Pulos, S. (1983). Children's cosmographies: Understanding the earth's shape and gravity. *Science Education, 67*(2), 205–221.

Solarte, L. J. I. (1984). Analysis of an earth science topic for understanding levels of demand. *Dissertation Abstracts International, 44*(7), 2108A.

Stepans, J., & Kuehn, C. (1985). What research says: Children's conceptions of weather. *Science and Children, 23,* 44–47.

Stevens, A. L., & Collins, A. (1980). Multiple conceptual models of a complex system. In R. E. Snow, P. A. Federico, & W. E. Montague (Eds.), *Aptitude, learning and instruction. Vol. 2: Cognitive process analyses of learning and problem solving.* Hillsdale, NJ: Lawrence Erlbaum.

Thomas, B. S. (1968). *An analysis of the effects of instructional methods upon selected outcomes of instruction in an interdisciplinary science unit.* Unpublished doctoral dissertation, Iowa University, Iowa City.

Tunley, A. T. (1986). A classroom study of science achievement in high school geology. *Journal of Geological Education, 34*(2), 110–113.

Ulerick, S. L. (1982). The integration of knowledge from instructional discourses in a college-level geology course. *Dissertation Abstracts International, 42*(11), 4783A.

Westerback, M. E., Gonzalez, C., & Primavera, L. H. (1984). Comparison of anxiety levels of students in introductory earth science and geology courses. *Journal of Research in Science Teaching, 21*(9), 913–929.

Wiley, D. A. (1984). A comparison of the effectiveness of geological conceptual development within two field trip types as compared with classroom instruction. *Dissertation Abstracts International, 45*(1), 143A.

· 10 ·

RESEARCH ON PROBLEM SOLVING: GENETICS

James Stewart

UNIVERSITY OF WISCONSIN–MADISON

Robert Hafner

WESTERN MICHIGAN UNIVERSITY

Understanding and problem solving are cognitively similar activities. Their similarity is underscored by the current focus of philosophers and sociologists of science (Darden, 1991; Kitcher, 1984, 1985; Latour, 1987; Nersessian, 1987) on analyses of the problem-solving aspects of scientific practice. This focus represents a shift from a retrospective analysis of the products of inquiry to a forward-looking viewpoint characteristic of the practicing scientist. Wimsatt (1979) summed up this viewpoint by asserting that: "We cannot have an adequate philosophy of science without putting a realistic model of the scientist as decision maker and problem solver back into our model of science" (p. 359).

Current with the increased study by philosophers of scientific practice and of scientists as problem solvers, psychological researchers have turned their attention to the problem-solving performance of students and scientists. In this chapter we review the research on problem-solving performance in the biological sciences and outline directions for its continuation.

In biology, classical genetics is the area in which most problem-solving research has been done. This review is divided into genetics research done (1) within a Piagetian perspective and 2) within an "expert-novice" perspective. The latter category is organized into research using (1) textbook problems that emphasize meiosis, (2) textbook problems that require solvers to reason from effects to causes, and (3) computer-generated problems that require effect-to-cause reasoning. In the computer-generated problem category, we make a distinction between situations in which the solver uses existing knowledge to solve problems (model-using problem solving) and situations in which the solver must modify existing explanatory models as a part of the problem solution (model-revising problem solving).

PIAGETIAN STUDIES

Piagetian research in the biological sciences has been extensive and has included studies of students' conceptions of scientific phenomena, as well as of the cognitive operations assumed to underlie scientific thinking. In genetics, one approach has been to search for correlations between Piagetian operational levels and problem-solving success (Gipson & Abraham, 1985; Gipson, Abraham, & Renner, 1989; Walker, Hendrix, & Mertens, 1980; Walker, Mertens, & Hendrix, 1979). Walker et al. (1979, 1980) began with the premise that many university students do not use formal operations and, therefore, genetics instruction emphasizing combinatorial, propositional, and hypothetico-deductive reasoning (all features of formal-operational thought) could increase the students' genetics problem-solving performance. Walker et al. (1980) developed a programmed learning approach designed to cause students mild disequilibrium as they encountered new genetics content. The intent was for students to resolve conflicts that led to disequilibrium and thus develop a better understanding of genetics. Using a non-equivalent group posttest-only study, the authors compared test scores of pupils taught by the programmed learning approach with those taught by a traditional lecture-laboratory method. A posttest consisting of 69 problem-oriented items was adminis-

The authors wish to thank Angelo Collins (Florida State University) for her helpful comments on an earlier draft of this chapter and Judith Van Kirk (University of Wisconsin) for comments on innumerable drafts.

tered to both groups, and the experimental group outperformed the control group.

Gipson et al. (1985, 1989) also assumed that the formal operations of proportional, combinatorial, and probabilistic reasoning are necessary for working with ratios in Punnett squares, forming gamete combinations, and determining the genotypes of zygotes. They examined the relationship between formal reasoning (as measured by standard Piagetian tasks) and genetics problem-solving success. Of secondary interest was the correlation of operational level with retention of genetics problem-solving ability. They used Pearson correlations, factor analyses, and analysis of variance (ANOVA) to examine the relationships among the performances of 71 college students on three Piagetian operations (proportional, probabilistic, and combinatorial reasoning) and a six-item genetics test covering zygote formation, the law of independent assortment, and the probability of gamete and zygote formation. Even though "specific identification of areas of formal thought related to corresponding areas in genetics were not conclusively shown by any of the statistical measures used" (p. 818), the authors concluded:

In considering overall results, one strong general conclusion emerges: We can say with a high degree of certainty that formal thought is necessary for retaining the ability to solve some types of Mendelian genetics problems, but we do not know enough about formal-operational thought to relate its characteristics directly to what we perceive to be the specific reasoning type necessary in solving genetics problems. Instead of specific reasoning types, the general ability of formal-operational thought seems to be necessary. (p. 818)

It is interesting that in the face of unexpected results, the authors did not consider non-Piagetian explanations. Rather, they postulated a more general correlation between formal reasoning (in an aggregate form) and academic task performance. Gipson et al. assume that students who fail to obtain correct solutions to genetics problems are unable to apply formal reasoning to the situation. However, it is equally conceivable that such students reason probabilistically (or combinatorially or proportionally) but do so with inappropriate genetics knowledge.

Kinnear (1983) examined the probabilistic reasoning of university biology students by creating two sets of problems (see Figure 10.1): those that did not require any discipline-specific knowledge (the "drug problem") and those in which the same probability concepts were embedded in a genetics context (the "horse problem"). She found that 69.3 percent of the students answered the drug problem correctly, whereas only 14.8 percent did so for the horse problem. Kinnear then extended the study to more students and more problems (six real-world and six genetics problems). The results were the same. For the most part, students were able to deal adequately with probability when solving real-world problems. They were much less able to do so with the genetics problems. Kinnear accounts for the differences in the following way:

It appears that the "discordant" students have two working concepts of chance—one that applies in the context of Mendelian genetics and focuses on classic ratios, applying them in a deterministic manner, and a

---

Problem 1

Measured dosages of a drug are packaged into capsules in a pharmaceutical factory. Each capsule is made of two halves that fit together. Bin 1 contains 10,000 halves, made up of equal numbers of red halves and white halves. Bin 2 contains 10,000 matching halves; 5000 of each color. A machine selects one half from bin 1, fills it with the drug and then seals it with a half from bin 2. An inspector watched this process and saw four capsules made. Answer the following by ticking ( ) the appropriate answer(s).

The four capsules made could have been

4 all-white	( )
3 all-white and 1 red or bicolored	( )
2 all-white and 2 red or bicolored	( )
1 all-white and 3 red or bicolored	( )
4 red or bicolored	( )

Problem 2

In horses, black coat color (B) is dominant to chestnut (b). A black stallion was mated with two chestnut mares, and each mating produced a black foal. Over subsequent years, these foals were mated and a total of four offspring were produced. Answer the following by ticking ( ) the appropriate answer(s).

The four offspring could have been

4 chestnut	( )
3 chestnut and 1 black	( )
2 chestnut and 2 black	( )
1 chestnut and 3 black	( )
4 black	( )

FIGURE 10.1. Problems Used by Kinnear (1983).

second working concept that applies in non-genetic "real world" situations . . . . Students are using a strategy which largely ignores the operation of chance when they are solving genetics problems. (p. 89)

Kinnear's results raise serious questions about the sufficiency of probabilistic reasoning as a basis for success in genetics problem solving. Challenges to the Piagetian claims about genetics problem solving have also been made by Smith and Good (Smith, 1986a; Smith & Good, 1983). They compared the problem-solving performance of 11 novices (undergraduate science and nonscience majors who had just completed their first genetics instruction) and 9 experts (genetics graduate students and genetics faculty) on a set of complex genetics problems with their performance on proportional, combinatorial, and probabilistic reasoning tasks. They claimed that:

Formal operational thought was clearly demonstrated to be an insufficient condition to determine problem-solving success. Successful manipulation of genetic combinations by non-formal subjects was also observed, suggesting that the formal operational schema of combinations may be an unnecessary condition to successfully solving the selected problems as well. (Smith & Good, 1983, p. 6)

TEXTBOOK PROBLEMS THAT EMPHASIZE MEIOSIS

Genetics problem-solving researchers at the University of Wisconsin have described the knowledge organization and problem-solving performance of students and geneticists. They have also developed a framework for problem-based teaching (Stewart & Van Kirk, 1990) that stems from their belief that, although the ability to solve problems has been used to indicate whether learning has taken place, problem solving as a means of learning has often been ignored. Stewart (1988) argued that problem solving may allow students to better understand the conceptual structure of genetics, discipline-independent problem-solving heuristics, genetics-specific problem-solving heuristics and algorithms, and the nature of science as an intellectual activity.

Wisconsin researchers first studied the procedural and conceptual knowledge of genetics and meiosis used by high school students to solve textbook problems that required them to reason from causes (known information on inheritance patterns, such as which variation of a trait is dominant) to effects (such as the prediction of offspring phenotype ratios). Because these problems can be solved by using algorithms, correct answers do not necessarily measure students' understanding of genetics (Stewart, 1983; Stewart & Dale, 1981, 1989; Stewart, Hafner & Dale, 1990).

In these studies, data were also gathered on students' ability to justify their solutions in terms of the mechanism of meiosis. They were asked to think out loud as they solved cause-to-effect problems and were interviewed after the problem-solving sessions. A large percentage of the students were able to obtain correct answers to the problems. However, the answers (both correct and incorrect) masked the fact that most students:

1. Obtained answers by applying taught algorithms, without fully understanding the underlying concepts.
2. Engaged in interesting and understandable thinking, albeit often with incorrect knowledge of chromosome-gene models.

Based on these conclusions, Stewart (1983) argued that traditional classroom assessments of knowledge (simply having students solve problems) do not provide adequate insight into what students know or do not know. At times this leads to overestimation of knowledge (when correct answers are taken as indicators of correct conceptual knowledge) and at other times to underestimation of thinking ability (when a student is able to manipulate an incorrect model to account for an answer). Overestimation of student knowledge is an obvious problem for teachers, but underestimation is significant as well.

Model building is at the heart of science, but students who do not obtain correct answers to problems are not recognized as engaging in this important activity.

Interesting examples of model building with incorrect knowledge can be seen in students' attempts to justify their solutions to problems with models of chromosome-gene behavior during meiosis (Stewart & Dale, 1989). Most students were able to use their knowledge, correct or not, to develop models of meiosis to explain their problem solutions. Whereas a correct chromosome-gene model for the problem solved had four chromosomes, most students described either one-chromosome, two-chromosome, or incorrect four-chromosome models (examples of these incorrect models are shown in Figures 10.2, 10.3, and 10.4). The significant feature of these alternative models is that they were functional. These students were able to use their models, which followed logically from their conceptual knowledge, to account for their problem solutions. Although few of these models corresponded to those that were presented during instruction, the students could show the genetic makeup of the parents using chromosomes and genes. In addition, they could use these chromosome-gene diagrams to determine possible gametes from the parents and possible offspring genetic makeups.

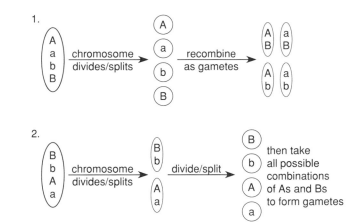

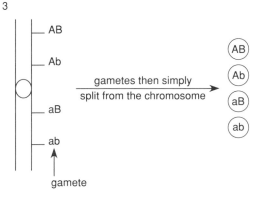

FIGURE 10.2. Students' One-Chromosome Models.

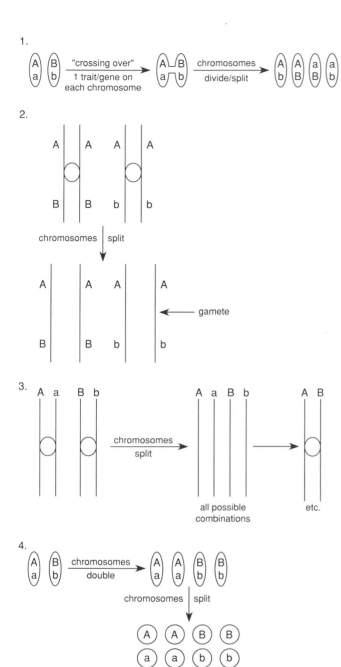

FIGURE 10.3. Students' Two-Chromosome Models.

Kindfield (Hildebrand, 1989; Kindfield, 1991a, 1991b, 1991c, 1991d) has extended the understanding of the knowledge of meiosis possessed by experts (geneticists), experienced novices (honors undergraduates who had completed a university genetics course), and inexperienced novices (undergraduates who had completed an introductory university biology course). The significance of her research stems from the detailed exploration of how individuals use their knowledge of meiosis in diverse situations and the role that diagrams play in problem solving. Kindfield (1991d) documented a number of misconceptions about meiosis (see Table 10.1) and, on the basis of this, made suggestions for improving instruction. Even though her focus was on university students, her results can be applied to understanding the difficulties high school students have in learning about meiosis.

In addition to detailing a wide range of specific misconceptions, Kindfield provided a glimpse of how successful (typically experts and advanced novices) and less successful (typically beginning novices) solvers use their knowledge of meiosis when solving problems. Successful solvers used diagrams of meiosis to organize and focus their reasoning as they solve problems; less successful solvers recalled diagrams that they were taught but did not use the diagrams to "think with." Successful solvers also used meiosis diagrams to recall and accumulate relevant knowledge as they were engaged in solving problems. In addition, they used diagrams they had drawn to check the accuracy of their reasoning.

Kindfield's research contains rich insights into how successful and less successful solvers think. These insights can be used to inform instruction in which it is recognized that "a robust understanding of meiosis includes a variety of pictorial skills that are virtually never addressed in instruction" (Kindfield, 1991b, p. 59).

TEXTBOOK PROBLEMS: EFFECT-TO-CAUSE REASONING

Two researchers who have studied textbook problem solving requiring effect-to-cause reasoning are Smith and Hackling. Each has used problems in which solvers were provided with phenotype data, including pedigree problems, and had to describe a genetic mechanism that could have produced the data.

Smith (Smith 1986a, 1986b, 1988a, 1988b, 1990; Smith & Good, 1983, 1984a, 1984b) focused his research on the following questions:

1. What comparisons can be made among solvers' problem-solving performance, background expertise, and problem-solving success?
2. What problem-solving behaviors are found in genetics that parallel those reported in other domains?

Using the constructs of successful and unsuccessful solvers (Smith recognized that novices may be successful and experts unsuccessful), he reported a set of "behavioral tendencies" that

While not giving students the same credit for alternative models of meiosis, Moll and Allen (1987) reported similar findings:

The misconceptions exhibited by many students ... revealed specific difficulties. Most difficult was relating a genetics problem statement to the organization of chromosomes and genes in a cell.... The most common difficulty in this respect was the tendency to place both alleles for a trait on one chromosome of a homologous pair. Deficiencies in students' understanding of chromosome movements during meiosis were also apparent. (p. 232)

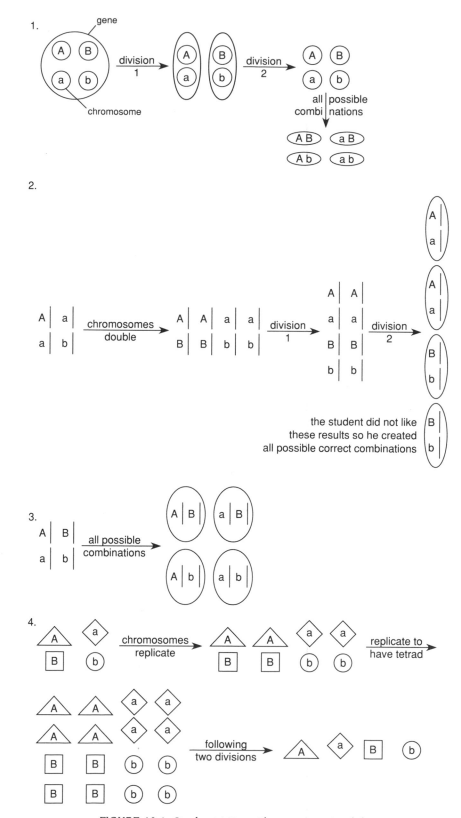

FIGURE 10.4. Students' Four-Chromosome Models.

TABLE 10.1. Meiosis Misconceptions Reported by Kindfield (1991b)

Chromosome Misconceptions

a. Sister chromatids of replicated chromosomes are not physically attached to one another at any time during the process of meiosis.

b. Single, rather than paired, replicated chromosomes align and separate during meiosis I; homologous chromatids then join in the resulting intermediate cells to form two-chromatid-plus-centromere entities.

c. Confusing chromosome structure with ploidy (chromosome number).

Process Misconceptions

a. Replication misconceptions: the common replication misconceptions involved its timing. This was part of a larger misconception concerning the relationship of meiosis to the cell cycle.

b. Crossing-over misconceptions: here, too, the timing of the process was not commonly understood. In addition, many of Kindfield's solvers did not recognize the role that the "presence and position of the centromere to crossing over" (p. 18). Also, solvers tended to think of chromosome pairing as part of the process of crossing over rather than as a prior, separate event.

c. Alignment and segregation misconceptions: the common misconception in this category involved the timing of the pairing of chromosomes. Solvers typically had single replicated chromosomes pair during meiosis I and homologous chromosomes pair and align during meiosis II rather than the reverse.

d. Replication was thought to be part of the process of meiosis rather than a temporally distinct process (another indication of how meiosis relates to the cell cycle).

e. Various meiotic events "could occur at more than one temporal position in the sequence of events that constitute meiosis rather than each event occurring once at a unique time." (p. 23)

f. Lack of recognition that each of the specific events of meiosis occurs independently of other events. Here it is thought that the solvers had some knowledge of the events of meiosis, but not of their timing, so they "mixed and matched" these components in convenient places in order to account for what they felt were the necessary gamete combinations that could result from a meiotic division.

successful, but not unsuccessful, solvers engaged in as they solved problems. The most significant among them were:

1. The perception of successful solvers that the task is not simply to arrive at an answer. Successful solvers recognize that they need to develop a solution that is both internally consistent and externally valid with respect to the larger body of genetics knowledge. For them, a complete solution involves a great deal of explanation and justification.
2. That successful solvers take a forward-working, or knowledge development, approach to the solution. Less successful solvers are more likely to use a backward-working (means-ends analysis) approach.

In addition to these more general heuristics, Smith reported on differences in the knowledge of genetics between his successful and unsuccessful students (see Table 10.2). Smith describes the knowledge organization of successful solvers as condition-action units that allow an appropriate action to be taken when an appropriate condition for that action is recognized. For example:

Condition: Does the mating of two unaffected individuals produce an affected individual? If so,

Action: Identify the trait as recessive (assuming complete penetrance) (Smith, 1988b, p. 414).

Smith (1988b) then used the pedigree problem data from the 20 subjects in his original study (Smith & Good, 1983, 1984a, 1984b), and five new subjects (three students and two genetics instructors) solving that problem, plus two additional ones, to extend his claims. He reported that solvers use one of two strategies: they look for cues in the pedigrees that suggest a particular inheritance mechanism, or they list a hypothesized

TABLE 10.2. Characteristics of Successful\Moderately Successful Subjects Compared with Unsuccessful Subjects

Unsuccessful Subjects Tendencies (1)
Successful or Moderately Successful Subjects Tendencies (2)

1. To perceive a problem as a task to recall patterns.
2. To perceive a problem as a task analysis and reason.

1. To attempt one-step solutions.
2. To break the problem into parts and solve it step by step.

1. To use a means-ends approach, often using trial and error.
2. To use a knowledge-development approach, not often using trial and error.

1. To produce unacceptable solutions.
2. To produce acceptable solutions.

1. To emphasize and use unimportant or irrelevant information.
2. Not to emphasize or use unimportant or irrelevant information.

1. Not to know as much genetic information (e.g., common genetic ratios and the conditions for using them).
2. To know more genetic information including ratios.

1. Not to spontaneously categorize problems.
2. To spontaneously categorize problems according to deep structure.

1. To not attempt to explain biological events in the problem information.
2. To offer accurate biological explanations of the event in the problem.

1. Not to know typical procedures for modeling the events in the problems (e.g., meiosis).
2. To know typical procedures for modeling the events in the problems.

TABLE 10.2. Continued

Unsuccessful Subjects Tendencies (1)
Successful or Moderately Successful Subjects Tendencies (2)

1. Not to be accurate and detailed in their bookkeeping.
2. To use accurate and detailed bookkeeping procedures.

1. Infrequently to recheck the problem statement after obtaining a solution.
2. Frequently to recheck the problem statement after obtaining a solution.

1. Not to make comments evidencing planning.
2. To make comments evidencing planning.

1. To use few general heuristics.
2. To use a broad range of general heuristics.

1. To consider symbol selection and definition as an abbreviation process.
2. To consider symbol selection and definition as a transformation process.

1. Infrequently to use the fact that the sum of all probabilities is 1.0.
2. To use the fact that the sum of all probabilities is 1.0 as a check on their accuracy.

1. Infrequently to check to see if the obtained solution seemed intuitively correct.
2. Frequently to check to see if the obtained solution seemed intuitively correct.

1. To more frequently check to ensure that the gene number was maintained in the procedure used.
2. Not to check to ensure that the gene number was maintained.

1. Frequently to make an overt check of whether or not certain work had been done.
2. Infrequently to make overt checks of whether or not certain work had been done.

Source: Data from Smith, 1984a.

mode of inheritance by assigning possible genotypes to individuals. Smith found that unsuccessful solvers:

1. Made decisions based on errors of logic. Successful subjects were aware of the difference between evidence that falsified a hypothesis and evidence that merely weakened it. Successful subjects also realized when data indicated that a particular inheritance pattern was or was not operating. The unsuccessful subjects were less able to rule in or rule out inheritance patterns.
2. Made decisions that ignored essential assumptions. Many of the problems contained stated assumptions that successful subjects tended to but unsuccessful subjects did not.
3. Made decisions that ignored earlier work. Unsuccessful subjects showed lack of consistency of thought throughout a problem. For example, an individual might have been la-

beled heterozygous at one point in the solving process and then, with no conflicting data, labeled homozygous later on. Slack and Stewart (1990) characterized this as lack of generational thinking, which leads to each new cross being treated as a new problem rather than part of a single problem involving successive generations.

Smith (1984b) described a developmental sequence for the emergence of expertise in genetics problem solving:

1. The trial-and-error stage was exhibited by unsuccessful subjects. Even though Smith calls this strategy trial and error, it requires relatively extensive knowledge of genotype, phenotype, sex-linkage, and cross-diagramming procedures. The difficulties for those who used this process were related to the interpretation of cross results. Smith hypothesized that the solver's conceptual genetics knowledge may not have been organized for solving problems that required reasoning from effects to causes.
2. In the second stage, solvers spent little time dealing with unfruitful options because they were able to discriminate between fruitful and unfruitful, based on their knowledge of inheritance patterns.
3. The final stage was represented by successful solvers who used powerful forward-working, knowledge-producing search strategies that were driven, often tacitly, by the logic of genetics.

Smith realized that this continuum cannot simply be used to characterize the performance of an individual. Experts or successful solvers, when faced with particularly challenging problems, often resort to more trial-and-error procedures.

In a more recent study, Smith (1988a) studied the problem-categorization schemes used by various subjects when sorting a group of problems. In physics, Chi, Feltovich, and Glaser (1981) claimed that experts categorize problems on the basis of deep structure (major organizing principles in a discipline), whereas novices categorize on surface structures (relatively unimportant features of a problem, such as that they all involve inclined planes). Thus different categorization schemes can be taken to represent differences in knowledge organization of experts and novices, or, in Smith's case, successful and unsuccessful solvers, and can be used to interpret differences in problem-solving performance. Smith had genetics faculty, genetics counselors, and university genetics students sort a set of 28 problems into categories based on how they would solve them. In addition, they were asked to solve four problems. Using multiple correlations and cluster analysis Smith analyzed the sort patterns by group to which the solver belonged (faculty, counselor, or student) and by whether they were successful (at least one of the four problems correct) or unsuccessful (no problems correct). He found that genetics counselors categorized problems more similarly to the students than they did to the genetics instructors. The counselors and students categorized on surface features, although the counselors focused more often "on the procedures to be used in the solution, e.g., drawing a pedigree (when none is given) as an aid to the solution process" (p. 10).

Hackling (Hackling 1984a, 1984b, 1986, 1990; Hackling & Lawrence, 1988) has also contributed to the understanding of problem solving in classical genetics. Employing a combination of paper-and-pencil tests and think-aloud protocols, he studied how high school students and teachers, university students and teachers, and genetics counselors solve pedigree problems. He was most concerned with the hypothesis-testing strategies that solvers use, their recognition of cues in pedigrees that lead to the generation and testing of hypotheses, and whether solvers take a confirmationist or falsificationist approach to hypothesis testing. He was also concerned with the efficiency of problem solvers and how it might be possible to use the strategies of experts (genetics lecturers and genetics counselors) to improve the efficiency of novices (Hackling, 1990).

In his early studies (Hackling 1984a, 1984b), three standard and not particularly challenging problems were presented to three high school biology teachers and two university biology lecturers. The subjects were asked to think aloud as they solved the problems. Hackling then analyzed transcripts of the think-aloud protocols to determine what "keys were recognized and used to hypothesize the most likely mode of inheritance . . . and the extent to which hypotheses were tested by assigning genotypes to individuals in the pedigrees" (p. 4).

From the data, Hackling produced a "Model of the Strategies Used by Experts to Solve Genetic Pedigree Problems." The salient features of the model are that experts:

1. Peruse the pedigree to identify key features indicative of the most likely mode of inheritance.
2. Generate the most plausible hypothesis and then test it by assigning genotypes to all the individuals in the pedigree.
3. Test some alternative hypotheses in an attempt to disprove them and confirm the initial hypothesis.
4. Search for any evidence that could be used to determine which of the two hypotheses was most probable.

With this model, Hackling examined students' (novices') performance on identical problems. As would be expected, experts (high school teachers and university lecturers) were more successful in obtaining correct answers. In addition, experts recognized and used more cues to identify potential modes of inheritance and to limit hypotheses.

The aim of Hackling's next study (1990) was to compare genetics counselors', genetics lecturers', advanced students', and novice students' use of genetics knowledge to generate and test hypotheses while solving pedigree problems. Each subject was presented with five problems. Two of the problems had a single mode of inheritance that was tenable (autosomal dominant or recessive), two had two tenable but a single most probable mode of inheritance (X-linked dominant or Y-linked recessive), and the last had several possible modes of inheritance. Think-aloud protocols were gathered from each subject for each problem and analyzed according to (1) cues critical to the solution of the problems, (2) the best interpretation of each of the cues, and (3) the consensus pattern of inheritance of the trait for each of the problems. For the most part, the two expert groups were more successful at identifying the correct mode of inheritance and produced more complete solutions than did the student groups.

Hackling also identified and analyzed the use of solution processes. One process was the use of cues in the pedigree to strengthen or weaken hypotheses about the mode of inheritance. The two expert groups utilized more cues than the two novice groups. A second solution process was the assigning of genotypes in the solution so that the subjects could test hypotheses about inheritance patterns. Hackling argued that the use of genotypes followed a sequence in the development of expertise. The first stage, used by novices, involved the construction of Punnett squares to represent particular matings in the pedigree. Once this was done, solvers checked to see if the ratio of phenotypes produced in the pedigree mating agreed with the phenotypic ratio predicted by the Punnett square, even though predicted phenotypic ratios are rarely realized in the small sample sizes represented in the pedigrees. The second stage, used by the more successful novice students and by advanced students, involved listing hypotheses by writing the genotypes, on an allele-by-allele basis, for every individual represented in the pedigree. If genotypes could be assigned to all individuals in the pedigree and were logically consistent with each other, the hypothesis was supported. The third stage, employed by successful advanced students and genetics lecturers, involved tracing the path of transmission of the trait down through the pedigree. These solvers thus attacked the problem at the level of genotypes rather than at the level of individual alleles.

Hackling (1990) then used a combination of a cue interpretation task and a pedigree genotype completion task to study pedigree cues, trying to determine which cues experts and novices were able to interpret correctly and which they misinterpreted, as well as the ability of novices to assign genotypes to autosomal and Y-linked modes of inheritance.

He was able to identify areas with implications for the teaching of genetics. Most significantly, he identified difficulties that novices have in identifying cues related to Y-linked dominance and recessiveness. Hackling also concluded that his earlier-mentioned paper-and-pencil test had as "high predictive validity for cue interpretation performance as pedigree problems."

Based on his research, Hackling (1990) designed instruction to teach novices the pedigree problem-solving knowledge and strategies that expert geneticists used. Compared with a control group, instructed students were able to obtain more correct answers and to provide more complete solutions to problems.

Results from Hackling's research program can be summarized as follows:

1. Successful solvers of pedigree problems test multiple inheritance pattern hypotheses while attempting to falsify alternative hypotheses. This success develops in four stages: use of Punnett squares, assigning genotypes on an allele-by-allele basis, an automated method of assigning genotypes, and cue interpretation. The more expert hypothesis-testing strategies appear to develop as proceduralized forms of earlier strategies, and solvers who use them are more conceptually driven and metacognitively aware than those who don't.

2. Optimal solutions are characterized by two features—extensive use of cues and sparing use of genotypes for testing hypotheses. This is particularly true of experts, who use a range of cues to evaluate alternative hypotheses at each of the decision-making points along a solution path.

3. Novices are far less successful at interpreting cues related to X linkage than cues related to dominant and recessive. They are also far less successful in assigning X-linked dominant genotypes to individuals in a pedigree than autosomal dominant genotypes. Most novice students lack the knowledge needed to test hypotheses related to X-linked modes of inheritance.

4. It is possible, through instruction, to teach students to use pedigree cues and other expert strategies, resulting in their correct identification of more modes of inheritance and production of more complete and conclusive solutions than a control group.

5. Instruction in pedigree analysis for novice genetics students should focus on the knowledge and skills necessary to:
   a. Assign autosomal and X-linkage genotypes to individuals.
   b. Use a multiple hypothesis-testing strategy with falsification of alternative hypotheses.
   c. Recognize and interpret critical cues that can be used to discriminate between dominant and recessive and between autosomal and X-linked modes of inheritance.
   d. Integrate pedigree cues into a decision-tree heuristic for efficient and systematic testing of inheritance hypotheses.
   e. Develop metacognitive awareness of the need to search the pedigree for cues to resolve specific decisions regarding alternative inheritance hypotheses.

Hackling's work holds promise for developing instruction to improve teaching and learning of genetics in high schools and universities. His emphasis on the best and most efficient solutions, however, raises some questions. Researchers or other external experts should judge what is "best" cautiously, as problems are not contextless. Solvers create a context that may differ from that assumed by the researcher. For example, on one simple problem in Hackling's 1990 study, more novices than genetics counselors obtained a correct solution. Based on their experience in clinical situations, the counselors unexpectedly transformed what was meant to be a straightforward problem, involving autosomal dominance, into something entirely different. From their clinical experience they expected nonviable offspring in the pedigree if autosomal dominance were operating. In the absence of such information, they transformed the problem into one involving multifactorial inheritance. Researchers who look for best solutions, where best has been determined on some a priori basis, may miscategorize individuals and not understand the thinking processes of the solvers.

A more theoretical issue is also raised when "best" solutions are emphasized. Should researchers make prescriptions for instruction that are couched in the language of optimality and efficiency? It is not obvious or necessary that efficiency and understanding are congruent. This was beautifully stated by an expert in Collins's (1986) study, who said that any cross is a good cross as long as something can be learned from it. By using efficiency as a standard, much of the "fits and starts" work

of expert problem solvers, their blind alleys and "productive mistakes," may be overlooked, to the detriment of novices struggling to develop expertise.

COMPUTER-GENERATED EFFECT-TO-CAUSE PROBLEMS

Many researchers have studied solvers as they solve effect-to-cause problems generated by microcomputers. Such problems cannot easily be solved by using algorithms, and they require solvers, at some point in a solution, to reason from effects (phenotype data) to causes (the genotype producing the phenotype data). Thus, instead of being given genetics patterns as part of the problem statement, the solver faces what a geneticist looking at a population faces—a group of organisms exhibiting various traits with diverse variations. Without appropriate underlying conceptual knowledge, these problems are not readily solved by the applications of algorithms, so they are more likely than typical textbook problems to promote the four learning outcomes mentioned earlier—understanding of the conceptual structure of genetics, of general problem-solving heuristics, of genetics-specific problem-solving heuristics, and of the nature of science as an intellectual activity (Stewart, 1988). In addition, computer-generated problems require that solvers generate as well as interpret data.

Researchers have used four different programs to generate problems. Jungck and Calley's (1984, 1985) GENETICS CONSTRUCTION KIT (GCK) has been used by researchers at the University of Wisconsin (Albright, 1987; Collins, 1986; Collins & Stewart, 1987; Collins, Stewart, & Slack, 1988; Hafner, 1991; Slack & Stewart, 1990). Kinnear's CATLAB, BIRDBREED, and KANGASAURUS have been used by Kinnear (1986; undated); Krajcik, Simmons, and Lunetta (1988); Peard (1983); and Simmons and Kinnear (1991). The following discussion focuses on the research done at the University of Wisconsin, with references to other studies where appropriate.

Geneticists' Problem Solving. Using GCK, Collins (1986) developed a model of Ph.D. geneticists' problem-solving performance. She reported that their knowledge was organized by problem-solving schemata composed of general genetics knowledge, as well as knowledge about particular genetics phenomena. These schemata included representations of phenotype-to-genotype mappings (which Collins called "abstract expression charts") for inheritance patterns that allowed the geneticists to move back and forth easily between the phenotype and genotype levels. Another feature of experts' schemata was abstract cross rules that they used to select crosses and to interpret the resulting data.

Geneticists also had a high-level, problem-solving agenda that interacted with their problem-solving schemata and consisted of a cycle of:

1. Redescribing field population phenotype data in terms of the number and names of traits and variations of each trait.
2. Using this redescription to generate a tentative inheritance pattern hypothesis.

3. Synthesizing a solution using the initial hypothesis to access an inheritance-pattern schema, which was then used to guide cross selection, hypotheses about crosses, and cross interpretations.
4. Redescribing the phenotype data from crosses.
5. Considering other phenomena that could produce the same data.
6. Checking the solution for accuracy.

The problem solving of geneticists was hypothesis driven, with an initial hypothesis formulated early in the problem-solving process, often in response to the data in the field collection. After the hypothesis generation came solution synthesis, in which the geneticists performed crosses, redescribed cross data, and interpreted cross data. They first tried to confirm their tentative inheritance pattern hypotheses in terms of developing problem-specific genotype-to-phenotype mappings. They used cross rules to explain data from crosses and to predict data for crosses about to be performed. At the beginning of hypothesis testing, abstract genotype entries of the expression chart were tentatively known while the phenotype entries were blank. If a cross produced data consistent with what could be expected given the legal cross rules for that inheritance pattern, the data were used to complete the expression chart. The following comment illustrates features of expert performance and schemata:

I've got four classes each of males and females [**data redescription**] so there is no reason not to think it is simple dominance [**invoking a schema**] so I'll cross the dw's (dumpy wing, white eye) with the sc's (shiny wing, cinnabar eye) and all the offspring are dw (dumpy wing, white eye), so if d (dumpy wing) and w (white eye) are dominant, the offspring are all heterozygotes [**use of a cross rule**] (Collins, 1986, p. 121)

In this example, redescription allowed the expert to identify cues to access an inheritance pattern schema, with its corresponding expression chart and cross rules. The expert then decided to use the cross rule "if one parent is homozygous dominant and the other parent is homozygous recessive, the offspring will be heterozygous and have a dominant phenotype." This cross rule was used to predict the distribution of variations among the offspring.

Once the experts had synthesized solutions, they assessed them for completeness and accuracy. In solution assessment, additional confirming evidence was often collected. Assessment sometimes included performing a chi-square test, although the experts instead usually compared the ratios by intuition. Experts also increased their confidence in the accuracy of their inheritance pattern and modifier hypotheses by doing additional crosses that could be explained or predicted from appropriate cross rules.

Despite having rich, accurate, well-developed schemata, geneticists did not solve all problems correctly. For example, they had difficulty solving multiple-allele problems that did not follow the ABO blood type pattern. In such situations they often overlooked cross data supporting the multiple-alleles hypothesis and assumed gene interaction as the inheritance pattern.

This example indicates the strength of the inheritance pattern schemata of experts. Once a schema was accessed, an expert tended not to abandon it until the problem space it defined was completely searched. In these situations, cues to other inheritance pattern schemata were regularly overlooked.

In a study of faculty and advanced graduate students, Simmons and Kinnear (1991) found that their subjects followed an agenda similar to that described by Collins. The stages of problem solving they found were "identifying the problem, representing the problem, reformulating the problem, planning the strategies, implementing the strategies, evaluating the strategies, and evaluating the solution in the context of the entire problem" (p. 4). One interesting feature of this study was the use of a data-gathering approach that had been reported previously by Krajcik et al. (1988). It involved sending the computer's video output directly to a videocassette recorder, so that the sequence of interactions that solvers had with the computer program could be replayed.

Student Problem Solving. In contrast to geneticists, students have neither well-organized schemata nor well-defined problem-solving agendas (Slack & Stewart, 1990). This difference is evident early in a solution, as students were less systematic than experts at data redescription. Rather than explicitly identifying the names and numbers of the traits and names and numbers of variations associated with each trait, students tended simply to read all of the information from the computer screen. They usually looked quickly at the initial population data, without generating hypotheses, before performing crosses. Slack and Stewart speculated that students do this because, even though they have learned about simple dominance and codominance, they have not organized that knowledge into inheritance-pattern schemata containing abstract expression charts and cross rules that would facilitate problem solving. Kinnear (undated) underscores this point. She gave a group of 68 college biology students two textbook problems (one involving simple dominance and the second autosomal linkage) and two analogous problems generated by CATLAB. The students were able to solve the paper-and-pencil problems with no perceived difficulty. Yet nearly a third of those who answered the textbook problems correctly "showed defects in their problems-solving strategies when faced with the computer-based problem" (p. 10). Kinnear postulated that the algorithms students learned to solve the textbook problems were insufficient and did not include appropriate or sufficient conceptual knowledge for success on the computer problems.

Slack and Stewart (1990) also reported that students did not use inheritance patterns to organize their problem solving. As a result, solution synthesis and hypothesis testing were less goal directed for them than for Collins' experts. There was little evidence of students attempting to confirm inheritance-pattern hypotheses by completing expression charts. Rather, their hypotheses were about individual crosses. In addition, their cross rules were often inaccurate or incomplete, resulting in unwarranted inferences about cross data. For example, a cross in which parents with like variations produced offspring of the same variation was frequently used by students as an indicator that all of the individuals were homozygous recessive. However, three other cross rules could explain the same data: both parents could have been homozygous dominant; one parent could have been homozygous dominant and the other heterozygous; or, less likely, both parents could have been heterozygous.

Students did not have the knowledge to identify features in the data that would provide access to an inheritance-pattern schema. In addition, they showed limited evidence of hypothesis testing within an inheritance-pattern schema. Sometimes they formulated tentative hypotheses after the initial data redescription, such as "There are more females than males, therefore this must be sex-linked," or "There are more red, so red must be dominant." Such conclusions are unwarranted, however. Most students explained the results of their crosses using incomplete or inaccurate cross rules; very few selected crosses carefully, predicting the results of a cross before actually performing it. One student actually said:

You know, I didn't expect, until I got down here to the end where I realized, on that last problem, that the results woulda been 1 to 1. But I think that was the first time that I, that I really thought out in advance what should happen.

In contrast to experts, students rarely confirmed their solutions. When they tried to, they (1) repeated crosses that they had used to determine solutions, using the same parents or parents from the same class of phenotypes, or (2) looked back at the crosses they had performed to make sure that the solutions were consistent with these data, ignoring the crosses they couldn't explain. This is an indication that students did not engage in genotypic thinking (solving the problem in terms of allele relationships) or generational thinking. That is, students, unlike geneticists, did not solve problems in terms of familial inheritance, including looking at patterns in the data over at least three generations.

From the research on computer-generated effect-to-cause problem solving, important differences can be seen between the approaches of students and geneticists. One primary difference is that geneticists' knowledge is highly integrated, organized into schemata consisting of both content and procedural knowledge. Their content knowledge consists of genotype-to-phenotype matches for a particular inheritance pattern, organized into mental expression charts that allowed them to move back and forth between genotypes and phenotypes. Their procedural knowledge includes cross rules used to test tentative hypotheses about inheritance patterns or crosses. Students' knowledge was less extensive and less highly structured. However, it is important to reiterate that students were able to describe and manipulate mental models of meiosis, a significant intellectual accomplishment, even though the models were often different from those their teachers expected.

Students also have "misconceptions" that interfere with their problem solving. Among the most significant that have been reported are that students:

1. Tend to view ratios as deterministic rather than probabilistic. (This was pointed out by most of the researchers previously mentioned and by Browning and Lehman, 1988.)

2. Confuse the dominant variation with the most frequent one. (This was pointed out by all of the researchers previously mentioned and by Peard, 1983, and Tolman, 1981, 1982.)
3. Did not think generationally. That is, they often treated a new cross as a new problem.
4. Did not think genotypically.
5. Because of alternative conceptions of how meiosis occurred, had difficulty justifying solutions to genetics problems in terms of meiosis.

EXTENDING PROBLEM-SOLVING RESEARCH

A motivation for extending the realm of problem-solving research comes from work by philosophers and sociologists of science who have pointed out the importance of adopting a forward-looking rather than a retrospective view of science (Latour, 1987). With this emphasis comes increased attention to "scientific discovery" (the generation and development of bodies of knowledge), the closely aligned increased interest in scientific problems as units for analysis, and, important for our purposes, increased attention to the generation and development of bodies of knowledge as well as their appraisal.

In this vein, Nickles (1981, 1987) has suggested that insight into the nature of the reasoning associated with the generation and development phases of scientific inquiry may derive from an improved account of problems:

Problem contexts include much more than empirical datum constraints on adequate problem solutions. Large conceptual problem contexts contain constraints of many kinds, many of them previous theoretical results, which function as consistency conditions, limit conditions, derivational requirements, and so on. There are general methodological demands and perhaps content-specific programmatic demands as well. . . . These constraints on an adequate problem solution are not just there in the problem situation: together they largely define the problem and give it structure. (Nickles, 1981, p. 36)

A problem thus consists of all the conditions or constraints on the solution plus the demand that the solution, an object satisfying the constraints, be found. Van Balen (1987) uses the problems that the Mendelian research program faced at the turn of the century as an illustration of this type of analysis.

By recognizing that problems are not merely a matter of explaining data, problems themselves, and not just the theories that solve them, can be considered conceptually deep and interesting. The constraints characterizing such problems constitute a rich supply of premises and content-specific heuristics for reasoning toward solutions, thus providing a guide to inquiry by restricting its path. At the same time, the constraints on a problem solution must be flexible if conceptual change is to be possible. For example, the constraint on the ontological status of the nonobservable Mendelian factors, as well as the constraint that every theory of heredity should have to account for the problem of development, had to be violated for the Mendelian research program to adopt the chromosome theory (Van Balen, 1987). The result is an "essential tension" between tradition and innovation that is rooted in the problem-solving process (Kuhn, 1962).

Furthermore, contemporary developments in history and philosophy of science and cognitive science suggest that the growth of knowledge in a discipline and the growth of knowledge in individuals share common principles and can be described in a common language. For example, in science education Duschl (1988) has argued for a strong link between psychology and epistemology. By arguing that teaching and learning need to be consistent with the spirit and character of scientific inquiry, both national commissions and professional societies have recommended giving increased emphasis to problem solving, as opposed to teaching science as a "rhetoric of conclusions" (Schwab, 1960) or as "final form science" (Duschl, 1990). Such a position is nicely summed up by Moore (1986) in his observation that: "There is wide acceptance that one important goal of education is to give students experience in solving problems. If this is accepted, then it is necessary to deal explicitly with problem solving and not just with solutions" (p. 587).

Underlying such recommendations is the explicit acknowledgment that concepts are important. Thus, accompanying the calls for problem-based learning is the belief that by situating the conceptual knowledge of a scientific domain in the context of its use in problem solving, students may develop a highly structured and functional understanding of the conceptual knowledge of that domain, general and domain-specific problem-solving strategies, and insight into the nature of science as an intellectual activity.

We feel it is important for problem-solving researchers to acknowledge that, in science, reasons to solve problems go beyond using existing knowledge to obtain solutions. Scientific problem solving also involves generating new knowledge, including new explanatory models. So too, in science classrooms students can solve problems not only to demonstrate their knowledge but also to invent and test new explanatory models. For example, solvers may learn new conceptual or procedural knowledge related to their existing models, leading to *model elaboration*. Or the *real* problem may be *model revising*, so that an existing model fits new and seemingly anomalous data. Such problems require that students work within the context of discovery. When revising models, students may learn new conceptual and procedural knowledge as well as developing understanding of how science is conducted.

In the following section we sketch a view of problem solving that we feel extends it to include **model-elaborating** and **model-revising** situations and hints at new ways of thinking about problem-solving research.

MODEL-ELABORATING PROBLEM SOLVING

One type of problem solving that involves learning is **model-elaborating** problem solving (Stewart & Hafner, 1991). While still bound by the concept of an existing model that includes objects, states those objects can exist in, and processes that act on the objects to cause changes of state, model elaboration includes interesting reasoning, and new insights may emerge from solving problems in which the data do not conflict with the existing models of solvers. Even though model elaboration does not involve the invention of new objects, processes, or states, it does entail sophisticated thinking and is an analogue of much scientific activity. It may result in:

1. Learning more efficient procedures for generating data
2. Developing within-model conceptual insights
3. Linking models because they share objects, processes, or states
4. Linking models to produce a larger framework

Kinnear (undated) quotes a student who had just solved a CATLAB problem involving two genetics phenomena that had been taught independent of each other:

I found this problem interesting as well as frustrating. The majority of the genetics problems on linkage we have done before, we do on the assumption that there is no gene interaction, that is, no epistatic relationships. I imagine, like me, most people would not even consider epistasis unless it was actually stated in the problem. Yet, gene interaction is a part of genetics crosses, and it appears that we only consider epistatic relationships as though they were a totally separate subject. (p. 11)

Little research has been done on how models are elaborated during problem solving. Important research questions in this area concern what students learn from solving problems, what metacognitive strategies they use to learn from solving, and what instructional strategies are likely to promote the development of knowledge via solving problems.

MODEL-REVISING PROBLEM SOLVING

Research is also needed on the thought processes of solvers who encounter data inconsistent with their existing models. This is **model-revising** problem solving, because the solver's primary task is to revise an existing explanatory model to account for anomalous data. The need to revise a model comes about when a solver recognizes that the empirical world fails to fit completely any models that have previously facilitated description, explanation, and prediction. This recognition may prompt solvers to invent objects or processes that did not exist in previous models or to posit new states for "old" objects.

For example, in genetics the simple dominance model leads one to expect that each trait will have two variations. Thus, a solver is able to recognize that three or more variations of a trait is inconsistent with the simple dominance model. With this recognition, the solver is in a situation that requires revising an existing model to account for new data.

The study of students who are faced with anomalous data and are thus expected to revise their models is important because problem solving is not completely circumscribed by what we have called model using and model elaborating. To be complete, any theory of problem solving should account for the thought processes of individuals faced with data that are inconsistent with their existing models. Important research questions on model revising and guidelines for the analysis of research data can be derived from Wimsatt's (1987) insight:

The primary virtue a model must have if we are to learn from its failures is that it, and the experimental and heuristic tools we have for analyzing it, are structured in such a way that we can localize its errors and attribute them to some parts, aspects, assumptions, or subcomponents of the model. If we can do this, then "piecemeal engineering" can improve the model by modifying its offending parts. (p. 30)

Hafner's (1991) research is an example of model-revising problem-solving research. He investigated high school students' model-revising problem solving in the context of an elective 9-week genetics class (Johnson & Stewart, 1991a, 1991b) by asking the following questions:

1. Do students use initial models to develop new models of increasing complexity and realism when they encounter anomalous phenomena?
2. If yes to 1, what is the nature of their model-revising process?
   a. How do successive models vary with respect to objects, states, and processes?
   b. What general and/or domain-specific heuristics do students use to construct successively more complex and refined models?
3. On what basis do solvers decide that a revised model is sufficient (that it has both predictive and explanatory power)?

Six students participated in his study. These students had received instruction on the models of simple dominance (Figure 10.5) and meiosis that facilitated solving realistic problems. Their instruction emphasized important objects (such as genes, alleles, traits, and variations), states that those objects can exist in (such as heterozygous, diploid), and processes that are responsible for those objects changing state (such as segregation and fertilization). The students then used these models to solve realistic monohybrid and dihybrid simple dominance problems generated by the strategic simulation GCK (Jungck & Calley, 1984). After this practice in model-using problem solving, the students encountered GCK problems with the following anomalous phenomena: multiple alleles, gene interaction, pleiotropy, sex linkage, and autosomal linkage. They were then asked to revise their initial models to solve these problems.

Hafner (1991) used three sources of data in his analysis of students' model-revising problem solving:

1. Computer printouts of the initial population produced for each problem, the individual organisms selected for a cross, and the offspring from each cross
2. "Think-aloud" protocols of students' as they solved problems
3. Written materials that resulted from the model-revising problem-solving process

Hafner combined these three types of data into units termed frames, which consisted of that portion of each of the three types of data used with one cross. Frames were created for every cross, for every problem, and for every solver and were

OBJECTS in the Simple Dominance Model	
No. of Traits	1
No. of Variations	2
No. of Genes	1
No. of Alleles/Gene in a Species	2
No. of Alleles/Gene in an Individual	2
No. of Possible Allele Combinations	3

STATES in the Simple Dominance Model
Type of Allele Combinations (1,1) and (2,2): homozygous
                             (1,2): heterozygous
Allele Combinations Can Map to Variations in the Following Ways:
    (1,1) to Variation A
    (1,2) to Variation A
    (2,2) to Variation B
Cross possibilities A x A > All A
                     A x B > All A or All B (depending on
                     A x B > 1/2 A and 1/2 B
                     B x B > All B

PROCESSES in the Simple Dominance Model
Segregation:
Separates one member of a pair of alleles of a parent during gamete production. Segregation changes the state of an object (an allele pair) from double to single.
Independent Assortment:
Assorts allele pairs so that members of the pair for one trait are distributed to gametes independently of the members of pairs for other traits.
Fertilization:
Joins gametes that have a single member of each allele pair so that a zygote has two members of a pair.

FIGURE 10.5. A Model of Simple Dominance.

then analyzed for significant student activity in each of two problem spaces: a space of experiments (in this case crosses performed) and a space of models. Insight into when model revision occurred; how successive models varied with respect to objects, states, and processes; and whether revised models represented an increase in complexity and realism over initial models was derived from an examination of the model space searched for each problem attempted. Examination of the portion of the experimental space (all possible crosses for a given problem type) used to test components of a revised model provided insight into the basis for students deciding a revised model was sufficient. Analysis of the sequence and nature of the interactions between the problem spaces afforded insight into the general and domain-specific strategies utilized in model-revising problem solving.

Model-revising problem solving is a complex and challenging endeavor, yet each student engaged in it successfully. Each of the six students used initial models to develop revised models of increased complexity and realism. Furthermore, all of the students produced revised models that represented the accepted view of the genetic phenomena operating in at least one of the problem types. Overall, revised models were developed for 30 of the 46 problems attempted. All 30 models represented an increase in complexity and 27 (90 percent) of the 30 models represented revisions compatible with accepted genetics theory. The model revisions consisted of:

1. Changes in number of objects; more specifically, the number of alleles
2. Changes in unstated biochemical processes: more than two genotypes mapping to a given phenotype; one gene determining the variations of more than one trait; more then one phenotype mapping to a given genotype; sex influence; and genotypes, rather than sex chromosomes, determining the sex of offspring
3. Changes in state—more specifically, the association or linkage of a gene with sex chromosomes

The revisions were characterized by their cumulative nature over time and problem types. Students utilized and built on revisions which had been developed for prior problem types.

Students used numerous heuristics in their model-revising problem solving. Among the more commonly used heuristics are those included in Table 10.3.

Twenty-five of the 46 problems attempted resulted in a revised model with which students were satisfied. Student satisfaction, and thus model sufficiency, was dependent on a high degree of fit between a model and the crosses performed. Overall, the revised models developed for these 25 problems accommodated 98 percent of the crosses performed. The students utilized their revised models, with which they were satisfied, to explain on average 44 percent and to predict on average 13 percent of the crosses they performed. Model sufficiency was more dependent on explanatory than on predictive value.

Using Hafner's study and Wimsatt's work on models, we believe that the following questions could guide research on model-revising problem solving:

1. Do solvers use oversimplified models as starting points to develop a series of models of increasing complexity and realism? If they do, what objects, processes, or states of existing models do they tinker with as they develop revised models?
2. What anomalous data are overlooked by solvers, and is it possible to account for differences in the oversights in terms of different models that different solvers use?
3. On what basis do solvers decide that a revised model is sufficient (that it has both predictive and explanatory power)?
4. How do different solvers (including experts and students) resolve the tension between tradition and innovation when faced with model-revising situations? What "constraints" does tradition place on innovation in problem solving?
5. Do solvers use two models to define the extremes of a continuum of cases in which the real case is presumed to lie?
6. Do solvers use false models to look for results that are true in all the models and therefore presumably independent of specific assumptions that vary from model to model?

7. Do solvers use incomplete models as templates for estimating the magnitude of parameters that are not included in their models?
8. Do solvers use incomplete models as templates to capture larger or otherwise more obvious effects that can then be "factored out" to detect phenomena which would otherwise be masked or be too small to be seen?
9. Do solvers consistently use general or domain-specific strategies to revise models?
10. Is it possible to improve model-revising problem-solving performance by making solvers aware of the ways in which constraints have limited their problem-solving flexibility or in which simple or false models may facilitate their model revising?

These questions are couched in terms of how individual solvers work to revise models, yet we feel that it is also important to study model elaborating and revising as they occur among groups of solvers (Stewart, Hafner, Johnson, & Finkel, 1992). Even though it is individuals who learn, we feel that the context for model revising should include groups, because science is a community activity. Thus, generating, testing, and persuading peers about the utility of a model should have a place in instruction and in research about model revising. Just as research has provided theoretical insight into model-using problem solving, so should related research on model-revising problem solving contribute to theories of human problem-solving performance and expertise.

The results of such research could influence the organization of instruction in science. Helping students develop model-revising skills would address the needs associated with the "context of discovery," which have been identified as weaknesses in precollege science education. Just as it has been possible to apply model-using problem solving research to prescribe instruction, so should research on model revising help students develop and use the powerful general and domain-specific strategies that may be identified in the model revising of experts or adept students. A longer-term consequence is that students might better understand certain aspects of the nature of science related to the context of discovery:

1. Models determine what will or will not be considered a problem.
2. Science process skills, even those as basic as observation, are not independent of the presuppositions of models. As Hanson (1958) observed, "there is more to seeing than meets the eye."
3. Inquiry is driven by a theoretical view that influences what data are generated and how they are interpreted. Models and theories "tell" scientists what is worth observing and recording.
4. It is best not to think of models or theories as right or wrong but rather as useful or not useful in exploring nature. In other words, they serve a heuristic function.
5. The claims made in science are tentative because science is not solely a process of justifying claims; but it also involves model and theory invention or revising.

TABLE 10.3. General and Domain-Specific Heuristics Used by Students When Revising Models

General Heuristics
Genetics-Specific Instantiations

1. Use existing models as templates to interpret features of the cross space search that: conform to the expectations of the model; do not conform to model expectations (anomaly recognition).

Use the codominance model as a template to interpret the six-variation (three-alleles) multiple-alleles problem type: to identify three variations as heterozygous (codominant) and three variations as homozygous; to recognize six variations for one trait and the cross type, parents with unlike phenotypic variations producing offspring with two new phenotypic variations in the ratio of 1:1, as anomalous.

2. Use existing models as templates to postulate additional causal factors (changes in objects, states, and processes) operating in a new problem type.

Use the six-variation (three-allele) multiple-alleles problem type model as a template to postulate one phenotype mapping to three genotypes (process change) in a four-variation (three-allele) multiple-alleles problem type.

3. Use limiting or defining relations within existing models to: postulate additional causal factors; explore the implications of a change in one component of a model for other components of that model.

After inferring that three variations in a six-variation multiple-alleles problem have homozygous genotypes, use the limiting relation between number of homozygotes and number of alleles to postulate the existence of three alleles. Then use the limiting relation between number of alleles and number of possible allele combinations to identify six possible genotypes.

4. Bracket the case of interest with cases on either side (interpolation).

Bracket a six-variation (three-allele) multiple-alleles problem with the following two models in an attempt at model-revising problem solving: two alleles and three alleles/gene/individual producing four possible genotypes; and three alleles and three alleles/gene/individual producing ten possible allele combinations.

5. Break problems of increased complexity into independent subproblems, applying existing models to each. Subsequently focus on the nature of the possible dependence between the subproblems.

Treat a six-variation (three-allele) multiple-alleles problem type as two independent codominance problems. Subsequently identify the nature of the dependence between those two problems (that variations are shared between the two subproblems and thus the codominance models must share an allele in common).

6. Generally match a revised model to data before making specific mappings.

In applying a six-variation (three-allele) problem type model to a five-variation (three-allele) multiple-alleles problem type, establish the relation between number of genotypes and phenotypes (six genotypes map to five variations and thus two of those genotypes will have to map to one of the variations) before mapping genotypes to specific phenotypes.

7. Use model revisions to determine the structure of the revised cross space and systematically explore that space, using the revised model to accommodate previously unsearched portions of the space.

A proposed model with three alleles and six possible allele combinations, each of which maps to its own genotype, implies a 21-cross possibility cross space with three new cross types phenotypically. Utilize that model to predict results for cross possibilities not previously performed.

8. Utilize or build on model revisions that have worked in the past. A model that can accommodate a large number of problem types, and thus provide explanatory unification within a discipline, is valued.

Utilize a six-variation (three-allele) multiple-alleles problem type model to reason about a four-variation (three-allele) multiple-alleles problem type. Build on or revise that model by adding the process change of having three genotypes map to one phenotypic variation.

9. Use a process of elimination in the search for a causal factor(s).

Perform repeat crosses using individuals with different genotypes. If the cross results are identical with respect to missing classes of offspring by sex, then assume genotypes are not causal with respect to that phenomenon.

SUMMARY

Research on problem-solving performance in science has made and will continue to make contributions to the improvement of instruction. For its potential to be realized, the research must reflect the range of problem-solving activities that scientists engage in and that philosophers of science have begun to

analyze. Thus, research should emphasize questions about what students learn from solving problems (model elaboration) and how they revise models in response to anomalies in data, in addition to questions related to model-using situations. We believe that the research of philosophers of science on models and problem solving includes insights to guide both problem-solving research and instructional development.

References

Albright, W. C. (1987). *A description of the performance of university students solving realistic genetics problems.* Unpublished master's thesis, University of Wisconsin–Madison.

Browning, M., & Lehman, J. (1988). Identification of student misconceptions in genetics problem solving via computer program. *Journal of Research in Science Teaching, 25*(9), 747–761.

Chi, M. T. H., Feltovich, P. J., & Glaser, R. (1981). Categorization and representation of physics problems by experts and novices. *Cognition and Instruction, 5*(2), 121–152.

Collins, A. (1986). *Strategic knowledge required for desired performance in solving transmission genetics problems.* Unpublished doctoral dissertation, University of Wisconsin–Madison.

Collins, A., & Stewart, J. (1987). *A description of the strategic knowledge of experts solving realistic genetics problems* (MENDEL Research Report No. 1). Madison: University of Wisconsin.

Collins, A., Stewart, J., & Slack, S. (1988). *A comparison of the problem solving performance of experts and novices on realistic genetics problems* (MENDEL Research Report No. 3). Madison: University of Wisconsin.

Dardin, L. (1991). *Theory change in science: Strategies from Mendelian genetics.* New York: Oxford University Press.

Duschl, R. A. (1988, April 9). *Scientific theory as schema: An epistemological perspective.* Paper presented at the annual meeting of the American Education Research Association, New Orleans.

Duschl, R. A. (1990). *Restructuring science education: The importance of theories and their development.* New York: Teachers College Press.

Gipson, M., & Abraham, M. (1985, April 15–18). *Relationships between formal-operational thought and conceptual difficulties in genetics problem solving.* Paper presented at the annual meeting of the National Association for Research in Science Teaching, French Lick Springs, IN.

Gipson, M., Abraham. M., & Renner, J. (1989). Relationships between formal-operational thought and conceptual difficulties in genetics problem solving. *Journal of Research in Science Teaching, 26*(9), 811–821.

Hackling, M. (1984a). A model of the strategies used by experts to solve genetic pedigree problems. *Proceedings of the 10th Annual Conference of the Western Australian Science Education Association* (pp. 31–40). Perth, W. Austr.

Hackling, M. (1984b). Expert and novice performance in solving genetic pedigree problems. *Proceedings of the Annual Conference of the Australian Association for Research in Education* (pp. 304–309).

Hackling, M. W. (1986, April). *Hypothesis testing strategies used in the solution of pedigree problems.* Paper presented at the annual meeting of the American Educational Research Association. San Francisco.

Hackling, M. (1990). *The development of expertise in genetic pedigree problem solving.* Unpublished Ph.D. dissertation, Murdoch University, Western Australia.

Hackling, M., & Lawrence, J. (1988). Expert and novice solutions of genetic pedigree problems. *Journal of Research in Science Teaching, 25*(7), 531–546.

Hafner, R. (1991). *High school student's model-revising problem solving in genetics.* Unpublished Ph.D. dissertation, University of Wisconsin–Madison.

Hanson, N. (1958). *Patterns of discovery.* London: Cambridge University Press.

Hildebrand, A. C. (1989). *Pictorial representations and understanding genetics: An expert/novice study of meiosis knowledge.* Unpublished doctoral dissertation, University of California, Berkeley. (UMI order number 9103719)

Johnson, S., & Stewart, J. (1991a). Using philosophy of science in curriculum development: An example from high school genetics. *International Journal of Science Education, 12*(3), 297–307.

Johnson, S., & Stewart, J. (1991b). Using philosophy of science in curriculum development: An example from high school genetics. In M. Matthews (Ed.), *History, philosophy, and science teaching: Selected readings* (pp. 201–212). New York: Teachers College Press.

Jungck, J. R., & Calley, J. N. (1984) GENETICS CONSTRUCTION KIT. COMPress Software, Wentworth, NH.

Jungck, J. R., & Calley, J. (1985). Strategic simulations and post-Socratic pedagogy: Constructing computer software to develop long-term inference through experimental inquiry. *American Biology Teacher, 47*, 11–15.

Kindfield, A. C. H. (1991a, April). *Constructing understanding of a basic biological process: Meiosis as an example.* Paper presented at the annual meeting of the National Association for Research in Science Teaching, Fontana, WI.

Kindfield, A. C. H. (1991b, April). *Understanding a basic biological process: Expert and novice models of meiosis.* Paper presented at the annual meeting of the National Association for Research in Science Teaching, Fontana, WI.

Kindfield, A. C. H. (1991c, April). *Biology diagrams: Tools to think with.* Paper presented at the annual meeting of the American Educational Research Association, Chicago.

Kindfield, A. C. H. (1991d). Confusing chromosome number and structure: A common student error. *Journal of Biological Education, 25*(3), pp. 193–200.

Kinnear, J. (1983). Identification of misconceptions in genetics and the use of computer simulations in their correction. In H. Helms & J. Novak (Eds.), *Proceedings of the International Seminar on Misconceptions in Science and Mathematics* (pp. 84–92). Ithaca, NY: Cornell University.

Kinnear, J. (1986, April). *Computer simulation and problem solving in genetics.* Paper presented at the meeting of the American Educational Research Association, San Francisco.

Kinnear, J. (undated). *Extending problem-solving tasks in genetics by use of computer simulations.* Unpublished manuscript. Lincoln Institute of Health Sciences, Carlton, Victoria, Australia.

Kitcher, P. (1984). 1953 and all that, a tale of two sciences. *Philosophical Review, XCIII*(3), 335–373.

Kitcher, P. (1985). Darwin's achievement. In N. Rescher (Ed.), *Reason and rationality in natural science* (pp. 127–189). Lanham, MD: University Press of America.

Krajcik, J., Simmons, P., & Lunetta, V. (1988). A research strategy for the dynamic study of students' concepts and problem solving strategies using science software. *Journal of Research in Science Teaching, 25*(2), 147–155.

Kuhn, T. (1962). *The structure of scientific revolutions.* Chicago: University of Chicago Press.

Latour, B. (1987). *Science in action.* Cambridge, MA: Harvard University Press.

Moll, M., & Allen, R. (1987). Student difficulties with Mendelian genetics problems. *American Biology Teacher, 49*(4), 229–233.

Moore, J. (1986). Science as a way of knowing—genetics. *American Zoologist, 26*, 583–747.

Nersessian, N. (Ed.). (1987). *The process of science.* Dordrecht: Martinus Nijhoff Publishers (Kluwer Academic).

Nickles, T. (1981). What is a problem that we may solve it? *Synthese, 47*, 85–118.

Nickles, T. (1987). Twixt method and madness. In N. J. Nersessian (Ed.), *The process of science* (pp. 41–67). Dordrecht: Martinus Nijhoff (Kluwer Academic).

Peard, T. (1983). The microcomputer in cognitive development research. In H. Helms & J. Novak (Eds.), *Proceedings of the International Seminar on Misconceptions in Science and Mathematics* (pp. 112–126). Ithaca, NY: Cornell University.

Schwab, J. (1960). The teaching of science as enquiry. In J. Schwab & P. Brandwein (Eds.), *The teaching of science* (pp. 1–103). Cambridge, MA: Harvard University Press.

Simmons, P., & Kinnear, J. (1991, April). *Early problem solving stages of subjects interacting with a genetics computer simulation.* Paper presented at the annual meeting of the National Association of Research in Science Teaching, Lake Geneva, WI.

Slack, S., & Stewart, J. (1990). High school students' problem solving on realistic genetics problems. *Journal of Research in Science Teaching, 27*(1), 55–67.

Smith, M. (1986a, April). *Problem solving in classical genetics: Successful and unsuccessful pedigree analysis.* Paper presented at the annual meeting of the American Educational Research Association, San Francisco.

Smith, M. (1986b, March 29). *Is formal thought required for solving classical genetics problems?* Paper presented at the annual convention of the Society of College Science Teachers and the National Science Teachers Association, San Francisco.

Smith, M. (1988a, April 13). *Expertise, mental representations, and problem-solving success: A study of the categorizations of classical genetics problems by biology faculty, genetic counselors, and students.* Paper presented at the annual meeting of the National Association for Research in Science Teaching, Lake of the Ozarks, MO.

Smith, M. (1988b). Successful and unsuccessful problem solving in classical genetic pedigrees. *Journal of Research in Science Teaching, 25*(6), 411–433.

Smith, M. (1990, April 18). Cell division: Student misconceptions and instructional implications. Paper presents at the annual meeting of the National Association for Research in Science Teaching, Boston.

Smith, M. & Good, R. (1983, April 5–8). *A comparative analysis of the performance of experts and novices while solving selected classical genetics problems.* Paper presented at the annual meeting of the National Association for Research in Science Teaching, Dallas.

Smith, M., & Good, R. (1984a). Problem solving and classical genetics: Successful vs. unsuccessful performance. *Journal of Research in Science Teaching, 21*(9), 895–912.

Smith, M., & Good, R. (1984b, April). *A proposed developmental sequence for problem-solving ability in classical genetics: The trial-and-error to deductive logic continuum.* Paper presented at the annual meeting of the National Association for Research in Science Teaching, New Orleans.

Stewart, J. (1983). Student problem solving in high school genetics. *Science Education, 67*, 523–540.

Stewart, J. (1988). Potential learning outcomes from solving genetics problems: A typology of problems. *Science Education, 72*(2), 237–254.

Stewart, J., & Dale, M. (1981). Solutions to genetics problems: Are they the same as correct answers? *Australian Science Teacher's Journal, 27*, 59–64.

Stewart, J., & Dale, M. (1989). High school students' understanding of chromosome/gene behavior during meiosis. *Science Education, 73*(4), 501–521.

Stewart, J., & Hafner, R. (1991). Extending the conception of problem in problem-solving research. *Science Education, 75*(1), 105–120.

Stewart, J., Hafner, R., & Dale, M. (1990). Students' alternate views of meiosis. *American Biology Teacher, 52*(4), 228–232.

Stewart, J., Hafner, R., Johnson, S., & Finkel, L. Using computers to facilitate learning science and learning about science. *Educational Psychologist 27*(3), 317–336.

Stewart, J., & Van Kirk, J. (1990). Understanding and problem solving in classical genetics. *International Journal of Science Education, 12*(5), 575–588.

Tolman, R. (1981). *An information processing model of genetics problem-solving.* Unpublished manuscript, Center for Educational Research and Evaluation, Louisville, CO.

Tolman, R. (1982). Difficulties in genetics problem solving. *American Biology Teacher, 44*(9), 525–527.

Van Balen, G. (1987). Conceptual tensions between theory and program: The chromosome theory and the Mendelian research program. *Biology and Philosophy, 2*, 435–461.

Walker, R., Mertens, R., & Hendrix, J. (1979). Formal operational reasoning patterns and scholastic achievement in genetics. *Journal of College Science Teaching, 8*(3), 156–160.

Walker, R. A., Hendrix, J. R., & Mertens, T. R. (1980). Sequenced instruction in genetics and Piagetian cognitive development. *American Biology Teacher, 42*(2), 104–108.

Wimsatt, W. C. (1979). Reduction and reductionism. In P. D. Asquith & H. E. Kyburg (Eds.), *Current research in the philosophy of science* (pp. 352–357). East Lansing, MI: Philosophy of Science Association.

Wimsatt, W. C. (1987). False models as means to truer theories. In M. Nitecki (Ed.), *Neutral models in biology* (pp. 23–55). London: Oxford University Press.

•11•

RESEARCH ON PROBLEM SOLVING: CHEMISTRY

Dorothy L. Gabel

INDIANA UNIVERSITY

Diane M. Bunce

THE CATHOLIC UNIVERSITY OF AMERICA

Problem solving in any subject area is a very complex human behavior. This is documented by the vast number of studies and articles that have appeared in research and teaching journals over the past 12 years, reflecting a renewed interest in how students solve problems. Problem solving has always been a stumbling block for students enrolled in chemistry courses, and most teachers of introductory courses recognize this (Silberman, 1981).

This review of research studies on problem solving in chemistry is organized according to factors that appear to influence students' success as problem solvers. The chapter is divided into three main sections that reflect the studies' principal areas of focus: (1) the nature of the problem and the underlying concepts on which the problem is based (as well as students' understanding of these concepts); (2) learner characteristics, that is, how students' aptitudes and attitudes relate to problem-solving success; and (3) learning environment, the contextual or environmental factors encountered by the problem solver that are external to the problem or the learner (these range from strategies that can be used on an individual basis to those that involve cooperative groups).

In the real world of problem solving, of course, these three factors overlap. Each plays an important part in an individual's success in solving a given problem and each must be accounted for or controlled in experimental research studies of problem solving. All problem-solving studies should take these three factors into account. However, most research studies are more limited in scope and concentrate on one or two of the factors simultaneously.

Many researchers whose studies are considered in this review use a constructivist or information-processing model of learning to explain their findings. Other research studies have no theoretical underpinnings. As a means of synthesizing the collection of studies in each section of the chapter, the authors assume a constructivist or information-processing view. From a constructivist point of view, individuals create or construct their own knowledge throughout life and continually modify it by relating new concepts to those already stored in long-term memory.

Problem solving has been defined in a variety of ways. The definition of a problem that is used in this chapter is that of Hayes (1981), who states that a problem exists when a person perceives a gap between where he or she is and where he or she wants to be but doesn't know how to cross the gap. This broad definition includes societal problems as well as what might be considered stereotypical exercises by expert problem solvers. However, the studies in this review focus exclusively on chemistry problems involving mathematical reasoning skills that students have difficulty solving. These appear to be the major types of problems that chemistry education researchers have examined. Unfortunately, this is a reflection of the practice of chemistry educators, who may not have recognized the importance of involving students in solving problems of a more global nature or may simply feel that the curriculum is already too dense to add more topics. It is in contrast to the study of other scientific disciplines such as the earth sciences, in which students examine environmental problems, or biology, in which students might solve problems related to overpopula-

The authors wish to thank Joseph S. Krajcik (University of Michigan) for his insightful comments on the first draft of this chapter.

tion. However, the situation in chemistry is beginning to change with the introduction of new curriculum projects emphasizing societal applications. In the future, chemistry problem-solving research may become broader in scope.

NATURE OF THE PROBLEM

In order to solve a chemistry problem in an acceptable manner, the problem solver must have both conceptual scientific knowledge and procedural knowledge. In addition, the successful problem solver is able to decode or translate the words given in the problem in a meaningful way. This has been called problem representation by Chi, Feltovitch, and Glaser (1981) and translation by Lee (1985). It requires the problem solver to create a cognitive structure corresponding to the problem. Studies by Greenbowe (1983) and Lee (1985) indicate that beginning chemistry students are generally unable to restructure or translate chemistry problems. Certain in-task variables may prevent the solver from creating this cognitive structure. For example, the vocabulary used in the problem, the format of the problem, or the number of variables involved might be a barrier to solving a given problem. The in-task factors and the conceptual knowledge required are important factors that are relevant to this discussion, and all of these require an awareness of learner characteristics. This section focuses on the two factors, in-task and conceptual knowledge. Procedural knowledge is discussed primarily in the section on learning environment.

In-Task Variables

The first step in successful problem solving is understanding the meaning of the problem. This requires knowing the vocabulary in which the problem is stated and its syntax. Experienced teachers know that rewording a complex problem using simpler vocabulary sometimes leads students to success. The vocabulary in a stated problem includes words that are not specific to science (such as *negligible*, *relevant*, and *contrast*), words that are used in everyday language but whose everyday meaning and scientific meaning are different (such as "the candy *melted* in my mouth," "the coffee is *weak*"), and scientific terms (such as *stoichiometry*, *molarity*, and *oxidation*). Cassels and Johnstone (1984, 1985), in a study of student understanding of science classroom vocabulary, found that when the key words in questions were simplified, substantially greater numbers of students were able to solve a problem correctly. For example, 7 percent more students could answer the question, "Which of the following is a choking gas?" than could answer, "Which . . . is a pungent gas?" If quantitative questions were changed from "least" to "most," success rates also increased. Successful answers increased by 26 percent when that substitution was made in this question: "Which of the following solutions of a salt in water is least concentrated?" Cassels and Johnstone (1984) reported similar results after changing negative forms to positive, after simplifying long and complex sentences, and after shifting from passive voice to active voice.

Cassels and Johnstone's (1985) research also produced an extensive list of rather common words that British students at various levels did not understand. Some of these are concept, contrast, displace, diversity, factor, fundamental, incident, negligible, relative, and spontaneous. If words such as these appear in chemistry problems, it is reasonable to assume that for some students they would be barriers to successful problem solution.

Gabel and Sherwood (1984) and Gabel and Samuel (1986) sought to determine whether the use of technical terms (i.e., mole and molarity) rather than the more underlying concepts involved in problems (i.e., working with collections of objects or concentration) was the cause of students' difficulties in problem solving. They compared how well students solved problems using familiar terms involving analogues such as dozens of oranges or sugar solutions rather than moles of an unfamiliar substance like sodium hydroxide. They concluded that for difficult problems requiring more than one step, such as limiting-reagent problems or in solution problems in which additional solute was added, students' difficulties were due to lack of conceptual understanding of the processes, rather than use of the technical terminology. In addition, Gabel and Sherwood determined that the vocabulary does make a difference. A simple substitution such as the numeral 1 for *one*, *a*, or *an* makes the problem easier to solve. They also found that students were more successful on problems involving multiplication rather than division and problems containing even multiples of numbers rather than fractional parts.

Other in-task variables found to be related to problem-solving success are the inclusion of irrelevant data and problems requiring implicit information. Falls and Voss (1985) found that the presence of in-task variables made problems difficult to solve for field-dependent students. Field-independent students were more capable of isolating the relevant information in a problem that contained both relevant and irrelevant data or required the use of implicit information. These students were also more successful in solving problems that contained a direct logic task, requiring the algebraic format $A = kB$.

Other in-task variables affecting problem difficulty are the format of the problem and the number of variables and other information involved. Staver (1986) studied the effect of problem format and the number of independent variables in Piaget's bending rods problem with eighth-grade science students. There were no differences in the correct problem solution according to problem format, in contrast to an earlier study on mealworms (Staver, 1984), but there were differences according to the number of variables involved. He found significant reduction in achievement when a problem contained four or five variables in contrast to when it contained two or three variables. Although no test was administered to measure the capacity of students' working memory, one reason postulated for the finding was the excessive task demand on working memory.

Whether the problem has a free-response or multiple-choice format is another factor of importance. Schmidt (1988) found that by cleverly devising foils (wrong answers) based on students' misconceptions, he could determine their thinking from their choice of the foil. Friedel and Maloney (1992) also use this method of determining students' thinking.

The relationship between working-memory capacity and academic performance in solving chemistry problems has been examined by Johnstone and El-Banna (1986) and by Opdenacker et al. (1990). Johnstone and El-Banna administered the Digit Backwards Test (DBT) and the Figural Intersection Test (FIT) to 206 secondary and 271 first-year university chemistry students to determine their capacity for memorization of data and their ability to perform thought operations on the data, respectively. They hypothesized that these functions that are required to solve problems are operations of working memory and that students of limited working memory would have difficulty solving complex problems that demanded large working-memory space. They found that this was the case. When the demand of the problem exceeded the working-memory space, most students were not successful. Interviews showed that students with limited working-memory capacity who were successful on more complex problems had been taught or devised for themselves strategies for performing beyond their measured capacity. An additional finding was that short-term memory space did not account for all of the variance; other factors must be involved.

In their two-part study, Opdenacker et al. (1990) confirmed the findings of Johnstone and El-Banna. They set out to test the hypothesis that a student can solve a chemistry problem successfully only if the capacity of working memory exceeds or equals the information-processing demand of the problem. The DBT was used to measure the capacity of working memory, and the information demand of chemistry problems was determined by three expert problem solvers. In their first experiment, they found few differences on several measures of chemistry achievement as a function of the problem demand and student working-memory capacity. Their explanation was that students of lower working-memory capacity may not have used that capacity but solved the problems using a schema that they memorized. In the second experiment, in which more care was taken in testing the students and in the problem demand determination, a moderate correlation (0.3) was found between the percentage of correct solutions and increased student memory capacity, when students were grouped according to working-memory capacity. The authors suggested that these findings must be interpreted in a wider context such as that proposed by Niaz (1987a).

A series of studies of chemistry problem solving by Niaz (1987a, 1987b, 1988a, 1988b, 1988c, 1989) included not only working-memory capacity as a dependent variable but also students' cognitive styles and formal-operational reasoning patterns. In all the studies, introductory college chemistry students were administered the Figural Intersection Test to measure the structural M space (a measure of functioning working memory capacity), the Group Imbedded Figures Test (a measure of field dependence or independence), a modified version of the Lawson Classroom Test of Formal Reasoning (a measure of developmental level), and, in all but one case (1988), the Raven Standard Progressive Matrices Test (a measure of the general level of intelligence).

Results of all five studies, which are discussed in more detail in the section on learner characteristics, showed that in terms of success in problem solving there appeared to be a relationship between the students' functioning working-memory capacity and the demand of the problems, especially for problems with a high M demand. Niaz recommended that science instructors carefully consider the effect of adding even a small amount of information to a problem on the students' success rate in solving the problem.

The overload-of-working-memory hypothesis has also been used by Sevenair and Burkett (1988) to explain why some students score less than the chance level on complex multiple-choice items involving problem solving. They postulated that when problems involve multiple steps and the distracter in the multiple-choice item is a correct answer to one of these steps, but not the final step, students with limited working-memory capacity choose the distracter rather than simply guess at the answer.

Summary. The studies reviewed in this section highlight the impact of problem presentation (in-task variables) on success. Although many factors may be the cause of students' lack of success, the overload of working memory contributes significantly (Johnstone, 1984). As problems become more complex and students must use more of their working memory to process numerous chunks of information, memory overload is likely to occur. However, long-term memory also has a role. Students' understanding of science concepts, how these concepts are networked in long-term memory, and the ease of transferability to working memory are important conditions leading to success or failure in problem solving. Students' conceptual understanding of science is considered in the next section.

The Need for Conceptual Knowledge in Problem Solving

Because one of the objectives of including problem solving in a chemistry course is to further students' conceptual knowledge of chemistry, important considerations are the level of conceptual knowledge that students have before solving chemistry problems, the level that is needed to solve the problems, and the level they have after solving problems. Although some students experience success on problems or exercises that are similar to those illustrated in the text or illustrated by the instructor in the classroom, if distant transfer is to occur—that is, transfer of skills and expertise to a new problem-solving domain that is distinctly different in its stimulus features from the original problem—rudimentary possession of the concepts underlying the problem is a necessity. Several researchers have examined problem solving and its relationship to students' conceptual understanding.

Use of Algorithms in Problem Solving. Evidence from studies by Nurrenbern (1979), Gabel (1981), Gabel, Sherwood, and Enochs (1984), Anamuah-Mensah (1986), Herron and Greenbowe (1986), Lythcott (1990), and Bunce and Gabel (1991) indicates that many students' conceptual understanding is inadequate to solve chemistry problems that require transfer and that students frequently solve even "exercises" using an algorithmic approach almost exclusively. In all of the studies, students were interviewed using the "think-aloud" technique to determine their level of conceptual understanding as related to problem solving.

In a study of high school students who were taught to use one of the four common strategies to solve chemistry problems (factor-label method, diagrams, analogies, or proportionality), Gabel (1981) found the majority of students relied on strictly algorithmic techniques. Their lack of conceptual understanding and hence lack of use of concepts in solving problems were shown by the fact that many more students could solve problems successfully than could answer the interview questions about the concepts involved. In addition, few students were able to solve problems requiring transfer, and in the majority of cases in which students used algorithms, there was no evidence that reasoning was used in the problem-solving process.

These results were corroborated by Herron and Greenbowe (1986) and Bunce et al. (1990). Both studies involved college chemistry students. Herron and Greenbowe found in a case study of "Sue" that even though she had the intellectual ability to solve problems, she did not use it effectively and represented problems in a manner inconsistent with the physical reality described. In the Bunce et al. study, students who were interviewed after solving chemistry problems stated that they didn't need or use any conceptual understanding when solving mathematical chemistry problems. Student after student made comments like, "I just think about the numbers and the chemical equation to see how ... each piece fits into it and the numbers." The researchers characterized this as the "Rolodex Approach" to chemistry problem solving, where the problem solver rotates the formulas in his or her mind and plugs numbers into the one that seems most appropriate.

The study by Anamuah-Mensah (1986) produced similar results. High school students in British Columbia were asked to verbalize their thinking using the think-aloud technique in calculating the concentration of a base after performing a titration experiment. Analysis of the transcripts of the taped interviews showed that about 80 percent of the students used a formula approach and 20 percent a proportional reasoning approach to solve the problems. Those using the formula approach made no attempt to demonstrate an understanding of the relationships among the variables in the formula they used. Even though students using the proportional reasoning approach showed some evidence of examining these relationships, the overall experiment showed that when the students were actually manipulating materials and examining the macroscopic properties of matter, little connection was made between conceptual understanding and mathematical problem solving. This was evident in the same experiment when students were asked to make predictions when different acids or bases were used in the titration. Although 27 percent of the students were 100 percent correct in making the three predictions, many had difficulty in giving a correct reason for their prediction.

The study by Lythcott (1990) of high school chemistry problem solving confirms these findings. Two groups of students were taught to solve stoichiometry problems in different ways. One group used an algorithmic approach, whereas the other group used a more conceptual approach. After the treatment, students in both groups were asked to solve problems aloud. They were then interviewed on the meaning of concepts involved in the problem, such as moles and mass, and also asked questions about the balanced equation and the particles involved. Although the algorithmic group performed slightly better on the number of perfectly solved problems, both the percentage of totally inadequate solutions and the number of students who had no idea about what to do were also from this group. She concluded that even "students who solved simple mass-mass problems correctly had woefully inadequate chemical knowledge."

Use of Conceptual Knowledge Base in Problem Solving. Several researchers who examined student knowledge using a variety of other techniques besides interviews provided additional evidence for the lack of a conceptual knowledge base for problem solving. Two systematic studies of stoichiometry (Griffiths, Pottle, & Whelan, 1983; Schmidt, 1990) showed that beginning students lack numerous concepts necessary for successful problem solving even though they have been presented in class. Schmidt studied the conceptual understanding of more than 6000 students in grades 10 to 13 through the use of carefully designed multiple-choice items. He found, for example, that students often equate amount with reacting mass or molar mass and reacting mass. Griffiths et al. determined students' conceptual background by using free-response items based on a learning hierarchy. Their findings confirm those of other researchers that students did not appear to understand the meaning of a balanced equation and were generally operating in an algorithmic rather than reasoning fashion.

Nurrenbern and Pickering (1987) and Sawrey (1990) studied the relationship between conceptual understanding and introductory college students' ability to solve chemistry problems by comparing students' achievement on both traditional and conceptual problems. The conceptual problems depicted molecules and atoms (as dots in circles representing space) of phenomena related to the traditional problem and assumed by most instructors to be understood by students who solve the traditional problems. The results show that over twice as many students solve the traditional problems correctly than solve the conceptual problems correctly. The Sawrey study, which had a larger sample size ($n = 285$ and 382, respectively), showed an even wider disparity between the two types of correct solutions, with about a 3:1 ratio for the gas-law problem and a 6:1 ratio for the stoichiometry problem.

In another study of college students, Pickering (1990), using some of the same problems as Sawrey (1990), examined whether there was a difference in the type of students who could solve conceptual versus traditional chemistry problems. Pickering found no differences in the organic chemistry grades between students who during the previous year were able to solve the conceptual chemistry problem correctly and those who could not and Pickering concluded that differences on the conceptual problems stemmed from students' lack of factual information.

Application of Conceptual Knowledge in Problem Solving.
Several studies provide additional insight into the relationship between concept knowledge and problem solving. Niaz and Robinson (1989, 1992) sought to determine whether certain student cognitive characteristics were related to the types of problems students could solve by using algorithms versus those that required conceptual understanding. The problems used in the study were similar to the traditional and conceptual problems used by Nurrenbern and Pickering (1987). Results indicated that formal-operational reasoning was necessary for solving the algorithmic problems but that no single cognitive characteristic was related to solving the conceptual problems.

The in-depth understanding of chemistry concepts required for solving problems involves more than knowledge of isolated concepts. Research indicates that problem solving includes making connections among the concepts involved in the problem solution. Gorodetsky and Hoz (1980) used the think-aloud technique to study the sequence of concepts used by tenth-grade students to solve a chemistry problem involving the boiling point of water of the Dead Sea. After identifying the concepts needed to solve the problem and their sequence, they found that both successful and unsuccessful problem solvers used the concepts listed in their profile analysis. However, the successful solvers made more connections between concepts, and the gap between frequency of connections for the two groups increased at higher levels of the concept profile. The researchers concluded that although the nonsolvers were using the concepts, they were unable to process them in the required mode, and their lack of ability to solve the problem stemmed from a partial or unclear perception of the concepts involved or from their lack of strategic knowledge.

Sumfleth (1988) reached a similar conclusion in his study of 16-year-old students in Germany who had studied chemistry for 2 years. From an analysis of an "explanation" test, an "achievement" test and a "connectivity" test, he concluded that even though students had a reasonable basic knowledge of chemical terms, they were unable to establish correlations between them and apply their knowledge in problem solving.

Although these studies indicate that students frequently do not use conceptual understanding in solving mathematical chemistry problems, they also provide evidence that students are limited in their ability to solve distant transfer problems without an in-depth understanding of relevant chemistry concepts. Hence, chemistry educators have always been interested in enhancing students' understanding of chemical concepts. Different methods have been proposed for doing this, such as making laboratory instruction more meaningful (Ward & Herron, 1980) or using Vee diagrams and concept mapping (Novak, 1984). Other proposed methods include teaching chemistry according to a spiral curriculum (Schmidkunz & Buttner, 1986) and making use of writing assignments (Horton, Fronk, & Walton, 1985; VanOrden, 1990).

Summary. The studies reviewed in this section show that many students do not understand the chemistry concepts involved in chemistry problems or are unable to apply the conceptual knowledge in solving the problems. In lieu of solving problems on the basis of conceptual understanding, they use algorithms and formulas to arrive at "correct answers." Because of the desirability of basing problem solutions on underlying chemical concepts and the close-knit relationship between meaningful problem solving and conceptual understanding, the next section provides a brief review of studies of students' understanding of concepts that are important in chemistry problem solving or that chemistry instructors might assume that students possess. In most cases, the researcher conducting the study did not relate conceptual understanding to problem solving. There has been no attempt to review studies of textbook analyses, even though the authors realize that textbooks may influence students' conceptual understanding.

Students' Acquisition of Conceptual Knowledge in Chemistry

There are many sources of the conceptual knowledge base necessary for solving chemistry problems. Pines and West (1986) found it useful to discriminate between two sources of knowledge: that spontaneously acquired from interactions with the environment and that acquired in a formal fashion through the intervention of the school. They indicate that when learning occurs, these two sources of knowledge that may be in conflict can interact in a positive way, resulting in conceptual exchange that leads to conceptual understanding. They may also not interact, resulting in long-term memory compartmentalization of conflicting views. The concepts which students acquire that are in conflict with scientific meaning are frequently labeled naive conceptions, intuitive conceptions, alternative conceptions, or even misconceptions.

Frequently, these terms are used in the research interchangeably. However, there may be some subtle differences among them. Knowledge acquired from interactions with the environment is perhaps best called an "intuitive conception," because this term identifies the source of the knowledge without implying a negative image of the learner. These intuitive conceptions are difficult to dispel because they seem natural to the learner and have been acquired over a long period of time. There are also some concepts acquired from formal school instruction that are in conflict with acceptable scientific meaning but are not intuitive. These may result when the formal presentation of a concept is inadequate or not illustrated in a variety of contexts. For example, children might think that everything boils at 100° C because they have been given no other examples but that of water boiling! They may also result because the presentation by the teacher contained conceptual errors or the learner was not paying attention. (Lawrenz [1986] found that elementary school teachers answered many of the National Assessment of Educational Progress test items for 17-year-olds incorrectly.) These are easier to dispel and perhaps might be better described as "misconceptions" or simply "conceptual errors." Because many researchers do not make any distinctions between the terms, the reader should be aware that they are sometimes used interchangeably.

The most comprehensive study of students' conceptual understanding of science over time was reported by Novak and Musonda (1991). Concepts of particular interest in the study of chemistry were the particulate nature of matter and the states of matter and its changes. Audio tutorial lessons containing an introduction to these and other science concepts were provided to 191 first- and second-grade children. A count of the "invalid notions" of 48 of these children (instructed) was compared to that of an uninstructed sample of the same size. Results indicated that the instructed group had significantly fewer nonscientific ideas than the uninstructed group at grades 2, 7, 10, and 12. Analysis of concept maps derived from interviews with 12 students from each group also revealed that students in the instructed group had a more coherent conceptual framework at each of the four grade levels tested. (Means were higher, although not significantly except at grade 7.) However, the research also showed the tenacity of intuitive conceptions and nonscientific knowledge. In some cases, there was a reversal to nonscientific explanations, and a number of students at grade 12 still held "invalid notions."

Because one of the objectives of schooling is the acquisition of concepts, several other studies have examined students' understanding of science concepts and whether the understanding is retained over a period of time. Unlike the study by Novak and Musonda (1991), these studies do not compare the same students over time, nor do they address or control the kind of instruction available to the students.

Several studies (Yager & Yager, 1984; deBuerger-Van der Borght & Mabille, 1989; & Kleinman, Griffin, & Kerner, 1987) examined student's knowledge of key words used in science or chemistry textbooks. Yager and Yager, using multiple-choice tests to determine students' definitions of eight science terms (volume, organism, motion, energy, molecule, cell, enzyme, and fossil), concluded that third-graders were not as familiar with the terms as were seventh- or eleventh-graders and that there was a general decline in students' understanding of the meaning of words from seventh to eleventh grade.

In a similar study in Belgium, deBuerger-Van der Borght and Mabille (1989) examined students' understanding of 20 words (carbon, three-dimensional, unit, capillary, element, energy, equilibrium, to identify, nucleus, mineral, unstable, organism, oxidize, model, structure, synthesis, system, acid, combustion, periodic), using a more open-ended procedure. Students were given a list of words and asked to write down all the meanings they associated with them. The authors concluded that although there was no regression in knowledge between 12- and 18-year-old students, new information given each year in the curriculum has a short life span and students generally did not link what was learned in science one year with what was learned the next.

A study of the differences in how undergraduate students, graduate students, and chemistry faculty members view certain science concepts was reported by Kleinman, Griffin, and Kerner (1987). They asked 10, 8, and 3 persons from the previously mentioned groups to tell them what images they had for each of 10 key words found in standard chemistry texts (bond, energy, equilibrium, functional group, mole, orbital, rate, resonance, solubility, and spontaneous process). Analysis of taped interviews indicated that there was an increase in both the level of abstraction and the number of images per person from undergraduate to faculty. Undergraduate students' primary mode of thinking was associative, whereas that of faculty involved using models.

Other studies (Cros, Chastrette, & Fayol, 1988; Hendley, 1990; Rowell, Dawson, & Lyndon, 1990) show that chemistry instruction does modify students' conceptions, but not to any great extent. In addition, sometimes it appears that instruction actually creates errors in scientific conceptions or "misconceptions."

The study by Cros et al. (1988), which was a follow-up to a previous study (Cros et al., 1986) in which students entering the university were tested on the constituents of matter and on their notions of acids and bases, sought to determine how their conceptions changed after 1 year of university chemistry study. They found that certain concepts, such as the Bohr model of the atom, were extremely difficult to dislodge. They concluded that although some progress had been made, it was "disappointingly low," and chemistry teaching needs to be modified to make concepts more accessible to students.

Hendley (1990) studied how the conceptions of tenth-grade high school honors chemistry students, who had been in special science classes for the gifted for several years, changed as a result of a chemistry course. He concluded, "Although the number of alternative conceptions generally decreased, alternative conceptions were still present, some of which were fairly common. Several alternative conceptions were developed during instruction."

The acquisition of conceptual knowledge has also been shown to be related to a person's belief system. Okebukola and Jegede (1990) found that ecocultural variables affect students' concept attainment in science. Students in class four in a secondary school in Nigeria had higher concept attainment if their reasoning patterns were empirical (versus magical). Concept acquisition was also affected by preference for cooperative versus competitive learning, living in a predominantly automated versus manual environment, and coming from a permissive versus authoritarian home environment.

Concept acquisition by students who are not taught in their native language becomes problematic in countries in which many native languages exist. Rutherford and Nkopodi (1990) compared the recognition of science concept definitions in English and North Sotho in South Africa. They found that concept definitions were recognized better in English than in the native language in very few cases. Because instruction was in English and students recognized science definitions in both languages, it appeared that students were internalizing the definitions rather than merely learning them by rote.

As the research studies in this section show, many students' understanding of science concepts is inadequate, misconceptions are firmly rooted, and chemistry and other science courses do little to improve conceptual understanding. The results of the study by Novak and Musonda (1991) are encouraging in that children who had a carefully controlled introduction to some rather complex scientific conceptions at a young age had a better conceptual framework in the later grades. Unfortunately, the correct conceptual understanding may have to be provided through technology, because many teachers have the same incorrect conceptions as the children.

Summary. In a previous section, it was shown that students make little use of their conceptual understanding in solving chemistry problems. This is understandable because the research just discussed has shown that students frequently lack the appropriate conceptual knowledge on which to base problem solving. If chemistry problem solving is to improve, students must solve problems in a logical manner by using appropriate concepts. Consequently, chemistry teachers need to be aware of students' understanding of the specific concepts that are critical for problem solving.

Students' Understanding of Specific Concepts in Chemistry

Many concepts in chemistry must be understood by students before they can even attempt to solve problems in a meaningful way. Some of these are mass, volume, mole, molecule, solid, liquid, gas, molecular mass, and molar volume. Other concepts important to specific types of frequently encountered problems are solution, concentration, density, temperature, standard temperature and pressure (STP), and equilibrium. Concepts such as isomer, nucleus, and atomic structure are not as critical for problem solving, unless the problem is directly related to them. Because concept development is so important in chemistry problem solving, research reports on those concepts of fundamental importance are reviewed in this section. Additional studies have been reviewed by Krajcik (1991). Concepts less directly related to problem solving have received less attention in this review.

Matter and Its States. Chemistry problems frequently contain terms such as "substance" or "solid," "liquid," or "gas." Teachers of introductory chemistry courses assume that students understand these and other fundamental concepts when they enter the course, and little time is generally spent in reviewing them. It is also assumed that students understand that matter is conserved. Research studies in this area have generally focused on children's lack of ability to classify substances as solids, liquids, and gases (Jones, Lynch, & Reesink, 1989; Stavy, 1991a) or their concepts about gases, changes in state, and conservation of mass or weight (Osborne & Cosgrove, 1983; Russell, Harlen, & Watt, 1989; Stavy, 1988; Stavy 1990a, 1990b, 1991b). However, some of these studies and several others show that naive conceptions held by children about matter during the elementary and middle school years persist and are identical to those held by students at the high school and even the college level. Furio Mas, Perez, and Harris (1987) reported that when given a question concerning the burning of paper in a sealed container, 54 percent of a sample of Spanish 17- and 18-year-olds, who had studied chemistry and physics as an elective subject, did not conserve weight. They commented that the students' conception of gases parallels that held by the early scientists in the history of chemistry. Stavy (1990a) showed that only 50 percent of seventh grade Israeli students understand that matter is conserved during evaporation. In another study, Stavy (1990b) tested Israeli students in grades 4 through 9 for their ability to conserve matter, property, and weight when a colorless liquid (acetone) evaporated in a closed container. She found that the ability to conserve increased with age but that at grade 9, although all students conserved matter, about 20 percent did not conserve weight. Stavy (1991b) found that students could be helped to overcome their misconceptions by using an analogue that they understood and could apply to the less familiar situation. Flick (1991) also found the use of analogies effective with children in grades 3, 4, and 5 in learning about particle explanations for changes of state. Another successful method of changing children's misconceptions about volume displacement was reported by Rowell et al. (1990). They found that students had fewer misconceptions if, after they were introduced to the idea that it was the volume of objects rather than their mass that affected water displacement, they were given additional opportunities to predict and observe the "old way" of thinking versus the "new way" of thinking. This was in contrast to asking students to work cooperatively after an initial introduction to the idea.

Students' tenacity of belief in their naive conceptions was illustrated by Shepherd and Renner (1982). They asked tenth- and twelfth-grade students to write a paragraph describing why solids always keep the same shape while gases and liquids do not. Analysis of the paragraphs indicated that no one in their 135-student sample got the answer 100 percent correct according to the kinetic molecular theory and that about 50 percent of the students had specific misconceptions about solids similar to those reported by Jones et al. (1989) and Stavy (1991a) for children.

Several of the aforementioned studies examined factors other than conservation as a part of their research on changes of state. Osborne and Cosgrove (1983) examined 12- to 17-year-old New Zealand students' conceptions about what happens to the chemical species of water as it boils, evaporates, and condenses. Although students' naive conceptions generally decreased with age, the 17-year-old students gave incorrect responses. About 40 percent thought that the bubbles in boiling water were made of hydrogen and oxygen. Almost 30 percent thought that water changed to oxygen and hydrogen when it evaporated, and 35 percent thought that hydrogen and oxygen combined to form condensation on the outside of a cold jar.

Bar and Travis (1991) conducted a similar research study with Israeli children of ages 6 to 14 involving three different ways to collect data. These were open-ended oral individual tests, a written multiple-choice test, and an open-ended written test. Results indicated that the testing format made a difference in the kinds of responses that students gave. Even though children frequently gave a correct response on an open-ended question, when scientific distracters were added in the multiple-choice format, such as heat or hydrogen/oxygen, students selected these. The researchers concluded that real understanding of scientific terms proceeds slowly. Students' understanding is frequently superficial, and alternative views are persistent.

Dibar Ure and Colinvaux (1989) examined answers to similar questions presented to adults (15–27 years old) of low socioeconomic status in Brazil. The majority of adults interviewed had misconceptions about the phenomena, although they did not use hydrogen and oxygen in their explanations. After an intervention consisting of more detailed explanations of the role of water vapor in air, they found that some persons modified their views but others did not. The persistence of misconceptions has also been shown by Bodner (1991), who used some of the same questions with first-year chemistry graduate students. He found that even after completing an undergraduate major in chemistry, naive conceptions persisted. For example, 20 percent of the graduate students thought that the bubbles in boiling water were air or oxygen and 5 percent thought that they were a hydrogen-oxygen mixture.

From the research reviewed thus far, it easy to understand why students do not feel it is necessary to use their conceptual understanding of chemistry to solve problems. Their conceptual understanding of change in state is often so incomplete that it would not aid them in attaining correct problem solutions.

Particulate Nature of Matter. Knowledge of the particulate nature of matter is also important in solving chemistry problems. Although some stoichiometry problems can be solved without considering the particle makeup of matter, molarity problems and gas-law problems can be understood more readily by making use of the kinetic molecular theory. Yet mounting evidence suggests that many chemistry students lack this conceptual framework that is so fundamental in the study of chemistry. Studies already reviewed that show students lacking in this area include Osborne and Cosgrove (1983), Sawrey (1990), Nurrenbern and Pickering (1987), Pickering (1990), Dibar Ure and Colinvaux (1989), Stavy (1988, 1991), and Shepherd and Renner (1982).

Many of these studies and several others show that students' conception of matter as a collection of moving particles is rudimentary and that instruction in this area is not as effective as might be expected. Novick and Nussbaum (1978) examined eighth-grade Israeli students' conception of the particle nature of matter. Eighth-grade students were selected because all children in Israel study the particle nature of matter in seventh grade. They found that "a significant portion of the sample failed to internalize important aspects of the particle model" (p. 278). Aspects of the particle model that were most difficult for students to conceive were the existence of empty space, the intrinsic motion of particles, and interactions between particles. Stavy (1988), who studied Israeli children's conception of gases at grades 7, 8, and 9, found that even though students had studied the particle nature of matter in grade 7, none of them used particles in their definition of gases during this grade. At grade 8, 25 percent used the particle theory in their definitions, and at grade 9, most of them did. Results of the previous two studies were corroborated in the United States by Berkheimer and Blakeslee (1988) and Mitchell and Kellington (1982). Berkheimer and Blakeslee, through extensive field testing of a National Science Foundation-funded curriculum development project produced materials on the particle nature of matter intended originally for the elementary school curriculum, concluded that sixth grade was the earliest grade level to begin the study of the particulate nature of matter. They also concluded that instruction at grade 6 should be limited to phase changes. Mitchell and Kellington found that seventh grade students in The Netherlands who had studied the particulate theory of matter as part of the Scottish curriculum had difficulty retaining the information by the end of the course.

Several other studies show that this lack of understanding of the particulate nature of matter persists in high school and university students. Novick and Nussbaum (1981) found only 35 percent of high school and university students used particles correctly to explain a vacuum. Shepherd and Renner (1982) found that none of the high school students in grades 10 and 12 used a completely correct explanation of differences among solids, liquids, and gases using particles. Brook, Briggs, and Driver (1984) examined answers to six written questions (300 responses per question) provided by 15-year-old students in England who had taken the national examination. They interviewed a smaller sample of students. Although more than half of the students used a particle answer in their explanation for every question, less than one in five applied ideas that were taught involving the particulate nature of matter. Stavridou and Solomonidou (1989) asked French students ranging in age from 8 to 17 to group a collection of changes (both chemical and physical) in different categories. Not only did few students use other than surface features to group the changes, but no one in the study used the particulate nature of matter to explain their classification scheme. Haidar and Abraham (1991), in a study of high school students' applied and theoretical knowledge of selected concepts based on the particulate theory, found that students use the particulate theory to explain theoretical questions more frequently than comparable applied questions. They found that 23 percent of the students continued to use general explanations even when cued to use particle explanations. Westbrook and Marek (1991) examined the understanding of diffusion by seventh- and tenth-grade students as well as college zoology students. They found that no student displayed a complete and sound understanding of the concept using a molecular interpretation. In this case, as in other studies such as those of BouJaoude (1991) on students' understanding of burning and Gabriela, Ribeiro, Costa Pereira, and Maskill (1990) on chemical reactions and spontaneity, students tend to explain chemical phenomena using mostly visual criteria related to macroscopic properties.

Portuguese students' knowledge of the particulate nature of matter was examined by Pereira and Pestana (1991), who explicitly asked students (13–18 years old) to write down a representation of water in its three states using a model. The students' understanding of the particulate nature was similar to that found by Gabel, Samuel, and Hunn (1987), who reported that students think the sizes of the particles increase with changes of state and have incorrect notions about relative spacing in the liquid and gaseous states.

Gabel et al. (1987) compared the conceptual understanding of the particulate nature of matter of prospective elementary teachers who had taken a chemistry course and those who had not. Although there were some significant differences between groups on certain aspects of the particulate model, both groups had an inadequate understanding. Rollnick and Rutherford (1990) also found preservice primary teachers in Africa to have an inadequate understanding of the particulate nature of matter. In a study of prospective teachers' conceptions about air, they found that teachers used explanations involving particles incorrectly.

Hesse and Anderson (1992) studied students' conceptions of chemical change. They sought to determine what knowledge is used by high school students when they describe chemical change, the role of conservation reasoning in their explanations, and the nature of the explanatory details. Only 1 of the 11 students interviewed in the study explained chemical change in terms of molecules and atoms. Most students used common analogies to explain change.

Gabel (1990) sought to determine whether an increased emphasis on the particulate nature of matter would lead to an increased understanding of the "symbolic" and "phenomena" levels of chemistry as well. She found a 10 percent increase in achievement on all three levels for students who had received the increased emphasis on the particulate nature of matter. However, the low overall test mean indicated that few high school students had a good grasp of any of the levels of chemistry.

The previous studies indicate that chemistry students do not have satisfactory knowledge of the kinetic molecular theory either before or after taking an introductory chemistry course. Another area in which students have difficulty is that of attributing the properties of a collection of atoms to a single atom of the substance. Ben-Zvi, Eylon, and Silberstein (1986) show that this is quite common among Israeli students (about 51 percent of tenth-grade students) and that instruction helps in overcoming this misunderstanding for some students. The research of de Vos and Verdonk (1987) indicates that 15-year-old chemistry students in The Netherlands have the same misunderstanding and that this and other nonscientific conceptions are difficult to dispel with instruction.

Griffiths and Preston (1989) examined grade 12 Newfoundland students' conceptions about fundamental characteristics of molecules and atoms. Fifty-two nonscientific conceptions were found. One-half of the sample who grossly overestimated molecular size believed that molecules of the same substance vary in size or have different shapes in different phases. They also believed that molecules change weight when a substance changes its phase and that atoms are "alive." Other studies pertaining to these and other areas have been reviewed by Gilbert and Watts (1983) and Andersson (1986, 1989).

Students' understanding of chemical formulas and equations are two other areas important in chemistry problem solving that are dependent on students' conceptualization of particles. Studies by Ben-Zvi, Eylon, and Silberstein (1982), Eylon, Ben-Zvi, and Silberstein (1982), Yarroch (1985), Friedel and Maloney (1992), and Maloney and Friedel (1991) indicate that high school and college chemistry students do not associate chemical formulas with the appropriate representation at the particle level. Students in all of these studies had difficulty in relating the subscript of the formula to the appropriate number of atoms when given particle drawings or when asked to represent the particles themselves. Yarroch showed that students do not make appropriate distinctions between the coefficients preceding the formula in an equation and the subscripts of the for-

mula. Moreover, among the high school students that he interviewed who had balanced equations correctly, more than 50 percent did not understand this relationship. Niaz and Lawson (1985) also found that nonmajor college students enrolled in chemistry find balancing equations difficult and that equation-balancing skill was related to their formal reasoning ability and mental capacity.

In conclusion, the studies reviewed in this section indicate that many students at the elementary, middle school, high school, and college levels have erroneous views about the particulate nature of matter, even after instruction. From the study by Novak and Musonda (1991) reviewed in the previous section, one might conclude that early instruction on these topics would facilitate a more scientific view. However, Berkheimer and Blakeslee (1988) found that this was difficult to implement by ordinary teachers at the lower grade levels. That matter is made up of particles, that the particles are in motion, and that a collection of particles has different properties than a single particle are not intuitive ideas. These and related ideas about the nature of matter are very abstract and therefore difficult for students to understand. Perhaps the approach proposed by Johnstone (1990), of teaching chemistry by relating physical phenomena to symbolic and particle representations simultaneously, would help secondary and college students dispel some of their nonscientific conceptions.

Mass, Volume, and Density. Mass, volume, and density are three chemistry concepts important in problem solving. Students' understanding of these is also linked to their knowledge of the particulate nature of matter. The following studies show that students generally lack a scientific understanding of these concepts. Because many of the studies reviewed in this section examined more than one of these concepts simultaneously, no attempt has been made to consider mass, volume, and density separately.

Mullet and Gervais (1990) and Rager and Stavy (1989) explored students' understanding of mass. Mullet and Gervais sought to determine whether French high school students (13–15 years-old) understood the difference between the concepts of weight and mass. Using the Information Integration Theory, they examined how students integrate information expressed in terms of density, volume, gravitation, and position to make judgments on the mass and weight of objects. They concluded that students have an intuitive concept of mass that is distinct from that of weight. This concept is not associated with the word "mass" but rather with "quantity of matter." Few differences between 13- and 15-year-old students were found.

Rager and Stavy (1989) studied 66 Israeli students' understanding of mass, volume, and number of particles through the use of interviews. The ninth- and tenth-grade students were asked questions about both static and dynamic systems. Typical questions for the static system held one of the three variables (mass, volume, number of particles) constant for three different materials (wood, aluminum, plastic) and asked about the other two variables. The dynamic system changes included expansion by heat, changes of state, and dissolving solids, liquids, and gases in water. Several general conclusions were reached: boys outperformed girls; student performance was best when dealing with solids, followed by liquids and gases; and for conservation-of-particle problems performance was higher for mass than for volume. The authors recommended that teaching of problem solving begin the study of moles with variables with which students are most familiar. Another recommendation included using solids, rather than liquids or gases, and having students solve mole-to-mass problems before problems involving numbers of particles.

Enochs and Gabel (1984) and Rowell and Dawson (1981) studied students' conception of volume. Enochs and Gabel gave elementary education majors a test that contained drawings of several objects and asked students to select from a list of possible methods those by which the volume and surface area could be determined. Even after instruction, a large percentage of students did not understand the concept of volume and were unable to distinguish surface area from volume. Most students (77 percent) recognized that the volume of a rectangular solid was obtained from the product of length, width, and height, but only 44 percent recognized that it could also be determined from the product of area and height. Most students used a "memorizing" mode to solve the problems rather than a "thinking" mode.

Rowell and Dawson (1981) sought to determine whether the conservation of volume for eighth-grade students would improve with conflict instructional strategies. In the pretest, they found that only about 25 percent of the students had mastery of the conservation-of-volume concept. Results of the study indicate that whole-class instruction using conflict instructional strategies is an effective way to improve students' conceptions of the conservation of volume.

Howe and Shayer (1981), Gennaro (1981), Hewson (1986), and Shepherd and Renner (1982) explored various aspects of students' conceptions of density. Howe and Shayer showed that gender differences exist in students' conceptions of volume and that these differences persist even after hands-on activities. In a sample of both American and British children, boys performed better than girls on volume and density tasks. After hands-on instruction that included working with centimeter cubes, plasticene, and so forth, both boys' and girls' conceptions of volume improved, but the difference in achievement between the sexes remained.

Hewson (1986) studied the conceptions of mass, volume, particles, and density of 40 high school students in the Qwa Qwa region of South Africa using interviews. She found that the scientific conception of volume was missing in most students, that the particle mass-weight concept was missing in all students, and that most students used alternative conceptions for mass-weight, volume, and density. She attributed this to their use of the terminology in everyday experiences and to the structure of their home language.

Gennaro found that ninth-grade students who had studied density for 2 weeks using a hands-on approach as part of an introductory physical science course showed a very low understanding of the concept. He concluded that about two-thirds of the students ($n = 333$) did not understand the density-displacement concept. This was confirmed by Baker (1988), who found that even after instruction, students did not demonstrate a good understanding of the density concept.

Shepherd and Renner (1982) studied tenth- and twelfth-grade students' conception of density in an applied setting when they asked them to explain why the temperature of water at the bottom of a lake never gets below 39° F. They found that only 10 percent of students gave an explanation that showed a sound or even partial understanding of the correct explanation involving the density of water reversal at that temperature.

These studies indicate the very abstract nature of the concepts of mass, volume, and density. The studies indicate that many students lack sufficient scientific understanding of the very concepts that form the basis for the study of matter. This lack of understanding explains why students have difficulty solving chemistry problems and why they resort to solving problems using formulas and other algorithmic modes.

Moles. Students' lack of a scientific conception of mass and volume undoubtedly affects their understanding of the concept of mole. Studies in this section show that students lack a scientific conception of the mole. Also included are research reports that indicate the strategies used by students to solve problems using moles, learning hierarchies associated with the mole concept, and effective methods for teaching the mole concept. A review of stoichiometry studies by Dierks (1981) considers these and other conceptual difficulties students have in solving chemistry problems.

Studies by Duncan and Johnstone (1973), Novick and Menis (1976), Lazonby, Morris, and Waddington (1982, 1985), Cervellati, Montuschi, Perugini, Grimellini-Tomasini, and Balandi (1982), Gabel and Sherwood (1984), Friedel, Gabel, and Samuel (1990), and de Jong (1990) focused on students' conception of the mole.

Duncan and Johnstone (1973) obtained 14- to 15-year-old Scottish students' conception of the mole by administering multiple-choice tests. They concluded that some of the major misunderstandings were that 1 mole of one substance always acts with 1 mole of another substance, balancing equations, and the concentration of solutions. Novick and Menis (1976) interviewed 29 Israeli 15-year-olds using 21 questions on the mole concept that had been formulated after giving a multiple-choice mole test to a larger group of students. The analysis of the data indicated three major erroneous conceptions held by the students: (1) a mole is a certain mass and not a certain number, (2) a mole is a certain number of particles of gas, and (3) a mole is a property of a molecule. Cervellati et al., using a multiple-choice test to determine how secondary school students (ages 15–17) in Bologna perceived the mole, showed that students' understanding improved with age but was generally low at all ages. Findings include that (1) students were not familiar with the mole as amount rather than mass; (2) 22.4 liters, although well known as a number, was not linked to pressure, temperature, or condition of state; (3) most students were familiar with the magnitude of Avogadro's number; and (4) students found solving stoichiometric problems difficult.

In an attempt to determine whether it was the "mole" concept per se that was causing high school students' difficulty in solving problems or whether it was working with mass, volume, and particles, Gabel and Sherwood (1984) examined student's use of these three attributes as applied to sugar and oranges. Tests administered by teachers on the mole concept were rewritten using the familiar substances of sugar and oranges in place of the chemical names like sodium hydroxide and using "dozen" in place of "mole." Results showed that the students' difficulty in problem solving is probably due to use of the term "mole" and other unfamiliar terms rather than their lack of understanding of volume, mass, and a collection of particles.

In a follow-up to this study, Friedel et al. (1990) sought to determine how well students recognized problems involving familiar objects such as oranges as being similar to chemistry problems. College students, enrolled in a chemistry course that had as its goal to remedy students' deficiencies before enrollment in a regular introductory course, were asked to match seven problems using moles and conventional chemicals with analogous problems using oranges and dozens. A test was administered at the end of a one-semester course during which half of the students had instruction using similar analogous problems. Results showed that although students receiving the analogue instruction were able to match a higher percentage of the chemistry problems with their analogues, the matching was lower than expected. Except for one problem that was well matched by both groups (about 90 percent), matching on the other problems ranged from 33 to 56 percent for the students without analogue instruction as compared with 42 to 63 percent for students with instruction.

These studies all indicate the lack of students' understanding about moles. Because the mole is a concept devised by scientists to aid in chemical calculations, students' erroneous or non-conceptions could hardly be called intuitive conceptions. They arise because of insufficient instruction or inappropriate teaching strategies.

Lazonby et al. (1982, 1985) and de Jong (1990) focused on the methodology used to obtain students' conceptions about the mole that have implications for teaching. Lazonby et al. (1982) identified students' lack of understanding of the molar interpretation of subscripts in chemical formulas and coefficients in equations through the use of interviews. They then tested eleventh-grade students in England using different test formats and found that it may not be students' lack of understanding of the mole concept that prevents them from solving problems correctly, but the way the problem is written. When problems are presented as a series of steps or as independent problems, instead of a single unit, students (especially girls) were more likely to solve the problem correctly. In the 1985 study, a larger sample of students were tested, and the results were basically the same except that there were no differences between boys and girls.

De Jong (1990) suggested that research on chemical calculations should take place in the classroom setting, where students are given problems to solve in a cooperative setting. From the written responses that students give to problems that are solved in groups as well as from interviews, much information can be gained about the strategies students use in solving problems. Using these methods, de Jong reported that students in The Netherlands generally fail to give coefficients in balanced chemical equations a molar interpretation and that in teaching the mole concept, it is better to begin with the comparison of

molar masses than to begin by presenting moles as a collection of particles.

Other research studies on how to teach the mole concept were conducted by Chiappetta and McBride (1980), Gower, Daniels, and Lloyd (1977), Rowell and Dawson (1980), Griffiths, Kass, and Cornish (1983), Vermont (1985), and Rawwas (1986). Gower et al. (1977) and Rowell and Dawson (1980) set out to determine the hierarchies among concepts that underlie the mole so that teachers might present the underlying concepts in an appropriate order. Gower et al. used a consistency ratio approach to determine the hierarchy, whereas Griffiths et al. compared several different approaches. Both concluded that data rarely confirm all steps in a hierarchy. An interesting finding in the Griffith et al. study corroborates the suggestions based on the findings of Rager and Stavy (1989) and those of de Jong (1990) that in teaching the mole concept, students will be more successful if molar mass is introduced before the mole is presented as a collection of particles.

Rawwas (1986), using the lower five elements of the Griffith-validated hierarchy, found that instruction with a high kinetic structure was generally more profitable for students than instruction with a moderate kinetic structure. Vermont found no differences in achievement on the mole concept using different instructional approaches. Those tested in her study included a cognitive learning and development strategy, the learning-cycle approach, and a lecture-laboratory format. Although Rowell and Dawson did not use a comparison group, they found that an instructional strategy making use of analogues and very systematic instruction was effective for teaching the mole concept to 89 percent of their sample of eleventh-grade students in South Australia. The instruction lasted 6 weeks and involved solving stoichiometric problems, including those involving gases. However, Chiappetta and McBride found that general remediation for teaching the mole concept to ninth-grade students who were unsuccessful in learning the concept during an initial presentation was ineffective, and Griffiths et al. found that specific remediation keyed to student errors was also unsuccessful.

The research studies reviewed in this section point out that students lack a strong conceptual understanding of the mole concept. Some studies have shown that this is not due to students' lack of understanding of mass, volume, and collections of particles but to the terminology itself. Because most chemistry problems make use of the mole concept, students need additional instruction in this area before they will be capable of solving more complex problems.

Molarity, Solutions, Acids, and Bases. Problems involving solutions are even more complex than the problems previously reviewed. These problems involve not only the concepts involved in the mole concept but also an understanding of the solution process, molarity, and frequently acids and bases. The studies in this section show that students at the middle school and high school levels do not have an adequate understanding of solutions. This, coupled with their lack of understanding of molarity and acids and bases, prevents many students from solving chemistry problems in a meaningful way.

General studies of early adolescents' conceptual understanding of solubility have been reported. Longden, Black, and

Solomon (1991) explored 11–12- and 13–14-year-old English students' notions of dissolving. Prietio, Blanco, and Rodriguez (1989) explored Spanish students' ideas of basic chemical aspects of solutions in grades 6 to 8. Both studies used written, open-ended questions in which students were asked to explain phenomena and to make drawings of what was occurring. Longden et al. found that only 20 percent of the 11–12-year-old students and 30 percent of the 13–14-year-old students understood dissolving on both macro- and microscopic levels. They also found that fewer students understood dissolving on the everyday (macro) level than on the particle (micro) level and concluded that many students do not connect the two levels. Prietio et al. concluded: (1) about three-fourths of the children believe in homogeneity of solutions, (2) about one-half of the students depicted solutions as continuous, (3) students generally use familiar rather than scientific terminology to describe the dissolution process even though they have had instruction using more scientific terms, and (4) examples of solutions generally refer only to solids dissolving in liquids.

Gennaro (1981) and Gabel and Samuel (1986) examined more quantitative aspects of solubility. Gennaro looked at ninth-grade students' understanding of solubility by administering a multiple-choice test after students completed study of the *Introductory Physical Science* chapter on "Solubility." He found that the percentage of students answering each of the five items correctly ranged from 21 to 82 percent, indicating a very low level of understanding.

Gabel and Samuel (1986) set out to determine whether students' difficulties in solving molarity problems were related to their misconceptions concerning solutions in general or to their inability to understand the mole concept. Using chemistry molarity problems and equivalent analogue problems involving dissolving lemonade, they tested 400 high school chemistry students using a multiple-choice format before and after instruction on molarity problems. Results indicated that for simple problems, students were able to solve both the analogue and molarity problems. However, when the problems became more complex, such as when the solvent was evaporated or additional solvent was added, fewer students (and in about the same proportion) could solve the analogue and chemistry problems correctly. Even after chemistry instruction, there was little improvement in students' ability to solve the more complex analogue problems, indicating that the students' difficulties were related more to not understanding solutions than to lack of understanding of the mole concept.

Students' ability to solve volumetric analysis problems depends on their understanding of acids and bases as well as mole and solution concepts. The conceptions of first-year university students in France were studied by Cros et al. (1986). They found that students could name several acids but were less adept at naming bases. Although students were somewhat familiar with the technical definitions of pH, they had little idea of the meaning as applied to everyday life. Ross and Munby (1991) reported similar findings with advanced twelfth-grade students in Ontario, Canada. They used multiple-choice tests to construct concept maps and interviews as their research instruments.

Schmidt (1991) also used multiple-choice tests to identify students' conceptions of neutralization. Tests were administered to 1500 German students (grades 11 to 13) to test their understanding of neutralization using multiple-choice items that were carefully constructed to reveal students' erroneous ideas. He found that many students applied the neutralization concept only for strong acids and bases and believed that most neutralization reactions went to completion. He attributes part of the difficulty to the ambiguous use of the word "neutral" in ordinary language and in a chemical context.

Nakhleh and Krajcik (1991) measured high school students' conceptions of acids and bases using concept maps. They found that students' conceptions of acids and bases were more successfully modified when they used microcomputer-based laboratory experiments rather than laboratory exercises involving chemical indicators or pH meters.

These studies on acids and bases indicate that students have only a rudimentary knowledge of acids and bases. This lack of understanding, coupled with their lack of knowledge about solutions and molarity, hinders them from successfully solving volumetric analysis problems.

Chemical Equilibrium. The concept of chemical equilibrium depends on numerous other chemical concepts. Hence, problems involving chemical equilibrium are generally quite complex, depend on students' understanding of stoichiometric relationships as well as solubility and concentration, and are among the most difficult problems to solve for beginning chemistry students. Studies that illustrate this complexity by comparing the problem-solving behaviors of experts and novices were conducted by Crosby (1988) and Camacho and Good (1989). Both conducted interviews using the think-aloud technique to establish differences in novice and expert behavior, and both report content and procedural weaknesses. Camacho and Good listed many of the stoichiometric errors of students and compared the behaviors of successful and unsuccessful problem solvers. Because they interviewed a wide range of experts and novices, they illustrate well how problem solving improves on the novice-to-expert continuum.

Other methods of obtaining data on student conceptions about equilibrium have been used in several studies. Bergquist and Heikkinen (1990) used more structured interviews to determine the conceptions of general college chemistry students. They found that many students did not understand the meaning of concentration and were confused by the scientific use of common terms.

Several studies have used word-association tasks to determine students' conceptions. Gorodetsky and Hoz (1985) used a free-sort task containing 21 chemical equilibrium concepts to construct a concept profile to determine the effectiveness of instruction of a one-semester introductory university chemistry course in Israel. This method, as well as using a knowledge test and a "misconceptions" test containing items similar to those developed by Wheeler and Kass (1978), was used by Gorodetsky and Gussarsky (1986) with twelfth-grade Israeli chemistry students. Results of the study not only unveiled students' conceptual errors about chemical equilibrium but also established the usefulness of the free-sort test, particularly with high-ability students. Two other studies, using 18 chemical equilibrium terms in a free-sort test and analyzing data using related coefficients, were reported by Gussarsky and Gorodetsky (1988, 1990).

Other research reports on equilibrium are associated with modifying instruction. Maskill and Cachapuz (1989) administered a word-association test containing six words to 14-year-old pupils in an English school at several times during the instruction. This was followed by individual administration of the test in which an interviewer asked for a brief explanation of the pupils' responses. The main purpose of their study was to develop an instrument that could be used for diagnostic purposes with students. Cachapuz and Maskill (1989) used a modification of the approach in which they administered the word-association task before and after students solved two equilibrium problems in an interview context.

Rahal (1986) devised a model (CHEM-PLAN) consisting of a network of interconnected nodes or circles representing subconcepts needed for solution of solubility-product problems to be used for instruction. Banerjee and Power (1991) devised a series of modules on chemical equilibrium that appeared to be successful in helping chemistry teachers in Australia develop an understanding of equilibrium and familiarizing them with the conceptual errors of students on the topic.

In summary, research on the understanding of chemical equilibrium has revealed that many students lack the conceptual understanding of fundamental concepts (such as concentration) on which the conceptual framework of equilibrium is based.

Heat, Temperature, and Thermodynamics. There have been many studies of students' conceptions of heat and temperature. Although many of these concepts are important for understanding science and constitute a fundamental knowledge domain, an in-depth understanding of them is not essential for solving many of the chemistry exercises or problems that appear in introductory chemistry textbooks. In this section, only a brief overview and listing of some of the studies in this area is given. Studies indicate that students' conceptions generally become more scientific over time and that certain types of instruction facilitate the change.

Early studies using interviews with or without demonstration tasks were conducted by Albert (1978), Erickson (1979, 1980), Frenkel and Strauss (1985), Appleton (1985), and Shayer and Wylam (1981) to determine children's conceptions of heat and temperature from ages 4 to 9 (Albert), 4 to 11 (Frenkel & Strauss), 8 to 12 (Appleton), and 10 to 13 (Erickson, Shayer, & Wylam). In the past few years, studies of older students' conceptions of heat, temperature, and more advanced thermodynamic topics have been examined using more sophisticated ways to collect data. Boyes and Stanisstreet (1990) determined 11- to 16-year-old students' views on energy sources using a series of questionnaires. They found that students' conceptions generally improved with age, although some misconceptions were more prevalent among older subjects. Sciarretta, Stilli, and Missoni (1990) used real-life phenomena and a paper-and-pencil format to determine the conceptions about heat and temperature of both middle school students and teachers. Concept maps were constructed to provide an individual profile for each subject. Nachmias, Stavy, and Avrams (1990) devised a microcomputer-

based diagnostic system for identifying students' conceptions about heat and temperature. The system provided profiles of students' understanding and could be used before and after instruction to judge its effectiveness for both individual students and the entire class. Rozier and Viennot (1991) used written questionnaires to determine the conceptual understanding of more complex topics of first-year university students in Paris with the purpose of making recommendations for appropriate instructional sequences.

Other studies have addressed teaching methodology more directly by observing the results of particular types of instruction. Stavy and Berkovitz (1980) found that using cognitive conflict with fourth-grade Israeli children enhanced their learning about temperature. Trumper (1990) used comparative events and an anthropomorphic framework to help Israeli 9- to 11-year-olds understand the energy concept. Cullen (1983) showed that instruction, including attempts to show overt linkages between entropy and other concepts in the chemistry course, was partially successful in helping students solve novel problems. Linn and Songer (1991) successfully modified the cognitive demands of a 13-week thermodynamic curriculum for eighth-grade students by actively involving them in predicting and reconciling results and having them use a heat-flow model of thermodynamics to integrate their experimental results. Students used real-time data collection utilizing the computer as a silent laboratory partner.

As in other areas, students generally lack a comprehensive understanding of the distinction between heat and temperature. Because most of the chemistry problems assigned to students in an introductory course do not require them to make this distinction, students have little opportunity to acquire the distinction, and chemistry teachers may be unaware that students lack this skill.

Other Concepts. Reports on students' understanding of other concepts related to problem solving, but not quite as central as those previously discussed, are as follows:

1. Atomic structure: Cros et al. (1986) and Cervellati and Perugini (1981).
2. Chemical bonding: Staver and Halsted (1985) and Peterson and Treagust (1989).
3. Electrochemistry: Hillman, Hudson, and McLean (1981) and Garnett, Garnett, and Treagust (1990).
4. Isomers: Widing (1985).
5. Metals: Biddulph and Osborne (1983).

Summary. From the aforementioned studies of specific science concepts, it is obvious that many students lack the conceptual base for meaningful problem solving. All problems included in an introductory course involve matter. Quantities of matter are expressed in terms of mass, volume, and particles. Additional modes of expression, including moles, density, and concentration, make the problems more complex. Adding more concepts to the problems (such as acids, bases, solubility, and equilibrium) adds more complexity. As a result, students who are unable to understand the chemical concepts adopt algorithmic methods to solve problems. The use of algorithms without understanding results in students' inability to solve the more complex problems and is probably responsible for the student perception that success in chemistry is not necessarily within their control.

Summary

The nature of the problem itself has been shown to be a primary factor in problem difficulty. The research has shown that both in-task variables, such as the vocabulary used in the problem and the problem structure itself, and the nature of the concepts around which the problem is built have important implications for the ease with which the student will solve the problem.

LEARNER CHARACTERISTICS

Another important consideration related to why some people succeed in solving problems while others do not involves inherent differences in the learners themselves. This section considers are studies that deal with experts versus novices and the developmental level of the learner, including proportional reasoning ability. Other factors include spatial ability, memory capacity, and prior knowledge.

Experts Versus Novices

In a search for a unified approach to problem-solving differences among learners, several researchers have examined the problem in terms of the expert-novice paradigm. Camacho and Good (1989), Heyworth (1988), Ploger (1987), and Reif (1983) reach consensus on several key issues. First, novices do not spend much, if any, time analyzing and restructuring the problem as do experts. Instead, they reach for a formula or algorithm and start plugging in numbers. This behavior is aggravated by the novice's lack of a cohesive knowledge structure on which to draw in developing or modifying mathematical relationships. Three suggestions for attending to these deficits are (1) increasing the underlying conceptual understanding of novices, (2) making explicit the actual steps taken by experts to solve problems (Crosby, 1988; Reif, 1983), and (3) helping construct explicit relationships among the chemical principles, laboratory investigations, and mathematical applications for a given topic (Bowen, 1990).

Rowe (1983) identified four different causes of mental lapses that novices experience during a typical lecture: (1) short-term memory overload, which occurs when too many ideas are presented in too short a time; (2) the idea presented doesn't match the existing mental structure of the learner; (3) the lecturer uses symbols different from those used in the text; and (4) something presented in lecture sets off a complementary chain of thoughts in the learner. These memory lapses can be partially overcome with note taking and pausing and encouraging students to share information after 10 minutes of lecture, according to Rowe. Differences between experts and novices

may be explained by Piagetian developmental level and other factors such as spatial ability, memory capacity, prior knowledge, and student attitude.

Developmental Level of the Learner

A major emphasis in the literature is the relationship among Piagetian levels of development and learners' problem-solving ability. Most studies are concerned with the mismatch between learners' developmental level and the cognitive demands of the problems. It has been established that many chemistry topics require formal-operational thought (Kavanaugh & Moomow, 1981; Niaz & Lawson, 1985). Such thought is characterized by abstract thinking and includes such topics as stoichiometry, limiting reagent, mole, gas laws, and balancing equations, to name a few. Yet according to the literature, from 30 to 70 percent of our chemistry students operate on the formal level in chemistry. Atwater and Alick (1990) mode a similar finding for African American students. The percentage of formal-operational students seems to depend on the type of problem used in the test (Thornton & Fuller, 1981). Wiseman (1981) reports a correlation of 0.76 between performance in high school science with developmental level. Similar correlations exist for introductory college chemistry and consumer chemistry courses (Wiseman, 1981). The conclusion of many researchers (Krajcik & Haney, 1987; Kavanaugh & Moomow, 1981) is that maximum learning occurs when the course taught matches the developmental level of the learner. It has been determined by Ward and Herron (1980) that concrete and formal students do equally well on concrete material, such as memorization of facts and figures, but formal students outperform their concrete counterparts on more abstract material.

Kletzly (1980) looked at the effects of teaching abstract high school chemistry topics using either traditional or Piagetian-based instruction. Traditional instruction included lecturing, review of notes, doing assigned problems, and performing a laboratory experiment. The Piagetian-based instruction followed large-group instruction with small-group, self-paced, and self-directed study. Laboratory experiments were also included. She found that both instructional methods were equally effective for formal students on the topic of atomic theory but that the Piagetian-based method was more effective for transitional-level students on the same topic.

Proportional Reasoning. When Piaget's development classification is examined more closely, the significance of the proportionality variable included in the classification becomes obvious. Proportional reasoning ability can be used to differentiate the performance levels within formal-operational groups (Krajcik & Haney, 1987) and as a predictor of success in more mathematical introductory chemistry courses (Bender & Milakofsky, 1982). Proportional reasoning is important not only in mathematical topics such as stoichiometry but also in other chemistry topics such as nomenclature and formula writing (Wheeler & Kass, 1977), although to a lesser degree (Thornton & Fuller, 1981). Wheeler and Kass looked at the relationship between proportionality and the chemistry achievement of 168 tenth-grade students. Proportionality accounted for 63.4 percent of the total variance of the chemistry achievement test. Wheeler and Kass comment that there is low transferability between general proportional reasoning and proportional reasoning in chemistry. They measured proportional reasoning in chemistry, specifically, and found that concrete students who solved problems successfully may have determined the correct answer by using additive rather than proportional reasoning techniques. The higher developmental-level requirement of some chemistry problems may be lowered, however, through the use of algorithms. Bender and Milakofsky (1982) call for the development of chemistry programs that deemphasize formal tasks in the beginning while attempts are made in the course to improve the cognitive functioning level of all students.

Proportional Reasoning and Algorithms. Students' proportional reasoning ability and the use they make of algorithms and heuristics in problem solving have been shown to be related in several studies. Some studies (Anamuah-Mensah, 1986; Anamuah-Mensah, Erickson, & Gaskell, 1987; Gabel & Sherwood, 1984) have shown that students use algorithms both with and without knowledge of proportional reasoning and conceptual understanding. Anamuah-Mensah (1986) found that in titration problems, when either a formula (algorithm) approach or a proportional reasoning approach (based on mole ratios of acid and base) could be used, students in both the medium- and high-ability groups chose the formula approach. This result reinforces the already demonstrated need for a variety of test items in chemistry, some of which test proportional reasoning and conceptual understanding in a qualitative fashion.

Anamuah-Mensah et al. (1987) found that teachers can help students switch to the use of algorithms with understanding by breaking complicated problems down into subproblems and solving the subproblems in a hierarchical fashion. Such explicit teaching helps students understand how to approach problem solving.

Gabel et al. (1984) have further shown that students with low proportional reasoning ability and high math anxiety learn the least from use of the factor-label method. In contrast, Wheeler and Kass (1977) stated that the factor-label strategy reduces the difficulties experienced by concrete thinkers. This seemingly contradictory finding may result from differing definitions of successful problem solving used in the two studies. Concrete thinkers may be able to use the factor-label method to solve problems if they view it as an algorithm. Such use, however, does not necessarily promote true learning as defined by conceptual understanding. The combination of conceptual understanding and proportional reasoning is still more strongly connected to successful problem solving than algorithmic use of the factor-label approach. However, Cervellati et al. (1982) report that in many textbooks little attention is paid to the development of a conceptual understanding of moles in stoichiometric calculations, offering support for the common practice of uncoupling conceptual understanding and mathematical manipulations. Even if conceptual understanding is separated from the use of proportional reasoning, successful use of proportional reasoning may develop with continued use during subsequent science courses (Raven, 1987).

Not all researchers believe that use of algorithms is sufficient for success in problem solving. Bodner (1987) states that algorithms are insufficient for students completing examination problems. They are, however, useful in solving routine exercises.

Bapat and Kiellerup (1981) looked at the problem-solving strategies used by high school and first-year college students. They report that 90 percent of the students studied used an analytical approach to problem solving rather than an intuitive one. No correlation was found between a subject's mental ability, personality profile, and chosen problem-solving strategy. Carter (1987) reported that for a small sample of college students ($n = 9$), students' beliefs do influence their selection of algorithms.

Spatial Ability, Memory Capacity, Prior Knowledge, and Student Attitude

Several other factors including spatial ability and memory capacity influence problem solving. They have been identified and investigated by Carter, LaRussa, and Bodner (1987), Niaz (1985), and Pribyl and Bodner (1987).

Niaz (1985) reports that student performance on a problem decreases as the memory demand increases but that the same items can be rewritten to reduce their M demand. Carter et al. (1987) and Pribyl and Bodner (1987) have shown that students with high spatial ability do better on problems that require mental manipulations of representations and outlining of multistep processes. Spatial ability is measured by a learner's ability to dissemble information from a visual field and restructure it (Bodner & McMillan, 1986). This ability is often referred to as a measure of field dependence. Field dependence has been shown to enable students to better understand the problem during the early stages of problem solving (Pribyl & Bodner, 1987).

Anamuah-Mensah (1990), Chandran, Treagust, and Tobin (1987), Niaz and Lawson (1985), Niaz and Robinson (1989), and Staver and Jacks (1988) have looked at the comparative effects of developmental level and other variables on problem solving. Niaz and Robinson (1989) and Niaz (1987) have differentiated between the types of chemistry problems and relevant learner variables. They contend that the Piagetian development level is the best single predictor in problems requiring mathematical manipulations and field dependence (M capacity) is the best predictor of conceptual understanding. Niaz and Lawson (1985) make a distinction between level of complexity of the same chemistry topic and learner variables. Their research shows that developmental level alone is important in predicting success of one-step equation-balancing problems, while developmental level and large M capacity are significant factors in multistep equation balancing. Chandran, Treagust, and Tobin (1987) and Anamuah-Mensah (1986), however, identify only developmental reasoning and former knowledge as statistically related to success in chemistry.

As indicated in the first section, the effect of previous knowledge and the quality of that knowledge significantly affect achievement. Gooding, Swift, Schell, Swift, and McCroskery

(1990), Greenbowe (1983), Pope and Gilbert (1983), and Yarroch (1985) concur that the better the learner's conceptual knowledge, the fewer mistakes he or she will make in solving problems. All call for restructuring of tests to reflect the importance of conceptual understanding in the study of chemistry, as well as mathematical problem solving. Pope and Gilbert (1983) suggest that teaching should also reflect this change by utilizing constructivist strategies such as role-playing, simulations, and discussions.

Braathen and Hewson (1988) conducted a case study of five students enrolled in a college chemistry tutorial program. They found that conceptual change was related to both the student's learning approach and the quantity and quality of prior knowledge.

Other variables such as student attitude toward the learning environment (Gooding et al., 1990) and learning style (Alcorn, 1985) have been shown to affect achievement significantly. Gooding et al. (1990) report that the two variables of previous achievement in chemistry and student attitude toward the class account for 70 percent of the variance in final examination scores. Alcorn (1985) measured problem-solving style in relation to the Myer-Briggs personality type. Results of this research show that the problem-solving style used by students is responsible for an increase in variance in college chemistry scores over that predicted using other measures.

Gender

The variable of gender differences has been explored by Squires (1977) and Staver and Halsted (1985). It has not been shown to be a consistently significant factor in problem-solving research.

Summary

In this section, the research has shown that learner characteristics are important factors in the way students solve problems. Experts solve problems using a more conceptual approach. Problem-solving ability is dependent on the developmental level of the learner (in particular, proportional reasoning ability), spatial ability, memory capacity, prior knowledge, attitude toward the learning environment, and personality orientation.

LEARNING ENVIRONMENT

The success of problem-solving activity is also dependent on the environment and context in which the activity takes place. Many environmental factors can affect the outcome, including whether (1) problem solving is an individual or group activity, (2) the process is taught explicitly or implicitly, (3) the conceptual foundation is elucidated through the use of analogies, (4) computer instruction is incorporated in the process, and (5) laboratory, science-focused math curricula, and remediation are used.

In a learning environment in which students are asked to assimilate 6000 to 6750 units of information contained in chemistry textbooks and lecturers add an additional 20 percent new material as indicated by Rowe (1983), students need help in structuring the learning environment. Manley's (1978) experiment with 80 classes of high school students enrolled in an Interdisciplinary Approaches to Chemistry course found that according to student opinion surveys, the learning *environment* is the most important variable in determining student attitudes toward chemistry. Environment is more important than the type of chemistry class in bringing about positive student opinion. Students in the top 25 percent of the study (with strong positive attitudes toward chemistry) viewed classes as more goal directed and showing less favoritism than students in the lowest quadrille. Attitude is important in both problem-solving success and the choice to pursue a career in a science-related field.

Social Interaction

Social interaction is a powerful variable in the learning process, but in American classrooms it has often been ignored. However, Nigeria, Okebukola, and Jegede (1990) investigated the effects of general environment, reasoning pattern, nature of the home environment, and goal structure of 128 secondary school students. The children in the study belonged to the Yoruba ethnic group, in which the extended family is the predominant social unit and the traditional authority of the father is paramount, resulting in a high degree of social conformity. Differences in science achievement between the more traditional rural student and the transitional urban student were examined. It was found that students whose goal structure was characterized by a preference for cooperative rather than individual, competitive work had higher achievement scores on a science concept test. Students from traditional, authoritative homes had lower achievement scores than students from less authoritative, transitional homes.

Several studies have examined the effects of cooperative learning groups on science achievement (Basili & Sanford, 1991; Grant, 1978; Robinson & Niaz, 1991; Tingle & Good, 1990). Grant's (1978) study of tenth-, eleventh-, and twelfth-grade high school students compared the quality of responses to science problems provided by students working individually and in groups of four. Some of the groups had extended experience with each other in group work in other courses. Other groups were formed for the purpose of this study. The groups with extended experience provided higher-quality answers than those from the newly formed groups. Both groups provided higher-quality answers than individual students. The difference in quality of answers was found not to be related to the intellectual levels of students.

Tingle and Good's (1990) study of 178 high school students in regular and honors chemistry classes did not show an advantage for cooperative group versus individual learning. In this study, groups were formed heterogeneously based on the proportional reasoning ability of students. An analysis of covariance based on the categorically independent variables of proportional reasoning ability, class type (honors or regular), and type of grouping (individual or cooperative) was performed with the continuous dependent measure of chemistry achievement. Student volunteers of varying proportional reasoning ability from both the treatment and control groups were videotaped while solving problems aloud. There was no statistical difference in achievement for students regardless of their class (honors or regular) or grouping (individual or cooperative). Furthermore, characteristics of successful problem solvers, as viewed on videotape, were the same in both individual and group settings. These characteristics included confidence, persistence, and evidence of a strong conceptual base.

Basili and Sanford (1991) studied the conceptual change strategies of 62 community college students involved in cooperative groups. Their analysis of the cooperative-operative learning groups revealed that peer discussion can help students clarify their science views and help make explicit the distinction between scientific and everyday words. Facilitating effective group work, however, involves explicitly teaching the behaviors involved in good group leadership.

Robinson and Niaz's (1991) study of 82 college students in two sections of an introductory chemistry course, although not a study of cooperative learning per se, did look at the effects of active student involvement in the classroom. Students were taught using either a lecture or an interactive discussion format. Interactive discussion was a combination of discussion of alternative approaches and error analysis of a problem solution between the members of the group and the instructor. Although different from cooperative learning groups with the teacher as coach or mentor, interactive discussion does *actively* involve students in the learning process. The treatment group performed significantly better than the control group on a 1-hour class examination. Even though fewer problems were covered in the treatment group, students with lower M capacity, as measured by the Figural Intersection Test, did significantly better than high-M-capacity students. Higher-formal-reasoning students, as measured by the Group Assessment of Logical Thinking Test (GALT), did better than lower-reasoning-level students in both the treatment and control groups.

Summary. These studies suggest that other learner attributes may have significant bearing on the effectiveness of cooperative learning. These include (1) previous cultural or academic experience with cooperative grouping, (2) depth and breath of the conceptual bases of students in the groups, (3) student learning style and developmental level, and (4) the skill of the teacher in the role of group mentor.

Heuristics and Explicit Problem-Solving Approaches

Polya (1945) was one of the first to suggest that an organized, general approach to problem solving be taught to students along with scientific or mathematical content. Polya's heuristic has four stages: (1) understanding the problem, (2) devising a plan, (3) carrying out the plan, and (4) looking back. This general heuristic can be applied in any problem-solving situation and has been elaborated and modified in chemistry teaching by several researchers (Bunce & Heikkinen, 1986; Bunce, Gabel, & Samuel, 1991; Bunce et al., 1990; Frank & Herron, 1987; Kramers-Pals, Lambrechts, & Wolff, 1982; Mettes, Pilot, Roossink, & Kramers-Pals, 1980; Reif, 1983; Stiff, 1988; Waddling, 1988).

Frank and Herron (1987) taught college chemistry students to solve problems using general heuristics. Treatment students had higher achievement scores than control students, but the increase was not significant. However, students trained in heuristics pursued the problem more persistently, checked work more often, made fewer math errors, and formed more generalizable representations of the problem.

Reif (1983) has stated that students must be taught *explicit processes* to achieve the performance in problem solving that experts demonstrate automatically. Reif describes three specific components of good problem solving: analyzing, planning a solution, and testing the result. The step included in Polya's model but not enumerated by Reif is the actual carrying out of the planned solution. The analysis stage of Reif's plan has two distinct parts: (1) *identification* of information to be found in the problem and gathering of other pertinent information for use in the solution and (2) *generation* of a theoretical description of the problem using the student's knowledge base. The similarity between Reif and Polya is striking. According to Reif, current teaching practices pay more attention to the *product* than the *process* of problem solving.

Waddling (1988) stressed the importance of student-designed networks in problem solving. Such networks have a twofold purpose: to provide students with a way to organize their solutions and to provide teachers with a representation of students' thinking patterns. Such networks can help guide teachers to the cause of student difficulties in solving problems.

Mettes et al. (1980) described a problem-solving approach with four steps: (1) analysis, (2) establishing whether the problem is a standard problem and, if not, breaking it into subproblems that will convert it to a standard problem, (3) executing operations necessary to solve nonroutine problems, and (4) checking and interpreting the answer. Applying this approach to problem solving, Kramers-Pals and Pilot (1988) suggest that tests should be designed to give credit for a student's progress to different points in the solution process, thus encouraging students to progress in the problem-solving process during the course.

Kramers-Pals et al. (1982) believe that teachers not only do not teach an organized approach to problem solving but also hinder students' progress by underestimating the difficulties students have with both the content and the problem-solving process. Kramers-Pals et al. pointed out that because almost all problems in a course are routine for teachers, they skip the transformation of a nonstandard problem into a standard problem. In addition, teachers take several steps at once in a problem's solution, so only part of their solution plan is obvious or written down for students to follow.

Bunce and Heikkinen (1986) employed an explicit problem-solving method as both a learning and teaching tool. The method, called PACE in this study and modified as the Explicit Method in subsequent studies, incorporated the following six steps: (1) identification of what is given; (2) identification of what is asked for in the problem; (3) recall of all pertinent information including rules, definitions, equations, and relationships; (4) construction of a schematic outline of the steps necessary to get from the given information to what is asked for; (5) mathematical solution including ratios or factor-label approaches; and (6) review of the problem statement, given information, requested solution, recall and overall plan. Such a visual framework for problem solving was also reported to be important for student conceptualization of both the content and the process of problem solving by Cameron (1985).

In the study by Bunce and Heikkinen (1986), half of the 200 college students enrolled in an introductory chemistry course were taught an explicit problem-solving approach. Although no significant differences in achievement for control and treatment students were detected, it became obvious that student use of such problem-solving techniques during tests required more time and space than normal.

A second study of college nursing chemistry students (Bunce & Gabel, 1991) showed that additional exposure to one stage of the Explicit Method of problem solving, namely categorization of the problem according to type, resulted in significant improvement in achievement in an unannounced testing situation. Categorization of problems was inherent in the recall section, but in this experiment it was stressed. Through analysis of think-aloud problem-solving interview tapes, student use of a Rolodex approach to problem solving became evident. The Rolodex approach is characterized by students mentally flipping through an imaginary file of formulas and trying to match units between the problem to be solved and the formula in the mental Rolodex. When a match is found, the formula is selected and immediately implemented without further analysis. Even though students appear to have a fairly good conceptual understanding of the chemistry involved in the problem, this knowledge is often not utilized in the formula retrieval and implementation phases of their problem solving. Raines (1984) and Ryan (1987) both reported that the ability to categorize a problem is essential for successful problem solving.

A more intensive study of 290 high school chemistry students' use of the Explicit Method of problem solving on stoichiometry problems was conducted by Bunce et al. (1990). In this study, three different levels of questions (bare-bones problems similar to end-of-chapter problems, those with extraneous information, and those involving a two-step stoichiometric reaction) were presented to students. The effect of using the Explicit Method was tested in three different testing situations (announced test, unannounced test, and final exam). A significant treatment effect was found across all testing situations. Furthermore, there was a differential significant treatment effect according to students' developmental level as measured by the GALT test. Students in the transitional phase showed a significant treatment effect on the announced and unannounced tests. They also showed a significant treatment effect, across all testing situations, for problems containing extraneous information.

Stiff (1988) described an explicit problem-solving strategy whereby teachers provide students with a narrative of how they solve problems and students use this model as they keep problem-solving journals. This method, called Problem Solving by Example, is considered by the author to be an intermediate step toward mastering problem-solving heuristics.

Summary. In summary, the studies in this section show that when chemistry students are taught to solve problems in a systematic manner, they are more successful. Strategies based on Polya's heuristics or variations thereof appear to facilitate students' ability to solve routine problems even though there is some evidence that students may be doing so using algorithms.

Analogies

In an attempt to help students understand the abstract concepts of chemistry, many researchers have searched for a link between students' past experiences and new chemistry concepts. In addition, researchers have attempted to provide concrete examples to help students understand abstract concepts. The use of analogies achieves both the linkage and concrete examples. An analogy expresses an abstract idea by grounding it in concrete experience. It facilitates the development of reasoning while new data are gathered. Research on the use of analogies in the teaching of chemistry problem solving has focused on the use of word analogies (Freidel et al., 1990; Gabel & Samuel, 1986; Gabel & Sherwood, 1984).

Gabel and Sherwood (1984) used word analogies of familiar tasks involving sugar and oranges with 322 high school students. The results show that the size of the object used in the analogy makes no difference in the problem's difficulty. Some students experienced difficulty with problem solving whether they used analogies or not because they did not understand scientific notation, division, and how to analyze two-step versus one-step problems. In another study, Gabel and Samuel (1986) used molarity problems and their analogue counterparts with 619 high school students. The researchers found that students had difficulty solving chemistry problems mainly because they did not understand the underlying chemical concepts. In some instances, the use of analogies may have helped conceptual understanding; however, sometimes students had the same difficulties understanding the analogue that they had with the chemical concept. For large numbers of students, problems were solved by the use of algorithms and students saw little or no relation between the chemical concept involved and the problem's solution. Student performance on certain types of problems may improve as the connection between the analogue and problem is better understood.

Freidel et al. (1990), working with preparatory college chemistry students, found no significant effect on achievement for students using analogies. However, once again, students had difficulty understanding the analogies used in the study and the authors point to the limited time allotted for the treatment as a factor in the nonsignificant results.

Summary. Results of the studies reviewed in this section indicate that analogies may not be as useful in teaching problem solving as one might think. If students are unfamiliar with the analogy or unable to make the connection between the analogy and the chemical phenomenon, analogies will not aid students in problem solving.

Laboratory, Mathematics, and Remediation

Teachers have traditionally tried to improve the problem-solving ability of students through teaching by example, providing better links between science and math courses, and providing more examples. In the past, these measures have met with mixed success, mainly because they addressed the symptoms and often missed the overriding issues. Lundeberg (1990) reported on a supplemental instructional session that helped students articulate their scientific conceptions and develop confidence and ability in problem solving. Ben-Zvi and Silberstein (1980) suggested using laboratory exercises to motivate students in learning quantitative concepts. Goldman (1974) looked at the effects of a more directed mathematics curriculum on science learning, and Chiappetta and McBride (1980) used problem-solving remediation to increase student achievement.

Lundeberg (1990) provided 42 1-hour sessions during the course of a school year to college-level students enrolled in chemistry for the allied health professions. Students attending the sessions, who were encouraged to think aloud while solving problems with mentors and to articulate their scientific conceptions, did significantly better in the course than students not attending the sessions.

Ben-Zvi and Silberstein (1980) taught difficult quantitative chemistry topics to tenth-grade students by introducing the topic with a demonstration, asking students to explain the demonstration, and then using the students' explanations as part of the teaching scheme. Students who were taught using this method did not score significantly better than the control group but they did report the topics to be more interesting and less difficult than the control group.

Although Dierks (1981) reported that student errors in stoichiometry calculations are related to students' inability to transpose chemistry statements into the language of mathematics, studies by Goldman (1974) and Chiappetta and McBride (1980) indicate that increased mathematics emphasis does not improve students' problem capability. Goldman looked at the effect of stressing science-situational problem-solving skills in a mathematics course with student achievement in high school chemistry. Although these students performed significantly better on the mathematics course posttest than students in the control group, there was no significant difference in achievement in the chemistry course between treatment and control groups. Chiappetta and McBride (1980) studied the effect of remediation on the achievement of 99 ninth-grade physical science students on the mole concept. Students who did not achieve at the 80 percent level initially were provided with remediation booklets and then retested. Second remediation booklets were provided if needed. No significant increase in achievement between treatment and control groups was demonstrated.

Summary. From these studies, it appears that an increased emphasis on mathematics does not necessarily improve students' problem-solving ability. Other factors such as conceptual understanding appear to be more related to their success in problem solving.

Science in Other Languages

Researchers have examined the effect of language on science conceptual understanding. Lynch, Chipman and Pachaury (1985) examined the effect of language (Hindi and English) on students' understanding of 16 concept words including mass, length, and volume. In the Hindi language, 12 of the 16 terms are used to describe science only; in English, this number drops to 5. This means that in the English language, most words do double duty as common usage and scientific vocabulary, but this is not the case in Hindi. The researchers found that in only 2 of the 16 cases did Hindi-speaking students have an advantage in concept word recognition over their English-speaking counterparts. Science occupies approximately the same proportion of the curriculum in the two school systems, but the sequencing and exposure to topics may differ. Although specific languages may appear to favor concept recognition, other variables must also be investigated. In other words, cultural and prior educational experiences, not language, are responsible for most of the differences between Hindi- and English-speaking high school students.

Summary

The learning environment encompasses many aspects of the learning experience. Through the use of cooperative learning situations, explicit problem-solving strategies, word analogies, and laboratory demonstration experience, students of varying developmental levels can experience increased success in chemistry. Perhaps a restructuring of the learning environment should be viewed as a set of variables that is well under the control of the teacher. Through careful planning, the teacher can help bring about a closer match between the developmental level of the student and the cognitive demands of the material.

Learner Characteristics and Learning Environment

Manifestation of learner characteristics is influenced by the learning environment. The learning environment can either compensate for or bring out deficiencies in the learner. Some curriculum approaches to the problem of learners' differing developmental levels and abilities involve careful organization or sequencing of materials.

Learning Cycle. One attempt to help students overcome the difficulties inherent in learning in a traditional classroom lecture involves use of the learning cycle (Ward & Herron, 1980). Here, through physical manipulation of materials and peer interaction, the discrepancies among the aptitudes of the learners in the class can be minimized. Teachers too, can help novice learners by either reducing the cognitive load or providing explicit strategies to help novices group elements of the concept into a meaningful framework links a learner's current knowledge with new knowledge (Johnstone, 1983). Presenting a topic at a simple, hands-on level and building on this basis, together with providing of immediate feedback for students, can help students restructure their cognitive framework (Kavanaugh & Moomaw, 1981). Niaz (1988c, 1989) suggests that teacher-led restructuring of the cognitive task to meet the level and limited memory capacity of novices will facilitate student success.

It has already been established that the development of an accurate and extensive conceptual base is essential for successful problem solving, especially solving novel or transfer problems. Several people have looked into the use of the learning cycle (Abraham & Renner, 1986; Jackman, Moellenberg, & Brabson, 1990; Vermont, 1985).

The learning cycle, based on Piagetian theory, divides learning into an *exploration* or data-gathering phase, *conceptual invention*, and *conceptual expansion* phases. The learning cycle was originally developed by Karplus for the Science Curriculum Improvement Study elementary curriculum project, in which the three components were called exploration, invention, and discovery. Each learning cycle begins with a student activity that provides a common set of experiences on which to build the concept. Because most curricula contain the three phases of the learning cycle in some order, Abraham and Renner (1986) decided to test the importance of the sequence of these phases with students of different developmental levels. They used several different sequences involving movement of the invention phase of the learning cycle to different positions in the learning sequence. Six classes of high school chemistry students took part in the study. Student achievement was measured on content achievement tests. In addition, classes were observed and tape recorded. One student from each class was randomly selected for an in-depth interview. The results of this study show that the position of the invention phase of the learning cycle is important for optimum learning, but the actual preferred sequence depends on whether the concept being taught is new or one with which students have had some previous experience. For new concepts, the optimum sequence includes invention as the middle phase. For review concepts, formal students do best when invention is the first phase, but concrete students achieve highest when invention is the last phase.

Vermont (1985) compared the effectiveness of three learning strategies on 60 community college chemistry students. The three learning strategies included the learning cycle, cognitive learning and development strategy, and a lecture-laboratory format. Data were analyzed, and all three methods were found to be equally effective in teaching the mole concept.

Jackman, Moellenberg, and Brabson (1990) investigated three different teaching methods in a general chemistry laboratory with 350 college students and their interaction with the conceptual systems of their students. The three teaching methods included traditional, learning-cycle, and computer-simulation techniques. The learning cycle and computer simulation were more effective than the traditional approach for all students. There was no significant interaction between the conceptual systems of the students and the teaching method employed.

Spiral Curriculum. Schmidkunz and Buttner (1986) discussed the advantages of introducing a spiral-curriculum model for the teaching of chemistry. A spiral curriculum includes fundamental knowledge of chemistry taught at different levels of understanding throughout the school experience. Such an approach allows a topic such as the atomic model to be introduced at several key points in the curriculum, such as in the topics of redox, chemical bonding, and acid-base chemistry. The curriculum is thus organized systematically, beginning with a simple explanation of phenomena and moving to more complex models. This approach provides flexibility of presentation and reinforcement of key chemical concepts in conjunction with a student's developmental growth.

Summary

Previous studies have shown that restructuring the learning environment so that it is more compatible with the learner's characteristics, such as developmental level, can have positive results. Both the learning cycle and the spiral curriculum provide opportunities for students of different developmental levels to investigate chemical concepts in a manner more in keeping with the learner's characteristics. Therefore, it is expected that their use would have positive effects on learners' achievement.

CONCLUSIONS

The diverse variables studied and results reported in the array of problem-solving research reports reviewed in this chapter confirm that chemistry problem solving is a very complex human activity. Research shows that successful problem solving is dependent on the nature of the problem itself. Problems requiring several steps for solution, the problem format, and the language used in the problems can affect problem difficulty. Learner characteristics—in particular, proportional reasoning ability, spatial visualization, and memory capacity—are factors that appear to be related to problem-solving ability. Although it is questionable whether these characteristics can be modified sufficiently in students to improve their problem-solving ability, the research has shown that if students are taught to be more systematic in their approach, they will be more successful problem solvers.

This review of problem-solving research also leads the reviewers to conclude that most of the problems given to students to solve in chemistry might be considered exercises rather than problems. Providing solutions even to these exercises is apparently difficult for students. There is overwhelming evidence that they solve the exercises by relying primarily on the use of algorithms rather than the meaning of the underlying concepts. Studies on concept learning indicate that students lack the conceptual understanding needed to solve problems in a meaningful way. An important implication for teaching is that problem solving could be made more meaningful and students might become more successful if problems were presented in such a way that students could see the relationship between the problem, the phenomena on which the problem is based, and the microscopic representation of that phenomena. An increased emphasis on understanding the underlying concepts on which problems are based is essential for improving meaningful problem solving. Improving students' mathematical skills without increasing their conceptual knowledge will have little, if any, effect on their problem-solving capability.

With regard to the research studies that are reviewed, it should be obvious to the reader that very few research studies address the intersection of all three areas of this review, namely nature of the problem, learner characteristics, and learning environment. At best, studies concentrate on an overlap of two areas. There are many practical reasons for this, including the fact that such research is difficult to conduct in the multivariable process of classroom teaching and learning. However, researchers should strive at least to control variables effectively in areas that cannot be directly investigated. Without effective control of such variables, significant outcomes may well be overshadowed by unexplained statistical error. The fact that prob-

lem-solving research often addresses only one area, such as the problem, learner, or environment, rather than the overlap of all three, may be responsible for the large number of research studies that fail to reach conclusive results. A new generation of research studies is needed that addresses, or at least controls, a larger number of variables in the problem-solving process. Of course, identifying the pertinent variables in the teaching-learning process is made easier if the researcher approaches the problem from a theoretical framework. Without such theoretical constructs, our research efforts will result in unfocused, narrow conclusions that have little generalizability. A theoretical framework, whether it can be classified as constructivist, information processing, Piagetian, or something else, can direct the researcher toward a meaningful exploration of the issues in chemistry problem solving.

References

Abraham, M. R., & Renner, J. W. (1986). The sequence of learning cycle activities in high school chemistry. *Journal of Research in Science Teaching, 23*(2), 121–143.

Albert, E. (1978). Development of the concept of heat in children. *Science Education, 62*(3), 389–399.

Alcorn, F. L. (1985). The relationships between problem solving style as measured by the Myers Briggs type indicator & achievement in college chemistry at the community college (Doctoral dissertation, Virginia Polytechnic Institute & State University, 1984). *Dissertation Abstracts International, 46*(4), 939A.

Anamuah-Mensah, J. (1986). Cognitive strategies used by chemistry students to solve volumetric analysis problems. *Journal of Research in Science Teaching, 23*(9), 759–769.

Anamuah-Mensah, J. (1990). Cognitive factors in chemistry achievement: some observations. *Journal of Research in Science Teaching, 27*(6), 607–609.

Anamuah-Mensah, J., Erickson, G., & Gaskell, J. (1987). Development and validation of a path analytic model of students' performance in chemistry. *Journal of Research in Science Teaching, 24*(8), 723–738.

Andersson, B. (1986). Pupils' explanations of some aspects of chemical reactions. *Science Education, 70*(5), 549–563.

Andersson, B. (1989). Pupils' conceptions of matter and its transformations (12–16). *Proceedings of the Second Jerusalem Convention on Education 1989* (pp. 21–49). Jerusalem: Hebrew University of Jerusalem.

Appleton, K. (1985). Children's ideas about temperature. In R. P. Tisher (Ed.), *Research in science education* (Vol. 15, pp. 122–126). Selections of papers from the annual conference of the Australian Science Education Research Association, Capricornia Institute, Rockhampton, Queensland, Australia. (ERIC Document Reproduction Service No. ED 267 974)

Atwater, M. M., & Alick, B. (1990). Cognitive development and problem solving of Afro-American students in chemistry. *Journal of Research in Science Teaching, 27*(2), 157–172.

Baker, C. A. (1988). A comparison of student-directed and teacher-directed modes of instruction for presentation of density to high school chemistry students (Doctoral dissertation, Purdue University, 1987). *Dissertation Abstracts International, 49*(5), 1107A.

Banerjee, A. C., & Power, C. N. (1991). The development of modules for the teaching of chemical equilibrium. *International Journal of Science Education, 13*(3), 355–362.

Bapat, J. B., & Kiellerup, D. M. (1981). *An investigation into the problem solving strategies of some H. S. C. and first year students.* Caulfield

Institute of Technology, Victoria, Australia. (ERIC Document Reproduction Service No. ED 215 880)

Bar, V., & Travis, A. S. (1991). Children's views concerning phase changes. *Journal of Research in Science Teaching, 28*(4), 363–382.

Basili, P. A., & Sanford, J. P. (1991). Conceptual change strategies and cooperative group work in chemistry. *Journal of Research in Science Teaching, 28*(4), 293–304.

Ben-Zvi, R., Eylon, B., & Silberstein, J. (1982, April). *Student conception of gas and solid difficulties to function in a multi-atomic context, part II.* Paper presented at the National Association of Research in Science Teaching Conference, Lake Geneva, WI.

Ben-Zvi, R., Eylon, B., & Silberstein, J. (1986). Is an atom of copper malleable? *Journal of Chemical Education, 63*, 64–66.

Ben-Zvi, R., & Silberstein, J. (1980). The use of motivational experiments in the teaching of quantitative concepts in chemistry. *Journal of Chemical Education, 57*(11), 792–794.

Bender, D. S., & Milakofsky, L. (1982). College chemistry and Piaget: the relationship of aptitude and achievement measures. *Journal of Research in Science Teaching, 19*(3), 205–216.

Berkheimer, G. D., & Blakeslee, T. D. (1988, April). *Using a new model of curriculum development to write a matter and molecules teaching unit.* Paper presented at the annual meeting of the National Association for Research on Science Teaching, Lake Ozark, MO.

Bergquist, W., & Heikkinen, H. (1990). Student ideas regarding chemical equilibrium. What written test answers do not reveal. *Journal of Chemical Education, 67*(12), 1000–1003.

Biddulph, F., & Osborne R. (1983). *Children's ideas about metals* (Working Paper No. 112). A Working Paper of the Learning in Science Project (Primary). Waikato University, Hamilton, New Zealand. (ERIC Document Reproduction Service No. ED 252 395)

Bodner, G. M. (1987). The role of algorithms in teaching problem solving. *Journal of Chemical Education, 64*(6), 513–514.

Bodner, G. M. (1991). I have found you an argument: the conceptual knowledge of beginning chemistry graduate students. *Chemical Education '91—Educational Possibilities.* Kenosha: University of Wisconsin–Kenosha.

Bodner, G. M., & McMillan, T. L. B. (1986). Cognitive restructuring as an early stage in problem solving. *Journal of Research in Science Teaching, 23*(8), 727–737.

BouJaoude, S. B. (1991). A study of the nature of students' understandings about the concept of burning. *Journal of Research in Science Teaching, 28*(8), 689–704.

Bowen, C. W. (1990). Representational systems used by graduate students while problem solving inorganic synthesis. *Journal of Research in Science Teaching, 27*(4), 351–370.

Boyes, E., & Stanisstreet, M. (1990). Pupils' ideas concerning energy sources. *International Journal of Science Education, 12*(5), 513–529.

Braathen, P. C., & Hewson, P. W. (1988, April). *A case study of prior knowledge, learning approach and conceptual change in an introductory college chemistry tutorial program.* Paper presented at the annual meeting of the National Association for Research in Science Teaching, Lake Ozark, MO. (ERIC Document Reproduction Service No. ED 292 687)

Brook, A., Briggs, H., & Driver, R. (1984). *Aspects of secondary students' understanding of the particulate nature of matter.* Children's Learning in Science Project, Leeds: Centre for Studies in Science and Mathematics Education, The University.

Bunce, D. M., Baxter, K., DeGennaro, A., Jackson, G., Lyman, J., Olive, M., & Yohe, B. (1990, December). *Teaching students to solve chemistry problems—A cooperative research project.* Paper presented at National Science Teachers Association Convention, Washington, DC.

Bunce, D. M., Gabel, D. L., & Samuel, K. B. (1991). Enhancing chemistry problem-solving achievement using problem categorization. *Journal of Research in Science Teaching, 28*(6), 505–521.

Bunce, D. M., & Heikkinen, H. (1986). The effects of an explicit problem-solving approach on mathematical chemistry achievement. *Journal of Research in Science Teaching, 23*, 11–20.

Camacho, M., & Good, R. (1989). Problem solving and chemical equilibrium: Successful versus unsuccessful performance. *Journal of Research in Science Teaching, 26*(3), 251–272.

Cameron, D. L. (1985). A pictorial framework to aid conceptualization of reaction stoichiometry. *Journal of Chemical Education, 62*(6), 510–511.

Carter, C. S. (1987). The role of beliefs in general chemistry problem solving (Doctoral dissertation, Purdue University). *Dissertation Abstracts International, 49*(5), 1107A.

Carter, C. S., LaRussa, M. A., & Bodner, G. M. (1987). A study of two measures of spatial ability as predictors of success in different levels of general chemistry. *Journal of Research in Science Teaching, 24*(7), 645–657.

Cassels, J. R. T., & Johnstone, A. H. (1984). The effect of language on student performance on multiple-choice tests in chemistry. *Journal of Chemical Education, 61*(7), 613–615.

Cassels, J. R. T., & Johnstone, A. H. (1985). *Words that matter in science.* London: Education Division, Royal Society of Chemistry, Burlington House, Piccadilly.

Cervellati, R., Montuschi, A., Perugini, D., Grimellini-Tomasini, N., & Balandi, B. P. (1982). Investigation of secondary school students' understanding of the mole concept in Italy. *Journal of Chemical Education, 59*(10), 852–856.

Cervellati, R., & Perugini, D. (1981). The understanding of the atomic orbital concept by Italian high school students. *Journal of Chemical Education, 58*(7), 568–569.

Chandran, S., Treagust, D. F., & Tobin, K. (1987). The role of cognitive factors in chemistry achievement. *Journal of Research in Science Teaching, 24*(2), 145–160.

Chi, M. T. H., Feltovitch, P. J., & Glaser, R. (1981). *Categorization and representation of physics problems by experts and novices* (Technical report no. 4). Pittsburgh, PA: Learning Research and Development Center, University of Pittsburgh.

Chiappetta, E. L., & McBride, J. W. (1980). Exploring the effects of general remediation on ninth-graders' achievement of the mole concept. *Science Education, 64*(5), 609–614.

Cros, D., Chastrette, M., & Fayol, M. (1988). Conceptions of second year university students of some fundamental notions of chemistry. *International Journal of Science Education, 10*(3), 331–336.

Cros, D., Maurin, M., Amouroux, R., Chastrette, M., Leber, J., & Fayol, M. (1986). Conceptions of first-year university students of the constituents of matter and the notions of acids and bases. *European Journal of Science Education, 8*, 305–313.

Crosby, G. L. (1988). Qualitative chemical equilibrium problem solving: College student conceptions (Doctoral dissertation, University of Maryland, College Park, 1987). *Dissertation Abstracts International, 48*(8), 2035A.

Cullen, J. F., Jr. (1983). Concept learning and problem solving: The use of the entropy concept in college chemistry. *Dissertation Abstracts International, 44*(6), 1747A.

deBueger-Van der Borght, C., & Mabille, A. (1989). The evolution in the meanings given by Belgian secondary school pupils to biological and chemical terms. *International Journal of Science Education, 11*(3), 347–362.

de Jong, O. (1990, June). Towards a more effective methodology for research on teaching and learning chemical calculations. *Empirical Research in Mathematics and Science Education* (pp. 92–105). Proceedings of the International Seminar, University of Dortmund, Germany.

de Vos, W., & Verdonk, A. H. (1987). A new road to reaction, part 4: The substance and its molecules. *Journal of Chemical Education, 64*(8), 692–694.

Dibar Ure, M. C., & Colinvaux, D. (1989). Developing adults' views on the phenomenon of change of physical state in water. *International Journal of Science Education, 11*(2), 153–160.

Dierks, W. (1981). Stoichiometric calculations: Known problems and proposed solutions at a chemistry-mathematics interface. *Studies in Science Education, 8*, 93–105.

Duncan, I. M., & Johnstone, A. H. (1973). The mole concept. *Educational Chemistry, 10*, 212–214.

Enochs, L. G., & Gabel, D. L. (1984). Preservice elementary teachers' conceptions of volume. *School Science and Mathematics, 84*(8), 670–680.

Erickson, G. L. (1979). Children's conceptions of heat and temperature. *Science Education, 63*(2), 221–230.

Erickson, G. L. (1980). Children's viewpoints of heat: A second look. *Science Education, 64*(3), 323–336.

Eylon, B., Ben-Zvi, R., & Silberstein, J. (1982, April). *Student conception of structure and process in chemistry.* Paper presented at the annual meeting of the National Association for Research in Science Teaching, Lake Genera, WI.

Falls, T. H., & Voss, B. (1985, April). *The ability of high school chemistry students to solve computational problems requiring proportional reasoning as affected by item in-task variables.* Paper presented at the National Association for Research in Science Teaching Conference, French Lick Springs, IN. (ERIC Document Reproduction Service No. ED 257 654)

Flick, L. (1991). Where concepts meet percepts: Stimulating analogical thought in children. *Science Education, 75*(2), 215–230.

Frank, D. V., & Herron, J. D. (1987, April). *Teaching problem solving to university general students.* Paper presented at the annual meeting of the National Association for Research in Science Teaching, Washington, DC.

Frenkel, P., & Strauss, S. (1985). *The development of the concept of temperature when assessed via three developmental models* (Working paper no. 46). Tel-Aviv University (Israel), Unit on Human Development and Education. (ERIC Document Reproduction Service No. ED 267 968)

Friedel, A. W., Gabel, D. L., & Samuel, J. (1990). Using analogs for chemistry problem solving: Does it increase understanding? *School Science and Mathematics, 90*(8), 674–682.

Friedel, A. W., & Maloney, D. P. (1992). An exploratory, classroom-based investigation of students' difficulties with subscripts in chemical formulas. *Science Education, 76*, 65–78.

Gabel, D. L. (1981, February). *Facilitating problem solving in high school chemistry*. Indiana University, School of Education, Bloomington. (ERIC Document Reproduction Service No. ED 210 192)

Gabel, D. L. (1990, June). Students' understanding of the particle nature of matter and its relation to problem solving. *Empirical research in mathematics and science education* (pp. 92–105). Proceedings of the International Seminar, University of Dortmund, Germany.

Gabel, D. L., & Samuel, K. V. (1986). High school students' ability to solve molarity problems and their analog counterparts. *Journal of Research in Science Teaching, 23*(2), 165–176.

Gabel, D. L., Samuel, K. V., & Hunn, D. (1987). Understanding the particulate nature of matter. *Journal of Chemical Education, 64*(8), 695–697.

Gabel, D., & Sherwood, R. D. (1984). Analyzing difficulties with mole-concept tasks by using familiar analog tasks. *Journal of Research in Science Teaching, 21*(8), 843–851.

Gabel, D. L., Sherwood, R. D., & Enochs, L. G. (1984). Problem-solving skills of high school chemistry students. *Journal of Research in Science Teaching, 21*(2), 221–233.

Gabriela T. C., Ribeiro, M., Costa Pereira, D. J. V., & Maskill, R. (1990). Reaction and spontaneity: The influence of meaning from everyday language on fourth year undergraduates' interpretations of some simple chemical phenomena. *International Journal of Science Education, 12*(4), 391–401.

Garnett, P. J., Garnett, P. J., & Treagust, D. F. (1990). Implications of research on students' understanding of electrochemistry for improving science curricula and classroom practice. *International Journal of Science Education, 12*(2), 147–156.

Gennaro, E. D. (1981). Assessing junior high students' understanding of density and solubility. *School Science and Mathematics, 81*(5), 399–404.

George, B., Wystrach, V. P., & Perkins, R. (1985). Why do students choose chemistry as a major? *Journal of Chemical Education, 62*(6), 501–503.

Gilbert, J. K., & Watts, M. D. (1983). Concepts, misconceptions and alternative conceptions: Changing perspectives in science education. *Studies in Science Education, 10*, 61–98.

Goldman, B. D. (1974). The effects of an experimental mathematics curriculum stressing problem-solving skills in science on student achievement in chemistry (Boston University School of Education). *Dissertation Abstracts International, 35*, 166A.

Gooding, C. T., Swift, J. N., Schell, R. E., Swift, P. R., & McCroskery, J. H. (1990). A causal analysis relating previous achievement, attitude, discourse and intervention to achievement in biology and chemistry. *Journal of Research in Science Teaching, 27*(8), 789–801.

Gorodetsky, M., & Gussarsky, E. (1986). Misconceptualization of the chemical equilibrium concept as revealed by different evaluation methods. *European Journal of Science Education, 8*(4), 427–441.

Gorodetsky, M., & Hoz, R. (1980). Use of concept profile analysis to identify difficulties in solving science problems. *Science Education, 64*(5), 671–678.

Gorodetsky, M., & Hoz, R. (1985). Changes in the group cognitive structure of some chemical equilibrium concepts following a university course in general chemistry. *Science Education, 69*(2), 185–199.

Gower, D. M., Daniels, J. J., & Lloyd, G. (1977). Hierarchies among the concepts which underlie the mole. *School Science Review, 59*, 285–299.

Grant, R. M. (1978). Group and individual problem solving: high school students (Doctoral dissertation, University of Oklahoma). *Dissertation Abstracts International, 39*(11), 6672A.

Greenbowe, T. J. (1983). An investigation of variables involved in chemistry problem solving (Doctoral dissertation, Purdue University). *Dissertation Abstracts International, 44*(12), 3651A.

Griffiths, A. K., Kass, H., & Cornish, A. G. (1983). Validation of a learning hierarchy for the mole concept. *Journal of Research in Science Teaching, 20*(7), 639–654.

Griffiths, A. K., Pottle, J., & Whelan, P. (1983, April). *Application of the learning hierarchy model to the identification of specific misconceptions for the two science concepts*. Paper presented at the annual meeting of the National Association for Research in Science Teaching, Dallas.

Griffiths, A. K., & Preston, K. R. (1989, March). *An investigation of grade 12 student's misconceptions relating to fundamental characteristics of molecules and atoms*. Paper presented at the 62nd conference of the National Association for Research in Science Teaching, San Francisco.

Gussarsky, E., & Gorodetsky, M. (1988). On the chemical equilibrium concept: Constrained word associations and conception. *Journal of Research in Science Teaching, 25*(5), 319–333.

Gussarsky, E., & Gorodetsky, M. (1990). On the concept "chemical equilibrium": The associative framework. *Journal of Research in Science Teaching, 27*(3), 197–204.

Haidar, A. H., & Abraham, M. R. (1991). A comparison of applied and theoretical knowledge of concepts based on the particulate nature of matter. *Journal of Research in Science Teaching, 28*(10), 919–938.

Hayes, J. R. (1981). *The complete problem solver*. Philadelphia: The Franklin Institute.

Hendley, C. T., III. (1990). *Student's conceptions in chemistry*. Unpublished master's thesis, University of Maryland, College Park.

Herron, J. D., & Greenbowe, T. J. (1986). What can we do about Sue: A case study of competence. *Journal of Chemical Education, 63*(6), 528–531.

Hesse, J. J., III, & Anderson, C. W. (1992). Students' conceptions of chemical change. *Journal of Research in Science Teaching, 29*(3), 277–299.

Hewson, M. G. (1986). The acquisition of scientific knowledge: Analysis and representation of student conceptions concerning density. *Science Education, 70*(2), 159–170.

Heyworth, R. M. (1988). Mental representation of knowledge for a topic in high school chemistry (Doctoral dissertation, Stanford University). *Dissertation Abstracts International, 49*(6), 1409A.

Hillman, R. A. H., Hudson, M. J., & McLean, I. C. (1981). A preliminary study of some of the learning and assessment difficulties in connection with O-level electrochemistry. *School Science Review, 63*(222), 157–163.

Horton, P. B., Fronk, R. H., & Walton, R. W. (1985). The effect of writing assignments on achievement in college general chemistry. *Journal of Research in Science Teaching, 22*(6), 535–541.

Howe, A. C., & Shayer, M. (1981). Sex related differences on a task of volume and density. *Journal of Research in Science Teaching, 18*(2), 169–175.

Jackman, L. E., Moellenberg, W. P., & Brabson, G. D. (1990). Effects of conceptual systems and instructional methods on general chemistry laboratory achievement. *Journal of Research in Science Teaching, 27*(7), 699–709.

Johnstone, A. H. (1983). Chemical education research, facts, findings, and consequences. *Journal of Chemical Education, 60*(11), 968–971.

Johnstone, A. H. (1984). New stars for the teacher to steer by? *Journal of Chemical Education, 61*(10), 847–849.

Johnstone, A. H. (1990, September). *Fashions, fads and facts in chemistry education*. Paper presented at the American Chemical Society Meeting, Washington, DC.

Johnstone, A. H., & El-Banna, H. (1986). Capacities, demands and processes—a predictive model for science education. *Education in Chemistry, 23,* 80–84.

Jones, B. L., Lynch, P. P., & Reesink, C. (1989). Children's understanding of the notions of solid and liquid in relation to some common substance. *International Journal of Science Education, 11*(4), 417–427.

Kavanaugh, R. D., & Moomaw, W. R. (1981). Inducing formal thought in introductory chemistry students. *Journal of Chemical Education, 58*(3), 263–265.

Kleinman, R. W., Griffin, H. C., & Kerner, N. K. (1987). Images in chemistry. *Journal of Chemical Education, 64*(9), 766–770.

Kletzly, N. E. (1980). *The effects of two methods of teaching abstract topics in high school chemistry* (Master's dissertation, Sam Houston State University). (ERIC Document Reproduction Service No. ED 199 063)

Krajcik, J. S. (1991). Developing students' understanding of chemical concepts. In S. M. Glynn, R. H. Yeany, & B. K. Britton (Eds.), *The psychology of learning science* (pp. 117–147). Hillsdale, NJ: Lawrence Erlbaum.

Krajcik, J. S., & Haney, R. E. (1987). Proportional reasoning and achievement in high school chemistry. *School Science and Mathematics, 87,* 25–32.

Kramers-Pals, H., Lambrechts, J., & Wolff, P. J. (1982). Recurrent difficulties: Solving quantitative problems. *Journal of Chemical Education, 59*(6), 509–513.

Kramers-Pals, H., & Pilot, A. (1988). Solving quantitative problems: Guidelines for teaching derived from research. *International Journal of Science Education, 10*(5), 511–522.

Lawrenz, F. (1986). Misconceptions of physical science concepts among elementary school teachers. *School Science and Mathematics, 86*(8), 654–660.

Lazonby, J. N., Morris, J. E., & Waddington, D. J. (1982). The muddlesome mole. *Education in Chemistry, 19,* 109–111.

Lazonby, J. N., Morris, J. E., & Waddington, D. J. (1985). The mole: Questioning format can make a difference. *Journal of Chemical Education, 62,* 60–61.

Lee, K. (1985). Cognitive variables in problem solving in chemistry. In R. P. Tisher (Ed.), *Research in science education: Vol. 15.* Selections of papers from the annual conference of the Australian Science Education Research Association, Capricornia Institute, Rockhampton, Queensland, Australia. (ERIC Document Reproduction Service No. ED 267 974)

Linn, M. C., & Songer, N. B. (1991). Teaching thermodynamics to middle school students: What are appropriate cognitive demands? *Journal of Research in Science Teaching, 28*(10), 885–918.

Longden, K., Black, P., & Solomon, J. (1991). Children's interpretation of dissolving. *International Journal of Science Education, 13,* 59–68.

Lundeberg, M. A. (1990). Supplemental instruction in chemistry. *Journal of Research in Science Teaching, 27*(2), 145–155.

Lynch, P. P., Chipman, H. H., & Pachaury, A. C. (1985). The language of science and the high school student: The recognition of concept definitions: A comparison between Hindi speaking students in India and English speaking students in Australia. *Journal of Research in Science Teaching, 22*(7), 675–686.

Lythcott, J. (1990). Problem solving and requisite knowledge of chemistry. *Journal of Chemical Education, 67*(3), 248–252.

Maloney, D. P., & Friedel, A. W. (1991). *Students' difficulties with subscripts in chemical formulas.* Paper presented at the annual meeting of the National Association for Research in Science Teaching, Fontana, WI.

Manley, B. L. (1978, March). *The relationship of the learning environment to student attitudes toward chemistry.* Paper presented at the annual meeting of the National Association for Research in Science Teaching, Toronto.

Maskill, R., & Cachapuz, A. F. C. (1989). Learning about the chemistry topic of equilibrium: The use of word association tests to detect developing conceptualizations. *International Journal of Science Education, 11,* 57–69.

Mettes, C. T. C. W., Pilot, A., Roossink, H. J., & Kramers-Pals, H. (1980). Teaching and learning problem solving in science. *Journal of Chemical Education, 57*(12), 882–885.

Mitchell, H., & Kellington, S. (1982). Learning difficulties associated with the particulate theory of matter in the Scottish integrated science courses. *European Journal of Science Education, 4,* 429–440.

Mullet, E., & Gervais, H. (1990). Distinction between the concepts of weight and mass in high school students. *International Journal of Science Education, 12*(2), 217–226.

Nachmias, R., Stavy, R., & Avrams, R. (1990). A microcomputer-based diagnostic system for identifying students' conception of heat and temperature. *International Journal of Science Education, 12*(2), 123–132.

Nakhleh, M. B., & Krajcik, J. S. (1991, April). The effect of level of information as presented by different technologies on students' understanding of acid, base and pH concepts. Paper presented at the Annual Meeting of the National Association for Research in Science Teaching, Lake Geneva, WI.

Niaz, M. (1985). Relation between M-space of students and M-demand of different items of general chemistry and its interpretation based on the neo-Piagetian theory of Pascual-Leone. *Journal of Chemical Education, 64*(6), 502–505.

Niaz, M. (1987). The role of cognitive factors in the teaching of science. *Research in Science and Technological Education, 5,* 7–16.

Niaz, M. (1988a). The information-processing demand of chemistry problems and its relation to Pascual-Leone's functional M-capacity. *International Journal of Science Education, 10*(2), 231–238.

Niaz, M. (1988b). Manipulation of M-demand of chemistry problems and its effect on student performance: A neo-Piagetian study. *Journal of Research in Science Teaching, 25*(8), 643–657.

Niaz, M. (1988c). Student performance in water pouring and balance beam tasks: Effect of manipulation of perceptual field factor. *Research in Science and Technological Education, 6,* 39–50.

Niaz, M. (1989). Relation between Pascual-Leone's structural and functional M-space and its effect on problem solving in chemistry. *Journal of International Chemical Education, 11,* 93–99.

Niaz, M., & Lawson, A. E. (1985). Balancing chemical equations: The role of developmental level and mental capacity. *Journal of Research in Science Teaching, 22,* 41–51.

Niaz, M., & Robinson, W. R. (1989, April). *Teaching algorithmic problem solving or conceptual understanding: Role of developmental level, mental capacity, and cognitive style.* Paper presented at the annual meeting of the National Association for Research in Science Teaching, Lake Geneva, WI.

Niaz, M., & Robinson, W. R. (1992). Manipulation of logical structure of chemistry problems and its effect on student performance. *Journal of Research in Science Teaching, 29*(3), 211–226.

Novak, J. D. (1984). Application of advances in learning theory and philosophy of science to the improvement of chemistry teaching. *Journal of Chemical Education, 61*(7), 607–612.

Novak, J. D., & Musonda, D. (1991). A twelve-year longitudinal study of science concept learning. *American Educational Research Journal, 28,* 117–153.

Novick, S., & Menis, J. (1976). A study of student perceptions of the mole concept. *Journal of Chemical Education, 53*(11), 720–722.

Novick, S., & Nussbaum, J. (1978). Junior high school pupils' understanding of the particulate nature of matter: An interview study. *Science Education, 62*(3), 273–281.

Novick, S., & Nussbaum, J. (1981). Pupils' understanding of the particulate nature of matter: A cross-age study. *Science Education, 65*(2), 187–196.

Nurrenbern, S. C. (1979). *Problem-solving behaviors of concrete and formal operational high school chemistry students when solving chemistry problems requiring Piagetian formal reasoning skills.* Unpublished doctoral dissertation, Purdue University, Lafayette, IN.

Nurrenbern, S. C., & Pickering, M. (1987). Concept learning versus problem solving: Is there a difference? *Journal of Chemical Education, 64*(6), 508–510.

Okebukola, P. A., & Jegede, O. J. (1990). Eco-cultural influences upon students' concept attainment in science. *Journal of Research in Science Teaching, 27*(7), 661–669.

Opdenacker, C., Fierens, H., Van Brabant, H., Sevenants, J., Spruyt, J., Slootmaekers, P. J., & Johnstone, A. H. (1990). Academic performance in solving chemistry problems related to student working memory capacity. *International Journal of Science Education, 12*(2), 177–185.

Osborne, R. J., & Cosgrove, M. M. (1983). Children's conceptions of the changes of state of water. *Journal of Research in Science Teaching, 20*(9), 825–838.

Padiglione, C., & Torracca, E. (1990). Logical processes in experimental contexts and chemistry teaching. *International Journal of Science Education, 12*(2), 187–194.

Peterson, R. F., & Treagust, D. F. (1989). Grade-12 students' misconceptions of covalent bonding and structure. *Journal of Chemical Education, 66*(6), 459–460.

Pereira, M. P., & Pestana, M. E. M. (1991). Pupils' representations of models of water. *International Journal of Science Education, 13*(3), 313–319.

Pickering, M. (1990). Further studies on concept learning versus problem solving. *Journal of Chemical Education, 67*(3), 254–255.

Pines, A. L., & West, L. H. T. (1986). Conceptual understanding and science learning: An interpretation of research within a source of knowledge framework. *Science Education, 70*(5), 583–604.

Ploger, D. H. (1987). *Expert and novice performance in biochemistry problem solving and explanation* (Doctoral dissertation, Rutgers, the State University of New Jersey, New Brunswick, 1986). *Dissertation Abstracts International, 47*(11), 4684B.

Polya, G. (1945). *How to solve it: A new aspect of mathematical method.* Princeton, NJ: Princeton University Press.

Pope, M., & Gilbert, J. (1983). Personal experience and the construction of knowledge in science. *Science Education, 67*(2), 193–203.

Pribyl, J. R., & Bodner, G. M. (1987). Spatial ability and its role in organic chemistry: A study of four organic courses. *Journal of Research in Science Teaching, 24*(3), 229–240.

Prieto, T., Blanco, A., & Rodriguez, A. (1989). The ideas of 11- to 14-year-old students about the nature of solutions. *International Journal of Science Education, 11*(4), 451–463.

Rager, T., & Stavy, R. (1989). Students' difficulties in understanding conceptions of the various dimensions of the quantity of matter. *Proceedings of the Second Jerusalem Convention on Education,* pp. 164–176. Jerusalem: Hebrew University of Jerusalem.

Rahal, T. M. (1986). A model for improving problem solving during chemistry instruction: The concept of solubility product constant (Doctoral dissertation, Columbia University Teachers College, 1985). *Dissertation Abstracts International, 46*(10), 2988A.

Raines, S. J. (1984). Problem-similarity recognition and problem-solving success in high school chemistry (Doctoral dissertation, Virginia Polytechnic Institute and State University). *Dissertation Abstracts International, 45*(3), 800A.

Raven, R. J. (1987). A study of the use of ratios in science problem solving. *Science Education, 71*(4), 565–570.

Rawwas, M. A. (1986). The effects of variation in sequential kinetic structure on acquisition of the mole concept in chemistry based on a hierarchical learning model using computer-based materials (Doctoral dissertation, Columbia University Teachers College, 1985). *Dissertation Abstracts International, 46*(9), 2650A.

Reif, F. (1983). How can chemists teach problem solving? *Journal of Chemical Education, 60*(11), 948–953.

Robinson, W. R., & Niaz, M. (1991). Performance based on instruction by lecture or by interaction and its relationship to cognitive variables. *International Journal of Science Education, 13*(2), 203–215.

Ross, B., & Munby, H. (1991). Concept mapping and misconceptions: A study of high-school students' understandings of acids and bases. *International Journal of Science Education, 13,* 11–23.

Rowe, M. B. (1983). Getting chemistry off the killer course list. *Journal of Chemical Education, 60*(11), 954–956.

Rowell, J. A., & Dawson, C. J. (1980). Mountain or mole hill: Can cognitive psychology reduce the dimensions of conceptual problems in classroom practice? *Science Education, 64*(5), 693–708.

Rowell, J. A., & Dawson, C. J. (1981). Volume, conservation and instruction: A classroom based Solomon Four Group Study of conflict. *Journal of Research in Science Teaching, 18*(6), 533–546.

Rowell, J. A., Dawson, C. J., & Lyndon, H. (1990). Changing misconceptions: A challenge to science educators. *International Journal of Science Education, 12*(2), 167–175.

Rozier, S., & Viennot, L. (1991). Students' reasonings in thermodynamics. *International Journal of Science Education, 13*(2), 159–170.

Russell, T., Harlen, W., & Watt, D. (1989). Children's ideas about evaporation. *International Journal of Science Education, 11*(5), 566–576.

Rutherford, M., & Nkopodi, N. (1990). A comparison of the recognition of some science concept definitions in English and North Sotho for second language English speakers. *International Journal of Science Education, 12*(4), 443–456.

Ryan, J. N. (1987). The name's the game in problem solving. *Journal of Chemical Education, 64*(6), 524–525.

Sawrey, B. A. (1990). Concept learning versus problem solving: Revisited. *Journal of Chemical Education, 67*(3), 253–254.

Schmidkunz, H., & Buttner, D. (1986). Teaching chemistry according to a spiral curriculum. *European Journal of Science Education, 8,* 9–16.

Schmidt, H. (1988, April). *Mind the red-herrings and deliberate distraction of pupil's strategies solving multiple choice questions in chemistry.* Paper presented at the annual meeting of the National Association for Research in Science Teaching, Lake of the Ozarks, MO. (ERIC Document Reproduction Service No. ED 291 577)

Schmidt, H. (1991). A label as a hidden persuader: Chemists' neutralization concept. *International Journal of Science Education, 13*(4), 459–471.

Schmidt, H. (1990). Secondary school students' strategies in stoichiometry. *International Journal of Science Education, 12*(4), 457–471.

Sciarretta, M. R., Stilli, R., & Vicentini Missoni, M. (1990). On the thermal properties of materials: Common sense knowledge of Italian students and teachers. *International Journal of Science Education, 12*(4), 369–379.

Sevenair, J. P., & Burkett, A. R. (1988). Difficulty and discrimination of multiple-choice questions: A counter-intuitive result. *Journal of Chemical Education, 65*(5), 441–442.

Shayer, M., & Wylam, H. (1981). The development of the concepts of heat and temperature in 10–13-year-olds. *Journal of Research in Science Teaching, 18*(5), 419–434.

Shepherd, D. L., & Renner, J. W. (1982). Student understandings and misunderstandings of states of matter and density changes. *School Science and Mathematics, 82*(8), 650–665.

Silberman, R. G. (1981). Problems with chemistry problems: Student perception and suggestions. *Journal of Chemical Education, 58*(12), 1036.

Squires, F. H. (1977). An analysis of sex differences and cognitive styles on science problem-solving situations (Doctoral dissertation, Ohio State University). *Dissertation Abstracts International, 38,* 2688A.

Staver, J. R. (1984). Effects of method and format on subjects' responses to a control of variables reasoning problem. *Journal of Research in Science Teaching, 21*(5), 517–526.

Staver, J. R. (1986). The effects of problem format, number of independent variables, and their interaction on student performance on a control of variables reasoning problem. *Journal of Research in Science Teaching, 23*(6), 533–542.

Staver, J. R., & Halsted, D. A. (1985). The effects of reasoning, use of models, sex type, and their interactions on posttest achievement in chemical bonding after constant instruction. *Journal of Research in Science Teaching, 22*(5), 437–447.

Staver, J. R., & Jacks, T. (1988). The influence of cognitive reasoning level, cognitive restructuring ability, disembedding ability, working memory capacity, and prior knowledge on students' performance on balancing equations by inspection. *Journal of Research in Science Teaching, 25*(9), 763–775.

Stavridou, H., & Solomonidou, C. (1989). Physical phenomena–chemical phenomena: Do pupils make the distinction? *International Journal of Science Education, 11,* 83–92.

Stavy, R. (1988). Children's conception of gas. *International Journal of Science Education, 10*(5), 553–560.

Stavy, R. (1990a). Children's conception of changes in the state of matter: From liquid (or solid) to gas. *Journal of Research in Science Teaching, 27*(3), 247–266.

Stavy, R. (1990b). Pupils' problems in understanding conservation of matter. *International Journal of Science Education, 12*(5), 501–512.

Stavy, R. (1991a). Children's ideas about matter. *School Science and Mathematics, 91*(6), 240–244.

Stavy, R. (1991b). Using analogy to overcome misconceptions about conservation of matter. *Journal of Research in Science Teaching, 28*(4), 305–313.

Stavy, R., & Berkovitz, B. (1980). Cognitive conflict as a basis for teaching quantitative aspects of the concept of temperature. *Science Education, 64*(5), 679–692.

Stiff, L. V. (1988). Problem solving by example. *School Science and Mathematics, 88*(8), 666–675.

Sumfleth, E. (1988). Knowledge of terms and problem-solving in chemistry. *International Journal of Science Education, 10,* 45–60.

Thornton, M. C., & Fuller, R. G. (1981). How do college students solve proportion problems? *Journal of Research in Science Teaching, 18*(4), 335–340.

Tingle, J. B., & Good, R. (1990). Effects of cooperative grouping on stoichiometric problem solving in high school chemistry. *Journal of Research in Science Teaching, 27*(7), 671–683.

Trumper, R. (1990). Being constructive: An alternative approach to the teaching of the energy concept—part one. *International Journal of Science Education, 12*(4), 343–354.

VanOrden, N. (1990). Is writing an effective way to learn chemical concepts? *Journal of Chemical Education, 67*(7), 583–585.

Vermont, D. F. (1985). Comparative effectiveness of instructional strategies on developing the chemical mole concept (Doctoral dissertation, University of Missouri, St. Louis, 1984). *Dissertation Abstracts International, 45*(8), 2473A.

Waddling, R. E. (1988). Pictorial problem-solving networks. *Journal of Chemical Education, 65*(3), 260–262.

Ward, C. R., & Herron, J. D. (1980). Helping students understand formal chemical concepts. *Journal of Research in Science Teaching, 17*(5), 387–400.

Westbrook, S. L., & Marek, E. A. (1991). A cross-age study of student understanding of the concept of diffusion. *Journal of Research in Science Teaching, 28*(8), 649–660.

Widing, R. W., Jr. (1985). Systematic analysis of chemistry topics (University of Illinois at Chicago, 1984). *Dissertation Abstracts International, 45*(7), 2057A.

Wheeler, A. E., & Kass, H. (1977, March). *Proportional reasoning in introductory high school chemistry.* Paper presented at the annual meeting of the National Association for Research in Science Teaching, Cincinnati.

Wheeler, A. E., & Kass, H. (1978). Student misconceptions in chemical equilibrium. *Science Education, 62*(2), 223–232.

Wiseman, F. L., Jr. (1981). The teaching of college chemistry. Role of student development level. *Journal of Chemical Education, 58*(6), 484–488.

Yager, R. E., & Yager, S. O. (1984). The effect of schooling upon understanding of selected science terms. *Journal of Research in Science Teaching, 22*(4), 359–364.

Yarroch, W. L. (1985). Student understanding of chemical equation balancing. *Journal of Research in Science Teaching, 22*(5), 449–459.

RESEARCH ON PROBLEM SOLVING: PHYSICS

David P. Maloney

INDIANA UNIVERSITY
PURDUE UNIVERSITY–FORT WAYNE

Three main areas of research on problem solving in physics are examined in this review. One area is detailed research on how individuals solve problems. The second comprises studies of pedagogical methods for improving students' problem-solving abilities. Research in these areas examines how subjects deal with standard physics problems—that is, the numerical tasks typically found at the ends of the chapters in introductory physics books. The third area consists of investigations into other types of problems, questions of transfer, what students learn from solving problems, and other issues. The chapter closes with a general discussion of problem solving and instructional issues.

How individuals solve physics problems has been investigated by a variety of researchers, including cognitive psychologists and physics educators; consequently, the studies are reported in a variety of journals. Most of the research on how individuals, often experts compared with novices, solve physics problems is found in the psychological and cognitive science literature. Most of the research on instructional aspects is found in either the physics education or science education literature. Research from the third area can be found in the science education, psychological, or cognitive science literature.

Although it does not involve direct investigation of problem solving another area of research that impinges on physics problem solving is the investigation of students' conceptual understanding. Because the conceptual research is reviewed elsewhere in this volume, no attempt is made to review it here. However, conceptual understanding enters into any consideration of problem solving in physics because the solver's knowledge base is a critical factor in how the solver proceeds.

Obviously, any examination of problem solving requires a clear definition of what constitutes a problem. Clement (1978) titled one of his investigations "Understanding Students' Causal Thinking from a Problem Solving Protocol," but the task assigned was to predict how certain aspects of a physical situation would affect the behavior of the system. Is this what we mean by problem solving? According to Hayes' (1981) definition of a problem—"whenever there is a gap between where you are now and where you want to be, and you don't know how to find a way to cross that gap, you have a problem" (p. i)—it could qualify. Likewise, Clement's task for the student would qualify as a problem under Newell and Simon's (1972) definition: "A person is confronted with a problem when he wants something and does not know immediately what series of actions he can perform to get it" (p. 72). But this is not what comes to mind when most physicists and science education or psychological researchers think of a physics problem.

For much of this review, the term "problem" will be taken to mean the kind of task that is usually found at the ends of the chapters in introductory college physics books. Typically, these tasks present a situation in which certain information is given, most often as numerical values for variables in the situation, and the value of one of the other possible variables is to be determined. These tasks are very specific and well defined, since only relevant variables are included and the unknown is explicitly identified. Several issues arise from restricting the term "problem" to these kinds of tasks, some of which are considered in the third section of the chapter.

HOW INDIVIDUALS SOLVE PHYSICS PROBLEMS

Human problem-solving abilities have been a subject of interest for centuries, but serious systematic research into these abilities is relatively recent. There are good reviews, such as

The author thanks Patricia Heller (University of Minnesota) and Alan Van Heuvelen (The Ohio State University) for their very productive comments on an earlier version of this chapter.

that by Mayer (1983), of the early research. Consequently, no attempt is made here to review that early research.

However, one early work on problem solving that is very important to the subsequent development of problem-solving research in mathematics and physics is Polya's book *How to Solve It*, originally published in 1945. Polya was a productive mathematician and teacher who wrote the book to provide systematic help for students trying to learn how to solve mathematics problems. The book was a result of his own experiences in solving problems, as well as his experiences as a teacher. It had two themes that have been recurrent in problem-solving research in physics: an overall plan of attack for solving problems, and the identification and use of heuristics in problem solving.

Polya (1945) developed a four-step general framework for problem solving: understanding the problem, devising a plan, carrying out the plan, and looking back. This framework, or variations thereof, are part of many studies of problem solving. For example, Reif, Larkin, and Brackett (1976) used a framework based on Polya's in their early work on problem solving. Similarly Schoenfeld (1978) made a concerted effort to teach students such a framework, as well as many of the heuristics Polya identified. These researchers subsequently investigated different issues and concerns, but they were clearly influenced in their early work by Polya, as were many others.

The dominant theoretical framework for more recent research on how individuals solve physics problems has been the information-processing framework. Information-processing research focuses on how an information processor, human or computer, takes a natural-language problem statement, translates it into an internal representation on which it can operate, carries out the appropriate operations, and then outputs the result. Such research seeks to determine things like what knowledge individuals bring to the task, what information in the problem they key on, and how they use these items of information. Obtaining this kind of detailed evidence requires intensive examination of how a few individuals solve problems. Consequently, most of the studies in this section employ think-aloud protocols of individuals solving problems. Frequently, the studies contrast the performance of novices, usually students who are taking or have just completed a college-level general physics course, with that of experts. The experts are usually experienced college physics teachers. Because of the nature of think-aloud protocols and their analysis, the studies in this section usually involve relatively small numbers of subjects.

In this first section, we review investigations aimed at determining how individuals solve physics problems. These studies do not involve any attempts to develop or test pedagogical approaches for improving student problem solving. Those studies are reviewed in a later section. This section is broken into four parts: information-processing/computer-modeling studies, studies of qualitative analysis and representations, investigations of problem solvers' knowledge structures, and investigations of students' study procedures. All of these studies share a focus on how individuals proceed when trying to solve problems or when trying to learn on their own how to solve problems.

Information-Processing/Computer-Modeling Studies

Information-processing studies of problem solving began in earnest with the publication of Newell and Simon's *Human Problem Solving* (1972). The early stages of information-processing work on problem solving concentrated on how humans solve puzzles, games (i.e., chess), and problems (i.e., logic and cryptarithmetic). Gradually, the domains of interest expanded to include areas such as physics.

Most information-processing studies employ a computer analogy or modeling in some way. There are basically two ways to do this. One way is to study how humans solve some particular type of problem and then construct a computer program that can reproduce their performance. The second is to construct a computer program based on theoretical perspectives of efficient problem solving, to solve a particular type of problem or problems in a particular domain, and then have several human subjects solve the problem(s). The performances of the computer program and human subjects are then compared. The first approach is the one usually taken by cognitive psychologists and cognitive science education researchers; the second is the preferred approach for artifical-intelligence researchers. Clearly, these approaches are complementary.

One of the early studies of problem solving in physics was that of Reif, Larkin, and Brackett (1976). They were interested in teaching general learning and problem-solving skills and used think-aloud protocols of students solving problems. These researchers found that students usually just start calculating something in a hapazard manner and lack any systematic strategy for guiding their activity. A simple four-step problem-solving strategy, based on Polya (1945), was taught to a group of students. The steps in the strategy were: (1) description, (2) planning, (3) implementation, and (4) checking. Students who received the problem-solving instruction showed improvement, but the study involved small numbers and a short time period.

Another early study used only one subject and a slightly different topic domain. Bhaskar and Simon (1977) reported on how a single, reasonably proficient subject solved a group of engineering thermodynamics problems. The authors' stated goal was to "extend the theory of human problem solving to semantically richer domains" (p. 193). Although engineering thermodynamics is not, strictly speaking, physics, it shares many problem-solving characteristics with physics. The major difference is that the engineering problems often require data from reference tables, whereas physics problems usually do not. The authors had their subject think aloud while solving six problems. They employed a protocol encoding system that they argued could also be considered a "weak theory of the problem-solving process" (p. 199). They concluded that the subject "follows a consistent pattern of approach that might be described as a form of means-ends analysis modified by his knowledge of the central role that the conservation of energy equation plays in such problems" (p. 214). The use of the general heuristic of means-ends analysis and the place of general principles in problem solver's efforts are two recurrent themes in problem-solving research. Means-ends analysis is a general heuristic process of identifying the goal, comparing the current state to the goal, and then carrying out available operations to reduce the difference between the current state and the goal state. General principles are usually employed by experts, but means-ends analysis, usually focusing on specific equations, is a common approach for novices, as the following study indicates.

Simon and Simon (1978) studied two subjects working one-dimensional kinematics problems. One subject was more experienced at working such problems than the other, but neither was what is normally termed an expert. The two subjects behaved in a similar way, in that both proceeded by reading the problem, finding an equation, plugging numbers into the equation, and finally solving the equation. Both subjects' procedures could be effectively modeled with similar production-system computer models. (A production-system model is one that is composed of a linked sequence of productions. A production is a condition-action statement—*if all of the forces acting on an object are known then carry out a vector addition of those forces to determine the magnitude and direction of the net force*.) But there was also a clear difference between the two subjects. The more experienced solver used a "working forward" approach in which he took the given information and plugged it into an equation, or a series of equations, until he found the unknown value. In contrast, the novice used a "working backward" approach in which she started with an equation that contained the unknown and tried to solve that. If one of the other items in that equation was missing, she looked for an equation involving that quantity and solved that. Then she used the value just found as one of the knowns in the first equation she had tried. This working-backward approach is an example of means-ends analysis, which we will find is a procedure frequently employed by novices.

Although the evidence was indirect, Simon and Simon (1978) claimed that another difference between the two solvers was that the more experienced solver employed "physical intuition" but the novice did not. The authors gave a definition for physical intuition, saying that it might be interpreted in the following way:

When a physical situation is described in words, a person may construct a perspicuous representation of that situation in memory. By a perspicuous representation we mean one that represents explicitly the main direct connections, especially causal connections, of the components of the situation. (p. 337)

Simon and Simon argue that the more experienced solver in their study was guided by physical intuition in his problem-solving efforts and that physical intuition was a major factor in why his problem-solving efforts were more efficient and effective. What Simon and Simon (1978) refer to as physical intuition is most likely the use of general physical principles, which seems to occur only after an individual acquires a broader knowledge base in physics than novices usually have. How would the situation of two individuals with similar background experience differ if they were presented with the same problem? The following study identifies one factor that can produce variations.

Simon and Simon (1979) report on another study of problem solving in which both subjects were experienced problem solvers. One of the subjects was told that the experimenters were interested in how students solve textbook problems, and the other tackled the problem without any information about context. The problem used in this study differed from standard textbook problems in that it provided a "cover story" about astronauts having an accident on the moon.

The subject who had been informed of the experimenters' interests worked the problem as a standard textbook problem; he extracted the standard problem and solved that, ignoring the cover story. In contrast, the other subject treated the problem in the context of the cover story; he carried out an extensive analysis to determine what the most pressing problem was and then performed a means-ends analysis to solve that problem. The results of this study support the claim that the representation of a problem the solver constructs will strongly control how the solver proceeds. This study also provides some support for the contention of Reusser (1988), explained later, that the context in which the problem occurs has a significant effect on how it is worked. Clearly, these two subjects were working the same task but proceeded in different ways because the contexts were different.

The idea of problem representation is a fundamental feature of current problem-solving research. A problem representation is an internal mental model of the situation that can be manipulated as the individual attempts to solve the problem. Most problems are presented to students in natural language, perhaps with a diagram to supplement the description. The student has to translate this natural-language statement into a form to which the physical relations can be applied (usually in the form of mathematical equations) to determine an answer. The result of the translation process is the solver's representation. Authors use different terms, such as internal model, original description, qualitative physical description, and mental model, for representations. Although there are some subtle differences among these terms in some situations, they need not concern us here.

Larkin (1979) carefully observed experts solve some problems. She identified the experts essential problem-solving strategies and then explicitly tried to teach these procedures to students. She found that experts conducted a qualitative analysis before starting any work with equations and that they had the main physical principles "chunked" together around important ideas. Larkin (1979) identified two functions of the qualitative analysis:

1) It reduces the chance of error because the qualitative analysis is easy to check both against the original problem situation and against the subsequently generated quantitative equations.
2) It provides a concise and easily-remembered description of the global features of the problem. This description can then be used as a guide when the details of the solution are worked out. (p. 286)

Larkin (1979) then tried an experiment in which she taught seven relations individually to 10 students. Five of the students received an additional hour of instruction on organizing the relations. The students who received the additional instruction were more successful in solving problems that required more than one relation for their solution. In a second, classroom-based experiment, Larkin (1979) taught students in a physics course a problem-solving "strategy" that incorporated features of the experts' approach. Preliminary observations indicated that these students used better general problem-solving procedures. The results of this study suggest that a definite problem-solving strategy can be useful. Nevertheless, how effective is such a strategy if the individual does not have an effective command of the domain knowledge to use with it? The next two studies emphasize the qualitative analysis aspect of a general problem-solving strategy.

Larkin and Reif (1979) reported on how two subjects (a novice and an experienced physics problem solver) solved five standard physics textbook problems in mechanics. The researchers developed two models to capture the main features of how the expert and novice solved the problems. These two models shared one common feature, the construction of an original description that involved only general knowledge and no specific physics concepts. The model of the novice went from the original description to a mathematical description, which was constructed using relevant physical principles, and then combined equations to eliminate undesired quantities. In contrast, the expert moved from the original description to a qualitative physical description and then to the mathematical description. Clearly, the major distinction between the two models is the expert's construction of the qualitative physical description—a "second stage" domain-specific representation. Construction of this domain-specific representation requires physics knowledge that novices either lack or do not possess in a form that allows them to access it effectively.

Larkin and Reif (1979) derived two instructional suggestions from their results. First, it is useful to help students learn to organize separate physical principles and relations into structured combinations which are applicable to a range of problems. Second, it seems to be important to teach students the value of approaching problems by successive descriptions that go from "more global to more detailed aspects of a problem." The value of using successive descriptions that identify the major features of the problem is supported by the next study.

In another early study, Larkin (1980a) described the problem-solving strategies of one highly skilled subject, which she modeled with a "program," called HIPLAN, that fully incorporated the major features of the process but did not try to capture the details of local procedures or the actual time segments of the process. The main feature of HIPLAN was that it was a "hierarchical planning model which uses extensively the strategy of planning by abstracting and solving a simpler problem, and using the resulting abstracted solution as a guide for more detailed work" (Larkin, 1980a, p. 287). The author contended that the results indicate the importance of being able to identify the major features of a problem and use that information to plan at a low level of detail before proceeding with a solution.

Larkin, McDermott, Simon, and Simon published two articles in 1980 that provided summaries of computer-modeling/information-processing research up to that time. One article, published in *Science* (1980a), brought this research to the notice of many active scientists. The second, published in *Cognitive Science* (1980b), described two computer programs that brought out the differences between means-ends analysis and knowledge development as strategies for problem solving.

Larkin et al. (1980a) reviewed the results of the research on developing computer models of expert and novice performance and what this research had revealed about problem-solving processes. They pointed out that the most obvious difference between experts and novices is the extensive knowledge the expert possesses and can access readily. They reviewed the research indicating that this knowledge is probably stored as "chunks," each of which is actually an associated group of items. For the expert, many of these chunks serve as perceptual patterns that are used to index domain-specific knowledge and strategies.

The authors considered the translation process by which the natural-language statement of a problem is converted to an internal representation. They described a program, STUDENT, that could translate algebra problems into equations. STUDENT did not have any semantic knowledge, so it had no way to determine whether the statements in the problem made sense in terms of real objects. Because the nature and properties of real objects are much more important for physics problems, such an approach would not work in physics, but STUDENT did provide insights into aspects of the translation process.

Before describing the next computer model, we need to define and describe a construct called a schema (plural schemata or schemas). Chi and Glaser (1985) define a schema as a "theoretical construct which describes the format of an organized body of knowledge in memory" (p. 241). They go on to say that "Researchers conceive of a schema as a modifiable information structure that represents generic concepts stored in memory" (p. 241). In other words, a schema is a model we have in memory for an object, situation, event, and so forth.

The next computer model, ISAAC, evaluated by Larkin et al. (1980a) has a set of schemata describing physical objects as a major component of the model. These schemata constituted ISAAC's semantic knowledge. Using its schemata, ISAAC translated from the problem statement to a physical representation that was then used to guide the construction of the equations. Both STUDENT and ISAAC show that going from a natural-language statement of a problem to an internal representation of the problem is a complex process. As we will see, these insights derived from computer-modeling research are strongly supported by studies of novices solving problems.

Larkin et al. (1980a) then described three differences between a novice and an expert working typical kinematics problems: the expert was much faster; the expert worked forward whereas the novice worked backward; and the novice mentioned each equation as it was used and explicitly substituted values into the equation, whereas the expert usually mentioned only the result of the substitution process. The authors then described how experts proceed when working dynamics problems in which the successive representations in the solution paths of the experts are important.

These authors pointed out that all the research indicates that an expert in any field must have considerable knowledge of the field. But this knowledge must also be indexed effectively so that it can be used when needed. The authors argued that the indexing is done through schemata. They finished with the hope that the understanding of how someone becomes an expert would lead to better ways to help people learn the necessary knowledge and problem-solving processes.

Larkin et al. (1980b) developed two computer models that solve physics problems in a manner similar to more and less competent human solvers. One model, which they called ME because it employed means-ends analysis extensively, was found to be a good match to the way novices solved kinematics and dynamics problems. The other model, called KD because it employed a knowledge-development approach, was a good match for the experts in their study. They had one expert and one novice solve 19 kinematics problems in think-aloud protocols and then had 11 experts and 11 novices solve two dynamics problems, again in a think-aloud framework.

Two major differences between the ME and KD models were identified. First, the order in which each model uses physical principles differs. The ME model essentially works backward from the unknown to the given information. The KD model proceeds in the opposite direction, working forward from the given information. The other difference between the two models is the way they connect information in the problem to variables in an equation. The ME model essentially writes an equation and then associates each term in the equation with a value from the problem. In contrast, in the KD model the values of the given quantities in the problem are essentially associated with each term of the equation as the equation is set up. The complexity of the models in this study helps to make clear the complex nature of physics problem solving. However, because these models involve very little physics knowledge, two obvious questions arise: How do solvers with different amounts of physics knowledge differ, and how does an individual's problem-solving efforts change as he or she acquires more physics knowledge? One attempt to answer the latter question is described in the following study.

Larkin (1981) described a set of computer models that demonstrate a form of learning. The most basic model (called ABLE) solves mechanics problems similarly to the way many novices proceed. Consider a novice trying to solve a problem. First, one has to remember a principle that might be applicable to the problem. Then one must determine whether the conditions for the application of the principle are indeed satisfied by the problem. Finally, because the principle will be accessed in its general form, one must identify the values to be inserted for each variable in the principle. The ABLE model uses a general algebraic means-ends strategy to identify a principle. It is not necessary to carry out the check on whether the principle applies, because all the principles it has are applicable to the problems presented to it. ABLE then carries out the third step by matching algebraic symbols in the problem statement with identical ones in the principle.

ABLE learns by essentially remembering each successful application of a principle as a chunk. The models that result from learning are called MORE ABLE models, and their solution procedures differ in three ways from the solution procedures of the ABLE model: (1) the order in which the principles are chosen is very different; (2) whenever a MORE ABLE model applies a principle it generates new information, which was not true for the ABLE model; and (3) the process of associating values with the variables in the principle, which ABLE had to carry out each time, is essentially eliminated. Larkin (1981) then compared the solution paths of her models with the work of novices and experts on the same problems. She found that the ABLE model did a good job of reproducing the solution paths of the novices and that MORE ABLE's solution paths were a good match for the experts.

The early information-processing/computer-modeling studies reviewed up to this point began to make it clear that solving standard physics problems was a complex and involved process that required a significant, well-organized knowledge base for competent, efficient performance. Two more recent computer studies have explored issues of transfer of problem-solving processes and learning.

Larkin, Reif, Carbonell, and Gugliotta (1988) developed a computer expert system (FERMI) that has several unique features. FERMI combines factual and strategic knowledge, of different levels of generality, into one system. These different kinds of knowledge—domain-specific factual knowledge, general factual knowledge, and general strategic methods (i.e., weak methods)—are organized into separate semantic heirarchies. This structuring gives FERMI significant power to solve problems in more than one domain. The different domains of problems in this study were pressure in liquids, center of mass, and electric circuits. Although these domains involve different entities, relations, and principles, they have some commonality beause they are all part of physics.

The common problem-solving methods that were examined in this study were decomposition and invariance. The authors also discussed three other general methods—constraint satisfaction, semantic interpretation of algebra, and simple reasoning by analogy—which they believe could be added to the system. This study provides insight into the value and use of weak methods in problem solving.

Elio and Scharf (1990) developed a computer model of the shift in the nature of knowledge organization and problem-solving strategy of novice and expert. Their program (EUREKA) starts out with only textbook-like knowledge and a means-ends strategy—a combination representative of what novices have to work with—and proceeds to learn by solving problems. In the process, EUREKA develops a set of problem schemas with associated solution strategies. EUREKA has three components: a bank of unorganized textbook knowledge, a problem solver, and a long-term memory that contains the problem schemas represented as a P-MOP (Problem-Memory Organization Packets) network. The unorganized textbook knowledge contains equations and concepts that a novice might learn from studying the textbook. However, this knowledge is not organized in any manner that would facilitate problem solving. The problem solver uses means-ends analysis but also has a "meta-strategy" that enables EUREKA to develop a focus of attention when the means-ends approach hits a dead end.

The P-MOP network, which forms EUREKA's long-term memory, is the major focus of the learning process. It consists of a set of constantly changing problem schemas. These schemas evolve, as EUREKA solves problems, to become more focused on physical entities and to discriminate better the strategies to employ. This enables EUREKA to shift from a concern for the unknown and use of means-ends analysis to a knowledge-development strategy using physical principles. This is a difference between novices and experts found in a number of studies.

The authors raise four issues identified by the behavior of EUREKA. First, EUREKA remembers everything it did as it tries to solve each problem. Clearly, humans do not remember everything they do when solving a problem, but what do they remember from any problem-solving experience? Second, suppose a novice and an expert are both given a problem that initially appears to fit one schema but actually fits a different schema when the full problem is read. Would the expert start off in the wrong direction and then correct his or her approach? Such a problem should not cause the novice any difficulty, because the theory indicates that the novice would not do anything until the whole problem has been read. Would things really work out this way? Third, EUREKA has a particular mechanism for dealing with situations in which the means-ends analysis is stymied. What do novice problem solvers do when they hit a dead end? Fourth, how do novices' problem prototypes change as they learn and what features are added to problem schemas?

Summary. The studies in this section share many similarities because they reflect the same theoretical perspective. These studies also share a methodology that involves intensive examination of how a small number of individuals solve problems and construction of computer models that exhibit similarities to the ways the novices and experts solve problems. Not surprisingly, common themes emerge from this work. One theme is the difference in knowledge base between novice and expert and the ways in which that difference affects how experts and novices solve problems. Because the experts have so much more knowledge in the domain, they are in a position to conduct a qualitative analysis, construct a rich and productive representation, and work forward from the given information. Novices, by contrast, start immediately with equations and begin with an equation that contains the unknown. If that equation cannot be solved, they must pick a second equation. This sequence is the reverse of the way experts proceed. Although these insights are useful and important, they provide only a start in understanding how the knowledge bases of novices and experts differ. Getting a better idea of the specific differences in the knowledge bases of novices and experts and learning how novices can build more expert-like knowledge bases are ongoing concerns, as a number of the subsequent studies show.

A second common theme of the studies in this section is the importance of the representation the solver constructs. The computer models differed in the quality of the representations they could construct, and those representations were critical to the solution. Building a computer model that could make the translation between a problem stated in natural language and a useful physics representation shows just how complex that process actually is. But the computer models cannot tell us what representations novice problem solvers actually construct and why they construct them.

The last two studies in this section share characteristics of earlier computer studies but are more complex and larger in scope than the earlier studies. These two studies deal with issues of knowledge organization as earlier studies did, but in these studies additional issues arise. The EUREKA model has a much more involved knowledge base than earlier models, and it "learns" by continually updating its memory. This updating is not simply an additive process; it also reorganizes the knowledge and priorities for using items. The FERMI model solves problems in several different areas, so it has specific domain knowledge, as did models in earlier studies, but it also has more general problem-solving knowledge that it can use in several domains. Obviously, this model addresses the issue of transfer of knowledge in the limited domains it explores. The issue of transfer is one of the oldest and most important issues in educational research.

Left unanswered by the studies in this section are several questions: What knowledge do novices typically use when faced with physics problems? How is the knowledge that a novice possesses organized in memory? How do alternative conceptions affect novices' representations? How can novices be efficiently helped to develop more expert-like knowledge structures?

Studies of Qualitative Analysis and Representations

Investigations into the ways people solve standard physics problems really started to increase around 1980. The Larkin et al. (1980a, 1980b) articles were a reflection of this increased activity. Subsequent work focused on further explicating the representational processes used by novices and experts and the knowledge structures of experts.

Larkin (1983) considered the role of problem representations in the process of solving physics problems. She contrasted naive representations, which involve the actual physical objects and examine real-time developments, with physical representations, which involve physics entities such as forces, energy, and electric fields. She argues that these physical representations differ in several other ways from naive representations. The physical representations are time independent because they incorporate the relevant physical principles, which are constraint relations that can be applied in any order. These representations provide redundant inference sources, so additional information can often be determined in several ways. Finally, these representations contain entities, such as forces, momenta, and velocity, that are "localized"; that is, the attributes of the entities are localized to the entities themselves and do not depend on the environment.

Larkin (1983) conducted some research to determine how physical representations function in problem solving. She had 11 novice and 11 expert subjects solve two easy problems and two hard problems. She found that the order in which the experts applied physical principles to the solution of the easy problems was consistent with drawing inferences from physical representations. In contrast, the novices showed no such sequencing, but rather displayed patterns consistent with means-ends analysis.

On the hard problems, the expert subjects used the physical

representation to construct schemas—for example, application of Newton's second law to an object—which had to be coordinated to solve the problem. The novice solvers attempted to write equations involving the unknown quantities; when the initial equation contained too many unknowns, they would write another equation containing one of the missing quantities in hopes of solving for that and substituting back in the first equation. Their solution attempts usually ended when they ran out of relations for missing quantities.

Larkin (1983) also had five experts solve a very hard problem. She found that the experts first attempted to construct a useful physical representation of the problem. Two of the experts immediately developed the most effective representation for the problem and proceeded to solve it. The other three experts ran into difficulty with their first representation, which they abandoned, and then tried a different one. If that one was also ineffective, they went on to a third one. For all three of these experts, however, the final representation was followed immediately by writing appropriate mathematical equations, which they proceeded to solve. In other words, these experts did not solve the problem until they had constructed a useful physical representation. This study makes a strong case for the importance of physical representations in problem solving.

Anzai and Yokoyama (1984) conducted an interesting, but not well known, investigation of novices' representations of three physics problems, as well as how readily the novices could be persuaded to change their representations. The authors identified three classes of what they called internal models: experiential models, false scientific models, and correct scientific models. (The authors use the term internal model, but it is essentially synonymous with internal representation.) Experiential models are what the name implies, models that employ knowledge derived from an individual's experience; they do not involve scientific concepts or terms. The two types of scientific models do explicitly involve scientific concepts, terms, and relations. The false scientific model incorrectly characterizes the problem information. The authors also introduce the idea of "semantic sensitivity," which describes how easily a problem solver can be enticed to shift from one internal model to another. The authors are interested in both how a model is generated and how it is modified.

Anzai and Yokoyama (1984) constructed three physics problems. In the yo-yo problem, a yo-yo, which had a string wound around the axle, was sitting on a horizontal surface. The question asked was, What will happen if you pull the string, assuming the disks may roll on the surface but never slide? The balance problem had a two-pan balance with a 1-kg mass on one pan and a container with 1 kg of water on the other. The container was said to be massless. A 1-kg ball was suspended from a string so that it was completely immersed in the water. The question was, When the ball is immersed in the water, what will happen to the balance? Finally, the pulley problem had two different mass blocks hanging from a string running over a pulley. The pulley was hung from a spring balance. The question was, Assuming the weight of the pulley is 0, what is the reading on the spring balance? In addition to these three problems, the authors constructed augmented versions that gave physics cues to how to solve the problems.

In their first study, Anzai and Yokoyama (1984) had two experts and one novice solve the problems. The two experts solved just the basic problems, while the novice solved the basic problems and then also solved the augmented problems. The results of this first study indicated that the experts generated a model that they used to solve the problem, whereas the novice tended to generate several models and compare them before deciding which to use to solve the problem. One of the experts also generated more than one model, but he generated additional models primarily to verify his solution based on the initial model. In other words, the reason for shifting between models differed for the expert and the novice; the novice shifted because he was unsure which model to use, but the expert shifted because he wanted additional support for his initial solution.

In their second study, Anzai and Yokoyama (1984) analyzed the protocols of 15 novices and 7 experts to construct a taxonomy of internal models for the three problems. In general, the novices constructed different models than the experts. The authors identified an average of seven models, from all three of their classes, for each problem. The novices' models for the basic problems were primarily either experiential or scientifically false. On the augmented problems, the novices were able to construct correct scientific models for the yo-yo and pulley problems but the balance problem was more difficult. This second study indicated that the number of models that subjects might construct for these problems is limited and that many novice models can be shifted by providing physics cues.

Anzai and Yokoyama (1984) then conducted a series of seven group investigations on how readily novices' models could be shifted. For all but one of these investigations the analysis was based on written answers and the percentage correct. The first two investigations dealt with the yo-yo problem, the third through the fifth with the balance problem, and the last two with the pulley problem.

In the first investigation, seven problems, the basic yo-yo problem and six variations, were given to 216 university freshman enrolled in a college physics course. The variations had additional cues, some pointing toward the correct scientific model and others neutral or providing irrelevant information. Cues pointing toward the correct model produced better performance. The second investigation was designed to determine what physical cues the novices' initial models were semantically sensitive to, that is, what cues were effective in getting the subjects to shift to a scientifically correct model. The 143 subjects, from the same population as in the first investigation, worked the original problem and then one of five variations. The different variations had different physical cues. The only variation that produced a significant shift was the augmented version used in the initial studies. This augmented version provided two pieces of information, whereas the other variations provided only one. The results demonstrate that both items of physical information were necessary to get the novices to shift their model.

The next three investigations, all of which used subjects from the same population as in the first investigation, involved the balance problem, which was the only one of the three for which the subjects in the protocol studies generated false scientific models. This result made the authors suspect that the false scientific models might be semantically insensitive to physical cue. Three variations on the original problem were constructed for this investigation. One of the three was the augmented version from the protocol studies, the second had an experiential cue rather than a physical cue, and the third had both physical and experiential cues. For this investigation 159 subjects completed one of the four problems. There were no differences in the subjects' performance on the four problems.

In the next investigation the authors sought more salient physical cues. They sought these cues in the action-reaction forces—the buoyant force the water exerts on the ball and the force the ball exerts on the water. They constructed a simplified version of the balance problem which had a single-pan balance and then constructed five variations that provided different physical cues, such as the reaction force the ball exerts on the water. The 148 subjects in this investigation worked the basic single-pan balance problem and then one of the five variations. There were no differences among the five groups, showing that none of these cues was effective in producing a shift to a correct scientific model. The authors suspected that the difficulty was that subjects were not considering the reaction force independently of the buoyant force but rather as part of a pair.

Consequently, in the third investigation with the balance problem the authors took the double-pan balance, inserted a spring balance into the string supporting the ball, and broke the task into two questions. The first question asked for the reading on the spring balance, and the second asked for what happens to the pan balance. Four variations were created with different physical cues and motivational instructions. The 200 subjects were divided into five groups, with each group working one of the five problems. About 86 percent of the subjects answered the first question about the reading on the spring scale correctly, but the percentage of correct answers dropped to around 56 percent for the second question. There were no significant differences for any of the different physical cues. This indicates that the subjects understood the effect of the buoyant force on the ball but did not understand the effect of the reaction force, as hypothesized.

The two investigations on the yo-yo problem indicated that experiential models constructed by novices could be shifted by physically relevant cues to correct scientific models. In contrast, the three investigations on the balance problem indicated that it can be difficult to provide effective cues to enable novices to shift from false scientific models to correct scientific models. From their results up to this point, Anzai and Yokoyama (1984) concluded that "semantic sensitivity to cues may depend on principles that the cues are related to, and also the knowledge the presently evoked model is based on" (p. 434). They used their two investigations into the pulley problem described next to explore the validity of this conclusion.

In the first investigation on the pulley problem, 10 subjects, who were either senior undergraduates or first-year or second-year graduate students, worked the original pulley problem and then a variation that supplied information about the tensions at the ends of the two strings, the accelerations of the objects, and the acceleration due to gravity. The subjects worked the problems in think-aloud protocols. Analysis of the protocols indicated that the majority of the subjects who mentioned acceleration first solved the second problem correctly even though they had worked the first problem incorrectly. In the second investigation on the pulley problem, the authors created six variations on the basic problem. The variations had different physical cues; half of the variations had cues about acceleration and the other half had cues about the tension. The 210 subjects from the population identified earlier were divided into seven groups; six groups worked the original problem and then one of the variations, the seventh group worked the original problem twice. The subjects who had the variations with the acceleration cues all showed significant improvement, but none of the groups with variations having tension cues improved significantly.

The Anzai and Yokoyama (1984) article makes a solid case for the relation between the problem solver's knowledge base, the representation he or she constructs, and information in the task environment. However, a word of caution about generalizing the conclusion derived from the investigations with the balance problems may be in order. Newton's third law is possibly the most difficult principle in physics for novices to understand and accept, so the results in the balance problem may be due in part to the difficulty of the principle involved (Boyle & Maloney, 1991; Brown, 1989; Maloney, 1984). Further evidence for the importance of the problem representation is presented in the following study.

Larkin (1985) again discusses the importance of problem representations in an article describing another computer model, called ATWOOD. The model is designed to take a verbal problem description and construct from it a representation involving entities and relations between them; that is, it builds semantic nets as its problem representations. The semantic nets constructed by ATWOOD were judged to be acceptable as long as they contained no errors and contained sufficient information to solve the problem. The nets constructed contained only everyday terms, not physics entities such as forces or energies. The author argued that these representations are similar to those a novice would construct but that an expert would be able to supplement such representations with specific physics knowledge.

To test this hypothesis, two subjects, one expert and one novice, were given abbreviated versions of 15 problems. The abbreviated versions were constructed by including only the terms ATWOOD used in successfully processing the 15 problems. The two subjects were instructed to reconstruct the original problems from the abbreviated versions. The expert was able to reconstruct 10 of the 15 problems, but the novice could reconstruct only 2. The author argued that these results indicate that the expert's physics knowledge enabled him to understand something that was basically unintelligible to the novice.

Up to this point, most of the studies have compared novices and experts, but there is clearly a fair amount of variation within each of these groups. The more interesting variation is among novices, because identifying its nature should provide useful information for designing instructional activities. The next two studies investigate some of these variations.

Finegold and Mass (1985) studied the differences between good and poor problem solvers. Their subjects were high school students who were taking an advanced placement physics course. They used four problems, two electric circuit problems and two mechanics problems. Subjects were asked to think aloud as they worked the problems. Subjects were rated as either good problem solvers (GPS) or poor problem solvers (PPS) by their teachers. Finegold and Mass randomly chose eight of each from the teachers' lists.

Finegold and Mass (1985) formulated eight hypotheses about the differences between the GPS and PPS. These hypotheses were based on the four-step solution framework identified by Polya (1945). Finegold and Mass (1985) found support for the following five of their eight hypotheses:

- Good solvers translate the problem statements more correctly and more exactly than do poor solvers.
- Good solvers plan their solutions more fully and in greater detail before carrying them out than do poor solvers who tend to solve without planning.
- Good solvers complete the solution to a problem in less time than do poor solvers.
- Good solvers spend relatively more time on translation and planning than do poor solvers.
- Good solvers make greater use of physical reasoning than do poor solvers. (p. 66)

The authors argue that, because any student who scored less than 60 percent in the advanced placement physics course was excluded from the study, all of the subjects in their study had the basic declarative knowledge necessary to solve these problems. Consequently, they argue that the differences cited are due to difficulties the poor solvers have in solving problems, not to any major deficiencies in their basic knowledge of the domains. This study is important because it is one of the few that tried to hold the novice knowledge base constant and look for differences in general problem-solving skills.

McMillan and Swadener (1991) studied how six novice volunteers solved an electrostatics problem. They were interested in the qualitative analysis students carried out while trying to determine the quantitative value asked for in the problem. Five of the six students were A or B students in a second-semester calculus-based physics course in which they were enrolled. The sixth student was a self-reported D student. The authors gave the subjects a common electrostatics problem involving two charges on a line. In this problem, the separation distance between the two charges, the sign and magnitude of one charge, and the distance from that charge to the point(s) where the electric field was zero were given, and the subjects had to determine the other charge. One feature of the problem was that there are actually two points where the field could be zero, depending on the sign of the unknown charge.

The researchers found that the five good students were all able to solve the problem to determine one of the two possible solutions. The D student was unable to solve the problem. However, none of the students used any qualitative reasoning in their solution efforts. The students, apparently without conscious consideration, chose one of the two possible locations where the electric field could be zero—this choice was equivalent to assuming that the charge was positive—and then proceeded to apply the relevant equations to obtain a numerical value for the unknown charge. When asked, the students were unable to explain their choice of the point for the zero-field location.

The essentially complete failure of these students to use any qualitative reasoning was the major finding of the study. This result is disturbing if our goal is that students learn to solve problems with understanding. As the authors comment, "A numerical solution to a problem in any situation is useless without some conceptual and interpretive knowledge of what that answer means" (McMillan & Swadener, 1991, p. 669). One question such results raise is what students are learning from working standard problems in physics courses.

Summary. The studies in this section make clear the critical importance of constructing an appropriate physical representation of a problem, as shown by the inability of the experts in Larkin's study (1983) to solve the very hard problem until they had an appropriate representation. Such a representation identifies the relevant physical quantities and provides a "map" to guide the solver. This construction process is primarily qualitative and obviously requires a well-organized knowledge base. The Anzai and Yokoyama (1984) study provides clear evidence for a strong relation between a problem solver's knowledge and the representation he or she constructs, as well as a relation between the solver's knowledge and task information. The Finegold and Mass (1985) and McMillan and Swadener (1991) studies provide strong evidence that poor problem solvers have difficulty, in large part, because they fail to or are unable to carry out the necessary qualitative analysis to construct an appropriate representation. This naturally leads to questions about what knowledge novices have and how they organize their knowledge.

Investigations of Problem Solver's Knowledge Structures

One early study, not strictly within the information-processing framework, deserves mention here. While not directly studying students problem-solving ability Thro (1978) did investigate the development of college students' associative knowledge structure over the span of one semester. Her study built on earlier work of Shavelson (1972) (Shavelson & Stanton, 1975), who employed word-association tasks to gain information about subjects' cognitive structure. Thro (1978) also used word-association tasks, from which she constructed a "relatedness coefficient" as a quantitative measure that could be related to students' performance on a 26-item multiple-choice posttest. The posttest investigations contained a variety of questions, some involving definitions, some simple applications, and some problem-solving ability. Based on the results of her analysis, Thro concluded, "Thus, the ability of subjects to solve problems was seen to vary directly with the establishment of related concepts in their cognitive structure" (Thro, 1978, p. 976).

One of the early information-processing investigations of novice knowledge structures was reported by Chi, Feltovich, and Glaser (1981). They contend that "The knowledge useful for a particular problem is indexed when a given physics problem is categorized as a specific type" (Chi et al., 1981, p. 122). Consequently, they conducted a series of studies exploring how novices and experts categorize standard mechanics problems.

Operating from this perspective, Chi et al. (1981) had eight experts, Ph.D. candidates in physics, and eight novices, undergraduates who had just studied mechanics, categorize a set of 24 problems from a college general physics book. The subjects were instructed to categorize the problems on the basis of the similarity of solution processes. The authors carried out a cluster analysis of the sortings of each group and then examined the four pairs of problems that each group considered most similar. On the basis of this analysis, they concluded that the novices categorized primarily on the basis of what the authors called surface structure—that is, literal objects in the situations, explicit physics terms, or the specific physical arrangement. The experts, in contrast, categorized primarily on the basis of physical principles, which the authors called deep structure.

In the second study in Chi et al. (1981). a set of 20 problems that combined both surface structure and deep structure features were constructed. Three subjects, a physics graduate student, a senior undergraduate physics major, and a student who had taken one course in mechanics, categorized this set of problems. The novice's categorization was primarily along the lines of the preidentified surface features, and the expert's categorization was essentially along the lines of the physical principles. The intermediate subject used a categorization scheme that, while based primarily on the physical principles, employed surface features as secondary factors. This study provided further support for the interpretation of the first study.

In their third study, Chi et al. (1981) presented a set of 20 category labels to two novices and two experts. The subjects were given 3 minutes to tell everything they could about problems from each category. The authors analyzed the protocol of one novice and one expert two different ways, by node-link structure and production rules. The expert's production rules contained actions that were explicit solution methods, whereas the novice's productions contained actions that were attempts to find explicit unknowns. In addition, both had productions involving conservation of energy, but this principle was part of the condition side for the novice and part of the action side for the expert. The fourth study reported by Chi et al. (1981) was an effort to determine what features of the problems subjects use in deciding how to categorize the problems. Subjects for this study were two experienced physics instructors and two novices who had earned an A in a basic mechanics course. The subjects were asked to read each problem and then think aloud about how they would solve it. For the two experts, the basic approach for solving the problem translated to determining what physical principle was involved. The novices either made global descriptions or started actually solving the problem. An additional difference was that the experts and novices mentioned different features of the problems as important. Other parts of the investigation carried out by these researchers are discussed next.

Chi, Glaser, and Rees (1982) describe a sequence of studies of the nature of novice and expert knowledge structures in relation to problem solving. In the first study they compare and contrast the problem-solving procedures of two experts and two novices. They found that quantitative differences were quite variable and individual, but there were identifiable qualitative differences. These differences were primarily in the area of the inferences drawn from the qualitative analyses. Both the novices and the experts carried out qualitative analyses, but the inferences drawn were different. The second study was the card-sorting task described more fully in Chi et al. (1981), in which the novices sorted primarily on the basis of the surface features (e.g., the objects in the problem or the explicit physics concepts mentioned) while the experts sorted on the basis of the physical principle they would use to solve the problem. The third study in Chi et al. (1982) was the same as the second study in Chi et al. (1981).

In the fourth study in the set, subjects initially sorted a set of 40 problems and were then asked to subdivide each category if they could and to combine their original categories if they could. Novices formed more categories initially, tended to produce subcategories that contained one or two problems, and had difficulty generating higher-level categories compared with the experts. In the fifth study, four experts and four novices reviewed a chapter on particle dynamics for 5 minutes and then summarized the chapter. The book was available for subjects' use while they were summarizing. The experts were found to have more complete knowledge than the novices, even though the novices had studied the material and had the book available if they needed it. In the sixth study, two novices and two experts were asked to elaborate on a set of 20 prototypical concepts that subjects in the earlier studies had used to describe their classifications. The main finding of this study was that "the knowledge common to subjects of both skill groups pertains to the physical configuration and its properties but that the expert has additional knowledge relevant to the solution procedures based on major physics laws" (Chi et al., 1982, p. 59).

The seventh and eighth studies were designed to determine differences in novices' and experts' abilities to identify key features of problems. Study seven had two experts and two novices identify explicitly the features in the problems that they would use to key their solution procedures. The novices and experts identified different features. The last study asked six novices and six experts to judge the difficulty of each of the 20 problems and to identify the cues they used to determine difficulty. This time there was strong overlap in the cues novices and experts used. However, novices were less accurate in their judgments of difficulty than experts. Even though the two groups used the same cues, their reasoning for using those cues and the way they used the cues were different, with the result that the novices were less able to identify the difficulty accurately.

Veldhuis (1986) replicated and extended the Chi et al. (1981) categorization study. He used 94 novices, 5 intermediate subjects, and 5 experts, and constructed problem sets that explicitly varied either surface or deep structure characteristics. He analyzed his results using cluster analysis (Veldhuis, 1990). His results supported the findings of Chi et al. (1981) with respect to the categorization of the experts, but his novices used both surface and deep structure in their categorizations.

In a study with methodological similarities to the Chi et al. (1982) study, de Jong and Ferguson-Hessler (1986) explored the cognitive structures of good and poor novice problem solvers. They cite two trends in the research on problem solving: studies of the solution processes employed by problem solvers and studies of the knowledge bases, including the organization of the knowledge bases, of the problem solvers. de Jong and Ferguson-Hessler (1986) argue that just having the knowledge is not sufficient; it must be organized in a useful manner. They argue that the Chi et. al. (1981) and Larkin (1979) investigations indicate that experts have problem schemata organized around physical principles whereas novices' problem schemata are organized around surface features.

De Jong and Ferguson-Hessler (1986) list three kinds of schema-specific knowledge they believe are needed for successful problem solving: declarative knowledge (principles, formulas, and concepts), procedural knowledge (e.g., when to use a particular relation), and knowledge of the characteristics of problem situations. But problem solvers also need strategic knowledge (Schoenfeld, 1978) to use schema-specific knowledge effectively.

The purpose of the de Jong and Ferguson-Hessler (1986) study was to test the hypothesis that good novice problem solvers have their knowledge organized according to problem schemata. The researchers constructed at least one declarative knowledge statement, one procedural knowledge statement, and one problem situation statement for each of the 12 problem types. A total of 65 knowledge statements resulted. Each statement was printed on a card, and subjects sorted index cards into "coherent" piles. That is, they were told that cards in each pile should be more closely related to each other than to cards in other piles. The subjects were enrolled in a course on electricity and magnetism. Results of an examination in this course were used as the measure of the subjects' problem-solving ability. Based on the subjects' card sorting, de Jong and Ferguson-Hessler (1986) constructed a complex value that they called a problem-type–centered (PC) measure. As described, the researchers had 12 problems with 65 knowledge elements, so they constructed a 65 by 65 symmetrical matrix in which each cell corresponded to a pair of knowledge elements. When the 12 problems were sorted into the 12 preidentified types, all of the elements in the matrix were ones. Subjects' matrices were constructed by entering a one in a cell when the elements were placed in the predetermined category and a zero in each cell if they were categorized differently. The entries were summed for each subject and the sums were divided by the value for the ideal sort to get the PC measure. They then calculated the correlation between PC and performance on the exam. They found significant correlations and concluded that their hypothesis was supported.

In a related article, Ferguson-Hessler and de Jong (1987) present a theoretical analysis of two knowledge structures, hierarchical and problem schemata. (The reason for choosing a hierarchical knowledge structure for comparison becomes clearer when the next set of studies is examined.) Their problem schemata are organized around principles or fundamental concepts such as Ampère's law. Each problem schema has elements of declarative knowledge and procedural knowledge, as well as characteristics of typical problem situations. The hierarchical organization they construct has matter and fields as the two highest-level concepts. The authors compare the two types of organization and argue that the two "do not contradict but supplement each other" (Ferguson-Hessler & de Jong, 1987, p. 495). They also contend that the problem schemata are an efficient way for novices in a first course to organize the knowledge to make it available for problem solving. They point out that the introductory-level course provides little opportunity for abstraction, which is integral to the hierarchical organization.

Ferguson-Hessler and de Jong (1987) briefly review their study of good and poor problem solvers (de Jong & Ferguson-Hessler, 1986) and present the idea that good novice problem solvers have their knowledge organized more closely along problem schemata lines than do poor problem solvers. They argue for instruction that explicitly encourages students to organize their knowledge around problem schemata. They contend that explicit use of problem schemata will help instructors make their tacit knowledge visible to the students and, consequently, will enable students to realize that characteristics of problem situations are used to determine which principles to apply.

The next three studies (Eylon & Reif, 1984; Heller & Reif, 1984; Reif & Heller, 1982) are a sequence on the usefulness of hierarchical knowledge organization for problem solving. The Reif and Heller (1982) article was a theoretical examination of how a particular knowledge structure can produce efficient problem solving. The second article (Heller & Reif, 1984) was an empirical investigation of the effectiveness of the problem-solving procedure identified in the theoretical analysis. The third study (Eylon & Reif, 1984) explored the knowledge-organization aspect of the theoretical analysis by comparing the effectiveness of a hierarchical organization to that of several other ways to organize a knowledge base. The theoretical analysis Reif and Heller (1982) carried out on the relation between a person's knowledge structure and his or her problem-solving procedures brings out the complexities involved in problem solving in physics. They adopt what they term a "prescriptive" approach to physics problem solving. That is, they try to ascertain what efficient and effective problem solving would be like without trying to match it to what human experts in the field actually do. They focus on three stages of the solution process: the description stage, the search for a solution, and the assessment of the solution.

The knowledge structure that they argue is best is a hierarchical one in which the top level contains information about the basic entities of interest in the domain, as well as the relations among those entities and the general principles that apply to the domain. The items in this top level are then elaborated at the lower levels of the structure. Each item in the knowledge structure must be accompanied by ancillary knowledge that enables the individual to interpret the concepts and principles and apply them where appropriate.

Reif and Heller (1982) then consider the actual construction of a solution. They argue that this is essentially a search process guided by some generally applicable methods such as decomposition, constraint satisfaction, and multiple levels of description. Each of these methods is itself a complex process that varies to some extent with the different situations to which it is applied. To apply the process effectively in any domain, the individual must have a reasonable understanding of the domain-specific knowledge.

Reif and Heller (1982) briefly consider the assessment of the solution stage and then consider instructional implications. They argue that "the cognitive mechanisms needed for effective scientific problem solving are complex and thus not easily learned from mere examples and practice. A potentially more effective instructional method would teach problem-solving skills explicitly" (p. 124). We return to this instructional issue later in the chapter.

Reif and Heller (1982) identified a group of general procedures used by successful problem solvers:

- Developing accurate and complete basic descriptions of the problem
- Developing accurate and complete theoretical descriptions of the problem
- Conducting exploratory analysis
- Employing effective executive, or metacognitive, processes
- Carrying out the solution plan effectively
- Evaluating the solution

Heller and Reif (1984) investigated the effectiveness of their prescriptive model of problem solving. They focused on the process of formulating a theoretical description—that is, a physical representation (Larkin, 1983)—for mechanics problems. They thoroughly pretested a set of directions derived from the theoretical model to be sure they were interpretable and executable. They also developed a set of directions, which they called the modified model, that was designed to be similar to the problem-solving procedures often found in general physics textbooks.

The subjects were 24 undergraduates who were in their second course of an introductory physics sequence. They were paid volunteers who had received a grade of B− or better in the first course. There were three groups, one using the full model, one using the modified model, and a control group who worked without external guidance, with eight subjects randomly assigned to each. Subjects completed a pretest consisting of three mechanics problems, which they solved individually without guidance. Subjects in the two experimental groups were then presented with the models as groups and practiced with the models for about a half-hour. Approximately 1 week later, subjects returned to work three more problems. Subjects in the control group worked the problems on their own. Subjects in the two treatment groups worked through the problems with external guidance.

The researchers identified and defined four performance measures to assess the quality of the subjects' solutions. The solutions generated by the subjects using the prescriptive model were significantly better than those of the control group on all four measures and significantly better than the solutions

of the subjects using the modified model on three of the four measures. These results strongly support the sufficiency and effectiveness of the prescriptive model used.

Eylon and Reif (1984) investigated the aspect of knowledge organization as a factor in effective problem solving. In two experiments, they presented subjects with knowledge organized in specific ways and then tested them on recall and problem-solving tasks. The problem-solving tasks were not standard physics problems but did require the subjects to use the knowledge actively. One of the knowledge organizations in each of the experiments was hierarchical, which the authors believed to be the most effective and useful organization.

In the first experiment, the hierarchical organization was compared with a single-level organization. There were three groups, one receiving the hierarchical treatment, one the single-level treatment, and the third the single-level treatment twice. There were 12 subjects in each treatment group. Subjects were also classified into three ability groups determined by performance on a prior physics test. The results showed that the subjects in the hierarchical treatment group performed significantly better on complex tasks involving appreciable information retrieval. Another finding was that the low-ability subjects benefited least from the hierarchical treatment.

The second experiment compared two different hierarchical organizations of the same information. Both dealt with a "fictitious universe of nuclear particles," but in one case the information was organized hierarchically by physical model and in the other it was organized hierarchically by historical development. Performance measures were designed to be either deductive, which was consistent with the physical model, or historical. Subjects performed better on the tasks that were consistent with the knowledge organization they had studied.

The three studies (Eylon & Reif, 1984; Heller & Reif, 1984; Reif & Heller, 1982) just reviewed may seem to stand in opposition to the two studies of de Jong and Ferguson-Hessler (1986) (Ferguson-Hessler & de Jong, 1987) on the issue of knowledge organization, but several factors should be considered before drawing such a conclusion. First, two different topic domains within physics were involved. The Reif group used mechanics as the domain of interest, whereas de Jong and Ferguson-Hessler worked in the area of electricity and magnetism. Because these two domains have different fundamental entities, principles, and relations, they may not be equally amenable to a particular structuring. Second, some of the items that have different names in the two sets of studies could be the same thing. For example, the individual knowledge items with their ancillary knowledge that make up what Reif and Heller call "basic functional knowledge components" are very similar to problem schemata. Third, the best and most efficient knowledge organization may be hierarchical, but nonetheless for novices problem schemata may be a better way to organize their knowledge, as suggested by Elio and Scharf (1990).

In contrast to the previous studies, which looked at large-scale knowledge structures, Robertson (1990) was interested in "cognitive structures" that enable students to understand specific concepts well enough to apply them to transfer problems. He defined transfer problems as those that are "structurally, but not conceptually unfamiliar to the solver." He used think-aloud protocols to investigate the relation between subjects' cognitive structures and their ability to solve problems involving Newton's second law. He focused on one aspect of understanding Newton's second law—the system concept. The system concept is basically the idea that the object (or objects) that is to be treated as the entity of interest, to which the vector summation of forces will be applied, must be explicitly identified. In many problems the choice of system of interest is not unique.

Robertson (1990) constructed a "system-concept index" by looking for both positive and negative indicators of an understanding of the need to identify a system explicitly. He found a significant correlation between this system-concept index and the subjects' scores on the transfer problems. A significant correlation was also found between the error of "transmitting a contact force" and an incorrect choice of system for the subjects in this study. Robertson concluded that students need to make connections between physical principles and everyday situations, but they also need to make connections between the various physical concepts to be able to employ them effectively.

Another of Robertson's (1990) implications is that the analytical technique he employed is specific for each problem. This again raises the issue of how generalizable the results of such research are. If this kind of analysis has to be carried out for each individual problem for each student, what insights does such research have for classroom teachers, who must work with groups of students? Robertson calls for studies employing detailed task analyses of problems under study.

Summary. The studies in this section explore several aspects of the relationship between an individual's knowledge base and his or her problem-solving behavior. Chi et al. (1981), Chi et al. (1982), Veldhuis (1986), and de Jong and Ferguson-Hessler (1986) employed card-sorting methodology to explore the knowledge bases of contrasting groups. Three of these studies compared experts and novices. These studies found differences—such as the experts' production rules containing explicit solution methods; the experts having more complete knowledge; the novices' knowledge being strongly focused on physical objects, configuration, and properties; and experts having the ability to use cues on problem difficulty more effectively—that help explain the experts' greater ability to solve problems. De Jong and Ferguson-Hessler (1986) studied good problem solvers and poor problem solvers, rather than experts and novices, and found that the good problem solvers had better problem schemata than the poor solvers. Reif and his colleagues explored the differences between hierarchical and other knowledge organizations. These researchers argue that a hierarchical organization is most effective for using the knowledge. Robertson (1990) found evidence that students need to connect the concepts, principles, and relationships they are studying to each other in order to be able to use them effectively. One question these studies raise is, how do students acquire and organize the appropriate knowledge? Are Ferguson-Hessler and de Jong (1987) correct that helping novices organize their knowledge around problem schemata is an effective and efficient way to assist the eventual development of a hierarchical organization? Problem schemata would seem to be a reasonable way for novices to relate physical principles to

everyday knowledge and to each other, which Robertson (1990) argues is an important step for students. The studies in the next section explore how students use existing textbook materials in trying to learn the relevant knowledge.

Investigations of Students' Study Procedures

Chi, Bassock, Lewis, Reimann, and Glaser (1989) studied how students use worked examples when they are trying to learn how to solve problems. Using think-aloud protocols, they studied what the students did while they were studying the worked examples and how the students used the worked examples when trying to solve later problems. Their major finding was that there was a difference between "good" and "poor" students in how the students used the examples. Good students studied the examples until they understood them, whereas poor students tended to "walk" through the examples without checking to see if they understood them. The good students could identify points where they did not understand an example and amplify on the example until they figured out the difficulty. In other words, the good students were significantly better at identifying comprehension failures and pinpointing their precise nature. Identification of such failures is critical for the student to know where he or she needs to revise his or her thinking.

The good students filled in many of the missing (i.e., implicit) steps in the worked example. Most of these missing steps involved the conditions for application of the principles and relations, that is, the procedural aspects of the relevant knowledge. The fact that the good students filled in these missing steps when studying the worked examples, whereas the poor students did not, means that the good students had developed sufficient understanding of the principles and relations to be able to use them effectively on new problems.

This latter conclusion is confirmed by the differences in how the good and poor students used the worked examples when they were working later problems. The good students tended to refer back to the worked examples when they needed a specific relation or when they wanted to compare and check what they had done. When they went back to a worked example, they usually referred to only a short segment of the example. In contrast, the poor students tended to go back and read major segments of the worked examples to try to find some procedure they could use in a wholesale manner on the problem they were trying to solve. In other words, there was evidence that the good students consulted the worked examples after they had formulated a plan for solving the problem, whereas the poor students consulted the worked examples in hopes of finding a plan they could use to solve the problem.

Ferguson-Hessler and de Jong (1990) also investigated the study processes of good and poor problem solvers. Their study is related to the Chi et al. (1989) study, but they used different techniques and worked in a different topic domain, electricity and magnetism. Nevertheless, they found results similar to those of Chi et al. (1989). Ferguson-Hessler and de Jong (1990) had students study a segment of text and, at selected points, think aloud about what they were doing. Subsequently, these students worked a test. Scores on that test were combined with scores on tests from the physics course in which the subjects were enrolled and with results from a test administered in the same course several weeks after the research. Five students who did well on all three exams and five who performed poorly on all three were the subjects for the analysis.

From a global perspective of how actively the students studied the text, both good and poor subjects were found to be similar. They differed in the nature of the study processes they employed. The good performers used more "deep processing" than the poor performers. Examples of deep processing are imposing structure not given in the text and making procedures and assumptions explicit. The poor performers were more likely to use the process Ferguson-Hessler and de Jong (1990) call "taking for granted" (p. 47) than the good performers. In addition the poor performers focused more strongly on declarative knowledge, while the good performers worked more on procedural and situational knowledge. The poor performers were also more likely to say "everything is clear" than the good performers.

Summary. The results of the two studies in this section, Chi et al. (1989) and Ferguson-Hessler and de Jong (1990), are consistent in showing that good performers are aware of when they do not understand something and make an effort to develop the necessary understanding. In contrast, poor performers are often unaware of the fact that they do not understand something, which means they are extremely unlikely to correct the deficiency. These studies indicate that good students work out the procedural aspects of the problems that are usually left implicit in textbook presentations, but the poor students need assistance in identifying and understanding these aspects. The fact that good and poor students also use worked examples differently when trying to solve problems is important information for instructors, because it alerts them to a behavior they can watch for and try to modify.

Section Summary

All of the studies in this section share a focus on individuals solving, or learning to solve, physics problems. The knowledge that novices and experts have and the ways they use that knowledge were explored, as were some of the differences between good and poor novices. The importance of thorough knowledge of concepts, principles, and relations, as well as effective organization of this knowledge, was demonstrated by many of the studies. This knowledge was shown to be critical for the construction of productive representations, without which a problem solver is reduced to employing general heuristics such as means-ends analysis. A definite problem-solving strategy was also shown to be important, but in a knowledge-rich domain such as physics it must also be coupled with an extensive, well-organized knowledge base.

Besides the importance of a well-organized knowledge base and general problem-solving strategies, the other thing that stands out from the studies in this section is the importance of problem representations. Novices have consistently been found

to generate different representations from experts. Novice representations involve physical objects, the arrangement and interactions of these objects, and the evolution of the systems over time. In contrast, experts' representations involve physical entities such as forces, energy, electric fields, and waves. A major obstacle novices face when trying to solve a physics problem is that of translating from the verbal statement of the problem to an accurate and effective physical representation of the problem. Because most novices lack the necessary knowledge base, including appropriate procedural knowledge, to apply the concepts, principles, and relations effectively to new situations, they must fall back on applying means-ends analysis to the mathematical relations they have available. Clearly, one of the primary tasks for converting novices to more expert physics problem solvers is to get them to learn the concepts, principles, and relations in such a way that they can use them to construct useful physical problem representations.

The majority of the studies examined up to now, while informative, are limited because they were not done in actual classrooms. Although these studies provide useful information about how individuals solve problems, they do not provide much information about how problem-solving skills can be taught, or developed, in large classes. For example, Heller and Reif (1984) demonstrated the effectiveness of the prescriptive model with individual subjects, but this technique could not be implemented in the same way in a large class. The studies in the following section take ideas from the research examined and try to apply those ideas to the classroom or other learning contexts.

PEDAGOGICAL METHODS FOR IMPROVING PROBLEM SOLVING

Although several of the studies reviewed earlier had instructional elements (Larkin, 1979; Larkin & Reif, 1979), they were small-scale investigations. The studies in this section involve instructional investigations with full classes actively engaged in learning physics or with students working with computers. There were several investigations of aspects of how students learn to solve problems that preceded the current information-processing and constructivist-oriented research. Because the theoretical frameworks, when they could be explicitly identified, varied greatly in these studies, it is difficult to draw coherent inferences from them.

One early study was carried out by Thorsland and Novak (1974). Although this study did not investigate the effectiveness of a particular pedagogical intervention, it did investigate students studying physics in a classroom context. Thorsland and Novak identified two "approaches" to problem solving, intuitive and analytic. The analytic approach involved a careful step-by-step analysis of the problem; in contrast, the intuitive approach involved an implicit "feel" for the situation. They studied a group of 70 students, in an introductory physics course, who received audiotutorial instruction on experiments and selected study activities. Twenty-five of the 70 students were randomly selected to participate in problem-solving interviews. Based on the interviews, the students were rated with respect to intuitive and analytic approach, and these ratings were used to identify four groups as high or low intuitive (I dimension) or analytic (A dimension). The groups were compared on several learning-related factors.

The authors contended that their hypothesis that "the A dimension is more highly related to scholastic ability than is the I dimension" (Thorsland & Novak, 1974, p. 249) was supported by a significant correlation between the A rating and SAT math and verbal scores. The authors concluded that the problem-solving approach of individuals could be reliably identified and is related to learning. However, this study had a major drawback.

Early in their presentation, Thorsland and Novak (1974) state that "students proceeded in the problem solving encounters in two distinct ways" (p. 246). These ways are identified as the analytic and intuitive approaches. They contend that "some students appeared to utilize both approaches successfully while others relied on only one approach" (p. 246), so they assigned both an analytic and an intuitive rating to each student. These two approaches are related to a theoretical framework derived from the work of Ausubel (1968). But the relation of the two approaches to the theoretical "differentiated cognitive structure" (Thorsland & Novak, 1974, p. 246) is not at all clear. For example, if the two approaches are directly linked to two different cognitive structures, how can the same individual use both approaches successfully?

This study is typical of early studies in that it is primarily statistical in nature. Little information is provided about the reasoning the subjects employed in the tasks. If one is not familiar with the theoretical framework, it can be difficult to know what the implications of the study are.

Mettes, Pilot, and Roossink (1981) report on efforts to formulate a general program of problem-solving instruction. They use as their theoretical framework a model of "stage-by-stage formation of mental actions" developed by Gal'perin (Talyzina, 1973). They developed what they called a program of actions and methods, which they used to train students in a chemical thermodynamics course. The actual instructional materials employed a prescriptive system of heuristics called a "systematic approach to problem solving," which has some similarities to the work of Reif and Heller (1982). The authors evaluated their program over a period of 5 years. They established three criteria for judging the effectiveness of the program: learning outcomes, time spent by students and teachers, and satisfaction with the course. The results for the last two times the course was offered on an experimental basis indicated that the program was superior to the control course structure. This study is interesting because of the attempt to develop a general instructional program, but the approach is difficult to evaluate because it is based on an obscure theoretical framework.

Another early study is reported in Richardson (1981). This study was based on some of the early information-processing work on problem solving in physics which indicated that experts conducted schema-based qualitative analyses of problems. An instructional program was developed that tried to provide students with a "core schema" for electrostatic problems, which they were encouraged to use when confronted with such problems. The author chose the topic domain of electrostatics because he wanted to avoid topics for which students have definite alternative conceptions. The treatment was found to have no significant effect on problem-solving ability. Two major reasons were advanced for the lack of effect: short treatment period and weak homework grading. The treatment consisted of lecture presentation of the core schema and practice during recitation sections. The practice was limited to three recitation periods, obviously a short intervention. Students' homework assignments were not closely graded; consequently, they received relatively little detailed feedback about their performance.

Despite its lack of effect, this study is interesting for several reasons. First, it is one of the few studies that has attempted to apply directly the findings of information-processing analysis of problem solving in physics. It did attempt to incorporate two of the recurring ideas from that research: importance of schemas in problem solving and qualitative analysis of problems before applying equations. Second, the identified reasons for lack of effect point out some important considerations about what is necessary to produce improved instruction in problem solving. Finally, the topic domain chosen for this study is one that has received significantly less attention than the topic of mechanics, so the study provides information about the generalizability of conclusions about problem-solving methods derived from mechanics studies.

Bascones and Novak (1985) tested the effectiveness of an Ausubelian approach to instruction on high school students' problem-solving ability. The Ausubelian approach had the content sequenced with the most inclusive and general principles and concepts introduced earlier to serve as conceptual anchors for later ideas. This approach was compared with a traditional instructional program. The authors established a 15-point scale of problem-solving skill, which had features such as: identify the unknown, draw a sketch, organize available data, design a method for solving the problem, and argue about the limitations of the answer. The authors considered that a score of 9 was the minimum to demonstrate successful problem solving. Students were tested on eight problems during the term. The students were also classified into one of three ability groups on the basis of scores on the Raven Progressive Matrices.

The results showed a significant difference between the Ausubelian group and the traditional group. However, neither group's mean score reached the value of 9 set up as the criterion for successful problem solving. Although little information was provided about the nature of the instructional intervention, as described it did have some similarities to the presentation of a heirachical knowledge organization as advocated by Eylon and Reif (1984). This study would thus indicate support for such an instructional approach.

Two things stand out when looking at these early studies. First, there were very few studies of any aspect of the effect or place of problem solving in learning physics. This reflects the almost unanimous belief among physics instructors of the period that solving problems was the way to learn physics. Few people questioned this belief. Second, these early instructional studies are very diverse. They use different theoretical bases, where they have an explicit theoretical basis; they focus on different features of problem solving; and they use vastly different methodologies. Where they do tend to be similar is in a strong statistical orientation. But these early studies do contain elements, such as interest in knowledge organization and problem-solving procedures, that are also a focus of later work.

Three classroom-based instructional studies have been reported since the research on problem solving in physics began in earnest. These studies address the development of students' problem-solving skills in the context of a college general physics course. They are important because they involve regular classroom-size samples in the context of a typical college physics curriculum.

Wright and Williams (1986) present an explicit problem-solving procedure, which is part of an Explicitly Structured Physics Instruction (ESPI) system, and some data supporting its efficacy. Their "WISE" strategy is similar to other systems such as those of Polya (1945), Schoenfeld (1978), and Heller, Keith, and Anderson (1992). WISE stands for What's happening (identify givens and unknown, draw a diagram, and identify the relevant physical principle), Isolate the unknown (select an equation, solve algebraically, and look for additional equations if one is insufficient), Substitute (plug in both numbers and units), and Evaluate (check the reasonableness of the answer).

The authors tested the ESPI system in general physics classes at a community college. A total of 136 students participated in the study. The authors found that the students who used the WISE strategy on most of their homework and test questions performed significantly better in the course than the students who failed to use the strategy. The better performance was determined by course grade. Students' comments were also strongly supportive of the value of the WISE strategy, with 134 saying it was somewhat or very helpful. The authors also found that the retention rate increased while the ESPI system was used. The authors present the advantages of a systematic approach to problem solving, but they also point out that students are reluctant, in some cases actively opposed, to adopting such a system because of the work involved. An obvious question is, Is it the instructor's job to train students to behave in a systematic manner, even if they are reluctant to adopt such an approach?

Heller, Keith, and Anderson (1992) studied the relationship between group and individual problem solving in an algebra-based general physics course. The sample consisted of 91 students who enrolled in a special section of the course. The students in this special section met 7 hours a week: 4 hours of lecture, 1 hour of discussion, and 2 hours of laboratory.

Heller et al. (1992) posed the problem-solving dilemma they were trying to address in the following way:

If the problems are simple enough to be solved moderately well using their novice strategy, then students see no reason to abandon this strategy—even if the prescribed strategy works as well or better. If the problems are complex enough so the novice strategy clearly fails, then students are initially unsuccessful at using the prescribed strategy, so they revert back to their novice strategy.

Students in the study were taught a five-step strategy: (1) visualize the problem, (2) describe the problem in physics terms, (3) plan a solution, (4) execute the plan, and (5) check and evaluate. The course instructor employed this strategy during the lectures and students practiced it during the discussion sessions. In the discussion sessions, students worked in cooperative groups most of the time.

To address the dilemma, the researchers constructed special problems for the students. These problems, designated "context rich" (Heller & Hollabaugh, 1992), were more complex than typical textbook problems. They contained extraneous information and/or missing relevant information that had to be obtained from some reference source, and they were presented in the context of a "real-world" cover story. Although the cover stories were sometimes far-fetched or humorous, they did provide a reason for solving the problem. In addition, the unknown variable to be found in these problems was often not explicitly identified.

The authors identified six characteristics of problems that affect their difficulty: problem context, problem cues, given information, explicitness of question, number of approaches that could be used, and memory load. The tests the students took were divided into two parts, a group segment consisting of one or two quantitative problems and an individual segment that had a qualitative item and two quantitative problems. When the quantitative problems were rated for difficulty on a scale using the preceding six characteristics, student performance was found to correlate with this rating.

The authors developed a scoring scale for evaluating the written problem solutions with respect to how well the student, or group, had generated a physics description, planned the solution, and carried out the plan. The authors matched problems from the individual test segment to the group problem on the six difficulty-related characteristics and compared the performance of the best student in each group with the group's performance. They found that the groups' solutions were consistently better than the solutions of the best individuals. They also determined which components of the problem-solving process were best done in groups. They found that the greatest differences between individual and group solutions were in the area of qualitative analysis. In other words, the groups were noticeably better at carrying out steps two and three of the five step problem-solving strategy.

Van Heuvelen (1991) reported on a different approach to physics instruction called Overview, Case Study (OCS). Because OCS is a full-scale instructional framework, it addresses more than just problem solving. Nonetheless, problem solving was one of the characteristics specifically investigated. An integral feature of the OCS approach is the explicit development of multiple representations, especially qualitative physics representations, of problems. Two other features are explicit hierarchical organization charts for the different topic areas of physics and an emphasis on active reasoning and peer interaction during the class periods. The results showed that students who had been in the OCS course were noticeably better at solving problems from the College Board Advanced Placement test in physics than students from a conventionally taught course.

A study by Dufresne, Gerace, Hardiman, and Mestre (1992), although not a classroom-based study, explored how novices' use of principles and relations changes when solving a sequence of problems with the aid of a computer. Dufresne et al. (1992), whose work draws heavily on the studies of Chi et al. (1981) and the analysis of Heller and Reif (1984), explored the effect of two different computer aids. One was called the EST, for Equation Sorting Tool, and the other was called the HAT, for Hierarchical Analysis Tool. In their first experiment, 14 subjects in each of three treatment conditions (HAT, EST, and control) worked 25 problems over a period of 3 weeks. All subjects had completed the first semester of a college physics course. The subjects had completed similarity judgment tasks, along the lines of those used by Chi et al. (1981), before the treatment, and the results of that pretest were used to establish three equal-size treatment groups with approximately equal ability. After the treatment, all of the subjects again completed the similarity judgment task. The group who had used the HAT had a statistically significant improvement on the posttest compared with the pretest; they were the only group to show an improvement and were significantly better than the other two groups on the posttest.

The second study reported by Dufresne et al. (1992) was designed to determine whether the improvement with the HAT was uniform for all subjects using the tool. In this experiment the similarity judgment pretest results were used to classify subjects on the basis of their most frequently used type of reasoning: primarily surface judgments, primarily physical principle judgments, or mixed. Fifteen subjects, five from each category, used the HAT, and 15 others, again five from each category, served as controls by solving the same problems without any aids. The main finding of this experiment was that the primarily surface feature and mixed subjects showed significantly greater improvement in physical principle use than the other subjects. In a third experiment, Dufresne et al. (1992) also checked to see whether the problem-solving ability of the subjects in the experiment just described changed as a result of the treatment. They found that the subjects using the HAT showed a significant improvement in problem-solving ability, considering both correct solution and application of correct principle, compared with the control subjects.

The Dufresne et al. (1992) study was not a direct examination of a pedagogical technique, but it does have pedagogical implications. The HAT certainly has potential as a pedagogical tool, considering the results it produced in the study. This is especially true when we realize that the subjects in this study were not given any feedback about their performance but nonetheless moved to more principle-based solution judgments as a result of using the HAT.

Section Summary

All of the studies in this section examine the effect of pedagogical interventions on the development of novices' problem-solving skills. The early studies reviewed do not provide much useful guidance because of their statistical character, obscure theoretical base, and/or short treatment. The more recent studies are much more promising and helpful. They show that getting students to adopt a definite problem-solving strategy, which includes conducting qualitative analyses of problems,

results in better problem solving. These studies also indicate that instruction that pays explicit attention to multiple representations and gets the students to interact with each other while engaged in problem-solving tasks is likely to be more productive than traditional lecture techniques.

ALTERNATIVE PERSPECTIVES ON PROBLEM SOLVING

Up to this point, the problems that have been used in most of the studies discussed have been the standard physics textbook problems described in the introduction. An interesting question is whether these tasks qualify as "problems," in the sense of the definitions given earlier by Newell and Simon (1972) and Hayes (1981), for the experts consulted in many of the studies. They do qualify as problems in this sense for the students who were subjects in the studies reviewed, but to most of the experts these tasks are more in the nature of exercises. One issue relating to the extensive use of these standard problems is whether they are the only type of problem that can or should be presented to students in their first formal physics course and, if not, what other forms problems could take for introductory physics courses. A related issue is whether these standard physics problems are a productive, or the most productive, way to help students learn physics. These are two of the issues considered in this section.

When talking about academic problem solving—especially in physics and chemistry—the problems are almost always well-defined, along the lines of the definition given by McCarthy (1956), tasks that involve mathematical manipulations. In contrast, in fields such as political science most of the problems are ill-defined (Voss, 1989). Ill-defined problems are not found in high school or college physics textbooks. But scientists engaged in scientific research do encounter ill-defined problems. This raises questions about whether the problem-solving skills needed to solve ill-defined scientific problems are similar to those used with well-defined problems and whether the problems students currently work in their general physics courses are really the best ones. Unfortunately, there has been little research on these issues. One of the few studies in this area is Clement's (1988) investigation of how scientists use analogies to solve an unfamiliar scientific problem. We return to these issues and unanswered questions after examining some alternative ideas about what should constitute a productive problem for helping students learn.

This section has four subsections, which, although variable in several ways, share a focus on other types of problems or other aspects of problem solving. The first subsection looks at studies that propose or use tasks that do not fit the standard physics problem model used up to this point. The second subsection argues for using carefully structured worked examples instead of many of the standard problem assignments. In the third subsection, two studies that examine physics problem solving outside the academic environment are reviewed. The fourth subsection looks at studies that examine issues of transfer of problem-solving procedures, effect of presentation context on problem solving, and development of process knowledge as a result of problem solving. All of the studies in this section present issues that must be considered to design effective instructional frameworks for helping students learn to solve physics problems.

Ideas About Different Kinds of Problems

Gil Pérez and Martinez Torregrosa (1983) propose an interesting alternative approach to problem solving. They argue that the currently popular process, which essentially involves linking given data to an unknown, is not consistent with scientific inquiry. Furthermore, they contend that "In fact, pupils in general do not learn how to solve problems but merely memorize solutions explained by the teacher as simple exercises in application" (Gil Pérez & Martinez Torregrosa, 1983, p. 447). They identify three characteristics of scientific methodology: the role of theoretical paradigms, the formulation of hypotheses and the design of experiments, and the collective, social nature of scientific development. Then they propose reformulating standard problems in a way that is more consistent with this view of scientific methodology. Their reformulation involves changing the basically numerical problems to more qualitative ones and incorporating a four-part "solution" process. The first part is to carry out a qualitative analysis of the situation and put forward hypotheses about the mechanisms at work. An interesting question raised by this perspective is whether such an approach would be more effective than the current one in helping students develop the appropriate conceptual understanding.

Garrett (1987) raises the issue of what constitutes a problem in science education. He makes a distinction between problems, for which a solution may not exist, and puzzles, for which solutions do exist. He argues that identifying of a problem is an important and creative skill that is not sufficiently developed in science education because most of the tasks assigned to students are actually puzzles. Garrett (1987) also contends that "Problem-solving can be regarded as an element of thinking but is probably more properly considered as a complex learning activity that involves thinking" (p. 127). Clearly, the issue of the nature of problems involves many different perspectives.

Garrett, Satterly, Gil Pérez, and Martinez Torregrosa (1990) argue that the standard physics problems found in most textbooks are not reflective of scientific reasoning and are actually the source of many students' difficulties. They conducted a study with preservice teachers to try to change the teachers' ideas of what constitutes a problem. They state that the research on the differences between good and poor problem solvers has "not made substantial contribution to the reduction of the high levels of failure encountered among pupils" (p. 1). In exploring the reasons for this result, the authors raise the questions: What is a problem? To what extent is what we teach in class representative of authentic problem solving? They propose that we should approach the solving of problems as investigations and they suggest modifying the standard problems into open-ended qualitative tasks.

Johsua and Dupin (1991) conducted a study to determine how and when class exercises become problems for students.

They followed tenth-grade French students who were studying electric circuits over the course of 1 year. The authors identified three types of exercises. Standard exercises were very similar in content and process to ones worked out in class. Innovative exercises were tasks for which the questions asked and the procedures that had to be used had not been taught in the same manner previously. Implicitly difficult exercises were those that initially appeared to be the same as standard exercises but proved to be unexpectedly difficult.

The authors also identified three groups of students in the course. "Good" students functioned effectively in all areas of the course. "Poor" students either did not study or try, or tried but had one or more conceptual difficulties or background deficiencies. The group of students labeled "average" was the most diverse, so the authors subdivided it into two subgroups: "average plus" students, who had no major conceptual difficulties but were not as quick at identifying and applying the appropriate strategies as the good students, and "average minus" students, who had conceptual problems but managed to function sufficiently to avoid major difficulties.

Johsua and Dupin (1991) present some interesting ideas about the implications of their results for instruction:

The present results confirm the need to concentrate effort on determining what causes problems for the student. We have shown here that this cannot be done by examining the problems themselves—they must be considered in the context of the work actually done in class. A "problem for the students" exists whenever the standard solving procedure previously provided by the teacher is not available or appropriate. The present study has indicated that a problem statement can very easily be outside such a limited framework so that what appears beforehand to be a minor deviation from the standard model can significantly perturb the solving conditions for the student. (p. 299)

The authors point out that any task perceived by the students as a "problem" is going to contain some element of newness for the students. Consequently, in an instructional setting we should "watch out for such 'newness' while in fact being as conformist as possible" (Johsua & Dupin, 1991, p. 299). Approaching instruction in this way clearly presents a dilemma. Should our instruction involve only tasks that are clearly established minor modifications of previously presented tasks? If so, how useful are such tasks for helping students learn to apply their knowledge in new contexts? Or should we accept that many students will find many of the assigned tasks more challenging than we might intend but have some assurance that students who can handle the tasks are learning to deal effectively with challenging new tasks? Johsua and Dupin (1991) argue that this really is not an either-or situation but that more time and effort should be devoted to helping the students become functional with the concepts.

This leads us back to the question of whether students think in conceptual terms when solving physics problems. Maloney and Siegler (1993) used problems that varied somewhat from the standard type of problem to explore what strategies (i.e., essentially what algebraic equations) novice subjects choose among and what keys the choices. They had two different groups of novice subjects, one composed of science or engineering majors and the other composed of nonscience majors.

All subjects were undergraduates at a selective private university. The problems used varied from standard physics problems in that they contained extraneous information, and the task in the problem was to compare five objects to detemine which had the largest value for a particular quantity and which had the smallest value. In contrast, in standard problems the task is to determine the specific value for one object or system. Two-thirds of the problems had the question phrased in everyday language, and the other third had the question phrased in explicit physics language, that is, asking about momentum and kinetic energy. Three different objects—bullets fired from rifles, medicine balls thrown by a machine, and toy cars shot from launchers—were used in the problem texts.

Maloney and Siegler (1993) found that the subjects had multiple strategies for these problems. The different objects in the different versions had no effect on which strategy subjects employed, which disagrees to some extent with the findings of Chi et al. (1981, 1982), but the disagreement may be due to the differences in the problems. The change from phrasing the question in everyday language to using explicit physics terms did affect the strategy choice on the problems involving kinetic energy but had significantly less effect on the momentum questions. The reason is that the subjects tended to use momentum on most of the problems phrased in everyday language, so no change was generated when the question explicitly asked about momentum, but explicitly asking about kinetic energy did produce a change. This change was stronger for science majors than for nonscience majors. This finding of the existence of multiple strategies and the shifting from one strategy to another based on cues in the problem is similar to the results of Anzai and Yokoyama (1984). But the fact that the strategies the subjects used in the Maloney and Siegler (1993) study were mathematical formulas adds evidence to other findings that students are not learning physics concepts but rather rote algorithmic procedures.

Summary. The studies just reviewed call into question the value of having students solve standard physics problems. Garrett et al. (1990) argue that these standard problems do not reflect the nature of scientific problem solving and so have limited value. They propose changing the nature of the problems assigned to students by making them more open-ended qualitative tasks that also involve such processes as problem identification. But how do we get students to learn to cope effectively with such open-ended tasks? Johsua and Dupin (1991) make a case for tasks that are similar to those used in the instructional context, because tasks containing an element of "newness" are problematic for students. The study by Maloney and Siegler (1993) is consistent with this perspective; they found that students, who had the appropriate algorithm in memory, failed to apply their physics knowledge to comparison tasks phrased in everyday terms rather than explicit physics terminology. Open-ended tasks involving problem identification are likely to have even stronger elements of "newness" than has been the case for traditional problems, so introduction of such tasks raises a number of questions about what, and how, students will actually learn from them. Although the studies in this section vary in several ways, they raise interesting questions

about what students learn by working standard physics problems and what types of tasks might be used to replace the standard problems.

The Value of Worked Examples

Sweller and his colleagues have also raised the question of how useful standard physics problems are for helping students learn physics. They argue, in a series of articles (Sweller, 1988; Sweller & Levine, 1982; Sweller, Mawer, & Ward, 1982; Ward & Sweller, 1990), that the typical problems assigned in physics courses are actually counterproductive for learning physics. Novices typically employ means-ends analysis in solving assigned problems, and this process makes such a demand on the cognitive resources that the student doesn't have sufficient resources left to devote to acquiring the relevant knowledge.

Sweller et al. (1982) investigated the development of students' strategies with practice on a restricted group of problems. They used two categories of constant-velocity kinematic problems: in one, the final speed and the time were given and the unknown was the distance traveled; in the other, the distance traveled and the time were given and the unknown was the final speed. Subjects were told to use only three equations in solving the problems.

In the first experiment, subjects worked a large number of identical problems of each type. The results of this experiment provided evidence for a transition from the use of a means-ends analysis–based strategy to a forward-working strategy more commonly used by experts in a field. The problems in this first experiment had specific goals, such as find the final speed.

Two groups of subjects were used in the second and third experiments. One group worked the set of problems with specific goals for the problems. The other group was told to "calculate the value of as many variables as you can." Significantly more of the subjects in the no-goal group developed forward-working strategies, and these subjects needed fewer steps to solve the problems in which they employed the forward-working strategy. Both of these characteristics are associated with experts' problem solving.

Ward and Sweller (1990) describe a sequence of investigations designed to test the relative effectiveness of worked example problems versus conventional homework problems. They used the topics of geometric optics and kinematics. Their basic argument is that having students work conventional problems is not the best way for the students to learn the relevant concepts and strategies. The authors' theoretical perspective is that the basic components of problem-solving skill are an effective set of schemas and the automation of relevant rules. The schemas enable expert problem solvers to recognize problems and associated solution methods. Rule automation takes place when an individual can use a rule (i.e., a mathematical procedure or physical principle) without needing to put much conscious effort into the application. The authors argue that someone learning a new topic needs to work on acquiring the relevant schemas. If applying appropriate rules in the solution of the problems encountered requires definite conscious effort, that will detract from the acquistion of the schema. For example, having to use means-ends analysis to solve a problem focuses the student's attention on the specific goal of the problem rather than the nature of the problem as a whole.

The investigations reported in the Ward and Sweller (1990) article were designed to determine what characteristics effective worked examples should have to promote scheme acquisition and learning. All five of the experiments reported were classroom-based studies using the same basic format. Students were instructed in the topic, given either worked examples or conventional problems for homework, and on the next day tested on the topic.

The first investigation compared worked examples to conventional problems in the domain of geometric optics. The group who received the worked examples were superior on the posttest on problems similar to the ones considered during instruction. The second experiment was designed to check the effect of worked examples versus standard problems on students' ability to solve problems structurally different from those used during the instruction phase. As was true in the first experiment, the students who received the worked examples were superior in solving these problems. At that point the authors contend:

We suggest the simplest explanation of these and previous findings is that: (a) schema acquisition and rule automation are essential to problem-solving skill; (b) efficient problem solving via a means-ends strategy requires a search for operators (rules) to reduce differences between problem states; (c) this search, although efficient from a problem-solving perspective, is an inefficient learning device, because it inappropriately directs attention and imposes a heavy cognitive load that interferes with learning; (d) with respect to schema acquisition and rule automation, worked examples can appropriately direct attention and reduce cognitive load. (Ward & Sweller, 1990, p. 17)

Experiments 3 through 5 were designed to determine the characteristics of worked examples that facilitate schema acquisition and rule automation. A previous study (Tarmizi & Sweller, 1988) indicated that the effectiveness of worked examples essentially disappeared if the examples required students to divide their attention between different sources of information. That is, if the worked example contained both equations and verbal descriptions that were at different places in the example but had to be integrated mentally to acquire the schema, it would be no more effective than having students work conventional problems. Experiment 3 was designed to test this idea.

Four types of problems from mechanics—linear motion, multiple-step linear motion, projectile motion, and collision—were used in this experiment. The worked examples in these areas had both verbal explanations and equations. The results showed that the worked examples were no more effective than conventional problems for the two types of linear-motion problems and were actually less effective for projectile motion and collision problems. This was hypothesized to be due to the effort students had to put into integrating the two different types of information, verbal and mathematical.

Experiment 4 was designed to test the latter hypothesis. Two kinds of worked examples were used in this experiment, the typical kind used in experiment 3 and one in which the two different types of information were integrated, by being spatially associated, in the example. These integrated worked examples were found to facilitate students' learning. The final experiment also tested the hypothesis about the critical characteristic of worked examples. In this experiment, the worked geometric optics examples used in experiment 1 were altered by inserting a verbal explanation into the diagrammatic presentation; which reduced their effectiveness. The results of this experiment further supported the hypothesis.

Summary. Sweller and his colleagues make a strong case that goal-specific problems are counterproductive for schema acquisition because too many of the students' cognitive resources are tied up in means-ends analysis, leaving insufficient resources to attend to what knowledge is being used and how it applies. These authors argue for the use of carefully constructed worked example problems to help students acquire useful problem schemas. But these worked examples must have the relevant information presented in an integrated manner so that students do not have to try to put different representations, such as diagrams, verbal descriptions, and equations, together into a coherent package.

Real-World Physics Problems

Essentially all of the problems in the studies examined so far have been academic, or textbook, problems. The next two studies expand the perspective somewhat by looking at some real-world physics problems. Two different physics domains are involved, mechanics and electronics.

Hegarty (1991) examined the knowledge and information-processing mechanisms involved in solving mechanics problems. She classified mechanics problems into two categories: everyday mechanics problems and formal problems. The latter are essentially the standard physics problems found in textbooks. Everyday mechanics problems involve such things as designing, operating, and troubleshooting machines. Hegarty also argued that there are three types of conceptual knowledge about mechanics. Intuitive conceptual knowledge is what people learn about motion and mechanical phenomena from observing the world around them. Practical knowledge is acquired from actively interacting with mechanical devices, including diagnosing what is wrong when the machine fails and repairing the deficiency. The third type of conceptual knowledge is theoretical and is usually acquired from formal instruction.

Hegarty (1991) contrasts conceptual knowledge with procedural knowledge, which she views as "mechanisms of problem solving." She looked at two general mechanisms—spatial mechanisms and mechanisms for accessing problem-relevant information. The spatial mechanisms serve two functions: they enable the problem solver to identify information by location, which greatly reduces the search process as described by Larkin and Simon (1987), and they help the solver infer the behavior of the system. This latter function enables the solver to construct a better mental model of the system. With regard to mechanisms for accessing problem-relevant information, Hegarty (1991) argues that novices have two sets of knowledge items—intuitive and theoretical—from which they must construct a problem representation. These two types of knowledge are not integrated, so the novice must choose one or the other. In contrast, experts have integrated their everyday and formal knowledge, so they develop representations by working within this integrated knowledge base.

Hegarty (1991) then argues that an individual's conceptual knowledge changes as he or she learns in a semantically rich domain such as mechanics. She argues that this change takes place in three ways: (1) the level of specificity of concepts broadens and is tied to underlying physical principles rather than to surface features, (2) concepts change from qualitative to quantitative, and (3) concepts are applied more consistently to appropriate situations. Clearly, mature understanding of the concepts in a domain is important for successful problem solving in the domain. As one of her conclusions, Hegarty (1991) states:

As in other problem-solving domains, success in solving mechanical problems is due not only to possession of the relevant domain knowledge, but also to the ability to access this knowledge as a problem demands. (p. 281)

This is by now a familiar theme.

In another study involving some real-world problem solving, Lesgold and Lajoie (1991) looked at troubleshooting electronic equipment. This activity involves both practical experience with the equipment and theoretical knowledge of how the system should behave. Clearly, such problems are quite different from the kinds of tasks students are given in academic settings, but the conceptual knowledge is physics, and it turns out that there are definite similarities to other types of problem solving. Lesgold and Lajoie (1991) report on the results of a large project called SHERLOCK, which involved many researchers over a number of years. The project employed a number of investigative techniques including diagnostic tests, actual troubleshooting tasks with real equipment, and observations of students working on a computer tutor developed as part of the project. Rather than contrasting the performance of experts and novices, most of the focus of the project was on contrasts among relatively inexperienced technicians who were still learning their job.

Although this kind of problem solving is different from that in most of the other studies in this chapter, there are also some similarities. In their conclusions, Lesgold and Lajoie (1991) point out that "expertise in complex technical domains is multifaceted" (p. 311). They identify three facets: conceptual knowledge, procedural knowledge, and experience with problems from the domain. All three of these facets have been found in other investigations of problem solving.

Lesgold and Lajoie (1991) make a couple of important points in their presentation. First, "Real world problem-solving skill is highly adaptive. It requires considerable knowledge, but not necessarily all of the knowledge that seems, from an academic perspective, to be important" (p. 298). This contrasts with the rigid approaches students often adapt to problem solving in

academic situations, as Reusser (1988) describes. The second point concerns motivational approaches to engender confidence and persistence in novice problem solvers. The authors comment that "we have encountered experiences in which people with low levels of job experience were paralyzed in the face of a task that they were capable of doing. They just did not realize that successful performance was possible" (p. 302). Certainly such difficulties are common for students studying physics. How do we help novices accurately estimate the difficulty of a problem and develop the persistence to stay with a problem they can solve? These issues are related to the difficulty Heller et al. (1992) identify in getting novices to use a general plan for problem solving. A general plan, a tendency to persist in the face of difficulty, and a habit of estimating problem difficulty are attitudes and skills that we hope students will transfer from one domain to another. The next two studies investigate another aspect of transfer.

Summary. The two studies just reviewed take physics problem solving beyond the academic situation by considering how people use physics knowledge to solve real-world problems. These two studies emphasize the need for procedural knowledge as well as conceptual knowledge, and they also highlight the importance of experience with problem solving in a complex domain such as physics.

Issues of Transfer, Context Effects, and Process Knowledge

One of the perennial questions in any research on learning is, "Will this knowledge or skill transfer?" Bassok and Holyoak (1989) investigated transfer of use of algebraic equations between algebraic contexts and physics contexts. They looked at how often subjects used the same solution procedure for isomorphic problems involving constant acceleration in a straight line and arithmetic-progression word problems. They conducted three experiments. In the first, subjects studied either the arithmetic-progression relation in an abstract algebra context or the constant-acceleration relation in a physics context. Subjects were then given a posttest that had problems from both the context they studied and the other context. Subjects who had studied in the algebra context transferred use of the relation to the physics problems significantly more frequently than the physics-context subjects transferred the relations they studied to the algebra problems. In a second experiment the researchers found evidence that there was nothing inherently more difficult about the physics context than any other learning context. In their third experiment they explored how the embedding context affected transfer. They employed three conditions differing in how specifically the introduction of the posttest tasks focused on motion problems. They found a small reduction in transfer when the embedding context was very specific to motion problems.

Bassok (1990) conducted a follow-up study in which she tried to identify the factors that facilitate transfer of mathematical problem-solving procedures that have been learned in content-specific situations. Specifically, she explored what factors in the two contexts, learning and transfer, facilitated spontaneous transfer of the use of an algebraic procedure. In the follow-up study she used financial contexts as well as physics contexts. She found that subjects in the financial context were better able to transfer what they learned to the algebra problems than the physics-trained subjects were. The type of variable that the tasks contained was the main factor affecting transfer. Bassok identified two types of variables: intensive and extensive. She argued that quantities such as "potatoes per day" are extensive quantities because the two quantities involved, potatoes and days, are normally separate and are molded into one quantity, the ratio, only in the context of the problem. In contrast, a quantity like speed is intensive because the two quantities it involves, distance and time, are combined into a single "conceptual entity."

Bassok (1990) found that transfer between situations in which the quantities were of the same type was relatively straightforward. In contrast, when the learning tasks and transfer tasks had different types of quantities, spontaneous transfer was blocked. Transfer could be produced in these situations by providing a hint, but even then it required some work to abstract the transferable features from the learned tasks and map them onto the transfer tasks.

Because transfer is one of the major goals of physics instruction, these studies are of interest. The results indicate that students should be able to transfer the mathematical procedures and skills they have learned in their mathematics courses to physics, and this is the common experience of physics instructors. Students who have learned their mathematics well seldom have difficulty with the mathematical facets of learning physics, and students who have not mastered the necessary mathematical skills usually have significant difficulty. Where the mathematically competent students often do have difficulty is in recognizing when to employ the various relations.

The question of transfer is especially important for problem solving because each new problem requires the problem solver to apply some knowledge in a new way, or in a new context. Students frequently have difficulty even transferring concepts and procedures from one part of physics to another. For example, students often study vector summation as an abstract mathematical procedure at the start of a general physics course. Then they have to use these ideas when summing forces in Newton's second law. But when the same ideas—application of Newton's second law and vector addition—arise again in the study of electrostatic fields and forces, many students essentially start over. Is this because they never really learn the ideas in the first place? Or is it because they don't recognize the electrostatic context as appropriate for the transfer of the earlier ideas? Or is transfer blocked by conceptual difficulties in understanding the new context? Does the lack of transfer indicate something about the effectiveness of standard problems for helping students learn the concepts?

Another question related to the transfer issue is how strongly the context in which something is learned affects what the individual learns and does with the knowledge. Reusser (1988) takes a broader view of classroom problem solving as both a cognitive task and a social-cognitive activity. He argues that both the nature and content of the task and the context in which the task is presented affect how students approach the problems. He conducted a study in which he presented one of four versions of a kinematic task involving the calculation of average speed for a cyclist who travels three equal-length segments at different speeds. Because the times required to travel the three segments vary, simply averaging the three speeds is an incorrect procedure, but it is also a popular one. Two versions of the task were constructed. In the simpler (S) version the three speeds were explicitly stated; in the complicated (C) version the initial speed was stated, but the other two speeds were expressed in terms of the initial speed or the first two speeds. In other words, the speed on the second and third segments of the trip had to be calculated in the C version. The problems were presented to the subjects with one of two statements designed to induce an expectation that the problem would either be difficult or relatively easy. The problems were administered to 68 junior high school students and 51 college students.

In analyzing the students' solutions, any subject who calculated, or attempted to calculate, the elapsed times for the three segments of the trip was considered correct. The results indicated that both the complexity of the task and the expectation of the level of difficulty affected performance. More subjects worked the simple task correctly, but more subjects worked the task, whether simple or complex, correctly when they expected the task to be of high difficulty. The reason for the latter is hypothesized to be that students would avoid the simple averaging trap when expecting the task to be difficult. Reusser (1988) then considered the context of classroom problem solving and described several other small-scale investigations on the effect of context, and violations of expected context, on students' problem-solving efforts. Reusser made several interesting observations about classroom problem solving. First

These examples show that there are factors in the whole classroom setting which can heavily impair the quality of comprehension in problem solving. The studies and observations highlight the difficulties that students of all ages have in both rejecting an ambiguous or apparently senseless or unsolvable task and in simply admitting that one is unable to come up with a sensible solution. (p. 328)

Second,

Also, students get almost no experience in solving ill-defined or unsolvable textbook problems. Almost every systematic dealing with ambiguity and unsolvability is factually excluded from textbooks, from curricula, and from the school setting where it even seems alien. (p. 328)

Third,

Sometimes it is easier to solve a problem than to understand it. (p. 328)

Reusser (1988) also pointed out that students often use information about whether the numerical values they are calculating come out to be "nice" numbers as a clue to whether they are proceeding correctly or not. These results emphasize the narrow scope of the problems students encounter in their courses as well as some of the unintended things students learn from doing such tasks.

One other study does not exactly fit into any of the earlier categories or perspectives. Mohapatra (1987) conducted a study to determine whether students' "process knowledge" of problem solving could be determined from think-aloud studies of students actually solving problems. The author assumed that students who had studied physics and worked problems had acquired some problem-solving process knowledge during the experience. Mohapatra (1987) identified three areas of problem-solving process knowledge: observation, which involved being able to use either classroom or life-centered experience to identify irrelevant information in a problem situation; identification, which involved being able to set up suitable subgoals for solving the problem; and analysis, which involved the ability to evaluate the solution to the problem.

Mohapatra (1987) conducted a study using two groups of students who were 1 year apart (17- or 18-year-olds) in their studies and, consequently, differed in the amount of problem-solving experience they had. The students worked a problem that had irrelevant information, involved two unknown quantities, could involve equations of motion or energy relations, and had an answer that had to be carefully interpreted. The more experienced students had better process knowledge in all three areas.

Summary. Three different issues, transfer of knowledge, context effects, and process knowledge, were explored in the studies in this subsection. Bassock (Bassock, 1990; Bassock & Holyoak, 1989) identifies the nature of the variables involved in the tasks as a critical feature affecting transfer. Reusser (1988) presents the idea that the context in which the task is presented strongly affects how students approach the task. He also points out that any task that is ambiguous or unsolvable is typically excluded from instructional situations. In contrast, in real-world situations the ability to identify ambiguity can be important and extremely difficult or unsolvable problems can be common. Mohapatra (1987) examined what he called process knowledge. He used tasks that had elements of ambiguity and high difficulty and found that more experienced problem solvers were better able to identify irrelevant information, set up appropriate subgoals, and evaluate the appropriateness of the solution.

Section Summary

This section reviews studies that have employed, explored, or advocated alternative types of tasks instead of the standard physics problems usually used in instructional settings. Several of the studies in this section argue that standard problems are not effective tools for helping students learn the relevant concepts. Garrett et al. (1990) argue that such problems do not adequately reflect the nature of real scientific investigations and, consequently, fail to focus the students' attention on the relevant concepts, principles, and relations. Sweller and his colleagues argue that the effort students have to devote to means-ends analysis to come up with the numerical goals of the problems detracts from the cognitive resources they need to learn the concepts. Johsua and Dupin (1991) argue that "new-

ness" in problems can distract students from learning. All three groups suggest the incorporation of more qualitative components in the problems students are assigned. Reusser (1988) goes even further, arguing for the inclusion of ambiguous and/or unsolvable problems. Some of the features encouraged by the authors in this section, such as emphasis on qualitative components and inclusion of irrelevant information, are incorporated in the instructional studies of Van Heuvelen (1991) and Heller et al. (1992) reviewed in the previous section. Those studies support the value of these ideas. Some of the other ideas and features proposed in the studies in this section remain more speculative.

Two studies of problem solving outside the academic setting were reviewed in this section. Hegarty (1991) compared mechanics problem solving in academic and real-world contexts. She argued that novices have separate banks of conceptual knowledge—one intuitive, acquired from real-world experience, and the other theoretical, acquired in formal instruction—that they have to access to construct a problem representation. In contrast, experts have their knowledge integrated. Lesgold and Lajoie (1991) also explored some real-world problem solving, but they were studying how electronic technicians develop skill in troubleshooting equipment. They emphasize the adaptive character of real-world problem solving and the importance of developing confidence in novices so they have the persistence necessary to solve the problems they encounter.

Comparing problem solving in academic settings to real-world problem solving brings out questions about transfer and the effect of context. Bassok (Bassok, 1990; Bassok & Holyoak, 1989) found one factor affecting transfer, the nature of the variables involved in the tasks. But there are obviously other factors that need to be identified and explored. Reusser (1988) investigated the effect of context on performance by providing the students with explicit expectations about difficulty and using two problems that differed in complexity. He found that expectation had a definite effect on the students' success. The studies reviewed in this section identify some important aspects of transferring knowledge and of the effect of context on problem solving, but much still needs to be investigated about these issues.

DISCUSSION

This review focuses on research on problem solving in physics. I now discuss some issues concerning instruction from my perspective as someone who is involved in both research in physics education and college-level physics instruction. These issues are approached from the dual perspectives of what we have learned from research and how that information can help us teach physics more effectively and more efficiently.

In the following quotation, chemical engineering thermodynamics can be replaced by physics. It provides a useful introduction to this discussion because it represented a real possibility at the point in time when the research reviewed in this chapter was beginning in earnest.

It might be questioned whether there is anything special to be said about problem solving in particular problem domains. It is not implausible to suppose that the processes a person uses to solve problems in chemical engineering thermodynamics are the same ones he uses to play chess, or even to compose music. According to this view, problems are solved in any domain by using common problem-solving processes that then draw upon specific knowledge of that domain—and the specific knowledge is simply the information to be found in a good textbook on the subject. (Bhaskar & Simon, 1977, p. 194)

We now have significantly better insights into the place of domain knowledge and general problem-solving procedures in solving physics problems, and they establish the critical importance of domain knowledge, as well as general problem-solving processes.

Research has supported a number of ideas about problem solving in physics. First, domain knowledge is an integral part of physics problem solving (Bhaskar & Simon, 1977; Chi et al., 1981; de Jong & Ferguson-Hessler, 1986; Heller et al., 1992; Larkin, 1980; Reif & Heller, 1982). Second, experts have extensive domain knowledge that is organized hierarchically with general physical principles at the top of the organization. In contrast, novices have significantly less domain knowledge and it is organized around surface features of situations (Chi et al., 1981; Chi et al., 1982; Elio & Scharf, 1990; Ferguson-Hessler & de Jong, 1987; Reif & Heller, 1982). Third, novices use means-ends analysis applied to the equations they have been taught as their primary problem-solving procedure (Larkin, 1981; Larkin et al., 1980a, 1980b; Simon & Simon, 1978). Fourth, carrying out a qualitative analysis and developing a good physical representation are important for problem solving with understanding (Finegold & Mass, 1985; Heller et al., 1992; Larkin, 1979, 1983, 1985; Larkin et al., 1980b). Fifth, procedural knowledge as well as declarative (propositional) knowledge is needed in order to know when and how to apply the declarative knowledge (Chi et al., 1989; de Jong & Ferguson-Hessler, 1986; Ferguson-Hessler & de Jong, 1990; Heller & Reif, 1984). Sixth, there are identifiable differences between good and poor novice problem solvers (Chi et al., 1989; Ferguson-Hessler & de Jong, 1990; Finegold & Mass, 1985; McMillan & Swadener, 1991). Finally, standard problems are often largely ineffective in helping novices learn physics concepts (Heller et al., 1992; McMillan & Swadener, 1991; Sweller, 1988; Sweller & Levine, 1982; Sweller et al., 1982; Ward & Sweller, 1990). Some of these ideas were common knowledge for physics instructors before research provided an empirical foundation for them, but others contradicted physics instructors' ideas.

We now have a significantly better understanding of how people solve physics problems than we did 15 years ago. But an interesting question is whether we have better ideas about how to help students learn how to solve problems. That is, we know more about what it takes to solve problems effectively, but do we know more about how to develop these problem-solving skills in students?

The early studies reviewed here are of limited value as guides for developing better instruction. This is primarily because those early studies were mainly statistical "snapshots" of large groups' initial and final states and were usually exploratory with little theoretical foundation. Statistical snapshots provide little information about the mechanisms involved in the transitions from initial to final state. Studies with weak theoretical foundations often provide only vague guidelines for improving instruction.

Early studies of problem solving (e.g., Bhaskar & Simon, 1977; Newell & Simon, 1972; Polya, 1945; Reif & Heller, 1982) established the importance of domain knowledge for effective problem solving in any specific topic domain. They also provided initial models of general problem-solving strategies that have been useful for subsequent studies.

The first results that are really useful from an instructional viewpoint came from the early information-processing research. The studies in this category have a theoretical framework that takes the computer as a model and analogy for the way humans think. (It is important to keep in mind that the extent to which research on human problem-solving skills is theory based is different from the extent to which research in science, such as physics, is theory based. Nonetheless, there is a difference between educational or psychological research that is exploratory and that which fits within a particular theoretical framework.)

This research focused on how an information processor, human or computer, could take a natural-language problem statement, translate it into an internal representation on which it could operate, carry out the appropriate operations, and then output the result. This research identified the importance of the translation phase and made clear the domain-specific nature of the operations that could be carried out on a representation.

Larkin (1983) provided clear evidence for the importance of appropriate representations to the solution of problems. Her expert subjects did not get to the point of writing an equation for the difficult problem until they had found an applicable representation. Heller and Reif (1984) showed that constraining novice subjects to an explicit prescription for developing a physical description could significantly improve their ability to solve problems. Thus, the importance of proper representations seems to be clear, but how do we get students to develop these representations?

Several studies (e.g., Johsua & Dupin, 1991; Larkin, 1981; Reif et al., 1976) conclude that novices have a strong tendency to employ means-ends analysis with the mathematical relations they know. Such an approach is different from one in which time and effort are devoted to developing a qualitative representation before any equations are used. How can instructors get students to alter their preferred method?

Wright and Williams (1986) provided evidence that adopting a systematic approach to problem solving was productive for students. In their study the students were strongly encouraged to adopt the systematic approach because use of that approach was part of their evaluation. This is one way to get students to learn such a method, but will this approach convince students of the value of the method sufficiently that they will continue to use it beyond the course? If not, how much experience is necessary to convince them of the usefulness of the method?

In Heller et al. (1992), the students were encouraged to use the prescribed method by seeing their instructors use it and by use of supplementary aids. In addition, the problems the students worked provided a reason for using the method. It would be interesting to find out if these students would still use parts of the method a year later.

Another outcome of the information-processing studies is the organization of domain knowledge. The importance of problem schemas is illustrated in the work of Chi et al. (1982), in which the novices used surface features, while the experts classified problems using problem schemas incorporating relevant physical principles and relations. Reif and Heller (1982) gave a detailed account of how the knowledge of one domain (mechanics) could be structured hierarchically and of the value of such a structure. The organization they present facilitates the use of problem schemas that involve the important principles and relations because those features are found in the higher levels of the organization. However, Ferguson-Hessler and de Jong (1987) argue that for novices, organizing knowledge around problem schemas, rather than an abstract hierarchical organization, is more efficient. This issue merits additional investigation.

How do we help students develop whichever knowledge structure (abstract hierarchical or problem schema) we believe is best? Do we need new approaches to instruction such as the Overview, Case Study approach (Van Heuvelen, 1991), or can modifications in the way we use problems in a traditional context accomplish the goal?

One theme that emerges from this review is that students have been expected to learn the conditions for application and the limitation on application of principles and relations on their own. Simon and Simon (1978) pointed out that students are usually taught principles and relations explicitly but conditions for and limitations on the application of these principles and relations are taught implicitly. Several of the studies reviewed indicate that, not surprisingly, many students fail to acquire this procedural knowledge. (Chi et al., 1989; Ferguson-Hessler & de Jong, 1990).

Chi et al. (1989) found that good problem solvers expanded worked examples with their own explanations of the conditions and consequences of actions that were left implicit in the examples. Heller and Reif (1984), in their prescriptive analysis, devote significant effort to making explicit the conditions for determining when contact forces are acting on an object. Ward and Sweller (1990) argue that effective worked examples need to integrate verbal presentations, equations, and diagrams rather than having different types of information that the students need to integrate on their own. All of these findings point to a need to study what constitutes good worked examples, that is, worked examples that will effectively and efficiently promote learning.

Another area of interest is the dichotomy between learning to solve physics problems and solving problems to learn physics. It is often assumed that students who can solve the problems at the ends of the chapters in physics books have, and are using, a solid understanding of concepts. The research on students' conceptual understanding (e.g., Halloun & Hestenes, 1985; Peters, 1982; Trowbridge & McDermott, 1980), has made it clear that such an assumption is overly optimistic. Two questions arise: What is the place of standard problems in physics instruction? What form should any new problems take?

A number of researchers question the use of standard problems in the traditional way. Sweller and his colleagues (Sweller, 1988; Sweller et al., 1982; Ward & Sweller, 1990) conclude that standard problems can actually be counterproductive. They believe standard problems encourage students to use means-ends analysis, which places a heavy load on their cognitive resources and leaves them few resources to devote to learning relevant problem schemas. Ward and Sweller (1990) suggest that worked examples may prove more useful than standard problems.

Garret et al. (1990) argue against the use of standard problems, but their suggested alternative is different. They contend that problems should be converted into exercises that more closely resemble the kinds of activities scientists engage in when doing scientific research. They suggest converting standard problems into open-ended tasks that require qualitative analysis and consideration of how variables could be measured in actual situations.

Before considering how to modify or replace standard problems, we should ask what students actually learn from working them. Most studies on problem solving have investigated how novices work problems and what knowledge they use when trying to solve problems, but I am not aware of any studies to determine specifically what students learn from solving standard problems. In spite of any shortcomings of standard problems, some of the things students learn from working them are things we want them to learn. Can students develop solid, thorough declarative knowledge bases before they are given problems to solve, or does solid understanding require that they attempt to apply the knowledge from the domain? This question relates to the Chi et al. (1989) and Larkin (1979) studies as well as others. The Chi et al. (1989) study implies that working with problem examples is an important part of learning the declarative knowledge in an area. Does that also mean that assigning problems for students to work on their own is an important part of learning? Other studies (e.g., the expert-novice studies) seem to imply that students need to have a solid knowledge base to solve some problems effectively.

How is this knowledge base to be established? How important is it to help students develop well-organized knowledge structures to enable them to solve problems?

Changing the numerical values from one problem to another is the smallest change that can be made in a problem and probably does not modify the problem sufficiently to cause a student to think of it as a new problem. However, changing any other feature of the problem, such as the object involved, the direction of motion, or the agent exerting a force, makes a new problem for at least some of the students. The student needs to have problem schemata that are organized around physical principles in order to ignore changes in surface structure of problems.

Reusser (1988) and Lesgold and Lajoie (1991) take these considerations further. Reusser's studies raise a number of questions. To try to wean students from using contextual cues, should we use ambiguous or unsolvable problems with novices? How would that affect students' confidence, persistence, and motivation? As Lesgold and Lajoie point out, students find it hard to estimate the difficulty of standard problems and to know when to persist; won't ambiguous and unsolvable problems aggravate the situation?

In addition, any improved instructional procedures derived from the research studies would require more time or effort for the students to learn. Consequently, it would be impossible to accomplish all of the suggested changes and still have the students learn the same amount of course material. This means we need to make some difficult decisions about what we want the students to learn. If good problem-solving ability is our goal, we may need to cover fewer topics. If only some "basic" level of problem-solving skill is required, that would allow the same number of topics to be examined. If one or two parts of the problem-solving process are considered most important, students could sometimes work on those components (e.g., planning or executing) and ignore the other parts of the process. Or students could learn to outline a solution path for a problem without doing all of the calculations. (This latter might actually be more difficult than completely solving the problem.)

Chi et al. (1989) and Ferguson-Hessler and de Jong (1990) investigate what beliefs the poor performers had about learning physics. Hammer (1989) suggests that "metaknowledge" or beliefs affect how students approach learning physics, including how they approach problem solving. If the student thinks of physics as a collection of formulas and rules without any real coherence, he or she will approach problem solving as a matter of finding the right formula. With such an approach there is little use for conceptual analysis of the problem situation.

Some poor performers may not feel any need to understand the material as long as they have procedures they feel they can use to work the problems. How would the approach of Heller et al. (1992) or techniques such as those suggested by Garrett et al. (1990) or Reusser (1988) affect these students? Would these students benefit from the worked examples proposed by Ward and Sweller (1990) more than by working problems?

Although the research to date is very informative, a number of issues are still in need of careful investigation. Among these are the place of general heuristics; the effect of alternative conceptions on problem solving; the place of problem solving in learning physics concepts, principles, and relations; "larger" cognitive issues such as epistemological beliefs; and noncognitive issues such as questions of attitude.

A number of researchers (Chi & Glaser, 1985; Schoenfeld, 1978; Simon & Simon, 1978) point out that heuristics are often important for solving problems. The question arises again, what do we expect students to learn from their problem-solving experiences? Do we expect them to learn the concepts, principles, and relations of physics, or do we expect them to learn general problem-solving skills such as heuristics like means-ends analysis, subgoaling, working backward, decomposing? Students are seldom given sufficient time on any one topic to learn both effectively. If learning heuristics is one of their goals, physics instructors need to be explicit about what the heuristics are and how they are used.

There is now a rich literature on the alternative conceptions students bring to the study of physics (see Chapter 5 in this volume). Snider (1989) has reviewed some of the alternative-conception literature, as well as some of the problem-solving literature, in terms of how problem solving could be used to help students learn to modify their conceptions. Clearly, this perspective considers problem solving as a vehicle for learning. But solving a problem in physics requires the construction of an effective representation, and that in turn requires accurate and well-organized domain knowledge. Often students have no real idea that they are using an inappropriate representation. The fact that they have failed to solve the problem will not tell them their conception needs to be modified.

The research reviewed in this chapter focuses on cognitive aspects of physics problem solving, but issues of motivation, persistence, and confidence, among others, are also relevant to improving instruction. What kinds of problems, presented in what ways, will support and foster these traits? Some ideas are starting to appear (e.g., Heller et al., 1992; Van Heuvelen, 1991) about how to modify our instructional approaches, but much needs to be done.

One of the things that this review makes clear is that physics instructors need to consider carefully what their goals for employing problem solving are in order to determine what can reasonably be accomplished. If we want to stress problem-solving skills, we need to make general problem-solving procedures, such as heuristics, an explicit part of our instruction and provide opportunities to practice these procedures. If we want to use problem solving to help students learn the concepts, principles, and relations, we may need to alter the form of the problems that are assigned. What the nature and form of such new problems should be is just being explored. ·

References

Anzai, Y., & Yokoyama, T. (1984). Internal models in physics problem solving. *Cognition and Instruction*, 1(4), 397–450.

Ausubel, D. P. (1968). *Educational psychology, A cognitive view*. New York: Holt, Rinehart & Winston.

Bascones, J., & Novak, J. D. (1985). Alternative instructional systems and the development of problem-solving skills in physics. *European Journal of Science Education*, 7(3), 253–261.

Bassok, M. (1990). Transfer of domain-specific problem-solving procedures. *Journal of Experimental Psychology: Learning, Memory, & Cognition*, 16(3), 522–533.

Bassok, M., & Holyoak, K. J. (1989). Interdomain transfer between isomorphic topics in algebra and physics. *Journal of Experimental Psychology: Learning, Memory and Cognition*, 15, 153–166.

Bhaskar, R., & Simon, H. A. (1977). Problem solving in semantically rich domains: An example from engineering thermodynamics. *Cognitive Science*, 1, 193–215.

Boyle, R. K., & Maloney, D. P. (1991). Effect of written text on usage of Newton's third law. *Journal of Research in Science Teaching*, 28(2), 123–140.

Brown, D. E. (1989). Students' concept of force: The importance of understanding Newton's third law. *Physics Education*, 24(6), 353–358.

Chi, M. T. H., Bassok, M., Lewis, M. W., Reimann, P., & Glaser, R. (1989). Self-explanations: How students study and use examples in learning to solve problems. *Cognitive Science, 13,* 145–182.

Chi, M. T. H., Feltovich, P. S., & Glaser, R. (1981). Categorization and representation of physics problems by experts and novices. *Cognitive Science, 5,* 121–152.

Chi, M. T. H., & Glaser, R. (1985). Problem-solving ability. In R. J. Sternberg (Ed.), *Human abilities: An information-processing approach* (pp. 227–250). New York: Freeman.

Chi, M. T. H., Glaser, R., & Rees, E. (1982). Expertise in problem solving. In R. J. Sternberg (Ed.), *Advances in the psychology of human intelligence* (Vol. 1; pp. 7–75). Hillsdale, NJ: Lawrence Erlbaum.

Clement, J. S. (1978). Mapping a student's causal conceptions from a problem solving protocol. In J. Lochhead & J. J. Clement (Eds.), *Cognitive process instruction* (pp. 133–146). Philadelphia: Franklin Institute Press.

Clement, J. (1988). Observed methods for generating analogies in scientific problem solving. *Cognitive Science, 12*(4), 563–586.

de Jong, T., & Ferguson-Hessler, M. G. M. (1986). Cognitive structures of good and poor novice problem solvers in physics. *Journal of Educational Psychology, 78*(4), 279–288.

Dufresne, R. J., Gerace, W. J., Thibodeau Hardiman, P., & Mestre, J. P. (1992). Constraining novices to perform expertlike problem analyses: Effects on schema acquisition. *Journal of the Learning Sciences, 2*(3), 307–331.

Elio, R., & Scharf, P. B. (1990). Modeling novice-to-expert shifts in problem-solving strategy and knowledge organization. *Cognitive Science, 14*(4), 579–639.

Eylon, B., & Reif, F. (1984) Effects of knowledge organization on task performance. *Cognition and Instruction, 1,* 5–44.

Ferguson-Hessler, M. G. M., & de Jong, T. (1987). On the quality of knowledge in the field of electricity and magnetism. *American Journal of Physics, 55*(6), 492–497.

Ferguson-Hessler, M. G. M., & de Jong, T. (1990). Studying physics texts: Differences in study processes between good and poor performers. *Cognition and Instruction, 7,* 41–54.

Finegold, M., & Mass, R. (1985). Differences in the process of solving physics problems between good physics problem solvers and poor physics problem solvers. *Research in Science and Technological Education, 3,* 59–67.

Garrett, R. M. (1987). Issues in science education: Problem-solving, creativity and originality. *International Journal of Science Education, 9*(2), 125–137.

Garrett, R. M., Satterly, D., Gil Pérez, D., & Martinez Torregrosa, J. (1990). Turning exercises into problems: An experimental study with teachers in training. *International Journal of Science Education, 12,* 1–12.

Gil Pérez, D., & Martinez Torregrosa, J. (1983). A model for problem-solving in accordance with scientific methology. *European Journal of Science Education, 5*(4), 447–455.

Halloun, I. A., & Hestenes, D. (1985). The initial knowledge state of college physics students. *American Journal of Physics, 53,* 1043–1055.

Hammer, D. (1989). Two approaches to learning physics. *The Physics Teacher, 27,* 664–670.

Hayes, J. R. (1981). *The complete problem solver.* Philadelphia: Franklin Institute Press.

Hegarty, M. (1991). Knowledge and processes in mechanical problem solving. In R. J. Sternberg & P. A. Frensch (Eds.), *Complex problem solving: Principles and mechanisms* (pp. 253–285). Hillsdale, NJ: Lawrence Erlbaum.

Heller, P., & Hollabaugh, M. (1992). Teaching problem solving through cooperative grouping. Part 2: Designing problems and structuring groups. *American Journal of Physics, 60*(7), 637–645.

Heller, P., Keith, R., & Anderson, S. (1992). Teaching problem solving through cooperative grouping. Part 1: Group versus individual problem solving. *American Journal of Physics, 60*(7), 627–636.

Heller, J. I., & Reif, F. (1984). Prescribing effective human problem-solving processes: Problem description in physics. *Cognition and Instruction, 1*(2), 177–216.

Johsua, S., & Dupin, J. J. (1991). In physics class, exercises can also cause problems. *International Journal of Science Education, 13*(3), 291–301.

Larkin, J. H. (1979, December). Processing information for effective problem solving. *Engineering Education,* 285–288.

Larkin, J. H. (1980a). Skilled problem solving in physics: A hierarchical planning model. *Journal of Structural Learning, 1,* 271–297.

Larkin, J. H. (1980b). Teaching problem solving in physics: The psychological laboratory and the practical classroom. In D. T. Tuma & F. Reif (Eds.), *Problem solving and education: Issues in teaching and research* (pp. 111–125). New York: Wiley.

Larkin, J. H. (1981). Enriching formal knowledge: A model for learning to solve textbook physics problems. In J. R. Anderson (Ed.), *Cognitive skills and their acquisition* (pp. 311–334). Hillsdale, NJ: Lawrence Erlbaum.

Larkin, J. H. (1983). The role of problem representation in physics. In D. Gentner & A. L. Stevens (Eds.), *Mental models* (pp. 75–98). Hillsdale, NJ: Lawrence Erlbaum.

Larkin, J. H. (1985). Understanding, problem representations, and skill in physics. In S. F. Chipman, J. W. Segal, & R. Glaser (Eds.), *Thinking and learning skills* (Vol. 2; pp. 141–159). Hillsdale, NJ: Lawrence Erlbaum.

Larkin, J. H., McDermott, J., Simon, D. P., & Simon, H. A. (1980a). Expert and novice performance in solving physics problems. *Science, 208,* 1335–1342.

Larkin, J. H., McDermott, J., Simon, D. P., & Simon, H. A. (1980b). Models of competence in solving physics problems. *Cognitive Science, 4,* 317–345.

Larkin, J., & Reif, F. (1979). Understanding and teaching problem solving in physics. *European Journal of Science Education, 1*(2), 191–203.

Larkin, J. H., Reif, F., Carbonell, J., & Gugliotta, A. (1988). FERMI: A flexible expert reasoner with multi-domain inferencing. *Cognitive Science, 12,* 101–138.

Larkin, J. H., & Simon, H. A. (1987). Why a diagram is (sometimes) worth ten thousand words. *Cognitive Science, 11,* 65–99.

Lesgold, A., & Lajoie, S. (1991). Complex problem solving in electronics. In R. J. Sternberg & P. A. Frensch (Eds.), *Complex problem solving: Principles and mechanisms* (pp. 287–316). Hillsdale, NJ: Lawrence Erlbaum.

Maloney, D. P. (1984). Rule-governed approaches to physics: Newton's third law. *Physics Education, 19,* 37–42.

Maloney, D. P., & Siegler, R. S. (1993). Conceptual competition in physics learning. *International Journal of Science Education, 15,* 283–295.

Mayer, R. E. (1983). *Thinking, problem solving, cognition.* New York: Freeman.

McCarthy, J. (1956). The inversion of functions defined by machines. In C. E. Shannon & J. McCarthy (Eds.), *Automata studies, annals of mathematics studies* (pp. 177–181). Princeton, NJ: Princeton University.

McMillan, C., & Swadener, M. (1991). Novice use of qualitative versus quantitative problem solving in electrostatics. *Journal of Research in Science Teaching, 28*(8), 661–670.

Mettes, C. T. C. W., Pilot, A., & Roossink, H. J. (1981). Linking factual and procedural knowledge in solving science problems: A case study in thermodynamics course. *Instructional Science, 10,* 333–361.

Mohapatra, J. K. (1987). Can problem-solving in physics give an indication of pupils' "process knowledge"? *International Journal of Science Education, 9*, 117–123.

Newell, A., & Simon, H. A. (1972). *Human problem solving*. Englewood Cliffs, NJ: Prentice Hall.

Peters, P. C. (1982). Even honors students have conceptual difficulties with physics. *American Journal of Physics, 50*, 501–508.

Polya, G. (1945). *How to solve it*. Garden City, NY: Doubleday.

Reif, F., & Heller, J. I. (1982). Knowledge structures and problem solving in physics. *Educational Psychologist, 17*(2), 102–127.

Reif, F., Larkin, J. H., & Brackett, G. C. (1976). Teaching general learning and problem solving skills. *American Journal of Physics, 44*(3), 212–217.

Reusser, K. (1988). Problem solving beyond the logic of things: Contextual effects on understanding and solving word problems. *Instructional Science, 17*(4), 309–338.

Richardson, J. J. (1981). *Problem solving instruction for physics*. Unpublished doctoral dissertation, University of Colorado, Denver.

Robertson, W. C. (1990). Detection of cognitive structure with protocol data: Predicting performance on physics transfer problems. *Cognitive Science, 14*(2), 253–280.

Schoenfeld, A. H. (1978). Can heuristics be taught? In J. Lochhead & J. J. Clement (Eds.), *Cognitive process instruction* (pp. 315–338). Philadelphia: Franklin Institute Press.

Shavelson, R. J. (1972). Some aspects of the correspondence between content structure and cognitive structure in physics instruction. *Journal of Educational Psychology, 63*(3), 225–234.

Shavelson, R. J., & Stanton, G. C. (1975). Construct validation: Methodology and application to three measures of cognitive structure. *Journal of Educational Measurements, 12*(2), 67–85.

Simon, D. P., & Simon, H. A. (1978). Individual differences in solving physics problems. In R. S. Siegler (Ed.), *Children's thinking: What develops?* (pp. 325–348). Hillsdale, NJ: Lawrence Erlbaum.

Simon, D. P., & Simon, H. A. (1979). A tale of two protocols. In J. Lochhead & J. S. Clement (Eds.), *Cognitive process instruction* (pp. 119–132). Philadelphia: Franklin Institute Press.

Snider, R. M. (1989). Using problem solving in physics classes to help overcome naive misconceptions. In D. Gabel (Ed.), *What research says to the science teacher: Problem solving* (pp. 51–65). Washington, DC: National Science Teachers Association.

Sweller, J. (1988). Cognitive load during problem solving: Effects on learning. *Cognitive Science, 12*, 257–285.

Sweller, J., & Levine, M. (1982). Effects of goal specificity on means-ends analysis and learning. *Journal of Experimental Psychology: Learning, Memory and Cognition, 8*(5), 463–474.

Sweller, J., Mawer, R., & Ward, M. (1982). Consequences of history: Cued and means-end strategies in problem solving. *American Journal of Psychology, 95*(3), 455–483.

Talyzina, N. F. (1973). Psychological bases of programmed instruction. *Instructional Science, 2*(3), 243–280.

Tarmizi, R., & Sweller, J. (1988). Guidance during mathematical problem solving. *Journal of Educational Psychology, 80*(4), 424–436.

Thorsland, M. N., & Novak, J. D. (1974). The identification and significance of intuitive and analytic problem solving approaches among college physics students. *Science Education, 58*(2), 245–265.

Thro, M. P. (1978). Relationships between associative and content structure of physics concepts. *Journal of Educational Psychology, 70*(6), 971–978.

Trowbridge, D. E., & McDermott, L. C. (1980). Investigation of student understanding of the concept of velocity in one dimension. *American Journal of Physics, 48*(12), 1020–1028.

Van Heuvelen, A. (1991). Overview, case study physics. *American Journal of Physics, 59*(10), 898–906.

Veldhuis, G. H. (1986). *Differences in the categorization of physics problems by novices and experts*. Unpublished doctoral dissertation, Iowa State University, Ames.

Veldhuis, G. H. (1990). The use of cluster analysis in categorization of physics problems. *Science Education, 74*, 105–118.

Voss, J. F. (1989). Problem solving and the educational process. In A. Lesgold & R. Glaser (Eds.), *Foundations for a psychology of education* (pp. 251–294). Hillsdale, NJ: Lawrence Erlbaum.

Ward, M., & Sweller, J. (1990). Structuring effective worked examples. *Cognition and Instruction, 7*, 1–39.

Wright, D. S., & Williams, C. D. (1986). A wise strategy for introductory physics. *Physics Teacher, 24*(4), 211–216.

Part

·IV·

CURRICULUM

·13·

RESEARCH ON GOALS FOR THE SCIENCE CURRICULUM

Rodger W. Bybee
BIOLOGICAL SCIENCES CURRICULUM STUDY

George E. DeBoer
COLGATE UNIVERSITY

Science education is in a period of significant reform, due in large measure to changes that are occurring in society. One critical aspect of the contemporary reform is a review and modification of the goals that form the structure of science curricula. The contemporary reform is not the first time in history that societal circumstances transformed the goals for science education. In this chapter historical review is used to clarify the goals of science education and the societal circumstances that have influenced a revision of science education's goals and subsequently influenced the science curriculum.

The purposes of this chapter are to (1) describe changes in specific goals at different times in history, (2) clarify different priorities of goals for the science curriculum, and (3) identify the emergence of goals that are important for today's science curriculum. This is an essay on the nature of progress in science education as shown in the historical changes in goals and their subsequent influences on the structure and function of science curricula.

BASIC QUESTIONS FOR THE SCIENCE CURRICULUM

At any one time, a variety of opinions exist in the goals of science teaching. Over time, changes in goals influence subsequent changes in science curricula and instructional tech-

niques. Historical review reveals a relatively small number of common goals that we have had for our students throughout history, and various configurations of these goals underlie all science programs.

What Science Should Be Learned?

Goal statements can be viewed in several different ways. One way is to focus on *what* is to be learned. Three such student outcome goals are (1) to acquire scientific knowledge, (2) to learn the processes or methodologies of the sciences, and (3) to understand the applications of science, especially the relationships between science and society and science-technology-society. Under this organization, the students should have some knowledge of the products of science, should have experience with and understand the methods of science, and should understand how science is a force in their world.

Science *knowledge* includes the range of accumulated observations and systematic information about the natural world. With specific reference to science curriculum, knowledge is often discussed using terms such as "facts," "concepts," "laws," "principles," and "theories." In recent educational literature, the term "cognitive domain" has been used in describing the goal of knowledge in teaching. In science education, the aim of understanding conceptual schemes (National Science Teachers Association, 1964) is another example of the knowledge goal.

The authors thank reviewer Paul Dehart Hurd (Stanford University).

357

Method is a manner of acting, a predisposition to behave, perform, and think in certain ways toward an object or objects of study. Of particular importance here are scientific methods as they have been variously described in the history of science education. One example of method as a stated goal of science teaching is the "processes of science" goal described in the 1964 National Science Teachers Association (NSTA) monograph *Theory into Action*. Another example is the list of science processes described by the American Association for the Advancement of Science (Livermore, 1964). Emphases on inquiry, discovery, and problem solving are further examples of the method goal of science education. We should point out that there has never been *a* scientific method. Certainly, the research methods in theoretical physics, observational astronomy, and field ecology vary considerably.

The societal *applications* goal is an attempt to relate science and society, especially through the technological advances that have had an impact on our social world. The science-technology-society (STS) emphasis of the 1980s and 1990s, the goal statements of NSTA dealing with scientific literacy, and the "societal challenges" dimension of the National Science Education Standards are all expressions of the societal-applications goal. The societal-applications goal implies that science curricula should have some connection to society and that students should develop understanding of science as it influences and is influenced by society.

Another way to view the goals of science teaching is to look at the ends to which the knowledge, method, and applications apply. The goals of science teaching then become (1) personal development, which includes aesthetic appreciation, intellectual development, and career awareness; (2) social efficiency and effectiveness, including the maintenance of a stable social order, economic productivity, and the preparation of citizens who feel comfortable in a technical world and understand issues such as environmental conservation, disease prevention, birth control, industrial development, and computer literacy; (3) the development of science itself, which involves the cultural transmission of scientific knowledge from one generation to the next so that each subsequent generation has a base from which new scientific discoveries can be made; and (4) national security, which includes the development of a strong, technically able military, an internationally competitive work force, and a citizenry that is sympathetic to the importance of science as a competitive force in this world.

Why Should Students Learn Science?

These four goals focus on the function or utility of the science content students should learn, that is, on the justification for including science in the school curriculum. There are many examples of how the goals that are viewed in these two ways interact. In the middle and late nineteenth century, scientific method, especially inductive reasoning, was taught as a means of disciplining the mind. In other words, the emphasis was on method as a means to personal development. Toward the end of the nineteenth and the beginning of the twentieth century, as the United States became more industrialized and urbanized, scientific method was looked to as a model of problem solving that could be applied to social problems. The emphasis on scientific method was justified in terms of a concern for the society at large. Individuals would understand and be able to solve problems that stemmed from urbanization and industrialization, problems of sanitation, birth control, housing, and crime. During the Cold War, the focus turned to national security and the training of a technically competent military. The emphasis was on a rigorous study of the structure of scientific knowledge to ensure the well-being of the nation. More recently, such social concerns as the environment, infectious disease, and birth control have led to a greater emphasis on the societal applications of science. Students learn the applications of science so that they may contribute to the welfare of the society. Coupled with this recent emphasis on the societal-applications goal has been concern about our competitiveness in international economic markets and a desire for U.S. students to demonstrate a mastery of basic scientific facts, at least to the extent of students in other countries. Thus, scientific knowledge is taught for reasons of national economic security. Through these examples, we can see the many ways in which the teaching of science knowledge, method, and applications can be justified, depending on the particular concerns of the historical moment.

BASIC GOALS FOR SCIENCE CURRICULUM

We have chosen to examine three major goal areas of science teaching and to combine the *what* and the *why* of science teaching in each category. These goals show us where the efforts of science educators have been directed over the years. The three goals are (1) personal and social development, (2) knowledge of scientific facts and principles, and (3) scientific methods and skills and their application. A brief explanation of what is included in each of these goals follows.

Personal and Social Development

Far and away, the most compelling reason for teaching science is the effect that it has on the development of individuals and the influence it has on the well-being and improvement of society. We place personal and social development under the same category because so often what is considered good for the individual is considered good for the society. Personal development includes such things as intellectual growth, personal satisfaction, career awareness, and building moral character. Social development includes the maintenance of public health, a productive economy, a stable and orderly society, a physically safe environment, and a safe and secure nation. At times throughout our history, personal development has taken precedence over social development, and sometimes societal concerns seem more important than issues of personal development. When these different emphases occur, we will point to those differences.

Knowledge of Scientific Facts and Principles

Throughout our educational history, the development of scientific understanding has been a primary goal of science education. In many programs, learning science facts and principles carries with it no particular rationale. Learning scientific knowledge is simply considered a basic element of science education that needs no justification. One specific reason why science has been taught is so that at least some of the youth of our schools will become scientists and eventually advance the frontiers of scientific knowledge. Knowledge of scientific principles has also been justified because it is thought to lead to intellectual development, personal satisfaction, and national security. Recently, we have emphasized greater scientific understanding in order to increase our position on international assessments.

Scientific Methods and Skills and Their Application

Another extremely important focus of science teaching over the years has been the development of both an *understanding* of the methods of science and *abilities* in applying those methods. As with scientific knowledge, the teaching of scientific method has been justified in a variety of ways, but educational efforts have been directed toward the acquisition of skill and understanding in this area itself, thus making it a major goal of science teaching. Scientific method has been taught as a way to develop the intellect, a general method for dealing with social problems, and a means for acquiring scientific knowledge. Unfortunately, the representation of this goal in science curricula has taken the form of *a* method, with steps such as defining the problem, making observations, forming a hypothesis, performing an experiment, and presenting a conclusion. Scientists and science teachers know this is not an accurate representation, but it continues in many textbooks.

As we look at what U.S. science educators have tried to accomplish at times and the rationales that they have offered, we must realize that what actually goes on in schools does not always match the highly publicized statements of key educational leaders or national organizations. The goals and rationales are important because they provide a direction and an argument for moving in a certain direction, but they should not be viewed as statements describing practice. Educational practice changes much more slowly than does the stated emphasis on one goal or another, or the justifications for moving in one direction instead of another. We intend to take the long historical view of science education goals because history shows that there is a remarkable stability in what we are trying to accomplish and even in why we are trying to accomplish it. There is a rich history that can provide a context for evaluating even our present goals, justifications, and practices. This history extends back at least to the middle of the nineteenth century, and most of the arguments made then are still useful to us today.

HISTORICAL PERSPECTIVES OF SCIENCE CURRICULA

The structure of the science curriculum is determined by the priority and emphasis (Roberts & Chastko, 1990) given to the respective goals. For example, if a teacher has decided that knowledge of conceptual schemes is the important goal of elementary science teaching, then other goals become subordinate to this goal. The latter goals may be included in the science curriculum and may be very important, but the curriculum that results from this arrangement of goals is quite different from a curriculum in which personal development is the primary goal and the content and methods chosen best suit the personal development goal. This scheme is one way to describe the difference in emphasis between two elementary science curriculum programs from the 1970s, *Science: A Process Approach* (SAPA) (AAAS, 1967; Gagné, 1967) and *Conceptually Oriented Program in Elementary Science* (COPES) (Barnard, 1971). The goal of SAPA is understanding the processes (methods) of science; the means for achieving this goal is to involve the student with the processes of observation, classification, and so on in the science activities. The program also has goals of knowledge and personal development, but these are secondary to the teaching of scientific method. COPES, however, has the primary goal of understanding the major conceptual schemes of science (knowledge), with method used as a means of achieving the knowledge goal. Furthermore, both programs are structured with recognition of the elementary students' developmental growth. Little recognition was given to societal and career development issues in other programs.

As seen in these examples, examination of the organization of goals is a way to describe differences in science curricula and science instruction. The priority and structure of goals are also meaningful ways to describe historical changes and contemporary trends in science teaching. For example, if two science curriculum projects separated by space or time are compared, the variations between programs can be explained by differences in the structured goals. It is then possible to analyze the identified differences and the factors influencing transformations in science education.

To understand present trends and transformations in science education, it is important to know what has happened in the past. This is certainly true for the goals of science teaching because of their important function in the underlying structure of science curricula. In general, an analysis of the history of science education reveals the conditions under which the goals of science education were formulated and the circumstances under which they have changed.

In the next sections, historical examples of goal statements and science programs are discussed to show how societal circumstances affect goals and, subsequently, the science curriculum. This discussion is not intended to be a comprehensive history of science education. Rather, selected historical examples are used to show the dynamics of changing goals and their influences on the structure of the science curriculum. Readers will find a more thorough discussion of the history of science education in *A History of Ideas in Science Education* (DeBoer, 1991).

EUROPEAN INFLUENCES ON THE SCIENCE CURRICULUM

In the next several pages, we focus on some of the forces that led to a changing conception of education in the United States and made it possible for science to become a major part of the school curriculum. We are particularly interested in some of the statements made by individual scientists during the nineteenth century who vigorously supported the inclusion of science in the curriculum. We are interested in the nineteenth century because it was a period of significant transition in thinking about education and a rich source of arguments about the appropriate form of education. During this time, some educators argued for a new approach to education, while others defended the traditional forms of education.

The nineteenth century saw profound changes in many of our social institutions. In education, a system that had its basis in the ecclesiastical scholasticism of the Middle Ages and the classical studies of the Renaissance slowly gave way to a system based on the practical concerns of the day. As the United States became more industrialized and scientific and technical knowledge took on new importance, the arguments for including science in the school curriculum became increasingly compelling.

Those who argued for the inclusion of science in the curriculum had to make the case that science was at least as important as the more traditional subjects. For the most part, the curriculum before the mid-nineteenth century included reading, writing, and arithmetic at the primary level and study of classical literature and languages at the upper levels. The approach to education was authoritarian, with little room for individual expression or independent thought. When science was taught, it was usually offered as a book study and often with a natural theology approach in which the natural world was seen as a revelation of God's unfolding plan for the world.

A number of European educators influenced the thinking of educators in the United States. As a group, they emphasized more child-centered approaches, independent thought, sense impressions as the basis of knowledge, and an active rather than a passive role for the learner. The approaches were nonauthoritarian and focused on the power of individuals to understand the world independently without divine revelation interpreted by ecclesiastical authority. From our vantage point, we can see that this influence not only changed the overall focus of education in the United States but also contributed greatly to the growth of science teaching as we know it today.

John Amos Comenius

One of the earliest individuals to advance a form of education that supported empirical investigation and practical studies was the seventeenth-century educator John Amos Comenius (1592–1671). His *Orbis Sensualium Pictus*, which appeared in 1658, was a popular illustrated textbook for children. Comenius has sometimes been referred to as the person who, through that book, first brought the study of science to the classroom. Although it was used largely as a means of building vocabulary and did not form a substantial science curriculum, the book did include many science topics (Underhill, 1941). Comenius believed that our ideas originate from experience and that children should be presented with materials from their natural environment. The textbook illustrations were representations of that natural world.

John Locke

Another person who influenced the direction of education in the United States and in Europe was the seventeenth-century British empiricist John Locke (1632–1704). Locke, like Comenius, believed that all our ideas come from experience and that ideas are true when they correspond to concrete, objective reality. Locke, too, emphasized the secular and practical, as opposed to religious aspects of education, and the power of individuals to derive meaning from their natural world.

Johann Heinrich Pestalozzi

In the early nineteenth century, one of the most powerful influences on the direction of education came from Johann Heinrich Pestalozzi. Pestalozzi was a Swiss educator who was influenced by the thinking of Jean Jacques Rousseau (1712–1781). Rousseau believed that the purpose of education was to prepare a child for life by developing the child's natural inborn capacities. According to Rousseau, a study of natural phenomena, in a manner consistent with the natural development of the mind of the child, would lead to growth in a way that the formal, authoritarian methods in use at the time would not.

Along the same lines as Rousseau, Pestalozzi advocated a nonauthoritarian approach to education. According to Pestalozzi, the goal of education was the development of independent self-activity. This goal would be accomplished by having students learn about the natural world through active investigation and experimentation and not memorized lessons taught by an authoritarian teacher. The teacher's responsibility was to determine each child's understanding at the various stages of life and to present that child with the materials that would lead to meaningful learning and the development of the child's mental faculties.

Pestalozzi also inspired the development of object lessons as an educational innovation in the early nineteenth century. One of the earliest books of object lessons was written by Elizabeth Mayo after she and her brother Charles returned from a visit with Pestalozzi in Switzerland. Each object lesson focused on the formal study of some natural object and became an early form of science instruction in the nineteenth century. We return to a more detailed discussion of object lessons when we discuss nineteenth-century teaching practices in the United States.

Friedrich Froebel

Friedrich Froebel was a follower of Pestalozzi who believed that education should begin early. His own schools for children in Germany, known as Kindergartens, accepted students as young as 3 years of age. According to Froebel, early-childhood education should include a combination of play, music, and physical activity. Froebel had profound respect for the individuality and dynamic nature of the child. He believed that education's purpose was to link the spirit of the child with the divine through a study of the natural world. In his schools, natural objects, story telling, and cooperative social activities were used constructively. Again, experience with the natural world and a faith in the power of the individual to generate meaning from first-hand experience formed the basis for educational practice.

Johann Friedrich Herbart

Another early nineteenth-century European educator who took sense perception as the beginning point in education was Johann Friedrich Herbart. Herbart wrote his *General Pedagogy* in 1806, and from 1810 to 1832 he held the chair of philosophy at the University of Konigsberg, a chair previously held by Immanual Kant. According to Herbart, the purpose of education was the development of mind through the acquisition of ideas. Education was meant not to exercise the mental faculties, but to create an organization of ideas that would enable a person to live a well-rounded, moral life. The notion that specific content was important was in contrast to the idea that a mind could be exercised and strengthened through the study of difficult material, regardless of its content. Herbart's influence led educators to view as important the selecting and organizing of content to create conceptual schemes in the minds of learners.

Herbart also spoke of the importance of the connectedness of ideas and the value of having learners discover the relationships between ideas instead of having those relationships presented to them directly. If teachers attempted to make the connections for the child, they would inevitably miss the mark, not knowing what the child already knew and where to make the connections most effectively. According to Herbart, if the learners discovered the relationships between ideas on their own, their understanding would be fuller and more meaningful.

Thomas Huxley

Huxley argued that education as it was practiced in the mid-nineteenth century was outmoded. The curriculum was dominated by basic skills at the primary level and by the study of Greek and Roman civilization and the Greek and Latin languages at the upper levels. According to Huxley, the humanists who controlled the curriculum failed to recognize the importance of contemporary civilization. Modern citizens needed an education that would prepare them to live in their own world, a world of science, industry, and technological development. The practical issues of their day were being ignored as they studied the truths of the ancient past. In Huxley's (1899) words:

There is one feature of the present state of the civilized world which separates it more widely from the Renascence, than the Renascence was separated from the middle ages. This distinctive character of our own times lies in the vast and constantly increasing part which is played by natural knowledge. Not only is our daily life shaped by it, not only does the prosperity of millions of men depend upon it, but our whole theory of life has long been influenced, consciously or unconsciously, by the general conception of the universe, which have been forced upon us by physical science. (p. 149)

According to Huxley, the study of science would not only enable a person to adapt to a rapidly changing world but also develop the mind of the individual in ways that other school subjects could not. Most psychologists and educators of the mid-nineteenth century believed that the mind was made up of discrete abilities or faculties that could be exercised through hard mental work. Thus, the memory faculty could be exercised through practice in memorization. The deductive faculty could be strengthened by means of such subjects as mathematics and grammar, where the applications are determined by the authority of basic rules. According to Huxley, the study of science would strengthen the inductive and observational faculties. Through an emphasis on direct observation of the natural world and the generation of principles from empirical data, science study could accomplish goals that were unique to the discipline.

For science teaching to accomplish these goals, Huxley believed it must not be taught as a book study. The power of science was in its ability to develop aspects of the mind untouched by other school subjects. The observational and inductive faculties would have to be strengthened through direct contact with the physical world, and students would have to be given the freedom to investigate on their own and formulate generalizations from sense data on their own. As Huxley said, "If scientific education is to be dealt with as mere bookwork, it will be better not to attempt it, but to stick to the Latin Grammar which makes no pretense to be anything but bookwork" (Huxley, 1899, p. 125). The teaching of verbal abstractions alone would be ineffective. The real world had to be experienced directly. Huxley (1899) said

In teaching him botany, he must handle the plants and dissect the flowers for himself; in teaching him physics and chemistry, you must not be solicitous to fill him with information, but you must be careful that what he learns he knows of his own knowledge. Don't be satisfied with telling him that a magnet attracts iron. Let him see that it does; let him feel the pull of the one upon the other for himself. And especially, tell him that it is his duty to doubt until he is compelled, by the absolute authority of Nature, to believe that which is written in books. (p. 127)

This argument for direct contact with the natural world, made by Huxley and others, contributed to the middle to late nineteenth-century fascination with the science laboratory as an essential part of science instruction.

As did Froebel, Rousseau, Pestalozzi, and others, scientists like Huxley placed confidence in the ability of children to generate meaning from the world around them. Huxley recommended that the study of science begin as early as possible. He recommended that the earliest course in science begin with an overview of the natural world so that children could satisfy their natural curiosity about the weather, the sky, the sea, the animals, and themselves.

Herbert Spencer

Herbert Spencer, like Thomas Huxley, argued that nineteenth-century youth were not learning those things that were most important for life in an increasingly industrialized society. Spencer said that the "vital knowledge" that had created the great industrialized nations of the world was being ignored as schools were "mumbling little else but dead formulas" (Spencer, ·1864, p. 54).

In his classic essay "What Knowledge Is of Most Worth?" (1864), Spencer evaluated the various goals of education, using as his measure of value the impact of different kinds of knowledge on human welfare. Spencer was interested in the utility of education for individual well-being. He classified human activity into several major categories and identified the knowledge that would be of most worth. Spencer's categories included the rearing and disciplining of offspring, maintaining social and political relations, self-preservation, and leisure activities "devoted to the gratification of the tastes and feeling" (Spencer, 1864, p. 32). In all of these categories of human activity, Spencer found a vital role for science study. Knowledge of the functions of the human body and their relation to good health was important for self-preservation. Earning a living, an indirect form of self-preservation, was dependent on a knowledge of the products and processes that formed a major part of the new industrial and agricultural economy. Knowledge of machines (the lever, wheel, and axle), the steam engine, smelting furnaces, gunpowder manufacturing, sugar refining, and agricultural production through courses in physics, chemistry, and biology would be useful in most people's lives because so many people were involved directly or indirectly in the production, preparation, and distribution of commodities.

Child rearing would be enhanced by parents having a more complete understanding of physiology and the laws of psychological development. Even aesthetic appreciation was dependent on a knowledge of science, according to Spencer. A person who studied science would have a better understanding of such concepts as physical form, proportion, and balance, and this would enhance both the appreciation of works of art and their production.

Finally, Spencer believed that the study of science would contribute to the development of a person's mental faculties. The ability to draw conclusions from observations of the natural world would come from direct contact with the physical environment. Like Huxley, Spencer was an advocate of the laboratory as a way to provide students contact with these objects and processes. Direct contact with the real world would provide the students with more precise mental images than verbal abstractions alone, and practice in drawing conclusions from observations would lead to the inductive faculty, which Spencer called "judgment." In Spencer's (1864) words:

No extent of acquaintance with the meaning of words, can give the power of forming correct inferences respecting causes and effects. The constant habit of drawing conclusions from data, and then of verifying those conclusions by observation and experiment, can alone give the power of judgment correctly. And that it necessitates this habit is one of the immense advantages of science. (p. 88)

In the mid–nineteenth century, a strong theme in European thinking about education ran counter to the traditional notions of religious authority. European educators, such as John Amos Comenius, John Locke, Johann Heinrich Pestalozzi, Freidrich Froebel, and Johann Friedrich Herbart, had a significant influence on American science education. As a group, Europeans emphasized new approaches to education. With their emphasis on independent thinking, sense data as the starting point in education, and respect for the individual's ability to generate meaning from their world, these ideas led the way for the inclusion of the natural sciences in the school curriculum.

Both Thomas Huxley and Herbert Spencer were influenced by the new thinking in education. Both of them argued for a changing conception of education that would place greater emphasis on a study of the natural world. Although the statements by Huxley and Spencer represent individual perspectives rather than goals agreed upon by the educational community, these perspectives are important for two reasons. The first reason is that during the nineteenth century there was no formal mechanism for debating and coming to a consensus on the goals of science teaching. One of the earliest examples of the latter is the 1893 report of the National Education Association's Committee of Ten. The second reason for including these statements is that both Huxley and Spencer, and many of their colleagues, spoke clearly and forcefully in defense of science teaching and thus had a significant influence on the goals for science teaching and the science curriculum.

EARLY AMERICAN INFLUENCE ON SCIENCE CURRICULUM

In this same period, the late 1800s, several Americans had an influence on science education. We have selected J. M. Rice and Charles W. Eliot as individuals who exemplified that period's view of what science should be in the curriculum and how it should be taught.

J. M. Rice

The contrast between teaching science for the acquisition of knowledge and teaching science for reasons of personal development was made clear in a series of articles that J. M. Rice wrote for *The Forum* in the early 1890s. Rice described the old approach, focusing on the acquisition of knowledge, as "mindless" and "mechanical." He said that mechanical schools, which followed the traditional practices of the past, saw as their function the "crowding into the memory of the child a certain number of cut-and-dried facts" (Rice, 1893, p. 20). Traditional practices included the use of textbooks instead of real-world experiences, harsh discipline, and mindless recitations that required no active thought on the part of the children.

In contrast, the new approach to education made use of many of the principles described by such educators as Pestalozzi and Herbart. Rice called these "scientific principles" of teaching. In schools that followed the new approach, the aim was "to lead the child to observe, to reason, and to acquire manual dexterity as well as to memorize facts—in a word, to develop the child naturally in all his faculties, intellectual, moral, and physical." (Rice, 1893, p. 21). Rice described what he saw in one school that had attempted to apply the modern approach.

I entered one of the rooms containing the youngest children at the time of the opening exercises. The scene I encountered was a glimpse of fairyland. I was in a room full of bright and happy children, whose eyes were directed toward the teacher.... She understood them, sympathized with and loved them, and did all in her power to make them happy. The window sills were filled with plants, and plants were scattered here and there throughout the room. (pp. 101–102)

After the children had sung a few little songs, the first lesson of the day was in order. This was a lesson in science; its subject was a flower. It began with the recitation of a poem. The object of introducing these poems into the plant and animal lessons is to inspire the child with love for the beautiful, with love for nature, and with sympathy for all living things. (p. 102)

Before the teacher endeavored to bring out the points to which she desired to direct the special attention of the class, the children were urged to make their own unaided observations and to express them. As each child was anxious to tell what he had observed in relation to the plant itself, what he otherwise knew of it, how it grew, where it grew, and perhaps some little incident that the flower recalled to him, the class was full of life and enthusiasm. (pp. 102–103)

The teacher . . . by her careful questioning led the children to observe the particular things to which she had decided to call their attention that morning. Her questions were not put to individual children, but to the whole class, so that every question might serve to set every pupil observing and thinking. (p. 104)

By the 1890s, there was considerable pressure to move away from what was thought to be the harshness, formality, and mindlessness of the traditional forms of education. As Rice noted, however, this type of education still dominated the schools of the day. Clearly, the pressure to change was present, and the nature-study approach to science teaching is an example of that, but rigorous study of subject matter for purposes of mental development or subject matter for its own sake was deeply ingrained in the thinking of American educators.

Charles W. Eliot

Another educator who observed the traditional practices that were so common in the schools of the late nineteenth century was Charles Eliot. Eliot was president of Harvard University from 1869 to 1895 and was an active participant in discussions concerning education at the elementary and secondary levels as well. Eliot strongly supported the use of the laboratory in science teaching as a way to develop one's powers of observation and inductive thought and the child-centered approaches that were being advocated by European educators. Eliot found, however, when he visited schools that science was taught from a book "as if it were grammar or history" (Eliot, 1898, p. 112) and that there was an excessive emphasis on memory and boring routine totally lacking in human interest.

In one school he observed a class in physiology, anatomy, and hygiene. I imagined that there might be a skeleton in that school, or a manikin, or a model of the brain, stomach, lungs, eye, ear, head, or arm, and that the children might be shown some of these beautiful organs. But no; there was nothing of the sort in the school house, and there never had been. Everything concerning that natural history subject was taught out of a little book; the children had nothing but flat figures of the things described, and were required to make them stand for the various members of the human body.... Observational teaching of the human body is, of course, a fascinating and profitable study for children, just as observational teaching in geography makes that subject one of the most charming in the world whether for children or adults. Until we can get the means of teaching scientific subjects properly, let us not teach them at all. (p. 190)

The belief of mental disciplinarians was that rigorous study of difficult materials would strengthen the will of the individual to work hard and persist at difficult tasks. It did not matter whether these materials were interesting or uninteresting. If uninteresting, the person would just have to be forced to learn them. For this reason, drudgery often characterized life in classrooms. Describing this condition in one of the schools that he observed, Eliot (1898) said:

The main characteristic of the instruction, as developed through those books—unless lightened by the personality of the teacher—is dullness, a complete lack of human interest, and a consequent lack in the child of the sense of increasing power. Nothing is so fatiguing as dull, hopeless effort, with the feeling that, do one's best, one cannot succeed. That is the condition of too many children in American schools—not the condition for half an hour, but the chronic condition day after day and month after month. Make the work interesting, and give the children the sense of success, and the stress which is now felt by them will be greatly diminished. (pp. 184–185)

Eliot was obviously influenced by the new, child-centered approaches to education that took account of a child's interests and ability to understand the world through direct observation. The dominant goal of science education, however, for Eliot and most other educators of the nineteenth century was personal development. This might be the development of the will through difficult memory exercises, or the development of the powers of inductive thought through observation and drawing inferences from those observations, but the focus was on the developing person. Some believed that the mind had to be shaped by rigorous, even if boring, effort. Others like Rousseau, Pestalozzi, and Herbart believed that there was a natural development of the human mind that needed simply to be brought into contact with the natural world. But individual human development was the clear focus of education up until the latter years of the nineteenth century. Eliot identifies a number of these personal development goals in the following statement, in which he summarizes his views on the major purposes of education.

Effective power in action is the true end of education, rather than the storing up of information, or the cultivation of faculties which are mainly receptive, discriminating, or critical. We are no longer content, in either school or college, with imparting a variety of useful and ornamental information, or with cultivating aesthetic taste or critical faculty in literature or art. We are not content with simply increasing our pupils' capacity for intellectual or sentimental enjoyment. All these good things we seek, to be sure; but they are no longer our main ends. The main object of education, nowadays, is to give the pupil the power of doing himself an endless variety of things which, uneducated, he could not do. An education which does not produce in the pupil the power of applying theory, or putting acquisitions into practice, and of personally using for productive ends his disciplined faculties, is an education which has missed its main end. (pp. 323–324)

To summarize, the European influence on education in the late 1800s was expressed by American educators such as J. M. Rice and Charles Eliot. These individuals emphasized the goal of personal development and criticized stress on the acquisition of knowledge. Individuals such as Rice and Eliot maintained that science could be used to enhance personal development and that processes such as observation and drawing inferences were effective means to the goal of development. The statements of some American educators in this period have echoed down through the decades. Writings by individuals such as Rice and Eliot did result in science being included in education programs; however, the goal of personal development has not been prominent.

NINETEENTH CENTURY: SCIENCE EDUCATION IN THE UNITED STATES

One of the major changes during the nineteenth century was a general movement to include science in school programs and to deemphasize literary studies. This was accomplished because of the efforts of scientists like Herbert Spencer and Thomas Huxley; European educators, such as John Comenius, John Locke, Johann Pestazolli, Friedrich Froebel, and Johann Herbart; and American educators such as J. M. Rice and Charles W. Eliot. In the next section, we review more specifically the three major goals of science education that we identified in the introduction to this chapter as they were developed during the nineteenth century.

The Goal of Scientific Knowledge

The dominant goal of science teaching during the early part of the nineteenth century was factual knowledge, the majority of which had to support theology. The educational theories of John Locke and Jean Jacques Rousseau were influential in the development of the didactic materials and goals of this period. Learning interesting subject matter was the goal of the educational model provided by the writings of these men (Underhill, 1941).

One of the first systematic approaches to elementary science, identifiable by about 1860, was object lessons. This approach was based on the pedagogical theories of Johann Pesta-lozzi. Although the main goal of this curriculum model was personal intellectual development, the knowledge goal was stressed as well. Materials from the environment were used to facilitate the development of the mental faculties. The uses of materials and the level of the lesson were to be governed primarily by the student's level of development. Because of the use of environmental materials, science was important in the object-lesson movement. The goal, however, often shifted from personal development to learning science facts as they related to the objects of study.

An overall curriculum plan was seldom developed for object teaching; thus, each lesson became an end in itself. Object lessons were then criticized for lack of order and direction and overemphasis on science facts. There is little evidence of object lessons ever being widely implemented. Lack of materials and understanding by teachers accounted for the decline of object lessons, but the curriculum model of science teaching represented by the arrangement of goals as seen in object lessons became the model for the nature-study movement.

The advent of the object-lesson movement produced a deemphasis in the acquisition of knowledge as the major goal of science education. Even though knowledge was not the dominant goal and its manner of presentation was different from that seen in the earlier part of the nineteenth century, it was still an important part of science teaching throughout the century.

The Goal of Scientific Methods

The influence of scientists such as Thomas Huxley and Herbert Spencer was significant in swaying sentiment toward the inclusion of science in school programs, and their use of modernization as justification for including science in the curriculum foreshadowed the social goal but, for the most part, they did not advocate teaching science in a social context. Rather, they primarily emphasized the knowledge goal and introduced the methods goal into science education. The methods of science (e.g., observation) were seen as an ideal means to enhance both intellectual development and "scientific habits of mind."

Arguments for basing science education on natural objects themselves, direct observation, and laboratory approaches are examples of the goal of scientific methods being translated to pedagogy. Scientists legitimized science in school programs and focused on the laboratory as a means of learning science concepts and developing the intellect. The methods goal incorporated to some extent both the knowledge and development goals.

The Goal of Personal-Social Development

Development as a goal of education is evident in *Emile* by Jean Jacques Rousseau (Boyd, 1956), written in 1762. The origins can be further traced to the humanist educators of the early Renaissance (Woodward, 1963). We see the effect Rousseau had on the ideas of educators such as Pestalozzi.

By about 1860, Johann Pestalozzi's ideas on child development were incorporated in the science program and became an important goal of science education. As we have mentioned, the pedagogical model of Pestalozzi's ideas was the object lesson. The assumption underlying object lessons was that an individual's faculties developed in a serial order with age, ranging from simple to complex, concrete to abstract. Object teaching was an attempt to align the curriculum and instruction with the natural development of the student.

Object teaching has for its purpose a thorough development of all the child's faculties and their proper employment in the acquisition of knowledge . . . A system of training. . . . Its purpose is not the attainment of facts . . . but the development, to vigorous and healthy action, of the child's powers of getting and using knowledge. . . . It does not signify the things about which something is taught, not that which is taught about them, so much as it seems the principles by which the teaching is performed, the purpose and manner of the teaching. (Calkins, 1890, p. 15)

As pointed out in the quotation, object lessons provided a method of teaching that was appropriate for the teaching of science. The method used all aspects of the environment to develop the students' faculties. Teachers complemented the students' sequential development through the use of objects from the environment. Psychological development was the primary goal of this instructional approach. The logical structure of the science curriculum and other goals, such as knowledge, were subordinate in this curriculum model of teaching elementary science. The lack of organization of the objects of study into subject areas, or logical bodies of knowledge, brought about major criticism of this model of teaching science.

During the same period, the late 1800s, the object-lesson model of teaching also influenced science curriculum at the secondary level through the development of the laboratory. Laboratory instruction was also supported by an emphasis on faculty psychology that advanced the notion that direct contact with the objects of the world would develop observational and inductive reasoning skills. The laboratory manual was used especially in biology (Hurd, 1961). These activities fulfilled the developmental "need" of the faculties of reasoning, observation, concentration, sense, perception, and the general discipline of a scientific mind.

Object lessons were never widely adopted in American schools despite the quantity of professional literature acclaiming the merits of this approach. The reasons for the decline of object lessons included lack of specialized methodology, the exclusion of textbooks to facilitate an understanding of the concepts, and the heavy demands made on the ability and knowledge of the teacher.

The important point to be made in this section is that object lessons of the period from 1860 to 1880 did represent a definite curriculum model for teaching science and the model had development as the primary goal. The arrangement of goals represented by the object-lesson model reemerged in the nature-study movement of the late nineteenth and early twentieth centuries. In the interval from about 1870 until the 1900s, the major goal of science education was the teaching of scientific knowledge.

Throughout the nineteenth century, the goal of personal intellectual development competed with the goal of learning science facts and information. The methods of science were also considered important, especially by scientists who encouraged the inclusion of science in the curriculum, but developing inductive thought and observational skill was largely justified as a way of developing the intellect, not a goal independent of the others.

The competition between the development and knowledge goals is evident in two curricular models that became popular toward the end of the nineteenth century. One of these emphasized the knowledge and the other the development goal. We will look at these two models to demonstrate the tension that had developed between these two major goals by the late nineteenth century.

By the end of the nineteenth century, there were two models of science curricula at the elementary level. One model was called nature study; the second was a subject matter, knowledge-oriented approach referred to as elementary science. By the mid–nineteenth century, the results of technological development had become evident and popular interest in science had emerged. Industrial expansion, with resulting emigration from rural to urban areas, supported two differing curriculum models for science at the elementary level. The post–Civil War depression of 1873 brought severe criticism of the public schools and demands for clarifying the goals of education. The educational journals had numerous articles stating a case for more science in the school program. With the changes in social and economic patterns toward an industrial-technological society, science was perceived as an important program in the schools. Changes in society gave impetus to a more practical, knowledge-oriented model of elementary school science. Several important educational leaders contributed to this model of elementary school science.

Elementary School Science

The model of science teaching known as elementary science was structured with knowledge as the primary goal. This model was influenced by the rise of industrialism in the United States and the need for individuals who were knowledgeable about the raw materials, processes, and products of the new industrial economy. F. W. Parker synthesized Pestalozzi's method, Froebel's view of the child, and Herbart's theory of learning and developed a curriculum model that represented a transition to a new scientific pedagogy. Parker's ideas also marked a new lack of dependence on direct importation of European educational ideas (Cremin, 1965). Parker's desire was to use science as a unifying theme in the elementary school; the goals were to understand the universe and to use scientific techniques as a means of problem solving (Underhill, 1941). Scientific knowledge and method were the basic goals in Parker's curriculum model of elementary science.

Wilbur S. Jackman worked with Francis Parker at the Cook County Normal School, now the School of Education of the University of Chicago. With Parker's support, Jackman developed an elementary science program that is similar to some contemporary programs of science instruction. The plan of science instruction formulated by Jackman was the first curriculum model to emphasize generalizations of science as the major organizing goals; although they were not clearly stated as such (Jackman, 1891). Jackman's program emphasized the understanding of significant concepts and emphasized observation and experimentation. In the context of this essay, Jackman's curriculum model had scientific knowledge as the primary goal and scientific method as the means of achieving that goal. Jackman was also influential in the nature-study movement (Jackman, 1904), which for a time overshadowed his work on the curriculum model of elementary science with scientific knowledge as the primary goal.

Another advocate of the knowledge goal was William T. Harris. Harris is generally credited with formulating the first substantial elementary school science curriculum (Harris, 1896). The Harris program emphasized science concepts and their relationship; the framework used in the organization was the specialized science disciplines. The Harris curriculum placed science in a subordinate role to the humanities, but his was still one of the first fully formulated science programs. The curriculum was organized with scientific knowledge as the goal, although he did include personal development as a goal. Harris advocated teaching science so that various branches of science and topics within those branches would relate to both the student's psychological needs and the environment. In Harris' program, there was recognition of the need for an interdisciplinary approach to science and functionality of content in terms of the needs of the student.

E. G. Howe later changed the Harris science curriculum into a program entitled *Systematic Science Teaching* (Howe, 1894). The Harris and Howe programs were predecessors of many contemporary programs; they included statements of learning objectives, teaching procedures, and required materials.

During this period, 1870 to 1900, the approach used in the elementary science program changed from the presentation of facts to the use of science generalizations as a basis for organization. With several transformations, this knowledge-oriented elementary science model has continued to the present time. Scientific method, as a means of acquiring knowledge, was emerging in the form of a greater emphasis on the laboratory and experimentation by the pupils. Influential people such as William Harris and Wilbur Jackman included methods such as observation and first-hand experience in their science programs as ways to acquire knowledge. The goals of personal and social development were evident, but not emphasized, in the science programs of this period. It should also be noted that in spite of the quantity of professional literature, little science was actually taught in the elementary school. The literature emphasized the value of science, but in practice most early programs were not implemented.

Nature Study

Nature study was similar to the earlier movement of object teaching in that personal-social development was the primary goal. Nature study was part of a larger movement that combined the influence of romantic literature (e.g., Emerson, Thoreau, Longfellow) and the principles of educational reform expressed in the writings of Comenius, Pestalozzi, Rousseau, and Froebel. Respectively, these two movements exalted individualism and emotionalism and focused attention on the child as a developing biological organism (Underhill, 1941).

In elementary school science education, the nature-study movement began during the depression of 1891 to 1893 in an attempt to slow the migration from the agricultural communities to New York City, migration that added to the relief rolls. In 1895, a joint conference was called by the Committee for the Promotion of Agriculture in New York State and the New York Association for Improving the Condition of the Poor. The objective of the conference was to examine the causes of the agricultural depression. Nature study was seen as a way of interesting children in farming and keeping them at home. As Lawrence Cremin stated, "Nature study was the great remedy for the alienation of man from the land and from his neighbor" (Cremin, 1964, p. 77).

The nature-study program was located at Cornell University under the direction of Liberty Hyde Bailey. Bailey (1903) and her associates at Cornell became instrumental in the organization and dissemination of materials that directly influenced the nature-study movement. Bailey described the function of nature study:

Nature study is a revolt from the teaching of mere science in the elementary grades. In teaching-practice, the work and the methods of the two [science and nature study] integrate . . . and as the high school and college are approached, nature study passes into science teaching, or gives way to it; but the ideals are distinct—they should be contrasted rather than compared. Nature study is not science. It is not facts. It is concerned with the child's outlook on the world. (pp. 4–5)

The nature-study curriculum model had interest as the motivating force within the child. As in the object-lesson movement, the educational theory underlying nature study stressed the importance of the child as a developing biological organism with motivational needs. As it evolved, nature study acquired an interdisciplinary theme. It included reading, writing, mathematics, drawing, and music as complements of study.

The spirit of nature study requires that the pupils be intelligently directed in the study of their immediate environment in its relations to themselves; that there shall be, under the natural stimulus of the desire to know, a constant effort at a rational interpretation of common things observed. . . . This plan . . . will lay a sound foundation for the expansive scientific study which gradually creates a world picture, and at the same time enables the student, by means of the microscope, the dissecting knife, and the alembic, to penetrate intelligently into its minute detail. (p. 9)

Wilbur Jackman, who was influential in promoting and developing the elementary science model, with scientific knowledge as the primary goal, was also influential in developing the nature-study model. Jackman's elementary school science curriculum stressed observation and experimentation as means of obtaining scientific concepts and understanding of science methods (Jackman, 1904). This combination of scientific knowledge and scientific method with the developmental goal could have provided a balance for science curriculum and instruction in the elementary school; however, for the most part, that did not happen. Jackman's model of science curriculum can be contrasted with the extreme romantic position of the larger nature-study movement and the knowledge emphasis in both elementary and secondary science.

High school teachers, operating from a model of science education with scientific knowledge as the primary goal and little recognition of personal-social development, voiced protest against nature study and especially against the goals of appreciation and interest:

High school teachers of science must protest against a mass of so-called nature study, more or less sentimental and worthless ... and must not be content simply with treating lightly this farcical science teaching, or passing it by with silent contempt. (Brownell, 1902, p. 253)

The nature-study model was faulted for not having the goals of science education structured in the same way as in the knowledge-based curriculum model, an argument that still continues.

This distinction between nature study and elementary science is made clear in Bailey's words:

Nature may be studied with either of two objects: to discover new truths for the purpose of increasing the sum of human knowledge; or to put the pupil in a sympathetic attitude toward nature for the purpose of increasing his joy of living. The first object, whether pursued in a technical or elementary way, is a science-teaching movement, and its professed purpose is to make investigators and specialists. The second object is a nature study movement, and its purpose is to enable everyone to live a richer life, whatever his business or profession may be. (Bailey, 1903, pp. 4–5)

To summarize, the basic differences between the nature-study model of science education and the elementary science model were that (1) nature study stressed facts primarily for their own sake unrelated to larger generalizations, whereas elementary science was concerned with facts in relations to generalizations, that is, the structure of science; (2) nature study stressed personal development, and elementary science did not; and (3) nature study incorporated aesthetics and appreciation, whereas elementary science stressed classification, experimentation, and an organized system of knowledge. The underlying goals of knowledge, methods, and personal-social development were present in both the elementary science programs and the nature-study movement. The primary difference was in structure and emphasis, and the different arrangements were at least in part a result of countervailing societal pressures, one toward industrialism and the other toward a rural, agrarian naturalism. Nature study declined for many reasons, but this model that stressed personal development and appreciation of the environment remained a subordinate model in science education over the years. In the early twentieth century, the nature-study movement lost prominence and the goal of science teaching at the elementary level continued its long hold on organized knowledge.

THE EARLY TWENTIETH CENTURY: A TIME OF TRANSITION

Committee of Ten

Before the late nineteenth century, most of the statements of goals for education were made by prominent individuals and did not have the backing of educational institutions or professional organizations. That changed in 1893 with the publication of the report of the Committee of Ten (National Education Association, 1893). This committee of the National Education Association (NEA) was formed in July 1892 at the NEA's annual meeting. The committee was charged with the responsibility of coordinating entrance requirements for college admission and thoroughly examining each of the major subjects that students were expected to take before being admitted to college. A separate conference was established for each major subject area.

There were nine subject area conferences in all, three of them in the sciences. One science conference was on natural history, including physiology, zoology, and botany; a second was on physics, chemistry, and astronomy; and a third was on geography, including physical geography, geology, and meteorology.

The Committee of Ten was headed by Charles Eliot, president of Harvard University. Members of the committee included William T. Harris, the U.S. Commissioner of Education, and eight others including college presidents and principals of secondary schools.

By the late nineteenth century, science was established as a subject in the school curriculum, but it was a relatively new subject and conferees took advantage of the opportunity to prove that science did indeed belong in the curriculum along with the older subjects such as Latin, Greek, and mathematics. They argued that science had disciplinary value, that is, that the study of science would help to develop the intellect, especially the observational and inductive faculties. Their arguments did not support either the informational or commercial value of science. The most important thing was for students to be placed in contact with the physical world and to draw conclusions from it. Thus, laboratory work played a major role in the science courses proposed by the science conferences.

Perhaps one of the most significant recommendations of the science conferences was the amount of time that should be devoted to science in the school curriculum. At a meeting of the three science conferences, the groups passed a joint resolution calling for 25 percent of the total school curriculum to be devoted to a study of science. As the Committee of Ten drew up sample courses of study to accommodate different student interests in the curriculum, they assigned considerable time to the study of science in each of the courses. Even the classical course that included the least amount of science required students to take physical geography, physics, and chemistry. The other sample courses (Latin-scientific, modern languages, and English) required students to take, in addition, varying amounts of zoology or botany; astronomy and meteorology; geology and physiography; and anatomy, physiology, and hygiene. In these recommended courses, approximately one-fifth of a student's time would be spent in the study of science.

Recommendations for such an extensive time commitment to science were not to be realized, however. In 1896, the Department of Secondary Education of the NEA formed a committee to discuss ways to implement the recommendations of the Committee of Ten. This Committee on College-Entrance Requirements submitted its final report at the annual meeting of the NEA in Los Angeles in 1899. The report endorsed the work of the Committee of Ten, and it recommended that the courses proposed by the Committee serve as norms for college admission and high school graduation. Rather than having students select a particular curriculum in advance, the Committee on College-Entrance Requirements recommended that a set of constants be the heart of everyone's course of study and that students complete their programs from a list of approved electives. The Committee recommended that students study English for 2 years, languages other than English for 4 years, mathematics for 2 years, history for 1 year, and science for 1 year. In addition, students would be required to choose six electives. Of the total 16 units, only 1 was required in science. Students could elect to take more science for high school graduation and college admission, but only 1 year of science study was required. The requirement for linguistic studies, however, was 6 units.

A Reorganization of Goals for the Science Curriculum

One of the assumptions of the Committee of Ten and the Committee on College-Entrance Requirements was that all students in high school, whether they were planning to attend college or not, would take the same subjects. They believed that as long as the subjects disciplined the intellect, they would be useful both as preparation for life and for college admission. During the coming years, however, it became more and more difficult to defend a traditional, disciplinary curriculum as preparation for life. More and more students were entering the schools looking for a practical curriculum that would enhance their opportunities in the work force. Commercial and industrial arts courses were being requested in addition to the traditional academic courses described by these committees.

At the same time, educational leaders began to promote goals having social utility over goals that were personally satisfying. In fact, proper balancing of the socially and personally beneficial goals became a subject of discussion among educators in the early twentieth century. Social utility became a concern among educators because of the rapid changes that were occurring in society. Chief among these were the social problems that stemmed from immigration, industrialization, and urbanization. Crime and poverty, poor sanitation, and the spread of infectious diseases were just a few of the problems facing society. Perhaps the schools could assist in educating citizens to deal effectively with these problems. In addition, it was thought that the schools could be used to Americanize the immigrants for life in a democratic society. A reorganized curriculum would teach the dominant values of the society and make individuals productive members of that society.

The Commission on the Reorganization of Secondary Education

In 1911, the Department of Secondary Education of the NEA formed the Committee on the Articulation of High School and College to continue to investigate the relationship between the school and college curriculum (National Education Association, 1911). The committee's report shows how quickly the mood in education had shifted from goals that focused on personal development (intellectual development and personal satisfaction) to ones that were socially relevant. The main recommendation was to reduce the requirement in foreign and classical languages from 4 to 2 years and to allow students to include vocational courses in their electives for graduation and college admission. In addition, a set of individual subject-matter committees was formed to examine the nature of school subjects and determine their usefulness to society. The parent committee became known as the Commission on the Reorganization of Secondary Education (CRSE), and in 1918 the CRSE issued its report entitled *Cardinal Principles of Secondary Education* (NEA, 1918). In 1920, the science committee submitted its report entitled *Reorganization of Science in Secondary Schools* (NEA, 1920).

The CRSE stated that all school subjects, both academic and vocational, should be defended in relation to seven goals or "cardinal principles" of education. The overall aim of education was to prepare youth for effectiveness in a social world, what they called "complete and worthy living for all youth" (NEA, 1918, p. 32). Preparation for living effectively in a democratic society would focus equally on the needs of the individual and the needs of society. Neither should be subordinated to the other. The seven goals of education were (1) health, (2) command of fundamental processes, (3) worthy home membership, (4) vocation, (5) citizenship, (6) worthy use of leisure, and (7) ethical character. Accomplishment of these goals would prepare good citizens, good family members, productive workers, and, most important, a stable society. The primary objective of education was to develop individuals who would be happy and contributing members of society.

In their final report, the science committee defended the presence of the sciences in the curriculum in terms of six of the seven cardinal principles. Only "command of fundamental processes" was omitted. With respect to the health goal, the committee said that science courses could help protect people from illness and help control disease by providing them with knowledge of public sanitation and personal hygiene. Worthy home membership would be enhanced by teaching people how to operate and repair any number of appliances and conveniences in the home. With respect to the vocational goal, the committee said, "In the field of vocational preparation, courses in shop physics, applied electricity, physics of the home, industrial and household chemistry, applied biological sciences, physiology, and hygiene will be of value to many students if properly adapted to their needs" (NEA, 1920, p. 13). Concerning the goal of citizenship, science courses could make individuals more appreciative of the role of scientists in society and better able to select technical experts for their special roles in society. Science courses could contribute to the worthy use of leisure by instilling in individuals an appreciation of the natural world and by developing an understanding of such science-based hobbies as photography. Finally, the study of science could assist in the development of ethical character "by establishing a more adequate conception of truth and a confidence in the laws of cause and effect" (NEA, 1920, p. 14).

In 1893, science was defended as a disciplinary study, one that would develop a person's mental faculties. By 1920, the study of science was being defended for its ability to develop individuals who would contribute to a stable, smoothly functioning society. Although this was an obvious change in the direction that the goal statements took, close examination of the statements of the CRSE showed that other goals were considered important as well. In addition to the goal of knowledge for social welfare (sanitation, technological applications, sex education), science-based avocational interests and career awareness were stressed (i.e., personal-social development goals), as was the goal of scientific method. Scientific method was not to be taught for purposes of mental discipline, however, but as a method of social problem solving. In addition, the basic principles of science, as organized by scientists, formed a central part of the science curriculum, regardless of their practical utility.

The Progressive Era

The years from approximately 1917 to 1957 represent the Progressive Era in American education (Cremin, 1964). The Association for the Advancement of Progressive Education, later renamed the Progressive Education Association, was organized in 1919 by Stanwood Cobb and held its final meeting in 1955. The organization's journal *Progressive Education* appeared in 1924 and published its last issue in 1957. For many educators this period reaffirmed ideas developed during the years preceding the publication of the cardinal principles. The Progressive Era was a period of confirmation of child-centered education, the importance of real-world applications, the social importance of knowledge, and the need to make school learning enjoyable and meaningful to the student.

As was the case in the first two decades of the twentieth century, the writing in science education during this period revised and gave a new priority to goals. Progressive educators argued against traditional methods and content in favor of content with greater social relevance and methods that would give students the tools to solve problems in their everyday worlds. Too many teachers were clinging to the old methods of instruction in which memory of facts and meaningless tasks dominated teaching and learning. Frustration was often expressed over the limited progress that had been made in moving toward the goals presented in the cardinal principles. These debates were repeated with calls for an end to traditional education and suggestions for an improved approach to education.

A major accomplishment of this period was the establishment of a definite sequence of courses that included general science in the first year of high school, followed by the specialized courses of biology, physics, and chemistry in the last 3 years. When the Committee of Ten issued its report, many courses were competing for a place in the high school curriculum. Zoology, botany, chemistry, physics, physical geography, geology, astronomy, agriculture, physiology, and hygiene were just some of the high school science offerings. In part because of the recommendations of the Committee of Ten that only some of these courses be considered electives, enrollments in subjects like geology and astronomy declined by 1910, while chemistry and physics became firmly entrenched in the school system. However, the establishment of biology as a single course encompassing zoology, botany, physiology, and hygiene made slow but steady progress through the early years of the twentieth century despite the Committee of Ten's recommendation that a separate course in either zoology or botany would provide better disciplinary training.

During the Progressive Era, there was considerable lack of agreement on the goals of science education. The debate was between the old and the new, between an emphasis on subject matter (i.e., the knowledge goals) and an emphasis on the application of subject matter to the lives of the students (i.e., the development goal). It was difficult, if not impossible, to blend the emphasis on knowledge of the subject, scientific method, and social and personal development in a way that was satisfying to a broad spectrum of educators. Groups tended to be polarized at the extremes of the various positions. Throughout the Progressive Era, however, there were strong, if not unified, commitments to personal-social applications and to the use of organizing themes as a way to structure scientific knowledge to make it more meaningful to the students. There was also a strong commitment to scientific method as a means of developing general problem-solving skills that could be used in the solution of social problems.

The Goal of Scientific Knowledge in the Progressive Era: A Summary

A survey by the NEA (1926) listed the common objectives of elementary science. The objectives included observation and purposeful activity, learning the names of simple objects, combining the facts and emotions, learning the use of scientific thinking, understanding simple concepts, and finally a reverence for nature. This mixture of objectives represents an amalgam of past history in elementary science and a recognition of scientific knowledge, method, and personal-social development as fundamental goals for elementary school science. With this new, more balanced list of goals, the time seemed right for the emergence of a new curriculum model in elementary school science, and it was developed by Gerald Craig at Columbia University.

The Craigian model for curriculum and instruction included modifications of the generalizations and ideas outlined earlier by Wilbur Jackman. The model presented by Craig (1927) in "Certain Techniques Used in Developing a Course of Study in Science" for the Horace Mann Elementary School had a major influence on the structure of contemporary science education.

In 1927, Gerald Craig published the results of several years of research into the important scientific ideas that might be included in the science curriculum. These were the scientific principles that appeared in children's questions. From these data, Craig identified a number of generalizations, principles, and concepts that could be used as the core of the elementary school science program (Craig, 1927). Although Craig's model included the developmental needs of children and experience with important scientific methods and social and democratic attitudes, he felt that scientific knowledge was the important goal of elementary school science.

In 1932, the National Society for the Study of Education (NSSE) published *A Program for Teaching Science* (NSSE, 1932). Craig was on the yearbook committee, and his influence was significant. For example, the shift of knowledge from facts to principles and generalizations was emphasized in the yearbook, and the transfer of Gerald Craig's curriculum model to the secondary level also was established by this book.

The yearbook committee stressed the use of principles and generalizations as the knowledge goal of science education in an attempt to deal the final blow to the nature-study movement (NSSE, 1932). Scientific principles and methods were clearly stated as the goals of science instruction. Experience, activity, and the "scientific method" were the means of achieving these ends. The report also stressed the importance of a continuous and unified science program. The authoritative influence of this yearbook perpetuated the curriculum model Craig established.

In 1932, Craig published a series of elementary textbooks that were structured on scientific generalizations. The books usually had major divisions of biology, earth, and physical science and included activities to complement the major scientific principles. These activities gave students experience that verified the scientific principle that was the main objective of the lesson. This was the predecessor of the science textbook series that dominated elementary school science until the mid–1960s. Many individuals referred to these as elementary science textbooks as "science readers" because reading was the major means of science instruction; and understanding scientific knowledge, in the form of generalizations, was the primary aim (Hurd, 1970, p. 3).

The pragmatic philosophy of John Dewey was also included in the elementary curriculum of the 1930s. Topics such as water supply, transportation, communications, heating, health, and conservation were included in the science program for elementary schools. An individual having knowledge and understanding of concepts related to these socially functional topics would be useful as a citizen in a democracy (Hurd, 1969).

In the 1947 NSSE Yearbook, *Science Education in American Schools*, the goals of science teaching remained similar to those of the previous decade, that is, understanding scientific principles and generalizations, developing an understanding of scientific methods and problem-solving skills, having an interest in and appreciation of the environment, and acquiring favorable social attitudes (NSSE, 1947).

In a 1957 publication entitled *Science in the Elementary Schools* (1957), Craig again reinforced his model of science curriculum. The legacy of Gerald Craig's model of science teaching had now continued in science education for more than 30 years. The structure of his model was finally transformed with the curriculum movement of the 1960s, but the textbook series designed in keeping with Craig's ideas are still used.

The Goal of Scientific Methods in the Progressive Era: A Summary

Although numerous nineteenth-century scientists had advocated the use of scientific processes as a way to acquire science knowledge, scientific methods were recognized by few as a goal of education in science until after the 1900s. The Baconian idea of the scientific method was expressed earlier by Herbert Spencer. Spencer stressed the importance of scientific knowledge and an application of the "scientific method" to raw data. If the scientific method was acquired by the student, it would provide a continuing means to knowledge even if knowledge expanded and changed (Spencer, 1966).

About 1890, Henry E. Armstrong devised a heuristic method for teaching science. Armstrong advocated "placing students as far as possible in the attitude of the discoverer ... [applying] methods which involve their finding out instead of being merely told about things (Armstrong, 1913, p. 236). Armstrong's emphasis on scientific method is evident in his own statements.

The facts must always be so presented ... that the process by which results are obtained is made sufficiently clear as well as the methods by which any conclusions based on the facts are deduced. And before any didactic teaching is entered upon to any considerable extent, a thorough course of heuristic training must have been arrived at and the power of using it acquired; scientific habits of mind, scientific ways of working, must become ingrained habits from which it is impossible to escape. And as a necessary corollary, subjects must be taught in such an order that those which can be treated heuristically shall be mainly attended to in the first instance. (p. 255)

The progression from Francis Bacon to Herbert Spencer to Henry Armstrong can be viewed as a transformation of Bacon's inductivist philosophy of science to the pedagogical application of his ideas. These pedagogical applications were widely developed during the Progressive Era.

Beginning in the early part of the twentieth century, there was an increasing concern about the methods of science being included in the science curriculum. This growing emphasis on method as a goal of science education was underscored by the need to solve social and economic problems and increased understanding of individual thinking in the process of scientific discovery (Underhill, 1941). One individual who was concerned with both of these problems was John Dewey.

Dewey's contributions to science education were numerous, but perhaps one of the most significant was his contention that the methods of science were at least as important as, if not more important, than scientific knowledge. In 1916, Dewey strongly advocated an experimental approach to science teaching (Dewey, 1944). The mediating psychological step between experience and scientific knowledge was the scientific method. He stated the goal of science teaching: "The end of science teaching is to make us aware of what constitutes the more effective use of mind, of intelligence" (Dewey, 1944, p. 120).

Dewey later stated the means to this end; note that "scientific method" is a means and end, logical, and psychological.

What is desired of the pupil is that starting from the ordinary unclassed material of experience he shall acquire command of the points of view, the ideas and methods, which *make* it physical or chemical or whatever ... the dynamic point of view [is] the really scientific one, or the understanding of *process* as the heart of the scientific attitude. (p. 122)

Thus, for Dewey, scientific methods were the means to scientific knowledge and also the means to understanding scientific method as a goal of science instruction.

When incorporated in the curriculum, Dewey's goals of reflective thinking became presentation of a *problem*, formation of a *hypothesis*, *experimenting* to collect data, and finally the formulation of a *conclusion* (Dewey, 1944).

Hurd points out, however, that this approach was weak, not in the goal but because of the difficulty of being implemented in science programs as a means to the goal. That is, experimenting with materials was not practical in the classroom during this early period (Hurd, 1969).

Scientific methods and attitudes important for living in a democracy were emerging as goals of science teaching. Hurd reports:

Many scientists at this time (1920–1930) expressed the opinion that the central purpose of education in science should be the development of an understanding of the nature of science, its methods, attitudes and cultural impact. As a result much educational research in the succeeding years was directed toward identifying the elements of the scientific method and scientific attitudes. (Hurd, 1961, p. 51)

Scientific method gained further support in the 1928 report entitled "On the Place of Science in Education." It was resolved that "Science as method is quite as important as science as subject matter and should receive as much attention in science instruction" (*On the Place of Science in Education*, 1928, p. 664). As the leaders in science education continued researching the role science could play in education, they identified attitudes: applying major scientific principles and concepts, using the scientific method in solving problems, and appreciating the habits of conservation (Central Association of Science and Mathematics Teachers, 1950).

The Thirty-First NSSE Yearbook, *A Program for Science Teaching* (NSSE, 1932), gave five methods of science as objectives for science education.

1. Conviction of universal basic cause and effect relations.
2. Sensitive curiosity concerning reasons for happenings.
3. Habit of delayed response, holding views tentatively for suitable reflection.
4. Habit of weighing evidence with respect to its pertinence, soundness, and adequacy.
5. Respect for another's point of view, openmindedness, and willingness to be convinced by evidence. (p. 56)

John Dewey's model of reflective thinking was one of the goals of the Progressive Education Association's (PEA) 1938 report *Science in General Education*. The process of reflective thinking was stated as having five phases:

1. A sense of perplexity, or of want, or of being thwarted, followed by identification of the problem.
2. 'Occurrence' of tentative hypothesis ...
3. Testing and elaborating hypotheses ...
4. Devising more and rigorous tests ...
5. Arriving at a satisfactory solution and acting upon it. (pp. 309–310)

As seen in this quotation, the PEA emphasized the methods of science as an aim of science education. Publication of *General Education in a Free Society* in 1945 gave further support to the aim of scientific methods. The Harvard Committee stressed the psychological process more than the logical. In an approach outlining the scientific method, they underscored solving problems in the natural environment (Harvard Committee, 1945). The report also made it clear that scientific knowledge stated as broad, integrative concepts and taught in an interdisciplinary approach was the preferable goal of science education.

Scientific methods were restated in the Forty-Sixth NSSE Yearbook, *Science Education in American Schools* (NSSE, 1947). Included were objectives such as problem-solving skills, instrumental skills, the scientific attitudes of openmindedness, honesty, and suspended judgment, appreciation of science, and interest in science (NSSE, 1947). The issue of attitudes was placed in the larger context of science instruction:

Instruction in science must take cognizance of the social impact of developments produced by science. It is not enough that they be understood in a technical or scientific sense; it is most important that these effects on attitudes and relationships of people be studied and understood. Science instruction has not only a great potential contribution to make but also a responsibility to help develop in our youth the qualities of mind and the attitudes that will be of greatest usefulness to them in meeting the pressing social and economic problems that face the world. (p. 1)

The aim of scientific method seen in various forms was firmly in the professional literature by the late 1940s. Scientific methods were researched and a problem-solving checklist was produced in 1956 (Obourn, 1956). Unfortunately, textbooks dominated the period from 1930 to 1960, and the role of scientific methods was sharply reduced by the imposed pedagogical parameters of a textbook.

The Fifty-Ninth NSSE Yearbook, *Rethinking Science Education*, (NSSE, 1960) also stressed the importance of inquiry as a characteristic of the scientific process:

One function of the elementary school has always been to help children learn a part of what they need to know from the world's storehouse of knowledge. In recent years this function has embraced more and more science. Scientific methods of investigation by which knowledge may be acquired and tested are now very much a part of culture. The elementary school should help children become more acquainted with these methods. (pp. 112–113)

In this statement, the yearbook committee makes a strong plea for the goal of scientific methods, which by this time was established in principle as a viable goal in science education but was not widely established in practice.

The Goal of Personal/Social Development in the Progressive Era: A Summary

In education, the overshadowing figure of the Progressive Era was John Dewey. His pragmatic philosophy reaffirmed for educators the thesis that education should center on the student and reflect the realities of the time. In the first chapter of *Experience and Education*, written in 1938, Dewey outlines the essential difference between traditional and progressive education. The knowledge model of curriculum in science education as we have discussed it was generally aligned with Dewey's concept of traditional education. Dewey characterized traditional education:

The subject matter of education consists of bodies of information and of skills that have been worked out in the past; therefore, the chief business of the school is to transmit them to the generation. (p. 17)

Dewey continues by suggesting two other factors that have contributed to traditional education: rules of conduct and the patterns of organization (schedules, bells, examination, etc.) within the school (Dewey, 1938).

With reference to this essay, Dewey's comments can be viewed as criticisms of the curriculum model that uses scientific knowledge as its primary goal. Dewey (1938) clearly states his position on knowledge, "We may reject knowledge of the past as the *end* of education and thereby only emphasize its importance as a *means*" (p. 23).

Underlying Dewey's philosophy is a concern for explaining events in terms of their consequences. Thinking is seen as an instrument of adaptation resulting from experiences that caused disequilibrium. Thus, there is a constant interaction between the changing individual and the changing environment, the result of which is personal development. Dewey uses the "scientific method" as a means of explaining the process of equilibration once the individual was at disequilibrium. To Dewey, personal-social development is the aim of education, and this goal is achieved through experience that by its nature includes the goals of scientific knowledge and method.

What does a progressive curriculum look like when these ideas are implemented? Dewey provided a short summary that contrasts his curriculum model with the knowledge model of the traditional school.

To imposition from above is opposed expression and cultivation of individuality; to external discipline is opposed free activity; to learning from texts and teachers, learning through experience; to acquisition of isolated skills and techniques by drill, is opposed acquisition of them as means of attaining ends which make direct vital appear; to preparation for a more or less remote future is opposed making the most of the opportunities of present life; to static aims and materials is opposed acquaintance with a changing world. (Dewey, 1938, p. 20)

During the Progressive Era, there was considerable tension between the traditional curriculum models that emphasized knowledge acquisition and newer progressive models that emphasized individual development and social relevance. The tension is clearly evident in the NSSE's Thirty-First Yearbook, in which contributors struggled with the relative importance of focusing on the subject matter or the needs of the learner. In the end, the Yearbook Committee found it difficult to give up the traditional emphasis on knowledge acquisition, even though they did give considerable recognition to individual experience and the needs of society.

Publications of the PEA, however, show a clearer vision of child-centered education and the social relevance of what was to be learned. *Science in General Education* (PEA, 1938), for example, has a distinctly different perspective from the NSSE Thirty-First Yearbook. The committee for the PEA recognized the nature of the individual: "the human organism is inherently characterized by tensions between itself and its environment, and that its activities represent the attempt to satisfy these tensions" (PEA, 1938, p. 17). Later the report (PEA, 1938) rejects the stimulus-response psychology of E. L. Thorndike.

This conception of the nature of individual behavior places an emphasis upon continuous reconstruction of experience of the learner as the essence of education. It implies a rather complete rejection of the stimulus-response concept of learning (the analysis of situations into fixed elements to which unit responses are learned one by one); on the contrary, emphasis is placed upon the total situation. Learning takes place through the reorganization of the individual's present behavior patterns (or wholes) into more inclusive, more adequate inclusive, more adequate patterns by bringing together and elements that have meaningful relations to the individuals' behavior pattern and to the situation at hand. (pp. 19–20)

The goal of personal-social development can be identified in the PEA's statement. It includes emphasis on the total experience, the continuous reconstruction of elements of knowledge, and a holistic approach to understanding learning. The PEA position is in marked contrast to the 1932 NSSE Yearbook, *A Program for Teaching Science*. Because of the philosophical and psychological orientation of the knowledge model, the Thirty-First NSSE Yearbook gave less consideration to personal needs. "No attention has been given in the foregoing paragraphs to the questions of whether learning activities are pupil-directed or teacher-directed. A discussion of this question is not necessary" (NSSE, 1932, p. 20). The philosophical position of the Thirty-First NSSE Yearbook was that experience was important because it provided the connections or associations; however, one did not have to consider the individual except as a receptor of the experience. Their conclusion then emphasizes a logical organization of knowledge with a logical system of experiences, a conclusion much different from the developmental goal expressed by the PEA's 1938 book, *Science in General Education*, which gives greater emphasis to the developmental goal:

As he deals with the situation, he learns physically ... he responds emotionally; he builds attitudes; he finds out what will work and what won't work in analogous situations; and he organizes on his level the meanings which the situation has for him—all of these responses being in progress at the same time. (pp. 20–21)

The PEA Committee on Science summarizes its position:

The individual is seen always in interaction with his environment, and it is necessary to take account of this fact in order to understand and deal helpfully with him. The multiple of needs characteristic of any individual grows out of his dynamic quality of human life; the individual is active and purposeful in the attempt to satisfy drives which take on meaning and character in interaction with his environment. It is his peculiar combination of needs, longings, desires, and his preferred ways of meeting them which constitute his uniqueness and which define the effective approach for him. Whatever goes on as he deals with his environment leaves its effect on him physically, emotionally, intellectually, he learns as a whole. (p. 22)

In summary, by the end of the Progressive Era, no consensus had been reached concerning the relative importance of the three major education goals. The traditional emphasis on knowledge acquisition persisted despite vigorous attempts to reform education along other lines, especially to make it more personally and socially relevant. The themes of personal and social relevance were very strong in some quarters, and where they were strongest they were sometimes seen as providing a curriculum that lacked rigor and was weak in content coverage. Scientific method was also discussed as a goal of science teaching, but primarily as a way of providing students with a model for solving the problems of society. Learning the scientific method had shifted from an earlier emphasis on mental discipline to a focus on social relevance.

CURRICULUM REFORM OF THE SIXTIES

By the mid–1950s, scientists began responding to the criticisms leveled at American education in general and science education in particular. With the support of their professional organizations and the financial backing of the National Science Foundation (NSF), scientists began discussing ways in which they could help renew the school science programs.

When the Soviet Union launched its earth-orbiting satellite *Sputnik* in October 1957, the U.S. government enthusiastically backed, with unprecedented financial support, a reform effort begun several years earlier. The involvement in science teaching by the federal government over the following two decades was the largest in our history, hence the title the "Golden Age." Curriculum in this era used an approach to science education that emphasized the logical structure of the disciplines and the processes of science.

The Goal of Scientific Knowledge

During the late 1950s and early 1960s, there were three significant changes in the goals of science education: (1) recognition of the personal-social developmental goal declined; (2) scientific knowledge was modified to emphasize understanding the structure of scientific disciplines and this goal be-

came the primary goal of science curricula, especially at the secondary level; and (3) scientific methods, now discussed as inquiry, discovery, and problem solving, became the means of achieving the knowledge goal and not a means of general problem solving to solve society's problems. These changes were brought about by pressure from two different sources, the scientists who directed the curriculum projects and the shortage of technically trained individuals in our industrial society.

In 1960, Jerome Bruner's book *The Process of Education* (Bruner, 1960) provided a vision that transformed the goals and structure of science education. Several of the new science projects were being developed before the publication of Bruner's book, but it was *The Process of Education* that symbolized the new model of science education. Bruner modified the knowledge goal from Gerald Craig's emphasis on generalizations and stressed the structure of scientific disciplines instead. Important knowledge for science teaching consisted of the concepts that formed the structure of a discipline. Bruner (1960) stated his aim:

The curriculum of a subject should be determined by the most fundamental understanding that can be achieved of the underlying principles that give structure to that subject. (p. 31)

In this statement, Bruner sets knowledge as the primary goal of the science curriculum. He follows with the idea that fundamental knowledge can be taught in some form to any child at any stage of development (Bruner, 1960). Here, Bruner directly criticized the notion of readiness by asserting children are always ready for knowledge if it is presented at an appropriate level of simplicity or complexity. Mastery of the structure of the discipline could be accomplished by utilizing a spiral curriculum in which concepts were first taught at an elementary level and then returned to again and again with progressively more complexity.

Bruner also discussed three processes involved in the act of learning: acquisition of new knowledge, modification of the knowledge to fit new tasks, and evaluation of the modified knowledge to see if it was adequate to the new task (Bruner, 1960). Bruner, like Dewey, was concerned with both the logical and psychological dimensions of knowing. Bruner also discussed the value of intuitive thinking, a concern generally not recognized by science educators.

The goal of personal-social development received relatively little emphasis during the reform movement of the 1960s. When development was discussed in *The Process of Education*, it concerned almost exclusively elementary school levels (Bruner, 1960). At the secondary level, the goal of personal-social development was largely neglected. By the mid-1960s, the social mood of the period had changed from concerns about the Soviet Union to domestic concerns, and the curriculum reform movement was subsequently being criticized. In *The Genius of American Education* (Cremin, 1965), Lawrence Cremin pointed out some of the considerations that would have to be faced if the reform movement was to achieve its maximum effect. Cremin's criticisms included lack of careful thought about what knowledge was to be taught to any child at any stage of development, the error of constructing curricula based on the logical structure of a discipline, seeming ambivalence over psychological questions of pedagogy, the piecemeal approach to the K–12 science program, and, finally, the class bias inherent in the materials (Cremin, 1965). It should be noted that these criticisms were made by a scholar outside the field of science education. In 1969, Paul DeHart Hurd, a science educator, addressed some of the issues of the reform movement of the 1960s in *New Directions in Teaching Secondary School Science* (Hurd, 1969).

Hurd pointed out that the reform movement had used scientists as the project directors and their method of developing curriculum materials was structured by the question, "What does it take to provide an accurate picture of my discipline?" Both the goals and the means to those goals are contained within the answer to this question. Broadly speaking, the goal is the conceptual schemes that form the discipline and the means is the process by which the scientists pursue knowledge about the basic concepts of their discipline. Hurd (1969) points out that only secondary consideration was given the student's personal development and the social conditions under which the student lives.

The goals for science teaching as they have emerged during the curriculum reform are not subject to neat statements that can be listed as an introduction to a course of study in the usual way. What is expressed is more a point of view or rationale for the teaching of science. This point of view lacks the societal orientation usually found in statements of educational goals. The "new" goals of science teaching are drawn entirely from the disciplines of science. Social problems, individual needs, life problems and other means used in the past to define educational goals are not considered. (p. 34)

This summary by Hurd generally characterizes the structure of science education as seen in the curriculum reform movement of the 1960s. Scientific knowledge was the primary goal and scientific methods the secondary goal and the means of achieving the knowledge goal. Personal-social development of the student was given only minor recognition and that at the elementary level. Hurd also points out some of the problems of the "structure of a discipline" design for materials and instruction. The social realities of the last half of the decade slowly began shifting the goals of science education to include personal-social development. This contention, a shift toward the developmental aim, is clearly evident in statements made by Hurd (1970).

The curriculum has not considered in any direct way the relation of science to the affairs of man, the actualities of life, and the human condition. (p. 13)

Hurd (1970) then states his goal of science teaching; it is clearly the personal/social goal:

to foster the emergence of an enlightened citizenry, capable of using the intellectual resources of science to create a favorable environment that will promote the development of man as a human being. (p. 14)

A decade of experience brought the "process of education" into perspective. The encounter of science curriculum materials with the personal, social, political, and economic realities of school and society magnified the omissions and problems discussed by both Cremin and Hurd. Even Bruner modified his position relative to the "process of education." Ten years later, he, too, gave greater emphasis to the developmental goals. First, he discussed the assumptions from which he and others at the Woods Hole meeting viewed education.

The movement of which *The Process of Education* was a part was based on a formula of faith: that learning was what students wanted to do, that they wanted to achieve an expertise in some particular subject matter. Their motivation was taken for granted. It accepted the tacit assumption that everybody who came to these curricula in the schools already had been the beneficiary of the middle class hidden curricula that taught them the analytic skills and launched them in the traditional intellectual use of mind. (Bruner, 1971, p. 24)

Addressing the question, "What would one do now—a decade later?" Bruner (1971) stated:

The issues would have to do with how one gives back initiative and a sense of potency, how one activates to tempt one to want to learn again. When that is accomplished then curriculum becomes an issue again—curriculum not as a subject but as an approach to learning and using knowledge. (p. 26)

Bruner ends with a statement indicating that he now has greater concern for the developmental goals, that is, concern for personal growth as well as the intellectual and attitudinal aspect of the student.

A decade later, we realize that *The Process of Education* was the beginning of a revolution, and one cannot know how far it will go. Reform of curriculum is not enough. Reform of the school is probably not enough. The issue is one of man's capacity for creating a culture, society, and technology that not only feed him but keep him caring and belonging. (p. 30)

Seen in historical perspective, we think Bruner is correct. Contemporary trends include greater emphasis on the developmental goals. The changes in the knowledge goal are toward major concepts that are not structured by a discipline, such as physics, chemistry, or biology, but structured by a multiple of disciplines. This change in the goal of scientific knowledge seems to have been influenced by contemporary problems that are interdisciplinary; the energy crisis and environmental pollution are examples.

The Goal of Scientific Methods

The Golden Age of curriculum reform firmly established scientific methods as a goal for science curriculum materials and instructional techniques. During this period, curriculum developers termed the methods goal inquiry "problem solving," "discovery teaching," and "the processes of science."

It is somewhat ironic that Jerome Bruner's *The Process of Education* (1960), the book that was so influential during the reform movement, had little mention of scientific method as it was finally seen in science curriculum materials. In his chapter on "Intuitive and Analytic Thinking," Bruner (1960) argued strongly for greater understanding of intuitive thinking as a complement to the rational problem-solving model of analytical thinking. Bruner later elaborated on his theme of intuitive thinking in *On Knowing: Essays for the Left Hand* (1969). Although first published in 1962, this book had little influence in science curriculum reform until the early 1970s (Samples, 1971, 1975).

Early in the 1960s, scientific method was discussed in the professional literature with a sense of balance with the knowledge goal of science education. For example, in 1962, J. Darrell Barnard stated:

A good science program gives the student an understanding of basic concepts and the processes of science or methods of critical thinking, and through it he learns that concept and process are inextricably related. (p. 4)

Probably the major catalyst of the period as far as establishing the goal of scientific methods was the NSTA's publication *Theory into Action* (1964). This publication listed five major items in the process of science and influenced the development of at least one major elementary curriculum project that had scientific method as the major goal (Gagné, 1967).

Examination of the structure of secondary curriculum projects of the 1960s indicates that scientific knowledge was generally to be achieved through the use of inquiry, problem solving, or critical thinking, all forms of the scientific method as a goal. During the nineteenth century, learning the scientific method was seen as a way to discipline the mind; in the first half of the twentieth century, as a tool for solving societal problems; and in the 1960s, as a way of learning the structure of a discipline as a scientist does.

The Goal of Personal/Social Development

The curriculum movement of the 1960s modified the goal of scientific knowledge and incorporated personal-social development, if at all, as part of either the scientific knowledge or scientific methods goal. At the secondary level, this change was especially true. The elementary level had a greater range of psychological counsel in the construction of curriculum materials; thus, in one form or another, the development goals were given more importance than at the secondary level. For example, *Elementary Science Study* (Hawkins, 1965) was based on a "romantic" view of development and *Science Curriculum Improvement Study* (Karplus, 1964) was generally an application of an interactionist approach to cognitive development. In general, however, science education was beginning to suffer from decreasing recognition of personal-social development as an integral goal. Paul DeHart Hurd indicates the status of the development goals in the period from 1945 to 1960. This same status continued into the 1970s.

Only minor attention had been given to theories and research in human learning in the past fifteen years (1945–1960). The characteristics of the time indicated the needs for a greater efficiency in learning and better teaching techniques. But there was not adequate psychological research available to develop the necessary educational hypotheses. (Hurd, 1961, p. 110)

In 1964, the NSTA publication *Theory Into Action* gave little recognition to the developmental goals. One of the major statements on learning science was:

The strategies of learning must be related to the conditions that will lead to an understanding of the conceptual structures of science and of the modes of scientific inquiry. (p. 9)

Clearly, the restructuring of science education was toward science knowledge and methods and away from personal-social development. According to Hurd (1969):

The modern movement in science curriculum development looks to the nature of science for answers to questions about instructional goals, the design of the curriculum and how science should be taught and learned. (p. 55)

The curriculum reform movement of the 1960s was, in part, a reaction to what was perceived to be an intellectually weak focus on personal-social relevance in the curriculum during the Progressive Era. Given that perception, issues of personal-social concern received little support in the curriculum models of the 1960s. The drive was toward intellectual rigor, mastery of the structure of the discipline in a way that modeled the way scientists themselves thought and created knowledge. Structured knowledge was the end, scientific inquiry was the means, and personal-social development was largely ignored. In the next section, we see how the goal of personal-social development was reinstated into the science curriculum. Rather than look at the three goals separately, as we have in previous sections, we focus on the developmental goal specifically and show how it was reintroduced with the curriculum and integrated within the other goals and through what became known as "a search for scientific literacy."

CONTEMPORARY REFORM

By the end of the 1960s, societal emphasis on science education had shifted from the "space race" to problems centering on the individual, the cities, and the environment. Examination of issues in these areas underscored the need for attention to personal concerns and a response to the deteriorating environment.

After a low point in the mid–1960s, developmental goals were again stressed during the early 1970s in response to social and economic conditions. Personal-social development re-emerged in the professional literature and curriculum materials of science education. For example, in *Designs for Progress in Science Education*, Butts (1969) discussed the concepts of identity and alienation that were forces influencing education and thus also important for science education.

Many educators who had been skeptical of the curriculum reformers' emphasis on the structure of the disciplines were quick to point out to the failure of the courses such as PSSC physics and BSCS biology to meet the new challenges of education. Society needed scientific literacy, not a scientific elite. To these critics, the science curriculum should be relevant to the lives of all students, not just those planning careers in science, and the methods of instruction should demonstrate concern for the diversity in ability and interest of students.

In response to the new emphasis, various forms of individualized science were discussed; for example, a new NSF project, *Environmental Studies* (Griffith, 1972), was designed for personal growth in science and an appreciation of the environment. Some science educators discussed humanistic approaches to science teaching (Bybee, 1974; Bybee & Welch, 1972; Rutherford, 1970). The Biological Sciences Curriculum Study (BSCS) (1973) began work on "Human Sciences: A Developmental Approach to Adolescent Education" that centered on the individual's development.

The developmental psychology of Jean Piaget also had a significant influence on science education over the previous 10 years. In *To Understand Is to Invent* (1973), Piaget discussed the educational implications of his theory, devoting a significant portion of the book to science teaching. He states:

It is apparent that reorganization would involve not only specialized training in the various sciences (mathematics, physics, chemistry, biology, etc.), but also more general questions, such as the role of preschool education (ages four to six); that of the true significance of active methods (which everyone talks about but which few educators effectively apply); that of the application of child and adolescent psychology; and that of the necessarily interdisciplinary nature at every level of the subjects taught. (p. 12)

In Piaget's statement, the goals of scientific knowledge, scientific method, and personal-social development are recognizable, although Piaget used "active methods" (i.e., experimentation and inquiry) to refer to the scientific method. What is the primary aim of Piaget's model? For Piaget (1973), the goal is full development of the human personality and strengthening respect for human rights and fundamental freedoms. Piaget points out that this goal cannot be achieved in the traditional atmosphere of intellectual and moral restraints. He states that a "lived" experience and freedom for investigation are necessary. In Piaget's ideas, the aforementioned goals of the progressive movement have been transformed.

The increased attention to student development and the renewed focus on the social relevance of the science curriculum were reminiscent of the fundamental principles of progressive education. The similarities between the goals of the two periods led Diane Ravitch (1983) to refer to the educational emphasis of the 1970s as the "New Progressivism." In science education, the term "scientific literacy" was used to describe an education in science for *all* youth that was personally relevant and that focused on socially important issues. Some science educators argued that all students should become scientifically literate if they were to deal effectively as adults with such social concerns as environmental degradation and resource depletion. For many science educators, scientific literacy became the slogan for the 1970s. The term lacked precision, and subsequently it was used to describe an assortment of educational goals.

Origins of Scientific Literacy

Paul DeHart Hurd (1958) was one of the first to discuss the idea of scientific literacy, in an article entitled "Science Literacy: Its Meaning for American Schools." Hurd used the term "science literacy" to describe an understanding of science and its applications to our social experience. Hurd argued that science and its applications in technology had become a dominant force in our society and that it was difficult to discuss human values, political and economic problems, or educational objectives without consideration of the role played by science. Hurd (1958) stated, "More than a casual acquaintance with scientific forces and phenomena is essential for effective citizenship today. Science instruction can no longer be regarded as an intellectual luxury for the select few" (p. 13).

In 1963, Robert Carlton, Executive Secretary of NSTA, asked a number of scientists and science educators to define the term "scientific literacy" and how they felt the terms should be used. In their definitions, most of the respondents focused on the goal of science knowledge in a broad range of scientific fields. Only a few spoke of the relationship between science and society as Hurd had in his 1958 article. Gerald Holton, professor of physics at Harvard University, described two facets of scientific literacy: "(1) some narrow-area contact—knowing and keeping up with at least one chosen, even though small, part of science, and (2) broad range-contact—trying to keep in touch with a variety of other scientific developments" (Carlton, 1963, p. 33). Hugh Odishaw, executive director of the Space Science Board of the National Academy of Sciences, said: "Scientific literacy can be defined as a comfortable familiarity with the development, methodology, achievements, and problems of the principal scientific disciplines. A thoughtful reading of some fifty books could establish such familiarity" (p. 34). Howard Meyerhoff, chairman of the Department of Geology at the University of Pennsylvania, said that scientific literacy did not imply "omniscience" but rather "familiarity with scientific methods and . . . sufficient knowledge in the several fields of science to understand reports of new discoveries and advances" (p. 34).

Still another conception of scientific literacy was that it was a set of skills and knowledge that enabled one to read and understand the science that was discussed in the popular media. Koelsche (1965) identified 175 science principles and 693 science vocabulary words in a sampling of newspapers and magazines. Koelsche believed that schools could provide students with a working knowledge of most of these terms and principles and that "the lists constitute desirable subject matter for scientific literacy" (p. 723).

In 1967, Milton Pella published a study in which 100 published science education articles were examined to determine what educators meant when they used the term "scientific literacy." He identified the six most frequently used referents:

1. Interrelations between science and society
2. Ethics of science
3. Nature of science
4. Conceptual knowledge
5. Science and technology
6. Science in the humanities

Pella then evaluated the extent to which the new NSF-supported science curricula addressed issues of scientific literacy. He found they failed to include adequate mention of technology, social application, or the relation between science and the humanities.

Norman Smith (1974), a retired National Aeronautics and Space Administration (NASA) aerospace research scientist, defined scientific literacy as an understanding of the events around us, the ability to verify the truth of claims made by laypersons in the popular media about science, and the ability to evaluate the relevance and importance of scientific developments and projects to the needs of society. According to Smith, television, radio, newspapers, magazines, and our own experiences inundate us with "opportunities to test our mettle as seekers and promoters of scientific literacy" (p. 34).

Although definitions of scientific literacy varied, to most people the concept implied a broad and functional understanding of science much like that emphasized by the Thirty-First NSSE Yearbook authors and not the "structure of the discipline" approach of the new curriculum projects. Science knowledge was to be used to answer important questions that people encountered in their everyday lives, not just to answer questions about theoretical science. The definitions indicated that science should be understood as an important social force, not simply as a set of abstract principles held by experts.

Hurd, especially, believed that more than anything else scientific literacy meant understanding the coherence of science and society. Speaking of ways to accomplish the goal of scientific literacy in contrast to the goals of the curriculum reformers, Hurd (1970) said that science had to be taught "in a wider context than the processes and concepts of which it is formed" (p. 109). In terms of our discussion in this essay, Hurd's argument was for greater recognition of personal-social goals in the science curriculum. Besides the intellectual side of science teaching, it was important that aesthetic and social values be taught. In fact, to Hurd, the social context of science was the *only* appropriate context for science teaching for general educational purposes.

The theme of scientific literacy gained additional prominence when the NSTA identified it as the most important goal of science education in its position statement "School Science Education for the 70s." The statement, approved by the NSTA Board of Directors in 1971, began, "The major goal of science education is to develop scientifically literate and personally concerned individuals with a high competence for rational thought and action" (NSTA, 1971, p. 47).

The choice of goals is based on the belief that achieving scientific literacy involves the development of attitudes, process skills and concepts, necessary to meet the more general goals of all education. (NSTA Committee on Curriculum Studies, 1971, p. 47)

The goals of scientific knowledge, method, and personal-social development were all considered important by this NSTA Committee, but the prominent goal was personal development even though it was related to scientific literacy. This NSTA document indicates a transformation in the structure of science education. Societal pressures had set a mood of the times different from that of the curriculum movement of the 1960s.

As the goal of scientific literacy was elaborated in the 1971 position statement, it became clear that NSTA was taking a very different stand on science education than the curriculum reformers of the Golden Age had taken a decade earlier. The scientifically literate person was described as one who "uses science concepts, process skills, and values in making everyday decisions as he interacts with other people and with his environment" and who "understands the interrelationships between science, technology and other facets of society, including social and economic development" (NSTA, 1971, pp. 47–48). Learning objectives were to be analyzed on the basis of their consistency with the nature of science but also on the basis of their "desirability and potential usefulness by thoughtful lay citizens" and their potential interest to students. In addition, science programs were to be evaluated for their student-centeredness and their ability to "foster student liking for science in general and for independent investigation in particular" (p. 49).

What was evident in the NSTA's position statement were the themes of social relevance, student interest, the relationships between science and other areas of the curriculum, the interdependence of science and technology, and the human aspects of the scientific enterprise. What was abandoned was the idea that the structures of the science disciplines and the processes of science should be studied largely for their own sake. The goals of education were to teach the aspects of science that would help students understand the world around them and to provide them with the tools for acquiring new science knowledge. The NSTA statement was a strong reaffirmation of the personal-social goals of education and the principles of progressive education elaborated by leaders in science education during the first half of the twentieth century.

The Science-Technology-Society Theme

Although the term "scientific literacy" continued to be used to describe a wide range of progressive educational goals throughout the 1970s and into the 1980s, the importance of the relationship between science and society aligned with the phrase science-technology-society, or STS. One of the first persons to propose formally that science education be oriented around an STS theme was James Gallagher in a 1971 article entitled "A Broader Base for Science Teaching." Gallagher argued that the curriculum projects of the 1960s took a limited view of science by focusing on the conceptual schemes and processes of science. Gallagher argued that learners should become familiar with the social interactions of science as well as scientific knowledge and processes:

For future citizens in a democracy, understanding the interrelationships of science, technology, and society may be as important as understanding the concepts and processes of science. An awareness of the interrelations between science, technology, and society may be a prerequisite to intelligent action on the part of a future electorate and their chosen leaders. (p. 337)

In 1982, the NSTA Board of Directors adopted a new position statement entitled "Science-Technology-Society: Science

Education for the 1980s," and by the mid–1980s the science education literature was filled with discussions of the STS theme. The close relationship between the STS and scientific-literacy themes was evident in the 1982 NSTA position statement. The following excerpts make this point:

Many of the problems we face today can be solved only by persons educated in the ideas and processes of science and technology. A scientific literacy is *basic* for living, working, and decision making in the 1980s and beyond. (NSTA, 1982)

The goal of science education during the 1980s is to develop scientifically literate individuals who understand how science, technology, and society influence one another and who are able to use this knowledge in their everyday decision-making. The scientifically literate person has a substantial knowledge base of facts, concepts, conceptual networks, and process skills which enable the individual to continue to learn and think logically. This individual both appreciates the value of science and technology in society and understands their limitations. (NSTA, 1982)

According to STS advocates, science education in the 1970s and 1980s was to be humanistic, value oriented, and relevant to a wide range of personal, societal, and environmental concerns.

Environmental Concerns

Perhaps one of the greatest forces behind the STS approach to science education in the 1970s and 1980s was the tremendous growth in environmental awareness during this time and the commitment of individuals to protect and preserve their life space. Issues of energy conservation, environmental pollution, resource use, and global problems such as ozone depletion and the greenhouse effect were concerns affecting all inhabitants of the earth and were intimately tied to a wide range of science fields and to technology. Virtually every science education writer who advocated scientific literacy within an STS context raised environmental-ecological concerns as potential subject matter for school science.

In 1979, Bybee argued that ecology would become an organizing theme for both society and for science education:

Our future support and growth will be based on ecology, which is a set of biological and psychological principles, and will become a social metaphor all of which will guide the formation of policy. In the short period of 20 years ecology has become a household word. Ecology will increasingly influence our ideas, values, and growth as we continue the transformation from a mechanistic-industrial-technical paradigm to an organismic-environmental-community paradigm. (p. 107)

Science education was considered especially important in this ecological society because the new society would have science at its core. Bybee (1979) suggested four goals of science education that were related to the new paradigm: (1) the full development and nurture of the individual; (2) the protection, conservation, and improvement of the environment; (3) appropriate use of natural resources; and (4) the development of a sense of community from the local to the international level (pp. 107–108). This ecological perspective, combined with

some elements of humanism, and the development of decision-making skills for informed citizenship became important themes of the STS movement in the 1970s and 1980s.

Besides being an important part of the STS movement, environmental education existed as a separate entity. Throughout the entire twentieth century and earlier, science education had included goals and curriculum models that incorporated a study of the natural environment, although none of these earlier efforts were in the context of the environment as a critical social issue. For example, earlier in the twentieth century, nature study, outdoor education, and conservation education taught students about the outdoor world and encouraged them to respect and preserve it, but courses such as these did not deal with the environment as a source of social problems to be solved. As Troy and Schwaab (1982) put it:

Man has been educating himself about the environment for a long time, perhaps since his earliest presence on earth. However, formal environmental education is considered a relatively new academic discipline. It may have experienced its true birth around the time of Earth Day, April, 1970. (p. 209)

The beginnings of the modern environmental movement in education can also be marked by the publication of the *Journal of Environmental Education* in 1969 and by congressional passage of the Environmental Education Act (Public Law 91–516) in 1970.

By the 1980s, however, environmental education had a tenuous status in the schools. Although Troy and Schwaab (1982) pointed out that most state education departments had developed environmental education courses or curricular materials, there were still many disagreements in the field about the definition of environmental education, especially whether it was to be considered a separate subject or to be integrated into the entire K–12 curriculum.

Regardless of the status of environmental education as an organized field of study, the environmental perspective did find its way into the science curriculum during the 1970s and 1980s, whether it was used to illustrate other science principles or as an organizing principle for the science curriculum. Comments by a number of science educators during this period pointed to that fact. For example, Moon and Brezinski (1974) said that environmental education was "fast becoming a major focal issue of this decade." They believed that the efforts that had been expanded during the curriculum reform movement were now being redirected into "more broadened interdisciplinary approaches to the biosphere" (p. 371).

A Conflict of Two Curriculum Models

During most of the 1970s and 1980s, there was little criticism of the STS approach to science education or the STS interpretation of scientific literacy. However, when Avi Hofstein and Robert Yager argued forcefully and unambiguously in 1982 that the science curriculum should be *organized* around social issues rather than around the concepts of the disciplines, a heated controversy arose. The controversy was based on the strong differences of opinion of educators concerning the relative importance of knowledge versus the personal-social development goals. Knowledge acquisition has long been the primary goal in science education, but educational reformers often point to its inadequacies and offer alternative models that emphasize other goals. In the debate that follows, personal-social development is offered as an alternative to the knowledge-acquisition goal.

Avi Hofstein and Robert Yager (1982) challenged the assumptions of disciplinary-based science educators when they said:

Rather than including information merely because it is basic to the discipline, the decisions should be made on the basis of relationship to real-life problems, thoughts, and interpretations. There needs to be a cultural validity to school science as well as scientific validity. Courses in science should be bounded not by discipline lines, but by a specific context, or relation to current issues and concerns. (p. 542)

Hofstein and Yager (1982) advocated a curriculum whose *logic* would be based on social issues and not the logic of the scientific disciplines. According to this view, content should be selected not because of its disciplinary value but because of its utility in helping students deal with real-world problems.

Hofstein and Yager said that the curriculum reformers of the 1960s were wrong to assume that "science would be inherently interesting to all students if it were presented in the way it is known to scientists" (p. 542). They argued that the discipline-centered curriculum projects that focused on an understanding of science for its own sake were motivating to only a small segment of the society and that an issues approach would appeal to a much wider group of students.

Hofstein and Yager (1982) illustrated the use of societal issues as organizers for the science curriculum with numerous examples. They used issues related to the personal-social needs of people.

Information is needed regarding controversial areas such as abortion and family planning; use of energy in homes; planning for proper food preparation and use; care and maintenance of the body; stress and mental health; life style and its effects upon others; pollution rights *and* responsibilities; disease prevention and cure. (p. 545)

In these examples, the reformulated goals and a number of central ideas of the new progressivism are clear. First, science was supposed to be presented in relation to important aspects of contemporary life. Second, science content was connected to the personal needs of individuals both because of the utility of such knowledge for the well-being of society at large and because of the motivational power of such personally relevant knowledge. Third, science was presented in a larger interdisciplinary context, not simply as an isolated body of knowledge but as part of the entire body of human knowledge that encompasses the arts, literature, mathematics, and the social sciences. Fourth, the students should understand the relationships between the scientific enterprise and other aspects of life in a democratic society. The effect of science on the quality of life, the ethical dilemmas raised by scientific discovery, and the importance of public support if science is to flourish should all have their place in the science education curriculum.

What was discomforting to opponents of this STS position on science literacy was not the emphasis on social relevance but rather the idea that science courses would be *organized* on the basis of contemporary social issues instead of scientific concepts. Kromhout and Good (1983) responded to the proposal to use societal issues as organizers of the science curriculum in an article entitled "Beware of Societal Issues as Organizers for Science Education." Although they did not object to the use of socially relevant problems to motivate students and lead them into a coherent study of science, they saw the use of societal issues as fundamental organizers of the science curriculum as part of a widespread flight from science that was occurring in the United States.

To illustrate their concerns, Kromhout and Good (1983) pointed to one science curriculum, the Individualized Science Instructional System (ISIS) program, that was based on a study of socially relevant issues. They suggested that the modules of the program were interesting and motivational and could teach basic analytical skills and terminology to students; however, "since the units are organized around a particular socially important problem, . . . they are independent of each other by design, and therefore are unable to convey any real understanding of the structural integrity of science" (p. 649). The authors argued that today's technological problems will fade and that the societal-issues approach will not provide students with skills for dealing with technological problems of the future. "Thus our position is that if the *organization* of science courses is to be dictated by social issues, it will be counterproductive in the struggle for scientific literacy" (p. 649). The authors also expressed their concerns that under such an organization of science, the "basics simply do not get taught," that an orderly systematic approach to content acquisition was needed if science was to be learned effectively, and that "societal issues normally do not provide that logical, orderly framework" (pp. 649–650).

In another article, Good, Herron, Lawson, and Renner (1985) suggested that science education was "the discipline devoted to discovering, developing, and evaluating improved methods and materials to teach science, i.e., the quest for knowledge, as well as the knowledge generated by that quest" (p. 140). This definition of science education was based on what these authors called a scientist's definition of science—that science is both the knowledge of the physical world and the processes of seeking that knowledge. Science education is involved with the methods and materials for passing this knowledge on to students. To Good and his colleagues, the science and society theme overemphasized the political and sociological aspects of the field and ignored theories of learning. Under the science and society approach, they argued, mathematics and foreign language education should likewise become the study of the impact of those subjects on society and the impact of society on those subjects. Continuing their argument, they said:

Physics education is not the study of the impact of nuclear energy on society, nor is it the impact of society on nuclear physics. Chemistry education is not the study of the impact of chemical dumping on society, nor is it the study of the impact of society on chemical dumpers or on the chemists who synthesized the chemicals in the first place. Biology education is not the study of impact of gene splicing on society, nor is it the study of the impact of society on genetic engineers. This, of course, is not to say that these sorts of issues are unimportant, yet surely they do not constitute the essence of science education. (p. 141)

The Debate over STS Continues

Debates such as the one over STS are not new in the history of science education. There was, for example, a similar debate at the turn of the century between proponents of the natural-study movement and the elementary school science movement. In midcentury there was the debate between progressive educators and traditional educators. Each side in the debate has a different priority for the same goals of science education, and they present different curriculum models based on their orientation.

As we approached the 1990s, the debates concerning issues-based versus discipline-based science curriculum subsided. The question of which of these should be the focus of school programs, however, remained unresolved. The argument that science literacy was best achieved by science instruction that was appropriate for all students and relevant to their lives as opposed to one geared primarily to the needs and interests of future scientists was widely discussed throughout the first half of the century. Early in the twentieth century, the reaction was to nineteenth-century disciplinary studies. In the 1980s, it was to the structure-of-the-discipline approach that characterized the curriculum reform movement of the 1950s and 1960s. In both instances, a major reason that was given for modifying traditional science instruction was the continuing decline in science enrollments. Because the majority of students had avoided taking traditional, disciplinary-based science courses, the solution in the 1980s, as before, was to reform and restructure the goals of science education and to offer science curricula that could help people in their everyday lives, that would allow them to make a contribution to the well-being of society, and that was interesting to the students.

Concurrent with the themes of personal relevance and social benefit, however, both in the 1980s and in the 1930s, were themes suggesting that the organized study of the disciplines of science and of the processes of scientific investigation were essential elements of any respectable program of science education. As the arguments made by Good and Yager illustrated, this issue represented an important and often emotional point of disagreement between science educators in the 1980s and one that was not easily resolved.

CURRICULUM MODELS IN SCIENCE EDUCATION: A LOOK AHEAD

Three major goals have shaped the curriculum and instructional practices in science education. These three goals have been repeated, with continuing variation, through the 200-year history of science education in the United States. Broadly defined, these goals are understanding scientific knowledge, understanding and using scientific methods, and promoting personal–social development.

The organization of curricula and instruction in science is determined by the way these goals are structured. For example, if an elementary science teacher has decided that scientific knowledge is the important goal of instruction, then the goals of scientific methods and personal–social development usually become subordinate to the primary goal in the science curriculum. The latter goals may be included in the science curriculum, and indeed they usually are. However, the curriculum is very different from one in which a science teacher uses personal-social development as the primary goal. Examination of the underlying goals of science education is a useful way to identify and describe differences in present curricula and a way to describe historical changes and contemporary trends in science education.

To this last point, the objective of this chapter is to describe historical changes in the goals of science education, to identify arrangements of goals that were relatively stable for a period of time, to describe changes in the arrangements of goals, and to discuss factors that have influenced changes in the goals of science education and the different arrangements of these goals. The following sections summarize answers to these questions. We then turn to the contemporary reform and review the changes in goals that are being proposed.

Historical Changes in the Goals of Science Education

The contemporary origins of science teaching in the schools have been traced to the eighteenth century. During the early period of science teaching, the goal of scientific knowledge was the presentation of facts and information concerning science, often in a theological context. In the nineteenth century, scientists such as Thomas Huxley, Herbert Spencer, and Charles Eliot shaped the goals of science education and the curriculum with their emphasis on scientific method for purposes of mental discipline and on the meaningful learning of science concepts. In the late nineteenth century, the goal of scientific knowledge changed from the presentation of facts to the use of broader scientific generalizations. The change of the knowledge goal at the turn of the century can be attributed to the simultaneous influence of science educators such as Wilbur Jackman and William Harris, the influence of prominent scientists, and the needs of an emerging industrial society.

The next major change in the goal of scientific knowledge occurred in the 1920s and was firmly established during the depression years. At this time, Gerald Craig refined the earlier work of Jackman and Harris. Understanding major scientific generalizations and important principles of science organized into units such as magnetism, fossils, and air became the knowledge goal at the elementary level. The "generalization" approach to the knowledge goal was a prominent model of science education for the next 30 years. The Thirty-First NSSE Yearbook added support to the idea of organizing science content around a small number of general themes at both the secondary and elementary levels.

In the late 1950s, the goal of scientific knowledge was further transformed into a focus on the major conceptual schemes of a specific science discipline (Bruner, 1961) or more generally the major *conceptual schemes* of science approach (NSTA, 1964). This change is usually associated with Jerome Bruner and was in response to perceived deficiencies in the educational system, particularly in science and mathematics as represented by a storage of scientists.

The goal of knowledge acquisition in science education has become increasingly encompassing and abstract with time as more is learned about our natural world. The greater the scientific knowledge, the more refined the goal of scientific knowledge has become in the literature and policies of science education. The scientific knowledge goal has had a long and steady tenure as *the* important goal of science teaching. In comparison with the goal of scientific methods, which has grown in importance in the history of science education, and the goal of personal-social development, which has been subject to recurrent ascendence and descendence, the scientific knowledge goal has been prominent and stable in the science curriculum throughout our history.

The second goal is scientific methods. Understanding and using scientific methods were goals of science education by the late 1800s in American schools. Terms such as "scientific method," "heuristics," and "scientific thinking" were first used to describe this goal. When the goal of scientific methods was actually used in the classroom, it was in the form of activities such as observation, description, and experimentation. During this time, scientific methods were seen as a way to discipline the mental faculties of observation and inductive thought and as a way to form a more complete understanding of the natural world. The science laboratory was a major innovation of the late nineteenth century.

In the early 1900s, John Dewey's Progressive Education brought the goal of scientific method into a prominent position in the science curriculum. The ascendence of this goal can be accounted for by Dewey's epistemology, which emphasized mental problem solving and the logical application of this goal to the scientific curriculum; that is, the student should have experience with the scientific method (Dewey, 1944). Including laboratories and demonstrations as part of instruction was a manifestation of this goal in the science curriculum.

During the period of curriculum reform in the 1960s, the goal of scientific methods was discussed using terminology such as inquiry, discovery, heuristics, problem solving, and the processes of science.

The major change in the goal of scientific methods was not in the use of terminology. The same terms that were used a hundred years ago are still used today. From the early history of science education, only minor changes have occurred in the terms used to describe this goal. The important change has been in the changing reasons for teaching scientific method. During the late nineteenth century, the emphasis on scientific method was for purposes of personal mental development. During the first half of the twentieth century, the focus was on a scientific method to be used to solve societal problems. In the 1960s, scientific method was defended as a way of acquiring scientific knowledge. The goal has had growing importance in science curriculum and instruction throughout our history. This importance was reaffirmed in the emphasis placed on laboratory investigations and active involvement with the natural world at both the elementary and secondary levels in the curriculum projects of the 1960s.

Personal-social development is the third major goal of science education. The developmental goals have not had a clearly defined terminology in the history of science education. The early literature discussed "development of faculties" and the natural "development of the child" as the goals toward which the teaching of elementary science should be directed. John Dewey and others in the Progressive Era attempted to center on development of the student as a whole person as the goal of education. Personal experience was the primary means by which teachers could achieve this goal. Jean Piaget presented his ideas and the implications of active methods and experience as means of developing students cognitively and socially.

The developmental goal has been changed and updated as psychologists have accumulated more knowledge about personal growth. Unlike the knowledge goal of science education, there has not been a distinct terminology representative of the changes. Throughout history, development has been a latent factor in the goals of science education. This goal, more than the other two goals, has vacillated in prominence. The vacillation often parallels changes in society, such as economic and environmental concerns. This goal was given a strong emphasis during the Progressive Era in reaction to the traditional disciplinary approach of the nineteenth century and again more recently as part of the STS and scientific-literacy movements.

The Organization of Goals—A Historical Summary

We have discussed historical changes in the individual goals of science education. The purpose of this section is to answer the questions: When were the different goals dominant and when were they subordinate in the structure of science curricula? When were the goals structured in an organization that was stable for a period of time? What changes have occurred in the organization of goals?

In the early history of science teaching in the United States, the dominant goal was factual information that was supportive of theological ideas. With time, there was a gradual decrease in religious indoctrination and an increase in utilitarian objectives for science teaching.

In 1860, the object-lesson movement established personal-social development as the important goal of science teaching; scientific knowledge was secondary, and scientific method was discussed but not prominent. In 1880, there was a dramatic shift toward scientific knowledge as the dominant goal; this resulted from severe criticism of the schools during the post–Civil War depression and increasing industrialization. At this time, a model of science education began evolving with the work of Parker, Jackman, and Harris. This model had scientific knowledge as the primary goal and continued in some form until about 1927. By about 1890, the nature-study movement had evolved from the earlier object-lesson movement. The dominant goal of nature study was personal-social development. Nature study was to be a countervailing force for the movement of people from rural areas to urban centers.

For about 30 years, there were these two models of elementary school science, one having knowledge as the dominant goal and the other having personal-social development as the dominant goal. Both models incorporated scientific methods as a goal, but it was subordinate to either scientific knowledge or personal-social development.

In 1927, Gerald Craig refined the knowledge model of science teaching by focusing on broader generalizations in science. This model continued to influence science teaching for about 30 years. Craig's model became predominant during the depression years of the 1930s and remained so until about 1960. In this model, scientific knowledge was the dominant goal and secondary recognition was given to scientific methods and personal-social development. This model also influenced secondary education, but the influence was primarily in updating the knowledge from numerous facts to generalizations as discussed in prior sections.

During the same years, the Progressive model of education advocated by John Dewey stressed personal-social development as the major goal. This movement in science education was basically a transformation of the nature-study concept. It maintained the goal of development but, perhaps more important, brought the goal of scientific method into a prominent position in the science curriculum. The two models of science education, one influenced by Craig's ideas and the other by Dewey's, remained until the 1960s.

In the curriculum reform movement of the early 1960s, there was a dramatic shift toward the scientific knowledge goal in science curriculum. This shift was triggered by the shortage of scientists, the apparent need to update the science and mathematics curriculum in American schools, and the primary use of scientists as designers of science curricula. The model adopted by the majority of curriculum projects was described by Bruner in *The Process of Education* (Bruner, 1960). The goal was scientific knowledge, and it was to be achieved through the methods of science. The goal of scientific methods was dominant in at least one curriculum project, *Science: A Process Approach* (Gagné, 1967). In fact, emphasis on scientific methods was the one unifying theme of all the materials produced during this period. The developmental goal was incorporated to some degree at the elementary level, but it was seldom overtly recognized at the secondary level.

In the mid–1960s, problems with our cities and with the natural environment began receiving attention in educational literature. Problems in urban centers could be classified as personal needs, for example, alienation, motivation, self-concept, and identity. Pollution, population, and conservation of natural resources were the growing problems concerning the natural environment. The factors described brought about reemergence of the personal-social developmental goals in the professional literature of science education and in some of the curriculum materials of the late 1960s and early 1970s.

LOOKING TOWARD THE YEAR 2000

In the 1980s, more than 300 reports called for a reform of education. There is no precedent in history for such widespread reform efforts in education. These reports made fairly consistent recommendations for improving science education. The particular goals and implied curriculum varied with the group making the recommendations. In this section, we discuss several reports and frameworks that are influencing the reform of science education. Those frameworks include the American Association for the Advancement of Science (AAAS) report *Science for All Americans* (1989), the NSTA project "Scope, Sequence, and Coordination" (Aldridge, 1989, 1992), and the National Center for Improving Science Education (NCISE) reports on middle-level education (Bybee et al., 1990) and secondary education (Champagne et al., 1991).

Science for All Americans

In the late 1980s, F. James Rutherford established Project 2061 at AAAS. Project 2061 takes a long-term, large-scale view of educational reform in the sciences. In the reform of science education developed by Project 2061, the aim of scientific literacy is the basis for restating and restructuring the goals of science education.

The primary core of the 2061 report *Science for All Americans* (AAAS, 1989) consists of recommendations on what understandings and habits of mind are essential for all citizens in a scientifically literate society. Scientific literacy—which embraces science, mathematics, and technology—is a central goal of science education. In preparing its recommendations, Project 2061 used the reports of five independent scientific panels. In addition, Project 2061 sought the advice of a large and diverse array of consultants and reviewers—scientists, engineers, mathematicians, historians, and educators. The process took over 3 years, involved hundreds of individuals, and culminated in the publication of *Science for All Americans* (AAAS, 1989).

Science for All Americans attempts to characterize scientific literacy. Thus, its recommendations are presented in the form of basic learning goals for American students. A premise of Project 2061 is that the schools do not need to teach more; they should teach less so that content can be taught better. The recommendations in *Science for All Americans* concerning the basic dimensions of scientific literacy are

1. Being familiar with the natural world and recognizing its diversity and its unity
2. Understanding concepts and principles of science
3. Being aware of some of the ways in which science, mathematics, and technology depend upon one another
4. Knowing that science, mathematics, and technology are human enterprises and knowing about their strengths and limitations
5. Developing a capacity for scientific ways of thinking
6. Using scientific knowledge and ways of thinking for individuals and social purposes

Science for All Americans covers an array of topics, many of which are common in school curricula. These include the structure of matter, the basic functions of cells, prevention of disease, communications technology, and different uses of numbers. However, the treatment of such topics differs from the traditional approaches in two ways. First, boundaries between traditional subject-matter categories are softened and connections are emphasized through the use of major common themes, such as systems, evolution, cycles, and energy. Transformations of energy, for example, occur in physical, biological, and technological systems, and evolutionary change appears in stars, organisms, and societies. Second, the amount of detail that students are expected to remember is less than in traditional science, mathematics, and technology courses. Major concepts and thinking skills are emphasized instead of specialized vocabulary and memorized procedures. The major ideas not only make sense at a simple level but also provide a lasting foundation for learning more science.

Recommendations in *Science for All Americans* also include some topics that are not common in school curricula. Among them are the nature of the scientific enterprise and how science, mathematics, and technology relate to one another and to the social system in general. The report also calls for some knowledge of the history of science and technology.

Science for All Americans (AAAS, 1989) is a clear and detailed statement on the knowledge, values, and skills that constitute science literacy. It offers a relatively balanced presentation of the three major goals discussed in this essay. Scientific knowledge is the dominant goal, and it is represented in the chapters on the physical setting, the living environment, the human organisms, human society, the designed world, and the mathematical world. It is also represented in a chapter on common themes, that is, systems, models, constancy, patterns of change, evolution, and scale. The processes of science and recognition of personal-social goals are also amply represented. The latter are presented as knowledge goals, but they certainly represent a contrast to contemporary textbook programs and to programs from the 1960s and 1970s.

Scope, Sequence, and Coordination

A second approach to the reform of secondary school science has been suggested by Aldridge (1989). In an analysis of extant school programs, Aldridge found deficiencies related to the scope, sequence, and coordination of programs. The deficiencies were revealed by comparison with science programs in other countries, specifically the Soviet Union and the People's Republic of China.

The Project on Scope, Sequence, and Coordination of Secondary School Science is an effort to restructure science teaching at the secondary level. It calls for elimination of the tracking of students, recommends that all students study science every year for 6 years, and advocates the study of science as carefully sequenced, well-coordinated instruction in physics, chemistry, biology, and earth-space science. As opposed to the traditional curriculum, in which science is taught in year-long and separate disciplines, the NSTA project provides for spacing the study of each of the sciences over several years. Research on the "spacing effect" indicates that students can learn and retain new material better if they study it in spaced intervals. In this way,

students can revisit a concept at successively higher levels of abstraction.

The Scope, Sequence, and Coordination reform effort also uses appropriate "sequencing" of instructional strategies, taking into account how students learn. In science, understanding develops from concrete experiences with a phenomenon before it is given a name or a symbol. Students need experience with a concept in several different contexts before it becomes part of their mental repertoire. With prior hands-on experience, students can come to understand the major concepts and processes of science. The practical components of this instruction should begin in the seventh grade with issues and phenomena of concern to students at a personal level and then move toward a more global scope in the upper grades. As they mature, students are able to generalize from concrete, direct experiences to more abstract and broader theoretical thinking. With a sequenced approach, students should no longer be expected to memorize facts and information. With practical applications, science should make sense and have meaning.

The third component of the Scope, Sequence, and Coordination project is the "coordination" of science concepts and topics. Earth-space science, biology, chemistry, and physics have significant features and processes in common. Coordination among these disciplines leads to awareness of the interdependence of the sciences and how the disciplines form a body of knowledge. Seeing a concept, law, or principle in the context of two or three different subjects helps to establish it firmly in the student's mind.

At first, students are introduced to the descriptive and phenomenological aspects of the sciences. Empirical and semiquantitative treatments are emphasized in the middle years, and emphasis on the most abstract and theoretical concepts and processes occurs in the later years. Computers and technology and practical applications are integrated directly into each course. Most important, students are taught science in such a way that they are able to understand and apply it, whether as scientists or citizens.

The Scope, Sequence, and Coordination project has taken the classical disciplines and reorganized the scientific knowledge and processes, but there is little emphasis on the goals of personal development and social issues. Other groups, however, have modified the Scope, Sequence, and Coordination framework to include the STS theme and other goals. Scope, Sequence, and Coordination is not really a curriculum project but rather an organization of existing courses in a way that matches the developmental levels of students to maximize understanding.

The National Center for Improving Science Education

Development of local school science programs can be enhanced by frameworks for curriculum, assessment, and staff development such as those produced by NCISE for the middle level (Bybee et al., 1990; Loucks-Horsley, 1990; Raizen et al., 1990) and the secondary level (Champagne et al., 1991).

The curriculum and instruction frameworks for the middle years and high school extend NCISE's proposed framework for the elementary years (Bybee et al., 1989; Loucks-Horsley, 1989; Raizen et al., 1989). The treatments of the recommended organizing concepts, however, are more complex. The organizing concepts detailed in the technical report for middle schools include cause and effect, change and conservation, diversity and variation, energy and matter, evolution and equilibrium, models and theories, probability and prediction, structure and function, systems and interaction, and time and scale. The concepts need not be independent units of study; they should at least link subjects, topics, and disciplines. Emphasis in the curriculum should include scientific habits of mind, such as willingness to modify explanations, cooperation in answering questions and solving problems, respecting reasons, reliance on data, and skepticism. Students also should develop skills in answering questions and solving problems, making decisions, and taking action. Content in the program should be related to the life and world of the students and provide a context for presenting new knowledge, skills, and attitudes. The focus of curriculum and instruction should be on depth study, not breadth of topics.

The NCISE frameworks present major conceptual themes as integrating principles for scientific knowledge and the processes of science. Scientific knowledge and processes should be learned in a personal-social context. The NCISE frameworks present a balanced curriculum model for the goals of science education.

At the time this chapter was being completed (mid–1992), the National Research Council of the National Academy of Sciences initiated the development of National Standards for Science Education. The results of this effort hold the promise of coordinating and organizing the variety of recommendations contained in the hundreds of reports stimulating contemporary reform.

New Science Curriculum

If the 1980s were characterized by reports, the 1990s must be characterized by new programs. Primarily with funding from the NSF, new science curricula are being developed and implemented. Some of the developers include Biological Sciences Curriculum Study (BSCS), Lawrence Hall of Science (LHS), Educational Development Center (EDC), and Technical Education Research Center (TERC). The new programs are available for school systems and contributing to the reform of education. The programs being developed in the 1990s represent diverse curriculum models and variations on the emphasis given to the goals of science education.

We should also note that many new science programs are being developed at the local level. As in the national programs, diverse models are based on variations of the goals of science education.

CONCLUSION

In the last decades of the twentieth century, there has been a public outburst calling for the reform of science education. Never in U.S. history have we witnessed so many groups, committees, and commissions with so much to recommend about a variety of aspects of science education. There have been state and local initiatives and national mandates. By the mid–1990s, the United States will have national standards for science curriculum, instruction, and assessment. However, writing reports about the reform and actually reforming science education are very different activities. The former requires that a small group agrees on a set of ideas and expresses those ideas clearly and with adequate justification. The latter requires that millions of school personnel in thousands of autonomous school districts change. For changes to occur in science education, school personnel must change their way of doing things, and the most important factors influencing the possibility of this occurring are the programs and practices currently in place and supported by the school system.

Reform of the educational system has widespread support. The president and governors have identified a goal for science education—the United States should be number one in science and mathematics in the world by the year 2000. This goal suggests that an international assessment will be used to see if we have achieved the goal, but it neither identifies the content of the assessment nor gives an orientation or context for the goals and curriculum as we have discussed them.

Scientific and technological literacy is the major purpose of K–12 science education. This purpose is for *all* students, not just those destined for careers in science and engineering. The curriculum for science education is inadequate to the challenge of achieving scientific and technological literacy by the year 2000. For this reason, many reports urge a review of school personnel and science programs. Increasing the scientific and technological literacy of students requires several fundamental changes in science curricula. First, the amount of information presented must be replaced by major conceptual themes that are learned in some depth. Second, at the secondary level, the rigid disciplinary boundaries of earth science, biology, chemistry, and physics should be softened and greater emphasis placed on connections between the sciences and among disciplines generally thought of as outside school science, such as technology, mathematics, ethics, and social studies (Confrey, 1990; Newmann, 1988).

Achieving the goal of scientific and technological literacy requires more than understanding major concepts and processes of science and technology. Indeed, there is a need for citizens to understand science and technology as an integral part of our society. Science and technology are enterprises that shape and are shaped by human thought and social actions. As mentioned earlier, aspects of this theme are discussed as STS (Bybee, 1987). The prevailing approach to STS is to center on science-related social problems, such as environmental pollution, resource use, and population growth. The recommendation we are making expands the STS theme to include some understanding of the nature and history of science and technology (BSCS, 1992a, 1992b). There is recent and substantial support for this recommendation, but few curriculum materials. Including the nature and history of science and technology provides opportunities to focus on topics that blur disciplinary boundaries and to show connections between fields such as science and social studies.

A substantial body of research on learning should be the basis for making instruction more effective. This research suggests that students learn by constructing their own meaning of the experiences they have (Driver & Oldham, 1986; Sachse, 1989; Watson & Konicek, 1990). A constructivist approach requires very different methods of science instruction at the elementary and secondary levels.

Not unrelated to the implications of research on learning theory is the age-old theme that science teaching should consist of experiences that exemplify the spirit, character, and nature of science and technology. Students should begin study with questions about the natural world (science) and problems about humans adapting (technology). They should be actively involved in the process of inquiry and problem solving. They should have opportunities to present their explanations for phenomena and solutions to problems and to compare their explanations and solutions to those concepts of science and technology. They should also have a chance to apply their understandings in new situations.

In short, the laboratory is an infrequent experience for science students, but it should be a central part of the student's experience in science education at all levels. Extensive use of the laboratory is consistent with our other recommendations, and it has widespread support (Costenson & Lawson, 1986).

In the next decade, the issue of equity must be addressed in science programs and by school personnel. For the past several decades, science educators at all levels have discussed the importance of changing science programs to enhance opportunities for historically underrepresented groups. Calls for scientific and technological literacy assume the inclusion of *all* Americans. Other justifications, if any are needed for this position, include the supply of future scientists and engineers, the changing demographics, and prerequisites for work. Research results, curricula recommendations, and practical suggestions are available to those developing science curricula for the secondary school (Atwater, 1986, 1989; Gardner, Mason, & Matyas, 1989; Linn & Hyde, 1989; Malcom, 1990; Oakes & The Rand Corporation, 1990). We should note that attempts to define a core curriculum for *all* students date at least from the Committee of Ten and have been made more or less continuously across the decades.

Reform of science education must be viewed as part of the general reform of education. Approaching the improvement of science education by changing textbooks, buying new computers, or adding a new course simply will not work. Fortunately, there is widespread educational reform that includes science education. The improvement of science education at the secondary level must be a part of the reconstruction of science education K–12 and include all courses and students, a staff development program, reform of science teacher preparation, and support from school administrators. This comprehensive or systemic recommendation is based on the research on implementation (Hall, 1989; Fullan, 1982) and research literature on school change and restructuring (Kloosterman, Matkin, & Ault, 1988; Roberts & Chastko, 1990; Tobin & Espinet, 1989; Yeany & Padilla, 1986).

Looking toward the year 2000, we see a system already in the process of reform. Although distinctly different from earlier reforms, this reform holds greater promise of accomplishing the goals of scientific and technological literacy for all Americans. Many of the most important goals of science education are in place. Indeed, they have been in place since the nineteenth century. In the past, there were many attempts to reform education in keeping with what educators thought was most important for students to learn. Throughout our history, the shifts in emphasis have been dramatic, and a balanced program has been difficult to achieve. In the current reform, most educators recognize the value of each of the major goals of science education and are creating programs that include all of them in a balanced way. This seems to be the key to reform. If balance in goal emphasis can be achieved and if new findings concerning the psychology of learning can be applied to the educational setting, genuine improvements in the science curriculum will result.

References

Aldridge, B. G. (1989). *Essential changes in secondary school science: Scope, sequence, and coordination* (unpublished paper). Washington, DC: National Science Teachers Association.

American Association for the Advancement of Science (AAAS). (1967). *Science: A process approach.* New York: Xerox Division, Ginn and Company.

American Association for the Advancement of Science (AAAS). (1989). *Science for all Americans: A project 2061 report on literacy goals in science, mathematics, and technology.* Washington, DC: Author.

Armstrong, H. E. (1913). *Teaching scientific method and other papers on education.* London: Macmillan.

Atwater, M. M. (1986). We are leaving our minority students behind. *Science Teacher, 5,* 54–58.

Atwater, M. M. (1989). Including multicultural education in science education: Definitions, competencies, and activities. *Journal of Science Teacher Education, 1,* 17–20.

Bailey, L. H. (1903). *The nature study idea.* New York: Doubleday.

Barnard, J. D. (1971). COPES: The new elementary science program. *Science and Children, 9,* 9–11.

Biological Sciences Curriculum Study (BSCS). (1973). *Human sciences: A developmental approach to adolescent education.* Boulder, CO: Author.

Biological Sciences Curriculum Study and Social Science Education Consortium. (1992a). *Teaching about the history and nature of science and technology: A curriculum framework.* Colorado Springs, CO: Author.

Biological Sciences Curriculum Study and Social Science Education Consortium. (1992b). *Teaching about the history and nature of science and technology: Background papers.* Colorado Springs, CO: Author.

Boyd, W. (Ed.). (1956). *The Emile of Jacques Rousseau.* New York: Teachers College Press.

Brownell, H. (1902). Science teaching preparatory for the high school. *School Science and Mathematics, 2,* 253.

Bruner, J. S. (1960). *The process of education.* New York: Vintage.

Bruner, J. S. (1961). The act of discovery. *Harvard Educational Review, 30,* 21–32.

Bruner, J. S. (1969). *On knowing: Essays for the left hand.* New York: Athenum.

Bruner, J. S. (1971). The process of education revisited. *Phi Delta Kappan, 5,* 17–21.

Butts, D. P. (Ed.). (1969). *Designs for progress in science education.* Washington, DC: National Science Teachers Association.

Bybee, R. W. (1974). *Personalizing science teaching.* Washington, DC: National Science Teachers Association.

Bybee, R. W. (1979). Science education and the emerging ecological society. *Science Education, 63,* 95–109.

Bybee, R. W. (1987). Science education and the science-technology-society (STS) theme. *Science Education, 71,* 667–683.

Bybee, R. W., & Welch, I. D. (1972). The third force: Humanistic psychology and science education. *Science Teacher, 39,* 18–22.

Bybee, R. W., Buchwald, C. E., Crissman, S., Heil, D. R., Kuerbis, P. J., Matsumoto, C., & McInerney, J. D. (1989). *Science and technology education for the elementary years: Frameworks for curriculum and instruction.* Washington, DC: National Center for Improving Science Education.

Bybee, R. W., Buchwald, C. E., Crissman, S., Heil, D. R., Kuerbis, P. J., Matsumoto, C., & McInerney, J. D. (1990). *Science and technology education for the middle years: Frameworks for curriculum and instruction.* Washington, DC: National Center for Improving Science Education.

Calkins, N. A. (1890). *Manual of object teaching with illustrative lessons on methods and the science of education.* New York: Harper.

Carlton, R. (1963). On scientific literacy. *NEA Journal, 52,* 33–35.

Central Association of Science and Mathematics Teachers. (1950). *A half century of science and mathematics teaching.* Oak Park, IL: Author.

Champagne, A. B. (1991). *Science and technology education for the high school years.* Washington, DC: National Center for Improving Science Education.

Confrey, J. (1990). A review of the research on student conceptions in mathematics, science, and programming. In C. B. Cazden (Ed.), *Review of research in education* (pp. 3–56). Washington, DC: American Educational Research Association.

Costenson, K., & Lawson, A. E. (1986). Why isn't inquiry used in more classrooms? *American Biology Teacher, 48,* 150–158.

Craig, G. S. (1927). Certain techniques used in developing a course study in science for the Horace Mann Elementary School. In *Contributions to education* (No. 276). Bureau of Publications, Columbia University. New York: Teachers College Press.

Craig, G. S. (1957). *Science in the elementary schools.* Washington, DC: National Education Association.

Cremin, L. (1964). *The transformation of the school.* New York: Vintage.

Cremin, L. (1965). *The genius of American education.* New York: Vintage.

DeBoer, G. E. (1991). *A history of ideas in science education: Implications for practice.* New York: Teachers College Press.

Dewey, J. (1938). *Experience and education.* New York: Collier.

Dewey, J. (1944). *Democracy and education.* New York: Free Press.

Driver, R., & Oldham, V. (1986). A constructivist approach to curriculum development in science. *Studies in Science Education, 13,* 105–122.

Eliot, C. W. (1898). *Educational reform.* New York: Century.

Environmental Education Act. (1970). *Statutes at Large, 84,* 1312–1315.

Fullan, M. (1982). *The meaning of educational change.* New York: Teachers College Press.

Gagné, R. M. (1967). *Science—A process approach–Purposes, accomplishments, expectations.* Washington, DC: AAAS Miscellaneous Publication.

Gardner, A. L., Mason, C. L., & Matyas, M. L. (1989). Equity, excellence,
and 'just plain good teaching'. *American Biology Teacher, 51,* 72–78.

Good, R., Herron, J., Lawson, A., & Renner, J. (1985). The domain of science education. *Science Education, 69,* 139–141.

Griffith, G. (1972). Environmental studies: A curriculum for people. *Science and Children, 9,* 18–21.

Hall, G. E. (1989). Changing practice in high schools: A process, not an event. In W. G. Rosen (Ed.), *High school biology: Today and tomorrow* (pp. 298–323). Washington, DC: National Academy Press.

Harris, W. T. (1896). Course of study from primary school to university. *NEA proceedings* (pp. 153–154). Washington, DC.

Harvard Committee. (1945). *General education in a free society.* Cambridge, MA: Harvard University Press.

Hawkins, D. (1965). Messing about in science. *Science and Children, 2,* 5–9.

Hofstein, A., & Yager, R. (1982). Societal issues as organizers for science education in the 80s. *School Science and Mathematics, 82,* 539–547.

Howe, E. G. (1894). *Systematic science teaching.* New York: Appleton.

Hurd, P. D. (1958). Science literacy: Its meaning for American schools. *Educational Leadership, 16,* 13–16.

Hurd, P. D. (1961). *Biology education in American schools, 1890–1960* (Biological Sciences Curriculum Study Bulletin No. 1). Washington, DC: American Institute of Biological Sciences.

Hurd, P. D. (1969). *New directions in teaching secondary school science.* Chicago: Rand-McNally.

Hurd, P. D. (1970). Scientific enlightenment for an age of science. *Science Teacher, 37,* 13.

Huxley, T. H. (1899). *Science and education.* New York: Appleton.

Jackman, W. (1891). *Nature study for the common schools.* New York: Holt.

Jackman, W. (1904). *Nature study. The Third Yearbook of the National Society for the Scientific Study of Education, Part II.* Chicago: University of Chicago Press.

Karplus, R. (1964). The Science Curriculum Improvement Study: A report to the Piaget Conference. *Journal of Research in Science Teaching, 2,* 236–240.

Kloosterman, P., Matkin, J., & Ault, P. C. (1988). Preparation and certification of teachers in mathematics and science. *Contemporary Education, 59,* 146–149.

Koelsche, C. (1965). Scientific literacy as related to the media of mass communication. *School Science and Mathematics, 65,* 719–724.

Kromhout, R., & Good, R. (1983). Beware of societal issues as organizers for science education. *School Science and Mathematics, 83,* 647–650.

Linn, M. C., & Hyde, J. S. (1989). Gender, mathematics, and science. *Educational Researcher, 18,* 17–19; 22–27.

Livermore, A. H. (1964). The process approach of the AAAS Commission on Science Education. *Journal of Research in Science Teaching, 2,* 271–282.

Loucks-Horsley, S., Carlson, M. O., Brink, L. H., Horwitz, P., Marsh, D. D., Pratt, H., Roy, K. R., & Worth, K. (1989). *Developing and supporting teachers for elementary school science education.* Washington, DC: National Center for Improving Science Education.

Malcom, S. M. (1990). Who will do science in the next century? *Scientific American, 262*(2), 112.

Moon, T. C., & Brezenski, B. (1974). Environmental education from a historical perspective. *School Science and Mathematics, 74,* 371–374.

National Education Association (NEA). (1893). *Report of the committee on secondary school studies.* Washington, DC: U.S. Government Printing Office.

National Education Association (NEA). (1911). Report of the committee of nine on the articulation of high school and college. *Journal of proceedings and addresses of the forty-ninth annual meeting.* Chicago: Author.

National Education Association (NEA). (1918). *Cardinal principles of secondary education: A report of the commission on the reorganization of secondary education* (U.S. Bureau of Education, Bulletin No. 35). Washington, DC: U.S. Government Printing Office.

National Education Association (NEA). (1920). *Reorganization of science in secondary schools: A report of the commission on the reorganization of secondary education* (U.S. Bureau of Education, Bulletin No. 26). Washington, DC: U.S. Government Printing Office.

National Education Association (NEA). (1926). *The nation at work on the public school curriculum. Fourth yearbook.* Washington, DC: Department of the Superintendent.

National Science Teachers Association (NSTA). (1964). *Theory into action.* Washington, DC: Author.

National Science Teachers Association (NSTA). (1982). *Science-technology-society: Science education for the 1980s.* A position statement. Washington, DC: Author.

National Science Teachers Association (NSTA). (1992a): *The content core: A guide for curriculum designers. Scope, sequence and coordination of secondary school science,* Vol. I. Washington, DC: Author.

National Science Teachers Association (NSTA). (1992b): *Relevent research. Scope, sequence and coordination,* Vol. II. Washington, DC: Author.

National Science Teachers Association Committee on Curriculum Studies. (1971, November). School science education for the 70s. *Science Teacher, 38,* 46–56.

National Society for the Study of Education (NSSE). (1932). *A program for teaching science. Thirty-first Yearbook of the National Society for the Study of Education,* 2 Parts. Bloomington, IL: Public School Publishing Company.

National Society for the Study of Education (NSSE). (1947). *Science education in American schools. Forty-sixth Yearbook of NSSE,* 2 parts. Chicago: University of Chicago Press.

National Society for the Study of Education (NSSE). (1960). *Rethinking science education: Fifty-ninth yearbook of the NSSE,* Part 1. Chicago: University of Chicago Press.

Newmann, F. M. (1988). Can depth replace coverage in the high school curriculum? *Phi Delta Kappan, 69,* 345–348.

Oakes, J., & The Rand Corporation. (1990). Opportunities, achievement, and choice: Women and minority students in science and mathematics. In C. B. Cazden (Ed.), *Review of research in education* (pp. 153–222). Washington, DC: American Educational Research Association.

Obourn, E. S. (1956). An analysis and check list on the problem solving objective. *Science Education, 40,* 388–392.

On the place of science in education. (1928). *School Science and Mathematics, 28,* 640–664.

Pella, M. (1967). Science literacy and the high school curriculum. *School Science and Mathematics, 67,* 346–356.

Piaget, J. (1973). *To understand is to invent.* New York: Grossman.

Progressive Education Association (PEA). (1938). *Science in general education.* New York: Appleton-Century-Crofts.

Raizen, S. A., Baron, J. B., Champagne, A. B., Haertel, E., Mullis, I. V. S., & Oakes, J. (1989). *Assessment in elementary school science education.* Washington, DC: National Center for Improving Science Education.

Raizen, S. A., Baron, J. B., Champagne, A. B., Haertel, E., Mullis, I. V. S., & Oakes, J. (1990). *Assessment in middle school science education.* Washington, DC: The National Center for Improving Science Education.

Ravitch, D. (1983). *The troubled crusade: American education.* New York: Basic Books.

Rice, J. (1893). *The public school system of the United States.* New York: Century.

Roberts, D. A., & Chastko, A. M. (1990). Absorption, refraction, reflection: An exploration of beginning science teacher thinking. *Science Education, 74,* 197–224.

Rutherford, F. J. (1970). A humanistic approach to science teaching. *NASSP Bulletin, 56,* 555–568.

Sachse, T. P. (1989). Making science happen. *Educational Leadership, 47*(3), 18–21.

Samples, R. (1971). Environmental studies. *Science Teacher, 38,* 36–37.

Samples, R. (1975). Educating for both sides of the human mind. *Science Teacher, 42,* 21–23.

Smith, N. (1974). The challenge of scientific literacy. *Science Teacher, 41,* 34–35.

Spencer, H. (1864). *Education: Intellectual, moral and physical.* New York: Appleton.

Spencer, H. (1966). What knowledge is of most worth. In A. M. Kazamias (Ed.), *Herbert Spencer on education* (pp. 121–159). New York: Teachers College Press.

Tobin, K., & Espinet, M. (1989). Impediments to change: Applications of coaching in high school science teaching. *Journal of Research in Science Teaching, 26,* 105–120.

Troy, T. D., & Schwaab, K. E. (1982). A decade of environmental education. *School Science and Mathematics, 82,* 209–216.

Underhill, O. E. (1941). *The origins and development of elementary school science.* Chicago: Scott, Foresman.

Watson, B., & Konicek, R. (1990). Teaching for conceptual change: Confronting children's experience. *Phi Delta Kappan, 71,* 680–685.

Woodward, W. H. (1963). *Vittorino da Feltre and other humanist educators.* New York: Teachers College Press.

Yeany, R. H., & Padilla, M. J. (1986). Training science teachers to utilize better teaching strategies: A research synthesis. *Journal of Research in Science Teaching, 23,* 85–95.

·14·

RESEARCH ON ASSESSMENT IN SCIENCE

Rodney L. Doran
SUNY AT BUFFALO

Frances Lawrenz
UNIVERSITY OF MINNESOTA

Stanley Helgeson
OHIO STATE UNIVERSITY

Assessment is one of those words that can be used in a great many contexts and can, therefore, have a variety of meanings for different people. Specifically, this inquiry addresses the role of assessment in science education. This chapter begins with a brief description of the term assessment and then examines the relationship of assessment to other concepts.

The authors believe that assessment can be defined as the collection of information, both quantitative and qualitative, obtained through various tests, observations, and many other techniques (e.g., checklists, inventories), that is used to determine individual, group, or program performance. Measurement is a term thought to be closely related to assessment. However, measurement is not quite as encompassing a term as assessment. Measurement has generally been defined as the process of testing, but it is also accepted as a more encompassing term that has included the use of observations, discussions, and so forth, as well as paper- and-pencil tests. The term testing is used to describe various teacher-made tests, as well as standardized forms of testing such as inventories, questionnaires, and checklists. Evaluation is the process of making carefully determined value judgments and decisions related to the issues and concerns a given assessment has focused on (e.g., student achievement, program quality). Figure 14.1 shows an interactive illustration of this terminology and the relationships between these assessment terms.

Several types or methods of assessment have been identified in relation to science education. Some of these methods of assessment are discussed for the purpose of illustrating some divisions or styles of assessment that are widely used, but again used with some definitional variations. Figure 14.2 shows some of these methods of assessment, grouped in several ways to show some of these variations and differences. This list is presented not to show consensus or acceptance of this interpretation of these terms, but rather to illustrate the magnitude and breadth of methods of assessment. It should also be noted that there are subdivisions within these methods. For example, an essay method used as a means of assessment can be further defined as either a restricted response or an extended response format. Another example of a further subdivision might be found in the use of a practical examination. This method of assessment can be further defined or described as consisting of an individual or station format. Educators have identified many other dimensions to help clarify and describe their assessment efforts. Learning theorists have described the outcomes of learning in terms of the cognitive, affective, and psychomotor domains for purposes of assessment. Each domain has been further subdivided into hierarchical taxonomies of interrelated concepts. Bloom, Engelhart, Furst, Hill, and Krathwohl (1956) have described the cognitive domain; Krathwohl, Bloom, and Masia (1964) have presented the affective domain; Lunetta and

The authors thank reviewer Victor L. Willson (Texas A & M University).

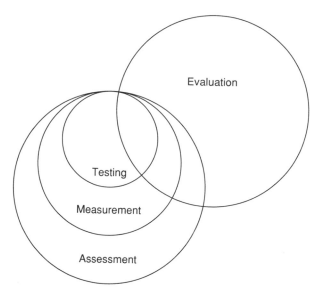

FIGURE 14.1. Diagram Illustrating Relationships Among Assessment Terms

Tamir (1979) have categorized the psychomotor domain in relationship to science laboratory skills.

Other researchers have proposed new categories for the learning domains, and some have modified the original categories. Notable in the science area has been the work by Klopfer

(1971) and Klinckman (1963) in this regard. The National Assessment of Educational Progress (NAEP, 1987a) designed a "collapsed" version of Bloom's taxonomy that utilized only three categories (knowing, using, and extending).

Another dimension of assessment is the timing of the assessment's use with respect to the instruction or intervention phase. The terms diagnostic, formative, and summative refer to assessment before, during, and after instruction.

Many educators have used these dimensions in planning their research and assessment activities. Quite often, this planning has resulted in the formation of a matrix (usually composed of two dimensions or sometimes three). Meng and Doran (1990) developed a matrix to help select appropriate assessment methods (e.g., written tests, practical tests) for various educational outcomes related to content, skills, and problem solving. A similar but more detailed matrix was developed by the Ontario Ministry of Education (1983) in order to match evaluation instruments with related learning behaviors. They used the nine-category Klopfer scheme and its divisions to identify Bloom's behavioral categories. They listed 12 specific assessment formats (e.g., multiple choice, checklist) within five categories of instrumentation based on selected response, created response, attitude instrument, lab tests, and observations.

The matrix in Figure 14.3 could be used by a science department. It maps out an assessment plan for all three domains of learning. For each domain, three subcategories are included; for example, Attitude, Interest, and Value are included for the affective domain. The other domains include several methods

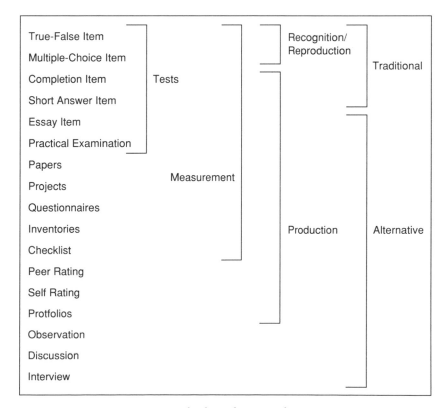

FIGURE 14.2. Methods and Types of Assessment

METHODS OF ASSESSMENT							
Outcomes	Written Tests	Inventories Questionnaires	Rating Scales	Group Practical	Individual Practical	Observation Discussion	Projects
Knowing							
Using							
Extending							
Attitude							
Interest							
Value							
Planning/ design							
Performing							
Analyzing/ interpreting							

* Possible
** Appropriate
*** Ideal

FIGURE 14.3. Matrix for Matching Format of Assessment with Instructional Outcomes

of assessment from written (paper and pencil) tests to student projects.

The major purpose of the matrix is to select methods of assessment that are most appropriate and efficient for the given outcomes. Meng and Doran (1990) described some of the advantages and disadvantages related to several methods of assessment:

Although it is currently chic to attack paper and pencil tests, they are still more than adequate for determining whether a student has mastered certain kinds of content, especially facts and terminology. Written tests, however, are not as effective in assessing whether students can apply that knowledge. (p. 43)

One form of planning matrix has been widely used by educators in designing tests for a particular course or unit of study. This matrix is often called a test grid or a table of specifications. The matrix in Figure 14.4 was used by Doran (1980) to illustrate this technique. Another example of a planning matrix has been suggested by Meng and Doran (1992) for use in guiding a sampling matrix for program assessment. Their example is shown in Figure 14.5.

In this example of multiple matrix sampling, the intent was to evaluate the achievement of a group of students on a large number of instruments. In order to not create a testing burden, each student was to complete one of five forms of the written tests, one of the two group practical tests, and one of the five individual practical tests. In this way, the teacher and curriculum developer could obtain data on a wide selection of cognitive and skill outcomes from the administration of these several tests.

One of the main foci of public interest in the United States during the 1980s and 1990s has been assessment of educational outcomes, especially in science and mathematics education. This interest was exemplified by the historic "education summit" that took place in late 1989. The president and the nation's governors declared at this summit that, for the first time in U.S. history, the time had come to establish clear, national performance goals necessary to make the United States internationally competitive. In February 1990, several important national education goals were adopted by the president and the governors. One of these goals specified that by the year 2000, students from the United States will be ranked first in the world in both mathematics and science achievement.

Before this pronouncement there had been a series of national reports on the state of education in the United States. A variety of data, in addition to the data presented in these reports, have helped to inform the general public. For example, the following studies and organizations have provided a great deal of data and information on the status of education: the National Council of Educational Statistics, the National Assess-

	Kinematics	Electricity & Light 20%		Magnetism
Atomic 20% & Nuclear 10%				
BEHAVIOR				
Application 50%	25%	10%	10%	5%
Comprehension 30%	15%	6%	6%	3%
Knowledge 10%	5%	2%	2%	1%
Analysis 10%	5%	2%	2%	1%

FIGURE 14.4. Blueprint for a Physics Test

Students	Written Tests					Group Practical		Individual Practical				
	I	II	III	IV	V	A	B	1	2	3	4	5
A	X					X		X				
B		X					X		X			
C			X			X				X		
D				X			X				X	
E					X	X						X
F	X						X	X				
G		X				X			X			
H			X				X			X		
I				X		X					X	
J					X	X						X
K	X					X		X				
L		X					X		X			
M			X			X				X		
N				X			X				X	
O					X	X						X

FIGURE 14.5. Program Assessment Chart

ment of Educational Progress (NAEP) studies, and Jon Miller's longitudinal studies of science literacy.

In the midst of this increasing interest in accountability, there has been a reassessment of the roles of curriculum, instruction, and assessment in our educational system. In the past, national interest has centered on instruction and curriculum and has somewhat neglected assessment. Now, although there is still considerable interest in instruction and curriculum, assessment has become a major interest. As discussed later in this chapter, the history of assessment is varied and offers insight into the changing needs and interests of our society. Today educational discussions are dominated by questions of assessment (e.g., Do teachers "teach to the test"? Is it bad if they do? Is it possible to do valid assessments? What do assessment results really mean? Can alternative assessment be done economically and efficiently?). The emerging bottom line is that assessment is only as good as the devices employed and the people charged with interpretation of the results.

The importance of assessment in science education is well recognized; thus, changes in assessment have also been found to be of significance. The goals for education have changed during recent years, as have techniques of assessment. As noted by writers at the Bank Street College of Education, "Altering assessment practices is likely to effect curriculum, teaching methods, and students' understanding of the meaning of their work" ("Alternative Assessment and Technology," 1992, p. 1). It is beyond the scope of this chapter to consider all of the ramifications of assessment and the accompanying changes in science education. Rather, consideration is given to a selection of topics that are viewed as indicators of current issues and concerns.

The purpose of this chapter is to provide an overview of assessment in science education, especially for grades K–12. Major issues in assessment are discussed, and current research findings are presented. In addition, current references are provided to assist the reader with alternative or more in-depth looks at these issues. The chapter begins with a history of science education assessment, including two major assessment movements, the NAEP and the International Association for the Evaluation of Educational Achievement (IEA). This historical background is followed by a brief look at the uses of assessment, followed by a more detailed description of various assessment categories and techniques. It is hoped that the information contained in this chapter will help clarify certain issues related to assessment and that the content presented will be useful to science teachers at all levels.

Reviews of Science Education Research

Reviews of science education research began in the 1920s with the first "Curtis Digest" (1926). This was followed by two more volumes (Curtis, 1931, 1939). These efforts were followed by a series of reviews sponsored by the Science Education Department of Teachers College, Columbia University (Boenig, 1969; Lawlor, 1970; Swift, 1969). The first "annual" review began with the *Review of Science Education Research* for 1972. These reviews were produced by the Educational Resources Information Clearinghouse (ERIC) for Science, Mathematics, and Environmental Education and distributed as a supplement to *Science Education*. Beginning in 1977, the annual reviews were included as a special issue of *Science Education*. Three articles (Voelker, Thompson, & VanDeman, 1980; Voelker & Wall, 1972, 1973) were found to be excellent guides to the many reviews available for science education researchers at those times.

In preparing this chapter, these annual reviews were examined, beginning with the 1978 annual ERIC review, for evidence of work that had been attempted in developing or revising assessment instruments related to science education. As each year's review was written by different people, the organization of each review was unique. It was difficult to locate research in assessment and then make comparisons across the years. However, the number of assessment instruments that were developed appeared to be relatively stable from 1978 to 1986. After 1986, the number of reported studies seemed to decline dramatically (Figure 14.6). Given the varying organization of the reviews, it was possible to miss some references to work on assessment that were located in other sections. Another dimension that was examined was the outcomes measured by the various assessment instruments. Again, the categories of the reviews varied, and it was necessary to develop new categories for analysis. The categories that were looked at and the number of citations found in each are shown in Figure 14.7. Again, these numbers should not be viewed as perfectly accurate, for reasons already cited. However, they should be viewed as a general indication of researcher interests. If the laboratory and process skills are viewed together, the cognitive and laboratory domains received approximately equal attention for instrument development. A somewhat surprising observation was the large number of projects that attempted to assess affective objectives.

Several introductions or summaries of the instrument development sections are worth citing here. In the 1978 annual sum-

1978	1979	1980	1981	1982	1983	1984	1985	1986	1987	1988	1989
19	18	14	14	9	13	15	10	10	5	2	5

FIGURE 14.6. Number of Articles, Reports, and Dissertations on Instrument Development

Student Characteristics	Cognitive Achievement	Affective Outcomes	Laboratory Skills	Process Skills
23	28	38	8	15

FIGURE 14.7. Outcomes Measured by Assessment Instruments (Annual Review 1978–1989)

mary of science education research, Gabel, Kagan, and Sherwood (1980) observed that "the progress of research in science education is critically related to the ability to measure relevant variables reliably. As the focus of research shifts with time, it is to be expected that new instruments will be developed to meet changing requirements" (p. 515).

As part of the 1979 annual summary of science education research, Butts (1981) reported:

In the more than 340 research studies reported in 1979, 18 described efforts to develop or improve the measurement tools researchers need. A hypothesis that an element of a filter is related to or is influencing the outcome variables must be tested with valid and reliable instrumentation. Since most of the independent and dependent variables in science education research required indirect measurement, the instrumentation needed to conduct research in science education becomes even more critical. While other studies did include the development of an instrument as part of their procedure, the studies described in this chapter include those for which measurement development was the primary purpose of the research. Of the 18 test development studies reported in 1979, 9 dealt with the measurement of achievement and attitudinal outcomes; 4 described measurement of elements of the student filter; and 4 focused on an emerging variable: teacher behavior in the learning context. To provide the information that a user needs, a study describing a measurement scheme should include:
a) An operational definition of the construct to be measured;
b) A discussion of the alternative means that could be used to measure that construct;
c) A description of the test strategy used including directions for its administration;
d) The estimation of the validity of the test in measuring the construct;
e) The estimate of the reliability of the test in measuring the construct; and
f) A description of the people for whom the test would be useful. (p. 422)

In the 1980 annual summary of research, Sipe and Farmer (1982) described the work on instrumentation as follows:

Progress toward generating a legacy of logically consistent measures of aptitude, performance and achievement can be reported. Diagnostic instrumentation and quality predictive measures seem to be in short supply. The linking of subscores on a variety of concepts to a more comprehensive instrument has met with some success. (p. 396)

As part of the 1981 annual summary of research, Voss (1983) summarized: "Not a great deal was added to the collection of research tests for science education.... One would agree with Wilson (1981)... many tests are available but that new areas need developing" (pp. 350–351).

Shymansky and Kyle (1988) summarized the research of 1986 which strove: to offer new and creative ways to assess the process of teaching, learning and schooling in science education. While we wholeheartedly agree with the recommendations of Schibeci, admonishing the indiscriminate development of instruments when more research could be fruitfully directed towards the validation of existing instruments, science educators must balance this recommendation with the need to develop new instruments to address new questions. (p. 399)

As part of the 1987 annual summary of research, Staver et al. (1989) stated that:

Not only do science educators need to become more systematic in assessing the validity of specific instruments, but additionally, they need to be more cognizant of the existence of other instruments which may be more applicable within a given research context. As an example, the development of yet another measure of process skills might be perfectly legitimate; however, if such a development was conducted without a prior examination of already existing instruments, then the new measure becomes an exercise in how to put an instrument together rather than a true generation of new knowledge. If indeed an instrument with acceptable validity exists and is applicable in the given research context, then it should be used in preference to the development of another instrument that must itself undergo a rigorous treatment of reliability and validity. (p. 336)

Welch (1985) commented on research effects to improve the techniques of measuring student learning. According to Welch,

Much work exists in this area but it is difficult to summarize its progress. Many tests have been developed and used to evaluate progress, predict future success in science, have been the focus of many dissertations.... Many student outcomes have been measured and researched. The influence of Bloom's taxonomy has been great and tests are now superior to those of 25 years ago. Further improvement will probably grow out of entirely new approaches and procedures for measuring outcomes. (p. 436)

Welch continued by discussing the assessment of student skills in science, which have occurred primarily at the lower grade levels. There have been isolated attempts to develop newer techniques, but the cost and noninclusion on national tests have delayed extensive development.

Uses of Assessment

The validity of any assessment is heavily dependent on the quality of the user. Good assessment devices can be misused, and poor devices can be stretched beyond their limitations. One way to look at these users has been to consider six levels of information aggregation: students and parents, teachers and instruction, districts and programs, states, national, and international. Figure 14.8 shows how these six user levels can be crossed by three levels of timing for assessment: beginning (i.e., diagnostic or baseline), middle (i.e., formative or process), and end (i.e., summative or outcome).

Although some timing levels might be found to be more common for some user levels, all users could probably make use of information from all times. For example, if one were to take the student user level across the three timing levels, the following types of assessment could possibly be observed. First, a student begins something; his or her existing understanding, feelings, and dispositions could be explored. This could be viewed as useful to the student, who would then be more aware of his or her current state of mind. Often students have not been made aware of how they think or feel about a given topic. Then, assessment for students at the middle level of timing would result in the student being made aware of how he or she has progressed in pursuing a particular goal. This would include information on what has happened, what things were done, what reactions occurred, what interactions took place, and so on. Lastly, assessment for the student at the third level of timing would include an examination of information related to the student's strengths and weaknesses related to obtaining a particular goal; what outcomes were reached in comparison to others pursuing the same goal or to preexisting standards; and what changes were made in relation to the student's initial understandings.

Another example, utilizing a simple accountability approach, exemplifies taking the end timing level through six user levels and could result in the following types of assessments. (Possible assessments for students were discussed in the preceding paragraph.) End-level assessments for teachers would focus on their capabilities and could include some descriptions of their teaching techniques, their classroom environment, their interaction patterns, and their effectiveness in facilitating student learning. End-level assessments for districts would focus on district- or programwide issues and could include specific information related to the numbers of teachers doing or having certain things; comparisons of instructional or curricular approaches; comparisons among schools or teachers; opinions of effectiveness from participants, parents, and the community;

and comparisons over time. End-level assessments for states would focus on statewide issues such as: Have state guidelines been followed? What percent of the teachers are teaching in areas they are certified in? How do school districts within the state differ? How have things changed over time? End-level assessments nationally would focus on issues of national priority, comparisons among states, longitudinal effects of national initiatives, comparisons among different federal funding mechanisms, and so forth. End-level international assessments would focus on comparisons among countries and, the effectiveness of educational strategies in addressing worldwide issues like climate control, among other things.

The important aspects of this grid are that assessment is viewed as a complex topic, and what is good for one level is not necessarily good for another. Sometimes, overlap occurs, but that is not always the case. Assessment that might have served admirably for accountability questions at the state level may not be useful to students or teachers. Care must be taken to design and select assessments that will serve the purpose at hand and to limit the use of assessment information to the purpose for which it was obtained.

Issues have also been identified related to the use of assessment that can cut across the cells of the grid. These issues have included types of comparisons, units of analysis, and types of analysis. All the issues are dependent on the purpose of the assessment, which can be defined by examining the cell of the grid in which it is found. Many of the issues are research design oriented, and they are reflective of the more detailed quantitative and qualitative domains.

The issue of types of comparisons has often been of interest in assessment activities. The two main types of comparisons that can be made are comparison of two things (e.g., individuals and programs, or groups and programs) on some measure or device or comparison of things with some predetermined standard. The first type of comparison is often normative. It is done, for example, in standardized tests where students are compared to the normative group that the developers of the standardized tests gave the test to initially. For these types of comparisons to be made and to be found useful, the comparison group must be one that makes sense for the given situation. Interpretation of the results is dependent on the relationship of the assessed population to the norm or comparison group. The second type of comparison is with a predetermined standard. An example of this type of comparison would be a criterion-

Levels of Timing	User Levels					
	Students and Parents	Teachers and Instruction	Districts and Programs	States	National	International
Beginning						
Middle						
End						

FIGURE 14.8. Matrix for Analyzing Assessment in Terms of Levels of Timing and User Levels

referenced test on which performance is compared to a specified ideal (e.g., use a graduated cylinder to measure the amount of a liquid within 0.5 ml).

The second across-grid issue that has been identified is choosing the appropriate unit of analysis for assessment activities. Questions in this area have included both how the assessment data should be grouped and at what level of aggregation the analysis should be run. Groups can be made up on the basis of race, gender, and so forth. The appropriateness of these groupings is related to the specific questions to be examined by the assessment and the sampling strategy selected for use. The unit of analysis used could be, for example, the individual, the classroom, the school, or the district. The choice of the appropriate unit of analysis is dependent on the purpose and theoretical basis for the assessment. For example, if the goals were to compare classroom learning environments, the classroom, rather than the individual learner, might be the appropriate level at which to do the analysis.

The third issue to be described is the type of analysis that should be conducted. Many different types of analysis have been developed, all of which have assumptions and criteria that must be met before they can be used validly. The decision of what type of analysis to be used, for either quantitative or qualitative data, must be carefully made and often requires an expert's advice.

Purposes of Assessment

Reasons given for assessing student learning in science are improvement of science instruction and programs; conveying expectations to students, parents, teachers, and administrators; monitoring the status of individuals, classes, districts, states, and the nation; and accountability (Raizen, Baron, Champagne, Haertel, Mullis, & Oakes, 1989). The areas of science to be assessed are generally considered to include knowledge of facts and concepts, science process skills, science thinking and problem-solving skills, skills needed to manipulate laboratory equipment, and the disposition to apply science knowledge and skills (Raizen, Baron, Champagne, Haertel, Mullis, & Oakes, 1991; Swain, 1985).

The impact of assessment cannot be overlooked. Shavelson, Carey, and Webb (1990) stated that "Developers of scholastic tests have become overseers of a very powerful instrument of education policy making: achievement tests" (p. 692). Mitchell (1992) noted that "Evaluation sends a message. It points to what is valued and ignores what is perceived to be unimportant" (p. vii). Because of the intense interest in science achievement in recent years, achievement test scores have influenced the behavior of superintendents, principals, teachers, and parents. It is widely accepted that achievement tests strongly influence and direct curriculum development ("Alternative Assessment and Technology," 1992). The power of achievement test scores has not stopped here. Political decisions at the highest levels of government have reflected concern about science scores; improvement of these scores was cited as one of the goals of the national education strategy presented by President Bush (U.S. Department of Education, 1991).

Summary for Introduction

This section began with some definitions of terms that are widely used in science assessment (i.e., assessment, measurement, testing, and evaluation). Types or methods of assessment were presented next, noting that a variety of dimensions such as method of data collection, the specific domains of learning, or the timing and purpose of assessment can affect the design, implementation, and outcomes of the assessment process.

Currently, much attention is focused on science assessment. This interest has been fueled by many national reports and international surveys. Assessment has also been a consistent but growing component of science education, as evidenced by the historical overview presented in this section of the many reviews of science education research.

Assessment can have a powerful impact on students, teachers, parents, school administrators, and the public. This section also identified a variety of uses and the types of comparisons, units of analysis, and types of analysis that can be made with various assessment projects. Care must be taken to interpret the results of assessment according to these parameters, which guided the development of the assessment plan.

Lastly, this section examined purposes or reasons for assessment in science education. Again, attention was given to the powerful impact that assessment can have in all areas of science education. The next section provides a brief history of science education assessments.

HISTORY OF SCIENCE ASSESSMENT

How to assess ability is not a new problem. Chinese officials conducted civil service examinations as early as 2000 B.C., and Greek scholars like Socrates used individualized questioning to assess student learning. However, Travers (1983) contended that before the mid-1800s no formal evaluation had been conducted in American education. Before 1837, political and religious beliefs dictated most educational choices and were opposed to the collection and use of evaluative information. Educational evaluation, as we know it today in the United States, probably began with Horace Mann and Henry Barnard. In a series of reports to the Board of Education for the Commonwealth of Massachusetts, Mann identified all educational concerns that had some sort of empirical support. Teacher competency and appropriate curriculum materials were included. The Boston Survey, conducted in 1845, was the first use of printed tests for wide-scale assessment of student achievement in several areas. These tests included definitions, geography, grammar, civil history, natural philosophy, astronomy, writing, and arithmetic (Travers, 1983). From 1895 to 1905, Joseph Rice carried out an assessment similar to the Boston Survey in a number of large school systems across the United States. He found negligible differences in spelling skills but large differences in arithmetic skills (Worthen & Sanders, 1973). These early tests were mainly criterion referenced and assessed the students' degree of mastery of predetermined topics. *Elements of Pedagogy* (White, 1886) stated that "there has been no change in school training in the past thirty years more marked or general than the use of written exercises" (p. 193). White then discussed the failings of written tests as opposed to recitation and the dangers inherent in teaching to the test.

In the late 1800s and early 1900s, measurement techniques for assessing human abilities flourished under the guidance of Edward Lee Thorndike. Thorndike was responsible for the emergence of a scientific approach to assessment. The interest in assessment was exemplified by the Civil Service Act of 1883, which created the civil service examination system. By 1900, half of all federal jobs were awarded on the basis of competitive examinations. The Stanford Binet scale was developed by Terman in 1916, and the army Alpha (for literates) and Beta (for nonliterates) tests were developed and used to help ascertain the needs of the soldiers required for World War I. The large Army testing programs encouraged people to believe that test results could be used as a basis for decision making and that large numbers of individuals could be tested efficiently. Lewis M. Terman and Robert M. Yerkes from the Army Psychological Division received a grant from the Rockefeller Foundation to develop a standardized intelligence test for the schools, which was used widely to classify students into ability groups for instruction.

In 1915, the school board of Gary, Indiana purchased an external evaluation examination to assess its educational program; it is the first school board known to have done so (Worthen & Sanders, 1987). At about the same time, several school systems created bureaus of research whose purpose was to work on large-scale assessments of student achievement (Madaus, Arasian, & Kellaghan, 1980).

The testing movement continued to spread with the development of the New York Board of Regents examinations and the Iowa tests in the early 1900s. School grading systems came to be mostly based on the results of teacher-made tests of student achievement. In addition, personality and interest profiles were developed by psychologists or researchers. The first of the Curtis Digests, which covered science education research from 1920 to 1925 and "those researches published before 1920 as seen in the author's judgment to give data of more or less permanent value" (p. ix), showed that science educators used a variety of assessment devices. These included students conducting experiments and writing essay responses to content questions or "writing up lessons," as well as surveys of students' science interests and analyses of curriculum content. The Second Curtis Digest covered research from 1925 to 1930. This research also included a variety of assessment devices related to science education but revealed a growth in the more standardized multiple-choice formats. As in 1920 to 1925, digest topics included studies of different teaching styles, student interests, and text or content analysis. A report of a study of the relative value of methods of scoring examinations revealed that the "new" examinations were of the completion, modified true-false, multiple-response, and modified multiple-response types.

The Progressive education movement in the 1930s fostered many new educational programs, as well as growth in the testing enterprise. The Buros' *Educational Psychological and Personality Tests of 1933 and 1934* was only 44 pages long. The 1938 edition had expanded to 400 pages and listed over 4000 tests (Haney, 1984).

Although many of the new educational programs of the 1930s were evaluated, most of the evaluation was informal and subject to criticism. The ensuing controversy led to the Eight Year Study, directed by Ralph Tyler (Smith & Tyler, 1942). The study employed a wide variety of assessment devices (e.g., tests, scales, inventories, questionnaires, check lists, and pupil logs) to gather information about achievement of curriculum objectives in 30 high schools. This project and later work on taxonomies of educational objectives (Bloom, Engelhart, Furst, Hill, & Krathwohl, 1956; Krathwohl et al., 1964) crystallized the objectives-based approach to educational evaluation, which is still popular today in the NAEP and in many individual state testing programs.

The Third Curtis Digest reviewed science education research from 1930 to 1937. Although the topics of this research were similar to those of previous years, movement toward a more quantitative assessment that reflected more statistical treatment of the data was reported. In 1937, the Committee on the Function of Science in General Education reported that the general technique for the construction of science evaluation instruments should follow the Tyler model of stating objectives, describing student behavior that indicates growth toward the objectives, and methods for recording behavior. Suggested methods included anecdotal records, observations by trained observers, questionnaires, interviews, student creative products, and student diaries or other activities. The committee recommended that less attention be given to paper-and-pencil tests and more attention be given to the other techniques.

School accreditation agencies also flourished in the 1930s, replacing the older system of school inspections. Teams of external educators were sent to review self-study reports of member institutions and to make their own observations (Worthen & Sanders, 1987).

Instead of the Curtis Digests, science education research for 1938 to 1957 was reported by three different authors, Robert W. Boenig (1969) for 1938 to 1947, J. Nathan Swift (1969) for 1948 to 1952, and Elizabeth Phelan Lawlor for 1953 to 1957. These texts contained articles that describe the construction and validation of science "tests," as well as more sophisticated statistical analyses and the use of other types of tests (e.g., IQ, mental rigidity) to assess student characteristics. Interestingly, one of the tests described was a paper-and-pencil test for the evaluation of laboratory instruction. *The Forty-Sixth Yearbook* (National Society for the Study of Education, 1947) recommended that evaluation be an integral part of teaching and learning of science in the elementary school. It suggested that an evaluational effort include the interpretation of behavior in relation to such basic objectives as skills, interests, appreciation, and attitudes, as well as to the functional understanding of facts, large conceptions, and basic principles of science.

Another significant period in the history of evaluation was begun in the United States in response to the launching of the Soviets' *Sputnik I* in 1957. Several new educational programs, especially in mathematics and science, were developed as a result of the launching.

The early 1960s witnessed three major educational evaluation efforts. The Coleman study focused on equality of opportunity for minority children. Project TALENT was a longitudinal study of the relationships among abilities and careers, and the

NAEP sponsored a yearly national assessment by subject area. The NAEP was one of the first examples of criterion-referenced testing. This type of testing was supported by the task analysis procedures developed by B. F. Skinner. The tasks were translated into behavioral objectives with predetermined standards for accomplishment of the task, often specifically stated.

The passage of the Elementary and Secondary Education Act (ESEA) in 1965, which included the Title 1 (later to be called Chapter 1) programs, enhanced these three evaluation efforts. The recipients of grants under Title 1 and Title 3 programs were required to provide an evaluation report on the use of their funds. In response to the pervasive need for evaluation engendered by the ESEA grants, much theoretical and practical work in evaluation took place at this time. New models for conceptualizing the evaluation process appeared, as well as new methodological approaches (Stufflebeam & Webster, 1980). New evaluation societies and corresponding publications were initiated. Many of these emerging evaluation models were products of a positivistic philosophy. As the field of evaluation matured, more differing viewpoints were incorporated (e.g., pragmatic, naturalistic, phenomenological approaches) leading to the quantitative-qualitative controversies of the 1980s and 1990s (Guba & Lincoln, 1988).

The 1970s were a time of social unrest and concern over fairness in testing. Standardized tests were criticized by the National Association for the Advancement of Colored People and by the National Association of Elementary School Principals, to name just two influential groups. This concern about test validity helped to promote statewide minimum competency testing (Baker & Stites, 1991). The 1970s also witnessed growth in cognitive psychology, assessments involving writing, and domain-referenced testing (Baker, 1974); in techniques like multiple matrix sampling (Sirotnik, 1970); and in item response theory (Lord, 1980).

The 1980s were a time of maturation and consolidation in evaluation and increasing interest in assessment. In 1981 *Standards for Evaluations of Educational Programs* (Joint Committee on Standards for Educational Evaluation, 1981) and *Personnel Evaluation Standards* (Joint Committee on Standards for Educational Evaluation, 1988) were published. Major issues were the political context of evaluation, paradigm boundaries, and the utilization of evaluation results. An additional concern raised by the report *A Nation at Risk* (National Commission on Excellence In Education, 1983) was that U.S. students performed more poorly than students from other nations in mathematics and science. Similarly, the results of several international assessments drew attention to the poorer performance of U.S. students. These concerns have led to an increase in local and state testing efforts and the use of tests to assess teachers' instruction. The task of education was viewed as raising test scores rather than educating children. The emphasis on test scores also seemed to further stress tracking of students (Oakes, 1985). In 1987, the Longitudinal Study of American Youth (LSAY) was begun by Jon Miller. Students were selected in the seventh and tenth grades, and they, their parents, and teachers completed a variety of instruments over several years. The purpose was to provide accurate and reliable estimates of the factors contributing to students' learning of science and mathematics. During this time, Miller enlarged his studies of

adult science literacy in the United States. Miller (1989) reported that only 1 in 20 American adults met a minimal definition of scientific literacy.

Another major data-gathering effort was conducted by the Council of Chief State School Officers. They worked to improve the quality and usefulness of data on science and mathematics education and to define and implement a set of indicators to use in improving science and mathematics education (Blank & Dalkilic, 1990).

In the 1990s attention was drawn to the development of policies and procedures for assessing progress in higher-level cognitive abilities (National Center for Improving Science Education, 1991). The interest in international comparisons was heightened by the *America 2000* report (U.S. Department of Education, 1991), which stated the federal government's goals for education by the year 2000. The report suggested that U.S. students should become first in the world in science and mathematics. Monitoring whether this goal is achieved has required continued federal support of international achievement testing.

Summary for History of Science Assessment

This brief history has detailed the growth of educational evaluation in the United States. Evaluation, like many other issues in education, is somewhat spiral in nature. Early educational evaluation in the United States was characterized by very site-specific, anecdotal types of assessment procedures. If comparisons were made at all, they were of a descriptive nature. The desire to be fair to all and to provide information that would be consistent across sites or persons has led to the development of a variety of "tests." Experimentalists and psychologists have led the assessment effort within a context of categorization of people, curricula, programs, and so forth. Quantitative methods and approaches were defined, and mathematical solutions to human questions seemed logical. With the proliferation of this type of assessment, however, came dissatisfaction and distrust. The numbers did not seem to represent and explain the issues involved, and only a few people knew or wanted to understand what these sophisticated statistical techniques were actually saying. The folk belief was that you could prove anything with statistics. The questionable validity of and lack of trust in quantitative approaches prompted the movement toward qualitative assessment and ultimately to the merging of these techniques, paradigm arguments notwithstanding. Also at issue was the notion of "authentic" assessment—actual performance of the desired outcome as opposed to measurement of an intermediate, predictive, skill (e.g., actually conducting an experiment as opposed to saying how you might conduct an experiment or choosing one of some given ways of conducting the experiment). The notion of assessment has undergone several iterations, with each offering unique perspectives and strengths and filling in the weaknesses of previous models.

The greatest public attention to science test scores has been generated by publicized results of tests that are external to the schools, such as National Assessment of Educational Progress (NAEP), the Second International Science Study (SISS), and the various state-mandated tests. This area, therefore, is the starting point for our consideration.

NATIONAL AND INTERNATIONAL ASSESSMENT

Surveys of the achievements of U.S. students at various grade levels or ages have been conducted in many content areas (including science) since the late 1960s as part of the NAEP. Other countries with active science assessment programs include Israel with its biology matriculation system and the Assessment of Performance Unit (APU) in the United Kingdom.

For over two decades the International Association for the Evaluation of Educational Achievement (IEA) has conducted cross-national surveys, including two science studies. More recently, the Educational Testing Service (ETS) has conducted surveys of science achievement involving a number of countries, called the International Assessment of Educational Progress (IAEP). As each of these studies has had considerable impact on assessment practices at all levels, they are reviewed in some detail.

As widely reported, achievement scores on national assessments of science have not been encouraging (O'Neil, 1991). "Trends for 9-, 13-, and 17-year-olds across five national science assessments conducted by NAEP from 1969 to 1986 reveal a pattern of initial declines followed by subsequent recovery at all three age groups. To date, however, the recoveries have not matched the declines" (Mullis & Jenkins, 1988, p. 5). Because the 1990 NEAP assessment was scored differently from earlier assessments, direct comparison of results was not useful. However, a few of the findings are noted as indicators. In this study, fewer than half of the seniors tested demonstrated the knowledge and reasoning abilities needed to use scientific knowledge to interpret data found in tables and graphs or to evaluate and design science experiments, nor did they have an in-depth knowledge of scientific information. Approximately two-thirds of the eighth-graders and approximately one-third of the fourth-graders were able to understand basic information in the physical sciences and basic ecological principles and exhibited a beginning ability to interpret experimental results (National Center for Education Statistics [NCES], 1992, p.3). More than half of the 17-year-olds tested in 1986 were deemed to be inadequately prepared either to perform competently jobs requiring technical skill or to benefit from specialized training. The 1990 results appeared only marginally better (Mullis & Jenkins, 1988; NCES, 1992).

Differences in science proficiency between racial and ethnic groups, although decreasing, still persisted. Although no gender difference appeared at grade 4 in the 1990 study, gender differences in science performance continued to appear and to increase at the higher grade levels (Hueftle, Rakow, & Welch, 1983; Mullis & Jenkins, 1988; NCES, 1992). Socioeconomic status was also related to science achievement. Students from advantaged urban areas outperformed students from disadvantaged urban areas. Several factors in the home environment were also found to be related to science proficiency. At all three age or grade levels, students who had access to more reading materials performed better than students who had fewer materials available; students who watched television for six or more hours a day had lower average proficiency than students who watched less; and students who had both parents living at home had higher average proficiency than those from single-parent families or those who lived apart from both (NCES, 1992).

By and large, these findings parallel the findings of international studies in which U.S. students took part. As a rule, science achievement in the United States was lower than in other countries (Jacobson, Doran, Chang, Humrich, & Keeves, 1987; Jacobson & Doran, 1988; Jacobson, Doran, Humrich, & Kanis, 1988; LaPointe, Askew, & Mead, 1992). It is interesting to note that the 1990 International Assessment of Educational Progress study found that, in nearly all of the 13-year-old populations, at least 10 percent of the students performed at least 15 points above the IAEP average and at least 10 percent performed at least 15 points below the average (LaPointe, Askew, & Mead, 1992). Gender differences, much like those found in the U.S. studies, also appeared worldwide. Only in biology in a few countries did females score higher than males (Jacobson & Doran, 1988). However, in science process laboratory skills, females did about as well as males. This suggests that teaching science through process tasks that emphasize manipulative activities might encourage females to study and succeed in science (Humrich, 1988; Jacobson, Doran, Humrich, & Kanis, 1988; Kanis, Doran, & Jacobson, 1990). In light of the relationship between television viewing and science proficiency, it was interesting to note that 13-year-olds in most countries were likely to spend much more time watching television (i.e., 2 to 4 hours per day) than doing homework (i.e., 1 hour or less per week in all countries except the Soviet Union) (LaPointe, et al., 1992). The school factor most strongly correlated with achievement was the opportunity to learn (Jacobson, Doran, Humrich, & Kanis, 1988). In countries that reported the highest science achievement scores there was relatively early specialization, and secondary school students studied the sciences for 3 or 4 years (Jacobson & Doran, 1988).

In an effort to determine whether science and mathematics education has improved since the publication of *A Nation At Risk* (National Commission on Excellence in Education, 1983), the Council of Chief State School Officers summarized data from national transcript studies, the NAEP data, and the State Indicators of Science and Mathematics (Blank & Engler, 1992). The report noted that students are now receiving more science instruction. From 1980 to 1987, 40 states increased their science requirements for graduation. The rates of course enrollments by course level indicated that the proportion of students advancing through the secondary science curriculum has been increasing. "The percentage of students taking first year biology by the time they graduated increased from 75 percent in 1982 to 95 percent in 1990, the percentage taking chemistry went from 31 percent to 45 percent, and the physics enrollment went from 14 percent to 20 percent" (p. 3). Gender and race or ethnicity differences in course taking have continued. Although science and math enrollments were below the recommendations of the Excellence Commission, these enrollments are increasing. States with higher requirements have more overall course taking in science and mathematics. They also reported slightly more upper-level courses being taken in these states.

Using symbolic representations	• Reading information from graphs, tables, and charts • Representing information as graphs, tables, and charts • Using scientific symbols and conventions
Using apparatus and measuring instruments	• Using measuring instruments • Estimating quantities • Following instructions for practical work
Using observations	• Using an identification key • Observing similarities and differences • Interpreting observations
Interpretation and application patterns in information	• Describing and using • Judging the applicability of a given generalization • Distinguishing degrees of inference • Applying science concepts • Generating alternative hypotheses
Design of investigations	• Assessing testable statements • Assessing experimental procedures • Devising and describing observations
Performing investigations	

FIGURE 14.9. Categories and Subcategories of the Assessment Framework Before 1980

Data from the NAEP indicated slight average increases in science achievement from 1982 to 1990. As each of these studies has had a considerable impact on assessment practices at all levels, they are reviewed in some detail.

Assessment of Performance Unit—Science Surveys

The Assessment of Performance Unit (APU) was set up in 1975 within the Department of Education and Science (DES) to promote the development of methods of assessing and monitoring of the achievement of children in school, and to identify the incidence of underachievement. The APU exists to provide objective information about national standards of children's achievement at all levels of ability. Such information can help inform decisions about where and how improvement can be made. (Johnson, 1989)

This quotation from the cover page of an APU report spells out the purpose of the APU and the likely uses for the results it has obtained. Dozens of researchers, teachers, and other specialists have worked on the many aspects of the APU Science Survey. Almost a quarter of a million individual students were tested with some of the APU instruments. Data were gathered annually for 5 years, from 1980 to 1984. The focus of this assessment was on students at ages 11, 13, and 15. The categories and subcategories for this assessment framework are shown in Figure 14.9.

This framework guided the early development of the assessment program (Assessment of Performance Unit, 1978), but was modified to the simple framework shown in Figure 14.10.

Reports of the procedure used, including sample items, and the results obtained were prepared for each of the age populations for each of the years surveyed. Results were available for each of the science categories and subcategories within each report. The reports on the 1984 surveys also included a review of the previous surveys. These APU survey reports are presented in Figure 14.11.

For those interested in the technical details of the APU science assessment surveys, an excellent report by Johnson (1989) is recommended. The APU has been known for two important procedures: domain sampling and performance assessment. The sampling done by the APU was with respect to student and school, as well as domain or content. They found that domain sampling was a workable strategy for most of the subcategories of the assessment framework but that more test questions per subcategory were necessary for satisfactory precision than ini-

Using graphs, tables, and charts
Making and interpreting observations
Interpreting presented information
Planning parts of investigations

Applying biology concepts
Applying chemistry concepts
Applying physics concepts

FIGURE 14.10. Domain-Sampled Subcategories in the Assessment Framework After 1980

1980 Surveys

Harlen, W., Black, P., Johnson, S. (1981). *Science in schools. Age 11: Report no. 1.* London: HMSO.
Schofield, B., Murphy, P., Johnson, S., Black, P. (1982). *Science in schools. Age 13: Report no. 1.* London: HMSO.
Driver, R., Gott, R., Johnson, S., Worsley, C., Wylie, F. (1982). *Science in schools. Age 15: Report no. 1.* London: HMSO.

1981 Surveys

Harlen, W., Black, P., Johnson, S., Palacio, D. (1983). *Science in schools. Age 11: Report no. 2.* London: APU.
Schofield, B., Black, P., Head, J., Murphy, P. (1984). *Science in schools. Age 13: Report no. 2.* London: APU.
Driver, R., Child, D., Gott, R., Head, J., Johnson, S., Worsley, C., Wylie, F. (1984). *Science in schools. Age 15: Report no. 2.* London: APU.

1982 Surveys

Harlen, W., Black, P., Johnson, S., Palacio, D., Russell, T. (1984). *Science in schools. Age 11: Report no. 3.* London: APU.
Gott, R., Davey, A., Gamble, R., Head, J., Khaligh, N., Murphy, P., Orgee, T., Schofield, B., Welford, G. (1985). *Science in schools. Ages 13 and 15: Report no. 3.* London: APU.

1983 Surveys

Harlen, W., Black, P., Khaligh, N., Palacio, D., Russell, T. (1985). *Science in schools. Age 11: Report no. 4.* London: APU.
Khaligh, N., Johnson, S., Murphy, P., Orgee, T., Schofield, B. (1986). *Science in schools. Age 13: Report no. 4.* London: APU.
Welford, G., Bell, J., Davey, A., Gamble, R., Gott, R. (1986). *Science in schools. Age 15: Report no. 4.* London: APU.

Review Reports—Incorporating 1984 Survey Results

Russell, T., Black, P., Harlen, W., Johnson, S., Palacio, D. (1988). *Science at age 11: A review of APU findings 1980-1984.* London: HMSO.
Schofield, B., Black, P., Bell, J., Johnson, S., Murphy, P., Qualter, A., Russell, T. (1988). *Science at age 13: A review of APU findings 1980-84.* London: HMSO.
Archenhold, W. F., Bell, J., Donnelly, J., Johnson, S., Welford, G. (1988). *Science at age 15: A review of APU findings 1980-84.* London: HMSO.

FIGURE 14.11. APU Survey Reports

tially anticipated. Johnson (1989) concluded that 60 questions per subcategory was a more appropriate number for use in a survey. Pupils were selected using a two-stage stratified sampling procedure, with schools being the first-stage unit. Typically, 12,000 to 16,000 students were tested in a single survey, and they were drawn from 500 to 600 schools. Approximately 300 schools were selected in England and about 100 were selected from both Wales and Northern Ireland. Subsets of the test questions were grouped into parallel subtests that were administered to random samples of students. In other words, questions were administered to pupils according to a multiple-matrix sampling scheme. The testing time for each pupil was limited to 45 minutes for 11-year-olds and 1 hour for 13- and 15-year-olds.

Although the APU assessment procedure included a wide variety of paper-and-pencil formats, a major contribution was their work in developing and validating performance assessment techniques *and* the implementation of practical testing on a national scale. Practical testing is by nature logistically complex. The task of achieving acceptable levels of validity and administrative reliability was a major one.

Two major formats were used for administering practical examinations: the "circus" or "station" format and the individually administered format. In the station format, 6 to 12 pupils were tested at one time, moving from one station to another on a timed basis. While at each station, the pupils worked independently, reading instructions, performing tasks, and recording results in a test booklet.

Individualized administration was clearly necessary for the "performing investigation" category, in which pupils were presented with a range of equipment and materials and required to solve a given problem by conducting a practical investigation. The APU researchers decided to provide to the schools the trained administrators, equipment, and materials necessary for practical testing. The administrators were practicing science teachers who were recruited for 2 days of training and 1 to 2 weeks of testing. Another set of science teachers were recruited and trained to mark or score the student test booklets. The traditional reliability indicator, average marker intercorrelation, was .95 or higher for most subcategories.

A national assessment program in England is described by Burstall (1986). The target audience for science consisted of 11-, 13-, and 15-year-olds. The overall assessment included both an oral and a practical component. The oral test was designed to determine the student's ability to communicate effectively and dealt with general rather than science-specific information. The science practical test included a script to guide the examiner's verbal assessment of student understanding and a check list to record observations of the student's performance. Burstall found the practical assessment to have the following major advantages: (1) because all questions were oral, poor readers were not penalized; (2) students could have asked for clarification of the test; (3) the pace and extent of the testing could be adjusted to suit the individual student; (4) students had the opportunity to retract or amend an answer; (5) assessors were permitted to prompt or to direct students toward an appropriate strategy; and (6) assessors were able to observe the method and problem-solving strategies of the student.

The British researchers believed that many science skills could not be completely and adequately assessed other than by a practical test. They thought that students would not necessarily do better on practical tasks; rather, they would perform differently because the tasks themselves are different. The form of presentation altered the starting point for students. Written performance can be very different from practical performance, reflecting the different skills or combinations of skills required. The experiences that the APU reported support the idea that it is possible to make valid and reliable measurements of the complex pupil activities involved in practical science.

Much can be learned from the APU experience and reports. Johnson (1989) acknowledges some implications for assessment outside the APU:

Quite clearly the first implication for any new assessment program is that of time allowance: time for planning, time for domain definition, time for question development, time for reflection on findings, time for dissemination. There is a real danger, as in all areas of activity, that the time needed will be underestimated. The costs of the original underestimation in the APU survey program have been spelt out in this report. Future program planners might learn from this experience. (p. 115)

Johnson continued,

the APU Science program has blazed a trail for innovative process and practical assessment on a very large scale. Many novel ideas have been incorporated in the test questions created for survey use.... It has proved possible to produce practically-based questions, and to administer them in an administratively reliable way on a nationwide basis. But the task is not an easy one, and the development time and effort involved are significant. (p. 115)

Another APU innovation was their

systematic computerized question-banking.... The system of question labelling, particularly the keyterms option with the thesaurus link, provides a very powerful system for question retrieval and, more importantly, for in-depth research into factors determining both question difficulty and pupil performance. The on-line availability of the wealth of question information currently available enables immediate links to the survey response database, and opens up wide research possibilities with minimum effort. (Johnson, 1989, p. 115)

However, limitations were uncovered by the APU work. Johnson observed that these limitations emerged from the experience of domain definition and from the variance component analyses.

In addition to the reports listed earlier, the APU developed a series of "short reports for teachers." These 40 to 50-page reports were organized with several foci: report of findings by age, reports describing the assessment framework and instruments, and reports on specific skills and content topics. Furthermore, the procedures and results of the APU surveys were reported in a wide variety of articles, chapters, and occasional papers, mostly in British or international journals. A summary of the APU accomplishments was authored by Black (1991), who was one of the key people on the APU team.

Since its last large-scale APU science survey in 1984, the APU has carried out a follow-up study of a cohort of pupils between the ages of 12 and 14 during the period 1987 to 1989. With the introduction of the National Curriculum in Science, the APU is no longer in existence. Since April 1989, the Schools Examination and Assessment Council (SEAC) has assumed responsibility for the assessment and evaluation of the National Curriculum. A new section, the Evaluation and Monitoring Unit (EMU), was formed within SEAC for these specific responsibilities. Furthermore, the EMU has assembled some reports that are representative of the culmination of the APU work in science. These reports can be found in a series entitled *Assessment Matters*. Taylor and Swatton (1990) wrote the first report on graphing in school science, and Strang (1990) composed the next report on measurement in school science. In addition, six other reports were made available in this series. Several of these reports concerned specific skills such as planning investigations, and observing in school science.

Israeli Matriculation Examination in Biology

Many countries have school exit or matriculation examinations for students completing their secondary schooling. In most cases, these exams are composed of essay and/or multiple-choice items that assess cognitive outcomes. Because Israel had a high school biology program that stressed laboratory and fieldwork, the Israeli matriculation exam in biology was multifaceted. The Israeli Biological Sciences Curriculum Study (BSCS) biology examination consists of a written examination, which counts for 60 percent of the total examination score, and a practical test, which counts for the rest. The resulting total examination score is then combined with the teacher's overall assessment grade to form the student's final grade (Tamir, 1974).

The practical examination, held in April, comprises the following three parts:

1. Plant or animal identification with the aid of a key
2. Oral examination on animals and plants, as related to their environment
3. Practical laboratory examination (2½ hours).

The schools are sent a list of materials and equipment to prepare and set up, 3 weeks before the scheduled examination date. Bottles, test tubes, thermometers, microscopes, and Pasteur pipettes are arranged at different stations in the school laboratory by the teacher and the laboratory technician. The examiners bring with them living materials such as *Daphnia* and yeast and certain chemicals such as adrenaline (epinephrine). The students are examined in groups of 10 to 18. Each student is assigned to a different working table, on which are all the necessary materials.

All of the materials and the examination paper are provided to the student in the laboratory. The student is requested at several points during the examination to call the examiner, who circles specific answers with red ink and hands over cue notes that direct the continuation of the investigation. These notes are used to avoid or reduce the need for discussion between the examiner and the examinee. The laboratory test is purported to measure a student's ability to solve novel problems in the laboratory.

The corresponding paper-and-pencil test is allocated 3 hours and consists of the following:

1. 30 multiple-choice items.
2. 3 multiple-choice items that have the correct answer marked; students must write an explanation for the marked answer.
3. 9 items based on data presented in a table or graph.
4. 9 essay questions.
5. 6 questions based on a previously unseen research report.

This multifaceted examination is based on the high school biology syllabus, which is prescribed by the Israeli Ministry of Education. Responsibility for the development of new science curricula, the organization of inservice education programs, and the design of the matriculation examination has been given to the Israeli Science Teacher Center (ISTC). The ISTC is expected to work closely with the Ministry of Education in providing these essential services.

IEA Studies in Science

Much attention has been focused on international comparisons, particularly in science and mathematics. A major source of data for these descriptions has been the IEA (International Association for the Evaluation of Educational Achievement) studies in science. The IEA is a multinational consortium of research institutes, universities, and individual researchers interested in educational assessment. Studies of educational achievement in an international context have been coordinated by the IEA in many fields. A small IEA staff is located in The Netherlands; however, each country must provide financial and professional support for collecting and analyzing their data for the international reports. Each country is strongly encouraged to prepare its own national reports, as well.

The First IEA Science Study (FISS) involved the collection of data during the 1969–1970 school year. The international report of the study was written by Comber and Keeves (1973). The report of U.S. involvement in FISS was part of a volume that also reported on IEA studies in mathematics, civics education, reading comprehension, French as a foreign language, and literature (Wolf, 1977). This collection of studies was called the Six Subject Survey. An article (Klopfer, 1973) described the FISS, including the populations sampled, the test framework, sample items, and some cross-national test results. The mean scores Klopfer reported for the U.S. sample did not compare well internationally.

The populations commonly surveyed by IEA studies are 10-year-olds (Population 1), 14-year-olds (Population 2), and students in the last year of secondary schools in their country (Population 3). Population 1 was chosen because it corresponded most clearly with the end of intact class instruction. In the second IEA studies, population 1 was broadened to include students in the modal grade level of 10-year-olds, because it was easier in most countries to sample by grade level rather than age. Population 2 was also broadened to include students in the modal grade level of 14-year-olds, to correspond to the end of compulsory schooling in most countries. Population 3 (called Population 4 in FISS) was chosen to demonstrate the cumulative effect of a country's science programs on students by the time they exited the secondary school.

The sampling of schools and students is important in all research. It is also viewed as critical in international studies. In IEA studies, researchers from all participating countries met and agreed on population definitions, excluded groups, appropriate strata for sampling, size of sample, and so forth. A sampling referee was put in charge of approving each country's sampling plan before data collection. In addition to sampling, a major concern faced by international surveyors is the relevance and quality of the tests used. In the IEA studies, each country was responsible for submitting an analysis of their country's science curriculum at the three population levels, in terms of content and objectives, the relative emphasis on the specific content topics, and the levels of objectives. A synthesis of each country's curricular analyses became the test grid for each population. Items were collected from previous IEA tests and other tests used in the participating countries. Items that best fit the test grid were assembled into trial test forms and pilot tested in some countries. Based on trial test data and relevance to the test grid, the final tests were assembled. In addition to these achievement tests (assessing cognitive outcomes), IEA has always administered additional questionnaires to students to provide data needed for comparison of samples and prediction of achievement. In addition to the student background questions (gender, family size, grades in science, etc.), IEA has had much success with a special question that probed for family support for education—the "Number of books in home" question. IEA studies have usually included an attitudinal scale designed to be related to the particular discipline being assessed and the Word Knowledge Test, a short inventory measuring verbal aptitude. For the science studies, a short test of mathematics concepts and skills was also made available for a country's use. The achievement tests required approximately two periods of administration time or about 2 hours. The other inventories and questionnaires are administered in a third period. It should be noted that schools in general resisted giving more of their class time for these kinds of assessment surveys.

In addition to the student tests, inventories, and questionnaires, the IEA surveys included teacher and school questionnaires. These were intended to help each country interpret or understand their achievement pattern and provided valuable information about school support for science programs, teacher's certification and experience, and other items. A construct that IEA pioneered has been referred to as the Opportunity to Learn (OTL). Teachers were asked with the OTL to rate the extent to which their students have studied the information and skill within each test question.

The Second International Science Study (SISS) involved data collection in the mid-1980s with the addition in the United States of advanced science samples and practical tests at grades 5 and 9 in 1986. As one of the purposes of a second study was to relate achievement to the first study, a considerable number of items were common to both FISS and SISS studies. These common items were called "bridge" items. For validation and comparison purposes, IEA administered a few items to two adjoining populations (e.g., 1 and 2, 2 and 3). These items are called "anchor" items because they are used across the different levels. A preliminary report of the findings of SISS was published in 1988 (IEA, 1988). A report of the curriculum analyses and other characteristics of the participating countries was edited by Rosier and Keeves (1991). The main international findings were compiled in a report by Postlethwaite and Wiley (1992). Ten countries participated in both the first and second IEA studies in science. A report documenting and explaining the change in achievement in science that occurred between 1970 and 1984 in those 10 countries was edited by Keeves (1992).

Results of the U.S. involvement in SISS have been presented in various reports and articles (Doran & Jacobson, 1991). Eleven dissertation projects were based on some aspect of the SISS–United States activities. Seven monographs have been written on specific elements of the study and distributed through the National Science Teachers Association (NSTA). These topics included curriculum analysis (Miller, 1988), examination of science performance of grade 5 students (Schneider, Muller, Doran, & Jacobson, 1990), examination of performance of grade 9 students (Schneider, Muller, Doran, & Jacobson, 1991), biology (Anderson, 1990), practical testing in science education (Kanis et al., 1990), performance of the advanced science student (Ferko, Jacobson, & Doran, 1991), and gender and science achievement (Humrich, 1992). An overall report of the U.S. involvement in the SISS study was compiled by Jacobson and Doran (1988). Numerous articles reporting on specific results about the U.S. survey have been written in both the United States and international professional journals.

One of the optional components of the FISS and SISS was the practical skills testing of students in Populations 1 and 2. As an optional part of FISS, England and Japan administered practical tests to students at the grade-9 level. Only England administered the FISS practical tests to students in grade 5. The practical test was devised by IEA researchers to measure students' abilities to carry out scientific experiments by requiring them to undertake a certain kind of measurement or experiment in a laboratory situation. This test was composed of five separate tasks: measuring the area of a leaflike figure by counting the number of squares, finding differences between photographs of two fruit flies, observing chemical changes when dilute sodium hydroxide is added to two unknown substances, measuring length and its error, and measuring electric current after constructing a circuit given in a diagram.

Scores on these laboratory practical tests were correlated with scores on a paper-and-pencil practical test and a paper-and-pencil achievement test. Based on the modest size of the coefficients obtained, Comber and Keeves (1973), concluded that laboratory practical tests measure some attributes distinct from those measured by the achievement exam and that these attributes are probed to a limited extent by the paper-and-pencil practical items. Kojima (1974) analyzed the responses of a sample ($N = 1934$) of Japanese ninth graders to this practical test. Kojima concluded that the practical tests measured some important aspects of science learning not measured by traditional paper-and-pencil tests. It was hoped that the practical tests developed for the FISS project would form a useful starting point for future work.

During the SISS, data were collected from students in grades 5 and 9 on a set of practical laboratory test items. Six countries participated in this optional dimension of the study (Hungary, Israel, Japan, Korea, Singapore, and the United States). These tests were based on science content and equipment found to be common in these respective grade levels in biology, chemistry, and physics. Furthermore, the skills assessed by the items were placed in the general categories of performing, investigating, and reasoning. Examples of specific science skills within these general categories are listed in Figure 14.12.

The six tasks that were used at the grade-9 level are shown in Figure 14.13. They were selected after trial testing in the participating countries and subsequent revision.

In the classes tested, half of the class responded to the three tasks labeled A and the other half responded to the B tasks. Each student was administered a set of three tasks in a 45-minute time period. Students were allowed 10 minutes per task. Additional time was needed to explain the testing procedure, clean the station, move to a new station, and collect test booklets. Detailed lists of equipment and materials were prepared to standardize as much as possible the objects and phenomena experienced. A person other than the students' science teacher administered and supervised the tasks according to specific directions. Lastly, a detailed scoring guide was prepared by the researchers to evaluate student responses.

According to Kanis (1991), the U.S. ninth-grade students were quite successful at performing tasks in which they were expected to observe, measure, and manipulate materials. Few differences were observed between the U.S. males and females in these science practical skills tests. This was in sharp contrast to the gender differences reported on the paper-and-pencil

Performing	To include	observing, measuring, and manipulating
Investigating	To include	planning and design of experiments
Reasoning	To include	interpreting data, formulating generalizations, and building and revising models

FIGURE 14.12. General Categories for Assessed Skills

achievement tests in science. The scores for males and females on the ninth-grade practical skills test are presented in Figure 14.14.

Generally speaking, the mean scores for males and females were not found to be different. Performance by males and females on some of the individual tasks varied somewhat: males performed better than females on tasks with a physical science content, and females performed better on items with a biological science content. For example, males did better than females on the item involving electric circuits, and females did better on the paper chromatography item. Similar results were found for the fifth-grade sample (Kanis, 1990).

The international report of this optional part of SISS prepared as a series of articles, edited by Doran and Tamir (1992). It was a special section in the December 1992 issue of *Studies in Educational Evaluation*.

A third IEA study in science is being organized (1991). This study will be combined with an assessment of mathematics, and it has scheduled a two-phase data collection procedure that has already been established for the 1994 and 1998. The plans for the science assessment component include some performance tasks as part of the required test package, as well as some additional performance tasks as optional components.

National Assessments of Education Progress

A highly significant event in assessment in the United States was the establishment of the NAEP. The act establishing the U.S. Department of Education specified that its purpose should include collecting information that would show the condition and progress of education. In 1963, Francis Keppel, the U.S. Commissioner of Education, and John Gardner, president of the Carnegie Corporation, sponsored two meetings to explore the questions of whether, and if so how, some kind of national testing program should be established (Greenbaum, Garet, & Solomon, 1977). The meetings led to the creation of the Exploratory Committee on Assessing the Progress of Education (ECAPE), chaired by Ralph Tyler. Later ECAPE became the Committee on Assessing the Progress of Education (CAPE), and in

Grade/Group/Task	Performance
9A1	By testing with battery-bulb apparatus, determine the circuit within a "black box". (Physics)
9A2	Using phenolphthalein and litmus paper, prepare and execute a plan to identify three solutions as being either acid, base, or neutral. (Chemistry)
9A3	Using iodine solution, prepare and execute a plan to determine the starch content of three unknown solutions. (Biology)
9B1	Using a spring scale and graduated cylinder, determine the densisty of a metal sinker. (Physics)
9B2	Explain movement rates and separation of water soluble dots in paper chromatography activity. (Biology)
9B3	Using a sugar test tape and iodine solution, identify three unknown solutions as to presence of starch and/or sugar. (Chemistry)

FIGURE 14.13. Six Tasks Used at the Ninth-Grade Level

Item	Ninth-Grade Males	Ninth-Grade Females
Subtest A		
A1	76.9%	70.7%
A2	47.0%	46.0%
A3	46.3%	44.0%
Mean (A)	56.1%	53.1%
Subtest B		
B1	44.6%	38.2%
B2	67.5%	71.8%
B3	44.0%	45.3%
Mean (B)	49.8%	49.0%

FIGURE 14.14. Manipulative Process Test for Males and Females (Percent Scoring Correctly)

1969 CAPE became a project of the Education Commission of the States and was renamed the National Assessment of Educational Progress (NAEP).

One of the first NAEP reports came out in 1975 (Johnson, 1975). In the foreword of that document, Tyler briefly discussed the reasons and objectives established for NAEP. Tyler pointed out that in 1974 more than $110 billion was spent by various U.S. educational institutions, with an average expenditure per pupil of $1000 and federal support constituting almost $13 billion. In view of these amounts, Tyler suggested that it was critically important to gather national-level data about educational achievement and to monitor changes related to educational achievement over the years. Tyler stated that although NAEP was designed to be a long-term project, its data were already an integral part of federal, state, and local needs. Tyler reported, "This is an impressive beginning for a vital program whose most promising years still lie ahead" (Johnson, 1975, p. xii).

Despite the achievements of NAEP in its early years, several concerns were raised along with suggestions for improvements. One concern was that NAEP would lead to a national curriculum or to conducting undesirable comparisons (e.g., among states). A set of essays gathered by Olson (1976) exemplifies some of the other criticisms of NAEP. He noted that many held concerns about the narrowness of NAEP's assessment in regard to less powerful groups like small communities, urban centers, and diverse cultural groups. The theme of a "top-down," powerful group imposition of mundane and narrow goals was pervasive in these essays. In addition to these reflections from individuals and groups, a comprehensive evaluation of NAEP was commissioned by the Carnegie Corporation. The evaluation was completed in 1973 but not published until 1975 because of extensive interactions with the NAEP staff about the findings. The report carefully documented the beginnings of NAEP, with many direct quotations from its earlier meetings. The report's epilogue stated that "the Assessment's (NAEP) consensual style obscures the fundamental conflicts in the goals and structure of American schooling and makes any fundamental resolution impossible. This indicates that the data collected will be of little immediate use in more closely approximating the nation's labyrinthine educational goals" (Greenbaum et al., 1977, p. 192). NAEP has attempted to address these concerns in subsequent surveys.

NAEP had gathered information from samples of the U.S. population at ages 9, 13, 17, and 26–35 about a variety of subject matter areas including science. The report of the first science assessment (Johnson, 1975) showed some interesting results. NAEP developed science objectives by working with teachers, scientists, educators, and laypersons throughout the country. The main objectives were to (1) know the fundamental facts and principles of science, (2) possess the abilities and skills needed to engage in the processes of science, (3) understand the investigative nature of science, and (4) have attitudes about and an appreciation of scientists, science, and the consequences of science that stem from adequate understandings. These objectives served as guides in the development of assessment tasks. These tasks were mostly multiple-choice questions, but other types of tasks were used as well. For example, there were tasks that used pictures, tapes, films, and practical everyday items. Also used were individual interviews, the manipulation of apparatus to solve a problem, and observations of individuals' problem-solving techniques. In science, in particular, individuals were asked to conduct a brief experiment. Attitude survey questions were also included. Tasks were administered to groups of 12 or less (sometimes individually), and instructions and exercises were presented on tape recordings.

As part of the 1972–1973 NAEP survey in science, subsamples of 9-, 13-, and 17-year-olds were administered practical skills tests. The tasks shown in Figure 14.15 were included in this study. The volume-of-rock activity focused on the displacement of water with a graduated cylinder. The task was individually administered with interaction between the administrator and the student. One interesting result was that almost as many students (42 percent) weighed the rock on the spring scale as measured the displacement of water in a graduated cylinder (50 percent) (NAEP, 1975). The researchers concluded that this activity was a good learning experience for the students as well as a test of their ability to demonstrate a simple procedure.

The water-temperature task was quite similar to that described earlier in the SISS section. Ten percent of the 17-year-olds admitted that they did not know how to use a thermometer. Only 35 percent of the 17-year-olds could explain verbally what the temperature of the mixture would be compared with the temperatures of the hot and cold water. Acceptable answers included such phrases as "halfway between hot and cold," or an actual midpoint temperature, or an indication that "it is an average."

A general conclusion in the NAEP report is that:

There is a wide gap between the ability of students to demonstrate procedures and their ability to successfully explain these procedures. At all three age levels, the percentage of students able to do things was much greater than the percentages of students able to explain how, why, or what they were doing. (NAEP, 1975, p. 17)

The science findings reported in 1975 are mirrored by many of those found more recently. Males achieved at higher levels than females, especially at the higher grade levels. The attitudes

Conducting a Simple Test:	Volume of rock
Conducting a Simple Experiment:	Circuit board
Conducting a Simple Experiment:	Water temperature
Applying Knowledge to Direct Observations:	Rock type
Applying Knowledge to Direct Observations:	Photo cell box
Demonstrating Principles Using a Model:	Rotation and revolution
Demonstrating Principles Using a Model:	Faulting and folding

FIGURE 14.15. Tasks Administered by the 1972–1973 NAEP Survey in Science

and curiosity of school-age blacks about science were roughly typical of Americans generally, but their mastery of scientific skills and knowledge was well below national levels. Adults performed less well than 17-year-olds in the aspects usually dealt with in school but performed better in the aspects that people pick up in daily living. Inner-city and rural children performed less well than suburban children.

The next science assessment was conducted in a similar fashion by NAEP in 1976–1977. Again, an extensive effort was made to verify or specify objectives and to develop assessment items or tasks, while maintaining some of the original items to assess change over time. The third assessment (NAEP, 1978) included for the first time a battery of items designed to assess student attitudes and perceptions of science, science teachers, and science courses. In 1981–1982, because of financial constraints at NAEP, the national science assessment of school-age students was conducted by the Science Assessment and Research Project (SARP), based at the University of Minnesota, through a grant from the National Science Foundation. The testing domain included science content, inquiry, science-technology-society, and science attitudes. The results of this assessment were presented and compared with the results from previous years (Hueftle et al., 1983). Highlights of the report included decreases in attitudes toward or interest in science, improvement on achievement items by 9-year-olds, no change for 13-year-olds, a decrease for 17-year olds, and improvements in the achievement of black students.

The 1986 NAEP science assessment was conducted by Educational Testing Service (Mullis & Jenkins, 1988) under contract to the U.S. Department of Education's National Center for Educational Statistics. This time students were sampled at grades 3, 7, and 11, and the areas assessed included science proficiency (i.e., nature of science, life science, chemistry, physics, and earth and space sciences), attitudes toward science, participation in science activities, students' home environments, and the amount and kinds of science instruction children had received. These new areas were added to provide background information and to help interpret the more traditional types of data collected. In the beginning of this report, Mullis and Jenkins put

the assessment data in the framework of other research and issues in science education. This combination allowed them to draw some conclusions about the nature of science education:

Evidence from NAEP and other sources indicates that both the content and structure of our school science curricula are generally incongruent with the ideals of the scientific enterprise. By neglecting the kinds of instructional activities that make purposeful connections between the study and practice of science, we fail to help students understand the true spirit of science, as described in these pages.

In limiting opportunities for true science learning, our nation is producing a generation of students who lack the intellectual skills necessary to assess the validity of evidence or the logic of arguments, and who are misinformed about the nature of scientific endeavors. The NAEP data support a growing body of literature urging fundamental reforms in science education—reforms in which students learn to use the tools of science to better understand the world that surrounds them. (Mullis & Jenkins, 1988, p. 17)

In 1986, the National Science Foundation awarded ETS a grant to determine the feasibility of testing "higher order thinking skills" (NAEP, 1987a). Thirty tasks were developed and pilot tested at grades 3, 7, and 11 in four skill areas: sorting and classifying, observing and formulating hypotheses, interpreting data, and designing and conducting an experiment. Many of the tasks used were adopted from those used in the APU evaluations. In addition, NAEP developed some tasks in which students interacted with a problem via a computer simulation.

The final report of this pilot study was written by Blumberg, Epstein, MacDonald, and Mullis (1986). A manual designed for use by science and mathematics coordinators and teachers, entitled *Learning by Doing* (NAEP, 1987b), was published in May 1987. This manual included a concise rationale for practical testing and numerous examples of tasks used. The equipment needed, the student test questions, a picture of a student doing the activity, and illustrative successful responses were presented in the NAEP report.

The tasks were classified as group activities, station activities, or complete experiments. The pilot study assessed about 1000 students in grades 3, 7, and 11, with about 100 to 300 responses obtained for each task. NAEP collected the pilot data primarily to assess the quality and grade-level appropriateness of the tasks rather than levels of student performance. It was planned that these tasks would be prototypes for use in future national assessments.

The NAEP researchers concluded that:

Although managing equipment and training administrators requires ingenuity and painstaking effort, conducting hands-on assessment is feasible and extremely worthwhile. The school administrators, teachers, students, and consultants were all very enthusiastic. The students found the materials engaging, and the school staff and consultants were more than supportive in encouraging further use of these kinds of tasks in both instruction and assessment. (NAEP, 1987b, p. 7)

The summary of the study reported the enthusiasm of everyone for the items, the vast amount of time needed in preparing and administering the items, as well as the costs involved. Overall, the report contended that performance assessment is feasible.

The 1990 NAEP (Jones, Mullis, Raizen, Weiss, & Weston, 1992) was conducted by ETS under contract to the National Center for Educational Statistics. Data were collected from fourth, eighth, and twelfth grades in four content areas: life sciences, physical sciences, earth and space sciences, and the nature of science. In addition, data were collected on student science experiences and instruction, and teachers of eighth-graders completed questionnaires about their background training and instructional approaches. The results were analyzed using item response theory (IRT) methods, allowing NAEP to describe performance across the grades and subpopulations on a 0 to 500 proficiency scale. The results of the assessment provided further information about students' lack of preparation in the sciences, their apparent disinclination to enroll in challenging science courses, and the comparatively low achievement of blacks and Hispanic students, females, economically disadvantaged students, and non-college bound students.

In addition to forming the basis for the specific assessment reports, the NAEP data have been made publicly available since 1979 and have been used in other research. For example, Linn, DeBenedictis, Delucchi, Harris, and Stage (1987) studied the gender effects of the "I don't know" option on the NAEP tests. Welch, Walberg, and Fraser (1986) used the NAEP data for 9-year-olds to test the validity of a model of educational productivity in identifying factors linked with achievement and attitudinal attainment. Rakow (1985) used the NAEP data for 17-year-olds to test the effectiveness of a model of educational productivity in predicting inquiry skills for males and females, as well as for whites and nonwhites. Napier and Riley (1985) used the NAEP data for 17-year-olds to test for effective determinants of science achievement. Yager and Penick (1986) used the NAEP assessment as a guide to preparing a more specific local assessment. NAEP spin-offs and other results were reported in instruction-oriented journals, as well as in research journals (e.g., Yager & Penick, 1989). NAEP has also periodically reexamined its own data to produce other reports. Most recently the report, *Trends in Academic Progress: Achievement of U.S. Students in Science 1969–70 to 1990, Mathematics 1973 to 1990,* *Reading 1971 to 1990, and Writing 1984 to 1990* was generated by NAEP (1992).

Today, NAEP is alive and well as part of our nation's proclivity to testing. Mandated testing alone has consumed $700 million to $900 million dollars and 20 million school days annually (Hudson, 1990). The 1990 and future assessments may include options that have not been available in the past. For example, states have been given the option of supplementing their samples so that state comparisons may be possible. Another possibility is an alternative to existing proficiency levels that would classify student achievement at each grade level as basic, proficient, or advanced. Also, the NAEP has been suggested as the vehicle of choice for state-level indicator systems and perhaps for identifying progress in international arenas (Hudson, 1990).

A parallel effort to NAEP has been the studies conducted by the International Assessment of Educational Progress (IAEP). The first IAEP was conducted in 1988 (LaPointe, Mead, & Phillips, 1989) and provided results on mathematics and science achievement of 13-year-olds from six countries: Canada, Ireland, Korea, Spain, the United Kingdom, and the United States. The IAEP was designed to collect and report data on what students know and can do, on the educational and cultural factors associated with achievement, and on students' attitudes, backgrounds, and classroom experiences. A second IAEP was conducted in 1991 (LaPointe et al., 1992). This time the assessment had four parts: a main assessment of 13-year-olds' performance in mathematics and science; an assessment of 9-year-olds' performance in mathematics and science; an experimental, performance-based assessment of 13-year-olds' ability to use equipment and materials to solve mathematics and science problems; and a short probe of the geography skills and knowledge of 13-year-olds. Twenty countries participated in the second IAEP. The performance of the U.S. students on both assessments was mediocre in comparison with that of students in other countries.

Summary for National and International Assessment

Interest in science assessment in the United States has been increased by national reports of declining scores and international studies that have generally revealed lower science achievement in the United States than in other countries. This section began with a general description of the growth of national and international assessment programs. NAEP has monitored achievement since the mid–1960s in this country. Great Britain's APU and Israel's biology matriculation system were presented as other examples of national science assessment programs. International assessment surveys (IEA and IAEP) were also described.

Each assessment effort was then described in greater detail, beginning with Britain's APU. This assessment program was recognized for implementing domain or content sampling and devising a procedure for measuring performance assessment. After noting that many science skills were not adequately measured by the more traditional paper-and-pencil assessments, APU developed and validated their own performance assessment techniques. They concluded that it was both possible and beneficial to implement a practical test on a national scale.

Israel's high school biology program provided another example of national assessment. This national biology program has stressed content as well as laboratory work and fieldwork. Recognizing that traditional paper-and-pencil tests cannot fully assess the practical skills component, the Israeli matriculation examination in biology was designed to be multifaceted with both a written component and practical tests.

This section next reviewed the IEA studies in science, which have allowed researchers to make international comparisons related to educational achievement. Two studies (FISS and SISS) sampled 10-year-olds, 14-year-olds, and students registered in the last year of their secondary schooling. The purpose of the IEA studies was to aid the participating countries in interpreting their students' achievement patterns. Several tests were assembled for use in the science studies, including a science content achievement test; a science practical skills test; student, school, and teacher background questionnaires; an attitudinal scale; a test of verbal aptitude; and tests of concepts and skills related to mathematics. A notable conclusion of these studies was that laboratory practical tests measure some attributes that are different from those measured by traditional achievement tests.

A brief historical overview of NAEP was then presented. A major goal of NAEP was to gather data and monitor changes for the purpose of examining the condition and progress of education in the United States. The NAEP science studies collected data in 1970, 1973, 1977, 1982, 1986, and 1990. The next NAEP surveys of science is scheduled for 1994. Over these studies, NAEP has portrayed achievement as declining or staying the same for their sampled groups of students at age 9, 13, and 17. These NAEP surveys also drew attention to lower achievement of blacks, Hispanics, females, economically disadvantaged students, and students not bound for college.

Lastly, two IAEP studies (LaPointe et al., 1989 & 1992), provided another example of international cooperation in the assessment of mathematical and science achievement in several countries. Data were collected on what students know and can do, various educational and cultural factors related to achievement, and students' attitudes, backgrounds, and classroom experiences. In both studies, the performance of students from the United States was mediocre compared with that of students from other countries.

SCIENCE ASSESSMENT IN THE STATES AND CANADIAN PROVINCES

Part of the recent publicity about educational accountability has resulted in increased assessment activity at the state level in the United States and the provincial level in Canada. A recent New York State Education Department report listed, for each state, traditional and alternative assessment practices (New York State Education Department, 1991). Some of their general conclusions were the following:

1. Twenty states use exit exams, most frequently tied to high school graduation.

2. Fifteen states test for minimum competencies.
3. Overall, 34 states have some form of performance-based activity.

The report also stated that the 34 states using performance-based tests are considering expanding these modes of assessment. Of the remaining 16 states, 9 are planning or developing performance assessment, 2 states have been at the discussing stage, and only 5 states have expressed no interest in using alternative assessment methods.

Because of the importance of state assessment programs for students and school districts, the activities of a number of states most active in science assessment are summarized. The Canadian provinces with innovative assessment programs are also included in this review.

State-Mandated Assessments

With the call for accountability in the past decades came an increase in general state-mandated assessment of student achievement. Most often this has been accomplished by means of various standardized tests (Raizen et al., 1991). Although testing is held by many to benefit education, the validity and value of traditional standardized testing have been increasingly debated among educators and concerned community members. Because of the many negative factors that have been attributed to the standardized tests, Herman and Golan (1992) conducted a study to discover more about the specific effects of standardized testing on teachers and classroom instruction. They surveyed more than 340 teachers in 48 different schools, and they were able to identify and define relationships between standardized testing and the teaching and learning process. Among other things, they found that teachers felt strong pressure, especially from district administrators and the media, to improve their students' standardized test scores. Administrators were reported to spend considerable time discussing with teachers ways to improve these test scores, and they provided teachers with materials to support or enhance students' test-taking skills. This finding is in agreement with one recently reported by Martens (1992) in a study conducted in the state of New York.

Herman and Golan (1992) also reported that standardized testing substantially influenced teachers' classroom planning. Teachers made sure that their instructional programs covered test objectives and many teachers looked at prior tests to ensure a good match. Teachers also adjusted curricular scope and sequence based on test content and students' prior performance. Furthermore, teachers devoted substantial time to test preparation activities, test-wiseness instruction, and practice tests. The results of these tests were, in the teachers' minds, of uncertain meaning and of uncertain value in school improvement. Teachers did not believe that standardized testing was helping schools to improve or that testing helped clarify school goals, provide useful feedback, or assess the most useful learning for students.

The negative attitudes of teachers toward standardized testing reported by Herman and Golan might be justified. Too often the tests have appeared to result in comparison of schools and districts, rather than in informed science program improvement. A documented outcome of standardized testing would seem to be the teaching of test taking. In fact, there has been some suggestion that testing may actually result in less science being taught. All too often program evaluations have been reduced to single scores, with little supporting evidence that these scores are related to the science program, instructional approach, or student understanding of science.

Attempts have been made, however, to include other facets in statewide assessment. Several states have been endeavoring to include some type of performance in their assessment effort, even if the attempts are not always as successful as one might like (Clark, 1992; Comfort, 1992; Dana & Nichols, 1992; Green, 1992; New York State Education Department, 1991). In addition, state assessment has the possibility of greater success because of the advent of computer-assisted assessment and performance-based assessment. It is anticipated that the computer will offer educators a greater opportunity to evaluate a wider range of outcomes with even more flexibility.

Assessment of Science in New York State

In few states is education tested as much as it is in New York State. Since the initiation of the first Regents eighth-grade examination in 1865, students in New York have been taking state tests for a variety of purposes (e.g., to receive credit for Regents courses, to meet requirements for high school graduation). A brief overall description of the state's testing program is given before describing in detail the state's assessment specifically as it relates to science.

After the 1865 beginning, the high school Regents course examinations followed in 1878. The state Pupil Evaluation Program (PEP) tests were used in New York in 1965 to determine who was eligible for federal funding for disadvantaged students. Competency tests were first administered during the 1979–1980 school year in the subject areas of reading, writing, and mathematics. Successful completion of these tests was established as a requirement for high school graduation for all students. The 1984 Regents Action Plan stated that program evaluation tests were to be developed for certain curricular areas. Science and social studies program evaluation tests were created and administered to assess how well school programs were functioning. Most recently, New York has established *A New Compact for Learning* (1990) to guide, coordinate, and influence the state's assessment plans, instruction, community involvement, and so forth.

Statewide science assessment in New York is begun at the fourth-grade level with the Elementary Science Program Evaluation Test (ESPET) (New York State Education Department, 1989). ESPET is designed to measure content, attitude, and skill outcomes contained in the New York State *Elementary Science Syllabus* (1985). A 45-item multiple-choice test has been developed to assess the content and some of the skill outcomes of the syllabus. The other required ESPET component, which is administered to grade-4 students in all public and nonpublic schools, is the Manipulative Skills Test (MST). These two re-

quired tests have been administered to approximately 200,000 grade-4 students during the month of May each year since 1969. Some of the psychometric properties of both of the ESPET required instruments (the multiple-choice test and the manipulative skills tests) have been described by Doran (1992a, 1992b). The optional instruments are the student attitude inventory and the four program environment surveys. These instruments are to be administered to students, teachers, parents, and school administrators and are described by Isenberg (1989).

The program environment surveys were based on a model of variables that could influence success in elementary school science. This model was presented in the *Guide for Program Evaluation* (New York State Education Department, 1988) and includes the following factors: school science program, community support, instructional planning, instructional technique, class engagement, class potential, and home support.

These factors were represented by items in one or more of the program environment surveys. Through data collected from these surveys, schools can examine how these factors may relate to patterns and trends of achievement in their school. The main focus of ESPET, as a whole, has been to identify how well each school's science program is enhancing pupils' achievement of content, attitude, and skill outcomes; what are the strengths and weaknesses within the program; and what are the factors that relate to these strengths and weaknesses. ESPET data is *not* to be used for evaluation of students or teachers but rather as an instrument to promote change.

Cognizant of the danger that test scores alone can lead to unfortunate inferences, the ESPET developers added a nonmandatory component to the instrument. The purpose of this component was to collect "information about student attitudes toward science, and about elements of the science program as perceived by students, administrators, teachers and parents/ guardians" (Martens, 1992, p. 2). These qualitative data were intended to help explain student performance on the required components of the test. Martens chronicled the events that preceded and followed the administration of the test in one relatively small, suburban district.

Martens (1992) found, among other things, that the assumptions held by the superintendent, the principal, the teachers, and the science curriculum director were rarely clear to all those involved (e.g., educational professionals, students, parents). She noted that these assumptions, instead of being shared, were, in fact, in direct conflict much of the time. Martens described, for example, the science director's attempts to improve the science program. Changes that the director and the teachers believed would improve the science program were found to affect negatively their students' scores on the ESPET. Shortly afterward, the test scores for every district in the county were published in the local paper. Resulting pressure from the administration to raise the district's scores led to behaviors much like those described by Herman and Golan (1992) and away from the changes in the science program that had been sought by the teachers and science director. Student scores in this district were found to improve the third year the tests were administered. This seemed to vindicate the actions taken to improve scores. Martens (1992) observed that it was highly questionable, however, whether these higher scores could be interpreted to mean improvement in the school's science program. She noted that rather than encouraging administrators to engage in a wide-ranging examination of the purpose, process, and structure of the program based on the test results, the pressure to improve their results puts educators in the position of having to rely on "quick-fix" activities that were often narrowly focused on the tests themselves. This study clearly supported that conclusion.

Another New York State science test used throughout the state is the Regents Competency Test in Science (RCT-Science), which is one of a set of competency tests that must be passed in order to qualify for high school graduation. The RCT-Science is composed of 70 multiple-choice items that are based on the outcomes from the Junior High School Blocks (the "curriculum"). The RCTs may be attempted by students at the end of ninth grade and repeated, if necessary, until passed. The RCT is perceived as a baseline criterion. The test may be waived by the student's successful completion of any Regents high school science course.

The best known New York science tests are the examinations associated with Regents syllabi in earth science, biology, chemistry, and physics. These courses are commonly taken by students in grades 9 through 12. Schools are allowed to vary the order in which these courses are experienced. Each of these exams is divided into several parts. Multiple-choice items that have been designed to assess core objectives constitute Part I. Part II is composed of several groups of items that represent elective objectives. Students choose the Part II groups of items that were taught at their school or that they feel best prepared to answer. The biology and physics Regents exams have a Part III section. Students are required to write responses and construct graphs and/or data tables in this section.

To be eligible for the final examination in each of the Regents science courses, students must have successfully completed 30 periods of laboratory work or fieldwork. During the laboratory or fieldwork, teachers must ascertain that each student has demonstrated attainment of a set of skills unique to each subject. In addition, for the earth science exam, students must have completed a laboratory performance exam, which is worth 10 points of the final exam total of 100 points. These laboratory exams are set up, administered, and scored by the classroom teacher with detailed guidelines provided by the State Education Department.

"Unlike most states, New York's state-level testing program contains no norm-referenced tests (NRTs); the tests in use are criterion-referenced tests (CRTs)" (New York State Education Department, 1991, p. 5). Student or program performance is with respect to learning outcomes specified in state documents, generally called syllabi. In these tests, each item is cross-referenced to specific content or skill objectives listed in the respective syllabus.

"In New York State, teachers are integral to test development. Committees of teachers plan, write questions, revise and assemble, as well as review the final versions of each test" (New York State Education Department, 1991, p. 5). A recent New York policy document has proposed changes in statewide assessment practices, to incorporate standards for mastery and excellence in addition to minimum competency. This document calls for each pupil in grades 4, 8, and 12 to be evaluated "by a three part assessment process that would include portfolio work, teacher evaluation of the students' desirable qualities, and core subject examinations to measure problem-solving skills and factual recall" (New York State Education Department, 1991, p. 5).

Science Assessment in Connecticut

With support from the National Science Foundation and under the guidance of Joan Baron, Connecticut has developed assessments that are called Enriched Performance Assessment Tasks. Baron (1990) noted that, such tasks have several characteristics in common in that they:

1. Are grounded in real world contexts;
2. Often take several days of combined in-class and out-of-class time;
3. Are broad in scope, frequently integrating several scientific principles and concepts;
4. Blend essential content with essential processes;
5. Present nonroutine, open-ended, and sometimes loosely structured problems that require students to define the problem and develop strategies for studying it;
6. Afford both multiple solutions and multiple methods.

Exemplars of how this assessment might appear in science classes were also provided:

1. In a biology class, students will be working in groups of three or four to design and conduct an experiment to determine the optimal salinity of water to be used to ship brine shrimp to a friend.
2. In a chemistry class, three or four students will be collaborating to design and conduct an experiment to determine which of two liquids is the regular soda pop and which is the diet version.
3. In an earth science class, students will be meeting in teams of three or four to decide whether a particular site would be appropriate for a nuclear power plant.
4. In a physics class, students will be assigned to groups of three or four to design and conduct an experiment to calculate the distance that a hot wheels car can jump between ramps when released from a given height on an inclined plane. (Baron, 1990, p. 127)

This work is part of a larger project designed to assess the extent to which students display knowledge and traits described in the Connecticut Core of Learning. The core of learning is based on a foundation of performance testing through the Connecticut Assessment of Educational Program (CAEP) and the Connecticut Mastery Test (CMT) program. Since 1989, CAEP has used performance testing in 11 subject areas, including science (Baron, Forgione, Rindone, Kraglanski, & Davey, 1989). The Connecticut science performance assessment in 1984–1985 included the use of science apparatus such as a triple-beam balance, graduated cylinders, a microscope, and an electrical circuit. This testing effort also required students to design and conduct an experiment. It was modeled after the APU work done in England. The tasks developed for use in Connecticut, in which students work individually and as part of a team, are generally composed of several stages or phases. These "assessments" can take several class periods and have almost every appearance of an instructional activity. In addition to some teacher evaluation, students assess themselves and their peers.

Science Assessment in California

The state of California is committed to use the power and impact of educational assessment as a large component in its attempt to reform education.

If we are to make effective educational reform, we must discover what students know and what they don't know, what they can do with their knowledge and skills and what they can't, how well they are being prepared for the next level of education and for life beyond the school walls. We need the right kinds of assessment to provide us with that information. (California Assessment Program, 1990, p. 1)

In this CAP report, the California researchers recognized that "assessment and curriculum development are interactive—linked in a circle of cause and effect" (CAP, 1990, p. 1). Tests are designed to assess the objectives of a curriculum, and curriculum and instruction are redesigned to improve the students' performance on the test. They concluded that if a test is not a full and complete measure of educational achievement, the pressure for higher test scores will distort what is taught and how it is taught. Large-scale testing in California and elsewhere has been dominated by multiple-choice tests, in which students darken the "bubble" of their choice. A California 1989 "Beyond the Bubble" Curriculum/Assessment Alignment Conference helped to lay out a plan for future assessments. The state has embarked on a path to develop and use new testing formats in all content areas and at all grade levels. These new testing formats are encouraged to include essay, open-ended, short-answer, performance, portfolio, oral, and integrated formats. The CAP direct writing assessment is a performance assessment that has been in place since 1987.

The California assessment work in science is designed to reflect the outcomes stated in the new *Science Framework Addendum* (1984). Developmental work in two formats has been conducted (performance task and open-ended questions) and assessment instruments were trial tested in 1989 and 1990. The work on performance tasks has focused on grade 6, with grade 12 established as the next level for development. Over 50,000 sixth-grade students from approximately 1000 schools have participated in the field testing of the grade 6 performance tasks.

The format of their performance tasks testing was five "stations" with one task per station. Students moved through the stations, spending approximately 10 minutes at each station, completing all five tasks in one testing session. The five stations required the students to demonstrate their skills in observing, classifying, sorting, detecting patterns, inferring, formulating hypotheses, and interpreting results. The science context of the five tasks were:

1. Testing objects for electrical conductivity;
2. Classifying a collection of leaves;
3. Testing and classifying a collection of rocks;
4. Estimating and measuring height, mass, volume, and temperature; and
5. Testing the pH of some water samples.

In each of these five tasks, students were asked to go beyond the specific activity results and apply what they had learned to the understanding of complex, natural phenomena.

Open-ended science questions were developed and field tested with sixth-grade students. The outcomes assessed by these questions were creating hypotheses, designing scientific hypotheses, and writing about social and ethical issues in science. Students could have responded to the questions in one of three ways. They could have interpreted or entered data on a chart, drawn a picture, or written a short paragraph. Each open-ended question required 10 to 15 minutes for the students to complete.

In the future, the science tests administered through CAP (for either program or school evaluation), as well as the Golden State Examination for high school science courses, which is a student evaluation, will consist of multiple-choice items, open-ended questions, and performance tasks.

Massachusetts Educational Assessment Program—Science

The Commonwealth of Massachusetts has conducted an assessment every 2 years of students at grades 4, 8, and 12 in four contents areas: reading, math, science, and social studies. Of the over 3000 items used in 1988, the large majority were in the multiple-choice format, but some were in an open-ended format that required students to write their responses. These open-ended questions were included in only one form of the test that was administered at each grade level. Only 1/12 of the fourth grade, 1/16 of the eighth grade, and 1/20 of the twelfth grade were asked to respond to open-ended questions. Approximately 10 questions were used in each content area, and they were distributed across the test booklets, so not all students received the same questions. The Massachusetts researchers expressed three reasons why they included open-ended questions:

1. They contribute to a more valid, comprehensive estimation of student achievement.
2. They are important to reinforce the value of thinking, reasoning, and solving problems.
3. They will act as a model for classroom testing.

They contended that learning in all subjects areas is a process of constructing meaning, of restructuring what we know and believe to accommodate more information and ideas. Learning involves recollection, questioning, comparing, and adapting—

the many kinds of intellectual activities that are an integral part of "critical thinking."

The Massachusetts open-ended questions in science were of two types: (1) those requiring students to use different aspects of the scientific approach to answer questions and (2) those probing student understanding of some fundamental concept in the field of life or physical science. They attempted to place all questions in the context of real problems. Their purpose was not only to report on how well students did on these tasks but also to illustrate activities for in-class instruction and evaluation. Some questions were used at one grade level, whereas others were used at several levels.

In the pamphlet *On Their Own* (Badger & Thomas, 1989), four open-ended questions were illustrated that investigated student inquiry skills at several places on a continuum from completely unstructured to a relatively defined situation. Generally, students were found to perform better on problems that are well defined and well structured. The four example items of "understanding science concepts" assessed the life-science concepts of ecosystem and plant growth and the physical-science concepts of energy and electrical conductivity.

For each illustrated item, some sample student responses were cited as demonstrating good reasoning, misunderstandings, as well as "on-topic gibberish." Badger and Thomas (1989) concluded that:

Taken as a whole, the results from these open-ended tasks, which required the application of scientific procedures—in contrast to the verbal knowledge of scientific procedures—were disappointing, particularly in light of student results on multiple-choice questions. Although students appear to know and recognize the rules and principles of scientific inquiry when presented as stated options, unstructured situations that demand an application of these principles seem to baffle them. Asking them to *do* rather than to *recognize*, made apparent their lack of actual understanding. (p. 39)

For their 1989 assessment in science, performance testing was added to the repertoire of instruments used. This testing and paper-and-pencil testing will be administered in alternate years. The importance of including these measures was rationalized in the report *Beyond Paper and Pencil: Background Summary* (Badger, Thomas, & McCormack, 1990a). Over 2000 pairs of fourth- and eighth-grade students were assessed on their ability to apply mathematical and science concepts. These students were enrolled in 150 fourth-grade and 75 eighth-grade schools selected by a stratified random sampling procedure. In addition to the performance testing, all students had completed a short, 30-minute, multiple-choice mathematics and science test several months before. Science teachers and curriculum coordinators were recruited and trained to administer these tasks. Sixty-five individuals completed the 2-day training session, which included working with the equipment, watching videotapes of students doing the task, and practice scoring. The test administrators were able to test 6 to 10 pairs of students in a single day. The students' written responses and the administrator's observations were part of the data collected from the testing. The three science tasks are described briefly in the following.

Animal/Leaf Key (Fourth and Eighth Grades)(Badger et al., 1990d). (1) Students were shown cards and asked to identify which illustrated an animal and to classify the remainder. (2) They were asked to sort a set of clown faces and describe their categorizations. (3) They were asked to sort a set of leaves and to describe their classifications. (Badger et al., 1990a).

Circuits (Fourth and Eighth Grades)(Badger et al., 1990b). (1) Given a battery, wire, and light bulb, students were asked to light a bulb. (2) They were asked to light a bulb using a circuit board and to describe the flow of electricity. (3) They were presented with four identical boxes and asked to use a set of materials to identify which contained a battery, a wire, a toothpick, and a resistor (Badger et al., 1990a).

Insulation (Eighth Grade)(Badger et al., 1990c). Students were asked to evaluate the insulating capacity of three hot drink cups and to come to a decision based on their findings. They were provided with the materials necessary to carry out and record such an investigation (Badger et al., 1990a).

These tasks ranged from very structured to open-ended. The administrators noted that when students did have prior experience with this kind of activity, their success rate was better with the tasks.

Rhode Island Science Laboratory Assessment

The Rhode Island Distinguished Merit Program (1991) is a method used to encourage and recognize excellence in academic achievement. This voluntary program is made available in a wide variety of subject areas to all seniors in Rhode Island's public and private high schools. Students are encouraged to participate, at no cost, in as many subject areas as they wish. Students who have met the standard of merit receive a certificate and a pin. Each area has a written test and, for those who score high on it, an audition, essay, or performance test, depending on the subject.

In science the written test is a 90-item multiple-choice test with a 1-hour time limit. The items are reflective of knowledge of scientific practices and interpretation of descriptions of phenomena presented to students in both text and graphic form (Rhode Island Department of Education, 1991). Those who have passed the written test participate in the second part of the examination, the laboratory test. Students select to do the laboratory test in one of three science subject areas: biology, chemistry, or physics. These tests were designed so that enrollment in an advanced placement course was not necessary.

Following are the problems that were given to the students to be solved in the three laboratory tests:

1. Biology: to determine the affect of time and concentration on the diffusion of potassium permanganate into potato cubes
2. Chemistry: to determine the formula of a blue hydrate
3. Physics: to determine how acceleration varies when a constant force is applied to different masses

In each of these laboratory tests, students were provided with material lists, procedures to follow, and directions or questions for interpretations and conclusions. The students were evaluated on their laboratory skills, handling of data, and a laboratory report. These three areas were weighted approximately equally in the three tests.

Michigan Educational Assessment Program (MEAP)

This state assessment program has been functioning since 1969. *The Michigan Educational Assessment Program Handbook* stated that the purpose of state assessment is to provide information on what our students are learning and doing compared with what they are expected to know and do (Michigan Department of Education, 1987). The first years of the program used standardized, norm-referenced tests to assess basic skills of all students in the fourth and seventh grades. In 1979, tenth-grade tests were added (Roeber, Donovan, & Cole, 1980). The current assessment tests were written by Michigan educators and are referenced to sets of performance objectives established for the areas of reading, mathematics, and science. These untimed tests have allowed students to work at their own rate. Each objective is measured by a set of three test items. Students must correctly answer two of the three items for each objective in order to be credited with attaining that objective. There are 30 objectives in science for grade 4, 31 objectives for grade 7, and 32 objectives for grade 10. These objectives are further divided into the following areas: life science; physical science; earth and space science; science processes; and science, technology, and society, with the heaviest emphasis at all grade levels on the science-processes category. All the items in the standard MEAP tests are constructed in multiple-choice format. In recent years, Michigan has begun to administer a hands-on performance task assessment for science skills to a small sample of students.

Low scores on the MEAP are an indication of educational needs. MEAP results can be used to assist in curriculum planning and in the allocation of state funds. Various reports have been generated from the data collected, including individual student reports, classroom listing reports, school summary reports, district summary reports, and feeder school reports. Sections of Michigan's *Handbook* have been designed specifically for use with teachers, parents, and the community. The objectives of MEAP are minimal, and districts are encouraged to develop their own lists of objectives and supplemental assessment instruments. MEAP has described four categories of achievement that are found to be similar to quartiles but refer to the percentage of students who attained a certain proportion of the objectives. For instance, category 3 is described by schools in which one-half to three-fourths of the objectives have been attained by the students. As schools have collected data for several years, schools are classified as improving, declining, or stable on the basis of these data. These ratings are done for each content area.

British Columbia Science Assessment

The British Columbia provincial assessment program was implemented in 1976, based on the principle that systematic collection and interpretation of comprehensive, objective information were essential components in the effective management of education (Province of British Columbia, 1986). Data from these provincial assessments were collected in 1978, 1982, 1986, and 1991. The first two assessments were conducted at grade levels 4, 8, and 12 to permit generalization about the outcomes of primary, intermediate, and secondary schooling, respectively. In 1986 and 1991, the grade level focus became 4, 7, and 10 to emphasize the outcomes of the primary, intermediate, and junior secondary levels (Erickson et al., 1993).

In the British Columbia assessments, all students in the public schools at the target grade levels are viewed as eligible to complete the student surveys. In May 1986, over 95,000 students (approximately equally divided across grades 4, 7, and 10) participated in the assessment. In order to survey as many content topics and objectives as possible, three parallel test forms were developed at each grade level. Each test form consisted of affective and achievement test items balanced by domain and objectives. Each student completed only one of these three parallel tests.

The affective component of the survey included items related to school science, science in society, careers in science, and special issues. The major domains addressed with the achievement tests were science processes, knowledge, higher-level thinking, technology and the nature of science, and safety.

The weighting of these domains at the three grade levels varied, with processes being more important at grade 4 and higher-level thinking more important at grade 10. Many of the items in the grade 4, 1986 survey had been used in the 1982 survey. A comparison over time was thus possible in some areas. In the science process area, there was a small gain of 3 percent over that time period, while there was a 4 percent gain on the knowledge items. The authors of the report concluded that:

These analyses show that, contrary to popular belief, student performance is improving over time in science, not decreasing . . . . What seems to be occurring is that societal expectations of students are increasing faster than student performance is increasing. This may, at least in part, explain the popular belief that students of today are not as good as students of yesteryear. (Province of British Columbia, 1986, p. 29)

As British Columbia prepares for another assessment, they are concerned with planning to expand their repertoire of instruments to include performance items. The performance tasks were of two administrative formats, consisting of stations and investigations. As the investigations were presented by an administrator to an individual student, there were only two investigative tasks. These were entitled Paper Towels and Magnets and were administered to a sample of students at the fourth-, seventh-, and tenth-grade levels. Detailed check lists and observation forms were produced for the administrators.

For the investigation and station tasks, student booklets, equipment lists, diagrams for station setup, and detailed directions for administration as well as scoring were developed after considerable trial testing. The British Columbia assessment team credited several sources as influencing their work, including work done in New York State and California, as well as the

NAEP and APU. The British Columbia performance assessment was based on the following six Science Skill Learning Dimensions:

1. Observation and classification of experience
2. Measurement
3. Use of apparatus and equipment
4. Communication
5. Planning experiments
6. Performing experiments

A total of 22 station tasks were assembled, emphasizing a variety of science process skills and the several content areas. At each of the grade levels, 12 station tasks were administered, with most being used at two or three grade levels. This plan allowed statements to be made about performance over the grade levels.

The 12 station tasks for each grade level were grouped into two sets of 6 tasks. Each student would work individually to complete one set of six tasks. The student was allowed to spend 6 to 7 minutes at each station and write her or his response in a booklet. For the investigation tasks, students worked in pairs to find an answer to an open-ended problem. According to Bartley (1991), the emphasis of this part of the assessment was on how students frame an operational question, control variables, use equipment, and deal with data they have generated. After the introduction and time to check familiarity with the material, the student pairs had 45 minutes to work on their investigation. The test administrators were teachers who were nominated by their district and trained by the British Columbia assessment staff.

The results from the 1991 performance testing in British Columbia were included in a separate report (Erickson et al., 1993). The tasks used in British Columbia were chosen because of their potential to assess one or more science skills. Tasks were then selected for further trial testing if they met the following criteria:

Durability: would the station remain stable as different students attempted it and could it be packed and transported to another site?
Cost: was the equipment available at a reasonable price?
Motivation: did the station appear interesting and likely to engage students and their teachers?
Content: what aspects of school science did the station address? Did it relate to the provincial curriculum?
Range of appropriateness: did the station appear to be suitable for more than one grade? Was it suitable for all three?
Instructions: did it appear possible to convey our intent for the station to the students in appropriate language? Was it possible to fit the instructions and the student response on one page?
Manipulative Skills: did it appear likely that the students would be able to make appropriate use of the equipment supplied?
Size: how big was the equipment? (some would be shipped over 1000 km);
Time: could the student complete the station in the time allowed? (Bartley, Carlisle, Erickson, & Meyer, 1991, p. 3)

From their pilot studies, they received positive feedback from the students and teachers. The students were motivated by doing the tasks, and teachers were impressed by strategies that students displayed while doing the tasks. Furthermore, some of the stations could be adapted for use by teachers in an instructional setting.

Manitoba Science Assessment Program

The province of Manitoba has conducted an assessment of a sample of students in their schools on a regular basis. Assessments have collected data during the school years 1980, 1986, and 1990. The purposes of these provincial assessments have been to:

1. provide benchmark indicators of student achievement in the Province of Manitoba;
2. obtain data on student achievement that will assist in curriculum and program improvement, both at the provincial and local levels;
3. assist school divisions in both student and system evaluation; and
4. help teachers improve their student evaluation skills. (Manitoba Department of Education, 1988)

The science assessments are based on the objectives of the Manitoba science curriculum, as well as the input of teachers who attend provincial meetings. Each assessment contained written tests and teacher surveys. Performance testing was included in the 1986 study. The results are then reported on a provincial basis only. No individual student, school, or school division is identified.

The focus of the 1980 assessment was on grade levels 5, 8, and 11, and the focus in 1986 was on grades 3, 6, and 9. The 1990 assessment focused on the high school science courses: Science 100, generally taken by grade-10 students; Biology 200, taken by grade 11 students; and Physics 300, generally taken by grade 12 students. These and all provincial tests are administered in the English and French languages and to students in French immersion programs. Approximately 10 percent of students in the targeted grade levels or science courses were selected to complete the provincial testing.

The following categories were used to construct the 1986 science assessment framework: content of science, processes of science, skills of science, problem solving and decision making, attitude toward science, science and society, and nature of science. The content areas for grades 3 and 6 were earth and space science, physical science, and life science, and for grade 9, environmental science, physics, and chemistry. The content and processes categories were most heavily emphasized at all grade levels. Some categories were not used at the lower grade levels. An interesting component of the assessments was the comparison testing, which consisted of a readministration of selected parts of the instruments used in the previous assessment.

These Manitoba tests were not designed for individual student diagnosis or grading, but rather to provide an indication of strengths and weaknesses in the program. The teachers were reminded that their personal grading of students should be based on a variety of observations and tests given throughout the school year. Very detailed reports of student achievement and attitudes by grade level, summaries of teacher surveys, and recommendations for improvement were included in the final reports.

The performance testing, which was administered only in 1986, consisted of two formats: station and individual. For each format, there were four to six tasks at each tested grade level. Several tasks were used at more than one grade level. These performance tasks were the sole measure of some objectives (e.g., using laboratory equipment, safety skills) and were one source of data, in addition to written tests, for other objectives. The following section was obtained from the summary and recommendations section regarding the grade 6 performance test:

The grade 6 students' ability to interpret data varied in accordance with the context of the performance task. Inferences and predictions were poorly done in the Station tasks, but were better for the Individual tasks. These students were relatively good at making and recording observations but evidenced weaknesses in the using of inquiry techniques, especially control of variables. They generally had difficulty with open-ended problems for which they had to design solutions. (Manitoba Department of Education, 1988, p. 154).

Ontario Assessment Instrument Pools (OAIP)

The Ontario Ministry of Education (1981, 1983, 1989) sponsored the development of collections of items for their high school science courses, entitled Ontario Assessment Instrument Pools. The first subject in which a pool was assembled was chemistry in 1981, followed by physics in 1983, and lastly with a draft document for biology in 1989. Although the format in each subject area varies somewhat, several volumes were developed that included multiple-choice items, other paper-and-pencil items, and a large collection of items assessing laboratory outcomes. The items were written to assess the outcomes outlined in provincial guidelines. The instruments in these pools were intended to supplement the evaluation procedures used by the teacher. The guidebooks to the pools suggested that specific instruments could be chosen based on their relevance to a given school program. These instruments could then be used for pre- and/or posttesting. In all the OAIP volumes, estimated difficulty levels were provided based on trial testing with students in Ontario classrooms.

The chemistry volume included a set of diagnostic instruments focusing on three key concepts in understanding chemistry: conservation of mass, solutions, and equilibrium. The chemistry laboratory instruments were developed along two models, in which students were provided with either clues or procedural steps if they needed help to continue with an investigation.

In physics, the laboratory instruments were of two types: laboratory-skills instrument and laboratory-experiment instrument. The laboratory-skills items were designed to assess student proficiency with individual skills, such as using specific instruments to make measurements. As these were often used

in a format with a number of different test stations arranged around a room and students moving on to another station after a set period of time, they have been called "bell ringers." The lab experiment test took longer to finish, approximately 40 minutes. It presented a scientific problem that required the students to design and implement a procedure for solving a problem and then report their results. In most cases the test experiment represented a slightly different situation than the instruction experiment, so the task required a modest extension or modification of work already done.

The biology pool was organized differently from physics and chemistry. It was broken down into several content units (e.g., genetics, ecology). Within each of these units, a wide variety of item types was presented to include the standard paper-and-pencil formats of multiple-choice and matching. There were four categories of laboratory items including laboratory or field data analysis, laboratory exercise, laboratory problems, and laboratory skills. Lastly, there were items in which students had to respond with short answer, draw and label, essay-restricted response, or essay-extended response. The instruments in both the biology and physics pools were also classified by behavior or objective level, which corresponded to the curriculum categories, as well as by content topic and curricular emphases. Detailed suggestions are provided for scoring or marking student responses for all types of formats for each item.

Summary for Science Assessment in the States and Canadian Provinces

The pressure for educational accountability has led to an increase in assessment activities at the state and provincial levels, related to science education. The many negative factors that have been associated with traditional standardized tests have led many to examine and include other approaches in their assessment efforts. Several states and provinces have been trying to include some type of performance testing in their overall assessment procedures, as well as strategies other than multiple-choice tests. Innovative programs from New York State, Connecticut, California, Massachusetts, Rhode Island, Michigan, British Columbia, Manitoba, and Ontario were presented in this section. Most programs combine a variety of assessment techniques, such as attitude inventories; criterion-referenced tests; performance assessment tasks; and new testing formats such as essay, open-ended, short-answer, portfolio, oral, and integrated approaches. Care must be taken to ensure that testing activities related to program evaluation are as comprehensive and complete a measure of educational achievement as possible.

PERFORMANCE ASSESSMENT OF SCIENCE LABORATORY AND INQUIRY SKILLS

This section reviews and describes work in the area of assessing the development of the critical science skills in modes in which students are required to perform in some way or actually "do science." The inclusion of laboratory activities in school science courses marked the beginning of doing science in the schools. It was invigorated through the development of activity- or inquiry-oriented curricula in the United States and other countries during the 1960s. A report entitled, *Testing in American Schools* (U.S. Congress, Office of Technology Assessment, 1992), described performance assessment as, "testing methods that require students to create an answer or a product that demonstrates their knowledge or skills. Performance assessment can take many different forms including writing an extended essay, conducting an experiment, presenting an oral argument, or assembling a portfolio of representative work" (p. 3). Even within the science field, a wide variety of performance assessments have been developed.

Much attention has been focused on achievement in science. With regard to assessment of achievement, Shavelson, et al. (1990) noted:

> Unfortunately, in an attempt to create achievement tests in science and other subjects that do not unduly favor one or another of the nearly infinite number of curricula in our country, the current technology produces tests that emphasize recall of facts and performance of isolated skills but tend not to measure students' conceptual understanding and problem-solving skills. Consequently, the current technology works against what many people value as education. (p. 697)

If science programs were designed only to acquire content, paper-and-pencil testing might suffice for determining how much and what students need to know. As more problem-solving and process skills are incorporated in the science program, different forms of assessment are required to measure these learning outcomes (Hein, 1987). A variety of assessment methods are suggested by Meng and Doran (1990). In addition to paper-and-pencil tests, there are practical tests, observations, discussions or interviews, practical tests in which students manipulate materials, as well as projects and written work in which the students can demonstrate investigative research or construct something. It is necessary, of course, that the type of assessment be appropriate for the learning outcomes being measured. Although paper-and-pencil tests are currently regarded with suspicion, they are thought to be adequate for measuring knowledge of certain kinds of content, such as facts and terminology. They do not, however, effectively measure whether or not students can apply their knowledge. Application of knowledge involving problem-solving and process skills is better assessed by means of practical tests or observations of student performance.

Alternative approaches to assessment offer both encouragement and opportunities to educators. The encouragement stems from the possibilities that are gained to develop a more complete perspective of student knowledge and understanding. The opportunities are derived from the work still to be done in determining the validity and reliability for most of these approaches. Most of these alternatives have been found to be more time consuming than traditional testing. The increased availability and flexibility of the computer and its related technology may sharply reduce the cost factor in both time and effort. The additional data-collecting, storing, and analysis capability made possible with the computer has enhanced the possibilities associated with the development of alternative forms of assessment.

As the science curriculum increasingly emphasizes process-oriented objectives and activities, it is necessary for the assessment techniques to be reflective of these objectives and activities. Doran (1990) has stated that these assessment procedures should include some situations in which students interact directly with materials, equipment, and objects. Such techniques will guide instruction in useful directions, as well as provide information about students' understanding of science concepts and their proficiency in various science skills. Perhaps the most powerful argument is that if teachers wish to encourage the development of practical skills, then any assessment system must be seen by the students to reflect this aim (Giddings, Hofstein, & Lunetta, 1992).

Science teachers who have wanted their students to develop problem-solving and laboratory skills have made the necessary effort to ensure that these kinds of learning find their way into their tests and other evaluation procedures. Teachers have understood that grading systems reflect the true goals of teachers and schools and that many students have geared their best efforts to activities that will be rewarded at test time (e.g., "Will it be on the test?").

Bryce and Robertson (1985) have observed that, when one moves away from the United States, with its dependence on multiple-choice and other forms of "objective" testing, it appears that science teachers place greater emphasis on the development of laboratory-related skills. For example, in Great Britain and Israel, laboratory skills are assessed more regularly, both by test items and by the evaluation of student behavior in the laboratory.

Tamir (1985) described five approaches for assessing science laboratory outcomes:

1. Continuous assessment by the science teacher based on systematic observations and records
2. Evaluation of laboratory reports made by students on the basis of their laboratory experiences
3. Individual student projects based on practical skills
4. Paper-and-pencil test items pertaining to laboratory experiences and laboratory-related issues
5. Practical examinations

In this section, attention is focused on what Tamir has labeled practical examinations. In the United States, this phrase would be interpreted as laboratory practical exams or science process skill assessment. Although these tasks may or may not have included reading or writing, they have always included some involvement of students with objects, materials, equipment, or organisms.

Process or inquiry skills have been included among the major objectives of elementary and secondary school science programs for decades. In the United States, prime instruments for this emphasis were the National Science Foundation (NSF)–supported curricula . Science—A Process Approach (SAPA) and the two other elementary curriculum projects, Science Curriculum Improvement Study (SCIS) and Elementary Science Study (ESS), provided excellent science-based activities for youngsters in elementary schools. The emphasis on process or skill objectives and the resultant material-centered activities (also called manipulative, practical, or "hands-on" activities) were extended with many commercially published textbook series and textbook-kit programs.

At the high school level, U.S. curricular efforts in each of the science fields have encouraged a laboratory-based instructional approach. For example, the Earth Science Curriculum Project (ESCP), Biological Sciences Curriculum Study (BSCS), Chemical Educational Materials Study (Chem Study), and Physical Science Study Committee (PSSC) have produced laboratory guides with extensive activities and experiments.

The impetus was continued by the National Science Teachers Association (NSTA) through their position statements for the 1970s and the 1980s. The major NSTA goal of science education for the 1970s was to develop scientifically literate citizens with intellectual resources, values, attitudes, and inquiry skills. One of the five fundamental goals of NSTA for the 1980s was to develop scientific and technological process and inquiry skills to promote the development of rational human beings.

During this decade of curriculum development and implementation, little was accomplished in the assessment of laboratory-related objectives, especially when this is compared with the testing of cognitive objectives that occurred. After examining the evaluation programs of these various science curriculum projects, Grobman (1970) concluded that there has been little testing that requires actual performance in a real situation or in a simulated situation that approaches reality. This condition was recognized by Tamir and Glassman (1971a, 1971b) as occurring with the BSCS curricula, even though lab-centered activities were a significant part of the BSCS programs. Grobman (1970) recognized that testing in this area is difficult and expensive, but because the long-run primary aims of projects generally involved doing something rather than writing about something, this area should not be neglected in the evaluation of curricula.

Hofstein and Lunetta (1982) reviewed the history, goals, and research findings regarding the laboratory as a medium of instruction in introductory science teaching. They considered the area of practical skills assessment a neglected aspect of research and argued that most past research studies on the educational effectiveness of laboratory work neglected important questions: What is the student really doing in the laboratory? What are the appropriate ways to measure his or her activity? Stannard (1982) identified a shift in emphasis in the previous 15 years by Australian teachers toward a more laboratory-centered approach to science education. He concluded, however, that there had not been a change in their methods of evaluating students' achievement to accommodate the curriculum shift.

In a review of practical assessment, Bryce and Robertson (1985) observed that many British teachers have placed great value in science as a practical subject, in that "The practical nature of the subject is commonly regarded as an important source of pupil motivation. Science is taught in laboratories and teachers spend a considerable amount of time in supervising

practical work. *Yet the bulk of science assessment is traditionally non-practical"* (p. 1).

Conceptual Framework for Skills

The assessment of science laboratory skills has been guided by a variety of frameworks. One of the most widely used was developed by Lunetta and Tamir (1979) and implemented by Lunetta, Hofstein, and Giddings (1981). It consists of four (and later five) major categories, each associated with several specific skills: (1) *planning and design* (e.g., formulating questions, predicting results, formulating hypotheses to be tested, designing experimental procedures); (2) *performance* (e.g., conducting the experiment, manipulating materials, making decisions about investigative technique, observing and reporting data); (3) *analysis and interpretation* (e.g., processing data, explaining relationships, developing generalizations, examining the accuracy of data, outlining assumptions and limitations, formulating new questions based on the investigation); and (4) *application* (e.g., making predictions for new situations, formulating hypotheses on the basis of investigative results, applying laboratory techniques to new problems).

Many people have collapsed the third and fourth categories, creating a three-category system that closely parallels the prelab–lab–postlab format of the laboratory investigations encouraged by BSCS, PSSC, and other curriculum projects. In the Second International Science Study (SISS) project, the labels for these three categories were investigating, performing, and reasoning (Doran & Tamir, 1992).

Others have prepared lengthy lists of specific skills found relevant to the science laboratory or the "doing of science." In the New York State Elementary Science Syllabus (1985), the following inquiry skills are presented as being relevant to science (listed alphabetically):

Classifying, creating models, formulating hypotheses, generalizing, identifying variables, inferring, interpreting data, making decisions, manipulating materials, measuring, observing, predicting, recording data, replicating, and using numbers.

For a high school chemistry project, Gardner (1990) listed the following 14 skills as appropriate to assess in the laboratory:

1. Following instructions
2. Selecting appropriate apparatus
3. Using appropriate manipulative skills
4. Conducting experiments safely
5. Making accurate observations and measurements
6. Recording results and observations accurately and clearly
7. Presenting experimental results clearly
8. Interpreting and evaluating observations and data
9. Drawing conclusions and making generalizations from experiments
10. Planning and organizing experimental investigations
11. Identifying problems

12. Evaluating methods and suggesting improvements
13. Communicating plans and results
14. Identifying and controlling variables

The Assessment of Performance Unit (APU) developed the following six-category scheme for constructing the assessment instruments and reporting the results for use with students at ages 11, 13, and 15 (Johnson, 1989).

1. Use of graphical and symbolic representations
   • reading from graphs, tables, and charts
   • representing as graphs, tables, and charts
2. Use of apparatus and measuring instruments
   • using instruments
   • estimating
   • following instructions
3. Observations
   • making and interpreting
4. Interpretations and application
   • interpreting presented information
   • applying science concepts from biology, chemistry, physics
5. Planning investigations
   • planning parts of investigations
   • planning entire investigations
6. Performing investigations
   • performing entire investigations

With the development of a detailed list of the practical-skills objectives of a science course within a suitable conceptual framework, the teacher or researcher should be in a much better position to make a valid assessment of a pupil's practical skills. The results of the APU suggested that the performance of these skills was very dependent on the content and context of the assessment situations. It was clear that "pure" laboratory skills were not being measured. It was important to recognize the role that context can play in solving science problems. It was just as clear that measuring some skills in a "practical" situation would produce different results than assessing those same skills in "paper-and-pencil" situations. It was not clear that one mode of assessment was always easier than the others. The expectations of the performance in different contexts and with different contents were found to be different. Looking at specific performance tasks made one realize that success was dependent on a student's ability to read and follow directions; perform various mathematical calculations; and knowing specific science terms, concepts, and principles. It also should be mentioned that many of these "science laboratory" skills can and should be experienced in other content areas. The skills of formulating hypotheses, interpreting data, identifying variables, and communicating results should be a part of many other school courses.

Types of Performance Assessments

Tamir (1985) defined practical examinations as those "tasks which require some manipulation of apparatus or some action

on material and which involve direct experience of the examinee with the materials or events at hand" (p. 4014). These examinations may be administered to students individually or in groups.

Individual performance exams involved a student who performed the required tasks and an examiner who observed and/or guided the performance and assigned marks (Tamir, 1985). Tamir also described three examples of slightly different formats for individual performance exams. In the first format, the examiner was to observe the student closely and monitor specific tasks, often utilizing a prepared check list. Usually the examiner did not ask questions or give directions but rather encouraged the examinee to follow written directions; hence, assessment was more confined to the observed behaviors.

Another mode is found to be similar to the clinical interview that was developed and used by Piaget. This technique is used by many researchers who are exploring the conceptions of youngsters about various natural phenomena. An example of this technique is provided by Nussbaum (1979), who probed the understanding children have of "Earth." Experienced administrators were able to sense appropriate questions to pose to youngsters within or beyond the prepared questions. The results of the examinations are often stated as direct quotations from participating individual children as follows:

Ruth (11 years old) told in detail how astronauts go on a space voyage and how they see from their space ship that the Earth is round. Only after probing with the aid of the concrete props was it found that she had believed that we live on a flat Earth. The round Earth is up in the sky and when astronauts go high enough they can photograph it and thus see that the Earth is round. (Tamir, 1985, p. 4014)

The third example of an individual performance examination was an oral exam based on concrete phenomena or materials. These oral exams were often administered as follow-ups to an investigation, project, or report. High school biology students in Israel who carried out an ecological project for several months were expected to bring a selection of organisms that they had studied to serve as objects for their oral examination. The following example with pine branches illustrates the relatively unstructured nature of this type of oral exam.

On the table lie a few pine tree branches with flowers and cones of different ages. The examinee is asked to suggest as many ways as she can to determine the age of a particular branch. She has to examine the branch to identify and point at the clues that may be used such as the stage of development of the ovulate cones, the bare patches created by the falling of the staminate flowers—one patch per year—and the pattern of new growth. (Tamir, 1985, p. 4015)

During the discussion, the student is expected to talk and demonstrate a general familiarity with the structures and their functions. Although the examination started with one specific question, it continued in accordance with the student's responses and the questions the examiner was led to ask based on student response.

The one (examinee)-to-one (examiner) practical examinations just described have their place as an important research tool and a means of evaluation, under special circumstances in school. However, they are too expensive to become a routine assessment procedure in most schools. Unlike the one-to-one examination in which assessment is based on direct observation and oral probing, group examinations are based on written responses. Three types of group-administered practical examinations are described next. All of these types of examinations have been used in the matriculation examinations in Israel, as well as in high schools and universities in the United Kingdom. They have also been used in other countries such as Japan, Hungary, and the United States to some extent. Many countries, however, have avoided them because of the difficulties in their design, administration, and scoring. The first example cited by Tamir (1985) involved the use of a dichotomous key to identify an object, such as a rock, mineral, or plant. The following description by Tamir is a type used in a high school biology exam in Israel:

A group of students, usually no more than 30, is seated one student per desk. A key, magnifying glass, needles, razor blade, pencil and paper, and a plant bearing flowers, fruit, and roots are provided. On some occasions students are asked to make a full description of the plant referring to the unique characteristics of roots, stem, arrangement of leaves, shape of leaves and their edges, arrangement of flowers, structure of flower, type of fruit. Following this description they proceed to the actual process of identification. In order to save time it was decided in Israel to skip the advance description and to proceed directly to the identification with the aid of the key. However, in order to keep track of the sequence used by the examinee and her ability to identify correctly the different features of the organism, the student is required to record the numbers of items which she followed as well as the descriptions in the text which fit the organism under examination. A detailed scoring key is used according to which the student loses a pre-agreed number of marks for each mistake. Half of the marks are assigned to correct identification of the family, genus and species, while the other 50 percent are assigned to the correct recording of sequence and features. The examination lasts 30 to 45 minutes and interrater reliability averages 85 percent of agreement. (Tamir, 1985, p. 4015)

The second and more common type of group performance exam has been called a "station" or "circuit" or "bell ringer" format. A number of stations are set up at tables around a room. Each station is provided with the equipment, materials, and instructions necessary for a particular task. The student is allowed a certain amount of time to work at each station before moving on to the next station. At the end of the allowed time (e.g., 2 minutes or 10 minutes), a bell is rung or an announcement is made indicating that it is time to move. The number of questions and complexity of the tasks at each station are required to be such that most students can complete the task in the designated time. As most group station exams are designed for 10 to 20 students, they must not require much detailed checking by the administrator. For instance, an administrator could *not* realistically check individually the position and focus of microscope slides for that many students. The primary role of the administrator is defined as monitoring the correct functioning of equipment, the on-task behavior of the students, and the replenishing or replacement of some materials. The administrator is expected to read directions and to announce the

times for each movement. At each station, the students are expected to read directions, interact with materials and record their answer in a test booklet. Several examples are presented to illustrate this versatile format.

The following tasks were among those assembled by Robinson (1969) for use with students in high school biology courses. The students were allowed 90 seconds at each station.

A. A hair is under the high power of a microscope.
   1. What is the diameter of the hair in microns?
   (Do not move the slide.)
B. Two test tubes each contain water and bromthymol blue, one with *Elodea*. A light, equidistant from each tube, is shining on the tubes.
   2. What is the "control"?
C. Two test tubes each contain bromthymol blue and a brass screw. One tube contains a piece of paper toweling and a meal worm.
   3. What is the experimental variable in this experiment?
D. Live yeast cells and Congo red stain are under microscope #1. Boiled yeast cells and Congo red stain under microscope #2.
   4. What change has occurred in the organisms from microscope #1 and #2 that is indicated by the staining?

Bathory (1986) conducted a study of 14-year-old Hungarian students' performance on some practical science tasks. The following was one of the tasks from the physics area. The pupils had 10 minutes to solve this problem.

Task. What different properties of an electric current can you demonstrate with the help of the following equipment? Sketch the electric circuits you construct and demonstrate their effects. Write down their effects in a table.
Equipment.

1 battery (4.5 V)	1 compass
1 measuring device	4 lengths of wire
1 lamp	1 freshly cut potato
1 glass of copper sulphate solution	2 identical electrodes

The third format of group performance assessment was the "complete investigation," like those initiated by Tamir (1974) in the Israeli matriculation examination in high school biology and used each year with approximately 3500 twelfth-grade students. Woolnough (1986) described why the APU included some "complete investigation" assessment tasks:

It [the Assessment of Performance Unit] has further shown that students, even those who do not do well on tests of specific practical skills, will often perform very competently on a problem to which they can relate. Such insights have caused many of us to ... emphasize and encourage the essence of genuine scientific activity through the assessment of whole investigations. (p. 195)

Bryce and Robertson (1985) recommended that the "stations format of practical test items is a worthwhile first strategy (for teachers) in gaining familiarity with what the test items reveal of pupils strengths and weaknesses. Thereafter teachers may successfully move to more integrated forms of assessment" (p. 20).

As these inquiry-oriented tests required one to two class periods of a student's time, they had to be developed and used carefully. With this in mind, Tamir (1976) cited some important considerations.

1. They should pose some real and intrinsically valuable problems before the students.
2. It should be possible to perform the task and conclude the investigation within a reasonable time limit (i.e., 2 hours).
3. The problems should be novel to the examinee, but the level of difficulty and the required skills should be compatible with the objectives of and experience provided by the curriculum.
4. Since every student will be able to perform just one full investigation, several different problems must be used simultaneously to ensure independent work within a group setting.
5. The student performing a complete investigation may encounter certain difficulties at various steps of his work. It is inconceivable that he should fail the whole examination just because, for instance, he made some incorrect observations. Therefore a procedure is needed for prompting—or providing certain leads during the examinations without damaging the standards of assessment.
6. Since the tests are based on open-ended problems, measures of divergence are needed, but accepted limits of this divergence must still be set.
7. When tests of this kind are used as external examinations, special logistic problems are to be solved. For example, although certain materials can be prepared by the schools, some materials and organisms must be brought by the examiners in order to prevent the examinees from obtaining clues regarding the tasks to be assigned during the examination.

There is evidence that practical work is sufficiently distinct from more cognitive aspects of science to merit direct assessment (Tamir, 1972). Thus, most science teachers would have accepted the claim of Kelly and Lister (1971) that practical work involved abilities that are both manual and intellectual in nature and then in some measure distinct from those used in nonpractical work. Abouseif and Lee (1965) concluded that practical tests revealed abilities different from those required for written tests and that different kinds of practical tests assessed different abilities. Tamir (1974) argued that the laboratory offered not only a unique mode of instruction but also a unique mode of assessment.

Researchers have consistently found low correlations between laboratory-based practical examinations, tests, or series of tests and paper-and-pencil tests (Ben-Zvi, Hofstein, Samuel, & Kempa, 1977; Head, 1966; Kelly & Lister, 1971; Kojima, 1974; Robinson, 1969; Wood & Ferguson, 1975). The evaluation of student *behavior* is especially important in that there appears to be low correlation between student success on laboratory practical tests and achievement on paper-and-pencil tests.

Exemplars of Performance Assessment

Although performance assessments have not been widely used in school, state, national, or international science testing programs, there is much current activity in this area. Doran (1990) reviewed several relevant efforts. The next few paragraphs describe some exemplars—some by content area, as many assessment questions focused on individual subjects, and some by project, as their purpose and focus will be made more discernible in this manner. The description begins with physics because some of the first work was done in this area.

There are numerous examples of efforts to assess the outcomes of science courses through laboratory performance tests (i.e., "lab practicals"). Notable among these has been the work done at the University of Minnesota physics department. Kruglak (1954a, 1954b, 1955) early focused on the development and validation of performance tests that were designed to evaluate the outcomes of physics that can be learned best in the laboratory. It was hoped that performance tests would improve motivation, improve outcome measurements, and expand the range in grades. Generally, the tests that were developed were employed during a 2-hour laboratory period in which students encountered eight short items and two long items. The Minnesota experimenters had access to several teaching assistants who were able to monitor the laboratory tests. The same test was used for several laboratory periods during each day of the week, with no reported advantage for students taking the test on the last day. This low level of "leaking over" was attributed to the low theoretical and high performance nature of the items involved in the assessment.

Kruglak (1955) also developed essay and multiple-choice tests to measure the same outcomes as the laboratory performance tests. Based on the results of these tests, he concluded that paper-and-pencil equivalent tests were at best only crude approximations to the unique ability of the actual performance test, where students dealt directly with laboratory materials and apparatus. Klopfer suggested that the techniques developed by Kruglak might be readily adaptable to science fields other than physics (1971).

Skills and processes that are developed and used by students to carry out laboratory experiments in high school chemistry were identified by Hearle (1974). He constructed a two-part test to measure their achievement. Part A was composed of manipulative items. These items were designed to measure students' ability to identify and use equipment in collecting data, stressing the perceptual and manipulative objectives of the chemistry curriculum. Students were required to go to a laboratory setting, perform the task described, and record which of the four choices available corresponded to their answer. The second part consisted of primarily cognitive-type items. The items in part B measured the ability to organize and summarize data, draw conclusions, and design further experiments. Most of the items in this part were accompanied by sketches, graphs, or presentations of data. Hearle's research was based on the observation that students who have done well in laboratory settings have not always done well on content examinations, and students who have learned manipulative skills have not necessarily learned cognitive-based skills.

A sophisticated performance test in chemistry was developed by Talesnick (1979). Over a single laboratory period, a student was expected to proceed in stages to solve a posed problem. At each stage, the student was required to check with the instructor before proceeding. If the student's work was judged to have insufficient information, the instructor would provide the students with helpful clues. However, the student's score for that particular stage would be lower than that of students who were able to complete the task without any clues or assistance. A problem used by Talesnick, for example, was the removal and verification of all the chromate ions from a solution of potassium chromate. Students were provided with a kit containing standard glassware and hardware. In this problem, the task was broken down into five procedural steps. A maximum of 15 points could be earned in the design phase and 10 points in the performance phase.

Marjorie Gardner directed a Laboratory Leadership Workshop (1987–89) that focused on laboratory-based instruction in high school chemistry. A monograph produced by the workshop, *Laboratory Assessment Builds Success* (Gardner, 1990), presents 18 laboratory activities that include several modes of assessment. These modes included laboratory practicals, performance checks, laboratory journals, and laboratory reports. Practical examinations in chemistry have been developed in several other countries as well (Ben-Zvi et al., 1977; Eglen & Kempa, 1974; Hearle, 1974; Tamir 1974).

Tamir and Glassman (1970, 1971b) developed the laboratory practical examination of the BSCS programs. During this examination, students working at separate tables are given detailed instructions and necessary equipment and materials to solve specific problems. Examiners are required to observe the students as they collect data and respond to questions about each problem during the 2-hour laboratory period. Experiments used in this testing program include the following: measurement of the rate of respiration, *Daphnia* alternation of activity, measurement of the rate of human respiration, grasshopper respiration, yeast fermentation, and water retention of a plant tissue. Tamir's scoring guide directed examiners to observe and then rate students' self-reliance and manipulative skills. Scores on observation, investigation, communication, and reasoning were obtained from students' written answers in a test book. Scores on both the practical and written exams were then used in determining a student's final grade on the laboratory exam. The relatively high levels of interrater agreement in scoring this practical exam were so encouraging that Tamir and Glassman suggested that teachers and schools use such examinations in their existing assessment procedures.

Testing programs at the state, national, and international levels have to varying degrees recognized the need for the assessment of students' ability to perform various manipulative tasks. The New York State Regents Earth Science Examination, for example, has included a 10-point laboratory practical exam constituting 10 percent of the students' final exam grade (New York State Education Department, 1992). Students are asked to perform five basic measurement tasks involving length, mass, volume, period, and angular displacement. Teachers are given detailed guidelines for selecting and assembling the required apparatus and materials in addition to precise criteria for scoring the students' responses.

The laboratory has been an indispensable part of the Introductory Physical Science (IPS) course, and success in the course is dependent on the students' ability to perform laboratory investigations skillfully and interpret them properly. Accordingly, Dodge (1970) claimed that it was appropriate to administer tests to measure achievement in the laboratory related to physical science objectives. He stressed that the emphasis of this testing should be placed on the actual techniques used and the reasoning involved in the investigation process. As with IPS laboratory activities, equipment and chemicals are readily available. Students are permitted access to their textbook and laboratory notebooks during the IPS lab tests. Because students are expected to do IPS laboratories in pairs, the same grouping is recommended by Dodge for the IPS lab tests. The teacher is required to observe student activity very closely and to take several notes to aid in the evaluation. The final evaluation is based on the scores obtained from the reasoning level of the students, the evidence offered by the students for their conclusions, as well as the observation of their performance by the teacher. One of the IPS laboratory tests, requiring four or five laboratory periods, is called the Sludge Test because the students must separate the components of a complex mixture and describe them as accurately and completely as possible. Suggestions about grading the laboratory test are given in the Teachers's Guide to the Laboratory Test. IPS has been one of the few science curriculum projects to develop instruments for assessing student ability to function in the laboratory setting as an integral part of the course curriculum.

In the belief that assessment techniques beyond the traditional paper-and-pencil tests were needed, Doran, Boorman, Chan, and Hejaily (1992a) conducted a study to develop and validate instruments needed to assess the level of laboratory skills of students completing high school science courses (i.e., biology, chemistry, and physics). More than 1000 students from 35 Ohio schools participated in the study. It was decided that the whole investigation format of laboratory practical testing would be the model used. In each of the science areas, tests were developed around six laboratory tasks (Doran, Boorman, Chan, & Hejaily, 1992b). The model used was one developed by Lunetta and Tamir (1979) and their colleagues, and it was composed of stages that are congruent with the prelab–lab–postlab format of many inquiry-oriented science programs (Lunetta, Hofstein, & Giddings, 1981). The three stages used were planning, performing, and reasoning. A two-part test was designed to account for the different kinds of skills needed in the planning-design stage and those required in the other two stages. The scoring system that was developed was general in the sense that it was applicable across the tasks in biology, chemistry, and physics. The scoring was subjective in that the test booklets were scored by raters. Thus, test reliability, interrater agreement, and correlation between raters had to be determined. Analyses of the data indicated that the reliabilities and correlations were sufficient to warrant further development and investigation.

Keys (1992) described a study in which the students' laboratory reports were analyzed for evidence of conceptual and procedural understandings in science. The eighth-grade students involved did two laboratory activities, one on the inclined plane and one on the action of levers. The students had been previously trained to write laboratory reports in full sentences using a specified structure that included statements related to the problem, hypothesis, materials, procedure, data-observations, conclusion, and discussion sections. Questioning prompts de-

veloped by the researcher were included in the report-writing stage to encourage the students to write more about the procedures and the concepts involved in each activity. The laboratory reports were then analyzed for conceptual and procedural understandings, through a scoring guide. No reliability was determined for the scoring guide, as this was intended to be further investigated and refined. Qualitative analyses of the students'

Item No.	Answer	Scoring	Item Value
1.	Test tube A—solution turned pink (purple-pink).* Test tube B—no change.** Test tube C—no change.**	1 point if observed changes are reported correctly. * required for point. ** not required for point.	1
2.	Test tube contains a basic solution.	1 point for correct conclusion.	1
3.	Dip a piece of blue and/or pink litmus paper (blue must be used; pink is optional) into test tube(s) B and/or C. Testing A is not required. Other procedures may be acceptable (e.g., combining equal amount from vial B or C with vial A).	2 points for complete plan. 1 point for partial plan.	2
4.	Test tube A—pink litmus paper turns blue. Test tube B—no color change with either paper. Test tube C—blue litmus paper turns pink. (NOTE: The observations from item 4 and explanation in item 5 may vary if "mixing" procedure was used instead of litmus paper.)	2 points for listing 2 and 3. 1 point for listing 2 or 3.	2
5.	Test tube A—contains base. Reason: pink litmus turned blue when dipped in it, or phenolphthalein turned solution pink.		
	Test tube—contains water.* Reason: No color change with either pink or blue litmus paper.	A. Identification 2 points for correct identification of 2 and 3 (1 point each) (1 is optional)	2
	Test tube C—contains an acid.* Reason: blue litmus paper turned pink.	A. Explanation 2 points for correct explanation of 2 and 3 (1 point each) (1 is optional)	2
		10 points total	

FIGURE 14.16. Scoring Guide for General Lab Exercises, Experiment 2

reports revealed many naive conceptions and instances in which students used the data they had collected to support their misconceptions. There was also a tendency for students who had collected reasonable data to draw conclusions that the data did not support. In some cases, students drew conclusions in a direction opposite to that which should have been inferred from the data. Others obtained unreasonable data but still arrived at correct conclusions. Another tendency was for some students to oversimplify the results in an attempt to make the drawing of conclusions more manageable. The author concluded that although there were still some problems to be solved with this approach to assessment, it was able to provide an additional option for researchers to pursue.

Many other testing options have been reported. The March 1992 issue of *Science Scope* is devoted to alternative forms of assessment. Similarly, a chapter in *Science Assessment in the Service of Reform* (Kulm & Malcom, 1991) included descriptions of assessment alternatives. Articles were included on performance-based assessment, portfolio assessment, group assessment involving team approaches, concept mapping, scoring rubrics techniques, dynamic assessment, and assessment for individual differences. Other useful resources for performance assessment at the elementary school level have been found in the studies of Ostlund (1992) and Meng and Doran (1992) and at the middle and senior high level in the work of Accongio and Doran (1992). These approaches, for the most part, are representative of still-emerging options whose validity and reliability need to be established, but they have offered ways to broaden assessment and improve the prospects for gaining a more complete picture of student understanding (Shavelson & Boxter, 1992).

Scoring of Performance Tests

Researchers working on performance assessment have developed a series of strategies that can produce a valid method of data collection. These strategies include written and verbal directions, detailed lists of equipment and material, and student test booklets or answer sheets. The foregoing procedures of collecting student written responses to laboratory tasks or activities have been developed following psychological and educational principles.

Similarly, procedures for reliable scoring of these student responses have been developed. As with the data collection, considerable effort in constructing the scoring criteria and in training the scorers is required. Two major formats of scoring are described, one using specific criteria for each question or item and the other utilizing general standards.

The specific scoring procedure is illustrated with a task from the SISS (Doran & Tamir, 1992). The number of points and the criteria for awarding each point value are associated with a specific response unique to a particular question (Figure 14.16). This form of scoring has commonly included lists of examples of acceptable responses, often the original student handwritten answers, which have earned each possible point value. The generic scoring procedures are used when some similar outcomes can be expected in a variety of contexts or with several different tasks. Often these scoring procedures are used with investigations conducted by high school students. Commonly, the outcomes are hypotheses, statement of dependent and independent variables, observation or measurement procedures, ways to record and organize data, and so forth. If the scorer is an experienced science teacher, these products can be screened across a number of contexts and tasks with relative ease. The following examples illustrate this technique.

The first example was used with the physics problem in the Rhode Island program (Rhode Island Department of Education, 1990–91). There are sets of expectations and a range of points that are possible within three phases of the laboratory experience: laboratory skills, handling of data, and the laboratory report (Figure 14.17).

Taking this scoring model one step further, Bock and Doran (1992) developed a scoring form with specific elements within each skill and descriptions of criteria for each point value of their performance (Figure 14.18).

For scoring the Israeli biology laboratory examination, Tamir (1974) used a scoring system with the following six skills with their respective weights:

Laboratory Skills

- Safety
- Massing and measuring techniques
- Proper handling of materials
- Proper handling of equipment
- Clean-up

Possible Points 15
Awarded Points ____

Handling of Data

- Labeling of data
- Significant figure
- Accuracy
- Units
- Utilization of data

Possible Points 15
Awarded Points ____

Laboratory Report

- Title
- Nature of investigation and purpose
- Data/calculation
- Conclusion
- Questions/further investigation

Possible Points 20
Awarded Points ____

Student ____________
Lab Judge ____________
Report Judge ____________ TOTAL POINTS ____

FIGURE 14.17. Scoring Form for Distinguished Merit Program in Physics

| SCORING FORM | | | | SCIENCE LABORATORY TEST | | | | | | |

SCORING FORM

School/Student ID No. _____

Date _____

Subject Area   B   C   P

PART A: Experiment Design

1. **Statement of hypothesis**
   - Dependent variable   _____
   - Independent variable   _____
   - Expected effect   _____
   - Directionality effect   _____
   - Experimentally feasible   _____

2. Procedure for investigation
   - Sequenced strategy or plan   _____
   - Detailed procedure   _____
   - Safety procedures   _____
   - Diagram of set-up   _____
   - List of materials   _____

3. Plan for recording and organizing observations/data
   - Appropriate dependent variables   _____
   - Appropriate independent variables   _____
   - Units included   _____
   - Concern for repeated trials   _____
   - Organization   _____

PART B: Experiment Report

4. Quality of observations/data
   - Qualitative description   _____
   - Accuracy of measurements/observations   _____
   - Completeness of data table   _____
   - Correct use of units   _____
   - Consistency of data   _____

5. Graph
   - Has an appropriate title   _____
   - Axes labelled with correct variables/units   _____
   - The scale of the graph is appropriate   _____
   - Points are plotted accurately   _____
   - The curve is appropriate for the data trend   _____

6. Calculations
   - Used all data available   _____
   - Correct use of units   _____
   - Accuracy of calculations   _____
   - Knowledge of relationships   _____
   - Correct substitution into relationship   _____

7. Forms a conclusion from the experiment
   - Correct use of variables   _____
   - Accurate relationship among variables   _____
   - Consistent with data   _____
   - Sources of error   _____
   - Consistent with scientific principle   _____

SCIENCE LABORATORY TEST

Reader ID No. _____

Time _____

Task   1   2   3   4   5   6

1. NR   0   1   2   3   4   5   NA

2. NR   0   1   2   3   4   5   NA

3. NR   0   1   2   3   4   5   NA

4. NR   0   1   2   3   4   5   NA

5. NR   0   1   2   3   4   5   NA

6. NR   0   1   2   3   4   5   NA

7. NR   0   1   2   3   4   5   NA

FIGURE 14.18. Scoring Form for Science Laboratory Test

Skill	Maximum Score
Manipulation	10
Self-Reliance	10
Observation	10
Investigation (Experimental Design)	25
Communication	15
Reasoning	30
Total	100

As an aid to teachers who developed their own laboratory practical examinations, Tamir, Nussinovitz, and Friedler (1982) designed a guide for the assessment of skills. They assembled 21 different skill categories ranging from problem formulation to application of outcomes to new situations. Each assessment task had to be analyzed to determine which skills were involved with the task. Following are the specific indicators of the various degrees of skill with a specified task, graphing, and their respective point values.

Score Points for Making Graphs	
Complete and accurate including title	7
Complete and accurate but no title	6
Dependent variable on x instead of y axis	5
Inadequate scaling	4
Inaccurate drawing of graph	3
Inaccurate or missing title	2
At least two of the above	1
No graph is presented	0

One could then collect the assessment guides, developed by Tamir, for the specific skills addressed in a task, creating a scoring system for this task. Tamir called the collection the Practical Test Assessment Inventory. Hargraves and Lynch (1988) used the PTAI with practical assessment in Tasmania. They found that teacher confidence in the reliability and validity of the marking process was considerably enhanced after using the inventory.

Researchers agree that an important issue for all scoring systems is reliability. The reliability of a scoring system must be established with each sample and administration. Because of its uniqueness, this is recognized as an especially important issue for performance assessment. Various methods and calculations have been identified as providing support for different kinds of reliability: within one test (i.e., internal consistency), across time (i.e., stability), across form (i.e., equivalency), and across raters (interrater agreement and correlation). These techniques have been widely used by different researchers with different tests.

On a 20-item or 20-station laboratory practical examination for high school biology, Robinson (1969) reported a reliability of .63 (calculated using the Hoyt ANOVA technique). He obtained a correlation of .33 between the laboratory practical examination scores and a 65-item paper-and-pencil test based on the content of the first semester of the science course. Robinson stated that this low correlation indicated that the two

evaluation instruments were sampling sufficiently different aspects of student understandings, and he concluded that further investigation of this laboratory performance test was warranted.

Tamir (1974) used two measures of reliability, Cronbach's alpha and interrater agreement, with a number of biology laboratory exams from 1971 and 1972 administrations. Tamir reported that when the individual raw scores in each of the six skills was used, the Cronbach reliability value ranged between .56 and .69 with a mean of .67. The interrater agreement was obtained with six examiners, in 1971 and 10 examiners in 1972. Each test was scored by two examiners, who were assigned to the tests at random. Each examiner scored between 30 and 100 papers. The correlation between the total raw scores given by each pair of examiners were computed. These values were regarded as measures of interrater reliability coefficients. The values ranged between .54 and .89 in 1971 and .57 and .87 in 1972. The mean interrater correlations were .82 and .79 in 1971 and 1972, respectively.

Baxter, Shavelson, Goldman, and Pine (1992) described:

a procedure-based scoring system for a performance assessment (an observed paper towels investigation) and a notebook surrogate completed by fifth-grade students varying in hands-on science experience. Results suggested interrater reliability of scores for observed performance and notebooks was adequate (>.80) with the reliability of the former higher. In contrast, interrater agreement on procedures was higher for observed hands-on performance (.92) than for notebooks (.66). Moreover, for the notebooks, the reliability of scores and agreement on procedures varied by students experience, but this was not so for observed performance. Both the observed-performance and notebook measures correlated less with traditional ability than did a multiple-choice science achievement test. The correlation between the two performance assessments and the multiple-choice test was only moderate (mean = .46), suggesting that different aspects of science achievement have been measured. Finally, the correlation between the observed-performance scores and the notebook scores was .83, suggesting that notebooks may provide a reasonable, albeit less reliable, surrogate for the observed hands-on performance of students. (p. 1)

Baxter and colleagues concluded that hands-on science investigations can be reliably scored. They found the relationship between the traditional test (i.e., multiple-choice items) and the performance measures to be moderate, at best. This suggested that each test measured different aspects of science achievement. The major contribution of their work was the development of a reliable scoring system that reflected student performance on science tasks. The utility of this scoring system was found to be in its applicability across measurement methods (i.e., hands-on and notebook), ease of training, ability to characterize the diversity of procedure, and the short time needed for scoring.

However, Shavelson, Baxter, and Pine (1992) found that intertask reliability was difficult to attain. In their research, some students performed well on one investigation while others performed well on other investigations. They suggested that one would have to watch students perform several investigations (perhaps between 10 and 20) to obtain an accurate picture of individual student science achievement.

Bryce and Robertson (1985) concluded from their work in Scotland and from the APU work in England and Wales that many of the basic skills were not acquired by the sampled populations, which was contrary to the expectations of many. Even in the case of measurement skills, which are usually close to the hearts of many science teachers, performance was poor. Although it was appealing and logical to propose that complete practical investigations be required for advanced science courses, Bryce and Robertson (1985) were reluctant to propose how their assessment could be validly conducted. Researchers are thus reminded to be cautious in making the leap from the position understood at the present to that of assessing pupil experiments. The latter is thought to be much more complex. Bryce and Robertson observed that there is little point in over-extrapolating from the success of carefully structured tasks for assessing practical skills to what might be the most valid forms of assessment for practical investigations. They encouraged teachers and researchers to remain optimistic about what pupils will be able to do, given appropriate experiences in practical science and appropriate assessment techniques designed to check their acquisition. Real progress will be made only when performance assessments are an integral part of all science programs.

Despite the growing support for performance assessment, concerns, cautions, and reservations have been raised. A number of these issues were consolidated in a New York State Education Department document (1992). Following are some concerns related to performance assessment in it that it could:

1. be untried and may not prove to better measurement of student ability;
2. be considerably higher in cost than standardized testing;
3. be identified by some psychometric experts as less reliable and less valid than traditional tests;
4. be just as vulnerable to being "taught to" as multiple choice–style testing;
5. be liable not to result in the better performance by those who fare poorly on the standardized tests since performance testing reflects reality;
6. be likely to reveal wider gaps in the achievement of disadvantaged students since changing the type of assessment will not change outcomes; and
7. necessitate that teachers will be required to teach in new ways consequently requiring professional development and re-education.

Much research is needed on performance assessment, despite the reasonable progress that has been made so far. To be accepted by education professionals, performance assessment, like other assessment modes, must meet the established standards for reliability and validity. Performance assessment has credible levels of internal consistency and interrater agreement based on a small sample of studies. Such research needs to be continued and replicated, and other modes such as test-retest reliability and reliability among alternative forms need to be explored.

The support for content validity of performance tasks has appeared to be strong and obvious. However, if this mode of assessment is to be fully accepted, researchers must establish additional support through criterion and construct-related validity. If some forms of performance assessment become accepted, they can act as the criteria for other measures. Currently, the evidence for "uniqueness" of the performance tasks has been for low to moderate correlation with standard (usually multiple-choice) achievement tests. Researchers need to investigate success on their performance tasks with measures of verbal and mathematical intelligence, reading ability, level of cognitive development, and similar other educational variables to get a more in-depth understanding of what these tasks are assessing.

In addition to the psychometric concerns, the economic issues cannot be ignored. The costs involved in the development, administration, scoring, and interpretation of performance assessment must be monitored. No assessment system is without its costs, but the relative costs-benefits of the several modes of assessment will be a critical issue over the next several years.

When an instructional or assessment strategy becomes commercially available, it is clear that the strategy has met with general acceptance. In 1992, two of the major testing companies advertised test lines or supplements that are performance based (The Psychological Corporation, 1992 & CTB Macmillan, McGraw-Hill, 1992). The performance expected from the students on these tests consists of short, written answers to questions presented through diagrams, tables, charts, and short written text passages. Such performance-based science tests are useful but hardly capture the full range of outcomes found in the most activity-based instructional programs. O'Neil observed that:

We believe that a substantial investment must be made to develop better assessment technologies. Assessment programs that increase the use of "performance tasks" (in which students actually conduct, for example, a science experiment), portfolios, or constructed-response items should be encouraged and funded at levels commensurate to the importance of the task. (p. 33)

Summary of Performance Assessment of Science Laboratory and Inquiry Skills

This section reviewed research related to the growth of performance assessment in science education. The need to assess the laboratory component of the inquiry-oriented curricula of the 1960's created interest in performance assessment. Traditional achievement testing and paper-and-pencil tests were designed to measure content acquisition. These types of tests emphasized recall of factual information, definitions of terminology, and content. They were found inadequate for measuring students' conceptual understanding or their ability to apply knowledge to new situations. Problem-solving and process skills required different forms of assessment, such as practical skills tests, observations, projects, and written work that can demonstrate investigative research. The type of assessment chosen must be appropriate for the learning outcomes being measured. Testing for practical skills was criticized as being expensive and difficult, but because one of the primary goals of science curricula involved doing something instead of just reading and writing about something, the need to develop this type of test was recognized.

Detailed lists of practical skill objectives were highlighted in this section. Researchers noted that the performance of these skills was very dependent on the content and context of the assessment situations. Success on specific performance tasks was dependent on one's ability to read and follow directions, ability to perform various mathematical calculations, and knowledge of specific science terms, concepts, and principles.

Several types of performance assessment were also presented in this section. Individual performance exams required a student to perform a given task, while an examiner observed, guided, and graded the student's performance. Strategies were also presented for establishing validity and reliability related to performance assessment. The need for constructing reliable scoring criteria and the training of scorers was addressed. The reliability of a scoring system must be established for each administration and sample. This section concluded that much research needs to be done on this new mode of assessment.

ASSESSMENT OF ATTITUDES IN SCIENCE EDUCATION

Interest in attitudes toward science is not new, but the subject is receiving increased attention in the overall assessment efforts at national, state, and district levels. Several major reviews of studies related to student attitudes toward science have been conducted in the past two decades. Ormerod and Duckworth (1975) summarized the results and implications of more than 500 attitude studies related to science. Gardner (1975) evaluated results of studies and instruments used and noted that it was possible to distinguish two broad categories: attitudes toward science and scientific attitudes. Gauld and Hukins (1980), in a review of scientific attitudes, identified as a major problem the lack of agreement about the meanings attributed to various terms that are used. Munby (1983a) examined the problems of assessment and instrumentation through a review of more than 50 instruments used to assess attitudes. Schibeci (1984) updated the research on attitudes toward science and presented general conclusions and issues from more than 200 studies. Two problems appeared consistently in these studies: lack of conceptual clarity in defining attitude toward science and difficulty with the instruments used to assess attitudes (Krynowsky, 1988). Drawing on the writing of several authors (Blosser, 1984; Gardner, 1975; Munby, 1983a), Germann (1988) stated:

First, the construct of attitude has been vague, inconsistent, and ambiguous. Second, research has often been conducted without a theoretical model of the relationship of attitude with other variables. Third, the attitude instruments themselves are judged to be immature and inadequate. (p. 689)

In their review of assessment instruments in science, Mayer and Richmond found that few instruments were submitted to repeated use and continual refinement. They also found extensive duplication of efforts, especially in assessment of attitudes toward science and scientists. Numerous attitude assessment instruments are available, and they contended that efforts should be directed toward the revision or refinement of these instruments. Based on the work of Gardner (1975), they suggest guidelines for the development and refinement of such instruments:

1. Specification of a clear theoretical construct to underlie the instrument. This construct should then guide the selection and development of items.
2. Avoidance of confusion between different theoretical constructs. If more than one is to be included in a single instrument, each should be identified and scored as a separate factor.
3. Elimination of defective items such as those that combine two or more understandings or perceptions.
4. Preliminary trial of the instrument on a population with characteristics approximating those of the population with which the instrument is to be used.
5. Provision in the design for filtering out influences of respondent knowledge about the scientific enterprise from attitudes toward it.
6. Refinement of the instrument for each factor or subscale such that reasonable internal consistency is obtained.
7. Determination of the stability of the instrument through the test-retest technique.
8. Use of factor analysis to validate factors empirically.

The authors contend that a few attitude instruments developed and refined in this manner will prove much more useful than the continued generation of unrelated instruments in isolated studies.

Munby (1983b) used conceptual analysis to investigate the validity of the Scientific Attitude Inventory (SAI), the most popular attitude instrument in science education research (Moore & Sutman, 1970). The SAI was designed for use with high school students. It is a 60-item Likert-type instrument with a 4-point response scale. Using conceptual perspectives developed in the study, Munby (1983b) examined the instrument itself and 30 studies in which it was used. He concluded that the instrument's validity was highly questionable. He noted that the field of measuring attitudes is replete with instruments, but that these instruments are used in rather cavalier fashion, without heed to their reliability and validity.

Based on the contention that validity testing of an instrument is dependent on there being a conceptual link between the construct, being measured and another construct, Munby, Kitto, and Wilson (1976) described the application of a multi-trait-multimethod model for determining validity. This model involved using two methods to measure the same trait. The correlation between the two measures should be significant because they have measured the same trait. Then a second trait, representing a construct coherent and converging to the first, is identified. Two more instruments are then selected that measure this second trait, with the intent that the scores of the first instruments should correlate significantly with the second pair of scores. Finally, the model required that the first two instruments be measures of a distinctive trait. That is, it was necessary that they discriminate from other measures, a third pair, of a trait conceptually understood to have no relationship with the first two traits.

Two examples of attitude instrument development based on both a theoretical foundation and empirical evidence are now provided. Krynowsky (1988) used the Azjen and Fishbein theory to guide the development of the Attitude Toward the Subject Science Scale (ATSS). A theoretical relationship between attitude and behavior was posited. This is interpreted to mean that a student's attitude towards a specific subject area in science is defined as a learned predisposition of an individual to respond in a consistently favorable or unfavorable way to performing behaviors related to the teaching-learning of the subject (Krynowsky, 1988). Thus, an attitude assessment involved students evaluating the prospective performance of these behaviors. The ATSS was developed, refined, and tested for reliability and validity. Test-retest reliability coefficients of .82 and .84 were obtained. Validity was established by two techniques. First, teachers of two science classes were asked to rank students in terms of most positive attitude toward the subject science, and the students' rank order was then correlated with the students' rank order on ATSS scores. Spearman rank-order scores were .79 ($n = 25$) and .65 ($n = 19$) for the two classes. The second approach compared student ATSS scores with scores obtained on a reliable scale of attitude toward the subject science, the School Science scale used in the British Columbia Assessment. The correlation of student scores was .70. The investigator intended that this example of using a theoretical-empirical basis for development would be pursued further by other science education researchers.

Drawing on Fishbein's view that beliefs and behavioral intentions are determinants of attitude, Germann (1988) developed the Attitude Toward Science in School Assessment (ATSSA) as a measure of a unidimensional concept of attitude. The ATSSA "was to measure a single dimension of a general attitude toward science, specifically, how students feel toward science as a subject in school" (p. 694). The instrument consisted of a Likert-type scale with five responses ranging from strongly agree to strongly disagree. The instrument was subjected to a series of pilot tests, in which student scores were correlated with their own estimates of attitudes and with teachers' estimates, and to principal-component factor analysis. The revised instrument, consisting of 14 items, was then field tested in four studies. In all four studies, Cronbach's alpha estimates of reliability were greater than .95. All 14 items were found to load on only one factor with consistent factor loadings in all four studies.

Discrimination was demonstrated by item-total correlations that ranged from .61 to .89. Attitude scores were compared from two classes, one of which was marked by poor discipline and inappropriate teacher behavior and the other by more experienced and skillful teaching. A t-test showed a significant difference between mean scores ($p = .001$), with the students in the more experienced class having more favorable attitudes. In a comparison of scores from two classes with equally skilled teachers, a nonsignificant difference was found. Germann noted that this evidence lacked experimental control, but it addressed the instrument's ability to discriminate. In applying the instrument to study the relationship between attitude and achievement, he found relatively low correlations and concluded that other factors were more important in the kinds of learning measured by achievement tests. However, achievement, including an evaluation of the consistency and quality of classwork as incorporated in a course grade, seemed to be more strongly correlated with attitude.

Summary for Assessment of Attitudes in Science Education

This section has provided evidence that ambiguity of terms and quality of instruments are two serious problems facing by those interested in assessing attitudes to science. The lack of a theoretical base has been identified in nearly all cases as a hindrance to assessment. Furthermore, the lack of empirical support for most of the existing instruments has exacerbated the situation. On the encouraging side, techniques for establishing the validity of an attitude assessment instrument have been detailed by Munby et al. (1976). Promising attempts have been made to establish a theoretical foundation for assessment by Krynowsky (1988) and Germann (1988). Taken in combination, the guidelines and foundation exist; it now remains to capitalize on them. However, a note of caution is in order. In a meta-analysis of 49 studies on attitudes toward science, Haladyna and Shaugnessy (1982) found that small differences existed in attitude between males and females; the effect of instructional programs on attitude were generally positive, but variable; and the teacher and the classroom environment played an important role in affecting attitude. As Germann pointed out, care must be taken in determining what factors are to be examined in relation to student attitudes.

Gender, family influence, home environment, self-concept, and peer pressure have all appeared to influence student attitudes (Gardner, 1975; Simpson & Troost, 1982). Willson (1983), in a meta-analysis of 43 studies for correlation between achievement and attitude, concluded that differences were small and that achievement in science seemed more highly related to interest in science than to psychologically scaled attitude. He suggested that science curricula should concentrate on achievement and let positive affect follow the improved achievement because the data suggested that successful achievement causes positive attitude.

COMPUTER APPLICATIONS IN ASSESSMENT

In summarizing the research findings on computer-based education, Waugh and Currier (1986) found that (1) groups experiencing some kind of computer-based education attained test scores that were on average between .25 and .44 standard deviations higher than their comparison groups, (2) there was evidence favoring the use of computer-based education with academically disadvantaged students, (3) long-term retention was no better for computer-based education than for other modes of instruction, (4) secondary students who experienced computer-based education had more positive attitudes toward computers than did their peers who did not experience computer-based education, and (5) significantly less time was required for computer-based education than for conventional instruction. It should be noted that many of the studies summarized relied heavily on drill and practice modes of instruction. Such programs depend on immediate feedback as a major function. Although this may not fit the common perception of assessment, it has been made clear that it does, in fact, function in such a manner and that the immediate feedback may well have a positive impact on learning.

A common use of computers in assessment has been to provide teachers with access to large banks of items for testing. These may range from specific topics, such as medical biochemistry (Aesche & Parslow, 1988) for instructors of a given course, to a test bank designed for state assessment (Willis, 1988), to a broad range of juried test items that teachers anywhere in the country may access and download onto their own computers (Dawson, 1987). Once the item banks have been put into place, the computer may be used to devise unique combinations of test items for each student and to use the results of those tests to develop remedial learning activities for each student. In each case, the computer can administer the quizzes, grade and record the results, and provide the student with immediate feedback (Dunkleberger, 1980). Use of the computer to file test questions, assemble examinations, handle all records, produce and grade tests, and guide students as to what should be done next enables testing to be done with an efficiency not possible for any teacher (Heikkinen & Dunkleberger, 1985; Vogel, 1985).

A program of regular and continuous computer-based student evaluation is described by Summers (1984) for an introductory college biology course. The course comprised independent study modules and was defined as mastery based with a 70 percent achievement score needed to pass each module. Ten-item multiple-choice quizzes on each module are completed at the student's convenience and are computer scored and recorded. Students may repeat quizzes until mastery is achieved, provided a 10-day time limit has not passed. The program had been in operation for four semesters, during which time 715 preprofessional or life-science students and 3473 nonmajors used it. Results indicated that 92 percent of all students attempt a quiz for any particular module. Among those who participated, an average of 90.9 percent were able to master the unit. Students who achieved mastery typically did so on the first or second attempt at the quiz. About two-thirds of the students who failed did so without making a second attempt; this was usually because they failed on the last day of the testing period and were not able to repeat the quiz. Questionnaires distributed at the end of the term reveal that responses were generally favorable. Two-thirds of the students reported that the computer testing program was useful in helping them budget their time. About 60 percent of the students felt that their final grade would be higher as a result of the computer testing program.

Leuba (1986) argued that machine-scored testing was appropriate, efficient, and effective in basic engineering sciences, especially in large classes. His arguments can be applied to computer-assisted testing, particularly since computers have become a viable component of much of engineering science

today. Leuba maintained that certain conditions were applicable. First, certain basic knowledge should be instilled in a student's "lifetime" memory, and such knowledge could then be explicitly tested for. Second, an upcoming test could be the stimulus for learning how to learn (i.e., practice in becoming proficient in new technical matter within a limited time). This condition was viewed as especially important, given the competing demands on students' time. Third, an impending test could stimulate students to sharpen their problem-solving skills, and the test should measure these problem-solving skills. Fourth, testing could promote learning. The test could be an impetus to study, and a well-designed test could also reinforce learning. By using machine-scored tests, students could be presented with four times as many questions as can be handled in the same time if hand-scoring techniques are used. Thus, a better sample of the student's universe of knowledge would be possible. If care is taken in designing the test, partial credit could be allotted even in a machine-scored approach. These conditions could be equally well met using computers, although it would admittedly require careful programming. In addition, the computer has an advantage of providing immediate feedback, further strengthening the reinforcement argument.

Another type of formative assessment is found in using computers to evaluate student data collected in laboratory exercises. Such checking of data and calculations has been repetitive, prone to error, and not cost effective when done by humans. Computers, on the other hand, have excelled at this type of task (Harrison & Pitre, 1983, 1988). Programs used in this way are designed to check for realistic values, a range of data, and values clearly outside acceptable limits. When incorrect answers are given, students may be asked to redo their calculations and submit revised figures (May, Murray, & Williams, 1985). The programs may also be created to accept answers within a certain range tentatively, but to suggest that students return to places of potential error and check their work (Harrison & Pitre, 1988).

As part of a project to integrate computer-generated homework into physical science college courses, Milkent and Roth (1989) used computer-generated problems as homework assignments and monitored student progress with computer-generated multiple-choice quizzes. They found that use of the computer-generated homework significantly reduced the effectiveness of ACT scores as predictors of course achievement. As a result of this homework approach, students had greater opportunities for achieving mastery and for minimizing the potential influence of entry-level aptitude and prior academic preparation. This was found in addition to the teacher advantages of an efficient system for homework management and freedom from bookkeeping procedures.

Incorporation of computers into science instruction has often taken the form of microcomputer-based laboratories (MBLs). Assessment has frequently been a part of such a system. However, in some cases this has meant simply presenting multiple-choice questions by means of the computer screen (Bross, 1986). If immediate feedback was not available, learning gains may not be accrued to such computer use. Increased ease of data collection and processing may still make this approach to testing of value to the instructor. A more useful approach might be that described by Browning and Lehman (1988) for identifying student misconceptions in genetics problem solving. Four computer programs were presented, and the students' responses were recorded and analyzed for evidence of misconceptions and difficulties in the problem-solving process. Three main problem areas were identified: difficulties with computational skills, difficulties in the determination of gametes, and inappropriate application of previous learning to new problems. Evaluation of this type seem to show considerable promise for remedial instruction and improved student learning.

Collins (1984) conducted a study to determine whether learning would be improved with computerized tests. The students ($n = 210$) were enrolled in a one-semester introductory biology course. Students in the computer section took computer-generated tests in addition to the tests taken by students in the other sections. The students taking the computer tests were given immediate feedback on their scores and then told which responses were correct and which were incorrect. In addition, the computer recorded student data on disk, allowing later analysis by the instructor. Collins concluded that computer testing led to enhanced learning, as indicated by higher scores on weekly in-class written tests, the midterm examination, the final examination, and final class marks.

Collins and Earle (1989–90) examined the effects of computer-based learning and computer-administered testing in an introductory biology class. They found that the greatest benefit was attained by those using the computer units in addition to attending regular lectures. Taking weekly computer-administered multiple-choice tests also appeared to benefit students of middle and upper ability but not students of lower ability. That the use of weekly computer tests can increase students' scores has reinforced a finding of an earlier study by Collins (1984). Although students benefited from using either the computer learning units or the computer tests, use of the two together did not result in even more gain, as might have been expected. Frequency of use of the units appeared to be a factor; the "frequent" user group achieved a much higher mean score and higher pass rate than the "infrequent" user group.

Llabre et al. (1987) studied the effect of a computer-administered test on test anxiety and performance. The sample included 26 male and 14 female college students enrolled in a developmental reading course. Subjects were randomly assigned to either a computerized or a paper-and-pencil testing situation. Both groups were administered a revised verion of the Test Anxiety Scale (TAS-R) and a sample of items from the California Short Form Test of Mental Maturity (CMM). Results of t-tests showed (1) a significant difference in anxiety level ($p < .05$), with the group taking the computerized test displaying higher levels of anxiety; and (2) a significance difference in CMM test performance ($p < .01$), with the computerized test group getting lower scores. The researchers concluded that computer-administered testing can potentially increase test anxiety and depress test performance for students who are relatively unfamiliar with computers.

The possibility that students were being disadvantaged by taking computer tests instead of written paper forms of the same tests was studied by Fletcher and Collins (1986–87). They found that students' mean scores on the computer-administered test and the written forms of the same test were roughly equivalent and concluded that the students were not disadvantaged by taking the computer tests. Most of the students favored the computer-administered tests and cited several major advantages: (1) immediacy of scoring; (2) immediate feedback on incorrect answers; (3) more convenient, straightforward, and easy to use; and (4) faster than written tests. Two major disadvantages were noted by the students: (1) not being able to review all their responses at the end of the test and make changes and (2) not being able to skip questions and come back to answer them later.

Jackson (1988) attempted to discover whether a computer could give any significant educational advantage to a student. He questioned whether the computer could improve student motivation during a testing situation by providing instant feedback and marking, thus improving understanding and leading to an enhanced score in a future test. The middle school science students who were tested by computer and given immediate feedback scored significantly higher in a later test on the same material than did students who were tested using the traditional paper-and-pencil method. An additional gain for the teacher was the ability to conduct further analyses, such as test item analysis, on the computer-recorded student data; such analyses could not be easily carried out without computer-administered testing.

Computerized adaptive testing has been emerging as a more efficient way to assess student knowledge and aptitudes. A unique characteristic of this technique has been that each examinee is given an individualized test comprising questions from a content-valid item bank. The adaptive algorithm selects questions that provide the most information about the examinee given his or her current estimated ability measure. After answering each question, the examinee's ability is reestimated. If the correct answer is given, the examinee's measure is increased and the next question is more difficult. If an incorrect response is submitted, the measure decreases and the next item administered is easier. This has resulted in a test that is tailored to the needs of each individual. The tests can be of various lengths depending on how far above or below the pass-fail threshold the examinee's performance has fallen. Thus, a test sufficiently long to determine the best decision can be presented with no wasted questions (Herb, 1992).

The following two reports concern studies of computer adaptive testing. Lunz (in Herb, 1992) reported a pilot study designed to examine the effectiveness of computerized adaptive testing for certification of students in five medical technology fields. It was found that 50 to 100 computerized questions provided the necessary pass-fail information, when compared with 109 written questions. Samples of 52 students from each of the five areas tested were surveyed about their experience. When asked if they thought future tests should be offered on the computer, 92 percent responded affirmatively. In addition, 82 percent said they would not have preferred the written test. The computerized test took 2 to 2½ hours to complete, compared with 4 hours for the written test. Other benefits of computerized testing included a shorter turnaround time for test results, improved security and data collection, and less chance of cheating because of the individualized nature of the exams.

Moe and Johnson (1988) investigated students' reactions to a computerized adaptive ability test and examined the practicability of this testing method in the classroom. The students in the study included 161 females and 154 males. They were fairly evenly divided among grades 8 through 12 and included a few college students as well. The subjects took a computerized version and a printed version of a standardized aptitude test battery and a survey assessing their reactions. Overall reactions to the computerized tests were overwhelmingly positive. Only 8.6 percent of the students were using a computer for the first time in taking the test. Analysis showed no difference in performance between first-time users and those who had prior experience. More than half of the students (51 percent) reported no difference in the amount of nervousness they felt in taking the computerized version compared with the printed version of the test. Females were more likely ($p < .05$) to report nervousness than males, but analysis of variance revealed no significant difference in performance between males and females.

The effects of microcomputer-administered diagnostic testing on both student achievement and attitudes were of concern to Waugh (1985). Students in one group were given the unit objectives and responded to a computer-administered diagnostic test consisting of one item per objective. The other group received the objectives and were assigned an out-of-class task of completing an objective-specific miniproject. The results showed that microcomputer-administered diagnostic testing could positively influence the immediate achievement of students in science. Evidence was not, however, supportive of the hypothesis that exposure to diagnostic testing might influence continuing achievement. The findings indicated that the microcomputer-administered diagnostic testing was successful in increasing student achievement in science by an average of 6 percent with no loss of positive attitude toward school, learning, or science. The evidence further indicated that diagnostic testing might have played a role in arousing student interest in microcomputers.

Using routines that allowed students to skip and return and to review questions, Ward, Hooper, and Hannafin (1989) examined the effects of computerized tests on college students' ($n = 50$) performance and attitudes. The students were randomly assigned to one of two testing conditions. In one condition students were tested in a traditional paper-and-pencil fashion. The students responded to a 25-item exam on mark-sense forms, which were then scanned and graded by computer. In the other test condition, the same 25 items were administered by computer. Students' responses were recorded and a score report was provided upon leaving the test site. A t-test revealed no significant difference in test performance. Scores on a Likert scale indicated that the computer-tested group had significantly higher ($p < .025$) anxiety scores, even though student performances did not differ significantly.

Student attitudes were also the focus of a study by Knight and Dunkleberger (1977) in a comparison of computer-managed self-paced instruction with teacher-managed group-paced instruction for ninth-grade students. The course consisted of large-group lectures (23 percent of the overall time), small-group seminars (46 percent of the time), and laboratory activities (31 percent of the time). The computer-managed self-paced group and the teacher-managed group-paced students received the same large-group lectures and small-group seminars. The computer group was allowed to self-pace through the laboratory activities, while the teacher-managed group followed a group pace. The computer served as an assessment and recordkeeping device for the computer-managed students. The quizzes were four-choice, multiple-choice questions and students received immediate feedback after completing each item. Although the different instructional approaches were applied only during the laboratory component of the course (31 percent), the positive reaction of the computer-managed self-paced group was sufficiently strong to effect a significant difference in attitudes toward the study of science.

The impact of an emerging technology, interactive video disk (IVD), was studied by Huang and Aloi (1991) in a first-year biology course. The IVD involved 17 menu-driven chapters integrating computer text with laser disk images and computer graphics. The students were organized into groups with intergroup competition in answering true-false, multiple-choice, and completion questions. The researchers compared, using an unpaired t-test, the proportion of students getting A, B, C, D, F, and W (withdraw) for 11 semesters prior to using IVD with the proportions during the 5 semesters following its use. They found that the proportion receiving A's increased significantly ($p < .005$) after use of the IVD. The percentage increases were: A, 6 percent before and 18 percent after; B, 21 percent before and 32 percent after; C, 20 percent before and 36 percent after; D, 10 percent before and 4 percent after; F's did not change. Retention of students was also increased. The proportion of W's was 33 percent before IVD use and 24 percent after. Thus, the use of IVD resulted in increased proportions of success at nearly all levels of achievement.

IVD was also used as a tool in assessing science teachers' knowledge of safety regulations in school laboratories for purposes of teacher certification by the Connecticut State Department of Education (Lomask, Jacobson, & Hafner, 1992). The program simulated a typical laboratory activity in a secondary school general science course and showed four students performing a simple laboratory experiment to identify unknown materials. The IVD assessment included two stages: stage one dealt with safety equipment and storage of chemicals, and stage two dealt with students' laboratory practices. The examinees are asked to assume the role of the laboratory teacher by viewing an IVD-simulated classroom. The teachers are then asked to identify safety violations and to suggest preventive or corrective measures. Subjects' responses are recorded for later analysis and scoring.

There appear to be several advantages of incorporating some form of computer assistance in assessment. Immediate feedback to the students has seemed to be a consistent factor in increased achievement. Ease of test taking, together with improved recordkeeping, has suggested the possibility of improved efficiency for both students and teachers. The availability of large test-item banks has made possible the offering of several intermediate quizzes with apparent achievement gains as a result. Such formative evaluation has served as both a diagnostic tool and a remediation device, indicating where corrections are needed. The data collection capability of computer testing has also permitted more extensive data analysis, especially in the area of test-item analysis, which in turn should yield more reliable and, presumably, more valid assessment. Two cautions must be noted, however. First, the simplicity of devising multiple-choice, true-false, matching, and other objective tests could lull the teacher into simply doing a better job of assessing low-level recall knowledge. Second, the linear nature of most computer testing has not allowed the student to go back and reflect on a particular item, nor to view the completed test as a whole to check for consistency of responses. The increased improvement and implementation of such emerging technologies as interactive video and hypermedia (Kumar, 1991) have shown high promise for overcoming both difficulties by providing opportunities for both improved levels of questions and increased flexibility in the testing process.

Summary for Computer Application in Assessment

This section presented research findings related to the use of the computer in various assessment procedures. Computer-based education has had a tendency to rely on drill and practice techniques, providing immediate feedback to the learner. Computers have given teachers access to large banks of items and questions for testing, and they can allow more efficient data collection and processing. Some studies have shown that students' anxiety levels were raised by taking a computer-based test rather than a traditional paper-and-pencil test. Computerized testing can allow a more individualized approach to testing. The research in this section indicates that there is a need to develop better assessment technologies to address the wider range of learning outcomes associated with problem-solving, activity oriented, decision-making science programs.

CLASSROOM ASSESSMENT

In an overview of assessment instruments in science, Mayer and Richmond (1982) noted that "evaluation efforts have typically focused on assessment of the degree of student growth in content knowledge, and, more recently, on student changes in science-process ability and in attitudes" (pp. 49–50). Mayer (1974) found that of 209 instruments developed in the United States, 119 instruments assessed level of scientific knowledge, 32 assessed level of achievement in the skills and processes of science, and 25 assessed affective objectives of science teaching. In selecting some of the instruments for more extensive review, Mayer and Richmond found that most of the instruments have been used infrequently and with small populations. Very few of the instruments seem to have been put through an extensive period of development and refinement. Noting that assessment of gains in student knowledge would continue to be a prime focus of evaluation efforts, Mayer and Richmond recognized that the content of science curricula was highly dependent on local conditions. Thus, evaluation instruments must be assembled for a specific curriculum. Although the instruments must be specific to the local curriculum, individual item banks may have wide evaluation utility. This is supportive of the argument for item banks that would also support domain-referenced testing. In this approach, items are randomly selected from a large bank (domain) that has been designed to reflect all the objectives of the curriculum being evaluated. The availability of constantly improving computer technology has made this kind of instrument development increasingly feasible. The range of assessment methods used by most elementary teachers was described by Raizen et al. (1989):

Although they have a wide repertoire of approaches to assessment to draw on, including keeping record of and observing childrens' performance in science, teachers usually model their own tests on the end-of-chapter quizzes in textbooks or on the items in standardized tests. (p. 719)

Hastings and Stuart (1983) examined the status of homemade achievement tests and compared the results to those of a similar study reported earlier by Anderson (1972). Six years (January 1975 to December 1980) of *Science Education* and *Journal of Research in Science Teaching* were reviewed. A total of 142 articles that contained reference to at least one written homemade achievement test were analyzed according to the categories used in the Anderson study. The findings showed that reliability estimates and the procedures for selecting test items were mentioned more frequently than in Anderson's study. There was little or no improvement in describing the relationship of test items to instruction. The majority of articles in both studies reported no empirical test development procedure. This is suggestive of problems with the establishment of validity. Although several articles reported that some form of content validity was established, it was difficult to determine what means were employed to establish it. Multiple-choice items were the most commonly reported in both studies, followed by completion items. It was suggested that reports of research involving homemade tests should give (1) an explanation of reasons for selecting test questions and (2) a complete description of the relationship between instruction and test questions.

Content validity as the primary assurance of the measurement accuracy of science assessment examinations was a concern of Yarroch (1991). Yarroch proposed item validity as an alternative procedure. Item validity was based on qualitative comparisons between (1) student answers to objective items, (2) clinical interviews with examinees to ascertain their knowledge and understanding of objective items, and (3) student answers to essay items prepared as an equivalent to the objective examination items. Tenth-grade students ($n = 93$) were administered the Michigan Educational Assessment Program pilot examination. Eight items were selected for analysis and 20 students were randomly selected from those whose scores were plus or minus 1 standard deviation about the mean. Student knowledge was categorized as correct for the correct reason, correct for the wrong reason, incorrect for the wrong reason, and incorrect for the correct reason (i.e., students did indeed lack the necessary knowledge). Items from the science assessment actually overestimated student understanding of the science content. Overestimation occurs when a student answers an item correctly, but for a reason other than that needed for an understanding of the content in question. There was little evidence that students incorrectly answered items but cited a correct reason, resulting in underestimation of knowledge. Essay items written to mirror the content objectives for four of the items were found to have much higher coefficients of validity. One alternative, then, would be to include essay questions in examinations. This is considered a costly and time-consuming choice. The other alternative would be the systematic evaluation and rewriting of items to better reflect examinee response to items. This could be accomplished by (1) design of items so that they more closely reflect the objective being measured; (2) selection of options, especially those that reflect common misconceptions; and (3) use of novel items not traditionally found on objective-style examinations. This might have included Likert-type items where a preference is asked for or multiple-tier items in which a student is asked to select an answer and a justification for the selected answer.

A post hoc analysis of teacher-made tests was conducted by Gullickson and Ellwein (1985) to determine how well practice coincided with recommended procedure. Based on survey responses from elementary and secondary teachers ($n = 323$), they found that more than 90 percent reported completing at least one course in educational measurement. The use of test statistics by these teachers was reported as follows:

Total test score for each student	85%
Range of test scores for the class	40%
Mean test score for the class	13%
Median test score for the class	12%
Standard deviation for the class	9%
Item difficulty	31%
Reliability of test	28%

There must be an inconsistency or misunderstanding with respect to the use of test reliability. In order to compute the reliability of a test, one must have first computed the mean and the standard deviation. Gullickson and Ellwein (1985) reexamined the data to determine how many of the teachers reporting use of the reliability information had also reported a total score, mean, and standard deviation. When analyzed this way, only 1.2 percent of the teachers met the criterion for verified use of reliability. The results of this study showed that elementary and secondary teachers do not analyze test results in the manner suggested by measurement courses. Teachers have claimed that the major purpose of tests is to provide feedback on student learning and the evaluation of instructional practice. Without systematic analysis of their tests, teachers have not had assurance that their tests function as desired.

It has been a common practice in multiple-choice tests to ask the examinee to select only the correct or best answer, although it is also possible to ask for the selection of more than one answer or the only choice that is not correct. Abu-Sayf (1979) investigated whether the abilities necessary for students in the course of multiple-choice testing were more favorably exhibited by one response approach or another. Two forms of a 40-item test were devised, with one form requesting the selection of a single best answer and the other form requesting the selection of the incorrect choice. The subjects were 82 eighth-graders, evenly divided between the two tests. The group taking the form requesting selection of the correct choice scored significantly higher than the group taking the form asking for the incorrect response. Analysis of errors indicated that it was harder for those who took the test calling for the incorrect choice to adjust from the conventional approach. Thus, the conventional multiple-choice testing approach appeared easier for both the test constructor and the examinees.

Item sequencing has long been a consideration in assessment, with easy items usually placed at the beginning of the test on the assumption that such an arrangement increases student motivation and produces more reliable tests. The effect of changes in item sequencing on student performance in a multiple-choice chemistry test was studied by Hodson (1984). Fifty items were arranged in three sequences of difficulty: random, easy to hard, and hard to easy. The tests were distributed randomly to 157 students (ages 16–19). The mean test score was significantly higher for the test sequenced easy to hard than for the hard-to-easy sequence. The item difficulty index was raised by placing easier items toward the beginning of the test and lowered by placing these items toward the end of the test. Test reliability was largely independent of item sequence. The tests sequenced hard to easy took the students slightly longer to complete and produced more negative postexamination attitudes.

The effect of question types in textual reading on retention of biology concepts was studied by Leonard and Lowery (1984). Although this was not, strictly speaking, an assessment study, the effect of question type seemed of value to those concerned with assessment instruments. The sample consisted of all 383 students enrolled in a one-semester course in general biology. The sample was randomly assorted into six approximately equal groups. A passage on the concept of multicellularity was read by five groups; the sixth group received no instructional treatment. One group received the text only, and four of the treatment groups had questions interspersed throughout the text at the beginning of selected paragraphs. The questions were one of four types: theoretical, factual (i.e., recall), valuing, or hypothesizing. A 20-item, multiple-choice test to assess recall of facts or concepts and applications of multicelluarlity was administered immediately after the reading, 2 weeks after the reading, and again 9 weeks after the reading. For the testing period immediately following the reading, none of the mean scores of the groups reading with questions differed significantly from the mean scores of the no-questions group. For the testing period that followed 2 weeks after the reading, all groups reading with questions had significantly lower mean scores than the reading-only group. At 9 weeks after the reading, only the rhetorical questions group scored lower than the no-questions group. As a check on whether reading the passage had an effect, scores of the reading groups were compared with those of the group that had no reading. The results of this study were found to be in opposition to those of most studies of the effects of text questions on learning or retention. However, Leonard and Lowery noted that in most reported cases, the test questions are placed either before or after the text material. In this study the questions were interspersed throughout the text. Clearly, this finding suggested further study.

Recognizing that much, perhaps most, of classroom assessment is conducted for grading purposes, Stiggins, Frisbie, and Griswold (1989) compared practices recommended by the measurement community with practices of 15 classroom teachers. The range of student characteristics to be assessed included achievement, ability, effort, attitude, interest, and personality. Methods of assessment included daily assignments, paper-and-pencil tests and quizzes, responses to oral questions during discussions, and performance assessments (i.e., observations of behavior and products). Information was gathered from each of the teachers on 19 specific dimensions of their grading practices. Included in these dimensions were student characteristics, measurement methods, and the manner in which the grade data were managed and reported. To determine the extent of adherence to recommendations, teachers were observed and/or interviewed in their classrooms. For issues related to grade data collection and management, recommended practice is indicated along with observed practice.

1. Recommendation: Grading methods should be clearly communicated to students.
   Practice: Expectations were clearly stated.
2. Recommendation: Achievement should be the sole ingredient in determining grades.
   Practice: Many teachers also incorporate effort.
3. Recommendation: Intelligence, cognitive ability, and aptitude should be considered when determining grades.
   Practice: Teachers were equally divided in their use of ability in assessment. High-ability students were frequently graded solely on achievement while lower-ability students' grades took into account perceived ability.
4. Recommendation: Attitude, interest, and personality should not be considered in determining grades.
   Practice: Attitude, interest, and personality were *not* considered.
5. Recommendation: Level of effort and seriousness should not be a part of grading criteria.
   Practice: Effort was given significant weight.
6. Recommendation: Methods of obtaining grading data: Daily assignment can serve two purposes—formative process, providing students with practice, in which case they should not be graded, or summative assessments to monitor learning, in which case they should.
   Practice: Daily assignments were used in determining grades, but not in any consistent way.
7. Recommendation: Oral questioning should *not* be used to determine grades.
   Practice: Teachers generally did *not* use oral data.

8. Recommendation: Performance assessment can be used in part for grading if certain rules of evidence are followed.
   Practice: About half of the teachers reported using performance measures. Although some of these measures seemed based on vague criteria and inadequate sampling of performance, their use was consistent with recommendations.
9. Recommendation: Amount of data to be collected for grading purposes should be adequate, about two or three high-quality measures in a quarter.
   Practice: Teachers "overgather" data. They tend to treat everything as summative. More time needs to be spent in the construction of quality testing instruments.
10. Recommendation: More time needs to be spent in the construction of quality testing instruments. Higher cognitive levels should be included.
    Practice: Teachers did not spend enough effort in developing well-balanced and challenging instruments.
11. Recommendation: For a challenging norm-referenced grading approach, each component score should be adjusted so that all distributions have equal variability (standard deviation) and the weighted components added to get a composite.
    Practice: Teachers failed to do this.

Several possible reasons have been noted for discrepancies between recommendations and practice. It is thought that there may not be a single best approach. What is felt to be best practice may be a matter of opinion or philosophical position, not established fact. Recommendations from measurement specialists may not be taken into account because of some of the practical constraints of the classroom. It is thought that teachers may be unaware of the recommendation or may lack the expertise to implement them. In any case, the issues involved are complex and need further exploration and research.

The assessment practices of 36 teachers were examined by Stiggins, Griswold, and Wikelund (1989) to determine the extent to which higher-order thinking skills were measured. Teacher assessment activities were recorded through interviews, observations in the teachers' classrooms, and analysis of their paper-and-pencil instruments. The results showed that the largest percentage of items on the written instruments tested recall of facts and information and that nearly half of the questions that were asked in the observed classrooms assessed recall of facts and information. Higher-order thinking skills such as inference, analysis, evaluation, and comparison were largely overlooked in these teachers' assessments. The results were very similar across grade levels and across content areas. Furthermore, teachers who lacked training in the teaching and assessing of higher-order thinking skills tended to assess them less often. Although this was not a surprising finding, it does suggest that more attention be given to these skills in teacher education programs.

Higher-order thinking skills were also the subject of a study by Lawrenz and Orton (1989), who surveyed a random sample of seventh- and eighth-grade science and mathematics teachers in Minnesota. Responses were received from 139 science teachers and 146 mathematics teachers. It was found that the mathematics teachers placed more emphasis on preparation for future study of mathematics and a systematic approach to solving problems than did science teachers. In contrast, the science teachers gave more emphasis to developing inquiry skills, learning what evidence is needed to constitute proof, learning about applications of science and mathematics in technology, learning how scientific explanations have changed throughout history, and developing curiosity about natural phenomena. With respect to categories of assessment items used, science teachers had considerable variety in their assessment categories. They placed more emphasis on class discussion, attendance, behavior, and projects than did mathematics teachers. The mathematics teachers gave more emphasis to homework and to classroom tests than did the science teachers. Science teachers were prone to use true-false, multiple-choice, and essay-type items. They also placed more emphasis on items that required definition of concepts, those that required students to explain their reasoning, and those that had more than one answer. Mathematics teachers placed more emphasis on items that required use of algorithms to solve problems, required recall of specific information, were minor variations on homework problems, had one particular answer, and required more than one step to find a solution. A major problem facing teachers attempting to teach higher-order thinking skills was identified. Although many teachers would like to get students to think about the subject matter in a critical way, many pressures existed. School administration was identified as having bureaucratic procedures to organize and control large numbers of students. At the same time, the public expectation has been that teachers will be concerned with educating and caring for the individual. Thus, the teachers are caught in the middle.

The message sent over the past four decades has been that assessment should serve an accountability purpose. Stiggins, Conklin, and Bridgeford (1986) contended that this measurement paradigm is too narrow and restrictive. Based on a summary of the research, they found that "the kind of measurement referenced under this paradigm represents only a small fraction of the assessments that take place in schools and that influence the quality of schooling and student learning" (p. 6). Some of the findings summarized are as follows:

1. More than 350 high school teachers taking part in a survey reported that 75 percent of the tests they used were teacher developed. More than 400 elementary teachers surveyed indicated they relied more heavily on curriculum-embedded tests (Herman & Dorr-Bremme, 1982).
2. As grade level increased, the weight given to objective tests and structured tests went up, while that given to published tests and spontaneous observations went down.
3. Math and science teachers tended to rely most on paper-and-pencil objective tests, whereas teachers focusing in areas

reflective of communication skills relied more on structured observations and professional judgments (Stiggins & Bridgeford, 1985).

The analysis of nearly 400 teacher-developed tests by Stiggins et al. (1986), including thousands of items, led to the following conclusions:

1. Teachers used short-answer questions most frequently in their test making.
2. Teachers, even English teachers, generally avoided essay questions.
3. Teachers used more matching items than multiple-choice or true-false items.
4. Teachers devised more test questions to sample knowledge of facts than any of the other behavioral categories studied.
5. When categories related to knowledge of terms, knowledge of facts, and knowledge of rules and principles are combined, almost 80 percent of the test questions reviewed focused on these areas.
6. Teachers developed few questions to test behaviors that could be classified as ability to make applications.
7. Comparison across school levels demonstrated that junior high school teachers used more questions to tap knowledge of terms, knowledge of facts, and knowledge of rules and principles than did elementary or senior high school teachers.

A survey of 228 classroom teachers on their use of performance assessment by Stiggins and Bridgeford (1985) found that teachers described their performance tests to be:

1. Equally divided between evaluations of processes (students performing as in speaking) and products created by students
2. Scored both holistically and analytically, but resulting in a single grade being assigned
3. Scored by the teacher rather than by the student or a colleague
4. Interpreted in criterion-referenced terms, in terms of a pre-established standard
5. Public and preannounced rather than unobtrusive assessments
6. Often used with little attention to assessment quality

Summary for Classroom Assessment

Classroom assessment is a complex task. Teachers are expected to measure dozens of student variables for a variety of reasons and at an incredible pace. Serious questions have been raised about the quality of classroom assessment. Failure to deal with this problem has led to the use of unsound measurement practices in the schools. Using unsound measurement results in poor decision making. This, in turn, could lead to inefficient instruction and failure to learn.

Classroom assessment is a major component of classroom teacher activity. "Teachers may spend as much as 20% or 30% of their time directly involved in assessment-related activities"

(Stiggins, 1988, p. 364). Teachers have not received much training in classroom assessment as part of their teacher education programs, nor has enough technical help been made available to them in their daily work. Most of the assessment research to date has tended to focus on large-scale, standardized tests; relatively little attention has been directed to teacher assessment in the classroom. As Stiggins (1988) noted, with the press of "nationalized standardized tests, statewide assessments, and of measurement driven instruction, we are failing to address the central issue in school assessment: insuring the quality and appropriate use of teacher-directed assessment of student achievement used every day in the classroom coast to coast" (p. 363).

OVERALL SUMMARY OF ASSESSMENT CHAPTER

Following the sections that provide an introduction and brief history of assessment, separate sections focus on a sample of topics chosen to illustrate areas of assessment that are active. These sections include state and provincial assessment efforts, national and international surveys, assessment of student attitudes, assessment utilizing computers, classroom assessment, and the assessment of performance or inquiry skills.

The introductory section describes the many perspectives and approaches to assessment. Assessment is described as the efforts to collect information, usually with the purpose of being used to help evaluate students, curricula, program, teachers, teaching, instructional innovations, or research treatments. Assessments are conducted formally or informally; can result in quantitative or qualitative information; can focus on cognitive, affective, or motor-skill outcomes; and can serve diagnostic, formative, and/or summative purposes. Assessment is organized, planned, conducted, and used by teachers, administrators, and researchers.

Much energy and attention is currently being focused on assessment, ostensibly to improve its validity, effectiveness, and usefulness. Some of these effects coincide with efforts to reform the science curriculum and instruction. Some of the reformers suggest that assessment should become more authentic and performance based and include modes (alternatives) other than those in current use (traditional). Ironically, many of the modes of assessment that are proposed were used in the past and were replaced as more public education evolved. As with all academic pursuits, an examination of the history of an issue is critical to understanding and analyzing current efforts.

Much attention has been given to results of national and international surveys of the science achievement of students at several stages of schooling. The main national survey has been that of the National Assessment of Educational Progress (NAEP), conducted six times between 1970 and 1990. The next NAEP survey in science is scheduled for 1994. The NAEP-Science Studies initially sampled students of ages 9, 13, and 17 and recently students at grades 4, 8, and 12. From these studies, NAEP has portrayed science achievement as declining or staying the same. These NAEP surveys have drawn attention to the lower achievement of blacks, Hispanics, females, economically disadvantaged students, and students not planning to attend college.

In the same time period, international surveys of science achievement have been coordinated by two groups: the International Association for the Evaluation of Educational Achievement (IEA) and the International Assessment Educational Program (IAEP). IEA is a consortium of nations and research institutes, and IAEP is an outgrowth of the NAEP. Despite variations in approaches, both international surveys reported results in which the American samples had lower mean scores than most of the other countries. These results and the decline or stable performance on the NAEP studies have fueled many of the current reform efforts.

Some of the science assessment activities in two other countries are also described—the work in Israel with the biology matriculation exam and the work of the Assessment of Performance Unit (APU) in the United Kingdom (England, Wales, and Northern Ireland). One of the unique aspects of the Israeli biology exam was the innovative practical testing (with organisms, objectives, and equipment). The APU used individual and group practical tests as well as paper-and-pencil tests to conduct several surveys of students at ages 11, 13, and 15. The Israeli and APU assessment strategies have been valuable models for the assessments conducted within a variety of state and provincial assessment systems and by individual researchers.

Recent pressure for educational accountability has led to an increase of assessment activities at the state or provincial level. Innovative assessment programs from New York Connecticut, California, Massachusetts, Rhode Island, Michigan, British Columbia, Ontario, and Manitoba are described in this section. Typically, these innovations (including performance testing and portfolios) were used as supplements to existing techniques.

A separate section presents a review of research on assessment via the performance or practical testing formats of inquiry or laboratory skills. This mode of assessment has been rationalized as more appropriate for some outcomes that are currently assessed by paper-pencil methods or not at all. Performance assessment can be conducted with individual students or with groups of students moving to several different places where collections of material are organized. The students are directed to perform certain skills or manipulations and commonly record their observations, measurements, calculations, and conclusion in a student booklet. Other performance assessments involve the students interacting with specially prepared computer programs or assembling a project or portfolio that demonstrates a set of skills and conceptual understandings. These performance assessment techniques have ample evidence of content validity and have been found to be capable of consistent scoring by trained raters. The major concerns are test reliability and variability across tasks.

Assessment of attitudes has been a strong interest of science educators for a long time. A number of excellent reviews summarize the development in the field and the continuing concerns. The lack of a theoretical basis for much of the research is cited repeatedly, as is the weak and incomplete empirical support provided in the research reports. Researchers have attempted to explain the relationship between attitudes and achievement, gender, instructional program, teacher variables, the classroom environment, and family and peer influences.

As computers and related technologies are being used in various ways in schools, their use in assessment activities seem logical and consistent. A few studies have examined student attitudes and anxiety levels related to computer-based testing. However, the greatest use of computers in assessment has been as a teacher resource for storing item pools, for creating tests from these pools, and as a data management system for determining student grades.

Various assessment activities are described as occupying approximately one-fourth of a teacher's teaching schedule. For this very complex task, teachers receive little preparation, either in preservice or inservice programs. Similarly, little research has focused on these critical elements of assessment.

This chapter attempts to review a large, complex field by focusing attention on a sampling of specific areas. In doing so, the authors recognize that some areas receive little or no attention. We hope that this brief, targeted review provides a road map to your further examination of this field.

References

Abouseif, A. A., & Lee, D. M. (1965). The evaluation of certain science practical tasks at the secondary school level. *British Journal of Educational Psychology, 35,* 41–49.

Abu-Sayf, F. K. (1979). Does multiple choice item format affect student assessment results in science? *Journal of Research in Science Teaching, 16,* 359–362.

Accongio, J. L., & Doran, R. L. (1992). The science test-key to reform in science education. *EPIC for Science , Mathematics, and Environmental Education,* Columbus, OH.

Aesche, D. W., & Parslow, G. R. (1988). Use of questions from the medical biochemistry question bank with the "q" instruction package. *Biochemical Education, 16*(1), 24–28.

Alternative assessment and technology. (1992). *News from the Center for Children and Technology, 1*(3), 1–5. Bank Street College of Education, NY.

Anderson, O. R. (1990). *The teaching and learning of biology in the United States.* Washington, DC: National Science Teachers Association.

Anderson, R. C. (1972). How to construct achievement tests to assess comprehension. *Review of Educational Research, 42,* 145–170.

Assessment of Performance Unit. (1978). *Science progress report 1977–78.* London: Elizabeth House.

Badger, E., & Thomas, B. (1989). *On their own: Student response to open-ended tests in science.* Quincy: Massachusetts Board of Education.

Badger, E., Thomas B., & McCormack, E. (1990a). *Beyond paper and pencil: Background summary.* Quincy: Massachusetts Educational Assessment Program.

Badger, E., Thomas B., & McCormack, E. (1990b). *Beyond paper and pencil: Circuits.* Quincy: Massachusetts Educational Assessment Program.

Badger, E., Thomas B., & McCormack, E. (1990c). *Beyond paper and pencil: Insulation.* Quincy: Massachusetts Educational Assessment Program.

Badger, E., Thomas B., & McCormack, E. (1990d). *Beyond paper and pencil: Animal/leaf key.* Quincy: Massachusetts Educational Assessment Program.

Baker, E. L. (1974). Beyond objectives: Domain-referenced test for evaluation and instructional improvement. *Educational Technology, 14*(6), 10–16.

Baker, E. L., & Stites, R. (1991). *Trends in testing in the United States.* Los Angeles: UCLA Center for Research on Evaluation, Standards and Student Testing.

Baron, J. (1990). Performance assessment: Blurring the edges among assessment, curriculum, and instruction. In A. Champagne, B. Lovitts, & B. Calinger (Eds.), *Assessment in the service of instruction.*

Washington, DC: American Association for the Advancement of Science.

Baron, J., Forgione, P., Rindone, D., Kraglanski, H., & Davey, B. (1989, March). *Toward a new generation of student outcome measures: Connecticut common core of learning assessment.* Paper presented at the annual meeting of the American Educational Research Association, San Francisco.

Bartley, T. (1991). *1991 British Columbia science assessment performance task, administration manual.* Victoria: Province of British Columbia.

Bartley, T., Carlisle, B., Erickson, G., & Meyer, K. (1991, November). *Performance assessment in British Columbia.* Paper presented at the National Science Teachers Association Regional Meeting, Vancouver, B. C.

Báthory, Z. (1986). The CTD science practical survey In Z. Báthory & J. Kádár-Fülöp (Eds.), *Educational evaluation studies in Hungary* (Vol. 9, pp. 165–174). *Evaluation in education: An international review series.* Oxford: Pergamon.

Baxter, G. P., Shavelson, R. J., Goldman, S. R., & Pine, J. (1992). Evaluation of procedure-based scoring for hands-on science assessment. *Journal of Educational Measurement, 29,* 1–18.

Ben-Zvi, R., Hofstein, A., Samuel, D., & Kempa, R. (1977). Modes of instruction in high school chemistry. *Journal of Research in Science Teaching, 14,* 433–440.

Black, P. J. (1991). APU science—the past and the future. *School Science Review, 72,* 13–28.

Blank, R. K., & Dalkilic, M. (1990). *State indicators of science and mathematics education 1990.* Washington, DC: Council of Chief State School Officers, State Education Assessments Center.

Blank, R. K., & Engler, P. (1992, January). Has science and mathematics education improved since *A nation at risk?* Science and Mathematics Indicators Project, Washington, DC: Council of Chief State School Officers.

Bloom, B. S., Engelhart, M. D., Furst, E. J., Hill, W. H., & Krathwohl, D. R. (1956). *Taxonomy of educational objectives: Handbook I: Cognitive domain.* New York: David McKay.

Blosser, P. E. (1984). Attitude research in science education. *Information Bulletin No. 1.* Columbus: ERIC Clearinghouse for Science, Mathematics, and Environmental Education, Ohio State University.

Blumberg, F., Epstein, M., MacDonald, W., & Mullis, I. (1986). *A pilot study of higher order thinking skills assessment techniques in science and mathematics.* Princeton, NJ: National Assessment for Educational Progress.

Bock, R. D., & Doran, R. L. (1992). *Scoring procedures for science laboratory tests.* Unpublished manuscript, State University of New York at Buffalo, Graduate School of Education, Amherst, NY.

Boenig, R. W. (1969). *Research in science education, 1938 through 1947*. New York: Teachers College Press.

Bross, T. R. (1986). The microcomputer–based science laboratory. *Journal of Computers in Mathematics and Science Teaching, 5*, 16–18.

Browning, M. E., & Lehman, J. (1988). Identification of student misconceptions in genetics problem solving via computer program. *Journal of Research in Science Teaching, 25*, 747–761.

Bryce, T. G. K., & Robertson, I. J. (1985). What can they do? A review of practical assessment in science. *Studies in Science Education, 12*, 1–24.

Burstall, C. (1986). Innovative forms of assessment: A United Kingdom approach. *Educational Measurement: Issues and Practice, 5*, 17–22.

Butts, D. P. (1981). A summary of research in science education—1979. *Science Education, 65*, 335–438.

California Assessment Program. (1990). *California: The state of assessment*. Sacramento: Author.

California State Department of Education. (1984). *Science framework addendum for California public schools, kindergarten and grades one through twelve*. Sacramento: Author.

Clark, D. (1992). Missouri's process skills approach, *Science Scope, 15*(6), 56.

Collins, M. A. J. (1984). Improved learning with computerized tests. *American Biology Teacher, 46*(3), 188–190.

Collins, M. A. J., & Earle, P. (1989–90). A comparison of the effects of computer–administered testing, computer–based learning and lecture approaches on learning in an introductory biology course. *Journal of Computers in Mathematics and Science Teaching, 9*(2), 77–84.

Comber, L. C., & Keeves, J. P. (1973). *Science education in nineteen countries: An empirical study*. New York: Wiley.

Comfort, K. (1992). Crime solvers in California. *Science Scope, 15*(6), 56–57.

Committee on the Function of Science in General Education. (1937). *Science in general education*. New York: Appleton–Century.

CTB Macmillan McGraw-Hill. (1992). *Performance assessment supplement to the California achievement tests* (5th ed.). Monterey, CA: Author.

Curtis, F. (1926). *A digest of investigation in the teaching of science in the elementary and secondary schools*. New York: Teacher's College Press.

Curtis, F. (1931). *Second digest of investigations in the teaching of science*. New York: Teachers College Press.

Curtis, F. (1939). *Third digest of investigations in the teaching of science*. New York: Teachers College Press.

Dana, T. M., & Nichols, S. (1992, March). *Assessing the state of science education reform in Florida*. Paper presented at the annual meeting of the National Association for Research in Science Teaching, Boston.

Dawson, G. (1987). 10,000 science questions by modem. *Science Teacher, 54*(3), 41–44.

Dodge, J. H. (1970). Testing-measuring student achievement. In *Introductory physical science—Physical science II—A progress report* (pp. 11–14). Newton, MA: Educational Development Center.

Doran, R. L. (1980). *Basic measurement and evaluation of science instruction*. Washington, DC: National Science Teachers Association.

Doran, R. L. (1990). What research says . . . about assessment. *Science and Children, 27*(8), 26–27.

Doran, R. L. (1992a). *An analysis of the ESPET manipulative skills test*. Unpublished manuscript, State University of New York at Buffalo, Graduate School of Education, Amherst, NY.

Doran, R. L. (1992b). *An analysis of the ESPET objective test*. Unpublished manuscript, State University of New York at Buffalo, Graduate School of Education, Amherst, NY.

Doran, R. L., Boorman, J., Chan, A., & Hejaily, N. (1992a, March). *Alternative assessment of high school laboratory skills*. Paper presented at the annual meeting of the National Association for Research in Science Teaching, Boston.

Doran, R., Boorman, J., Chan, A., & Hejaily, N. (1992b). Successful laboratory assessment. *Science Teacher, 59*(4), 22–27.

Doran, R. L., & Jacobson, W. J. (1991). Science achievement in the United States and sixteen other countries. In S. K. Majunder, L. M. Rosenfeld, P. A. Rubba, E. W. Miller, & R. F. Schmalz (Eds.), *Science education in the United States: Issues, crises, and priorities*. Easton: Pennsylvania Academy of Sciences.

Doran, R. L., & Tamir, P. (1992). An international assessment of science practical skills. *Studies in Educational Evaluation, 18*(3), 263–406.

Dunkleberger, G. E. (1980). Applying 20/20 hindsight to self–pacing. *Science Teacher, 47*(8), 32–34.

Eglen, J. R., & Kempa, R. F. (1974). Assessing manipulative skills in practical chemistry. *School Science Review, 56*, 261–273.

Erickson, G., Bartley, A., Blake, L., Carlisle, R., Meyer, K., & Stavy, R. (1993). *British Columbia assessment of science 1991 technical report: Student performance component*. Victoria: Ministry of Education.

Ferko, A., Jacobson, W. J., & Doran, R. L. (1991). *Advanced science student performance*. Washington, DC: National Science Teachers Association.

Fletcher, P., & Collins, M. A. J. (1986–87). Computer–administered versus written tests—advantages and disadvantages. *Journal of Computers in Mathematics and Science Teaching, 6*(2), 38–43.

Gabel, D. L., Kagan, M. H., & Sherwood, R. D. (1980). A summary of research in science education—1978. *Science Education, 64*, 425–568.

Gardner, M. (1990). *Laboratory assessment builds success (LABS)*. Berkeley: Lawrence Hall of Science.

Gardner, P. L. (1975). Attitudes to science: A review. *Studies in Science Education, 2*, 1–41.

Gauld, C. F., & Hukins, A. A. (1980). Scientific attitudes: A review. *Studies in Science Education, 7*, 129–161.

Germann, P. J. (1988). Development of the attitude toward science in school assessment and its use to investigate the relationship between science achievement and attitude toward science in school. *Journal of Research in Science Teaching, 25*, 689–703.

Giddings, G. J., Hofstein, A., & Lunetta, V. N. (1992). Assessment and evaluation in the science laboratory. In B. Woolnough (Ed.), *Practical science*. London: Open University Press.

Green, B. (1992). Statewide assessment in Texas. *Science Scope, 15*(6), 58.

Greenbaum, W., Garet, M., & Solomon, E. (1977). *Measuring educational progress: A study of the National Assessment*. New York: McGraw-Hill.

Grobman, H. (1970). *Developmental curriculum projects: Decision points and processes: A study of similarities and differences in methods of producing developmental curricula*. Itasca, IL: Peacock.

Guba, E. G., & Lincoln, Y. S. (1988). Do inquiry paradigms imply methodologies? In D. M. Fetterman (Ed.), *Qualitative approaches to evaluation in education: The silent scientific revolution*. New York: Praeger.

Gullickson, A., & Ellwein, M. (1985). Post hoc analysis of teacher-made tests: The goodness of fit between prescription and practice. *Educational Measurement: Issues and Practice, 4*(1), 15–18.

Haladyna, T., & Shaughnessy, J. (1982). Attitudes toward science: A quantitative synthesis. *Science Education, 66*, 547–563.

Haney, W. (1984). Testing reasoning and reasoning about testing. *Review of Educational Research, 54*(4), 597–654.

Hargraves, N., & Lynch, P. (1988). Experimenting with assessment: The practical examination in HSC biology revisited. *Australian Science Teachers Journal, 34,* 30–37.

Harrison, D., & Pitre, J. (1983). Computerized lab test on error analysis. *Physics Teacher, 21,* 588–589.

Harrison, D., & Pitre, J. (1988). Computerized pre–checking of laboratory data. *Physics Teacher, 26,* 156.

Hastings, J., & Stuart, J. (1983). An analysis of research studies in which "homemade" achievement instruments were utilized. *Journal of Research in Science Teaching, 20,* 697–703.

Head, J. (1966). An objectively marked practical examination in practical biology. *School Science Review, 68*(164), 85–95.

Hearle, R. (1974, April). *The development of an instrument to evaluate chemistry laboratory skills.* Paper presented to the annual conference of the National Association for Research in Science Teaching, Chicago.

Heikkinen, H., & Dunkleberger, G. E. (1985). On disk with mastery learning. *Science Teacher, 52*(7), 26–28.

Hein, G. E. (1987). The right test for hands-on learning? *Science and Children, 25*(2), 8–12.

Herb, A. (1992, February 17). Computer adaptive testing. *ADVANCE* (Newsletter of the Medical Laboratory Professionals).

Herman, J. L., & Dorr-Bremme, D. (1982, April). *Assessing students: Teachers' routine practices and reasoning.* Paper presented at the annual meeting of the American Educational Research Association, New York.

Herman, J., & Golan, S. (1992). *Effects of standardized testing on teachers and learning—another look* (CSE Technical Report 334). Los Angeles: National Center for Research on Evaluation, Standards and Student Testing, University of California.

Hodson, D. (1984). The effect of changes in item sequence on student performance in a multiple-choice chemistry test. *Journal of Research in Science Teaching, 21,* 489–496.

Hofstein, A., & Lunetta, V. (1982). The role of the laboratory in science teaching: Neglected aspects of research. *Review of Educational Research, 52,* 201–217.

Huang, S. D., & Aloi, J. (1991). The impact of using interactive–video in teaching general biology. *American Biology Teacher, 53,* 281–284.

Hudson, L. (1990). National initiatives for assessing science education. In A. Champagne, B. Lovitts, & B. Calinger (Eds.), *Assessment in the service of instruction.* Washington, DC: American Association for the Advancement of Science.

Hueftle, S. J., Rakow, S. J., & Welch, W. W. (1983). *Images of science: A summary of results from the 1981–82 national assessment in science.* Minneapolis: University of Minnesota, Minnesota Research and Evaluation Center.

Humrich, E. (1988, April). *Sex differences in the second IEA science study—U.S. results in an international context.* Paper presented at the annual meeting of the National Association for Research in Science Teaching, Lake of the Ozarks, MO.

Humrich, E. (1992). *Sex differences in science attitude and achievement.* Washington, DC: National Science Teachers Association.

International Association for the Evaluation of Educational Achievement (IEA). (1988). *Science achievement in seventeen countries: A preliminary report.* New York: Pergamon.

Isenberg, S. (1989). New York—on the cutting edge. *Science and Children, 26*(7), 28–29.

Jackson, B. (1988). A comparison between computer–based and traditional assessment tests, and their effects on pupil learning and scoring. *School Science Review, 69,* 809–815.

Jacobson, W. J., & Doran, R. L. (1988). *Science achievement in the United States and sixteen countries: A report to the public.* Washington, DC: National Science Teachers Association.

Jacobson, W. J., Doran, R. L., Chang, E., Humrich, E., & Keeves, J. (1987). *The second IEA science study—U.S. second IEA science study.* New York: Teachers College, Columbia University.

Jacobson, W. J., Doran, R. L., Humrich, E., & Kanis, I. (1988). *International science report card.* New York: Teachers College, Columbia University.

Johnson, S. (1975). *Update on education: A digest of the national assessment of educational progress.* Denver: Educational Commission of the States.

Johnson, S. (1989). *National assessment: The APU science approach.* London: Department of Education and Science.

Joint Committee on Standards for Educational Evaluation. (1981). *Standards for evaluations of educational programs, projects, and materials.* New York: McGraw–Hill.

Joint Committee on Standards for Educational Evaluation. (1988). *The personnel evaluation standards.* Newbury Park, CA: Sage.

Jones, L., Mullis, D., Raizen, S., Weiss, I., & Weston, E. (1992). *The 1990 science report card: NAEP's assessment of fourth, eighth and twelfth graders* (Library of Congress No. 92–60173). Washington, DC: Educational Testing Service National Center for Educational Statistics.

Kanis, I. B. (1990). What is the current status of laboratory skills of our grade five students? *Science Teacher Bulletin, 54,* 20–29.

Kanis, I. B. (1991). Ninth grade lab skills: An assessment. *Science Teacher, 58,* 29–33.

Kanis, I. B., Doran, R. L., & Jacobson, W. J. (1990). *Assessing science process laboratory skills at the elementary and middle/junior high levels.* Washington, DC: National Science Teachers Association.

Keeves, J. P. (1992). *The IEA study of science III: Changes in science education and achievement 1970 to 1984.* Oxford: Pergamon.

Kelly, P., & Lister, R. (1971). Assessing practical ability in Nuffield. In B. Bloom, J. Hastings, & G. Madaus (Eds.), *Handbook for the formative and summative evaluation of student learning.* New York: McGraw-Hill.

Keys, C. (1992, March). *Evaluating students' written reports for evidence of conceptual and procedural understandings in science.* Poster session presented at the annual meeting of the National Association for Research in Science Teaching, Boston.

Klinckman, E. (1963). The BSCS grid for test analysis. *BSCS Newsletter, 19,* 17–21.

Klopfer, L. E. (1971). Evaluation of learning in science. In B. Bloom, J. Hastings, & G. Madaus (Eds.), *Handbook for the formative and summative evaluation of student learning.* New York: McGraw-Hill.

Klopfer, L. E. (1973). Evaluation of science achievement and science test development in an international context: The IEA study in science. *Science Education, 57,* 387–403.

Knight, C. W., & Dunkleberger, G. E. (1977). The influence of computer–managed self–paced instruction on science attitudes of students. *Journal of Research in Science Teaching, 14,* 551–555.

Kojima, S. (1974). IEA science study in Japan with special reference to the practical test. *Comparative Education Review, 18*(2), 262–267.

Krathwohl, D. R., Bloom, B. S., & Masia, B. B. (1964). *Taxonomy of educational objectives: Handbook II: Affective domain.* New York: David McKay.

Kruglak, H. (1954a). Measurement of laboratory achievement, part I. *American Journal of Physics, 22,* 442–463.

Kruglak, H. (1954b). The measurement of laboratory achievement, part II. Paper-pencil laboratory achievement tests. *American Journal of Physics, 22,* 452–463.

Kruglak, H. (1955). Measurement of laboratory achievement, part III. Paper-pencil analogy of laboratory performance tests. *American Journal of Physics, 23,* 82–87.

Krynowsky, B. (1988). Science education assessment instruments: Problems in assessing student attitude in science education: A partial solution. *Science Education, 72,* 575–584.

Kulm, G., & Malcolm, S. M., Eds. (1991). *Science assessment in the service of reform.* Washington, DC: American Association for the Advancement of Science.

Kumar, D. (1991, September). *A note on hyperequation.* Paper presented at the National Center for Science Teaching and Learning, Ohio State University, Columbus.

LaPointe, A., Askew, J., & Mead, N. (1992). *Learning science: IAEP.* Princeton, NJ: Educational Testing Service.

LaPointe, A., Mead, N., & Phillips, G. (1989). *A world of differences. An international assessment of mathematics and science.* Princeton, NJ: Educational Testing Service.

Lawlor, E. P. (1969). *Research in science education, 1953 through 1957.* New York: Teachers College Press.

Lawrenz, F., & Orton, R. (1989). A comparison of critical thinking related teaching practices of seventh and eighth grade science and mathematics students. *School Science and Mathematics, 89,* 361–372.

Lawson, A. P., Castenson, K., & Cisneros, R. (1986). A summary of research in science education—1984. *Science Education, 70*(3), 189–346.

Leonard, W., & Lowery, L. (1984). The effects of question types in textual reading upon retention of biology concepts. *Journal of Research in Science Teaching, 21,* 377–384.

Leuba, R. J. (1986a). Machine-scored testing, part I: Purposes, principles, and practices. *Engineering Education, 77*(2), 89–95.

Linn, M., DeBenedictis, T., Delucchi, K., Harris, A., & Stage, E. (1987). Gender differences in national assessment of educational progress science items: What does "I don't know" really mean? *Journal of Research in Science Teaching, 24*(3), 267–278.

Llabre, M., Clements, N., Fitzhugh, K., Lancelotta, G., Mazzagatti, R., & Quinones, N. (1987). The effect of computer-administered testing on test anxiety and performance. *Journal of Educational Computing Research, 3,* 429–433.

Lomask, M., Johnson, L., & Hafner, L. (1992, March). *Interactive videodisc as a tool for assessing science teachers' knowledge of safety regulations in school laboratories.* Paper presented at the annual meeting of the National Association for Research in Science Teaching, Boston.

Lord, F. (1980). *Applications of item response theory to practical testing problems.* Hillsdale, NJ: Lawrence Erlbaum.

Lunetta, V., Hofstein, A., & Giddings, G. (1981). Evaluating science laboratory skills. *The Science Teacher, 48*(1), 22–25.

Lunetta, V., & Tamir, P. (1979). Matching lab activities with teaching goals. *The Science Teacher, 46*(5), 22–24.

Madaus, G., Arasian, P., & Kellaghan, T. (1980). *School effectiveness.* New York: McGraw-Hill.

Manitoba Department of Education. (1981). *Manitoba science assessment program 1980—Final report.* Winnipeg: Author.

Manitoba Department of Education. (1988). *Manitoba science assessment 1986—Final report.* Winnipeg: Author.

Manitoba Department of Education. (1990). *Manitoba science assessment 1990—Preliminary report.* Winnipeg: Author.

Martens, M. (1992, March). *Effects of state-mandated testing on local science programs: Case study of a suburban elementary school.* Paper presented at the annual meeting of the National Association for Research in Science Teaching, Boston.

May, P., Murray, K., & Williams, D. (1985). CAPE: A computer program to assist with practical assessment. *Journal of Chemical Education, 62*(4), 310–312.

Mayer, V. J. (1974). *Unpublished evaluation instruments in science education: A handbook.* Columbus: ERIC Clearinghouse for Science, Mathematics, and Environmental Education, The Ohio State University.

Mayer, V. J., & Richmond, J. (1982). An overview of assessment instruments in science. *Science Education, 66*(1), 49–66.

McKerson, R. S. (1989). New directions in educational assessment. *Educational Researcher, 18*(9).

Meng, E., & Doran, R. L. (1990). What research says about appropriate methods of assessment. *The Science Teacher, 28*(1), 42–45.

Meng, E., & Doran, R. L. (1992). Improving instruction and learning through evaluation-elementary school science. *ERIC for Science, Mathematics, and Environmental Education,* Columbus, OH.

Michigan Department of Education. (1987). *Michigan educational assessment program handbook.* East Lansing: Author.

Milkent, M. M., & Roth, W. M. (1989). Enhancing student achievement through computer-generated homework. *Journal of Research in Science Teaching, 26,* 567–573.

Miller, J. K. (1988). *An analysis of science curricula in the United States.* Washington, DC: National Science Teachers Association.

Miller, J. K. (1989). *Who is scientifically literate?* Paper presented at American Association for Advancement of Science. Science Education Forum, Washington, DC.

Mitchell, R. (1992). *Testing for learning—How new approaches to evaluation can improve American schools.* New York: Free Press.

Moe, K. C., & Johnson, M. F. (1988). Participants' reactions to computerized testing. *Journal of Educational Computing Research, 4,* 79–86.

Moore, R. W., & Sutman, F. X. (1970). The development, field test and validation of an inventory of scientific attitudes. *Journal of Research in Science Teaching, 7,* 85–94.

Mullis, I., & Jenkins, L. B. (1988). *The science report card: Elements of risk and recovery. Trends and achievement based on the 1986 national assessment.* Princeton, NJ: Educational Testing Service.

Munby, H. (1983a). *An investigation into the measurement of attitudes in science education.* Columbus: SMEAC Information Reference Center, Ohio State University. (ERIC Document Reproduction Service No. ED 237 347)

Munby, H. (1983b). Thirty studies involving the "scientific attitude inventory": What confidence can we have in this instrument? *Journal of Research in Science Teaching, 20,* 141–162.

Munby, H., Kitto, R. J., & Wilson, R. J. (1976). Validating constructs in science education research: The construct "view of science." *Science Education, 60,* 313–321.

Napier, J., & Riley, J. (1985). Relationship between affective determinants and achievement in science for seventeen-year-olds. *Journal of Research in Science Teaching, 22,* 364–383.

National Assessment of Educational Progress. (1975). *Selected results from the national assessments of science: Scientific principles and procedures* (Report No. 04–5–02). Princeton, NJ: Author.

National Assessment of Educational Progress. (1978). *The national assessment in sciences: Changes in achievement, 1969–72.* Denver: Educational Commission of the States.

National Assessment of Educational Progress. (1987a). *Science objectives, 1985–86 assessment.* Princeton, NJ: Educational Testing Service.

National Assessment of Educational Progress. (1987b). *Learning by doing—a manual for teaching and assessing higher order skills in science and mathematics* (Report No. 17, HOS–80). Princeton, NJ: Educational Testing Services.

National Assessment of Educational Progress. (1992). *Trends in academic progress: Achievement of U.S. students in science 1969–70 to 1990, mathematics 1973 to 1990, reading 1971 to 1990, and writing 1984 to 1990.* Princeton, NJ: Educational Testing Service.

National Center for Education Statistics. (1992). *The 1990 science report card: NAEP'S assessment of fourth, eighth, and twelfth graders.* Washington, DC: U.S. Department of Education.

National Center for Improving Science Education. (1991). *The high stakes of high school science.* Andover, MA: The NETWORK.

National Commission on Excellence in Education. (1983). *A nation at risk.* Washington, DC: Author.

National Society for the Study of Education. (1947). *The forty-sixth yearbook: Part 1: Science education in American schools.* Chicago: University of Chicago Press.

New York State Education Department. (1985). *Elementary science syllabus.* Albany: Author.

New York State Education Department. (1988). *Guide for program evaluation.* Albany: Author.

New York State Education Department. (1989). *New York State program evaluation test in science.* Albany: Author.

New York State Education Department. (1990). *A new compact for learning.* Albany: Author.

New York State Education Department. (1991). *Student assessment: A review of current practice and trends in the United States and selected countries.* Albany: Author.

New York State Education Department. (1992). *Regents high school examination: Earth science—performance test manual.* Albany, NY: Author.

Nussbaum, J. (1979). Children's conceptions of earth as a cosmic body: A cross age study. *Science Education, 63,* 83–93.

Oakes, J. (1985). *Keeping track: How schools structure inequality.* New Haven, CT: Yale University Press.

Olson, P. (1976). *A view of power: Four essays on the national assessment of educational progress.* University of North Dakota. Grand Forks: North Dakota Study Group on Evaluation.

O'Neil, J. (1991). *Raising our sights: Improving United States achievement in mathematics and science.* Alexandria, VA: Association for Supervision in Curriculum Development.

Ontario Ministry of Education. (1981). *Ontario assessment instrument pool—chemistry.* Toronto: Author.

Ontario Ministry of Education. (1983). *Ontario assessment instrument pool—physics.* Toronto: Author.

Ontario Ministry of Education. (1989). *Ontario assessment instrument pool—Biology—draft.* Toronto: Author.

Ormerod, M. B., & Duckworth, D. (1975). *Pupils' attitudes to science: A review of research.* Windsor, England: NFER.

Ostlund, K. L. (1992). Science process skills—assessing hands-on student performance. Menlo Park, CA: Addison-Wesley.

Postlethwaite, T. N., & Wiley, D. E. (1992). *The IEA study of science II: Science achievement in twenty-three countries.* Oxford: Pergamon.

Province of British Columbia. (1986). *British Columbia science assessment 1986—Summary report.* Victoria: Author.

Raizen, S. A., Baron, J. B., Champagne, A. B., Haertel, E., Mullis, I. V. S., & Oakes, J. (1989). *Assessment in elementary school science.* The National Center for Improving Science Education; Andover, MA: The NETWORK.

Raizen, S. A., Baron, J. B., Champagne, A. B., Haertel, E., Mullis, I. V. S., & Oakes, J. (1991). *Assessment in science education: The middle years.* The National Center for Improving Science Education; Andover, MA: The NETWORK.

Rakow, S. (1985). Prediction of the science inquiry skill of seventeen-year-olds: A test of the model of educational productivity. *Journal of Research in Science Teaching, 22,* 289–302.

Rhode Island Department of Education. (1990–91). *Rhode Island distinguished merit program handbook.* Providence: Author.

Robinson, J. (1969). Evaluating laboratory work in high school biology. *American Biology Teacher, 31*(3), 236–240.

Roeber, E., Donovan, D., & Cole, R. (1980). Telling the statewide testing story—and living to tell it again. *Phi Delta Kappan, 62,* 273–274.

Rosier, M. J., & Keeves, J., Eds. (1991). *The IEA study of science I: Science education and curricula in twenty-three countries.* Oxford: Pergamon.

Schibeci, R. A. (1984). Attitudes to science: An update. *Studies in Science Education, 11,* 26–59.

Schneider, A., Muller, E., Doran, R., & Jacobson, W. (1990). *The second international science study, items and results: Population 1—grade 5.* Washington, DC: National Science Teachers Association.

Schneider, A. E., Muller, E. W., Doran, R. L., & Jacobson, W. J. (1991). *The second international science study, items and results: Population 2—grade 9.* Washington, DC: National Science Teachers Association.

Science Scope. (1992, March). Special issue on assessment, *15*(6).

Shavelson, R. J., & Baxter, G. P. (1992). What we've learned about assessing hands on science. *Educational Leadership, 49*(8), 20–25.

Shavelson, R. J., Baxter, G. P., & Pine, J. (1992). Performance assessments: Political rhetoric and measurement reality. *Educational Researcher, 21*(4), 22–27.

Shavelson, R. J., Carey, N. B., & Webb, N. M. (1990). Indicators of science achievement: Options for a powerful policy instrument. *Phi Delta Kappan, 71,* 692–697.

Shymansky, J. A., & Kyle, W. C., Jr. (1988). A summary of research in science education—1986. *Science Education, 72,* 249–402.

Simpson, R. D., & Troost, K. M. (1982). Influences on commitment to and learning of science among adolescent students. *Science Education, 66,* 763–782.

Sipe, C., & Farmer, W. A. (1982). A summary of research in science education—1980. *Science Education, 66,* 299–494.

Sirotnik, K. (1970). An investigation of the context effect in matrix sampling. *Journal of Educational Measurement, 7,* 199–207.

Smith, E. R., & Tyler, R. W. (1942). *Appraising and recording student progress.* New York: Harper & Row.

Stannard, P. (1982). Evaluating laboratory performance. *Queensland Science Teacher, 41–63.*

Staver, J. R., Enochs, L. G., Koeppe, O. J., McGrath, D., McLellan, H., Oliver, J. S., Scharmann, L. C., & Wright, E. L. (1989). A summary of research in science education—1987. *Science Education, 73*(3), 243–389.

Stiggins, R. J. (1988). Revitalizing classroom assessment: The highest instructional priority. *Phi Delta Kappan, 69,* 363–368.

Stiggins, R. J., & Bridgeford, N. (1985). The ecology of classroom assessment. *Journal of Educational Measurement, 22*(4), 271–286.

Stiggins, R. J., Conklin, N., & Bridgeford, N. (1986). Classroom assessment: A key to effective education. *Educational Measurement: Issues and Practice, 5*(2), 5–17.

Stiggins, R. J., Frisbie, D., & Griswold, P. (1989). Inside high school grading practices: Building a research agenda. *Educational Measurement: Issues and Practice, 8,* 5–14.

Stiggins, R. J., Griswold M., & Wikelund, K. (1989). Measuring thinking skills through classroom assessment. *Journal of Educational Measurement, 26,* 233–246.

Strang, J. (1990). *Measurement in school science—assessment matters No. 2.* London: School Examination and Assessment Council.

Stufflebeam, D. L., & Webster, W. J. (1980). An analysis of alternative approaches to evaluation. *Educational Evaluation and Policy Analysis, 2*(3), 5–20.

Summers, G. (1984). Testor. *Journal of College Science Teaching, 13,* 356–358.

Swain, J. (1985). Toward a framework for assessment. *School Science Review, 67,* 139–159.

Swift, J. N. (1969). *Research in science education, 1948 through 1952.* New York: Teachers College Press.

Talesnick, I. (1979). *Chemistry laboratory achievement test.* Kingston, Ontario: Queen's University.

Tamir, P. (1972). The practical mode—a distinct mode of performance in biology. *Journal of Biological Education, 6*, 175–182.

Tamir, P. (1974). An inquiry oriented laboratory examination. *Journal of Educational Measurement, 11*, 25–33.

Tamir, P. (1976). "Invitations to inquiry" and teacher training. *American Biology Teacher, 38*, 50–52.

Tamir, P. (1985). Practical examinations. In T. Husen & N. Postlewaite (Eds.), *International encyclopedia of educational research*. London: Pergamon.

Tamir, P., & Glassman, S. (1970). A practical examination for BSCS students. *Journal of Research in Science Teaching, 7*, 107–112.

Tamir, P., & Glassman, S. (1971a). A practical examination for BSCS students: A progress report. *Journal of Research in Science Teaching, 8*, 307–315.

Tamir, P., & Glassman, S. (1971b). Laboratory tests for BSCS students. *BSCS Newsletter, 42*, 90–113.

Tamir, P., Nussinovitz, R., & Friedler, Y. (1982). The design and use of the practical test assessment inventory. *Journal of Biological Education, 16*, 42–50.

Taylor, R. M., & Swatton, P. (1990). *Graph work in school science—a booklet for teachers*. London: School Examination and Assessment Council.

The Psychological Corporation. (1992). *Goals: A performance-based measure of science*. San Antonio, TX: Author.

Travers, R. M. W. (1983). *How research changed American schools*. Kalamazoo, MI: Mythos Press.

U.S. Congress, Office of Technology Assessment. (1992). *Testing in American schools: Asking the right questions*. Washington, DC: U.S. Government Printing Office.

U.S. Department of Education. (1991). *America 2000: An education strategy*. Washington, DC: Author.

Voelker, A., Thompson, T., & Van Deman, B. (1980). Research reviews in science education: An update. *Science Education, 64*, 569–578.

Voelker, A., & Wall, C. (1972). Research review in science education. *Science Education, 56*, 487–501.

Voelker, A., & Wall, C. (1973). Research and development in science education: A bibliographic history. *Science Education, 57*, 263–270.

Vogel, G. (1985). A microcomputer–based system for filing test questions and assembling examinations. *Journal of Chemical Education, 62*, 1024–1027.

Voss, B. (1983). A summary of research in science education—1981. *Science Education, 67*, 283–420.

Ward, T. J., Jr., Hooper, S. R., & Hannafin, K. M. (1989). The effect of computerized tests on the performance and attitudes of college students. *Journal of Educational Computing Research, 5*, 327–333.

Waugh, M. L. (1985). Effects of microcomputer–administered diagnostic testing on immediate and continuing science achievement and attitudes. *Journal of Research in Science Teaching, 22*, 793–805.

Waugh, M. L., & Currier, D. (1986). Computer–based education: What we need to know. *Journal of Computers in Mathematics and Science Teaching, 5*(3), 13–15.

Welch, W. W. (1985). Research in science education: Review and recommendations. *Science Education, 69*, 421–448.

Welch, W. W., Walberg, H., & Fraser, B. (1986). Predicting elementary science learning using national assessments data. *Journal of Research in Science Teaching, 23*, 699–706.

White, E. E. (1886). *The elements of pedagogy*. New York: American Book Company.

Willis, J. A. (1988). Learning outcome testing: A statewide approach to accountability. *Technological Horizons in Education, 16*, 69–73.

Willson, V. L. (1983). A meta-analysis of the relationship between science achievement and science attitude: Kindergarten through college. *Journal of Research in Science Teaching, 20*, 839–850.

Wilson, J. T. (1983). Toward a disciplined study of testing in science education. *Science Education, 65*(3), 259–270.

Wolf, R. M. (1977). *Achievement in America*. New York: Teachers College Press.

Wood, K., & Ferguson, C. M. (1975). Teacher assessment of practical skills in A level chemistry. *School Science Review, 56*, 605–608.

Woolnough, B. E. (1986). The assessment of practical work. *Physics Education, 21*, 195–196.

Worthen, B. R., & Sanders, J. (1973). *Educational evaluation: Theory and practice*. Worthington, OH: Charles Jones.

Worthen, B. R., & Sanders, J. (1987). *Educational evaluation: Alternative approaches and practical guidelines*. New York: Longman.

Yager, R., & Penick, J. (1986). Perceptions of four age groups toward science classes, teachers, and the value of science. *Science Education, 70*, 355–363.

Yager, R., & Penick, J. (1989). An exemplary program payoff. *Science Teacher, 56*, 54–56.

Yarroch, W. L. (1991). The implications of content versus item validity on science tests. *Journal of Research in Science Teaching, 28*, 619–629.

·15·

RESEARCH ON THE HISTORY AND PHILOSOPHY OF SCIENCE

Richard A. Duschl

UNIVERSITY OF PITTSBURGH

The renewed scholarly interest in history and philosophy of science (HPS) and science teaching in the last several years highlights but a portion of the activities in this domain of science education research. In 1989 the First International Conference on History and Philosophy of Science and Science Teaching was held at Florida State University. In 1991, a new international scholarly journal, *Science & Education: Contributions from History, Philosophy and Sociology of Science and Mathematics*, was initiated to broaden and sustain the interfield dialogue that has begun. The first three issues of this new journal contained some of the papers delivered at the Second International Conference on History and Philosophy of Science and Science Teaching held at Queen's University, Kingston, Ontario. But these recent and important events were preceded by many important HPS and science education developments since the 1960s.

This chapter takes as its focus the application of HPS to science education curricula and instruction. Several organizational and analytical schemes are followed. It is shown that the early (1950s and 1960s) application of HPS to science education was principally to the design of curriculum (inquiry into inquiry). Although a concern for curriculum issues has continued, researchers have begun to consider how HPS can be applied to instructional, learning, and assessment strategies. Another trend in HPS that parallels the shift in educational application is the shift by philosophers away from the practice of examining only the justification of knowledge toward explaining development or generation of knowledge. It is the dual application of knowledge-justification frameworks and knowledge-development frameworks that has reawakened the application of HPS to science education. This double-edged application of HPS provides researchers of science teaching, science teacher education, and science learning with frameworks for analyzing and investigating the dynamic processes associated with the growth of scientific knowledge.

This review examines major developments between HPS and science education in the 40-year time frame 1950 to 1990. Let me state up front that the preparation of this review has greatly benefited from the reading of HPS and science teaching reviews by Michael Matthews (1989, 1990) and Derek Hodson(1985). However, this review differs from those of Matthews and of Hodson in that it seeks to place HPS and science education within the schema of a shift from curriculum only to curriculum and instruction applications and a shift from a predominant concern about the justification of knowledge to a more balanced concern between the justification and the development of knowledge. What emerges is a perspective on science and about science education that stresses restructuring and replacement in the growth of scientific knowledge. In turn, the application of this perspective of science to learning and teaching and to curriculum and instruction holds both promise and problems for science education researchers.

The chapter is organized into four parts: an overview of the major developments in philosophy of science between 1950 and 1990 and a brief discussion of salient educational issues; selected important developments in the history and philosophy of science and science education during the 1950s and 1960s; selected important developments in the 1970s and 1980s; and a summary section.

The author thanks reviewer Fred N. Finley (University of Minnesota).

AN OVERVIEW OF PHILOSOPHY OF SCIENCE

Traditional Views

Philosophers, in general, seek to understand the uniquely human process of thinking and the factors that influence thought processes (i.e., ethics, logic). Philosophers of science focus on understanding matters related to the growth of knowledge and on issues that establish the rational and justified bases of these knowledge claims. It is, simply put, an analysis of the relationship between evidence and explanations. More exactly, one activity of philosophers of science is to account for and explain the processes associated with scientific inquiry, which, in turn, help shape and determine the form of knowledge. This branch of philosophy of science is called epistemology. Another activity of philosophers of science is the investigation of the rationality and truthfulness of scientific knowledge claims. On what grounds can the products of science (theories, principles, laws, etc.) claim to be real, or are they merely instruments that one employs to serve a particular purpose? The realist-instrumentalist debate divides many philosophers. Nagle (1961), in his chapter "The Cognitive Status of Theories," argues that the distinction is basically a semantic one. More contemporary discussion of this important philosophical problem can be found in Giere (1988) and Suppe (1989), who embrace realist versions of science, and Van Fraassen (1980), who embraces an instrumentalist or antirealist version of science. This branch of philosophy of science is called ontology.

Modern philosophy of science that gave rise to the "classical approach" or the "received view" began with the start of the twentieth century in Vienna, Austria, with a group of philosophers who came to be known as the "Vienna Circle." Prominent members of the Circle included Mach, Carnap, and Reichenbach. It is beyond the scope of this chapter to provide a detailed review of the developments in philosophy of science during the first half of the twentieth century. What follows is, as the section heading asserts, an overview.

The end of the nineteenth century found scientists engaging in the development of many enormously abstract theoretical positions—for example, evolution, natural selection, thermodynamics, kinetic-molecular theory, existence of the atom, and the physics of atomic particles. What confidence could we have in these new theoretical speculations? The group of scientists and philosophers of science in and around Vienna said very little. They began to take issue with the confidence one could have in theoretical statements built from or based on unobservable data. More exactly, the criticism and discussion focused on whether science could establish a solid foundation on such ephemeral theoretical speculation. Their concerns were both epistemological and ontological. The reaction was to promote a philosophy of science—positivism—that stressed the importance of science proceeding from observable evidence to accurate predictions.

Thus, great care was taken to develop logically consistent rules outlining how theoretical statements can be derived from observational statements. The intent was to create a single set of rules to guide the practice of theory justification. For the philosophers of the Vienna Circle, the objective was to develop one singular form for judging all theoretical statements in science. Crucial elements of this form were empirical observations and logic. The emphasis on empirical observations sought to develop a neutral observational language, while the emphasis on logic was an appeal to the need for the reduction of language to axioms. Carnap (1939) provided a sense of the intention of philosophers at the time:

Any physical theory, and likewise the whole of physics, can ... be presented in the form of an interpreted system, consisting of a specific calculus (axiom systems) and a system of semantical rules for its interpretation. (p. 202)

From the reliance on empiricism and logic come the labels we associate with this type of philosophy, namely logical positivism and logical empiricism.

One important element of logical empiricism, then, is the separation of observations from theories. This appeal for a neutral observational language is known as the observational-theoretical distinction. It was an argument for empirically grounding and preserving the rationality of scientific knowledge claims. We still find the influence of this view of the nature of science in science programs that emphasize that students observe or "discover" concepts without any consideration or understanding of the background knowledge needed for seeing and discovering. See Hanson (1958, Chapter 1) for a discussion of what is wrong with the observational-theoretical distinction. Grandy (1973) is an edited volume of developments on this question. For a discussion of the role of observations in science education, see Duschl (1985a), Norris (1982, 1984, 1985), and Willson (1987). Norris and Willson have opposing perspectives on the nature and role of observation in science classrooms and their point-counterpoint dialogue is informative.

A second important element of logical empiricism is the role of logic. For logical empiricists a theory of science was considered a strong theory if and only if its theoretical statements could be logically justified by observational statements. Recall that the motivation was to establish confidence in the knowledge claims being generated by scientists. From the very beginning, however, the Vienna Circle's attempt to bolster the need for empirical observations, to make theory dependent on observation, and to set a singular logical form for science inquiry was met with resistance. Scientists, in particular, weren't cooperating. The speculative theoretical thought of the 1800s was not just continuing, it was expanding. Theoretical speculation was dictating observation. Recall that the first several decades of the twentieth century were a time when some very speculative theories emerged in science. For example, in physics there were Einstein's theory of relativity and the collective theories of quantum mechanics; in geology there was Alfred Wegener's theory of continental drift; and in biology, the great synthesis of Mendelian genetics with Darwinian natural selection took place.

Indeed, theoretical speculation was alive and well and contributing to the growth of science. Attempts to hold such speculation down were feeble and constant revisions were made to the philosophy of empiricism—revisions that, over time, had to account for the increased role theoretical statements have in the growth of scientific knowledge (Grandy, 1973). From the beginning of the twentieth century to the post–World War II years, continual reconsideration of the role of observation and theory in science resulted in several brands of empiricist philosophy. In chronological order, the major empiricist philosophies were positivism, logical positivism, and hypothetico-deductivism (Losee, 1980). It is the last of these that science teachers would immediately recognize as the standard scientific method:

1. Select a hypothesis.
2. Conduct observations.
3. Collect data.
4. Test hypothesis.
5. Reject or accept hypothesis.

It is a method for testing hypotheses. The hypothetico-deductive method of justifying scientific knowledge claims had one very important characteristic in common with the early twentieth-century philosophies—observations were still considered to be mutually exclusive of theories and logic could be applied to the testing of theories; logic, though, had nothing to do with the discovery of theories or other types of scientific knowledge claims.

The relations between observation and theory and between the testing and discovery of knowledge claims became two of the most important issues in philosophy of science in the 1950s and 1960s. The question of the rationality of science emerged as a serious issue during the 1960s with the release of Kuhn's *Structures*. But the debate between realist perspectives of science and instrumentalist or antirealist perspectives of science was already established (Nagel, 1961). More recent discussions of the topic of rationality can be found in Laudan (1977) and van Fraassen (1980). As philosophers struggled with answers to these and other questions about the growth and form of scientific knowledge and the rationality of scientific knowledge, two new academic disciplines were emerging that would come to have a significant impact on the discipline of philosophy of science. The new disciplines were the history of science and the sociology of science.

The Emergence of History and Sociology of Science

The credit for establishing history of science as a scholarly pursuit in the United States goes to George Sarton. "When George Sarton launched his ISIS in 1912, it remained to be shown that the history of science was a discipline. By the time he relinquished his editorship, some four decades later, the demonstration had been given and the discipline established" (Thackray, 1985, p. 467). It was the collective work of scholars in this discipline that began to influence philosophers of science.

Besides laying the foundation for making history of science a scholarly activity, Sarton firmly established a style of doing history of science. For Sarton, good history of science meant going well beyond merely cataloguing the chronological order of the successes of science. Sarton put in place a set of guidelines for doing history of science that sought to characterize and understand the choices scientists made in the pursuit of scientific explanations and the conditions, social-political or otherwise, under which the choices were made. Kuhn writes, "I was drawn ... to history of science by a totally unanticipated fascination with the reconstruction of old scientific ideas and of the processes by which they were transformed to more recent ones" (Kuhn, 1984, p. 31). The key word is reconstruction, and thus the lines of reasoning not taken or the experiments dropped were as important to the historian as those ultimately selected. As a result, much more information was included in these histories. Following World War II, when the world was settling back into its normal paces, many scholars turned to the study of history of science. The impact these collections of works had on philosophy of science was, in a sense, as revolutionary as the revolutions of science they were reporting.

Sarton also had an impact on the development of sociology of science as a scholarly discipline (Cohen, 1988; Merton, 1985; Shapin, 1988). Robert Merton was a student of Sarton's at Harvard in the 1930s and Merton's dissertation *Science, Technology, and Society in Seventeenth Century England* is considered a classic study on the social forces that affect science (Merton, 1938/1970). From this particular work emerged the "Merton Thesis," which is a theory about the social and cultural relations of science. Steve Shapin's synopsis of the Merton Thesis in the *Dictionary of the History of Science* (Bynum, Browe, & Porter, 1981) states that there are two critical elements of the thesis:

1. The quantitative analysis of papers presented to the Royal Academy of Science of London, demonstrating the emphasis placed on the scientific topics directly or indirectly related to capitalist and military technical requirements, and
2. The argument for positive links between Puritan forms of religion and the institutionalization of science.

Over time, what the collective critical histories and sociologies of science discovered was that the growth of scientific knowledge was not an activity that grew without disruption, upheaval, or alteration to central ideas. Close scrutiny of historical events in science indicated that science was better characterized as a discipline in which dynamic change and alteration were the rule rather than the exception. The view of science as an inductively logical process—a process of moving from empirical fact to the development of scientific theory—was not supported by these historical studies either.

The new methods for writing history of science and the new findings from such analyses provided historians and philosophers of science with evidence that the rigid form of science advocated by logical empiricists and logical positivists just did not exist. Rather, these histories of science reveal that all aspects of science—its standards, meanings of terms, application of methods, and theoretical forms—progress through stages of development (Thackray, 1984). With its national studies; discipline studies; science and religion studies; science, medicine, and technology studies; philosophy, psychology, and sociology of science studies; and great people studies, history of science has had a significant impact on philosophy of science. As Shapere (1984) asserts, these historical and sociological studies provided very new perspectives for philosophers of science. Based on an analysis of the work by historians of science, he claims philosophers of science discovered three things about the nature of science:

1. The standards used to assess the adequacy of scientific theories and explanations can change from one generation of scientists to another.
2. The standard used to judge theories at one time are not better or more correct than the standard used at another time.
3. The standard used to assess scientific explanations are closely linked to the then-current beliefs of the scientific community.

It became increasingly apparent, now more than ever with the aid of history and sociology of science, that what was observed, measured, evaluated, or hypothesized in science was done with strong theoretical commitments. In other words, not only did the observational-theoretical distinction not exist in science, if anything, it seemed that theory determined observation. Norwood Hanson (1958) put it this way: what we see is determined by what we know. This idea about the impact of prior knowledge on what is learned has, of course, carried over to our contemporary view of how people learn science and explains the interest researchers have in studying the role of misconceptions. The emphasis on theoretical commitments and the fact that such commitments change gave rise to a new philosophy of science.

The New Philosophy of Science

In the 1950s, philosophers of science began to take seriously the evidence presented by historians of science. One example of a historical case study of the development of scientific knowledge that had an impact on philosophers of science is Fleck's *Genesis and Development of a Scientific Fact* (1935/1979). Thomas Kuhn writes in the forward to the English translation of the book,

To the best of my recollection, I first read the book during the year 1949 or early in 1950. At that time I was a member of the Harvard Society of Fellows, trying both to prepare myself for the transition from research physics to history of science and also, simultaneously, to explore a revelation that had come to me two or three years before. The revelation was the role played in scientific development by the occasional noncumulative episodes that I have since labeled scientific revolutions. . . . I immediately recognized [from the title of Fleck's book that it] was likely to speak to my concerns. Acquaintance with Fleck's text soon confirmed that intuition. (Fleck, 1979, pp. vii–viii)

Like Kuhn, many other philosophers of science were examining the scholarly papers being produced by historians of science. And, as mentioned earlier, the strict final-form version of scientific theories as purely logical entities was not being supported by these historical studies of the growth of scientific knowledge. Thus, the consideration of historical data about the

growth of scientific knowledge forced philosophers to call into question the "classical approach" or "received view" of the nature of science. Examples of important and influential writings at this time were the works of Toulmin (1953), Wittgenstein (1953), Kuhn (1957), Hanson (1958), and Polyani (1958).

We find in these writers the roots of the position that scientific knowledge is affected by the theoretical perspectives held by the investigator or shared among a community of investigators. Philosophically, this problem is referred to as the observational-theoretical distinction, and the position being developed is that it is not possible to distinguish observational terms in science (e.g., red the color) from theoretical terms in science (e.g., red the wavelength of electromagnetic radiation). Thus, a neutral observational language does not exist. What we see is in a large sense determined by what we know (Hanson, 1958). The HPS evidence clearly demonstrated that the background knowledge and standards of investigators are critical to the inferences made about acceptance of evidence, what counts as an observation, the appropriate design of experiments, and so forth. Most important, the standards themselves can change by the same mechanisms. Shapere (1987) has referred to this restructuring of standards as a learning-how-to-learn-about-nature process among scientists.

The philosophical shift toward considering the influence of theories, paradigms, and social forces raised questions about the rationality of science. That is, to what extent, if any, could it be established that science, as a way of knowing, was indeed an objective and rational way of knowing? Kuhn's thesis that science had periods of turmoil and revolution prompted some philosophers to argue science had no method (e.g., Feyerabend, 1978). Historical accounts of science that provided pictures of scientific inquiry being less than objective and rational led Stephen Brush (1974) to ask, "Should the history of science be X-rated?" These are not trivial issues for science educators, and they throw us headlong into the realist-instrumentalist debate.

Briefly, the realist-instrumentalist controversy concerns the truth of scientific knowledge claims. Are scientific theories approximations of what "truly" exists in the world? Are revisions of models, theories, and explanations progressive in the sense that they are better approximations of the actual structure of nature? Or are scientific theories instruments or inventions that are used by scientists and discarded when they no longer provide an adequate account of nature? In either case, it is clear that the activity of science is best conceived as one in which replacement, substitution, and even outright abandonment of knowledge claims, in lieu of additive growth, are the more accurate description of the growth of scientific knowledge.

It is within this new philosophy of science that we find the roots of science educators' theory of conceptual change teaching (Posner, Strike, Hewson, & Gertzog, 1982), cognitive psychologists' schema theory (Rumelhart, 1980), and educational psychologists' domain-specific knowledge (Glaser, 1984). The challenge being made by philosophers at this time was that science could not be viewed strictly as a cumulation of knowledge claims. Indeed, a great deal of historical evidence suggested that science was a discipline that from time to time found itself in a state of flux about precisely what constituted the core knowledge and core activities of investigation within a field of inquiry.

Nor was the activity of science to any great extent normative. Here again, works from history of science and sociology of science show that there is a great deal of diversity of scientific practice both between and within content domains. Thus, what emerges is a direct challenge to the idea that a common scientific method exists. In spite of the fact that most science textbooks and curricula speak of the existence of a scientific method (state a hypothesis, conduct observations, collect data, test hypothesis), it is generally agreed among philosophers of science that the hypothetico-deductive (H-D) method is *not* an accurate portrayal of the growth of scientific knowledge. Nickles (1991) writes,

The H-D method has come under severe attack in the past twenty years. Three lines of attack on the traditional H-D are: the attacks on the theoretical/observational distinction and related foundational moves that underwrote the traditional H-D method; Glymour's (1980) technical attack on the logic of the H-D method and the . . . Dorling-Nickles defense of generative methodology, harking back to the 17th century. (p. 369)

Glymour's attack is an argument from logic that shows it is possible always to have two separate and contradictory solution paths for the test of a single hypothesis. It is referred to as the "bootstrapping" approach to theory testing. The position of Dorling (1973) and Nickles (1987) is that the H-D method is not an accurate portrayal of events from the history of science. It is more often the case that an artificial story of the personal history of scientific inquiry is constructed. In such a story the individual scientist pieces together the line of argument after the conclusion is in sight.

The dismantling of strict empiricism and positivism left many unanswered problems for philosophers. In his contribution to the 1979 Philosophy of Science Association Critical Problems Conference, Suppe (1979) comments:

Today virtually every significant part of the Positivistic view points has been found wanting and has been rejected by philosophy of science. . . . The dominant thrust of contemporary work in the philosophy of science is the development of new views of science which proceed from, and tend to depend heavily on, close examination of actual practice and products(e.g., theories, explanations) of science. . . . Among the important impacts of this philosophical revolution has been a radical change in what are viewed as the crucial problems in philosophy of science. . . . The situation today . . . is that a new problem—the growth of scientific knowledge, including the dynamics of theorizing—has emerged as the central problem in philosophy of science, and . . . the focus of these problems now has become, roughly, understanding how reduction, explanation, induction and confirmation, theory structure, etc. can, do, or could contribute to the growth of scientific knowledge in actual practice. (Suppe, 1979, pp. 317–319)

What, then, are the essential characteristics of knowledge growth and development in science? On what grounds can we establish the rationality of knowledge claims? Similarly, on what grounds can we assert that one knowledge claim is "better" than another? Contemporary philosophers of science like Laudan, Van Fraassen, Giere, and Suppe, among others, have begun to address these issues and extend the work of Kuhn, Lakatos, and the other contributors to the new philosophy of science. The examination of "the growth of scientific knowledge in actual practice" (Suppe, 1979, p. 19) encouraged the extension of philosophical analysis into domains of science not previously considered—the geological and biological sciences. Out of such analyses, as opposed to analyses in the physical sciences, a very different image of the growth of scientific knowledge emerges.

A Cognitive Twist to the New Philosophy of Science

The emphasis on practice and the quest to preserve and understand the rationality of science has led to the emergence of many alternative epistemologies. In recent years, the dialogue about scientific discovery and the growth of scientific knowledge has moved into the realm of cognitive science. This interfield application has led to a position that there is indeed a structure that can be applied to scientific discovery (Giere, 1988, 1992; Nersessian, 1987, 1989, 1992; Nowak & Thagard, 1992; Thagard, 1988). For a point-counterpoint argument concerning the development of a philosophy of science based on cognitive and cultural dimensions, see the attack by Siegel (1989) and the defense by Giere (1988).

The evidence and format for establishing the case for scientific discovery have increasingly examined the cognitive status of knowledge claims (cf. Laudan, 1984) and the cognitive strategies scientists employ to make discoveries or inventions (cf. Nersessian, 1987, 1991). Hence, what has begun to emerge from this line of philosophical research is the use of interpretive procedures from cognitive science to interpret philosophy of science. One example of this line of inquiry is the research program of David Gooding. Gooding (1990) has carefully examined Michael Faraday's notebooks and developed a notation to map experiment as an active process in a real-world environment and to display the human, cognitive element too often written out of most narrative accounts of scientific discovery. In a similar vein, Nersessian (1987) has carefully studied the development of the electromagnetic field theory from Faraday to Maxwell and argues that specific cognitive tasks are a component of scientific discovery. Thagard (1988, 1992) has advanced a computational theory of philosophy of science—Explanatory Coherence (ECHO)—that employs computer programs to construct models of explanatory coherence to track knowledge growth and development in science.

Another example is Laudan (1984), who argues for a triadic network model to explain scientific knowledge growth and development. In this model commitments to theories, to methods, and to aims can, and do, function independently of one another to bring about change in scientific knowledge. Shapere (1984) maintains that the meanings of concepts and the standards of what counts as an observation or theory have themselves undergone meaning changes as scientists have learned how to learn about nature. His recommendation is that one examine the role of observation in practice rather than from some distant philosophical perspective. From this perspective of practice, then, Shapere posits that over time scientific observations have relied less and less on sense perception observations and more and more on the theories of science, which have altered what counts as an observational event.

Common among these philosophers is the use of historical and contemporary cases of science in practice to base their philosophical positions. What differentiates these philosophers from Kuhn, Lakatos, and other early voices of the new philosophy of science is the extent or degree to which the process of growth in scientific knowledge is detailed. Laudan (1984), for example, draws a sharp distinction between Kuhn's holistic gestalt shift-type brand of scientific change and his own piecemeal reticulated model of scientific change. The examination of practice as it is reported by historians, philosophers, and sociologists of science combines to suggest that the activity of science is a great deal more complicated than the positivists thought. It still involves evidence and observations, but it raises questions about what counts as evidence and what constitutes an observational event. The answers to such issues and questions are increasingly being determined by complex sets of background knowledge and theoretical commitments.

Summary

The educational importance of these developments in history, philosophy, and sociology of science is, of course, the effect it has had on our conceptualization of curriculum design or "what to teach" and of instructional practices or "how to teach." The 40-year period 1950 to 1990 is, in part, a story of transition. The important information to extract from the previous overview is that during the years leading up to the 1950s, philosophical inquiry placed an emphasis on logical arguments and analytical frameworks that sought to portray the justification or testing of scientific knowledge. Brody and Grandy (1989) refer to this brand of philosophy as the "classical approach," while Suppe (1977) labels it the "received view." Both of these texts are recommended for individuals seeking a scholarly historical grounding in twentieth-century philosophy of science. The Brody and Grandy (1989) work is an edited volume of readings in the philosophy of science organized into four parts: theories; explanation and causality; confirmation of scientific hypotheses; selected problems of particular sciences. The readings are organized to reflect the development of philosophical ideas in each of the four parts. Suppe's (1977) is also an edited volume of papers presented at a symposium on the structure of scientific theories. The volume begins with a extensive introduction that many consider to be one of the best analyses of events in the philosophy of science during the first half of the twentieth century. The afterword to the book is also an excellent comment on developments in philosophy of science since the 1950s. For individuals seeking an overview of this period of philosophy, shorter and more readable reviews can be found in Bechtel, (1988), Brown (1977, Part 1), Brush (1988, Chapter 11), Duschl (1990, Chapters 2 & 3), and Losee (1980).

This classical approach or received view of philosophy, known as empiricism, logical empiricism, and logical positivism, has had and in many respects continues to have an effect on the teaching of science during this 40-year period. But since the 1950s philosophers of science have used findings of historical and sociological studies and cognitive research to guide a restructuring of our conception of the growth of scientific knowledge. This new conception recognizes the important role of theory in science and further recognizes that an adequate philosophy of science will be one that explains both the structure and the mechanics of restructuring scientific knowledge claims. Thus, the context of knowledge development has risen in importance as a philosophical problem domain and resides beside the established context of knowledge justification.

The fundamental premises for a leading theory of conceptual change learning advanced by Posner et al. (1982) are taken from Kuhn's (1962) and Lakatos' (1970) epistemological arguments that the growth of scientific knowledge develops through cyclic periods of consensus (Kuhn's normal science episodes) and dissensus (Kuhn's revolutionary science episodes) among practitioners. Since the publication of this important paper, Hewson and Thornley (1989), Pintrich, Marx, and Boyle (1993), and Strike and Posner (1992) have elaborated on the conditions required for effecting conceptual change in learners. Nonetheless, it is the 1982 paper that appears most frequently in the literature and it remains a position of practical importance. Supported by studies in the history of science, the application of this revisionary conceptual change view of science has clear implications for understanding and for teaching science (Anderson & Smith, 1987; Tisher & White, 1985). The problem, however, according to Laudan and other philosophers interested in context of development issues, is not with recognizing that science has times when there is agreement or consensus among participants and times when there is disagreement or dissensus among participants. The more important philosophical problem is establishing an accurate account—a mechanism if you will—of how scientists move from a period of consensus to one of dissensus or vice versa. For this reason, among others, philosophers of science have taken a keen interest in the development and restructuring of scientific knowledge claims.

In the reform of science education in the 1950s and 1960s, scientists sought to devise curriculum standards that would prepare a new generation of scientists. The approach was decidedly "science for scientists." Thus, the curriculum approaches portrayed science in a logical-positivistic way and sustained the idea that a scientific method exists. In practice, the label of science as inquiry applied only to Schwab's stable version. The attempt to move science instruction from a text-based program to an investigation-based program was successful, but at a cost. The cost was making science in the minds of learners a subject that could be mastered by only a few. The focus on the H-D method of inquiry in these programs was challenged early on by educational researchers (Easley, 1959) as to whether the emphasis on teaching students how to operate as scientists was a significant educational objective.

Today we have adopted the call for a "science for all" (Rutherford & Ahlgren, 1990). We have adopted perspectives about science education that ask for a wider distribution of involvement among learners in the curriculum (Shymansky & Kyle, 1991). The authoritarian role of the sciences as a rhetoric of conclusions or final-form presentations of scientific knowledge persists in the minds of teachers and students to this day. In the early 1960s, however, a number of science educators sought to change the positivistic image of science and approach to science education by applying HPS.

The next section examines in more detail the impact of this revolution in HPS on early (1950–1960) and later (1970–1980) reforms in science education. In Kuhnian terms, this 40-year period marks science education's revolutionary period. It is a revolution, though, that has yet to end. Science educators have yet to reach a consensus about "what counts" as the standards and exemplars of educational practice in science classrooms. What is emerging is a clear sense that considerations from HPS can provide bases for describing, analyzing, and guiding the various and sundry processes associated with the growth of scientific knowledge in the teaching and learning of the sciences.

SCIENCE EDUCATION AND HISTORY AND PHILOSOPHY OF SCIENCE IN THE 1950s AND 1960s

The 1950s and 1960s are extremely interesting for science educators to consider. In addition to the dynamic curriculum reform efforts at the time and the changes in HPS just outlined, it was during these two decades that radical developments in the disciplines of cognitive psychology and artificial intelligence began to take place. Although developments in these disciplines were certainly known by scientists who directed or participated in the initial National Science Foundation (NSF) curriculum reform program (e.g., Bruner, 1960), these developments were, for all intents and purposes, effectively ignored in the design and implementation of NSF precollege science programs (Connelly, 1972; Duschl, 1985b, 1990; Holton, 1978; Jones, 1977; Roberts, 1980/1984; Welch, 1979).

In the 1950s a very important paper was being researched and written by Joseph Schwab (1960). It asked a simple but at the same time complex question, "What do scientists' do?" The foundation of the paper was an examination of some 2000 scientific research papers, contemporary as well as historical. One of the contributions is his explanation of the role of inquiry (Schwab's convention was to use the spelling enquiry) in science and in science education. The second contribution is his reflections on the structure of scientific knowledge into syntactical (rules for knowing) and semantic (meanings of knowing) categories (Schwab, 1962a).

Stable and Fluid Enquiry

It is interesting for science educators and students of science education to note that while Thomas Kuhn was developing his ideas about normal and revolutionary periods of science during the 1950s, Joseph Schwab was developing his thesis that scientific investigations employ two distinct modes of inquiry—stable inquiry and fluid inquiry. Readers of Kuhn's *Structures* and Schwabs's "The teaching of science as inquiry" (1962b) will be struck at the similarities that exist between normal science and

stable inquiry and revolutionary science and fluid inquiry. Each scholar, Kuhn working principally with the physical sciences and Schwab working principally with the biological sciences, arrives at the same general set of conclusions about the character of scientific investigations. That is, there are normal or stable times when members of the scientific community are in agreement about the background knowledge and critical problems within their respective domains of inquiry, which, in turn, establishes the important questions and activities within the domain. During normal or stable times, scientific activities turn to the refinement of established knowledge claims.

But the review of historical documents and actual practice of scientists revealed that there are also times when members of a scientific community are in disagreement about the appropriate background knowledge and critical problems that should guide the design of investigations and the evaluation of evidence and knowledge claims. During these periods of flux, the activities among scientists are considered to be revolutionary and fluid.

The teaching of science as inquiry is based on the general proposition that science curricula should reflect the nature of science. It represents an early HPS application to science education. Two implications of this position for classroom practice are:

1. Having the learner acquire knowledge and science experiences through investigations employing procedures similar to what scientists themselves employ, and
2. Making learners aware that the knowledge being acquired from the investigation is subject to change.

In many of his writings, Schwab stresses the need to avoid a "rhetoric of conclusions" in science education programs, inasmuch as the knowledge base of science is one that changes.

However, an examination and analysis of early NSF curriculum project efforts reveals that the majority of activities and investigations are of the type that support a final-form version of scientific knowledge (Duschl, 1988, 1990; Nadeau & Desautels, 1984; Russell, 1981) and thus also support a stable or normal view of scientific inquiry. The application of Schwab's ideas about teaching science as an "inquiry into inquiry" was successful on only one of his two fronts. Research since the 1960s on students' understanding of the tentative nature of science revealed that the students maintain an authoritarian perception of science as the quest for truth and final-form statements about the world (Carey & Strauss, 1968, 1970; Cooley & Klopfer, 1963; Kimball, 1967; Klopfer & Cooley, 1963; Miller, 1963; Rubba, 1977). A summary of 30 years of research on students' and teachers' conceptions of the nature of science can be found in Lederman (1992).

The concept of making science instruction an inquiry into inquiry was (Schwab, 1964) and still is a good idea (Kyle, 1980; Welch, Klopfer, Aikenhead, & Robinson, 1981). The inquiry approach in science firmly established the role of the laboratory and the doing of science by children. The problem is that the meaning of inquiry in the 1950s and 1960s was limited to learning how to test knowledge claims, thereby focusing on stable inquiry.

Science instruction has been and continues to be dominated by the teaching of facts, hypotheses, and theories for the contribution each makes to establishing modern knowledge. Schwab (1964) calls it the "rhetoric of conclusions," Duschl (1990) calls it "final form science." How scientific knowledge came to exist is, as the logical empiricists proposed, treated as a nonissue. The consequence is that an incomplete picture of science is presented to students. Learners are given instructional tasks designed to teach what is known by science. A large part of this instruction involves teaching students processes that justify what we know. Teaching about the "what" without teaching about the "how" runs the risk of making science instruction incomplete. Kilbourn (1982) has coined the term "epistemological flatness" to describe science curriculum materials or instruction strategies that do not give a complete picture of the concepts being taught. Too often, he argues, science instruction is taken out of context and presented without the critical background material necessary to understand the meanings or transitions of science. It is important to consider the historical context. Others, as Ziman (1984) contends, include the political and social contexts in which science and technology functions. The intermingling of science with technological advancements and social conditions cannot be denied (BSCS, 1984; Ziman, 1984). For an interesting examination of how science education might function in a sociocultural context, see Kelly, Carlsen, and Cunningham (1993) and Pintrich et al. (1993).

What is presently missing in our science curricula are instructional units that teach about the other face of science—the how. One example of teaching about the how is Aikenhead's (1991) high school science textbook *Logical Reasoning in Science and Technology*. He applies a number of philosophical frameworks that demonstrate the "how" in the development and evaluation of scientific arguments. Another curriculum effort that is embracing the how is the BSCS Middle School project. Here an emphasis has been placed on the construction and reconstruction of models that explain scientific evidence.

The 13 processes of science adopted for the Science—A Process Approach (S-APA) curriculum directed by the American Association for the Advancement of Science and still prevalent today in science curricula are representative of the final form brand of science presented in most curricula.

Basic Processes	Integrated Processes
Observing	Controlling variables
Using space/time relationships	Interpreting data
Classifying	Formulating hypotheses
Using numbers	Defining operationally
Measuring	Experimenting
Communicating	
Predicting	
Inferring	

Originally designed for implementation in the elementary grades, the processes of science are now typically partitioned between basic processes (grades K–5) and integrated processes (6–9). The process approach is a carryover of the view of science defined by empiricism, in which science begins with observations and then proceeds logically to the formulation of scientific theories. It is a view of science provided by practitioners (e.g., physicists) working with neat problems that can be precisely measured and with hypotheses that could be tested with true controlled experiments. Missing from this perspective is explicit attention to several important processes of science, such as modeling, explaining, and evaluating. Although the original design team for S-APA intended to include models as a separate process and subsume explaining and evaluating into other processes, the result has been exclusion of these important scientific activities in most K–8 science curricula.

The significance of models, explanation, and evaluation is that each represents a product of scientific inquiry determined by the theoretical commitments an individual or a community of individuals—in this case scientists—adopts in constructing a personal world view. That is, what counts as an explanation for you, or one scientific community, may not be satisfactory to me, or another scientific community, and vice versa. What make an explanation unacceptable or what make it appear to be correct are the criteria we employ in our investigation and evaluation processes. The criteria scientists employ determine the standards for measurement, for selecting important research questions, for designing an experiment, for accepting the outcomes of an investigation. The adoption of criteria plays an important role in the growth of science. And, as Kuhn and Schwab point out, an important aspect of science is the process through which the criteria for "what counts" change. This latter fluid inquiry focus was missing from the majority of the 1960s science education reform efforts. Finley (1983) presents an excellent discussion on the reinterpretation of scientific processes in light of the "new philosophy of science."

As I stated earlier, one can take the position that the bulk of activity of the early NSF, and in Britain the Nuffield, curriculum efforts to reform science education was of the kind that focused on understanding the stable inquiry aspects of science. Science as it is understood by scientists who are practicing the discipline was the logical extension of Schwab's theses that science education focus on what scientists do. But the representation of work by individual scientists taken from, say, an internalist perspective is quite different from a representation of a community of scientists taken from, say, an externalist perspective. Historical studies of science are typically either internalist or externalist reports. The internal view of scientific developments presents scientific knowledge growth and development within a framework that is independent of social, political, religious, economic, or other forces external to the discipline. The perspective examines restructuring of knowledge on the basis of conceptual frameworks, methods, and evidence for or against a theory. The external view of scientific developments, on the other hand, embraces such social forces as viable elements to consider in a reconstruction of the growth of scientific knowledge. See Shapin (1991) and Brush (1991) for reviews of the debate that exists among historians of science concerning the internalist-externalist writing of the history of science. See Nickels (1991) for a philosopher's perspective on the role of history of science in philosophy of science. The "new philosophy of science" has sought to provide epistemological accounts of knowledge growth and development that embrace both the internal and external elements of science.

Early HPS Curriculum Efforts

Unfortunately for science education, conclusions and views about the new philosophy of science were unavailable to science educators and scientists preparing new curriculum materials for precollege science programs in the 1950s. Similarly, neither were the new views emerging from philosophy of science fully considered by individuals working on NSF science curriculum projects. The result is that at the same time that science education was adopting new perspectives for teaching science as an inquiry into inquiry, the basic definitions of inquiry were being altered.

There exist, however, examples of curriculum efforts and theoretical frameworks that sought to embrace the fluid inquiry ideas suggested by historians and philosophers of science. Schwab's Biological Sciences Curriculum Study (BSCS) *Biology Teacher's Handbook* (1965) with its Invitations to Inquiry is a well-known example of a science curriculum effort that seeks to have students understand both stable inquiry and fluid inquiry. Klopfer's (1964–66) adaptation and extension of the *Harvard Case Studies in Experimental Science* (Conant, 1957) for use in high school science programs is another example of an effort that asks students to explore the forces involved in the fluid nature of scientific inquiry. Nine *History of Science Cases* were written by Klopfer, with each unit consisting of a case booklet for students and a teachers' guide containing extensive commentary on science content and on the ideas about science and scientists developed in the case. Also included was a unit test designed to assess the extent of student's mastery of the science content and ideas about the scientific enterprise. The test is an extension and application of the work Klopfer has done in the design and validation of the Test on Understanding Science instrument for assessing students' understanding of the nature of science (Cooley & Klopfer, 1963; Klopfer & Cooley, 1963).

Among the NSF curriculum efforts, Schwab's Invitations to Inquiry and the Harvard Project Physics (Rutherford, Holton, & Watson, 1970) were the only ones to consider seriously the application of history of science to high school science. It is interesting and important to note that these curriculum efforts by Schwab, Klopfer, Rutherford, Watson, and Holton were but a part of the larger activities taking place at their respective campuses, University of Chicago and Harvard University. For example, Harvard Project Physics grew out of the larger history of science program that existed at the time, which included Thomas Kuhn as one of its members. Other infusions of history of science into science education at Harvard were by Holton and Brush (1952) and Cohen and Watson (1952). Others (Arons, 1965; Kemble, 1966) were also engaged in the application of history of science to science education, but nothing approached what was taking place in Cambridge, MA.

A benchmark for examining the inclusion of history and philosophy of science and science education during the 1960s is a National Association for Research in Science Teaching symposium in Chicago in 1968. Three papers, presented by Leopold Klopfer, James T. Robinson, and F. Michael Connelly, were accompanied by insightful commentary by Arnold Lahti, Marshall D. Herron, and J.W. George Ivany, respectively. The full papers and commentaries are published together in Volume 6, Issue 1 of the *Journal of Research in Science Teaching*. One clear message at this point is that the issue of how philosophy of science and history of science related to matters of science education was alive. Another message is that the application of HPS to science education was focusing primarily on the design of curriculum and not the design of instructional practices.

Klopfer's work has already been outlined. His symposium paper takes the position via example, Humphry Davy's visit to France, that an examination of the development of knowledge in science—fluid inquiry—involves, when analyzed from a historical view, forces external to the enterprise of science. Recall the earlier discussion of the external-internal debate about doing the history of science. Thus, for Klopfer there is certainly much more to scientific knowledge than "knowing" the facts. An examination of the items on his "Test On Understanding Science" (Klopfer & Cooley, 1963) makes this claim apparent.

James T. Robinson is the author of the then and now influential book *The Nature of Science and Science Teaching* (1968), which, like his paper, advocated an infusion of philosophy of science into science teacher education and science education practice. His principal thesis is that the view of science held by a teacher affects the discourse and dialogue that occur in the classroom. More exactly, employing Schwab's (1962b) labels, he distinguishes between a classroom language of science that presents science as a "rhetoric of conclusions" rather than as a "narrative of enquiry." Thus, Robinson was committed to teachers having a thorough introduction to the new philosophy of science that was emphasizing the fluid nature or revolutionary nature of scientific knowledge. But he did little to carve out a precise path that science teacher educators or science teachers could follow in order to implement his innovative ideas.

The third presenter at the symposium was F. Michael Connelly, who has as much as anyone else advanced curriculum recommendations, along with some instructional recommendations for the fluid-inquiry representation of science. He was a student of Schwab and, not surprisingly, extended his mentor's ideas. Like Schwab Klopfer, Connelly was more directly concerned with matters of curriculum. By this I mean that the practical application of concepts from philosophy of science and from history of science had more to do with what was being taught than with how teaching took place. The avoidance of a rhetoric of conclusions had more to do with what was selected as the topic of instruction (e.g., Invitations to Inquiry or History of Science Cases) than with applying an epistemically sound framework to guide instructional strategies by teachers or learning strategies by students. The emphasis on matters of curriculum by Connelly is seen in his review of Robinson's book (Connelly, 1969), and it is also developed in the Patterns of Enquiry Project. But the Patterns of Enquiry Project represents a turning point in the application of HPS in science instruction as well as science curriculum.

Summary

The relations between observation and theory and between the testing and discovery of knowledge claims became two of the most important issues in philosophy of science in the 1950s and 1960s. The new methods for writing history of science and the new findings from such analyses provided historians and philosophers of science with evidence that the rigid form of science advocated by logical empiricists and logical positivists did not exist. Rather, these histories of science reveal that all aspects of science—its standards, meanings of terms, application of methods, and theoretical forms—progress through stages of development.

With its national studies; discipline studies; science and religion studies; science, medicine, and technology studies; philosophy, psychology, and sociology of science studies; and great-people studies, history of science has had a significant impact on philosophy of science. These historical studies provided very new perspectives for philosophers of science. It became increasingly apparent, now more than ever with the aid of history and sociology of science, that what was observed, measured, evaluated, or hypothesized in science was done so with strong theoretical commitments. In other words, not only did the observational-theoretical distinction not exist in science, if anything it seemed that theory determined observation. Norwood Hanson (1958) put it this way: what we see is determined by what we know. This same general opinion has carried over into our views of how children learn science and the importance of addressing misconceptions and faulty prior knowledge; what is being learned is affected by the existing knowledge base of the learner.

Unfortunately for science education, conclusions and views such as these were unavailable to science educators and scientists preparing new curriculum materials for precollege science programs in the 1950s. Similarly, neither were the new views emerging from philosophy of science fully considered by individuals working on NSF science curriculum projects. The result is that while science education was adopting new perspectives for teaching science as an inquiry into inquiry, the basic definitions of inquiry were being altered. HPS efforts in the 1960s did not significantly affect science education practice, but they did provide a sound foundation for subsequent work in HPS and science teaching.

HISTORY AND PHILOSOPHY OF SCIENCE AND THE EPISTEMIC TURN—HPS AND SCIENCE EDUCATION IN 1970s and 1980s

The Patterns of Enquiry Project (Connelly, Finegold, Clipsham, & Wahlstrom, 1977) is an important pivotal contribution to science education theory and practice and to the application of HPS to science education. The Project efforts expanded Schwab's ideas of teaching science as inquiry by examining the role or meaning of inquiry with regard to three factors in an instructional setting—subject matter, the student, the teacher.

Thus, the project team developed positions about inquiry as logic, inquiry as mode of learning, and inquiry as instructional methodology.

The Project was but part of the larger efforts taking place at the Ontario Institute for Studies in Education (OISE) and the University of Toronto under the leadership of Connelly and colleagues, notably Douglas Roberts. Concerted efforts were being made to determine precisely how elements of philosophy could be used to guide the analysis of educational practice. Roberts and Russell (1975) present a summary report of this line of research and an overview of the methodological argument for the procedures used by researchers to apply philosophical analysis to the examination of practice. Whereas Connelly's work is an extension of Schwab's thinking, the application of philosophical methods to educational thought and argument is an extension of the work of Israel Scheffler (1960). Noteworthy examples of such extensional efforts include Kilbourn's (1974) application of Pepper's world views to analyze curriculum materials, Munby's (1973) application of Scheffler's conditions of knowledge to analyze the intellectual conditions of teaching, Russell's (1976) application of Toulmin's argument patterns to the analysis of teachers' views about the nature of science and nature of teaching, and Robert's (1982) critique of science education research methodology. Developed versions of this collection of work can be found in Munby, Orpwood, and Russell (1984).

There existed at OISE, then, a major effort to apply HPS and philosophy in general to educational matters. To demonstrate this, let's return to the discussion of the Patterns of Enquiry Project. A core element of the curriculum model—Patterns of Enquiry—is to articulate a learning situation that "demonstrates to students the degree of legitimate doubt ordinarily attached to scientific knowledge claims" (Connelly et al., 1977, p. 6). The instructional goal is to provide students with the inquiry skills, most important content, and habits of mind "so as to be able to assess the status of knowledge claims" (p. 18). Central to this goal is teaching the students how to recognize the "guiding conceptions" of the scientific research. Connelly et al. (1977) refer to the guiding conceptions of a research project or report as the set of assumptions scientists make about the appropriate background knowledge to consider. In this way, it frames the way scientists view the world and thus influences all the other steps of inquiry.

The language is subtly different but the intent in the Patterns of Enquiry is similar to that of philosophers who sought to dismantle the observational-theoretical distinction. It is an attempt to answer the question of what counts. The Patterns of Enquiry Project is a progressive application of new philosophy of science to the design of curriculum. The Project efforts also sought to affect the nature of instruction in the classroom. Through the transcription and analysis of classroom dialogue, the Project staff were able to make recommendations for how teachers ought to guide students in the academic work of the classroom. One specific application is the Patterns of Enquiry question-asking strategy, which involves three levels of questions: structural questions, the parts of the enquiry; functional questions, the relationships among the parts; and evaluative questions, the adequacy of the knowledge claims made in the enquiry.

The Patterns of Enquiry Project effort to affect the instructional environment as well as curriculum materials represents a significant turn of events in the application of HPS to science education and science teaching. We find here the beginnings of specific recommendations about how to monitor and effect changes in the learning environment. The turn to philosophy became a research program for a group of science education researchers associated directly or indirectly with this Project working out of the Ontario Institute for Studies in Education, Toronto.

Another effort to merge HPS and science education was that of Michael Martin (1972/1986). Following in the tradition of James Robinson, he sought "to bridge the gulf that now exists between philosophy of science and science education" (p. vii). His text is an important contribution because it argues via examples how the new philosophical conceptions about observation, explanation, and definition can and should alter practice and the goals of science education. Thus, Martin (1972/1986) writes, "(o)ne of the major theses of this book is that the study of philosophy of science can help science educators in their thinking about science education and in their educational practice" (p. 5).

The analysis of contemporary philosophy of science for science education practiced by Martin (1971, 1972/1986) and the combined Canadian efforts of applying contemporary philosophical analysis to study educational practice (Roberts & Russell, 1975; Munby, Orpwood, & Russell, 1984) demonstrate the lack of scholarly attention being given to foundational aspects of science education. The number of individuals participating in this line of research within the community of science education was small. At the time, during the 1970s and early 1980s, this line of research was also out of the mainstream for science education research. The importance of this work, then and now, is that it (1) focuses educational researchers' attention on the new philosophy of science and (2) begins the application of HPS to matters of instructional practice. The next two sections address these two developments.

The New Philosophy of Science—Critical Developments in the 1970s and 1980s

The work in history and philosophy of science and science education suffered from the lack of analysis by philosophers and by historians. A small number of individuals were engaged in the enterprise, but too often it was not enough to influence practice, in the sense of becoming a dominant mode for formatting instruction. Harvard Project Physics has always had a secondary place in physics behind problem-solving math-oriented programs. Yet to be clearly established were the benefits that could be incurred from an application of the history and philosophy of science (Russell, 1981). As long as the arguments for its inclusion resided in a more accurate view of the nature of science, a view that was rapidly changing and not shared by practicing scientists themselves, it should come as no surprise

that HPS was perceived as having little more to offer to science education than providing anecdotes.

Philosophers themselves are partly to blame because they were proceeding through a "Kuhnian revolution" over the demise of the observational-theoretical distinction and hypothetico-deductive models of science. There was, and still is, little agreement about the nature of theory restructuring in science. Theory generation and restructuring are generally believed to be a central issue for philosophers of science. However, recall the earlier discussion about realist-instrumentalist debates among philosophers and externalist-internalist debates among historians. There is, as Duschl, Hamilton, and Grandy (1990) assert, little consensus among historians and philosophers of science about what constitutes the appropriate image of scientific inquiry and the growth of scientific knowledge.

The dismantling of the received view and the positivistic traditions in philosophy of science involved a much larger cast than simply Thomas Kuhn. Although his role is critical because it began to turn philosophers' attention to issues surrounding the growth of scientific knowledge, his was one of many voices that have very different opinions. During the 1970s we find in the philosophy of science literature a number of efforts to articulate the dynamics of the growth of knowledge in practice. Prominent among them are Lakatos'(1970) idea of the role of research programs, Laudan's (1977) research traditions and the role of rationality in science, and Holton's (1978) thematic analysis of science.

In brief, Laudan (1977) extends Lakatos (1970), who extends Kuhn (1962), and all owe a measure of insight to Fleck's (1979) *Genesis and Development of a Scientific Fact*. The emerging issue in HPS is the rational reconstruction of scientific knowledge. In other words, the recent effort among philosophers has been to articulate in detail a context of discovery for scientific knowledge, with scientific knowledge being identified primarily as theories and explanations. (See the earlier Cognitive Twist subsection).

The point is that perspectives about and details for describing and understanding the nature of science have progressed far past Kuhn. Kuhn's work was influential and vital but it was so steep in generalities that it threatened the role of rationality in science. Kuhn did indeed alter the landscape of philosophy of science, but it was others who provided the detailed maps of that landscape. Yet, from the proceedings from the first and second international conferences (Hegbert, 1990a, 1990b; Hills, 1992), one could get the impression that the new philosophy of science began and ended with Kuhn. In these volumes, few applications of HPS to science education focus on the contributions of contemporary philosophers, or for that matter of contemporary psychologists. Many of the justifications for adopting a view about the nature of science as being developmental rest on the arguments of Kuhn and of Piaget. As important and influential as these two scholars are, they each neglected to provide the details.

It has become an agenda of philosophers over the past three decades to articulate precisely what is meant by theory-ladenness and to describe and explain the structure and restructuring processes of scientific theories (e.g., Giere, 1989; Salmon, 1984; Shapere, 1984; Suppe, 1989; van Fraassen, 1980).

The search for details about the growth of scientific knowledge led to two principal lines of philosophical inquiry during the 1970s and 1980s. One is the Semantic Conception of Theories (SCT) and the other is Naturalized Philosophy (NP). Proponents of the semantic conception of theories include Stegmuller (1976), Suppe (1989), Suppes (1967), and Van Fraassen (1980). Proponents of the naturalized philosophy include Giere (1985, 1988), Laudan (1984), Quine (1969), and Shapere (1984).

Some important distinctions between the two camps are that STC maintains that theories "consist of mathematical structures . . . which are propounded as standing in some representational relationship to actual and physically possible phenomena" (Suppe, 1979, p. 320). It follows that the STC position is that the essential features of theories are nonlinguistic claims. NP embraces the idea that theories are linguistic in nature and therefore are best represented as cognitive devices. The NP school is "concerned with investigating how theories are employed and function in actual scientific contexts (historical and contemporary)" (Suppe, 1979, p. 320). In other words, for NP the meanings of terms and mechanisms that bring about changes or evolution of the meanings of terms are considered central to establishing a philosophy that explains the growth of scientific knowledge. Giere (1985) provides a good overview of the dynamics of NP. Another important distinguishing feature of NP is the appeal to theories in the disciplines of biology and geology. Attempts to document and explain the growth of knowledge in these scientific fields have prompted a broader consideration of the important theoretical structures employed by scientists. Given the emphasis on the meanings of scientific terms and the conception of theories as cognitive devices, it should come as no surprise that the principal application of HPS to contemporary educational research and practice has developed from NP positions.

Cognitive science and philosophy of science

When we consider the educational arguments made by Schwab some 30 years ago, that science education needs to be about teaching both faces or characterizations of the nature of science (e.g., science as a process of justifying what we know, or stable enquiry, and science as a process of discovering new knowledge or how we've come to know, or fluid enquiry), we begin to understand how the application of cognitive science to philosophical knowledge and of philosophical knowledge to educational thought and practice can be helpful. In both cognitive science and philosophy of science, the turn has been toward understanding in much more detail the dynamics of knowledge growth and development. The examination and development of the cognitive and the epistemic nature of knowledge growth have begun to provide a new level of resolution about fluid inquiry. The similarity of the two domains with regard to the emphasis each has placed on understanding not only the structure of knowledge but also the conditions that bring about a restructuring of knowledge has led Derek Hodson (1988) to claim that a type of harmony finally exists between the psychological and epistemological foundations of science education.

The application of knowledge restructuring and conceptual change processes to the teaching and learning of science has become the focus of HPS and science teaching as we move toward the twenty-first century. In addressing the topic of research on conceptual change learning in science, Susan Carey succinctly summarizes the symbiotic relationship among the relevant disciplines:

If my diagnosis of the problem that developmental psychologists face is correct, then at least we know what we are up against—the fundamental problems of induction, epistemology, and philosophy of science. We ignore the work in these fields at our peril. (Carey, 1985b, p. 514)

and,

I hope to provide a feel for the complexity of the issues, to show that progress is being made, and to suggest that success will require the collaboration of cognitive scientists and science educators, who together must be aware of the understanding of science provided by both historians and philosophers of science. (Carey, 1986, p. 1125)

It was argued earlier that the extant perspectives on the application of philosophy of science to education by researchers interested in studying students' epistemologies are informed principally by early philosophical models of theory restructuring. Specifically, I am referring to the models of knowledge growth proposed by Thomas Kuhn (1962) and Imre Lakatos (1970) and summarily adopted by education researchers (Anderson & Smith, 1987; Cobb, Wood, & Yackel, 1991; Linn, 1986; Posner et al., 1982) and cognitive scientists (Carey, 1986; Carey, et al., 1989; Chi, 1992; Kuhn, Amshel, & O'Loughlin, 1988; Vosniadou & Brewer, 1987).

One consequence of the application of the epistemologies of Kuhn and Lakatos to science learning research is the categorization of the mechanism of knowledge change into two fundamental types—weak restructuring and radical restructuring (e.g., Carey, 1985; Chi. 1992; Vosniadou & Brewer, 1987). The weak-restructuring idea is taken from Kuhn's idea of alterations during "normal science" that function within the same exemplars and Lakatos' idea of changes to the soft core of a research program. Here the schematic framework of the learner is acceptable. Strong restructuring, on the other hand, is akin to Kuhn's "revolutionary science," in which new exemplars are adopted, and to Lakatos' idea of the abandonment of a research program's hard core of theoretical commitments. Here the schematic framework of the learner is unacceptable and must be replaced.

I want to point out that exemplars is used in place of paradigms in order to distance the debate from Kuhn's critics that science does not function at two distinct levels of activities. The exemplars represent the shared examples among scientists and one aspect of the disciplinary matrix Kuhn argues scientists employ in the pursuit of knowledge. The replacement of exemplar for paradigm is a move that Kuhn makes in his postscript

(p. 187) to the second edition of *The Structure of Scientific Revolutions*.

Posner et al. (1982) argue that this philosophical view of the growth of scientific knowledge can be used to develop an epistemological framework for the science classroom. Their position suggests that if learners are to change their commitments to scientific ideas or schematic frameworks, then four conditions must be met:

1. Existing ideas must be found to be unsatisfactory.
2. The new idea must be intelligible; it must appear to be both coherent and internally consistent.
3. The new idea is plausible.
4. The new idea is fruitful; it is preferable to the old viewpoint on the grounds of perceived elegance, parsimony, and/or usefulness.

Scientific thinking is grounded in the particulars of a domain (Glaser, 1984). Thus, declarative or domain-specific knowledge related to principles, laws, theories and generalizations of science must be taught, alongside the procedural or generic strategic knowledge of the domain. Within the context of normal scientific developments or weak restructurings, there is a small amount of procedural knowledge to be acquired about the fine-tuning of theories and the adjustment of conceptual relations. But if we are to produce radical restructuring of concepts, the personal correlate of Kuhn's revolutionary science, it seems that we must also teach the procedural knowledge involved in the evaluation of theory, evidence, observation, and data. In short, we must be prepared to engage in teaching that fosters changes in learners' epistemologies as Carey et al. (1989) and Strike and Posner (1992) suggest. The reform recommendations suggesting that the science curriculum include topics that foster meaning making among students have yet to explore fully the specific procedures for implementing a curriculum of this type. Newer epistemological models that articulate details about theory development must be sought to serve as a guide.

A review of the cognitive science literature reveals that many researchers and scholars have found it prudent to join psychological concepts focusing on schema theory with philosophical concepts addressing theory development (Carey, 1985a, 1985b; Chi, 1992; Giere, 1992; Rissland, 1985; Vosniadou & Brewer, 1987). The process of theory development by scientists has often been compared to the development and acquisition of an individual's knowledge of the world (Kitchener, 1986, 1987, 1992; Krupa, Selman, & Jaquette, 1985; Piaget, 1967). It is important to remember that it is with respect to changes in theory that conceptual change is taken from philosophy of science and applied to education and to learning in science. The framework in which these "common" terms—schemata, schematic frameworks, theory, conceptual change—are to be addressed concerns the activities associated with the growth of knowledge or knowledge restructuring; that is, weak and radical restructuring of learners' schematic frameworks. Thus, Carey's (1986) request for science educators to consider the contributions that principles and concepts from the history and philosophy of science can make to researchers attempting to interpret and understand the conceptual development of young learners is welcomed.

Making the focus of science education, then, a study of the various dynamics involved in the growth of scientific knowledge, development, and justification was a theme among the papers presented at the First International Conference on History and Philosophy of Science and Science Teaching. For example, Winchester (1990) addresses the role of thought experiments in conceptual revision; Stewart and Hafner (1991) apply epistemic criteria to extend the conception of problem in problem-solving research and teaching; Siegel (1989) examines linkages between critical thinking and epistemic claims about the rationality of science; Nersessian (1989) demonstrates how cognitive strategies mined from the history of science can be applied to conceptual change teaching; Benson (1989), Collins (1990), Cobb et al. (1991), King (1991), and Gallagher (1991) apply philosophical tenets to examine teacher practice; and Abell (1989) analyzes how elementary science textbooks portray the new philosophy of science. Applications of HPS to the development of instructional models include Wandersee's (1990) use of historical vignettes; Hodson's (1990) model for curriculum planning; Shahn's (1990) description of a foundations of science course for nonmajors; Nussbaum's (1989) analysis of conceptual-change classrooms; and Arons' (1988), Cushing's (1989), Stinner's (1990), and Kenealy's (1989) recommendations for modifying introductory physics education. A full listing of the papers with references for this conference is presented in Table 15.1.

Summary

The general direction of HPS and science education is one that asks us to look at the growth of science more closely. It asks that we employ history of science in ways that examine the paths taken and not taken in the pursuit of scientific knowledge. It asks that we consider epistemic issues in the structure of scientific theories and processes and dynamics that invoke a restructuring of the same scientific theories. It is clear that the application of HPS to science education is in concert with the position that science education needs to be concerned with students' depth of knowledge about science. Given this consensus that the learning and doing of science demand knowledge restructuring and the pedagogical position that knowledge restructuring requires the development of a disposition in the learner to examine the status of knowledge claims, both personal and scientific, then an important set of research questions and issues in science education are those that seek to understand the personal epistemologies that students employ in the construction and the reconstruction of scientific knowledge.

Whereas applications of HPS to science education in the 1960s and 1970s sought ways to restructure the curriculum materials of what was taught, the current applications of HPS to science education are seeking ways to restructure instructional practice and teacher decision making.

TABLE 15.1. Publications From the First International Conference on HPS and Science
Teaching Held at Florida State University, November 6–10, 1989

Author	Title	Key Words	Pages
	Synthese, 80(1), 1989		
Siegel, H.	The rationality of science, critical thinking, and science education	Rationality of science; aims of education	9–42
Silverman, M. P.	Two sides of wonder: philosophical keys to the motivation of science learning	Context of science learning; student motivation	43–62
Jordan, T.	Themes and schemes: a philosophical approach to interdisciplinary science teaching	Integrated science; thematic basis for science courses	63–80
Eger, M.	The "interests" of science and the problems of education	Influence of PS on educative processes	81–106
Benson, G. D.	The misrepresentation of science by philosophers and teachers of science	Links between interpretations	107–120
von Glasersfeld, E.	Cognition, construction of knowledge, and teaching	Adaptive function of cognition; Piaget's scheme theory; communication processes; social interaction	121–140
Rowell, J. A.	Piagetian epistemology: equilibrations and the teaching of science	Piagetian equilibration and application to science teaching	141–162
Nersessian, N. J.	Conceptual change in science and science education	Conceptual change: nature of, reasoning in, need for cognitive model of	163–184
Matthews, M. R.	HPS and science teaching: a bibliography	Bibliography of HPS works	185–195
	Educational Philosophy and Theory, 20(2), 1988		
Rohrlich, F.	Four philosophical issues essential for good science teaching	Discovery/invention distinction; belief justification, metaphysics	1–6
Newton, D. P.	Relevance and science education	Ambiguity of idea of relevance	7–12
Arons, A. B.	Historical and philosophical perspectives attainable in introductory physics courses	Examples of integrating HPS into physics	13–23
Sloep, P. B., & van der Steen, W. J.	A natural alliance of teaching and philosophy of science	Philosophy in/of biology	24–32
Birch, C.	Whitehead and science education	Process thought; Whitehead	33–41
Shahn, E.	On science literacy	Scientific literacy: definition, relation to cognitive development, role of language, past approaches	42–52
Hodson, D.	Experiments in science and in science teaching	Role, nature, pitfalls of experimentation; scientists vs. science educators	53–66
Matthews, M. R.	A role for history and philosophy in science teaching	Curriculum, learning psychology, and science teaching	67–81
	Interchange, 20(2), 1989		
Matthews, M. R.	A role for history and philosophy in science teaching	Curriculum goals; integrating HPS into curriculum	3–15
Stinner, A.	Science, humanities, and society—the Snow-Leavis controversy	Scientific/humanistic confrontation	16–23
Selley, N. J.	The philosophy of school science	Philosophical assumptions in textbooks	24–32
Jacoby, B. A., & Spargo, P. E.	Ptolemy revived? The existence of a mild instrumentalism in some selected British, American and South African high school physical science textbooks	Instrumentalism, realism	33–53
Cushing, J. T.	A tough act—history, philosophy, and introductory physics (an American perspective)	HPS and goals of introductory science courses	54–59
Brush, S. G.	History of science and science education	Curriculum; integration of HS with technical content	60–70
Brackenridge, J. B.	Education in science, history of science, and the textbook—necessary vs. sufficient conditions	Role of texts in school science; humanities/science distinction	71–80
Kauffman, G. B.	History in the chemistry curriculum	Advantages/disadvantages of HPS in undergraduate chemistry	81–94
Tamir, P.	History and philosophy of science and biological education in Israel	HPS in secondary school biology	95–98
Matthews, M. R.	History, philosophy, and science teaching—a bibliography	Bibliography of HPS works	99–111
Dublin, M.	Joseph J. Schwab—a memoir and a tribute	Biographical, historical, anecdotal account	112–115

TABLE 15.1. *Continued*

Author	Title	Key Words	Pages
Studies in Philosophy and Education, 10(1), 1990			
Pitt, J. C.	The myth of science education	Integrated science teaching; images of science	7–18
Garrison, J. W., & Bentley, M. L.	Science education, conceptual change, and breaking with everyday experience	Conceptual development, cognitive psychology, many worlds thesis	19–36
Otte, M.	Arithmetics and geometry: some remarks on the concept of complementarity	Complementarity, intuitionism, axiomatic proof	37–62
Ginev, D.	Towards a new image of science: science teaching and nonanalytical philosophy of science	Didactical approach	63–72
Winchester, I.	Thought experiments and conceptual revision	Thought experiments, rationalism, empiricism, a priori reasoning	73–80
Ruse, M.	Making use of creationism: a case study for the philosophy of science classroom	Creationism as a case study	81–92
Matthews, M. R.	History, philosophy, and science teaching: what can be done in an undergraduate course?	HPS in science education courses	93–98
International Journal of Science Education, 24(1), 1990			
Duschl, R. A., Hamilton, R., & Grandy, R.	Psychology and epistemology: match or mismatch when applied to science education?	Development of science, conceptual change/restructuring	230–243
Stinner, A.	Philosophy, thought experiments and large context problems in the secondary school physics course	Contexts of inquiry, large context problems	244–257
Settle, T.	How to avoid implying that physicalism is true: A problem for teachers of science	Physicalism; metaphysical schemes	258–264
Steinberg, M. S., Brown, D. E., & Clement, J.	Genius is not immune to persistent misconceptions: conceptual difficulties impeding Isaac Newton and contemporary physics students	Misconceptions, qualitative conceptions, conceptual struggles	265–273
Jenkins, E. W.	History of science in British schools: retrospect and prospect	British national curriculum; role of laboratory work, experimentation	274–281
Krasilchik, M.	The scientists: an experiment in science teaching	Curriculum about scientists and their experiments	282–287
Götschl, J.	Philosophical and scientific conceptions of nature and the place of responsibility	Placement of responsibility in philosophical and scientific conceptions	288–296
Johnson, S. K., & Stewart, J.	Using philosophy of science in curriculum development: an example from high school genetics	Student construction of conceptual models	297–307
Nielsen, H., & Thomsen, P. V.	History and philosophy of science in physics education	Curriculum implementation; HPS in secondary school physics	308–316
Matthews, M. R.	Ernst Mach and contemporary science education reforms	Contributions to comtemporary science education	317–325
Science Education, 75(1), 1990			
Wheatley, G.	Constructive perspectives on mathematics and science learning	Constructivism, science and mathematics, language	9–22
Cobb, P. et al.	Analogies from philosophy and sociology of science for understanding classroom life	Epistemological development of mathematics education	23–44
Sequeira, M., & Leite, L.	Alternative conceptions and history of science in physics teacher education	Student conceptions of mechanics, conceptual change, physics teacher education	45–56
Tamir, P., & Zohar, A.	Anthropomorphism and teleology in reasoning about biological phenomena	Anthropomorphism, teleology, biological phenomena	57–68
Kinnear, J. F.	Using historical perspectives to enrich the teaching of linkage in genetics	Linkage in genetics	69–86
Ray, C.	Breaking free from dogma: philosophical prejudices in science education	Role of diversity and debate in determining philosophy of science education	87–94
Solomon, J.	Teaching about the nature of science in the British National Curriculum	Nature of science; British national curriculum	95–104
Steward, J., & Hafner, B.	Extending the conception of "problem" in problem-solving research	Context of discovery, context of justification	105–120
Gallagher, J.	Prospective and practising secondary school science teachers' knowledge and beliefs	Textbooks, teacher education, teacher knowledge and beliefs	121–134
King, B.	Beginning teachers' knowledge of and attitudes toward issues in the history	Research, beginning teachers, knowledge and attitudes	135–142

TABLE 15.1. *Continued*

Author	Title	Key Words	Pages
Bybee, R. W., Powell, J. C., Ellis, J. D., Giese, J. R., Parisi, L., & Singleton, L.	Integrating the history and nature of science and technology in science and social studies curriculum	Integration of science and social studies	143–156

The History and Philosophy of Science in Science Teaching. (1989) D. E. Herget (Ed.), Tallahassee, FL: Florida State University

Author	Title	Key Words	Pages
Abell, S. K.	The nature of science as portrayed to preservice elementary teachers via methods textbooks	Teacher education; scientific discourse	1–14
Akeroyd, F. M.	Philosophy of science in a national curriculum	Teacher education; science education policy	15–22
Bakker, G. R., & Clark, L.	The concept of explanation: teaching the philosophy of science to science majors	College science	23–29
Brandon, E. P.	Subverting common sense: textbooks and scientific theory	Conceptual change teaching	30–40
Carter, C.	Scientific knowledge, school science, and socialization in to science: issues in teacher education	Teacher education; sociology of science	41–51
Chazan, D.	Instructional implications of a research project on students' understandings of the differences between empirical verification and mathematical proof	Philosophy of mathematics; arguments and proofs	52–60
Collins, A.	Assessing biology teachers: understanding the nature of science and its influence on the practice of teaching	Teacher education; portfolio assessment	61–70
Confrey, J., & Smith, E.	Alternative representations of ratio: the Greek concept of anthyphairesis and modern decimal notation	History of mathematics; mathematics education	71–82
Cossman, G. W.	A comparison of the image of science found in two future oriented guideline documents for science education	Science education policy; social studies of science	83–105
Dagher, Z. R., & Cossman, G. W.	The nature of verbal explanations given by science teachers	Scientific discourse; classroom discourse	106–112
Davson-Galle, P.	History and philosophy of science—mixture or compound? The dangers of a prevailing view of philosophy of science for science education	Context of discovery; context of justification	113–127
Dell'Aquila, J. R.	Physical science in cultural context	College science	128–131
Edmondson, K. M.	College students' conceptions of the nature of scientific knowledge	College science; epistemology	132–142
Eger, M.	Rationality and objectivity in a historical approach: A response to Harvey Siegel	Critical thinking; history of science	143–153
Felder, D. W.	Is instructional design a science?	Curriculum theory	154–160
Fleury, S. & Swift, J. N.	Students as peasants: are we developing thinkers rather than memorizers?	Scientific literacy; classroom discourse	161–165
Glasson, G. E. & Garrison, J. W.	Hypothetico-deductive thought: On the horns of a dilemma?	Piagetian learning theory	166–169
Gruender, C. D.	Some philosophical reflections on constructivism	Philosophy of education	170–176
Hostetler, K.	The rationality of ethical inquiry in science: lessons from philosophy of science	Science-technology-society; curriculum theory	177–184
Jegede, O.	Toward a philosophical basis for science education of the 1990's: An African viewpoint	Science-technology-society; curriculum theory	185–198
Johanningmeier, E. V.	How educational researchers have defined educational research	Educational research	199–208
Kenealy, P.	Telling a coherent "story": a role for the history and philosophy of science in a physical science course	College science; elementary teacher education	209–220
Lochhead, J. & Defresne, R.	Helping students understand difficult science concepts through the use of dialogues with history	Scientific discourse	221–229
Lucas, A. M.	Public understanding of science and the "official" English model of science	Curriculum theory; science education policy	230–241
Martin, J. R.	What should science educators do about the gender bias in science?	Feminist theory; curriculum theory	242–255

TABLE 15.1. *Continued*

Author	Title	Key Words	Pages
McClain, B. R.	A teacher's analysis of the absence of the history and philosophy of science in the State of Florida's student performance standards	Science education policy	256–258
McFadden, C. P.	Redefining the school curriculum	Curriculum theory; science-technology-society	259–270
Mermelstein, E.	Clarification of conservation of liquid quantity and liquid volume	Learning theory	271–277
Nussbuam, J.	Classroom conceptual change: philosophical perspectives	Conceptual change	278–291
Robin, N. & Ohlsson, S.	Impetus then and now: a detailed comparison between Jean Buridan and a single comtemporary subject	Reasoning about scientific explanation	292–305
Sánchez, L.	On the implicit use of history of science in science education	Curriculum design	306–312
Sassower, R.	Collavoration as a pedagogical device	Cooperative learning; sociology of science	313–321
Schilk, J. M., Driscoll, S. E., & Carter, C. S.	Problem-solving and construction of scientific knowledge: a case study in epistemology	College science; conceptual change	322–331
Solomon, J.	The retrial of Galileo	Curriculum design	332–338
Stein, F. and AAAS Staff	Project 2061: Education for a changing future	Science education policy	339–343
Thompson, M. & de Zengotita, T.	Science literacy and the language game of science	Classroom discourse	344–352
Wallenmaier, T. E.	Dewey and science education	Philosophy of education	353–358

More History and Philosophy of Science in Science Teaching. (1990) D. E. Herget (Ed.), Tallahassee: Florida State University

Author	Title	Key Words	Pages
Espinet, M.	Newton's and Goethe's processes of inquiry: Two alternative ways of knowing in science	Applying HS to SE, conceptual change	3–12
McCarty, L. P.	Philosophy, science and education: the Deweyan perspective	Philosophy of education; Dewey, Hegel, Habermas	13–20
Briscoe, C.	John Dewey's philosophy of science: its relationship to science education	Philosophy of education; SE practice	21–27
Bailin, S.	Creativity, discovery and science education: Kuhn and Feyerabend revisited	Concept of creativity; context of discovery; scientific method	28–39
Woolnough, B. E.	Towards a holistic view of science education (or the whole is greater than the sum of its parts, and different)	Scientific method; curriculum theory	40–54
Cordero, A.	Science and the limits of the scientific understanding of things	Scientific objectivity	57–65
Davson-Galle, P.	Is realistic, context-bound-normative philosophy of science possible of desirable?	Ontology	66–70
Levy, T.	Science, hermeneutics, and the notion of rationality	Scientific discourse; Habermas	71–78
Hamburger, A. I.	Epistemological and historical studies of physics concepts for science teaching	Physics education; thermodynamics; social context of knowledge	79–85
Chandler, M.	Philosophy of gravity: intuitions of four-dimensional curved spacetime	Scientific discourse; observation theoretical distinction	86–101
Rowell, J. A., Dawson, C. J., & Lyndon, H.	Changing misconceptions: a challenge to science educators	Conceptual change	102–110
Jenkins, E. S.	Benjamin Banneker and science education	African American scientists	113–121
Carter, C.	Gender and equity issues in science classrooms: values and curricular discourse	Scientific discourse; feminist theory	122–132
Gilmer, P. J.	What it's really like to be a woman in science	Feminist perspective of science	133–135
Hills, G. L., & McAndrews, B.	Inquity, comfort, and community	Elementary teacher education	136–155
Go ab-Meyer, & Ruijgrok, T. W.	Science teaching: ethics in the classroom	Science-technology-society	156–162
Allchin, D. (Ed.)	Teachers' views of prospects for the history and philosophy of science in science teaching	Reflections on the 1st International Conference on HPS	165–171
Shapiro, B. L.	"That's not true, it doesn't make sense": one elementary student's views on the worthiness of scientific ideas	Scientific discourse; conceptual change	172–183

TABLE 15.1. *Continued*

Author	Title	Key Words	Pages
Taylor, P.	The influence of teacher beliefs on constructivist teaching practices	Conceptual change; classroom restructuring; mathematics ed.	184–201
Jungwirth, E., & Dreyfus, A.	Identification and acceptance of a posteriori causal assertions invalidated by faulty enquiry methodology: an international study of curricular expectation and reality	Biology education; enquiry	202–211
Pessoa de Carvalho, A. M.	The influence of the history of momentum and its conservation on the teaching of mechanics in high schools	Physics education; history of science	212–220
Davson-Galle, P.	Philosophy of science done in the "philosophy for children" manner in lower secondary schools	Philosophy for and by children	223–230
Gruender, C. D.	Uses of Galilean drama	Science, religion, and society	231–235
Désautels, J., & Larochelle, M.	A constructivist pedagogical strategy: the epistemological disturbance (experiment and preliminary results)	Conceptual change; curriculum; college teaching	236–257
Machold, D. K.	The historical objections to the special theory of relativity and a method of instruction for overcoming these difficulties	Physics education	258–265
Lühl, J.	The history of atomic theory with its societal and philosophical implications in chemistry class	Science-technology-society; chemistry education	266–274
Wandersee, J. H.	On the value and use of the history of science in teaching today's science: constructing historical vignettes	History of science; conceptual change	277–283
Wesley, W. G.	Theory in elementary physics	Physics education; Einstein on theory; Gowin's Vee	284–291
Hodson, D.	Making the implicit explicit: A curriculum planning model for enhancing children's understanding of science	Knowledge about science; curriculum theory	292–310
Shahn, E.	Foundations of science: a lab course for non-science majors	College science; history of science	311–352
Smith, H. A., Pilch, D. R., & Welch, G. R.	Holism vs. reductionism: a model for introducing philosophy into the biology curriculum	College teaching; biology education	353–372
Herget, D. E.	Micro chaos	Chaos theory; microcomputers	373–389
Ignatz, M.	A misconception propagated in many of the current physics textbooks about the "hunter-monkey" demonstration	Physics education	390–394

HISTORY AND PHILOSOPHY OF SCIENCE AND SCIENCE EDUCATION RESEARCH AND PRACTICE TODAY

Although there may be little agreement about the exact role philosophy of science should have in education, HPS and science education have benefited from this exploration. The detailed analyses of theory building and the development of a richer understanding of the role of theories in the growth of scientific knowledge have proved useful for both cognitive psychologists (Carey, 1986; Carey et al., 1989; Kuhn et al., 1988; Ohlsson, 1992) examining scientific reasoning and science educators (Duschl & Gitomer, 1991; Hodson, 1988, 1992; Posner et al., 1982; Russell, 1983) examining the dynamics of science learning environments.

Cognitive science research on children's understanding of science suggests that students construct explanations at very early ages and are guided by epistemological models. Of course, these theories are often incomplete (e.g., White & Frederiksen, 1987), incoherent (e.g., Ranney & Thagard, 1988), and misguided (e.g. Caramazza, McCloskey, & Green, 1981). Often, these student theories recapitulate the historical development of scientific thought (e.g., Nersessian, 1989; Nussbaum, 1983; Thagard, 1990, 1992). The consensus among science education researchers is that instructional activities ought to provide opportunities for students' current conceptions to be confronted and challenged and through a set of teacher-guided interactions, for theories to be restructured. Linn (1986), Novak (1977), Novak and Gowin (1984), Resnick (1983), Finley (1983), Anderson and Smith (1986), and Krupa et al. (1985), among others, each speak to the effect of a learner's prior knowledge on subsequent learning. This collective body of research implies that learners, as Carey (1986) asserts, develop their cognitive abilities by a process of progressively changing conceptual schemes.

However, arguments that the structures of scientific theories and cognitive schemas have much in common (Carey, 1986; Giere, 1988) must consider both the structural characteristics and the developmental characteristics of these conceptual frameworks. From a structural perspective, scientific theories and cognitive schemas may indeed have much in common. Theories can be thought of as composed of facts, principles, and lawlike statements that are molded together by accepted methodological and axiological practices. Cognitive schemas, from a structural perspective, can also be thought of as made up of concepts and propositional statements governed by rules and values that guide synthesis.

A problem occurs, though, when it is necessary—and it is—to describe the mechanisms for change in the structure of theories or in the structure of conceptual schemas as each relates to the growth of knowledge. As I see it, this is a central issue in the application of epistemology to science education. Furthermore, research on the dynamics of conceptual change has added another important element to the task—the learner's epistemological framework. Strike and Posner (1992), Schauble, Klopfer, and Raghavan (1990), Kuhn et al. (1988) and Carey et al. (1989) have acquired evidence that a learner's epistemological framework is a factor in effecting changes in knowledge representation. At the level of the classroom, it translates into what learners consider to be evidence for or against an emerging scientific explanation. The debates about constructivism and models of conceptual change learning environments need to focus on the role of evidence and what it is that counts (exemplars) or does not count (anomalous data) as evidence in the knowledge-restructuring domain. It is an important issue for bringing about knowledge restructuring because the epistemology of the learner becomes yet another element that must change.

We need also keep alive the lessons learned from previous HPS and science education research. The work carried out at OISE, for example, reminds us of the importance of having frameworks for teachers and students to employ in meaning making and reasoning. The HPS frameworks and stories then become much more: they become tools to guide and judge the development of individuals' knowledge growth and sense making in science—tools for teachers as well as learners. Thus, there are implications for teacher education programs. They become the guidelines for engaging, doing, and reviewing science. The new philosophies of science have come out of an analysis of actual practice and, in turn, they can and should inform practice. The challenge for science education researchers interested in applying HPS to science education rests squarely in the domain of providing details about this practice.

In one sense, we are asking the same question Schwab asked decades before: What is it that scientists do? But a focus on the development of theories and explanations moves us beyond the behaviors and into the cognitive dimensions of that question. Contemporary HPS and forward-looking applications of HPS to science education are exploring the details and consequences of both the behavioral and cognitive dimensions of science as it is practiced.

References

Abell, S. K. (1989). The nature of science as portrayed to preservice elementary teachers via methods textbooks. In D. E. Herget (Ed.), *The history and philosophy of science in science teaching* (pp. 1–14). Tallahassee: Florida State University

Aikenhead, G. S. (1991). *Logical reasoning in science and technology.* Toronto: Wiley.

Anderson, C. W., & Smith, E. (1987). Teaching science. In V. Koehler (Ed.), *The educator's handbook: A research perspective.* New York: Longman.

Arons, A. B. (1965). *Development of concepts of physics.* Reading, MA: Addison-Wesley.

Arons, A. B. (1988). Historical and philosophical perspectives attainable in introductory physics courses. *Educational Philosophy and Theory, 20*(2), 13–23.

Bechtel, W. (1988). *Philosophy of science: An overview for cognitive science.* Hillsdale, NJ: Lawrence Erlbaum Assoc.

Benson, G. D. (1989). The misrepresentation of science by philosophers and teachers of science. *Synthese, 80,* 107–120.

Biological Sciences Curriculum Study. (1984). *Science and technology: Investigating human dimensions.* Middle School Science Series. Dubuque, IA: Kendall Hunt.

Brody, B., & Grandy, R., Eds. (1989). *Readings in the philosophy of science* (2nd ed.). New York: Prentice Hall.

Brown, H. I. (1977). *Perception, theory and commitment: The new philosophy of science.* Chicago: Precedent Publishing.

Bruner, J. (1960). *The process of education.* Cambridge, MA: Harvard University Press.

Brush, S. (1974). Should the history of science be x-rated? *Science, 183*(4130), 1164–1172.

Brush, S. (1988). *The history of modern science: A guide to the second scientific revolution, 1800–1950.* Ames: Iowa State University Press.

Brush, S. (1991). Should scientists write history of science? *Critical problems and research frontiers in history of science and history of technology* (pp. 67–91). Conference proceedings, Madison, WI.

Bynum, W. F., Browe, E. J., & Porter, R., Eds. (1981). *Dictionary of the history of science.* Princeton, NJ: Princeton University Press.

Caramazza, A., McCloskey, M., & Green, B. (1981). Naive beliefs in "sophisticated" subjects: Misconceptions about trajectories of objects. *Cognition, 9,* 117–123.

Carey, S. (1985a). *Conceptual change in childhood.* Cambridge, MA: Bradford Books, MIT Press.

Carey, S. (1985b) Are children fundamentally different kinds of thinkers and learners than adults? In S. Chipman, J. Segal, & R. Glaser (Eds.), *Thinking and learning skills* (Vol. 2, (pp. 485–517). Hillsdale, NJ: Lawrence Erlbaum.

Carey, S. (1986). Cognitive science and science education. *American Psychologist, 41*(10), 1123–1130.

Carey, S., Evans, R., Honda, M., Jay, E., & Unger, C. (1989). 'An experiment is when you try it and see if it works': A study of junior high students' understanding of the construction of scientific knowledge. *International Journal of Science Education, 11,* 514–529.

Carey, R. L., & Strauss, N. G. (1968). An analysis of the understanding of the nature of science by prospective secondary science teachers. *Science Education, 58,* 358–363.

Carey, R. L., & Strauss, N. G. (1970). An analysis of the experienced science teachers' understanding of the nature of science. *School Science and Mathematics, 70*(5), 366–376.

Carnap, R. (1939). Foundations of logic and mathematics. In O. Neurath,

R. Carnap, & C. Morris (Eds.), *International encyclopedia of unified science* (Vol. I, Part 3). Chicago: University of Chicago Press.

Chi, M. (1992). Conceptual change across ontological categories: Examples from learning and discovery in science. In R. Giere (Ed.), *Cognitive models of science: Minnesota studies in the philosophy of science*. Minneapolis: University of Minnesota Press.

Cobb, P., Wood, T., & Yackel, E. (1991). Analogies from the philosophy and sociology of science for understanding classroom life. *Science Education, 75*, 23–44.

Cohen, I. B. (1988). The publication of *Science, technology and society*: Circumstances and consequences. *ISIS, 79*(299), 571–581.

Cohen, I. B., & Watson, F. G. (1952). *General education in science*. Cambridge, MA: Harvard University Press.

Collins, A. (1990). Assessing biology teachers: Understanding the nature of science and its influence on the practice of teaching. In D. E. Herget (Ed.), *The history and philosophy of science in science teaching* (pp. 61–70). Tallahassee: Florida State University.

Conant, J. B. (1957). *Harvard case histories in experimental science* (Vols. 1 & 2). Cambridge, MA: Harvard University Press.

Connelly, F. M. (1969). Philosophy of science and science curriculum. *Journal of Research in Science Teaching, 6*, 108–113.

Connelly, F. M. (1972). The functions of curriculum development. *Interchange, 3*(2–3), 161–77.

Connelly, F. M., Finegold, M., Clipsham, J., & Wahlstrom, M. W. (1977). *Scientific enquiry and the teaching of science: Patterns of Enquiry Project*. Toronto: OISE Press.

Cooley, W. W., & Klopfer, L. E. (1963). The evaluation of specific educational innovations. *Journal of Research in Science Teaching, 1*, 73–80.

Cushing, J. T. (1989). A tough act: History, philosophy, and introductory physics. *Interchange, 20* (2), 54–59.

Dorling, J. (1973). Demonstrative induction: Its significant role in the history of physics. *Philosophy of Science, 40*, 360–372.

Duschl, R. (1985a). The changing concept of scientific observation. In R. W. Bybee (Ed.), *Science technology society: 1985 yearbook of the National Science Teachers Association* (pp. 60–69). Washington, DC: National Science Teachers Association.

Duschl, R. (1985b). Science education and philosophy of science Twenty-five years of mutually exclusive development. *School Science and Mathematics, 85*(7), 541–555.

Duschl, R. (1988). Abandoning the scientistic legacy of science education. *Science Education, 72*, 51–62.

Duschl, R. (1990). *Restructuring science education: The importance of theories and their development*. New York: Teachers College Press.

Duschl, R., & Gitomer, D. H. (1991). Epistemological perspectives on conceptual change: Implications for educational practice. *Journal of Research in Science Teaching, 28*(9), 839–858.

Duschl, R., Hamilton, R., & Grandy, R. (1990). Psychology and epistemology: Match or mismatch when applied to science education? *International Journal of Science Education, 12*(3), 230–243.

Easley, J. A. (1959). The Physical Science Study Committee and educational theory. *Harvard Educational Review, 29*, 4–11.

Feyerabend, P. K. (1978). *Against method*. London: New Left Books.

Finley, F. (1983). Science processes. *Journal of Research in Science Teaching, 20*, 47–54.

Fleck, L. (1979). *Genesis and development of a scientific fact*. Chicago: University of Chicago Press.

Gallagher, J. J. (1991). Perspective and practicing secondary school science teachers' knowledge and beliefs about the philosophy of science. *Science Education, 75*, 121–134.

Giere, R. N. (1985). Philosophy of science naturalized. *Philosophy of Science, 52*, 331–356.

Giere, R. N. (1988). *Explaining science: A cognitive approach*. Chicago: University of Chicago Press.

Giere, R. N. (1989). Scientific rationality as instrument rationality. *Studies in History and Philosophy of Science, 20*(3), 377–385.

Giere, R. N. (1992). *Cognitive models of science: Minnesota studies in the philosophy of science*. Minneapolis, MN: University of Minnesota Press.

Glaser, R. (1984). Education and thinking: The role of knowledge. *American Psychologist, 39*(2), 93–104.

Glymour, C. (1980). *Theory and evidence*. Princeton, NJ: Princeton University Press.

Gooding, D. (1990). Mapping experiment as a learning process: How the first electromagnetic motor was invented. *Science, Technology, & Human Values, 15*(2), 165–201.

Grandy, R., Ed. (1973). *Theories and observation in science*. Englewood Cliffs, NJ: Prentice Hall.

Hanson, N. (1958). *Patterns of discovery*. Cambridge: Cambridge University Press.

Hegbert, D. E. (1990a). *The history and philosophy of science in science teaching: Proceedings of the 1st International Conference on the History and Philosophy of Science and Science Teaching* (Vol. 1). Tallahassee: Florida State University.

Hegbert, D. E. (1990b). *More history and philosophy of science in science teaching: Proceedings of the 1st International Conference on the History and Philosophy of Science and Science Teaching* (Vol. 2). Tallahassee: Florida State University.

Hewson, R., & Thornley, R. (1989). The conditions of conceptual change in the classroom. *International Journal of Science Education, 11*, 541–553.

Hills, S. (1992). *The history and philosophy of science in science education: Proceedings of the 2nd International Conference on the History and Philosophy of Science in Science Teaching* (Vols. 1 & 2). Kingston, Ontario: Queen's University.

Hodson, D. (1985). Philosophy of science, science and science education. *Studies in Science Education, 12*, 25–57.

Hodson, D. (1988). Toward a philosophically more valid science curriculum. *Science Education, 72*, 19–40.

Hodson, D. (1990). Making the implicit explicit: A curriculum planning model for enhancing children's understanding of science. In D. E. Herget (Ed.), *More history and philosophy of science in science teaching* (pp. 292–310). Tallahassee: Florida State University.

Hodson, D. (1992). Assessment of practical work: Some considerations in philosophy of science. *Science and Education, 1*, 115–144.

Holton, G. (1978). *The scientific imagination: Case studies*. Cambridge: Cambridge University Press.

Holton, G., & Brush, S. (1952). *Introduction to concepts and theories in physical science* (2nd ed.). Reading, MA: Addison-Wesley.

Jones, H. (1977). The past, present, and future of science education before, during, and after the year of the golden-fleeced MACOS. In G. Hall (Ed.), *Science teacher education: Vantage Point 1976, AETS Yearbook* (pp. 189–213). Columbus, OH: ERIC Clearinghouse Science, Mathematics, and Environmental Education, College of Education, Ohio State University.

Kelly, G., Carlsen, W., & Cunningham, C. (1993). Science education in sociocultural context: Perspectives from the sociology of science. *Science Education, 77*(2), 207–220.

Kemble, E. C. (1966). *Physical science: Its structure and development*. Cambridge, MA: MIT Press.

Kenealy, P. (1989). Telling a coherent "story": A role for the history and philosophy of science in a physical science course. In D. E. Herget (Ed.), *The history and philosophy of science in science teaching* (pp. 209–220). Tallahassee: Florida State University.

Kilbourn, B. S. (1974). *Identifying world views projected by science teaching materials: A case study using Pepper's World hypotheses to analyze a biology textbook*. (Doctoral thesis, University of Toronto). *Dissertation Abstracts International, 37*(5), 2763A.

Kilbourn, B. S. (1982). Curriculum materials, teaching, and potential outcomes for students: A qualitative analysis. *Journal of Research in Science Teaching, 19*(8), 675–688.

Kimball, M. E. (1967). Understanding the nature of science: A comparison of scientists and science teachers. *Journal of Research in Science Teaching, 5*(2), 110–120.

King, B. B. (1991). Beginning teachers' knowledge of and attitude toward history and philosophy of science. *Science Education, 75*, 135–142.

Kitchener, R. (1986). *Piaget's theory of knowledge: Genetic epistemology and scientific reason.* New Haven, CT: Yale University Press.

Kitchener, R. (1987). Genetic epistemology: Equilibration and the rationality of scientific change. *Studies in the History and Philosophy of Science, 18*, 339–366.

Kitchener, R. (1992). Piaget's genetic epistemology: Epistemological implications for science education. In R. Duschl & R. Hamilton (Eds.), *Philosophy of science, cognitive psychology, and educational theory and practice* (pp. 116–146). Albany: SUNY Press.

Klopfer, L. (1964–1966). *History of science cases.* Chicago: Science Research Associates.

Klopfer, L., & Cooley, W. (1963). Test on understanding science. Princeton, NJ: Educational Testing Service.

Krupa, M., Selman, R., & Jaquette, D. (1985). The development of science explanations in children and adolescents: A structural approach. In S. Chipman, J. Segal, & R. Glaser (Eds.), *Thinking and learning skills* (Vol. 2, pp. 427–455). Hillsdale, NJ: Lawrence Erlbaum.

Kuhn, D., Amshel, E., & O'Loughlin, M. (1988). *The development of scientific thinking skills.* Orlando, FL: Academic Press.

Kuhn, T. (1957). *The Copernican revolution.* Cambridge, MA: Harvard University Press.

Kuhn, T. (1962). *The structure of scientific revolutions* (2nd ed.). Chicago: University of Chicago Press.

Kuhn, T. (1984). Professionalization reflected in tranquility. *ISIS, 45*(276), 29–33.

Kyle, W. C., Jr. (1980). The distinction between inquiry and scientific inquiry and why high school students should be cognizant of the distinction. *Journal of Research in Science Teaching, 17*(2), 123–130.

Lakatos, I. (1970). Falsification and the methodology of scientific research programs. In I. Lakatos & A. Musgrave (Eds.), *Criticism and the growth of knowledge* (pp. 91–196). London: Cambridge University Press.

Laudan, L. (1977). *Progress and its problems: Toward a theory of scientific growth.* Berkeley: University of California Press.

Laudan, L. (1984). *Science and values: An essay on the aims of science and their role in scientific debate.* Berkeley: University of California Press.

Lederman, E. (1992). Research on students' and teachers' conceptions of the nature of science: A review of the research. *Journal of Research in Science Teaching, 29*(4), 327–330.

Linn, M. (1986). Science. In R. Dillon & R. Sternberg (Eds.), *Cognition and instruction* (pp. 155–204). New York: Academic Press.

Losee, J. (1980). *A historical introduction to the philosophy of science* (2nd ed.). Oxford: Oxford University Press.

Martin, M. (1971). The relevance of philosophy of science for science education. *Boston Studies in the Philosophy of Science, 32*, 293–300.

Martin, M. (1972/1986). *Concepts of science education: A philosophical analysis.* Glenview, IL: Scott, Foresman. (Republication with University Press of America, Lanham, MD.)

Matthews, M. (1989). A brief review. *Synthese, 80*, 1–8.

Matthews, M. (1990). History, philosophy and science teaching: A rapprochement. *Studies in Science Education, 18*, 25–51.

Merton, R. (1938/1970). *Science, technology and society in seventeenth century England* (2nd ed.). New York: Harper & Row.

Merton, R. (1985). George Sarton: Episodic recollections by an unruly apprentice. *ISIS, 76*(284), 470–486.

Miller, P. E. (1963). A comparison of the abilities of secondary teachers and students of biology to understand science. *Iowa Academy of Science, 70*, 510–513.

Munby, A. H. (1973). The provision made for selected intellectual consequences by science teaching: Derivation and application of an analytical scheme. (Doctoral dissertation, University of Toronto). *Dissertation Abstracts International, 34*(10), 6364A.

Munby, H., Orpwood, G., & Russell, T. (1984). Seeing curriculum in a new light: Essays from science education. Lanham, MD: University Press of America.

Nadeau, R., & Desautels, J. (1984). *Epistemology and the teaching of science.* Ottawa: Science Council of Canada.

Nagel, E. (1961). *The structure of science: Problems in the logic of scientific explanation.* New York: Harcourt, Brace, & World.

Nersessian, N., Ed. (1987). *The process of science: Contemporary philosophical approaches to understanding scientific practice.* Dorcrecht: Martinus Nijhoff.

Nersessian, N. (1989). Conceptual change in science and in science education. *Synthese, 80*, 163–183.

Nersessian, N. (1991). The cognitive sciences and the history of science. *Critical problems and research frontiers in history of science and history of technology* (pp. 92–115). Conference proceedings, Madison, WI.

Nersessian, N. (1992). Constructing and instructing : The role of "abstraction techniques" in creating and learning physics. In R. Duschl & R. Hamilton (Eds.), *Philosophy of science, cognitive psychology, and educational theory and practice* (pp. 48–68). Albany: SUNY Press.

Nickles, T. (1987). From natural philosophy to metaphilosophy of science. In R. Kargon & P. Achinstein (Eds.), *Kelvin's Baltimore lectures and modern theoretical physics: Historical and philosophical perspectives* (pp. 507–541). Cambridge, MA: MIT Press.

Nickles, T. (1991). *Critical problems and research frontiers in history of science and history of technology* (pp. 349–373). Conference proceedings, Madison, WI.

Norris, S. P. (1982). A concept of observation statements. In D. R. DeNicola (Ed.), *Philosophy of education, 1981.* Normal, IL: Philosophy of Education Society.

Norris, S. P. (1984). Defining observational competence. *Science Education, 68*, 129–142.

Norris, S. P. (1985). The philosophical basis of observation in science and science education. *Journal of Research in Science Teaching, 22*(9), 817–833.

Novak, J. D. (1977). *A theory of education.* Ithaca, NY: Cornell University Press.

Novak, J., & Gowin, R. (1984). *Learning how to learn.* Cambridge: Cambridge University Press.

Nowak, T., & Thagard, P. (1992). Newton, Descartes, and explanatory coherence. In R. Duschl & R. Hamilton (Eds.), *Philosophy of science, cognitive psychology, and educational theory and practice* (pp. 69–115). Albany: SUNY Press.

Nussbaum, J. (1983). Classroom conceptual change: The lesson to be learned from the history of science. In H. Helm & J. Novak (Eds.), *Proceedings of the International Seminar on Misconceptions in Science and Mathematics.* Ithaca, NY: Department of Education, Cornell University.

Nussbaum, J. (1989). Classroom conceptual change: Philosophical perspectives. *International Journal of Science Education, 11*(5), 530–540.

Ohlsson, S. (1992). The cognitive skill of theory articulation: A neglected aspect of science education. *Science and Education, 1,* 181–192.

Piaget, J. (1967). *Six psychological studies.* New York: Random House.

Pintrich, P. R., Marx, R. W., & Boyle, R. A. (1993). Beyond cold conceptual change: The role of motivational beliefs and classroom contextual factors in the process of conceptual change. *Review of Educational Research, 63,* 167–199.

Polanyi, M. (1958). *Personal knowledge: Towards a post-critical philosophy.* Chicago: University of Chicago Press.

Posner, G., Strike, K., Hewson, P., & Gertzog, W. (1982). Accommodation of a scientific conception: Toward a theory of conceptual change. *Science Education, 66*(2), 211–227.

Quine, W. V. O. (1969). Epistemology naturalized. In *Ontological relativity and other essays.* New York: Columbia University Press.

Ranney M., & Thagard, P. (1988). *Explanatory coherence and belief revision in naive physics* (Cognitive Science Laboratory Report 31). Princeton, NJ: Princeton University.

Resnick, L. (1983). Mathematics and science learning: A new conception. *Science, 220,* 477–478.

Rissland, E. (1985). The structure of knowledge in complex domains. In S. Chipman J. Segal and R. Glaser, (Eds.), *Thinking and learning skills* (Vol. 2, pp. 107–125). Hillsdale, NJ: Lawrence Erlbaum.

Roberts, D. (1980/1984). Theory, curriculum development, and the unique events of practice. In H. Munby, G. Orpwood, & T. Russell (Eds.), *Seeing curriculum in a new light: Essays from science education* (pp. 65–87). Lanham, MD: University Press of America. (Reprint of 1980 copyright held by OISE.)

Roberts, D. A. (1982). Developing the concept of "curriculum emphases" in science education. *Science Education, 66*(2), 243–260.

Roberts, D., & Russell, T. (1975). An alternative approach to science education research: Drawing from philosophical analysis to examine practice. *Curriculum Theory Network, 5*(2), 107–125.

Robinson, J. T. (1968). *The nature of science and science teaching.* Belmont, CA: Wadsworth.

Rubba, P. (1977). *The development, field testing, and validation of an instrument to assist secondary school students' understanding of the nature of scientific knowledge.* Unpublished doctoral dissertation, Indiana University, Bloomington.

Rumelhart, D. E. (1980). Schemata: The building blocks of cognition. In R. Spiro, B. Bruce, & W. Brewer (Eds.), *Theoretical issues in reading comprehension.* Hillsdale, NJ: Lawrence Erlbaum.

Russell, T. (1976). On the provision made for development of views of science and teaching in science teacher education. (Doctoral dissertation, University of Toronto). *Dissertation Abstracts International, 39*(3), 1496A.

Russell, T. (1981). What history of science, how much and why? *Science Education, 65,* 51–64.

Russell, T. (1983). Analyzing arguments in science classroom discourse: Can teachers questions distort scientific authority? *Journal of Research in Science Teaching, 20,* 27–46.

Rutherford, F. J., & Ahlgren, A. (1990). *Science for all Americans.* New York: Oxford University Press.

Rutherford, F. J., Holton, G., & Watson, F. G. (1970). *Harvard Project Physics.* New York: Holt, Rinehart, & Winston.

Salmon, W. C. (1984). *Scientific explanation and the causal structure of the world.* Princeton, NJ: Princeton University Press.

Schauble, L., Klopfer, L. & Raghavan, K. (1990). Students' transition form an engineering model to a science model of experimentation. *Journal of Research in Science Teaching, 28*(9), 859–882.

Scheffler, I. (1960). *The language of education.* Springfield, IL: Charles C Thomas.

Schwab, J. J. (1960). What do scientists do? *Behavioral Science, 5,* 1–27.

Schwab, J. J. (1962a). The concept of the structure of a discipline. *Educational Record, 43,* 197–205.

Schwab, J. J. (1962b). The teaching of science as enquiry. In J. J. Schwab & P. F. Brandwein (Eds.), *The teaching of science* (pp. 3–103). Cambridge: Harvard University Press.

Schwab, J. J. (1964). The structure of the natural sciences. In G. W. Ford & L. Pugno (Eds.), *The structure of knowledge and the curriculum* (pp. 31–49). Chicago: Rand-McNally.

Schwab, J. J. (1965). *The biology teacher's handbook.* American Institute of Biological Sciences, Biological Sciences Curriculum Study. New York: Wiley.

Shahn, E. (1990). Foundations of science: A lab course for nonscience majors. In D. E. Herget (Ed.), *More history and philosophy of science in science teaching* (pp. 311–352). Tallahassee: Florida State University.

Shapere, D. (1984) . Reason and the search for knowledge: Investigation in the philosophy of science. Dordrecht, The Netherlands: Reidel.

Shapere, D. (1987). Method in the philosophy of science and epistemology: How to inquire about inquiry and knowledge. In N. Nersessian (Ed.), *The process of science* (pp. 1–38). Dordrecht: Martinus Nijhoff.

Shapin, S. (1988) Understanding the Merton thesis. *ISIS, 79*(299), 594–605.

Shapin, S. (1991). Discipline and bounding: The history and sociology of science as seen through the externalism-internalism debate. *Critical problems and research frontiers in history of science and history of technology* (pp. 203–228). Conference proceedings, Madison, WI.

Shymansky, J. A., & Kyle, W. C., Jr. (1991). *Establishing a research agenda: The critical issues of science curriculum reform.* Washington, DC: National Association for Research in Science Teaching.

Siegel, H. (1989). Philosophy of science naturalized? Some problems with Giere's naturalism. *Studies in the History of Philosophy of Science, 20,* 365–375.

Stegmuller, W. (1976). *The structure and dynamics of theories.* New York: Springer-Verlag.

Stewart, J., & Hafner, R. (1991). Extending the conception of "problem" in problem solving research. *Science Education, 75,* 105–120.

Stinner, A. (1990). Philosophy, thought experiments and large context problems in the secondary school physics course. *International Journal of Science Education, 12*(3), 244–257.

Strike, K., & Posner, G. (1992). A revisionist theory of conceptual change. In R. Duschl & R. Hamilton (Eds.), *Philosophy of science, cognitive psychology, and educational theory and practice* (pp. 147–176). Albany: SUNY Press.

Suppe, F., Ed. (1977). *The structure of scientific theories* (2nd ed.). Champagne-Urbana: University of Illinois Press.

Suppe, F. (1979). Theory structure. In P. Asquith & H. E. Kyburg, Jr. (Eds.), *Current research in philosophy of science—Proceedings of the P.S.A. critical research problems conference* (pp. 317–338). East Lansing, MI: Philosophy of Science Association.

Suppe, F. (1989). *The semantic conception of theories and scientific realism.* Champagne-Urbana: University of Illinois Press.

Suppes, P. (1967). What is a scientific theory? In S. Morgenbesser (Ed.), *Philosophy of science today.* New York: Basic Books.

Thackray, A. (1984). Sarton, science, and history. *ISIS, 75,* 7–10.

Thackray, A. (1985). An end and a beginning. *ISIS, 76,* 467–469.

Thagard, P. (1988). *Computational philosophy of science.* Cambridge, MA: MIT Press.

Thagard, P. (1990). The conceptual structure of the chemical revolution. *Philosophy of Science, 57*(2), 183–209.

Thagard, P. (1992). *Conceptual revolutions.* Ewing, NJ: Princeton University Press.

Tisher, R. P., & White, R. T. (1985). Research on natural sciences. In M. C. Wittrock (Ed.), *Handbook of research on teaching* (3rd ed.) (pp. 874–905). New York: American Educational Research Association.

Toulmin, S. E. (1953). *The philosophy of science: An introduction.* London: Hutchinson.

van Fraassen, B. C. (1980). *The scientific image.* London: Clarendon.

Vosniadou, S., & Brewer, W. F. (1987). Theories of knowledge restructuring in development. *Review of Educational Research, 57,* 51–67.

Wandersee, J. H. (1990). On the value and use of the history of science in teaching today's science: Constructing historical vignettes. In D. E. Herget (Ed.), *More history and philosophy of science in science teaching* (pp. 277–283). Tallahassee: Florida State University.

Welch, W. (1979). Twenty years of science education development: A look back. *Review of Research in Education, 7,* 282–306.

Welch, W. W., Klopfer, L. E., Aikenhead, G. S., & Robinson, J. T. (1981). The role of inquiry in science education: Analysis and recommendations. *Science Education, 65,* 33–50.

White, B., & Frederiksen, J. (1987). *Progressions of qualitative models as a foundation for intelligent learning environments* (BBN Report No. 6277). Cambridge, MA: Bolt, Beranck and Newman, Inc.

Willson, V. (1987). Theory-building and theory-confirming observation in science and science education. *Journal of Research in Science Teaching, 24,* 279–286.

Winchester, I. (1990). Thought experiments and conceptual revision in science. *Studies in Philosophy and Education, 10,* 73–80.

Wittgenstein, L. (1953). *The philosophical investigations.* Translated by G. E. M. Anscombe. New York: Macmillan.

Ziman, J. M. (1984). *An introduction to science studies: The philosophical and social aspects of science and technology.* Cambridge: Cambridge University Press.

·16·

RESEARCH ON THE USES OF TECHNOLOGY IN SCIENCE EDUCATION

Carl F. Berger, Casey R. Lu, Sharolyn J. Belzer, and Burton E. Voss

THE UNIVERSITY OF MICHIGAN

Research reviews are normally limited to the study of research that has occurred. However, the research currently being carried out in instructional technology, particularly that involving science teaching and learning, may be undergoing a paradigm shift. In this review, then, we present traditional research, current research, and a projection of future research that is suggested by research in progress. In each section, we propose a basic question as a conceptual organizer. These questions are either from the background of literature or from our own thoughts about research directions. In either case, the use of a conceptual organizer allows us to state "up front" our philosophical view of the underlying research concerns presented in each following section.

The first section emphasizes the completed traditional research. We define traditional research as that which emphasizes comparison studies, that is, studies that compare results of instruction with and without computer assistance or involvement. This research is most mature and includes research that contributes to meta-analytic studies and the report of those meta-analyses. The organizing question of this section is, "Does the use of instructional technology provide more efficient learning than learning without the use of instructional technology?"

The second section contains studies that emphasize a shift toward new uses of instructional technology. They include studies where comparison studies are not appropriate because there are often no analogues for instruction outside the use of technology. In such studies, comparison research would be misleading at best and irresponsible at worst. Irresponsible because it would assume that the instruction should be carried out without the use of technology in instances where it is neither technologically possible nor educationally acceptable to

do so. The organizing question for this section is, Do media influence student learning or do media merely influence the way learning is delivered?

The third section includes a report of research under way with few final results available. Results of this research are most often taken from conference presentations rather than journal articles. It is the most risky section to speculate on as some of the research may never appear in easily accessible research journals or, more problematic, some of the ideas presented as harbingers of the future may never come to pass. Nevertheless, we include these ideas as we believe that they may be indicators of a paradigm shift, a fundamental shift in the direction of research and the way we carry out our research in the use of technology in science education. The organizing question for this section is, What are new and promising avenues of research that may indicate a paradigm shift of the use of instructional technology in science education? We hope that the last section further presents interesting suggestions for alternatives in research studies and techniques.

TRADITIONAL COMPUTER-BASED EDUCATION (CBE)

Background

Does the use of instructional technology provide more efficient learning than learning without the use of instructional technology? Efficient learning is best described as more learning in the same amount of time or the same learning in less

The authors thank reviewer Robert D. Sherwood (Vanderbilt University).

466

time. The following research reviews may provide an answer to this question.

Microcomputers are now an essential part of our information-based society. They occur regularly in homes, offices, businesses, universities, and public and nonpublic school classrooms, to name just a few locations. The number of computers being used for instruction in American public and nonpublic schools increased from one or two thousand to more than 1 million between spring 1980 and spring 1985 (Becker, 1986). The Office of Technology Assessment (1982) predicted that by the year 1994 there would be more than 4 million microcomputers in the schools. Clearly, the computer has become an integral part of the U.S. educational system, yet the impact of computer-based education on learning is not fully understood. Mitchell and Montague (1986) found that although science teachers in Texas used computers in their classroom, very few used them for other than word processing and reported that they would have little use for computers in the laboratory (indicating that they were unaware of the powerful use of microcomputer-based laboratories) (Mitchell & Montague, 1986).

Studies on effectiveness of CBE are explored first by examining the work of researchers who have reviewed CBE research over the past several years, including both descriptive and meta-analytic review studies; second, attention is focused on reviews of the effects of CBE on learning in the science domain; and, third, exemplary CBE programs from the science field are discussed.

Studies of Computer-Based Education

Studies investigating the effects of computer-based education on learning (usually achievement measures) have been reviewed periodically starting in the early 1970s. These reviews usually include CBE studies on the effects of computer-assisted instruction (CAI), generally referring to drill or practice and tutorials; computer-managed instruction (CMI), generally referring to computer evaluation of student test performance, guiding students to appropriate instructional resources, and record keeping; computer-simulated experimentation (CSE); and microcomputer-based laboratories (MBL), referring to an interface between a computer and a data-collecting device. Roblyer, Castine, and King (1988) reviewed 26 of these past reviews (Aiello & Wolfe, 1980; Bangert-Drowns, Kulik & Kulik, 1985; Blume, 1984; Burns & Bozeman, 1981; Carter, 1984; Crosby, 1983; Edwards, Norton, Taylor, Weiss, & Dusseldorp, 1975; Florida Department of Education, 1980; Glass, 1982; Hasselbring, 1987; Jamsion, Suppes, & Wells, 1974; Kulik, Bangert, & Williams, 1983; Kulik, Kulik, & Cohen, 1980; Kulik, Kulik, & Schwalb, 1986; Kulik, Kulik, & Bangert-Drowns, 1985; Lawton & Gerschner, 1982; Niemiec, Samson, Weinstein, & Walberg, 1987; Niemiec & Walberg, 1985; Okey, 1985; Orlansky, 1983; Roblyer & King, 1983; Samson, Niemiec, Weinstein, & Walberg, 1985; Thomas, 1979, Vinsonhalcr & Bass, 1972; Willett, Yamashita, & Anderson, 1983), then added their own meta-analysis of the effects of microcomputer-based education on learning from 1980 to 1987. The Roblyer et al. (1988) review examined studies that were descriptive "box score" type re-

views and reviews that used the meta-analysis techniques, pioneered by Glass (1976, 1977), that report effect sizes. Box scores generally provide a narrative review of studies, counting and comparing the number of studies with positive effects versus negative effects versus no effects, hence the name box score. Box-score reviews provide qualitative insights into the impact of a given type of treatment.

Effect sizes, however, are a quantitative measure of the impact a new method would have over an old one. Effect sizes are calculated by subtracting the mean score achieved by the control group from the mean score achieved by the treatment group. The result is divided by a measure of the spread of scores achieved by the two groups, usually the pooled standard deviation. Some authors use the standard deviation of the control group in place of the pooled standard deviation. This statistical procedure provides a measure of the impact of a treatment in standard-deviation units and can be converted to percentiles. For example, an average effect size of 0.25 would translate into an effect on the performance of the treatment group so as to place their average performance at the 60th percentile, while the control group average performance would be at the 50th percentile. Cohen (1977) has suggested that the magnitude of effect sizes be interpreted as follows: ES of 0.2 or less = small effect; ES of 0.5 to 0.6 = medium effect; ES of 0.8 or more = large effect.

Roblyer et al.'s (1988) comprehensive review of box-score and meta-analytic reviews reported the following trends across the 26 earlier review studies:

1. Nearly all reviews found that CBE provided some benefit over other instructional methods.
2. There was a reduction in the amount of learning time required for CBE, but most of the significant findings for time saving were reported for subjects in higher grades. Kulik et al. (1985) reported that there was an average time saving of 32 percent for the CBE group.
3. Attitudes were positively affected by CBE; however, the largest average effect size of 0.62 was for attitude toward computers, with a lower average effect size of 0.22 reported for attitude toward subjects and school learning (Bangert-Drowns et al., 1985; Florida Department of Education, 1980; Kulik et al., 1980, 1983, 1986).
4. There were no clear trends indicating whether drill-practice or tutorial CBE was more effective at increasing achievement. In the areas of mathematics (Burns & Bozeman, 1981), reading and language (Roblyer & King, 1983), and secondary level education (Samson et al., 1985) better results were reported for tutorials than for drill-practice CBE. Niemiec et al. (1987) and Burns and Bozeman (1981) found drill and practice more effective than tutorials at the elementary level. Other studies were inconclusive as to which type of CBE was more effective.
5. Supplemental use of CBE was more effective than total replacement of the classroom teacher by CBE (Crosby, 1983; Edwards et al., 1975; Okey, 1985).
6. There was some indication that CBE may be especially effective for use with lower ability learners (Bangert-Drowns et al., 1985; Edwards et al., 1975; Jamison et al., 1974; Kulik,

Kulik, & Bangert-Drowns, 1985; Niemiec et al., 1987; Samson et al., 1985). However, Roblyer et al. cautioned that the software tested was often specifically designed for this group of learners. Thus, finding that CBE was especially effective with these learners may have been due to the specific software used and not CBE in general.

7. CAI (drill-practice and tutorial) was more effective at lower grade levels, and computer-managed instruction and computer simulations were more effective at higher grade levels (Kulik et al., 1985, 1986).

8. The magnitude of the effect of CBE in different content areas was not clear. In mathematics and reading it appeared that CBE was effective at increasing achievement (mathematics drill-practice average ES = 0.34; mathematics tutorial average ES = 0.44 [Burns & Bozeman, 1981]; reading overall average ES = 0.35 [Roblyer & King, 1983]).

Aiello and Wolfe (1980), Willett et al. (1983), Wise and Okey (1983), and Wise (1988) reviewed the science field with different results. Aiello and Wolfe found an overall positive average effective size for CBE in science of 0.42. Willett et al. reported a large effect size of 1.45 for computer-simulated experiments (this was based on only one study) and much lower average effect sizes of 0.01 and 0.05 for CAI (drill-practice and tutorials) and CMI, respectively. Wise and Okey reported an overall average effect size of 0.82 for CBE in math and science. Wise (1988) reported an overall average effect size of 0.34 for science CBE. There was not enough quantitative work on CBE effects on learning in other subject areas to evaluate them quantitatively.

Findings of the Roblyer et al. (1988) Review of Research on Microcomputer-Based Education from 1980 to 1987

As part of their comprehensive review of the effects of CBE on learning, Roblyer et al. (1988) conducted a meta-analysis of the effects of CBE from 1980 to 1987 to focus more specifically on the effects of the microcomputer. Earlier reviews had included research on CBE that was mostly mainframe computer based. The inherent constraints of mainframe-based CBE, such as the large physical size of the computing system, need for users to have access and be connected to a mainframe system by direct hard-wired lines or phone lines, need for terminals, and so forth, certainly influenced learning outcomes compared with microcomputer-based CBE. The introduction of affordable microcomputers greatly facilitated the task of providing CBE to a large number of elementary, secondary, and postsecondary learners, and this has become the predominant medium for delivering CBE. Thus, an investigation into the effects of microcomputer-based CBE is highly appropriate.

Roblyer's group selected research studies according to the following criteria: statistical information such as means and standard deviations, as well as necessary information on sample sizes, had to be reported; studies had to have a control group; only studies available after 1980, mostly microcomputer-based were included; and the studies had to be free from severe methodological problems. Using these criteria, Roblyer et al.

conducted Dialog searches of ERIC and *Dissertation Abstracts* data bases; conducted by-hand searches of well-known instructional computing and research journals; 1980–1987 proceedings and programs from research and technology conferences; contacted well-known authors of recent non-meta-analysis reviews and requested recent unpublished reports that these authors used for their own review work; and, they searched bibliographies of articles and reports, generated by these procedures, for further studies. This search procedure resulted in the inclusion of 38 studies and 44 dissertations in their meta-analysis. Outlier studies were identified by using tests for homogeneity, and effect sizes were calculated with and without including outlier studies. Effect sizes with and without outliers were reported. Roblyer et al.'s findings are shown in Table 16.1.

General trends that emerged from this meta-analysis of the effects of CBE (microcomputer-based) on learning from 1980 to 1987 are as follows:

1. Attitudes toward both school or subject and self were significantly and positively affected by CBE. This finding had not previously been supported by reviews of CBE. It should be noted that there were only three studies reporting measures of attitude toward self (self-image and self-confidence); thus these results must be treated as tentative findings. However, improving learner self-image and self-confidence is an important goal for education, as attitude toward self can play a significant role in learner motivation (Pintrich, 1986, 1987). Similar results demonstrating the power of computers to motivate have been reported by Johnson (1982) and Cox and Berger (1981).

2. CBE was found to be effective at increasing achievement levels of treatment groups over those of control groups in the subject areas of mathematics, reading-language, and science. This finding confirms the positive average effect sizes reported for these subject areas in past reviews. Roblyer et al.'s positive findings in the area of science are from studies of the simulation form of CBE only and should not be generalized to other types of CBE.

The researchers could find only four studies from the science domain between the years 1980 and 1987 that could be used in their meta-analysis (i.e., studies that had control and treatment groups, reported means and standard deviations, were free from major methodological flaws, etc.). This indicates that there is still a great need for sound experimental and quasi-experimental research on the effects of different types of CBE in science. As noted previously, more quantitative research is also needed in other subject areas before quantitative reviews can address the effect of CBE across more disciplines.

3. The results from the analysis of the type of CBE application (drill-practice, tutorial, simulation, problem solving) that CBE was most effective at increasing achievement were inconclusive. Only in the subject areas of mathematics and reading were enough studies found to make statistical comparisons. In mathematics, drill-practice and tutorial seemed to be equally effective, as there was no significant difference between their average effect sizes. In reading, tutorials pro-

TABLE 16.1 Summary of Effect Sizes (d) and Percentages of Outliers Across Groups (Numbers in parentheses indicate number of studies in each category, original and reduced sets)

	Original Set d Groups**	Reduced Set d	Percentage	Difference of Outliers	Difference from 0*
ACHIEVEMENT					
Elementary level	0.32 (44)	0.29 (37)	16	—	—
Secondary level	0.19 (22)	0.19 (19)	14	—	—
College/adult level	0.57 (10)	0.66 (9)	10	—	—(H)
Mathematics	0.36 (35)	0.37 (31)	11	—	
Science	0.49 (4)	0.64 (3)	25	—	
Reading/language	0.26 (33)	0.30 (24)	27	—	
Cognitive skills	0.35 (14)	0.28 (13)	7	—	
Math drill/practice	0.25 (17)	0.20 (16)	6	—	
Math tutorial	0.22 (5)	0.30 (3)	40	—	
Math: other applications	0.20 (8)	0.38 (7)	13	—	
Reading drill/practice	0.26 (18)	0.25 (11)	39	—	
Reading tutorial	0.41 (2)		0	—	
Reading: other applications		0.32 (5)		0	—
Math computation	0.24 (9)		0	—	
Math concepts/problem solving		0.26 (4)		0	—
Reading comprehension	0.21 (7)		0	—	
Reading vocabulary	0.31 (4)	−0.20 (2)	50	—	
Reading word analysis	0.55 (2)		0	—	
Language usage/grammar	0.30 (4)	0.28 (3)	25	—	
English as second language	−0.05 (3)	−0.15(2)	33		
Word processing	0.23 (8)		0	—	
Problem solving with Logo	0.39 (11)		0	—	—(H)
Problem solving with CAI	0.20 (4)	0.01 (3)	25		
Creativity	0.73 (3)		0	—	
Low-achieving students	0.45 (15)	0.36 (14)	7	—	
Regular students	0.32 (29)	0.22 (27)	10	—	
Females	0.17 (11)	0.12 (10)	9		
Males	0.19 (11)	0.31 (10)	9	—	
ATTITUDE					
Self	0.25 (3)		0	—	
School/subject matter	0.28 (12)		0	—	
Computer medium	0.06 (4)	0.19 (3)	25		
Females	0.05 (5)	0.18 (4)	20	—	
Males	0.29 (5)		0	—	

* $p < .05$ that d is 0 after outliers were removed.

** $p < .05$ that d is the same among groups after outliers were removed.

H = highest d among groups.

Source: "Assessing the impact of computer-based instruction "by Roblyer et al., 1988, Computers in the Schools, 5, p. 121. Copyright 1988 by Haworth Press, Inc. Adapted by permission.

duced the highest average effect size. In the area of science, only simulation-type CBE studies were located and they found a large positive effect size on achievement. Thus, as reported in the earlier reviews of CBE and in Roblyer et al.'s meta-analysis, the type of CBE that is most effective for a given learning situation is still equivocal.

4. College and adult learners had a significantly higher average effect size than that of other grade levels, which is a result contrary to findings from the past reviews, in which college adult learners seemed to benefit *least* from CBE, with younger learners gaining the most. Thus, it appears that CBE can now be effective for older as well as younger learners' needs. This could be due to the availability of improved software that is better matched to college-adult learners, such as software that allows more learner control of the learning process and pace of instruction. Secondary-level CBE had a significantly lower average effect size than both college-adult and elementary level CBE.

5. Roblyer et al.'s finding that CBE is effective for lower-level learners also supports the findings of the earlier reviews. However, the difference in average effect sizes between lower-level and regular-level learners was not significant. It should be noted, again, that this higher average effect size for lower-level learners could be due to the design of some of the software (specifically designed for slower learners) and not to CBE in general.

Although this research seems exhaustive, the results can be phrased in a few sentences. If we define learning efficiency as learning more in the same amount of time, the studies show confusing trends. If, however, we define learning efficiency as the ratio of amount of learning to the time taken to learn (LE = L/T), then the results are clear. CBE can save learning time and can increase learning; that is, students can learn the same amount in less time and can learn more in the same amount of time. Other major findings from these reviews can be summarized as follows: CBE can be effective for all ages of learners; it has a significant positive effect on achievement; and it may have a positive effect on attitudes toward computers, subject being taught, and self. Before we conclude that CBE is a panacea for the learning of students of all ages, develops positive attitudes, and increases achievement, note that CBE is generally more effective as a supplement to than as a complete replacement for instruction. CBE is most often related to replacement or supplementation of presentation or questioning strategies. Equally, the effect of CBE in the content areas is still unclear and the most effective type of CBE is still unclear. We still have much to research and learn.

Effects of Computer-Based Education on Learning in the Science Domain

This section concentrates on the effects of CBE in the science domain. Several reviews have reported findings specifically from the science subject areas (Aiello & Wolfe, 1980; Kulik & Kulik, 1986; Kulik, Kulik, Bangert-Drowns, 1984; Kulik et al., 1983; Roblyer et al., 1988; Willett et al., 1983; Wise, 1988; Wise & Okey, 1983). Their findings are summarized in Table 16.2.

Aiello and Wolfe (1980) conducted a meta-analysis on individualized instruction in science (audiotutorial instruction, CAI, personalized system of instruction, programmed instruction, and a combination category that included studies that did not fit exactly into any of the first four categories). Their results indicated that CAI and personalized system of instruction (PSI)

were the most effective forms of individualized instruction, with overall average effect sizes for achievement of 0.42 for both CAI ($n = 14$) and PSI ($n = 28$). They had enough effect sizes to break their analysis down into specific subject areas, and they report an effect size for CAI of 0.36 in biology ($n = 1$) and average effect sizes of 0.52 for chemistry ($n = 8$) and 0.23 for physics ($n = 4$).

Willett et al. (1983) meta-analyzed the effects of instructional systems in science teaching (audiotutorial, computer-linked, contracts for learning, departmentalized elementary school, individualized instruction, mastery learning, media-based instruction, PSI, programmed learning, self-directed study, use of original source papers in the teaching of science, and team teaching). They found that CBE produced an overall average effect size of 0.13 ($n = 14$) but that CAI's and CMI's average effect sizes were only 0.01 ($n = 5$) and 0.05 ($n = 8$), respectively. CSE produced a huge effect size of 1.45 ($n = 1$), heavily biasing the overall average effect size in the positive direction. They found that mastery learning and PSI were the most effective instructional systems, with average effect sizes of 0.60 ($n = 13$) and 0.64 ($n = 15$), respectively.

Wise and Okey (1983) reported meta-analysis results on CBE's effect on achievement in math and science at the elementary and secondary levels in the same year that Willett et al. reported their meta-analysis of instructional systems. Wise and Okey found a large positive overall average effect of 0.82 ($n = 12$) for CBE. The difference between Wise and Okey's results versus Willett et al.'s results (ES of 0.82 versus 0.13) was presumably due to methodological differences or the inclusion of mathematics CBE studies by Wise and Okey.

Kulik et al. (1983, 1984) and Kulik and Kulik (1986) performed several meta-analyses on the effects of CBE on learning across subject domains. They reported an effect size of 0.36 ($n = 1$) for elementary-level science CBE and average effect sizes of 0.31 ($n = 11$) for secondary-level science CBE and 0.15 ($n = 44$) for college-level hard sciences, engineering, mathematics, and agriculture CBE.

As discussed earlier, Roblyer et al. (1988) reported results of a meta-analysis of the effects of CBE on learning across subject

TABLE 16.2. Summary of Findings from Reviews of the Effects of CBE on Achievement in Science

Reviewer(s)	Year	Grade Level	Number of Studies	Type of CBE	Overall Average Effect Sizes***
Aiello & Wolfe	1980	Secondary and college science	11	CAI	0.42
Willet et al.	1983	Elementary and secondary science	14	CAI, CMI, CSE	0.13
Wise & Okey	1983	Elementary and secondary math and science	12		0.82
Kulik et al.	1983	Secondary science	11		0.31
Kulik et al.	1984	Elementary science	1	CMI	0.36
Kulik & Kulik	1986	College (hard sciences*)	44	CAI, CMI, CEI**	0.15
Roblyer et al.	1988	Secondary and college science	4(3)*	CSE	0.49 (0.64)
Wise	1988	Elementary, secondary, and college science	26	MBL, CAI, CMI, CSE, VDBL****	0.34

* Hard sciences = the hard sciences plus engineering, mathematics, and agriculture.

** Computer-enhanced instruction: computer serving as a tool (word processing, graphing data in science classes, etc.) and simulation device.

*** Numbers in parentheses are for analysis with outliers removed.

**** Microcomputer-based laboratories, tutorials, diagnostic testing, simulations, and video disk–based lessons.

areas. They found an average effect size of 0.44 ($n = 4$) (ES = 0.64; $n = 3$ when one outlier study was removed from the analysis) for simulation CBE in science. Probably the most interesting finding in the science domain was that only four studies from 1980 to 1987 were allowed into their meta-analysis after searching ERIC and *Dissertation Abstracts* data bases, searching current computer instruction and educational research journals, contacting non–meta-analysis reviewers personally for further references, and examining bibliographies. Although they did miss some studies (e.g., Bunderson, Baillio, Olsen, Lipson, & Fisher, 1984), the fact that a fairly comprehensive search turned up only four sound quantitative studies in science CBE indicates a great need for more quantitative research in the science domain.

Finally, Wise (1988) conducted a science domain-specific meta-analysis. He was able to locate 26 studies from 1982 to 1988, 22 more than Roblyer et al., but still a relatively small number of quantitative studies. Wise's criteria for considering a study CBE, that "teachers used microcomputers in some way to deliver science instruction" (p. 107), were fairly loose and simple video disk systems were considered CBE in his study. His less stringent criteria for allowing studies into his meta-analysis probably account for the greater number of science CBE studies located by Wise.

Wise reported an overall average effect size for CBE of 0.34 ($n = 51$). When he looked at the effect of different types of CBE, he found that microcomputer-based labs (MBL) produced the highest average effect size of 0.76 ($n = 6$), followed by tutorials with an average effect size of 0.45 ($n = 7$). Video disk–based lessons produced an average effect size of 0.40 ($n = 11$), followed by microcomputer-based diagnostic testing and microcomputer-based simulations with average effect sizes of 0.28 ($n = 3$) and 0.18 ($n = 24$), respectively. All of the effect sizes were significantly different from zero. Wise's finding of a low average effect size for computer-simulated experimentation is somewhat surprising, as previous reviews generally reported that CSE produced high average effect sizes (Kulik & Kulik, 1986; Kulik et al., 1983; Roblyer et al., 1988; Willett et al., 1983). Wise also looked at average effect sizes for different science curricular areas. He found that CBE in physical science produced the highest average effect size of 0.45 ($n = 16$). One science-technology-society (STS) CBE study was located, which produced an effect size of 0.42. Multiple science areas CBE produced an average effect size of 0.36 ($n = 4$), science process skills CBE produced an average effect size of 0.33 ($n = 14$), and biology CBE produced an average effect size of 0.22 ($n = 16$). Wise found no significant differences in the magnitude of the average effect sizes across these groups. He also looked at grade level and CBE effects and found average effect sizes of 0.12 ($n = 4$) for K–4, 0.24 ($n = 11$) for college level, 0.39 ($n = 20$) for grades 5–8, and 0.40 ($n = 16$) for grades 9–12. There were no significant differences across these groups either. Finally, Wise reported that most studies he reviewed indicated increased motivation and interest on the part of the students using CBE.

Wise noted that computer-based tutorials are one form of CBE that have produced consistent positive average effects across reviews. Tutorials produced the second highest average effect size in his study (ES = 0.45); Burns and Bozeman (1981) reported an average effect size for tutorials of 0.44; Kulik et al. (1983) reported an average effect size for tutorials of 0.36; and Niemiec and Walberg (1985) reported an average effect size of 0.34 for tutorials. Meta-analyses provide an overall average measure of how computers affect learning in science, but they cannot provide some of the specific information indicating how certain programs have affected the learning process and the specific contexts in which CBE programs were successful. Therefore, we highlight a case study that illustrates the effectiveness of the learning-efficiency CBE model.

Several CBE programs, which can be considered exemplary, have been utilized in the science field. This discussion of exemplary projects is not meant to be comprehensive and many exemplary CBE projects in the science domain are not discussed here (for more examples of exemplary CBE programs see "Higher Education Software Awards" information materials produced by the EDUCOM, Washington, DC).

An Exemplary Drill-Practice and Tutorial Curriculum Project

Kleinsmith's (University of Michigan Biology Department) national award-winning computer-based Biology Study Center (1988 NCRIPTAL/EDUCOM Best Curriculum Innovation Award) is an example of how drill-practice and tutorial CBE can be effectively utilized to help learners strengthen their conceptual understanding of biological concepts, as well as learn critical thinking (Johnston & Kleinsmith, 1987). Kleinsmith undertook the computer-based Biology Study Center project to help alleviate the extreme disparity in the level of understanding across students when lecture alone is used to deliver instruction.

Kleinsmith developed drill-practice programs, which allow learners to self-test their understanding of concepts learned in lecture, laboratory, and text readings, and animated tutorials, which help learners add new concepts to their knowledge schemata. Drill-practice software usually connotes fact-recall questions that require only lower-level thinking on the part of the learner. The Biology Study Center's drill-practice questions are rarely of this type. Instead, they are extremely well-designed questions requiring learners to apply concepts and analyze and synthesize information, often requiring learners to manipulate and transform information before a solution can be found. By practicing with analytic or problem-solving–type questions, and aided by the computer programs' guiding feedback, many more learners become proficient at solving higher-level biology questions than before the Biology Study Center was available. The use of specific learning strategies is not explicitly discussed at any time during the introductory biology course; however, the computer programs do provide models of successful problem-solving strategies. Presumably, introductory biology students are able to understand and incorporate these strategies into their own schemata, through use of the Biology Study Center programs, and then apply these strategies during exams and quizzes. It is probable that the "infinitely patient tutor" is a major factor in this cognitive accomplishment.

The questions and answers are randomized each time the students run a particular program, thus removing the possibility of memorizing an answer pattern and allowing an almost infinite number of combinations of questions with which to self-test their knowledge. Feedback is provided for all answers, correct or incorrect. When an incorrect answer is selected, the feedback explains the error in logic and may provide a clue but does not give the correct answer. Students have repeatedly reported learning *more* from the incorrect-answer feedback screens than from the correct ones. They often study by investigating incorrect answers.

In addition to drill-practice programs, animated tutorials were needed for teaching introductory biology, because many biological concepts involve processes and substances moving and interacting in time and space. It is difficult to teach these dynamic concepts using static textbooks, overhead projector diagrams, and verbal descriptions. The computer is better equipped to tutor learners about these concepts because it can present learner-controlled animation. Consequently, Kleinsmith produced animated tutorials that teach the concepts of protein synthesis, the *lac* operon (an example of genetic regulation), and mitosis and meiosis. These concepts are difficult for learners to grasp because they are complex and dynamic, with processes occurring in time and space. Before the Biology Study Center's animated tutorials were available, learners had to mentally translate static diagrams into dynamic biological systems, as well as understand what each biological component was and where complex interactions were taking place. The computer uses simple animation to move shapes, which represent biological components, around the screen in the appropriate sequence and through intricate and vitally important interactions, thus representing a dynamic biological system in action. This greatly eases the cognitive load that would normally be required of the learner in order to understand the dynamic nature of these concepts. Students have control over the speed of the animation and may jump backward or forward in the program, and there are self-quizzes covering each concept. These tutorials make the task of understanding surmountable for a greater number of introductory biology learners.

To assess the impact of the Biology Study Center software, 469 introductory biology students, attending The University of Michigan during the fall 1986 semester, were given a self-report questionnaire asking them to estimate the amount of time they spent using the problem sets, to rate the problem sets, and to provide suggestions for improvement. Students were also asked for their grade level and their overall grade point average (GPA) at The University of Michigan. Teaching assistants provided student exam scores, average quiz scores, section numbers, and teaching assistant for each section.

The results of the analysis revealed that the achievement of the overall class had gone up from an average of 65.6, for 1979–1983, to an average of 81.2 for the 1986 exam. The amount of time spent using the study center software, along with GPA, was a significant predictor of exam success.

Average exam scores of underprepared minority students went up even more dramatically than the overall class average score, jumping from 48.0 for 1979–1983, to an average score of 80.3 for the 1986 exam. In other words, the achievement gap between the overall class average exam score and the average score of underprepared minority students had disappeared.

Student attitudes toward the study center were also very favorable. Ninety-seven percent of the students used the study center software and 85 percent of the class ranked it "1," which corresponded to "very favorable" on the questionnaire. No student ranked the software as "not valuable" on the questionnaire. Kleinsmith also reported subjective personal observations indicating that students were no longer as concerned with the issue of "what will be on the exam" but were truly interested in understanding the biological topics being taught in the class. Kleinsmith's work reinforces the findings of Berger (1984) on his research using microcomputers to gather, analyze, and display data. Berger concluded that the area was one of the fastest growing and had the greatest potential for student learning of science laboratory technique as well as concepts (Berger, 1984b).

It is interesting to note that Kleinsmith's project, recognized as exemplary, would not fit the category of a carefully designed study. Because of its lack of randomized assignment to a control or experimental group and the use of historical data for comparison, Kleinsmith's study would not be included in the previous meta-analysis section. Yet the results of the study fit very well into the findings of the previous sections and indicate that carefully designed CBE is very effective for students with diverse learning styles and cultural backgrounds.

CURRENT RESEARCH IN TECHNOLOGY

Background

Whereas Kleinsmith's work highlights the value of well-designed drill and practice CBE, efforts have shifted in the kinds of questions asked and answered in more current research. In this section the organizing question is, Why should people be involved in research on educational technology? Do media influence student learning? Or, as Clark (1983) has suggested, do media merely influence the way learning is delivered?

Kozma (1991) argued that learners, actively working with a medium, construct knowledge and that the medium and the methods can cause more or different learning depending on the kind of medium being utilized by the learner. He further argued that it is possible to provide a theoretical framework, wherein the learner is seen as "actively collaborating with the medium to construct knowledge" (p. 179), to refute Clark's contention that no further research should be done on media (instructional technology) until a novel theory is proposed.

Mokros and Tinker of the Technical Education Research Centers (TERC) (1987) further argued that the generation and analysis of real data by students provided for in MBL causes visualization of science concepts that cannot be achieved in any other form. The students' attempt to reproduce a velocity graph while moving in front of a motion sensor and watching the real data automatically gathered is a profound influence, whether they are in the first grade or at the university.

We propose that there are strong bases for supporting Kozma's theoretical framework. Aside from the theoretical aspects of learning, there are practical and pragmatic concerns for including instructional technology in schools and, thus, for research on the use of technology in the science curriculum. As we shall see, the paradigm shifts highlighted by Collins (1991) support this viewpoint.

In his article "Learning with Media," Kozma (1991) reviewed research literature related to learning with books, learning with television, learning with computers, and learning with multimedia. He summarized studies that examined the processing capabilities of the computer and that showed how they can influence the mental representations and cognitive processes of learners.

> The transformation capabilities of the computer connected the symbolic expressions of graphs to the real-world phenomena they represent. Computers also have the capability of creating dynamic, symbolic representations of non concrete, formal constructs that are frequently missing in the mental models of novices. More importantly, they are able to proceduralize the relationships between these objects. Learners can manipulate these representations within computer microworlds to work out differences between their incomplete, inaccurate mental models and the formal principles represented in the system. (Kozma, 1991, p. 199)

Thus, the computer is described as a powerful tool for assisting learning. Kozma (1991) concluded that "Clark (1983) calls for a moratorium on media research, but this article provides a rationale for additional research on media" (p. 205).

Salomon, Perkins, and Globerson (1991) presented their views on how human intelligence can be extended with intelligent technologies. Their main points were:

1. Computers "potentially allow a learner to function at a level that transcends the limitations of his or her cognitive system" (p. 4). Effects with technology can redefine and enhance performance as students work in partnership with intelligent technologies—those that undertake a significant part of the cognitive processing that would otherwise have to be managed by the person.
2. The "effects *of* technology can occur when partnership *with* a technology leaves a cognitive residue, equipping people with thinking skills and strategies that reorganize and enhance their performance even away from the technology in question" (p. 8). Learners' use of technology can have spinoff effects on learning and thinking when one approaches a new problem without the aid of the technology. Learners' previous work with the technology might affect how the learners approach the new problem.
3. Learners must be in a state of "mindfulness" for technology to work. Mindfulness is employment of nonautomatic, effortful, and thus metacognitively guided processes (Salomon & Globerson, 1987).

Collins (1991), writing in the *Phi Delta Kappan* on the role of computer technology in restructuring schools, states, "In a society where most work is becoming computer-based, 'school-work' cannot forever resist the change" (p. 28). Collins bases his belief on instructional changes in the reform movements that are taking place. The shift toward constructivist teaching facilitates the rationale for adding educational technology systems in schools. Collins further presents eight shifts in instruction in schools that have adopted computers:

1. A shift from whole-class to small-group instruction.
2. A shift from lecture and recitation to coaching.
3. A shift from working with better students to working with weaker students.
4. A shift toward more engaged students.
5. A shift from assessment based on test performance to assessment based on products, progress, and effort.
6. A shift from a competitive to a cooperative social structure.
7. A shift from all students learning the same thing to different students learning different things.
8. A shift from the primacy of verbal thinking to the integration of visual and verbal thinking.

Thus, technology can enhance major changes in instruction in the schools. Such enhancement can be seen in the following areas.

Microcomputer-Based Laboratories

Microcomputer-based laboratories are systems that couple a data-gathering device, an analog-to-digital converter, with a microcomputer. The microcomputer is then used to record the incoming information from a sensor (such as a temperature probe) and help with organization of that information, often producing graphs from the data (e.g., Friedler, Nachmias, & Linn, 1990; Mokros & Tinker, 1987; Technological Education Research Center, 1984; Tinker, 1985). The use of MBL during science lessons has produced large average positive effect sizes (Wise, 1988).

Adams and Shrum (1990) studied the effects of MBL and conventional laboratory graphing exercises on acquisition of graph construction and graph interpretation skills in tenth grade learners ($n = 20$). Adams and Shrum found that conventional laboratory graphing exercises led to high achievement on graph construction ($p = 0.05$), while MBL appears to lead to higher achievement on graph interpretation (not statistically significant with the small group). The researchers recommend that conventional graphing exercises be coupled with MBL exercises.

Beichner (1990) studied the effect of VideoGraph software (simulating an MBL exercise) compared with traditional laboratory. The subjects consisted of 165 high school physics students and 72 college physics students randomly assigned to four different treatment groups within buildings or institutions. The treatments consisted of (1) traditional laboratory with students viewing an object thrown from one end of the room to another forming an arc; (2) traditional laboratory with students examining and making graphs from stroboscopic photographic representations of the projectile motion; (3) students working with

VideoGraph animation of the projectile motion with students able to start, stop, and replay the motion; and (4) students working with static computer representations of projectile motion paralleling the content of the static stroboscopic laboratory experiment. Beichner did not find any significant achievement differences across the four treatment group. The gains from pretest to posttest were very small, on the order of +1 point out of 24 total points. Beichner believes these results indicate that a real motion event *must* be experienced by learners coupled with the simultaneous graphing associated with other MBL experiments (see, for example, the study of Brassell, 1987) or the significant achievement gains reported by others will not occur. The design of this study was excellent: large N, research based on theory, and random assignment of subjects. Unfortunately, the results were marginal. Why was there only a one- to two-point gain from pretest to posttest for all treatment groups? Beichner notes that this particular kinematics experiment was very simple and most students were probably familiar with the event (a projectile describing an arc in space). Therefore, this motion concept may have been too simple for advanced high school and college students, not requiring an MBL experience for learners to be able to connect the motion event to a graphic representation. It is possible that a more complex and less familiar motion event would turn up significant effects for the VideoGraphic treatment. As this simulation of MBL is not MBL as defined previously, it would be interesting to compare a VideoGraphic treatment with an actual MBL experiment .

Krajcik and Laymann (1989) posited that the overall effectiveness of MBL will depend on teachers' understanding of how to use the new technique, their personal knowledge of the concepts involved, and their knowledge of how to help students link their experiences with the concepts.

Nakhleh and Krajcik (1992) investigated how different levels of information presented by different technologies affected secondary school students' understanding of acid, base, and pH concepts. They found that students using MBL had a large positive shift in their concept map scores, which indicated greater differentiation and integration of their knowledge of acids and bases and provided some evidence that the MBL affected their understanding of chemical concepts. Although students had a positive shift in concept map scores, the use of computers and MBL does not guarantee that thin conceptions (conceptions that, while correct, are very limited in conceptual richness) and misconceptions will be eliminated. Krajcik (1991) analyzed several MBL studies and suggested that the effectiveness of MBL and level of concept richness are connected to the instructional sequence surrounding the MBL activity.

Wise (1986) and Berger (1987a) studied the misconceptions and thin conceptions of students using microcomputers and found that microcomputers are not a panacea to cure misconceptions or thin conceptions. They voiced concerns that use of poorly designed simulations can create misconceptions as well as modify them.

Berger (1987a) reported similar findings with the misconceptions and thin conceptions of teachers using microcomputers. He found many instances of the misuse of computers that even confronted the concepts of science by students.

An Exemplary Microcomputer Based Laboratory Study

Brassell (1987) conducted research on MBL by asking the following questions: How will a very brief treatment with an MBL kinematics unit affect the ability of students to translate between a physical event and the graphic representation of it? How will real-time graphing, as opposed to delayed graphing, of data affect achievement?

She used an ultrasonic motion detector interfaced with a microcomputer to gather and graph data generated by students walking through the detector's ultrasonic field. On day one, high school physics students ($n = 75$) were given an orientation to the study and a pretest that required them to translate between a verbal description of a physical event and the graphic representation (for an example of the type of questions asked see Mokros & Tinker, 1987).

Students present on the second day were randomly assigned to one of four treatments: (1) standard microcomputer-based learning laboratory (graph of data constructed as data come in, real-time graphing), (2) delayed microcomputer-based learning laboratory (20-second delay before data are graphed by microcomputer), (3) pencil-and-paper activity control group, or (4) test-only group. The MBL exercise or pencil-and-paper activities were then carried out. On the third day, the posttest (not the same as the pretest) was administered.

The randomization of the students failed to produce equivalent groups, so analysis of covariance was used to equate the groups statistically. Analysis of covariance results indicated significant positive effects of the standard MBL exercise on graphing achievement scores compared with the other treatments ($F_{3,68} = 6.59, p < .001$). The standard MBL exercises had the strongest positive effect on achievement for distance graphing. The velocity subtest scores were not significantly different across the four treatment groups, although the standard MBL treatment produced the highest scores of all treatments.

The standard MBL treatment also significantly decreased the number of graphing errors on the distance subtest compared to the other treatments. A short delay between time of data collection and graphing (about 20 seconds) caused inhibition of the positive effect of the standard MBL exercise (increased achievement scores and reduction of graphing errors). Brassell proposed that limitations of learners' short-term memories may have been the cause of the inhibition. Standard MBL eliminated the cognitive task of having to hold the actual data-generating event in short-term memory. Instead, learners using standard MBL could simultaneously observe the events and process the information into graphic form as it came in (thanks to the computer's graphing software), enabling them to make meaningful connections in their knowledge schemata. Brassell also noted two other possible explanations for the inhibition of increased achievement by the 20-second delay: (1) lack of motivation for remembering the data-generating event and (2) lack of learners understanding that they should remember or think about the event that had just occurred.

Brassell concluded that real-time graphing MBL activities had a significant positive effect on the achievement scores of distance graphing for high school physics students. The nonsignificant difference on velocity achievement scores may have been due to the experimental design.

Although this experiment was short in duration and straightforward in design, Brassell found significant positive effects for her treatment. In addition, her experiment used statistical matching of students (ANCOVA), which allowed clearer attribution of the variance.

Brassell conducted her study with senior high school physics students. These students were among the higher-achieving students in the school, which limits the external validity of this study. Would the same significant and positive effects result for general track students, younger students, or older students? This study should be repeated with middle school and college-level learners of varying abilities.

Finally, the researcher's failure to find a significant difference between the treatment groups on the velocity subtest may be remedied by increasing the sample size. Also, the difficulty with activating the detector at the same time as the student began moving could be remedied by use of a photogate. The problems with this particular study are mostly methodological and can be remedied.

Brassell's motion detector experiment could be a candidate to determine whether there is a significant difference in favor of the complete MBL procedure, as Beichner believes would be the case.

Interactive Video Disk Projects

The possibility of producing very high quality computer-based science instruction through the combination of microcomputers with video disk systems is also being investigated. This combination of a video disk, capable of storing 108,000 photographic frames, and a microcomputer offers a form of CBE with fantastic possibilities for education. The video disk system allows presentation of high-quality visual images, while the computer makes interactive instruction possible. Efforts by Leonard (1989, 1992) and Bunderson et al. (1984) have shown that this combination is technically possible, albeit expensive, for teaching science concepts.

An Exemplary Interactive Video Disk Project

The WICAT Systems project, funded by the National Science Foundation (NSF) and produced by Bunderson et al. (1984), is probably one of the most instructionally complete and theoretically sound pieces of biological CBE to be designed and studied recently. Bunderson et al.'s group produced an interactive video disk system, "Development of Living Things," which was developed through a series of three formative and summative stages over a 6-year period. The purpose was to design and evaluate an "intelligent videodisk" system in biology. They based their instructional design on current cognitive learning theory, incorporating the constructivist learning theory into the instructional design, and they included rudimentary attempts at explicitly teaching learning strategies to students during the interactive video disk instruction.

The final summative evaluation consisted of the testing of one complete interactive video disk lesson that was used as *part* of a whole semester of instruction in biology. The final CBE program was delivered using text, still images, and moving images and allowed student interaction. Students controlled the pace of the lesson and could branch ahead or backward as needed to form the proper links with their own concept networks. There were practice items and practice tests for each lesson. Learning strategies were modeled, upon user request, by presenting example solutions that gave step-by-step procedures explaining how to approach and solve a given problem.

The interactive video disk lesson was tested for instructional effectiveness against lecture groups at Brigham Young University (BYU) and Brookhaven Community College (BCC). Subjects were assigned randomly to each treatment group—interactive video disk program or lecture—for one complete lesson on specific topics in biology. The topics covered in this study were DNA structure and function and the transcription and translation phases of protein synthesis. Instruction for the interactive video disk treatment groups was carried out at one point during the term by the interactive video disk program; at the same time, these topics were covered in lecture for the lecture treatment groups. Final assessment of student performance was done with a pencil-and-paper exam and questionnaires. When the data were analyzed, the researchers found that the interactive video group did significantly better on test scores (objective and short answer; overall effect size for BCC learners = 0.72; overall effect size for BYU learners = 0.57) and spent about 30 to 40 percent *less* time learning the material (viewing the lesson and studying versus going to lecture and studying) than the lecture group. This time saving is in agreement with those reported by Kulik et al. (1985) for CBE across learners and subjects. Bunderson et al. concluded: (1) The interactive video disk treatment group consistently outperformed the lecture group on tests of biology achievement, (2) the interactive video disk group spent significantly less time (30 to 40 percent less) learning the material than the lecture group, and (3) the productivity ratios (achievement gain/learning time) showed a twofold increase for interactive video disk instruction compared with traditional instruction.

This is an example of an in-depth quantitative CBE study. The researchers were able to use a random treatment–control group design. They found that there was no significant difference between the groups on the pretest measures. Although the evaluation was done in-house, which could result in biases favoring the effects of their product, it does not appear that this was a problem. (The researchers had strong records in instructional evaluation before their affiliation with WICAT Systems.)

Conclusion

From the preceeding review of research, we can reasonably conclude that Kozma's call for more rather than less research on the topics in this review can be supported. Not only are the basic structures in place for testing a constructivist theory, but the tools necessary to carry out the task are starting to appear. Cognitive and motivational aspects of learning are being emphasized to a greater extent in the development of computer-based instruction. Additional studies are needed to solve questions. For example, how can the student link ideas together to create a flexible knowledge structure? How much control should a student have over his or her own learning, and how much guidance should be given or withheld from a student in order to optimize learning and intrinsic interest? In the next section, we present several interesting and insightful research projects and procedures that may start to answer these questions.

NEW RESEARCH ON TECHNOLOGY IN SCIENCE EDUCATION

What are new and promising avenues of research that may indicate a paradigm shift of the use of instructional technology in science education? Indicators of a paradigm shift may be found not only in the kind of instructional technology being developed to solve problems in science education but also in the way such research is carried out. The following research reviews provide a potpourri of what we believe are new and interesting projects, some of which may lead to a true paradigm shift in this arena.

In this section, we examine the following genera of projects; contextual-based problem solving using video disk, problem solving at a distance using telecommunications, science teaching simulations, hypermedia, microworlds, and expert tutoring systems. We also examine research techniques, including automatic logging of student actions with instructional technology and hypermedia techniques. We allocate more discussion to hypermedia and an example of a hypermedia tool, Hyper-Card™, as we believe it offers a unique opportunity for research in science education.

Background

CBE has been in existence for more than two decades and, as a result of hundreds of studies on CBE, some of its effects on learning are beginning to come into focus. There is no question that certain forms of CBE have raised achievement levels in various subject areas relative to a traditional teaching approach (e.g., Kulik et al., 1980; Kulik et al., 1983; Kulik et al., 1986; Roblyer et al., 1988; Vinsonhaler & Bass, 1972; Wise, 1988). As with any other new genera of research, strong changes in the kinds of studies have occurred over time. In this case, early studies focused on drill and practice CBE using mainframe computers with limited graphics abilities and requiring strong programming efforts. Later studies have provided more innovative drill and practice programs; have moved to more and more simulation and MBL; and involve the addition of sophisticated graphics, video disk and high-level programming techniques such as HyperCard™. It would be easy to dismiss drill and practice as an early and simple example of the use of technol-

ogy, but when used effectively even the most direct drill and practice can be very effective. It is clear, however, that the future of instructional technology lies with the ability to use multimedia to provide supporting instruction experiences for a wide variety of students with diverse backgrounds not only of culture but also of learning styles.

Furthermore, studies comparing computer-based learning with non-computer-based learning do not necessarily provide a link to or understanding of what is going on while students are learning using instructional technology. Even the best pre-post and randomized designs cannot answer such questions. By accompanying a detailed analysis of students' interactions with the computer (often by keeping log files of keystrokes and mouse moves), a start to analyze student learning during the experience with technology can be inferred. Such research started early in the 1980s and continues today (Berger, 1982; Berger, Newman, & Prince, 1981; Dershimer, Berger, & Jackson, 1991). The examination of the use of science processes with microcomputers is one such area. Berger and his colleagues have shown powerful learning curves and teaching efficiency (that is, more learning in a shorter time) as first-year college students learn chemistry (Zitzewitz & Berger, 1985). Starting with a discussion of how log files could be used to describe the learning of the process of estimation in a gamelike environment in 1981, they have shown learning with the use of microcomputers provides strong insights into learning in alternative modes (Berger, 1987b; Berger & Jackson, 1989, 1990; Berger & Pintrich, 1986; Berger, Pintrich, & Stemmer, 1987; Cox & Berger, 1985; Edwards, Jackson, & Berger, 1990; Jackson, Edwards, & Berger, 1990). Later work has focused on the use of hypermedia to take advantage of such models (Dershimer & Rasmussen, 1990).

CBE has the potential to aid learners in learning how to learn and has been used successfully for teaching problem-solving skills and cognitive strategies (see Table 16.1). Friedler et al. (1990) have shown that problem-solving skills or scientific reasoning can be improved (as measured by a "scientific reasoning test") via CBE, such as by use of Discover a Science Project software by SunBurst combined with MBL activities on heating and cooling.

In addition to the use of instructional technology in the delivering of instruction, Berger (1982) and Fisher (1990) proposed that there are compelling examples of the use of technology to research the learning that may be going on while students use mediated instruction. Their use of automated event recording, development of concept maps, and semantic networks indicates that the technology can be used to assess a level of understanding and navigation through a body of knowledge that provides for a rich base of inference for student learning.

Callman, Faletti, and Fisher (1985) developed a prototype computer-based semantic network in molecular biology that uses concept mapping to introduce knowledge structures to learners. Callman's group has constructed an elaborate semantic network of microbiology concepts contained in computer files. Concepts are arranged hierarchically and are cross-linked to all other relevant concepts in the network. In a game format, students find a path from concept X to concept Y. The computer recreates the path taken by a student and reports the shortest and longest paths taken by students. This prototype program is just beginning to make use of concept mapping to help learners think about knowledge structures in the biology domain. The authors plan to continue to refine their program and report again on their progress.

The Jasper series, developed by the Cognition and Technology group at Vanderbilt (1990), is another good example of technology-based problem-solving instruction. In this professionally produced video disk–based learning environment, students and teachers share a common "macro-context," the adventures of Jasper Woodbury, which are shown in approximately 15-minute video segments. The task for the learners is to solve the problem that Jasper faces. For example, in the first episode, Jasper has traveled down a river and purchased a boat. He is faced with the problem of being able to get home before dark, given the amount of gasoline he has in the boat and the time of day. The students are engaged in a complex problem-solving task (requiring the generation of approximately 15 subgoals) in which they attempt to figure out when Jasper must leave in order to get home before dark and whether he can even make it with the amount of gas he has in the tank. The learning is anchored in a real-life situation, and the learners and teacher can quickly scan the video disk for "embedded data" necessary to solve Jasper's problem. Eventually, the Cognition and Technology Group plans to develop 6 to 10 Jasper adventures and accompanying computer data bases, which will provide the capability for the addition of new problems. With these video disk–based adventures and their accompanying data bases, students will have the opportunity to learn math, history, science, geography, and other subjects in a highly motivating problem-solving macro–context-based arena.

These studies are representative of the infant status of CBE that couples content instruction with explicit instruction about learning strategies and metacognition. This type of CBE is still in the developmental stage and future research should focus attention on this area. Bunderson et al. (1984) made an early attempt at overtly coupling the teaching of learning strategies with content delivery when they explicitly modeled learning strategies for the biology domain, during the interactive video disk instruction. Perhaps this was one reason that their program was so successful at improving achievement and decreasing learning time. Coupling comprehensive instruction in learning how to learn in a given content area with the presentation of content and procedural information could be highly effective at increasing learning (Resnick, 1987). CBE projects that present in-depth instruction on learning (including instruction about metacognition) in a content area, rather than simply teaching subjects a few learning strategies and testing to see if the treatment subjects use these strategies significantly more than a control group on a posttest, will be a major research undertaking but should be pursued.

A Prototype of a Drill-Practice and Tutorial Curriculum Project

The interactive video disk system developed by Bunderson et al. is certainly an impressive and important accomplishment.

However, that particular type of CBE will not reach the majority of public school systems, especially urban centers that are financially burdened, for years to come, if ever. Standard microcomputer-based education, including drill-practice, tutorials, simulations, problem-solving programs, MBL, and telecommunications, are already operating in American school systems. The effects of many, if not most, of these classroom-based CBE systems have not been, and are not going to be, tested. Instead, their merit is simply accepted as given. This could lead to poor or inappropriate instruction, wasted funds and time, and, most important, detrimental effects on young learners' education.

Baird (1988) suggests that truly successful CBE programs, at the national level, will require the following:

1. Involvement of practicing classroom teachers in the design and implementation of CBE programs
2. Design contributions by programmers, cognitive scientists, and teacher trainers
3. Major funding of design and implementation; at least over a 5-year period
4. Grass-roots communication among the design team and teacher practitioners
5. Low-cost products readily available to teachers on the most common hardware

Kleinsmith has undertaken a CBE curriculum development project, known as BioNet, with high school biology teachers that happens to follow Baird's guidelines (1–5) almost to the letter (1988). Public school teachers have written hundreds of questions covering their entire biology textbook. The questions were programmed by expert programmers and are currently being reviewed for correct content and educational value by public school teachers, former public school teachers, cognitive scientists, and teacher trainers (suggestions 1, 2, and 4). Kleinsmith has secured major funding for hardware, software, and consulting or troubleshooting for 3 years to date and plans to continue with the projects until they can become self-sustaining (suggestion 3). The software is running on both local area networks and stand-alone personal computers with dual floppy disk drives (suggestion 5). The effects of the BioNet computer programs on achievement and attitudes of high school biology students are currently being investigated.

The first generation of computer-based education has demonstrated that CBE can produce positive achievement gains; it is effective for all levels of learners; tutorial-based activities tend to dominate the field, are in place, and are effective; and student motivation to work with computers is high. The next generation deals with more complex operations of computers and with more powerful thinking on the part of learners. There is less research on the use of complex systems such as hypermedia, microworlds, and expert tutoring systems, but what is available is explored.

Contextual-Based Problem Solving Using Video Disk

Sherwood (1991), working with the Cognition and Technology Group at Vanderbilt, argued that technology, through anchored instruction, may provide the rich environment of experts that would be lacking in traditional classroom environments. Furthermore, the problems occurring in real situations seen through video "documentaries" provide an authentic base for hypothesis generation and solution.

The Cognition and Technology Group at Vanderbilt University (1990) initiated the concept of anchored instruction—that is, instruction situated in an environment that permits sustained exploration of students and teachers and enables them to understand the kinds of problems and opportunities that experts in various areas encounter and the knowledge that these experts use as tools. In-depth video disk–based problem-solving lessons, such as the Jasper series, are the kinds of contexts used for their anchored instruction. Their rationale for this technology is:

1. Instruction via video disk–based lessons allows relatively easy access to rich information, relatively easy ability for teachers to communicate problem contexts to students that are motivating and complex yet ultimately solvable by students.
2. Video disk technology can make it easier for students to develop confidence, skills, and knowledge necessary to solve problems and become independent thinkers and learners.
3. Video disk–based instruction is presented in visual format. This may allow students to (1) develop pattern recognition skills and (2) experience the learning situation for themselves rather than text that presents the author's interpretation of events; (3) it is dynamic, visual, and spatial, allowing learners more easily to form rich mental models of the problem situation, which is especially important for low-achieving learners and learners with little knowledge of the area (Bransford, Kinzer, Risko, Rowe, & Vye, 1989; Johnson, 1987); and (4) it allows random access, so students and teachers can go back over video disk information and explore problem situations from many angles.

Telecommunications

A relative newcomer to CBE involves collaborative science study via telecommunications. Telecommunication projects such as WaterNet, developed by Berger and Wolfe at the University of Michigan and funded by the Department of Education; MIX (McGraw-Hill Information Exchange); and Kids Network developed by the Technical Education Research Centers and funded by the NSF and the National Geographic Society, hold great promise for increasing the quality of science education in the near future (Lenk, 1988).

WaterNet is a telecommunication-based water pollution study involving high schools in the United States, West Germany, and Australia. Through sharing of data collected from their local sites, it is hoped that students will attain a critical understanding of the problems of water pollution and develop an interest in solving these social problems.

MIX involves elementary school students experimenting on the growth of corn seeds at different locations in the United States and Canada. They will share data and findings with each other, as well as being able to contact experts using telecommunication.

Kids Network is a telecommunication-based science curriculum for learners in fourth through sixth grades in the United States, Canada, Israel, and Argentina. The Kids Network participants will focus on different topics, such as acid rain, land use, weather, and health, every 6 weeks. They will gather data from their local sites and then share their data and findings with other school involved in Kids Network.

These telecommunication projects allow learners to collect and analyze data on local phenomena and share their data and findings with students around the world. They can also have access to experts and huge data bases, allowing students and teachers to open up their minds to the science classrooms of the world. These exciting and important new arenas of CBE are currently being implemented, but their effects on learning have not yet been thoroughly tested.

Science Teaching Simulations

Two studies that captured the use of the computer in teacher education were reported by Kimball (1974) and Shyu (1987). Kimball (1974) developed and designed a computer-simulated classroom, called Teach, which could serve as a teaching aid in a university methods classroom.

The development of the Teach computer simulation had three fundamental steps. First, a list of classroom management strategies was developed by the researcher and evaluated by experienced teachers. Second, a computer program was written that could generate classrooms of fictitious students with pretest characteristics. Finally, a series of assumptions were made allowing constants to be calculated that would produce a lifelike interaction between the strategies and the mythical classroom of students. During the computer simulation, the user has an opportunity to work with the whole class and with individual students. Discipline problems and possible ways to handle discipline problems were also included.

Kimball limited his study to a one-on-one teaching station because the data had to be entered into a terminal, sent to the central mainframe, and then returned to the teletype at the terminal. When the Apple computer's memory and processing capacities became larger, the program was adapted to the microcomputer by Shyu so that the Teach simulation can now provide simulated teaching experiences to teams of methods students competing against one another. Shyu also reformatted the Teach simulation by validating and updating the strategies with "honors" science teachers. Both studies showed increased gains in student teachers' awareness and use of teaching strategies (Voss, 1990).

Hypermedia

Hypermedia can best be described as a computer-based environment for thinking and communication. Hypermedia, of which Hypertext is an example and HyperCard™ is an implementation, consists of audiovisual, as well as text and graphic, capabilities and access to video disk, data collection, and analysis devices. Hypertext refers to text triggers or key words within the text that are designated by different type styles and that can be "clicked" on and opened when further information is necessary. HyperCard™ (and similar applications, like Course of Action™ or Toolbook® is a computer program that includes an extensive data base capability, an organizer of graphic images (still images or animations), and a programming language (Hypertalk). A variety of media, each of which reinforces the others, offers benefits to the learner. Interactive multimedia allows students to learn in many different ways—through seeing, hearing, reading, writing, and doing. The medium provides students with the opportunity to organize their thoughts, and create new thoughts. "By encountering the same material in a variety of modalities, students grasp the richness and depth of the material. They also extend and refine their own capabilities, becoming better viewers, creators, and critics" (Friedlander, 1989, p. 38).

HyperCard™ Features

"HyperCard™ is a piece of (computer) software that provides the user with access to decks of cards (representing screens) containing text and graphic elements which can be easily rearranged into a variety of combinations producing exciting new programs" (Gray, 1989, p. 39). HyperCard™ contains different levels of complexity from a read-only level to an authoring level at which the user can create, modify, or invent new applications from the existing stacks. HyperCard™ uses a stack-of-cards metaphor, whereas other applications, such as Toolbook®, use a page-chapter metaphor. In HyperCard™, for example, information stored on several related cards can be grouped into stacks. Programmable buttons can be used to link cards to one another or to access visuals and auditory overlays, still images, or animations (termed hypermedia). A point-click-zoom interface allows the user to make selections of content to cover, applications to use, and paths to explore through the stacks during instruction. The system keeps track of the user's path, allowing easy return from exploratory side paths. Many programs feature a global map that shows the entire hyperdocument and allows navigation within it and a local map that presents a view centered on a single document, displaying the links to its nearest neighbors in the web. Nonsequential links between one document and another allow users to follow paths in any direction without losing their original context (when students finish exploring a tangential path, they can return to the original path from which they made an excursion). Stacks are designed to be explored in greater and greater depth as students proceed through them. By giving students a tool that allows them immediately to gratify their intellectual curiosity through exploration, hypermedia turns students into active learners rather than passive receivers of knowledge. Students must take the initiative in selecting the paths they will explore and in clicking into the stacks—the software demands that the student make choices—"or everything just stops" (McCarthy, 1989, p. 31). Icons representing intuitive functions can be used

to guide navigation—when students click on a movie camera, animations begin; when they click on a forward arrow they move to the next stack; when they click on a help icon they may see a map overlay showing them where they have been and in what direction they are headed or providing more explicit instructions to address a problem related to the part of the hyperdocument in which they are engaged.

Marsh and Kumar (1992) reviewed recent findings on Hypermedia and made the following points on uses of it. Hypermedia allows:

1. Learner construction of knowledge, active learning, and learner control of learning; learners may explore in a nonlinear fashion.
2. Knowledge exploration without having to leave the medium (area) in which one is working. For example, if you are reading a textbook and wish to check one of the references, you must go to the library and look up that reference. Some or all of those references will be included in the hypermedia.
3. Easy integration with STS instruction because the lateral links necessary for connecting science concepts with societal issues are available.
4. Presentation of knowledge structures, which are omni directional in nature, so the learner can observe what the knowledge structure of a field might look like.

Construction of Knowledge

Although extensive research on hypermedia is not yet available, the utility of this medium for teaching science is undeniable—the combination of flexibility and a sizable storage capacity allows users to access and organize information in entirely new ways. The integration of HyperCard™ capabilities with graphics, video disk, real-time data collection devices, and data base resources makes the program a useful educational tool. Special attributes of the program can address the need to teach science students to construct their own flexible knowledge base, to analyze continuous changes in scientific theory, and to evaluate evidence with a skeptical intelligence. Fostering flexibility in the acquisition and structuring of knowledge is a desirable way to educate people capable of looking beyond the existing paradigms to new and yet undiscovered ideas.

Relating Ideas and Making Unconstrained Connections

HyperCard™ requires students to organize and make connections between their ideas. In order to learn (acquire and use knowledge), students must relate ideas to one another and to what they already know. Stacks of related cards allow students to make new connections and put ideas together in unconstrained ways. HyperCard™ makes it possible to find, store, and create connections more richly and flexibly than ever before. When students are required to make their own connections among information in the stacks, they are constructing their own understanding of the material. "Hypermedia gives the student power over the medium: the power to explore a body of information without being constrained by the author's view of how it all fits together, the power to follow an idea as far as one's imagination, and the medium, will allow" (Underwood, 1989, p. 8). Where students are required to make their own connections among the information in the stacks, they are participating in the construction of meaning and building their own knowledge base. The flexibility and open-endedness of hypermedia is noteworthy—the connections between pieces of information are dynamic and can be modified. Uninhibited exploration should enhance flexible thought, as well as originality, especially when students can so easily follow alternative paths and consider other possible explanations, combinations of ideas, or even new paradigms.

Nonlinear Navigation—Appropriate Path Choice

Hypermedia environments present texts and other applications nonlinearly, requiring users to decide what information to read and in what order to read it. "Appropriate" path choice (exposes students to information in a particular sequence) can enhance flexible and original thought, leading students to structure the information differently and even make unusual connections. However, some students may equate the hypermedium with textbooks (reading cards in sequence from the beginning to the end), not taking advantage of the full flexibility of the software (capabilities to relate new ideas or move through the medium in a nonlinear manner). Little research has investigated what strategies students adopt in navigating through the informational network, what factors influence choices of the sequence of their instruction, and what impact the choice of paths through the network has on student learning. Charney (1989) has proposed that presenting hypertext information in a poor order can impair student learning; this prediction is supported by empirical evidence that inexperienced readers of scientific articles and texts make poor selections of what to attend to or ignore important information altogether. Charney is testing this prediction by randomly assigning subjects to conditions (for reading information in a Hyper-Card™ network) that vary in randomness of reading and in the degree of student-planned or instructor-guided sequencing. If used properly, hypermedia could have a large impact on learning, but educators must make sure that students comprehend a larger, more global picture of what they are studying.

Problems Limiting the Utility of Hypermedia

The temptation is to make the system so free and interactive that users have complete control at every moment. While this is a praiseworthy goal, users can often feel bewildered and overwhelmed by choices and uncertainty. (Friedlander, 1989, p. 36)

The danger lies in getting caught up in the delightful concept of empowering students to do their own learning. Students should be the

navigators but what you don't want is just a lot of random learning about a given subject. You want a presentation that has some built in direction; something that leads the students from one important piece of information to another. (McCarthy, 1989, p. 31)

Because of the complexity and variability of the medium, students may have a tendency to lose their sense of direction and can experience a cognitive burden when information is too abundant and many decisions must be made. Such problems are being addressed by researchers such as Marchionini (1989, 1990). Two problems—disorientation and cognitive overload—may reduce student learning, limiting the utility of hypermedia. When students become overwhelmed in a hypermedia environment, they fail to benefit from the flexible features of the system that can allow them to make unusual connections. The additional effort and concentration required to maintain several tasks or paths simultaneously may result in a cognitive overload for the student. Several strategies are currently being implemented to alleviate the problem. Concept maps can be accessed intermittently to give the learner a sense of direction during instruction. Information can be filtered so that the learner is presented with a manageable level of complexity and detail and can shift the view or the detail suppression while navigating through the document. Such features, under the control of the student, can be accessed as necessary. Another problem associated with cognitive overload is that of "informational myopia"; the learner must decide whether following side paths is worth the distraction.

These problems are not new with hypertext, nor are they mere byproducts of computer-supported work. People who think for a living—writers, scientists, artists, designers, etc.—must contend with the fact that the brain can create ideas faster than the hand can write them or the mouth can speak them. There is always a balance between refining the current idea, returning to a previous idea to refine it, and attending to any of the vague 'proto-ideas' which are hovering at the edge of consciousness. Hypertext simply offers a sufficiently sophisticated 'pencil' to begin to engage the richness, variety, and interrelatedness of creative thought. (Conklin, 1987, p. 40)

To the extent possible, it is necessary to give our students realistic problems that reflect the complexity and uncertainty in science. Problems of informational myopia within hypermedia environments should be correlated with how quickly divergent thinking can occur and should reflect flexible and original thinking patterns. HyperCard™, with its flexible features (particularly stacks permitting rapid, unconstrained interrelation of ideas), may be an ideal medium for balancing the creation and refinement of ideas.

Networked Hypermedia—Collaborative Flexible Knowledge Construction

If use of hypermedia is completely undirected, students may construct faulty connections (misconceptions) that are never evaluated by the program (because it does not have an evaluative capability). However, as a solution to this problem, hypermedia can be used to encourage and enhance collaborative thought. Networked hypermedia provides a vehicle by which students can not only generate and store connections but also share the connections they have discovered with other students and the instructor (Campbell, 1989). The networking strategy can enhance flexible knowledge construction through collaboration with other students. In this way, the relationships among concepts and student "knowledge" can be subjected to the scrutiny of colleagues, just as scientific evidence and theory are subject to scrutiny by the scientific community.

The thinking process does not build new ideas one at a time, starting with nothing and turning out each idea as a finished pearl. Thinking seems rather to proceed on several fronts at once, developing and rejecting ideas at different levels and on different points in parallel, each idea depending on and contributing to the others. (Conklin, 1987, p. 32)

Students can build on a consensus of the class's connections and continue (throughout their academic and scientific careers) to access and modify their existing conceptual framework as their knowledge base grows and new evidence causes existing theory to be overturned. Students can contribute to the learning of their colleagues by posting their ideas for discussion. Critical thought and doubt are enhanced as students realize that justification for their ideas is crucial in determining how uncertain their position is and whether their "knowledge" should be tested further, or perhaps changed.

Hypermedia environments could store original documents as text and provide scaffolded questions that help students become more analytical about what they read and more independent in their analysis. Analysis and critiquing of original scientific papers have also been employed in university-level science courses to teach students to evaluate critically hypotheses, methods, data, and analyses and to confirm that data support the conclusions drawn. Even introductory students can and should learn to formulate and evaluate scientific arguments early in their science coursework so that as they learn the accepted scientific theory, they can critically evaluate it. Papers should be chosen carefully for this purpose and efforts should be made to involve students in collaboratively defining difficult terminology and identifying problems addressed in the research (Woodhull-McNeal, 1989). Students need to develop critical evaluative skills if they are to critique their own research as well as that research of their colleagues. Students should develop a skepticism that permits them to look beyond what is in print; they should realize that published articles are not always well supported and may on careful reading be found to have problems in methodology, interpretation, or analysis (Woodhull-McNeal, 1989; Janners, 1988). Hypermedia environments should facilitate the development of students' critical evaluation skills, particularly if students work collaboratively.

Prototypes of Hypermedia Applications

The application of hypermedia is widespread. It is being used to teach everything from foreign languages to Shakespeare and *The Grapes of Wrath*. It is also being used to teach science, particularly at the college level. For example, Hyper-Card™-based instruction is being developed at Stanford, to integrate the physiology of rapidly changing interactions between human body systems by using video disk to access complex audiovisual imagery with overlaid graphics and text (Friedlander, 1989).

Rohwedder (1990) has brought together several computer-aided environmental education studies that represent a sampling of the potential, problems, and promise of computers in general and specifically in the field of environmental education. Four major applications addressed in this publication are hypermedia (including interactive video disks), simulation and modeling, interactive software, and telecommunications. Examples of each of these applications and how they are invaluable in furthering environmental education are discussed.

Dersheimer and Rasmussen (1990) have, in conjunction with the Office of Instructional Technology (OIT) at The University of Michigan, designed a Supercard program (similar to HyperCard™) called Seeing Through Chemistry. This program uses several different media to reinforce the same concepts and principles so that students begin to integrate concepts into a robust conceptual framework that represents more than a collection of fragmented ideas. The program features a trail map that assists students in navigating through the hyperdocument and indicates the extent of guidance required by the student on each section. The program operates at several levels, permitting students to explore freely (browse), constructing their own understanding of the domain, and at other levels asks students inquiry-focused questions at three progressive levels of guidance. Inquiry-directed instruction permits students to jump reiteratively back and forth between their written explanation for a question and the stacks, without intervention. Assisted inquiry instruction (with hints) suggests areas in the stacks that will provide essential information. Guided instruction more directly guides the student through the sequence of stacks to consider in linking concepts, without giving students the answer. This multimedia learning environment is designed to place the burden of error correction on the student; the typed explanations to inquiry-focused questions are not evaluated by the computer but are turned in to the instructor.

Cordts, Beloin, and Gibson (1989) at Cornell University developed a HyperCard™-based program entitled A Tutorial in Recombinant DNA Technology. The program won an award in the 1989 EDUCOM/NCRIPTAL Software Awards Competition (University of Michigan); it had noteworthy animation and still-image features for use in student-directed learning. The hyperdocument included (1) a compulsory Help section for first-time users, which presented instructions for navigating through the program; (2) a Refresher Course, which reviewed essential background knowledge; (3) a quiz on background material so that students knew whether review was necessary; (4) the body of material (Recombinant DNA, Applications of recombinant technology, Safety and legal issues, and Environmental implications); (5) a bibliography citing references beyond the scope of the tutorial; and (6) an index, which permitted rapid movement between various parts of the tutorial. Each of these sections was represented in a map intended to assist students in navigating through the hyperdocument. Navigational concerns were addressed through features like an easily accessible help section, the map, and index features, which allowed students to move about in the hyperdocument. Point-and-click images permitted students to control their navigation—moving forward, backward, back to "Square 1" (the beginning of the instructional sequence), or into the animation segments. The tutorial was designed to resemble a book with a linear flow of information and division into "chapters" and "subchapters" that would be intuitively familiar to any user (Cordts et al., 1989). Novel hypertext features were also included in the program. "Pop-up" definitions, which, when clicked on, move students to the portion of the tutorial where a relevant explanation for the term exists, and "closer-look cards," which move the students to more detailed information without losing their main path in the tutorial, were included to capture the interest and enhance the understanding of a diverse range of college students.

Hallada (1990), a chemistry instructor at the University of Michigan, has developed a ChemTutor exam question and answer data base for introductory chemistry college students. The HyperCard™ program allows students to focus on areas in which they are having difficulty. Multiple-choice problems are randomly generated for students. When the student selects an answer, the program indicates whether or not the answer was correct. Wrong answers represent common misunderstandings and mistakes made by students; thus, further explanation is given when the answer is incorrect. The extent of guidance given depends on the needs and responses of each student. However, if students have questions beyond the scope of the answer options and explanations provided, teacher assistance is crucial. In addition, Hallada has developed a system for keeping records on each student, including information on academic and class standing, the amount of time each student spends using the problem-solving program, which answers they choose, and how many times they choose to enter a problem domain. Considerable effort was put forth in pioneering a networked system that could accommodate multiple users simultaneously. The data base offers students an opportunity to practice, repeatedly if necessary, solving problems of examination difficulty. The results of such efforts are impressive. Hallada reports strong increases in grade point average of students, particularly minorities, of over 15 percent (Hallada, M., personal communication, December 1, 1992).

Microworlds

Another kind of computer-assisted learning is the simulation microworld environment. These environments are computer laboratories simulating real-world phenomena that vary in complexity depending on the number of variables and the extent of interaction between the variables. Such environments usually allow students to discover principles on their own; the expertise of an instructor may be required to facilitate productive and meaningful learning. Students are usually expected to ask questions, interpret results, and draw conclusions on their own, with limited guidance from the microworld environment. Experiments can be scaffolded so that the increase in complexity is gradual, incorporating a greater number of interactive variables that more readily reflect real-world problems. This scaffolding can help students to learn fundamental principles that, with some or much modification, may be applied to more complicated systems. Microworld environments give students the opportunity to explore and discover scientific principles by actively constructing meaning (students work to put knowledge together and link ideas). These environments engage students beyond discovering a "fact"; students are "challenged to evaluate three structures of their knowledge: methodology (strategies), justification (how do you know that you know), and the perception of which components are under investigation (what is the real problem here)" (Peterson, Jungck, Sharpe, & Finzer, 1987, p. 113). Although microworlds should never replace laboratories, they can help students to think more critically, evaluating, not merely accepting, theory and methods of scientific investigation, and to gain practical experience design. "A user should be able to pose problems of greater and greater sophistication as his expertise grows. Problem posing is intimately tied to developing the ability to perceive critically" (Peterson et al., 1987, p. 115). Microworlds provide students with the power to explore realistic and scientifically relevant problems that sustain student interest; the microworld environment is designed to require students to construct and reconstruct meaning as they frequently encounter unanticipated consequences and evidence that counters their understanding. A microworld that is well designed will provide multiple views of the same knowledge, enabling the learner to build a better understanding of a complex system. "The combination of multiple views, multiply posed problems, and multiple strategies provides the robust richness of microworlds. No optimal path exists through the combinatorial complexity" (Peterson et al., 1987, p. 115).

Testing Hypotheses for Complex Problems with Multiple Solutions

Students must be able to explore, to solve problems with many solutions, to be wrong, and to engage in debates in which they are forced to justify their reasoning. If appropriately designed, microworld environments can teach science students how to think as a scientist would, with conditional reliance on existing information, based on the best available and carefully scrutinized evidence. Peterson and colleagues (1987), who have designed microworld environments, suggest that the process of scientific investigation is initiated from tentatively accepted theories and seeks to understand a phenomenon where the "goals are fuzzy, very fuzzy, when compared to the concise statements that will ultimately be published in the scientific literature" (p. 113). Students working in microworld environments assume the role of a novice scientist, developing and testing hypotheses; articulating a defense for their problem statement and design; entertaining several hypotheses simultaneously; testing and evaluating the evidence for each; and, where possible, ruling out alternative hypotheses. Students must be able to argue persuasively for their position with a qualified degree of confidence, in a manner that permits questioning, critical analysis, and possible acceptance of hypotheses and conclusions by the student's peers (based on available evidence and justification). Strategically, complex problems involving several solutions require students to evaluate the evidence and justify the best possible solution and experimental design; therefore, doubt and flexible and original thought are likely to be fostered. Through experience in the microworld environment, students should come to understand that scientific theories have withstood the challenge of testing and retesting by the scientific establishment and that theories are open to challenge and can be modified, as necessary. The real world, which is presumably what we are addressing in academia, presents scientists with complex, gray, ill-defined problems that have a large number of solutions, depending on the assumptions that are made. Scientists must decide what information is needed to work toward a solution (Zoller, 1987).

Gray, ill-defined problems do not confine or restrict student ideas but permit a fluid divergence in thought; assigning such problems forces students to deal actively and intimately with problems, to recognize that realistic problems are complicated and their solutions often numerous. Students should acquire a deeper understanding in considering what information to use in solving problems and how to justify their solutions. Students can become more original and flexible thinkers as they explore realistic problems, beyond discovering what is already "known" (maybe just to the student), and imagine possible alternatives. Students can become more skeptical as they consider what information is and is not necessary for solving problems and how their solutions are justified (Pinet, 1989).

Prototypes of Microworlds

As early as 1979, Berger and colleagues adapted a Power Company Minicomputer Simulation for use with the then-developing microcomputers. In the simulation, Berger found that students developed a more balanced view of energy use in terms of the supply, cost, and hazards of gas, oil, coal, and nuclear power. By actually simulating the running of a power company the student learned, in a far more holistic way, that the classroom presentation of energy concerns and issues was realistic (Berger, 1979). Other studies and programs added quickly to the use of simulation and microworlds, as evidenced by the work of Berger (1984a), Choi and Gennaro (1987), and Wells and Berger (1986).

Microworlds developed by Peterson and Campbell (1985) have the following features: (1) The programs use a graphic rather than a textual interface. Tutorials, text, questions, and answers are nearly absent, leaving the student to pose problems and design experiments. Manipulations necessary to effect changes in experimental design involve a point-and-click interface. (2) Student performance is evaluated by the instructor; no tutoring function is available, so students who require remediation must consult the instructor. (3) Experiments reflect reality, in that no endpoints to hypothesizing and experimentation exist—scientific inquiry is a continuous process.

A Cardiovascular Systems and Dynamics simulation-interactive video disk program allows medical students to explore a classic experiment encountered in their text, which is too complicated for them to perform in the laboratory (Peterson & Campbell, 1985). Students make and test hypotheses on the cardiovascular or respiratory systems in normal and diseased patients, with isolation of other systems. When each system is understood in isolation, simulated experiments are then complicated by interactions between the systems or interactions with drugs administered to the "patient."

Another program, the Genetics Construction Kit (Jungck & Calley, 1986), uses open-ended problem-solving simulations in genetics. It requires students to discover how many inheritance patterns are present in a given sample population, in much the same way as Mendel did—by performing crosses and observing the phenotypic traits of offspring. The simulation involves a great deal of uncertainty in genotype because some phenotypes may not be expressed in the original sample or through several generations of crosses.

Krajcik and Peters developed a dynamic display of particle animation in the "Molecular Velocities" computer microworld. Students used visual and nonvisual versions of the software. Krajcik found that although students may shift in their understanding of the dynamic nature of particle model, much depends on initial conceptions (Krajcik, 1991).

The autonomous classroom computer environments for learning (ACCEL) project, developed by Linn and colleagues (1988), investigated how precollege students, enrolled in programming courses, could increase their learning outcomes. These environments were similar to microworld learning environments. Expert techniques for solving programming problems were analyzed in order to improve instruction, identify ways to communicate to students the key techniques used by experts, and determine whether expert techniques were applicable to the learning activities of these students. Students learned to program through unguided discovery, primarily getting feedback when they ran their programs. Initial findings suggest that students need practice to acquire a set of well-defined algorithms and procedures typically used by experts in solving new problems. Students learned significantly more about programming when they wrote and ran programs themselves than when they read expert solutions to new programming problems.

Science Quest (1991), under development at Florida State University, is a middle and junior high school science program with an inquiry approach that combines the processing power of the computer with the visual capabilities of video disk. Video disk technology allows the students to explore a wide variety of environments. They are then involved in problem-solving situations that require them to do research using computer-based tools and video resources.

Expert Tutoring Systems (Artificial Intelligence)

Intelligent tutoring systems (ITS) are a third type of computer-assisted learning; they provide "adaptive" and supportive learning environments intended to facilitate active learning. In most cases, the expert tutoring environment is structured so that the student has the opportunity to make self-discoveries as much as possible. Coaching, characterized as "guided discovery learning," "assumes a constructivist position, in which the student constructs new knowledge from their existing knowledge.

In this theory, the notion of misconception or 'bug' plays a central role. Ideally, a student's bug will cause an erroneous result that he will notice" (Burton & Brown, 1982, p. 80). The bug is constructive if the student has enough information to determine what caused the error and can fix it. The bug is nonconstructive if the student does not have sufficient information to correct the error. The significant challenge in providing students with the appropriate tutorial guidance is to predict when students require additional information and what kind of information they require to transform nonconstructive bugs into constructive ones.

Considerations in System Design

Whenever possible, guidance should be minimized, and students should not simply be given answers. Sleeman and Brown (1982), who have designed expert tutoring systems, stated that

[Expert tutoring] systems attempt to combine the problem-solving experience and motivation of 'discovery' learning with the effective guidance of tutorial interactions. These two objectives are often in conflict since, to tutor well, the system must constrain the student's instructional paths and exercises to those whose answers and likely mistakes can be completely specified ahead of time. (p. 1)

To meet the instructional needs of students, the system must be fast, robust, and capable of handling inconsistent and incomplete information from students.

To overcome these limitations, the system must have its own problem-solving expertise, its own diagnostic or student modeling capacities and its own explanatory capabilities. In order to orchestrate these reasoning capabilities it must also have explicit control or tutorial strategies specifying when to interrupt a student's problem solving activity, what to say and how best to say it. (Sleeman & Brown, 1982, p. 1)

The coach must ensure that guidance given to the student is both relevant and memorable. Student wait time must be minimized; the coach may provide initial information for the student to ponder, while the system accumulates additional evidence based on the moves the student is making. If a student asks for help, expert tutoring systems can provide several levels of guidance. At the first level, the coach identifies the student's weakness. If a skill is required to make the optimal move, the coach suggests that the student consider that issue. At a second level, if the student is still floundering, the coach provides the student with a list of possible moves or outcomes. At the third level of guidance, the coach selects the optimal move and explains why this option is the best. Finally, at the fourth level, the coach fully describes how to make the optimal move, by example or in a step-by-step fashion. Obviously, the tutor progresses through these levels only as necessary because each level provides more explanation, permitting the student to discover less and less. When students are robbed of the opportunity to discover for themselves or to make mistakes, they are likely to lack the cognitive skills necessary for confronting problems with unknown characteristics or uncertain outcomes. Strategically, the coach should not interfere too much—mistakes are valuable as well as necessary. "While the student is making mistakes in the (tutorial) environment he is also experiencing the idea of learning from his mistakes and discovering the means to recover from his mistakes. If the Coach immediately points out the student's errors, there is a real danger that the student will never develop the necessary skills for examining his own behavior and looking for the causes of his own mistakes" (Burton & Brown, 1982, p. 80). However, there are times when interruption is necessary because students fixate on ineffective strategies or lack the necessary knowledge to solve a problem. "While a student's incorrect decisions sometimes lead to erroneous results that he can immediately detect, they often produce symptoms that are beyond his ability to recognize" (Burton & Brown, 1982, p. 79). The opportunity to self-discover and correct mistakes should enhance critical thinking among students.

Research suggests that student errors occur with a striking uniformity. "When students fail to make conceptual changes conventionally, they are forced to make do with what they already know in a new situation, where their knowledge is inadequate and inappropriate. By describing necessary conceptual changes, one can both predict that errors will occur and what those errors are likely to be" (Matz, 1982, p. 36). For the expert tutoring system to give advice to the student as problems develop, it is necessary to construct a model of the student's knowledge and learning attributes. To minimize guidance given to the student and interference during the learning process, the tutor constructs these models by developing hypotheses and making inferences from a sequence of moves made by the student. Because the coach obtains information indirectly, no particular answer is certain evidence for skills or knowledge that the student may be lacking. The coach must consider the sequence of worked examples and problems encountered by the student in developing hypotheses. Over time, the tutoring system may be able to detect weaknesses in the students' knowledge base by examining progress and referencing prior responses. Hypotheses are posed to determine whether the student possesses skills or pieces of knowledge that would have been used by the expert in an identical situation. "The Coach looks for an Issue in which the student is lacking and which is required for the Expert's better moves. Once an Issue has been determined, the Coach can present an explanation of that Issue together with a better move that illustrates the Issue. In this way, the student can see the usefulness of the Issue at a time when he will be most receptive to the idea presented—immediately after he has attempted a problem whose solution requires the Issue" (Burton & Brown, 1982, pp. 83–84). The coach increases the chance that the student will remember the criticism by choosing to illustrate the issue using examples in which the move being demonstrated is dramatically superior to the move made by the student.

Prototypes of Expert Tutoring Systems

To improve students' scientific inquiry skills, Shute and Boner (1986) designed the prototypes for several intelligent tutoring systems within microworld environments. Laboratory microworld environments in the domains of microeconomics, light refraction, and electrical circuits were created so that students could interactively learn the rules and regularities of each domain through observation and discovery. As students acquire new knowledge through the use of interrogation and inferential skills, they can formulate hypotheses, conduct experiments to test the hypotheses, and record their results. As students explore the domain, their inquiry strategies are compared to those used by expert and nonexpert learners. This project focuses on (1) whether students more easily manipulate variables and develop experiments across microworld environments, using inquiry skills and scientific behaviors, and (2) how well students learn targeted concepts. An "inquiry coach" has been partially designed to intervene when students' use of strategies could be more systematic and effective, relative to those of experts.

Lepper and Malone's (1987) findings on the cognitive and motivational consequences of computer-based learning environments influenced the design of three interactive expert tutoring environments developed by Reiser, Cohen, Hamid, and Kimberg (1993). Lepper and Malone (1987) argue that (1) an optimal level of challenge should make the task challenging and yet interesting without frustrating the student; (2) the task should include discrepant events that conflict with the learner's prior knowledge, exposing it as incomplete and inconsistent; and (3) the learning environment should allow the student to choose how to approach a task and to control the direction of the interaction, without being confused or unsure of where to begin. Reiser et al. (1993) empirically investigated the cognitive and motivational consequences of guided tutoring and discovery learning by using an intelligent tutoring system designed to help students understand how to construct and debug LISP (a computer programming language) programs. Reiser and colleagues suggest that although it is beneficial for students to learn by discovery, this may not be ideal when discovery learning is completely unassisted and exploratory in nature. The extent of guidance given to students can range from intervening and supplying a complete explanation (defeating opportunities for discovery learning) to "tutoring without talking," where the problem domain can be altered when necessary to facilitate the learning process. An expert instructional system, which can interpret students' understanding and thinking patterns as they construct a solution, is capable of providing careful guidance during problem solving (Anderson, Boyle, & Reiser, 1985). A discovery learning environment that allows students to construct their own knowledge and explore the problem-solving domain can be designed using the same intelligent tutoring system but allowing students limited access to the tutoring capability. By varying the amount and type of guidance that students receive, cognitive and motivational consequences of the learning environments can be directly assessed. Reiser and colleagues investigated three interactive learning environments that range from an interventionist tutoring system to a free-discovery learning environment: (1) Tutored (guided) learning—monitors student problem solving and intervenes whenever students make errors, providing complete explanatory feedback designed to help students understand and fix the error. (2) Discovery learning, assigned problems

(identical to problems in the first treatment)—enables students to articulate and test their hypotheses, checking their reasoning and making predictions as they proceed. The system does not interrupt when the student makes errors unless guidance is requested by the student; the system allows students to discover and repair their own errors (students find out whether their solution satisfies the problem specifications when it is complete and submitted for evaluation). When errors are evident (following evaluation of the completed program), the tutor provides feedback through example, giving the student the opportunity to modify the program. (3) Discovery learning, no assigned problems (optional list of problems assigned to students in other treatments is provided)—students are permitted complete control over their own learning. Students are free to explore by building whatever programs they desire and solving problems of their own choice; the system facilitates student articulation and tests of hypotheses. The tutor offers no instructional guidance when errors are made and programs have been completed; feedback is provided only upon student request. Undergraduates with little knowledge of computer programming and no knowledge of LISP were placed in one of the treatment groups according to their SAT math score (low SAT 420–630; high SAT 640–800), which has been shown to be the best predictor of success in learning to program. The treatment groups had equivalent average SAT math scores.

Results reported by Reiser et al. (1993) "have provided perhaps the first direct evidence that students' judgments of their own abilities and their resulting interest in the domain will be affected by the style of their interaction with the learning situation." Overall, discovery learning environments appear to provide greater gains for the high-ability students and greater losses for the low-ability students, suggesting that both guided and discovery learning environments should be utilized and that student ability should be considered in designing instruction. However, caution in using discovery learning is necessary even for higher-ability students. Unsupervised, students can invent misconceptions that guided-instruction students never develop, and the danger exists that students will fail to discover important aspects of the domain. Instructors should be cognizant of the demands that unassisted discovery instruction places on students, particularly those of low ability. With the freedom to learn by discovery (where problems are and are not assigned), successful students (typically those with high SAT math scores) take advantage of the learning situation, managing their own learning, learning to recognize and repair their own errors, and thereby acquiring better error-management skills than guided-learning students. The freedom to fail is helpful in allowing successful students to explore available options and develop necessary cognitive skills. "To facilitate learning, students must be encouraged to ask questions, to predict answers, to be wrong, and to explain and thereby profit from mistakes" (Reiser et al., 1993). Infrequent interruption and guidance from the tutor and a sense of control and challenge cultivate the intrinsic interest of successful students, causing them to attribute their success to their ability. However, without guidance from the tutor, students with low SAT scores are less likely to acquire the cognitive skills necessary to debug programs and detect their own errors. When less successful students engage in discovery learning situations and fail, they often attribute their failures to lack of ability, and these negative attributions are most likely reflected in poor self-esteem and lowered intrinsic interest in subsequent learning. Guided learning environments appear to provide a scaffold on which students with lower abilities can, through frequent interventions, continue solving difficult problems that they could otherwise not handle, preferably developing the skills that would permit them to transfer to discovery learning environments when ready. Guidance, provided to students by an expert tutor, involves predetermining when and how often to interrupt the student, what to say, and how best to say it; the optimal amount of guidance given is likely to vary with individual students' characteristics (e.g., academic ability, intellectual curiosity, degree of independent thinking, etc.).

CONCLUSION

We started this section with the question, What are new and promising avenues of research that may indicate a paradigm shift of the use of instructional technology in science education? We then outlined several instructional projects that have the possibility for new avenues of research in both content and method. What are the capabilities of applications like hypermedia, microworlds, expert tutoring systems, and telecommunication features that allow students to share, exchange, and even reconsider what they have learned? What are the capabilities of data base systems that permit students to retrieve information and make flexible connections between related ideas? What are the capabilities of microcomputer simulation environments that give students the opportunity to make hypotheses, test them, observe the results, and come to conclusions? If there are common themes or shifts of research methods in these questions, we may be better able to understand and categorize these new directions. We speculate on the following themes.

Common Themes

The instructional projects we have outlined have some common features that may help to indicate a future research agenda in technology in science instruction.

Decline in comparison studies. There is a decided shift away from comparison studies. Fewer studies raise the question, Would the use of technology offer greater learning than the same amount of time placed on current nontechnological instruction? We cite three major reasons for this shift.

1. The use of technology is so common that to compare its use to current nontechnological practice would be equivalent to comparing instruction with books to the same amount of instruction without books 20 years ago. Books are such an integral part of teaching and learning that it is inappropriate to do such a comparison, however interesting. The same

comparison studies with technology are rapidly becoming inappropriate as well.

2. Such comparison studies often cite time spent on instruction as a cogent reason for comparison. By citing time as a major variable, they ignore three major points. First, time and learning efficiency are only partially related. The real variable is learning efficiency, that is, the amount learned divided by the time spent. Second, by measuring only before and at the end of the study they ignore the increase in learning over a shorter period of time that can occur with a student-driven technological study. It is as if two different learning curves were operating: the "standard one," in which the learning starts slowly and rises steadily to the end of the instructional period, and the "technological one," in which the learning rises rapidly and peaks at a mastery level early. Thus, by not measuring at several times and measuring only at the end of the instructional period, both groups appear to have gone through the same learning process. Third, students do not spend the same amount of time on traditional instruction. Several researchers cite, either anecdotally or by design, that students work at well-constructed technological instruction for hours and are on task. Such times on task may be obtainable under research conditions for traditional instruction but not under standard everyday instruction.

3. It is inappropriate to use a comparison technique. The use of technology provides an instructional approach that cannot be matched by traditional nontechnological instruction. The ability to see graphs develop in real time, to see multiple representations of phenomena from macro to micro, cannot be duplicated even with the best lecture or reading. Thus, it is even less appropriate to try to design a comparison study.

Increases in Qualitative and Quantitative Understanding of What Is Going on During the Instruction. Several studies report using techniques that allow in-depth analysis of how students are constructing knowledge as they work, indicating a new research approach. One such promising technique outlined by Nakhleh and Krajcik (1992), involves videotaping students by connecting the video output of the monitor to the video input of the videocassette recorder (VCR) and recording the students' comments or situated interviewing on the audio portion of the VCR. Using this technique, the work of students as they carry out instruction can be analyzed later.

Increase in the Use of Studies to Reach and Work on the Needs of Underserved Populations. Of every 4000 seventh-graders in school (in the United States) today (in 1991), only six will ultimately receive a Ph.D. in science or engineering—of these six, only one will be a female. By the year 2000 minority students will account for 40 percent of our elementary and secondary school population. Yet only 4 percent of undergraduate science and engineering degrees are awarded to minorities (Massey, 1991).

The variety of features available in technology-based instruction can enhance the learning environment for a diverse range of students—if an appropriate learning environment can be determined and provided for each student (Dershimer & Berger, 1992).

Increase in the Attempts to Understand How Students See the Relationship of Concepts and Build the Links Among Those Concepts. The use of knowledge networks is also increasing, highlighted by the work of Fisher (1990) and Goldsmith, Johnson, and Acton (1991). Concept maps can be made and analyzed to determine student understanding.

Increase in the Involvement of Students to Determine Their Own Learning. New technological instructional methods are appearing that offer students minimal guidance. These methods can discern and give guidance only as it is needed, so that students can discover as much as possible on their own. Such techniques should foster the development of critical, original, and flexible thought. Microcomputers can be used by individual students to facilitate independent learning; some students can learn in one way and other students in a different way. Such instructional diversity is most difficult to achieve in a classroom setting, where the entire class is usually doing the same thing.

Summary

We have listed four promising research directions. Individual projects already exist that highlight both these research directions and the particular kind of technology that can be beneficial in such directions. Interactive video disk, CAI (developed from appropriate learning theory), and MBL are promising research projects, particularly for developing higher thinking skills for secondary school students. If we freely speculate on the future of technology in science education, we can imagine a scenario in which students are truly empowered to learn, mediated instruction supports the instructor rather than replaces him or her, and technology is an answer (not the answer) to providing instruction to a diverse range of students with varying learning needs.

Finally, we believe the new techniques that technology can provide to carry out research will someday be as common as the use of a statistical package. However, these packages can no more replace the careful design of research or the quantitative or qualitative analysis of research than such statistical packages do today. Clear thinking and care with conception and design may never be replaced by even the finest of artificial intelligence. Yet working with technology will provide new and exciting avenues and will lead to many more questions than answers in the future, just as it does now.

References

Adams, D. D., & Shrum, J. W. (1990). The effects of microcomputer-based laboratory exercises on the acquisition of line graph construction and interpretation skills by high school biology students. *Journal of Research in Science Teaching, 27*(8), 777–787.

Aiello, N. C., & Wolfe, L. M. (1980, April). *A meta-analysis of individualized instruction in science.* Paper presented at the annual meeting of the American Educational Research Association, Boston. (ERIC Document Reproduction Service No. ED 190 404)

Anderson, J. R., Boyle, C. F., & Reiser, B. J. (1985). Intelligent tutoring systems. *Science, 228,* 456–462.

Baird, W. E. (1988). Status of use: Microcomputers and science teaching. In J. D. Ellis (Ed.), *1988 AETS Yearbook* (pp. 85–104). Columbus: The Ohio State University.

Bangert-Drowns, R. L., Kulik, J. A., & Kulik, C. L. C. (1985). Effectiveness of computer-based education in secondary schools. *Journal of Computer-Based Instruction, 12*(3), 59–68.

Becker, H. J. (1986). *Instructional uses of school computers: Reports from the 1985 national survey* (Issue no. 1). Baltimore: Johns Hopkins University, Center for Social Organization of Schools.

Beichner, R. J. (1990). The effect of simultaneous motion presentation and graph generation in a kinematics lab. *Journal of Research in Science Teaching, 27*(8), 803–815.

Berger, C. (1979). *Power company program.* Ann Arbor: University of Michigan.

Berger, C. (1982). Attainment of skill in using science processes. I: Instrumentation, methodology and analysis. *Journal of Research in Science Teaching, 19*(3), 249–260.

Berger, C. (1984a). Learning more than facts: Microcomputer simulation in the science classroom. In D. Peterson (Ed.), *Intelligent schoolhouse: Readings on computers and learning.* Reston, VA: Reston.

Berger, C. (1984b, March). *Research using microcomputers in gathering displaying and analyzing laboratory data.* Paper presented at the meeting of the National Association for Research in Science Teaching, New Orleans.

Berger, C. (1987a, April). *Misconceptions and thin conceptions of teachers using microcomputer-based laboratories.* Paper presented at the meeting of the American Educational Research Association, Washington, DC.

Berger, C. (1987b, April). *Student estimation on linear and logarithmic scales.* Paper presented at the meeting of the National Association for Research in Science Teaching, Washington, DC.

Berger, C., & Jackson, D. F. (1989, April). *Student as grapher: Computer-assisted thinking tools.* Paper presented at the meeting of the American Educational Research Association, San Francisco.

Berger, C. F., & Jackson, D. F. (1990, April). *Using technology to interpret large-scale complexity: The use of scientific sequence analysis algorithms in research on computer-assisted problem solving.* Paper presented at the meeting of the National Association for Research in Science Teaching, Atlanta.

Berger, C., Newman, R., & Prince, J. (1981, March). *Applicants of scientific and mathematical processes using microcomputers. Part I: Estimation: The effect of information on the determination of linear distance.* Paper presented at the meeting of the National Association for Research in Science Teaching, New York.

Berger, C., & Pintrich, P. (1986). Attainment of skill in using science processes. II: Grade and task effects. *Journal of Research in Science Teaching, 23*(8), 739–747.

Berger, C., Pintrich, P., & Stemmer, P. (1987). Cognitive consequences of student estimation on linear and logarithmic scales. *Journal of Research in Science Teaching, 24*(5), 437–450.

Blume, G. W. (1984, April). *A review of research on the effects of computer programming on mathematical problem solving*. Paper presented at the annual meeting of the American Educational Research Association, New Orleans.

Bransford, J., Kinzer, C., Risko, V., Rowe, D., & Vye, N. (1989). Designing invitations to thinking: Some initial thoughts. In S. McCormick & J. Zutell (Eds.), *Cognitive and social perspectives for literacy and instruction* (pp. 35–54). Chicago: National Reading Conference.

Brassell, H. (1987). The effect of real-time laboratory graphing on learning graphic representations of distance and velocity. *Journal of Research in Science Teaching, 24*(4), 385–395.

Bunderson, C. V., Baillio, B., Olsen, J. B., Lipson, J. I., & Fisher, K. M. (1984). Instructional effectiveness of an intelligent videodisc in biology. *Machine Mediated Learning, 1*(2), 175–215.

Burns, P. K., & Bozeman, W. C. (1981). Computer-assisted instruction and mathematics achievement: Is there a relationship? *Educational Technology, 21*(10), 32–39.

Burton, R. R., & Brown, J. S. (1982). An investigation of computer coaching for informal learning activities. In D. Sleeman & J. S. Brown (Eds.), *Intelligent tutoring systems* (pp. 79–98). New York: Academic Press.

Callman, J. L., Faletti, J., & Fisher, K. M. (1985, March). *III. Computer-based semantic network in molecular biology: A demonstration*. Paper presented at the annual meeting of the National Association for Research in Science Teaching, San Francisco. (ERIC Document Reproduction Service No. ED 268 011)

Campbell, R. (1989). (I learned it) through the grapevine: Hypermedia at work in the classroom. *American Libraries, 20*(3), 200–205.

Carter, J. (1984). Instructional learner feedback: A literature review with implications for software development. *Computing Teacher, 12*(2), 53–55.

Charney, D. (1989). Comprehending non-linear text: The role of discourse cues and reading strategies. *Hypertext '87 proceedings*. New York: Association for Computing Machinery.

Choi, B. S., & Gennaro, E. (1987). The effectiveness of using computer simulated experiments on junior high students' understanding of the volume displacement concept. *Journal of Research in Science Teaching, 24*(6), 539–552.

Clark, R. E. (1983). Reconsidering research on learning from media. *Review of Educational Research, 53*(4), 445–459.

Cognition and Technology Group, Vanderbilt. (1990). Anchored instruction and its relationship to situated cognition. *Educational Researcher, 19*(6), 2–10.

Cohen, J. (1977). *Statistical power analysis for the behavioral sciences* (rev. ed.). New York: Academic Press.

Collins, A. (1991). The role of computer technology in restructuring schools. *Phi Delta Kappan, 73*, 28–36.

Conklin, J. (1987). Hypertext: An introduction and survey. *Computer, 20*, 17–41.

Cordts, M. L., Beloin, R., & Gibson, J. (1989). *A tutorial in recombinant DNA technology*. Unpublished manuscript, Cornell University, Ithaca, NY.

Cox, D. A., & Berger, C. (1981). Microcomputers are motivating. *Science and Children, 19*, 28–29.

Cox, D. A., & Berger, C. (1985). The importance of group size in the use of problem-solving skills on a microcomputer. *Journal of Educational Computing Research, 1*(4), 459–468.

Crosby, H. M. (1983). *The effectiveness of the instructional use of computers for students with mild learning problems: A review of the literature and research* (Concepts and issues paper). Urbana-Champaign: University of Illinois. (ERIC Document Reproduction Service No. ED 249 688)

Dershimer, C., & Berger, C. (1992, April). *Characterizing student interactions with a hypermedia learning environment*. Paper presented at the meeting of the American Education Research Association, San Francisco.

Dershimer, C., Berger, C., & Jackson, D. (1991, April). *Designing Hyper-Media for concept development: Formative evaluation through analysis of log files*. Paper presented at the meeting of the National Association for Research in Science Teaching, Lake Geneva, WI.

Dershimer, C., & Rasmussen, P. (1990). *Seeing through chemistry*. Computer multimedia package. Ann Arbor: Office of Instructional Technology, University of Michigan.

Edwards, B., Jackson, D., & Berger, C. (1990, April). *Teaching the design and interpretation of graphs through computer-aided graphical data analysis*. Paper presented at the meeting of the National Association for Research in Science Teaching, Atlanta.

Edwards, J., Norton, S., Taylor, S., Weiss, M., & Dusseldorp, R. (1975). How effective is CAI? A review of the research. *Educational Leadership, 32*, 147–153.

Fisher, K. M. (1990). Semantic networking: The new kid on the block. Special Issue: Perspectives on concept mapping. *Journal of Research in Science Teaching, 27*(10), 1001–1018.

Florida Department of Education. (1980). *More hands for teachers: Report of the Commissioners Advisory Committee on Instructional Computing*. Tallahassee, FL: Office of Educational Technology.

Friedlander, L. (1989). Moving images into the classroom: Multimedia in higher education. *Laserdisk Professional, 2*(4), 33–38.

Friedler, Y., Nachmias, R., & Linn, M. C. (1990). Learning scientific reasoning skills in microcomputer-based laboratories. *Journal of Research in Science Teaching, 27*(2), 173–191.

Glass, G. V. (1976). Primary, secondary, and meta-analysis of research. *Educational Researcher, 5*(10), 3–8.

Glass, G. V. (1977). Integrating findings: The meta-analysis of research. In L. Shulman (Ed.), *Review of research in education* (pp. 349–379). Itasca, IL: Peacock.

Goldsmith, T. E., Johnson, P. J., & Acton, W. H. (1991). Assessing structural knowledge. *Journal of Educational Psychology, 83*, 88–96.

Gray, R. A. (1989). HyperCard™ utilization in education: Possibilities and challenges. *Tech Trends, 34*, 39–40.

Hallada, M. (1990). ChemTutor (Macintosh computer program.) University of Michigan, Ann Arbor.

Hasselbring, T. S. (1987). Effective computer use in special education: What does the research tell us? In *Effectiveness of microcomputers in education* (Funder Forum, pp. 23–41). Cupertino, CA: Apple Computer.

Hawisher, G. E. (1986). Studies in word processing. *Computers and Composition, 4*, 8–31.

Jackson, D., Edwards, B., & Berger, C. (1990, April). *The design of software tools for meaningful learning by experience: Flexibility and feedback*. Paper presented at the meeting of the National Association for Research in Science Teaching, Atlanta.

Jamison, D., Suppes, P., & Wells, S. (1974). The effectiveness of alternative instructional media: A survey. *Review of Educational Research, 44*, 1–67.

Janners, M. Y. (1988). Inquiry, investigation, and communication in the student-directed laboratory. *Journal of College Science Teaching, 18*, 32–35.

Johnson, J. P. (1982). Can computers close the educational equity gap? (use of classroom computers with minority students). *Perspectives, 14*, 20–25.

Johnson, R. (1987). *The ability to retell a story: Effects of adult mediation in videodisc context on children's story recall and comprehension*. Unpublished doctoral dissertation, Vanderbilt University, Nashville, TN.

Johnston, J. J., & Kleinsmith, L. J. (1987). *Computers in higher education: Computer-Based tutorials in introductory biology*. Ann Arbor: University of Michigan, Institute for Social Research.

Jungck, J., & Calley, J. (1986). *Genetics: Strategic simulations in Mendelian genetics.* Wentworth, NH: COMPress.

Kimball, R. Q. (1974). *An experimental study of the use of a computer-based classroom simulation in teacher education.* Unpublished doctoral dissertation, University of Michigan, Ann Arbor.

Kleinsmith, L. J. (1987). A computer-based biology study center: Preliminary assessment of impact. *Academic Computing, 2*(3), 32–33, 49–50, 67.

Kleinsmith, L. J. (Ed.). (1989). *BioNet.* Ann Arbor, MI: University of Michigan.

Kozma, R. B. (1991). Learning with media. *Review of Educational Research, 61,* 179–211.

Krajcik, J. S. (1991). Developing students' understanding of chemical concepts. In S. H. Glynn, R. H. Yeany, & B. K. Britton (Eds.), *The psychology of learning science* (p. 117–147). Hillsdale, NJ: Lawrence Erlbaum.

Krajcik, J. S., & Layman, J. W. (1989, April). *Middle school teachers conceptions of heat and temperature: Personal and teaching knowledge.* Paper presented at the meeting of the National Association for Research in Science Teaching, San Francisco.

Kulik, C. L. C., & Kulik, J. A. (1986). Effectiveness of computer-based education in college. *AEDS Journal, 19,* 81–108.

Kulik, C. L. C., Kulik, J. A., & Bangert-Drowns, R. L. (1984, April). *Effects of computer-based education in elementary school pupils.* Paper presented at the annual meeting of the American Educational Research Association, New Orleans. (ERIC Document Reproduction Service No. ED 244 616)

Kulik, C. L. C., Kulik, J. A., & Schwalb, B. J. (1986). Effectiveness of computer-based adult education: A meta-analysis. *Journal of Educational Computing Research, 2,* 235–252.

Kulik, J. A., Bangert, R. L., & Williams, G. W. (1983). Effects of computer-based teaching on secondary school students. *Journal of Educational Psychology, 75,* 19–26.

Kulik, J. A., Kulik, C. L. C., & Bangert-Drowns, R. L. (1985). Effectiveness of computer-based education in elementary schools. *Computers in Human Behavior, 1,* 59–74.

Kulik, J. A., Kulik, C. L. C., & Cohen, P. A. (1980). Effectiveness of computer-based college teaching: A meta-analysis of findings. *Review of Educational Research, 50*(4), 525–544.

Lenk, C. (1988). Doing science through telecommunications. In J. D. Ellis (Ed.), *1988 AETS Yearbook* (pp. 25–34). Columbus: The Ohio State University.

Leonard, W. H. (1989). A comparison of student reactions to biology instruction by interactive videodisc or conventional laboratory. *Journal of Research in Science Teaching, 26*(2), 95–104.

Leonard, W. H. (1992). A comparison of student performance following instruction by interactive videodisc versus conventional laboratory. *Journal of Research in Science Teaching, 29,* 93–102.

Lepper, M., & Malone, T. T. (1987). Computer-based education. In R. E. Snow & M. J. Farr (Eds.), *Aptitude, learning and instruction, Vol. 3: Cognitive and affective process analyses.* Hillsdale, NJ: Lawrence Erlbaum.

Linn, M. (1988). *Autonomous classroom computer environments for learning.* Progress report and annotated bibliography. Washington, DC: National Science Foundation. (ERIC Document Reproduction Service No. ED 305 903)

Marchionini, G. (1989). Information-seeking strategies of novices using a full-text electronic encyclopedia. *Journal of American Society for Information Science, 40,* 54–66.

Marchionini, G. (1990, April). *Self-directed learning through hypermedia: Assessing the process.* Paper presented at the meeting of the American Education Research Association, Boston.

Marsh, E. J., & Kumar, D. D. (1992). Hypermedia: A conceptual framework for science education and review of recent findings. *Journal of Educational Multimedia and Hypermedia, 1,* 25–37.

Massey, W. E. (1991). *Educational technology: Computer-based instruction.* Washington, DC: U.S. Government Printing Office.

Matz, M. (1982). Towards a process model for high school algebra errors. In D. Sleeman & J. S. Brown (Eds.), *Intelligent tutoring systems* (pp. 25–50). New York: Academic Press.

McCarthy, R. (1989). Multimedia: What the excitement's all about. *Electronic Learning, 8*(8), 26–31.

Mitchell, K. L., & Montague, E. J. (1986, March). *Computer utilization in Texas secondary science classrooms.* Paper presented at the meeting of the National Association for Research in Science Teaching, San Francisco.

Mokros, J. R., & Tinker, R. F. (1987). The impact of microcomputer-based labs on children's ability to interpret graphs. *Journal of Research in Science Teaching, 24*(4), 369–383.

Nakhleh, M., & Krajcik, J. (1992, March). *A protocol analysis of the effect of technology on students' actions, verbal commentary, and thought processes during the performance of acid-based titrations.* Paper presented at the National Association for Research in Science Teaching, Boston.

Niemiec, R., Samson, G., Weinstein, T., & Walberg, H. J. (1987). The effects of computer-based instruction in elementary schools: A quantitative synthesis. *Journal of Research on Computing in Education, 20*(2), 85–103.

Niemiec, R. P., & Walberg, H. J. (1985). Computers and achievement in the elementary schools: A quantitative synthesis. *Journal of Educational Computing Research, 1*(4), 435–440.

Office of Technology Assessment. (1982). *Information technology and its impact on American education.* Washington, DC: U.S. Government Printing Office.

Okey, J. R. (1985, April). *The effectiveness of computer-based education: A review.* Paper presented at the annual meeting of the National Association for Research in Science Teaching, French Lick Springs, IN. (ERIC Document Reproduction Service No. ED 257 677)

Orlansky, J. (1983). Effectiveness of CAI: A different finding. *Electronic Learning, 3,* 58, 60.

Peterson, N. S., & Campbell, K. B. (1985). Teaching cardiovascular integrations with computer laboratories. *Psychologist, 28,* 159–169.

Peterson, N. S., Jungck, J. R., Sharpe, D. M., & Finzer, W. F. (1987). A design approach to science simulated laboratories: Learning via the construction of meaning. *Machine-Mediated Learning, 2*(1&2), 111–127.

Pinet, P. R. (1989). Understanding the language of argument and the methods of science. *Journal of Geological Education, 37*(3), 197–201.

Pintrich, P. R. (1986, April). *Motivation and learning strategies interactions with achievement.* Paper presented at the American Educational Research Association Convention, San Francisco.

Pintrich, P. R. (1987, April). *Motivated learning strategies in the college classroom.* Paper presented at the American Educational Association Convention, Washington, DC.

Reiser, B. J., Cohen, W. A., Hamid, A., & Kimberg, D. Y. (1993). *Cognitive and motivational consequences of tutoring and discovery learning.* Evanston, IL: Northwestern University.

Resnick, L. B. (1987). *Education and learning to think.* Washington, DC: National Academy Press.

Roblyer, M. D., Castine, W. H., & King, F. J. (1988). Assessing the impact of computer-based instruction. *Computers in the Schools, 5,* 1–149.

Roblyer, M. D., & King, F. J. (1983, April). *Reasonable expectations for computer-based instruction in basic reading skills.* Paper presented at the annual conference of the Association for Educational Communications and Technology, New Orleans.

Rohwedder, W. J. (Ed.). (1990). Computer-aided environmental education. *Monographs in environmental education and environmental studies* (Vol. 2). (ERIC Document Reproduction Service No. ED 328 441)

Salomon, G., & Globerson, T. (1987). Skill may not be enough: The role of mindfulness in learning and transfer. *International Journal of Educational Research, 11*, 623–638.

Salomon, G., Perkins, D. N., & Globerson, T. (1991). Partners in cognition: Extending human intelligence with intelligent technologies. *Educational Researcher, 20*(3), 2–9.

Samson, G. E., Niemiec, R., Weinstein, T., & Walberg, H. J. (1985, April). *Effects of computer-based instruction on secondary school achievement: A qualitative synthesis.* Paper presented at the American Educational Research Association, Chicago.

Science Quest. (1991). For further information write Houghton Mifflin Company, Department 91, Lebanon, NH.

Sherwood, R. D. (1991, April) *The development and preliminary evaluation of anchored instruction environments for developing mathematical and scientific thinking.* Paper presented at the National Association for Research in Science Teaching, Lake Geneva, WI.

Shyu, Y. J. (1987). *The use of a microcomputer-based classroom simulator in the preparation of science teachers.* Unpublished doctoral dissertation, University of Michigan, Ann Arbor.

Sleeman, D., & Brown, J. S. (1982). *Intelligent tutoring systems.* New York: Academic Press.

Technological Education Research Center. (1984). *Science and mathematics software opportunities and needs (SAMSON) project: Final report.* Report to the Office of Research and Educational Development. Cambridge, MA: U.S. Department of Education by the Technical Education Research Center.

Thomas, D. B. (1979). The effectiveness of computer-assisted instruction in secondary schools. *AEDS Journal, 12*(3), 103–116.

Tinker, R. F. (1985). How to turn your computer into a science lab. *Classroom computer learning, 5*(6), 26–29.

Underwood, J. (1989). HyperCard™ and interactive video. *Calico Journal, 6*(3), 7–20.

Vinsonhaler, J. F., & Bass, R. K. (1972). A summary of ten major studies on CAI drill and practice. *Educational Technology, 12*(7), 29–32.

Voss, B. E. (1990, February). *A microcomputer-based game that simulates several aspects of classroom management.* Paper presented at the American Association for the Advancement of Science Meeting, New Orleans.

Wells, G., & Berger, C. (1986). Teacher/student-developed spreadsheet simulations: A population growth example. *Journal of Computers in Mathematics and Science Teaching, 5*(2), 34–40.

Willett, J. B., Yamashita, J. M., & Anderson, R. D. (1983). A meta-analysis of instructional systems applied in science teaching. *Journal of Research in Science Teaching, 20*(5), 405–417.

Wise, K. C. (1988). The effects of using computing technologies in science instruction: A synthesis of classroom-based research. In J. D. Ellis (Ed.), *1988 AETS Yearbook* (pp. 105–118). Columbus: The Ohio State University.

Wise, K. C., & Okey, J. R. (1983, April). *The impact of microcomputer-based instruction on student achievement.* Paper presented at the annual meeting of the National Association for Research in Science Teaching, Dallas.

Wise, M. (1986, April). *Misconceptions of students using MBL.* Paper presented at the meeting of the American Educational Research Association, San Francisco.

Woodhull-McNeal, A. (1989). Teaching introductory science as inquiry. *College Teaching, 37*, 3–7.

Zitzewitz, B., & Berger, C. (1985). Applications of mathematical learning models to student performance on general chemistry: Microcomputer drill and practice programs. *Journal of Research in Science Teaching, 22*(9), 775–791.

Zoller, U. (1987). The fostering of question asking capability. *Journal of Chemical Education, 64*(6), 510–512.

Part

·V·

CONTEXT

·17·

RESEARCH ON CLASSROOM AND
SCHOOL CLIMATE

Barry J. Fraser

CURTIN UNIVERSITY OF TECHNOLOGY

Science educators often speak of a classroom's or school's climate, environment, atmosphere, tone, ethos, or ambience and consider it to be both important in its own right and influential in terms of student learning. Despite the fact that the educational environment is a somewhat subtle concept, remarkable progress has been made during the last two decades in conceptualizing it, assessing it, and researching its determinants and effects. Although important educational-climate work has been undertaken by researchers interested in a variety of school subject areas, science education researchers have led the world in terms of developing, validating, and applying environment assessment instruments.

Many questions of interest to teachers, educational researchers, curriculum developers, and policymakers in science education can be asked about classroom and school environment. Does a classroom's environment affect student learning and attitudes, and does a school's environment affect teacher job satisfaction and effectiveness? What is the impact of a new curriculum or teaching method on a classroom's environment? Can teachers conveniently assess the climates of their own classrooms and schools, and can they change these environments? What are some of the determinants of classroom and school environment? Is there a discrepancy between actual and preferred classroom environment, as perceived by students, and does this discrepancy matter in terms of student outcomes? Do teachers and their students perceive the same classroom environments similarly, and do school administrators and teachers see the same school environment similarly? These questions represent the thrust of the work on science education environments during the past 25 years and the main areas considered in this chapter.

Traditionally, research and evaluation in science education have tended to rely heavily and sometimes exclusively on the assessment of academic achievement and other valued learning outcomes. Although few responsible educators would dispute the worth of outcome measures, these measures cannot give a complete picture of the educational process. Because students spend up to 15,000 hours at school by the time they finish senior high school (Rutter, Maughan, Mortimore, Ouston, & Smith, 1979), they have a large stake in what happens to them at school and their reactions to and perceptions of their school experiences are significant. This chapter is devoted to one approach to conceptualizing, assessing, and investigating what happens to students during their schooling. In particular, the main focus is on students' and teachers' perceptions of important social and psychological aspects of the learning environments of classrooms and schools.

Educational environments can be considered as the social-psychological contexts or determinants of learning. Teaching is a determinant of learning, but it is more deliberate than aspects of the educational environment. Teaching, however, affects the environment and is in turn affected by it. Indeed, research on environments usually assumes that students, curricula, and other internal and external factors, as well as the teacher, affect the environment. Whereas the environment most often is measured by obtaining students' and teachers' perceptions of the classroom or school, research on teaching more often employs behavioral observations and case studies by people other than teachers and students.

Measures of environments are typically more like measures of motivation than measures of ability or achievement. They do not require demonstrations of performance but involve judg-

The authors thanks reviewer Bruce E. Perry (Miami University).

ments of psychological or social-psychological states of classes or schools. They often require participant ratings of such things as goal direction, democracy, and satisfaction. A distinctive research tradition has examined the correlation of environmental properties with causal antecedents and causal consequences. Such work is typically descriptive, multivariate, and correlational in its quest to study the relations among environmental and other variables as they naturally occur. The typical study of teaching, by contrast, assigns students or classes to alternative methods of teaching and attempts to gauge the relative effectiveness in terms of student outcomes.

The approach described here, which defines classroom or school environment in terms of the shared perceptions of the students and teachers in that environment, has the dual advantage of characterizing the setting through the eyes of the actual participants and capturing data that the observer could miss or consider unimportant. Students have a good vantage point to make judgments about classrooms because they have encountered many different learning environments and have enough time in a class to form accurate impressions. Also, even if teachers are inconsistent in their day-to-day behavior, they usually project a consistent image of the long-standing attributes of classroom environment.

This chapter falls into five main parts. First, an introductory section provides background information about the fields of school and classroom environment (including alternative assessment approaches, a historical perspective on past work, the distinction between school and classroom environment, and the unit-of-analysis question). Second, a section is devoted to instruments for assessing perceptions of classroom environment. Third, assessment instruments for school environment are considered. Fourth, an overview is given of several lines of past research involving environment assessments in science classrooms (including associations between outcomes and environment, the use of environment dimensions as criterion variables, and person-environment fit studies of whether students achieve better in their preferred environment). Fifth, recent research in which quantitative and qualitative methods were combined to advantage within the same classroom environment study is described. Sixth, teachers' use of classroom and school environment instruments in practical attempts to improve their own classrooms and schools is considered. Seventh, current trends and future desirable directions in research on educational environments are identified.

BACKGROUND

This introductory section sets the scene for the remainder of the chapter by raising four important issues that recur in subsequent sections: the method of assessing educational environments in terms of students' and teachers' perceptions is compared with alternative approaches, and the relative merits of perceptual measures are weighed; a historical perspective is taken on past work that has influenced the ways of conceptualizing, assessing, and investigating educational environments; the distinction between school-level and classroom-level envi-

ronment is considered; and the important issue of choosing an appropriate level or unit of analysis for classroom environment work is discussed.

Approaches to Studying Educational Environments

The use of students' and teachers' perceptions can be contrasted with the method of direct observation, which typically involves an external observer in systematic coding of classroom communication and events according to some category scheme (e.g., Dunkin & Biddle, 1974; Rosenshine & Furst, 1973). The distinction between the "objective" approach of directly observing the environment and the "subjective" approach based on milieu inhabitants' apprehension of the environment is widely recognized in the psychological literature (see Jessor & Jessor, 1973). In particular, Murray (1938) introduced the term *alpha press* to describe the environment as assessed by a detached observer and the term *beta press* to describe the environment as perceived by milieu inhabitants.

Rosenshine (1970) makes the distinction between *low-inference* and *high-inference* measures of classroom environment. Low-inference measures tap specific explicit phenomena (e.g., the number of student questions), whereas high-inference measures require a judgment about the meaning of classroom events (e.g., the degree of teacher friendliness). Compared with low-inference measures, high-inference measures are involved more with the psychological significance of classroom events for students and teachers. Whereas it has been common for classroom observation schemes to focus on low-inference variables, perceptual measures have tended to focus on high-inference variables.

Fraser and Walberg (1981) outline some advantages of student perceptual measures over observational techniques. First, paper-and-pencil perceptual measures are more economical than classroom observation techniques that involve the expense of trained outside observers. Second, perceptual measures are based on students' experiences over many lessons, whereas observational data are usually restricted to a small number of lessons. Third, perceptual measures involve the pooled judgments of all students in a class, whereas observation techniques typically involve only a single observer. Fourth, students' perceptions, because they are the determinants of student behavior more than the real situation, can be more important than observed behaviors. Fifth, perceptual measures of classroom environment typically have been found to account for considerably more variance in student learning outcomes than directly observed variables.

Another approach to studying educational environments involves application of the techniques of naturalistic inquiry, ethnography, and case study, which are well illustrated by the vivid descriptions of school settings found in popular books such as *To Sir With Love* and *Thirty-Six Children*. Some of the other approaches to conceptualizing and assessing human environments delineated by Moos (1973) include ecological dimensions (e.g., meteorological and geographic dimensions as well as the physical design and architectural features reviewed by Weinstein, 1979) or behavior settings, which are conceptualized as naturally occurring ecological units concerned with molar behavior and the ecological context in which it occurs (e.g., Barker & Gump, 1964). In another approach (e.g., Astin & Holland, 1961), the character of an environment is assumed to depend on the nature of its members, and the dominant features of an environment are considered to depend on its members' typical characteristics.

Historical Perspectives

It is now more than two decades since Herbert Walberg and Rudolf Moos began their seminal independent programs of research that form the starting points for the work reviewed in this chapter. More than 20 years ago, Walberg began developing earlier versions of the widely used Learning Environment Inventory as part of the research and evaluation activities of Harvard Project Physics (see Anderson & Walberg, 1968; Walberg, 1968; Walberg & Anderson, 1968a, 1968b). Also, over two decades ago, Moos began developing the first of his social climate scales, including those for use in psychiatric hospitals (Moos & Houts, 1968) and correctional institutions (Moos, 1968), which ultimately resulted in the development of the widely known Classroom Environment Scale (Moos & Trickett, 1974, 1987).

How the important pioneering work of Walberg and Moos on perceptions of classroom environment developed into major research programs and spawned a great deal of other research is reflected in numerous comprehensive literature overviews. These include books (Fraser, 1986a; Fraser & Walberg, 1991; Moos, 1979a; van der Sijde & van de Grift, in press; Walberg, 1979), monographs (Fraser, 1981b; Fraser & Fisher, 1983a), a guest-edited journal issue (Fraser, 1980b), an annotated bibliography (Moos & Spinrad, 1984), and several state-of-the-art literature reviews (Anderson & Walberg, 1974; Chavez, 1984; Fraser, 1986b, 1989b, 1991; Randhawa & Fu, 1973; Walberg, 1976; Walberg & Haertel, 1980) including special-purpose reviews with an emphasis on classroom environment work in science education (Fraser & Walberg, 1981), in Australia (Fraser, 1981a), and in Germany (Dreesman, 1982; Wolf, 1983). As well, the American Educational Research Association established a successful Special Interest Group (SIG) on the Study of Learning Environments in 1984, and this group sponsors an annual monograph (e.g., Fraser, 1986c, 1987b, 1988; Waxman & Ellett, 1990).

Although this chapter focuses predominantly on the work on educational environments over the previous two decades, it is fully acknowledged that this research builds on and has been influenced by two areas of earlier work. First, the influence of the momentous theoretical, conceptual, and measurement foundations laid half a century ago by pioneers like Lewin and Murray and their followers (e.g., Pace & Stern) is recognized. Second, Chavez (1984) observes that research involving assessments of perceptions of classroom environment epitomized in the work of Walberg and Moos was also influenced by prior work involving low-inference, direct observational methods of measuring classroom climate.

One fruitful way to think about educational life is in terms of Lewin's (1936) early but seminal work on field theory. Lewin recognized that both the environment and its interaction with personal characteristics of the individual are potent determinants of human behavior (see von Saldern, 1984). The familiar Lewinian formula, $B = f(P,E)$, was first enunciated largely for didactic reasons to stress the need for new research strategies in which behavior is considered a function of the person and the environment. Murray (1938) was the first worker to follow Lewin's approach by proposing a needs-press model, which allows the analogous representation of person and environment in common terms. Personal needs refer to motivational personality characteristics representing tendencies to move in the direction of certain goals; environmental press provides an external situational counterpart that supports or frustrates the expression of internalized personality needs. Needs-press theory has been popularized and elucidated in Pace and Stern's (1958) widely cited article and in Stern's (1970) comprehensive book.

Although the work described in this chapter clearly has some historical antecedents in the work of Lewin, Murray, and others, earlier writings neither focus sharply on educational settings nor provide empirical evidence to support linkages between climate and educational outcomes. Moreover, the epic work of Pace and Stern (1958), although involving high-inference measures of educational environments, focused on higher-education institutions rather than high schools or elementary schools; it also assessed the environment of the whole college and did not specifically consider the environment of individual classrooms. Consequently, the present chapter has a distinctive focus.

School-Level Versus Classroom-Level Environment

Various writers have found it useful to distinguish classroom or classroom-level environment from school or school-level environment, which involves psychosocial aspects of the climate of whole schools (Anderson, 1982; Fraser & Rentoul, 1982; Genn, 1984). Nevertheless, despite their simultaneous development and logical linkages, the fields of classroom-level and school-level environment have remained remarkably independent. It is common for workers in one field to have little cognizance of the other field and for different theoretical and conceptual foundations to underpin the two areas. Although the focus of past research in science education has been primarily on classroom-level environment, it is also acknowledged that it would be desirable to break away from the existing tradition of independence of the two fields of school and classroom environment and for there to be a confluence of the two areas.

A common way to view the school environment is to consider it as something distinct from and more global than classroom environment. For example, whereas classroom climate might involve relationships between the teacher and his or her students or among students, school climate might involve relationships between teachers and their teaching colleagues, head of department, and school principal. Similarly, whereas classroom environment is usually measured in terms of either student or teacher perceptions, school environment is usually (but not exclusively) assessed in terms of teacher perceptions.

School climate research owes much in theory, instrumentation, and methodology to earlier work on organizational climate in business contexts (Anderson, 1982). This is clearly illustrated by the fact that two widely used instruments in school environment research, namely Halpin and Croft's (1963) Organizational Climate Description Questionnaire (OCDQ) and Stern's (1970) College Characteristic Index (CCI), relied heavily on previous work in business organizations. Consequently, one feature of school-level environment work that distinguishes it from classroom-level environment research is that the former has tended to be associated with the field of educational administration and to rest on the assumption that schools can be viewed as formal organizations (Thomas, 1976). Another distinguishing feature is that, whereas classroom-level research has been concentrated on secondary and elementary schools rather than higher education, a sizable proportion of school-level environment research has involved the climate of higher-education institutions.

Level of Analysis: Private and Consensual Press

Murray's distinction between alpha press (the environment as observed by an external observer) and beta press (the environment as perceived by milieu inhabitants) has been extended by Stern, Stein, and Bloom (1956), who distinguish between the idiosyncratic view that each person has of the environment (*private* beta press) and the shared view that members of a group hold about the environment (*consensual* beta press). Private and consensual beta press could differ from each other, and both could differ from the detached view of alpha press of a trained nonparticipant observer. In designing classroom environment studies, science education researchers must decide whether their analyses will involve perception scores from individual students (private press) or whether these will be combined to obtain the average of the environment scores of all students in the same class (consensual press).

A growing body of literature acknowledges the importance and consequences of the choice of level or unit of statistical analysis and considers the hierarchical analysis and multilevel analysis of data (Burstein, 1978; Burstein, Linn, & Capell, 1978; Cheung, Keeves, Sellin, & Tsoi, 1990; Corno, Mitman, & Hedges, 1981; Cronbach, 1976; Cronbach & Snow, 1977; Cronbach & Webb, 1975; Goldstein, 1987; Larkin & Keeves, 1984; Lincoln & Zeitz, 1980; Raudenbush, 1988; Raudenbush & Bryk, 1986). The choice of unit of analysis is important for a number of reasons. First, measures having the same operational definition can have different substantive interpretations with different levels of aggregation. Second, relationships obtained using one unit of analysis could differ in magnitude and even in sign from relationships obtained using another unit (Robinson, 1950). Third, the use of certain units of analysis (e.g., individuals when classes are the primary sampling units) violates the requirement of independence of observations and calls into question the results of statistical significance tests because an unjustifiably small estimate of the sampling error is used (Peckham, Glass, & Hopkins, 1969; Ross, 1978). One solution to this dilemma followed in recent research (Ross, 1978) is to use the individual as the unit of analysis but to employ the jackknife technique (Mosteller & Tukey, 1977) or the bootstrap technique (Luecht & Smith, 1989) to adjust significance levels to allow for nonindependence of observations. Fourth, the use of different units of analysis involves the testing of conceptually different hypotheses (Burstein et al., 1978; Cronbach, 1976).

For example, in a study of the effects of classroom environment on some student outcome measure, use of the individual as the unit of analysis (i.e., a between-student analysis) involves substantive questions about the relationship between individuals' outcomes and their environment scores when class membership is disregarded. Use of the deviation of a student's score from the class mean as the unit of analysis (i.e., a pooled within-class analysis) involves substantive questions about whether the amount by which a student's classroom environment score differs from that of his or her classmates is related to how much his or her outcome performance differs from the class mean. Use of the class mean as the unit of analysis (i.e., a between-class analysis) asks whether the relationship between class means on the outcome measure varies with the average environment perceptions of the students within a class.

Although the unit-of-analysis problem has received considerable attention in the context of testing hypotheses using existing classroom environment instruments, Sirotnik (1980) considers it ironic that concerns about analytic units have been virtually nonexistent at the stage of developing and empirically investigating the dimensionality of new instruments. Because of the central importance of the unit-of-analysis problem in classroom environment research, subsequent sections of this chapter return to this problem. For example, separate validation information for the individual and the class as the unit of analysis is reported, and the research reviews consider the level of statistical analysis used in different studies.

INSTRUMENTS FOR ASSESSING CLASSROOM ENVIRONMENT

This section clarifies the background and nature of several instruments commonly used in prior research in science education to assess perceptions of classroom learning environment; instruments for assessing school-level environment are discussed in a separate section later. The instruments considered are the Learning Environment Inventory (LEI), Classroom Environment Scale (CES), Individualized Classroom Environment Questionnaire (ICEQ), My Class Inventory (MCI), College and University Classroom Environment Inventory (CUCEI), and Science Laboratory Environment Inventory (SLEI). Each instrument is suitable for convenient group administration, can be scored either by hand or by computer, and has been shown to be reliable in extensive field trials. Each of these instruments is considered in a separate subsection. In addition, separate subsections are devoted to preferred forms of scales; some economical short forms of the ICEQ, CES, and MCI; hand scoring procedures; and scale validation.

Table 17.1 shows the name of each scale contained in the LEI, CES, ICEQ, MCI, CUCEI, and SLEI. The table summarizes

TABLE 17.1. Overview of Scales Contained in Six Classroom Environment Instruments (LEI, CES, ICEQ, MCI, CUCEI, and SLEI)

Instrument	Level	Items per Scale	Scales Classified According to Moos's Scheme		
			Relationship Dimensions	Personal Development Dimensions	System Maintenance and Change Dimensions
Learning Environment Inventory (LEI)	Secondary	7	Cohesiveness Friction Favoritism Cliqueness Satisfaction Apathy	Speed Difficulty Competitiveness	Diversity Formality Material environment Goal direction Disorganization Democracy
Classroom Environment Scale (CES)	Secondary	10	Involvement Affiliation Teacher support	Task orientation Competition	Order and organization Rule clarity Teacher control Innovation
Individualized Classroom Environment Questionnaire (ICEQ)	Secondary	10	Personalization Participation	Independence Investigation	Differentiation
My Class Inventory (MCI)	Elementary	6–9	Cohesiveness Friction Satisfaction	Difficulty Competitiveness	
College and University Classroom Environment Inventory (CUCEI)	Higher education	7	Personalization Involvement Student cohesiveness Satisfaction	Task orientation	Innovation Individualization
Science Laboratory Environment Inventory (SLEI)	Senior secondary, higher education	6 or 7	Student cohesiveness	Open-endedness Integration	Rule clarity Material environment

the level (elementary, secondary, higher education) for which each instrument is suited, the number of items contained in each scale, and the classification of each scale according to Moos's (1974) scheme for classifying human environments. Moos's three basic types of dimension are Relationship Dimensions (which identify the nature and intensity of personal relationships within the environment and assess the extent to which people are involved in the environment and support and help each other), Personal Development Dimensions (which assess basic directions along which personal growth and self-enhancement tend to occur), and System Maintenance and System Change Dimensions (which involve the extent to which the environment is orderly, clear in expectations, maintains control, and is responsive to change).

Learning Environment Inventory

The development and validation of a preliminary version of the Learning Environment Inventory (LEI) began in the late 1960s in conjunction with the evaluation and research related to Harvard Project Physics (Anderson & Walberg, 1974; Fraser, Anderson, & Walberg, 1982). The LEI is a 15-scale expansion and improvement of the Classroom Climate Questionnaire. In selecting the 15 climate dimensions, an attempt was made to include as scales only concepts previously identified as good predictors of learning, concepts considered relevant to social psychological theory and research, concepts similar to those found useful in theory and research in education, or concepts intuitively judged relevant to the social psychology of the classroom.

The name of each of the 15 LEI scales is listed in Table 17.1 and each has a commonsense meaning. The final version of the LEI contains a total of 105 statements (or seven per scale) descriptive of typical school classes. The respondent expresses degree of agreement or disagreement with each statement on a four-point scale with response alternatives of strongly disagree, disagree, agree, and strongly agree. The scoring direction (or polarity) is reversed for some items. A typical item contained in the Cohesiveness scale is "All students know each other very well." An item from the Speed scale is "The pace of the class is rushed."

Classroom Environment Scale

The Classroom Environmental Scale (CES) was developed by Rudolf Moos at Stanford University (Fisher & Fraser, 1983d; Moos & Trickett, 1974, 1987; Trickett & Moos, 1973) and grew out of a comprehensive program of research involving perceptual measures of a variety of human environments, including psychiatric hospitals, prisons, university residences, and work milieus (Moos, 1974). The original version of the CES consisted of 242 items representing 13 conceptual dimensions. After trials of the items in 22 classrooms and subsequent item analysis, the number of items was reduced to 208. This item pool was administered in 45 classrooms and modified to form the final 90-item version of the CES. These items were evaluated statistically according to whether they discriminated significantly between the perceptions of students in different classrooms and whether they correlated highly with their scale scores.

Moos and Trickett's (1974, 1987) final published version of the CES contains nine scales with 10 items of true-false response format in each scale. Published materials include a test manual, a questionnaire, an answer sheet, and a transparent hand scoring key. Typical items in the CES are "The teacher takes a personal interest in the students" (Teacher Support) and "There is a clear set of rules for students to follow" (Rule Clarity).

Individualized Classroom Environment Questionnaire

The Individualized Classroom Environment Questionnaire (ICEQ) differs from other classroom environment scales in that it assesses the dimensions (e.g., Personalization, Participation) that distinguish individualized classrooms from conventional ones. The initial development of the long-form ICEQ (Rentoul & Fraser, 1979) was guided by several criteria. First, dimensions chosen characterized the classroom learning environment described in individualized curriculum materials and in the literature of individualized education, including open and inquiry-based classrooms. Second, extensive interviewing of teachers and secondary school students ensured that the ICEQ's dimensions and individual items were considered salient by teachers and students. Third, items were written and subsequently modified after receiving reactions sought from selected experts, teachers, and junior high school students. Fourth, data collected during field testing were subjected to item analyses to identify items whose removal would enhance scale statistics.

The final published version of the ICEQ (Fraser, 1990) contains 50 items, with an equal number of items belonging to each of the five scales. Each item is responded to on a five-point scale with the alternatives of almost never, seldom, sometimes, often, and very often. The scoring direction is reversed for many of the items. Typical items are "The teacher considers students' feelings" (Personalization) and "Different students use different books, equipment, and materials" (Differentiation). The progressive copyright arrangements for the published form of the ICEQ permit an unlimited number of copies of the questionnaires and response sheets to be made.

My Class Inventory

The LEI has been simplified to form the My Class Inventory (MCI) for use among children in the 8- to 12-year age range (Fisher & Fraser, 1981; Fraser et al., 1982; Fraser & O'Brien, 1985). Although the MCI was developed originally for use at the elementary school level, it has also been found useful with students in junior high school, especially those who might experience reading difficulties with the LEI. The MCI differs from the LEI in four important ways. First, in order to minimize fatigue among younger children, the MCI contains only five of the LEI's original 15 scales. Second, item wording has been simplified to enhance readability. Third, the LEI's four-point response format has been reduced to a two-point (yes-no) response format in the MCI. Fourth, students answer on the questionnaire itself instead of a separate response sheet to avoid errors in transferring responses from one place to another.

The final form of the MCI contains 38 items (6 for Cohesiveness, 8 for Friction, 8 for Difficulty, 9 for Satisfaction, and 7 for Competitiveness). Typical items contained in the MCI are "Children are always fighting with each other" (Friction) and "Children seem to like the class" (Satisfaction). It can be seen that the reading level of these MCI items is well suited to students at the elementary school level.

College and University Classroom Environment Inventory

Although some notable prior work has focused on the institutional-level or school-level environment in colleges and universities (e.g., Halpin & Croft, 1963; Pace & Stern, 1958; Stern, 1970), surprisingly little work has been done in higher education classrooms that is parallel to the traditions of classroom environment research at the secondary and elementary school levels. One likely explanation for this shortage is simply the unavailability of a suitable instrument, and the College and University Classroom Environment Inventory (CUCEI) was developed to fill this void (Fraser & Treagust, 1986; Fraser, Treagust, & Dennis, 1986). The CUCEI is intended for use in small classes (say up to 30 students) sometimes referred to as "seminars"; it is not suited to lectures or laboratory classes.

The initial development of the CUCEI involved examining the scales and items in the LEI, CES, and ICEQ to identify concepts and ideas relevant to higher-education settings. An initial pool of items was developed and then modified, first after subjecting items to the scrutiny of colleagues and then after performing item analyses on data collected during field trials. The final form of the CUCEI contains seven seven-item scales. Each item has four responses (strongly agree, agree, disagree, strongly disagree) and polarity is reversed for approximately half of the items. Typical items are "Activities in this class are clearly and carefully planned" (Task Orientation) and "Teaching approaches allow students to proceed at their own pace" (Individualization).

Appendix 17.A contains a complete copy of the actual form of the CUCEI, together with information about scoring.

Science Laboratory Environment Inventory

Because of the importance and uniqueness of laboratory settings in science education, a new instrument specifically suited to assessing the environment of science laboratory classes at the senior high school or higher-education levels was developed in collaboration with colleagues from various countries (Fraser, Giddings, & McRobbie, in press). This new questionnaire, the Science Laboratory Environment Inventory (SLEI), has the five scales (each with approximately seven items) listed in Table 17.1, and the five response alternatives are almost never, seldom, sometimes, often, and very often. Typical items are "The teacher is concerned about students' safety during laboratory sessions" (Teacher Supportiveness), and "We know the results that we are supposed to get before we commence a laboratory activity" (open-endedness).

Notably, the SLEI was field tested and validated simultaneously with a sample of more than 5447 students in 269 classes in six different countries (the United States, Canada, England, Israel, Australia, and Nigeria) to obtain comprehensive information about its cross-national validity and usefulness. The merits of cross-national studies such as these are discussed by Fraser, Giddings, and McRobbie (1991).

Preferred Forms of Scales

A distinctive feature of most of the instruments in Table 17.1 is that they have not only a form to measure perceptions of actual classroom environment but also a form to measure perceptions of preferred classroom environment. The preferred (or ideal) forms are concerned with goals and value orientations and measure perceptions of the classroom environment ideally liked or preferred. Although item wording is identical or similar for actual and preferred forms, instructions for answering them are different. For example, an item in the actual form such as "There is a clear set of rules for students to follow" would be changed in the preferred form to "There would be a clear set of rules for students to follow." Having different actual and preferred forms has enabled these instruments to be used for a range of research applications, which are discussed later. Although the LEI and MCI were originally designed to measure actual environment, Fraser and Deer (1983) and Fraser and O'Brien (1985) have used a preferred form of the MCI successfully with elementary school classes.

Short Forms of ICEQ, MCI, and CES

Although the long forms of classroom environment instruments have been used successfully for a variety of purposes, some researchers have expressed a preference for a more rapid and economical instrument. Similarly, some teachers using these scales for local, school-based applications have reported that they would like instruments to take less time to administer and score. Consequently, short forms of the ICEQ, MCI, and CES were developed (Fraser, 1982a; Fraser & Fisher, 1983b) to satisfy three main criteria. First, the total number of items in each instrument was reduced to approximately 25 to provide greater economy in testing and scoring time. Second, the short forms were designed to be amenable to easy hand scoring. Third, although most existing classroom environment instruments were developed to provide adequate reliability for the assessment of the perceptions of individual students, the majority of applications involve averaging the perceptions of students within a class to obtain class means. Consequently, it was decided that the short forms would be developed to have adequate reliability for uses involving the assessment of class means. Use of the long form of these instruments, however, is still recommended for applications involving the individual student as the unit of analysis.

The development of the short forms was based largely on the results of several item analyses performed on data obtained by administering the long form of each instrument to a large sample of science students. In particular, the internal consistency of each scale was maximized by selecting items with large item-remainder correlations (i.e., correlations between item score and total score on the rest of the scale), and discriminant validity was enhanced by including an item only if the correlation with its *a priori* assigned scale was smaller than the correlation with any other items in the battery. In addition, the development of the short forms was based on logical considerations including face validity and an attempt to achieve a balance of items with positive and negative scoring directions (both within each scale and within each instrument as a whole). Nevertheless, because the long forms of some scales have an imbalance in the number of items with positive and negative polarity, this imbalance tended to be maintained in the short forms of these scales.

The application of these criteria led to the development of short forms of the ICEQ and the MCI that each consisted of 25 items divided equally among the five scales making up the long form of the instrument. The long form of the CES consisted of 90 items and this was reduced considerably to form a short version with 24 items divided equally among six of the original nine scales. Furthermore, the development of this short form was guided by the fact that Trickett and Moos (1973) had recommended a short four-item version of each of the CES's nine scales. In fact, the present short form consists of five scales that are identical to those recommended by Trickett and Moos (namely, Involvement, Affiliation, Teacher Support, Order and Organization, and Rule Clarity) and a sixth scale (Task Orientation) that contains two of the four items recommended.

In order to clarify the nature of the short forms, a copy of the actual short form of the MCI is shown in Appendix B. Unlike the long form, the short form of the MCI does not make use of a separate answer sheet because all items and space for responding fit on a single page.

Hand Scoring Procedures

Appendix 17.B illustrates typical hand scoring procedures for the MCI. First, inclusion of the letter *R* in the Teacher Use Only column identifies items that must be scored in the reverse direction. Second, items are arranged in blocks and in cyclic order so that all items from the same scale are found in the same position in each block. For example, the first item in each block of five items in the MCI belongs to the Satisfaction scale (see Appendix 17.B). Items in Appendix 17.B *without* the letter *R* are scored by allocating a score of *3* for the response *Yes* and *1* for the response *No*. Underlined items *with* the letter *R* are scored in the reverse manner. Omitted or invalidly answered items are scored *2*.

To obtain scale totals, the five item scores for each scale are added. The first, second, third, fourth, and fifth items in each block of five, respectively, measure Satisfaction, Friction, Competitiveness, Difficulty, and Cohesiveness. For example, the total Satisfaction score is obtained by adding scores for items 1, 6, 11, 16, and 21. Scale totals can be written in the spaces provided at the bottom of the questionnaire. Appendix 17.B illustrates how these scoring procedures were used to obtain a total of 10

for the Satisfaction scale and a total of 12 for the Cohesiveness scale.

Validation of Scales

This subsection reports typical validation data for some classroom environment scales. Table 17.2 provides a summary of a limited amount of statistical information for the six instruments (the LEI, CES, ICEQ, MCI, CUCEI, and SLEI) considered previously. Attention is restricted to the student actual form and to use of the individual student as the unit of analysis. Table 17.2 provides information about each scale's internal consistency reliability (alpha coefficient) and discriminant validity (using the mean correlation of a scale with the other scales in the same instrument as a convenient index), and the ability to differentiate between the perceptions of students in different

TABLE 17.2. Internal Consistency (Alpha Reliability), Discriminant Validity (Mean Correlation of a Scale with Other Scales), and ANOVA Results for Class Membership Differences (Eta^2 Statistic and Significance Level) for Student Actual Form of Six Instruments Using Individual as Unit of Analysis

Scale	Alpha Reliability	Mean Correlation with Other Scales	ANOVA Results Eta^2	Scale	Alpha Reliability	Mean Correlation with Other Scales	ANOVA Results Eta^2
Learning Environment Inventory (LEI)				*Individualized Classroom Environment Questionnaire* (ICEQ)			
	(N = 1048 students)	(N = 149 classes)			(N = 1849 students)		
Cohesiveness	0.69	0.14	—	Personalization	0.79	0.28	0.31*
Diversity	0.54	0.16	—	Participation	0.70	0.27	0.21*
Formality	0.76	0.18	—	Independence	0.68	0.07	0.30*
Speed	0.70	0.17	—	Investigation	0.71	0.21	0.20*
Material environment	0.56	0.24	—	Differentiation	0.76	0.10	0.43*
Friction	0.72	0.36	—	*My Class Inventory* (MCI)			
Goal direction	0.85	0.37	—		(N = 2305 students)		
Favoritism	0.78	0.32	—	Cohesiveness	0.67	0.20	0.21*
Difficulty	0.64	0.16	—	Friction	0.67	0.26	0.31*
Apathy	0.82	0.39	—	Difficulty	0.62	0.14	0.18*
Democracy	0.67	0.34	—	Satisfaction	0.78	0.23	0.30*
Cliqueness	0.65	0.33	—	Competitiveness	0.71	0.10	0.19*
Satisfaction	0.79	0.39	—	*College and University Classroom Environment Inventory* (CUCEI)			
Disorganization	0.82	0.40	—		(N = 372 students)		
Competitiveness	0.78	0.08	—	Personalization	0.75	0.46	0.35*
Classroom Environment Scale (CES)				Involvement	0.70	0.47	0.40*
	(N = 1083 students)			Student cohesiveness	0.90	0.45	0.47*
Involvement	0.70	0.40	0.29*	Satisfaction	0.88	0.45	0.32*
Affiliation	0.60	0.24	0.21*	Task orientation	0.75	0.38	0.43*
Teacher support	0.72	0.29	0.34*	Innovation	0.81	0.46	0.41*
Task orientation	0.58	0.23	0.25*	Individualization	0.78	0.34	0.46*
Competition	0.51	0.09	0.18*	*Science Laboratory Environment Inventory* (SLEI)			
Order and organization	0.75	0.29	0.43*		(N = 3727 students)		
Rule clarity	0.63	0.29	0.21*	Student cohesiveness	0.77	0.34	0.21*
Teacher control	0.60	0.16	0.27*	Open-endedness	0.70	0.07	0.19*
Innovation	0.52	0.19	0.26*	Integration	0.83	0.37	0.23*
				Rule clarity	0.75	0.33	0.21*
				Material environment	0.75	0.37	0.21*

*$p < .01$.

TABLE 17.3. Internal Consistency (Alpha Reliability) and Discriminant Validity (Mean Correlation of a Scale with Other Four Scales) for Two Units of Analysis for ICEQ

Scale	Unit of Analysis	Alpha Reliability				Mean Correlation with Other Scales			
		Student Actual (N = 1849 & 150)[a]	Student Preferred (N = 1858 & 150)[a]	Teacher Actual (N = 90)	Teacher Preferred (N = 34)	Student Actual (N = 1849 & 150)[a]	Student Preferred (N = 1858 & 150)[a]	Teacher Actual (N = 90)	Teacher Preferred (N = 34)
Personalization	Individual	0.79	0.74	0.79	0.74	0.28	0.31	0.32	0.29
	Class	0.90	0.86			0.31	0.35		
Participation	Individual	0.70	0.67	0.79	0.82	0.27	0.29	0.39	0.34
	Class	0.80	0.75			0.32	0.32		
Independence	Individual	0.68	0.70	0.83	0.86	0.07	0.12	0.23	0.25
	Class	0.78	0.79			0.16	0.17		
Investigation	Individual	0.71	0.75	0.80	0.90	0.21	0.27	0.34	0.33
	Class	0.77	0.83			0.29	0.31		
Differentiation	Individual	0.76	0.75	0.85	0.81	0.10	0.16	0.29	0.16
	Class	0.91	0.92			0.19	0.20		

[a] The sample sizes shown are the number of individual students and classes, respectively.

classrooms (significance level and eta^2 statistic from ANOVAs). Statistics are based on 1048 students for the LEI, except for discriminant validity data, which are based on 149 class means (Fraser et al., 1982); 1083 students for the CES (Fisher & Fraser, 1983c); 1849 students for the ICEQ (Fraser, 1990); 2035 students for the MCI (Fisher & Fraser, 1981); 3727 senior high school students for the SLEI (Fraser et al., in press); and 372 students for the CUCEI (Fraser & Treagust, 1986). Generally, the data in Table 17.2 suggest that the actual form of each scale of each instrument has adequate internal consistency reliability and discriminant validity (although each instrument appears to assess somewhat overlapping aspects) and has the ability to differentiate between classrooms (although no data are available for the LEI for this characteristic).

Table 17.3 illustrates the reporting of more comprehensive validation information for one instrument, the ICEQ. This table shows reliability and discriminant validity data separately for students and teachers, separately for actual and preferred forms, and separately using the individual and class mean as the unit of analysis for the student statistics. The sample consists of 1849 students in 150 junior high school classes in Australia for the student actual form, 1858 students in the same 150 classes

for the student preferred form, 90 teachers of some of the same classes for the teacher actual form, and 34 teachers of some of the same classes for the teacher preferred form. Overall, Table 17.3 suggests that the ICEQ has adequate internal consistency reliability and discriminant validity for use with students or teachers, in its actual or preferred form, and using either the individual student or the class mean as the unit of analysis.

An especially noteworthy feature of the development and validation of the Science Laboratory Environment Inventory (SLEI) was that it was field tested simultaneously at both the senior high school and university levels in the United States, Canada, England, Australia, Israel, and Nigeria. During the application of item analysis techniques to the data, an item was retained in the instrument only if it displayed satisfactory statistical characteristics for each of the 12 subsamples (i.e., university or senior high school level in six countries). The total sample consisted of 3401 senior high school students in 183 laboratory classes and 1242 university students in 42 laboratory classses. Table 17.4 shows the Cronbach alpha reliabilities obtained for each of the 12 subsamples for both the actual and preferred forms of a 52-item version of the SLEI.

TABLE 17.4. Cronbach Alpha Reliability Coefficient for Actual and Preferred Form of 34-Item Version of SLEI for the 12 Subsamples

Scale	No. of Items	Form	Alpha Reliability Coefficient											
			USA		Canada		Australia		England		Israel		Nigeria	
			Sch	Uni	Sch	Uni	Sch	Uni	Sch	Uni	Sch	Uni	Sch	Uni
Student cohesiveness	7	Actual	0.81	0.76	0.75	0.83	0.78	0.81	0.77	0.83	0.69	0.78	0.56	0.56
		Preferred	0.73	0.74	0.63	0.70	0.72	0.69	0.57	0.61	0.74	0.72	0.51	0.52
Open-endedness	6	Actual	0.75	0.70	0.60	0.55	0.69	0.66	0.78	0.58	0.54	0.57	0.49	0.54
		Preferred	0.61	0.63	0.54	0.56	0.60	0.60	0.54	0.56	0.62	0.62	0.43	0.53
Integration	7	Actual	0.84	0.91	0.80	0.88	0.81	0.86	0.65	0.92	0.89	0.89	0.68	0.74
		Preferred	0.79	0.82	0.73	0.82	0.80	0.76	0.80	0.84	0.84	0.85	0.72	0.62
Rule clarity	7	Actual	0.76	0.74	0.76	0.81	0.72	0.69	0.84	0.76	0.71	0.54	0.61	0.62
		Preferred	0.71	0.68	0.65	0.69	0.68	0.58	0.68	0.65	0.75	0.53	0.51	0.54
Material environment	7	Actual	0.74	0.72	0.79	0.72	0.74	0.67	0.83	0.73	0.56	0.57	0.71	0.76
		Preferred	0.70	0.68	0.70	0.65	0.73	0.59	0.61	0.54	0.74	0.69	0.71	0.70
Sample size			885	719	282	323	1875	298	108	106	359	104	218	170

INSTRUMENTS FOR ASSESSING SCHOOL ENVIRONMENT

Over the past decade or so, the concept of school environment has appeared in the educational literature with increasing frequency. In fact, together with curriculum, resources, outcomes, and leadership, environment is considered to make a major contribution to the effectiveness of a school (Creemers, Peters, & Reynolds, 1989). However, in contrast to work on classroom-level environment, relatively little work has been directed toward helping science teachers assess and improve the environments of their own schools. Practical constraints inhibiting teachers' use of the school environment instruments include difficult access to instruments, the fact that many existing instruments lack economy of testing and scoring time, and the unavailability of case studies of teachers' successful attempts at improving school environments. Consequently, the subsections below describe the development and validation of two instruments, the Work Environment Scale (WES) and the School-Level Environment Questionnaire (SLEQ), which are of

widespread usefulness in measuring teachers' perceptions of psychosocial dimensions of the school environment.

One example of an earlier school environment instrument is the College Characteristics Index (CCI) (Pace & Stern, 1958), which measures student or staff perceptions of 30 environment characteristics. Each of these 30 variables (e.g., Affiliation, Aggression, Deference, Impulsiveness, Order) was based on Murray's (1938) taxonomy and paralleled a needs scale in the Stern et al. (1956) Activities Index. That is, each Activities Index scale corresponded to behavioral manifestations of a needs variable, while the parallel CCI scale corresponded to environmental press conditions likely to facilitate or impede their expression. The original CCI was adapted by Stern (1970) to form the High School Characteristics Index (HSCI), which is suitable for use at the high school level.

Probably the most widely used instrument measuring school-level environment is Halpin and Croft's Organizational Climate Description Questionnaire (OCDQ) (Halpin & Croft, 1963). In fact, Thomas (1976) noted that the OCDQ has been used in more than 200 studies in at least eight countries and that the instrument achieved something of bandwagon status in

TABLE 17.5. Description of Scales in the WES and their Classification According to Moos's Scheme

Scale Name	Description of Scale	Sample Item	Scoring Direction[a]	Moos's General Category
Involvement	The extent to which teachers are concerned and committed to their jobs.	Teachers put quite a lot of effort into what they do.	+	Relationship
Peer cohesion	The extent to which teachers are friendly and supportive of each other.	Teachers go out of their way to help a new teacher feel comfortable.	+	
Staff support	The extent to which the senior staff is supportive of teachers and encourages teachers to be supportive of each other.	Senior staff often criticize teachers over minor things.	−	
Autonomy	The extent to which teachers are encouraged to be self-sufficient and to make their own decisions.	Teachers can use their own initiative to do things.	+	Personal development
Task orientation	The extent of emphasis on planning and efficiency.	There is a lot of time wasted because of inefficiencies.	−	
Work pressure	The extent to which the pressure of work dominates the job milieu.	It is very hard to keep up with your work load.	+	System maintenance
Clarity	The extent to which teachers know what to expect in their daily routines and how explicitly school rules and policies are communicated.	Teachers are often confused about exactly what they are supposed to do.	−	
Control	The extent to which the school administration uses rules and pressures to keep teachers under control.	Teachers are expected to conform rather strictly to the rules and customs.	+	
Innovation	The extent to which variety, change and new approaches are emphasized in the school.	This place would be one of the first to try out a new idea.	+	
Physical comfort	The extent to which the physical surroundings contribute to a pleasant work environment.	The colors and decorations make the place warm and cheerful to work in.	+	

[a] Items designated (+) are scored by allocating 3 and 1, respectively, for the responses of true and false. Items designated (−) are scored in the reverse manner. Omitted or invalid responses are given a score of 2.

TABLE 17.6. Internal Consistency (Alpha Reliability) and Discriminant Validity (Mean Correlation with Other Scales) for Actual and Preferred Forms of WES for Two Units of Analysis

Scale	Unit of Analysis	Alpha Reliability				Mean Correlation with Other Scales			
		Moos Sample	Science Teachers Actual	New Sample Actual	New Sample Preferred	Moos Sample	Science Teachers Actual	New Sample Actual	New Sample Preferred
Involvement	Indiv	0.85	0.85	0.76	0.74	0.40	0.41	0.35	0.41
	School			0.93	0.91			0.48	0.50
Peer cohesion	Indiv	0.70	0.60	0.72	0.69	0.37	0.33	0.31	0.39
	School			0.95	0.90			0.46	0.54
Staff support	Indiv	0.78	0.66	0.71	0.68	0.25	0.29	0.30	0.36
	School			0.96	0.93			0.46	0.50
Autonomy	Indiv	0.76	0.61	0.60	0.55	0.35	0.27	0.32	0.33
	School			0.88	0.87			0.48	0.49
Task orientation	Indiv	0.78	0.78	0.70	0.60	0.31	0.34	0.33	0.34
	School			0.97	0.91			0.49	0.49
Work pressure	Indiv	0.84	0.74	0.79	0.70	0.21	0.16	0.11	0.28
	School			0.96	0.95			0.27	0.38
Clarity	Indiv	0.82	0.73	0.70	0.72	0.57	0.33	0.30	0.41
	School			0.90	0.95			0.41	0.54
Control	Indiv	0.77	0.64	0.64	0.62	0.18	0.18	0.17	0.11
	School			0.91	0.90			0.29	0.27
Innovation	Indiv	0.91	0.84	0.84	0.74	0.34	0.29	0.29	0.35
	School			0.98	0.95			0.44	0.47
Physical comfort	Indiv	0.83	0.70	0.71	0.72	0.24	0.26	0.23	0.36
	School			0.93	0.94			0.38	0.45
Sample sizes	Indiv	624	114	599	543	624	114	599	543
	School		35	34	34		35	34	34

research in the field of educational administration. The final version of the OCDQ measures teacher perceptions of eight factor-analytically derived dimensions. Four of these dimensions pertain to teachers' behavior and are called Disengagement, Hindrance, Esprit (i.e., morale), and Intimacy; the other four dimensions pertain to the principal's behavior and are called Aloofness, Production Emphasis, Thrust, and Consideration. Furthermore, Halpin and Croft suggested a method by which profiles of OCDQ scores can be used to classify schools into six climate types: open, autonomous, controlled, familiar, paternal, and closed. The OCDQ formed the basis for the development of some new factor-analytic school environment scales for use in secondary schools in England (Finlayson, 1973) and Australia (Deer, 1980).

Work Environment Scale

The Work Environment Scale (WES) (Moos, 1981) was designed for use in any work milieu rather than specifically in schools. However, its 10 dimensions are well suited to describing salient features of the teacher's school environment, even though it has had little usage specifically in school settings. To improve the WES's face validity for use in schools, the word "people" was changed to "teachers," the word "supervisor" to "senior staff," and the word "employee" to "teacher" (Fisher & Fraser, 1983c; Fraser, Docker, & Fisher, 1988).

Of the WES's 10 scales, three measure Relationship Dimensions (Involvement, Peer Cohesion, Staff Support), two measure Personal Development Dimensions (Autonomy, Task Orientation), and five measure System Maintenance and System Change Dimensions (Work Pressure, Clarity, Control, Innovation, Physical Comfort). The WES consists of 90 items of true-false response format, with an equal number of items in each scale. The WES is described in more detail in Table 17.5 which provides a scale description and a sample item for each scale and shows each scale's classification according to Moos's scheme.

Moos (1981) reported validation data for the original form of the WES based on its administration to a sample of 624 employees and supervisors in a broad range of work groups (e.g., salesmen, nurses, drivers, maintenance workers) in the United States. Table 17.6 summarizes Moos's results for this sample for each scale's internal consistency (alpha reliability coefficient) and discriminant validity (using the convenient index of the mean correlation of a scale with the other nine scales).

The WES was used for the first time specifically with school teachers in a study conducted among Australian science teachers (Fisher & Fraser, 1983c). The slightly modified version of the WES was administered to a sample of 114 science teachers in a representative sample of 35 secondary schools in Tasmania. Validation data for this sample also are included in Table 17.6.

Further validation data were generated in a more comprehensive study in Tasmania (Docker, Fraser, & Fisher, 1989) with a sample that included elementary schools as well as secondary schools. Furthermore, whereas the previous study involved only the actual form of the WES, the new sample responded to both an actual form and a preferred form. The total sample consisted of 34 schools, with 599 teachers responding to the actual form of the WES and 543 teachers responding to the preferred form. Because applications of school environment instruments could involve the school mean rather than the individual teacher as the unit of analysis, validation data in Table 17.6 are reported for both units for this sample.

Overall, the data in Table 17.6 indicate that the WES scales display satisfactory internal consistency and discriminant validity in either the actual or preferred form and with either the individual teacher or the school mean as the unit of analysis. The WES measures distinct, although somewhat overlapping, aspects of school environment.

Whether the actual form of each WES scale is capable of differentiating between the perceptions of teachers in different schools was explored for the sample of 599 teachers in 34 schools. A one-way analysis of variance (ANOVA) was performed for each scale, with school membership as the main effect. It was found that each scale differentiated significantly ($p < .001$) between schools and that the eta^2 statistic (an estimate of the proportion of variance in WES scores attributable to school membership) ranged from 0.18 for Autonomy to 0.40 for Innovation or Physical Comfort.

School-Level Environment Questionnaire

The School-Level Environment Questionnaire (SLEQ) was designed especially to assess school teachers' perceptions of psychosocial dimensions of the environment of the school. A careful review of potential strengths and problems associated with existing school environment instruments, including the WES, suggested that the SLEQ should satisfy the following six criteria (Rentoul & Fraser, 1983).

1. Relevant literature was consulted and dimensions included in the SLEQ were chosen to characterize important aspects of the school environment, such as relationships among teachers and between teachers and students and the organizational structure (e.g., decision making).
2. Dimensions chosen for the SLEQ provided coverage of Moos's three general categories of dimensions—Relationship, Personal Development, and System Maintenance and System Change.
3. Extensive interviewing ensured that the SLEQ's dimensions and individual items covered aspects of the school environment perceived to be salient by teachers.
4. Only material that was specifically relevant to the school was included.
5. As a number of good measures of classroom environment already exist, the SLEQ was designed to provide a measure of school-level environment that had minimal overlap with these existing measures of classroom-level environment.
6. In developing the SLEQ, an attempt was made to achieve economy by developing an instrument with a relatively small number of reliable scales, each containing a fairly small number of items.

It was found that these criteria could be satisfied with an instrument consisting of seven scales, with two measuring Relationship Dimensions (Student Support, Affiliation), one measuring the Personal Development Dimension (Professional Interest), and four measuring System Maintenance and System Change Dimensions (Staff Freedom, originally named Formalization; Participatory Decision Making, originally named Centralization; Innovation; and Resource Adequacy). To complete the view of the school environment, a scale named Work Pressure (based on the WES) was added recently to the latter dimension.

The SLEQ consists of 56 items, with each of the eight scales being assessed by 7 items. Each item is scored on a five-point scale with the responses of strongly agree, agree, not sure, disagree, and strongly disagree. Table 17.7 further clarifies the nature of the SLEQ by providing a scale description and sample item for each scale and showing each scale's classification according to Moos's scheme. As well, Table 17.7 provides information about the method and direction of scoring of SLEQ items.

In addition to an actual form that assesses perceptions of what a school's work environment is actually like, the SLEQ has a preferred form. An item such as "Teachers are encouraged to be innovative in this school" in the actual form would be changed to "Teachers would be encouraged to be innovative in this school" in the preferred form. Appendix 17.C contains a copy of all items in the latest version of the actual form of the SLEQ.

Validation data are available for the SLEQ for three samples and include information about each scale's internal consistency (Cronbach alpha reliability) and discriminant validity (mean correlation of a scale with the other seven scales). The first sample in Table 17.8 consisted of 83 teachers from 19 coeducational government schools (7 elementary and 12 secondary) in the Sydney metropolitan area. The second sample consisted of 34 secondary school teachers, each in a different government high school in New South Wales. The third sample consisted of 109 teachers in 10 elementary and secondary schools in Tasmania. The teachers in the third sample are the only ones who responded to the preferred form as well as the actual form. The recently added Work Pressure scale was not in the form of the questionnaire that was administered to these samples; hence no validation statistics for this scale are included.

Table 17.8 shows that the alpha coefficient for different SLEQ scales ranged from 0.70 to 0.91 for the first sample, from 0.68 to 0.91 for the second sample, from 0.64 to 0.85 for the actual form for the third sample, and from 0.64 to 0.81 for the preferred form for the third sample. These values suggest that each SLEQ scale displays satisfactory internal consistency for a scale composed of only seven items. The values of the mean correlation of a scale with the other scales shown in Table 17.8 range from 0.17 to 0.38 for the first sample, 0.05 to 0.29 for the second sample, 0.10 to 0.42 for the actual form for the third sample, and 0.28 to 0.44 for the preferred form for the third sample. These values indicate satisfactory discriminant validity

TABLE 17.7. Description of Scales in the SLEQ and their Classification According to Moos's Scheme

Scale Name	Description of Scale	Sample Item	Scoring Direction[a]	Moos's General Category
Student support	There is good rapport between teachers and students and students behave in a responsible self-disciplined manner.	There are many disruptive, difficult students in the school.	−	Relationship
Affiliation	Teachers can obtain assistance, advice and encouragement and are made to feel accepted by colleagues.	I feel that I could rely on my colleagues for assistance if I should need it.	+	
Professional interest	Teachers discuss professional matters, show interest in their work and seek further professional development.	Teachers frequently discuss teaching methods and strategies with each other.	+	Personal development or goal orientation
Staff freedom	Teachers are free of set rules, guidelines and procedures, and of supervision to ensure rule compliance.	I am often supervised to ensure that I follow directions correctly.	−	System maintenance and system change
Participatory decision making	Teachers have the opportunity to participate in decision making.	Teachers are frequently asked to participate in decisions concerning administrative policies and procedures.	+	
Innovation	The school is in favor of planned change and experimentation, and fosters classroom openness and individualization.	Teachers are encouraged to be innovative in this school.	+	
Resource adequacy	Support personnel, facilities, finance, equipment and resources are suitable and adequate.	The supply of equipment and resources is inadequate.	−	
Work pressure	The extent to which work pressure dominates the school environment.	Teachers have to work long hours to keep up with the workload.	+	

[a] Items designated (+) are scored by allocating 5, 4, 3, 2, 1, respectively, for the responses strongly agree, agree, not sure, disagree, strongly disagree. Items designated (−) are scored in the reverse manner. Omitted or invalid responses are given a score of 3.

TABLE 17.8. Internal Consistency (Alpha Reliability) and Discriminant Validity (Mean Correlation of Scale with Other Scales) for Each SLEQ Scale for Three Samples[a]

Scale	Number of Items	Form	Alpha Reliability			Mean Correlation with Other Scales		
			Sample 1	Sample 2	Sample 3	Sample 1	Sample 2	Sample 3
Student support	7	Actual	0.70	0.79	0.85	0.19	0.19	0.10
		Preferred			0.81			0.31
Affiliation	7	Actual	0.87	0.85	0.84	0.34	0.18	0.38
		Preferred			0.77			0.42
Professional interest	7	Actual	0.86	0.81	0.81	0.29	0.29	0.36
		Preferred			0.77			0.43
Staff freedom	7	Actual	0.73	0.68	0.64	0.31	0.05	0.30
		Preferred			0.76			0.30
Participatory decision making	7	Acutal	0.80	0.69	0.82	0.34	0.22	0.34
		Preferred			0.74			0.28
Innovation	7	Actual	0.84	0.78	0.81	0.38	0.22	0.42
		Preferred			0.77			0.31
Resource adequacy	7	Actual	0.81	0.80	0.65	0.22	0.19	0.35
		Preferred			0.64			0.44
Sample size			83	34	109	83	34	109
			19	34	10	19	34	10

[a] Note: No validation data for the new Work Pressure Scale are yet available.

and suggest that the SLEQ measures distinct although somewhat overlapping aspects of school environment.

The ability of each SLEQ scale to differentiate between the perceptions of teachers in different schools was explored for the actual form for the sample of 109 teachers in 10 schools described in Table 17.8. A one-way ANOVA revealed that each SLEQ scale differentiated significantly ($p < .001$) between schools and that the eta^2 statistic (the proportion of variance attributable to school membership) ranged from 0.16 to 0.40 for different scales.

RESEARCH INVOLVING EDUCATIONAL ENVIRONMENT INSTRUMENTS

In order to illustrate the range of possible uses of educational environment scales in science education research, past studies employing various instruments are reviewed briefly in this section. Three types of research considered involved (1) associations between student outcomes and environment, (2) use of environment dimensions as criterion variables (including the evaluation of educational innovations and investigations of differences between students' and teachers' perceptions of the same classrooms), and (3) investigations of whether students achieve better when in their preferred environments. Later sections focus on combining qualitative and quantitative methods in environment research and teachers' practical attempts to improve their classroom and school climates.

Associations Between Student Outcomes and Environment

The strongest tradition in past classroom environment research has involved investigation of associations between students' cognitive and affective learning outcomes and their perceptions of psychosocial characteristics of their classrooms (Haertel, Walberg, & Haertel, 1981). Numerous research programs have shown that student perceptions account for appreciable amounts of variance in learning outcomes, often beyond that attributable to background student characteristics. The practical implication of this research is that student outcomes might be improved by creating classroom environments found empirically to be conducive to learning.

Overview of Studies. Table 17.9 provides a broad overview of the comprehensive set of past studies in which the effects of classroom environment on science student outcomes were investigated. The only studies included in the table are those whose sample consisted wholly or partly of science classes at the secondary or higher education levels or of elementary school classes in which students take all their subjects, including science, with the same teacher and in the same room. Table 17.9 excludes studies that involved nonscience subject areas such as mathematics (O'Reilly, 1975) and social studies (Cort, 1979; Fraser, Pearse, & Azmi, 1982). Studies are grouped according to whether they involved use of the LEI, CES, ICEQ,

MCI, or other instruments. Also, research in developing countries is grouped together. This table shows that studies of associations between outcome measures and classroom environment perceptions have involved a variety of cognitive and affective outcome measures, a variety of classroom environment instruments, and a variety of samples (ranging across numerous countries and grade levels).

Meta-Analysis of Studies. The findings from prior research are highlighted in the results of an ambitious meta-analysis involving 734 correlations from a collection of 12 studies of 10 data sets from 823 classes in eight subject areas containing 17,805 students in four nations (Haertel, Walberg, & Haertel, 1981). Learning posttest scores and regression-adjusted gains were consistently and strongly associated with cognitive and affective learning outcomes, although correlations were generally higher in samples of older students and in studies employing collectivities such as classes and schools (in contrast to individual students) as the units of statistical analysis. In particular, better achievement on a variety of outcome measures was found consistently in classes perceived as having greater Cohesiveness, Satisfaction, and Goal Direction and less Disorganization and Friction. Other meta-analyses synthesized by Fraser, Walberg, Welch, and Hattie (1987) provide further evidence supporting the link between educational environments and student outcomes.

Cooperative Learning. Among the various lines of programmatic research on classroom environment, the work on the relative effectiveness of cooperative, competitive, and individualistic goal structures stands out because of the volume of studies completed (Johnson & Johnson, 1987, 1991; Johnson, Johnson, Johnson Holubec, & Roy, 1986). Although many past studies of student achievement found that cooperative learning is more successful than either competitive or individualistic learning, the evidence is not always consistent. The generally positive effect of cooperative learning approaches on student achievement is illustrated by the findings of a comprehensive meta-analysis involving 122 studies (Johnson et al., 1981), but not even this impressive synthesis is totally conclusive and generalizable. For instance, a large proportion of these studies involved group outcomes (e.g., the group's ability to solve problems) rather than the conventional individual student outcome, which is afforded so much importance in elementary and secondary schooling.

Slavin (1983a, 1983b) makes a significant contribution to our understanding of circumstances under which cooperation is most likely to lead to improved achievement. This research synthesis was restricted to 41 studies that were conducted in regular elementary and secondary schools, involved use of cooperative or other methods for at least 2 weeks, and were methodologically adequate (e.g., in terms of control for initial differences in experimental and control treatments). Overall, Slavin's review of studies of cognitive achievement supports the efficacy of cooperative learning in that 63 percent of these studies favored cooperative methods, and only one study's results favored the control group. Slavin's review also illuminates some of the conditions under which cooperative learning was

TABLE 17.9. Studies of Associations Between Student Outcomes
and Classroom Environment

Study	Outcome Measures	Sample
Studies involving LEI		
Anderson & Walberg (1968); Walberg & Anderson (1968a); Anderson (1970); Walberg (1969b, 1969c, 1972)	Selected from: achievement; understanding of nature of science; science processes; participation in physics activities; science interest; attitudes	Various samples (maximum of 144 classes) of senior high school physics students mainly in USA, but with some in Canada
Walberg & Anderson (1972)	Examination results	1600 grade 10 and 11 students in various subject areas in 64 classes in Montreal, Canada
Lawrenz (1976)	Science attitudes	238 senior high school science classes in midwest USA
Fraser (1978, 1979a)	Inquiry skills; attitudes; understanding of nature of science	531 students in 20 grade 7 science classes in Melbourne, Australia
Power & Tisher (1975, 1979)	Achievement; attitudes; satisfaction	315 junior high school students in 20 science classes in Melbourne, Australia
Hofstein et al. (1979)	Attitudes	400 grade 11 students in 12 chemistry classes in Israel
Haladyna, Olsen, & Shaughnessy (1982); Haladyna, Shaughnessy, & Redsun (1982a, 1982b); Haladyna, Shaughnessy, & Shaughnessy (1983)	Attitudes	5804 science, mathematics and social studies students in 277 grade 4, 7, and 9 classes in Oregon, USA
Studies involving CES		
Trickett & Moos (1974)	Satisfaction and mood criteria	608 students in 18 classes in USA
Moos & Moos (1978)	Absences; grades	19 high school classes in one school in USA
Moos (1979a)	Indexes of student reactions	241 secondary school classes in various subject areas
Fisher & Fraser (1983b) (See study reported in detail in this chapter)	Inquiry skills; attitudes	116 grade 8 and 9 science classes throughout Tasmania, Australia
Galluzi et al. (1980)	Psychological outcomes	414 grade 5 students in USA
Humphrey (1984)	Self-control	750 grade 4 and 5 children in 36 classes in USA
Keyser & Barling (1981)	Academic self-efficacy beliefs	504 grade 6 children in South Africa
Studies involving ICEQ		
Rentoul & Fraser (1980)	Inquiry skills; enjoyment	285 junior high school students in 15 science and social science classes in Sydney, Australia
Wierstra (1984)	Attitudes; achievement	398 15–16-year-old students in 9 classes in The Netherlands
Wierstra et al. (1987)	Attitudes; achievement	1105 secondary school students in 66 classes involved in Dutch option of Second International Science Study
Fraser (1981c); Fraser & Butts (1982)	Attitudes	Maximum of 712 students in 30 junior high school science classes in Sydney, Australia
Fraser, Nash, & Fisher (1983)	Anxiety	116 grade 8 and 9 science classes throughout Tasmania, Australia
Fraser & Fisher (1982b)	Inquiry skills; attitudes	116 grade 8 and 9 science classes throughout Tasmania, Australia
Studies involving MCI		
Fraser & Fisher (1982a, 1982c)	Inquiry skills; understanding of nature of science; attitudes	2305 grade 7 science students in 100 classes in Tasmania, Australia
Payne et al. (1974–75); Ellett et al. (1977); Ellett & Walberg (1979)	Achievement; school attendance	6151 grade 4 students in 89 schools in Georgia, USA

TABLE 17.9. Continued

Study	Outcome Measures	Sample
Fraser & O'Brien (1985)	Word knowledge; comprehension	758 grade 3 students in 32 classes in Sydney, Australia
Lawrenz (1988)	Energy knowledge; two energy attitude scales	Approximately 1000 grade 4 and 7 students in 34 classes in Arizona, USA
Studies involving other instruments		
Kelly (1980)	Achievement	41,657 students in 1735 schools in 14 developed countries involved in an IEA science study
Johnson et al. (1981); Johnson et al. (1986); Slavin (1983a, 1983b)	Different studies included: achievement; cross-ethnic relationships; cross-handicap relationships	Various samples involved in studies of cooperative learning strategies in various subjects, especially in USA
Fraser & Treagust (1986)	Satisfaction; locus of control	372 higher education students in 34 classes in various subject areas
Talton (1983)	Attitude; achievement	1456 grade 10 biology students in 70 classes in 4 schools in North Carolina
Perkins (1978)	Basic skills	3703 grade 4 students in 42 elementary schools in a SE state in USA
Brookover & Schneider (1975); Brookover et al. (1978, 1979)	Achievement	8078 grade 4 and 5 students in Michigan, USA
Gardner (1974, 1976)	Attitudes	1014 grade 11 physics students in 58 classses in Melbourne, Australia
Payne et al. (1974–75); Ellett & Walberg (1979)	Achievement	3350 elementary and 3613 secondary students in various subject areas and 1200 teachers in Georgia, USA
Wubbels et al. (1988) Brekelmans et al. (1990)	Achievement; attitudes	1105 secondary school students in 66 classes involved in Dutch option of Second International Science Study
Giddings & Fraser (1990)	Attitudes	4643 senior high school and university students in 225 laboratory classes in Australia, USA, England, Canada, and Israel
Studies in developing countries		
Walberg, Singh, & Rasher (1977)	Achievement	3000 grade 10 science and social science students in 150 classes in Rajasthan, India
Schibeci, Rideng, & Fraser (1987)	Attitudes	250 grade 11 biology students in six classes in Indonesia
Paige (1978, 1979)	Achievement; individual modernity	1621 grade 6 students in 60 schools in East Java, Indonesia
Holsinger (1972, 1973)	Information learning; individual modernity	2533 grade 3–5 students in 90 classes in Brazil
Persaud (1976)	Noncognitive outcomes including social development and aspiration levels	1277 grade 3 and 6 students in 18 schools in Jamaica
Chatiyanonda (1978)	Attitudes	989 grade 12 physics students in 31 classes in or near Bangkok, Thailand

most effective. First, cooperative learning methods were more effective than control methods only when *group rewards* were provided based on group members' achievement. Second, cooperative learning methods were superior to alternatives only when there was *individual accountability;* no study in which group members worked together to produce a single group product found positive achievement effects. Based on these findings, Slavin concluded that the effects of cooperative learning on achievement are primarily *motivational.* That is, whereas there is little evidence that working in small groups

per se is more or less effective than studying individually, it appears that working with others to achieve a group goal creates peer norms supporting learning and that these increase student motivation to achieve and help one another.

In addition to its focus on student achievement, past research on cooperative, competitive, and individualistic methods (see Johnson et al., 1986; Slavin, 1983a) has involved cross-ethnic relationships, cross-handicap relationships, and various other noncognitive outcomes. A number of prior studies consistently indicate that, when students work in ethnically mixed cooperative learning groups, positive relationships are developed between students of different races or ethnicities. Another set of studies focuses on relationships between academically handicapped and normal-progress students and shows that cooperative learning can overcome this barrier to friendship and interaction. Also, reviews of other research generally support the positive impact of cooperative learning strategies on noncognitive outcomes, including self-esteem, peer support for achievement, internal locus of control, time on task, liking of class and classmates, and cooperativeness (Johnson et al., 1986; Slavin, 1983a).

Educational Productivity Research. In another line of research, psychosocial learning environment has been incorporated as one factor in a multifactor psychological model of educational productivity (Walberg, 1981). This theory, which is based on an economic model of agricultural, industrial, and national productivity, holds that learning is a multiplicative, diminishing-returns function of student age, ability, and motivation; of quality and quantity of instruction; and of the psychosocial environments of the home, the classroom, the peer group, and the mass media. Because the function is multiplicative, it can be argued in principle that any factor at a zero point will result in zero learning; thus either zero motivation or zero time for instruction will result in zero learning. Moreover, it will do less good to raise a factor that already is high than to improve a factor that is currently the main constraint to learning. In contrast to much past research in education, which considered only two or three factors in a single study, the model provides a comprehensive approach to statistically controlled, multivariate research on productive factors in schooling.

Empirical probes of the educational productivity model were carried out by extensive research syntheses involving the correlations of learning with the factors in the model (Fraser et al., 1987; Walberg, 1984, 1986). Later, secondary analyses of large data bases collected as part of the National Assessment of Educational Achievement (Walberg, 1986) and National Assessment of Educational Progress (Fraser, Welch, & Walberg, 1986; Walberg, Fraser, & Welch, 1986) confirmed the importance of these factors as predictors of achievement when mutually controlled for one another. In particular, these educational productivity analyses showed that classroom and school environment was a strong predictor of both achievement and attitudes even when a comprehensive set of other factors was held constant.

A Case Study. Fisher and Fraser's (1983b) study reported in Table 17.10 illustrates some of the methodological complexity of rigorous studies of the effects of classroom environment on student outcomes. This study used the data base from the sample of science classes in Tasmania. It consisted of a representative group of 116 grade 8 and 9 classes, each with a different teacher, in 33 different schools. Approximately equal numbers of schools were in country and suburban areas, and approximately equal numbers of boys and girls made up the sample. Although the sample was not chosen randomly, it was selected to be as representative as possible of the population of schools.

Three cognitive and six affective measures were administered at both the beginning and end of the same school year, and classroom environment was assessed by administering the CES at midyear. The three cognitive outcomes were measured by the *Test of Enquiry Skills* (Fraser, 1979b), and the six attitude measures each consisted of 10 items of Likert format selected from the *Test of Science-Related Attitudes* (Fraser, 1981d). In addition, information was gathered about the general ability of the students using a version of the Otis test. In order to permit comparison with results from methodologically diverse past studies, data were analyzed in six different ways (simple, multiple, and canonical correlation analyses performed separately for raw posttest scores and residual posttest scores adjusted for corresponding pretest and general ability).

It has been common in prior research to perform a conservative test of outcome-environment relationships by controlling statistically certain student characteristics, especially corresponding pretest and general ability. That is, for simplicity, learning environment dimensions have been considered useful predictors of student learning outcomes only if they accounted for variance different from that attributable to well-established predictors such as pretest and general ability (Walberg & Haertel, 1980). Although conservative analyses in which student characteristics are controlled have the merit that they do not overestimate the variance component attributable to environment, they might well underestimate the importance of the environment component because any variance shared by environment and student characteristics is removed. For this reason, all analyses (simple, multiple, canonical correlation) were performed twice, once using raw posttest scores as the criterion variables and once using residual posttest scores adjusted for corresponding pretest and general ability.

Table 17.10 shows the results of the six types of analyses. The first pair of analyses are the least complex as they involve simple correlations between class means on the nine environment scales and class means on each of the nine outcome posttests (using either raw scores or residual scores). A major advantage of these simple correlational analyses is that they furnish data to other workers interested in associations between particular environment variables and particular outcomes. The results in Table 17.10 show that the number of significant outcome-environment correlations ($p < .05$) was 27 for the analyses involving raw posttest scores (about seven times that expected by chance) and 18 for the analyses using residual posttest scores (about four times that expected by chance).

The second pair of analyses reported in Table 17.10 consisted of a multiple correlation analysis involving the set of nine environment scales performed separately for each outcome using either raw or residual criterion scores. The multiple correlation provides a more parsimonious picture of the joint influence of correlated environment dimensions on outcomes and reduces the Type I error rate associated with simple correlational analyses. These analyses are likely to be of particular relevance to people interested in particular outcome measures. Table 17.10 shows that the multiple correlation between raw outcome scores and the set of classroom environment scales ranged from 0.30 to 0.51 and was significantly greater than zero ($p < .05$) for seven of the nine outcomes. As expected, multiple correlations were smaller for analyses involving residual

TABLE 17.10. Simple, Multiple, and Canonical Correlations Between Classroom Environment Dimensions and Learning Outcomes (Using Raw Outcome Scores and Residual Scores Adjusted for Corresponding Pretest and General Ability)

Learning Outcomes	Raw Scores/Residuals[a]	Simple Correlation									Multiple Correlation	Beta Weights for Significant Individual Environment Predictors[b]
		Innov	Affil	Teach Supp	Task Orien	Comp	Order & Org	Rule Clar	Teach Contr	Innov		
Social implications of science	Raw scores	.22*	.16	.16	.25**	.20*	.30**	.24*	.02	.06	.38*	.34* (Order & Org)
	Residuals	.27**	.24*	.15	.25**	.14	.33**	.26**	.03	.09	.39*	.36* (Order & Org)
Enjoyment of science lessons	Raw scores	.42**	.20*	.27**	.17	.13	.45**	.25*	−.02	.20*	.49**	.43** (Order & Org)
	Residuals	.36**	.27**	.16	.22*	.02	.40**	.20*	.05	.03	.44**	.38* (Order & Org)
Attitude to normality of scientists	Raw scores	.12	.10	.07	.23*	.03	.16	.10	.08	−.20*	.39*	.37* (Teach Supp); −.37** (Innov)
	Residuals	.17	.11	.15	.07	−.04	.10	.18	.08	−.04	.31	
Attitude to inquiry	Raw scores	.11	.18	.05	.18	.07	.10	.23*	.07	.03	.38*	.25* (Rule Clar)
	Residuals	.10	.18	.04	.13	.05	.09	.23*	.09	.01	.29	
Adoption of scientific attitudes	Raw scores	.07	.29**	.10	.25**	.14	.17	.06	−.04	−.13	.44**	.27* (Affil); −.26* (Innov)
	Residuals	.16	.26**	.21*	.07	−.01	.18	.15	−.02	.06	.31	
Leisure interest in science	Raw scores	.28**	.22*	.11	.25**	.08	.41**	.25**	.04	.20*	.51**	.56** (Order & Org); .31** (Innov)
	Residuals	.30**	.12	.11	.12	−.11	.35**	.21*	.00	.23**	.49**	.45** (Order & Org); .32* (Innov)
Comprehension of science reading	Raw scores	.02	.13	−.03	.15	.05	.13	−.05	−.04	−.13	.30	
	Residuals	.11	.13	.00	.03	.03	.17	.06	.06	.01	.27	
Design of experimental procedures	Raw scores	−.08	.03	−.06	.22*	.18	.11	.01	.05	−.20*	.37	
	Residuals	−.05	−.05	−.02	−.05	.09	.05	.09	.12	−.05	.28	
Conclusions and generalizations	Raw scores	.08	.17	.06	.31**	.15	.22*	−.02	.04	−.20*	.47**	.35* (Teach Supp); .27* (Task Orien); −.28* (Innov)
	Residuals	.18	.12	.07	.14	.07	.26**	.07	.12	−.02	.38**	.35* (Order & Org)
Canonical correlations	Raw scores						.67**,	.54*				
	Residuals						.62**					

* p < .05, **p < .01.

[a] Residual scores have been adjusted for performance on the corresponding pretest and general ability.

[b] Beta weights are shown for those individual predictors for which, first, the corresponding block of nine environment scales had a significant multiple correlation and, second, the b weight was significantly different from zero.

scores, but their magnitudes still ranged from 0.27 to 0.49 with four of these being statistically significant.

In order to interpret which individual classroom environment scales were making the largest contribution to explaining variance in learning outcomes, b and beta weights were examined for the regression equations for which the multiple correlation for the whole block of nine environment scales was significantly greater than zero ($p < .05$). The right-hand side of Table 17.10 lists the magnitude of the beta weight for the individual environment scales whose b weights were significantly different from zero ($p < .05$) and for which the corresponding block of environment scales also had a significant multiple correlation. This requirement that the multiple correlation for the whole block of environment scales should meet the .05 significance criterion protects against an inflated experimentwise Type I error rate. This table shows that the number of significant relationships for individual environment variables was 11 for raw criterion scores and 5 for residual criterion scores. Examples of the results for residual scores are that Social Implications of Science scores were higher in classes perceived as having greater Order and Organization, and Leisure Interest in Science scores were higher in classes perceived as having greater Order and Organization and Innovation.

Although use of multiple correlation analyses overcomes the problems of collinearity between environment scales, collinearity between outcome measures could still give rise to an inflated experimentwise Type I error rate. Canonical analysis, however, can provide a parsimonious picture of relationships between a domain of correlated learning outcomes and a domain of correlated environment dimensions. Consequently, two canonical analyses were conducted (one involving raw outcome scores and one involving residual scores) using the class mean as the unit of analysis. The bottom of Table 17.10 shows that both canonical analyses yielded at least one significant canonical correlation. Two significant canonical correlations of 0.67 ($p < .01$) and 0.54 ($p < .05$), respectively, were found between environment scales and raw posttest scores, and one significant canonical correlation of 0.62 ($p < .01$) was found between environment scales and residual posttest scores.

In order to interpret the results of the canonical analyses, the magnitudes and signs of the structure coefficients (i.e., simple correlations of a canonical variate with its constituent variables) associated with each significant canonical variate were examined. The first significant canonical correlation for the analysis involving raw scores was readily interpretable. It indicated that attitude scores on the Enjoyment of Science Lessons and Leisure Interest in Science scales were higher in classes perceived as having greater Order and Organization and Innovation. The interpretation of the second significant canonical correlation for the analysis of raw scores was less straightforward, but it suggested that cognitive outcome scores on the Conclusions and Generalizations scale tended to occur in classes perceived as having more Teacher Support and less Innovation. The straightforward interpretation of the significant canonical correlation for residual scores was that, with corresponding pretest scores and general ability controlled, Leisure Interest in Science scores were greater in classrooms perceived as having greater Order and Organization.

The separate methods of analysis yielded consistent support for the existence of outcome-environment relationships and led to no major conflicts when explicating the specific form of such relationships in terms of particular outcomes and environment dimensions. However, as expected, the interpretation for individual variables varied somewhat with the presence or absence of control for student background characteristics (i.e., the raw scores vs. residuals analyses) and with the extent to which collinearity among variables was allowed for (i.e., simple, multiple, or canonical correlational analyses). Nevertheless, the study has some important tentative implications for educators wishing to enhance science students' achievement of particular outcomes by creating classroom environments found empirically to be conducive to achievement. For example, the finding that Order and Organization seems to have a positive influence on student achievement of a variety of aims is likely to be useful to practitioners.

Use of Environment Perceptions as Criterion Variables

Table 17.11 overviews studies in which classroom environment dimensions were employed as dependent variables for a wide range of purposes. The past studies are organized by three themes: (1) evaluation of educational innovations, (2) differences between student and teacher perceptions of actual and preferred environment, and (3) studies involving other independent variables. Table 17.11 is restricted to studies involving samples consisting wholly or partly of science classes (including elementary classes in which students take all of their subjects with the same teacher). Studies involving students in other subject areas, such as social science (Baba & Fraser, 1983; Cort, 1979), are excluded.

Evaluation of Educational Innovations. One promising but largely neglected use of classroom environment instruments is as a source of process criteria in curriculum evaluation (Fraser, 1981b; Fraser, Williamson, & Tobin, 1987; Walberg, 1974). For example, as many curricula attempt to achieve more individualization, the ICEQ provides a useful tool for monitoring changes in student perceptions of five important aspects of individualization. When the ICEQ was used in evaluating a project aimed at promoting individualized learning approaches, it was found that students in the school perceived their classes as significantly more individualized on a number of ICEQ scales than did a comparison group of students (Fraser, 1980a). Another study involving an evaluation of the Australian Science Education Project (ASEP) revealed that, in comparison with a control group, students in ASEP classes perceived their classrooms as more satisfying and individualized and having a better material environment (Fraser, 1979a). The significance of the ASEP evaluation and Welch and Walberg's (1972) evaluation of Harvard Project Physics is that classroom environment variables differentiated revealingly between curricula, even when various achievement outcome measures showed negligible differences. Clearly, there is scope in science education for teachers and researchers to include classroom environment measures more

TABLE 17.11. Studies Using Classroom Environment Perceptions as
Criterion Variables

Study	Instrument	Independent Variable
Curriculum evaluation studies		
Anderson et al. (1969); Welch & Walberg (1972)	LEI	Use of Harvard Project Physics
Fraser (1976, 1979a); Tisher & Power (1976, 1978); Power & Tisher (1979); Northfield (1976);	LEI CAQ (Steele et al., 1971)	Use of Australian Science Project
Kuhlemeier (1983); Wierstra (1984); Wierstra et al. (1987)	ICEQ	Use of new Dutch physics curriculum
Levin (1980)	LEI	Use of individualized curriculum
Ainley (1978)	Locally developed	Standard of science facilities
Eash (1990)	OCIW	NSF summer workshop
Differences between student-teacher and actual-preferred forms		
Fisher & Fraser (1983a); Moos (1979a)	CES	Student actual vs. student preferred; student actual vs. teacher actual
Fraser (1982b)	ICEQ	As above
Fraser (1984)	MCI	As above
Fraser & Treagust (1986)	CUCEI	As above
Raviv et al. (1990)	CES	Student vs. teacher actual and preferred
Wubbels et al. (1991)	QTI	Student actual vs. teacher actual
Hofstein & Lazarowitz (1986)	LEI	Student actual vs. student preferred
Other studies involving environment dimensions as criterion variables		
Trickett et al. (1976, 1982)	CES	Single-sex vs. coeducational schools; independent vs. public schools
Fraser & Rentoul (1982)	ICEQ	School-level environment
Ellett & Masters (1978)	MCI	School-level environment
Lawrenz & Welch (1983)	LEI	Sex of science teacher
Walberg (1968)	LEI	Teacher personality
Walberg & Anderson (1968b)	LEI	"Achieving" vs. "creative" classes
Walberg (1969a) Anderson & Walberg (1972)	LEI	Class size
Walberg & Ahlgren (1970)	LEI	Various variables
Shaw & Mackinnon (1973); Randhawa & Michayluk (1975); Welch (1979)	LEI	Grade level
Anderson (1971); Steele et al. (1974); Kuert (1979); Welch (1979)	LEI CAQ (Steele et al., 1971)	Differences between school subjects
Hearn & Moos (1978)	CES	Differences between school subjects classified according to Holland's occupational classification
Randhawa & Michayluk (1975); Hofstein et al. (1980); Sharan & Yaakoby (1981)	LEI	Type of school
Trickett (1978)	CES	Type of public school (urban, rural, suburban, vocational, alternative)
Moos (1979a, 1980)	CES	Differences in overall context, achitectural characteristics, organizational characteristics; teacher characteristics; aggregate student characteristics
Walberg et al. (1972)	LEI	Student sex and socioeconomic status; school enrollment
Ellett et al. (1978)	MCI	Teacher competency
Lawrenz & Munch (1984)	MCI	Grouping students in laboratory on formal reasoning ability

TABLE 17.11. Continued

Study	Instrument	Independent Variable
Harty & Hassan (1983)	CES	Teacher control ideology
Rentoul & Fraser (1981)	ICEQ	Changes in beginning teachers' preferences for individualization
Owens & Straton (1980); Owens (1981)	Locally developed	Sex differences in classroom environment preferences
Byrne et al. (1986)	MCI, CES, ICEQ	Sex differences in classroom environment perceptions
Thistlewaite (1962); Astin (1965); Genn (1981)	CCI (Stern, 1970)	College environment as perceived by students following different specialisms
Costello (1988)	CES	Ability grouping

frequently in their evaluations of new curricula and teaching approaches.

In an evaluative study of some alternative high schools for adult learners, teachers' perceptions of school environment were used in conjunction with students' perceptions of classroom environment (Fraser et al., 1987). Taken together, the evidence supported the existence of a positive educational climate at this school and attested to the usefulness of including both school and classroom environment measures in the same study.

Differences Between Student and Teacher Perceptions of Actual and Preferred Environment. Having different actual and preferred forms of educational environment instruments that can be used with either teachers or students permits investigation of differences between students' and teachers' perceptions of the same actual classroom environment and differences between the actual environment and that preferred by students or teachers. This research into differences between forms was reported by Fisher and Fraser (1983a) using a sample of 116 classes in Tasmania for comparisons of student actual with student preferred scores. For the comparison of student actual with teacher actual form, a subsample of 56 of the teachers of these classes was available for contrast with the student class means for the corresponding 56 classes. The results of this study are depicted in Figure 17.1, which shows simplified plots of statistically significant differences between forms. Figure 17.1 clearly shows that students preferred a more positive classroom environment than was actually present for all five ICEQ dimensions and that teachers perceived a more positive classroom environment than did their students in the same classrooms on four of the ICEQ's dimensions. These interesting results replicate patterns in other studies in school classrooms in the United States (Moos, 1979a), Israel (Hofstein & Lazarowitz, 1986), The Netherlands (Wubbels, Brekelmans, & Hooymayers, 1991), and Australia (Fraser, 1982b; Fraser & O'Brien, 1985); in science laboratory classes at the senior high school and university levels in a cross-national study involving six countries (Giddings & Fraser, 1990); and in other settings such as hospital wards and work milieus (e.g., Moos, 1974, 1979b) and kindergartens (Jorde-Bloom, 1991). These studies show that students and teachers are likely to differ in the way they perceive the actual environment of the same classrooms and that the environment preferred by students commonly falls short of that actually present in classrooms.

Studies Involving Other Independent Variables. The third group of studies in Table 17.11 shows that other workers have used classroom environment dimensions as criterion variables in research into how the classroom environment varies with such factors as teacher personality, class size, grade level, subject matter, nature of the school-level environment, and type of school. For example, larger class sizes were associated with greater classroom Formality and less Cohesiveness (Anderson & Walberg, 1972; Walberg, 1969a).

In an interesting study of students' preferences for different types of classroom environments, girls were found to prefer cooperation more than boys, and boys preferred both competition and individualization more than girls (Owens & Straton, 1980). In a similar study, Byrne, Hattie, and Fraser (1986) found that boys preferred friction, competitiveness, and differentiation more than girls, whereas girls preferred teacher structure, personalization, and participation more than boys.

Person-Environment Fit Studies of Whether Students Achieve Better in Their Preferred Environment

Whereas past research has concentrated on investigations of associations between student outcomes and the nature of the actual environment, having both actual and preferred forms of educational environment instruments permits exploration of whether students achieve better when there is higher similarity between the actual classroom environment and that preferred by students. Such research is an example of what is referred to as person-environment fit research (Hunt, 1975). In fact, science education studies have extended prior research in a new direction by using a person-environment interactional framework in classroom environment research (Fraser & Fisher, 1983c, 1983d). The purpose of this research was to see whether student outcomes depended not only on the nature of the actual classroom environment but also on the match between students' preferences and the actual environment.

One person-environment fit study in science education involved using the ICEQ with a large sample consisting of the 116 classes described previously (Fraser & Fisher, 1983c). The class

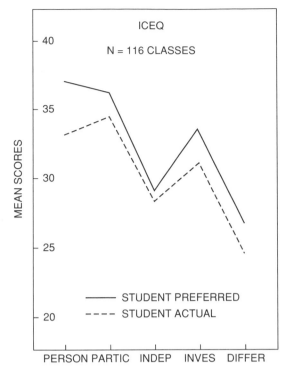

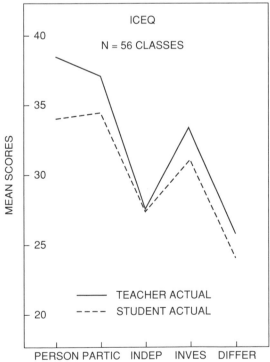

FIGURE 17.1. Simplified Plots of Significant Differences Between Forms of ICEQ.

was employed as the unit of analysis. A total of 29 variables was used in exploring relationships between achievement, actual environment, and actual-preferred interaction (i.e., person-environment fit). Student achievement was measured at the be-

ginning and end of the same school year using six affective and three cognitive outcome measures. The ICEQ was administered at midyear to obtain students' perceptions of five dimensions of actual classroom individualization and five dimensions of preferred classroom individualization. As preferred classroom individualization per se was not of interest, data from the ICEQ were used to provide five actual individualization variables and to generate five new variables indicating the congruence or interaction between actual and preferred individualization. In addition, the student background characteristic of general ability was measured in the study using a version of the Otis test. The basic design of the study, then, involved prediction of posttest achievement from pretest performance, general ability, the five actual individualization variables, and the five variables indicating actual-preferred interaction.

This study provided many methodological improvements over prior research (Fraser & Fisher, 1983c). In particular, it measured the person and the environment as sets of commensurate and continuous variables, it controlled for student background characteristics and actual environment when studying the effect of actual-preferred interaction, and it reduced the overall Type I error rate by ensuring that individual interactions were interpreted only when the block of all interactions was associated with a significant amount of criterion variance. Furthermore, regression surface analysis provided a powerful multivariate method of statistical analysis that enabled person-environment interactions to be represented as the products of continuous variables.

A regression analysis was conducted for the actual-preferred interaction for each of the five ICEQ scales for any outcome for which the block of interactions accounted for a significant increment in outcome variance. To satisfy the requirement that student background characteristics be controlled, each of these analyses was carried out using residual posttest criterion scores that had been adjusted for corresponding pretest and general ability. Also, to meet the condition that an interaction term should account for a significant increment in criterion variance over and above that explainable by the corresponding actual environment variable, each regression equation included an actual environment term in addition to an actual-preferred interaction. Consequently, the form of each of the 20 regression equations was

$$Y_{res} = a + b_1 A_i + b_2(A_i \times P_i)$$

where Y_{res} represents residual outcome scores (adjusted for corresponding pretest and general ability), a is the regression constant, b_1 is the raw regression coefficient for the ith continuous actual environment variable, and b_2 is the raw regression coefficient for the interaction formed by taking the product of the ith continuous actual environment variable and the ith continuous preferred variable.

One of the cases for which the outlined conditions were satisfied was for the Social Implications of Science outcome and the Personalization scale. In this case the regression equation was

$$Y_{res} = -0.3150 - 0.1171A + 0.0035(A \times P)$$

Because actual-preferred interactions had been formed by taking the products of continuous actual and preferred scores (in order to enhance statistical power), the two-dimensional plots conventionally used with ANOVA results were inappropriate. Instead, the interpretations of the significant interactions were based on three-dimensional regression surfaces that permitted actual and preferred scores to be represented as continuous variables. In each of these plots, the vertical axis represented residual posttest scores, one horizontal axis represented continuous scores on an actual environment scale, and the other horizontal axis represented continuous scores on the corresponding preferred environment scale. Each regression surface was plotted using values ranging from a minimum of 2 standard deviations (for class means) below the mean for the actual and preferred scales to a maximum of 2 standard deviations above the mean. Figure 17.2 shows the regression surface for the case involving Social Implications of Science and Personalization.

Figure 17.2 shows that the interpretation of the actual-preferred interaction for Personalization and Social Implications of Science was that the relationship between residual Social Implications scores and actual Personalization was negative for classes with preferred Personalization scores 2 standard deviations below the mean, was approximately zero for classes with preferred Personalization scores 1 standard deviation below the mean, and was positive for classes with preferred Personalization scores at or above the mean. That is, residual Social Implications scores increased with increasing amounts of actual Personalization for classes preferring high levels of actual Personalization but decreased with increasing actual Personalization for classes preferring low levels of actual Personalization. This finding, together with others from the same study, suggests that actual-preferred congruence (or person-environment fit) could be as important as individualization in predicting student achievement of important affective and cognitive aims.

The research has interesting practical implications, but one must be careful to ensure that the implications drawn are consistent with the unit of statistical analysis used. It cannot be assumed that an individual student's achievement would be improved by moving him or her to a classroom that matched his or her preferences. Rather, the practical implication of these findings for teachers is that class achievement of certain outcomes might be enhanced by attempting to change the actual classroom environment in ways that make it more congruent with that preferred by the class. Finally, although the reader is cautioned against generalizing the present findings from the class level to the individual level, it is noteworthy that a previous study (Fraser & Rentoul, 1980) in which the individual was the unit of analysis suggested that the effects of classroom environment on individual student cognitive achievement were also mediated by individual student preferences for classroom environment.

COMBINING QUANTITATIVE AND QUALITATIVE METHODS IN STUDIES OF EDUCATIONAL ENVIRONMENTS

For a number of years, workers in various areas of educational research, especially educational evaluation, have claimed that there are merits in moving beyond the customary practice of choosing either quantitative *or* qualitative methods and, instead, combining quantitative *and* qualitative methods (Cook & Reichardt, 1979; Firestone, 1987; Fraser, 1988; Howe, 1988; Smith & Fraser, 1980). In the field of classroom environment, research involving qualitative case study methods (Rutter et al., 1979; Stake & Easley, 1978) has provided rich insights into classroom life and the use of quantitative methods, involving assessment of student and teacher perceptions as described

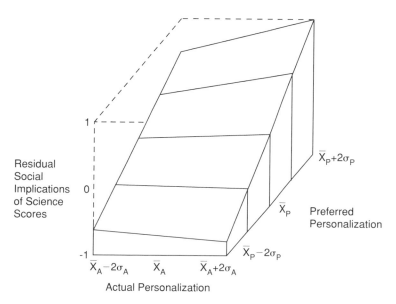

FIGURE 17.2. Regression Surface for an Actual-Preferred Interaction.

previously, has clearly advanced our understanding of classrooms. To date, however, only limited progress has been made toward the goal of combining quantitative and qualitative methods within the same study in research on classroom learning environments (see Fraser & Tobin, 1991). The fruitfulness of a confluence of qualitative and quantitative methods is illustrated next by reporting details of three recent studies in which ethnographic methods and classroom environment questionnaires were used together to advantage. The three studies, respectively, focused on target students, exemplary teachers, and higher-level cognitive learning.

Target Students

In many classroom environment studies using the class mean as the unit of analysis, it is implicitly assumed that different students within a given class experience more or less the same learning environment. In the following research, however, it was hypothesized that there are certain students, referred to as "target" students (Tobin, 1987b; Tobin & Gallagher, 1987a; Tobin & Malone, 1989), who experience a different learning environment from other students in that they are involved in a markedly greater proportion of the whole-class interactions with the teacher than are their classmates. It was hypothesized that teachers rely on this set of students to provide responses to questions that act as a bridge to new content areas or provide a basis for illustrating points that need to be taught. As a consequence, nontarget students are able to engage only in a covert way for a significant proportion of class time. Among its other purposes, this study's aims were to identify the presence and roles of target students in science classes and to compare the learning environment perceptions of target and nontarget students.

Past research (Tobin & Gallagher, 1987b) suggests that science teachers tend to spend a high proportion of time in whole-class interactive settings. In these settings, some evidence of differential engagement was reported by Sadker and Sadker (1985), who found that, in approximately half of all classrooms, a few "salient" students received more than three times their share of classroom interactions. They also noted that 25 percent of the students in all classes did not participate in any classroom interactions.

Twelve male and three female science teachers from two coeducational high schools, called Private School and Public School in this chapter, in the metropolitan area of Perth participated in the study. Each teacher was observed teaching several lessons to three or four different classes by a team of four researchers.

The qualitative component of the investigation involved data based on field notes from observation of approximately 200 grade 8 to 12 science lessons, written self-report data on student engagement, questionnaire data from teachers, and interview data from teachers and students. The quantitative component of the study involved the assessment of grade 8 students' perceptions of classroom environment using the short form of the Classroom Environment Scale (CES) (Fraser, 1986a), which assesses student perceptions of Involvement, Affiliation,

Teacher Support, Task Orientation, Order and Organization, and Rule Clarity.

Data collection occurred in three distinct phases. In the first phase, nine teachers from Private School were observed for 6 weeks. During the initial 3 weeks, five grade 8 classes were observed during a general science class on the topic of liquids. Two observers visited each class for a total of approximately 8 to 12 lessons during that time. In a second 3-week interval, additional observations were conducted in grade 9 general science classes and in grade 11 elective studies in chemistry, physical science, biology, and human biology. These data enabled hypotheses formulated on the basis of the observations of grade 8 classes to be tested for other grade levels and science topics.

The second phase of the study continued for 12 weeks and consisted of observations of six volunteer teachers from Public School. Observers tracked a teacher for approximately half a day, and observations occurred each day of the week. The third phase of the study involved formal interviews with all 15 teachers and 86 students using standard protocols. To achieve equal samples of male and female students and a wide spectrum of students, a stratified random sampling procedure was used to select students from each science class at Private School. Students from Public School were not involved in the formal interviews, which were tape recorded and transcribed at a later time.

At the conclusion of the classroom observations, teachers completed a questionnaire in which they were asked to nominate the target students from their class and to describe how these students were involved in class. Teachers were also asked to provide the names of students who did not contribute to classroom discussion or who were not called on to contribute. The five nontracked grade 8 classes at Private School formed a convenient sample for examining the characteristics of target students. The results of the study of target students are summarized as three assertions.

Assertion 1: In whole-class settings, a small group of target students dominated verbal interactions.

The observations revealed a trend in most classes for whole-class interaction to involve a set of approximately three to seven target students. The first type of target student consisted of those selected by the teacher to respond to questions because, in the opinion of the teacher, they could contribute a response to facilitate learning and content coverage. In the interviews, the teacher described these students as high achievers. This trend was confirmed in analyses involving the science achievement and formal reasoning ability of target students (Tobin & Gallagher, 1987a). The second type of target student consisted of pupils who projected themselves into whole-class interactions by raising their hands or by calling out to respond to teacher questions, to ask questions, or to evaluate the responses of others. Of course, the two types of target student were not mutually exclusive. In each class, there were some target students who volunteered and who were also called on by the teacher.

Of 59 questions asked in a grade 11 biology class, 47 were

answered by a group of only seven students. Similarly, in a grade 11 human biology class, a group of seven students answered 46 of 53 questions (with one student answering as many as 20 of the questions).

Without exception, teachers involved in the study were able to identify the target students in their class and the students who rarely participated in whole-class interactions. Many teachers, however, were content to have verbal interactions dominated by a relatively small proportion of the students because they believed that target students provided correct or partially correct answers on which the teacher could build and from which other students would learn. These teachers seemed little concerned about the inequity of providing some students with more opportunities to interact than others. In particular, inequity was a serious problem because the majority of target students in grades 8 to 10 were boys, irrespective of the teacher's gender. For example, in two grade 8 classes taught by a female, all 15 target students were boys. Of a total of 21 target students in several grade 8 classes at Private School, there were 17 boys and only 4 girls.

Assertion 2: Teachers tended to direct higher-level cognitive questions to target students.

In all classes, lower-level cognitive questions tended to be asked almost at random, whereas higher-level cognitive questions tended to be directed to target students. In an interview, one teacher noted that he would ask one or two students to respond but would then ask a more able student to complete the answer or provide the correct answer. That is, students perceived by teachers as more able were nurtured with higher-level cognitive questions, they received positive feedback for their responses to questions, and they generally received more teacher attention. The target students received opportunities to develop answers to higher-level cognitive questions and to "think on their feet" as they provided a response. Ironically, students who were able to give satisfactory answers received practice in responding, while those perceived as being unable to make a worthwhile contribution were denied similar opportunities and challenges. The differential involvement of target and nontarget students over a number of years might result in differences in reasoning ability, achievement, motivation, and self-esteem.

Assertion 3: Target students held more favorable perceptions of the learning environment than nonparticipant students.

When the classroom environment perceptions of 21 target students were contrasted with those of 22 nonparticipants from the five grade 8 classes, statistically significant differences were obtained for Involvement and Rule Clarity. Target students tended to perceive more Involvement and greater Rule Clarity in their classes than did students from the nonparticipant group. These results are consistent with classroom observations suggesting that different learning environments existed for different students. In whole-class interactive settings, target

students tended to be very involved whereas nonparticipants tended to be quite passive. Similar trends were observed in small groups and in individualized activities. During group and individualized activities, some students appeared to work in a sustained and independent manner, whereas others tended to participate in a relatively passive manner, engage intermittently, or be off task in a disruptive way.

The results for Rule Clarity were similar to those for Involvement. Target students perceived more Rule Clarity than nonparticipant students. The observations indicated that the rules were applied differently for some students. For example, most target students were able to call out responses to teacher questions and could elaborate at length while responding, while other students were reprimanded for calling out an answer. Inconsistencies were also observed in the way in which student movement and noise were controlled in each of the activity settings. Some students were permitted to move about and to speak with peers, while others were disciplined for similar behavior. Differences in the application of rules undoubtedly reflected the teachers' expectations for different learners in the class. It appears that these differences could have resulted in different students perceiving somewhat different learning environments within the one classroom.

It is important that science teachers become more aware of the target student phenomenon and understand how it leads to inequitable treatment of students. Possibly the easiest area to change would be the arrangement of students in groups for instruction. In particular, teachers should consider using the whole-class interactive activity setting for a smaller proportion of the time; the use of small groups for discussion is one alternative. Similarly, individualized activities that promote higher cognitive engagement should be considered as an alternative to the relatively covert types of engagement that are possible in whole-class settings.

The learning-environment results could have implications for research involving the class mean as the unit as analysis. If discrete perceived learning environments exist within a class, it is relevant to ask what the class average actually represents. Further consideration should be directed to this issue and additional research should be conducted to ascertain whether stable differences exist between the perceptions of target and nonparticipant students and between other groups (e.g., males and females).

Exemplary Teachers

The literature contains many reports and research findings that highlight problems and shortcomings in the teaching and learning of science (National Commission on Excellence in Education, 1983; Stake & Easley, 1978; Tobin & Gallagher, 1987b). In order to provide a refreshing alternative, a series of case studies of exemplary practice in science teaching was initiated to provide a focus on the successful and positive facets of schooling (Tobin & Fraser, 1987, 1990, 1991). It was assumed that much could be learned from case studies of exemplary practice that would stimulate and guide improvements in science education (Penick & Yager, 1983).

In addition to some case studies of seven mathematics teachers, there were seven science case studies involving 15 exemplary science teachers from schools in the metropolitan area of Perth. The exemplary teachers were identified by asking key educators in Western Australia, including teachers, State Education personnel, and university staff, to nominate outstanding science teachers. Five different research teams examined physics, chemistry, biology, and general science at the high school level. At the elementary school level, two teams examined science taught in grades 6 and 7 and grades 3 and 5, respectively.

The study utilized an interpretive research methodology (Erickson, 1986) and primarily qualitative data were obtained by direct observation of at least eight lessons by participant observers, interviews with teachers and students, and examination of curriculum materials, tests, and student work. Altogether, approximately 300 hours of intensive observation of science classrooms were involved. Throughout the study, meetings of the research team were held to facilitate discussion of administrative matters and substantive issues related to interpretation. Analysis and interpretation occurred within three teams: the researchers involved in each case study (i.e., one or two researchers in most case studies), the nine researchers involved in the seven science case studies, and the 14 researchers involved in the 12 science and mathematics case studies. At regular intervals during the data-collection phase, field notes were discussed by the researchers and assertions consistent with the observations were formulated. Subsequent observations were used as a basis for refuting, revising, or accepting assertions.

In addition to the qualitative information, quantitative information was gathered by assessing student perceptions of the psychosocial learning environment (Fraser, 1986a) with a variety of questionnaires selected from Fraser and Fisher (1983a). The measures provided a quantified picture of life in the classrooms of exemplary teachers as perceived by students and enabled comparisons between the learning environments in classes taught by exemplary and nonexemplary science teachers. The findings of this study are summarized as the following four assertions.

Assertion 4: Exemplary teachers used management strategies that facilitated sustained student engagement.

A distinctive feature of the classes of the exemplary science teachers was a high level of managerial efficiency. For example, when Tobin (1987a) compared the teaching performance of two exemplary high school science teachers with that of colleagues teaching in the same schools, the main feature differentiating the exemplary teachers from their colleagues was the management of their classes. The exemplary teachers had well-ordered classes with a relaxed atmosphere characterized by pleasant interactions with students and subtle use of humor. In an important sense, management was the key to success, because the exemplary teachers were able to concentrate on teaching and learning rather than on controlling student behavior. Although different styles were used by different exemplary teachers in establishing and maintaining an environment conducive to learning, in all case studies the crucial link between management, teaching, and learning was highlighted.

Assertion 5: Exemplary teachers used strategies designed to increase student understanding of science.

Most of the exemplary teachers had a concern for assisting students to learn with understanding. As a consequence, the teachers set up activities in which students could have overt involvement in the academic tasks. In elementary grades, the activities were based on the use of materials to solve problems, and in high school grades teachers often used concrete exemplars for abstract concepts. However, the key to teaching with understanding was the verbal interaction that enabled teachers to monitor student understanding of science concepts. The exemplary teachers were effective in a range of verbal strategies, including asking questions to stimulate thinking, probing student responses for clarification and elaboration, and providing explanations to give students additional information.

Assertion 6: Exemplary teachers utilized strategies that encouraged students to participate actively in learning activities.

Sanford (1987) explains how "safety nets" can be used by teachers to allow students to participate without undue embarrassment in front of the teacher and their peers. In the classes of most exemplary science teachers, safety nets were used to encourage involvement from all students. Although involvement was maximized, the cognitive level of the activities was kept appropriately high. Teachers appeared able to make it safe for students to engage in whole-class, small-group, and individualized activities and maintain a focus on meaningful learning. For example, an exemplary biology teacher always treated students and their contributions with the utmost respect and endeavored to work from a given answer to the understanding that he wanted a student to construct. After his explanations, he offered all students a chance to request further explanation or clarification and encouraged questions. When students were unable to respond to a question, this teacher usually persisted by rephrasing the original question or asking supplementary questions until the student could contribute.

Assertion 7: Exemplary teachers maintained favorable classroom learning environments.

In an attempt to make meaningful interpretations of the learning environment data, the actual environments of exemplary teachers' classes were compared (1) with the perceived environment of comparison groups of classes from past research (2) with the classroom environment preferred by the exemplary teachers' students, and (3) with the perceived classroom environment of nonexemplary teachers of the same grade levels in the same school. Overall, the results provide considerable evidence that exemplary and nonexemplary science teachers can be differentiated in terms of the psychosocial environments of their classrooms as seen through their students' eyes and that exemplary teachers typically create classroom environments that are markedly more favorable than those of nonexemplary teachers (Fraser & Tobin, 1989).

For example, the grade 11 class of an exemplary biology teacher (Tobin, Treagust, & Fraser, 1988) was contrasted with a comparison group in terms of the six dimensions of the short form of the Classroom Environment Scale. Relative to the comparison group of the 116 junior high school science classes, the exemplary teacher's students perceived their classroom environments more favorably, especially in terms of higher levels of Involvement, Teacher Support, and Order and Organization. All differences were greater than 1 standard deviation for class means and some differences exceeded 2 standard deviations.

There is much to be learned from exemplary teachers that can be of benefit to others. Perhaps the most fruitful area concerns the activities and strategies used to teach specific areas of content. Detailed case studies describing activities in terms of teacher and student involvement in learning tasks can serve as the content of teacher education courses. As well, the finding that students perceived exemplary teachers' classes favorably provides a valuable validity check on the process of identifying the exemplary teachers through peer nomination. That is, teachers nominated as exemplary by their peers were also seen as very different from nonexemplary teachers by the students.

Higher-Level Cognitive Learning

Two high school science teachers were studied by six researchers using interpretive research methods over a period of 12 weeks to obtain insights into problems teachers encounter as they endeavored to provide students with an environment conducive to higher-level cognitive learning (Fraser, Rennie, & Tobin 1990; Tobin & Fraser, 1989; Tobin, Kahle, & Fraser, 1990). Because the acquisition of higher-level cognitive outcomes has been elusive, we chose a school with a record of emphasizing student learning. This school had a good reputation, was in a fashionable suburb, was innovative, had commenced with considerable optimism about catering for the needs of individuals, and had an open area design that facilitated team teaching and self-paced learning. The curriculum enabled students to learn at their own rates and study science topics in which they were interested. Workbooks provided students with independence, and staff concentrated on enhancing self-esteem and motivation to learn content in a meaningful and integrated manner. Students were given considerable autonomy in selecting when and for how long to study particular topics.

Two "above average" teachers, referred to as Peter and Sandra for the purposes of this chapter, participated in the study. Peter's class contained 11 males and 20 females, and Sandra's class contained 13 males and 18 females. During the first 5-week period, both teachers taught a topic on vertebrates; during the second 5-week interval, they taught a topic on nuclear energy. Two weeks separated the topics. A team of six researchers visited the two classes at various times throughout the study and continued to interact with the teachers for 1 year after the classroom observations.

Participant observer data-collecting strategies were employed predominantly (Erickson, 1986). These involved observing classrooms, interviewing teachers and students on a daily basis, working with students during class time, obtaining written responses to specific questions, and examining student notebooks and test papers. Feedback from the teachers on the

written reports of the study was used as another data source. In addition to the primarily qualitative information obtained by participant observation, quantitative data were obtained by administering instruments assessing students' perceptions of their classroom environments.

As soon as possible after each lesson, data were compiled into written field notes and circulated among members of the research team. Regular team meetings were used to discuss the data and their interpretation and to guide the collection of additional information to support or refute assertions formulated, modified, or rejected throughout the study.

Selected scales from the Individualized Classroom Environment Questionnaire (Fraser, 1990) and the Classroom Environment Scale (Moos & Trickett, 1987) were used. In fact, an important feature of the design of the study was that these classroom environment dimensions were selected *after* a certain amount of fieldwork had been done, so only dimensions considered to be salient for the present research were chosen for inclusion. The four scales used were Personalization, Participation, Order and Organization, and Task Orientation.

Assertion 8: Teachers conceptualized their roles in terms of metaphors that influenced the way in which they taught.

During interviews, Peter described his teaching role in terms of two metaphors: the teacher as *Entertainer* and the teacher as *Captain of the Ship*. The observations and interviews suggest that these metaphors influenced the way in which Peter perceived his role and the way in which he taught. When Peter was entertaining the class, he was humorous, interactive, and amenable to student noise and risque behavior. As Captain of the Ship, Peter was assertive and businesslike, emphasized whole-class activities in order to maintain control of a teacher-centered and teacher-paced learning environment, and was particularly severe on students who stepped out of line and scolded them in a strong voice.

Sandra conceptualized the teacher as a *Resource* and this metaphor influenced her teaching style during both topics. Sandra made herself available to students and assisted them to complete and understand the work. Few whole-class activities were conducted in the 10 weeks of instruction and, when they did occur, they were designed to clarify the schedule or provide details related to the administration of the program. Of particular note was Sandra's noninitiating, responsive role as she moved from one individual to the other throughout each lesson. In particular, extension and elaboration were left to other resources such as the textbook, the workbook, or reference books. Sandra endeavored to provide all students with equal access to her time during activities and was untiring in her efforts to share the teacher resource among the student consumers.

Assertion 9: Teacher beliefs had a major impact on the way in which the curriculum was implemented.

Peter's beliefs were influenced by his own experiences as a high school student when he discovered that learning was meaningful when it was related to personal life experiences.

Also, Peter emphasized the need to interact with students and encourage them to interact with one another, and he was not enthusiastic about the workbook approach, which reduced the amount of time he had for interacting with students. But, despite Peter's belief that students learn best as a result of interacting with peers, he regarded "lock-step", whole-class teaching as an effective means of monitoring student understanding.

Sandra's style of teaching in both topics was consistent with her belief that children learn best when they do things for themselves. Describing her ideal class, Sandra said that, if students could move through the activities efficiently, achieve the objectives, and be accurate in their work, she would arrange for them to have small-group discussions in which they could go over the work to ensure that they really understood it. In accordance with a belief that students should have opportunities to develop understandings about science, Sandra frequently provided students with direct experiences in the form of teacher demonstrations and laboratory activities. Sandra's teaching and her responses to interview questions indicated that she firmly believed that students should progress at their own rates in their science activities. Thus, students were organized into groups so that they could collaborate on their work and help one another learn.

Assertion 10: Knowledge limitations of teachers resulted in an emphasis on students learning facts and completing workbook exercises rather than learning with understanding.

Peter regarded himself as "in field" in terms of his content knowledge when he taught vertebrates and "out of field" when he taught nuclear energy, and his approach to teaching the two topics reflected his perceived level of expertise. When he taught in the vertebrates topic, he appeared confident and was not reliant on the student text as a source of knowledge. But Peter's teaching was focused on acquisition of lower-level cognitive outcomes, and he made various errors of fact. Peter taught in a less confident and less expansive manner in the nuclear energy topic, which he had not taught for 5 years. In this case, he relied on the student text for his knowledge and avoided explanations of key concepts such as nuclear instability and half-life. Instead, he focused on the social aspects of nuclear energy in his whole-class activities and provided students with much more time to engage in individualized activities. Consequently, in the nuclear energy topic, students were much more reliant on workbooks and textbooks than during the vertebrates topic. Peter was aware of his knowledge limitations in physics.

The situation in Sandra's class was quite different. Sandra appeared to have adequate knowledge to teach both topics. She indicated that the knowledge she constructed during her formal degree work was sufficient for her to teach biology and chemistry. Sandra had developed the knowledge needed to teach vertebrates by teaching it on earlier occasions and by teaching similar topics such as human biology in grades 11 and 12. The knowledge needed to teach nuclear energy to grade 10 students was constructed partially during a first-year physics course at university and partially while teaching the topic on other occasions.

Assertion 11: The student-perceived learning environment of the classes was related to teachers' knowledge and beliefs.

Figure 17.3 depicts profiles of mean perceived classroom environment scores obtained by averaging the individual scale scores of the 31 students in Peter's class and the 31 students in Sandra's class. These profiles have been constructed separately for the responses given by students during the vertebrates and the nuclear energy topics. This figure shows that, despite the existence of some small but systematic changes in classroom environment between the two topics, overall there is remarkable consistency between the shapes of the teachers' profiles for the two different topics. The two greatest student-perceived differences between the teachers for both topics were that Sandra's class had considerably more Personalization and less Order and Organization than Peter's class. Moreover, differences were statistically significant at the 0.01 level of confidence only for these two scales and for both topics.

Students' perceptions of the learning environment within each class are consistent with the observers' field records of the patterns of learning activities and engagement in each classroom. The high level of Personalization perceived in Sandra's classroom matches the large proportion of time spent in small-group activities during which she constantly moved about the classroom interacting with students. Furthermore, when Sandra offered desists, they were often private and she was never heard to use sarcasm or personal criticism in her interactions with students. The lower level of Personalization perceived in Peter's class is associated partly with the larger amount of time spent in the whole-class mode and the generally public nature of Peter's interactions with students. He spent much less time than Sandra dealing with students in quiet, small-group situations.

The second significant difference between the learning environments was the lower level of Order and Organization in

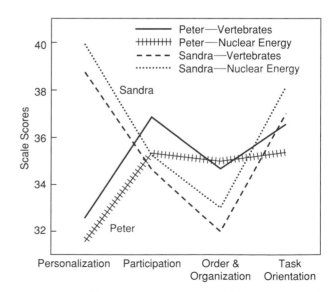

FIGURE 17.3. Classroom Environment Profiles for Two Teachers for Two Topics.

Sandra's class than in Peter's class. Sandra's class was observed to be noisier than Peter's and the high levels of off-task behavior, most of which was social, are consistent with the students' perceptions of a less orderly class. The physical arrangement of the classroom also contributed to the different levels of off-task behavior. To make it easy to collaborate in groups, Sandra's students sat around in tables formed by two desks. Unfortunately, this method of seating not only encouraged social interaction but also hindered effective scanning of the class for management purposes. As a result, many students with their backs to Sandra were able to carry on with their social agenda even during her whole-class presentations. In contrast, Peter's classroom had the desks in rows facing the front of the room, where Peter spent about half of the lesson time using the whole-class instructional mode. This seating arrangement facilitated management scanning, and Peter quickly targeted potential noisemakers for effective public desists. As a result, Peter's class was managed more effectively than Sandra's in terms of the proportion of student-engaged time.

The differences in the classroom environments created by the two teachers can also be considered in terms of the *teaching metaphors* adopted. For Personalization for Peter, for example, some students were confused by being treated in a depersonalized way as crew members during whole-class activities (Captain of the Ship role) but treated in a very friendly way during individual activities (Entertainer role). Moreover, only some students liked their personal interactions with Peter during individualized activities because it was not uncommon for Peter to interact with boys in a "macho" way and with girls in a sexist way. Consequently, Peter's class on average perceived a relatively low level of Personalization. Although Peter was keen about covering the content and being entertaining, he did not attempt to enhance classroom Personalization as a way to aid student understanding of the content. Similarly, the very high level of Personalization perceived in Sandra's class is consistent with her metaphor of the teacher as Resource. Her teaching approach almost exclusively involved individualized work (about 75 percent) and she devoted great amounts of energy to moving around the class to give students individual help.

The low Order and Organization perceived in Sandra's class is linked with her commitment not to use whole-class teaching. Although she appreciated that Order and Organization would probably improve in whole-class situations, her beliefs led her to concentrate on individualized approaches. In particular, Sandra's time was monopolized by a group of demanding and keen girls and another group of disruptive boys, whom she tried to control through proximity desists. Of course, with so much of her time devoted to these two groups, there was a tendency for the other students in the class to be off task and for the average class level of perceived Order and Organization to be low. In fact, Peter's management metaphor, especially his role as Captain of the Ship, resulted in levels of perceived Order and Organization that were higher than in Sandra's class.

The differences in each teacher's classroom environment between the vertebrates and nuclear energy topics, although relatively small, are interesting and consistent (see Figure 17.3). For example, a higher mean for student perceptions of Task Orientation in Sandra's class than in Peter's class during the nuclear energy topic can be tied in with the findings, based on

qualitative data, that Peter's content knowledge was weaker for nuclear energy than for vertebrates, and this led him to require students to use the workbooks to a greater extent for nuclear energy than for vertebrates. Figure 17.3 shows a decrease in student perceptions of Task Orientation in Peter's class from vertebrates to nuclear energy. In contrast, our qualitative methods strongly suggested that Sandra had good content knowledge in both topics, that she arranged the class in the same way for both topics, and that her confidence was high in both topics.

Furthermore, Sandra was interested in and concerned about the feedback she received on student perceptions of the learning environment during the vertebrates topic, and she was determined to change her classroom behavior in ways that would lead to improvements in the classroom environment. Peter dismissed the classroom environment information for vertebrates as irrelevant and, in all likelihood, made no attempt to change his classroom behavior. These observations are clearly reflected in the profiles in Figure 17.3, which show that Peter's classroom environment was less favorable for the second topic than the first on all dimensions except Order and Organization (for which differences were negligible) but that a small improvement occurred between the two testing occasions for all environment dimensions in Sandra's class.

Assertion 12: Teacher expectations of and attitudes toward individuals were reflected in individual students' perceptions of the learning environment.

Whereas the class means of actual environment scores depicted in Figure 17.3 furnish a useful overall picture of classroom environment, they do not provide information about how an individual student perceives his or her learning environment or about a student's preferred environment. Consequently, in this section, the learning environment profiles of some individual students are discussed and integrated with other information gathered using participant observation methods.

Two students, Jenny and Helen, were chosen for discussion because Peter interacted with them so differently and because they were representative of two clusters formed in a cluster analysis using the perceived learning environment scores. Students within a given cluster had relatively similar perceived environment scores, but the different clusters varied from each other in terms of the pattern of mean environment scores. Jenny is representative of a cluster with scores consistent across the scales, but the scores of Helen's group were more variable. Their scores on the learning environment perceived and preferred scales as measured during the vertebrates topic are graphed in Figure 17.4.

Both Jenny and Helen achieved at a little above the class average, were involved in frequent interactions with Peter, and enjoyed their science class and the company of other students. Jenny was nearly always on task. She paid close attention during whole-class activities, often responding to Peter's questions. Her class notes were complete, well presented, and consistently graded A by Peter. In an interview, Peter described Jenny as "mature, confident, bright, beautifully presented" and on another occasion as "pretty close to being one of the brighter kids in that class ... in the top 10 if not the top five." Jenny and

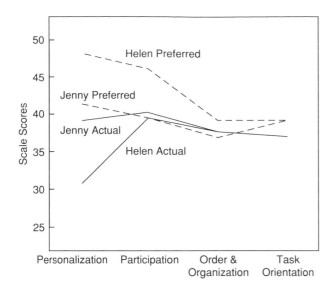

FIGURE 17.4. Classroom Environment Profiles for Two Students.

Peter seemed to get along well together, and Jenny's enjoyment of her class was reflected in her scores on the learning environment scales. She scored very highly on each scale, and the differences between her perceptions of the environment and her preferred environment were negligible.

Helen did not work as well as Jenny in class. She was off task about 20 percent of the time, compared to Jenny's 10 percent, and her classwork was somewhat incomplete, with some parts done well and others not. Her work was graded C for vertebrates and B for nuclear energy. Peter considered Helen to be "good but lazy," suggested that she had fallen out of favor, and sometimes made public statements in class that embarassed her. In fact, Helen did seem to be capable, as she performed at the fully formal level on a Piagetian Separation of Variables Task administered by one of the researchers. On the other hand, Jenny was rated on the same task at the concrete level. Both attained similar test scores for each topic—Jenny scored 76 and 82 on the two topic tests, respectively, and Helen scored 75 on each test.

Helen's perceptions of classroom Participation, Order and Organization, and Task Orientation were similar to Jenny's; it was only on the Personalization scale that there was a marked difference. Comparison of perceived and preferred environment scores accentuate the contrast between Helen and Jenny. Whereas the perceived-preferred difference for Jenny was 2 points on the Personalization scale, for Helen it was as large as 17 points. This contrast is especially interesting in view of the fact that, during the observation period, Helen and Jenny had the same number of interactions with Peter. However, about three-quarters of Helen's interactions with Peter were in a small-group situation.

When this study commenced, we held the view that the major problems in high school science education were associated with the use of whole-class activities for such a large proportion of the time. However, this study suggests that imple-

menting activities with a better balance between small-group and individualized instruction is no guarantee of success. Although Peter was able to manage student behavior in a variety of activity structures (as Captain of the Ship and as Entertainer), he did not have a sufficient repertoire of discipline-specific pedagogical knowledge to facilitate learning adequately in either topic. In contrast, despite Sandra's strong background in science, she lacked the pedagogical knowledge needed to manage student behavior effectively, so students did not gain maximum benefit from her content knowledge. These findings highlight the importance of two aspects of management. If teachers are to be successful facilitators of learning, they must manage both student behavior and the cues to initiate and sustain the cognitive processes associated with learning. Discipline-specific pedagogical knowledge and pedagogical knowledge together are seen as crucial ingredients of successful teaching. Neither is sufficient alone, and each is required if students are to attain the elusive goal of higher-level cognitive learning.

Student perceptions of the learning environment data show that student mind frames are influenced by teacher mind frames through classroom practices. The data were important for two reasons. First, the quantitative data enabled us to use statistical analyses to support assertions based on our qualitative information. Second, the findings showed that teachers dealt with students in an inequitable way on the basis of beliefs that were unsubstantiated. Different students perceived the learning environment of the same class differently partly because the teacher treated them differently within the same classroom. Researchers need to acknowledge these differences and develop new instruments to provide quantitative insights into the extent and nature of the different learning environments. Qualitative analyses should be used in conjunction with these instruments to provide salient insights into aspects of the environment that are not captured quantitatively.

Conclusion

There is little doubt that, in just over two decades, the field of classroom learning environment has progressed enormously and that research involving qualitative methods and research involving quantitative methods have made outstanding contributions to this overall progress. To date, however, research involving both qualitative and quantitative methods in the same study has been the exception rather than the rule. This section has illustrated the fruitfulness of a confluence of qualitative and quantitative research traditions in classroom environment research in three case studies of investigations that used qualitative ethnographic methods in conjunction with quantitative information obtained from classroom environment questionnaires. Clearly, combining of qualitative and quantitative methods is a desirable future direction for research on learning environments.

An important methodological point to emerge from the study of higher-level learning and the study of target students was that certain subgroups of students within a class appeared to perceive different subenvironments because of the teacher's differential treatment of students. This is most vividly illustrated by the way in which target students perceived significantly greater levels of involvement and rule clarity than nontarget students, which was consistent with our classroom observation that the teacher directed more questions at target students and allowed them (and not other students) to call out answers without being asked. These findings bring into question the common practice in past classroom environment research of using the class mean as the unit of analysis and highlight the need for more research on the differences in perceptions of individuals or subgroups within the class. However, existing classroom environment questionnaires seek students' perceptions of the class as whole, and there is a need to devise new questionnaires that are suited to assessing the different subenvironments that might exist for different students or subgroups within the same class. Already, one classroom environment questionnaire has been designed to have separate "class" and "personal" forms (Fraser et al., in press). Instruments such as these should be used in conjunction with ethnographic methods to enhance our understanding of the within-class subenvironments that might exist for different students.

TEACHERS' ATTEMPTS TO IMPROVE CLASSROOM AND SCHOOL ENVIRONMENTS

Although much research has been conducted on educational environments, surprisingly little has been done to help science teachers improve the environments of their own classrooms or schools. Consequently, the purpose of this section is to report how feedback information based on student or teacher perceptions was employed as a basis for reflection on, discussion of, and systematic attempts to improve classroom and school environments. The basic logic underlying the approach has been described by Fraser (1981e). It involves (1)

using assessments of student perceptions of both their actual and preferred environment to identify discrepancies between the actual environment and that preferred by students and (2) implementing strategies aimed at reducing existing discrepancies. This approach can be justified partly in terms of the person-environment fit research described previously, which suggests that students achieve better when in their preferred classroom environment. The proposed methods have been applied successfully in previous studies at the elementary (Fraser & Deer, 1982), secondary (Fraser, Seddon, & Eagleson, 1982), and higher-education levels (De Young, 1977).

Improving Classroom Environments

The attempt at improving classroom environments described next (Fraser & Fisher, 1986) made use of the short 24-item version of the CES discussed previously. The class involved in the study consisted of 22 grade 9 boys and girls of mixed ability studying science at a government school in Tasmania. The procedure followed by the teacher of the class incorporated the following five fundamental steps:

1. *Assessment.* The CES was administered to all students in the class. The preferred form was answered first; the actual form was administered in the same time slot 1 week later.
2. *Feedback.* The teacher was provided with feedback information derived from student responses in the form of the profiles shown in Figure 17.5, representing the class means of students' actual and preferred environment scores. These profiles permitted ready identification of the changes in classroom environment needed to reduce major differences between the nature of the actual environment and the preferred environment as currently perceived by students. Figure 17.5 shows that the interpretation of the larger differ-

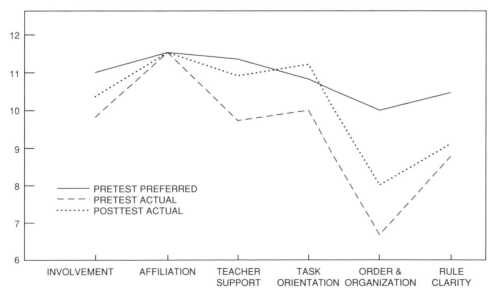

FIGURE 17.5. Pretest Actual, Pretest Preferred, and Posttest Actual Means for the CES.

ences was that students would prefer less Friction, less Competitiveness, and more Cohesiveness.

3. *Reflection and Discussion.* The teacher engaged in private reflection and informal discussion about the profiles to provide a basis for a decision about whether an attempt would be made to change the environment in terms of some of the CES's dimensions. The main criteria used for selection of dimensions for change were that there should be a sizable actual-preferred difference on that variable and that the teacher should feel concerned about this difference and want to try to reduce it. These considerations led the teacher to decide to introduce an intervention aimed at increasing the levels of Teacher Support and Order and Organization in the class.

4. *Intervention.* The teacher introduced an intervention of approximately 2 months' duration in an attempt to change the classroom environment. This intervention consisted of a variety of strategies that originated during discussions between teachers or were suggested by examining ideas contained in individual CES items. For example, strategies used to enhance Teacher Support involved the teacher moving around the class more to mix with students, providing assistance to students, and talking with them more than previously. Strategies used to increase Order and Organization involved taking considerable care with distribution and collection of materials during activities and ensuring that students worked more quietly.

5. *Reassessment.* The student actual form of the scales was readministered at the end of the intervention to see whether students perceived their classroom environments differently from before.

The results are summarized graphically in Figure 17.5, in which a dotted line indicates the class mean score for students' perceptions of actual environment on each of the CES's five scales at the time of posttesting. Figure 17.5 clearly shows that some change in actual environment occurred during the intervention. When tests of statistical significance were performed, it was found that pretest-posttest differences were significant ($p < .05$) only for Teacher Support, Task Orientation, and Order and Organization. These findings are noteworthy because two of the dimensions on which appreciable changes were recorded were those on which the teacher had attempted to promote change. (There also appears to be a side effect in that the intervention could have resulted in the classroom becoming more task oriented than the students would have preferred.)

Although the second administration of the environment scales marked the end of this teacher's attempt at changing a classroom, it might have been thought of as simply the beginning of another cycle. That is, the five steps outlined earlier could be repeated cyclically one or more times until changes in classroom environment reached the desired levels. Overall, this case study and other previous ones suggest the potential usefulness of science teachers employing classroom environment instruments to provide meaningful information about their classrooms and a tangible basis to guide improvements in classroom environments.

Improving School Environments

The method used for improving school environments was based on techniques similar to those used successfully in the past for improving classroom-level environments and described in a previous section. First, the actual and preferred forms of a school environment instrument were administered to teachers. Second, teachers considered feedback information derived from scoring the instrument and summarized as profiles of mean school scores. Third, teachers found private reflection and discussion with peers and researchers about the profiles led to decisions about which dimensions, if any, were to be the targets for change attempts. Fourth, teachers introduced various strategies, typically over a period of several months, aimed at improving selected dimensions of school environment. Fifth, the actual form of the instrument was readministered at the end of the intervention period.

Fraser et al. (1988) described the use of the Work Environment Scale (WES) as part of teacher development activities in Tasmania. In particular, they reported a case study of a school change attempt in an elementary school with a staff of 24 teachers. After pretesting with both the actual and preferred forms of the WES, mean scale scores were calculated and pretest actual and pretest preferred profiles were fed back to the school staff. The areas in which sizable differences between actual and preferred scores were evident were Peer Cohesion, Clarity, Innovation, and Physical Comfort. Consequently, following a staff meeting and considerable discussion, priorities for action were accepted (e.g. strategies for improving Clarity included making

TABLE 17.12. Priorities for Action in Improving School Environment

SLEQ Dimension	Priorities for Action
Resource adequacy	Conduct a survey of resources in the school
	Develop a plan of attack—immediate, intermediary and long term
	Check and repair already existing equipment
	Develop a plan for increased sharing of resources
Innovation	Conduct staff meetings in individual classrooms. These meetings should be rotated between elementary and infant rooms. Time should be given for the class teacher to comment on organization, display, problems, etc.
	Free teachers with particular skills to help in other rooms (drama, computers, science)
	Adopt a whole-school theme
	Attempt to "spot the innovator" (particularly by senior staff)
Work pressure	Have less staff meetings
	Use recess breaks for minor discussions
	Draw on the community for assistance with coaching sporting teams
	Provide opportunities for discussion about meeting the individual needs of children

available to teachers details about the amount of money for excursions, cooking, petty cash, and so forth and a simple information sheet about resource rooms). An intervention consisting of the accepted actions was implemented for approximately 10 weeks. At the end of this time, the actual form of the WES was administered to teachers again to determine whether there were any changes in the work environment as perceived by teachers. The results indicated that sizable changes did occur in three of the priority areas of Peer Cohesion, Clarity, and Physical Comfort.

Recently, the new form of the School-Level Environment Questionnaire (SLEQ; Fisher & Fraser, 1991a) incorporating the Work Pressure scale was used in similar school improvement studies using the same basic strategy. Reported next are details of a case study in an elementary school of 15 teachers. After pretesting with both the actual and preferred forms of the SLEQ, mean scale scores were calculated and pretest actual and pretest preferred profiles were fed back to the school staff. The results are depicted (Figure 17.6) in the form in which they were presented to a meeting of the school. Although there were sizable differences between actual and preferred scores

on a number of dimensions, the areas determined by the staff for initial improvement were Resource Adequacy, Work Pressure, and Innovation. Other dimensions were to be targets for a second round of change attempts.

Next, the staff was divided randomly into small groups to discuss the areas in which actual-preferred discrepancies were largest. These groups were asked to consider those areas and to make suggestions for improvement. The groups then were called together and group session leaders presented a report to the whole staff. Points were discussed at some length and the priorities for action listed in Table 17.12 were accepted.

An intervention consisting of the actions listed in Table 17.12 was implemented for approximately 10 weeks. At the end of this time, the actual form of the SLEQ was administered to teachers for a second time to determine whether there had been any changes in the work environment as perceived by teachers. The results of this assessment are depicted in Figure 17.7.

On examining the profiles in Figure 17.7, it can be seen that sizable changes did occur in two of the priority areas. Resource Adequacy increased 2.5 raw score points (about two-thirds of a

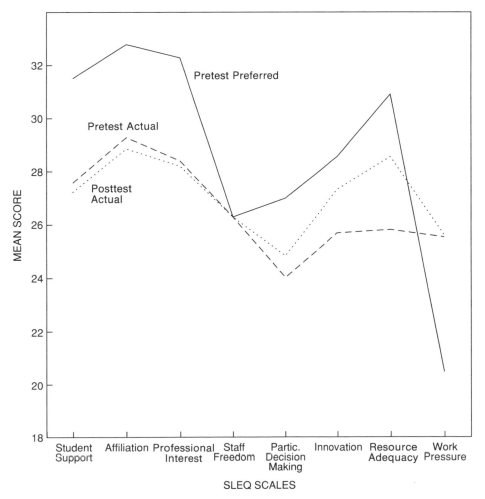

FIGURE 17.6. Pretest Actual, Pretest Preferred, and Posttest Actual Means for the SLEQ.

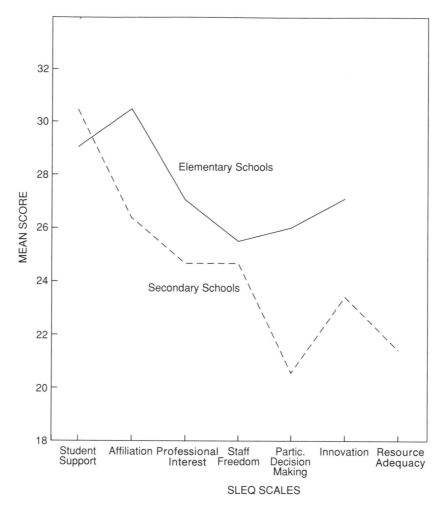

FIGURE 17.7. Comparison of the School Environment Profiles of Elementary and Secondary Schools.

standard deviation) and Innovation increased 1.7 raw score points (about half a standard deviation). The use of t-tests for dependent samples revealed that each of these differences was statistically significant ($p < .05$). However, the level of Work Pressure did not change. Nevertheless, the staff was pleased when presented with the results because they indicated that the concentrated effort in the other two chosen areas had been worthwhile.

RECENT TRENDS AND DESIRABLE FUTURE DIRECTIONS

Although the major thrust of this chapter has been to synthesize the previous two decades of accomplishments in research on classroom and school environment, the focus in this section is to describe several new lines of research that hold promise and are desirable directions for future research efforts on educational environments.

School-Level Environments

Although it was noted earlier that science education researchers have paid substantially more attention to classroom environment research than to school environment research, desirable future directions include a greater emphasis on the school-level environment and the integration of classroom and school climate variables within the same study. The possibility of greater attention to school climate variables has been facilitated greatly by the availability of two convenient instruments, the WES and the SLEQ (Fisher & Fraser, 1991a), which are described elsewhere in this chapter.

An interesting recent line of research with school environment instruments has involved examination of differences in the climates of different types of schools. Docker et al. (1989) reported the use of the WES with the sample of 599 teachers, described in Table 17.6, in investigating differences between the environments of various school types. When profiles of WES scale means were sketched for the various school types, reasonable similarity was found for preferred environment scales.

That is, there was a fair degree of agreement among teachers in different types of schools about what they would prefer their school environments to be like. In contrast, teachers' perceptions of their actual school environments varied markedly, with the climate in elementary schools emerging as more favorable than the environment of high schools on most of the WES scales. For example, elementary schools were viewed as having greater Involvement, Staff Support, Autonomy, Task Orientation, Clarity, Innovation, and Physical Comfort and less Work Pressure.

Recently the SLEQ has been used in exploring differences between the climates of elementary and high schools. The sample consisted of the 109 teachers in 10 schools in Tasmania comprising the third sample described in Table 17.8. Differences between the two types of schools were tested statistically for each SLEQ scale. The first step involved the performance of a one-way MANOVA in which the set of environment scales constituted the dependent variables and the type of school constituted the main effect. Because the multivariate test using Wilks' lambda criterion was statistically significant ($p < .01$), the univariate ANOVA results were examined for each of seven scales individually. The findings are summarized in Figure 17.7.

The profiles in Figure 17.7 reveal some clear general patterns of differences in the favorableness of the school environments in the types of schools. The most striking pattern is that the climate in elementary schools emerged as more favorable than the environment of high schools on most of the SLEQ scales. In particular, compared with high school teachers, elementary school teachers perceived their school climates considerably more favorably in terms of greater Affiliation, Professional Interest, Staff Freedom, Participatory Decision Making, Innovation, and Resource Adequacy. Differences were greater than 1 standard deviation for Affiliation, Participatory Decision Making, and Resource Adequacy.

It is hoped that educational researchers and teachers will make use of the widely applicable and extensively validated WES and SLEQ in assessing the important concept of school environment and in pursuing research and practical applications related to school-level environment analogous to those previously reported in this chapter. For example, assessments involving the WES or SLEQ could form the basis for studies of the effects of the school environment on such outcomes as teacher job satisfaction, student achievement, or morale. Further investigations might be made of the links between students' perceptions of classroom-level environment and teachers' perceptions of school-level environment (Fraser & Rentoul, 1982). The WES and SLEQ are likely to provide a useful source of criteria in the evaluation of innovative or alternative educational provisions (Fraser et al., 1987). In particular, teachers might use assessments of their perceptions of actual and preferred school work environments as a basis for discussing improvements in their school work settings that would reduce actual-preferred discrepancies (see the case study reported in the previous section). Through instruments such as the WES and SLEQ, it is possible to tap important but subtle aspects of teachers' professional lives (e.g., Staff Freedom, Professional Interest, Affiliation, Innovation, and Work Pressure). It is hoped that assessments of actual and preferred school environments,

as seen through the eyes of teachers themselves, will provide a useful foundation on which teachers can base attempts to improve the quality of their school settings and professional lives.

Constructivist Learning Environments

Traditionally, teachers have conceived their roles to be concerned with revealing or transmitting the logical structures of their knowledge and directing students through rational inquiry toward discovering the predetermined universal truths expressed in the form of laws, principles, rules, and algorithms. Recently, developments in history, philosophy, and sociology have provided educators with a better understanding of the nature of knowledge development. At the level of the individual learner, there has been a realization that meaningful learning is a cognitive process of making sense, or purposeful problem solving, of the experiential world of the individual in relation to the totality of the individual's already constructed knowledge. Because the individual belongs to a world populated by significant others, the sense-making process involves active negotiation and consensus building for the duration of the individual's lifetime, regardless of the learning context (von Glasersfeld, 1981, 1988). Educators are actively researching and negotiating the pedagogical applications and implications of a constructivist epistemology as a viable alternative to the positivist epistemology that shaped much of school teaching of the twentieth century.

Although classroom environment research has focused on the assessment and improvement of teaching and learning, it has done so largely within the context of the traditional, dominant epistemology underpinning the established classroom environment. Consequently, a new learning environment instrument is needed to help researchers assess the degree to which a particular classroom's environment is consistent with a constructivist epistemology and to help teachers reflect on their epistemological assumptions and reshape their teaching practice. The Constructivist Learning Environment Survey (CLES) was developed to meet this need and to assess the four scales of Autonomy, Prior Knowledge, Collaboration, and Reflection (Taylor & Fraser, 1991).

The CLES not only extends the field of learning environment research by providing a new assessment tool but also permits a confluence of the work of researchers who investigate constructivist teaching-learning approaches with the work of learning environment researchers. For example, researchers could make use of the CLES in monitoring the effectiveness of preservice-inservice attempts to change teaching-learning styles to a more constructivist approach; evaluating the impact of constructivist teaching approaches on student outcomes; guiding teacher-as-researcher attempts to reflect on and improve classroom environments; reducing the amount of classroom observation needed in studies of constructivist teaching-learning (through collection of information from students via the CLES); complementing qualitative information in constructing richer case studies that also include quantitative information based on student perceptions obtained with the CLES; and investigating the relationship between teacher cognition and teaching practice.

Personal Forms of Scales

Fraser and Tobin (1991) point out that there is potentially a major problem with nearly all existing classroom environment instruments when they are used to identify differences between subgroups within a classroom (e.g., boys and girls) or to construct case studies of individual students. The problem is that items are worded in such a way that they elicit an individual student's perceptions of the class as a whole, as distinct from that student's perceptions of his or her role within the classroom. Consequently, Fraser et al. (in press) have developed a Personal Form of the Science Laboratory Environment Inventory (SLEI) that parallels its Class Form. Table 17.13 shows how the wording of items differs in the Class Form and the Personal Form of the actual version of the SLEI. In addition to making this generally useful Personal Form available to researchers interested in individual or subgroup perceptions, this work makes it possible to investigate differences between the Class and Personal forms in terms of class means and the magnitudes of associations between achievement and environment.

Interpersonal Teacher Behavior

A line of research, which originated in The Netherlands, focuses on the nature and quality of interpersonal relationships between teachers and students (Créton, Hermans & Wubbels, 1990; Wubbels, Brekelmans, & Hermans, 1988; Wubbels et al., 1991). Drawing on Leary's (1957) theoretical model of proximity (cooperation-opposition) and influence (dominance-submission), the Questionnaire of Teacher Interaction (QTI) was developed to assess student perceptions of eight behavior aspects (leadership, helpful-friendly, understanding, student responsibility and freedom, uncertain, dissatisfied, admonishing, and strict). One interesting application of the QTI was in the development of an eightfold typology of teaching, namely directive, authoritative, tolerant and authoritative, tolerant, uncertain-tolerant, uncertain-aggressive, repressive, and drudging. In other applications, student perceptions of interpersonal teacher behavior explained cognitive and affective outcomes (Brekelmans, Wubbels, & Créton, 1990), and a mismatch was found between teachers' and students' perceptions of teacher behavior.

Although research with the QTI began at the senior high school level in The Netherlands, comparative work has been completed at the same grade levels in the United States (Wubbels & Levy, 1989). Replication is also under way in Australia at senior high school grades, and this line of work is being extended, using a modified version of the QTI, to elementary schools in the United States, Singapore, and Australia.

School Psychology

Given the school psychologist's changing role, the classroom psychosocial environment is a good example of an area that furnishes a number of ideas, techniques, and research findings that could be valuable in school psychology (Burden & Hornby, 1988; Fraser, 1987a, 1987c). Traditionally, school psychologists have tended to concentrate heavily and sometimes exclusively on their roles in assessing and enhancing academic achievement and other valued learning outcomes. The field of classroom environment provides an opportunity for school psychologists and teachers to become sensitized to subtle but important aspects of classroom life.

Burden and Fraser (1991) report the way in which classroom environment instruments were used to advantage in a variety of applications in school psychology in England. For example, one study in comprehensive schools in Bristol showed that the Individualized Classroom Environment Questionnaire can be valuable in helping teachers change their classroom interactive styles. In other studies, teachers used discrepancies between students' perceptions of actual and preferred environment as an effective basis to guide improvements in their classrooms. This research is also noteworthy because it represents one of the very few studies in the classroom environment tradition ever undertaken in Britain.

Links Between Educational Environments

Although most individual studies of educational environments in the past have tended to focus on a single environment, researchers have also considered the links between and joint influence of two or more environments. For example, Marjoribanks (1991) shows how the environments of the home and school interact and codetermine school achievement, and Moos (1991) illustrates the links between school, home, and parents' work environments. Earlier work by Fraser and Rentoul (1982) revealed some interesting associations between school-level and classroom-level environment.

TABLE 17.13. Differences in the Wording of Items in the Class Form and Personal Form

Scale	Class Form	Personal Form
Teacher supportiveness	The teacher/instructor is friendly toward students.	The teacher/instructor is friendly toward me.
	Students in this laboratory class get to know each other well.	I get to know students in this laboratory class well.
Open-endedness	In our laboratory sessions, different students do different experiments.	In my laboratory sessions, I do different experiments than some of the other students.
	There is opportunity for students to pursue their own science interests in this laboratory class.	There is opportunity for me to pursue my own science interests in this laboratory class.

Cross-National Studies

Reviews of past research on educational environments attest to its international character and indicate that research that began in one country has sometimes been replicated in other countries. However, it is rare for learning environment researchers from the outset to design studies that involve the cross-national development, validation, and application of instruments for assessing educational environments. Giddings and Fraser (1990) and Fraser et al. (in press), however, describe a study in which the Science Laboratory Environment Inventory was validated and applied simultaneously in six countries (the United States, Canada, England, Australia, Israel, and Nigeria). This study involved 5447 students in 269 classes at the senior high school and higher-education levels. In addition, Fraser et al. (1991) provide a useful overview of the major advantages of cross-national research and some timely warnings about the pitfalls that need to be avoided in future cross-national studies.

Changes in Classroom Environment during the Transition from Elementary to High Schools

Although there is considerable interest in the effects on early adolescents of a move from elementary school to the larger, less personal environment of the junior high school at this time of life, little attention has been focused specifically on the classroom environment. Several studies revealed a deterioration in the classroom environment when students moved from generally smaller elementary schools to larger, departmentally organized lower secondary schools (Feldlaufer, Midgley, & Eccles, 1988; Midgley, Eccles, & Feldlaufer, 1991; Midgley, Feldlaufer, & Eccles, 1989). Several of these studies suggest that the decline could be attributable to less positive student relations with teachers and reduced student opportunities for decision making in the classroom. However, there was also evidence that some students moved to a more facilitative environment and that this had a positive effect on the beliefs and attitudes of young adolescents.

Incorporating Learning Environment Ideas into Teacher Education

Improvement of preservice and inservice teacher education programs requires new ideas that will help teachers become more reflective and retrospective about their teaching. Although the thriving field of psychosocial learning environment furnishes a number of ideas and techniques which could be extremely valuable in teacher education programs, surprisingly little progress has been made in incorporating these ideas in teacher education (Fraser, 1989a). Consequently, Fisher and Fraser (1991a) decided to review some examples of successful past and current attempts to include learning environment work in teacher education programs and make suggestions about how teacher education can be improved through the input of ideas from learning environment research.

In particular, Fisher and Fraser (1991a) report some case studies of how classroom and school environment work has been used in preservice and inservice teacher education to (1) sensitize teachers to subtle but important aspects or classroom life, (2) illustrate the usefulness of including classroom and school environment assessments as part of a teacher's overall evaluation-monitoring activities, (3) show how assessment of classroom and school environment can be used to facilitate practical improvements in classrooms and schools, and (4) provide a valuable source of feedback about teaching performance for the formative and summative evaluation of student teaching (Dushl & Waxman, 1991). It appears that information on student perceptions of the classroom learning environment during preservice teachers' field experience is a useful addition to the information obtained from university supervisors, school-based cooperating teachers, and student teacher self-evaluation. Overall, learning environment ideas can promote a reflective or teacher-as-researcher stance among preservice teachers or in school-based professional development activities (Munby, 1989; Schon, 1983).

Créton et al. (1990) have used a systems communication perspective (Watzlawick, 1978; Wubbels, Créton, & Holvast, 1988) to provide guidance on how the teacher education programs can be changed to improve interpersonal teacher behavior in the classroom. Within this perspective, problems in classroom communication between teacher and students are seen as characteristics of the classroom system formed by students and teacher. In order to map interpersonal behavior, Wubbels, Brekelmans, and Hooymayers (1990) proposed an eight-sector model based on a proximity dimension (cooperation–submission) and a influence dimension (dominance–submission), and developed the Questionnaire on Teacher Interaction (see previous section) to assess student and teacher perceptions of interpersonal behavior.

Once research using the QTI had suggested what types of interpersonal teacher behavior are desirable, Créton et al. (1990) made some suggestions about how to promote these through teacher education. For example, using peer coaching, modeling, simulations, and so forth, an attempt is made to not only to increase teachers' behavioral repertoires but also to help teachers use their current behavioral repertoires to select the skills most suited to specific situations. Changing escaling spirals of breakdown in communication are given special attention (e.g., disorderly behavior among students leads to corrective behavior from teachers, which is reacted to negatively by students, who react more aggressively). Wubbels et al. (1990) show that the solution is not more of the same, but rather use of more friendly behavior. Because beginning teachers have particular problems in these areas, teacher training programs need to emphasize setting standards and maintaining norms strictly, but without too much of the admonishing and angry behaviors that can cause a negative spiral.

Teacher Assessment

An innovative teacher assessment system called the Louisiana STAR (System for Teaching and Learning Assessment and

Review) specifically includes learning environment dimensions in a set of four performance dimensions (Ellett, Loup, & Chauvin, 1989). The other three performance dimensions are Preparation, Planning, and Evaluation (e.g., teaching methods, homework, assessment), Classroom and Behavior Management (e.g., student engagement, monitoring student behavior), and Enhancement of Learning (e.g., content accuracy, thinking skills, pace, feedback). Teachers who are effective in terms of the psychosocial learning environment dimension "actively encourage positive interpersonal relationships within a classroom environment in which students feel comfortable and accepted. The teacher, through verbal and non-verbal behaviors, models enthusiasm and interest in learning, includes all students in learning activities and encourages active involvement" (p. 60).

With the STAR, multiple observers complete an assessment in 45 minutes by focusing on preparation and planning in addition to in-class performance, on student learning as well as teaching behavior, on higher-level as well as lower-level student learning, and on differential provision for different children.

DISCUSSION AND CONCLUSION

The major purpose of this chapter on perceptions of psychosocial characteristics of classroom and school environments has been to make this exciting research in science education more accessible to wider audiences. In portraying prior work, attention has been given to instruments for assessing classroom and school environments (including some interesting new instruments), several lines of previous research (e.g., associations between outcomes and environment, use of environment dimensions in evaluating educational innovations, person-environment fit studies of whether students achieve better in their preferred environment), recent classroom environment studies that have combined quantitative and qualitative methods, teachers' use of classroom environment perceptions in guiding practical attempts to improve their own classrooms and schools, and new lines of research that suggest desirable future directions for the field. The areas of current research discussed were school-level environment, school psychology, constructivist learning environments, personal forms of scales, interactional teacher behavior, links between different educational environments, cross-national studies, incorporating educational environment ideas into teacher education programs, and changes in environment in the transition from elementary to high school. Given the ready availability of instruments, the salience of educational environments, the impact of educational environments on student outcomes and teacher satisfaction, and the potential of environmental assessments in guiding educational improvement, it seems crucial that science education researchers and science teachers make more frequent use of classroom and school environment instruments for a variety of purposes.

It has been assumed in this chapter that having a positive classroom environment is an educationally desirable end in its own right. Moreover, the comprehensive evidence presented here clearly establishes that the nature of the classroom environment has a potent influence on how well students achieve a range of desired educational outcomes. Consequently, science educators need not feel that they must choose between striving to achieve constructive classroom environments and attempting to enhance student achievement of cognitive and affective aims. Rather, constructive educational climates can be viewed as both means to valuable ends and worthy ends in their own right.

Most prior classroom and school environment research has been correlational in nature. That is, studies have investigated associations between outcomes and actual environment or actual-preferred congruence in naturally occurring settings. Consequently, causal conclusions strictly cannot be drawn. Desirable in future research, then, are experimental studies in which the environment is deliberately changed in specific ways to establish more clearly the causal effects of these changes on students' outcomes.

In this chapter, more attention has been devoted to reporting past research uses of classroom and school environment instruments than to describing science teachers' uses of these instruments for a variety of practical, school-based purposes. This balance is an accurate reflection of the fact that educational environment instruments have tended to be used more by researchers than by teachers. But this chapter helps to pave the way for much greater involvement of teachers by making economical hand-scorable instruments readily accessible and reporting promising case studies of applications in which classroom or school environment assessments have been used successfully to guide improvements in environments.

Typically, student outcomes have been studied using quantitative approaches based on educational measurement traditions, whereas classroom processes or environment have usually involved qualitative approaches including informal observations and interviews. This chapter illustrates that classroom climate is susceptible to quantitative study. Admittedly, quantitative measures have well-known limitations; but so too do qualitative approaches. Rather than claiming that quantitative methods are superior to qualitative ones in the study of classroom environments, the intention has been to make a potentially useful tradition of quantitative assessment of classroom climate readily accessible so that studies might benefit from the use of a range of quantitative and qualitative approaches.

Although the focus of this chapter has been more on classroom-level rather than school-level environment, school-level environment work is also important in science education and has been given some prominence here. Promising recent work has combined the use of classroom and school environment measures to advantage within the one study (Fraser & Rentoul, 1982; Fraser et al., 1987), has used school climate scales to reveal interesting differences between elementary and secondary schools (Docker et al., 1989), and has applied the methods of improving classroom-level environments described in this publication to the improvement of school-level environments (Fraser et al., 1988).

In most prior classroom environment studies, researchers adopted either qualitative or quantitative methods, but seldom both. Therefore, from a methodological perspective, the combination of quantitative measures of classroom environment obtained from questionnaires with a range of qualitative data-gathering techniques in the two studies reported here is noteworthy for several reasons. First, the complementarity of qualitative observational data and quantitative classroom environment data added to the richness of the data base. Second, the use of classroom environment questionnaires provided an important source of students' views of their classrooms. Third, through a triangulation of qualitative and quantitative classroom climate information, greater credibility could be placed in findings because patterns emerged consistently from data obtained using a range of different data-collection methods. Overall, the studies described here illustrate the considerable advantages of incorporating both qualitative and quantitative methods in future classroom environment research in science education.

One potentially fruitful line for future research is the identification of typologies of classroom environments in different subject areas at different grade levels. Moos (1978) reports an example of this line of research in a cluster analysis with the responses to the Classroom Environment Scale of a sample of 200 junior high and high school classes. The five distinctive clusters that emerged were labeled as control oriented, innovation oriented, affiliation oriented, task oriented, and competition oriented. One of the most striking aspects of the results was the identification of 23 percent classrooms almost exclusively oriented toward teacher control of student behavior. More recently, Wubbels et al. (1991) reported a similar attempt to generate clusters based on student responses to the Questionnaire on Teacher Interaction. Taxonomies of classrooms such as the one developed by Moos and Wubbels could help educators understand how dimensions of classroom environment moderate each other, how they are linked with specific outcomes, and which students might benefit more from one type of classroom psychosocial environment than another.

In just over 20 years, as this chapter shows, older instruments have been more widely used and cross-validated in various countries, preferred forms and personal forms have been developed to augment the original actual forms, short and hand-scorable forms have been designed for the convenience of teachers, and new instruments have been developed to fill gaps (e.g., for use in higher-education classrooms or science laboratory classes). As this chapter is being written, workers around the world are continuing to translate and adapt instruments for use in different countries, to develop new instruments for settings not ideally catered for with existing questionnaires (e.g., computer-assisted instruction, preschool classrooms, constructivist classrooms), and to use the instruments in settings (e.g., various special education classes) in which they have not been used previously. For example, a Braille form of My Class Inventory was prepared recently (Genn, 1988).

Already in various countries the topic of classroom and school environment is being included in preservice and inservice courses for science teachers and is gaining attention from school psychologists. Given the potential usefulness of including educational environment topics in science teacher education programs (Fraser, 1989a; Fisher & Fraser, 1991b) and incorporating classroom and school environment assessments in the work of school psychologists (Burden & Fraser, 1991; Burden & Hornby, 1988; Fraser, 1987a, 1987c), it is probable that the use of educational climate measures in these two areas will continue to grow. For example, because school psychologists and teachers have sometimes tended to concentrate almost exclusively on their roles in assessing and enhancing academic achievement, the field of classroom and school environment provides an opportunity for them to become sensitized to many important, subtle aspects of school life. Also, past experience in using educational environment assessments suggests several important ways (e.g., in evaluating innovations) in which climate scales might be used to advantage by school psychologists.

APPENDIX 17.A

Directions

The purpose of this questionnaire is to find out your opinions about the class you are attending right *now*.

This questionnaire is designed for use in gathering opinions about *small classes* at universities or colleges (sometimes referred to as *seminars* or *tutorials*). It is *not* suitable for the rating of lectures or laboratory classes. This form of the questionnaire assesses your opinion about what this class is *actually like*. Indicate your opinion about each questionnaire statement by circling:

SA if you STRONGLY AGREE   that it describes what this class is actually like.

A if you AGREE   that it describes what this class is actually like.

D if you DISAGREE   that it describes what this class is actually like.

SD if you STRONGLY DISAGREE   that it describes what this class is actually like.

All responses should be given on the separate Response Sheet.

1. The instructor considers students' feelings.
2. The instructor talks rather than listens.
3. The class is made up of individuals who don't know each other well.
4. The students look forward to coming to classes.
5. Students know exactly what has to be done in our class.
6. New ideas are seldom tried out in this class.
7. All students in the class are expected to do the same work, in the same way, and in the same time.

8. The instructor talks individually with students.
9. Students put effort into what they do in classes.
10. Each student knows the other members of the class by their first names.
11. Students are dissatisfied with what is done in the class.
12. Getting a certain amount of work done is important in this class.
13. New and different ways of teaching are seldom used in the class.
14. Students are generally allowed to work at their own pace.

15. The instructor goes out of his/her way to help students.
16. Students "clock watch" in this class.
17. Friendships are made among students in this class.
18. After the class, the students have a sense of satisfaction.
19. The group often gets sidetracked instead of sticking to the point.
20. The instructor thinks up innovative activities for students.
21. Students have a say in how class time is spent.

22. The instructor helps each student who is having trouble with the work.
23. Students in this class pay attention to what others are saying.

24. Students do not have much chance to get to know each other in this class.
25. Classes are a waste of time.
26. This is a disorganized class.
27. Teaching approaches in this class are characterized by innovation and variety.
28. Students are allowed to choose activities and how they will work.
29. The instructor seldom moves around the classroom to talk with students.
30. Students seldom present their work to the class.
31. It takes a long time to get to know everybody by his/her first name in this class.
32. Classes are boring.
33. Class assignments are clear so that everyone knows what to do.
34. The seating in this class is arranged in the same way each week.
35. Teaching approaches allow students to proceed at their own pace.

36. The instructor is not interested in students' problems.
37. There are opportunities for students to express opinions in this class.
38. Students in this class get to know each other well.
39. Students enjoy going to this class.
40. This class seldom starts on time.
41. The instructor often thinks of unusual class activities.
42. There is little opportunity for a student to pursue his/her particular interest in this class.

43. The instructor is unfriendly and inconsiderate toward students.
44. The instructor dominates class discussions.
45. Students in this class are not very interested in getting to know other students.
46. Classes are interesting.
47. Activities in this class are clearly and carefully planned.
48. Students seem to do the same type of activities in every class.
49. It is the instructor who decides what will be done in our class.

Scoring

Items whose numbers are underlined are scored 1, 2, 4, and 5, respectively, for the responses SA, A, D, and SD. The other items are scored in the reverse manner. Omitted or invalidly answered items are scored 3.

To obtain the total score for each scale, add the scores for seven items in each scale. The items are arranged in cyclic order so that the first, second, third, fourth, fifth, sixth, and seventh item in each block, respectively, assesses Personalization, Involvement, Student Cohesiveness, Satisfaction, Task Orientation, Innovation, and Individualization.

College and University Classroom Environment Inventory (CUCEI)

APPENDIX 17.B

Directions

This is not a test. The questions inside are to find out what your class is *actually like*. Draw a circle around

    Yes   if you AGREE with the sentence
    No   if you DON'T AGREE with the sentence

Please answer all questions. If you change your mind about an answer, cross it out and circle the new answer.

Do not forget to write your name and other details on the top of the next page.

Name _____   *Class* _____

School _____

Remember you are describing your *actual* classroom	Circle Your Answer	Teacher Use Only
1. The pupils enjoy their schoolwork in my class.	(Yes) No	_3_
2. Children are always fighting with each other.	Yes No	_____
3. Children often race to see who can finish first.	Yes No	_____
4. In our class, the work is hard to do.	Yes No	_____
5. In my class, everybody is my friend.	Yes (No)	_1_
6. Some pupils are not happy in class.	(Yes) No	R _1_
7. Some of the children in our class are mean.	Yes No	_____
8. Most children want their work to be better than their friend's work.	Yes No	_____
9. Most children can do their schoolwork without help.	Yes No	R _____
10. Some people in my class are not my friends.	Yes (No)	R _3_
11. Children seem to like the class.	Yes (No)	_1_
12. Many children in our class like to fight.	Yes No	_____
13. Some pupils feel bad when they do not do as well as the others.	Yes No	_____
14. Only the smart pupils can do their work.	Yes No	_____
15. All pupils in my class are close friends.	(Yes)(No)	_2_
16. Some of the pupils do not like the class.	Yes (No)	R _3_
17. Certain pupils always want to have their own way.	Yes No	_____
18. Some pupils always try to do their work better than the others.	Yes No	_____
19. Schoolwork is hard to do.	Yes No	_____
20. All of the pupils in my class like one another.	(Yes) No	_3_
21. The class is fun.	Yes No	_2_
22. Children in our class fight a lot.	Yes No	_____
23. A few children in my class want to be first all of the time.	Yes No	_____
24. Most of the pupils in my class know how to do their work.	Yes No	R _____
25. Children in our class like each other as friends.	(Yes) No	_3_

S _10_   F ____   Cm ____   D ____   Ch _12_

My Class Inventory

Directions

There are 56 items in this questionnaire. They are statements about the school in which you work and your working environment.

Think about how well the statements describe your school environment.

Indicate your answer by circling:

SD   if you STRONGLY DISAGREE with the statement.
D   if you DISAGREE with the statement.
N   if you neither agree nor disagree with the statement or are not sure.
A   if you AGREE with the statement.
SA   if you STRONGLY AGREE with the statement.

If you change your mind about a response, cross out the old answer and circle the new choice.

1. There are many disruptive, difficult students in the school.
2. I seldom receive encouragement from colleagues.
3. Teachers frequently discuss teaching methods and strategies with each other.
4. I am often supervised to ensure that I follow directions correctly.
5. Decisions about the running of the school are usually made by the principal or a small group of teachers.
6. It is very difficult to change anything in this school.
7. The school or department library includes an adequate selection of books and periodicals.
8. There is constant pressure to keep working.

9. Most students are helpful and cooperative to teachers.
10. I feel accepted by other teachers.
11. Teachers avoid talking with each other about teaching and learning.
12. I am not expected to conform to a particular teaching style.
13. I have to refer even small matters to a senior member of staff for a final answer.
14. Teachers are encouraged to be innovative in this school.
15. The supply of equipment and resources is inadequate.
16. Teachers have to work long hours to complete all their work.

17. Most students are pleasant and friendly to teachers.
18. I am ignored by other teachers.
19. Professional matters are seldom discussed during staff meetings.
20. It is considered very important that I closely follow syllabuses and lesson plans.
21. Action can usually be taken without gaining the approval of the subject department head or a senior member of the staff.
22. There is a great deal of resistance to proposals for curriculum change.
23. Video equipment, tapes, and films are readily available and accessible.
24. Teachers do not have to work very hard in this school.

25. There are many noisy, badly behaved students.
26. I feel that I could rely on my colleagues for assistance if I should need it.
27. Many teachers attend inservice and other professional development courses.

28. There are a few rules and regulations that I am expected to follow.
29. Teachers are frequently asked to participate in decisions concerning administrative policies and procedures.
30. Most teachers like the idea of change.
31. Adequate duplicating facilities and services are available to teachers.
32. There is no time for teachers to relax.

33. Students get along well with teachers.
34. My colleagues seldom take notice of my professional views and opinions.
35. Teachers show little interest in what is happening in other schools.
36. I am allowed to do almost as I please in the classroom.
37. I am encouraged to make the decisions without reference to a senior member of staff.
38. New courses or curriculum materials are seldom implemented in the school.
39. Tape recorders and cassettes are seldom available when needed.
40. You can take it easy and still get the work done.

41. Most students are well-mannered and respectful to the school staff.
42. I feel that I have many friends among my colleagues at this school.
43. Teachers are keen to learn from their colleagues.
44. My classes are expected to use prescribed textbooks and prescribed resource materials.
45. I must ask my subject department head or senior member of staff before I do most things.
46. There is much experimentation with different teaching approaches.
47. Facilities are inadequate for catering for a variety of classroom activities and learning groups of different sizes.
48. Seldom are there deadlines to be met.

49. Very strict discipline is needed to control many of the students.
50. I often feel lonely and left out of things in the staffroom.
51. Teachers show considerable interest in the professional activities of their colleagues.
52. I am expected to maintain very strict control in the classroom.
53. I have very little say in the running of the school.
54. New and different ideas are always being tried out in this school.
55. Projectors for filmstrips, transparencies, and films are usually available when needed.
56. It is hard to keep up with your work load.

Scoring

Underlined items are scored 1, 2, 3, 4, and 5, respectively, for the responses SA, A, N, D, and SD. All other items are scored in the reverse manner. Invalid or omitted items are scored 3.

Scale scores are obtained by adding items scores for each of the seven items in a scale. Items are arranged in cyclic order so that the first, second, third, fourth, fifth, sixth, seventh, and eighth item, respectively, in each block measures, Student Support, Affiliation, Professional Interest, Staff Freedom, Participatory Decision Making, Innovation, and Research Adequacy.

School-Level Environment Questionnaire (SLEQ)

References

Ainley, J. G. (1978). Science facilities and variety in science teaching. *Research in Science Education, 8,* 99–109.

Anderson, C. S. (1982). The search for school climate: A review of the research. *Review of Educational Research, 52,* 368–420.

Anderson, G. J. (1970). Effects of classroom social climate on individual learning. *American Educational Research Journal, 7,* 135–152.

Anderson, G. J. (1971). Effects of course content and teacher sex on the social climate of learning. *American Educational Research Journal, 8,* 649–663.

Anderson, G. J., & Walberg, H. J. (1968). Classroom climate and group learning. *International Journal of Educational Sciences, 2,* 175–180.

Anderson, G. J., & Walberg, H. J. (1972). Class size and the social environment of learning: A replication. *Alberta Journal of Educational Research, 18,* 277–286.

Anderson, G. J., & Walberg, H. J. (1974). Learning environments. In H. J. Walberg (Ed.), *Evaluating educational performance: A sourcebook of methods, instruments, and examples.* Berkeley, CA: McCutchan.

Anderson, G. J., Walberg, H. J., & Welch, W. W. (1969). Curriculum effects on the social climate of learning: A new representation of discriminant functions. *American Educational Research Journal, 6,* 315–328.

Astin, A. W. (1965). Classroom environment in different fields of study. *Journal of Educational Psychology, 56,* 275–282.

Astin, A. W., & Holland, J. L. (1961). The environmental assessment technique: A way to measure college environments. *Journal of Educational Psychology, 52,* 308–316.

Baba, T. L., & Fraser, B. J. (1983). Student attitudes to UNDP Social Science Curriculum in Fiji—personal and environmental influences. *International Review of Education, 29,* 465–483.

Barker, R. G., & Gump P. V. (1964). *Big school small school: High school size and student behavior.* Stanford, CA: Stanford University Press.

Brekelmans, M., Wubbels, T., & Créton, H. (1990). A study of student perceptions of physics teacher behavior. *Journal of Research in Science Teaching, 27,* 335–350.

Brookover, W. B., Beady, C., Flood, P., Schweitzer, J., & Wisenbaker, J. (1979). *School social systems and student achievement: Schools can make a difference.* New York: Praeger.

Brookover, W. B., & Schneider, J. (1975). Academic environments and elementary school achievement. *Journal of Research and Development in Education, 9,* 82–91.

Brookover, W. B., Schweitzer, J. H., Schneider, J. M., Beady, C. H., Flood, P. K., & Wisenbaker, J. M. (1978). Elementary school social climate and school achievement. *American Educational Research Journal, 15,* 301–318.

Burden, R., & Fraser, B. J. (1991, April). *Use of classroom environment assessments in school psychology: A British perspective.* Paper presented at annual meeting of American Educational Research Association, Chicago.

Burden, R., & Hornby, T. A. (1988). Assessing classroom ethos: Some recent promising developments for the systems oriented educational psychologist. *Educational Psychology in Practice, 5,* 17–22.

Burstein, L. (1978). Assessing differences between grouped and individual-level regression coefficients: Alternative approaches. *Sociological Methods and Research, 7,* 5–28.

Burstein, L., Linn, R. L., & Capell, F. J. (1978). Analyzing multilevel data in the presence of heterogeneous within-class regressions. *Journal of Educational Statistics, 3,* 347–383.

Byrne, D. B., Hattie, J. A., & Fraser, B. J. (1986). Student perceptions of preferred classroom learning environment. *Journal of Educational Research, 81,* 10–18.

Chatiyanonda, S. (1978). *An evaluation of the IPST physics curriculum in Thailand.* Unpublished doctoral thesis, Monash University.

Chavez, R. C. (1984). The use of high inference measures to study classroom climates: A review. *Review of Educational Research, 54,* 237–261.

Cheung, K. C., Keeves, J. P., Sellin, N., & Tsoi, S. C. (1990). The analysis of multilevel data in educational research: Studies of problems and their solutions. *International Journal of Educational Research, 14*(3), 215–319 (whole issue).

Cook, T. D., & Reichardt, C. S. (Eds.). (1979). *Qualitative and quantitative methods in evaluation research.* Beverly Hills, CA: Sage.

Corno, L., Mitman, A., & Hedges, L. (1981). The influence of direct instruction on student self-appraisals: A hierarchical analysis of treatment and aptitude-treatment interaction effects. *American Educational Research Journal, 18,* 39–61.

Cort, H. R., Jr. (1979). A social studies evaluation. In H. J. Walberg (Ed.), *Educational environments and effects: Evaluation, policy, and productivity* (pp. 235–257). Berkeley, CA: McCutchan.

Costello, R. W. (1988). Relationships among ability grouping, classroom climate and academic achievement in mathematics and English classes. In B. J. Fraser (Ed.), *The study of learning environments* (Vol. 3; pp. 60–67). Perth: Curtin University of Technology.

Creemers, B., Peters, T., & Reynolds, D. (Eds.). (1989). *School effectiveness and school improvement.* Lisse, The Netherlands: Swets & Zweitlinger.

Créton, H., Hermans, J., & Wubbels, T. (1990). Improving interpersonal teacher behavior in the classroom: A systems communication perspective. *South Pacific Journal of Teacher Education, 18,* 85–94.

Cronbach, L. J. (with assistance of Deken, J. E., and Webb, N.). (1976). *Research on classrooms and schools: Formulation of questions, design, and analysis.* Unpublished paper of Stanford Evaluation Consortium, Stanford University.

Cronbach, L. J., & Snow, R. E. (1977). *Aptitudes and instructional methods: A handbook for research on interactions.* New York: Irvington.

Cronbach, L. J., & Webb, N. (1975). Between-class and within-class effects in a reported Aptitude × Treatment interaction: Reanalysis of a study by G. L. Anderson. *Journal of Educational Psychology, 67,* 717–724.

Deer, C. E. (1980). Measuring organizational climate in secondary schools. *Australian Journal of Education, 24,* 26–43.

DeYoung, A. J. (1977). Classroom climate and class success: A case study at the university level. *Journal of Educational Research, 70,* 252–257.

Docker, J. G., Fraser, B. J., & Fisher, D. L. (1989). Differences in the psychosocial work environment of different types of schools. *Journal of Research in Childhood Education, 4,* 5–17.

Dreesman, H. (1982). Classroom climate: Contributions from a European country. *Studies in Educational Evaluation, 8,* 53–64.

Dunkin, M. J., & Biddle, B. J. (1974). *The study of teaching.* New York: Holt, Rinehart & Winston.

Duschl, R. A., & Waxman, H. C. (1991). Influencing the learning environment of student teaching. In B. J. Fraser & H. J. Walberg (Eds.), *Educational environments: Evaluation, antecedents, and consequences.* (pp. 255–270). London: Pergamon.

Eash, M. J. (1990, *April*). *Changing learning environments in high school science: An evaluation of the results of an NSF workshop.* Paper presented at annual meeting of American Educational Research Association, Boston.

Ellett, C. D., Loup, K. S., & Chauvin, S. W. (1989). *System for Teaching and Learning Assessment and Review (STAR): Annotated guide to teaching and learning.* Baton Rouge: Louisiana Teaching Internship and Teacher Evaluation Projects, College of Education, Louisiana State University.

Ellett, C. D., & Masters, J. A. (1978, August). *Learning environment perception: Teacher and student relations.* Paper presented at Annual Convention of American Psychological Association, Toronto.

Ellettt, C. D., Masters, J. A., & Pool, J. E. (1978, January). *The incremental validity of teacher and student perceptions of school environment characteristics.* Paper presented at annual meeting of Georgia Educational Research Association, Atlanta.

Ellett, C. D., Payne, D. A., Masters, J. A., & Pool, J. E. (1977, September). *The relationship between teacher and student perceptions of school environment dimensions and school outcome variables.* Paper presented at annual meeting of Southeastern Psychological Association, Miami.

Ellett, C. D., & Walberg, H. J. (1979). Principals' competency, environment, and outcomes. In H. J. Walberg (Ed.), *Educational environments and effects: Evaluation, policy, and productivity.* Berkeley, CA: McCutchan.

Erickson, F. (1986). Qualitative research on teaching. In M. C. Wittrock (Ed.), *Handbook of research on teaching* (3rd ed.; pp. 119–161). New York: Macmillan.

Feldlaufer, H., Midgley, C., & Eccles, J. S. (1988). Student, teacher, and observer perceptions of the classroom environment before and after the transition to junior high school. *Journal of Early Adolescence, 8,* 133–156.

Finlayson, D. S. (1973). The school perceptions of teachers of differential status. *Research in Education, 9,* 83–92.

Firestone, W. A. (1987). Meaning in method: The rhetoric of quantitative and qualitative research. *Educational Researcher, 16*(7), 16–21.

Fisher, D. L., & Fraser, B. J. (1981). Validity and use of My Class Inventory. *Science Education, 65,* 145–156.

Fisher, D. L., & Fraser, B. J. (1983a). A comparison of actual and preferred classroom environment as perceived by science teachers and students. *Journal of Research in Science Teaching, 20,* 55–61.

Fisher, D. L., & Fraser, B. J. (1983b, April). *Use of Classroom Environment Scale in investigating effects of psychosocial milieu on science students' outcomes.* Paper presented at annual meeting of National Association for Research in Science Teaching, Dallas.

Fisher, D. L., & Fraser, B. J. (1983c). Use of WES to assess science teachers' perceptions of school environment. *European Journal of Science Education, 5,* 231–233.

Fisher, D. L., & Fraser, B. J. (1983d). Validity and use of Classroom Environment Scale. *Educational Evaluation and Policy Analysis, 5,* 261–271.

Fisher, D. L., & Fraser, B. J. (1991a, April). *Incorporating learning environment ideas into teacher education: An Australian perspective.* Paper presented at annual meeting of American Educational Research Association, Chicago.

Fisher, D. L., & Fraser, B. J. (1991b). School climate and teacher professional development. *South Pacific Journal of Teacher Education, 19,* 17–32.

Fraser, B. J. (1976). Pupil perceptions of the climate of ASEP classrooms. *Australian Science Teachers Journal, 22*(3), 127–128.

Fraser, B. J. (1978). Environmental factors affecting attitude toward different sources of scientific information. *Journal of Research in Science Teaching, 15,* 491–497.

Fraser, B. J. (1979a). Evaluation of a science-based curriculum. In H. J. Walberg (Ed.), *Educational environments and effects: Evaluation, policy, and productivity* (pp. 218–234). Berkeley, CA: McCutchan.

Fraser, B. J. (1979b). *Test of enquiry skills.* Melbourne: Australian Council for Educational Research.

Fraser, B. J. (1980a). *Criterion validity of individualized classroom environment questionnaire.* Report to Education Research and Development Committee, Canberra. (ERIC Document Reproduction Service No. ED 214 961)

Fraser, B. J. (1980b). Guest editor's introduction: Classroom environment research in the 1970's and 1980's. *Studies in Educational Evaluation, 6,* 221–223.

Fraser, B. J. (1981a). Australian research on classroom environment: State of the art. *Australian Journal of Education, 25,* 238–268.

Fraser, B. J. (1981b). *Learning environment in curriculum evaluation: A review.* Evaluation in Education series. Oxford: Pergamon.

Fraser, B. J. (1981c). Predictive validity of an individualized classroom environment questionnaire. *Alberta Journal of Educational Research, 27,* 240–251.

Fraser, B. J. (1981d). *Test of science-related attitudes.* Melbourne: Australian Council for Educational Research.

Fraser, B. J. (1981e). Using environmental assessments to make better classrooms. *Journal of Curriculum Studies, 13,* 131–144.

Fraser, B. J. (1982a). Development of short forms of several classroom environment scales. *Journal of Educational Measurement, 19,* 221–227.

Fraser, B. J. (1982b). Differences between student and teacher perceptions of actual and preferred classroom learning environment. *Educational Evaluation and Policy Analysis, 4,* 511–519.

Fraser, B. J. (1984). Differences between preferred and actual classroom environment as perceived by primary students and teachers. *British Journal of Educational Psychology, 54,* 336–339.

Fraser, B. J. (1986a). *Classroom environment.* London: Croom Helm.

Fraser, B. J. (1986b). Determinants of classroom psychosocial environment: A review. *Journal of Research in Childhood Education, 1,* 5–19.

Fraser, B. J. (Ed.). (1986c). *The study of learning environments* (Vol. 1). Salem, OR: Assessment Research.

Fraser, B. J. (1987a). Classroom learning environment and effective schooling. *Professional School Psychology, 2,* 25–41.

Fraser, B. J. (Ed.). (1987b). *The study of learning environments* (Vol. 2). Perth: Curtin University of Technology.

Fraser, B. J. (1987c). Use of classroom environment assessments in school psychology. *School Psychology International, 8,* 205–219.

Fraser, B. J. (Ed.). (1988). *The study of learning environments* (Vol. 3). Perth: Curtin University of Technology.

Fraser, B. J. (1989a). Improving science teacher education programs through inclusion of research on classroom psychosocial environments. In J. Barufaldi & M. Druver (Eds.), *Improving preservice/inservice science teacher education: Future perspectives* (pp. 25–53). 1987 Yearbook of the Association for the Education of Teachers of Science. Columbus, OH: SMEAC Information Reference Center and ERIC Clearinghouse for Science, Mathematics and Environmental Education.

Fraser, B. J. (1989b). Twenty years of classroom climate work: Progress and prospect. *Journal of Curriculum Studies, 21,* 307–327.

Fraser, B. J. (1990). *Individualised Classroom Environment Questionnaire.* Melbourne: Australian Council for Educational Research.

Fraser, B. J. (1991). Two decades of classroom environment research. In B. J. Fraser & H. J. Walberg (Eds.), *Educational environments: Evaluation, antecedents, and consequences* (pp. 3–27). London: Pergamon.

Fraser, B. J., Anderson, G. J., & Walberg, H. J. (1982). *Assessment of learning environments: Manual for Learning Environment Inventory (LEI) and My Class Inventory (MCI)* (third version). Perth: Western Australian Institute of Technology.

Fraser, B. J., & Butts, W. L. (1982). Relationship between perceived levels of classroom individualization and science-related attitudes. *Journal of Research in Science Teaching, 19,* 143–154.

Fraser, B. J., & Deer, C. E. (1983). Improving classrooms through use of information about learning environment. *Curriculum Perspectives, 3*(2), 44–46.

Fraser, B. J., Docker, J. G., & Fisher, D. L. (1988). Assessing and improving school climate. *Evaluation and Research in Education, 2*(3), 109–122.

Fraser, B. J., & Fisher, D. L. (1982a). Effects of classroom psychosocial environment on student learning. *British Journal of Educational Psychology, 52,* 374–377.

Fraser, B. J., & Fisher, D. L. (1982b). Predicting students' outcomes from their perceptions of classroom psychosocial environment. *American Educational Research Journal, 19,* 498–518.

Fraser, B. J., & Fisher, D. L. (1982c). Predictive validity of My Class Inventory. *Studies in Educational Evaluation, 8,* 129–140.

Fraser, B. J., & Fisher, D. L. (1983a). *Assessment of classroom psychosocial environment: Workshop manual.* Research Seminar and Workshop Series, Perth: Western Australian Institute of Technology.

Fraser, B. J., & Fisher, D. L. (1983b). Development and validation of short forms of some instruments measuring student perceptions of actual and preferred classroom learning environment. *Science Education, 67,* 115–131.

Fraser, B. J., & Fisher, D. L. (1983c). Student achievement as a function of person-environment fit: A regression surface analysis. *British Journal of Educational Psychology, 53,* 89–99.

Fraser, B. J., & Fisher, D. L. (1983d). Use of actual and preferred classroom environment scales in person-environment fit research. *Journal of Educational Psychology, 75,* 303–313.

Fraser, B. J., & Fisher, D. L. (1986). Using short forms of classroom climate instruments to assess and improve classroom psychosocial environment. *Journal of Research in Science Teaching, 5,* 387–413.

Fraser, B. J., Giddings, G. J., & McRobbie, C. J. (in press). Evolution and validation of a personal form of an instrument for assessing science laboratory classroom environments. *Journal of Research in Science Teaching.*

Fraser, B. J., Nash, R., & Fisher, D. L. (1983). Anxiety in science classrooms: Its measurement and relationship to classroom environment. *Research in Science and Technological Education, 1,* 201–208.

Fraser, B. J., & O'Brien, P. (1985). Student and teacher perceptions of the environment of elementary-school classrooms. *Elementary School Journal, 85,* 567–580.

Fraser, B. J., Pearse, R., & Azmi (1982). A study of Indonesian students' perceptions of classroom psychosocial environment. *International Review of Education, 28,* 337–355.

Fraser, B. J., Rennie, L. J., & Tobin, K. (1990). The learning environment as a focus in a study of higher-level cognitive learning. *International Journal of Science Education, 12,* 531–548.

Fraser, B. J., & Rentoul, A. J. (1980). Person-environment fit in open classrooms. *Journal of Educational Research, 73,* 159–167.

Fraser, B. J., & Rentoul, A. J. (1982). Relationship between school-level and classroom-level environment. *Alberta Journal of Educational Research, 28,* 212–225.

Fraser, B. J., Seddon, T., & Eagleson, J. (1982). Use of student perceptions in facilitating improvement in classroom environment. *Australian Journal of Teacher Education, 7,* 31–42.

Fraser, B. J., & Tobin, K. (1989). Student perceptions of psychosocial environments in classrooms of exemplary science teachers. *International Journal of Science Education, 11,* 14–34.

Fraser, B. J., & Tobin, K. (1991). Combining qualitative and quantitative methods in classroom environment research. In B. J. Fraser & H. J. Walberg (Eds.), *Educational environments: Evaluation, antecedents, consequences* (pp. 271–292). London: Pergamon.

Fraser, B. J., & Treagust, D. F. (1986). Validity and use of an instrument for assessing classroom psychosocial environment in higher education. *Higher Education, 15,* 37–57.

Fraser, B. J., Treagust, D. F., & Dennis, N. C. (1986). Development of an instrument for assessing classroom psychosocial environment at universities and colleges. *Studies in Higher Education, 11,* 43–54.

Fraser, B. J., & Walberg, H. J. (1981). Psychosocial learning environment in science classrooms: A review of research. *Studies in Science Education, 8,* 67–92.

Fraser, B. J., & Walberg, H. J. (Eds.). (1991). *Educational environments: Evaluation, antecedents, and consequences.* London: Pergamon.

Fraser, B. J., Walberg, H. J., Welch, W. W., & Hattie, J. A. (1987). Syntheses of educational productivity research. *International Journal of Educational Research, 11*(2), 145–252 (whole issue).

Fraser, B. J., Welch, W. W., & Walberg, H. J. (1986). Using secondary analysis of national assessment data to identify predictors of junior high school students' outcomes. *Alberta Journal of Educational Research, 32,* 37–50.

Fraser, B. J., Williamson, J., & Lake, J. (1988, April). *Combining quantitative and qualitative methods in the evaluation of an alternative to conventional high schools.* Paper presented at annual meeting of American Educational Research Association, New Orleans.

Fraser, B. J., Williamson, J. C., & Tobin, K. (1987). Use of classroom and school climate scales in evaluating alternative high schools. *Teaching and Teacher Education, 3,* 219–231.

Galluzzi, E. G., Kirby, E. A., & Zucker, K. B. (1980). Students' and teachers' perceptions of classroom environment and self and others-concepts. *Psychological Reports, 46,* 747–753.

Gardner, P. L. (1974). Pupil personality, teacher behaviour and attitudes to a physics course. In P. W. Musgrave (Ed.), *Contemporary Studies in the Curriculum* (pp. 173–199). Sydney: Angus & Robertson.

Gardner, P. L. (1976). Attitudes toward physics: Personal and environmental influences. *Journal of Research in Science Teaching, 13,* 111–125.

Genn, J. M. (1981). The climates of teaching and learning that Australian university teachers establish in their undergraduate classes. In R. Wellard (Ed.), *Research and development in higher education* (Vol. 4; pp. 78–89). Sydney: Higher Education Research and Development Society of Australasia.

Genn, J. M. (1984). Research into the climates of Australian schools, colleges and universities: Contributions and potential of need-press theory. *Australian Journal of Education, 28,* 227–248.

Genn, J. M. (1988). Review of "Classroom Environment." *South Pacific Journal of Teacher Education, 16,* 57–60.

Giddings, G. J., & Fraser, B. J. (1990, April). *Cross-national development, validation and use of an instrument for assessing the environment of science laboratory classes.* Paper presented at annual meeting of Annual Educational Research Association, Boston.

Goldstein, H. (1987). *Multilevel models in educational and social research.* London: Charles Griffin.

Haertel, G. D., Walberg, H. J., & Haertel, E. H. (1981). Socio-psychological environments and learning: A quantitative synthesis. *British Educational Research Journal, 7,* 27–36.

Haladyna, T., Olsen, R., & Shaughnessy, J. (1982). Relations of student, teacher and learning environment variables to attitudes toward science. *Science Education, 66,* 671–688.

Haladyna T., Shaughnessy, J., & Redsun, A. (1982a). Correlates of attitudes toward social studies. *Theory and Research in Social Education, 10,* 1–26.

Haladyna, T., Shaughnessy, J., & Redsun, A. (1982b). Relations of student, teacher, and learning environment variables to attitudes toward social studies. *Journal of Social Studies Research, 6,* 36–44.

Haladyna, T., Shaughnessy, J., & Shaughnessy, J. M. (1983). A causal analysis of attitude toward mathematics. *Journal of Research in Mathematics Education, 14,* 19–29.

Halpin, A. W., & Croft, D. B. (1963). *Organizational climate of schools.* Chicago: Midwest Administration Center, University of Chicago.

Harty, H., & Hassan, H. A. (1983). Student control ideology and the science classroom environment in urban secondary schools of Sudan. *Journal of Research in Science Teaching, 20,* 851–859.

Hearn, J. C., & Moos, R. H. (1978). Subject matter and classroom climate: A test of Holland's environmental propositions. *American Educational Research Journal, 15,* 111–124.

Hofstein, A., Gluzman, R., Ben-Zvi, R., & Samuel, D. (1979). Classroom learning environment and attitudes towards chemistry. *Studies in Educational Evaluation, 5,* 231–236.

Hofstein, A., Gluzman, R., Ben-Zvi, R., & Samuel, D. (1980). A comparative study of chemistry students' perceptions of the learning environment in high schools and vocational schools. *Journal of Research in Science Teaching, 17,* 547–552.

Hofstein, A., & Lazarowitz, R. (1986). A comparison of the actual and preferred classroom learning environment in biology and chemistry as perceived by high school students. *Journal of Research in Science Teaching, 23,* 189–199.

Holsinger, D. B. (1972). *The elementary school as an early socializer of modern values: A Brazilian study.* Unpublished doctoral dissertation, Stanford University.

Holsinger, D. B. (1973). The elementary school as modernizer: A Brazilian study. *International Journal of Comparative Sociology, 14,* 180–202.

Howe, K. (1988). Against the quantitative-qualitative incompatibility thesis: Or dogmas die hard. *Educational Researcher, 17*(8), 10–16.

Humphrey, L. L. (1984). Children's self-control in relation to perceived social environment. *Journal of Personality and Social Psychology, 46,* 178–188.

Hunt, D. E. (1975). Person-environment interaction: A challenge found wanting before it was tried. *Review of Educational Research, 45,* 209–230.

Jessor, R., & Jessor, S. L. (1973). The perceived environment in behavioral science. *American Behavioral Scientist, 16,* 801–828.

Johnson, D. W., & Johnson, R. T. (1987). *Learning together and alone: Cooperative, competitive, and individualistic learning* (2nd ed.). Englewood Cliffs, NJ: Prentice Hill.

Johnson, D. W., & Johnson, R. T. (1991). Cooperative learning and classroom and school climate. In B. J. Fraser & H. J. Walberg (Eds.), *Educational environments: Evaluation, antecedents, and consequences* (pp. 55–74). London: Pergamon.

Johnson, D. W., Johnson, R. T., Johnson Holubec, E., & Roy, P. (1986). *Circles of learning: Cooperation in the classroom* (2nd ed.). Alexandria, VA: Association for Supervision and Curriculum Development.

Johnson, D. W., Maruyama, G., Johnson R., Nelson, D., & Skon, L. (1981). Effects of cooperative, competitive, and individualistic goal structures on achievement: A meta-analysis. *Psychological Bulletin, 89,* 47–62.

Jorde-Bloom, P. (1991). Organizational climate in child care settings. In B. J. Fraser & H. J. Walberg (Eds.), *Educational environments: Evaluation, antecedents, and consequences* (pp. 161–175). London: Pergamon.

Kelly, A. (1980). Exploration and authority in science learning environments: An international study. *European Journal of Science Education, 2,* 161–174.

Keyser, V., & Barling, J. (1981). Determinants of children's self-efficacy beliefs in an academic environment. *Cognitive Therapy and Research, 5,* 29–40.

Kuert, W. P. (1979). Curricular structure. In H. J. Walberg (Ed.), *Educational environments and effects: Evaluation, policy, and productivity* (pp. 180–199). Berkeley, CA: McCutchan.

Kuhlemeier, H. (1983). Vergelijkend onderzoek naar de perceptie van de leeromgeving in PLON-onderwijs en regulier natuurkunde-onderwijs. *Tijdschrift Voor Onderwijsresearch, 8,* 1–16. (An educational research journal published in The Netherlands.)

Larkin, A. I., & Keeves, J. P. (1984). *The class size question: A study at different levels of analysis.* Melbourne: Australian Council for Educational Research.

Lawrenz, F. (1976). The prediction of student attitude toward science from student perception of the classroom learning environment. *Journal of Research in Science Teaching, 13,* 509–515.

Lawrenz, F. (1988). Prediction of student energy knowledge and attitudes. *School Science and Mathematics, 88,* 543–549.

Lawrenz, F., & Munch, T. W. (1984). The effect of grouping of laboratory students on selected educational outcomes. *Journal of Research in Science Teaching, 21,* 699–708.

Lawrenz, F. P., & Welch, W. W. (1983). Student perceptions of science classes taught by males and females. *Journal of Research in Science Teaching, 20,* 655–662.

Leary, T. F. (1957). *An interpersonal diagnosis of personality.* New York: Ronald.

Levin, T. (1980). Classroom climate as criterion in evaluating individualized instruction in Israel. *Studies in Educational Evaluation, 6,* 291–292.

Lewin, K. (1936). *Principles of topological psychology.* New York: McGraw.

Lincoln, J. R., & Zeitz, G. (1980). Organizational properties from aggregate data: Separating individual and structural effects. *American Sociological Review, 45,* 391–408.

Luecht, R. M., & Smith, P. L. (1989, April). *The effects of bootstrapping strategies on the estimation of variance components.* Paper presented at annual meeting of American Educational Research Association, San Francisco.

Marjoribanks, K. (1991). Families, schools, and students' educational outcomes. In B. J. Fraser & H. J. Walberg (Eds.), *Educational environments: Evaluation, antecedents, and consequences* (pp. 75–91). London: Pergamon.

Midgley, C., Eccles, J. S., & Feldlaufer, H. (1991). Classroom environment and the transition to junior high school. In B. J. Fraser & H. J. Walberg (Eds.), *Educational environments: Evaluation, antecedents, and consequences* (pp. 113–139). London: Pergamon.

Midgley, C., Feldlaufer, H., & Eccles, J. S. (1989). Student/teacher relations and attitudes toward mathematics before and after the transition to junior high school. *Child Development, 60*(4), 981–982.

Moos, R. H. (1968). The assessment of the social climates of correctional institutions. *Journal of Research in Crime and Delinquency, 5,* 174–188.

Moos, R. H. (1973). Conceptualizations of human environments. *American Psychologist, 28,* 652–665.

Moos, R. H. (1974). *The Social Climate Scales: An overview.* Palo Alto, CA: Consulting Psychologists Press.

Moos, R. H. (1978). A typology of junior high and high school classrooms. *American Educational Research Journal, 15,* 53–66.

Moos, R. H. (1979a). *Evaluating educational environments: Procedures, measures, findings and policy implications.* San Francisco: Jossey-Bass.

Moos, R. H. (1979b). Improving social settings by climate measurement and feedback. In R. F. Munoz, L. R. Snowden, & J. G. Kelly (Eds.), *Social and psychological research in community settings* (pp. 145–170). San Francisco: Jossey-Bass.

Moos, R. H. (1980). Evaluating classroom learning environments. *Studies in Educational Evaluation, 6,* 239–252.

Moos, R. H. (1981). *Manual for work environment scale.* Palo Alto, CA: Consulting Psychologist Press.

Moos, R. H. (1991). Connections between school, work, and family settings. In B. J. Fraser & H. J. Walberg (Eds.), *Educational environments: Evaluation, antecedents, and consequences* (pp. 29–53). London: Pergamon.

Moos, R. H., & Houts, P. S. (1968). The assessment of the social atmospheres of psychiatric wards. *Journal of Abnormal Psychology, 73,* 595–604.

Moos, R. H., & Moos, B. S. (1978). Classroom social climate and student absences and grades. *Journal of Educational Psychology, 70,* 263–269.

Moos, R. H., & Spinrad, S. (1984). *The Social Climate Scales: Annotated bibliography 1979–1983.* Palo Alto, CA: Consulting Psychologists Press.

Moos, R. H., & Trickett, E. J. (1974). *Classroom Environment Scale manual* (1st ed.). Palo Alto, CA: Consulting Psychologists Press.

Moos, R. H., & Trickett, E. J. (1987). *Classroom Environment Scale manual* (2nd ed.). Palo Alto, CA: Consulting Psychologists Press.

Mosteller, F., & Tukey, J. W. (1977). *Data analysis and regression: A second course in statistics.* Reading, MA: Addison-Wesley.

Munby, H. (1989, April). *Reflection-in-action and reflection-on-action.* Paper presented at annual meeting of American Educational Research Association, San Francisco.

Murray, H. A. (1938). *Explorations in personality.* New York: Oxford University Press.

National Commission on Excellence in Education. (1983). *A nation at risk: The imperative for educational reform.* Washington, DC: U.S. Government Printing Office.

Northfield, J. R. (1976). The effect of varying the mode of presentation of ASEP materials on pupils' perceptions of the classroom. *Research in Science Education, 6,* 63–71.

O'Reilly, R. (1975). Classroom climate and achievement in secondary school mathematics classes. *Alberta Journal of Educational Research, 21,* 241–248.

Owens, L. (1981, November). *The cooperative, competitive, and individualized learning preferences of primary and secondary teachers in Sydney.* Paper presented at the annual meeting of the Australian Association for Research in Education, Adelaide, South Australia.

Owens, L. C., & Straton, R. G. (1980). The development of a cooperative, competitive and individualized learning preference scale for students. *British Journal of Educational Psychology, 50,* 147–161.

Pace, C. R., & Stern, G. G. (1958). An approach to the measurement of psychological characteristics of college environments. *Journal of Educational Psychology, 49,* 269–277.

Paige, R. M. (1978). *The impact of classroom learning environment on academic achievement and individual modernity in East Java, Indonesia.* Unpublished doctoral dissertation, Stanford University.

Paige, R. M. (1979). The learning of modern culture: Formal education and psychosocial modernity in East Java, Indonesia. *International Journal of Intercultural Relations, 3,* 333–364.

Payne, D. A., Ellett, C. A., Perkins, M. L., Klein, A. E., & Shellinberger, S. (1974–75). *The verification and validation of principal competencies and performance indicators.* Unpublished final report of Results Oriented Management in Education (R.O.M.E.) Project, University of Georgia.

Peckham, P. D., Glass, G. V., & Hopkins, K. D. (1969). The experimental unit in statistical analysis: Comparative experiments with intact groups. *Journal of Special Education, 3,* 337–349.

Penick, J. E., & Yager, R. E. (1983). The search for excellence in science education. *Phi Delta Kappan, 64,* 621–623.

Perkins, M. L. (1978, April). *Predicting student performance from teachers' perceptions of the school environment.* Paper presented at the annual meeting of the American Educational Research Association, Toronto.

Persaud, G. (1976). *School authority pattern and students' social development in selected primary schools in Jamaica.* Unpublished doctoral dissertation, Stanford University.

Power, C. N., & Tisher, R. P. (1975, November). *Variations in the environment of self-paced science classrooms: Their nature, determinants, and effects.* Paper presented at the annual conference of the Australian Association for Research in Education, Adelaide, South Australia.

Power, C. N., & Tisher, R. P. (1979). A self-paced environment. In H. J. Walberg (Ed.), *Educational environments and effects: Evaluation, policy, and productivity* (pp. 200–217). Berkeley, CA: McCutchan.

Randhawa, B. S., & Fu, L. L. W. (1973). Assessment and effect of some classroom environment variables. *Review of Educational Research, 43,* 303–321.

Randhawa, B. S., & Michayluk, J. O. (1975). Learning environment in rural and urban classrooms. *American Educational Research Journal, 12,* 265–279.

Raudenbush, S. (1988). Educational applications of hierarchical linear models: A review. *Journal of Educational Statistics, 13*(2), 85–116.

Raudenbush, S., & Bryk, A. S. (1986). A hierarchical model of studying school effects. *Sociology of Education, 59,* 1–17.

Raviv, A., Raviv, A., & Reisel, E. (1990). Teacher and students: Two different perspectives?! Measuring social climate in the classroom. *American Educational Research Journal, 27,* 141–157.

Rentoul, A. J., & Fraser, B. J. (1979). Conceptualization of enquiry-based or open classrooms learning environments. *Journal of Curriculum Studies, 11,* 233–245.

Rentoul, A. J., & Fraser, B. J. (1980). Predicting learning from classroom individualization and actual-preferred congruence. *Studies in Educational Evaluation, 6,* 265–277.

Rentoul, A. J., & Fraser, B. J. (1981). Changes in beginning teachers' attitudes towards individualized teaching approaches during the first year of teaching. *Australian Journal of Teacher Education, 6,* 1–13.

Rentoul, A. J., & Fraser, B. J. (1983). Development of a school-level environment questionnaire. *Journal of Educational Administration, 21,* 21–39.

Robinson, W. S. (1950). Ecological correlations and the behavior of individuals. *American Sociological Review, 15,* 351–357.

Rosenshine, B. (1970). Evaluation of classroom instruction. *Review of Educational Research, 40,* 279–300.

Rosenshine, B., & Furst, N. (1973). The use of direct observation to study teaching. In R. M. W. Travers (Ed.), *Second handbook of research on teaching* (pp. 122–183). Chicago: Rand-McNally.

Ross, K. N. (1978). *Sample design for educational survey research.* (Evaluation in Education series.) Oxford: Pergamon.

Rutter, M., Maughan, B., Mortimore, P., Ouston, J., & Smith, A. (1979). *Fifteen thousand hours: Secondary schools and their effects on children.* Cambridge, MA: Harvard University Press.

Sadker, D., & Sadker, M. (1985). Is the O.K. classroom O.K.? *Phi Delta Kappan, 55,* 358–361.

Sanford, J. P. (1987). Management of science classroom tasks and effects on students' learning opportunities. *Journal of Research in Science Teaching, 24,* 249–265.

Schibeci, R. A., Rideng, I. M., & Fraser, B. J. (1987). Effects of classroom environment on science attitudes: A cross-cultural replication in Indonesia. *International Journal of Science Education, 9,* 169–186.

Schon, D. A. (1983). *The reflective practitioner: How professionals think in action.* New York: Basic Books.

Sharan, S., & Yaakoby, D. (1981). Classroom learning environment of city and kibbutz biology classrooms in Israel. *European Journal of Science Education, 3,* 321–328.

Shaw, A. R., & Mackinnon, P. (1973). *Evaluation of the learning environment.* Unpublished paper, Lord Elgin High School, Burlington, Ontario.

Sirotnik, K. A. (1980). Psychometric implications of the unit-of-analysis problem (with examples from the measurement of organizational climate). *Journal of Educational Measurement, 17,* 245–282.

Slavin, R. E. (1983a). *Cooperative learning.* New York: Longman.

Slavin, R. E. (1983b). When does cooperative learning increase student achievement? *Psychological Bulletin, 94,* 429–445.

Smith, D. L., & Fraser, B. J. (1980). Towards a confluence of quantitative and qualitative approaches in curriculum evaluation. *Journal of Curriculum Studies, 12,* 367–370.

Stake, R. E., & Easley, J. A., Jr. (1978). *Case studies in science education.* Urbana: University of Illinois.

Steele, J. M., House, E. R., & Kerins, T. (1971). An instrument for assessing instructional climate through low-inference student judgments. *American Educational Research Journal, 8,* 447–466.

Steele, J. M., Walberg, H. J., & House, E. R. (1974). Subject areas and cognitive press. *Journal of Educational Psychology, 66,* 363–366.

Stern, G. G. (1970). *People in context: Measuring person-environment congruence in education and industry.* New York: Wiley.

Stern, G. G., Stein, M. I., & Bloom, B. S. (1956). *Methods in personality assessment.* Glencoe, IL: Free Press.

Talton, E. L. (1983, April). *Peer and classroom influences on attitudes toward science and achievement in science among tenth grade biology students.* Paper presented at the annual meeting of the National Association for Research in Science Teaching, Dallas.

Taylor, P. C., & Fraser, B. J. (1991, April). *Development of an instrument for assessing constructivist learning environments.* Paper presented at the annual meeting of the American Educational Research Association, Chicago.

Thistlethwaite, D. L. (1962). Fields of study and development of motivation to seek advanced training. *Journal of Educational Psychology, 53,* 53–64.

Thomas, A. R. (1976). The organizational climate of schools. *International Review of Education, 22,* 441–463.

Tisher, R. P., & Power, C. N. (1976). Variations between ASEP and conventional learning environments. *Australian Science Teachers Journal, 22,* 35–39.

Tisher, R. P., & Power, C. N. (1978). The learning environment associated with an Australian curriculum innovation. *Journal of Curriculum Studies, 10,* 169–184.

Tobin, K. (1987a). A comparison of exemplary and non-exemplary teachers of science and mathematics. In K. Tobin & B. J. Fraser (Eds.), *Exemplary practice in science and mathematics education* (pp. 15–27). Perth: Curtin University of Technology.

Tobin, K. (1987b). Target student involvement in high school science. *International Journal of Science Education, 10,* 317–330.

Tobin, K., & Fraser, B. J. (Eds.). (1987). *Exemplary practice in science and mathematics education.* Perth: Curtin University of Technology.

Tobin, K., & Fraser, B. J. (1989). Barriers to higher-level cognitive learning in high school science. *Science Education, 73,* 659–682.

Tobin, K., & Fraser, B. J. (1990). What does it mean to be an exemplary science teacher? *Journal of Research in Science Teaching, 27,* 3–25.

Tobin, K., & Fraser, B. J. (1991). Learning from investigations of exemplary teachers. In H. C. Waxman & H. J. Walberg (Eds.), *Contemporary research on teaching* (pp. 217–236). Yearbook of the National Society for the Study of Education (NSSE). Berkeley, CA: McCutchan.

Tobin, K., & Gallagher, J. J. (1987a). The role of target students in the science classroom. *Journal of Research in Science Teaching, 24,* 61–75.

Tobin, K., & Gallagher, J. J. (1987b). What happens in high school science classrooms? *Journal of Curriculum Studies, 19,* 549–560.

Tobin, K., Kahle, J. B., & Fraser, B. J. (Eds.). (1990). *Windows into science classes: Problems associated with higher-level cognitive learning.* London: Falmer.

Tobin, K., & Malone, J. (1989). Differential student participation in whole-class activities. *Australian Journal of Education, 33,* 320–331.

Tobin, K., Treagust, D. F., & Fraser, B. J. (1988). An investigation of exemplary biology teaching. *American Biology Teacher, 50,* 142–147.

Trickett, E. J. (1978). Toward a social-ecological conception of adolescent socialization: Normative data on contrasting types of public school classrooms. *Child Development, 49,* 408–414.

Trickett, E. J., & Moos, R. H. (1973). Social environment of junior high and high school classrooms. *Journal of Educational Psychology, 65,* 93–102.

Trickett, E. J., & Moos, R. H. (1974). Personal correlates of contrasting environments: Student satisfactions in high school classrooms. *American Journal of Community Psychology, 2,* 1–12.

Trickett, P. K., Pendry, C., & Trickett, E. J. (1976). *A study of women's secondary education: The experience and effects of attending independent secondary schools.* Unpublished report, Department of Psychology, Yale University.

Trickett, E. J., Trickett, P. K., Castro, J. J., & Schaffner, P. (1982). The independent school experience: Aspects of normative environments of single sex and coed secondary schools. *Journal of Educational Psychology, 74,* 374–381.

van der Sijde, P. C., & van de Grift, W. (Eds.). (in press). *School en Klasklimaat.* Almere, The Netherlands: Versluys.

von Glasersfeld, E. (1981). The concepts of adaption and viability in a radical constructivist theory of knowledge. In I. E. Sigel, D. M. Brodzinsky, & R. M. Golinkoff (Eds.), *New directions in Piagetian theory and practice* (pp. 87–95). Hillsdale, NJ: Lawrence Erlbaum.

von Glasersfeld, E. (1988). The reluctance to change a way of thinking. *Irish Journal of Psychology, 9*(1), 83–90.

von Saldern, M. (1984). Kurt Lewin's influence on social emotional climate research in Germany and the United States. In E. Stivers & S. Wheelan (Eds.), *Proceedings of the First International Kurt Lewin Conference* (pp. 30–39). Berlin, Germany: Springer-Verlag.

Walberg H. J. (1968). Teacher personality and classroom climate. *Psychology in the Schools, 5,* 163–169.

Walberg, H. J. (1969a). Class size and the social environment of learning. *Human Relations, 22,* 465–475.

Walberg, H. J. (1969b). Predicting class learning: An approach to the class as a social system. *American Educational Research Journal, 6,* 529–542.

Walberg, H. J. (1969c). The social environment as mediator of classroom learning. *Journal of Educational Psychology, 60,* 443–448.

Walberg, H. J. (1972). Social environment and individual learning: A test of the Bloom model. *Journal of Educational Psychology, 63,* 69–73.

Walberg, H. J. (1974). Educational process evaluation. In M. W. Apple, M. J. Subkoviak, & H. S. Lufler, Jr. (Eds.), *Educational evaluation: Analysis and responsibility* (pp. 237–268). Berkeley, CA: McCutchan.

Walberg, H. J. (1976). The psychology of learning environments: Behavioral, structural, or perceptual? *Review of Research in Education, 4,* 142–178.

Walberg, H. J. (Ed.). (1979). *Educational environments and effects: Evaluation, policy, and productivity.* Berkeley, CA: McCutchan.

Walberg, H. J. (1981). A psychological theory of educational productivity. In F. H. Farley & N. J. Gordon (Eds.), *Psychology and education: The state of the union.* (pp. 81–108). Berkeley, CA: McCutchan.

Walberg, H. J. (1984). Improving the productivity of America's schools. *Educational Leadership, 41,* 19–30.

Walberg, H. J. (1986). Synthesis of research on teaching. In M. C. Wittrock (Ed.), *Handbook of research on teaching* (3rd. ed.; pp. 214–229). Washington, DC: American Educational Research Association.

Walberg, H. J., & Ahlgren, A. (1970). Predictors of the social environment of learning. *American Educational Research Journal, 7,* 153–167.

Walberg, H. J., & Anderson, G. J. (1968a). Classroom climate and individual learning. *Journal of Educational Psychology, 59,* 414–419.

Walberg, H. J., & Anderson, G. J. (1968b). The achievement-creativity dimension of classroom climate. *Journal of Creative Behavior, 2,* 281–291.

Walberg, H. J., & Anderson, G. J. (1972). Properties of the achieving urban class. *Journal of Educational Psychology, 63,* 381–385.

Walberg, H. J., Fraser, B. J., & Welch, W. W. (1986). A test of a model of educational productivity among senior high school students. *Journal of Educational Research, 79,* 133–139.

Walberg, H. J., & Haertel, G. D. (1980). Validity and use of educational environment assessments. *Studies in Educational Evaluation, 6*(3), 225–238.

Walberg, H. J., Singh, R., & Rasher, S. P. (1977). Predictive validity of student perceptions: A cross-cultural replication. *American Educational Research Journal, 14,* 45–49.

Walberg, H. J., Sorenson, J., & Fishbach, T. (1972). Ecological correlates of ambience in the learning environment. *American Educational Research Journal, 9,* 139–148.

Watzlawick, P. (1978). *The language of change.* New York: Basic Books.

Waxman, H. C., & Ellett, C. D. (Eds.). (1990). *The study of learning environments* (Vol. 4). Houston, TX: University of Houston.

Weinstein, C. S. (1979). The physical environment of the school: A review of research. *Review of Educational Research, 49,* 577–610.

Welch, W. W. (1979). Curricular and longitudinal effects on learning environments. In H. J. Walberg (Ed.), *Educational environments and effects: Evaluation, policy, and productivity* (pp. 167–179). Berkeley, CA: McCutchan.

Welch, W. W., & Walberg, H. J. (1972). A national experiment in curriculum evaluation. *American Educational Research Journal, 9,* 373–383.

Wierstra, R. (1984). A study on classroom environment and on cognitive and affective outcomes of the PLON-curriculum. *Studies in Educational Evaluation, 10,* 273–282.

Wierstra, R. F. A., Jorg, T. G. D., & Wubbells, T. (1987). Contextual and individually perceived learning environment in curriculum evaluation. In B. J. Fraser (Ed.), *The study of learning environments* (Vol. 2; pp. 31–41). Perth: Curtin University of Technology.

Wolf, B. (1983). On the assessment of the learning environment. *Studies in Educational Evaluation, 9,* 253–265.

Wubbels, T., Brekelmans, M., & Hermans, J. (1988). Teacher behavior: An important aspect of the learning environment? In B. J. Fraser (Ed.), *The study of learning environments* (Vol. 3; pp. 10–25). Perth: Curtin University of Technology.

Wubbels, T., Brekelmans, M., & Hooymayers, H. (1991). Interpersonal teacher behavior in the classroom. In B. J. Fraser & H. J. Walberg (Eds.), *Educational environments: Evaluation, antecedents, and consequences* (pp. 141–160). London: Pergamon.

Wubbels, T., Créton, H. A., & Holvast, A. J. C. D. (1988). Undesirable classroom situations: A systems communication perspective. *Interchange, 19*(2), 25–40.

Wubbels, T., & Levy, J. (1989, March). *A comparison of Dutch and American interpersonal behavior.* Paper presented at the annual meeting of the American Educational Research Association, San Francisco. (ERIC ED 307311)

·18·

RESEARCH ON GENDER ISSUES
IN THE CLASSROOM

Jane Butler Kahle

MIAMI UNIVERSITY

Judith Meece

UNIVERSITY OF NORTH CAROLINA

GENDER, MATHEMATICS, AND SCIENCE

Recently, concerns about the numbers and contributions of women in science have been voiced by various sectors of our society. Some concerns are pragmatic in origin, reflecting the changing nature of the work force; others are philosophical, mirroring issues about equity and equality. Regardless of the reason for one's concern, the basic issue remains: girls and, later, women are underrepresented in scientific and technological courses and careers. Nationwide, women constitute only 16 percent of all employed scientists and engineers (Vetter, 1990a, 1990b). Of employed women scientists and engineers in 1986, roughly 5 percent were African American, 5 percent were Asian American, 3 percent were Hispanic, and less than 1 percent were Native American (National Science Foundation [NSF], 1990). Women are also underrepresented among college students preparing for careers in science and engineering. Recent data indicate that U.S. women earn 30 percent of the bachelor's degrees in natural science and engineering and 21 percent of the doctorates in those fields (Vetter, 1990a, 1990b). Although women's representation in science and engineering has been increasing since 1980, much remains to be done in both recruiting and retaining women in the scientific work force (Brush, 1991; Vetter, 1987).

During the past two decades, both research and intervention programs have addressed the underrepresentation of girls in the scientific "pipeline." In this chapter we examine the factors underlying the differential participation of boys and girls in school science; then, we discuss interventions directed at increasing the participation of girls and women in science; and, last, we analyze where further progress is needed. As we began our research, we became aware of three limitations of our work. First, we write as though all girls and women are the same; that is, the nature of the data ignores the effects of economics and of race or ethnicity on individual choices. Although we are acutely aware that different forces affect the choices of Appalachian girls in Tennessee, Chicano girls in Texas, and Asian American girls in California, there is little direct evidence in the literature. Furthermore, we question whether conclusions based on predominately white respondents can be generalized to girls and women of color. Available data are seldom disaggregated by race or ethnicity and gender or by economic level and gender. A comprehensive review of the literature on girls' schooling (Wellesley College Center for Research on Women [WCCRW], 1992) has found that researchers seldom break down data by race or ethnicity and sex. In addition, the interactions between those two variables are rarely studied. Second, we focus primarily on the educational factors that affect girls in schools and in other educational settings. Although family, social, and cultural influences reinforce what happens in schools, we have chosen to address primarily factors that teach-

The authors thank reviewer Nancy Kreinberg (University of California at Berkeley).

ers and administrators can directly affect. Last, we focus on elementary and secondary education, because those are the years during which girls enter the mythical pipeline that leads to scientific and technical careers. According to Berryman (1983), the quantitative talent pool emerges by grade 9 and is essentially complete by grade 12. After high school essentially all changes in the pipeline are due to emigration from, not immigration to, the pool.

Girls and Mathematics

Because competency in mathematics is a prerequisite for entrance and persistence in scientific and technical fields, the study of mathematics was defined early as the "critical filter" (Sells, 1980). Studies in the 1970s showed that differences in Scholastic Aptitude Test-Math (SAT-M) scores between boys and girls could be partly attributed to different enrollment patterns in high school mathematics (Pallas & Alexander, 1983). Therefore, initial efforts to encourage girls in science and engineering focused on mathematics. Colleagues at the Lawrence Hall of Science were successful in identifying barriers, including high math anxiety, low parental expectations, and competitive classroom climates, to girls' participation in mathematics.

The outcome of the early intervention programs was that more girls began to enroll in algebra, geometry, and trigonometry. Today, there are large disparities in the numbers of girls and boys only in precalculus and calculus courses, but the percentage of girls enrolled in those courses in the United States has remained fairly stable during the past 15 years. For example, females constituted 36 percent of all precalculus and calculus students in 1978, 45 percent in 1982, and 39 percent in 1986 (Linn & Hyde, 1989). The increase percentage, found in 1982, reflects the period of most concern and effort toward retaining girls in mathematics.

As discussed later, several studies indicate that the gender gap in mathematics achievement has narrowed in recent years (Hyde, Fennema, & Lamon, 1990; Linn & Hyde, 1989; WCCRW, 1992). This research further suggests that when girls enroll in the same number and kinds of mathematics courses that boys do, gender differences in test scores, although not eliminated, are substantially smaller (Kimball, 1989). For example, the 1990 National Assessment of Educational Progress (NAEP) found few gender differences in the average mathematics proficiency of students at age 9. Gender differences begin to occur by age 13 and increase substantially by age 17, especially at the upper proficiency levels (Dossey, Mullis, Lindquist, & Chambers, 1988; Educational Testing Service [ETS], 1991). Those disparities occur after different course-taking patterns in mathematics are evident (Oakes, 1990).

Two decades of intervention programs were fairly successful in improving girls' participation and achievement in mathematics. Workshops and programs focused on reducing math anxiety, increasing self-confidence, and improving enrollment of girls also included strategies for enhancing their mathematics knowledge and skills. That is, both math attitudes and problem-solving skills were included in intervention efforts. However, a different pattern appeared in the workshops, projects, and research that focused on girls in science during the 1980s.

Girls and Science

Several years after the initial intervention programs in math, researchers began to examine the disparities in the science achievement of girls and boys. The work was instigated by the findings of the 1976–1977 NAEP Second Survey of Science. Although girls' achievement levels trailed those of boys at all levels, possible explanations for the differences were found among responses to other types of items. For the first time, questions concerning children's attitudes toward science and scientists, their desire to work in a scientific field, and their opinions of science classes were included in the survey. Responses indicated that the gender issues in science were more complicated than in math, where achievement differences could be traced to number and type of courses taken.

Results of the Second IEA Science Study (SISS) in the United States showed differences in science achievement favoring boys in biology, chemistry, and physics for all grades levels (5 to 12) (IEA, 1988). Few gender differences occurred on items assessing fifth- and ninth-grade students' abilities to perform scientific processes (Humrich, 1988). Similarly, the 1986 NAEP reported few gender differences at age 9, but boys outperformed girls in science achievement and the gender gap increased as students progressed in school. By grade 11, the areas of largest male advantage are physics, chemistry, earth science, and space science (Mullis & Jenkins, 1988). Among college-bound students, scores on science achievement tests have been consistently lower for women than men throughout the 1980s. Discrepancies in achievement scores range from 29 points for biology to 56 points for physics (NSF, 1990).

Gender differences also exist in levels of science proficiency. By age 17, about one-half of the boys but only one-third of the girls surveyed by NAEP demonstrate an ability to analyze and interpret scientific data (Mullis & Jenkins, 1988). In addition, trends in science achievement from 1970 to 1986 indicate that gender gaps in science performance have increased slightly for 9-year-olds, have more than doubled for 13-year-olds, and have narrowed slightly for 17-year-olds (Mullis & Jenkins, 1988).

Although gender differences in enrollment patterns for science generally begin in high school, the 1986 NAEP data revealed no significant gender differences in the numbers of male and female students enrolled in earth sciences and chemistry (Mullis & Jenkins, 1988). Slightly more females (90 percent) than males (87 percent) completed biology courses, but substantially more males (25 percent) than females (15 percent) completed physics (Nelson, Weiss, & Capper, 1990). Other evidence indicates that less than one-third of the girls who take chemistry continue to take physics, whereas more than one-half of the boys do. Nationally, only 35 percent of college-bound women take physics, compared to 51 percent of the men (Czujko & Bernstein, 1989).

Results from the 1986 NAEP further indicate that gender differences in science achievement cannot be explained by differential course-taking patterns for girls and boys. For example, gender differences in students' performance on the life science subscale are consistent for students who have or who have not enrolled in a biology course (Mullis & Jenkins, 1988). Similarly, course enrollment patterns have no effect on the magnitude of the gender effects found for the chemistry and physics subscales of the NAEP Science Survey. In fact, the gender gap is greater among students who had taken chemistry, suggesting that chemistry courses differentially benefit males.

In summary, an analysis of the research on mathematics and science achievement reveals several important discrepancies. First, the gender gap is closing in mathematics achievement but not in science achievement, especially during the middle school years. Second, the gender gap in science achievement increases from age 9 to 13, although most boys and girls are enrolled in similar courses during those years. In mathematics, gender differences in achievement do not occur with any consistency until the late high school years, when girls begin to take fewer mathematics courses than boys. Other analyses suggest attitudinal differences are evident at a younger age in science than in mathematics (Mullis & Jenkins, 1988). Boys and girls also have vastly different science-related experiences outside school that contribute to the gender gap in science achievement (Kahle & Lakes, 1983; Linn, 1990; Sjoberg & Imsen, 1988).

SOURCES OF INFLUENCE INDIVIDUAL, SOCIOCULTURAL, FAMILY, AND EDUCATIONAL VARIABLES

In this section, we discuss the factors that contribute to the differential achievement and participation of boys and girls in science. This section focuses on the role of individual, school, home, and sociocultural variables. We briefly summarize the contributions of each source and highlight directions for future research or intervention.

Individual Variables

Kahle and Danzl-Tauer (1991) discuss the evidence for a range of individual variables that affect girls in science. Specifically, they examine the evidence that genetic variables affect the differential achievement rates of girls and boys in science. They conclude that the distribution of female scores on tests of spatial ability contradicts the one expected if that trait were genetically passed by a recessive gene on the X chromosome. Furthermore, Linn and Peterson's (1985) review of a decade of research on possible biological factors concludes that there is "no noncontroversial evidence for genetic explanations of gender differences in cognitive and psychosocial factors" (p. 53). Our review of individual variables addresses both cognitive and attitudinal variables, but we do not imply that cognitive differences are due to biological traits.

Cognitive Abilities

Researchers (Benbow & Stanley, 1980; Maccoby & Jacklin, 1974) have argued that gender differences in students' spatial and mathematical abilities contribute to the differential achievement of boys and girls in science. Although gender differences appear in some measures of cognitive abilities, their influence on students' achievement in science has not been well established. Research suggests that spatial visualization tasks, which require reasoning about spatially presented information, are most closely associated with science achievement (Fennema & Sherman, 1977). Recently, meta-analysis, a technique that allows researchers to combine and interpret the results of multiple, individual studies, has provided useful information about the persistent gender gap reported for spatial visualization tasks. Results of meta-analysis are reported in terms of effect size representing the difference between groups. They are expressed as standard deviation units. An effect size of 0.2 or less is considered small, near 0.5 is considered medium, and 0.8 and higher are designated large. In the early 1980s, Linn and Peterson analyzed results from 172 studies and concluded that gender differences on spatial visualization tasks were minimal (Linn & Peterson, 1985). Further evidence suggests that gender differences on those tasks have declined in recent years. For example, Linn and Hyde (1989) reported that the effect size for gender was close to one-third of a standard deviation ($-.30$) for studies published before 1974, whereas a synthesis of subsequent research yielded a substantially smaller effect size ($-.13$). A difference of that small magnitude is likely to be highly responsive to training, and researchers have shown that instructional interventions can eliminate gender differences on some types of spatial perception tasks (Conner, Schackman, & Serbin, 1978; Liben & Golbeck, 1980, 1984; Whyte, 1986).

Similarly, research on gender differences in mathematical abilities does not yield a consistent pattern of findings. In a 1985 review, Stage, Kreinberg, Eccles, and Becker (1985) concluded that (1) girls occasionally outperform boys on tests of computational skills, (2) boys and girls perform similarly on tests of algebra and basic mathematical knowledge, and (3) high school boys score slightly higher on tests of mathematics reasoning and problem solving (primarily word problems). Two studies concluded that age-related gender differences in students' mathematics achievement levels are primarily a function of the cognitive level of the test. Girls are superior in computation during the elementary and middle school years, but no differences are found in high school populations. There were no gender differences in students' understanding of mathematical concepts at any of the ages examined. Also, the gender difference was essentially zero for tests of mathematical reasoning administered to elementary and middle school students. However, among high school and college students, there was a moderate gender difference favoring males. Tests of problem-solving skills thus appear to show the most dramatic age-related gender differences (Hyde, Fennema, Ryan, Frost, & Hopp, 1990b; Hyde, Fernandez, & Lamon, 1990a).

Several studies have reported large and consistent gender differences favoring boys in SAT-M achievement scores. The research of Hyde et al. (1990a) suggests that those gender differences may be partly attributed to the content of the test and the selectivity of the sample. First, the SAT-M contains a mixture of problem-solving and computation items, which, according to that study should favor males. Second, because this test is generally given to college-bound high school students, the selectivity of the sample is likely to increase the magnitude of gender differences. Evidence suggests that males taking the SAT-M tend to be drawn from a more select sample than are the female takers. For example, Hyde and Linn (1988) point out that the males tend to be more advantaged than the females in terms of parental income, father's education, and attendance in private school. Also, a larger number of females take the SAT, which is likely to increase the representativeness of that subsample. Hyde and her colleagues (1990a, 1990b) showed that when SAT data were eliminated from the analysis of gender differences in mathematics achievement, the effect size for gender dropped from .40 to .15.

These results are consistent with an analysis of data from the National Longitudinal Study of the high school class of 1972 (NLS-72) by Adelman (1991). Adelman argues that the highly touted gender difference in the SAT or American College Test (ACT) quantitative scores are largely erased if one compares female and male performance by standard deviation units rather than by mean scores. Because the women taking those tests have a wider range of abilities than the men, analysis of their scores result in a larger standard deviation than analysis of the scores of men. According to Adelman, the standard deviation unit, contrasted with mean score, takes into account the considerable variation in the backgrounds of the large populations that take the SAT or ACT. Therefore, it provides a fairer comparison of groups, in this case females and males. Women in the NLS-72 sample who had a solid background in college preparatory mathematics and/or science performed as well as their male peers on the SAT and ACT tests, taken over the period of the study. Adelman also notes that a comparison of the mean SAT or ACT quantitative scores of men and women is not very revealing, because more women than men come from the lowest socioeconomic quartile. Few comparisons of women's and men's performance levels on the SAT-M adjust for socioeconomic differences. Yet, differential mean achievement levels on the SAT and ACT are frequently cited as evidence of a general difference in math ability between boys and girls.

When gender differences in mathematics achievement are measured using course grades and overall grade point average in mathematics, a different pattern emerges. Some studies indicate no gender differences in students' grades in mathematics for the elementary school years (Entwistle & Baker, 1983) or for the middle and high school years (Eccles, 1983, 1984; Meece, Wigfield, & Eccles, 1990), whereas other studies (de Wolf, 1981; Pallas & Alexander, 1983) report gender differences favoring girls. Within gifted and talented populations, where boys generally outperform girls on standardized mathematics tests, girls also receive higher math grades than boys (Benbow & Stanley, 1983). Adelman (1991) reports that a higher percentage of the NLS-72 women than men were at the top of their high school graduating classes, even among students who had taken four semesters of a precollegiate curriculum in science or mathematics. Although fewer women than men may take a college preparatory curriculum in mathematics or science, the women who do tend to perform better than men in science classes.

Female superiority in high school math and science grades

is often ignored or dismissed in research on science achievement. Linn (1990) has argued that course grades are more reflective than test scores of the types of sustained problem-solving effort that is required for most careers in mathematics and science. Thus, to the extent that boys and girls do differ in their ability to perform spatial mathematical tasks, this difference alone does not sufficiently explain gender differences in science achievement.

Attitudinal Variables

Numerous studies have examined the influence of attitudinal variables on students' achievement and participation in science. This area of research assumes that girls generally develop a set of attitudes and beliefs that do not promote high levels of achievement and participation in science. For example, compared with boys, girls often have less confidence in their academic abilities, lower achievement expectations, less interest in challenging achievement activities, and more debilitating causal attribution patterns (Dweck, 1986). Research suggests that these gender differences are especially marked in achievement areas such as mathematics and science, which are sex typed as a "male domain" (Kelly, 1985; Lenny, 1977; McHugh, Fisher, & Frieze, 1982).

Consistent with this analysis, several large-scale studies have reported gender differences in students' self-concepts, interests, and values related to science. Similar to earlier science assessments (Hueftle, Rakow, & Welch, 1983; Kahle & Lakes, 1983), analysis of the 1986 NAEP data revealed a sex-differentiated pattern in students' attitudes toward science (Mullis & Jenkins, 1988; Nelson et al., 1990). Gender differences in students' feelings toward science were quite small in the third grade, but more girls (66%) than boys (59%) reported that science interested them. By the seventh and eleventh grades, gender differences favoring boys were evident in students' perceptions of both the difficulty and enjoyment value of science. There was little difference between boys' and girls' perceptions regarding the importance of science. However, fewer girls (34 percent) than boys (40 percent) believed that science would be important to their life's work.

Similarly, Zimmer and Bennett (1987) analyzed data from the 1986 California Statewide Assessment of Attitudes and Achievement involving 200,000 eighth-grade students. Boys reported greater liking for and enjoyment of science than did girls. They were also more likely to report benefiting from science activities inside and outside school, but fewer girls than boys were likely to agree with the statement that science was a more appropriate activity for boys.

Simpson and Oliver (1990) completed a longitudinal study of more than 4500 students in grades 6 to 10 from 178 science classes. The results indicated that boys score higher than girls on measures of science anxiety, self-concept, and attitudes toward science. Girls had more positive perceptions than boys of their achievement motivation and of their science classrooms. The results also showed that students' attitudes toward science become less positive from sixth through eighth grade; then they remain constant through the tenth grade.

Gender differences in students' attitudes toward science also appear among high-ability students. For example, Benbow and Minor (1986) examined gender differences in the science attitudes of more than 1900 high school students from the Study of Mathematically Precocious Youth (SMPY). These students were eligible to participate if, as seventh- or eighth-graders, they scored in the upper 5 percent nationally on the SAT-M test. Girls comprised 39 percent of the sample. At the end of high school, students were asked to rate their liking for biology, chemistry, physics as well as science in general. There was a gender difference favoring girls in biology and favoring boys in chemistry and physics among SMPY students. The effect size for physics was the largest. A greater number of boys (57 percent) than girls (47 percent) chose at least one science as one of their three favorite classes in high school. In addition, approximately 49 percent of the boys and 45 percent of the girls in this sample of highly able students indicated that they had considered possible careers in science.

In general, the gender differences reported in recent studies tend to be small in magnitude and consistent with previous reviews (Fleming & Malone, 1983; Halalyna & Shaughnessey, 1982; Steinkamp & Maehr, 1983, 1984; Wilson, 1983). Studies prior to 1980 report small gender effect sizes of .20 of a standard deviation or less for most measures of science attitudes. Findings also reveal that gender differences are not homogeneous across measures, age groups, and content areas, which makes it difficult to draw any general conclusions.

Overall, gender differences in science attitudes are larger for measures that assess self-concepts of science ability than those that focus on interest, importance, and enjoyment (Steinkamp & Maehr, 1984). There are also important age-related trends in gender differences. Some research (Wilson, 1983) suggests that gender differences in science attitudes are significant for studies of elementary and secondary students, whereas other studies (Steinkamp & Maehr, 1983, 1984) indicate the gender differences are greatest for junior high school students. In addition, the magnitude of the gender difference may depend on the ability level of the student and the content area assessed. Other evidence suggests that the geographic location of the school moderates gender effects (Matyas, 1984; Simpson & Oliver, 1985), and, as described later, gender differences in science attitudes are influenced by the student's socioeconomic background as well (Steinkamp & Maehr, 1984).

One study of the academic self-concepts, attributions, and achievements of high-ability science students showed that gender and race or ethnicity interacted to adversely affect Anglo, compared with Asian American, girls (Campbell, 1991). An analysis of Asian American and Anglo American boys and girls who were Westinghouse Talent Search winners found that Anglo American girls scored significantly lower than all other groups on a mix of variables called "technical orientation." Those variables included high SAT-M scores, high scores on assessments of mathematics and science self-concepts, high scores on degree of persistence to a scientific college major, and moderate assessments of the amount of personal effort needed to excel in science and math. Campbell postulated that broad socialization of Anglo American girls limited both their educational aspira-

tions and their aggressiveness. He concluded that with their low scores on the technical orientation variables, it was not surprising that, compared with Anglo American males and Asian American students, significantly more Anglo American females choose nontechnical college majors.

Results of studies that have investigated the relationship of affective variables to achievement or course enrollment in science are also mixed. Some studies report significant positive correlations (Simpson & Oliver, 1985; Steinkamp & Maehr, 1983, 1984; Wilson, 1983), while others report negative relations (Baker, 1985; Benbow & Minor, 1986; Fleming & Malone, 1983). The strength of the relationship found also varies across age groups and measures. In sum, available evidence suggests that gender differences in affective variables cannot adequately explain the underrepresentation of women in science.

Sociocultural Variables

Our discussion of sociocultural influences is based on the premise that science in Western European culture is defined and portrayed as a masculine enterprise (Bleier, 1984; Harding, 1986; Keller, 1985, 1986). Furthermore, there is increasing evidence that gender identification is developed at a very early age (Huston, 1983). The sociocultural stereotype of science as masculine interacts with a child's gender identification to affect who studies science.

Mead and Metreaux (1957) first analyzed the gender stereotyping of science by asking children to describe scientists. The verbal descriptions, collected over 30 years ago, still hold today and provide a vivid picture of a typical scientist. Furthermore, they indicate that most children stereotype scientists as men. More recently, researchers have asked children to draw pictures of scientists. Their pictures may be analyzed, according to indicators that have been described by Chambers (1983). The indicators allow researchers to rank the drawings from stereotypic to nonstereotypic (Kahle, 1989; Mason, Kahle, & Gardner, 1991). Analysis of thousands of drawings, collected in the United States, Australia, New Zealand, Ireland, Canada, and Norway, suggest a stereotyping of science and of scientists as masculine (cold, hard, analytical) (Chambers, 1983; Kahle, 1989; Maoldomhnaigh & Hunt, 1988; Mason et al., 1991, Schibeci, 1986; Schibeci & Sorensen, 1983).

According to Kelly (1985), "[T]he masculinity of science is often the prime reason that girls tend to avoid the subject at school" (p. 133). Her thesis is that science as an intellectual domain is perceived as masculine and that that perception discourages girls from expressing interest in science, from achieving well in science, and from continuing to study science. She notes three ways in which science is perceived as masculine: the numbers who practice and who are rewarded in science; the way in which science is packaged in curricular and instructional materials; and the way in which science is practiced in schools, including student-teacher and student-student interactions. She describes how the overwhelming number of men who both do and teach science combine with the ways in which science is packaged and practiced to produce a masculine image. She argues that that image, largely defined by sociocultural influences, is reproduced in schools, where it discourages girls from active participation in science.

Other researchers have shown that the stereotyping of science as masculine affects children's expressed interest in specific science topics and later in their science course selections (Baker, 1990; Shroyer, Powell, & Backe, 1991). Baker (1990) states that gender differences in course-taking patterns are established as early as kindergarten. She states, "Socialization acts as a filter for what students remember about their science instruction in terms of topics and pedagogy and in terms of instructional preferences. These early memories and preferences seem to influence subsequent course taking behavior" (p. 7). In a study designed to help curriculum developers produce a curriculum that would be of equal interest to boys and to girls, Shroyer et al. (1991) found that by grade 6 girls and boys reflected gender stereotyping of science in their topic choices (girls choose life science topics, while boys select physical science ones). Furthermore, there were gender differences in the reasons they gave for their choices. Girls more frequently responded that their selections were based on what "they should know." Boys, on the other hand, indicated that their choices were based upon what "they wanted to know." Their reasons suggest that both sexes consider science more appropriate for boys.

When older students are asked to rate school subjects on a masculine to feminine scale, the gender stereotyping of certain subjects is evident. Furthermore, their rankings are verified by sex-specific enrollment patterns in various courses. When British secondary school students were asked to arrange their subject options on a continuum from masculine to feminine, they ranked woodworking and physics, closely followed by mathematics and chemistry, as masculine subjects. History and biology received neutral rankings, while English, French, typing, and cooking were seen as feminine subjects (Weinreich-Haste, 1981). The course-taking patterns in American schools, discussed earlier, reinforce the rankings made by the British children and lead to continued gender stereotyping of specific courses. In addition, Garratt (1986) suggests that the gender stereotyping of subjects may be extended to the suitability of certain subjects for academically able children. He explains his premise in the following way.

Biology is perhaps perceived as being relevant to girls of all abilities, but only appropriate for boys of average ability. Conversely, physics may be seen as suitable for a broad ability band of boys, but only for girls of higher ability. (p. 68)

Furthermore, when girls are given the choice to "vote with their feet" both in and out of school, they select subjects other than science, according to a recent report of Girls Inc. (1991). In describing a large project designed to provide out-of-school, hands-on science activities for girls in Girls' Clubs, the authors note the following:

Another indication of girls' ambivalence about science and technology was their participation. They did not sign up or show up for science and computer classes as much as for crafts, cooking and woodworking. Similarly, the number of girls dwindled on the second and third visits to the science museum, despite their enthusiasm during the first visit. (p. 32)

Different participation rates that lead to the gender stereotyping of science as masculine continue in colleges and universities. In 1974, a study used characteristics and enrollment patterns to place academic fields along a continuum from masculine to feminine, as shown in Figure 18.1. Again, we find that students rank subfields in the biological sciences as neutral but that they stereotype fields in the physical sciences as well as mathematics as masculine.

Examination of actual transcripts, possible with the NLS-72 sample, also revealed gender stereotyping of subjects at colleges and universities. Transcripts were analyzed by the total time (as measured in credits) that men and women graduates spent in various courses. The analysis revealed a trichotomous pattern. Some courses were found to be *common courses;* that is, men and women spent approximately equal amounts of time in them. In the sciences, only general biology was a common course. Other courses in which women, compared with men, spent more time were labeled *women's courses*. Among the sciences, only anatomy and physiology fit that category. Last, courses in which men, compared with women, spent a disproportionate amount of time were labeled *men's courses*. Calculus, general physics, organic chemistry, electrical engineering, geology, and college algebra were all designated as men's courses (Adelman, 1991). These course-taking patterns, reflecting gender-stereotyped careers, further reinforce the masculine image of physical science and mathematics.

Enrollment	Field	Characteristics
High Male	Engineering Physics Geology Chemistry Mathematics Biochemistry	Masculine
	Microbiology Physiology Zoology Botany	Neutral
High Female	Liberal Arts Social Science Nursing	Feminine

FIGURE 18.1. Enrollment patterns of men and women students in science fields and sex-stereotyping of fields. Adapted from Kahle, J. B. (1986). *Equitable science education: A discrepancy model* (p. 19). Perth, Western Australia: Science & Mathematics Education Center, Curtin University of Technology.

Although biology has been seen over the years as a neutral course by both high school and college students, a detailed analysis of the college transcripts of the NLS-72 sample revealed that there are two curricula in the biological sciences. As shown in Figure 18.2, the male curriculum includes more analytical courses such as genetics and cellular biology and the female curriculum contains more descriptive courses such as microbiology and ethology. As Adelman (1991) explains, some of that dichotomy occurs because the premedical curriculum was taken more frequently by men than by women in the late 1970s. On the other hand, women's course selection may have been influenced by the nursing curriculum, prevalent in the 1970s. However, he concludes that neither the analytical frameworks involving women's more intimate and connected ways of approaching scientific knowledge (Gilligan, 1982; Keller, 1985) nor the research tradition that distinguishes between descriptive and empirical modes of thought (Bar-haim & Wilkes, 1989) explains the wholesale difference in course-taking patterns found between men and women in this large sample.

The masculine image of science is also reinforced by the sociocultural influences that work to marginalize women in the scientific enterprise. Both Watson (1981) and Keller (1986) have provided individual portraits of women scientists that attest to their lack of recognition and resources. In addition, Rossiter (1982) provides a historical examination of the marginalization of women in science through the nineteenth and twentieth centuries (until 1940). When women scientists are marginalized, their work is underrecognized and underrewarded and the gender stereotype of science as masculine is perpetuated.

Courses	Men		Women	
	BS degree	All	BS degree	All
"Men's courses"				
Biochemistry	7.9%	4.2%	6.5%	3.5%
Botany (general)	9.8	6.7	7.9	5.0
Cellular Biology	5.2	2.8	3.7	2.0
Ecology	11.2	6.7	7.5	4.0
Genetics	11.8	6.4	9.2	4.7
"Women's courses"				
Microbiology	9.1	5.6	13.1	10.7
Physiology	9.1	5.6	11.6	6.6
Bacteriology	2.0	1.0	2.6	1.6
Pharmacology	1.5	0.9	2.8	2.6
Ethology	1.7	0.8	1.9	0.9

FIGURE 18.2. Sex differentiated enrollment patterns in college science careers. Adapted from Adelman, C. (1991). *Women at thirtysomething: Paradoxes of attainment* (p. 12). Washington, DC: Office of Educational Research and Improvement, U.S. Department of Education.

Home and Family Variables

Although sociocultural variables may be attributed to many forces, they are often reinforced in the home and family environment. In this section, we examine that reinforcement and, to the extent that the data allow, the differential effects it has on girls' and boys' participation and achievement in science.

Reports of science achievement have focused on the influence of demographic characteristics of families such as ethnicity, socioeconomic status, and parental education. Family background variables are assumed to have an indirect effect on science achievement through their influence on the availability of economic resources, the quality of the home environment, the level of parents' educational and occupational aspirations, and the quality of the schools attended. For example, national studies indicate that Anglo and Asian American students score higher on tests of science achievement than African American and Hispanic students throughout elementary and secondary school (Hueftle et al., 1983; Mullis & Jenkins, 1988; Vetter & Babco, 1989). However, despite their lower level of achievement, African American students report more positive attitudes toward science than do Anglo American and Hispanic students (Mullis & Jenkins, 1988). In addition, both parental education and socioeconomic status are positively linked to students' achievement levels and participation rates in science (Mullis & Jenkins, 1988; Schibeci & Riley, 1986; Simpson & Oliver, 1990; Ware & Lee, 1988).

To date, few studies have examined whether family characteristics have a differential impact on girls' and boys' achievement in science. However, one study reports that gender differences in science attitudes favor girls in low socioeconomic families. Steinkamp and Maehr (1984) find that girls tend to have more positive motivational orientations than boys in economically disadvantaged samples, whereas boys have more positive attitudes than girls in upper middle-class families. Steinkamp and Maehr argue that in low socioeconomic groups, boys may show less interest in science because school activities in general are defined as feminine in that subgroup. This research suggests a need to examine more carefully how gender differences in science attitudes and achievement vary within different ethnic and socioeconomic groups.

In addition, little research has focused on socialization experiences in the home that may be more directly linked to gender differences in science achievement. Gender role socialization within the family can occur in many different ways. Considerable evidence suggests that parents structure the social and physical environment for boys and girls differently (Huston, 1983). Parents tend to buy more scientific games and toys for their sons than for their daughters (Astin, 1974; Casserly, 1980; Hilton & Berglund, 1974; Maccoby & Jacklin, 1974). Studies also show that boys are more likely than girls to play with toys that encourage manipulation, construction, or movement through space (Tracy, 1987).

Similarly, research on science achievement indicates that boys are more likely than girls to report that they had participated in a science activity at home, and students with more of these experiences tend to have higher science proficiency scores (Kahle & Lakes, 1983; Mullis & Jenkins, 1988). For example, Kahle and Lakes (1983) reported that 9-year-old boys, compared with girls, had more opportunities to use scientific equipment, to perform science experiments, and to take science-related field trips. Mullis and Jenkins (1988) also reported that the number of girls who have "tried to fix something mechanical" declined during the middle school years, while the number of boys doing those activities rose dramatically. On the basis of this evidence, several researchers have concluded that sex-typed play activities in the home may contribute to gender differences in science achievement (Kahle & Lakes, 1983; Tracy, 1987).

Parents also tend to have gender stereotypic behavioral expectations for their sons and daughters that have important long-term implications for achievement. For example, Hoffman (1977) found that parents stress occupational success, ambition, and intelligence as valued characteristics for their sons, whereas the desired qualities for their daughters were being kind, well mannered, and attractive, as well as having a good marriage and becoming a good parent. Researchers have also argued that early independence training may foster gender-differentiated achievement patterns, with parents encouraging and reinforcing girls' bids for help and interpersonal closeness but pushing boys toward self-reliance and achievement (Block, 1978; Hoffman, 1972). Campbell and Connolly (1987) report that high-achieving Anglo American females avoid contests like the Westinghouse Talent Search and that that avoidance is supported by their parents, who discourage teenage daughters from traveling to take advantage of research opportunities. They do not find evidence for that overprotective behavior among the Asian American parents studied.

Often, parents have different educational and vocational aspirations for their sons than they do for their daughters. For example, Adelman (1991) reported that 16 percent of the men and 9 percent of the NLS-72 women indicated that their fathers expected them to attain an advanced graduate degree. The pattern was similar for students' perceptions of their mothers' achievement expectations. In a study of mathematically gifted adolescents, Brody and Fox (1980) asked parents to list careers they would most like to see their child pursue. Approximately two-thirds of the mens' parents listed a mathematical or scientific career as their first choice; only one-third of the womens' parents did so.

Research has shown further that parents encourage their sons and daughters to take different academic courses in high school. Studies of adolescents found that both parents were more likely to encourage their sons than their daughters to take advanced mathematics, chemistry, and physics (Eccles, 1989; Eccles (Parsons), Adler, & Kaczala, 1982). Interestingly, mothers appeared to have a stronger impact than fathers on the achievement patterns of their children, especially on their daughters' achievement. This study also showed that parental advice is one of the most important influences on high school course enrollment decisions.

It is clear that parents can influence their children's achievement in science in a variety of indirect and direct ways. Much of the research suggests that parents treat their sons and daughters in ways that have important implications for their academic interests, skills, and attainment. In general, the socialization

experiences girls receive in the home are not likely to encourage achievement in science.

There is, however, one notable exception to this pattern. Findings indicate that children of employed mothers tend to hold less gender-stereotypic views for their children's personal characteristics, educational attainment, and occupational roles. The effects of maternal employment are particularly marked for girls, who tend to have higher educational aspirations, higher self-esteem, and be more career oriented than girls with non-employed mothers (Eccles & Hoffman, 1986). With a larger percentage of mothers in the work force than before, many of the socialization practices found in earlier research studies may no longer hold. Additional information is needed to understand how changing gender roles have affected children's socialization in the family and their educational attainment.

Educational Variables

In the past, much about the practice of science in schools was reported externally by researchers who feared that they would bias their data if they were involved directly in collecting them. Usually, scales and surveys were developed and tested independently, and either teachers or colleagues not directly involved with the study were asked to collect the data. The fear was that even the presence of the researcher in the classroom would bias the experiment and the resulting data. This practice seriously limited our ability to understand the teaching and learning process.

Recently, there has been a major change in the direction of science education research. Researchers realize that the richest data are collected on site, that understanding a complex process requires many sources of data, and that a long-term commitment is required to interpret adequately the teaching-learning process. Furthermore, objectivity has been balanced with the need for a joint, cooperative analysis. Researchers have learned that the interpretations of teachers and students are valid components of analysis. In addition, action research that informs and involves the subjects (teachers and/or students) in the purpose and plan of the study has been used (Whyte, 1986). In other cases, researchers and teachers have formed collaborative teams in order to investigate the complexities of the teaching-learning process (Tobin & Garnett, 1987; Tobin, Kahle, & Fraser, 1990). Therefore, many recent studies describe the processes of teaching and learning science through direct classroom observation. We shall report studies of teacher expectations, teaching behaviors, and classroom environments by leaning heavily on more descriptive studies.

Although differential teacher expectations for boys and girls in math are well documented and have been identified as a contributor to the different attribution and confidence patterns of boys and girls in math (Eccles et al., 1983; Fennema, 1990; Meece, Parsons, Kaczala, Goff, & Futterman, 1982), fewer studies have been done in science. Furthermore, there is evidence that collection of expectancy data by questionnaires and interviews may be unrealistic today. Recently, the results of a questionnaire, used to assess teacher opinions about the relevance, interest, and aptitudes of boys and girls in science, were questioned after the researchers observed the classrooms. Written survey responses suggested that teachers held similar expectations for the success of both girls and boys in science. However, behaviors suggesting differential expectations for girls and boys were recorded during classroom observations of lessons in electricity and simple machines. The researchers concluded that the survey responses reflected an awareness of equity; however, that awareness had not changed teacher actions (Kahle, Anderson, & Damnjanovic, 1991).

Two studies, one in the United States and the other in Australia, assessed teacher expectations for girls and boys for different areas of science (Kahle et al., 1991; Rennie & Parker, 1987). Teachers were queried about the relevance of science topics for boys and girls, about boys' and girls' self-confidence in science areas, about boys' and girls' performance levels across science topics, and about boys' and girls' levels of interest in selected topics. Results showed that both U.S. and Australian teachers considered all areas of science relevant for all children. Although U.S. teachers thought that girls were less confident than boys in all areas of science, Australian teachers perceived that difference only in physical science. U.S. teachers responded that both boys and girls performed less well in physical and earth than in biological science. Australian teachers, on the other hand, believed that girls performed significantly better than boys did in the biological sciences. When teachers were queried about boys' and girls' interest in science, Australians again separated the children by sex only in physical science. They thought that girls were significantly less interested than boys in physical science. However, the U.S. sample responded that girls were significantly less interested than boys in all areas of science.

In the past two decades, research on teacher behavior has centered on ethnographic and case studies. The richness of qualitative studies in science classrooms was verified and amplified by Stake and Easley's (1978) study of science classrooms in 11 schools. In addition, studies by Brophy and Good (1974) and others provided reliable and valid ways to code behavior. However, only recently have researchers coded data and analyzed observations for possible gender differences. Before Kahle and Lakes' (1983) analysis of 9-year-old girls' actual and desired experiences in science lessons, it was assumed that children in elementary classrooms had similar opportunities to do science.

A fair amount of research in the past 10 years has examined gender differences in classroom interaction patterns (Brophy, 1985; Eccles & Blumenfeld, 1985; Sadker & Sadker, 1985). This research assumes that gender differences in students' attitudes and achievement stem from differential treatment of boys and girls in the classroom. Research on science classroom environments provides partial support for this hypothesis (Jones & Wheatley, 1990; Kahle, 1990; Morse & Handley, 1985; Tobin & Garnett, 1987). Compared with girls, boys are more likely to initiate teacher interactions, to volunteer to answer teacher questions, to call out answers, and to receive praise, criticism, or feedback to prolong teacher interactions. These classroom interaction patterns result in greater opportunities for boys than girls to learn in science and may reflect favorable achievement expectations for boys.

Another classroom factor that differentially affects girls and boys is the type of instruction. Gender differences in classroom interactions are most pronounced in whole-class activities. Tobin and Garnett (1987) report that whole-class activities tend to be dominated by high-achieving boys. Research consistently shows that most girls prefer and take a more active role in cooperative, rather than competitive, learning activities (Baker, 1990; Eccles, 1989; Johnson & Johnson, 1987; Kahle, 1990; Smail, 1985). However, some evidence suggests that cooperative learning may not be effective in increasing girls' participation and achievement in science. In mixed-sex small groups, girls are less likely to receive help from boys, and girls are less likely than boys to assume a leadership role. In a dissertation study, Rogg (1990) did not find achievement gains associated with increased cooperative group work for either girls or boys in science.

Although gender differences in classroom interaction patterns are well documented, few studies have directly examined the influence of these experiences on students' attitudes and achievement in science. In a study of high school algebra classes, Koehler (1990) found no relation between the differential treatment of boys and girls in these classes and their achievement levels, even though boys received more favorable treatment. Similarly, Eccles (Parsons), Kaczala, and Meece (1982) reported that only a few classroom interaction patterns in junior high school mathematics classes related to students' self-concepts of math ability, and the relations differed for boys and girls. Studies also indicate that gender differences in students' attitudes and achievement may be largely context specific because gender effects are not observed in all classroom settings (Brophy, 1985; Eccles & Blumenfeld, 1985; Eccles (Parsons) et al., 1982). Additional research is needed to clarify the influence of classroom interaction patterns on gender differences in science attitudes and achievement.

INTERVENTION PROGRAMS

Research studies and practical intervention programs concerning the entrance, retention, and achievement of girls in mathematics indicated probable areas for fruitful programs in science. In the United States, intervention workshops and studies in science were initiated in the early 1980s and were often funded by the Women's Educational Equity Act (U.S. Department of Education) or by the Education and Human Resource Directorate of the National Science Foundation. During that time, projects were also conducted in the United Kingdom and Australia. Most intervention programs focused on schools and most were targeted at (1) demasculizing and demystifying science, usually by exposure to role models and career information; (2) improving girls' self-confidence and self-perceptions of their ability to do science; (3) implementing teaching strategies that actively involved girls in science lessons; and (4) developing girls' skills of doing science. This section reviews a series of intervention studies that attempted to identify specific factors affecting girls' participation and progress in science as well as to assess the effects of those factors.

Two studies (Kahle, 1985, 1987) identified factors that increased the interest levels as well as the retention rates of girls in science courses. The first focused on determining which, if any, teaching strategies and teacher behaviors were successful in encouraging girls to remain in science. The study involved eight sites across the United States and approximately 400 high school biology students. The researchers found that teachers who had a high proportion of girls continuing to enroll in high school chemistry and physics used specific teaching practices. For example, compared with a national sample (Weiss, 1978), they emphasized laboratory work and discussion groups, they quizzed their students weekly, they stressed creativity and basic skills, and they used numerous printed resources rather than relying solely on one textbook. The teachers, mostly females, also provided their students with career information and informal academic counseling. They all had attractive classrooms, decorated with posters and projects, and kept live plants and animals in their laboratories. The researchers hypothesized that the identified teaching strategies contributed to more girls continuing to take elective science courses (Kahle, 1985).

The study also showed that rural, compared with urban and suburban, high school students had fewer opportunities to work with the materials of science (both in and out of the classroom) and to experience science-related field trips, as well as less access to role models and career information. Therefore, a subsequent study focused on rural youth in biology classes taught by male teachers. Intervention strategies included access to career information, contact with female role models in science, experience with a variety of science activities and field trips, and use of laboratory exercises that included two- and three-dimensional representations, hypothesized to develop spatial skills (Kahle, 1987).

The project included two types of intervention conditions, "limited" intervention (teacher-requested activities and materials) and "full" intervention (researchers provided materials and assistance at regular intervals). In addition, an equivalent school, teacher, and students served as a control setting. After the intervention period, boys, compared with girls in all three groups, expressed more confidence in their scientific ability and had had more extracurricular science experiences. Girls in all groups revealed more positive attitudes toward women in science than boys did. Overall student responses to the attitude, interest, and aspiration questionnaires closely reflected the grades received. However, important differences were found in self-reported enrollment plans. More than 90 percent of the students in both intervention programs planned to enroll in chemistry, an optional science class elected by only 48 percent of the control school girls and, at that time, by 40 percent of high school girls nationally. The project had affected girls' retention rates in science but not their negative attitudes or levels of achievement.

Kelly and her colleagues in the United Kingdom also reported mixed findings from their 4 year action research project (Kelly, Whyte, & Smail, 1984; Smail, 1985; Whyte, 1986). Their program was also school based and focused on career information and role models. Because 10 compulsory, comprehensive, lower schools (more than 2000 children, ages 11 to 14) were involved, different course-taking options could be related to

students' subject selection when they entered noncompulsory, upper school. Interventions included presence of role models, curriculum units based on girls' expressed interests, and career information. The effect of those interventions on students' subject choices were inconsistent across schools. However, the enrollment of girls in one technical trade course (an option in one school) eliminated differences in girls' and boys' scores on one measure of spatial ability (Kelly et al., 1984). Kelly and her colleagues concluded that multiple factors influenced girls' choices and that further study was needed to separate and identify specific ones. However, they recommended "tinkering opportunities" and technical trade courses as a way to provide needed science-related experiences for girls.

By the mid–1980s, the results of intervention programs suggested that multiple factors were involved in increasing the participation of girls. They also indicated that girls' achievement levels in science were not affected by access to role models, career information, or interest-based curriculum. However, taking specific courses and using certain teaching strategies seemed to improve girls' retention rates in optional science courses (Mason & Kahle, 1989). Therefore, recent studies have tried to identify and assess factors that affect girls' achievement in science.

An Australian intervention program for teachers of upper primary school (fifth grade) concluded that science lessons based on girls' expressed areas of interest served to restrict their knowledge about and options in science (Parker & Rennie, 1986). However, when teachers interceded to involve girls more directly in lessons on electricity, both their competence and confidence increased. They found a slight, but consistent, tendency for students in classes taught by teachers who had participated in an equity workshop, compared with a control group of students, to perceive girls as competent with electricity. Furthermore, 85 percent of the girls in the experimental group (teachers' equity workshop), compared with 70 percent in the control group (teachers' science skill workshop), expressed confidence that they could be electricians (Rennie & Parker, 1987). Those findings were supported by the results of a U.S. study that used similar materials and intervention strategies. Kahle et al. (1991) reported that significantly more girls than boys, whose teachers had participated in an equity workshop, thought they could become electricians.

Another study assessed the effects of specific types of curriculum to improve retention rates and achievement levels of girls in school science. The study involved 10 rural, male biology teachers, who were introduced to quantitative laboratory activities that included spatial-visual exercises as well as the use of cooperative group learning (Danzl-Tauer, 1990). After the intervention program, most of the teachers increased the amount of time spent in individual and small-group activities as well as with hands-on, manipulative materials designed to develop students' spatial and quantitative skills. Interviews with teachers suggested that that change was due to the availability of quantitative and skill activities that were appropriate for alternative classroom interaction modes. By regressing classroom strategies against student variables, Danzl-Tauer found that the use of more interactive activities in an individual format was related positively to girls' enjoyment of science and gains in science achievement. However, she found a negative relationship between time spent in small groups and girls' attitudes concerning self-confidence in science ability and usefulness of science. Finally, she reported that the time spent in whole-class activities seemed unrelated to any of the student variables in the study. She cautioned that the negative relationship between use of small-group activities and attitude might have occurred because the small groups were not cooperative groups. She did not find a relationship between future enrollment in science and math and any teacher variable (interaction mode or number of activities used). Although she concluded that gender differences in enrollment patterns were a function of the schools attended rather than of the science classes and teachers, she did identify a factor (use of manipulative and quantitative materials) related to improved achievement of girls in science.

In sum, intervention programs have been fairly successful in identifying specific factors that influence girls' self-confidence and retention in science courses (Matyas & Malcom, 1991). However, they have been less successful in identifying specific factors that contribute to the continued and growing achievement gap between girls and boys in science. Recent work (Danzl-Tauer, 1990; Kahle et al., 1991; Klanin & Fensham, 1987; Rennie & Parker, 1987) indicates that emphasis on the skills of science may further enhance girls' retention and, with wide implementation, result in improved achievement levels.

SUMMARY AND IMPLICATIONS

Our review of the research on girls and science has presented a range of variables found to affect their retention and performance. We began by reviewing studies in mathematics because math is seen as the cognitive area that affects girls' achievement levels in science and because competence in mathematics is clearly a primary factor affecting girls' entrance, retention, and success in science courses and careers. Furthermore, research studies and intervention projects addressed the issue of girls in mathematics about a decade before such work was done in science. Therefore, a substantial body of knowledge as well as theoretical models exists in mathematics. Although mathematics was initially seen as the "critical filter" to girls' participation in science, mathematics should now be viewed as the "critical key" for understanding the continued underrepresentation and underachievement of girls and women in science.

Several different sources of gender differences in science participation and achievement, ranging from cognitive abilities to sociocultural stereotyping of science as masculine, have been identified. Researchers have gone into science classrooms to delineate differential participation patterns that, in addition to teachers' differential expectations for girls and boys, contribute to two types of experiences in school science. Those experiences are accumulative, and they compound the difference in out-of-school science activities that have been documented for boys and girls. The result is that many girls do not have equal opportunities to learn science, and their continued lag in achievement reflects that lack of opportunity.

Brickhouse (in press) maintains that the myth of objectivity, value-free inquiry, scientific method, and rationality that surrounds science presents students with a false image of science, and, thereby, limits their participation in it. Although she concurs that school science has a masculine image, she suggests three ways in which that image may be changed. First, she recommends that studying science within its historical and sociological context will allow students to see science as part of their culture and to realize that cultures other than the Western European tradition have developed sophisticated sciences. Second, she proposes that students examine current or historical scientific controversies about which scientists have not agreed. Third, she suggests teaching science and its uses together so that students can see science in relation to the world. She hypothesizes that these changes in the way school science is conceptualized and taught will change the masculine image of science, fostered by the sociocultural milieu (Brickhouse, in press).

In addition, because of the gender stereotyping of science as masculine in Western European culture, many adolescent girls experience gender-role conflict if they show an interest in science. Indirect evidence of the negative effect of this stereotyping is found in a study of girls' attitudes and achievement in science in Thailand. The authors show that chemistry is a neutral subject in that country and suggest that that perception enhances both girls' participation and achievement (Klanin & Fensham, 1987). Furthermore, girls in Western European culture hold more stereotypic views of science and scientists than they do of mathematics and mathematicians. Our analysis of a series of drawings from American and Australian high school and college students indicates that proportionately more female mathematicians are drawn and that female mathematicians are depicted with fewer masculine characteristics.

Family concerns clearly play a larger role in women's than in men's educational decisions. For example, Adelman (1991) reports that less than 8 percent of the NSL–72 women who chose not to go to college did so because they did not feel qualified for further education. Only 3 percent cited discouragement from parents and teachers, but women were more likely than men to indicate that future marriage and family plans influenced their decisions to continue their education. Similarly, other evidence suggests that many capable young women do not perceive careers in science as compatible with their family goals (Frieze & Hanusa, 1984; Matyas, 1986; Ware & Lee, 1988).

A series of intervention studies has focused on identifying factors that differentially affect girls and boys in science. Intervention projects by both Kahle (1987) and Kelly et al. (1984) and their colleagues have been conducted in science classrooms and have tried to affect directly the attitudes and retention levels of girls. Other projects have worked with teachers, assessing change in teacher behaviors or in students' self-confidence levels (Kahle et al., 1991; Rennie & Parker, 1987). However, in spite of a decade of work, the gender gap in achievement continues to grow in science. There is some evidence (Linn, DeBenedictis, Delucchi, Harris, & Stage, 1987; Murphy & Qualter 1986; Humrich, 1988) that some of that gap is due to the nature of the tests as well as to the types of test items used to assess achievement in science. For example, males on average score higher on objective tests, whereas females as a group score better on essay tests. Furthermore, many test items contain references to games, sports, and other activities that are based on boys' interests (baseball averages, motorcycle mileage, automobile engines). On the other hand, Humrich (1988) has found that girls are favored by tests that assess the processes of science. Linn and her colleagues have identified another component of multiple-choice, objective tests that affects results. That is, when the correct response is unknown, girls more frequently select the "I don't know" option than boys do. Therefore, boys have a 20 to 25 percent chance, depending on whether four or five responses are listed, of receiving credit for a correct guess (Linn et al., 1987). However, these types of evidence do not explain the size, or the consistency, of the achievement difference for girls and boys in science.

A careful review of the projects and programs that were designed to increase the numbers of girls in mathematics indicates that those interventions also provided instruction in mathematics. However, in science, the skills needed to do science were not included in most intervention programs. Indeed, Parker and Rennie (1986) argued that by focusing on curricula that were motivational for girls, some intervention projects actually served to limit girls' options as well as their preparation for advanced physical science classes. Because no one explanation has received unequivocal support, it is clear that a comprehensive approach is needed to understand gender differences in science achievement. The research, with few exceptions (Simpson & Oliver, 1990), has focused on the influence of single, rather than multiple, variables. As a result, intervention programs have been developed that focus singly on cognitive skills, attitudes, or the social ecology of science learning.

The lack of a theoretical model that integrates psychological and sociocultural variables has limited research on gender differences in science achievement. Two models have been developed in mathematics, one by Eccles and her colleagues and the other by Fennema and her research group. Although there are similarities in the models, there are also differences. Eccles and her colleagues (Eccles et al., 1983) have developed an academic choice model that has been used successfully to explain gender differences in mathematics achievement and to guide intervention projects (Eccles et al., 1983; Meece et al., 1982). Figure 18.3 summarizes the major components of the model.

The model is based on expectancy-value theories of achievement motivation. Accordingly, students' decisions to persist and to excel at a particular course of study are linked to their expectations for success and the subjective value of the achievement area. These perceptions are shaped over time by experiences with related activities, encouragement from others, cultural norms, and opportunity structures that exist in the culture. Research has shown that the variables included in the Eccles model explain a substantial amount of the variance in students' course-enrollment plans (Eccles, 1984; Eccles et al., 1983; Meece et al., 1990). More important, the model has identified several important areas for intervention.

The second model, called the autonomous learning behavior (ALB) model, attempts to explain causation of gender differences in achievement on mathematical tasks of high cognitive complexity. Fennema and her associates describe autonomous

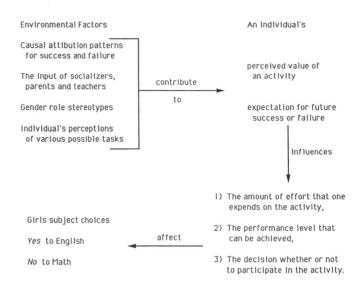

FIGURE 18.3. Eccles' academic choice model. Adapted from Eccles, J. (1989). Bringing young women to math and science. In M. Crawford & M. Gentry (Eds.), *Gender and thought: Psychological perspectives* (p. 38). New York: Springer Verlag.

learning behaviors as those that are used by students who increasingly assume control of the learning process. The model, shown in Figure 18.4, has been the basis of extensive research (Fennema & Petersen, 1985; Fennema & Sherman, 1977; Leder, 1984; Meyer & Koehler, 1990). It has been used to investigate the confluence of external and internal influences on successful achievement in mathematics. Models like the ones developed by Eccles and Fennema and their associates could be used to guide and integrate research on gender differences in science achievement.

Much of the research reviewed here is based on a deficit model that implies that girls must lack some cognitive skill,

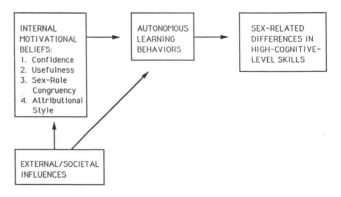

FIGURE 18.4. Fennema's autonomous learning behavior model. Adapted from Meyer, M., & Koehler, M. S. (1990). Internal influences on gender differences in mathematics. In E. Fennema & G. Leder (Eds.), *Mathematics and Gender* (p. 70). New York: Teachers College Press.

personal characteristic, or experience that which prevents them from achieving as well as boys. Eccles (1987) has argued that research characterizing females as deficient males perpetuates a distorted view of female achievement. It does not help to identify factors that uniquely influence women's educational and occupational choices. Furthermore, theoretical models that focus on women's deficiencies do little to address problems in the workplace that make it difficult for women to combine work and family roles.

Last, research on gender differences in science achievement has given very little attention to individual differences within female samples. The research examined in this review indicates that ability level, ethnicity, and socioeconomic status may moderate gender differences. By the end of the 1990s, nearly one-third of the youth in the United States will be members of an ethnic or racial minority group. Currently, one out of five children lives in poverty. With these changing population demographics, researchers can no longer continue to study science achievement patterns for either girls or boys in white, middle-class samples.

References

Adelman, C. (1991). *Women at thirtysomething: Paradoxes of attainment*. Washington, DC: Office of Educational Research and Improvement, U. S. Department of Education.

Astin, H. S. (1974). Sex differences in mathematical and science precocity. In J. C. Stanley, D. P. Keating, & L. H. Fox (Eds.), *Mathematical talent: Discovery, description and development*. Baltimore: Johns Hopkins University Press.

Baker, D. (1985). Predictive value of attitude, cognitive ability, and personality to science achievement in the middle school. *Journal of Research in Science Teaching, 22*(2), 103–113.

Baker, D. (1990, April). *Gender differences in science: Where they start and where they go*. Paper presented at the meeting of the National Association for Research in Science Teaching, Atlanta.

Bar-haim, G., & Wilkes, J. M. (1989). A cognitive interpretation of the marginality underrepresentation of women in science. *Journal of Higher Education, 60*(4), 373–387.

Benbow, C., & Minor, L. (1986). Mathematically talented males and females and achievement in high school sciences. *American Educational Research Journal, 23*(3), 393–414.

Benbow, C. P., & Stanley, J. C. (1980). Sex differences in mathematical ability: Fact or artifact. *Science, 210*, 1262–1264.

Benbow, C. P., & Stanley, J. C. (1983). Sex differences in mathematical reasoning ability: More facts. *Science, 222*, 1029–1031.

Berryman, S. E. (1983). *Who will do science?* Washington, DC: Rockefeller Foundation.

Bleier, R. (1984). *Science and gender: A critique of biology and its theories on women*. New York: Pergamon.

Block, J. H. (1978). Another look at sex differentiation in the socialization behavior of mothers and fathers. In J. A. Sherman & F. L. Denmark (Eds.), *Psychology of women: Future directions of research*. New York: Psychological Dimensions.

Brickhouse, N. (in press). Bringing in the outsiders: Transforming the sciences. *Journal of Curriculum Studies*.

Brody, L., & Fox, L. H. (1980). An accelerative intervention program for mathematically gifted girls. In L. H. Fox, L. Brody, & K. Tobin (Eds.), *Women and the mathematical mystique* (pp. 164–192). Baltimore: Johns Hopkins University Press.

Brophy, J. (1985). Interactions of male and female students with male and female teachers. In L. C. Wilkinson & C. B. Marrett (Eds.), *Gender influences in classroom interaction*. New York: Academic Press.

Brophy, J., & Good, T. L. (1974). *Teacher student relationships: Causes and consequences*. New York: Rinehart & Winston.

Brush, S. G. (1991). Women in science and engineering. *American Scientist, 79*(5), 404–419.

Campbell, J. R. (1991). The roots of gender in equity in technical areas. *Journal of Research in Science Teaching, 28*(3), 251–264.

Campbell, J. R., & Connolly, C. (1987). Deciphering the effects of socialization. *Journal of Educational Equity and Leadership, 7*(3), 208–222.

Casserly, P. L. (1980). Factors affecting female participation in advanced placement programs in mathematics, chemistry, and physics. In L. H. Fox, L. Brody, & K. Tobin (Eds.), *Women and the mathematical mystique* (pp. 138–163). Baltimore: Johns Hopkins University Press.

Chambers, D. W. (1983). Stereotypic images of the scientist: The draw-a-scientist test. *Science Education, 67*(2), 255–265.

Conner, J. M., Schackman, M., & Serbin, L. A. (1978). Sex-related differences in response to practice on a visual-spatial test and generalization to a related test. *Child Development, 49*, 24–29.

Czujko, R., & Bernstein, D. (1989). *Who takes science? A report on student coursework in high school science and mathematics* (AIP Publication No. R–345). New York: American Institute of Physics.

Danzl-Tauer, L. (1990). *The relationship between intervention, equity, and excellence in rural high school biology classrooms*. Unpublished doctoral dissertation, Purdue University, West Lafayette, IN.

de Wolf, V. A. (1981). High school mathematics preparation and sex differences in quantitative abilities. *Psychology of Women Quarterly, 5*(4), 555–567.

Dossey, L. A., Mullis, I. V. S., Lindquist, M. M., & Chambers, D. L. (1988). *The mathematic report card. Are we measuring up?* (Trends and achievements based on the 1986 National Assessment Report No. 17-M–01). Princeton, NJ: Educational Testing Service. In M. C. Linn & J. S. Hyde (1989). Gender, mathematics, and science. *Educational Researcher, 18*(8), 17–19, 22–27.

Dweck, C. S. (1986). Motivational processes affecting learning. *American Psychologist, 41*(10), 1040–1048.

Eccles, J. (1983). Expectancies, values, and academic behaviors. In J. T. Spence (Ed.), *Achievement and achievement motives* (pp. 75–146). San Francisco: Freeman.

Eccles, J. (1984). Sex differences in achievement patterns. *Nebraska Symposium on Motivation, 32*, 97–132.

Eccles, J. (1987). Gender roles and women's achievement-related decisions. *Psychology of Women Quarterly, 11*(2), 135–172.

Eccles, J. (1989). Bringing young women to math and science. In M. Crawford & M. Gentry (Eds.), *Gender and thought: Psychological perspectives*. New York: Springer-Verlag.

Eccles, J., Adler, T. F., Futterman, R., Goff, S. B., Kaczala, C. M., Meece, J. L., & Midgley, C. M. (1983). Expectancies, values and academic behaviors. In J. T. Spence (Ed.), *Achievement and achievement motivates*. San Francisco: Freeman.

Eccles (Parsons), J., Adler, T. F., & Kaczala, C. (1982). Socialization of achievement attitudes and beliefs: Parental influences. *Child Development, 53*(2), 310–321.

Eccles, J., & Blumenfeld, P. C. (1985). Classroom experiences and student gender: Are there differences and do they matter? In L. C.

Wilkinson & C. B. Marrett (Eds.), *Gender influences in classroom interactions*. New York: Academic Press.

Eccles, J., & Hoffman, L. W. (1986). Sex roles, socialization, and occupational behavior. In H. W. Stevenson & A. E. Siegel (Eds.), *Research in child development research and social policy* (Vol. 1). Chicago: University of Chicago Press.

Eccles (Parsons), J., Kaczala, C. M., & Meece, J. L. (1982). Socialization of achievement attitudes and beliefs: Classroom influences. *Child Development, 53*(2), 322–339.

Educational Testing Service (ETS). (1991). *Trends in academic progress: Achievement of American students in science, 1970–90, mathematics, 1973–90, reading, 1971–1990, and writing, 1984–90.* Washington, DC: National Center for Education Statistics, Office of Educational Research and Improvement.

Entwisle, D. R., & Baker, D. P. (1983). Gender and young children's expectations for performance in arithmetic. *Developmental Psychology, 19*(2), 200–209.

Fennema, E. (1990). Justice, equity, and mathematics education. In E. Fennema & G. C. Leder (Eds.), *Mathematics and gender* (pp. 1–9). New York: Teachers College Press.

Fennema, E., & Peterson, P. (1985). Autonomous learning behavior: A possible explanation of gender-related differences in mathematics. In L. C. Wilkinson & C. B. Marrett (Eds.), *Gender-related differences in classroom interaction* (pp. 17–35). New York: Academic Press.

Fennema, E., & Sherman, L. (1977). Sex-related differences in mathematics, achievement, spatial visualization, and sociocultural factors. *American Educational Research Journal, 14*, 51–71.

Fleming, M. L., & Malone, M. R. (1983). The relationship of student characteristics and student performance in science as viewed by meta-analysis research. *Journal of Research in Science Teaching, 20*(5), 481–495.

Frieze, I. H., & Hanusa, B. H. (1984). Women scientists: Overcoming barriers. In M. W. Steinkamp & M. L. Maehr (Eds.), *Advances in motivation and achievement: Women in science* (Vol. 2) (pp. 139–164). Greenwich, CT: JAI Press.

Garratt, L. (1986). Gender differences in relation to science choice at A-level. *Educational Review, 38*, 67–76.

Gilligan, C. (1982). *In a different voice: Psychological theory and women's development.* Cambridge, MA: Harvard University Press.

Girls Incorporated. (1991). *The explorer's pass: A report on case studies of girls and math, science and technology.* New York: Author.

Halalyna, T., & Shaughnessey, L. (1982). Attitudes toward science: A quantitative synthesis. *Science Education, 66*(4), 547–563.

Harding, J. (1986). *Perspectives on gender and science.* Philadelphia: Falmer.

Hilton, T. L., & Berglund, G. W. (1974). Sex differences in mathematics achievement: A longitudinal study. *Journal of Educational Research, 67*(5), 231–237.

Hoffman, L. W. (1972). Early childhood experiences and women's achievement motives. *Journal of Social Issues, 28*(2), 129–155.

Hoffman, L. W. (1977). Changes in family roles, socialization, and sex differences. *American Psychologist, 32*(8), 644–657.

Hueftle, S. J., Rakow, S. J., & Welch, W. W. (1983). *Images of science: A summary of results from the 1981–82 National Assessment in Science.* Minneapolis: Minnesota Research and Evaluation Center.

Humrich, E. (1988, April). *Sex differences in the second IEA science study—U.S. results in an international context.* Paper presented to the annual meeting of the National Association for Research in Science Teaching, Lake of the Ozarks, MO. (ERIC Document Reproduction Service No. ED 292 649)

Huston, A. C. (1983). Sex-typing. In E. M. Hetherington (Ed.), *Handbook of child psychology. Vol. 4: Socialization, personality, and social development* (pp. 387–468). New York: Wiley.

Hyde, J. S., Fennema, E., & Lamon, S. J. (1990a). Gender differences in mathematics performance: A meta-analysis. *Psychological Bulletin, 107*(2), 139–155.

Hyde, J. S., Fennema, E., Ryan, M., Frost, L., & Hopp, C. (1990b). Gender comparisons of mathematics attitudes and affect: A meta-analysis. *Psychology of Women Quarterly, 14*(3), 299–324.

Hyde, J. S., & Linn, M. C. (1988). Gender differences in verbal ability. A meta-analysis. *Psychological Bulletin, 104*, 53–69.

International Association for the Evaluation of Educational Achievement (IEA). (1988). *Science achievement in seventeen countries. A preliminary report.* Elmsford, NY: Pergamon.

Johnson, R. T., & Johnson, D. W. (1987). Cooperative learning and the achievement and socialization crisis in science and math classroom. In A. B. Champagne & L. E. Hornig (Eds.), *Students and science learning.* Washington, DC: American Association for the Advancement of Science.

Jones, G., & Wheatley, J. (1990). Gender differences in teacher-student interactions in science classrooms. *Journal of Research in Science Teaching, 27*(9), 861–874.

Kahle, J. B. (1985). Retention of girls in science: Case studies of secondary teachers. In J. B. Kahle (Ed.), *Women in science: A report from the field.* Phildelphia: Falmer.

Kahle, J. B. (1987). SCORES: A project for change? *International Journal of Science Education, 9*, 325–333.

Kahle, J. B. (1989). Images of scientists: Gender issues in science classrooms. *What research says to the science and mathematics teacher* (No. 4). Perth, W. Austr.: Key Centre for Science and Mathematics Education, Curtin University of Technology.

Kahle, J. B. (1990). Real students take chemistry and physics: Gender issues. In K. Tobin, J. B. Kahle, & B. J. Fraser (Eds.), *Windows into science classrooms: Problems associated with higher-level cognitive learning* (pp. 92–134). New York: Falmer.

Kahle, J. B., Anderson, A., & Damnjanovic, A. (1991). A comparison of elementary teacher attitudes and skills in teaching science in Australia and the United States. *Research in Science Education, 21*, 208–216.

Kahle, J. B., & Danzl-Tauer, L. (1991). The underutilized majority: The participation of women in science. In S. K. Majumdar, L. M. Rosenfeld, P. A. Rubba, E. W. Miller, & R. F. Schalz (Eds.), *Science education in the United States: Issues, crisis, and priorities.* Philadelphia: Pennsylvania Academy of Science Press.

Kahle, J. B., & Lakes, M. K. (1983). The myth of equality in science classrooms. *Journal of Research in Science Teaching, 20*(2), 131–140.

Keller, E. F. (1985). *Reflections on gender and science.* New Haven, CT: Yale University Press.

Keller, E. F. (1986). How gender matters: Or, why it's so hard for us to count past two. In J. Harding (Ed.), *Perspectives on gender and science* (pp. 168–183). London: Falmer.

Kelly, A. (1985). The construction of masculine science. *British Journal of Sociology of Education, 6*(2), 133–153.

Kelly, A., Whyte, J., & Smail, B. (1984). *Final report of the GIST project.* Manchester: Department of Sociology, Manchester University.

Kimball, M. M. (1989). A new perspective on women's math achievement. *Psychological Bulletin, 105*(2), 198–214.

Klanin, S., & Fensham, P. J. (1987). Activity based learning in chemistry. *International Journal of Science Education, 9*(2), 217–227.

Koehler, M. S. (1990). Classrooms, teachers, and gender differences in mathematics. In E. Fennema & G. C. Leder (Eds.), *Mathematics and gender* (pp. 128–148). New York: Teachers College Press.

Leder, G. C. (1984). Sex differences in attributions of success and failure. *Psychological Reports, 54*, 57–58.

Lenny, E. (1977). Women's self-confidence in achievement settings. *Psychological Bulletin, 84,* 1–13.

Liben, L. S., & Golbeck, S. L. (1980). Sex differences in performance on Piagetian spatial tasks: Differences in competence or performance? *Child Development, 51*(2), 594–597.

Liben, L. S., & Golbeck, S. L. (1984). Performance on Piagetian horizontality and verticality tasks: Sex related differences in knowledge of relevant physical phenomena. *Developmental Psychology, 20*(4), 595–606.

Linn, M. C. (1990, July). *Gender, mathematics and science: Trends and recommendations.* Paper prepared for the Council of Chief State Officers Summer Institute, Mystic, CT.

Linn, M. C., DeBenedictis, T., Delucchi, K., Harris, A., & Stage, E. (1987). Gender differences in national assessment of educational progress science items: What does "I don't know" really mean? *Journal of Research in Science Teaching, 24*(3), 267–278.

Linn, M. C., & Hyde, J. S. (1989). Gender, mathematics, and science. *Educational Researcher, 18*(8), 17–19, 22–27.

Linn, M. C., & Peterson, A. C. (1985). Facts and assumptions about the nature of sex differences. In S. S. Klein (Ed.), *Handbook for achieving sex equity through education* (pp. 53–77). Baltimore: Johns Hopkins University Press.

Maccoby, M. E., & Jacklin, C. N. (1974). *The psychology of sex differences.* Palo Alto, CA: Stanford University Press.

Maoldomhnaigh, M. O., & Hunt, A. (1988). Some factors affecting the image of the scientist drawn by older primary school pupils. *Research in Science and Technological Education, 6*(12), 159–166.

Mason, C. L., & Kahle, J. B. (1989). Student attitudes toward science and science related careers: A program designed to promote a stimulating gender-free environment. *Journal of Research in Science Teaching, 26,* 25–39.

Mason, C. L., Kahle, J. B., & Gardner, A. L. (1991). Draw-a-scientist test: Future implications. *School Science and Mathematics, 91*(5), 193–198.

Matyas, M. L. (1984, April). *Science career interests, attitudes, abilities and anxiety among secondary school students: The effects of gender, race/ethnicity, and school/type location.* Paper presented at the annual meeting of the National Association in Science Teaching, New Orleans.

Matyas, M. L. (1986, May). *Persistence in science-oriented majors: Factors related to attrition among male and female students.* Paper presented at the American Educational Research Association, San Francisco.

Matyas, M. L., & Malcom, S. M. (Eds.). (1991). *Investing in human potential: Science and engineering at the crossroads.* Washington, DC: American Association for the Advancement of Science.

McHugh, M. C., Fisher, J., & Frieze, I. (1982). Effect of situational factors on the self-attributions of females and males. *Sex Roles: A Journal of Research, 8*(4), 389–397.

Mead, M., & Metreaux, R. (1957). The image of the scientist among high school children. *Science, 126,* 384–389.

Meece, J. L., Parsons, J. E., Kaczala, C. M., Goff, S. B., & Futterman, R. (1982). Sex differences in math achievement: Toward a model of academic choice. *Psychological Bulletin, 91*(2), 324–348.

Meece, J. L., Wigfield, A., & Eccles, J. S. (1990). Predictors of math anxiety and its influence on young adolescents' course enrollment intentions and performance in mathematics. *Journal of Educational Psychology, 82,* 60–70.

Meyer, M. R., & Koehler, M. S. (1990). Internal influences on gender differences in mathematics. In E. Fennema & G. C. Leder (Eds.), *Mathematics and Gender* (pp. 60–95). New York: Teachers College Press.

Morse, L. W., & Handley, H. M. (1985). Listening to adolescents: Gender differences in science classroom interaction. In L. C. Wilkinson &

C. B. Marrett (Eds.), *Gender influences in classroom interactions.* New York: Academic Press.

Mullis, I. V. S., & Jenkins, L. B. (1988). *The science report card: Elements of risk and recovery.* Princeton, NJ: Educational Testing Service.

Murphy, P., & Qualter, A. (1986). Science for all? But do the children think so? In M. Dahms, L. Dirckinck-Holmfeld, K. G. Hanson, A. Kolmos, & J. Nielsen (Eds.), *Women challenge technology* (Vol. 1) (pp. 189–210). Elishore, Den: Centertrykkeriet, University of Aalborg.

National Science Foundation (NSF). (1990). *Women and minorities in science and engineering* (NSF 90-301). Washington, DC: Author.

Nelson, B., Weiss, I., & Capper, J. (1990). *Science and mathematics education briefing book* (Vol. II). Washington, DC: National Science Teachers Association.

Oakes, J. (1990). Opportunities, achievement, and choice: Women and minority students in science and math. In C. Cazden (Ed.), *Review of research in education* (Vol. 16). Washington, DC: The American Educational Research Association.

Pallas, A. M., & Alexander, K. L. (1983). Sex differences in quantitative SAT performance: New evidence on the differential coursework hypothesis. *American Educational Journal, 20*(2), 165–182.

Parker, L. H., & Rennie, L. J. (1986). Sex-stereotyped attitudes about science: Can they be changed? *European Journal of Science Education, 8,* 173–183.

Rennie, L. J., & Parker, L. H. (1987). Detecting & accounting for gender differences in mixed-sex and single-sex groupings in science lessons. *Educational Review, 39,* 65–73.

Rogg, S. R. J. (1990). *Toward the evaluation of the small instructional group in the secondary biology classroom.* Unpublished doctoral dissertation, Purdue University, West Lafayette, IN.

Rossiter, M. W. (1982). *Women scientists in America: Struggles and strategies to 1940.* Baltimore: Johns Hopkins University Press.

Sadker, D., & Sadker, M. (1985). Is the ok classroom, ok? *Phi Delta Kappan, 66*(5), 358–361.

Schibeci, R. A. (1986). Images of science and scientists and science education. *Science Education, 70*(2), 139–149.

Schibeci, R. A., & Riley, J. P. (1986). Influence of students' background and perceptions on science attitudes and achievement. *Journal of Research in Science Teaching, 23*(3), 177–187.

Schibeci, R. A., & Sorenson, I. (1983). Elementary school children's perceptions of scientists. *School Science and Mathematics, 83,* 15–20.

Sells, L. W. (1980). The mathematics filter and the education of women and minorities. In L. Fox, L. Brady, & K. Tobin (Eds.), *Women and the mathematic mystique* (pp. 66–75). Baltimore: Johns Hopkins University Press.

Shroyer, M. G., Powell, J. C., & Backe, K. A. (1991, April). *Gender differences: What do kids tell us and what should we do in the curriculum to make a difference?* Paper presented at the annual meeting of the National Association for Research in Science Teaching, Lake Geneva, WI.

Simpson, R. D., & Oliver, J. S. (1985). Attitude toward science achievement motivation profiles of male and female students in grades 6 through 10. *Science Education, 69*(4), 511–526.

Simpson, R. D., & Oliver, J. S. (1990). A summary of major influences on attitude toward achievement in science among adolescent students. *Science Education, 74,* 1–18.

Sjoberg, S., & Imsen, G. (1988). Gender and science education I. In P. Fensham (Ed.), *Development and dilemmas in science education* (pp. 218–248). London: Falmer.

Smail, B. (1985). An attempt to move mountains: The 'girls into science and technology' (GIST) project. *Journal of Curriculum Studies, 17*(3), 351–354.

Stage, E. K., Kreinberg, N., Eccles, J., & Becker, J. R. (1985). Increasing the participation and achievement of girls and women in mathematics, science, and engineering. In S. S. Klein (Ed.), *Handbook for achieving sex equity through education* (pp. 237–268). Baltimore: Johns Hopkins University Press.

Stake, R. A., & Easley, J. A. (1978). *Case studies in science education* (2 vols.). Washington, DC: U.S. Government Printing Office.

Steinkamp, M. W., & Maehr, M. L. (1983). Affect, ability, and science achievement: A quantitative synthesis of correctional research. *Review of Educational Research, 53*(3), 369–396.

Steinkamp, M. W., & Maehr, M. L. (1984). Gender differences in motivational orientations toward achievement in school science: A quantitative synthesis. *American Educational Research Journal, 21*, 39–59.

Tobin, K., & Garnett, P. (1987). Gender related differences in science activities. *Science Education, 71*, 91–103.

Tobin, K., Kahle, J. B., & Fraser, B. J. (Eds.). (1990). *Windows into science classrooms: Problems associated with higher level cognitive learning*. New York: Falmer.

Tracy, D. (1987). Toys, spatial ability, and science and mathematics achievement: Are they related? *Sex Roles: A Journal of Research, 17*, 115–138.

Vetter, B. M. (1987). Women's progress. *MOSAIC, 1*, 2–9.

Vetter, B. M. (1990a). *Who is in the pipeline? Science, math and engineering education*. Washington, DC: Commission on Professionals in Science and Technology.

Vetter, B. M. (1990b). *Women in science and engineering: An illustrated report*. Washington, DC: Commission on Professionals in Science and Technology.

Vetter, B. M., & Babco, E. L. (1989). *Professional women and minorities: A manpower data resource service*. Washington, DC: Commission on Professionals in Science and Technology.

Ware, N. C., & Lee, V. E. (1988). Sex differences in choice of college science majors. *American Educational Research Journal, 25*, 593–614.

Watson, J. D. (1981). *The double helix: A personal account of the discovery of the structure of DNA*. London: Weidenfeld & Nicolson.

Weinreich-Haste, H. (1981). The image of science. In A. Kelly (Ed.), *The missing half. Girls and science education*. Manchester: Manchester University Press.

Weiss, I. R. (1978). *Report of the 1977 National Survey of Science, Mathematics and Social Studies Education* (SE 78–72). Washington, DC: National Science Foundation.

Wellesley College Center for Research on Women (WCCRW). (1992). *How schools shortchange girls*. Washington, DC: American Association of University Women Educational Foundation.

Whyte, J. (1986). *Girls into science and technology*. London: Routledge & Kegan Paul.

Wilson, V. L. (1983). A meta-analysis of the relationship between science achievement and science attitude: Kindergarten through college. *Journal of Research in Science Teaching, 20*(9), 839–850.

Zimmer, L. K., & Bennett, S. M. (1987, March). *Gender differences on the California Statewide Assessment of Attitudes and Achievement in Science*. Paper presented at the annual meeting of the American Educational Research Association, Washington, DC.

RESEARCH ON CULTURAL DIVERSITY IN THE CLASSROOM

Mary M. Atwater
UNIVERSITY OF GEORGIA

Important facets of learning and teaching for African American, Asian American, Hispanic, and Native American students are the foci of this chapter. The philosophy of the author is that all school stakeholders must collaborate to develop an environment in which all of these students can be successful in science classrooms. The problem is not the student (a deficient-model philosophy). All students bring knowledge and skills that can help them be successful in science classrooms. We must build on that knowledge and those skills so that all students can understand the scientific principles, laws, and theories; use science processes to help them think critically; and develop values that aid them in their decision making.

This chapter cites research in four main categories: (1) ethnic and cultural diversity in science classrooms; (2) cognitive and affective student factors related to ethnicity, culture, and science learning; (3) classroom environments conducive to science learning; and (4) teacher knowledge, affective characteristics, teaching methods, and curriculum. The author reports on successful science programs for culturally diverse students.

A note about general terms for grouping people: There is diversity among African Americans, Asian Americans, Hispanics, Native Americans, and Americans catagorized as European or White. Most data on science learning and careers do not identify differences within the diverse groups; that is, statistics are given for large groupings. The author of this chapter wishes all people to recognize that diversity within groups certainly exists, but general terms are used unless more specific data are known. Occasionally, data refer to urban versus rural or low versus high economic conditions, which may blend culture and ethnicity.

ETHNIC AND CULTURAL DIVERSITY IN SCIENCE CLASSROOMS

The 1990 census counted 248,709,873 people in the United States and divided the populations as follows: African Americans (11.8 percent), Asian Americans (2.8 percent), Hispanics (9.0 percent), Native Americans (0.7 percent), and Whites (75.6 percent) (1990 Census of Population and Housing, 1992).

Student Populations

Since 1954, the science classrooms in the United States have been diversifying because of desegregation efforts, immigration, and higher birth rates of some ethnic groups. Before that time, students from several cultural and ethnic groups were segregated. In 1972, the K–12 student enrollment of African American, Asian American, Hispanic, and Native American groups in the nation's public schools was 21.7 percent (Hilliard, 1984; McNett, 1983). By 1984, that figure had increased to 30.4 percent (*The Condition of Education*, 1987). The student population in 23 of the 25 largest school districts in the United States is dominated by these diverse groups of students, with 75 percent being African American (Hilliard, 1984; McNett, 1983).

Enrollment in kindergarten through grade 8 was expected to increase from 33.3 million in 1989 to 33.8 million in 1990 (National Center for Education Statistics, 1990). By the year 2000, African Americans, Asian Americans, Hispanics, and Native Americans are expected to compose 33 percent of the K–12

The author thanks reviewer Steven J. Rakow (University of Houston–Clear Lake).

enrollment (McNett, 1983). According to 1990 census data, states such as Florida, New Mexico, and Utah have experienced changes in population and diversity since the 1980 census. Florida had a 32.7 percent increase in its population with these major increases: African American (31 percent), Asian American (171.9 percent), Hispanic (83.4 percent), and Native American (88.7 percent) ("The 1990 Census, State by State," 1991). Not only has the ethnic diversity been increasing in the United States, the language diversity has also.

Hispanics. In 1990, California, Texas, New York, and Florida had 1 million Hispanics in their population. (The group referred to as Hispanics consists of Mexican American or Chicano, Puerto Rican, Cuban, and other Spanish or Hispanic groups.) More than 80 percent of the Hispanic population reside in the following ten states: Texas, New York, Florida, Illinois, New Jersey, Arizona, New Mexico, Colorado, and Massachusetts (U.S. Department of Commerce, 1991).

Asians and Pacific Islanders. The Asian and Pacific Island population has doubled in the last 10 years. Chinese and Filipino people are among the largest groups in the Asian population with 1.6 and 1.4 million, respectively. Japanese, Asian Indians, and Koreans follow with a total population of 800,000 (U.S. Department of Commerce, 1991). The five largest Pacific Island groups include Hawaiian, Somonoan, Guamanian, Tongon, and Figian. These groups differ in culture, population size, and geographic distribution in the United States. Populations vary greatly from state to state. Sixty-three percent of the residents of Honolulu are Asian Americans. There is a 55 percent chance that any two randomly selected people in Honolulu are different ethnically or culturally ("Analysis Puts a Number on," 1991). However, the population of Kentucky is composed of only 0.48 percent Asian Americans.

Languages. The headline in an Atlanta newspapers read "Dekalb melting pot is bubbling. Asian, Hispanic influx enriches diversity, but stirs anxiety, too" (Scroggins, 1992a). In an elementary school in the Dekalb school district, 13 different languages are spoken on the playground. Along a 6-mile stretch of road in the county, one could purchase a car, attend a church, see a doctor, and lease a movie using the Vietnamese, Chinese, Korean, or Spanish language. The ethnic makeup of this southeastern county has changed from predominantly White to a more diverse community. The new percentages for the county's population are Whites (45 percent), Hispanics (24 percent), African Americans (19 percent), and Asian Americans (12 percent). Sixty percent of the population in the southeastern part of this county identified itself as either Hispanic or Asian.

Enrollment figures for 1988 showed that 2 million school-age students (5 percent of the school population) were categorized as limited-English-proficient (LEP) students. These students have not mastered English well enough for it to be the language for classroom instruction ("Staffing the Multilingually Impacted Schools," 1990). More than 60 percent of LEP students are from Spanish-speaking families; 30 percent of the federally funded programs serve students who speak Native American

languages, Southeast Asian languages, Asian languages, Arabic, Tagalog, and Urdu (Schmidt, 1991).

The only high school in the southeastern part of the United States that has established a special English laboratory for youths who are illiterate in their own language is in Georgia (Scroggins, 1992b). In some school districts on the West Coast and in the Northeast, 40 different first languages are spoken (Banks, 1989a). In 1984, approximately 25 percent of the public school teachers had LEP students. Only 3.3 percent of these students possessed the academic preparation or language skills to be instructed in English (*The Condition of Bilingual Education in the Nation*, 1988).

Economic Status. The student population in many of the large city-school districts is disproportionately poor. In 1959, 26.9 percent of children under 18 lived below the poverty level; in 1985, 20.1 percent were below the poverty level (Office of Educational Research and Improvement, 1987). The latter resource also lists the percentages of children in three different populations living below the poverty line in 1959 and 1985.

	1959	1985
Hispanic children	33.1%	39.6%
African American children	65.5%	43.1%
White children	20%	15.6%

Among children aged 5 or younger who lived in poverty in 1988, nearly 50 percent lived in families headed by mothers who began having babies when they were teenagers. Fifty-nine percent of the women who received Aid to Families with Dependent Children (AFDC) payments in 1988 were 19 or younger at the birth of their first child (Moore, 1990).

High School Dropout Rates. Data from *High School and Beyond* show that 17 percent of the 1980 high school sophomore cohort who were from low-socioeconomic-status families dropped out, compared with 9 percent and 5 percent from middle- and high-socioeconomic-status families (Peng & Takai, 1983).

High school completion rates of African American, Hispanic, and White students differ. Shown below are high school graduation rates of 18- and 19-year-old students in 1974 and 1985 from *The Condition of Education* (1987).

	1974	1987
African American students	55.8%	62.8%
Hispanic students	48.9%	49.8%
White students	76.2%	76.7%

According to Haukoos and Chandayot (1988), as many as 60 percent of Native Americans who live on some reservations will drop out of school before twelfth grade. Many Native Americans fall ½ to 1 ½ years behind by mid-elementary school. Some of the high school students are 2 to 2 ½ years behind. The percentages of tenth-graders in 1980 who later dropped out of school were 8 percent for Asian Americans, 15 percent for Whites, 22 percent for African Americans, 28 percent for Hispanics, and 36 percent for Native Americans (Indian Nations At Risk Task Force, 1991).

K–12 Teachers

In 1985, 86 percent of the public and private school teachers were White and 10 percent were African American. Sixty-eight percent of K–12 teachers were female (*The Condition of Education,* 1987). Most of the K–12 teachers in America are White and female.

Summary. Diversity in the classroom is both a challenge and an opportunity. It is a challenge because our knowledge and skills for effectively teaching diverse student populations has not kept up with the influx of students we need to serve. It is an opportunity to teach about the joy of experiencing a variety of cultures and languages and of respecting all people. It is necessary to provide K–12 science teachers with the knowledge and skills to teach all students science.

COGNITIVE AND AFFECTIVE STUDENT FACTORS RELATED TO ETHNICITY, CULTURE, AND SCIENCE LEARNING

With increasing diversity in classrooms, many acknowledge that there is a diversity in the ways students learn. Students learn at different rates and focus on different aspects of science. In addition, students bring with them a conceptual knowledge base of science to which they add new knowledge. They develop their science understanding through sensual impressions, language, brain organization, social environment, and instruction (Duit, 1991). However, students often do not alter their understanding of science concepts. This section focuses on cognitive and affective student factors related to ethnicity, culture, and science learning.

Ways of Knowing

Epistemology, the study of the nature and method of knowledge, is part of the exploration of philosophy and psychology. Learning has been defined in a variety of ways. According to the constructivist framework, learning, the process of acquiring new knowledge, is active and complex. This process is the result of an active interaction of key cognitive processes, such as perception, imagery, organization, and elaboration (Glynn, Yeany, & Britton, 1991, p. 5). It is a means of refining existing knowledge in which conceptual changes take place, rather than simple gaining of new knowledge. Students' prior knowledge, expectations, and preconceptions serve as filters for the information that is focused on. In science classes, the ideal way for students to understand science concepts includes students challenging the new concept, grappling with it, attempting to make meaning of it, and eventually integrating it with what they already known.

Many believe it is important for students to generate and utilize their ideas (Wittrock, 1985). According to Templin and Ebbs (1992), students need the opportunity to test their ideas critically, discuss them with others, and have them evaluated by others. However, some cultures in the United States do not require the student learner to discuss ideas with others.

Even within a specific culture, there are basic patterns of beliefs and values that define how people act, judge, decide, and solve problems in their own life and in the world; these systems vary across cultures (Anderson, 1988). Students, regardless of ethnicity or class, construct their own knowledge socially (Novak, 1985; Strike & Posner, 1985b). Hence, "cognitive abilities" are socially transmitted, socially constrained, socially nurtured, and socially encouraged" (Day, French, & Hall, 1985). World views vary from culture to culture.

Non-Western and Western World Views. According to Crichlow, Goodwin, Shakes, and Swartz (1990), "a wide range of resurgent cultures are asserting their privilege to represent themselves, their histories, realities, and voices in their own cultural tones" (p. 101). This may involve how a certain kind of knowledge, concept, or symbolic image is created. This emancipatory approach provides an opportunity for both the teacher and student to engage in further ways of knowing, thinking, and being.

Anderson (1988) has delineated some fundamental dimensions of non-Western and Western world views. People with non-Western world views, such as African Americans, Chinese Americans, Mexican Americans, Native Americans, and many European American females, emphasize group cooperation, value group achievement, believe that people should be in harmony with nature, use time relatively, accept affective expressions, embrace the extended-family concept, think holistically, embrace religion, accept world views of other cultures, and are socially oriented. Individuals with Western views, such as many European American males, Western acculturated African Americans, Japanese Americans, and Puerto Ricans tend to emphasize individual competition, value individual achievement, believe that people must conquer and dominate nature, follow rigid time schedules, devalue affective expressions, cling to the nuclear-family concept, think dualistically, separate religion from other parts of their culture, believe that their world views are better, and are very task oriented.

Native Americans emphasize a close relationship with nature, rather than control over nature (Caduto & Bruchac, 1989). To some, a tree or a rock is like a brother or sister. Many Native Americans view the relationship between people and nature as highly spiritual; thus some explanations for natural phenomena may be supernatural (Robbins, 1983).

Conceptual Change. According to Posner and his colleagues (Posner, Strike, Hewson, & Gertzog, 1982; Strike & Posner, 1985a), the following conditions must exist for conceptual change to occur in learners:

1. Learners must be displeased with their personal theories about the phenomenon under study.
2. Learners must have a minimal understanding of the new conception related to the phenomenon.
3. The new conception about the phenomenon must seem believable when introduced.
4. The new conception about the phenomenon must have explanatory and predictive power.

If students' prior explanations are still perceived as better than the alternative ones, then students will maintain prior beliefs and frameworks.

Summary. In order to have a scientifically literate society, it is necessary for all students to understand science concepts. Many factors influence science learning. Some are related to the student, others to the teachers. Both cognitive and affective characteristics of students determine how and what they learn. Ways of knowing, styles of learning, and prior knowledge and skills interact.

Styles of Learning

Cognitive style has been characterized by Saracho (1988) as a way of processing information and by Bagley (1988) as "the manner in which individuals perceive, interpret, organize and think about themselves in relation to their environment" (p. 143). Cognitive styles are represented on continua in processing information; the cognitive style of each individual student is found on the continuum (Driscoll, 1987). The term cognitive style has a myriad of meanings depending on the frame of reference. Thus, field dependence-independence, analytical-relational, reflection-impulsivity, differentiation-remote association, holistic and analytical perceptual strategy, and educational cognitive styles are discussed below.

Field Dependence—Field Independence. Oakes (1990) interpreted findings on the effects cognitive abilities and attitudes might have on ethnic differences in achievement levels and persistence in science. Field dependence is believed to be highly influenced by the context in which knowledge and skills are embedded. She cautioned about use of the results of studies in this area, because few studies controlled for the confounding effects of social class on the language-achievement relationship. However, it is important to investigate the findings in the area of cognitive and perceptual styles as they relate to different ethnic groups found in the classroom.

Originally, field-independent learners were identified as those who easily perceived a hidden figure on the Embedded Figures Test; field-sensitive (or field-dependent) learners found this task difficult because of the obscuring design (Castenada & Gray, 1974; Witkin, 1978). Field-sensitive learners are able to restructure a field on their own, whereas field-independent learners respond to the dominant properties of the field (Witkin, 1978). Field-sensitive learners do better in social situations.

Witkin maintained that field sensitive-independent cognitive style is relative to context; therefore individuals develop cognitive styles that are adaptive to their environment. Berry (1976) proposed that survival skills in a particular environment favor the earlier development of specific spatial cognitive abilities over other cognitive abilities. These adaptations include forming modes of behavior that are compatible with the cognitive style and influencing the behavior of others in a manner that accommodates the cognitive style so that good interpersonal relationships develop. Field-sensitive learners are influenced by opinions of others when making decisions and their performance is affected more by approval or disapproval of authority figures. Field-independent learners are not readily persuaded by opinions of others and their performance on tasks is less swayed by beliefs of authority figures (Ramirez, Casteneda, & Herold, 1974).

Findings by Holtzmann, Goldsmith, and Barrera (1974) support a relationship between field-dependent or field-independent style and science domain. They found that field-independent learners were much more likely to choose science and mathematics careers. In addition, Cross (1976) and Douglass (1976) have suggested that type of cognitive style affects how efficiently students learn and solve problems in a particular environment. Field-independent learners are more successful with inductive materials and open-ended problem-solving activities, although field-dependent (field-sensitive) learners experience more success with deductive materials because they are more global in their thinking.

Ramirez and Price-Williams (1974) found that fourth-grade African American and Mexican American students tended to be more field sensitive in their cognitive style and that White students were more field independent. The more acculturated Mexican American students are, the more their cognitive style resembled that of White mainstreamed students (Ramirez, Casteneda, and Herold, 1974).

Bagley (1988) summarized conclusions from the cross-cultural studies on field independence-dependence in the following manner:

1. Skill in disembedding figures is to some degree related to schooling experiences.
2. Cultural and social variables that influence cognitive styles rapidly vanish in Jamaicans and Gujeratis (from India) who migrated to metropolitan areas.
3. Japanese children are extremely field independent compared with other groups, including White children; this finding contradicts Witkin's original theory.
4. In all the studies of different cultural and migratory status reviewed by Bagley, females were more field sensitive than males.
5. The influences on field dependence or field independence are more culture specific than cross-cultural.

If one of the variables in the model of career selection is cognitive style, then the paucity of ethnic groups in the scientific profession can be explained. In a study of undergraduate students enrolled in an introductory biology, mathematics, and social science classes, Kahle (1982) found the following:

1. The normal distribution of scores on the Group Embedded Test was not found for this sample of undergraduate students.
2. Generally, the highest percentage of field-sensitive students came from small, rural communities or from urban cities.
3. The only selected major that had a greater percent of field-independent learners than field-dependent learners was the natural sciences. Twenty-three percent of the students selecting science as a college major were field independent as compared to 14 percent of the field-dependent students.
4. Approximately equal percentages (75 percent) of field-independent and field-sensitive students possessed positive attitudes toward biology.

Olstad, Juarez, Davenport, and Haury (1981) maintained that some of the differences in academic performance of ethnic groups in mathematics and science can be explained by cognitive styles. Hadfield, Martin, and Wooden (1992) stated that Navajos are taught to look at the big picture first in their culture. Because analytic skills, abstract and impersonal orientations, and working independently are valued in science and mathematics, field-sensitive students are incompatible with preferred school practice. Witkin (1978) and Anderson (1988) observed that field-sensitive students were less likely to do well in the sciences because of the present way of teaching the discipline.

Analytical and Relational. Cohen (1969) identified two other types of cognitive styles: analytical and relational. Analytical thinkers possess both breadth and depth of general information, are capable of analytic abstraction, and can extract salient information from embedding contexts. Relational thinkers have a difficult time extracting relevant information from text and performing analytic abstractions. She discovered that a larger percentage of relational thinkers came from low-income environments. She agreed with Cohen that analytical-relational style is the result of different social environments that stimulate, reinforce, and make functional one style of conceptual organization. Analytical thinkers prefer to function in formal groups, whereas relational thinkers like the shared-function style found in cooperative groups.

Reflection-Impulsivity. Fennema and Behr (1980) have developed the reflection-impulsivity cognitive style, a continuum on speed and accuracy consistency with which alternative hypotheses are expressed and the information processed. The Matching Figures Test is used to measure reflection-impulsivity (Witkin & Goodenough, 1981). Impulsive learners tend to pursue the first answer that comes to mind, while reflective learners reflect longer on the different possibilities for answers and problem solutions (Driscoll, 1987).

Differentiation—Remote Association. Morse and Gordon (1974) and Ba-Haim and Wilkes (1989) studied the differentiation—remote association continuum. Differentiation refers to the ability to recognize and formulate research problems. Ba-Haim and Wilkes (1989) described high differentiators as those who perceive their surrounding environment as a series of

distinct parts, recognize the unique qualities in people, and find it easy to make subtle distinctions even on subjective criteria. Because they are open to differences and sensitive to nuances and inconsistencies, they are able to formulate fruitful research questions. Low differentiators are limited by perceptions and view the environment as homogeneous. It is more difficult for them to accept the unanticipated as fact and focus less attention on explaining discrepancies in a theory.

Remote association alludes to the ability to generate and evaluate a myriad of alternative solutions to a given problem. High remote learners fashion optimal or appropriate solutions to problems by utilizing available materials, linking them with ingenuity, and bringing in new elements not normally connected with the problem. Bar-Haim and Wiles (1989) have applied this construct to specific areas of science. Individuals who are both high differentiators and remote learners are good integrators; they are able to identify problems and solve them. People who score low in both differentiation and remote association are the most restricted and can be viewed as the technicians of science. Individuals who score high on differentiation and low on remote association are good problem finders, whereas those scoring high on remote association and low on differentiation are good problem solvers.

Holistic and Analytical. Anderson (1988) described holistic and analytical thinking based on the works of Cohen (1969), Cooper (1980), and Vygotsky (1978). Holistic thinkers focus on descriptive abstractions, determine word meaning by content, rarely utilize synonyms and make comparisons, often employ relational and institutional classification, and are inclined to use the word "you" to reflect group identity and pull the reader into the reading. On the other hand, analytical thinkers focus on analytic abstractions, concentrate on the formal meanings of words, commonly use synonyms, often use many kinds of generalization comparisons, classify hierarchically, can easily adopt a third-person viewpoint, attempt to be very objective, and are able to reflect identities separate from what is going on.

Perceptual Strategy. Perceptual strategy or modality represents the method by which a student gathers and translates information from the environment. Sometimes, perceptual style and cognitive style are used interchangeably. However, perceptual modality, strategy, or style is usually measured by a self-report instrument that focuses on the manner in which students prefer (1) gathering and receiving information, (2) working together, and (3) expressing themselves. Specifically, students indicate whether they favor gathering linguistic and numerical information in learning by the use of their auditory senses or visual senses or by doing. Also, their preferences for group or individual study and for written or oral expression are delineated. Haukoos and Satterfied (1986) found that college biology students of the Sicangu Lakota of the Western Teton Sioux were more visual-linguistic than auditory-linguistic and preferred not to express themselves orally as compared with the White biology student referent group. The two groups in this study differed socioeconomically as well.

Fritz (1990) found that styles of learning center on the how but not the why, and they do not address aptitude. However, cognitive and perceptual styles do provide additional insights into understanding students in a science classroom.

Summary. Educators are reminded that the best ways to teach children depend on the individual students' cognitive style. The more traditional ways of teaching science may not be as effective with underrepresented groups in science as are some of the active, relational, and holistic approaches. However, further investigations are needed in this area of cognitive and perceptual styles.

Prior Knowledge and Skills

Some factors that affect the technical education of Hispanic undergraduate students include poor mathematical preparation, poor language skills, and conceptual difficulties (Burns, Gerace, Mestre, & Robinson, 1982). Atwater (1981) found that African American students who were successful in engineering and science programs had a strong background in the sciences and mathematics. Puerto Rican students majoring in the technical fields scored significantly lower on language proficiency and algebra tests. They had difficulties in solving algebra word problems—specifically, translating verbal statements into algebraic equations. Alick and Atwater (1988) found this also to be the case for African American students enrolled in general chemistry.

A significant correlation was found between mathematics achievement and language proficiency measures for Hispanic students (Mestre, 1981). In this study, it appears that Hispanic students had not reached a point where they felt proficient in reading and solving technical problems. Duran (1979) found that the performance of Puerto Rican college students on deductive reasoning problems was significantly related to reading comprehension skills in the language of the problems and that the pattern was similar across English and Spanish. Cummins (1979) proposed there is a threshold level of linguistic competence a bilingual student must reach in order to develop further cognitively.

Bilingual Programs. According to Mason and Barba (1992), few Hispanic students enrolled in bilingual programs receive science instruction in their primary instructional level or at their appropriate grade level. They have summarized the research findings on the science instruction of LEP students. These students receive science instruction from bilingual classroom aides rather than from certified science teachers. The aides rely on drill-and-practice science activities and other activities that focus on low-level thinking skills. The students use science textbooks that are not in their primary language. Some students (about 18 percent in California) have access to science textbooks written in Spanish, but not to laboratory manuals, dictionaries, encyclopedias, atlases, computer software, or audiovisual materials in Spanish. Therefore, these students fail to understand the science concepts or to develop problem-solving and inquiry skills. Mason and Barba (1992) recommended that actual science objects, models, and line drawings be presented to bilingual students in both Spanish and English so that the

students are able to form science schema in either Spanish, English, or both languages.

Sutman, Allen, and Shoemaker (1986) summarized the finding of several biology teaching studies. Although only 9 of 32 technical words such as *recombinant* and *episomal* created problems for LEP students, 45 of 53 common nontechnical words caused them problems. Researchers suggested that technical words gave these students fewer problems because the words are used internationally and older students are able to find good translations and explanations for scientific terms in bilingual technical dictionaries. When nontechnical words are used by science teachers, they should be written on the chalkboard and their meanings explained. Science teachers can help students in bilingual and multilingual classrooms by adapting their science activities using dialogues, cloze exercises, crossword puzzles, group activities, and hands-on activities.

Summary. Because difficulty in understanding science concepts is so often tied to mathematical and language knowledge, it is important for educators to be aware of students' knowledge and to work with them accordingly. When science is taught in the primary language of students and when they have access to bilingual visuals, science success is more apt to be within their reach.

Student Beliefs, Attitudes, and Values Toward Science Learning

Beliefs, attitudes, values, locus of control, achievement motivation, and career expectations influence the kind and amount of science learning occurring in schools. The influence of these factors is different for various ethnic and cultural groups. Social and economic influences, not the genetics of the students, are responsible.

Attitudes. Wiggins, Atwater, and Gardner (1992), found that about 20 percent of urban African American middle school students had a positive attitude toward science, and only 8 percent had a negative attitude. The large majority of the students had neutral attitudes toward science. The 20 percent with positive attitudes toward science had positive attitudes toward their science curricula and science teachers but negative attitudes toward their schools. The 8 percent of students with negative attitudes toward science had negative attitudes toward their science teachers, science curricula, science classroom climates, science classroom physical environment, and schools. The two groups of middle school students differed significantly on their attitudes toward their science teachers and the science curricula.

Positive attitudes toward science of 75 sixth-grade students correlated significantly with attitudes toward science curriculum and science teacher, science self-concept, achievement motivation, and family attitude toward science (Gardner, Atwater, & Wiggins, 1992). The same trend was noted for 92 seventh-graders; however attitude toward science correlated significantly with attitudes toward science teacher, science self-concept, and achievement motivation for eighth-graders. Attitude toward science teachers accounted for 16 percent of the variance in attitudes toward science for the eighth-graders.

Gender. Jacobowitz (1983) found that the relationship between science self-concept and sex-role concept of African American students may be stronger than the relationship between science self-concept and science achievement for urban junior high students. In addition, significant differences were found between the male and female mean scores on science self-concept; the males had higher scores than the females.

Using the 1981–1982 National Assessment in Science data, Rakow (1985a) found that 9-year-old White males had the most positive attitude toward science and 9-year-old African American females had the least positive attitude toward science. Hispanic females had the least positive attitudes toward science, and White males had the most positive attitudes toward science of any of the groups. Rakow (1985a) believed that this link was gender based rather than ethnic based.

Family Pressure in the Asian Culture. Shih (1988) stated that family pressures add a dimension to some Asian American success in the sciences. In most Asian cultures, a person's social identity is closely affiliated with the kin group. The children are taught that they are responsible for the honor or shame of each family member; therefore success or failure is shared with the family. Success is the necessary; failure is feared. When the family is very tightly knit, parents with high expectations for both school achievement and career goals provide additional pressures for Asian American students to succeed.

Peer Pressure in the African American Community. Fordham and Ogbu (1986) presented the argument that peer pressure not to act White is one major reason African American students do not reach their full potential in school. Successful African American males found ways to offset their academic success so that they would not be rejected by their peers for obtaining good science grades. They would participate in sports, clown around, and exchange academic assistance for inclusion. Some African American females maintained a low school profile in order to be included. They selectively skipped classes, halted their academic performance, and refused to participate in academic clubs and competitions. Fordham (1993) found that achieving African American females became silent in their classes and became the least visible of all the classroom students.

Locus of Control and Achievement Motivation

Some teachers motivate students with extrinsic rewards such as team scores, privileges, or extra credit (Kagan, 1980; Slavin, 1987, 1988); other teachers believe that students can be motivated by the concern for all group members (Aronson, 1978; Lickona, 1980). Based on the causal model of Templin and Ebbs (1992), teacher-student relationship, as well as socioeconomic level, has a large direct effect on locus of control of eighth-graders.

Templin and Ebbs (1992) proposed a causal path model for academic risk taking based on a sample of eighth-graders from the National Educational Longitudinal Survey–Base 1988 data set. Academic risk-taking construct was measured in relation to fear of asking questions. It was found that academic risk taking was related to locus of control; teacher-student relationship had a direct effect on locus of control.

Locus of control had an indirect effect on women's choice of field of study. Those with more internal orientation were more likely to persist in a quantitative field such as the natural sciences (Ethington & Wolfe, 1988). African American females were more likely to persist in science and mathematics than White females. Urban middle school students who have positive attitudes toward science and sixth-grade students with negative attitudes toward science were very motivated to achieve in school (Wiggins et al., 1992).

Summary. Attitudes toward science vary according to students' feelings about their science teachers and the science curricula. In junior high school, male students scored higher than female students in science self-concept, an attitude that continues through high school. Family pressure, peer pressure, and locus of control are also significant factors affecting attitudes toward science.

Career Expectations Related to Science

African American students' achievement is affected by the existence of a "job ceiling" or barriers in the opportunity structure (Ogbu, 1978). African American students perceive that the careers and the opportunities available to them are limited, especially in the sciences. Harper (1989) found significant negative correlation between an early career selection and science students' intended highest degree. The longer these African American students delayed in selecting a major, the greater the chance that they would elect to pursue a doctorate in the sciences or medicine.

Using regression analysis, Thomas (1986) found that having an interest in science hobbies during childhood, holding childhood aspirations of being a scientist, and receiving encouragement to major in science were positively related to students' interest in high school science.

Burns et al. (1982) surveyed seventh-, eighth-, and ninthgrade Hispanic and White students to discover their career plans. When Hispanic students were asked to describe what engineers do, 52 percent of them had no idea at all, whereas 40 percent of the White students were able to provide a description. Only 26 percent of the Hispanic students were willing to guess, while 53 percent of the White students were willing to guess. Four times as many Hispanic students as White students intended to enter a vocational rather than an academic program after high school graduation. Differences were found among the parental education levels of the two groups. It was suggested that because of the lower parental education level of Hispanic students, the students had little opportunity to associate with friends or relatives in the technical professions such as engineering, computer science, and physics; thus they possessed little information on these careers.

Gender in Science Careers. Women from underrepresented groups have special problems that White males and, to a lesser degree White females do not have. In 1975, a conference convened by American Association for the Advancement of Science (AAAS) highlighted the visibility of women from underrepresented groups in the sciences in its conference proceedings, *The Double Bind: The Price of Being a Minority Woman in Science.* The following perspective was shared in this document:

There seems therefore to be a range of costs to the individual in the attainment of a professional science career. The more an individual resembles the "typical scientist" the lower are his costs. Each factor of deviation from the norm raises the costs so that, as a group, minority women must pay a tremendous price for a career in science. This "differentness" of the minority woman in science may not only be a factor in the scientific community but also in the contest of her culture. The tremendous personal cost that results from the combined effect of being a scientist, a woman and a member of a minority racial or ethnic group was frequently alluded to in the conference discussions. The toll of foregone social and personal activity, highly valued in traditionally defined cultural roles, was for many severe.... The feeling of differentness, which for most of the conferees began to develop as early as their interest in science, was reinforced continually by the recurrent experience of being the only woman in so many situations. (*The Double Bind: The Price of Being a Minority*, cited in Malcom, 1989, p. 15).

Ethington and Wolfe (1988) established for females in the 1980 sophomore cohort of *High School and Beyond* that the number of mathematics and science courses taken in high school was the chief determinant in their model of choice of fields of study. African American women were more likely to select a quantitative field of study than any other female group. However, women from higher socioeconomic levels were less likely to enroll in an undergraduate quantitative major, such as the sciences.

Jacobowitz (1983) found that gender was the strongest predictor of science career preferences of urban eighth-graders. In a study of urban middle school students, Wiggins et al. (1992) found that 57 percent of the sixth-graders, 53 percent of the seventh-graders, and 47 percent of the eighth-graders planned to enter a career that involved science or mathematics. Yet, 50 percent of these students planned to take only the required science courses for graduation or were undecided. A high percentage of these middle school students did not see a connection between course-taking pattern and career options.

Intervention for Females. Malcom (1989) has recommended interventions that might be effective in promoting participation in science and engineering careers by African American, Asian American, Hispanics, and Native American females. Her suggestions for successful programs for these females include:

1. Targeting these females at an early age
2. A continuous focus on rigorous preparation in science and mathematics
3. Promotion of hands-on involvement with such activities as science fairs and projects
4. Associations with role models who are like the students

5. Procurement of appropriate career information
6. Early experiences in science research
7. Direct discussions about gender and ethnic specific issues such as the combination of marriage, family, and career responsibilities and conflicts of cultures
8. Opportunities for early work experiences in science-related careers

Adult role models and encouragement seem to influence some students' career decisions. Malcom, Hall, and Brown (1976) and Young and Young (1974) found that exposure to, along with interaction with, professional role models in the sciences and engineering was critical for the recruitment and retention of students' science interest and participation. African American students reported they received less encouragement to pursue science college majors and careers than did White students. Thomas (1986) found that encouragement to major in science was a very important factor in influencing students' interest in high school science courses.

Pallone, Hurley, and Rickard (1973) surveyed a group of Puerto Rican, African American, and White females to identify the key influences on their career choices. Mothers, school counselors, relatives not of the immediate family, persons employed in the field in question, teachers, peers, their fathers, their brother or sister, and neighbors or family friends were the key influences of Puerto Rican females. The major difference among the different females was the importance of school counselors. African American and White females ranked the counselor fourth and eighth, respectively, while the Puerto Rican females ranked the counselors second. At least in this study, counselors had more influence on Puerto Rican females than they did on African American or White students.

Socioeconomics and Careers. Zuckerman (see Pearson, 1986) proposed that the general social characteristics of African Americans differ from chracteristics of those who pursue science careers. In general, scientists come from well-educated families typically headed by professional or managerial fathers. Therefore, she asserted, African Americans are underrepresented in the sciences because of their socioeconomic level. A study by Harper (1989) found that few students who were enrolled in a graduate program in the sciences or medical school or who had successfully completed graduate school or a medical program had parents with science-related careers. In a study involving 60 Hispanic undergraduate students majoring in technical fields, equivalent percentages of the Hispanic and White undergraduates had relatives or friends with occupations in fields similar to those that the students had selected. Approximately one-third of each group did not know anyone in the field they had selected (Burns et al., 1982). However, there were differences between family income levels, with more Hispanic students in the lower income levels.

Webb (1992) iterated that children start to become engineers, scientists, and scientifically literate at an early stage in life. He maintained that preschool programs should be available to the 14 million children who live in poverty, of whom one-third are African American, Asian American, Hispanic, and Native American. These students drop out of college-track mathematics and science courses early, suffer from chronic shortages of good science teachers, and experience little or no hands-on science. The effect is "lower achievement scores for the Nation as whole, but personal hardship for those students who have not been taught the skills they need to take full part in our advanced industrial society" (p. 29).

Summary. Early interest and exposure, such as involvement in a science hobby, is one predictor in the choice of a science career. Others include role models and associations and family careers and education levels. Women, especially from underrepresented groups, can benefit from early intervention programs that prepare them in science and math.

Classroom Environment Conducive to Science Learning

Interactions among different ethnic groups of students play an important part in classroom environment. Douglass, Kahle, and Everhart (1980) found that urban students' opportunities for, and the nature of, contact with other ethnic groups accounted for 43 percent of the variance of unfriendly ethnic contacts. Other variences included family and neighborhood background (10 percent), perceptions and attitudes toward other ethnic groups (4 percent), and personal characteristics (2 percent). In this study, Latino and White students reported the most friendly interactions with each other and the least avoidance behavior among the various ethnic groups in that setting. African Americans recounted the least friendly interactions and the most avoidance behavior. Unfriendly interactions were most significantly related to the amount of anger students felt toward other ethnic students and relationships they had with students of their own ethnic group. In order to curtail avoidance behaviors in the classroom, opportunities for different ethnic students to work together in groups must be provided.

Farrell (1984) proposed that different social and emotional climates were needed for both field-sensitive (dependent) and field-independent students and teachers. Field-sensitive learners liked to work in groups to achieve a common goal, asked for aid from others in ambiguous situations, more openly expressed their feelings toward the teacher, and were interested in the personal experiences of their teachers. Field-sensitive teachers expressed more approval and warmth and encouraged group achievement by students. Field-independent learners preferred to work alone, were less likely to request help from others in ambiguous situations, needed less physical contact with teachers, and limited their interactions to the immediate tasks. Field-independent teachers maintained more formal relationships with students, focused on structural objectives, and encouraged individual student achievement. When the same kinds of learners are taught by the same kinds of science teachers, the social and emotional climates of the classrooms are compatible. When a mismatch occurs, the social and emotional climates are not supportive of science learning.

Hmong Adult Cooperative Learning. Hvitfeldt (1986) proposed that Hmong adults perceive the world and interact with others in ways that differ significantly from those of many White adults. The Hmong is the largest ethnic group in Laos; many have immigrated to this country. Hmong adults are successful with cooperative learning. Individual students continually monitor verbally and nonverbally the progress of the other members of the group. Unsure students seek assurance from others. Assistance is given freely by surer students to the unsure, even if not requested, and it is always accepted. When individuals were praised by the teacher, it was obvious that they did not want to be singled out. Classroom achievement was always perceived as a group accomplishment, not an individual accomplishment. This behavior did not extend to the family. Decisions in the family were made by the head of the household. If Hmong adult learners behave in this manner, one would expect their children to behave similarly in science classrooms.

Class Size. Urban middle school students have a less than positive attitude toward the physical environment of their science classrooms (Wiggins et al., 1992). Powell (1990) suggested that African American performance in the sciences can be improved if the educational environment is not crowded and noisy. The classrooms should be quiet, pleasant, and orderly; instruction should be given to a small group of students so that the teacher can focus on those students and provide constant feedback to them.

Safety. Maslow (1968) identified safety as one of the basic needs. It is important that students do not feel they will be victimized by other students. Sometimes, teachers are unaware of the breaches of safety. Marzano (1992) interviewed a secondary teacher who was positive that his students felt safe in his classroom. When students completed a questionnaire anonymously, he discovered that some of his students were demanding payment for protection.

Assessment and Evaluation Instruments

According to Singh (1987), the assessment debate in science centers around underachievement. It is also partly in response to complaints from employers that high school graduates lack the skills for employment. Science process skills are considered important, and many standardized and teacher-made tests do not assess these skills.

Racism and Testing. Samuda (1986) discussed structural, technical, and scientific racism as they relate to assessment. Structural racism includes the mind-set that each student must be judged according to the same standards, procedures, and values regardless of cultural or class differences. Technical racism implies the use of correlations and citations of constructs, content, and concurrent validity as statistical evidence to justify the labeling of African American, Hispanic, and Native American and the placement consequences of test results. Scientific racism resides in the theory that African Americans, Hispanics, and

Native Americans and some others are of inferior mental stock. Samuda found the following underlying assumptions of some test makers:

1. There is a commonality of experience shared by the test takers.
2. The educational opportunities in the home and neighborhood are similar for all test takers.
3. Every test taker has equal facility with the language used in class instruction.
4. The syntax and word usage are familiar to all the test takers and sociocultural, economic, and linguistic differences can be ignored.

None of these underlying assumptions are valid, of course.

Haukoos and Satterfield (1986) have suggested that with Native American students, evaluation should be viewed by the students as constructive, not punitive. Sometimes, Native American students are unaware of the purposes of testing and the significance of test results. Deyhle (1987) found that second-grade Navajo students defined tests in relation to placement of the desks during testing, rules of cheating and copying, location of the test materials on various chalkboards, and teachers who had to be listened to during the test. When the second-grade Navajos were asked about the importance of tests, they were unable to answer (Deyhle, 1987). However, after four additional years of testing, these students' responses were similar to those of white students who understood the importance of testing.

Ability Grouping. Norm-referenced assessment usually determines ability grouping. This practice has an integral function in the ethnic and socioeconomic segregation and imbalance in classrooms. It helps determine the labeling, the lowered self-concepts, and the lowered teacher expectations and perceptions, and it leads to poor teaching of students found in the lower ability groups. Samuda (1986, p. 51) stated that

there are still teachers practising in the field who need to be reminded that norm-referenced assessment is not very useful for minorities because tests of aptitude, intelligence and achievement comprise attitudes and patterns of behavior more typical of the mainstream Anglo- or Euro-Canadian middle-class groups and do not necessary reflect the cultural pattern of ethnic minorities; and that standardized test are for the most part normed on middle-class children and even when they do include minorities they do not properly represent the spectrum. Tests of ability are heavily loaded culturally and typify the performance of the mainstream pupil while discriminating against minorities.

In a study by Rakow (1985b), only 17.6 percent of variance in science inquiry skills of 17-year-old non-White students could be explained by ability, quantity of instruction, quality of instruction, classroom or motivation, and home environment; 31.7 percent of the variance of White students' science inquiry skills could be explained by these five variables. Ability accounted for 20.7 percent and 6.4 percent of the variance in science inquiry skills for White students and non-White students, respectively.

Perceived Importance of Science Education. In the Gallup Poll in 1984, teachers and the public differed in rating the importance of an understanding of science and scientific facts. Seventeen percent of the teachers and 45 percent of the public believed scientific literacy was an important educational goal (*The Condition of Education,* 1987).

Summary. Classroom environment is a variable in science learning. The environment includes interactions among groups of students, with unfriendly interactions detrimental to learning. Other factors are an emotional match with the teacher, workable class size, and safety. Testing can have biased results when it does not take into account cultural and class differences. Is it possible to develop nonbiased assessment instruments? The author maintains that it is impossible. The best science teachers can do is to use a variety of assessment techniques to determine their students' understanding of science.

TEACHER KNOWLEDGE, AFFECTIVE CHARACTERISTICS, TEACHING METHODS, AND CURRICULUM

This section deals with teacher knowledge; beliefs, values, attitudes, and expectations; verbal and nonverbal communication patterns; establishment of power in the classroom; teaching methods and strategies; and curriculum factors. A teacher's knowledge of science concepts is critical in student learning. However, the teachers' knowledge of their students also affects science learning. If teachers do not interact successfully with their students, learning takes place, but not the kind that teachers might desire. A poor understanding of students might prevent the students from developing an understanding of science concepts, principles, and theories.

Knowledge about History, Culture, and Contributions

Many science teachers are ill-prepared to be successful in some classrooms because they have limited knowledge about the history and culture of their students. Even when teachers confront negative attitudes about cultures different from their own, they still need knowledge about the history, culture, and science contributions of the various ethnic and cultural groups (Pearson, 1986). Some teachers may not personally know scientists or engineers from other cultures. Bennett (1990) advocates multiple historical perspectives. Past and current world events must be viewed from different perspectives. Environmental problems must be understood from multiple perspectives, because there is usually more than one solution to a problem.

Many first-year teachers are ill-prepared to meet the unique needs of Native American students because they arrive with little or no targeted education for teaching Native American students. They may spend the first year attempting to understand their students and discovering reasons why students do not respond to their teaching as did students during a previous teaching experience. They ask such questions as "Why don't they like me?" and make such comments as "My students are too quiet and don't seem to understand the lessons" (Allen & Seumptewa, 1988). Many of these teachers, when they are not successful, leave the reservation; others who are successful stay for a time, but eventually leave. Thus, the Native American science precollege teaching force suffers from the revolving-door syndrome.

Ethnic and Cultural Content in Curriculum. Banks (1989b) described four approaches for including ethnic and cultural content in the curriculum. The first approach adds contributions, heroes, celebrations, and people to the curriculum on special days. This approach provides a superficial understanding of different cultures. The second, additive, approach attaches content, themes, and perspectives to the curriculum without altering its structure; thus it enforces the idea that many ethnic groups should not be an integral part of the United States. The third approach, transformation, changes the curriculum so that students are able to get multiple viewpoints about an event or finding. The final approach, decision making, allows students to identify their own social or environmental problems so that they can clarify their values and understanding of the problems. Few science teachers know about the many science contributions made by African Americans, Asian Americans, Hispanics, and Native Americans. This information is not readily available to either classroom teachers or preservice teachers. If a few posters and calendars are all that is within the financial means of science teachers, they will be unable even to take the first approach or aspire to a higher level.

Teacher Beliefs, Values, Attitudes, and Expectations

According to Brophy and Evertson (1981), beliefs are statements considered to be true or false, regardless of whether they are or not. They define expectations as explicit or implicit cognitive predictions with varying degrees of strength and certainty. Attitudes are affective or emotional reactions to objects or people. There is a human need for consistency; thus people are predisposed to consider information selectively so that their beliefs, expectations, and attitudes will be congruous. Science teachers might even use defense mechanisms to avoid becoming aware of information that conflicts with their beliefs if there is a need to maintain their false impressions (Bem, 1970; Festinger, 1957). When impressions and perceptions are accurate, teachers' beliefs develop systematically so that expectations and attitudes formed through experiences accurately reflect the experiences.

Patterns of Teacher Interactions. Power (1972) studied contrasting patterns of teacher interactions in four eighth-grade science classes. He identified three different groups of students based on interaction patterns. (1) The rejection group was composed of students who were low achievers and felt alienated from school and the teachers. They sat at the back of the classroom and were rarely called on. Whenever the science teachers did interact with these students, they were most likely to be praised. Power explained that teachers believed that this behav-

ior compensated for the rejection these students suffered. (2) The concern group was composed of dependent, sensitive, low-ability, low-achievement students who had very positive attitudes toward science. The science teachers gave these students easier tasks and did not set high goals for them. (3) The "success syndrome" students were high-ability, high-achievement, high-status, participating students who attempted to answer difficult questions and volunteered for difficult assignments. Power did not report any teacher favoritism toward this group of students as some other authors did (Silberman, 1969).

According to Pulos, Stage, and Karplus (1982), environmental expectations along with self-perception may cause African American students to profit less from mathematics instruction than they might otherwise. These expectations, in turn, lead teachers to have lower expectations for these students, thereby reinforcing the students' attitudes.

Teachers with Varying Cultural Backgrounds. Dempsey (1981) measured the attitudes of Cuban American and White elementary teachers concerning females and males. He stated that it is generally accepted in the European culture that males are hostile and aggressive and females are passive and conforming in school situations. In some Latin American cultures, the male is viewed as emotional, earthy, and courteous, while the female is seen as controlled, spiritual, and assertive. White teachers described males as more giggly, silent, bossy, steady, open, vexing, factual, rational, bad, ugly, independent, and obtrusive. Cuban American teachers described the males differently; they were more morose, dependent, talkative, shy, protective, creative, pleasing, emotional, sweet, pretty, and imaginative. However, White elementary teachers found the female students to be more virile, obtrusive, mischievous, sharing, straightforward, careless, dependent, quiet, and cowardly. Cuban American elementary teachers described the females as more prissy, shy, malicious, sneaky, independent, loud, brave, and selfish. Dempsey advocated that teachers from different culture backgrounds must be made aware of possible conflicts that might arise from their expectations of students who come from different backgrounds than their own.

Teacher Expectations. Based on a review of 31 articles, Stegemiller (1989) concluded that teacher expectations are based on social class, attractiveness, ethnicity, and perhaps gender. Furthermore, Rodney, Perry, Parson, and Hrynuik (1986) concluded from a study of 662 junior high students that teachers are sensitive to the cultural values of their students in setting both normative and cognitive expectations. Research by Morrison (1990) with 22 preschool classes suggested that some teachers may be unaware of the differential treatment given in their classes.

The Parents' Role. Many teachers believe that parents of African American and Hispanic students do not care about their children's educational achievement. Based on a study by the Hispanic Policy Development Project, Olson (1990) reported that low-income Hispanic parents are deeply concerned about their children's education but do not understand the role that parents are to play in the educational process of children. Most teachers assume that parents have some responsibility for their children's education; the parents are expected to prepare children for school, teach them basic skills, and reinforce the learning that takes place in the classroom. Many parents do not believe they have the right to question teachers because the teachers are seen as the experts (Olson, 1990).

Attitudes Toward Physical Characteristics. According to Bennett (1990), research on prejudices demonstrates that skin color influences people's perceptions and judgments of each other. Unfortunately, lighter-skinned people were more preferred than those with darker skins by White and some African Americans and Hispanics. Even though the leaders of the 1960s civil rights movement attempted to help others to recognize that Black is beautiful, studies on teacher interactions support the idea that teachers expect more from White or lighter-skinned students (Bennett & Harris, 1982; Gay, 1974; Rist, 1970; *Teachers and students*, 1973). It is important, through preservice teacher education programs and inservice programs, to change the belief, value, and attitude systems of teachers of science who tolerate or promote academic racism (Boyer, 1983).

Summary. Teachers need to understand and appreciate the backgrounds of their students, especially when cultural or ethnic differences exist. What teachers think of their students and how they respond to students play a critical role in how students perceive themselves. Teachers must protect students from acquiring the debilitating self-doubt that can prevent them from reaching their potential.

Verbal and Nonverbal Communication Patterns

According to Anderson (1988), students from underrepresented groups in science and rural White students with a minimal degree of acculturation display communication styles that differ from those of nonrural White students. Longstreet (1978) outlined potential sources of misunderstanding in communication patterns related to verbal and nonverbal communication and orientation modes. These aspects of verbal communication include grammar, semantics, phonology (sound, pitch, rhythm, and tempo of words), and discussion modes. Important nonverbal patterns include body language, touching, and body positions.

Some ethnic groups use extensive imagery in their thinking, speaking, and writing. They are very visual, perceive thought to be living and holistic, create affective images based in the culture, extensively utilize expressions with concrete emotional words and metaphors, focus on the message, and introduce self into objective analysis of events. White groups are more apt to use notions or theoretical statements; perceive thought as mentalistic, devitalized, and static; minimize affective imaginary; rarely use expressions of concrete emotional words and metaphors; believe that the medium is most important because it communicates the message; contemplate actions prior to attempts to motivate; and remove self from their speaking and writing (Anderson, 1988).

Patterns of Speech. Speech patterns that reflect cognitive styles of groups are means of effective interaction within group communities and the larger society. The standard English communication style is one of many communication styles; it is more functional for those with an Anglo-European framework, but it should not be perceived as the best (Anderson, 1988).

Eye Contact. Patterns of nonverbal communication, such as eye contact, differs among ethnic groups. According to Bennett (1990), direct eye contact signifies that one is listening to the speaker and speaking honest, truthful words. When Whites and African Americans interact, they look directly at each other. Among Navajos and some Asian Americans, lowering the eyes is a sign of respect. Usually, first-generation Asian Americans practice this publically; however, teachers quickly attempt to teach these Asian students that the lowering of the eyes is not an acceptable practice in the United States.

Personal Space. The amount of personal space varies from culture to culture. People in high-context cultures such as China require less personal space and do more touching than people from low-context cultures (Bennett, 1990). In some cultures, people are more apt to greet each other with hugs and kisses; other cultures may behave more formally.

Summary. Means and Knapp (1991) made several recommendations for teaching students who are regarded as less likely to succeed in school. Dialogue should be the central medium for teaching and learning, because a dialogue is an interchange in which two individuals are full-fledged participants and both significantly influence the nature of the exchange. Boyer (1983) outlined 10 critical dimensions of multicultural teaching, learning, and service delivery. He maintained that a different understanding about the manner in which groups of people communicate needs to be acquired. Until this occurs, he proposes that education will remain stagnant, in the dark ages, and that some science teachers will not act on egalitarian values associated with pluralism.

Establishment of Power in the Classroom

Science teachers must be committed to empowering their students in science. Students need the capability to act with results in the area of science. Using a refinement of McLaren's (1989) definition of empowerment, it is the process by which students learn to use critically science knowledge that is outside their immediate experience to broaden their understanding of science, themselves, and the world and to realize the prospects for reforming the accepted assumptions about the way people live.

According to Sleeter (1991), empowering relationships differ from helping relationships. These "benevolent helping models" (Sleeter, 1991, p. 4) tend to maintain the status quo and disable students of oppressed groups. Examples of helping models used by science teachers are admonishing students with low science grades to work harder, describing America as a place of equal opportunity where one can get ahead through hard work, describing "at-risk" students as in need of experts to diagnose their needs and letting the experts prescribe treatments, and having classroom management philosophies that tell students how they ought to behave in science classrooms and school.

Some Native American students have characterized science and mathematics teachers as "hard, tough, unyielding, and difficult" (Green, Brown, & Long 1978). Cummins (1986) identified four areas for science teachers to consider in empowering relationships: (1) incorporation of students' culture and language in the teaching of science, (2) collaborative participation of the community in schools and science classrooms, (3) orientation of science pedagogy toward reciprocal interaction, and (4) advocacy rather than legitimacy of failure as a goal for science assessment. However, science teachers do not empower or disempower their students, they merely create the situations under which their students can empower themselves. Students' power involves them in judging more effectively when to use their science knowledge and skills.

Teaching Methods and Strategies

Cooperative learning is a teaching strategy that science teachers must be adept at utilizing in their classrooms. According to Sapon-Shevin and Schniedewind (1991), cooperative learning not only assists students who are members of underrepresented groups in the sciences in working together and appreciating each other but also enables students to learn how to act collectively on various issues. When science classrooms are structured around competition, students learn that some students are winners, others are losers, "that students who are different are not capable or worthy, and that success is a scarce commodity, not available to all, and accessible only by pleasing the teacher" (Sapon-Shevin & Schniedewind, 1991, p. 160). Johnson and Johnson (1975) listed three conditions in which competition can be motivating. One of these conditions is that all students must perceive themselves as having an equal probability of winning. In many science classrooms, students and teachers are aware that this is not true.

Studies with Hawaiian and Native American students have revealed that they will not engage in group contingencies and reward systems that place one student above another or would embarrass any of the students (Gallimore & Howard, 1968; Kohn, 1986). Mason and Barba (1992) recommended that bilingual students be matched with monolingual Spanish-speaking students in science classrooms. This arrangement allows for peer tutoring, especially in the students' primary language.

Peer Proofing. According to Fordham (1991), schools practice "peer-proofing" strategies. Peer-proofing African American students' school success has come to mean sequestering them from other African American students and changing their cultural and ethnic identity so that they no longer view themselves as African American people. Fordham wants peer proofing African Americans adolescents' academic competition must come to mean immersing them in the fictive kinship system (a system in which a sense of peoplehood or collective social identity has

developed among individuals who are not related by blood or marriage, but have complementary social and economic relationships) or collective ethos of the Black community, thereby combining their desires to be academically successful and their ethnic identification. The science curricula should be structured so the students feel responsible for each other and not in one-on-one competition against each other.

Learning by doing encourages divergent thinking in which students are required "to produce their own answers, not to choose from alternatives given to them" (Guilford, 1967, p. 138). Bryant (1988) urged teachers of all grade levels to acknowledge the intrinsic worth and intellectual capabilities of African American students to achieve at high levels of excellence. He stated that teachers must examine their attitudes and the pedagogical skills they use for positive impacts on African American science learning and eliminate attitudes that inhibit science learning.

Low Verbal Expressive Students. Haukoos and Satterfied (1986) suggested classroom changes for highly visual, low verbal expressive students such as Western Teton Sioux. More discussion and less lecture should be utilized by the teacher. Even discussion sessions should be saturated with more photographic slides and visuals with graphics, organisms, and natural settings. Peer and teacher small study groups should be used. When talking, the teacher should sit at eye level in front of the desk.

The Pressure of High Expectations. Asian-Americans, especially Japanese and Chinese Americans, have been referred to as the "model minority" (Shih, 1988; Takaki, 1989). However, many Asian American students need teachers to consider their mental and social needs in a classroom. Not all Asian-American students excel in computer, mathematics, and science classes.

Curriculum

Allen and Seumptewa (1988) stated there is no one appropriate science curriculum for the over 400 Native American nations, which have different cultures. They recommended that curriculum developers look primarily at universally accepted activities that are culturally sensitive for the numerous nations. To help Native American students interpret abstract science concepts, they should be taught in the forms of story situations that include foods, places, and events from the students' environment (Allen & Seumptewa, 1988; Johnson, 1975). Green et al. (1978) recommended discovering how the cultures use science and mathematics concepts in such areas as astronomy, cooking, and crafts and including these in the science curriculum.

Haukoos (1989) taught an inservice workshop for elementary teachers who taught Lakotan students. The goals of the workshop were:

1. To have dialogue among the participants about the European-type science and the Native American culture
2. To present the history of science from the Western perspec-

tive and demonstrate heuristic strategies used by all to obtain knowledge about their environment

3. To describe the science inquiry and problem-solving strategies and help the teachers to acquire those skills by doing certain elementary science activities
4. To develop elementary science activities that would promote Lakotan interest in science and be compatible with the Lakotan culture, and to try the activities with their students
5. To interview community friends and elders and share the findings of these interviews with the other participants

According to the participants in this workshop, there are varying degrees of consensus on the use of oral traditions in written materials. The following titles of the elementary science lessons reflect some of these traditions: Food Preservation, Fruit Drying, The Four Seasons as a Whole Year, Measuring Time Through Nature, Weather; Hot and Cold Air on the Plains; and Tin 'psila.

Duran and Weffer (1992) found that curriculum was the second most important predictor of science achievement for talented immigrant Mexican students. More coursework in science resulted in higher achievement for these students. The goal of the science intervention program was to reinforce concepts in biology, chemistry, and physics and to introduce college-level laboratory methods. When students participated in the program for twelfth-graders, the participation doubled the effect of curriculum on achievement. This was not surprising because immigrant students, even more than other students, depend on school coursework for knowledge of a particular subject area.

In order to maximize the opportunities for all students in science instruction, teachers must be alert to biases in their science curriculum materials. Detections of the following biases are crucial when teaching African American, Asian American, Hispanic, and Native American students (Atwater, 1993):

1. Invisibility of certain groups of people, which conveys to students that certain groups of people are less appreciated in the sciences
2. Stereotyping or assigning very rigid roles and attributes to groups of people in the sciences, which influences powerful people such as science teachers to discriminate against students in their classrooms who belong to these groups
3. Imbalance, selectivity, and unreality of a scientific issue so that students fail to become competent in thinking critically and solving problems
4. Fragmentation, isolation, and separation of scientific issues related to certain groups from the main science curricula so that students believe that science contributions of these people are nonexistent or insignificant

In addition, science learning materials should (1) reflect fully and accurately the reality of the U.S. ethnic and cultural diversity, (2) recognize the universality of human experiences and interdependence of all people and communities, and (3) present historical data and controversial science issues in a culturally unbiased manner (Ijaz, 1986). Bey (1992) has urged school leaders that one way to improve learning and teaching is

to generate an interest among teachers to utilize curriculum materials that are representative of the students backgrounds.

Martinez and Ortiz de Montellano (1983) have described culturally relevant student projects in agriculture, architecture, astronomy, botany, geology, and nutrition. They discussed the Mayan calendar, Native American buildings, and structures used for astronomical purposes and illustrated 12 figures on astronomy and timekeeping.

Mears (1986) challenged teachers to help students gain different perspectives. For example, the scientific basis for mapping is a fruitful source because maps contain ethnocentric biases. Comparing the concepts on which the Mercator and other traditionally used projections are based with those which the newer Peters' projections and the Buckminster Fuller Dymaxion maps are constructed is an enlightening activity for students studying earth science, geography, or social studies.

Textbook Representation. Powell and Garcia (1985) evaluated seven contemporary elementary science textbook series for depiction of African Americans, Asian Americans, and Native Americans. It was found that 75 percent of the adults represented in the science series were White; adults from underrepresented groups in the sciences rarely appeared in the same illustration, usually were shown in roles or activities dealing with parental or familial situations and in such occupational roles as teachers and mechanical workers, and appeared less often in science-related careers. Thirty-five percent of the children illustrated in the science series were from underrepresented groups in the sciences. Eighteen percent of illustrations that were coded as appendages pictured underrepresented groups in the sciences. Powell and Garcia concluded that the seven science textbook series did not provide their readers with a broad, comprehensive, and balanced view of science in society.

Lynch (1983) suggested some questions to ask about multicultural curriculum materials. The questions related to areas that have not been discussed are:

1. What is the "cultural identity" of the students and community from which they come? What languages and religions are represented? What cultural values and customs are evident? What businesses, professions, jobs, careers, and lifestyles are characterized by the community?
2. What should be the major components of the science curriculum and how should these reflect the community of which the students are a part? How can the community be involved in the science curriculum?

For example, sports is a domain of interest for many African American males. Topics in human physiology or physics of motion and matter can be organized around sports so that students' interests are piqued.

SUMMARY AND IMPLICATIONS

According to Goldberg (1984), successful schools whose student populations are predominantly African-Americans or Hispanics have the following characteristics:

1. The professional staff believe that its students can learn.
2. A consensus on school goals has been reached; these goals foster academic skills but do not exclude art and physical education.
3. High expectations for both student and staff performance are present.
4. Teachers hold the belief that if students fail, they have failed.
5. A well-developed support system for teachers is in place; this system includes administrative help so that instructional time is protected.
6. There is a collegial pressure among the teachers to work hard.
7. The goal of staff development activities is to improve the instructional program.
8. When new conditions emerge such as an influx of different kinds of students, the staff views it as an opportunity to do novel things.
9. Regular assessment focuses on student progress and instructional improvement; results are regularly reported to the parents.

In the final report of the Indian Nations at Risk Task Force (1991), 16 barriers that disempower and disenfranchise Native American students were identified. These barriers include an unfriendly school climate that fails to promote appropriate academic, social, cultural, and spiritual development among Native American students; curricula presented from a purely western or European point of view; low expectations and relegation of Native American students to low-ability tracks, and overt and subtle racism in schools with a lack of multicultural focus. Mc-Bay (1989) has forwarded nine recommendations for improving education for African Americans, Asian Americans, Hispanics, and Native Americans. Recommendations are as follows:

1. Early interventions in preschool and parent education, along with day care and nutrition, are required for lasting effects.
2. School systems must be restructured so that the education of all students is encouraged. Flexibility, decentralization, and accountability are needed.
3. Curricula that are sensitive to ethnic and cultural identity must be developed and utilized.
4. The best teachers must be available to the students who need them the most.
5. The role of the school must be changed, for it must also function as a social services provider.
6. Better coordination between existing community programs and services and schools must be achieved.
7. Revitalization of the community resources must occur so that role models and support are available.
8. Since there are leaks in the educational pipeline at transition points, students must be exposed early to higher levels of education.
9. Delineate the incentives for all students to participate in higher education.

In addition, Allen & Seumptewa (1988) have identified two strategies proposed by Native American Science Education Association to increase the number of Native Americans who could major in the sciences. First, in order to improve the quality of Native American science education before and during high school, preservice and inservice teachers must develop the knowledge and skills necessary for them to be successful in teaching Native American students. Second, the science curriculum must be sensitive to needs of the Native American students if the attrition of Native American students from the pool is to decrease.

Haberman (1991) described good teaching for urban students, which is good teaching for all students, in the following ways. He said good teaching is occurring if students are involved in

1. Solving problems that are of vital interest to them;
2. Explaining human differences, seeing major concepts, big ideas, and general principles, rather than accumulating isolated facts;
3. Planning what they will be doing;
4. Applying ideals such as fairness, equity, or justice to their world;
5. Doing an experiment and constructing things;
6. Reflecting on real-life experiences;
7. Working in heterogeneous groups;
8. Thinking about an idea in a way that questions common-sense or a widely accepted assumption, that relates new ideas to ones learned previously, or that applies an idea to the problems of living;
9. Redoing, polishing, or perfecting their own works;
10. Accessing information through technology; and
11. Reflecting on their own lives and how they come to believe and feel as they do.

Simon (1987) suggested, "If we do not give youth a sense of how to 'make it' within existing realities, all too often we doom them to social marginality: yet another high-minded way of perpetuating the structural inequalities in society" (p. 375). According to Takata (1991), empowered students collectively form a powerfully inventive and imaginative source from which innovations are developed and social changes guided. The science classrooms of the twenty-first century must help students to be empowered. Otherwise, many of our young adults will lack scientific literacy, and our entire community will lose the contributions of these students in the fields of engineering, science, and mathematics.

References

Alick, B., & Atwater, M. M. (1988). Problem-solving strategies and success of Afro-American science majors. *School Science and Mathematics, 88*(8), 659–665.

Allen, G. G., & Seumptewa, O. (1988). The need for strengthening Native American science and mathematics education. *Journal of College Science Teaching, 55*, 364–369.

Analysis puts a number on population mix. (1991, April). *USA Today*, 10A.

Anderson, J. A. (1988). Cognitive styles and multicultural populations. *Journal of Teacher Education, 39*(1), 2–9.

Aronson, E. (1978). *The jigsaw classroom.* Beverly Hills, CA: Addison-Wesley.

Atwater, M. M. (1981). The influence of cognitive and affective variables on the success of Black undergraduate students in science and engineering curricula at a predominantly White university (Doctoral dissertation, North Carolina State University, 1980). *Dissertation Abstracts International, 41*, 5044A.

Atwater, M. M. (1993). Multicultural science education: Assumptions and alternate views. *The Science Teacher, 60*(3), 32–38.

Bagley, C. (1988). Cognitive style and cultural adaptation in Blackfoot, Japanese, Jamaican, Italian, and Anglo-Celtic children in Canada. In G. K. Verma & C. Bagley (Eds.), *Cross-cultural studies of personality, attitudes and cognition* (pp. 143–159). New York: St. Martin's.

Banks, J. A. (1989a, April). Fostering language and cultural literacy in the schools. Paper presented at the Conference on Public Policy Issues in Bilingual Education, Washington, DC.

Banks, J. A. (1989b). Integrating the curriculum with ethnic content: Approaches and guidelines. In J. A. Banks & C. A. McGee Banks (Eds.), *Multicultural education: Issues and perspectives* (pp.189–207). Needam Heights, MA: Allyn & Bacon.

Bar-Haim, G., & Wilkes, J. M. (1989). A cognitive interpretation of the marginality and underrepresentation of women in science. *Journal of Higher Education, 60*, 371–387.

Bem, D. (1970). *Beliefs, attitudes, and human affairs.* Belmont, CA: Brooks/Cole.

Bennett, C. I. (1990). *Comprehensive multicultural education: Theory and practice.* Boston: Allyn & Bacon.

Bennett, C. I., & Harris, J. J., III. (1982). Suspensions and expulsion of male and Black students: A study of the causes of disproportionality. *Urban Education, 16*(4), 399–423.

Berry, B. (1976). *Human ecology and cognitive style: Comparative studies in cultural and psychological adaptation.* New York: Wiley.

Bey, T. M. (1992). Multicultural teacher development: Diversifying approaches for improvement. *Kappa Delta P. Record, 28*(2), 59–62.

Boyer, J. (1983). The ten most critical dimensions of cross-racial, cross-ethnic teaching and learning. *Educational Considerations, 10*(3), 2–4.

Brophy, J. E., & Evertson, C. M. (1981). *Student characteristics and teaching.* New York: Longman.

Bryant, N. (1988). Sons, daughters, where are your books? *Journal of College Science Teaching, 55*, 344–347.

Burns, M., Gerace, W., Mestre, J., & Robinson, H. (1982). The current status of Hispanic technical professionals: How can we improve recruitment and retention. *Integrated Education, 20*, 49–55.

Caduto, M. J., & Bruchac, J. (1989). *Keepers of the earth: Native American stories and environmental activities for children.* Golden, CO: Fulcrum, Inc.

Castaneda. A., & Gray, T. (1974). Bicognitive processes in multicultural education. *Educational Leadership, 32*, 203–207.

1990 census of population and housing summary population and housing characteristics, United States. (1992). Washington, DC: U.S. Government Printing Office.

The 1990 census, state by state. (1991, March). *The New York Times*, p. A22L.

Cohen, R. A. (1969). Conceptual styles, culture conflict and nonverbal tests of intelligences. *American Anthropologist, 71*, 828–856.

Collins, A., Hawkins, J. & Carver, S. M. (1987). A cognitive apprenticeship for disadvantaged students. In B. Means, C. Chelemer, & M. S. Knapp (Eds.), *Teaching advanced skills to at-risk students: Views from theory and practice* (pp. 216–254). San Francisco: Jossey-Bass.

The condition of bilingual education in the nation. (1988). Washington, DC: Office of Bilingual Education and Minority Language Affairs.

The condition of education, a statistical report. (1987). Washington, DC: Office of Educational Research and Improvement, U.S. Department of Education.

Cooper, G. (1980). Everyone does not think alike. *English Journal, 60*, 45–50.

Crichlow, W., Goodwin, S., Shakes, G., & Swartz, E. (1990). Multicultural ways of knowing: Implications for practice. *Journal of Education, 172*(2), 101–117.

Cross, K. P. (1976). Beyond education for all—toward education for each. *College Board Review, 99*, 5–10.

Cummins, J. (1979). Linguistic interdependence and the educational development of bilingual children. *Review of Educational Research, 49*, 222–251.

Cummins, J. (1986). Empowering minority students: A framework for intervention. *Harvard Educational Review, 56*, 18–36.

Day, J. D., French, L. A., & Hall, L. K. (1985). Social influences on cognitive development . In D. L. Forrest-Pressley, G. E. MacKinnon, & T. G. Waller (Eds.), *Metacognition, cognition, and human performance: Vol 1.* Miami, FL. (ERIC Document Reproduction Service No. ED 213 684)

Dempsey, A. D. (1981). *Anglo-American and Cuban-American teachers' perceptions of elementary school boys and girls* (Report No. SP-019-765). Miami, FL. (ERIC Document Reproduction Service No. ED 213 684)

Deyhle, D. (1987). Learning failure: Tests as gatekeepers and the culturally different child. In H. T. Trueba (Ed.), *Success or failure? Learning and the language minority student* (pp. 85–108). Cambridge: Newbury House.

Douglass, C., Kahle, J. B., & Everhart, J. S. (1980). Interracial behavior patterns in an inner city biology laboratory. *School Science and Mathematics, 80*(5), 413–422.

Douglass, C. B. (1976). *The effect of instructional sequence and cognitive style on achievement of high school biology students.* Unpublished doctoral dissertation, Purdue University, West Lafayette, IN.

Driscoll, M. (1987). *Research within reach: Secondary school mathematics.* Reston, VA: National Council of Teachers of Mathematics.

Duit, R. (1991). Students' conceptual frameworks: Consequences for learning science. In S. M. Glynn, R. H. Yeany, & B. K. Britton (Eds.), *The psychology of learning science* (pp. 65–85). Hillsdale, NJ: Lawrence Erlbaum.

Duran, B. J., & Weffer, R. E. (1992). Immigrants' aspirations, high school process, and academic outcomes. *American Educational Research Journal, 29*(1), 163–181.

Duran, R. P. (1979). *Logical reasoning skills of Puerto Rican bilinguals* (NIE-G-78-0135). Washington, DC: National Institute of Education.

Ethington, C. A., & Wolfe, L. M. (1988). Women's selection of quantitative undergraduate field of study: Direct and indirect influences. *American Educational Research Journal, 25*(2), 157–175.

Farrell, M. L. (1984, December). A review of research on styles: Implications for the development and implementation of CAI for bilingual students. In C. Bilotta (Ed.), *Conference proceedings: Delivering academic excellence to culturally diverse populations* (pp. 37–43). Teaneck, NJ: Fairleigh Dickinson University.

Fennema, E., & Behr, M. J (1980). Individual differences and the learning of mathematics. In R. J. Shuway (Ed.), *Research in mathematics education* (pp. 324–355). Reston, VA: National Council of Teachers of Mathematics.

Festinger, L. A.(1957). *A theory of cognitive dissonance.* Evanston, IL: Row, Peterson.

Fordham, S. (1991). Peer-proofing academic competition among black adolescents: "Acting White Black American style." In C. E. Sleeter

(Ed.), *Empowerment through multicultural education* (pp. 69–93). Albany: State University of New York Press.

Fordham, S. (1993). "Those loud Black girls": (Black) women, silence, and gender passing in the academy. *Anthropology and Education Quarterly, 24*(1), 3–32.

Fordham, S., & Ogbu, J. (1986). Black students' school success: Coping with the 'burden of acting White'. *Urban Review, 18*(3), 176–206.

Frank, B. M. (1986). Making math work for African Americans from a practitioner's perspective. In Mathematical Sciences Education Board (Ed.), *Making mathematics work for minorities: A compendium of papers prepared for the regional workshop.* Washington, DC: Mathematical Sciences Education Board.

Fritz, B. (1990). Contemporary issues in student preferences for style. Unpublished manuscript.

Gallimore, R. & Howard, A. (1968). *Studies in a Hawaiian community: Na Makanaka o Nanakuli* (Pacific Anthropological Record 1). Honolulu: Hawaiian Bishop Museum, Department of Anthropology.

Gardner, C., Atwater, M. M., & Wiggins, J. (1992, March). *Affective characteristics of urban African American middle school students with highly positive attitudes toward science.* Paper presented at the annual meeting of the National Association for Research in Science Teaching, Boston.

Gay, G. (1974). *Differential dyadic interactions of Black and White teachers with Black and White pupils in recently desegregated social studies classrooms: A function of teacher and pupil ethnicity.* (OE Report No. 2F113). Washington, DC: U.S. Civil Rights Commission.

Glynn, S. M., Yeany, R. H., & Britton, B. K. (1991). A constructive view of learning science. In S. M. Glynn, R. H. Yeany, & B. K. Britton (Eds.), *The psychology of learning science* (pp. 3–19). Hillsdale, NJ: Lawrence Erlbaum.

Goldberg, M. S. (1984). A report that changed history. In C. Bilotta (Ed.), *Conference proceedings : Delivering academic excellence to culturally diverse populations* (pp. 1–9). Teaneck, NJ: Bilingual Education Skills and Training Center , Fairleigh Dickinson University.

Green, R., Brown, J. W., & Long, R. (1978, February). *Report and recommendations: Conference on Mathematics in American Indian Education* (A report of the Project on Native Americans in Science). Washington, DC: Office of Opportunities in Science, American Association for the Advancement of Science.

Guilford, J. P. (1967). *The nature of human intelligence.* New York: McGraw-Hill.

Haberman, M. (1991). The pedagogy of poverty versus good teaching. *Phi Delta Kappan, 73*(4), 290–294.

Hadfield, O. D., Martin, J. D., & Wooden, S. (1992). Mathematics anxiety and learning style of the Navajo middle school student. *School Science and Mathematics, 92,* 171–175.

Harper, C. (1989). *The impact of societal and institutional factors on the production of Black graduate students in the sciences.* Unpublished master's thesis, Atlanta University, Atlanta, GA.

Haukoos, G. D. (1989, April). *Developing culturally sensitive science curricula for elementary teachers of Native American children.* Paper presented at the national convention of the National Science Teachers Association, Seattle.

Haukoos, G. D. & Chandayot, P. (1988). *A cross-cultural study of attitude toward science and related influential factors.* (Report no. SE048924). Lake of the Ozarks, MO: National Association for Research in Science Teaching. (ERIC Document Reproduction Service No. ED 291 591)

Haukoos, G. D., & Satterfield, R. (1986). Learning styles of minority students (Native Americans) and their application in developing a culturally sensitive classroom. *Community/Junior College Quarterly, 10,* 193–201.

Hilliard, A. G. (1984). *Saving the African American child.* A report of the National Alliance of Black School Educators, Inc., Washington, DC: 14(11).

Holtzman, E., Goldsmith, R., & Barrera, C. (1979). *Field-dependence and field-independence: Educational implications for bilingual education.* Austin, TX: Dissemination and Assessment Center for Bilingual Education.

Hvitfeldt, C. (1986). Traditional culture, perceptual style, and learning: The classroom behavior of Hmong adults. *Adult Education Quarterly, 36*(2), 65–67.

Ijaz, A. (1986). Guidelines for curriculum writing in a multicultural milieu. In R. J. Samuda & S. L. Kong (Eds). *Multicultural education: Programmes and methods* (pp. 109–114). Toronto: Intercultural Social Sciences Publication.

Indian Nations At Risk Task Force. (1991). *Indian nations at risk: An educational strategy for action.* Washington, DC: U.S. Department of Education.

Jacobowitz, T. (1983). Relationship of sex, achievement, and science self-concept to the science career preferences of black students. *Journal of Research in Science Teaching, 20*(7), 621–628.

Johnson, D., & Johnson, R. (1975). *Learning together and along: Cooperation, competition and individualization.* Englewood Cliffs, NJ: Prentice Hall.

Johnson, W. N. (1975). *Teaching mathematics in a multicultural setting: Some considerations when teachers and students are of differing cultural backgrounds.* Murray, KY: Murray State University (ERIC Document Reproduction Service No. ED 183 414)

Kagan, S. (1980). Cooperation-competition, culture and structural bias in classrooms. In S. Sharan, P. Hare, C. Webb, and R. Hertz-Lazaowitz (Eds.), *Cooperation in education* (p. 210). Provo, UT: Brigham Young University Press.

Kahle, J. (1982). Factors affecting minority participation and success in science. In R. E. Yager (Ed.), *What research says to the science teacher:* (Vol. 4, pp. 80–95). Washington, DC: National Science Teachers Association.

Kohn, A. (1986). *No contest: The case against competition.* Boston: Houghton Mifflin.

Lickona, T. (1980). Beyond justice: A curriculum for cooperation. In D. Cochrane & M. Manley-osmir (Eds.), *Development of moral reasoning* (pp. 108–144). New York: Praeger.

Longstreet, W. (1978). *Aspects of ethnicity: Understanding differences in pluralistic classrooms.* New York: Teachers College Press.

Lynch, J. (1983). *The multicultural curriculum.* London: Batsford Academic and Educational.

Malcom, S. (1989). Increasing the participation of black women in science and technology. *Sage, 6*(2), 15–17.

Malcom, S., Hall, P. Q., & Brown, J. W. (1976). *The double bind: The price of being a minority woman in science.* Washington, DC: American Association for the Advancement of Science.

Martinez, D., & Ortiz de Montellano, G. (1983, May). *Teaching culturally relevant science.* Paper presented at the Michigan Annual Hispanic Education Conference, Dearborn.

Marzano, R. J. (1992). *A different kind of classroom.* Alexandria, VA: Association for Supervision and Curriculum Development.

Maslow, A. H. (1968). *Toward a psychology of being.* New York: Van Nostrand Reinhold.

Mason, C. L., & Barba, R. H. (1992). Equal opportunity science. *Science Teacher, 59*(5), 23–26.

McBay, S. M. (1989). Improving education for minorities. *Issues in science and technology, 5*(4), 41–47.

McLaren, P. (1989). *Life in schools.* New York: Longman.

McNett, I. (1983). *Demographic imperatives: Implications for educational policy.* Washington, DC: American Council of Education.

Means, B., & Knapp, M. (1991). Cognitive approaches to teaching advanced skills to educationally disadvantaged students. *Phi Delta Kappan, 73*(4), 282–289.

Mears, T. (1986). Multicultural and antiracist approaches to the teaching of science in schools. In J. Gundara, C. Jones, & K. Kimberly (Eds.), *Racism, diversity, and education* (pp. 154–166). London: Hodder & Stoughton.

Mestre, J. P. (1981). Predicting academic achievement among bilingual Hispanic college technical students. *Educational and Psychological Measurement, 41*, 1255–1264.

Moore, K. A. (1990, November). *TSFACTS at a glance*. Michigan: Charles Stewart Mott Foundation.

Morrison, J. W. (1990). *Compensatory preschool teachers' interaction patterns with the classroom minority*. (Report No. PS 018 650). New York: Syracuse University. (ERIC Document Reproduction Service No. ED 317 271)

Morse, E. V., & Gordon, G. (1974). Cognitive skills: A determinant of scientists' local-cosmopolitan orientation. *Academy of Management Journal, 17*(4), 709–723.

National Center for Education Statistics. (1990). *Targeted forecast, elementary and secondary enrollment: fall 1988 to fall 1994*. Washington, DC: U.S. Department of Education.

Novak, J. D. (1985). Metalearning and metaknowledge strategies to help students to learn how to learn. In L. H. T. West & A. L. Pines (Eds.), *Cognitive structures and conceptual change* (pp. 189–209). Orlando, FL: Academic Press.

Oakes, J. (1990). Opportunities, achievement, and choice: Women and minority students in science and mathematics. In C. B. Cazden (Ed.), *Review of research in education*: Volume 16 (pp. 153–221). Washington, DC: American Educational Research Association.

Office of Educational Research and Improvement. (1987). *Digest of Education Statistics 1987*. Washington, DC: U.S. Department of Education.

Ogbu, J. (1978). *Minority education and caste: The American system in cross-cultural perspective*. New York: Academic Press.

Olson, L. (1990, May). Misreading said to hamper Hispanics' role in school. *Education Week, 1*, 2.

Olstad, R. G., Juarez, J. R., Davenport, L. J., & Haury, D. L. (1981). *Inhibitors to achievement in science and mathematics by ethnic minorities*. Seattle: University of Washington. (ERIC Document Reproduction Service No. ED 223 404)

Pallone, N. J., Hurley, R. B., & Rickard, F. S. (1973). Further data on key influencers of occupational expectation among minority youths. *Journal of Counselling Psychology, 20*, 484–486.

Pearson, W. J., Jr. (1986). Black American participation in American science: Winning some battles but losing the war. *Journal of Educational Equity and Leadership, 6*(1), 45–59.

Peng, S. S. & Takai, R. T. (1983). *High school dropouts: Descriptive information from high school and beyond*. Washington, DC: National Center for Education Statistics.

Posner, G. J., Strike, K. A., Hewson, P. W., & Gertzog, W. A. (1982). Accommodation of a scientific conception: Toward a theory of conceptual change. *Science Education, 66*, 211–227.

Powell, L. (1990). Factors associated with the underrepresentation of African Americans in mathematics and science. *Journal of Negro Education, 59*(3), 292–298.

Powell, R. R., & Garcia, J. (1985). The portrayal of minority and women in selected elementary series. *Journal of Research in Science Teaching, 22*(6), 519–533.

Power, C. (1972). The unintentional effects of science teaching. *Australian Science Teachers Journal, 18*(4), 103–111.

Pulos, S., Stage, E. K., & Karplus, R. (1982). Setting effects in mathematical reasoning of early adolescents: Findings from three urban schools. *Journal of Early Adolescence, 2*, 39–59.

Rakow, S. J. (1985a). Minority students in science: Perspectives from the 1981–1982 National Assessment in Science. *Urban Education, 20*(1), 103–113.

Rakow, S. J. (1985b). Prediction of the science inquiry skill of seventeen-

year-olds: A test of the model of educational productivity. *Journal of Research in Science Teaching, 22*(4), 289–302.

Ramirez, M., III, Castaneda, A., & Herold, P. L. (1974). The relationship of acculturation to cognitive style among Mexican Americans. *Journal of Cross-Cultural Psychology, 5*, 424–433.

Ramirez, M., III, & Price-Williams, D. R. (1974). Cognitive styles of children of three ethnic groups in the United States. *Journal of Cross-Cultural Psychology, 5*, 212–219.

Rist, R. (1970). Student social class and teacher expectations: The self-fulfilling prophecy in ghetto education. *Harvard Education Review, 40*, 411–451.

Robbins, R. (1983). John Dewey's philosophy and American Indian: A brief discussion of how it could work. *Journal of American Indian Education, 22*(3), 1–9.

Rodney, C. A., Perry, R. P., Parson, K., & Hrynuik, S. (1986). Effects of ethnicity and sex on teachers' expectations of junior high school students. *Sociology of Education, 59*, 58–67.

Samuda, R. J. (1986). The role of psychometry in multicultural education: Implications and consequences. In R. J. Samuda & S. L. Kong (Eds.), *Multicultural education: Programmes and methods* (pp. 47–98). Toronto: Intercultural Social Science Publication.

Sapon-Shevin, M., & Schniedewind, N. (1991). Cooperative as empowering pedagogy. In C. E. Sleeter (Ed.), *Empowerment through multicultural education* (pp. 159–178). Albany: State University of New York Press.

Saracho, O. N. (1988). Cognitive styles: Implications for the preparation of early childhood teachers. *Early Child Development and Care, 38*, 1–11.

Schmidt, P. (1991, February). Three types of bilingual education effective, E. D. study concludes. *Education Week, 1*, 23.

Scroggins, D. (1992a, January 12). Dekalb melting pot is bubbling. *Atlanta Journal*, A1, A8.

Scroggins, D. (1992b, January 12). 'We've got them all' at Cross Key High. *Atlanta Journal*, A9.

Simon, R. I. (1987). Disability as the basis for a social movement: Advocacy and the politics of definition. *Journal of Social Issues, 44*, 159–172.

Singh, B. (1987). Graded assessments: Hijacking 'process'. In D. Gill & L. Levidow (Eds.), *Antiracist science teaching* (pp. 210–269). London: Free Association Books.

Slavin, R. (1987). Cooperative and the cooperative school. *Educational Leadership, 45*, 7–13.

Slavin, R. (1988). Cooperative and student achievement. *Educational Leadership, 45*, 31–33.

Shih, F. H. (1988). Asian-American students: The myth of a model minority. *Journal of College Science Teaching, 55*, 356–369.

Silberman, M. (1976). Behavioral expression of teachers' attitudes toward elementary school students. *Journal of Educational Psychology, 60*, 402–407.

Sleeter, C. E. (1991). Introduction: Multicultural education and empowerment. In C. E. Sleeter (Ed.), *Empowerment through multicultural education* (pp. 1–23). Albany: State University of New York Press.

Staffing the multilingually impacted schools of the 1990's. (1990, May). In *Proceedings of the National Forum on Personnel Needs for Districts with Changing Demographics*. Washington, DC: Office of Bilingual Education and Minority Affairs, U.S. Department of Education.

Stegemiller, H. A. (1989). *An annotated bibliography of the literature dealing with the contributing factors of teacher expectations on student performance*. (Report N. SP 031 604). South Bend: Indiana University at South Bend. (ERIC Document Reproduction and Service No. ED 313 323)

Strike, K. A., & Posner, G. J. (1985a). A conceptual change view of learning and understanding. In L. H. T. West & A. L. Pines (Eds.),

Cognitive structure and conceptual change (pp. 211–231). Orlando, FL: Academic Press.

Strike, K. A., & Posner, G. J. (1985b). Relationships of attitude toward classroom environment with attitude toward and achievement in science among tenth grade biology students. *Journal of Research in Science Teaching, 24*(6), 507–525.

Sutman, F. X., Allen, V. F., & Shoemaker, F. (1986). *Learning English through science*. Washington, DC: National Science Teachers Association.

Takaki, R. (1989, May). *The fourth iron cage: Race and political economy in the 1990's*. Paper presented at the Green Bay Colloquium on Ethnicity and Public Policy, Green Bay, WI.

Takata, S. R. (1991). Who is empowering whom? The social construction of empowerment. In C. E. Sleeter (Ed.), *Empowerment through multicultural education* (pp. 251–271). Albany: State University of New York Press.

Teachers and students. Report V: Mexican-American education study: Differences in teacher interaction with Mexican-American and Anglo students. (1973). Washington, DC: U.S. Government Printing Office.

Templin, M., & Ebbs, C. (1992, March). *The meeting effects of teacher-student relationship on academic risk taking*. Paper presented at the meeting of National Association for Research in Science Teaching, Boston.

Thomas, G. E. (1986). Cultivating the interest of women and minorities in high school mathematics and science. *Science Education, 70*(1), 31–43.

U.S. Department of Commerce. (1991, December). Asian and Pacific islander population doubled. *Census and You* (Vol. 26, No. 12, p. 3). Washington, DC: U.S. Government Printing Office.

Vygotsky, L. S. (1978). *Mind in society: The development of higher psychological processes*. Cambridge, MA: Harvard University Press.

Webb, M. R. (1990). Nurturing talent for science. In T. Lopushinsky (Ed.), *Proceeding of a Forum on the needs and problems of the underrepresented college student, with possible solutions* (pp. 27–31). Atlanta: Society for College Science Teachers.

Wiggins, J., Atwater, M. M., & Gardner, C. (1992, March). *A descriptive study of urban middle school students with high and low attitudes toward science*. Paper presented at the annual meeting of the National Association for Research in Science Teaching, Boston.

Witkin, H. A. (1978). Cognitive styles in personal and cultural adaptation. Worchester, MA: Clark University Press.

Witkin, H. A., & Goodenough, D. R. (1981). *Cognitive styles: Essence and origins*. New York: International Universities Press.

Wittrock, M. C. (1985). Learning science by generating new conceptions from old ideas. In L. H. T. West & A. L. Pines (Eds.), *Cognitive structure and conceptual change* (pp. 259–266). Orlando, FL: Academic Press.

Young, H. A., & Young, B. H. (1974). *Scientists in the black perspective*. Louisville, KY: Lincoln Foundation.

Zuckerman, H. (1971). *Women and Blacks in American science: The principle of the double penalty*. Paper presented at the symposium on Women and Minority Groups in American Science and Engineering, California Institute of Technology, Pasadena.

NAME INDEX

SUBJECT INDEX